Fundamentals of Power Plant Engineering

(Conventional and Non-Conventional/Renewal)

–An Innovative Approach

Second Revised Edition

By the same Author

(1) Steam & Gas Turbines, Turbomachines and Power Plant Engineering
(2) Fundamentals of Engineering Thermodynamics
(3) Applied Thermodynamics
(4) Heat and Mass Transfer
(5) Internal Combustion Engines, Pollution and E-vehicles

Fundamentals of Power Plant Engineering

(Conventional and Non-Conventional/Renewal) –An Innovative Approach

Second Revised Edition

Prof. (Dr.) R. Yadav

Ph.D, F.E.I., M.I.S.T.E

Emeritus Professor, M.N.N.I.T., Allahabad (UP)

Former Professor & Head

M.N.R. Engg. College (now M.N.N.I.T.), Allahabad

Former Principal

Regional Institute of Technology (now N.I.T.), Jamshedpur

Former Dean, CET, AAI-DU, Allahabad

Former Director General, MIT, Moradabad

Prof. (Dr.) Sanjay

B.E., M.E., Ph.D.

Professor of Mech. Engg.

N.I.T. Jamshedpur

and

Er. Rajay

B. Tech. (IT BHU, now, IIT BHU)

Manager, B.H.E.L., Haridwar

I.K. International Pvt. Ltd.

NEW DELHI

Published by
I.K. International Pvt. Ltd.
4435-36/7, Ansari Road, Daryaganj
New Delhi-110 002 (India)
E-mail: info@ikinternational.com
Website: www.ikbooks.com

ISBN: 978-93-90620-90-6

Published by Krishan Makhijani for I.K. International Pvt. Ltd., 4435-36/7, Ansari Road, Daryaganj, New Delhi–110 002

Printed at Shree Maitrey Printech Pvt. Ltd., Noida

Dedicated to
My Late Parents
Elder Brother and Sister-in-Law

Preface to the Second Edition

Creativity lies in the minds of creators. The present form of the book titled **"Fundamentals of Power Plant Engineering (Conventional and Non-Conventional/Renewal)** is the result of **4 Ds (Desire, Direction, Dedication and Discipline)**, wide experience as a teacher, researcher and practitioner of the authors. The book contains 18 chapters covering wide spectrum of conventional part and brief aspect of non-conventional part. The contents of the book cover full syllabus on the subject of Power Plant Engineering of all Indian engineering institutes and universities. The book is highly suitable for all competitive examinations like Civil Services, Engineering Services and PSUs of central and state governments.

The purpose of the book is to meet the following objectives:

- to present a thorough treatment of conventional and non-conventional power plants from working, design and application point of view, through an innovative approach;
- to provide a brief idea of energy scenario and various terminologies used in energy technology;
- to prepare students to understand the working principles, control aspects, operation and maintenance of power plants;
- to give emphasis on design aspect of power plants; and
- to develop creativity and engineering judgement among students.

The book contains 18 chapters covered in about more than 800 pages. The special features of the book include the following aspects:

- A large number of typical and real life solved problems in each chapter with diagrammatic approach have been included.
- An exhaustive treatment of theory and design aspects with diagrammatic presentation especially in conventional part has been attempted throughout the book for in-depth knowledge of students.
- The language is simple and lucid to facilitate one time reading to understand the matter.
- **The beauty lies in Exercise portion (unsolved part) which includes many innovative features like** *Things to Think About, Short and Long Theoretical Questions, Software Based Questions, Design and Open Ended Questions and Multiple-Choice Questions.*

- A summary of formulae has been given where needed at the end of each chapter for quick look.
- Non-conventional part (Chapter 18) has been enlarged.
- Printing mistakes in the first edition has been rectified.

The questions given in *Things to Think About* enthuse students to think about the physics involved in the chapter concerned. *Short and Long Theoretical Questions* help in developing the core and comprehensive knowledge of the subject while *Numerical Questions* enhance the analytical capability in the subject. *Software Based Questions* offer a platform to students to develop the software skills in engineering problems related to power plants by developing model algorithm, studying parametric analysis and finally designing. *The Design and Open Ended Questions* offer a great platform for students concerning design experience involving creativity, engineering judgement, formulation of design task statements, application of realistic constraints consideration of alternative and preparation of exhaustive report. These questions are such that some of them lead the students upto the *High Research Level. Multiple-Choice Questions* put some check on the understanding and comprehension of the subject. The simple lucid language, depth, richness, emphasis on fundamentals and design, creativity, innovative approach and judgement enhancement capabilities are the beauty and strength of the book. We hope that with all type of riches, the book will serve the students and teachers in a big way.

While writing this book authors have consulted a large number of books and research papers which are listed in references and also elsewhere in the book. We acknowledge all the authors whom we have referred. We will appreciate if any mistake and error found in this book is brought to the notice of authors.

Finally, the authors thank Meera Yadav, wife of the main author and mother of two co-authors who has been a consistent source of encouragement, pillar of strength, serving the role of wife and mother in an unprecedented manner and endured certain hardships due to preoccupation in this venture.

In the last, we pray to the Almighty who has given us health and wisdom to believe in ***4Ts of Vedas (Tapsya, Tyag, Tejashwa, Tanmayata)*** which has helped to complete this cherished task.

R Yadav
Sanjay
Rajay

Acknowledgements

Sincere acknowledgements and thanks are due to various authors of books and research papers listed in References and elsewhere which were consulted in preparation of this book.

Contents

CHAPTER

1

Introduction – Energy and its Various Aspects

OVERVIEW

Energy is the primary requisite to sustain the Biosphere on this beautiful Blue Planet called the Earth. Every living organism uses energy to survive and grow. The sun, directly or indirectly, is the source of the all energy available on the earth. Our energy choices and decisions impact earth's natural system, so it is essential to choose carefully. This chapter deals with introductory aspects of conventional and non-conventional power plants discussing energy reserves, energy scenario, various energy conversion technologies, various nomenclatures used in energy technologies, energy, conservation and aspects of sustainable energy options.

1.1 INTRODUCTION TO THE WORLD OF ENERGY

Every living creature on the earth needs input of energy. Energy is the basic requirement for economic development. Energy sector of the national economy-agriculture, industry, transport, commercial and domestic needs energy. Thus, energy is the life line and driving force of civilization and utmost care should be taken for its efficient production and utilization. The word "energy" itself is derived from the Greek word **'en-ergon'**, which means 'in-work or work content'.

A systematic study of various forms of energy and energy transformation involving human experience and observation is termed ***energy science****. The applied part of energy science useful to human society, nation and individual is known an* ***energy technology****.*

History shows that a country's energy consumption correlates with its degree of industrialization and standard of living of its population. Fuel resources used for energy consumption in a given country have evolved over a period of time driven by facts such as available fuel resources, economics, technical advancement, political conditions and the degree of awareness of environmental impact. Energy security is imperative for sustained growth of economy. *Like per capita income, per capita energy consumption is now another index to know the standard of living of the people of a particular country.* It is worth mentioning that with 4.6% of world's population, USA consumes about 24% of world energy production while India having 16.5% of world population consumes about 3% of world's energy. *The present annual per capita electrical energy consumption in USA is 12,985 kWh and that in India is just 1180 kWh (March 2019 data) and one can differentiate the quality of life of people of both countries.*

Energy in the form of electricity is easy to produce, transport, use and control. The electric generation in India in January 2018, was around 334.4 GW. At present 70% of energy production in the world is from fossil fuels (coals, oil and gas). Hydroelectric, nuclear power and other non-conventional energy resources such as solar, wind, biomass, geothermal, ocean, etc., contribute about 30% of energy production.

Since the deposits of fossil fuels are depleted day by day, scientists and engineers are engaged in searching out new energy resources, especially non-conventional and renewal and developing energy converting devices for new resources. The increasing awareness of the exponential growth of energy consumption, the rapid depletion of our natural resources, the lagging development of new energy resources and technological advancements and the growing public and institutional demands for energy and material conservation and environment protection have forced the various countries to look forward to a more *rational plan for energy economy.* As a result, many government and private agencies have initiated broad studies of projected energy consumption and the possibility of conserving energy while *enhancing the quality* of environment. *Due to greenhouse effect caused by the emission of* CO_2 *owing to burning of fossil fuels*, the earth's temperature is increasing and as a result many low lying areas near the sea will be drowned owing to melting of glaciers, if the earth's temperature increases by 4°C.

Efficient use of energy could be achieved on the basis of a genuine energy strategy and the future of energy should be linked to the more efficiency, less vulnerable and environmentally ***sustainable source of energy****. Energy available from fossil fuels is called* ***Black Energy*** *and energy available from renewal sources is called* ***Green Energy****.*

The demand for energy and electricity increases steadily. The International Energy Agency (IEA) has predicted that the energy and electricity demands in the world will increase by 1.7% and 2.4% per year respectively, from 2000 to 2030. In 2030, the IEA has forecasted that fossil fuels will still account for the largest part of the energy demand and most of the new power generating capacity will be natural gas-fired combined gas/steam cycles as natural gas is most environment friendly among fossil fuels. This chapter deals with energy resources, classification and energy scenario of the world and India.

1.2 ENERGY CHARACTERISTICS

Energy is defined as the capacity to do work and the rate of doing work is called power. Energy has the following characteristics.

(i) Energy is available in different forms.
(ii) Energy can neither be created nor destroyed.
(iii) Energy can be transformed from one form to another.
(iv) Energy can be stored.
(v) Energy can be transported from one place to another.
(vi) Energy is associated with a potential. It is well-known natural law that free flow of energy takes place only from a higher potential to a lower potential.
(vii) The energy is measured in N.m or joule (J) and the power is measured in watt. The energy density is expressed as J/kg.
(viii) Energy is measured with reference to a datum (ground level, earth's atmospheric, etc.).
(xi) The transfer of heat energy takes place due to temperature gradient while other form of energy due to difference in other intensive properties.
(x) The energy is classified in two categories—**high grade energy**, such as electrical or mechanical energy and **low grade energy** such as heat energy. *The low grade energy is limited by second law of thermodynamics.*

1.3 FORMS OF ENERGY

There are various forms of energy as given below.

(i) Electrical energy
(ii) Mechanical energy including shaft energy, kinetic energy, potential energy, flow energy, etc.
(iii) Chemical energy
(iv) Heat energy
(v) Nuclear energy.

1.4 CLASSIFICATION OF ENERGY RESOURCES

Energy resources are those resources which have energy in any form. They are classified on the basis of many aspects as given below.

1.4.1 Based on Usability of Energy

(a) Primary Resources. Resources available in nature in raw form are known as primary energy resources, e.g., fossil fuels (coal, oil and gas), uranium, hydrogen energy, etc. Generally, this form of energy cannot be used as such. These are in the form of deposits and hence are located, explored, extracted, processed and are converted to a form as per requirement of consumers. The energy yield ratio is generally used to know that efficiency of energy extraction process which is defined as

$$\text{Energy yield ratio} = \frac{\text{Energy received as raw energy source}}{\text{Energy spent to obtain raw energy source}}$$

It is to be noted that only those resources for which the energy yield ratio is very high, are considered worth for exploration.

(b) Intermediate Resources. These are obtained from primary energy resources one or more steps of transformation and is used as **vehicle of energy**, such a synthetic gas, coal, oil, etc.

(c) Secondary Resources. *The form of energy which is finally supplied to consumer for utilisation is called secondary or usable energy,* e.g., electrical energy, thermal energy, in the form of hot water or steam and mechanical energy.

1.4.2 Based on Traditional Applications

(a) Conventional. *Energy resources, which have been traditionally in use before and around oil crisis of 1973,* are known as conventional energy resources, e.g., fossil fuels, nuclear and hydro-resources.

(b) Non-Conventional. *Energy resources which are considered for large scale use after the oil crisis of 1973, are known as non-conventional energy resources,* e.g., solar, wind, biomass, tidal, geothermal, fuel cells, etc.

Table 1.1 gives a glimpse of classification of various power plants grouped in the conventional and non-conventional categories.

Table 1.1 Classification of power plants

Power plant

Conventional
- Steam engine power plant (obsolete)
- Steam turbine power plant
- Gas turbine power plant
- Hydroelectric power plant
- Nuclear power plant
- Gas/steam combined cycle plant
- Rankine organic power plant
- Binary vapour power plant
- Free piston gas turbine

Non-Conventional
- Thermoelectric*
- Therminic*
- Fuel cell*
- Photovoltaic cell
- MHD*
- Fusion
- Biomass
- Geothermal
- Hydrogen energy
- Wind energy
- Ocean thermal energy conversion
- Wave energy
- Tidal wave
- Electric storage battery
- Electro gas dynamic

Some the non-conventional power plants marked by asterisk (*) are also called **direct energy conversion systems** and the rest are **indirect energy conversion systems**. In the indirect energy conversion, the chemical energy is first converted into heat energy and then into mechanical and electrical energy. In the direct energy conversion 'one form of energy is directly converted into electrical energy'.

1.4.3 Based on Long-term Availability

(a) Non-renewable. Energy resource which are finite and cannot be renewed after their consumption are known as non-renewable resources, e.g., fossil fuels, uranium, etc.

(b) Renewable. Energy resources, which are renewed by nature again and again and their supply is not affected by their rate of consumption are known as renewable resources, e.g., solar, wind, biomass, ocean (thermal, tidal and wave), geothermal, hydro, etc. Most of these alternative sources are the manifestation of solar energy as shown in Fig. 1.1.

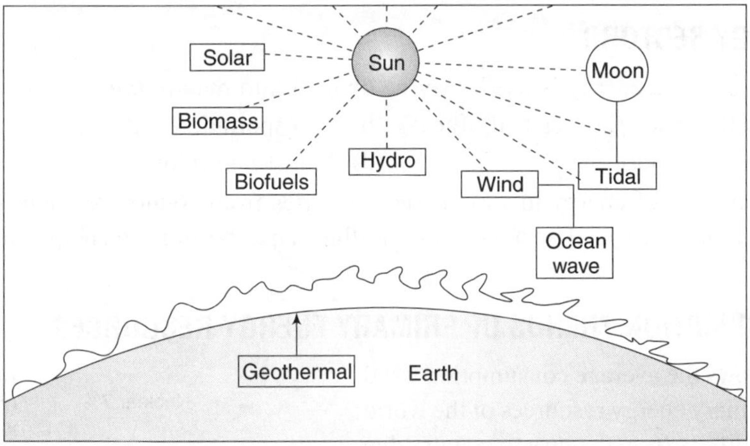

Figure 1.1 Renewable sources of energy

1.4.4 Based on Commercial Applications

(a) Commercial Energy Resource. The energy resources which are used in commercial applications are called commercial energy resources, e.g., electricity, petrol, gas, etc. The economy of a country depends upon the consumption of commercial energy resources.

(b) Non-commercial Energy Resource. The energy resources derived from nature and used directly without passing through commercial outlet are known as non-commercial energy resource, e.g., wood, animal dung, cake, crop residues, etc.

1.4.5 Based on Origin of Resources

The energy resources based on their origin are given below:

(a) Fossil fuels energy (coal, oil and gas)
(b) Nuclear energy
(c) Hydro energy
(d) Solar energy
(e) Wind energy
(f) Biomass energy
(g) Geothermal energy
(h) Ocean thermal tidal and wave energy.

At present 70% of world energy demand is met by fossil fuels. To safeguard our planet earth from greenhouse gas effect, it is very essential to switch over to the environment-friendly energy resources (i.e., renewable) such as solar, wind, biomass, etc.

The world had been affected first ***in 1973 by the oil crisis created by OPEC*** (Organisation of Petroleum Exporting Countries) which was established in 1960. The oil prices in 1973 shot up fourfold causing severe energy crises world over. **Two more "oil shocks" jolted the world in 1979 and 1990. On the first oil shock of 1973, the world took this very seriously** and then the use of alternative source of energy was felt and among them, renewal was the most important due to **eco-friendly in nature**.

1.5 ENERGY SECTORS

Energy sectors where energy is used, may be divided into mainly *four sectors* named as—

(a) Domestic sector (houses and offices) (b) Transportation sector

(c) Agriculture sector (d) Industrial sector

The consumption of energy in various sectors varies from country to country. In general, transportation and industrial sectors consume the major portion of energy in most of the countries.

1.6 CONSUMPTION TRENDS OF PRIMARY ENERGY RESOURCES

Figure 1.2 shows the average consumption trend of various primary energy resources of the world but the trend differs from country to country due to varying available energy resources. The trend shows that about 80% of world's energy supply comes mainly from fossil fuels whereas in India, the share of fossil fuels is around 68%.

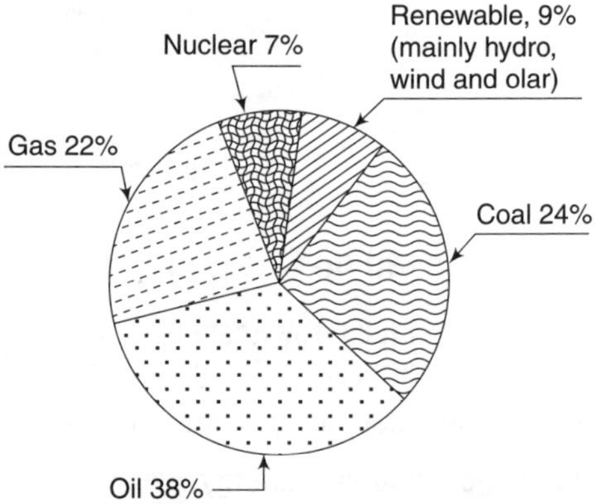

Figure 1.2 Percentage consumption of various primary energy resources

1.7 ENERGY CHAIN

In general (not always), we cannot use the energy from primary energy resources directly. For example, electric motor cannot be driven directly from coal, oil or uranium. The energy available from primary energy source is known as **raw energy**. This energy undergoes various forms of transformation before being utilised finally by consumers. The sequence of energy transformation between primary and secondary energy (usable energy say fuel) is called energy chain or energy route.

These transformations are given below and are shown in Fig. 1.3. At present, about 30% to 40% world's energy supply is met through electrical energy route.

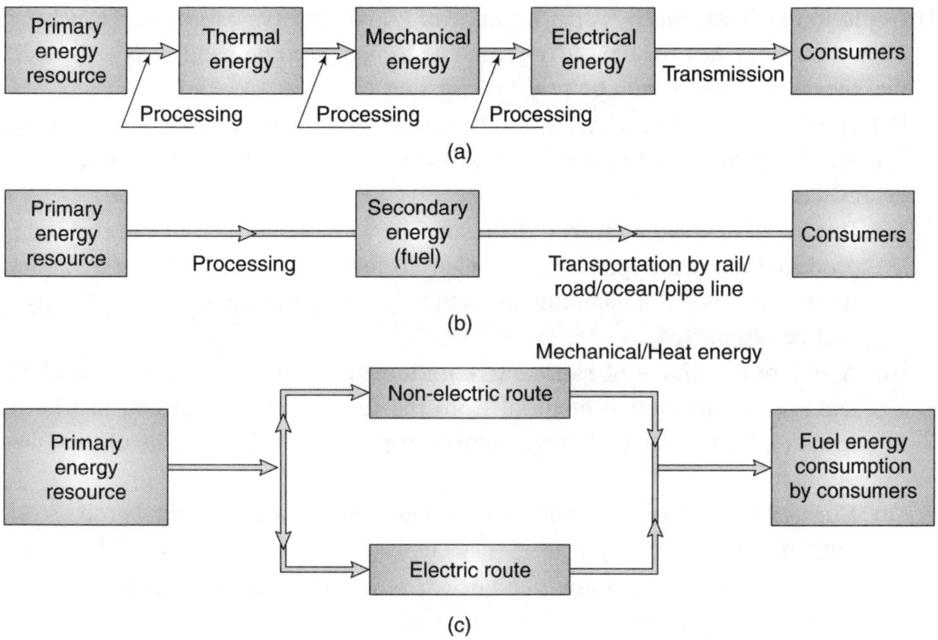

Figure 1.3 Energy chain

1.8 ADVANTAGES AND DISADVANTAGES OF CONVENTIONAL ENERGY SOURCES

1.8.1 Advantages

(i) Cost. At present, it is cheap as reserves are sufficient to last for about next 50 years. The approximate cost of energy from different sources at present is given as follows: ₹3.20 per kWh from gas, ₹3.30 per kWh from coal, ₹3.50 per kWh from diesel ₹2.20 kWh from hydro resources, ₹2.80 per kWh from nuclear sources ₹4.0 per kWh.

(ii) Security. As per the known reserve of fossil fuels, it may last for about next 50 years. By storing certain quantity, the energy availability can be ensured for a certain period.

(iii) Large energy demand from plant. The conventional energy resource can meet a large demand more than 3000 MW from a single plant.

(iv) Convenience. It is very convenient to use.

1.8.2 Disadvantages

(i) **Generation of pollutants.** Fossil fuels generate pollutants such as CO_2, CO, NOx, SOx, particulate matters, smoke and heat. These pollutants are so harmful that they degrade the environment, cause serious health problems, greenhouse gas effect, acid rain, etc.

(ii) Depletion of fossil fuels. As per estimate of known reserves, fossil fuels such as coal, oil and gas may last for next fifty years due to increased demand. It is very necessary that these fossil fuels may be reserved for future.

(iii) Safety of nuclear plants. A fool-proof safety of nuclear plants is a must. However, it is still a controversial subject. The following major problems arise due to nuclear energy.

(a) Radioactive waste material disposal. The waste material (nuclear fuel rods, etc.) produced in nuclear plant has a dangerous level of radioactivity for a long period. Its safe disposal is a daunting task and very costly failing which the locality/space will be radioactive.

(b) Accidental leakage of radioactive material. There is a possibility of accidental leakage of radioactive material from the reactor as it happened in Chernobyl, former USSR in 1986. A large number of persons died and some of them became disabled.

(c) Limited resources of uranium fuel. The uranium fuel reserves in the world are limited. The processing is also very costly.

(d) Sophisticated technology. The nuclear power plant technology is very sophisticated and only a few countries have acquired this.

Though, USSR, France, Germany, UK and USA bank upon nuclear energy ranging from 20% to 60% for their requirements but Sweden decided to ban it.

(iv) Problems associated with hydroelectric plants. Though, hydroelectric plants are the cleanest but the large hydro-reservoirs cause the following problems:

(a) Deforestation. Due to submergence of large land areas into water, deforestation starts.

(b) Ecological disturbance. It causes ecological disturbance leading to earthquakes.

(c) Dislocation of large people. It causes dislocation of large people which creates rehabilitation problems on a large scale.

1.9 ENERGY DENSITIES (HEATING VALUES OF VARIOUS FUELS)

Energy density or heating value of a fuel is defined as the amount of heat released per kg or per m^3 of fuel. Table 1.2 shows energy densities (heating values) of various fuels.

Table 1.2 Energy density (heating values) of various fuels

Primary resources (fuels)		*Energy density (MJ/kg)*
Coal	Anthracite	32–34
	Bituminous	26–30
	Coke	29

Contd.

Table 1.2 Contd...

Brown Coal	Lignite (old)	16–24
	Lignite (new)	10–14
	Peat	8–9
Crude Petroleum		45
Petrol		51–52
Diesel		45–46
Natural gas		50 (42 MJ/m^3)
Methane (85%)		45 (38 MJ/m^3)
Propane		50 (45 MJ/m^3)
Hydrogen		142 (112 MJ/m^3)
Wood		10-11
Natural Uranium		0.25-0.3 × 10^6
Enriched Uranium		2.6-3 × 10^6
U^{233}		83 × 10^6
U^{235}		82 × 10^6
Pu^{239}		81 × 10^6

1.10 SALIENT FEATURES OF NON-CONVENTIONAL ENERGY RESOURCES

World over, there is a great concern for environmental pollution due to the ever increasing use of fossil fuels and rapid depletion of these resources. This led to the development of alternative sources of energy, which are renewable and environment friendly. *Now there is a need of* ***sustainable energy*** *which ensures constant supply of energy without or very less environmental pollution.* Following are the merits and demerits of non-conventional energy sources.

1.10.1 Merits

(i) **Plenty availability.** These are available in plenty in nature and are free of cost.

(ii) **Loss or no pollution.** These produce no or very less pollution and nearly, they are environment friendly.

(iii) **Inexhaustible.** These resources are inexhaustible.

(iv) **Low gestation period.** They have low gestation period.

1.10.2 Demerits

(i) **High harnessing cost.** Though non-conventional energy resources are available freely in nature but the cost of harnessing energy is high as these are available in dilute forms. The cost of solar cells has come down drastically and comparable to thermal energy.

(ii) **Low power density.** Table 1.3 shows the power densities of various energy resources. Obviously, the power density of non-conventional energy resources is very low. **It means, for getting 1 kW, a large area is needed.**

Table 1.3 Power densities of various energy sources

Energy	*Power density (MJ/m²)*
Non-conventional	
(i) Wind	< 3
(ii) Solar	< 0.2
(iii) Tidal	< 0.002
(iv) Wave	< 1.00
(v) Biomass	< 0.0002
(vi) Geothermal energy	< 0.00006
Conventional	
(i) Gas and through hot plate	100
(ii) Coal (1 cm thickness)	500
(iii) Electrical energy through power cable (1 sq.m of surface area)	1000

(iii) Uncertainty of availability. The energy flow depends upon various natural phenomena, such as cloudy day, low air wind velocity, etc.

(iv) Transporting difficulty. There is a difficulty in transporting this form of energy.

1.11 ENERGY PARAMETERS

In order to conserve fuels, it is necessary to adopt various measures for maximizing economic development with minimum energy consumption. This may be possible with the help of some energy parameters as explained below.

1.11.1 Energy Intensity

It is defined as energy consumption per unit of GNP (Gross National Product). For any country, it is desirable to have a lower value of energy intensity signifying less energy consumption and at the same time achieving higher production. Table 1.4 shows the energy intensity over the four years.

Table 1.4 Energy intensity over four years

Parameters	*Years*			
	2015	*2014*	*2013*	*2009*
Energy intensity (MJ per dollar of GDP)	5	5	5	6
Change, %	−4.62%	−0.57%	−4.06%	1.52%

Table 1.4 shows that energy intensity has marginally fallen over the year. The downward trend will only continue when the per unit energy consumption for the production of energy intensive raw materials like steel and aluminium, etc., is gradually reduced. *It is worth to note that the developed countries have reduced energy intensity resulting in less energy consump-*

tion and at the same time achieving higher production. It is astonishing that India's energy intensity is 9.5 times higher than Japan and double that of USA.

It is, therefore, necessary that using the tools of energy conservation and efficiency that energy intensity may be reduced.

1.11.2 Energy-GDP Elasticity

Energy-GDP elasticity is defined as the percentage growth in energy requirement for one per cent growth in GDP. The lower the value of elasticity the higher is the overall efficiency. **It is worth to note that the value of elasticity for developed countries ranges from 0.8 to 1.0 whereas for India it is around 1.2.** There are various ways to reduce the value of elasticity. These include high efficiency energy systems, less energy consuming machines, reducing T&D losses, energy conservation measures, etc.

1.12 ENVIRONMENTAL ASPECTS OF ENERGY

The following issues are involved in the environmental aspects of energy:

(a) Trade-off between energy and environment; (b) Ecology; (c) Greenhouse effect; (d) Global warming and its consequences; (e) Pollution; (f) Green power—a sustainable energy option. Green energy is the need of hour.

1.12.1 Trade-off between Energy and Environment

Literally things surrounding us is environment. Air, soil, water and plants are the main constituents of the environment and the nature has provided these to human beings in clean form. But due to manmade reasons, their quality is being degraded and it has come to such a stage that there is a great concern for them. The major reasons for the degradation of environment is the activities concerning energy generation and its utilisation. During energy conversion, using fossil fuels, there is a heat rejection to the surroundings along with pollutants which degrade the environment. The lives of all creatures depend upon the clean environment so every steps should be taken to preserve environment.

In view of the above, while meeting the increased demand, every effort should be made to adopt various measures to minimise the degradation of environment. *Thus, there must be a trade-off between energy and environment.* In order to create public awareness about conservation of environment, every year, **5th June** is observed as **"World Environment Day"**.

1.12.2 Ecology

Ecology is defined as the relationship between biotic (human beings, animals, birds, plants and vegetation, etc.) and abiotic environment (earth surface, water bodies and atmosphere). Under normal circumstances, the nature is equipped with self cleaning capability and recycles or renews its resources through various processes such as water cycle, carbon cycle, etc. and thus maintains a state of equilibrium. It is worth to note that the ecological balance gets disturbed when the human activities harming the environment exceed the limit.

1.12.3 Greenhouse Effect

The CO_2 emissions from various manmade activities such as power plants, transport, etc., produce a effect similar to greenhouse effect. *A greenhouse is an enclosure having transparent glass panels or sheets to grow plants especially in a very cold climate. The same glass panes behave differently for incoming visible (short waves) radiation and outgoing infrared (long waves) radiation.* These glass panels behaving as transparent body for incoming solar radiation allow entry of sunlight in house/chamber but behave *as opaque body for reflected infrared radiation from the earth's surface which prevents the exit of heat from the house.* Thus, greenhouse maintains a controlled and warm environment inside of the house for growth of plants.

The emissions of CO_2 from manmade activities and the natural CO_2 present in atmosphere form an envelope around the globe in the atmosphere and behaves similar to glass panes and forms a big global greenhouse. As a result, the heat in the form of radiation emitted from the earth is prevented by this CO_2 envelope, which in turn, warms the earth. This phenomenon created by CO_2 envelope is known as greenhouse effect. More the CO_2 emission more will be warming of the earth which is known as global warming.

It is worth to note that the normal concentration of CO_2 in the atmosphere is around 0.03% which maintains an average-surface temperature of 15°C that is hospitable for life. In the absence of this layer, the earth would be a frozen planet at about –25°C. But if the concentration of CO_2 increases from this level obviously the temperature of the earth will increase. Apart from CO_2, other gases coming out from manmade activities such as CH_4 NO_x, etc., are called **greenhouse gases.** The main culprit for global warming is the developed countries which contribute 82% of total greenhouse gas emission of the world, so first they should go on fast track to reduce CO_2 emission drastically.

1.12.4 Global Warming and its Consequences

As mentioned above global warming is caused mainly by the emission of excessive CO_2 due to burning of fossil fuels, wood and agricultural products. In recent years, the rate of global warming has increased to the extent of 0.2°C per decade. This global warming has led to the melting of polar snowcaps resulting in the rise of sea level which will submerge the low lying areas and small islands, It is worth to note that already the ice cover in Arctic is down by 15% and in Antarctic by 25%. As per estimate, the earth temperature will increase by 4°C by the end of 21st century that will melt the major portion of ice in Arctic and Antarctic, which in turn, will submerge vast areas adjoining seashore. *There is a prediction that the global warming will result in more frequent and severe heat waves, more intense tropical cyclones, change in rainfall patterns, more floods and spread of epidemics.*

1.12.5 Pollution

As mentioned above, emission of harmful gases from the burning of fossil fuels has led to the pollution of the environment. Table 1.5 shows the comparison its emission among fossil fuels. The pollution may be indoor and outdoor.

Table 1.5 Comparison of emission (kg/TJ)

Pollutants	*Natural gas (kg/TJ)*	*Oil (kg/TJ)*	*Coal (kg/TJ)*
NO_x	43	142	359
SO_x	0.3	430	731
Particulates	2	36	1333

Indoor Pollution. In rural areas, a large number of conventional *chulhas* employing wood, agricultural products or coal are used which produce CO_2. This creates indoor air pollution. As per estimate, about 5 lakh children and women die from diseases every year caused by indoor air pollution. In order to reduce indoor air pollution, there is a need of high quality *chulhas* or shifting the LPG in rural areas.

Outdoor Pollution. It is mainly caused due to burning of fossil fuels in power plants, industries, transports, etc. Coal and oil produce more pollution than gaseous fuels.

It is worthwhile to discuss the harmful effect of various pollutants.

The various pollutants are CO_2, CO, SO_x, NO_x and particulate matter. The effects of every one are explained below:

(a) The effects of CO_2. CO_2, in general, is not considered a toxic gas. The presence of CO_2 in a certain percentage is essential for photosynthesis and production of essential oxygen and organic matter in nature. However, increase in (Fig. 1.1) concentration of CO_2 in atmosphere leads to global warming and its effects have been discussed above. It is worth to note that CO_2 in air during 1949 was less than 312 ppm and presently it is 350 ppm.

(b) The Effects of CO. Incomplete burning of fuel (carbon) is the cause of CO formation. It is highly harmful for human beings as it seriously impacts the oxygen dependent tissues in the body such as brain, heart and skeleton musceles. It is worth to note that CO concentration of 100 ppm causes headache, 500 ppm causes collapse and 1000 ppm is fatal. One can imagine the consequences of smoking as 400–450 ppm of CO are inhaled by smokers.

(c) The Effects of SO_x. In natural air, SO_2 is not present. It is the manmade actions which produce SO_2 due to burning of fossil fuels containing sulphur. The contribution of SO_2 from various sources is as follows:

- Power plants 70%
- Industry 15%
- Motor vehicles 8%
- Solid waste disposal 5%
- Other 2%

There is a further oxidation of SO_2 into SO_3, which in turn, forms sulphuric acid (H_2SO_4) when absorbed in water. The following are the harmful effects of the presence of SOx in air.

(i) It causes respiratory diseases like asthma and irritates eyes and respiratory track.

(ii) It causes acid rain (H_2SO_4) harming to agriculture, forest, vegetation, soil, stones and buildings.

(iii) It causes corrosion of metals, deterioration of electric contacts, paper, textile, building, stone, etc.

There is a safe limit of SO_x present upto 80 μg/m³.

(d) The Effect of NO_x. A large number of nitrogen oxides are generally referred to as NO_x which include N_2O, NO, NO_2, N_2O_3, etc. NO_x in the atmosphere is produced by two methods, one *is natural biological reactions and the other manmade.* 80% of NOx is produced by natural causes and 20% is produced by manmade causes due to combustion process in air at high flame temperature. The contributions of NO_x by manmade causes are as follows:

- Motor vehicles 7%
- Power plants 4%
- Industries 7%
- Solid water 2%

The following are the harmful effects of NOx:

(i) It causes respiratory and cardiovascular illnesses.

(ii) The tissues needing oxygen are deprived.

(iii) It forms acid in lungs and is more dangerous than CO.

The safe limit of NO_x is 100 mμg/m³ but air of most of the metropolitan cities has crossed this limit.

In **KYOTO protocol**, it was decided to reduce and control the greenhouse gas emissions by various technological advances.

1.12.6 Green Power—A Sustainable Energy Option

The sustainable energy or green power is the need of the hour. *It refers to energy produced and utilised without harming the environment through the pollution and ecological imbalance.* Renewable energy such as solar, wind, tidal and small hydro plant come in the category of green power. If we remove/separate the harmful gases due to combustion of fossil fuels that may also come in the category of green power but **cleaning process is very costly**.

1.13 WORLD ENERGY SCENARIO

The world is passing through energy crisis and global warming. There is a vast gap in the consumption of energy between developed, developing and poor countries and so is the quality of life.

The present annual primary energy consumption in the world is about 12×10^3 kWh/year. (405 Exajoules). About 40% of this energy is consumed in transportation and 60% by industries, domestic and social consumers. World energy demand is increasing at an average annual growth rate of about 3.5 to 5% and it is difficult to meet.

Table 1.6 shows the world total energy consumption by region (1990–2015). The world's average lies at 2.1 kW and if it rises to 3 kW by 2100, the per capita world energy consumption will rise about 3 times above the present value.

Table 1.6 World total energy consumption by region in quadrillion (10^{15}) Btu.

Region	1990	1993	2010	2015
OECD				
North America	100.2	104.2	129.8	135.6
United States	84.4	87.3	104.7	108
Canada	11.0	11.7	16.5	17.6
Mexico	4.9	5.1	6.6	9.9
OECD Europe	62.0	62.6	77.5	81.5
OECD Pacific	22.9	24.0	32.6	34.3
Japan	18.1	19.0	26.2	27.2
Other OECD	4.8	5.0	6.3	6.7
Total OECD	185.1	190.7	239.8	251.4
Non-OECD				
EF/FSD	75.3	60.9	74.0	79.7
Former Soviet Union	58.7	51.6	55.5	59.6
Eastern Euro	16.6	13.6	18.5	20.1
Non-OECD Asia	51.6	60.3	126.6	150.8
China	27.0	31.6	66.7	81.8
India	8.0	9.1	19.5	22.5
Other Asia	16.6	19.6	40.4	46.6
Middle East	11.3	12.2	18.1	19.9
Africa	10.6	10.3	14.5	16.0
Central and South America	14	19.7	22.1	24.6
Total Non-OECD	162.7	158.3	255.3	291.0
Total World	**347.8**	**349.1**	**495.1**	**542.23**

Modern life is much dependent on electric power. The per capita consumption of electricity is an index of prosperity of a nation. Table 1.7 gives the per capita electricity consumption of selected countries.

Table 1.7 Per capita electrical energy consumption (kWh/year)

Country	*Per capita energy consumption (kWh/year) March 2009*
Norway	23,000
Austraila	19,500
Canada	18,400
USA	15,000
Sweden	13,800
UK	8300

Contd.

Table 1.7 Contd...

Japan	9000
Brazil	2250
China	1600
India	700

1.13.1 Present Availability of Conventional Sources

(i) Fossil fuels. Fossil fuels such as coal, oil and gas are formed in several parts of the earth at varying depths, during several million years by slow decomposition and chemical actions of buried organic matter under favourable pressure, heat, and bacterial marine environment. The known reserves of these are not going to last for ever. Presently, the world is passing through the peak period of fossil age and decline will start after a few centuries. For example, coal will decline after 250 years while oil and gas will decline after 50 years and 70 years respectively.

Table 1.8 shows the world resources in fossil fuels. The projected demand by the end of 21st century could be 1.17×10^{15} kWh which is not possible to meet by conventional sources.

Table 1.8 World resources in fossil fuels

Fuel	*Consumption kWh*
Coal	5.27×10^{15}
Natural gas	0.618×10^{15}
Oil	0.996×10^{15}
Uranium	23.4×10^{15}
Thorium	14.0×10^{15}
Total	44.28×10^{15}

Table 1.9 shows the known energy sources available in India and the world.

Table 1.9 World energy sources [1 TWh = 10^9 kWh]

Energy Source	*World*	*India*
Coal (10^9 t)	7500	190
Lignite (10^5 t)	2000	2.064
Water power (TWh/years)	5000	218.5
Oil (10^6 t)	256000	140
Natural gas (109 m^3)	15000	63.7
Uranium (10^3 t)	5000	73
Thorium (10^3 t)	2700	500

(ii) Hydro Resources. *Among renewable sources of energy, hydro power is the most established source of electric power due to water cycle on the earth.* The early generation of electricity from hydro was made possible in 1880. From then world over, a large number hydro generating stations have been installed. As a result of huge capital requirement and strong environmental concern, only about 20% hydro potential energy has been tapped so far. It is worth to note that worldwide installed generating capacity of hydropower is about 6,30,000 MW which accounts for about 23% of world total installed electric power generation Norway topped the list in deriving 90% of required electric power from hydro resource. It is worth to note that there is a hydropower plant located in Brazil at Itaipu which generates 12,000 MW through a dam of 7.7 km long built over river Pirana. The ecological imbalance due to big hydropower is very high.

(iii) Nuclear Resources. Nuclear fuels such as U^{235}, U^{233} and Pu^{239} are used in nuclear reactor in which nuclear fission takes place releasing a large amount of heat and converted into mechanical and electrical energy through turbine and generator. Among all nuclear fuels, only U^{235} is found in nature (0.71%) and U^{233} and Pu^{239} are produced from fertile material like Th^{232} and U^{238} respectively in fast breeder reactor. Figure 1.4 (a) and (b) shows these reactions.

From the known reserves in the world, the uranium reserve is very small and probably will last for 50 years. The content of natural uranium in uranium is just 0.1 to 0.5% and as a result, the fuel processing is a costly affair. Further, major sources of uranium are available in limited countries such as in USA, Canada and Australia.

By the end of 2020, there were about 440 nuclear power plants working all over the world which generated about 10% of world's electricity. In France, the percentage of electricity generated by nuclear energy is about 70%. USSR also produces 40% of electricity by nuclear energy.

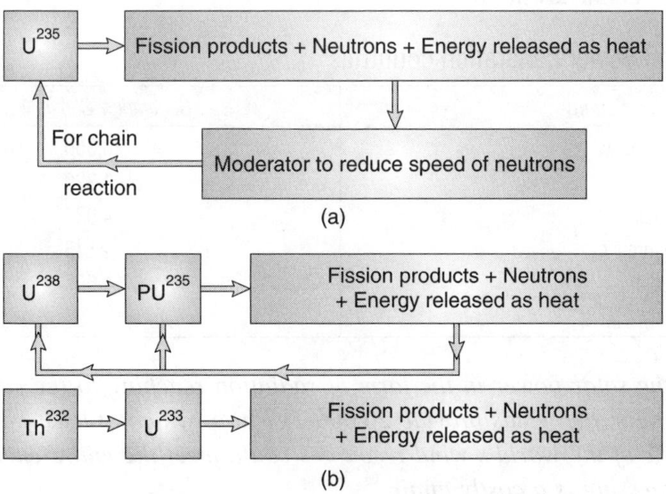

Figure 1.4 (a) Fission reaction of U^{235}. (b) Breeder reaction of U^{238} and Th^{232}

Nuclear fusion reaction has a lot more potential as vast resources are available and radioactivity effects are also very less, but commercially it has not been successful as yet. If commercial nuclear fusion is a reality, then it will be the main source of energy to the world.

1.13.2 Availability of Non-Conventional (Renewal) Sources

For sustainable energy, non-conventional energy resources are the option. Though, their contribution is less at present but in future, the major development will take place in this field.

(i) Solar Energy. Solar energy is a renewable form of inexhaustible energy for all living creatures on the earth. This energy can be utilised by using thermal and photovoltaic conversion systems. Solar radiation, known as solar constant, in about is 1367 W/m^2 outside the atmosphere but the average energy received on the earth is about 800 W/m^2. It is worth to note that the earth continuously intercepts solar power of 178 billion MW daily which is about 10,000 times the worlds demand of electricity but so far it has been harnessed in a very small percentage. It is estimated that if all the buildings of the world are covered with solar PV panels, the whole requirement of world's energy will be met. But, the solar PV power is till date comparable in tariff. *At present, the capital cost of solar PV system is ₹5.5 crore/MW as compared to ₹4 crore/MW for coal based power plant. It is expected that with further research in PV system, the cost will come down.*

Solar energy is already being used economically in many countries including India both for domestic and commercial purposes such as water heating, water distillation, refrigeration, drying, electricity generation, etc.

(ii) Wind Energy. Wind power has been used in many countries to sail vessels, pump water, and grind wheat and corn. At the turn of the present century, nearly 30,000 house mills capable of producing a total of 100 MW and 3000 industrial wind mills generating an another 100 MW will be in operation in Denmark. Table 1.10 shows power generating capacity by wind turbine in selected countries in 2010.

Table 1.10 Wind power generation countries

Country	*Generation Capacity (MW)*
Germany	10,650
USA	1,329
Spain	4,039
Denmark	2,515
India	1,507
Italy	755

About 2% of the solar power in the form of radiation reaching to the earth is converted into wind. Every year, the winds provide 26,000 TkWh, of which at least one-third is recoverable. A mere 1% of the world's wind resources could meet the entire energy needs of the world but its harnessing is a costly affair.

(iii) Biomass Energy. Energy resources available from animals and vegetation fall under the category of biomass energy. These are renewable energy sources and it is supposed that biomass energy resources do not produce any pollution as CO_2 released due to combustion of plants is absorbed by plants which are renewable in nature. These are very important sources of energy for developing countries. The biomass energy resources include the following:

(a) Trees (wood leaves and forest industry waste)
(b) Plants (stem leaves and roots)
(c) Algae and other vegetation from ocean and lakes
(d) Urban waste (municipal and industrial waste)
(e) Rural waste (agricultural and animal waste, crop residues).

It is worth to note that plants and trees grow absorbing solar energy through photosynthesis process which is estimated to be around 2×10^{18} kJ. Further, biomass resources may be used to produce biofuels such as biogas (methane), producer gas, ethanol, biodiesel, etc. World over there are millions of biogas plants especially in China and India.

(iv) Geothermal Energy. Geothermal energy is primarily energy from the earth's own interior in the form of internal energy. Vast natural heat stores lie beneath the earth's crust. It is recoverable in the form of steam or hot water. These are used for electricity generation, space heating, cooking and medicinal purposes. The main locations of geothermal energy are USA, Italy, Japan and New Zealand.

By the year 2020, word's total installed power generating capacity from geothermal resources was about 15,608 MWe and direct thermal use installed capacity was around 17,000 MW. The growth rate of geothermal power is around 7%. *It is worth to note that island of Hawaii generates 25% of its electricity from geothermal resources.* The oldest geothermal electric power station, located at Lordasello, in Italy and commissioned in 1904 is presently producing 460 MW. In EL-Salvador is producing 30% of total electricity requirement by geothermal energy. It is estimated that if the earth's crust is cooled by 1°C, the whole world will be getting energy for 20,000 to 50,000 years but nobody knows about the ecological imbalance due to this.

(v) Ocean Thermal Energy Conversion. The ocean thermal energy conversion plant converts the thermal energy (due to difference in temperature between surface and deep sea) acquired from solar radiation into electrical energy. The system has been developed now. Its potential is very high but there is a chance of disturbance of eco system.

(vi) Ocean Wave Energy. It is also in an infant stage. Some small prototype devices have been tested. The resource is more concentrated in deep sea which is difficult to harness. The estimated potential of this energy is around 2×10^6 MW.

(vii) Ocean Tidal Energy. *Tides are the joint action of sun and moon on the earth.* The potential is estimated to be very high and of the order of 1.20×10^3 MW. There are only a few working plants world over. The first and the largest tidal plant built in 1996 by France at the mouth of La Rance river near St Malo on the Brittany coast produces 240 MW. A 400 MW plant is located at Kislaya Cuba. In Canada, there is a 20 MW plant at Nova Scotio.

1.14 ENERGY SCENARIO IN INDIA

1.14.1 General Aspects Concerning Population, Per Capita Energy Consumption and Electricity Generation and its Planning

Energy consumption differs greatly between developed and developing countries. India being a developing country is struggling hard to meet the demand and the shortfall between demand and supply is huge (up to 30%). India's per capita consumption of total energy is 638 KGOE (kilogram of oil equivalent). The annual per capita electrical energy consumption is just 700 kWh which is less than 1/6th of **global average level of 3680 kWh** and below even half of **Asian average of 1250 kWh**. Population wise, India is 1.36 billion people, which is 1/6th of world population and consumes just 1/6th of **global average electricity**, one can imagine the quality of life of Indian people. Table 1.11 shows basic data of India with energy consumption.

At present, the total installed capacity of electric power generation in India is 334.4 GW by January, 2018 out of which, thermal (coal and gas) contribution is 68.9% hydro contribution is 24%, nuclear 2.7% and non-conventional is 4.5%. The electricity generation planning during past few 5-years plans are given in Table 1.12. The national goal of electric generation by 2018 as 3,44,002 MW to ensure reliable and quality power to all citizens. India's electric power sector has a annual growth rate of 6% while the demand is increasing at 9%. The gap between demand and supply is huge and the annual peak shortfall is about 20%. This has resulted in load shedding, under frequent tripping and blackouts. But due to Covid-19 the situation will change.

Table 1.11 Basic data of India including energy consumption (2019)

S.No.	*Parameters*	*Data*
1.	Population (March 19)	1.36 billion (17% of world population)
2.	Economic Category	Moderate growth in industry and agriculture
3.	Urbanisation	34%
4.	Rural	66% (6 lacs villages)
5.	GDP per capita (ppp) GDP (PPP USD Trillion) 2018 GDP (absolute, USD Trillion) 2019	USD 7144 11.4 2.97 (nominal)
6.	GDP per capita 2019 in USD	USD 2044 while China 10,000 and USA 65,112
7.	Crop land	0.3% of word's crop land
8.	Drinking water reserve	2.5% of world reserve
9.	Share of industry in GDP	27%
10.	Share of agriculture in GDP	16%
11.	Share of services in GDP	57%

Contd.

Table 1.11 Contd...

12.	Growth GDP per capita in 2019-20	4.2%
13.	Household saving (% GDP) 19	9%
14.	No. of airports (2019)	126 while in China 200
15.	Road investment (BUSS) 9	100 while China 117
16.	Largest company by market cap (USD Billion) 19	RIL 122.8 while Petro China 148.53
17.	Largest Bank by market cap (USD Billion) 19	SBI 48 while ICBC China 1910
18.	Oil imports to total exports	60%
19.	Energy consumption Total annual per capita (total) (i) Traditional energy (ii) Commercial energy	 637.40 KGOE 52% 48%
20.	Sector-wise energy consumption: Industry Transport Domestic Other	 40% 9.5% 47.3% 3.2%
21.	Electricity generation (2019): Thermal (coal and gas) Renewal Nuclear	360,788 MW while China 668200 GWh 79.8% 17.3% 2.9%
22.	Coal, oil and gas reserves	Coal 7% of world's reserve and oil and gas: 0.5% of world's reserve
23.	Electrical consumption kWh per capita (2019) Global average Asian average	1181 kWh, while in USA; 12,071 kWh 2674 kWh 2355 kWh

Table 1.12 Energy Planning in India under 5-year plans

5 year plan	*Demand*	*Installed capacity*	*Shortage*
6th (1980–85)	52,000 MW	47,000 MW	5,000 MW
7th (1985–90)	75,000 MW	65,000 MW	10,000 MW
8th (1990–92, 1992–97)	1,05,000 MW	85,000 MW	20,000 MW
9th (1997–2002)	1,32,000 MW	1,10,000 MW	22,000 MW
10th (2002–07)	1,46,000 MW	1,32,000 MW	14,000 MW
09-01-2019	3,72,693 MW	3,60,788 MW	11,905 MW

For uniform development of India electrification of rural areas is very important. Out of 6 lacs villages, 1 lac villages are yet to be electrified. Now about 98% electrified.

1.14.2 Availability of Commercial Primary Energy Resources

1.14.2.1 Geographical Distribution

Geographical distribution of available primary commercial energy sources in India is quite uneven. About 70% of the total coal reserves are located in the eastern region. Crude oil and natural gas are found in west, south and north-eastern parts. About 77% of the hydro-potential is located in northern and north-eastern regions. Geographical distribution of various energy resources in India is shown in Table 1.13.

Table 1.13 Primary commercial energy resources in India (Planning Commission)

Region of India	*Coal (Bt)*	*Lignite (Bt)*	*Crude oil (Mt)*	*Natural gas (B.c.m)*	*Hydropower (TWh)*
Northern	1.06	2.51	0.03	0.00	225.0
Western	56.90	1.87	516.47	516.42	31.4
Southern	15.46	30.38	45.84	80.94	61.8
Eastern	146.67	0.00	2.19	0.29	42.5
North-eastern	0.89	0.00	166.17	152.00	239.3
Total	220.98	34.76	733.7	749.65	600.00

Bt: Billion tonnes. B.c.m.: Billion cubic meters, TWh: Trillion Watt hours; Mt: Million tonnes

1.14.2.2 Primary Commercial Energy Production and Demand

The primary commercial energy demand grew almost threefold at an annual rate of 65% between 1981 and 2001 to reach 314.7 Mt of oil equivalent (MTOE). For the next decade will be the highest.

The average annual world economic growth in the period 2000–2020 is projected at 3.2% while the energy growth rate is estimated to be 2.1% per annum. Table 1.14 shows the production, consumption and demand in India.

Table 1.14 Primary commercial energy production and demand in India

Sources of Energy	*Unit*	*Production Year 2006-07*	*Demand year*	
			2006-07	*2011-12*
Coal	Mt	405.00	460.5	620.00
Lignite	Mt	55.96	57.79	81.54
Crude oil	Mt	33.97	134.5	172.47
Natural gas	B.c.m	37.62	47.45	64.00
Hydro power	BKWh	103.49	148.08	215.66
Nuclear power	BKWh	19.30	23.15	54.74

The difference in demand and supply can be met by imports, efficient energy management and exploitation of renewables. But at present, demand and production have increased.

1.14.2.3 Conventional Resources

(i) Fossil Fuels. India has vast reserves of coal, about 221 billion tonnes (Table 1.11). As per estimate, Indian reserves is 7% of world's reserves, 3rd largest in the world after China and United States. With the present mining rate of consumption, India has enough coal for 250 years. The worst aspect is that Indian coal has high ash content (25 to 50%), low heat value (3000 to 4000 kJ/kg) and low sulphur content (1%). Crude oil and natural gas reserves (Table 1.11) are 733.7 MT and 749.65 b.c.m. respectively and this has been estimated to be 0.5% of world reserves. As per present consumption, known oil and gas reserves could last only for 40 and 50 years respectively.

Indian oil and gas reserves are insufficient even for transport. *It is worthy to note that only 37% of requirement of oil and gas is met by domestic production and the balance 63% is met by imports costing huge expenditure.* The oil reserves of Assam and Bombay High have started giving less production. However, new findings of gas and oil reserves in Rajasthan and Krishna–Godavari basin off the Vishakhapatnam Coast may bring new hopes, if the actual total production starts.

(ii) Hydro Resources. India is endowed with an enormous hydropower potential which is assessed by Central Electricity Authority as 1,48,701 MW. Hydropower potential developed till 2019 was 156 TWh, which is only *about 38.71% of total potential.*

Table 1.15 gives the status of Hydroelectric Potential Development (as of 31.03.2008).

Table 1.15 Status of hydroelectric potential development (as of 31.03.2008 excluding less than 15 MW capacity) (More capacity has been added after 2008)

Region	*Identified capacity (MW)*	*Capacity Developed (MW)*	*Capacity under construction (MW)*	*Capacity yet to be developed (MW)*
Northern	53,395	11,070	7,959	34,366
Western	8,928	5,172	980	2,776
Southern	16,458	8,964	867	6,627
Eastern	10,949	2,384	1,012	7,553
North-eastern	58,971	1,095	2,934	54,942
All India	**141,971**	**28,685**	**13,752**	**1,06264**
All India	on 31, March, 2020	46,000 MW	–	–

India has a big potential (15,000 MW). Small (1 MW to 25 MW) hydel plants in Himachal Pradesh, Jammu & Kashmir, Uttarakhand, Haryana, Sikkim, West Bengal and Arunachal Pradesh. These projects can be set up on rivers and streams where generated power can be supplied to nearby villages. The installed capacity of such projects was 705 MW at the end of 2004. Hydropower is environment friendly. It involves no fuel cost and is renewable. Table 1.16 gives the list of some hydroelectric power plants in India generating above 100 MW power.

Table 1.16 List of hydropower plants generating above 100 MW in India

S.No.	Name of the power plant	Capacity, MW
1.	Mackhund	114.75
2.	Upper Sileru	120.00
3.	Bhakra, Kotla, Ganguwal	1204.60
4.	Sabirigiri	3000.0
5.	Gandhi Sagar	115.0
6.	Rana Pratap Sagar	129.0
7.	Koyan	540.0
8.	Bhira	132.00
9.	Jog	120.00
10.	Sharavati	534.0
11.	Hirakund (1 + 1)	270.0
12.	Kundah	445.0
13.	Mettur	240.0
14.	Periyar	140.0
15.	Rihand	300.0

(iii) Nuclear Resources. The condition of nuclear energy in meeting the total energy is just 2.7%. At present India's installed nuclear capacity for power generation is 3900 MWe. There are 14 operating nuclear reactors, out of which 12 are indigenous **Pressurized Heavy Water Reactor (PHWR)** based on natural uranium fuel which is mined from own reserves of about 78,000 tonnes. Two reactors at Tarapur are **Boiling Water Reactors (BWR)** dependent on enriched uranium which is imported from US, France, China and Russia. Further, there are two research reactors, namely, **Circus and Dhruva** which are mainly used to convert plutonium for nuclear weapons.

At present eight more reactors are under construction, out of which six are **PHWRs**. The other two are 1000 MWe **VVER** Russian reactors at Kundankulan (Tamil Nadu). These eight reactors will add a capacity of 3960 MWe which are now operational since 2009.

The founder of the country's atomic energy programme, Dr. Homi Jehangir Bhabha outlined a three stage programme for nuclear energy development in India as far back as 1955. These stages are discussed below:

Stage 1. PHWR reactors will be installed using natural uranium (0.7/U^{235} and 99.23% U^{238}) as fuel and heavy water as moderator and coolant.

Stage 2. Fast Breeder Reactors (FBR) will be installed in 2nd stage using spent fuel of PHWR. This FBR technology will be the basis for the generation of 5,00,000 MWe which is likely to provide energy security to the country.

Stage 3. In the third stage, reactors shall be built to use India's abundant reserve of thorium (about 4,50,000 tonnes) which is about 60% of world's thorium reserves. Thorium is available in the form of Monazite (ore) in sand beaches of Kerala. Thorium is a fertile material which is converted into a fissionable material U^{233} in a FBR. The advance Heavy Water Reactor (AHWR) will form the first stage of the third stage. The AHWR will use naturally available thorium as fuel and convert it to U^{233} which will then undergo fission to generate electricity. Enough progress in R&D activity has been made to develop the final phase of 3rd stage with the development of Compact High Temperature Reactor (CHTR) which will generate fission energy at 1000°C.

Now Nuclear Treaty with USA has been signed. By 2022, the contribution of nuclear energy will reach from 2.9% to 10%.

1.14.2.4 Non-conventional Energy Resources

Being located in tropical region. India is blessed with abundance of renewable energy resources such as solar, wind, biomass, hydro, geothermal, etc. As mentioned earlier, the main focus world over is towards renewable energy for energy security, so India is also striving hard to harness these energy resources through its various programmes. It is worth to note that India has potential to generate more than 10^5 MW from non-conventional energy resources.

At the end of 1999 the total contribution of renewable energy was 2414 MW. As of 31 March 2020, 35.86% of India's installed electricity generation capacity is from renewal sources, generating 21.22% of total utility electricity in the country. The country is aiming to achieve more ambitious target of 57% of the total electricity capacity from renewal sources by 2027 as per Central Electricity Authority's strategy blueprint. According to 2027 blueprint, India aims to have 275 GW from renewal energy, 72 GW of hydroelectricity, 15 GW of nuclear energy and nearly 100 GW from other zero emission sources. Government of India has planned to go for renewable resource in a big way to achieve 10% electricity generation by 2022, however, this is difficult to achieve.

Government of India has also set a target for installation of Rooftop Solar Projects (RTPs) of 40 GW by 2022 including installation on rooftops of houses.

As of October 2019 of the 175 GW interim target, 83 GW is already operational, 29 GW is under installation, 30 GW is under bidding and remaining 43 GW is under planning. 175 GW interim target is 100 GW of solar, 60 GW of wind, 10 GW of biomass and 5 GW of small hydro.

India has strong manufacturing base in wind power with 20 manufactures of 53 different wind turbine models of international quality up to 3 MW is size with exports to Europe, the U.S. and other countries. Wind or solar PV panel with four hour battery/storage systems is already cost-competitive without subsidy, as a source of dispatchable generation compared with new coal and new gas turbine plants in India.

Renewal energy overview and targets are as of 30 June, 2019 (Source: GOI site):

- Wind power: 36,368 MW (45.2% of total renewal)
- Solar power: 29,549 MW (36.7% of total renewal)

- Biomass power: 9896 MW (12.2% of total renewal)
- Small hydropower: 4604 MW (5.7% of total renewal)
- Waste-to-power 136 MW (0.2% of total renewal)

The installed grid interactive renewal power capacity (excluding large hydropower) as of 31 March 2020, is given in Table 1.17.

Table 1.17 Installed renewal grid power on 31 March 2020

Sources	*Total Installed Capacity (MW)*	*2022 Target (MW)*
Wind power	37,693.75	60,000
Solar power	34,627.82	100,000
Biomass power (Biomass and gasification & bagasse cogeneration)	9,875.71	10,000 (Including waste to power generation)
Small hydropower	4830.4	5,000
Total	**87,027.68**	**175,000**

Source: GOI Site

Table 1.18 The installed grid power capacity from all sources in India as of 31 March 2020.

Sources	*Capacity in MW*	*Percentage of Total (%)*
Coal	205,134.5	55.43%
Renewal except large hydro	87,027.68	23.51%
Large hydro	45,699.22	12.35%
Gas	24,955.36	6.74%
Nuclear	6,780.00	1.83%
Diesel	509.71	0.14%
Total	**370,106.46**	**100%**

Source: GOI site

1. Wind Energy. Government has a very big dream to harness wind energy on a very large scale. The programme was started in 1983–84. India has achieved a milestone by acquiring 5th rank in the world among countries having large installed capacity wind generator after Denmark, USA, Spain and Germany. The leading company of wind generator in India is Suzlon and it is providing expertise to other countries such as Turkey, Egypt, Philippines and Vietnam, etc. By 30 December, 2022, India could be able to generate just 60,000 MW. Most of the wind generators are located in coastal areas such Tamil Nadu, Gujarat and Maharashtra.

2. Solar Energy. India being a tropical country receives enough solar energy which can be harnessed. As per estimate India receives solar energy equivalent to 5000 trillion kWh per year which is more than the total energy consumption. The daily global radiation is around 5 kWh/m^2 per day and the bright aspect with India is that sun shine during 2300 to 3200 hours per year in its most parts. As a result, government of India has embarked upon to harness solar energy at a rapid rate, 100 GW by 2022.

There are three methods to harness solar energy as given below.

(a) Solar photovoltaic system programme; (b) Solar thermal energy system programme; (c) Space based solar power (new concept).

(a) Solar Photovoltaic System Programme. Solar photovoltaic (SPV) system is a highly developed and commercialized. Now it is so common that in remote villages also, the system has been installed. Indian government is promoting solar energy in the form of solar lanterns, home lighting systems, street lighting systems, solar water pumps and power plants. The total off-grid SPV installed in India was around 919.5 MW on 31 July 2019 and as a result India ranks fourth in the world in harnessing SPV system after Japan, USA and Germany. 100 kW and 200 kW SPV systems have been installed in some parts of the country.

(b) Solar Thermal System Programme. Solar thermal power plant is harnessed by three methods namely **solar concentrated system, distributed system** and **solar pond.** In India, now solar energy for water heating, cooking, drying and space heating through various schemes is in common use in order to promote the solar water heating system. Bangalore has been declared as a **solar thermal city.**

It is worthy to note that a project of 140 MW is under consideration under Integrated Solar Combined Cycle (ISCC) in Jodhpur district in Rajasthan. 35 MW will be generated by solar energy while rest 105 MW is to be generated by fossil fuel (regasified liquefied natural gas). A 100 MW solar power plant using solar concentrated system is under development. Reliance ADAG is going to install 500 MW of solar energy plants (PV and thermal systems) in India.

3. Biomass Energy. *India, being an agricultural country, has a large quantity of biomass in the form of dry waste agro-residues, fuel wood, twigs, etc. and wet waste like cattle dung, organic effluents, sugarcane bagasse, banana stem, etc.* It is worth to note that potential for generation of electric power from agro-residue and plantation is 16,881 MW while potential through cogeneration using bagasse is 5000 MW and that from waste is 2700 MW. **Energy farming** has a great potential in India through which bio-diesel is extracted from some plants (**Jatropha**, etc.). A large quantity of bio-diesel is being produced in India and *it can be mixed with fossil diesel from 10% to 20%.*

4. Small Hydro Resources. There are three types of hydro power plants under the banner of small hydro resources. They are small (less than 25 MW), mini (less than 1 MW) and micro (less than 100 kW). India has a potential of 15,000 MW out of which 2344.67 MW are being generated through more than 500 plants.

5. Geothermal Energy Resources. *Geothermal energy resources in India have a potential of about 10,000 MW, mainly in the form of hot water and very less in the form of steam.* Geothermal resources suitable for power generation have been located at Tattapani in Chhattisgarh and Puga Valley in Ladakh, Jammu & Kashmir. A 300 kW power plant is being installed at Tattapani. Hot water resources are mostly located in hilly region such are Badrinath, Kedarnath and other Himalayan regions but they are not utilized for power generation. However, these hot water springs are utilized for heating purposes at some locations.

6. Ocean Tidal Energy Resources. *India has reasonable potential from ocean tidal energy resources upto the magnitude of 9000 MW.* However, up till now India has not built any tidal power plant. Government of India has identified three sites for the development of tidal power plant, namely:

(a) Gulf of Kutch: Potential = 10^3 MW, with tidal range = 7.5 m
(b) Gulf of Cambay (Khambat): Potential = 7,000 MW, tidal range = 10.8 m
(c) Sundarban: Potential = 10^3 MW, tidal range = 3.9 m

Ministry of Non-conventional Energy Sources and West Bengal Renewable Energy Development Agency are jointly setting up 3.6 MW capacity tidal power plant at Durgadauxi creek in Sundarbans area of West Bengal and it is hoped that India will see such a power plant within five years.

7. Ocean Wave Resources. The average potential of ocean waves on annual basis for Indian coasts has been estimated at around 0.02 MW/m of wavefront. This is a reasonable potential to harness energy. A 150 kW pilot plant has been installed at Vizhingum harbour near Thiruvananthapuram, Kerala.

8. Ocean Thermal Energy (OTEC) Resources. *Ocean thermal energy works on the principle of temperature difference between upper surface temperature and deep sea.* India, being a tropical country possesses enough OTEC resource. A proposal is under consideration to install a OTEC power plant at Minicoy Island of Lakshadweep.

9. Fuel Cell and Hydrogen Energy. Fuel cell is coming up in a big way to meet the shortage of electrical energy. Hybrid system up to 40 MW plant in USA using hydrogen or natural gas in fuel cell will produce 80% of electrical energy from fuel cell and 20% energy from gas turbine using the exhaust of fuel cell. Hydrogen energy is supposed to a be clean energy. Both fuel cells and hydrogen energy are suited for stationary and portable power generation in transport. Many E-vehicles are working on hydrogen fuel-cell system in Japan.

In view of the enormous potential of fuel cells and hydrogen energy, Government of India has established a **National Hydrgoen Energy Board** to look after its various aspects such as *production, storage, delivery, application, safety issues,* etc.

1.15 ENERGY CONSERVATION

Energy is the backbone of modern civilization. The gap between energy demand and supply is widening. In order to reduce this gap, one of the novel ways is to conserve energy. *Energy conservation means reduction in energy consumption but without making any sacrifice of quality or quantity of production.* In other words, increasing the production from a given amount of energy input by reducing losses/wastage and maximizing the efficieny in the purpose of energy conservation. **Fourteenth December** is observed as **National Energy Conservation Day** and also **World's Energy Day.** The most important slogan in this respect is **"Energy Saved is Energy Generated"**.

1.15.1 Energy Audit

It is a buzzword in the field of energy conservation. It is an official, technical survey of energy consumption of an organization/process/plant equipment, aimed at reduction of energy consumption without affecting productivity and suggests methods for energy conservation and reduction in energy cost.

The various aspects of energy conservation include:

- Economic aspects
- Environmental aspects
- Depletion of non-renewable energy assets.

1.15.2 Principle of Energy Conservation

The following are some general principles of energy conservation.

(a) Recycling of waste
(b) Modernisation of technology
(c) Waste heat utilization (combined cycle)
(d) Proper housekeeping
(e) Judicious use of proper type of energy
(f) Judicious use of proper type of fuel
(g) Cogeneration
(h) Training of manpower
(i) Adopting daylight saving time

1.16 A BRIEF IDEA OF WORKING OF VARIOUS TYPES OF POWER PLANTS

A brief idea of working principle of various types of power plants with their simple sketches are discussed below.

A. CONVENTIONAL

1.16.1 Hydel Power Plant

Hydel power plant works on the principle of conversion of head (potential) energy of water into mechanical energy and then electrical energy. In hydel power plants, water is stored to a very high head by creating a dam on the river. This high head water passes through water turbines which generate mechanical power and generator mounted on the turbine shaft generates electrical energy (Fig. 1.5). Hydraulic efficiency ranges from 80 to 90%. It takes about 10 years to complete a hydel power plant.

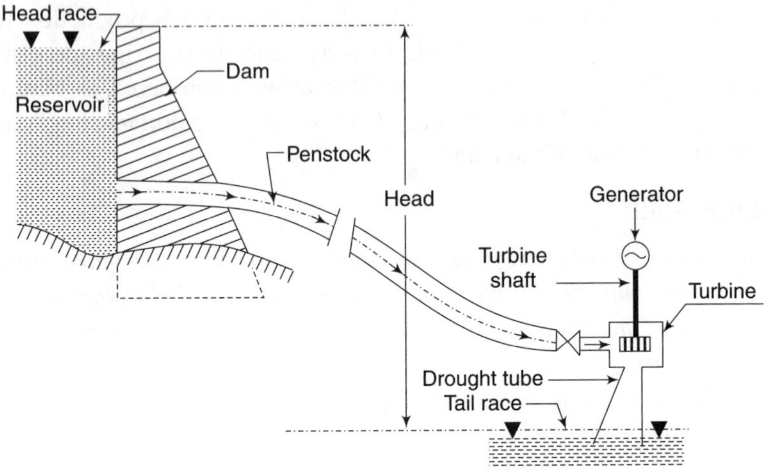

Figure 1.5 Schematic diagram of hydel power plant

1.16.2 Steam Engine

In steam engine (Fig. 1.6), now obsolete, steam generated in boiler expands in steam engine which develops power. The thermal efficiency is very low, around 15%.

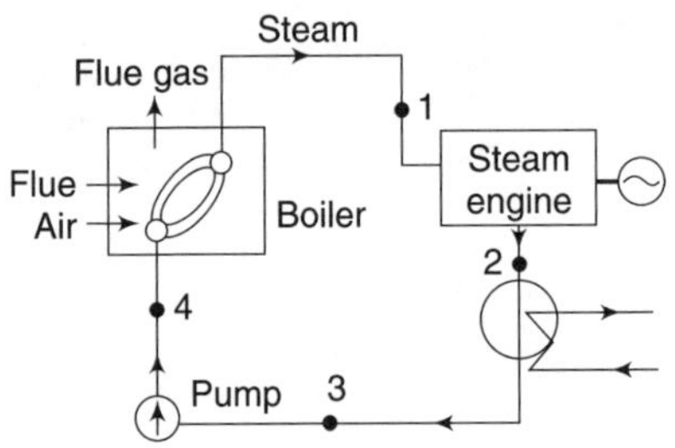

Figure 1.6 Steam engine power plant

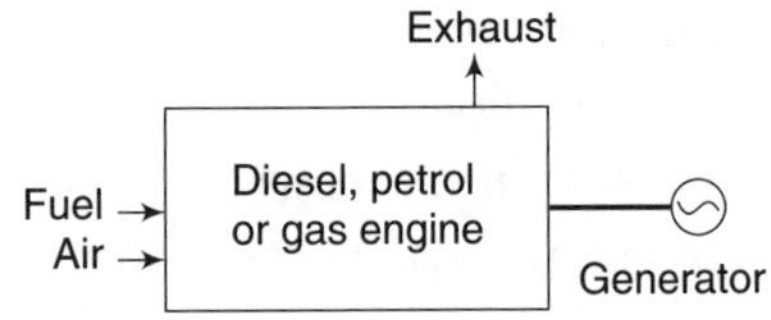

Figure 1.7 I.C. engine power plant

1.16.3 Diesel, Petrol and Gas Engine Power Plants

These are reciprocating internal combustion engines in which chemical energy of fuel is converted into heat energy due to combustion which gets converted into mechanical power and then electrical power (Fig. 1.7). The efficiency is around 32%.

1.16.4 Steam Turbine Power Plant

Steam power plant works on the principle of conversion of chemical energy of fuel into heat energy in the form of steam which expands through steam turbine and produces mechanical work. The mechanical power so produced is converted into electrical energy through electrical generator. The main elements of a steam power plant consists of boiler, steam turbine, con-

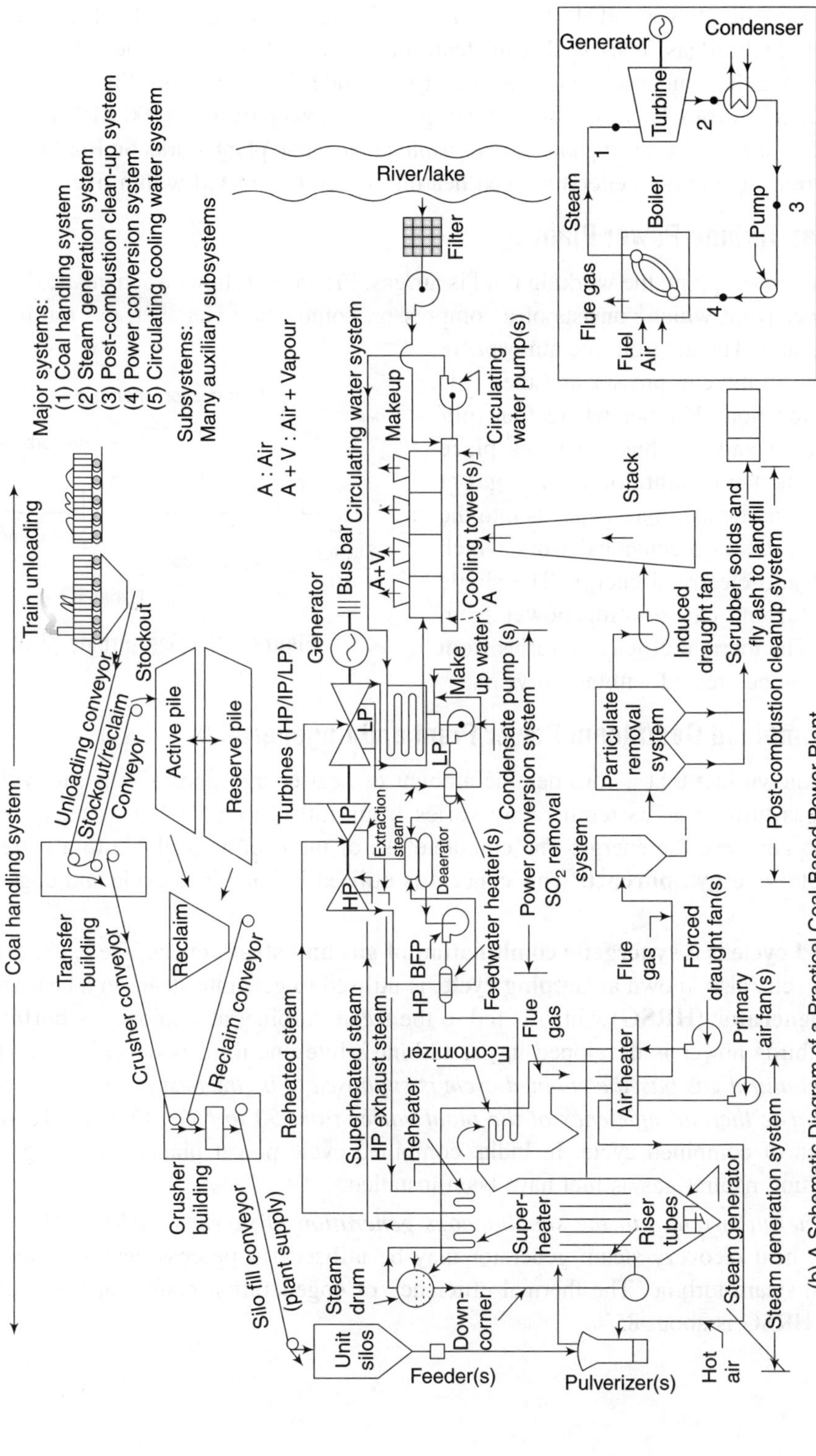

(b) A Schematic Diagram of a Practical Coal Based Power Plant

Figure 1.8

(a) Steam Power Plant

Figure 1.8(a)

denser, pump, cooling tower and electrical generator. Steam produced in boiler due to burning of fossil fuel (coal/oil/gas) is expanded in steam turbine and does work. The exhaust steam gets condensed in the condenser and is fed back to the boiler by feed pump (Fig. 1.8a, b). The thermal efficiency varies from 35–40%. It can generate power from 1-1100 MW through a single unit. *It is the most versatile and tested prime mover.* The plant works on Rankine cycle modified to reheating and regenerative feed heating. It can be erected within five years.

1.16.5 Gas Turbine Power Plant

In gas turbine power plant, the working fluid is air/gas. Figure 1.9 shows a schematic diagram of a gas power plant which consists of a compressor, combustion chamber, gas turbine and electric generator. The air from the atmosphere is compressed in the compressor and then it is led to a combustion chamber where fuel (oil/gas) is injected and combustion takes place which raises the temperature of air. The gas at high pressure and temperature expands into the turbine and produces mechanical work which is converted into electrical energy. The single unit of gas turbine can develop power from 1–350 MW. The thermal efficiency varies from 35–47%. It can be erected within two years.

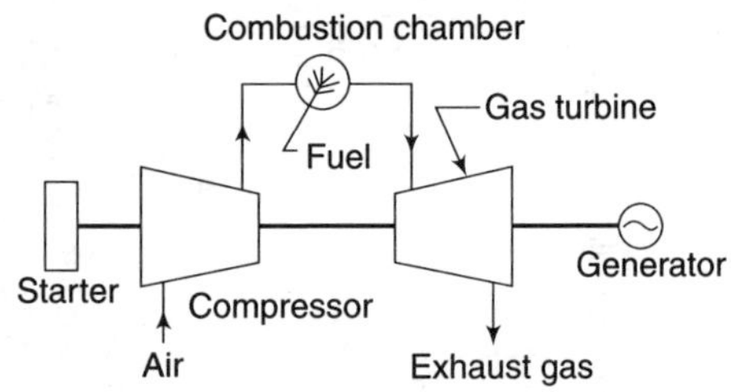

Figure 1.9 Gas turbine plant

1.16.6 Combined Gas/Steam Power Plant and Cogeneration

It is a well-known fact that a considerable amount of heat energy goes as a waste with the exhaust of gas turbine as its temperature varies from 450°C to 600°C. This energy must be utilised to conserve the energy. The complete use of the energy available to a system is called the **total energy approach**. This concept is utilised in combined cycle and cogeneration plant.

Combined cycle is a synergetic combination of gas and steam cycle. The exhaust from gas turbine cycle, also known as **topping cycle,** is utilised to generate steam in a heat **recovery steam generator (HRSG)** which is fed to the steam turbine, also known as **bottoming cycle.** Two-thirds power is developed in gas turbine while one-third is developed in steam turbine. *The natural gas based combined cycle is supposed to be the best prime mover of the 21st century. The thermal efficiency of the plant varies from 50 to 60%.* Figure 1.10 shows simple sketch of combined cycle. In India, combined cycle power plants producing about 4000 MW using natural gas as fuel have been installed.

The cogeneration refers to the simultaneous generation of power and heat. The steam generated in heat recovery steam generator may be utilised for process heating instead of expansion in steam turbine. The thermal efficiency of cogeneration plant consisting of gas turbine and HRSG is about 85%.

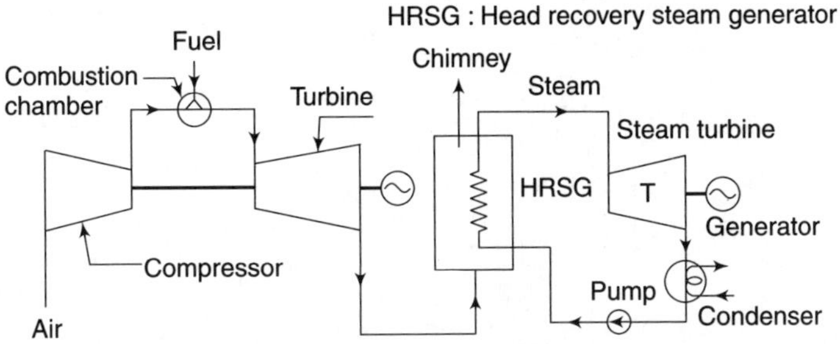

Figure 1.10 Combined gas/steam cycle power plant

1.16.7 Binary Vapour Cycle Power Plant

In a binary vapour cycle as the name implies, two working fluids, namely, mercury and steam are commonly used (Fig. 1.11). This also comes under the category of combined cycle. Mercury vapour from the mercury boiler expands in mercury turbine. The heat rejected in mercury condenser is utilised to generate steam which expands in steam turbine. The thermal efficiency of binary vapour cycle is more than simple steam cycle.

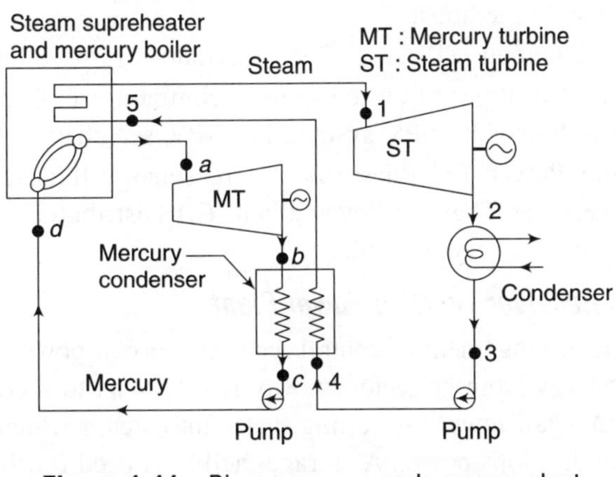

Figure 1.11 Binary vapour cycle power plant

1.16.8 Nuclear Power Plant

In the nuclear power plant, the boiler of the conventional steam power plant is replaced by nuclear reactor. The fuel used in the nuclear reactor is U-235 or isotopes. The nuclear reaction in the reactor involves splitting the nucleus (fission reaction) and is accompanied by a large amount of release of energy which may be expressed by famous formula $E = mc^2$ where m is the loss in mass and c is velocity of light. The heat so released is used to generate steam for expansion in steam turbine (Fig. 1.12). The cycle works on Rankine cycle.

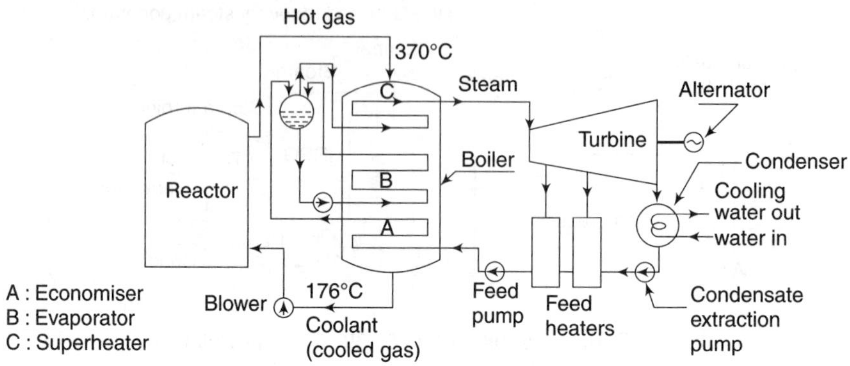

Figure 1.12 Nuclear power plant

B. NON-CONVENTIONAL POWER PLANT

1.16.9 Solar Energy Power Plant

Solar energy is a renewable form of energy and going to prove as one of the main attractive sources of energy to mankind without any harmful emission. The solar energy is harnessed by the following possible methods:

(a) Solar photovoltaic cell (SPV); (b) Solar thermal; (c) Space based solar power

Now both first two types of systems are in common use: in heating by solar thermal and lighting of homes by SPV system. For SPV, see chapter 18.

(b) **Solar Thermal Power.** To harness solar thermal energy, three technologies are available (i) Central Receiver Thermal Power Plant; (ii) Distributed Collector Solar Thermal Power plant, and (iii) Solar Pond.

1.16.9.1 Central Receiver Thermal Power Plant

Figure 1.13 shows an arrangement of central receiver thermal power plant. It optically transmits radiant solar energy from collector known as heliostats to a centrally located receiver (acting as boiler) on a tall tower converting water into steam which is fed to conventional steam turbine which develops power. A storage facility is used for the period when sun rays are not available at nights and cloudy days. It is capable to generate 100 MW by this system by generating steam at a pressure of 90 bars and temperature 510°C.

1.16.9.2 Distributed Collector Solar Thermal Plant

It converts radiant solar energy to heat at each local reflector site known as distributed collector system. Figure 1.14. shows the use of **a two axis tracking paraboloidal dish collector coupled to a steam Rankine cycle**. A provision of storage of thermal energy is provided by using a part of steam generated which is a must for all time operation. Solar pond is less common (see chapter 18).

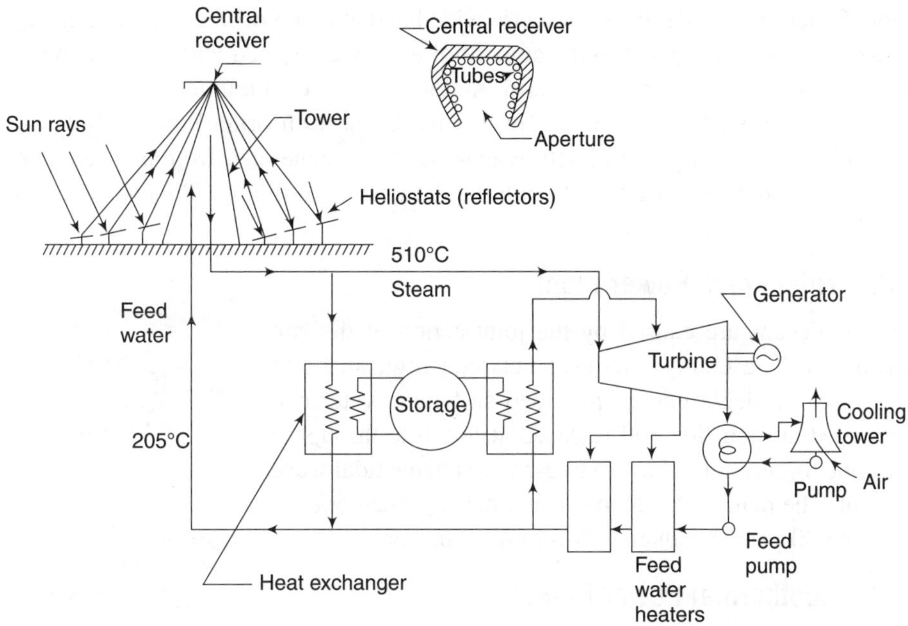

Figure 1.13 Central receiver solar thermal power plant

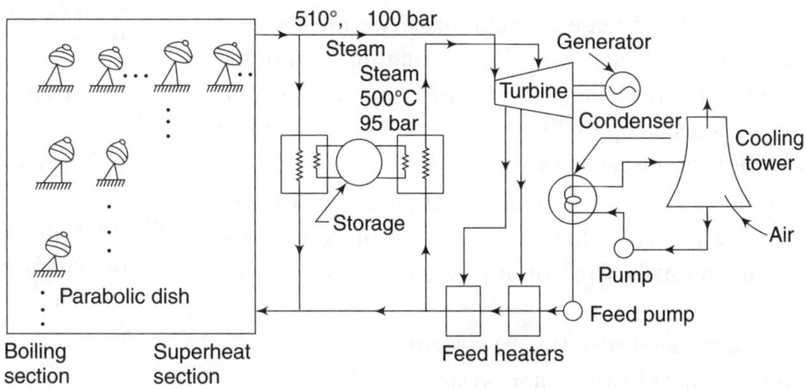

Figure 1.14 Distributed parabolic dish collector solar thermal power plant

1.16.9.3 Space Based Solar Power (SBSP)—A New Concept

Though solar energy is one of the most popular renewable and clean energy sources, the cost of tapping the same has always been comparatively more because of its diluted form, seasonable variation and non-availability for more than half of the day. SBSP overcomes this issue by placing the solar collector panels in a geostationary orbit in space and have the solar power beamed to the earth before converting it to useful electricity. Solar energy could be beamed to the earth as radio waves or coherent light (LASER beam). Just to understand the enormity of energy availability, a kilometre wide band at a geo-orbit experience enough solar flax in one year to nearly equal to the amount of energy contained within all known recoverable conventional oil reserves on the earth. And it is 99 per cent uninterrupted throughout the year. The research work is yet to start in this field. The California Power Utility agreed to buy 200 MW of electricity from 2016, which was to be generated from a space based solar station. SBSP is so promising that it could be the answer to the world's energy crisis if it is developed and commercialised.

1.16.10 Tidal Wave Power Plant

Tides in the oceans are caused by the joint action of the sun and the moon. The change in level between high and low tides produces available energy provided sufficient water can be impounded at high tide and released at low tide through a water turbine. Figure 1.15 shows a two-basin scheme tidal wave power plant. The plant at **La Rana in France** has been operating for some 30 years producing 240 MW of power.

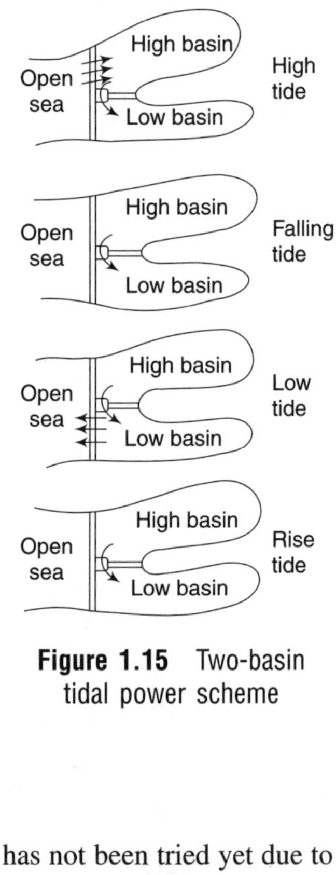

Figure 1.15 Two-basin tidal power scheme

1.16.11 Geothermal Power Plant

The heat of the earth's interior is the most intensive energy available to the mankind. Geothermal resources are derived from the distribution of temperature and thermal energy beneath earth's interior. It is has been estimated that if we merely cool the rocky crust under dryland by 1°C (ignoring input from the sun and the magma core) there would be enough energy to supply the world's needs up to 20,000 to 50,000 years. But this may create ecological imbalances.

Depending upon the type of geothermal energy resources, three systems are prevalent for the utilisation of geothermal energy. All of them finally work on Rankine cycle using steam or any vapour.

(i) Vapour dominated (dry steam) system
(ii) Liquid dominated (wet water) system
(iii) Dry (hot rock) geothermal formation (least exploited). It has not been tried yet due to the possibility of ecological imbalance.

1.16.11.1 Vapour Dominated System

In vapour dominated system the dry steam is extracted from steam well and is fed to the steam turbine (Fig. 1.16). Generally, the pressure and temperature of steam is around 8 bars and 200°C.

1.16.11.2 Liquid Dominated Hot Water System

Figure 1.17 shows a closed system geothermal power plant using hot water as the main heat source and a secondary fluid (isobutane) driving turbogenerator. In this plant, hot water is

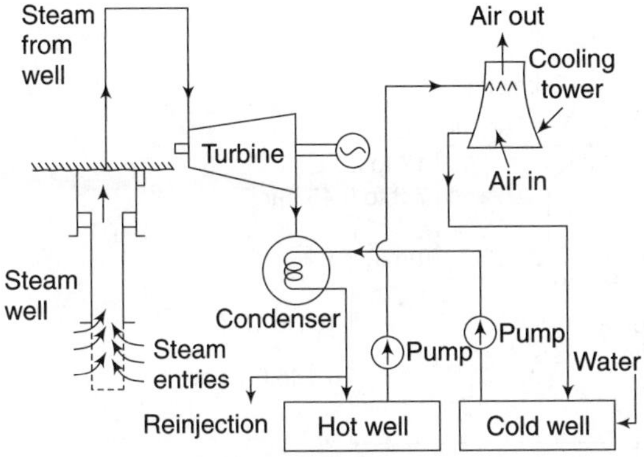

Figure 1.16 Geothermal power plant using dry steam

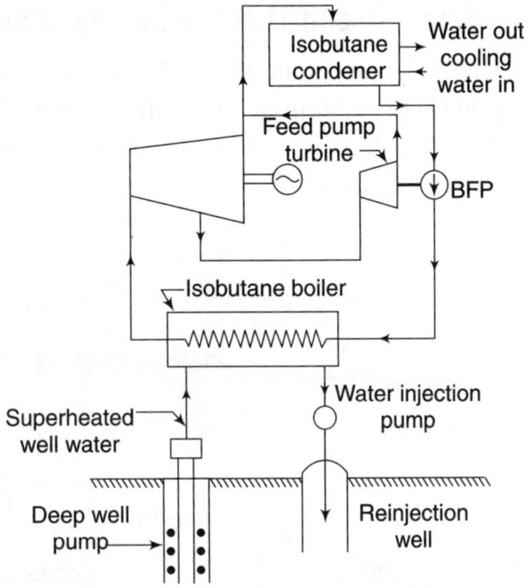

Figure 1.17 Geothelmal secondary vapour turbine cycle

pumped from the surface by a deep-well pump and in a heat exchanger an another fluid (isobutane) vapours are formed which expands in vapour turbine. In order to maintain the ground level water, the extracted water, is reinjected into well.

1.16.12 Wind Energy Power Plant

As discussed earlier, the wind energy resource on the earth, is enormous, however, some part of the world, it is more and somewhere less. Wind energy was used since 150 BC by Chinese. It may be used to pump water or to generate electricity. A wind mill or turbine is used to generate electrical energy (Fig. 1.18). The wind speed 6 m/s or above is desirable for a longer period of the day. The efficiency is around 20 to 30%.

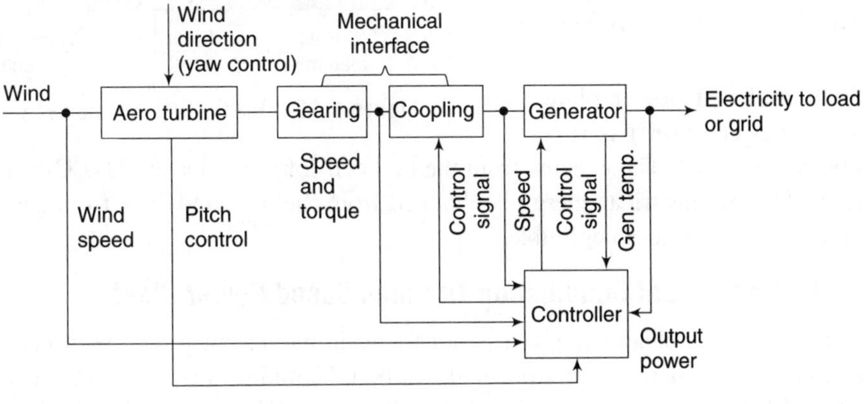

Figure 1.18 Wind electric system

1.16.13 Ocean Thermal Energy Conversion (OTEC) Power Plant

There is a great potential in OTEC to meet energy demand. Figure 1.19 shows a closed Ranking Cycle OTEC system in which the working fluid ammonia is vaporized by heat exchange with water and expands in NH_3 turbine and develops power (Fig. 1.19). Its conversion efficiency is very low about 3%.

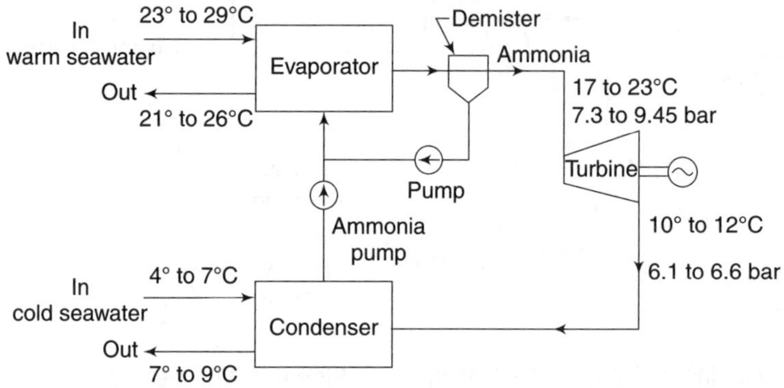

Figure 1.19 OTEC power plant

1.16.14 Fuel Cells Power Plant

A fuel cell is an electrochemical device which converts the chemical energy of a conventional fuel such as hydrogen, hydrocarbon, etc., directly into electrical energy. Figure 1.20 shows such H_2–O_2 fuel cell. It consists of an anode (non-consumable), a cathode then an electrolyte (invariant), fuel and oxidant supply system. The conversion efficiency is very high.

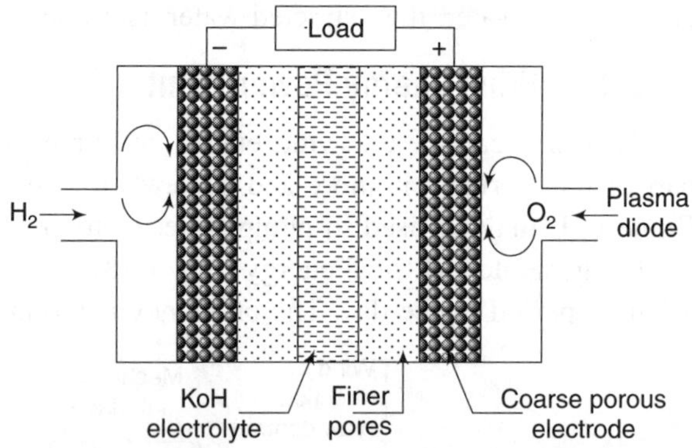

Figure 1.20 H_2–O_2 fuel cell power plant

A **hybrid fuel cell** power plant consisting of **fuel cell and gas turbine** has been developed. The exhaust from the fuel cell (of the order of 1000°C) expands in a gas turbine. 80% of electrical energy is obtained from fuel cell and 20% from gas turbine. The pollution is more or less negligible.

1.16.15 Hydrogen Fuel Combustion Turbines Based Power Plant

Hydrogen as a fuel is a clean energy. It is not found in nature but it can be produced from electrolysis process from water or from hydrocarbon. Thinking this, government of Japan, **Under a WE-NET programme,** is going to install a 400 MW plant based on **hydrogen fuel**

combustion turbine in 2020, which will be yielding **70% efficiency**. H_2 will be obtained from water through electrolysis process and O_2 will be obtained from air through liquefaction process. Figure 1.21 shows the schematic diagram of hydrogen fueled combustion turbine cycle which works on Rankine cycle. The design and production are going on. If this plant becomes successful, world's energy problem may be easier to solve without any pollution.

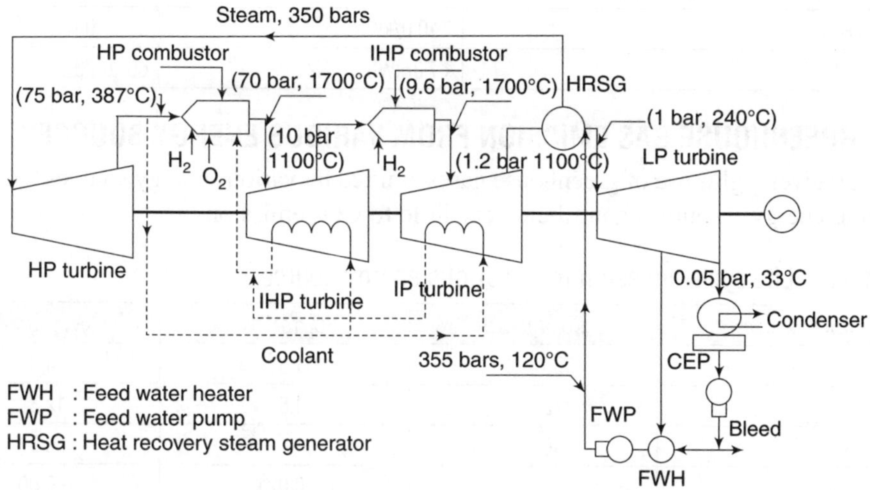

Figure 1.21 Hydrogen fuel combustion turbine based power plant

1.17 COST OF ELECTRICITY PRODUCTION WITH CONVERSION EFFICIENCY

Market penetration of any new technology depends upon its comparative economic and financial advantages. The cost details (based on 2019) of various energy sources are given below along with their conversion efficiencies in Table 1.19.

Table 1.19 Cost and conversion efficiency of various resources (based on 2018)

S.No.	Energy Resources	Cost/kWh	Conversion efficiency (%)
1.	Thermal	₹3.30/kWh + Ecological cost ₹3.20/kWh + Ecological cost	35 to 40 gas/steam plant 50 to 60 combined cycle
2.	Large hydro	₹2.20/kWh + Ecological cost	80 to 90
3.	Nuclear	₹2.80/kWh + Ecological cost	32 to 35
4.	Geothermal	₹4.30/kWh + Ecological cost	28 to 30
5.	Small hydro	₹3.50/kWh + Ecological cost	60 to 70
6.	Wind	₹3.30/kWh	20 to 30
7.	Biomass	₹3.50/kWh	30 to 32
8.	Biogas	₹3.0/kWh	30 to 32
9.	Solar thermal	₹4.50/kWh	30 to 32

Contd.

Table 1.18 Contd...

10.	Solar photovoltaic	₹4.5/kWh	70 to 75
11.	Fuel cell	₹10.0/kWh	75 to 80
12.	Hybrid fuel cell	₹10.50/kWh	70 to 72
13.	Hybrid fuel combustion turbine	₹8.0/kWh	70 to 72
14.	OTEC	₹10.0/kWh	03 to 04
15.	Tidal	₹7.0/kWh	60 to 70

1.18 GREENHOUSE GAS EMISSION FROM VARIOUS ENERGY SOURCES

Table 1.20 gives a glimpse of greenhouse gases emitted by various energy sources. Obviously, renewable energy resources contribute very little to CO_2 emission.

Table 1.20 Life cycle emission from various energy sources

Energy Sources	*CO_2 (g/kWh)*	*SO_2 (g/kWh)*	*NO_x (g/kWh)*
Coal	955	11.8	4.3
Diesel	772	1.6	12.3
Natural gas	430	–	0.5
Large hydro	11.6	0.021	0.006
Small hydro	09	0.03	0.07
Wind	09	0.09	0.06
Solar photovoltaic	167	0.34	0.30
Solar thermoelectric	38	0.27	0.13
Biomass	27	0.16	2.5
Geothermal	09	0.02	0.28

1.19 KYOTO PROTOCOL

Kyoto protocol refers to the agreement intended to limit the world's carbon emission which was signed by 84 countries in Kyoto, Japan in 1997. The Kyoto protocol calls upon 38 developed countries to cut their greenhouse gases (GHGs) emission between the years 2008 and 2012 to levels 5% below their 1990 levels. However, some developed countries, were allowed to have higher levels than 1990 levels such as Austria (105%), Iceland (110%), Norway (101%), New Zealand, Ukraine and Russian Federation (100%). **The worst aspect is that the USA, which creates 37% of global emissions highest in the world, refused to ratify the Kyoto protocol.**

In is worth to note that the International Agreement provides three other options to developed countries to earn **"Carbon Credits"** towards reduction in GHG emissions as given below:

- Clean Development Mechanism (CDM)

- Joint Implementation (JI)
- International Emission Trading (IET)

(i) **CDM.** It provides an opportunity to industrialized countries to earn carbon credits by investing in clean energy production or reducing emission projects in developing countries.

(ii) **EI.** It is an another approach for countries to earn carbon credits by investing in emission reduction projects in developed countries in agreement with Kyoto protocol.

(iii) **IET.** It allows buying and selling of carbon credits among the developed countries who are members in Kyoto protocol.

Kyoto protocol is a boon for the mankind if it is implemented in letter and sprit to solve the environmental challenges of global warming posed by emissions of greenhouse gases. It is to be noted that the dominant warming occurred in continental land mass between **40° and 70°N** latitudes and hence a special emphasis is to be given in this part of world.

Kyoto protocol was to become the international law for environmental agreement with two conditions, namely:

(a) 55 countries had to approve it by passing through their national legislation.
(b) Approving countries necessarily had to be industrial, so as to account for 55% of their global emission in 1990.

17th Feb. 2005 was the red latter day in the history of the world when Russia accorded the approval of **Kyoto Global Warming Pact.** According to this pact, the following measures have to be taken.

(i) Improving energy efficiency
(ii) Maximum use of renewable energy
(iii) Carbon dioxide sequestration
(iv) Passing out tax exemption and subsidies on all GHG emitting commercial energy sources.

As per this international law, it is hoped that global warming may be reduced. Now, Paris Agreement has also taken place.

1.20 SUSTAINABLE ENERGY OPTION FOR INDIAN ECONOMY

India has a population of about 136 crore people and GDP about USD 2.97 trillion (2019 year). With this population and average GDP, India cannot afford to follow the high energy consumption pattern of the West which resulted in global warming. For sustainable energy option government of Inida framed a National Electricity Policy (NEP) in 2005. The following are the guidelines of NEP:

(i) To increase per capita energy consumption from the present figure at 700 kWh/year to 1000 kWh/year by 2012 using efficient technologies to reduce pollution.
(ii) To add 1,50,000 MW extra electricity generation capacity during the next 10 years from the present value of 1,51,000 MW.

(iii) To accelerate development of renewable energy sources like solar photovoltaic, solar thermal, wind, biomass and small hydro in a big way.
(iv) To adopt biofuels up to 10% in energy mix to some fuels.
(v) To encourage captive power plants to supply their surplus to the grid and rural areas locally as distributed generation.
(vi) To cater for suitable margins and returns for investors in electricity sector.
(vii) To harness the untapped estimated hydropower of 1,50,000 MW at average load factor from the present level of 30,200 MW (by 2012).
(viii) To install thorium based nuclear reactors for generating 40,000 MW electricity and also from other nuclear fuels so as to bring the nuclear power contribution from 2.7% to 10% in the power grid. Nuclear power does not contribute to CO_2 emissions but leakage of radiation is a problem.
(ix) Now, many guidelines have been revised.

Now, there is a Paris Agreement to reduce CO_2 emission.

PROBLEMS

Things to Think About

1.1. What do you mean by primary and secondary energy sources?
1.2. Differentiate between conventional, and non-conventional energy sources.
1.3. What is the relation between energy production and environment?
1.4. What is greenhouse effect?
1.5. What are the measures that should be taken to stop global warming?

Short and Long Theoretical Questions

1.1. List the percentage share of various energy sources which contribute to energy demand.
1.2. What do you mean by renewable energy sources?
1.3. What are the advantages and disadvantages of conventional energy sources.
1.4. What do you understand by energy chain?
1.5. What is the importance of energy conservation?
1.6. Discuss the energy scenario of the world.
1.7. Discuss the Indian energy scenario.
1.8. What do you understand by Kyoto protocol
1.9. How the renewable energy sources may prove to be a boon in combating global warming?
1.10. Discuss in detail the causes of global warming.
1.11. What are the prospects of wind, solar and biomass in meeting energy demand.
1.12. What is the present status of nuclear energy and what are its future prospects?
1.13. With the help of simple sketch discuss the working principle of all conventional and non-conventional power plants in brief.

1.14. Discuss the various aspects of energy conservation.
1.15. Discuss the principle of energy conservation.
1.16. Classify the energy resources.
1.17. What do you know about oil crisis of 1973?
1.18. List of cost of electricity production from sources and their energy conversion efficiency.
1.19. Discuss sustainable energy option for Indian economy.

Software Based Question

1.1. Develop a software which can predict the cost of electricity production from all possible energy sources.

Design and Open Ended Question

1.1. Prepare a comprehensive report concerning electricity generation from all possible sources which do not or least produces greenhouse gases covering percentage of GHG, cost of electricity and conversion efficiency.

Multiple-Choice Questions

1.1. The per capita electrical energy consumption in India in 2019 was
(a) 400 kWh/year (b) 700 kWh/year (c) 880 kWh/year (d) 1180 kWh/year

1.2. The hydraulic efficiency of big hydel power plan•t is around
(a) 50 to 60% (b) 60 to 70% (c) 70 to 80% (d) 80 to 90%

1.3. The conversion efficiency of combined cycle plant ranges from
(a) 30 to 40% (b) 40 to 50% (c) 50 to 60% (d) 60 to 70%

1.4. CO_2 emission from coal in g/kWh is around
(a) 855 (b) 955 (c) 1055 (d) 1155

1.5. The thermal efficiency of coal based modern power plant is around
(a) 30 to 35% (b) 35 to 40% (c) 40 to 45% (d) 45 to 50%

1.6. By energy conservation the percentage of energy saved is
(a) 10% (b) 20% (c) 30% (d) 40%

Answers to Multiple-Choice Questions

1.1 (d) 1.2 (d) 1.3 (c) 1.4 (b) 1.5 (b) 1.6 (a)

CHAPTER

2

Variable Load Problem and Power Plant Economics

OVERVIEW

The electrical load depends upon the consumer's demand and so it will be variable. The essence of supplying a variable load influences the characteristics and method of use of power equipment. It is very essential to select a suitable generating unit which is reliable, efficient, less polluting, and cheap so that the cost of electricity generated becomes cheap. This chapter deals with variable load problem and power plant economics.

2.1 THE BASIC INFORMATION FOR SELECTION OF GENERATING UNITS

For **establishing a power plant** at a suitable location, ***two basic parameters*** are decided and they are—

(a) *Total power output to be installed* (MW_{inst}); and (b) *size of the generating units*

For **estimating the total installed capacity,** the following **information** is gathered:

(i) First demand (kW_{max}); (ii) Growth of demand; (iii) Required reserved capacity.

For selecting the ***size of generating units,*** **following information** should be gathered.

(i) Load variation (i.e., load curve) during 24 hours (summer, winter, weekdays, holidays).
(ii) Total capacity of units connected to the electrical grid.
(iii) Minimum start-up and shut-down periods of the units.
(iv) Planning of maintenance programme.
(v) Variation of plant efficiency vs. size of the unit.
(vi) Variation of price and space demand per kW vs. size of the unit.

2.2 LOAD CURVE

In order to estimate expected maximum load, the most useful tool is the construction of hypothetical load curve, hourly load variation in a metropolitan area as shown in Fig. 2.1.

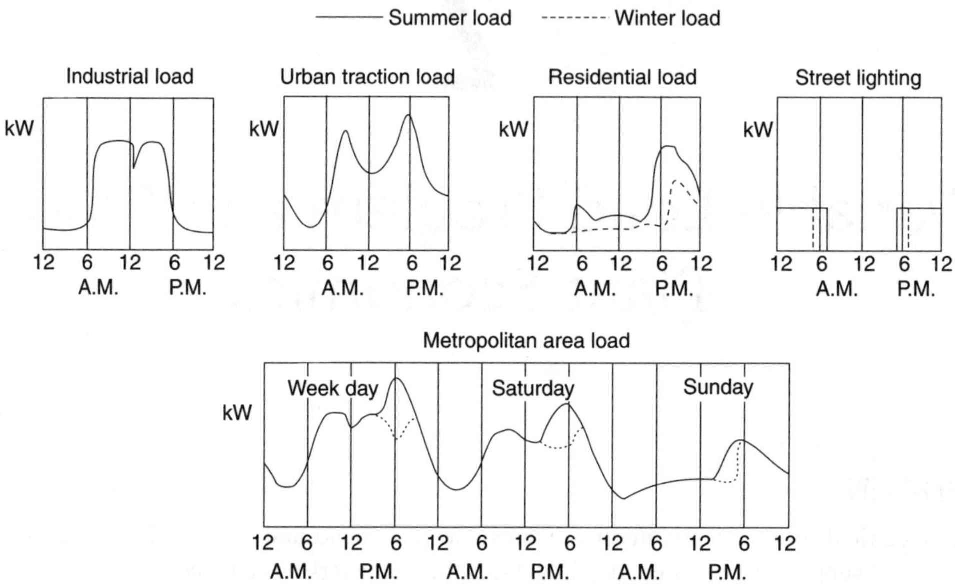

Figure 2.1 Different types of load curves

A consumer of electrical power will use the power as and when required, and hence the load will always be changing with time. *A curve showing the load demand (variations) of consumer with respect to time is known as load curve.* The load curve may be daily, weekly, monthly, and yearly for industrial, urban, residential, street and metropolitan. The loads of industrialized countries, as shown in Fig. 2.3(b) in the West in winter and summer are different. The winter load is more. The energy consumption of the consumer is given by an expression

$$\text{kWh} = E = \int_0^{24} (\text{kW})\, dt \quad \text{...(2.1)}$$

2.3 CONNECTED LOAD

It is the sum of ratings in kW of equipment installed in the consumer's premises. It may be residential, commercial, industrial, municipal load, etc.

2.4 DEMAND FACTOR

It is expressed as the ratio of maximum demand to the connected load. The maximum value of demand factor is unity.

2.5 AVERAGE LOAD

The average load is obtained by dividing the area under the load curve (expressed in kWh) by the time period considered (say 24 hours). The average load is shown in Fig. 2.2.

Thus, average load

$$= \frac{\text{Area under the load curve}}{24(\text{h})} \quad \text{...(2.2)}$$

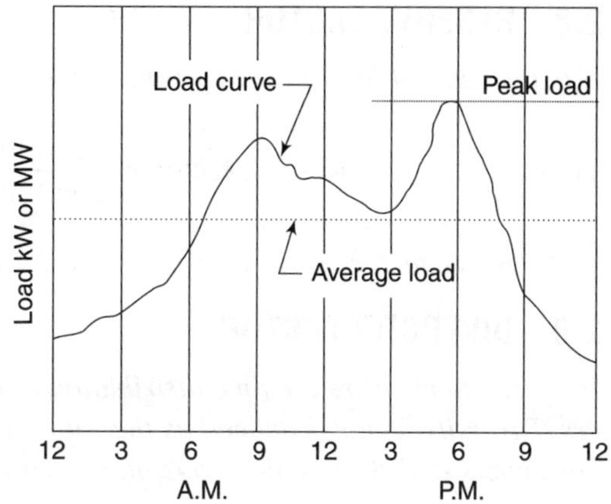

Figure 2.2 Load curve, average load and peak load

2.6 LOAD FACTOR

It is defined as the ratio of average load for a given period to the peak or maximum load during the same period determined by the consumer. It is always less than unity. Thus,

$$\text{Load factor} = \frac{\text{Average load over a given time interval}}{\text{Peak load during the same time}} \quad \text{...(2.3)}$$

$$= \frac{\text{kWh}_{\text{average}} \text{ in a year}}{\text{kW}_{\text{max}} \times 8760}; \text{ as one year} = 8760 \text{ h}$$

If a plant has a peak load of 2000 MW in a year but the average annual load is 900 MW, the annual load factor will be 0.45. This means that most of its capacity remains unutilized for the major part of the year, so the cost per kWh would be high and consumer will feel the punch. Thus, *the load factor refers to the economic health of the plant.* A high load factor is, in general, an indication of balanced load curve with small load changes.

2.7 CAPACITY FACTOR OR PLANT FACTOR

It is defined as the ratio of actual power produced in a period (say one year) to the rated capacity. It may also be expressed as average load to the plant rated capacity.

$$\text{Capacity or plant factor} = \frac{\text{Actual power generated in a year}}{\text{Rated capacity of the plant}}$$

$$= \frac{\text{kWh generated in a year}}{\text{kW}_{\text{installed}} \times 24 \times 365} = \frac{\text{kWh}_{\text{gen}} \text{ (year)}}{\text{kW}_{\text{installed}} \times 8760} \quad \text{...(2.4)}$$

The capacity factor shows how near the plants run to its full rating. In India, NTPC plants gives high plant factor even up to 95%.

2.8 RESERVE FACTOR

It is the ratio of load factor to capacity factor.

Thus, $$\text{Reserve factor} = \frac{\text{Load factors}}{\text{Capacity factors}} = \frac{kW_{inst}}{kW_{max}} \quad ...(2.5)$$

A reserve factor 1.2 to 1.25 is generally selected.

2.9 DIVERSITY FACTOR

It is used to measure the time distribution of maximum demands for similar types of consumers. Diversity factor is defined as the ratio of sum of the maximum demands of the individual consumers and the simultaneous maximum demand of the whole group during a particular time. Thus,

$$\text{Diversity factor} = \frac{\text{Sum of maximum demands of individual customer}}{\text{Simultaneous maximum demand of the group}} \quad ...(2.6)$$

It is shown in Fig. 2.3(a) in which a, b, c represent different loads and d represents total load. Thus, diversity factor = $\frac{a + b + c}{d}$.

Figure 2.3(b) shows that load curves of an industrial country.

The peak load is in winter. Maintenance is done during off-season.

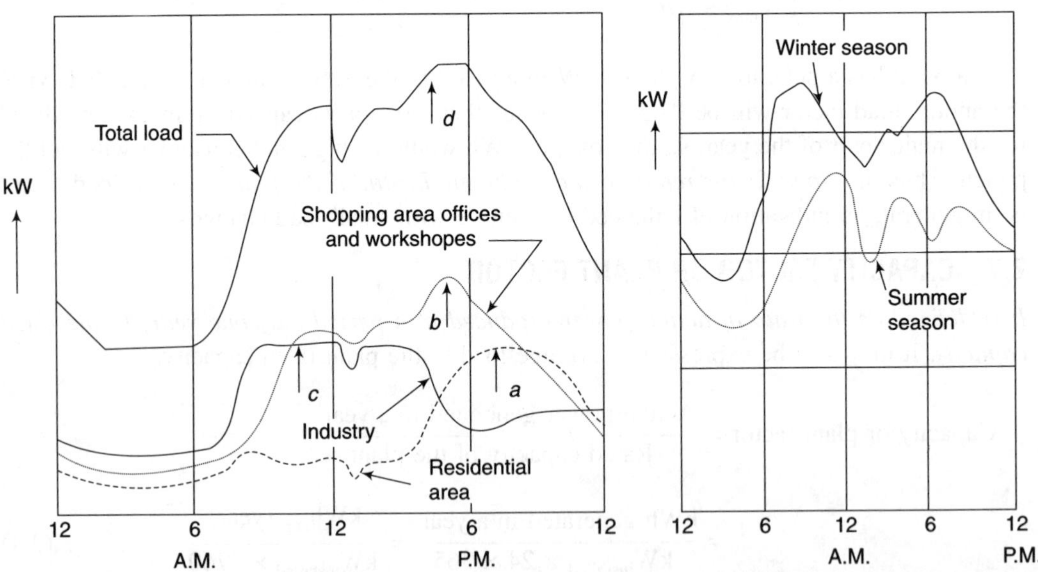

(a) Diversity of Loads (b) Load Curve of an Indutrialized Country

Figure 2.3

2.10 PLANT USE FACTOR

It is expressed as the ratio of energy produced in a given time to maximum possible energy that could have been produced during actual number of hours of operation. Thus,

$$\text{Plant use factor} = \frac{\text{Annual energy produced}}{\text{Capacity of plant} \times \text{Operating hours in a year}} \qquad ...(2.7)$$

If the operating hours in one year, then plant use factor is equal to the capacity factor. The cost of electricity produced reduces with the increase in the size of the unit.

2.11 CONSTRUCTION OF LOAD-DURATION CURVES

The construction of load-duration curves for typical days which indicate the power demands at different intervals of time is essential to obtain an economical operation of the various types of generating units connected to a grid and also for planning for fixing tariffs. As mentioned earlier, the loads may be residential, commercial, industrial, municipal, irrigation, traction, etc., and their load curves are given in Fig. 2.1.

Figure 2.4 shows the construction of load-duration curves from load curves. *It rearranges the load elements of a chronological curve in the descending order of the magnitude with*

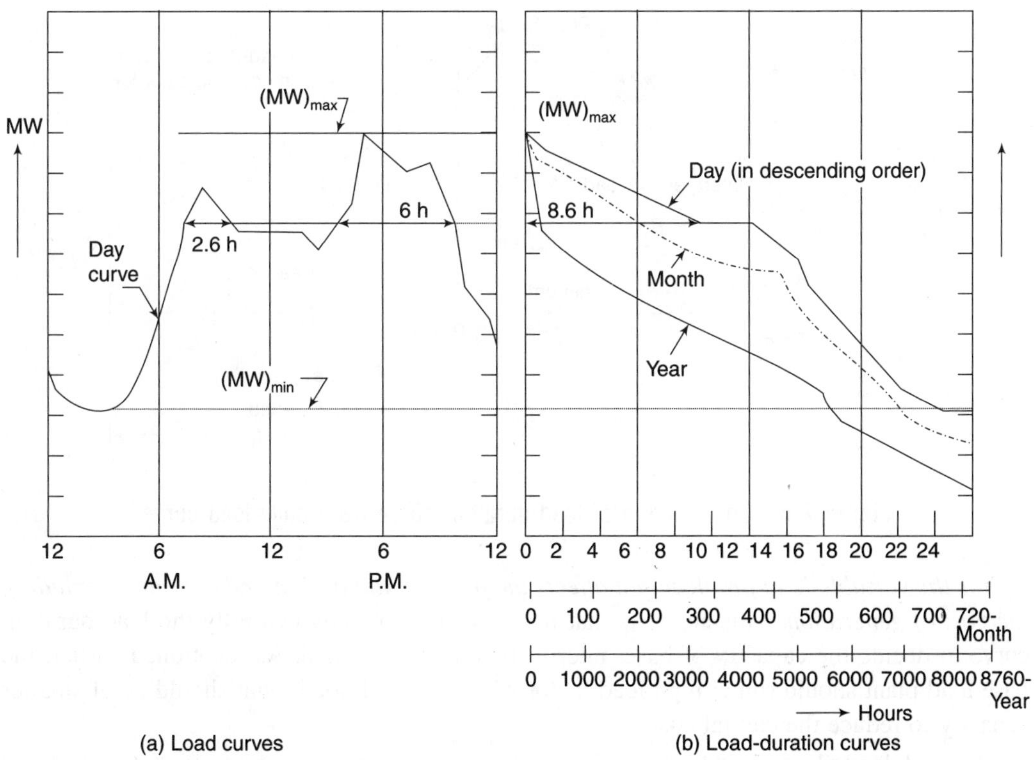

Figure 2.4 Construction of load-duration curve from chronological load curve of a day

time. The area under the curves kWh = $\int_0^{24}$ kW dt is the energy requirement during the day. The load duration curve depicts for how many hours a certain load has been required in the course of the day. From the compilation of load hours on various days, the load-duration curves for daily, one month (720 h) or one year (8760 h) can be obtained.

The load-duration curve constructed from daily load curve is given in Fig. 2.5. The daily load curve is usually divided in three parts namely ***base load, intermediate load*** and ***peak load***. The ***base load*** is the *load below which the demand never falls and is supplied 100% of the time.* Generally, the ***peak load*** occurs for *small period (say 15% of the daytime)*. The remaining load region represents the ***intermediate load***.

The base load plants are loaded very heavily and so its operation costs are of much significance for the whole plant. A high capital cost of base load plant may be permissible if its operating costs are low. Intermediate load plants are somewhat smaller in size than base load plants. As peaking load plants are needed for a smaller fraction of time, so the fuel costs are not important but minimum capital costs are the criterea for their selection.

A judicious planning of all three types of generation, namely, peak, intermediate and base, the maximum economy of the plant can be achieved so that cost per kWh charged to the consumer should be the minimum.

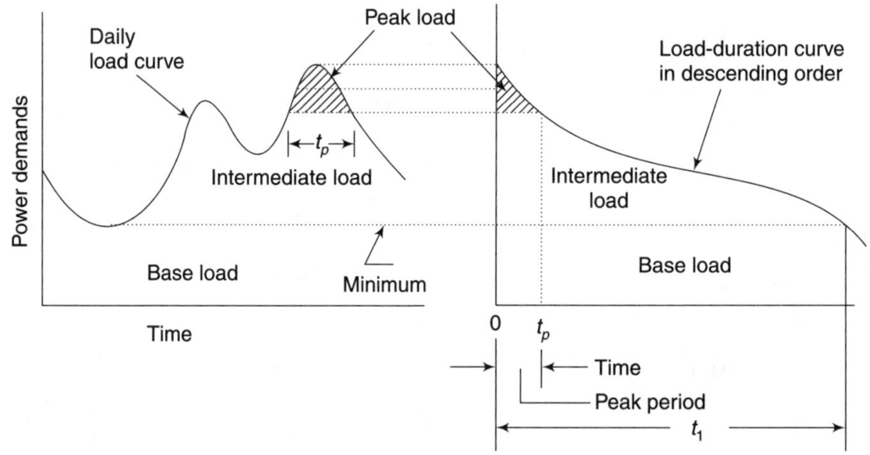

Figure 2.5 Construction of load-duration curve from daily load curve

It is the variable load problem that affects the power plant design and operation including the cost of generation. This suggests that it is essential to study carefully the load-duration curve to decide the capacity of base, intermediate and peak load. As mentioned earlier, the base load plant should run at high load factor while the peak load plant should be of smaller capacity to reduce the capital cost.

The peak load plant could be gas turbine unit, compressed air energy system, diesel engine or pumped hydrosystem depending upon the scope of availability and location.

2.12 EFFECT OF VARIABLE LOAD ON POWER PLANT DESIGN AND OPERATION

The electrical load depends upon the consumer's demand and so it will be variable. The essence of supplying a variable load influences the characteristics and method of use of power plant equipment. *The generation of power must be regulated according to the demand (i.e., load) and thus speed governing is essential to achieve the same. Another requirement of a power plant is a quick response to the load.*

Among all the variable load problems, the major problem is the generator that must be able to take variable load as quickly as possible without the change in voltage and frequency. The generator is coupled with the prime mover (turbine or engine). When the load on the generator increases, the speed of the rotor (i.e., prime mover) decreases, which in turn, decreases the frequency of electric supply. With the decrease in the speed of the prime mover due to the increase in load on the generator, the governing system must act, admitting more steam in the case of steam/nuclear power plant, more fuel in the case of gas turbine and more water in the case of hydel power plant to bring back the speed to the normal and take up the load. *The purpose of governing (discussed separately in subsequent chapters in each case) is to maintain the speed of turbine/engine constant irrespective of the load.*

The raw materials used in thermal power plant are fuel, air and water. In order to produce variable power from thermal power plant as per demand, it is necessary to vary the raw materials correspondingly. With an increase in load on the plant, the speed governor admits more steam and maintains the turbine speed. The governor response to this point follows rapidly the change of load but beyond this point the change is not so rapid. Since the steam generator operates with unbalance between heat transfer and steam demand long enough which yields a definite decrease of steam pressure. It is very difficult to secure good combustion and steady steam pressure with fluctuating steam demand, because efficient combustion requires good coordination of so many services. There is a certain time lag in combustion and heat transfer that is not true with electrical generators.

Due to complex problem associated with steam power plant for supplying variable loads, it is always desirable to allow the steam power plant to operate as a base load and gas turbine, diesel or hydel should meet fluctuating demand. *In general, to meet the peaking loads, reserve capacity is required which should be as small as possible.*

2.13 LOCATIONS OF POWER PLANT

1. Thermal Power Plant. The following factors are considered while locating the conventional base load thermal power plants:

(i) Availability of cooling water used in condenser. If cooling towers are used, the possibility of adequate make-up water has to be found out. Make-up water (treated) is also needed in steam generation. The source of water could be sea, lake or river.

(ii) Main wind direction and water currents in cooling water source in order to minimize air and water pollution.

(iii) Availability of fuel and the costs of fuel transport either by water, rail, road or pipe connections. The fuel may be coal, oil or gas.

(iv) Distance from the centre of gravity load demand to determine the transmission and distribution costs including their losses.

(v) Costs of land including space for extension, maintenance workshop and storage yard.

(vi) Character of the soil to determine costs of foundation.

(vii) Disposal of ash for coal-fired power plant.

(viii) Accommodation for staff if the plant is situated far from town.

(ix) Rail and road connection to the site.

(x) Security consideration around the site.

(xi) Environmental regulation.

2. Hydroelectric Power Plant. All the above requirements of thermal power plants except the above given terms (iii) and (vii) are also valid for hydroelectric power plant. Over and over these, there are two more factors:

(xii) Availability of water and water head is one of first and foremost requirements of hydraulic power plant.

(xiii) Danger of earthquake.

3. Nuclear Power Plants. In addition to all requirements of thermal power plants, two more factors are considered:

(xiv) Danger of earthquake.

(xv) Density of population in the vicinity to avoid radiation hazard if any.

2.14 POWER PLANT ECONOMICS

A power plant should not only be reliable and safe but must supply regular electricity at minimum cost to consumer. The **cost per** kWh_{net} is determined by following factors:

(i) *Fixed cost (FC).* It covers mainly interest, depreciation, insurance, taxes depending upon the capital invested, i.e., construction cost of the plant including the cost of the land and transmission cost.

(ii) *Operation and maintenance (O and M).* It covers salaries and wages, maintenance of equipment, repairs including spare parts, lubricating oil, chemicals and miscellaneous expenses.

(iii) *Fuel cost (C_f).* It depends upon the amount of electricity generated by the plant.

(iv) Net kWh_{net} of electricity sent out per year by power plant including transmission and distribution losses.

The **total annual production cost (C_{tap})** in a power plant is expressed as

$$C_{tap} = (FC + O\&M + C_f) = \left(\frac{I + D + T}{100}\right) C_C + (W + R + M) + C_f \quad ...(2.8)$$

where I is the interest rate (%), D is the depreciation (%), T is the taxes and insurance (%), C_c stands for construction cost, W is wages and salaries, R is repairs and maintenance, M stands for miscellaneous and C_f stands for fuel cost. Transmission and distribution (T&D)

loss is very high in India, around 21% whereas in western countries, it is around 8 to 10%. This loss is also taken care by transmission efficiency. (See Example 2.17).

The annual **amount of electricity sent** out by a power plant (kWh_{net}) is expressed as

$$kWh_{net} = kW_{inst} \times 8760 \times \left(1 - \frac{kW_{aux}}{100}\right) \times F_{pc} \qquad ...(2.9)$$

where, kW_{inst} is the rated (i.e., installed) output of generations; kW_{aux} is the power consumed by auxiliaries (%), F_{pc} is the plant capacity factor and 8760 = 24 × 365 hours/year.

It is worthy to note that in order to calculate the electrical power cost to a consumer, in addition to the production cost, the transmission and distribution cost, administrative expenses and return or profit on the investment are to be considered.

Thus, *consumer cost* will be

$$C_{consumer} = C_{tap} + C_{td} + C_{adm} + C_{profit} \qquad ...(2.10)$$

where C_{td} is the transmission and distribution cost, C_{adm} is the administrative expenses and C_{profit} is the profit cost.

Reliability of a power plant is governed by ***forced outage rate*** which is defined as:

$$\text{Forced outage rate} = \frac{\text{Forced outage hours in a year}}{\text{(Service hours + forced outage hours) in a year}}$$

Ideally, forced outage should be nil as it raises operation and maintenance costs. A proven reliability of equipment and regular preventive maintenance are essential to guarantee regular electric supply and better overall economy of the power plant.

The following factors ***reduce the cost*** of power generation

(i) Selection of quality equipment of longer life and proper capacities
(ii) Achievement of high load factor by the power station
(iii) Efficient plant having minimum cost per kW
(iv) Proper maintenance
(v) Proper supervision
(vi) Selection of simple design plant which does not need highly skilled personnel
(vii) Policy of total quality management and total quality people.

The cost to the consumers has a continuous upward trend due to monetary inflation, rising fuel prices and environmental regulation imposed by government agency.

2.15 CONSTRUCTION COST (C_C)

The following expenses are considered while calculating the construction costs.

1. Planning and design costs
2. Land costs
3. Preparation of site including road/rail connection costs
4. Building and machinery foundation costs
5. Plant equipment including transport to site costs

6. Erection and testing of equipment costs
7. Interest during construction for payments made before commissioning of plant
8. Administrative work costs.

There exists an economy of scale as shown in Fig. 2.6. *It is obvious from the result that the construction cost per kW decreases with size for all types of power plant, however the effect is a function of the equipment.* The cheapest is the jet type gas turbine plant without recuperator while the costliest is the nuclear plant.

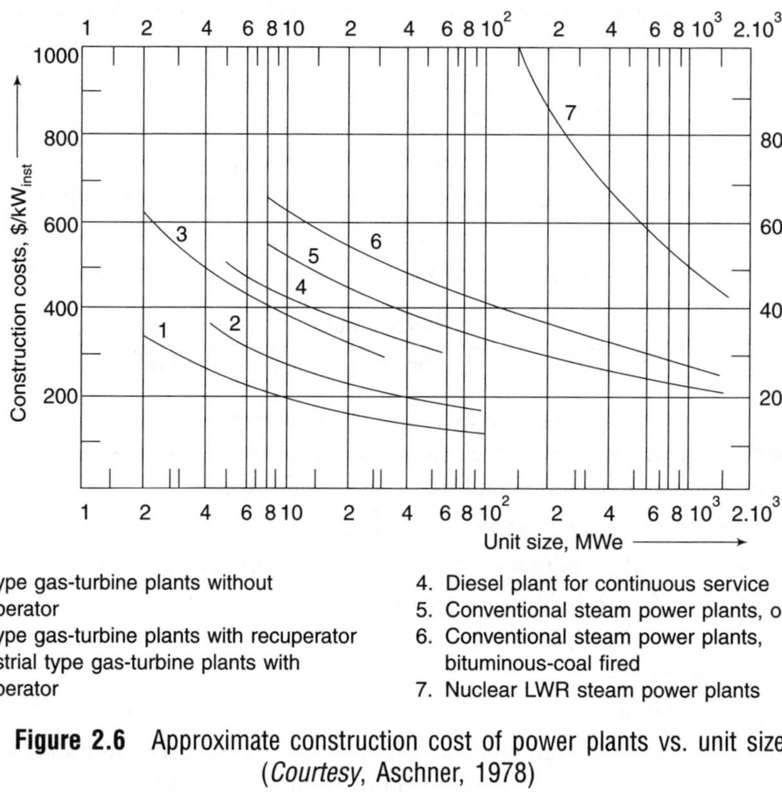

1. Jet type gas-turbine plants without recuperator
2. Jet type gas-turbine plants with recuperator
3. Industrial type gas-turbine plants with recuperator
4. Diesel plant for continuous service
5. Conventional steam power plants, oil fired
6. Conventional steam power plants, bituminous-coal fired
7. Nuclear LWR steam power plants

Figure 2.6 Approximate construction cost of power plants vs. unit size (*Courtesy,* Aschner, 1978)

2.16 FIXED COSTS (FC)

Fixed costs (FC) comprise of annual costs for interest (I), depreciation (D), taxes and insurance on property (T).

2.16.1 Interest Rates (I)

Various factors such as general financial conditions at the time of construction and source of funding (i.e., international, national, public, private or bond issues) influence the interest rates. Further, inflation (i.e., decrease in the value of the money) gives rise to the concept of real and charged interest rate. For example, if the real interest is fixed at 10% and a price inflation of 5% per year is forecast then the charged interest rate becomes 15%.

2.16.2 Depreciation Rates (D)

Any power plant and the equipment used in the station will have a certain period of useful life. The equipment becomes obsolete or loses its efficiency after years of use and thus needs replacement. In some cases, the equipment have to be changed, even fairly new, if more efficient and reliable equipment are available in the market. So for replacement of equipment, some money has to be put aside annually which is known as *depreciation fund.*

Depreciation rates (D) are calculated based on the life expectancy of equipment and the method used for calculating the depreciation. The following two methods are prevalent to calculate the depreciation fund.

(1) Straight-line method (2) Sinking-fund method

1. Straight-line Method. It is based on the assumption that depreciation is considered uniformly every year according to straight-line law. There is no interest on money saved. The depreciation charge every year is expressed as

$$D = \left(\frac{C_{\text{eqp}} - C_s}{N}\right) \quad \text{...(2.11)}$$

where C_{eqp} is the capital cost of equipment, C_s is the salvage after N useful years, and N is useful years.

Straight-line method is illustrated in Fig. 2.7. The straight-line (BE) indicates depreciation and the slope of the line indicates the rate at which the depreciation of the plant takes place. From Fig. 2.7, it is obvious that the value of equipment which is 100% initially, depreciates say to only 15% salvage value at the end of its 100% useful years of life. The depreciation at the end of 50% of its useful is FG and the total depreciation at the end of 100% of useful life is CE. The straight-line (AD) automatically will represent the depreciation reserve to be accumulated so that at the end of useful life, the total is DJ which is equal to the total depreciation (CE) signifying that 85% of the value of the equipment will come from accumulated reserve and 15% will come from salvage value.

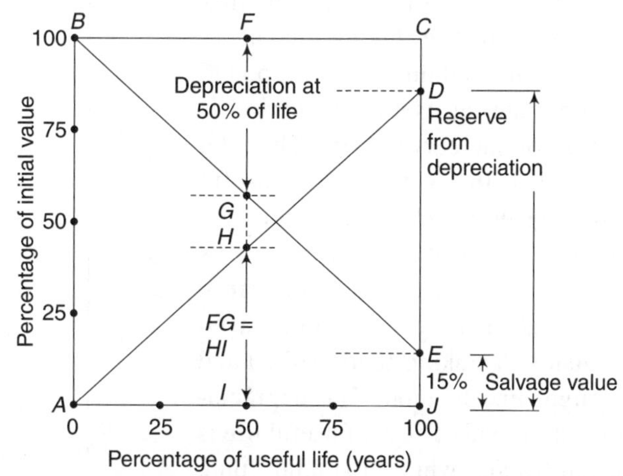

Figure 2.7 Straight-line method of depreciation

2. Sinking-Fund Method. Sinking-fund method is modeled on the assumption that a sum of money is set aside every year for N years and invested to earn compound interest.

Let P = annual deposit, i = interest compounded annually when the deposit is invested, N = useful years.

At the end of the first year, an amount P is deposited and after (N–1) years this will be worth $P(1 + i)^{N-1}$. At the end of second year, amount P is deposited and after (N–2) years, this will be worth $P(1 + i)^{N-2}$. At the end of (N–1) years, the amount P is deposited and this will be worth $P(1 + i)$.

At the end of Nth year, an amount P is deposited and this will be worth of amount P. Thus, at the end of N years, the sum of the amounts deposited will be

$$S = P(1 + i)^{N-1} + P(1 + i)^{N-2} + \ldots. P(1 + i) + \ldots. + P$$

$$= P + P(1 + i) + P(1 + i)^2 + \ldots. + P(1 + i)^{N-1}$$

The above equation is a geometrical progression with a common ratio of $r = 1 + i$.

Therefore, the sum is expressed as

$$S = \frac{P[(1 + i)^N - 1]}{i} \quad \text{...(2.12)}$$

The sum is equal to $C'_{eqp} = C_{eqp} - C_{sal}$ = Capital cost – Salvage value

$$\therefore \quad P = \frac{C'_{eqp} \cdot i}{(1 + i)^N - 1} \quad \text{...(2.13)}$$

Figure 2.8 illustrates the sinking-fund method. The sinking fund accumulated at a certain rate of compound interest during useful life period is shown by the curve, AHD. The total amount at the end of 100% useful life is given by JD which represents the initial cost of the plant minus its salvage value. There is a difference in the rate at which the sum is saved annually in sinking-fund method and straight-line method. The depreciation at the end of 50% of useful life is given by FG, while in sinking-fund, it is given by HI.

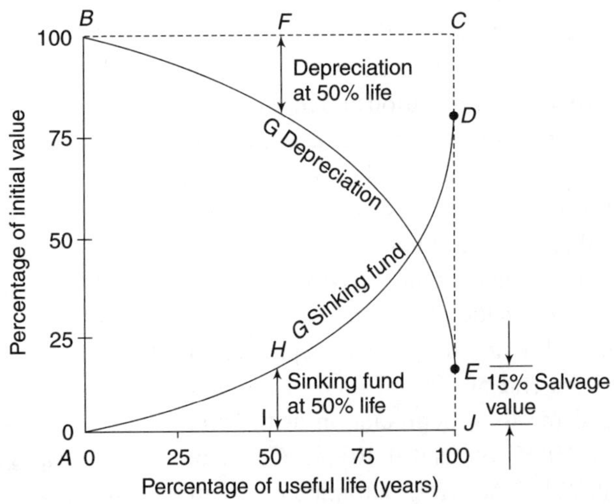

Figure 2.8 Sinking-fund method of deprection

2.17 FUEL COST (C_f)

The amount of electricity produced determines the fuel consumption. It is a well-known fact that the efficiency of power plant is the highest at rated load. At lower (i.e., part) loads, the plant efficiency decreases and so the fuel consumption increases. In other words, the cost of the electricity generation is the minimum at the rated load and it increases with the decrease in load. It is the calorific value of fuel, availability and transportation charges that also affect the cost of fuel.

2.18 PRESENT-WORTH (PW) CONCEPT

A huge sum of money is needed in setting up a power plant, and this amount is supposed to be on loan. The amount spent in setting up the plant is to be paid back with interest. In the evaluation of payment schemes (such as cash flow schemes, etc.) for financing a power plant project, the influence on interest on the value of funds to be spent on a later date has given rise to the present-worth concept.

PW is defined as the value of a sum of money at the present time that with compound interest will have a specified value at a certain time in the future. The present worth (PW) of a payment S to be made after N years is expressed as $\text{PW} = \dfrac{S}{(1+i)^N}$ ...(2.14)

It is obvious from the above expression that PW will be less than S, and it will be governed by factor $F_{PW} = (1 + i)^{-N}$. For example, if $i = 10\%$, and $N = 5$ years

$$\text{PW} = 0.621\ S$$

In many situations, equal amounts of annual expenses are required. Under such condition, the present worth will be calculated by substituting the value of S from Eq. (2.12).

Thus,
$$\text{PW} = \frac{S}{(1+i)^N} = \frac{P[(1+i)^N - 1]}{i} \times \frac{1}{(1+i)^N} \quad \text{...(2.15)}$$

where P is the payment every year. If $i = 10\%$, $N = 20$ years.

then
$$\text{PW} = \frac{P[(1+0.1)^{20} - 1]}{0.1} \times \frac{1}{1.1^{20}} = 8.51\ P.$$

2.19 PLANT NET HEAT RATE (PNHR), INCREMENTAL HEAT RATE (IHR) AND PLANT NET EFFICIENCY (η_{plant})

Mainly two parameters determine the performance of electricity generating plants and they are

Plant net heat rate (PNHR) expressed in kJ/kW

$$= \frac{\text{Heat input } (I)}{\text{Net work output } (L)} = \frac{3600}{\eta_{th}};\ \text{kJ/kWh} \quad \text{...(2.16)}$$

$$\text{Plant efficiency, } \eta_{plant} = (\eta_{th}) = \frac{\text{Net work output}}{\text{Heat input}} \quad \text{...(2.17)}$$

Table 2.1 depicts the typical performance of various generating systems running at rated loads. Sometimes, it is essential to generate additional amount of energy at any given load. This aspect is taken care by incremental heat rate (IHR). *The incremental heat rate (IHR) is defined as the ratio of incremental heat input to incremental net kW output.*

Thus,
$$\text{IHR} = \frac{dI}{dL} \quad \text{...(2.18)}$$

Table 2.1 Typical performance of generating system

S.N.	*Generating system*	*Unit size (MW)*	*Plant efficiency (%)*
1.	Steam	200–1000	30–40
2.	Nuclear (BWR/PWR)	500–1100	31–35
3.	Gas turbine	50–300	32–37
4.	Combined gas/steam cycle	300–600	46–60
5.	Diesel engine	10–30	27–30
6.	Hydel	50–700	70–90

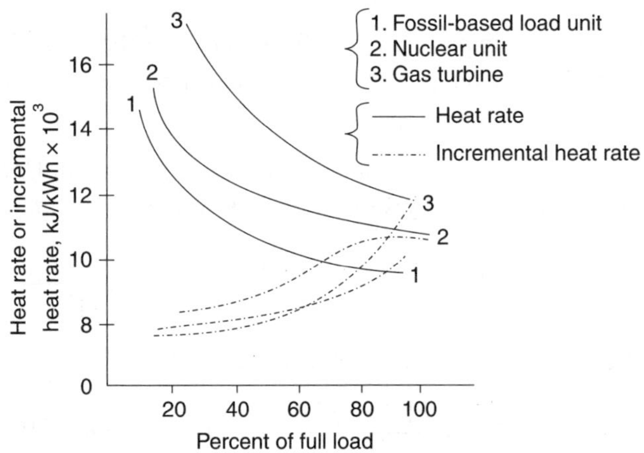

Figure 2.9 Plant net heat rate and incremental heat rate vs. load

Figure 2.9 shows the variation of PNHR and IHR with load. It is obvious from the results that plant net heat rate decreases with the increase in load while incremental heat rate increases with load.

2.20 ECONOMIC SCHEDULING PRINCIPLE FOR LOAD DISTRIBUTION

In any power station, there are more than one generating units. Generally, the number of units are 3, 5 or 7 so that each unit may be put for maintenance and overhauling as per schedule without affecting the power supply.

A major problem confronting the power engineers is to distribute properly the load among the generating units. In the event of improper distribution of load, the plant efficiency will go down as a whole. As per the economic scheduling principle, the load is so distributed that at any moment, all generating units will have the same incremental heat rate. With the increase of system load, the incremental heat rate of each unit will increase. As an example, economic scheduling principle is illustrated by assuming two generating units in the network.

Let I_c = combined input to units 1 and 2

L_c = combined output of units 1 and 2

I_1, I_2 = input of units 1 and 2 respectively

L_1, L_2 = outputs of units 1 and 2 respectively

The combined input, I_c, is a function of either L_1 or L_2 for a given combined output L_c.

For the minimum value of combined input, I_c the ratio of incremental input to incremental output by unit 1 should be zero, i.e.,

$$dI_c/dL_1 = 0 \quad \text{...(2.19)}$$

But $I_c = I_1 + I_2$, so

$$\therefore \quad dI_1/dL_1 + dI_2/dL_1 = 0 \quad \text{...(2.20)}$$

We have $$\frac{dI_2}{dL_1} = \left(\frac{dI_2}{dL_2}\right) \times \left(\frac{dL_2}{dL_1}\right)$$

As $$L_c = L_1 + L_2$$

so $$dL_2/dL_1 = -1$$

or $$\frac{dI_2}{dL_1} = -\frac{dI_2}{dL_2} \quad \text{...(2.21)}$$

Combining Eqs. (2.20) and (2.21), we have

$$\frac{dI_1}{dL_1} = \frac{dI_2}{dL_2} \quad (2.22)$$

From Eq. (2.22), it is clear that the combined input, I_c is a minimum only if the incremental heat rate of each two unit is the same. The same principle will be applicable to the analysis of multiple units.

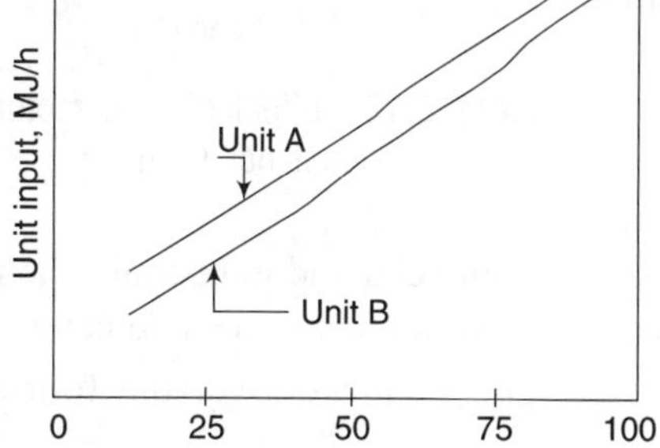

Figure 2.10 Input-output curves of two generating units in an utility

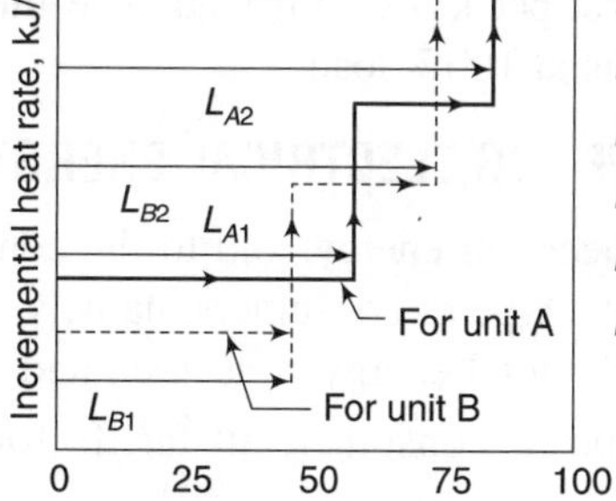

dL	dI
$L_{B1} + 0$	$I_{B1} + 0$
$L_{B1} + L_{A1}$	$I_{B1} + I_{A1}$
$L_{B2} + L_{A1}$	$I_{B2} + I_{A1}$
$L_{B2} + L_{A2}$	$I_{B2} + I_{A2}$
$L_{B2} + L_{A3}$	$I_{B2} + I_{A3}$
$L_{B3} + L_{A3}$	$I_{B3} + I_{A3}$

Figure 2.11 Incremental heat rate curves for two generating units in an utility

The input-output curves of two units operating in parallel and supplying to a common load is shown in Fig. 2.10. The corresponding incremental heat rate curves are depicted in Fig. 2.11. The principle of equal incremental heat rate is applied in the distribution of load to achieve the maximum fuel economy. The unit (turbine) *B*, starting from zero picks up the load say upto L_{B1}, while unit (turbine) *A* remains at zero load. When combined load, I_c is greater than L_{B1}, the unit *A* starts to pick up load until it reaches say L_{A1}. The second valve of turbine starts to open at the combined load of $(L_{A1} + L_{B1})$. With the increase in load, the unit (turbine) *B* picks up the additional load, while unit (turbine) *A* remains at the load L_{A1}. If the combined load is more than $(L_{A1} + L_{B2})$, the unit (turbine) *A* starts to open the second valve and later the third valve. It is worth to note that at the combined load $(L_{A3} + L_{B2})$, the unit (turbine) *A* would be at full load L_{A3} and the unit (turbine) *B* at the load, L_{B2}. With the increase in combined load beyond this level, the unit (turbine) *B* starts to open the last valve to meet the additional demand. At this level, both unit will be sharing full 100% load.

It is essential that the generating capacity of multiple units in a utility system must be greater than the system load. This difference is known as ***spinning reserve*** which is a function of type of utility. It is worth to remind that the principle of equal incremental heat rate is to be applied in the distribution of load among multiple unit in utility to minimise the system fuel cost.

2.21 EFFECT OF LOAD FACTOR ON COST PER kWh

In any utility system, the fixed costs are a function of the capacity of the plant while *fuel and operating costs are governed by the total energy produced by the station annually.* The fixed costs are invariant irrespective of whether the station is running or idle but the fuel and operating cost depend upon the load. It is obvious that power station will be fully utilized and maximum possible energy is produced when it operates at 100% load factor. At load factor less than 100%, less electricity is generated which means higher cost per kWh. Figure 2.12 shows the variation of cost per kWh with load. The minimum cost per kWh is at 100% load.

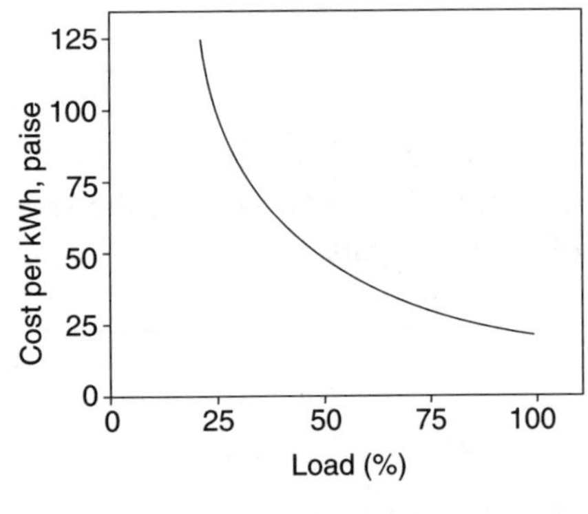

Figure 2.12 Effect of load factor on cost per kWh

2.22 TARIFF FOR ELECTRICAL ENERGY

The rates of electrical energy sold to the consumers depend upon the type of consumers, namely, domestic, commercial and industrial. The rates depend upon the total energy consumed and the load factor of the consumers demand.

For any type of consumer, all tariffs for electrical energy must cover the following items:

(i) Recovery of capital cost invested

(ii) Recovery of running costs such as operation cost, maintenance cost, maintaining the equipment cost, billing cost and many others

(iii) Reasonable profit on the invested capital.

General Rate Form. The general type of tariff can be represented by the following expression

$$z = ax + by + c \qquad ...(2.23)$$

where, z = Total amount of bill for the period considered

x = Maximum demand in kW

y = Energy consumed in kWh during the period considered

a = Rate per kW of maximum demand

b = Energy rate per kWh

c = Constant amount charged to the consumer during each period.

From the general rate form, the following charging rates are possible:

(i) Flat Demand Rate. This is expressed as

$$z = ax \qquad ...(2.24)$$

(ii) Straight Line Meter Rate. This is given by

$$z = by \qquad ...(2.25)$$

(iii) Black Meter Rate. In this case, the rate is expressed as

$$z = \Sigma b_i y_i = b_1 y_1 + b_2 y_2 + ... + b_n y_n \qquad ...(2.26)$$

(iv) Hopkinson Demand Rate of Two-part Tariff. This method charges the consumer according to his maximum demand and energy consumption. This is expressed as

$$z = a + by \qquad ...(2.27)$$

This method requires two meters to record the maximum demand and energy consumption by consumers. This method is generally used for industrial customer.

(v) Doherty Rate or Three Part Tariff. According to this method, the customer has to pay some fixed amount in addition to the charge for the maximum demand and energy consumed. This is expressed as

$$z = ax + by + c \qquad ...(2.28)$$

Example 2.1 The following loads are supplied to the consumer by a power station.

Time in hours	0–6	6–8	8–10	10–12	12–16	16–20	20–22	22–24
Load in MW	40	80	90	100	75	120	100	80

Work out the following: (a) Draw the load curve, (b) Determine the load factor of plant, (c) If load factor of a standby equipment of 30 MW capacity and takes all the loads above 90 MW, calculate the load factor, (d) Calculate plant use factor of standby plant.

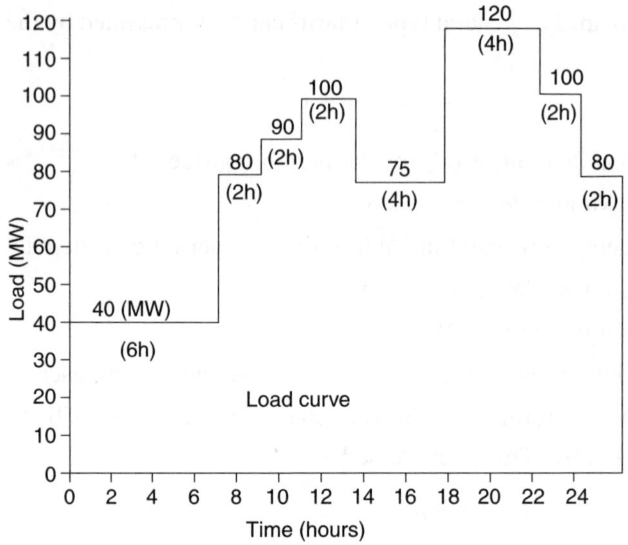

Figure 2.13 Example 2.1

Solution (a) Figure 2.13 shows the load curve based on data supplied.

Energy generated by the plant = Area under the load curve = E

$= 40 \times 6 + 80 \times 2 + 90 \times 2 + 100 \times 2 + 75 \times 4 + 120 \times 4 + 100 \times 2 + 80 \times 2 = 1920$ MWh

$$\text{Average load} = \frac{1920}{24} = 80 \text{ MW}$$

(b) $\text{Load factor} = \dfrac{\text{Average load}}{\text{Peak load}} = \dfrac{80}{120} = 0.6666 = 66.66\%$ ***Ans.***

(c) The energy generated by standby plant of 30 MW above 90 MW is

$$= 10 \times 2 + 30 \times 4 + 10 \times 2 = 160 \text{ MWh}$$

The period of operation of standby unit = 2 + 4 + 2 = 8 h

$\text{Average load of the standby plant} = \dfrac{160}{8} = 20 \text{ MW}$

$\therefore$ $\text{Load factor of standby plant} = \dfrac{20}{30} = 0.6666 = 66.66\%$ ***Ans.***

(d) The use factor of standby plant is given by

$\text{Use factor} = \dfrac{\text{Energy generated}}{\text{Plant capacity} \times \text{Operating hours}} = \dfrac{160 \times 10^3}{30 \times 10^3 \times 8} = 66.66\%$ ***Ans.***

Example 2.2 The rated capacity of a power plant is 400 MW. The peak load on the plant is 350 MW. The various consumer groups having maximum demands of 120 MW. 100 MW, 80 MW and 90 MW are connected to the power plant. The annual load factor is 0.8. Calculate

(a) the average load, (b) the capacity factor, (c) the energy supplied per year, (d) the reserve factor, (e) the demand factor and (f) the diversity factor.

Solution The load factor is given by

$$\text{Load factor} = \frac{\text{Average load}}{\text{Peak load}} = 0.8 = \frac{\text{Average load}}{350}$$

(a) Average load = 350 × 0.8 = 280 MW ***Ans.***

(b) $\text{Capacity factor} = \frac{\text{Average load}}{\text{Rated capacity}} = \frac{280}{400} = 0.7$ ***Ans.***

(c) Energy supplied per year = Average load × 8760 h

$$= 280 \times 8760 = 2452.8 \times 10^3 \text{ MWh}$$ ***Ans.***

(d) $\text{Reserve factor} = \frac{\text{Load factor}}{\text{Capacity factor}} = \frac{0.8}{0.7} = 1.142$ ***Ans.***

(e) $\text{Demand factor} = \frac{\text{Maximum demand}}{\text{Connected demand}} = \frac{350}{120 + 100 + 80 + 90} = 0.8974$ ***Ans.***

(f) $\text{Diversity factor} = \frac{\text{Sum of the individual maximum demands}}{\text{Simultaneous maximum demands}}$

$$= \frac{120 + 100 + 80 + 90}{350} = 1.114$$ ***Ans.***

Example 2.3 The peak load of a power plant is 400 MW having annual load factor 0.85, capacity factor 0.8, and use factor 0.82. Calculate (a) the annual energy production, (b) the reserve capacity over and above the peak load, (c) hours during which the plant is not in service per year and (d) the turnover of the plant if the energy is sold at ₹2 per unit.

Solution (a) $\text{Load factor} = \frac{\text{Average load}}{\text{Peak load}} = 0.85$

or Average load = 0.85 × 400 = 340 MW

$$\text{Annual energy production} = 340 \times 10^3 \times 8760 = 2978.4 \times 10^6 \text{ kWh}$$ ***Ans.***

(b) $\text{Capacity factor} = \frac{\text{Average load}}{\text{Plant capacity}} = 0.8$

or $\text{Plant capacity} = \frac{340}{0.8} = 425 \text{ MW}$

Reserve capacity over and above peak load = 425 – 400 = 25 MW ***Ans.***

(c) $\text{Use factor} = \frac{\text{Energy generated per year}}{\text{Plant capacity} \times \text{Hours in operation}} = 0.82$

$$\text{or hours in operation} = \frac{2978.4 \times 10^6}{425 \times 10^3 \times 0.82} = 8546.3 \text{ hours}$$

Hence, hours not in service = 8760 – 8546.3 = 213.7 h ***Ans.***

(d) Turnover = $2978.4 \times 10^6 \times 2$ = ₹5956.8×10^6 ***Ans.***

Example 2.4 The following data relate to a thermal power plant: Capacity of the plant = 350 MW, maximum load = 320 MW, Annual load factor = 0.7, Coal consumption = 1.4 kg/kWh, Cost of coal = ₹800.0 per tonne.

Calculate—(a) the capacity factor of the plant and (b) the net revenue earned if the energy is sold at ₹2/kWh, (c) the annual profit earned if all expenses except fuel come out to be 80% of the net revenue earned.

Solution Average load = Load factor × Peak load = 0.7 × 320 = 224 MW

$$\text{(a) Capacity factor} = \frac{\text{Average load}}{\text{Capacity of plant}} = \frac{224}{350} = 0.64$$ ***Ans.***

(b) Energy generated per year = $224 \times 8760 = 1962.136 \times 10^3$ MWh

Coal required per year = $1.4 \times 1962.24 \times 10^3 \times 10^3 = 2747.136 \times 10^6$ kg

= 2747.136×10^3 tonne

Cost of coal per year = $2747.136 \times 10^3 \times 800$ = ₹2197.70×10^6

Cost of energy sold = $2 \times 1962.24 \times 10^3 \times 10^3$ = ₹3924.48×10^6

Net revenue earned by the plant

= ₹3924.48×10^6 – ₹2197.708×10^6 = ₹1726.7×10^6

= ₹1726.7 crore ***Ans.***

(c) Profit = (1 – 0.8) × 1726.67 = ₹345.33 crore ***Ans.***

Example 2.5 A power plant is to be established with a maximum demand of 2000 MW and load factor of 0.8. Choice is to be made out of steam power plant, nuclear plant and hydel power plant. Find the overall cost per kWh in case of each system with the following data.

	Cost	*Steam power plant (A)*	*Hydel power plant (B)*	*Nuclear power plant (C)*
1.	Capital cost per MW installed	₹3 crore	₹4 crore	₹5 crore
2.	Interest	7%	6%	5%
3.	Depreciation	6%	4%	5%
4.	Operating cost including fuel per kWh	80 paise	30 paise	50 paise
5.	Transmission and distribution cost per kWh	5 paise	6 paise	5 paise

Solution **A. Steam Power Plant**

Capital cost = ₹3 × 2000 crore = 6000 crore = ₹6 × 10^{10}

Interest = ₹0.07 × 6 × 10^{10} = ₹0.42 × 10^{10}

Annual fixed cost = Interest + Depreciation = ₹0.78 × 10^{10}

Energy generated per year = Average load × 8760

= Load factor × Maximum demand × 8760

= 0.8 × 2000 × 10^3 × 8760 = 1.4016 × 10^{10} kWh

Overall cost/kWh = fixed cost kWh + operating cost/kWh + Transmission and distribution cost/kWh

$$= \frac{0.78 \times 10^{10}}{1.4016 \times 10^{10}} + 0.80 + 0.05 = ₹1.406$$ ***Ans.***

B. Hydel Power Plant

Capital cost = ₹4 × 2000 = ₹8000 crore = ₹8 × 10^{10}

Interest = ₹0.06 × 8 × 10^{10} = ₹0.48 × 10^{10}

Depreciation = ₹0.04 × 8 × 10^{10} = ₹0.32 × 10^{10}

Annual fixed cost = Interest + Depreciation = ₹0.8 × 10^{10}

Energy generated per year = 14016 × 10^6 kWh

$$\text{Overall cost/kWh} = \frac{0.8 \times 10^{10}}{1.4016 \times 10^{10}} + 0.3 + 0.06 = ₹0.93 = 93 \text{ paise}$$ ***Ans.***

C. Nuclear Power Plant

Capital cost = ₹5 × 2000 = ₹10000 crore = ₹10 × 10^{10}

Interest = ₹0.05 × 10 × 10^{10} = ₹0.5 × 10^{10}

Depreciation = ₹0.05 × 10 × 10^{10} = ₹0.5 × 10^{10}

Annual fixed cost = Interest + Depreciation = ₹1 × 10^{10}

Energy generated per year = 14016 × 10^6 kWh

$$\text{Overall cost/kWh} = \frac{1 \times 10^{10}}{1.4016 \times 10^{10}} + 0.5 + 0.05 = ₹1.26$$ ***Ans.***

Example 2.6 The following data related to a power plant: Capacity of the plant = 600 MW. Peak load = 570 MW. Capital cost = 1200 crore, Rate of interest and depreciation = 12%, Annual cost of fuel, salaries and taxation = ₹50 crore, Load factor = 0.7

Calculate (a) the cost of generation per kWh, (b) the saving in cost per kWh if the load factor is raised to 0.8 and (c) capacity factor.

Solution $\text{Load factor} = 0.7 = \frac{\text{Average load}}{\text{Peak load}}$ or Average load = 570 × 0.7 = 399 MW

Energy generated per year = $399 \times 10^3 \times 8760 = 349.524 \times 10^7$ kWh

Fixed cost (interest and depreciation) = 0.12×1200 = 144 crore = ₹144×10^7

Total annual cost = Fixed cost + Annual cost of fuel, salaries, etc.

$$= ₹144 \times 10^7 + ₹50 \times 10^7 = ₹194 \times 10^7$$

(a) Cost per kWh = $\dfrac{₹194 \times 10^7}{349.524 \times 10^7}$ = ₹0.55 = 55 paise ***Ans.***

(b) When the load factor is raised to 0.8,

Average load = 570×0.8 = 456 MW

Energy generated per year = $456 \times 10^3 \times 8760 = 3994.56 \times 10^6$ kWh

Fixed cost will remain the same, hence

Cost per kWh = $\dfrac{194 \times 10^7}{3994.56 \times 10^6}$ = ₹0.48 = 48 paise/kWh

Saving in cost per kWh due to increased load factor = 55 – 48 = 7 paise ***Ans.***

Example 2.7 The following data relate to a power plant having 350 MW stalled capacity: Capacity cost (installed) = ₹30,000/kW, Interest and depreciation = 12%, Annual capacity factor = 65%, Annual load factor = 70%, Annual running charge = ₹40 crore, Energy consumed by the power plant auxiliaries = 7%. Find (a) the peak load, (b) the reserve capacity and (c) the cost of power generation per kWh.

Solution (a) $\dfrac{\text{Load factor}}{\text{Capacity factor}} = \dfrac{\text{Average load}}{\text{Peak load}} \times \dfrac{\text{Capacity of plant}}{\text{Average load}}$

or $\dfrac{0.7}{0.65} = \dfrac{350}{\text{Peak load}}$

or Peak load = $\dfrac{350 \times 0.65}{0.7}$ = 325 MW ***Ans.***

(b) Reserve capacity = 350 – 325 = 25 MW ***Ans.***

(c) Average load = Load factor × Peak load = 0.7×325 = 227.5 MW

Energy generated per year = $227.5 \times 10^3 \times 8760 = 199.29 \times 10^7$ kWh

Net energy delivered = $(1 - 0.07) \times 199.29 \times 10^7 = 185.33 \times 10^7$ kWh

Fixed cost (interest and depreciation) = ₹$0.12 \times 30{,}000 \times 350 \times 10^3$ = ₹126×10^7

Total annual cost = Fixed cost + Running cost

$$= ₹126 \times 10^7 + ₹40 \times 10^7 = ₹166 \times 10^7$$

Cost per kWh = $\dfrac{166 \times 10^7}{185.33 \times 10^7}$ = ₹0.89 = 89 paise ***Ans.***

Example 2.8 A 2000 MW thermal power station supplied power to a system having maximum and minimum demand of 1900 MW and 1200 MW respectively in a year. By assuming the load duration curve to be a straight line over a year, calculate (a) the load factor, (b) the capacity factor and (c) the reserve factor.

Solution Refer to Fig. 2.14 Ex. 2.8. The straight line variation of load duration curve is shown in Fig. 2.14.

The energy generated and supplied per year

$$= 1200 \times 8760 \times 10^3 + \frac{1}{2}(1900 - 1200) \times 8760 \times 10^3 = 1357.8 \times 10^7 \text{ kWh}$$

$$\text{Average load} = \frac{1357.8 \times 10^7}{8760} = 1.55 \times 10^6 \text{ kW} = 1550 \text{ MW}$$

$$\text{(a) Load factor} = \frac{\text{Average load}}{\text{Peak load}} = \frac{1550}{1900} = 0.815$$ ***Ans.***

$$\text{(b) Capacity factor} = \frac{\text{Average load}}{\text{Capacity}} = \frac{1550}{2000} = 0.775$$ ***Ans.***

$$\text{(c) Reserve factor} = \frac{\text{Load factor}}{\text{Capacity factor}} = \frac{0.815}{0.775} = 1.0156$$ ***Ans.***

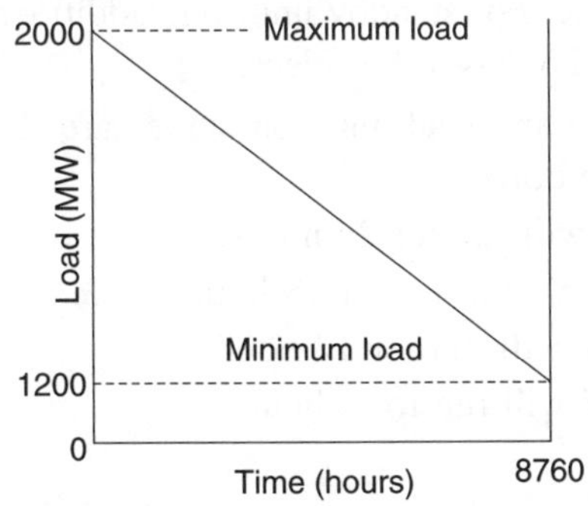

Figure 2.14 Example 2.8

Example 2.9 A power station is to be built to meet the following electrical load requirements.

Time (hours)	0–6	6–12	12–14	14–18	18–20	20–24
Load (MW)	120	360	240	400	200	150

Workout the following: (a) select the suitable generating units, (b) draw the load curve, (c) draw the load duration curve, (d) calculate the load factor and (e) calculate the capacity of plant and the plant capacity factor.

Solution Refer to Fig. 2.15. (a) Ex. 2.19. In order to supply the electrical power to meet the consumer demand, three generating units, each of 120 MW, and one generating unit of

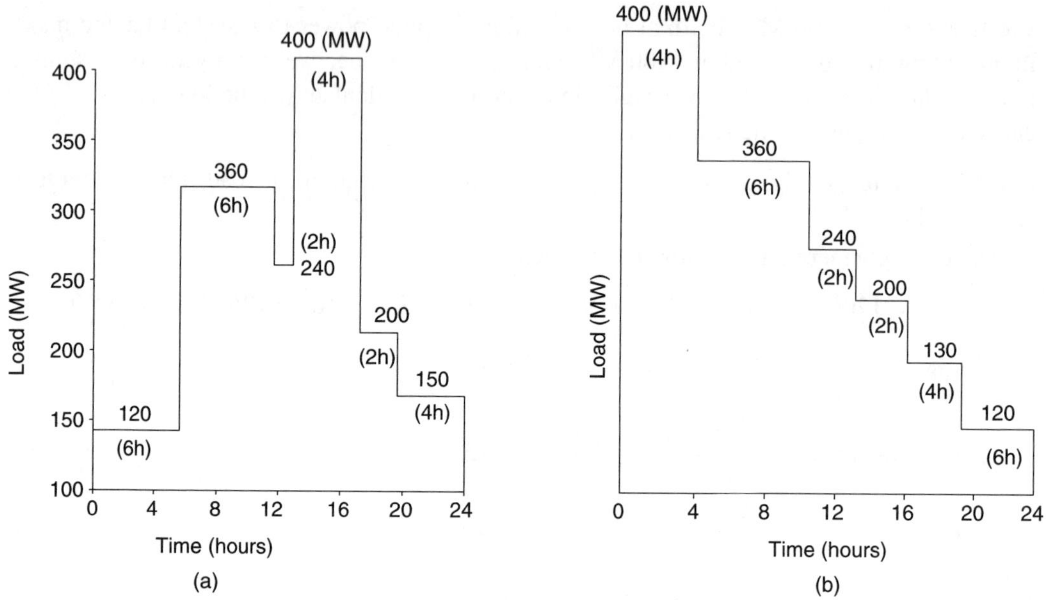

Figure 2.15 Example 2.9

40 MW will serve the purpose. For standby unit, one additional unit of capacity equivalent to the largest unit, i.e., 120 MW has to be selected.

(b) and (c). The load curve and load duration curves are shown in Fig. 2.15 from which the following conclusions are drawn

- One unit of 120 MW will run for 24 hours.
- Second unit of 120 MW will run for 18 hours.
- Third unit of 120 MW will run for 12 hours.
- Fourth unit of 40 MW will run for 4 hours.

(d) The energy generated

$$= 120 \times 6 + 360 \times 6 + 240 \times 2 + 400 \times 4 + 200 \times 2 + 150 \times 4 = 5960 \text{ MWh}$$

$$\text{Average load} = \frac{5960 \times 10^3}{24} = 248.33 \times 10^3 \text{ kW}$$

$$\text{Peak load} = 400 \text{ MW} = 400 \times 10^3 \text{ kW}$$

$$\therefore \quad \text{Load factor} = \frac{\text{Average load}}{\text{Peak load}} = \frac{248.33 \times 10^3}{400 \times 10^3} = 0.62$$ ***Ans.***

(c) Plant capacity = 120 × 4 + 40 × 1 = 520 MW ***Ans.***

$$\text{Plant capacity factor} = \frac{\text{Energy generated}}{\text{Capacity} \times \text{Operating time}}$$

$$= \frac{5960 \times 10^3 \text{ kWh}}{520 \times 10^3 \times 24 \text{ kWh}} = 0.478$$ ***Ans.***

Example 2.10 A 120 MW capacity power plant supplied the following loads with different load factors:

A. Domestic consumers with a maximum demand of 75 MW at a load factor of 30%.

B. Industrial load with a maximum demand of 40 MW at a load factor of 50%.

C. Street light load with a maximum of 5 MW at a load factor of 20%.

Calculate overall cost of electricity produced per kWh for each type of consumer for the following data: Capital cost of the plant = ₹30,000 per kW, Annual rate of interest and depreciation on capital = 12%. Total running cost = ₹20 crore.

Solution Energy supplied per year to all the three types of consumers

$$= [75 \times 0.3 + 40 \times 0.5 + 5 \times 0.2] \times 8760 \times 10^3$$

$$= 38.106 \times 10^7 \text{ kWh}$$

$$\text{Operating cost per kWh} = \frac{20 \times 10^7}{38.106 \times 10^7} = 0.52 = 52 \text{ paise for all three cases}$$

$$\text{Capital cost} = ₹30{,}000 \times 120 \times 10^3 = ₹36 \times 10^8$$

$$\text{Fixed cost per year} = ₹36 \times 10^8 \times 0.12 = ₹4.32 \times 10^8$$

$$\text{Fixed cost per kW} = \frac{4.32 \times 10^8}{120 \times 10^3} = ₹3600.00 \text{ for all three cases}$$

(A) For domestic consumers

Total cost = Fixed cost + Operating cost

$$= ₹3600 \times 75 \times 10^3 + ₹0.52 \times 75 \times 10^3 \times 0.3 \times 8760$$

$$= ₹387405 \times 10^3$$

$$\text{Overall cost per kWh} = \frac{37840 \times 10^3}{75 \times 10^3 \times 0.3 \times 8760} = ₹1.91$$ ***Ans.***

(B) For industrial consumers

$$= ₹3600 \times 40 \times 10^3 + ₹0.52 \times 40 \times 10^3 \times 0.5 \times 8760 = ₹235104 \times 10^3$$

$$\text{Overall cost per kWh} = \frac{235104 \times 10^3}{40 \times 10^3 \times 0.5 \times 8760} = ₹1.34 = ₹1.34 \text{ paise}$$ ***Ans.***

(C) For streetlight load

$$\text{Total cost} = ₹3600 \times 5.0 \times 10^3 + ₹0.52 \times 5 \times 10^3 \times 0.2 \times 8760 = ₹22555 \times 10^3$$

$$\text{Overall cost per kWh} = \frac{22555 \times 10^3}{5 \times 10^3 \times 0.2 \times 8760} = ₹2.57$$ ***Ans.***

Example 2.11 The working life of a power plant is 30 years while its capital cost is 90 crores. The salvage value is 10% of the capital cost. Using sinking-fund method, calculate the amount of money to be saved annually for replacement if the yearly rate of compound interest is 5%.

Solution The annual deposit by sinking-fund method is given by

$$P = \frac{C'_{\text{eqp}} \cdot i}{(1 + i)^N - 1}$$

Here, C'_{eqp} = Capital cost – Salvage value

$$= ₹90 \times 10^7 (1 - 0.1) = ₹81 \times 10^7$$

$$i = 05\% \text{ and } N = 30 \text{ years}$$

So, $$P = \frac{81 \times 10^7 \times 0.05}{(1.05)^{30} - 1} = ₹1.219 \times 10^7 = ₹1.219 \text{ crore}$$ ***Ans.***

Example 2.12 In establishing a power plant, an initial investment of ₹8000 crore has been made. After 5 years, a further of ₹2400 crore have to be paid and from 6th year to 15th year ₹500 crore per year are to be paid. The salvage value of the plant after an useful life of 30 years will be ₹1000 crore. Calculate the present worth of these payments at the time of commission if the interest rate is 10%.

Solution The present worth is given by

$$\text{PW} = \frac{S}{(1 + i)^N}$$

and $$S = \frac{P(1 + i)^N - 1}{i}$$

∴ $$\text{PW} = 8000 + \frac{2400}{(1 + 0.1)^5} + \frac{500[1.1^{15} - 1]}{0.1(1+0.1)^{15}} - \frac{500[1.1^6 - 1]}{0.1(1+ 0.1)^6} - \frac{1000}{(1 + 0.1)^{30}}$$

$$= 8000 + 1490.21 + 3803.03 - 217.76 - 57.3$$

$$= ₹13,018.18 \text{ crore}$$ ***Ans.***

Example 2.13 In a power station, there are two coal-fired generating units A and B and their incremental heat rate is given by

$$(IR)_A = 0.482 \times 10^{-7} L_A^4 - 0.91 \times 10^{-4} L_A^3 + 0.69 \times 10^{-1} L_A^2 - 0.22 \times 10 L_A + 9900$$

$$(IR)_B = 0.96 \times 10^{-7} L_B^4 - 0.78 \times 10^{-4} L_B^3 + 0.27 \times 10^{-1} L_B^2 - 0.22 \times 10 L_B + 9000$$

where IR is in kJ/kWh and L is in MW. If the combined output of these two units is 1000 MW, calculate the incremental heat rate.

Solution The combined output of the unit A and $B = L_C = L_A + L_B$

where L_A is the load of unit A and L_B is the load of unit B

Here, $$L_A + L_B = 1000$$

For economic scheduling principle,

$$(IR)_A = (IR)_B$$

Solving the three equations, namely, for $(IR)_A$, $(IR)_B$ and L_C by any numerical method such as Newton–Raphson, we get

$$L_A = 733.2 \text{ MW}, \quad L_B = 266.8 \text{ MW}, \quad (IR)_A = (IR)_B = 9294 \text{ kJ/kWh} \qquad \textbf{\textit{Ans.}}$$

Example 2.14 A thermal power plant has fuel and operating costs 50 paise per kWh generated. The fixed costs of the power plants are 3000 per kW of installed capacity. Calculate the cost of electrical energy per kWh at load factors of 100%, 80%, 60% and 40%. Comment on the results.

Solution The fixed costs per kWh is given by

$$\text{Fixed cost/kWh} = ₹\frac{3000}{8760} = ₹0.34 = 34 \text{ paise}$$

It is obvious that at 100% load factor, 1 kW plant running for 1 hour will produce 1 kWh. Similarly, at 80% load factor, it will produce 0.8 kWh. The costs per kWh for different load factors are tabulated as follows.

Load factor (%)	*Energy produced (kWh)*	*Fixed cost per hour (paise)*	*Fuel and operating cost (paise)*	*Total cost per hour (paise)*	*Cost per kWh (paise)*
100	1	34	50	85	84
80	0.8	34	40	74	32
60	0.6	34	30	64	106
40	0.4	34	20	54	135

Comment: It is obvious from the result that as the plant load factor decreases, the cost per kWh increases.

Example. 2.15 In a power plant, there are two generating units *A* and *B* whose incremental fuel costs are given by the expressions

$$\frac{dF_A}{dP_A} = 0.07\, P_A + 40 \qquad \text{and} \qquad \frac{dF_B}{dP_B} = 0.085\, P_B + 20$$

where *F* is the fuel cost in rupees per hour and *P* is the power output in MW. Calculate (a) the economic loading of two units if the total load supplied by the power plants is 400 MW, and (b) the loss in fuel cost per hour in the event equally sharing of the load by the two generating units.

Solution (a) As per data given: $P_A + P_B = 400$ ...(1)

For economical loadings of two units, there should be an equal incremental heat rates in both units, thus

$$\frac{dF_A}{dP_A} = \frac{dF_B}{dP_B}$$

Thus, $0.07\, P_A + 40 = 0.085\, P_B + 20$ ...(2)

From Eqs. (1) and (2), we have

$$P_A = 90.33 \text{ MW and } P_B = 309.67 \text{ MW} \qquad \textbf{\textit{Ans.}}$$

(b) In case of equal sharing of load by both units, the increase in cost of fuel for unit A

$$\int \frac{dF_A}{dP_A} = \int_{90.33}^{200} (0.07 P_A + 40) dP_A = \left[0.07 \frac{P_A^2}{2} + 40 P_A \right]_{90.33}^{200}$$

$$= 1114.41 + 4386.8 = ₹5501.21$$

The increase in cost of fuel for the generating unit B

$$\int \frac{dF_B}{dP_B} = \int_{309.67}^{200} (0.085 P_B + 20) dP_B = \left[0.085 \frac{P_B^2}{2} + 20 P_B \right]_{309.67}^{200}$$

$$= -2375.55 - 2193.4 = -₹4568.98$$

∴ The net increase in fuel cost on account of departure from economic distribution of load in units A and B is

$$₹5501.21 - 4568.98 = ₹932.23 \text{ per hour} \qquad \textbf{\textit{Ans.}}$$

Example. 2.16 Two steam turbines, each of 30 MW capacity take a total load of 45 MW. The steam consumption rates for two turbines A and B are given by the following expressions.

$$S_A = 2000 + 10\, L_A - 0.0001\, L_A^2$$

and

$$S_B = 1000 + 7\, L_A - 0.00005\, L_B^2$$

where L represents the load in kW and S represents steam consumption in kg/h.

Calculate the most economical loading when the total load taken by both units is 50 MW.

Solution As per data: $L_A + L_B = 45 \times 10^3$ kW ...(1)

For the most economical loading, the required condition is given by

$$\frac{dS_A}{dL_A} = \frac{dS_B}{dL_B}$$

or

$$10 - 0.0002\, L_A = 7 - 0.0001\, L_B \qquad ...(2)$$

From Eqs. (1) and (2), we have

$$L_A = 25000 \text{ kW} = 25 \text{ MW and } L_B = 20 \text{ MW} \qquad \textbf{\textit{Ans.}}$$

Example 2.17 The following data relate to a 500 MW thermal power plant.

Capital cost of the planet = ₹3 crore per MW, Insurance, interest and depreciation = 10% of capital cost, Capital cost of primary and secondary distribution = 30 crore, Insurance, interest and depreciation on the capital cost of primary and secondary distribution = 5% capital cost, Plant maintenance cost = ₹2 crore per year, Maintenance cost of primary and secondary distribution = 0.2 crore, Salary and wages = 10 crore per year, Consumption of coal = 1 kg/kWh, Cost of coal = ₹800 per tonne, Dividend to stock holders = ₹3 crore per year, Energy loss in transmission and distribution = 10%, Diversity factor = 1.6, Peak demand = 450 MW, Load Factor = 80%.

Calculate a two-part tariff and find the average cost/kWh.

Solution Average load = Peak load × Load factor = 450 × 0.8 = 360 MW

Energy generated per year = $360 \times 10^3 \times 8760 = 31536 \times 10^5$ kWh

Capital cost of the plant = ₹$3 \times 10^7 \times 500$ = ₹15×10^9

Insurance, interest and depreciation cost of plant per year

$$= 0.1 \times 15 \times 10^9 = ₹150 \times 10^7$$

Insurance, interest and depreciation of plant for secondary distribution per year

$$= 0.05 \times 30 \times 10^7 = ₹1.5 \times 10^7$$

Total fixed cost = Insurance + Interest + Depreciation + Dividend to stock holders cost

$$= ₹150 \times 10^7 + 1.5 \times 10^7 + 3 \times 10^7 = ₹154.5 \times 10^7$$

Sum of the individual maximum demand

$$= \text{Maximum demand} \times \text{Diversity factor} = 450 \times 10^3 \times 1.6 = 720 \times 10^3 \text{ kW}$$

$$\therefore \text{ Fixed cost per kW} = \frac{154.5 \times 10^7}{720 \times 10^3} = ₹2145.83$$

Total variable cost = All maintenance cost + Salaries and wages + Fuel cost

$$= ₹0.2 \times 10^7 + 2 \times 10^7 + 10 \times 10^7 + \frac{800 \times 1 \times 31536 \times 10^5}{1000}$$

$$= ₹12.2 \times 10^7 + 252.28 \times 10^7 = ₹264.48 \times 10^7 \qquad \textit{Ans.}$$

Energy transmitted = Energy generated × Transmission efficiency

$$= 31536 \times 10^5 \times 0.9 = 28382.4 \times 10^5 \text{ kWh}$$

$$\text{Cost for energy consumption per kWh} = \frac{264.48 \times 10^7}{28382.4 \times 10^5} = ₹0.93 = 93 \text{ paise} \qquad \textit{Ans.}$$

Total cost = Fixed costs + Variable costs

$$= ₹154.5 \times 10^7 + 264.48 \times 10^7 = ₹418.98 \times 10^7$$

$$\text{Average cost of supply/kWh} = ₹\frac{418.98 \times 10^7}{28382.4 \times 10^5} = ₹1.47 \qquad \textit{Ans.}$$

PROBLEMS

Things to Think About

2.1 What is cause of variable load problem?

2.2 Is it possible to run the plant at design load all the time? If yes, state methods.

Short and Long Theoretical Questions

2.1 Discuss various terminologies used concerning loads in power plants such as connected load, maximum demand, demand factor, average load, load factor, diversity factor, plant capacity factor and plant use factor.

2.2 What are the two basic parameters while planning for a power plant?

2.3 How is the total installed capacity of a power plant decided?

2.4 How is the load-duration curve constructed?

2.5 What do you mean by base load and peaking load? What is the reason that the base load plants are loaded heavily?

2.6 List the considerations for selecting the suitable site for a thermal hydraulic and nuclear power plants.

2.7 What are the elements that contribute to the cost of electricity? What is T&D loss?

2.8 What do you mean by the forced outage rate?

2.9 How can the cost of power generation be reduced?

2.10 What are fixed costs?

2.11 What do you mean by depreciation? Discuss the various methods of determination of depreciation.

2.12 What is the present worth concept? How is it estimated?

2.13 What is incremental heat rate?

2.14 Discuss the principle of economic scheduling applied to power plant economics.

2.15 Prove that for generating units running in parallel, the combined energy input will be the minimum if the incremental heat rate of unit 1 is equal to that of unit 2.

2.16 What is spinning reserve?

2.17 Discuss the effect of variable loads on the plant operation.

2.18 Discuss estimation of tariffs for electrical energy.

Numerical Questions

2.1 The following table shows the demand with respect to time

Time hours	0–6	6–8	8–12	12–14	14–18	18–22	22–24
Load (MW)	40	50	60	50	70	80	40

Draw the load curve and calculate the load factor. In the event of loads above 60 MW taken by a standby unit of 20 MW capacity, calculate the load factor and use factor of standby unit.

[*Ans.* 0.713, 0.751 and 0.601]

2.2 A 30 MW capacity of power plant has to meet the maximum annual demand of 25 MW. It supplies loads with a maximum demand of 10 MW, 8.5 MW, 5 MW and 4.5 MW. The load factor is 0.45. Calculate

(a) the average load.

(b) the energy supplied per year,

(c) the diversity factor, and

(d) the demand factor. [*Ans.* (a) 11.25 MW, (b) 98.54 × 10^6 kWh, (c) 1.10, (d) 0.89]

2.3 The annual load factor, capacity factor and use factor of a central power station are 70%, 50%, and 55% respectively. The maximum demand is 250 MW. Calculate (a) the annual energy production and revenue earned if the cost of energy is ₹2.5/kWh.

(a) The reserve capacity over and above the peak load.

(b) The hours per year the power station will not be in service to the customer.

2.4 The following data relate to a power plant of capacity 80 MW. Calculate the generating cost/kWh. Capital cost = ₹160 crore; Annual wage and taxation = ₹3.6 crore; Interest and depreciation = 10% of the capital cost; Annual load factor = 0.45. [*Ans.* 72 paise]

2.5 The following data related to a power plant: Installed capacity = 2000 MW; Capital cost = ₹4000 crore; Annual cost of fuel, taxes and salaries = ₹600 crore; Rate of interest = 5% of the capital cost; Rate of depreciation = 6% of the capital cost; Annual load factor = 0.8; Capacity factor = 0.7; Energy consumed in running auxiliaries 3% of total energy generated. Calculate the

(a) reserve capacity, and

(b) the cost of power generation per kWh.

2.6 The following data pertain to a power plant–

Load factor = 70%; Capacity factor = 50%;

Use factor = 60%; Peak load = 20 MW. Calculate

(a) the annual energy generated;

(b) the reserve capacity over and above the peak load and

(c) the hours per year during which the plant will note in service to the consumer.

[*Ans.* (a) 122.7 × 10^6 kWh. (b) 8 MW. (c) 2060 h]

2.7 In a power station there are two generating units A and B, whose incremental fuel costs are given by

$$\frac{dF_A}{dP_A} = 0.06\,P_A + 11.4; \qquad \frac{dF_B}{dP_B} = 0.07\,P_B + 10$$

Where P is power in MW and F is fuel cost in rupees/h. Calculate (a) the economic loading of the two units when the total load to be supplied by the power station is 150 MW and (b) the net increase in fuel cost per hour if the load is equally shared by the two units.

[*Ans.* (a) P_A = 70 MW, P_B = 80 MB (b) ₹1.62 per hour]

2.8 The following data relate to a power plant: Capacity of plant = 500 MW; Capital cost = ₹ 30,000 per kW installed; Interest and depreciation = 10% on the capital; Fuel consumption = 1.2 kg/kWh; Fuel cost = 800 per tonne; Salaries, wages, repair and maintenance = 300 × 10^6 per year; Peak load = 450 MW; Load factor = 0.75. Calculate the cost per kWh.

2.9 A power plant of 10 MW capacity has the following daily load

Time (hours)	6–8	8–12	12–1	1–5	5–7	7–9	9–11	11–6
Load (kW)	3,500	8,000	3,000	7,500	8,500	10,000	4,500	2,000

(a) Draw the load curve and load duration curve

(b) Select the size and number of generating units

(c) Draw up the operating schedule of the units

(d) Reserve capacity

(e) Load factor, capacity factor and plant use factor.

Software Based Questions

2.1 Develop a software for calculating the tariff for electric energy consumption by various types of consumers.

2.2 Develop a software for calculating depreciation rate for a different power plants.

2.3 Develop a software for economic scheduling of a power plant.

Design and Open Ended Question

2.1 Prepare a comprehensive report on solving the world problem of sustainable energy which should be pollution free as far as possible involving green energy.

Multiple-Choice Questions

2.1 Plant factor is defined as

(a) $\dfrac{\text{kWh generated in a year}}{\text{kW}_{\text{installed}}}$ (b) $\dfrac{\text{kWh generated}}{\text{kW}_{\text{installed}}}$

(c) $\dfrac{\text{kWh generated in a year}}{\text{kW}_{\text{installed}} \times 24 \times 365}$ (d) $\dfrac{\text{kWh generated}}{\text{kW}_{\text{installed}} \times 24 \times 365}$

2.2 In a steam power plant, the variable load problem may be solved by

(a) Hydel base load and gas or thermal as fluctuating load

(b) Gas base load and thermal or hydraulic as fluctuating load

(c) Thermal as and gas as base load and hydraulic as fluctuating load

(d) Thermal as base load and gas or hydraulic as fluctuating load

2.3 The total annual production cost (C_{tap}) in a power plant is given by

(a) $C_{tap} = \dfrac{(I + D)C}{100} + (W + R) + C_f$ (b) $C_{tap} = \dfrac{(I + D + T)C}{100} + (W + R + M) + C_f$

(c) $C_{tap} = \dfrac{(I + D + T)C}{100} + (W + R)$ (d) $C_{tap} = \dfrac{(I + D + T)C}{100} + (W + R + M) + C_f$

2.4 The plant efficiency of steam power plant ranges from

(a) 20 to 30% (b) 30 to 40% (c) 40 to 50% (d) 50 to 60%

2.5 The plant efficiency of a hydroelectric plant ranges from
(a) 30 to 60% (b) 40 to 60% (c) 50 to 70% (d) 70 to 90%

2.6 The plant efficiency of a combined cycle plant ranges from
(a) 30 to 40% (b) 40 to 50% (c) 50 to 60% (d) 60 to 70%

2.7 The plant efficiency of a gas turbine plant ranges from
(a) 20 to 30% (b) 25 to 35% (c) 35 to 45% (d) 45 to 60%

2.8 The plant efficiency of a nuclear plant ranges from
(a) 15 to 25% (b) 20 to 30% (c) 25 to 30% (d) 30 to 35%

Answers to Multiple-Choice Questions

2.1 (c) 2.2 (d) 2.3 (d) 2.4 (b) 2.5 (d) 2.6 (c) 2.7 (c) 2.8 (d)

CHAPTER

3

Steam Power Plant Cycles

OVERVIEW

A power cycle continuously converts heat (energy released by burning of fossil fuels or energy released from nuclear reaction) into work (shaft work) in which a working fluid repeatedly performs a succession of processes. In the *vapour power cycle,* the working fluid is water or binary fluids which undergoes a change of phase. *The modern, steam power plant uses the Rankine cycle, modified to include superheating, regenerating feed water heating and reheating.*

The coal based steam power plant (Fig. 3.1) consists of the following systems, and components:

(1) Coal handling system consisting of coal unloading, stock, conveyor, crusher, silos, feeders, pulverisers, etc.

(2) Steam generation system consisting of high pressure boiler equipped with superheating, reheating, economiser and preheating.

(3) Post-combustion clean-up system involving SO_2 removal, fly ash removal, draught system, stack, etc.

(4) Power conversion system involving steam turbine, generator, feed pump, etc.

(5) Circulating cooling water system consisting of condenser and circulating water pump, cooling tower, etc.

(6) Feed water treatment system.

(The discussion of steam power plant has been covered in chapters 3 to 9).

This chapter deals with all the elements of modern steam power plant cycles.*

---

*For more details also see the author's book **Steam & Gas Turbines, Turbomachines and Power Plant Engineering.**

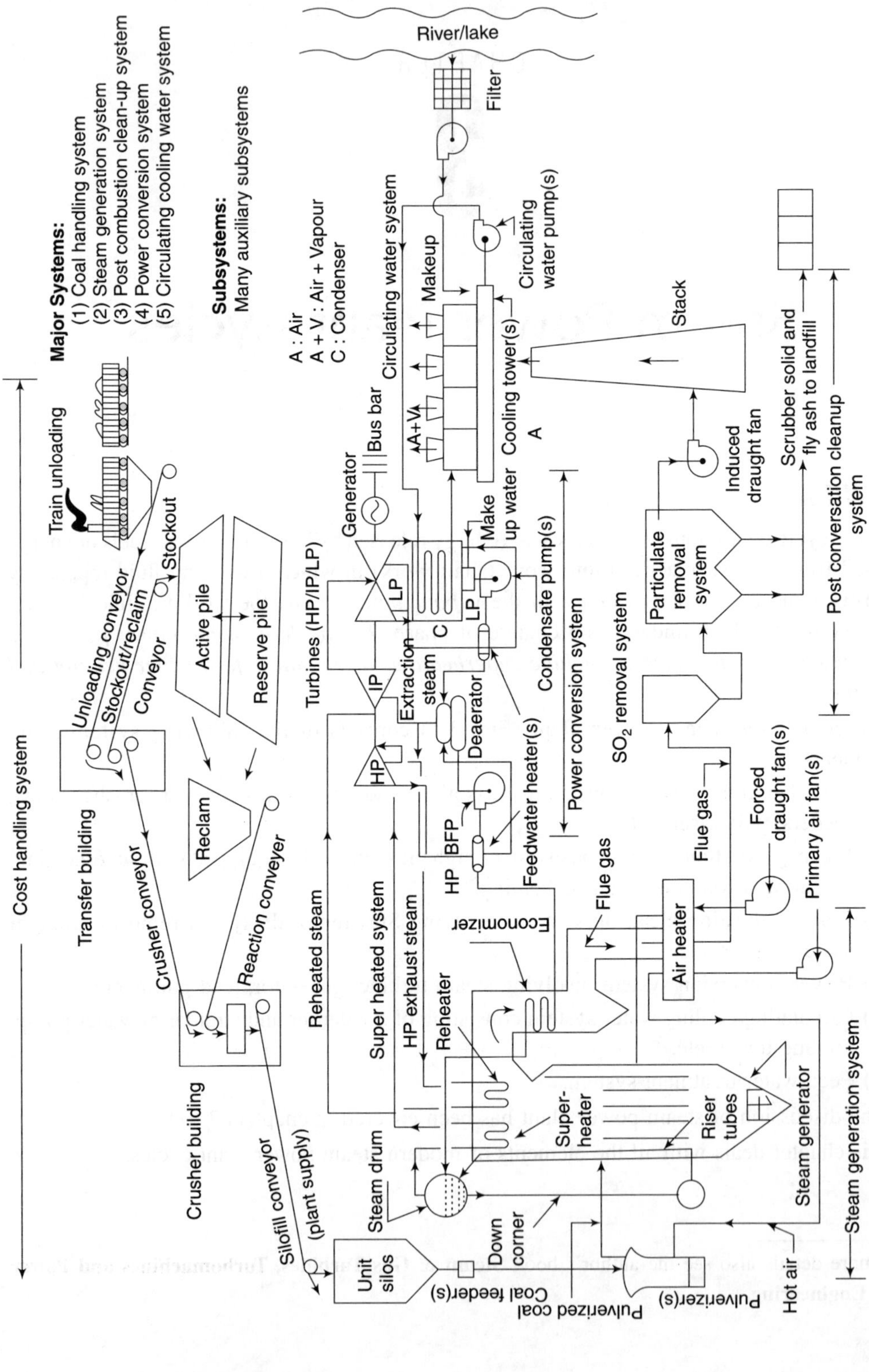

Figure 3.1 Detailed diagramatic arrangement of coal based steam power plant

3.1 SIMPLE RANKINE CYCLE

The theoretical basic cycle for the simple steam turbine power plant is the Rankine cycle which is closed one. A diagrammatic arrangement of a steam turbine power plant and representation of its working processes on *p-v*, *T-s* and *h-s* diagrams are shown in Fig. 3.1 and 3.2 respectively.

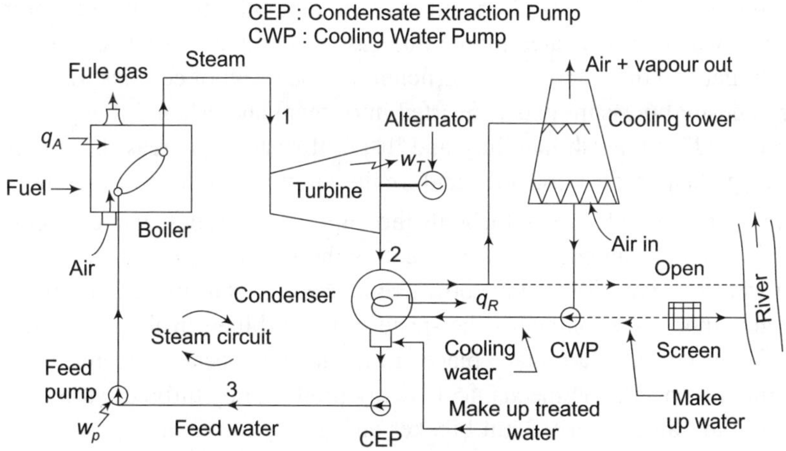

Figure 3.2 Simplified diagramatic arrangement of steam power plant

There are four fluid circuits in a steam power plant (Fig. 3.2). These are as follows:

(i) Steam circuit (closed) (ii) Cooling water circuit (closed or open)

(iii) Cooling air circuit (open) (iv) Combustion gas circuit (open)

A steam power plant consists of four fundamental elements, 1) Boiler, 2) Steam turbine, 3) Condenser, and 4) Feed pump. Many other elements, such as coal handling system, cooling tower, feed heaters, ash handling system, etc., are given in subsequent chapters.

1. *Boiler.* In the boiler plant, the working fluid water receives heat due to the combustion of fuel and is converted into steam at constant pressure. Its efficiency is about 90%.
2. *Steam turbine.* In the steam turbine, the steam from the boiler expands (i.e., steam does work by reducing its pressure, temperature and heat content) and thus performs mechanical work. It is a rotodynamic machine. Its isentropic efficiency is about 86 to 88%.
3. *Condenser.* In the condenser, the exhaust steam from the turbine gives up heat on condensation to the cooling water which cannot be converted into work and must be rejected to restore the initial condition of the working fluid. The condenser enables the exhaust steam to be used as the working fluid of the boiler again and again. It also increases the output of the turbine due to vacuum created inside the condenser. About 50% to 60% of the heat energy input is rejected in the condenser. The performance of condenser is judged by condenser and vacuum efficiencies.

4. *Feed Pump.* The feed pump (centrifugal type), pumps the feed water coming out of from the condenser to the boiler at the desired pressure. It is either motor or turbine driven. It consumes about 2 to 2.5% of the power output.

3.1.1 Working Principle

In the case of coal-fired boiler, the coal is supplied from the coal storage to the boiler through the coal handling plant. The atmospheric air is fed to the boiler through an air preheater (heat recovery device) where air is heated by the flue gases coming out as a waste heat. The heated air enters the boiler and thus increases its efficiency. As a result of combustion, water supplied in the boiler at desired pressure gets converted into steam and ash and flue gases are formed. The ash is removed by the ash handling and disposal system whereas the flue gases passes through the air preheater, dust collector and finally chimney to the atmosphere.

The steam so generated passes through the superheater tubes and gets converted into superheated steam. This superheated steam enters the turbine through the steam stop valve (SSV) and governor valve (GV). The stop valve is used for starting and stopping the turbine whereas the governor valve maintains the speed of the turbine sensibly constant irrespective of the load. Expansion of steam takes places in the turbine because steam does work in the turbine. Alternator converts the mechanical energy produced by turbine into electrical energy which is fed to the transformer, circuit breaker and finally to the bus bar.

The exhaust steam from the turbine is condensed in the condenser (shell and tube type heat exchanger) due to exchange of heat with cooling water. Condenser is equipped with a vacuum pump to extract any air which may be present due to leakage through joints and gases released upon condensation.

The condensate is extracted by a condensate extraction pump and led to L.P., feed heater (direct contact type) where feed water is heated with steam bled from the turbine. Heated feed water is pumped back to the boiler through H.P. feed heater (surface type).

The cooling water is supplied to the condenser by a circulating water pump through a closed circuit. The heated water is cooled in a cooling tower. Some quantity of cooling water in the form of water vapour is carried away by the air hence make-up cooling water to the condenser is supplied from the river or lake or ocean through a filter. If the source of cooling water is an ocean then there is a need of desalination plant. If the source of cooling water (i.e., river or lake, etc.) is very vast, the cooling tower can be dispensed with and the hot water is led to the river or lake as the case may be, as an open system.

Due to leakage of steam from the turbine, some quantity of steam gets lost. Hence, make-up water, well treated through a water treatment plant is generally added up in the well of condenser.

3.1.2 Thermodynamic Analysis

The above processes are shown in *p-v*, *T-s* and *h-s* diagrams of Rankine Cycle. The various processes are discussed as follows.

Process 3-4: The water from the hot-well or the surge tank which is at low pressure is pumped into the boiler at high pressure, p_1. Here pumping process 3-4 is isentropic.

Process 4-5: As the water enters the boiler, water is first heated up to the saturation temperature or evaporation temperature T_1 called sensible heating and during this process the state point moves along curve 4-5. The heat supplied during this process is $h_{f5} - h_{f4}$ and is represented by the area *L*-3-4-5-*M*, i.e., *sensible heat of water.*

Process 5-1: At constant pressure p_1 and temperature T_1, water is completely evaporated into steam. The heat supplied in this process is equal to $h_1 - h_{f5}$ and is represented by *M*-5-1-*N*, i.e., latent heat of vaporization. The state point 1 shows the dry and saturated condition of steam.

Process 1-2: It shows isentropic expansion of steam in the turbine from pressure p_1 to p_2.

Process 2-3: At constant pressure p_2 and temperature T_2, the exhaust steam from the turbine is condensed in the condenser.

It is possible that the steam leaving the boiler may be dry and saturated, wet or superheated and so the corresponding cycle on the *T-s* and *h-s* diagrams are 1-2-3-4-1, 1′-2′-3-4-1 or 1″-2″-3-4-1″ respectively.

The thermal efficiency of the Rankine cycle can be obtained by applying the first law of thermodynamics separately for each process. Let us assume that 1 kg of the working fluid flows through the various elements of the cycle.

From first law of thermodynamics for flow process.

$$\delta q - \delta w = de = dh + \Delta \text{KE} + \Delta \text{PE}$$

Since ΔKE and ΔPE are usually negligible, hence

$$\delta q - \delta w = dh$$

For boiler process 4-1: Process 4-1 is the combined process of 4-5 and 5-1 in the boiler. In this process, the work done is zero because no work is done in the boiler.

The heat supplied to the fluid is given by

$$q_{\text{boiler}} = \int_4^1 dh = h_1 - h_{f4} = (h_1 - h_{f3}) - (h_{f4} - h_{f3})$$

Here, h_{f4} is the sensible heat of water (fluid) in the boiler at state 4.

For turbine process 1-2: In this process, heat supplied is zero because the process is isentropic expansion and hence,

$$w_{\text{turbine}} = \int_1^2 -dh = \int_2^1 dh = h_1 - h_2$$

For condenser process 2-3: In this process, no work is done and so from the first law of thermodynamics

$$q_{\text{condenser}} = h_2 - h_{f3} = h_2 - h_{f2}$$

Here h_{f3} is the enthalpy of water of state 3 corresponding to pressure p_2, i.e., $hf_2 = hf_3$.

For feed pump process 3-4: In this process, no heat is supplied because the process is isentropic; hence from first law of thermodynamics

$$w_{\text{pump}} = h_{f4} - h_{f3} = v_{f2}\,(p_1 - p_2)$$

Here, v_{f2} = sp. volume of water in m^3/kg corresponding to final pressure of p_2, kN/m^2

p_1 = initial pressure of steam in kN/m^2

p_2 = final pressure of steam in kN/m^2

∴ Heat supplied = $q_{\text{boiler}} = h_1 - h_{f3} - v_{f2}\,(p_1 - p_2)$

So net work done is = $w_{\text{net}} = w_{\text{turbine}} - w_{\text{pump}}$

$$= (h_1 - h_2) - (h_{f4} - h_{f3}) = (h_1 - h_2) - v_{f2}\,(p_1 - p_2)$$

∴ Thermal efficiency = $\eta_{\text{th}} = \dfrac{\text{Net work}}{\text{Heat supplied}} = \dfrac{w_{\text{net}}}{q_{\text{boiler}}}$

or $$(\eta_{\text{th}})_{\text{Rankine}} = \frac{(h_1 - h_2) - (h_{f4} - h_{f3})}{(h_1 - h_{f3}) - (h_{f4} - h_{f3})} = \frac{(h_1 - h_2) - w_p}{(h_1 - h_{f3}) - w_p} \quad ...(3.1)$$

Compared to turbine work, pump work $(w)_{\text{pump}}$ is very small and it may be neglected because the specific volume of water is very small. Hence, we have

$$w_{\text{pump}} \approx 0, \quad \text{so} \quad h_{f4} = h_{f3}$$

Therefore, $$(\eta_{th})_{\text{Rankine}} = \frac{(h_1 - h_2)}{(h_1 - h_{f3})} = \frac{h_1 - h_2}{h_1 - h_{f2}} \quad ...(3.2)$$

The thermal efficiency of the Rankine cycle may be also expressed as (in the form of areas). Referring to Fig. 3.2.

$$(\eta_{\text{th}})_{\text{Rankine}} = \frac{\text{Area 1 2 3 4 5 1}}{\text{Area L 3 5 1 NL}}$$

It is to be noted that if the condensate is pumped back to the boiler from the state point 3′ instead of 3, then it forms the *Carnot cycle*. So, 1-2-3′-5-1 is Carnot cycle. We know that Carnot cycle has more efficiency than Rankine cycle and this is also obvious from Fig. 3.3 with the same temperature limits. The reason is that the average temperature at which heat is added in the Rankine cycle lies between T_4 and T_5 (or T_1) and is thus less than the constant temperature T_1 at which heat is added to the Carnot cycle. *Though the efficiency of the Carnot cycle is more than Rankine, still we*

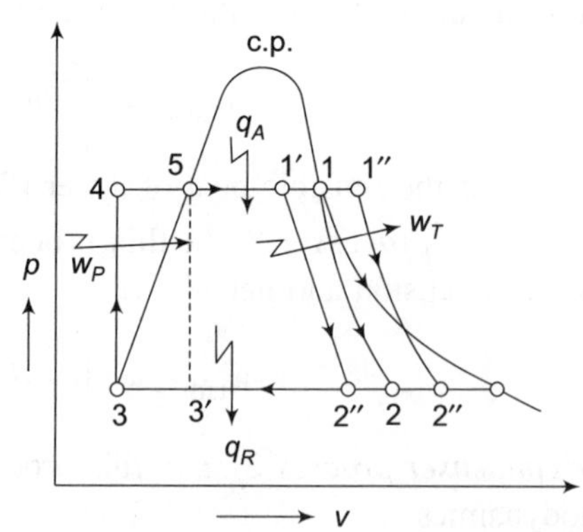

Figure 3.3 *p-v* diagram of Rankine cycle

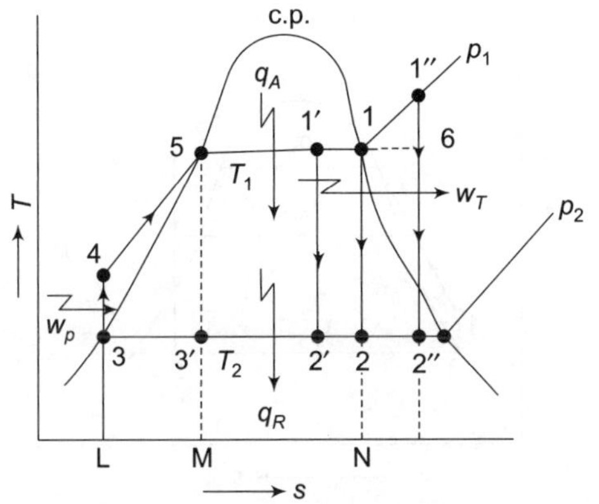

Figure 3.4 *T-s* diagram of Rankine cycle

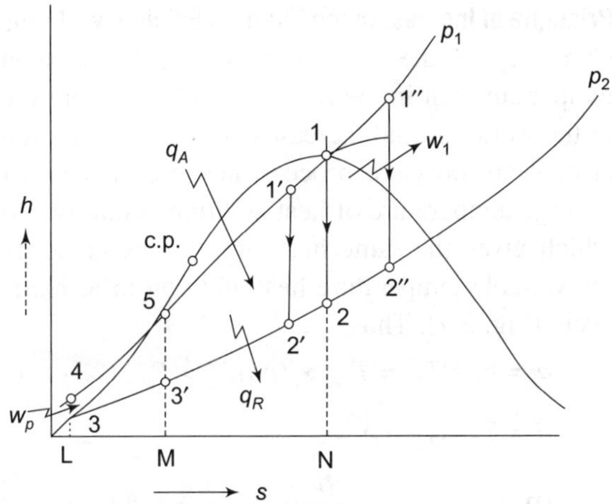

Figure 3.5 *h-s* diagram of Rankine cycle. Supercritical boiler pressure

take the Rankine cycle as the theoretical cycle for the steam power plant because of two main facts as follows.

Rankine cycle working on supercritical boiler pressure of steam is shown in Fig. 3.5.

3.1.3 Limitations of Carnot Cycle as Vapour Cycle

(a) In the Carnot cycle, the condensate is pumped to boiler from state point 3′ (Fig. 3.4) where the condensate is in the form of water and its vapour to state point 5 where the water is saturated. *It is very difficult to build a pump which can handle such a mixture of water and its vapour at state point 3′ and deliver saturated liquid at state point 5. Practically, it is very easy to completely condensate the vapor up to state point 3 and handle only liquid in the pump as is done in the Rankine cycle.*

(b) In the Carnot cycle, superheating of steam is a great problem. Superheating has to be done at constant temperature along path 1–6 but during this process pressure drops. This means that the heat will transfer to the vapour in the expansion process. In actual practice, this is very difficult to achieve. But in the Rankine cycle, the vapour is superheated at constant pressure from 1 to 1″ without any difficulty.

From the above facts, it is clear that the Rankine cycle is possible in actual practice and not Carnot cycle so the former is taken as theoretical cycle for the steam power plant.

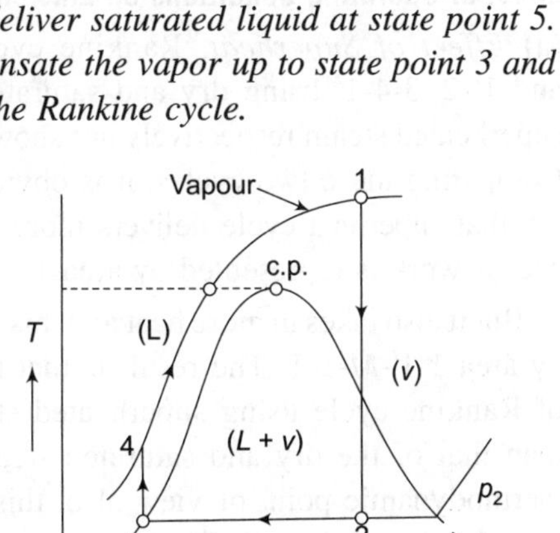

Figure 3.6 Rankine cycle with supercritical boiler pressure

Principle of Increasing the Thermal Efficiency. In the Carnot cycle, the heat addition is at the constant temperature but in the Rankine cycle, temperature of the working fluid increases during heat addition at constant pressure process, hence a concept of average temperature of heat addition is introduced which gives the same heat supplied as given by the variable temperature heat addition in Rankine cycle (Fig. 3.7). Thus,

$$q_A = h_1 - h_{f4} = T_{ave} \times (\Delta s)_{ave} = T_{ave} (s_2 - s_3)$$

$$q_R = T_2 (s_2 - s_3)$$

$$\therefore \quad (\eta_{th})_{Rankine} = 1 - \frac{q_R}{q_A} = 1 - \frac{T_2}{T_{ave}} = f(T_{ave})$$

only as T_2 is fixed.

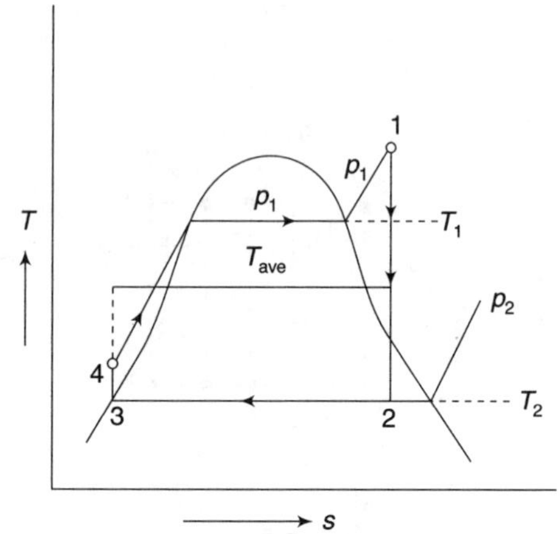

Figure 3.7 Concept of average temperature of heat addition

Thus, higher the average temperature of heat additions, the higher will be the Rankine efficiency.

Based on the concept of average temperature of heat addition in the Rankine cycle and the outcome of Carnot cycle, following two principles are established to increase the thermal efficiency of Rankine cycle.

(a) *By increasing the average temperature at which heat is added to the cycle.*

(b) *By decreasing the average temperature at which heat is rejected from the cycle.*

Effect of Operating Conditions on Efficiency

(a) *Effect of Superheat.* Rankine cycle 1-2-3-4-1 and 1′-2′-3-4-1′ using dry and saturated steam and superheated steam respectively are shown in Fig. 3.8. Comparing these two cycles, it is obvious from Fig. 3.8 that superheat cycle delivers more work and the excess work is represented by area 1-1′-2′-2-1.

But it also takes in more heat and this is represented by area 1-1′-*M*-*L*-1. The result is that the efficiency of Rankine cycle using superheated steam is more than that of the dry and saturated steam. From the thermodynamic point of view also, this result could have been anticipated. Due to the superheating of steam, the average temperature of heat addition to the cycle increases, so there should be an increase in the thermal efficiency compared to the cycle using dry and saturated steam.

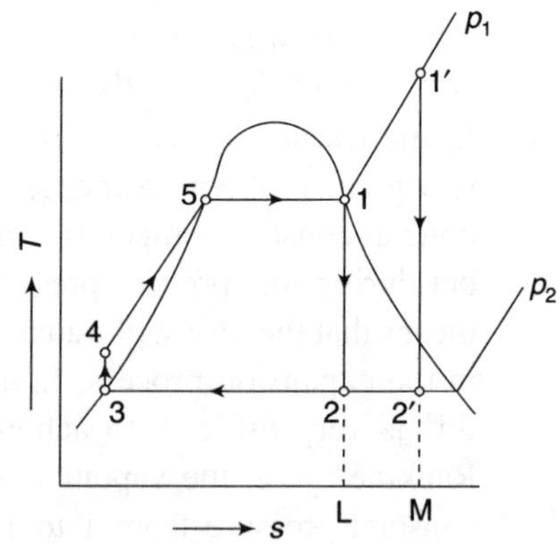

Figure 3.8 Effect of superheat on Rankine cycle

(b) *Effect of Maximum Pressure.* In Fig. 3.9, two cycles 1-2-3-4-1 and 1′-2′-3-4′-1′ have the same maximum temperature $T_1' = T_1$ but different maximum pressure $p'_1 > p_1$. The condenser pressure is the same in both the cases. It is obvious from Fig. 3.9 that due to the increase in the maximum pressure from pressure p_1 to p'_1, the net work increases by the area shown by the horizontal hatching line and decreases by the area shown by cross hatching. After drawing the diagram true to scale, it has been found that these areas are nearly equal, so the net work is nearly the same, but the heat rejected decreases by the area 2-2′-*L*-*M*. The thermal efficiency is given by

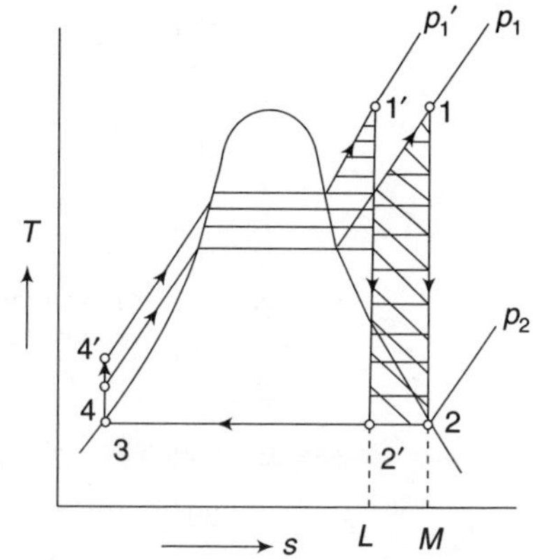

Figure 3.9 Effect of maximum pressure on rankine cycle

$$\eta_{th} = 1 - \frac{q_R}{q_A} = 1 - \frac{\text{Heat rejected}}{\text{Heat added}}$$

Since heat rejection is reduced in the case of increasing maximum pressure, so thermal efficiency also increases. From the second law of thermodynamics it can be anticipated that the thermal efficiency increases with increase in maximum pressure because the average temperature of heat addition increases with increase of maximum pressure.

(c) *Effect of Exhaust Pressure.* In Fig. 3.10, two cycles 1-2-3-4-1 and 1-2′-3′-4′-1 have the same maximum pressure and temperature but different exhaust pressures $p_2 < p_2'$. It is obvious from Fig. 3.10 that if the exhaust pressure is reduced from p_2 to p_2', the net work is increased by area 2-2′-3′-4′-4-3-2- and the heat supplied to steam increases by the area *L*-3′-4-4′-*M*-*L*.

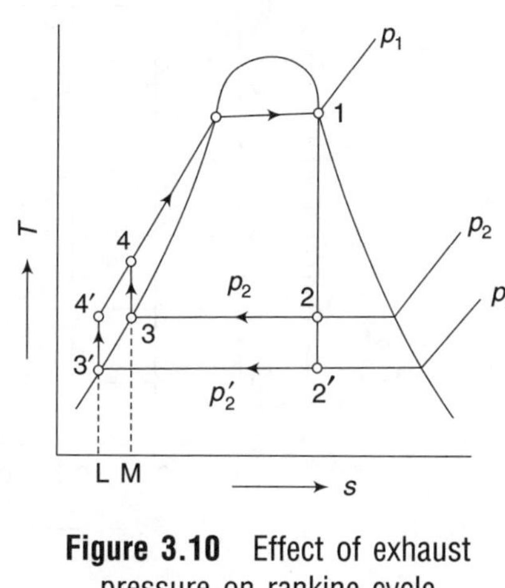

Figure 3.10 Effect of exhaust pressure on rankine cycle

Drawing the diagram to scale it has been found that these two areas are nearly equal, so the net result is an increase in the cycle thermal efficiency. From the second law of thermodynamics, it can be anticipated that the thermal efficiency increases with decrease in exhaust pressure because the average temperature of heat rejection decreases with decrease of exhaust pressure.

Performance Curves: Figure 3.11 shows effect of initial, i.e., throttle temperatures and initial pressures on the Rankine thermal efficiency. At a particular value of initial pressure except very high pressure, the thermal efficiency increases approximately linearly as the temperature increases. At a very high pressure in the beginning, the increase in thermal efficiency is very fast at the start and after that it increases slowly.

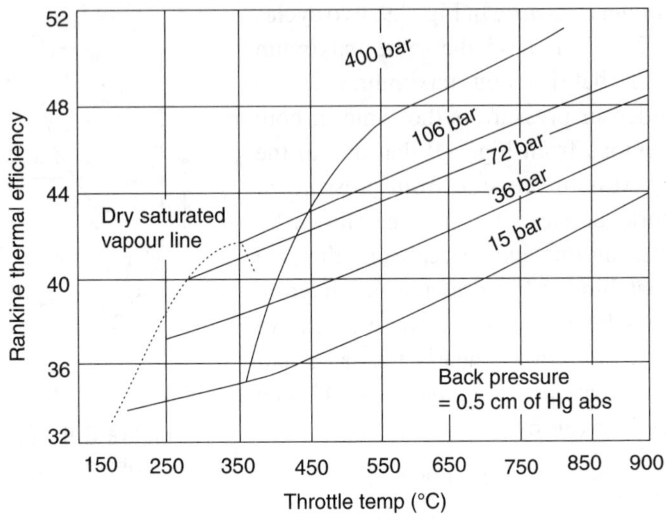

Figure 3.11 Effect of throttle temperature

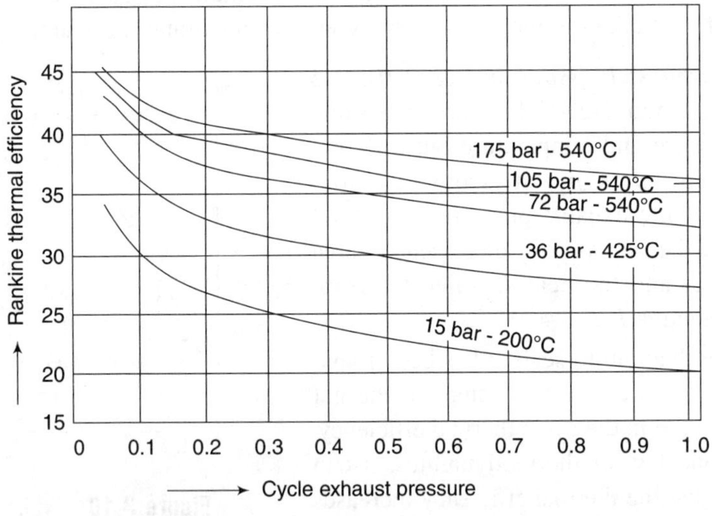

Figure 3.12 Effect of back and initial pressure

The effect of back pressure on thermal efficiency is shown in Fig. 3.12. As the back pressure decreases, the thermal efficiency increases, it is also obvious that as the initial pressure increases the thermal effiency increases.

Figure 3.13 shows the variation of cycle efficiency and specific steam consumption with boiler pressure and temperature.

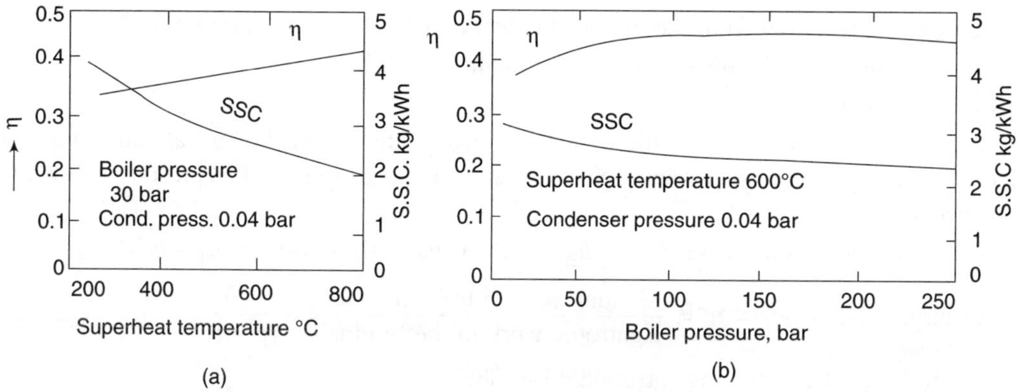

Figure 3.13 Variation of cycle efficiency and SSC with boiler pressure and temperature

3.1.4 Methods of Increasing the Thermal Efficiency

Based on the above two principles, the methods of increasing the thermal efficiency are as follows:

(i) By increasing the superheating temperature of steam
(ii) By increasing the maximum pressure of steam
(iii) By reducing the exhaust pressure of steam
(iv) By regenerative feed heating
(v) By reheating of steam
(vi) By water extraction
(vii) By using binary-vapour

Generally, a modern steam turbine power plant is a combination of all the above except term (vii). In some case term (vii) is also used.

3.1.5 Deviation of Actual Cycle from Theoretical Cycle (Internally Irreversible Cycle)

Actual steam turbine power plant cycle deviates from the theoretical, i.e., ideal cycle because losses occur in various components like boiler, turbine, pump, condenser and piping. Internal irreversibility of Rankine cycle is caused by fluid friction, throttling and mixing. These losses are shown in Fig. 3.14. We will now discuss each of them briefly.

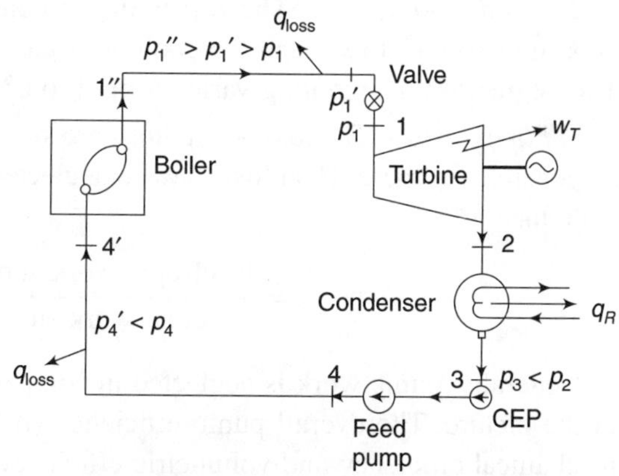

Figure 3.14 Losses in Rankine cycle

1. *Turbine losses.* The losses in turbine arises mainly due to:

(i) friction resulting from the flow of the working fluid through the turbine,

(ii) heat loss from the turbine to the surrounding, and

(iii) leakage loss.

The effect of friction is to increase the entropy and it can be visualized from Fig. 3.15. Process 1-2 represents isentropic expansion while process 1-2′ represents the actual expansion.

The isentropic turbine work = $h_1 - h_2$ and actual turbine work = $h_1 - h_2'$

$$\text{So turbine efficiency} = \eta_t = \frac{\text{Actual work in the turbine}}{\text{Isentropic work in the turbine}} = \frac{h_1 - h_2'}{h_1 - h_2} \quad \text{...(3.3)}$$

The turbine efficiency ranges from 85% to 88%.

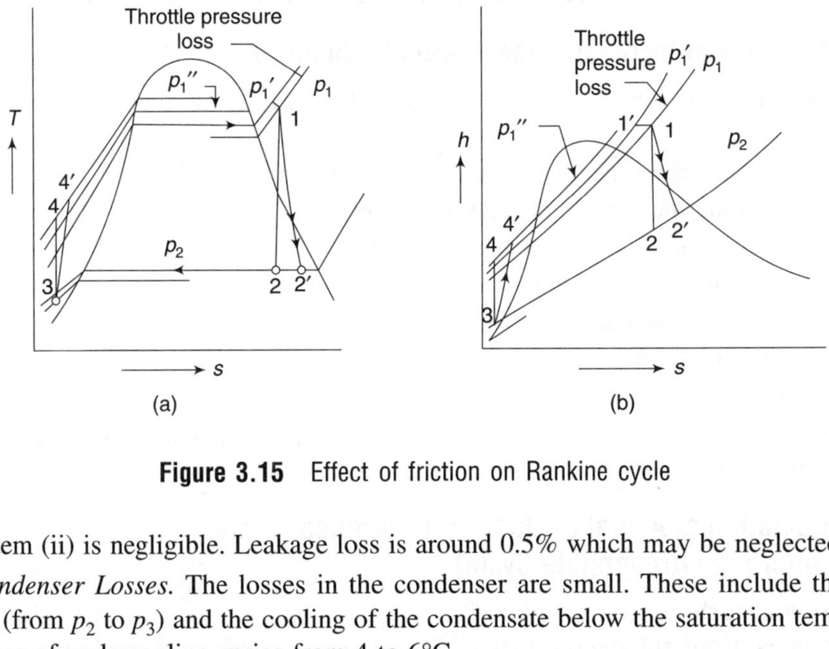

Figure 3.15 Effect of friction on Rankine cycle

The item (ii) is negligible. Leakage loss is around 0.5% which may be neglected.

2. *Condenser Losses.* The losses in the condenser are small. These include the loss in pressure (from p_2 to p_3) and the cooling of the condensate below the saturation temperature. The degree of undercooling varies from 4 to 6°C.

3. *Pump Losses.* The losses in pumps are similar to losses in turbines and friction is mainly responsible for these. Heat loss may be neglected because it is very low. The pump efficiency is defined as

$$\eta_{pump} = \frac{\text{Isentropic work supplied}}{\text{Actual work supplied}} = \frac{h_{f4} - h_{f3}}{h'_{f4} - h_{f3}} \quad \text{...(3.4)}$$

Generally, pump work is neglected in comparison to turbine, so pump losses do not come in the picture. The overall pump efficiency (η_p) which is a product of hydraulic efficiency, mechanical efficiency and volumetric efficiency ranges from 80% to 90%.

4. *Boiler Losses.* The losses in the boiler include the heat loss to the surroundings and heat carried away by the flue gases. Boiler efficiency ranges from 88% to 92%. Boiler losses cannot be depicted in *T-s* and *h-s* charts but it affects the overall thermal efficiency of the plant. Due to inefficiency of boiler, fuel requirement will be more.

5. *Piping Losses.* There are mainly two types of losses in steam pipes. Firstly, there is a pressure drop of steam (from p_1'' to p_1') due to friction between the pipe wall and steam. Secondly, there is a heat loss from the pipe to the surroundings. Due to these two losses the state of steam entering the turbine is at a lower pressure and temperature and greater entropy as compared to the state leaving the boiler. So the effect is that the net work of the cycle is reduced and thus the efficiency is lower.

6. *Regulating Valve Losses.* When the steam passes through regulating valves, there is a pressure drop (from p_1' to p_1) in the steam of constant enthalpy ($h_1' = h_1$) due to throttling action. As a result, the net work of the cycle is reduced and thus the thermal efficiency drops. About 5% higher pressure is supplied to boiler to take care of pressure loss.

3.1.6 Externally Irreversible Rankine Cycle

It is the temperature differences between the combustion gases and the working fluid on the source side and the temperature differences between the condensing working fluid and the condenser cooling water on the sink side cause external irreversibility in the Rankine cycle (Fig. 3.16). The products of combustion, i.e., flue gases get cooled from *a* to *d* and the working fluid (water) temperature rises from 4 to 1 in counter-flow heat exchanger (steam generator). The minimum temperature differences sbetween the two fluids are *c*-5 and *a*-1, and the points where these occur are called *pinch points*. Generally, *a*-1 is known as approach temperature difference. The pinch point plays an important part in deciding the steam generator efficiency. Too small a pinch point difference causes a lower external thermal irreversibility and an increase in surface area means more expensive steam generator, while a too large pinch point difference results in small and inexpensive steam generator resulting in reduced plant efficiency. This suggests that an optimum pinch point is desirable taking care efficiency and cost.

In the condenser side, line *e-f* represents the rise in temperature of cooling water while condensation of steam takes place along 2-3.

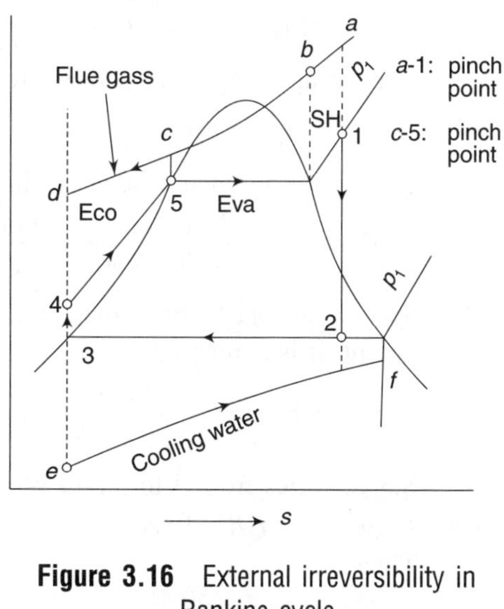

Figure 3.16 External irreversibility in Rankine cycle

3.1.7 Efficiencies

(i) *Isentropic cycle efficiency* $= \eta_c = (\eta_{th})_{Rankine} = \dfrac{\Delta h_{isen}}{q_A}$ ...(3.5) where Δh_{isen} = Isentropic heat drop in the turbine, kJ/kg of steam

q_A = Heat supplied in the cycle = $(h_1 - h_{f2}) = \dot{m}_f \times (LCV)_f$, kJ/kg

$\dot{m}_f$ = Mass flow rate of fuel, kg/kg of steam

$(LCV)_f$ = Lower calorific value of fuel, kJ/kg of fuel

(ii) Thermal efficiency of the turbine plant = $\eta_{th} = \dfrac{\Delta h_{coupa}}{q_A}$ ...(3.6)

where Δh_{coupa} = Actual coupling work considering all type of losses up to coupling.

(iii) *Thermal efficiency ratio* = $\eta_r = \dfrac{\eta_{th}}{\eta_c} = \dfrac{\Delta h_{coupa}}{q_A} \times \dfrac{q_A}{\Delta h_{isen}}$

or $$\eta_r = \frac{\Delta h_{coupa}}{\Delta h_{isen}} = \eta_t \times \eta_{mech} \quad (3.7)$$

So we see that the thermal efficiency ratio is the ratio of actual coupling work to the isentropic heat drop based on per kg of steam.

(iv) *Overall thermal efficiency* (η_{oth}). The overall thermal effficiency takes care of all the losses in various elements of steam power plant. It is given by

$$\eta_{oth} = \frac{\text{G.O.}}{q_{Aa}} = \frac{[w_{t,isen} \times \eta_t \times \eta_m \times \eta_{generator} - w_{p,isen}/\eta_p] \times \eta_{boiler}}{(q_A)} \quad ...(3.8)$$

where η_p = Overall pump efficiency = $\eta_h \times \eta_m \times \eta_v$

$(q_A)_a = [(h_1 - h_{f2}) - w_p/\eta_p] = \dot{m}_f \times (LCV)_f / \eta_{boiler} = q_A/\eta_{boiler}$

Neglecting pump work and where $q_A = \dot{m}_f \times (LCV)_f$

$$\eta_{oth} \approx (\eta_{th})_{Rankine} \times \eta_{boiler} \times \eta_{turb} \times \eta_{mech} \times \eta_{generator} \quad ...(3.9)$$

The overall thermal efficiency varies from 35 to 40%.

(v) Work ratio = $\dfrac{\text{Net work}}{\text{Turbine work}} = \dfrac{w_{net}}{w_t}$ ...(3.10)

(vi) *Specific steam consumption (S.S.C.):* It is defined as the steam consumption per power output. It is expressed as

$$\text{S.S.C.} = \frac{3600}{w_{net}\,(\text{kWh/kg})},\ \text{kg/kWh} = \frac{3600\ \dot{m}_s\ (\text{kg/s})}{W_{net}\,(\text{kW})} \quad ...(3.11)$$

S.S.C. varies from 3 to 5 kg/kWh.

(vii) *Heat Rate (HR):* It is defined as the amount of heat supplied per power output. Thus,

$$\text{H.R.} = \frac{3600}{\eta_{th}} = \frac{3600 \cdot q_A}{w_{net}};\ \text{kJ/kWh} \quad ...(3.12)$$

(viii) Specific circulating cooling water consumption (SCWC) = $\dfrac{m_w(\text{kg/h})}{W_{net}(\text{kW})}$; kg/kWh ...(3.13)

SCWC varies from; 120 to 200 kg/kWh and make-up water will be 2 to 10%.

3.1.8 Requirements of an Ideal Working Fluid

There are several vapours which have physical properties suitable for working fluid. They are steam, mercury vapour, sulphur dioxide, diphenyl oxide and certain hydrocarbons. The following are the requirements for an ideal working fluid:

(i) Ample amount should be available at low cost.
(ii) Critical temperature should be higher than metallurgical limits.
(iii) Reasonable saturation pressure at maximum temperature of the cycle from metallurgical point of view of boiler.
(iv) Steep saturated vapour line to minimise moisture problem in expansion of steam in the turbine.
(v) It should wet the boiler surfaces enveloping it and should be chemically stable at the maximum temperature of the boiler.
(vi) Saturation pressure at the minimum temperature of the cycle should be higher than atmospheric, otherwise the maintenance of the condenser will be costly.
(vii) Low liquid specific heat so that most of heat is added at the maximum temperature.
(viii) Considerable decrease in volume upon condensation.
(ix) Non-toxic and non-corrosive.
(x) Freezing point should be much below the normal atmospheric pressure.

Among all types of working fluids, water satisfies the maximum requirements. Its general abundancy and low cost is of prime importance due to which it is selected as the working fluid in steam turbine power plants.

3.1.9 Critical Temperature and Pressure of Water

At all temperatures above the critical, it is impossible to liquefy water vapour by using pressure, no matter how great the pressure is employed. The critical temperature of water is 374°C. At this temperature, the vapour is called critical pressure and its value is 221.2 bars.

Example 3.1 A steam turbine receives steam at pressure 20 bars and superheated to 88.6°C. The exhaust pressure is 0.07 bar and the expansion of strem takes place isentropically. Using steam table only, calculate the following:

(a) Heat supplied, assuming that the feed pump supplies water to the boiler at 20 bars; (b) Heat rejected; (c) Net work done; (d) Work done by the turbine; (e) Thermal efficiency and (f) Theoretical steam consumption.

If the actual steam consumption is 5 kg/kWh, what is the efficiency ratio of the turbine and actual power output at turbine coupling.

Solution Refer to Fig. 3.17 Example 3.1. From steam table, t_1 at 20 bars = 212.4°C.

So, initial temperature of steam = 212.4 + 88.6 = 301°C, h_1 = 3025 kJ/kg

Now $s_1 = s_2 = x_2\, s_{g2} + (1 - x_2)s_{f2}$ or $6.768 = 8.274\, x_2 + (1 - x_2)\, 0.559$

or $x_2 = 0.8047$

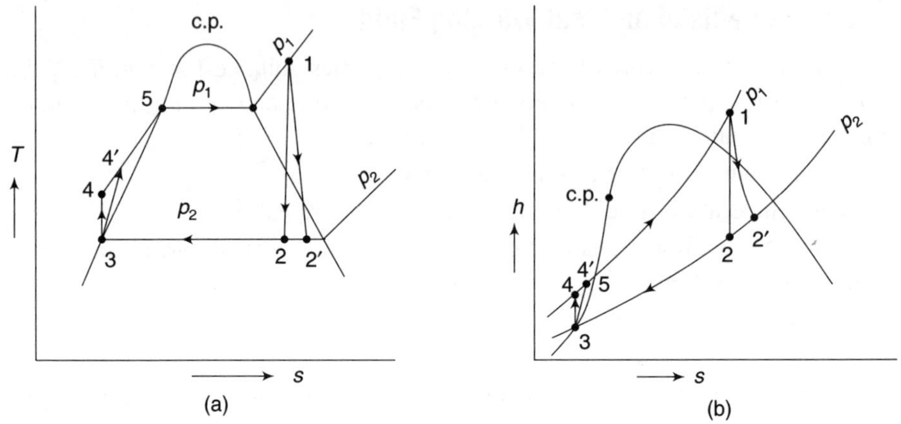

Figure 3.17 Example 3.1

Again, $h_2 = h_{f2} + x_2\, h_{fg} = 163 + 0.8047 \times 2409 = 2101.52$ kJ/kg

From steam table, $h_{f3} = 163$ kJ/kg

Pump work $= v_{f2}\,(p_1 - p_2)10^2 = 0.001(20 - 0.07)10^2 = 1.993$ kJ/kg

(a) Heat supplied $= h_1 - h_{f3}$ – Pump work $= 3025 - 163 - 1.993 = 2860$ kJ/kg ***Ans.***

(b) Heat rejected $= h_2 - h_{f3} = 2101.52 - 163 = 1938.52$ kJ/kg

(c) Net work done = Turbine work – Pump work

$$= (h_1 - h_2) - 1.993 = 3025 - 2101.52 - 1.993 = 921.847 \text{ kJ/kg}$$ ***Ans.***

(d) Work done by the turbine $= h_1 - h_2 = 3025 - 2101.52 = 923.48$ kJ/kg. ***Ans.***

(e) Thermal efficiency $= \dfrac{\text{Net work done}}{\text{Heat supplied}} = \dfrac{923.48}{2860} = 32.28\%$ ***Ans.***

(f) Theoretical steam consumption $= \dfrac{3600}{\text{Net work done}} = \dfrac{3600}{923.48} = 3.898$ kg/kWh ***Ans.***

$$\text{Efficiency ratio} = \frac{(\Delta h)_{\text{coupa}}}{(\Delta h)_{\text{isen}}}$$

$$\eta_r = \frac{\text{Theoretical steam consumption in kg/kWh}}{\text{Actual steam consumption in kg/kWh}} = \frac{3.898}{5} = 77.96\%$$ ***Ans.***

Actual power output at turbine coupling

$$w_{\text{neta}} = w_{t\ \text{isen}} \times \eta_r - w_p = 923.48 \times 0.7796 - 1.993 = 717.95 \text{ kJ/kg}$$ ***Ans.***

Example 3.2 The following are the mean observation made during test on a steam turbine—Duration of test = 2 hours, Power developed at (turbine coupling) = 4500 kW, Total mass of steam condensed = 50,000 kg. Steam pressure at stop valve = 14 bars, Steam temperature at stop valve = 300°C, Barometeric height = 76 cm of Hg. Vacuum = 72.5 cm of Hg.

Calculate the thermal efficiency of the turbine and also its efficiency ratio. Assume that the condensate is returned to the boiler plant at the exhaust saturation temperature. Also calculate the fuel cost per kWh generated in steam power plant if $\eta_b = 0.92$, L.C.V. of coal = 29000 kJ/kg. of steam, cost of coal = ₹1800.00 per tonne and 5% of steam generated is used to cover losses and auxiliary power.

Solution Refer to Fig. 3.17 also for Example 3.2. Absolute exhaust pressure = 76 – 72.5 = 3.5 cm of Hg.

$$p_2 = \frac{3.5 \times 1.013}{76} = 0.04665 \text{ bar}$$

From Mollier chart: $h_1 = 3038$ kJ/kg (at 14 bar 300°C),

$h_2 = 2105$ kJ/kg (at 0.04665) bar

From steam table: $h_{f3} = 133$ kJ/kg (at 0.04665 bar)

Neglecting the pump work

Heat supplied: $h_1 - h_{f3} = h_1 - h_{f2} = 3038 - 133 = 2905$ kJ/kg

Theoretical work done by turbine = $h_1 - h_2 = 3038 - 2105 = 933$ kJ/kg

$$\text{Actual steam consumption} = \frac{50{,}000}{2 \times 3600} = 6.944 \text{ kg/s}$$

$$\text{Actual thermal efficiency} = \frac{\text{Actual work done}}{\text{Actual heat supplied}} = \frac{4500 \times 0.92}{2905 \times 6.944} = 20.52\% \qquad \textit{Ans.}$$

$$\text{Efficiency ratio} = \frac{(\Delta_h)_{\text{coup}}}{(\Delta_h)_{\text{isen}}} = \frac{4500}{933 \times 6.944} = 69.45\% \qquad \textit{Ans.}$$

$$\dot{m}_f(\text{kg/kWh}) \times LCV(\text{kJ/kg}) \times \eta_b = \dot{m}_{sa}\ (\text{kg/kWh}) \times q_{Aa}\ (\text{kJ/kg})$$

or

$$\dot{m}_f \times 29000 \times 0.92 = \frac{6.944 \times 1.05 \times 3600 \times 2905}{4500}$$

or

$$\dot{m}_f = \frac{6.944 \times 1.05 \times 3600 \times 2905}{4500 \times 0.92 \times 29000} = 0.635 \text{ kg/kWh}$$

∴

$$\text{Fuel cost} = ₹\ \frac{1800 \times 0.635}{1000} = ₹1.143/\text{kWh}. \qquad \textit{Ans.}$$

This may be treated as generation cost only based on fuel cost. The total generation and distribution cost will include many other costs such as wages, depreciation, maintenance, distribution, etc.

3.2 BINARY-VAPOUR CYCLE

The main aim of the steam turbine designer is to achieve higher thermal efficiency of the turbine plant. We have already seen that the thermal efficiency of the Rankine cycle is increased by (a) *increasing the average temperature at which heat is added to the cycle and* (b) *decreasing the average temperature at which heat is rejected by the cycle. Metallurgical problems in the boiler and steam turbine limit the maximum temperature but at a lower pressure.* At present this maximum temperature limit is about 650°C. The cooling medium in the condenser is a limit to the minimum temperature of the cycle. This temperature is about 35°C, i.e., the temperature of natural water. If steam is used as the working fluid in Rankine cycle operating over the above temperature range, i.e., from 650°C to 35°C the difficulties arise. These are:

(i) Since the critical temperature of water is 374°C so for obtaining the maximum temperature of 650°C, the steam has to be superheated and this is only possible at very high pressure. The metallurgical problems do not allow high pressure and high temperature, and also, the average temperature of heat addition reduces thus reducing efficiency. This high temperature is therefore desirable at lower pressure.

(ii) High initial pressure causes excessive moisture in the low pressure stages.

(iii) As the pressure increases the latent heat of vaporization decreases.

Water also has disadvantages at the minimum temperature of the cycle. Condenser operates at vacuum because the saturation pressure corresponding to 35°C is 0.05622 bar which is much lower than atmospheric pressure. So it is essential to minimise air leakage in the condenser. It is also desirable that the back pressure is somewhat above the atmospheric pressure.

We observe, therefore, that all the requirements for an ideal working fluid are not met by water or any single working fluid. Two working fluids are therefore used in the cycle so that most of the requirements are fulfilled. Such a cycle is called *Binary-Vapour cycle* (also called combined cycle) and in this mercury and steam are commonly used as working fluids. The following are the advantages and disadvantages of mercury as a binary vapour cycle fluid.

3.2.1 Advantages of Mercury

(i) Moderate vapour pressure at high temperature. At 540°C, the saturation pressure is less than 14 bars.

(ii) Very stable.

(iii) Liquid mercury has a high density which is advantageous in the separation of the liquid from vapour in boiler and in feeding back to the boiler under hydrostatic head rather than under the pumping head.

(iv) Specific heat of the liquid is only 0.13, so the liquid line is steep, making the cycle closer to Carnot cycle.

(v) Specific enthalpy of mercury vapour is low thus resulting in lower jet velocities in mercury turbines.

(vi) Thermal efficiency is higher than Rankine steam cycle.

3.2.2 Disadvantages of Mercury

(i) High cost and limited supply.

(ii) Toxic in nature.

(iii) Pervasive, thus tends to leak through joints, cracks, etc.

(iv) Latent heat is very low, requiring a large amount of mercury for the same heat utilisation.

(v) The whole boiler operates under high temperature and not just the superheater.

(vi) It does not wet surfaces which cause poor heat transfer. This can, of course, be overcome by chemical treatment.

Hence, it is clear that neither mercury nor any other vapour can replace steam in all the ranges of working. However, mercury is really superior in the high temperature range.

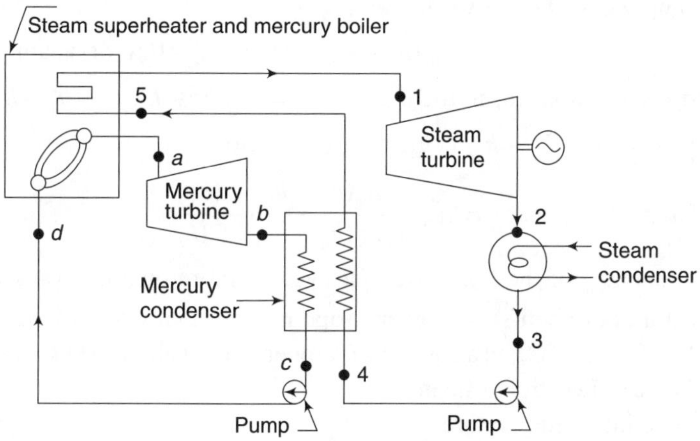

Figure 3.18 Binary-vapour cycle

The schematic diagram of a binary vapour (mercury-steam) cycle is shown in Fig. 3.18 and its corresponding *T-s* diagram is shown in Fig. 3.19.

In the brinay-vapour (mercury-steam) cycle, there are two different circuits, one for steam and another for mercury. Mercury vapour, dry and saturated, enters the mercury turbine at state point *a* and expands to *b*. The exhaust mercury vapour is condensed in mercury condenser to state point *c*. The condensate is pumped back to the boiler by the mercury pump. The heat rejected in the mercury condenser is used to vaporize fed water fed back by the feed water pump into steam at state point 5. Thus, we can say that the mercury condenser also acts as a boiler for feed water. The heat transfer from the exhaust mercury vapour to feed water is possible because there is a large difference in temperature between these two. The saturated steam coming out from the mercury condenser is superheated to state point 1 in the superheated and expanded to state point 2 in the steam turbine and condensed in the condenser to state point 3. Thus, 1-2-3-4-5-1 and *a-b-c-d-a* show the steam and mercury cycles respectively.

To calculate the thermal efficiency of the binary vapour cycle the process is as follows:

Let m_m = Mass of mercury circulated per kg of steam.

Assume 1 kg of steam entering the steam turbine.

In mercury condenser, heat given by mercury-vapour is received by feed water. Thus,

$$m_m(h_b - h_{mc}) = 1 \times (h_5 - h_{f4}) \quad \text{or} \quad m_m = \frac{(h_5 - h_{f4})}{(h_b - h_{mc})} \qquad ...(3.14)$$

Net work per kg of steam flow in the cycle

= (Mercury turbine work-mercury pump work) + (Steam turbine work-water pump work)

$= m_m(h_a - h_b) - m_m(h_{md} - h_{mc}) + (h_1 - h_2) - (h_{f4} - h_{f2})$ kJ/kg of steam

Neglecting pump work, the net work

$$w_{net} = m_m(h_a - h_b) + (h_1 - h_2) \text{ kJ/kg of steam} \qquad ...(3.15)$$

Heat supplied per kg of steam in the cycle = $q_A = m_m(h_a - h_{md}) + (h_1 - h_5)$

$\approx m_m (h_a - h_{mc}) + (h_1 - h_5)$ neglecting pump work ...(3.16)

$$\text{Thermal efficiency} = \eta_{th} = \frac{m_m(h_a - h_b) + (h_1 - h_2)}{m_m(h_a - h_{mc}) + (h_1 - h_5)} \qquad ...(3.17)$$

The thermal efficiency of the binary-vapour cycle is more than the Rankine steam cycle.

Frass proposed a potassium steam binary-vapour cycle that operates with a turbine inlet temperature of 830°C to 890°C and a condenser temperature of about 600°C. This topping cycle is superimposed on a conventional steam cycle with a turbine inlet temperature of 590°C by transforming the waste heat rejected from the condensing potassium vapour to the boiling water of steam cycle. This cycle is similar to the mercury-steam binary vapour cycle (employed in seven plants in USA). However, the operating temperature of these plants was limited to 500°C. But the potassium-steam binary cycle employing higher temperature made the mercury-vapour cycle obsolete.

The results obtained by Frass indicates that potassium-steam binary vapour cycle would give an overall thermal efficiency of about 54% an compared to only 40% from a conventional steam cycle. Binary-vapour cycle is also called combined cycle.

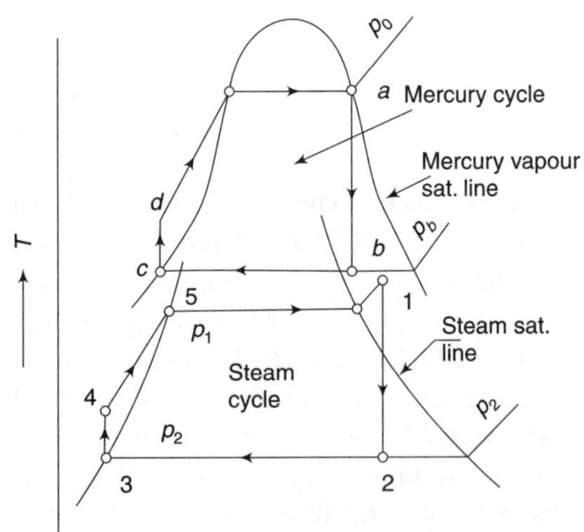

Figure 3.19 T-s diagram for binary-vapour cycle

3.3 REGENERATIVE FEED HEATING CYCLES

Regenerative feed heating cycle is the modified form of Rankine cycle with an aim to achieve higher thermal efficiency. It is a practical approach towards approaching Carnot cycle based on the philosophy of increasing average temperature at which heat is added to the cycle. All the steam power plants with single unit producing 30 MW and above employ regenerative feed heating arrangement.

3.3.1 Basic Principle of Regenerative Feed Heating

In the simple (basic) Rankine cycle 1-2-3-4-5-1 shown in Fig. 3.20, the working fluid at state point 4 is in the liquid phase and it is heated in the boiler from state point 4 to state point 5 due to sensible heating and from 5 to 1 due to vaporization process. Thus, the average temperature of heat addition to the simple Rankine cycle, $(T_{ave})_{Rank}$ lies between state points 4 and 5 which is lower than the average temperature of heat addition (i.e., at maximum temperature T_1) to the Carnot cycle 1-2-3′-5-1 in which it is represented by process 5-1 only. In other words, $(T_{ave})_{Carnot}$ is greater than $(T_{ave})_{Rank}$. That is why the efficiency of simple Rankine cycle is lower than the Carnot cycle between the same temperature limits. In another way, it can be said, that the irreversible mixing of the cold condensate with hot boiler water reduces the thermal efficiency of simple Rankine cycle. Since increasing the maximum temperature and pressure and lowering the minimum temperature and pressure of the cycle have practical limitations, any further improvement of cycle efficiency is possible only by eliminating or minimising the irreversibilities (i.e., the sensible heating) in the simple Rankine cycle if we are aiming to achieve the efficiency near or equal to Carnot cycle efficiency through some arrangement. Such an arrangement is possible by regenerative feed heating, *the name regenerative feed heating is given because the expanding steam in the turbine is used to heat the feed water* thus increasing average temperature of heat addition. The increase in efficiency due to regenerative feed heating may also be explained by the fact that the heat rejected in the condenser is reduced due to bled off steam.

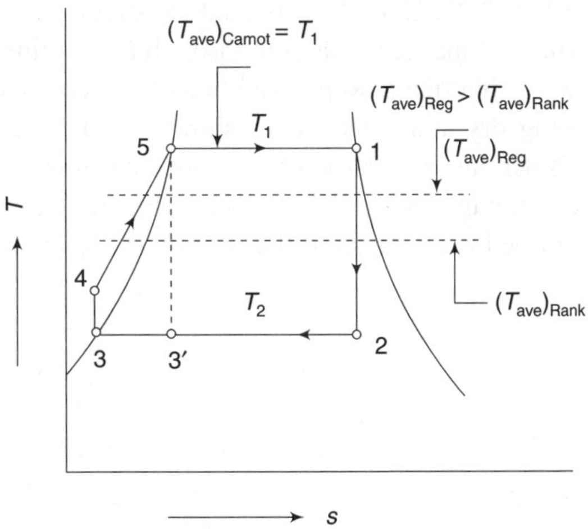

Figure 3.20 Rankine cycle on T-s diagram showing average temperature

The aim of regenerative feed heating cycle is to supply the feed water to the boiler at a temperature much higher than the temperature of the condensate in the condenser by heating the feed water in one or more feed heater with the help of steam bled off during expansion from stream turbine, thereby increasing the average temperature of heat addition in the cycle, which in turn, increases the thermal efficiency of regenerative feed heating cycle.

The feed water is heated either in a ***direct contact (dc) feed heater or in a surface (indirect) heater due to condensation*** of extracted steam, i.e., the heat is given up by the extracted steam which is received by the feed water. In this way, the average temperature of heat addition to the cycle is increased, which increases the thermal efficiency of the cycle. This type of system has become a standard practice in steam power plants both on land and, in the more powerful installations. *In practice, the appreciable cost of heaters, pipes, valves and accessories put a limit to the number of heaters employed* (Table 3.1). In large units, the temperature rise of feed water per heater is normally about 20°C to 35°C, and three to ten heaters are installed. The main points to be noted are that the bled steam must not be live steam otherwise no work from that live steam would be available nor must it be completely expanded, for then it would be useless for feed heating. For the selection and design of closed feed heaters, reference [6] may be consulted.

3.3.2 Most Ideal Regenerative Feed Heating Cycle

The aim of most ideal regenerative feed heating cycle is to achieve the efficiency equal to efficiency of Carnot cycle by supplying the feed water to boiler at saturation temperature, T_1 of boiler pressure, p_1, i.e., state 4 (Fig. 3.21). This *concept* may be visualized in *two ways.*

First Concept. The most ideal regenerative feed heating cycle consist of an *infinite number of heaters arranged between cylinders and the feed-water passing through the heaters* as shown in Fig. 3.21. Here, heat is made available for feed heating by changing the condition of the whole of the steam passing through the turbine, i.e., by a reduction of the dryness fraction of expanding steam. Representation of such a cycle is shown on T-s diagram, the initial condition being dry and saturated as shown in Fig. 3.22. Due to the infinite number of heaters, the temperature rise in each heater becomes indefinitely small and the process becomes reversible. The expansion line 1-2′ becomes smooth and parallel to the liquid line because the heat that will be bled is equal to the heat added to the feed water between condenser temperature and

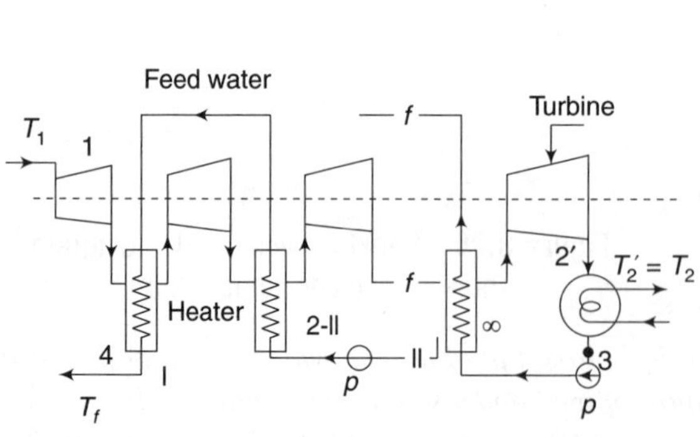

Figure 3.21 Most ideal regenerative feed heating arrangement having infinite number of feed heaters

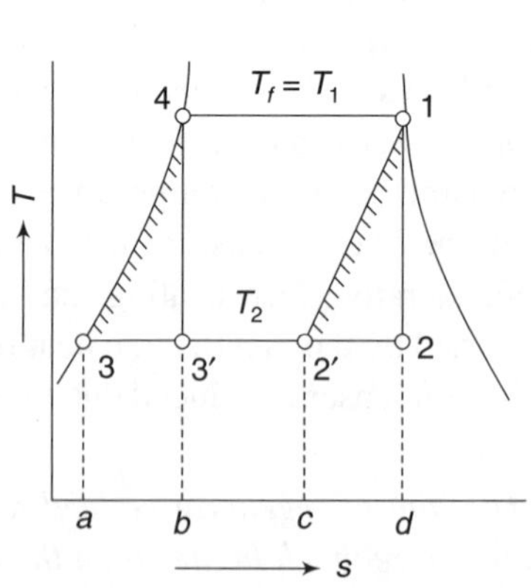

Figure 3.22 T-s diagram for infinite number of feed heaters. (Without pump work)

feed water temperature. Due to the infinite number of heaters the feed water is heated to boiler saturation temperature, i.e., $T_{fw} = T_4 = T_1$. In this way, the efficiency of an ideal regenerative cycle is equal to that of Carnot (i.e., area) 1 – 2 – 3′ – 4 – 1 = area 1 – 2′ – 3 – 4 – 1). It is, of course, assumed that the feed water is heated up to the saturation temperature in each heater. It is also assumed that there is no pressure or heat loss in the extract line, no under-cooling of condensate below condenser saturation temperature and expansion is isentropic.

It is to be noted that the efficiency of ideal regenerative cycle equals to that of Carnot only if the turbine uses saturated steam; the two are not equivalent when the steam is initially superheated because all the heat is received at high temperature in Carnot cycle while this is not true in regenerative cycle in which the steam is initially superheated since the steam entering the first heater should be saturated; the final feed water temperature T_{fw} will be less than highest temperature in Carnot cycle.

Second concept. This concept may also be visualized from the hollow casing arrangement shown in Fig. 3.23. The process is represented on *T-s* diagram as shown in Fig. 3.22. Dry and saturated steam enters the turbine at point 1 and expands isentropically to the condenser pressure at state point 2. The condensate from the condenser enters the hollow casing as saturated water at state point 3. The condensate or feed water receives heat from the steam on the other side through perfectly conducting portion on its way to the high pressure side of the turbine. Theoretically, it may be assumed that the temperature of water and steam tube are the same on both sides of the perfect heat conducting partition, so that it is a reversible heat transfer. The feed water is gradually heated along the path 3–4 and the steam is gradually losing the same amount of heat during expansion process, resulting in a decrease in entropy. Such a feed heating is called regenerative feed heating. Considering the casing as a part of the system, the heat added along the process 3–4 cancels with heat rejected along process 1–2′, so line 1–2′ is parallel to 3–4. By this way the efficiency of ideal regenerative cycle is equal to that of Carnot. In the Rankine cycle, the expansion process follows the path 1–2′. Mathematically also,

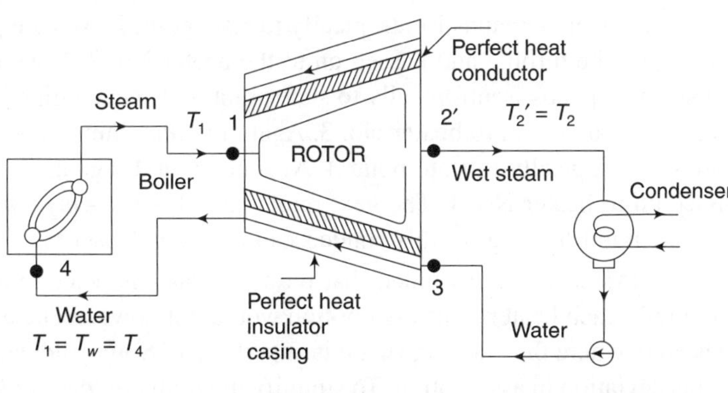

Figure 3.23 Most ideal regenerative feed heating system with hollow casing

Heat added = $q_A = T_1(s_1 - s_4)$, Heat rejected = $q_R = T_2(s_2' - s_3)$

Hence, net work = $w_{net} = q_A - q_R = T_1(s_1 - s_4) - T_2(s_2' - s_3)$

Since $(s_1 - s_4) = (s_2' - s_3)$, so $w_{net} = (T_1 - T_2)(s_1 - s_4)$ ...(3.18)

Hence, most ideal = $\eta_{Reg} = \dfrac{w_{net}}{q_A} = \dfrac{(T_1 - T_2)(s_1 - s_4)}{T_1(s_1 - s_4)} = \dfrac{T_1 - T_2}{T_1}$ ...(3.19)

Equation (3.19) shows the efficiency of most ideal regenerative cycle approaches the efficiency of Carnot cycle. As a thermodynamic principle, it may be said that since the average temperature of heat addition in the regenerative cycle is increased, so the thermal efficiency is increased. Sometimes, regenerative feel heating is called the Carnotization of Rankine cycle.

The *most ideal regenerative feed heating cycle is not practical for the following reasons:*

(i) It is difficult to put an infinite number of heaters or to pass feed water through hollow turbine casing. Economically also, it is not feasible.

(ii) Since the feed heating process changes the condition of whole of steam, it results in excessive moisture in the low-pressure stage of the turbine. This is not all desirable.

For these reasons, the practical regenerative feed heating cycle consists of only 3 to 10 heaters. (Table 3.1).

3.3.3 Regenerative Feed Heating Cycles and their Representation of Ideal Process on *T-s* and *h-s* Diagrams

For the analysis of regenerative feed heating cycle, we shall consider the three systems shown in Fig. 3.24. Figure 3.27 shows the representation of the process on *T-s* and *h-s* charts. Here, in each case, four feed heaters are shown. 1 kg of the steam at pressure p_0 and temperature t_0 enters the turbine at state point 0 and expands to the state point 1. At state point 1, m_1 kg of steam is bled off from the turbine and passed on the heater No. 1 for feed heating. The remaining $(1 - m_1)$ kg of steam expands isentropically to state point 2. At state point 2, m_2 kg of steam is bled off from the turbine and passed on to the heater No. 2. Now, the remaining $(1 - m_1 - m_2)$ kg of steam expands isentropically to state point 3. At state point 3, again, m_3 kg of steam is extracted and passed on to heater No. 3. Again the remaining $(1 - m_1 - m_2 - m_3)$ kg of steam expands isentropically to state point 4. At state point 4, again m_4 kg of steam is extracted and passed on to heater No. 4. The next remaining $(1 - m_1 - m_2 - m_3 - m_4)$ kg of steam expands to state point 5 and after this condenses in the condenser.

In each of the three systems, it is assumed that the feed water is heated up to the steam saturation temperature in each heater. This assumption would not however, be actually true in the case of those heaters where the entering steam is superheated. Some allowance, therefore, may be given for this deviation in assumption. To simplify the problem, *perfect heat-exchange between the extracted steam and the feed water is assumed in each heater though in actual practice it is not so.*

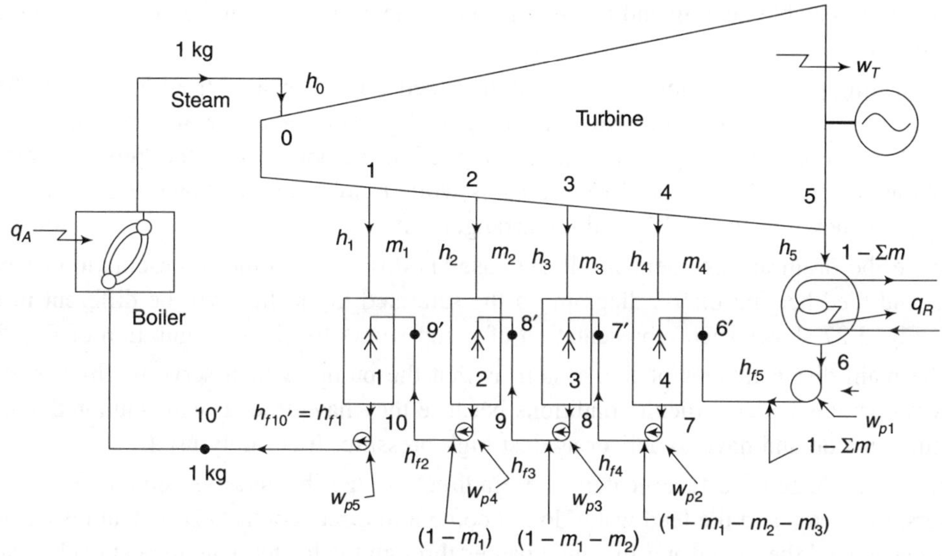

Figure 3.24 Regenerative feed heating system with four direct contact heaters, drain fed to HP side

Let the enthalpies of steam and feed water be represented by h and h_f respectively. Let m_1, m_2, m_3 and m_4 kg/kg of steam be extracted from stations 1, 2, 3 and 4 having pressures p_1, p_2, p_3 and p_4 respectively. In each heater, the quantity of bled steam and its enthalpy entering, the amount of feed water and drain and their enthalpies entering and coming out are shown. It will be seen that the enthalpy of feed water will be found corresponding to the respective pressure at which the steam is extracted.

In each of the three cases, the feed heating arrangement is different, so it is discussed separately as follows.

In Fig. 3.24, four direct contact heaters are shown. In each heater, steam is condensed in the feed water thus gives heat to feed water $(1 - m_1 - m_2 - m_3 - m_4)$ kg of feed water from condenser is pumped to heater No. 4 from state point 6 to 6′ where m_4 kg of steam is mixed and condenses from state point 4 to 7 and $(1 - m_1 - m_2 - m_3 - m_4)$ kg of water is heated from state point 6′–7. Now after mixing of feed water and steam $(1 - m_1 - m_2 - m_3)$ kg of feed water at state point 7 is pumped to heater No. 3 at state point 7′. In this heater, m_3 kg of bled steam condenses from point 3 to 8 and $(1 - m_1 - m_2 - m_3)$ kg of feed water is heated from state point 7′ to 8. At state point 8, $(1 - m_1 - m_2)$ kg of feed water is pumped to heater No. 2 at state point 8′. In this heater, m_2 kg of bled steam condenses from point 2 to 9 thus heating $(1 - m_1 - m_2)$ kg of feed water from state point 8′ to 9. At state point 9, $(1 - m_1)$ kg of feed water is pumped to heater No. 1 at state point 9′. In this heater, m_1 kg of steam is condensed from point 1 to 10, thus heating $(1 - m_1)$ kg of feed water from 9′ to 10. At state point 10, 1 kg of steam is pumped to boiler pressure at state point 10′ and feed water is heated from

10′ to 11 as sensible heating and from 11 to 0 as evaporation and supherheat in the boiler as shown in Fig. 3.27.

Neglecting pumping action, we may say that feed water is heated from h_{f3} to h_{f4} in heater No. 4, from h_{f4} to h_{f3} in heater No. 3 from h_{f3} to h_{f2} in heater No. 2 and from h_{f2} to h_{f1} in heater No. 1. From $h_{f1} = h_{f10}'$ to h_0 heat is supplied in the boiler. Note that representation on *h-s* shown in Fig. 3.28 is valid for these cases while representation shown Fig. 3.29 is valid for system shown in Fig. 3.25 for drain arrangement.

Since the drain arrangement in all the cases is different, so the representation of pump work and feed heating on *T-s* diagram on the saturated liquid line will be different in each case, Fig. 3.27 shows such representation of pump works for the configuration of Fig. 3.24.

The main disadvantages of this system is that the pumps which serve the high pressure heaters operates under difficult conditions because they have to deal with saturated water at the suction side and have to deliver against high pressure. It is rarely used.

In Fig. 3.25, four surface heaters (i.e., indirect contact heaters) are employed. In these heaters, the steam and the feed water do not come into direct contact. This steam is extracted into heaters and the feed water in a pipe passes through the heater. Due to heat exchange, the steam is condensed by giving up heat to the feed water in the heater. The condensed steam, i.e., the heater drain water is extracted by a drain pump and discharged into the feed pipe immediately after the heater. In this case also, the drain pumps suffer from the same practical difficulties as that shown in Fig. 3.24.

In Fig. 3.26, a regenerative feed heating cycle having four surface heaters integral with drain cooler is shown but the drain system is different from that shown in Fig. 3.25. In this case, it is assumed that a part of the heating surface is arranged for drain cooling. The direction of

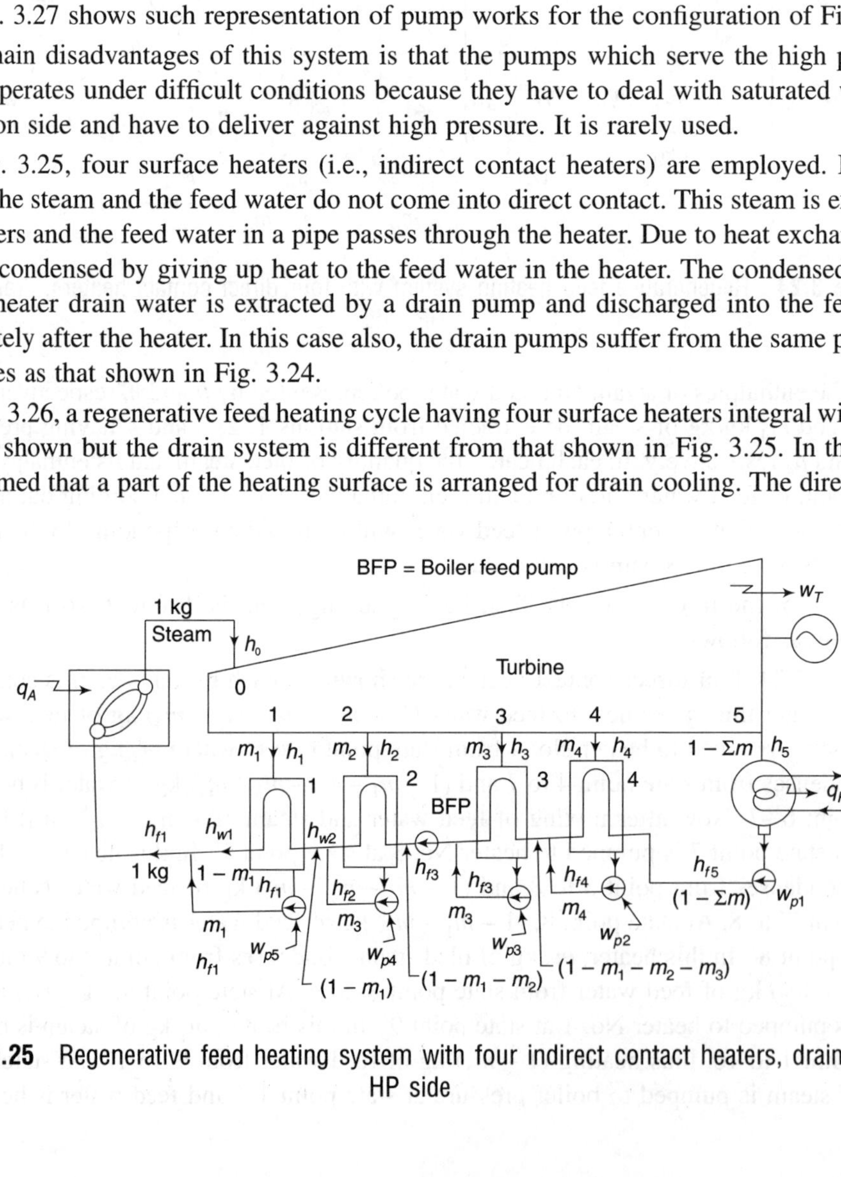

Figure 3.25 Regenerative feed heating system with four indirect contact heaters, drain fed to HP side

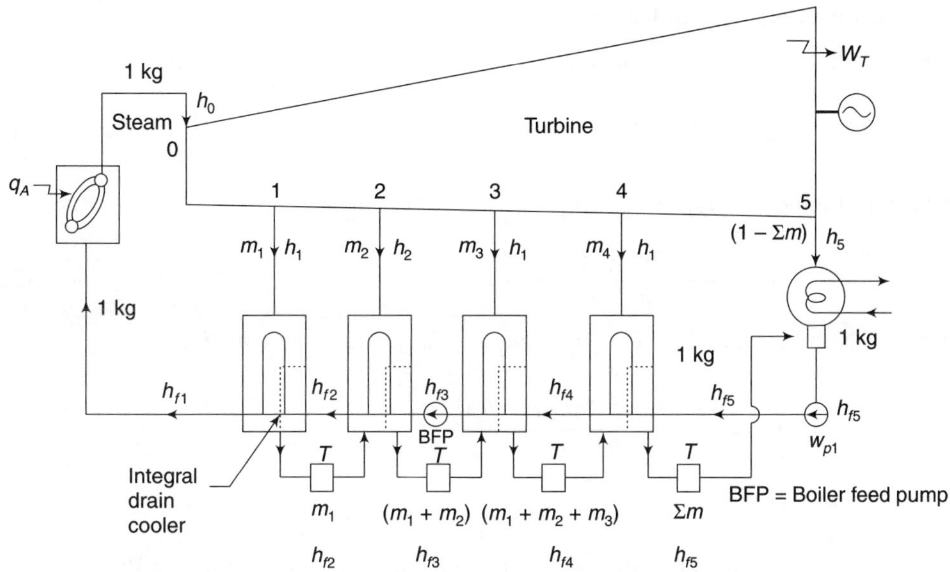

Figure 3.26 Regenerative feed heating system with four indirect contact heaters, drain fed to HP side

flow of the hot drain water is arranged such that it flows counter to the incoming feed water, and it is assumed that the drain water is cooled to the saturation temperature existing in the next heater on the low pressure side. On the drain system live steam traps T are arranged to avoid the escape of steam with the drain water. From the first heater, the feed water is led to the hot well and then fed back to the boiler. Here, 1 kg of feed water is flowing through the main feed pipe.

Note: For plain surface heater (without integral drain coolers), the enthalpies of drains leaving heaters 1, 2, 3 and 4 will be h_{f1}, h_{f2}, h_{f3} and h_{f4} respectively (Fig. 3.26).

The position of boiler feed pump (BFP) in this system is very important because the heaters that follow it must be designed for very high pressure, and *it should be placed at a convenient point in the feed system so that there is no risk of evaporation of feed water taking place due to pumping action.* Since the heaters are specially designed for high pressure they are more costly. Generally, BFP is located below deaerator so that priming problem is avoided.

3.3.3.1 Calculation of Amount of Bled Steam

To calculate the amount of bled steam m_1, m_2, m_3 and m_4 at each station, 1 kg of steam entering the turbine is assumed and energy balance over heater using control volume is applied, i.e., enthalpy entering and enthalpy leaving equated at each heater. Refer to Fig. 3.24 and Fig. 3.27 for writing enthalpy balance equation with control volume approach.

Heat entering = Heat leaving

At heater No. 1: $m_1 h_1 + (1 - m_1) h_{f2} = h_{f1}$

or $$m_1 = \frac{h_{f1} - h_{f2}}{h_1 - h_{f2}} \qquad ...(3.20)$$

At heater Number 2: $m_2 h_2 + (1 - m_1 - m_2) h_{f3} = (1 - m_1) h_{f2}$

or $$m_2 = \frac{(1 - m_1)(h_{f2} - h_{f3})}{h_2 - h_{f2}} \qquad ...(3.21)$$

At heater Number 3: $m_3 h_3 + (1 - m_1 - m_2 - m_3) h_{f4} = (1 - m_1 - m_2) h_{f3}$

or $$m_3 = \frac{(1 - m_1 - m_2)(h_{f3} - h_{f4})}{h_3 - h_{f4}} \qquad ...(3.22)$$

At heater Number 4:

$$m_4 h_4 + (1 - m_1 - m_2 - m_3 - m_4)\, hf_5 = (1 - m_1 - m_2 - m_3) h_{f4}$$

or $$m_4 = \frac{(1 - m_1 - m_2 - m_3)(h_{f4} - h_{f5})}{(h_4 - h_{f5})} \qquad ...(3.23)$$

In the other two systems, values of m_1, m_2, m_3 and m_4 will be calculated also by writing enthalpy balance equation.

Neglecting pump work:

The heat supplied externally in the cycle $= q_A = (h_0 - h_{f1})$ kJ/kg ...(3.24)

The heat rejected in the cycle $= q_R = (1 - m_1 - m_2 - m_3 - m_4)(h_5 - h_{f5})$

$$= (1 - \Sigma m)(h_5 - h_{f5}) \text{ kJ/kg} \qquad ...(3.25)$$

Therefore, the work done in the cycle per kg of steam

$$= \text{Turbine work per kg of steam} = w_{net} = w_T$$

$$w_T = (h_0 - h_{f1}) - (1 - \Sigma m)(h_5 - h_{f5}) \text{ kJ/kg} \qquad ...(3.26)$$

The work done by the turbine may also be calculated by summing up the products of the steam flow and the corresponding heat drop in turbine stages.

$$w_T = (h_0 - h_1) + (1 - m_1)(h_1 - h_2) + (1 - m_1 - m_2)(h_2 - h_3)$$
$$+ (1 - m_1 - m_2 - m_3)(h_3 - h_4) + (1 - \Sigma m)(h_4 - h_5)$$

The thermal efficiency of the regenerative cycle is

$$(\eta_{th})_{Reg} = \frac{\text{Work done}}{\text{Heat supplied}} = \frac{(h_0 - h_{f1}) - (1 - \Sigma m)(h_5 - h_{f5})}{h_0 - h_{f1}}$$

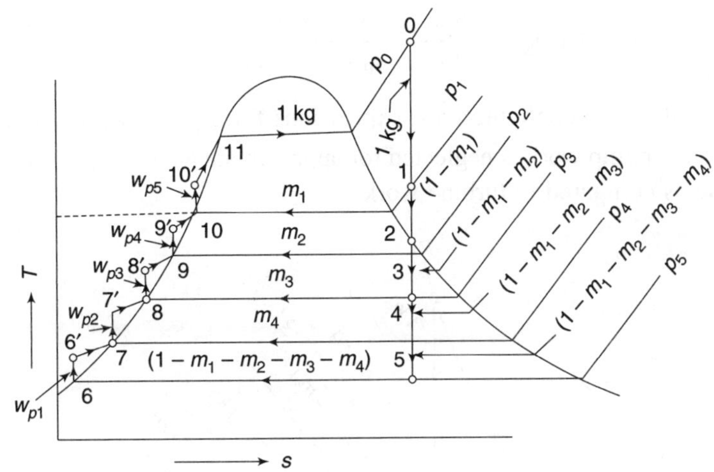

Figure 3.27 *T-s* Diagram with pump work, valid for Fig. 3.24 only

or

$$(\eta_{th})_{Reg} = 1 - \frac{(1 - \Sigma_m)(h_5 - h_{f5})}{h_0 - h_{f1}} \quad ...(3.27)$$

The thermal efficiency of these simple Rankine cycle is given by the expression

$$(\eta_{th})_{Rank} = 1 - \frac{h_5 - h_{f5}}{h_0 - h_{f5}} \quad ...(3.28)$$

Comparing these two equations we see that $(1 - \Sigma m)(h_5 - h_{f5})$ is less than $(h_5 - h_{f5})$ because $(1 - \Sigma m)$ is always less than unity. The value of $(h_0 - h_{f1})$ is less than $(h_0 - h_{f5})$ because of higher value of h_{f1} than h_{fs}. In the regenerative feed heating system, the ratio of heat rejected to heat supplied is less than that of the simple Rankine cycle. Thermal efficiency thus increases in the case of regenerative feed heating cycle.

The increase in thermal efficiency in the case of regenerative feed heating cycle, may be also realised on the basis of average temperature of heat addition. As shown in Fig. 3.27, the average temperature of heat addition in the regenerative feed heating system is more, so thermal efficiency is more.

Considering Pump Work: Refer to Fig. 3.27

$$\text{Net work in cycle} = w_{net} = w_T - \Sigma w_p \text{ kJ/kg}$$

where

$$\Sigma w_p \approx v_{wb} (p_0 - p_b) \times 10^2 \text{ kJ/kg.}$$

$$\text{Heat supplied} = q_A = h_0 - h'_{f10} = h_0 - h_{f0} - w_{p5} = h_0 - h_{f1} - w_{p5}$$

where

$$w_{p5} = v_{f10} (p_0 - p_1) \times 10^2 \text{ kJ/kg}$$

$$(\eta_{th})_{Reg} = \frac{w_{net}}{q_A} = \frac{w_T - \Sigma w_p}{h_0 - h_{f1} - w_{p5}} \quad ...(3.29)$$

Similarly, for other two configurations, expressions for m_1, m_2, m_3, m_4 may be derived.

Note: Generally, pump work is neglected for approximate analysis because it is very small and is negligible as compared to turbine work.

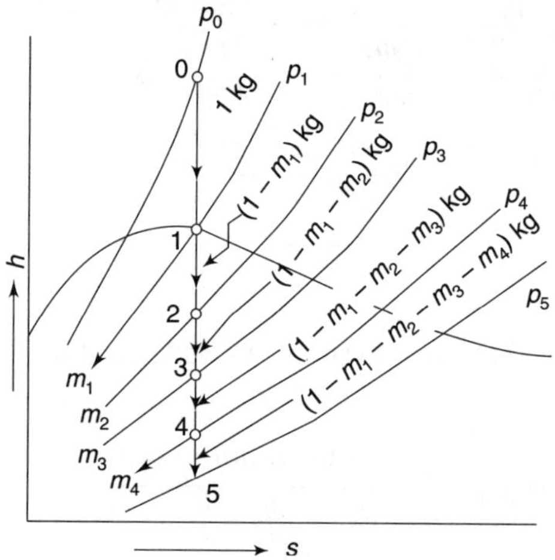

Figure 3.28 *h-s* diagram for all three systems (Figs. 3.24, 3.25, 3.26)

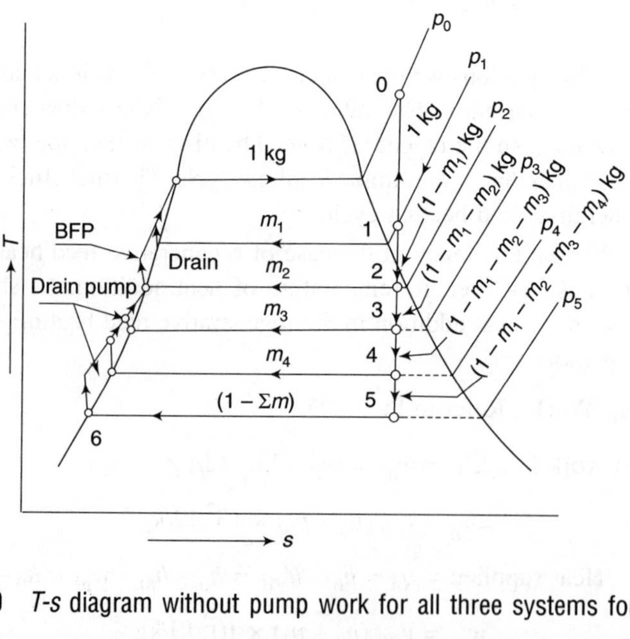

Figure 3.29 *T-s* diagram without pump work for all three systems for Fig. 3.25

3.3.3.2 Law of Efficiency Gain

It is worth to discuss the law of efficiency gain due to regenerative feed heating. If $(\Delta t)_0 = t_{boiler,\ sat} - t_{cond}$ and $(\Delta t)_{fw}$ is temperature rise of feed heaters, then

$n = 0$ (heater), $(\Delta t)_{fw} = 0$

$n = 1$ (heater), $(\Delta t)_{fw} = 1/2\ (\Delta t)_0$ Gain $= 1/6\ (\Delta t)_0$

$n = 2$ (heater), $(\Delta t)_{fw} = 2/3\ (\Delta t)_0$ Gain $= 1/12\ (\Delta t)_0$

$n = 3$ (heater), $(\Delta t)_{fw} = 3/4\ (\Delta t)_0$ Gain $= 1/20\ (\Delta t)_0$

$n = 4$ (heater), $(\Delta t)_{fw} = 4/5\ (\Delta t)_0$

Since the cycle efficiency is proportional to $(\Delta t)_{fw}$, the gain in efficiency follows law *of diminishing return with the increase in number of heaters.*

3.3.4 Representation of Actual Process on *T-s* and *h-s* Diagrams in Regenerative Cycle

In actual regenerative cycle, the expansion of steam cannot be isentropic. The actual condition of steam will be 1′, 2,′ 3,′ 4 and 5′ in the expansion process. (Fig. 3.30). So the new enthalpies of steam entering the heaters No 1, 2, 3 and 4 will be h_1', h_2', h_3' and h_4' respectively. With these enthalpies m_1, m_2, m_3 and m_4 will be calculated for all three systems.

Neglecting pump work:

The actual work done in the cycle = Actual turbine work $= q_A - q_R$

or
$$w_{Ta} = (h_0 - h_1') + (1 - m_1)(h_1' - h_2') + (1 - m_1 - m_2)(h_2' - h_3') + (1 - m_1 - m_2 - m_3)(h_3' - h_4') + (1 - \Sigma m)(h_4' - h_5')$$

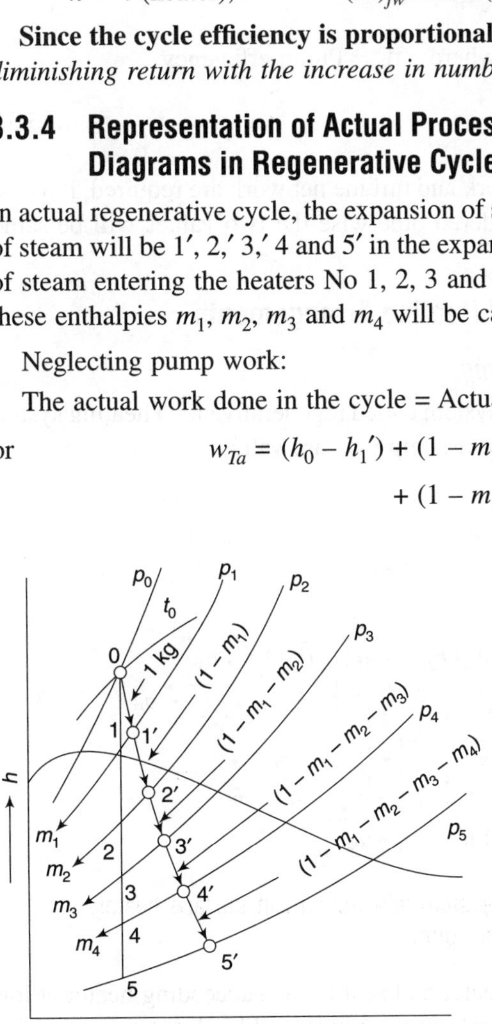

Figure 3.30(a) Actual process on *h-s* diagram without pump work

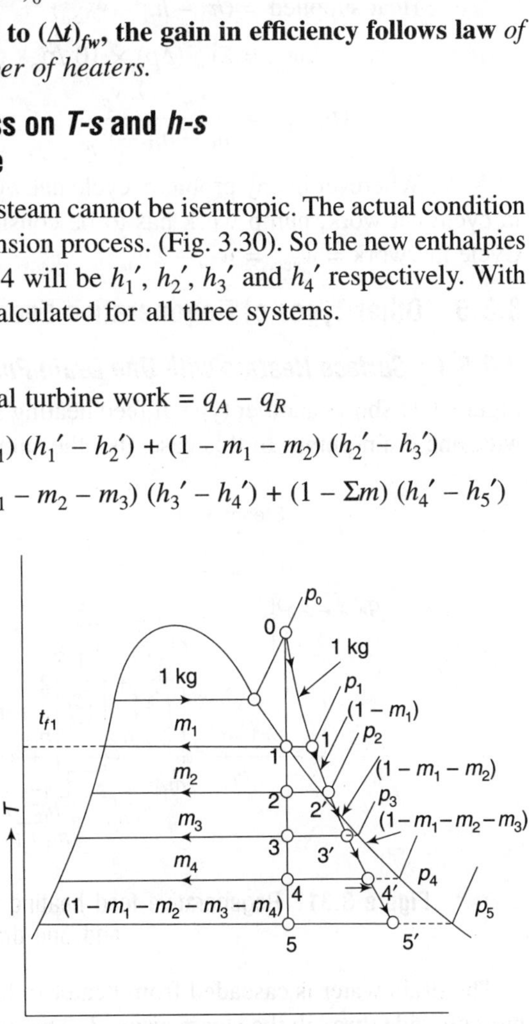

Figure 3.30(b) Actual process on *T-s* diagram without pump work

Here $h_0 - h_1' = \eta_t (h_0 - h_1)$ and $h_1' - h_2' = \eta_t (h_1' - h_2)$

Heat supplied $= q_{Aa} = (h_0' - h_{f1})$ kJ/kg

$$q_R = (1 - \Sigma m)(h_5' - h_{f5})$$

$$(\eta_{th})_{Reg} = \frac{\text{Actual work done}}{\text{Actual heat supplied}} = \frac{(h_0 - h_{f1}) - (1 - \Sigma m)(h_s' - h_{f5})}{(h_0 - h_{f1})} \quad ...(3.30)$$

Considering pump work:

$$\text{Net work} = w_{net,a} = w_{Ta} - \Sigma w_{pa}$$

$$\text{Heat supplied} = (h_0 - h_{f1} - w_{p5a})$$

$$\Sigma w_{pa} = \Sigma v_w (\Delta p) \times 10^2/\eta_p \quad \text{where} \quad \eta_p = \text{Pump efficiency}$$

$$(\eta_{th})_{Reg} = \frac{w_{Ta} - \Sigma w_{pa}}{h_0 - h_{f1} - w_{p5a}} \quad ...(3.31)$$

Note: Wherever in any problem, cycle net work and turbine net work are required, it means in cycle net work, pump work has to be considered otherwise the two values will be same. Cycle net work $= w_{neta} = w_{Ta} - \Sigma w_{pa}$.

3.3.5 Other Types of Regenerative Feed Heating Arrangements

3.3.5.1 Surface Heaters with One Drain Pump

Figure 3.31 shows another type of feed heating system called regenerative feed heating system with one drain pump. In this case also the same assumptions are valid.

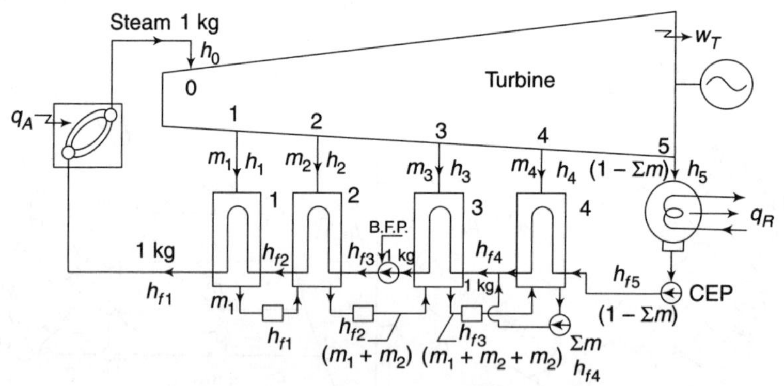

Figure 3.31 Regenerative feed heating system with four plain surface heaters and one drain pump

The drain water is cascaded from heater to heater and sent to the succeeding heater at low pressure side through the steam traps, *T*. The whole of the condensed bled steam is extracted from heater No. 4. By a drain pump, the drain is discharged to the main feed line on the outlet side of the heater No. 4 on high pressure side.

The various amounts of steam extracted are: Using energy balance,

At heater No 1: $m_1 h_1 + h_{f2} = h_{f1} + m_1 h_{f1}$

or $$m_1 = \frac{h_{f1} - h_{f2}}{h_1 - h_{f1}} \quad ...(3.32)$$

At heater No. 2: $m_2 h_2 + h_{f3} + m_1 h_{f1} = (m_1 + m_2)h_{f2} + h_{f2}$

or $$m_2 = \frac{(h_{f2} - h_{f3}) - m_1(h_{f1} - h_{f2})}{(h_2 - h_{f2})} \quad ...(3.33)$$

At heater No. 3: $m_3h_3 + (m_1 + m_2)h_{f2} + h_{f4} = (m_1 + m_2 + m_3)h_{f3} + h_{f3}$

or $$m_3 = \frac{(h_{f3} - h_{f4}) - (m_1 + m_2)(h_{f2} - h_{f3})}{h_3 - h_{f3}} \quad ...(3.34)$$

At heater No. 4: $m_4h_4 + (1 - m_1 - m_2 - m_3 - m_4)\, h_{f5} + (m_1 + m_2 + m_3)h_{f3} = h_{f4}$

or $$m_4 = \frac{(h_{f4} - h_{f5}) - (m_3 - m_2 - m_3)(h_{f3} - h_{fw5})}{h_4 - h_{f5}} \quad ...(3.35)$$

The expression for thermal efficiency will be on the line as discussed above.

3.3.5.2 Plain Surface Heaters with One Drain Pump and One Drain Cooler

Figure 3.32. shows an alternative system which is called regenerative feed heating with one drain pump and one drain cooler.

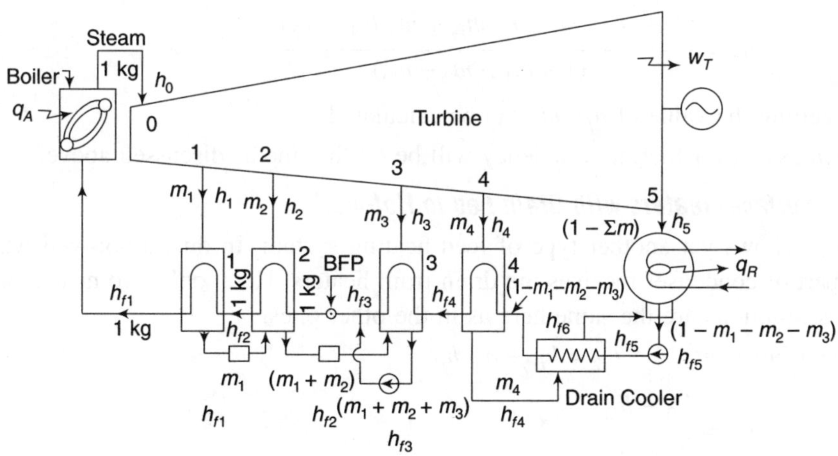

Figure 3.32 With drain surface heaters, one drain pump and one drain cooler

The condensed heating steam m_1 is cascaded from heater No. 1 to heater No. 2. From the heater No. 2, the condensed steam $(m_1 + m_2)$ is cascaded to heater No. 3. From the heater

No. 3 the condensed steam $(m_1 + m_2 + m_3)$ is extracted by a drain pump and delivered to the main feed pipe immediately after the heater, i.e., on the delivery side of the heater No. 3.

The drain from heater No. 4 is passed through a drain cooler to transfer some of the heat to the feed water and after cooling to condenser temperature it goes to the lower part of the condenser or hot-well.

At heater No 1: $m_1\, h_1 + h_{f2} = h_{f1} + m_1\, h_{f1}$

or $$m_1 = \frac{h_{f1} - h_{f2}}{h_1 - h_{f1}} \qquad ...(3.36)$$

At heater No. 2: $m_2h_2 + h_{f3} + m_1h_{f1} = (m_1 + m_2)h_{f2} + h_{f2}$

or $$m_2 = \frac{(h_{f2} - h_{f3}) - m_1\,(h_{f1} - h_{f2})}{h_2 - h_{f2}} \qquad ...(3.37)$$

At heater No. 3: $m_3h_3 + (m_1 + m_2)h_{f2} + h_{f4}\,(m_1 - m_2 - m_3)h_{f4} = h_{f3}$

or $$m_3 = \frac{(h_{f3} - h_{f4}) - (m_1 + m_2)(h_{f2} - h_{f3})}{h_3 - h_{f4}} \qquad ...(3.38)$$

At heater No. 4: $m_4h_4 + (1 - m_1 - m_2 - m_3)\, h_{f6} = (1 - m_1 - m_2 - m_3)h_{f4} + m_4 \cdot h_{f4}$

or $$m_4 = \frac{(1 - m_1 - m_2 - m_3)h_{f4} - h_{f6})}{h_4 - h_{f4}} \qquad ...(3.39)$$

At drain cooler: $(1 - m_1 - m_2 - m_3)\, h_{f5} + m_4\, h_{f4} = (1 - m_1 - m_2 - m_3)h_{f6} + m_4h_{f5}$

$$h_{f6} = \frac{(1 - m_1 - m_2 - m_3)h_{f5} + m_4(h_{f4} - h_{f5})}{(1 - m_1 - m_2 - m_3)} \qquad ...(3.40)$$

After getting the value of h_{f6}, m_4 can be calculated.

The expression for thermal efficiency will be on the line as discussed above.

3.3.5.3 Surface Heaters with Drain Fed to Hot-well

Figure 3.33 shows yet another type of feed heating system. In this, a hot-well which is an integral part of condenser receives the drain from heaters. Here, only two heaters are shown and the assumptions are the same here as in the other ones.

At heater No 1: $m_1\, h_1 + h_{f2} = h_{f1} + m_1\, h_{f1}$

or $$m_1 = \frac{h_{f1} - h_{f2}}{h_1 - h_{f1}} \qquad ...(3.41)$$

At heater No. 2: $m_1h_{f1} + m_2h_2 + h_{f4} = h_{f2}\,(m_1 + m_2)h_{f2}$

or $$m_2 = \frac{m_1h_{f2} + h_{f2} - h_{f4} - m_1h_{f1}}{(h_2 - h_{f3})} \qquad ...(3.42)$$

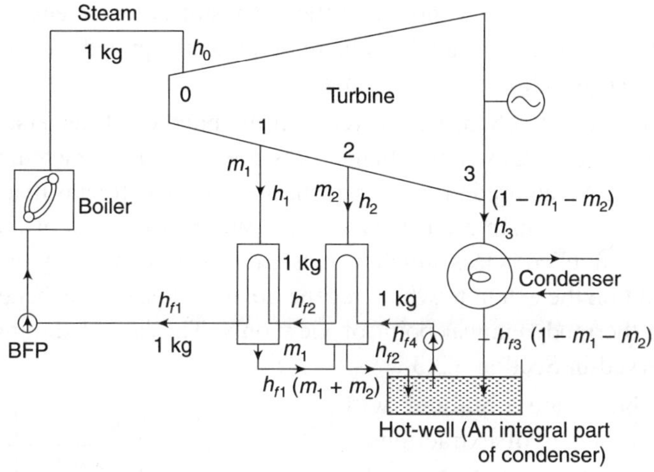

Figure 3.33 Heaters with plain surface heaters, drain fed to hot-well

At hot-well: $h_{f4} = (m_1 + m_2)h_{f2} + (1 - m_1 - m_2)h_{f3}$

Putting the value of h_{f4}, we can calculate m_2.

The expression for thermal efficiency will be on the above line.

3.3.6 Optimum Feed Water Temperature and Saving in Heat Rate

There is an optimum feed water temperature by which only the regenerative feed heating can be carried out very efficiently. The criterion governing the number of heaters is one of economic justifications. To understand the optimum feed water temperature, let us first define heat rate. *Heat rate is defined as the heat supplied per unit of work obtained, expressed generally as* kJ/kWh. *Sometimes, it is also termed heat consumption.*

$$\text{Heat rate} = \frac{\text{Heat supplied} \times 3600}{\text{Heat supplied} - \text{Heat rejected}}\text{; kJ/kWh} \quad ...(3.43)$$

$$\text{Heat rate} = \frac{3600}{\text{Thermal efficiency}}\text{; kJ/kWh} \quad ...(3.44)$$

3.3.6.1 Gain in Thermal Efficiency due to Regenerative Feed Heating

It is obvious from Eq. (3.43) that an increase in heat supplied reduces the heat rate and thereby increases efficiency, provided the heat rejected remains constant. But due to extraction of steam for regenerative feed water heating, a slightly lower amount of steam passes through the low pressure stages of the turbine thereby reducing the output. To maintain the original power level output, the throttle flow has to be increased. Since the extra steam due to increase in throttle flow is extracted in the middle of the expansion and does not reach the condenser, the heat rejected remains practically constant. When throttle flow is increased, the heat to be supplied per unit mass of steam in boiler is reduced due to feed heating by bled steam. When this extraction is less, i.e., the number of heaters is less, this reduction is relatively small as compared to increase in throttle flow and the net result is to increase the heat supplied with heat rejected remaining constant. This increase in heat supplied reduces heat rate hence efficiency is increased which is our main aim.

A further increase in extraction, i.e., increase in the number of heaters, causes relatively less increase in throttle flow, as steam at higher pressure stages is extracted, which has higher enthalpy and therefore, is required in smaller quantities. The net effect is that the heat supplied in boiler per unit mass of steam reduces at a faster rate than the rate of increase in throttle flow. Hence the heat supplied is less which increases the heat rate thereby reducing efficiency. This imposes a limit on the extent to which regenerative feed heating is beneficial. The above result is based on thermodynamical point of view only. The law of diminishing return has already been observed in Section 3.3.3.2.

Reduction in heat rate obtained with various numbers of stages of extractions is shown in Fig. 3.34. These curves show that the improvement in efficiency due to stage extraction increases as the final feed water temperature increases up to a certain point, after that it decreases. It is obvious, that as the number of stages of extraction increases, the most efficient final feed water temperature increases. Also, as the number of heaters increases the reduction in heat rate is increased, i.e., the gain in thermal efficiency increases.

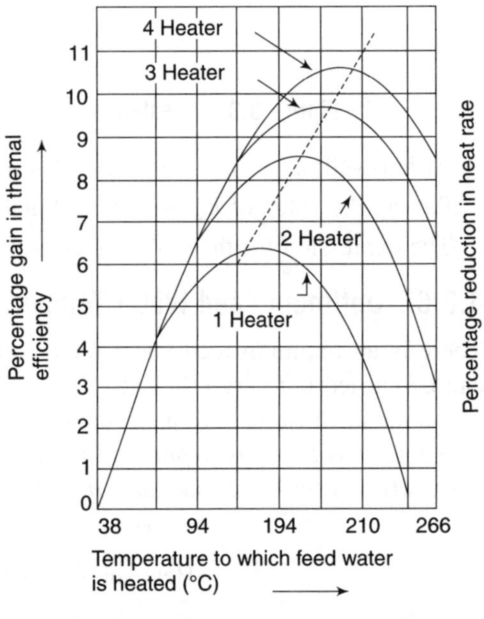

Figure 3.34 Gain in thermal efficiency due to regenerative feed heating

Figure 3.35 shows the effect of throttle pressure on temperature to which the feed water should be heated to obtain the maximum gain with stage extraction. It is seen from these curves that the most efficient final feed water temperature increases quite rapidly as heaters are added for the first four heaters while any further extraction above four affects the efficient final feed temperature only slightly. The higher the boiler pressure or lower the condenser pressure, the better will be gain affected by the regenerative cycle.

From the above discussion, the following facts are summarised.

(i) The thermal efficiency improves with each additional heater but the incremental gain with each becomes progressively smaller.

(ii) For a given initial steam temperature the decrease in heat rate (kJ/kWh) is greater with the higher initial pressure.

(iii) For a given initial steam pressure, the decrease in heat rate is slightly smaller with a higher initial temperature.

(iv) The optimum final feed water temperature increases with initial steam pressure.

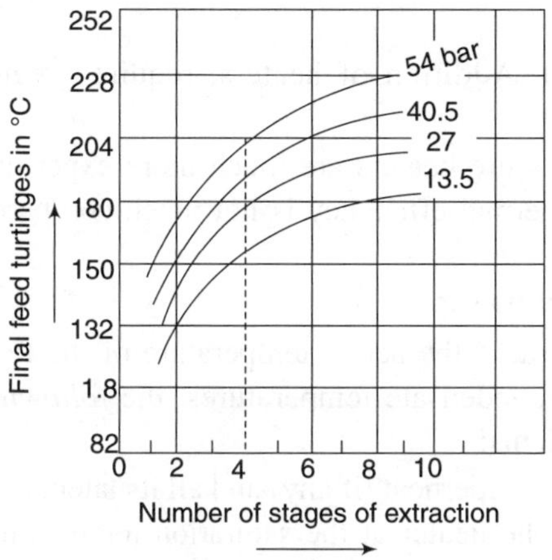

Figure 3.35 Final feed water temperature for maximum gain with various number of stages and steam pressure and temperature

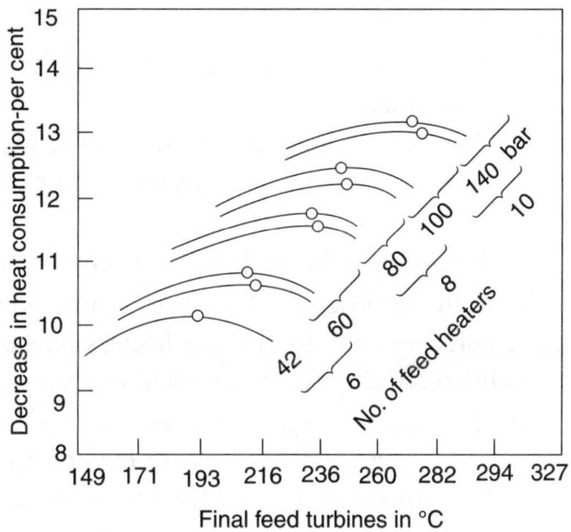

Figure 3.36 Decrease in heat consumption

3.3.7 Advantages and Disadvantages of Regenerative Feed Heating Cycle over Simple Rankine Cycle

Advantages:

(i) Irreversibility in the feed water heating process in the boiler is minimised, i.e., heating process in boiler trends to become reversible.

(ii) Average temperature of heat addition to the cycle is increased due to high temperature of feed water entering the boiler, thus increasing the thermal efficiency. As the numbers of heaters increases, the thermal efficiency increases as shown in Figs. 3.35 and 3.36.

(iii) Temperature ranges in the boiler are reduced, thus reducing thermal stresses set up in the boiler.

(iv) The efficiency with which the extracted steam rejects heat to feed water is appreciably high, near about 98 per cent. So the extraction of steam is advantageous because most of the extraction is at low pressure stages of the turbine where the steam expands with relatively low efficiency. The extracted quantity is limited by the feed water temperature.

(v) Less amount of steam is passed through the low pressure stages, so blade height will be less resulting in reduction in the cost of L.P. turbine.

(vi) Improved turbine drainage due to many extractions reduces erosion due to moisture.

(vii) It requires a reduced condenser capacity (i.e., condenser size) i.e., less costly condenser.

(viii) Heat rate is reduced.

Disadvantages:

(i) It increases the complications of the plant. Addition of heaters, require greater maintenance.

(ii) The heaters and pipings are costly. High pressure heaters are much more expensive than the low pressure heaters. The gain in thermal efficiency is not much compared to the heavier costs.

(iii) It requires a larger boiler capacity for a given power.

Note. In the absence of precise information regarding the actual temperature of the feed water entering and leaving the heaters and of the condensate temperatures, the *following assumption* should always be made in solving problems:

(i) The bled steam just condenses, i.e., gives up its superheat (if any) and all its latent heat only. The condensed steam therefore leaves the heater at the saturation temperature corresponding to the bleeding pressure.

(ii) Feed water is heated to saturation temperature at the pressure of the bled steam. And either.

(iii) (a) the condensed bled steam is returned to the system by separate pumps at a point down stream from the heater, i.e., drain is fed to the high pressure side; or

(b) the condensed bled steam discharges into the hot-well, thereby arising the hot-well temperature; or

(c) the condensed bled steam is cascaded down and discharges finally into the hot-well of the condenser with or without drain cooler.

3.3.8 Feed Heaters

As mentioned earlier, there are two types of feed heaters in common use, namely: (a) Direct contact heater, and (b) Indirect contact (surface) heater.

Basically, these are heat exchanges. Figure 3.37 shows diagrammatically the temperature rise of the feed water in the various types of feed heaters. The thermal performance of feed heater may be accessed by the difference in temperature between the water leaving the heater and the saturation temperature corresponding to the heater steam pressure; this difference is termed terminal ***temperature difference*** (TTD). Direct contact heaters result in zero TTD. For a plain condensing (indirect contact) heater, TTD is about 6°C. It is desirable to have TTD equal to zero (i.e., feed water is heated to saturation temperature corresponding to the pressure of steam) or even small negative value. This is possible only with the use of desuperheater when the heater is supplied with highly superheated steam. In desuperheater, the temperature of the steam is reduced to near saturation temperature before exhausting it to the main heater for condensation. This is done because of the fact that the heat transfer coefficient from the superheated steam to the feed water is about ten times less than that of saturated steam. The desuperheater may be a separate heat exchanges for reducing the steam to near saturation temperature before exhausting it to the main heater for condensation. Alternatively, it may integral with the main heater and consist of baffling around the tubes carrying the feed water to the outlet of the heater. Separate drain cooler (i.e., it cools condensed steam and thus

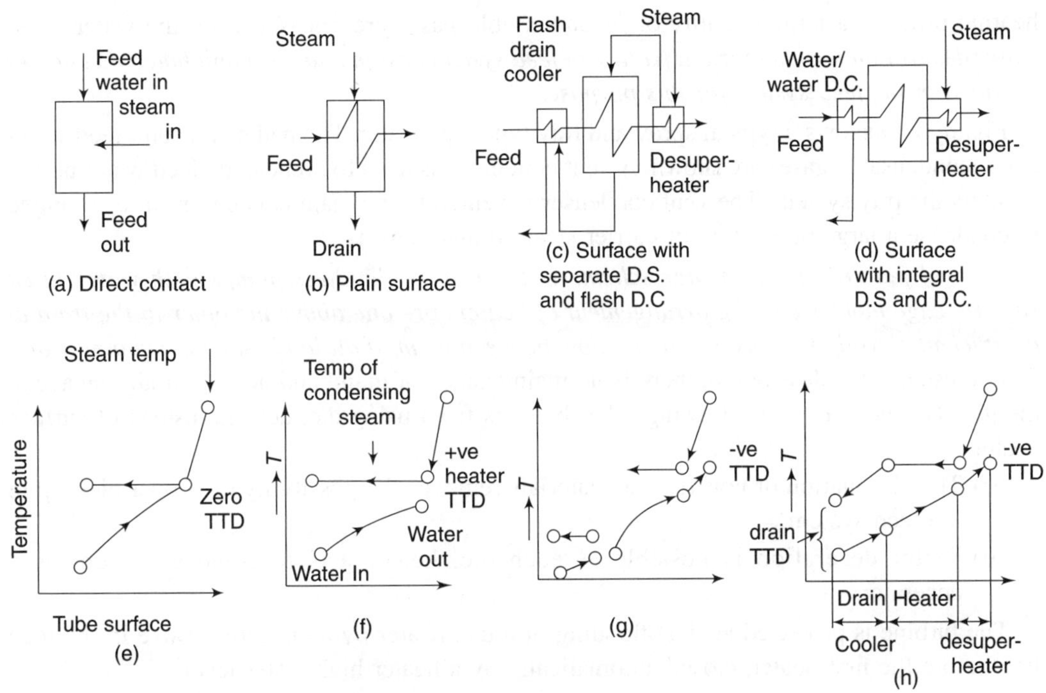

Figure 3.37 Temperature rise in feed heaters

heats up feed water) are in common use but these are also incorporated with h.p. heaters. *The inclusion of desuperheating and drain cooling sections within the shell complicates feed heater design and construction but this design reduces capital cost and so it is in current use.* Again, to access the thermal performance of a drain cooler, the tool used is the difference in temperature between the feed water entering and drain leaving (TTD). For drain cooler, TTD is equal to about 8°C. The flash type drain cooler is one in which drains entering it flash into steam and are then condensed. Integral drain cooler is water to water type heat exchanger.

3.3.9 Direct Contact (DC) Heaters

In this type of heater, there is a direct mixing of bled steam and feed water; the steam gives up its latent heat and thus condenses and as a result the feed water gets heated. In order to help this mixing process to take place in as short a time and in as small a volume as possible, the feed water entering at the top of the heater is atomised by spray nozzles and the resulting droplets fall through the steam atmosphere and cascade over perforated plates or tray stages (Fig. 3.38).

This type of heater is more efficient than the surface type since there is no metal barrier to the transfer of heat between water and steam. The feed water can be heated to the saturation temperature corresponding to the pressure of heating steam, i.e., TTD is zero. During the heating process, a large quantity of incondensable gases are released from the water. *Feed water deaeration is an important feature of feed system design and each unit had a d.c. heater installed in the feed system for this purpose.*

Figure 3.39 shows a typical spray and tray type d.c. heater. The paths of steam, feed water and incondensable gases are shown. A vent condenser is used to preheat the feed water before it enters the tray system. The vent condenser is vented to the main condenser but is arranged to condense a large proportion of carrier steam drawn into it.

A novel feature of d.c. heater scheme is the use of glandless pumps with water filled squirrel cage motors and the arrangement of heaters are one above the other in the train at a sufficient elevation to ensure gravity flow between them at all loads saving pumping work. Due to use of glandless pump, there is no maintenance of gland and no risk of air leakage at the glands. There are the following other benefits from using d.c. heaters instead of surface heaters:

(i) The elimination of non-ferrous materials removes the possibility of copper pick-up in the feed system.

(ii) Better deaeration is possible as each d.c. heater acts to some an extent as a deaerator.

The turbine is protected against flooding of a d.c. heater by an isolating valve in the feed line before the first heater, closed automatically by a heater high water level.

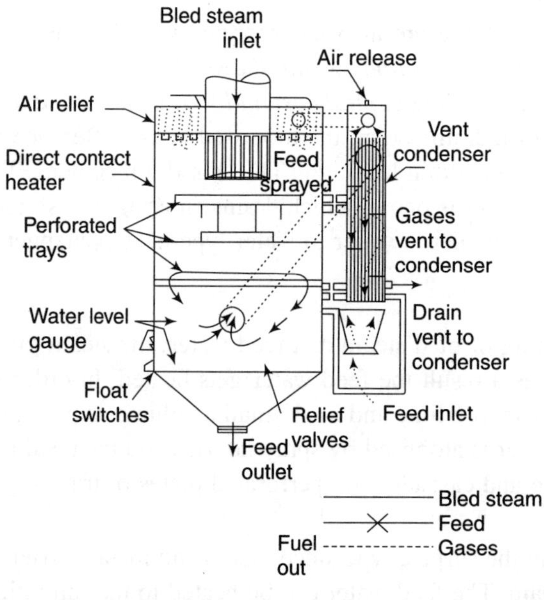

Figure 3.38 D.C. heater

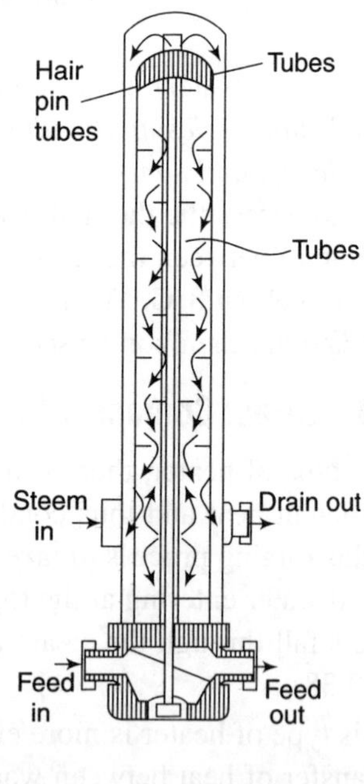

Figure 3.39 Surface heater

Particularly, during periods of changing load, the risk of flooding may be further reduced by designing the system to operate at full load with no stored water in the heaters. The use of water seal prevents the steam flow between heaters under all operating conditions.

3.3.10 Surface Heaters

In the surface type of feed heater, the feed water passes through *hairpin tubes* in two, four of occasionally six flow (three hairpin tubes in series), while the steam is passed over the tubes. Baffles are provided to direct the flow of steam through the tube nest. Baffles also act as support plates and allow thermal expansion of tubes.

The condensate is drained from the bottom of the heater shell to be cascaded to the next lower heater train, or to be pumped forward ahead of the heater into the feed line towards high pressure side. Pumping forward (high pressure side) mixes the drains with the feed water at the outlet temperature of the heater and is, therefore, a more efficient method of drains heat utilisation. The advantages lies in requiring a pump for each heater, its maintenance and the risk of air leakage at the glands.

Figure 3.39 shows a h.p. heater of surface type. *The current practice on large high pressure systems is to install desuperheating and drain cooling section within the shell of the heater.* Although the inclusion of desuperheating and drain cooling sections within the shell complicates the heater design and construction, but it reduces the overall capital cost. In such a heater, there are three zones, namely, *desuperheating, condensing, and drain cooling*.

3.3.10.1 Desuperheating Zone

This zone is the last feed heater zone through which the superheated bled steam enters and the feed water passes before leaving the heater. Figure 3.40 shows temperature/heat transferred diagram for a typical high pressure heater. It is obvious from the diagram that when the feed water enters the zone it has been heated in the condensing zone to within a few degree of the saturation temperature, corresponding to the bled steam pressure at the entry to the feed heater. Although, there is an addition of only a small percentage of the total heat transferred in the heater by desuperheating zone, this small rise in feed water temperature is very valuable in terms of thermal economy. The heat transferred in this zone is by convection process as the superheated steam may be treated to behave as a normal gas.

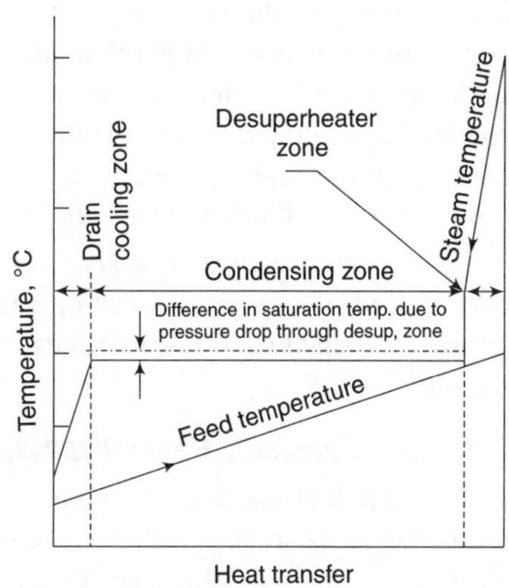

Figure 3.40 Temperature distribution in surface heater integral with draincooler and desuperheater

The normal practice is to allow the steam to leave the desuperheating zone with some residual superheat of about 27.8°C. The temperature gradient between the bulk steam and the feed water temperature is shown in the top of the figure. Since the steam leaves with some residual superheat, so the tube wall temperature is above saturation temperature of the bled steam and no condensation takes place on the tube wall under normal conditions. That is why there is no problem of formation of droplets of condensed steam and being carried into the condensing zone by the relatively high velocity steam and thus causing impingement attack.

3.3.10.2 Condensing Zone

The major heating of feed water takes place in the condensing zone as the overall heat transfer coefficient in this zone is very high. The main design requirement of large heaters is to provide good steam distribution to the condensing surface with minimum pressure loss. Under normal load running conditions, the ingress of non-condensable gases is unlikely, however, there is a likelihood of its accumulation after shut-down during two-shift operation. Under this condition, continuous venting is necessary to help the steam distribution and to clear any non-condensable gases accumulated. The use of diaphragm plates serves a dual purpose, in addition to supporting the tube nest, they also prevent the accumulation of condensate from running down the outsides of the tubes and forming a thick film which impairs the heat transfer coefficient.

3.3.10.3 Drain Cooling Zone

The drain cooling zone provides a fairly high heat transfer rate provided the drains are not reheated by heat transferred through the shrouds around the drain cooler convection one. The reheating of the drain is possible if there is no desuperheating section. In general, the desuperheating zone is larger than the drain cooling zone and it is possible to have the heater shell flooded with condensate above the drain cooler inlet. By this way, a stagnant pool of condensate is formed around the drain cooler shrouds causing them to follow the temperature gradient in the drain cooler. As a result of this, the problem of drain reheat is eliminated provided the shell drain valves are kept closed.

The present practice is to keep the high pressure heater upside down. This reduces the cost of feed pipe work and has operational advantage that it permits drainage of the tubes during prolonged shut-down periods and thus reducing feed-water contamination and metal pick-up.

3.3.10.4 Automatic Heater Bypassing

There is a risk of flooding of high pressure surface heater owing to bursting of tubes, leakage tube/tube-plate joints or drain blockage. Precautions must be taken to prevent water from a flooded heater shell being passed back into the turbine through the bled steam pipes. In order to circumvent this problem, an automatic bypass system around the high pressure heater or group of heater is installed, so that, when a dangerously high water level occurs in a heater shell, the unit is automatically isolated

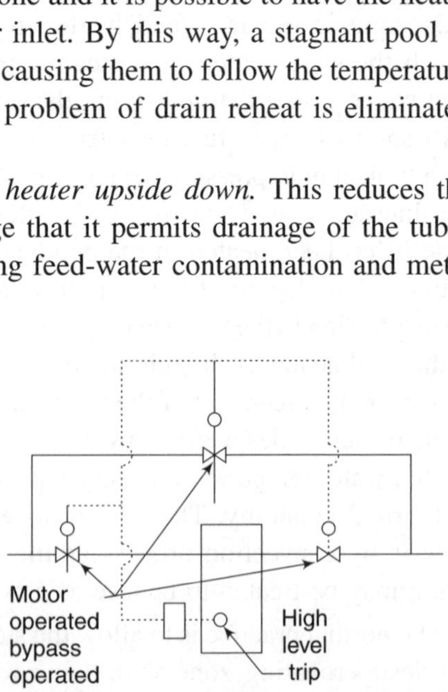

Figure 3.41 Feed heater bypass system

on the feed side from the feed pump (Fig. 3.41). By this way, any further ingress of water into the heater shell is prevented.

3.3.10.5 Tube Material

The tube material used for the heating surface in L.P. heaters is usually non-ferrous and copper and brass are used extensively because of their satisfactory heat transfer. The 90/100 grade of cupro-nickel alloy is in service in some *h-p* heaters with feed water temperature up to 224°C but steel is used for high temperature.

3.3.11 Deaerators

There are corrosive effects of water which contains oxygen dissolved from the atmosphere and thus produces detrimental effects on boiler tubes, steam pipes and other items of the plant. Therefore, one requirement of a modern feed water system is that it must supply water, which has an almost zero oxygen content, to the feed pumps for feeding the boiler. In general, for boilers operating at a pressure in excess of 70 bars, the feed water should have an oxygen content of less than 0.007 mg/kg which is possible by physical deaeration but it is possible to reduce this value to 0.007 mg/kg by chemical dosing. *Since a feed system is closed after the boiler feed pump, all physical deaeration must be carried out before the feed pump suction.*

The amount of oxygen that can be retained in solution is proportional to the partial gas pressure and depends inversely on the temperature of the water. It is to be noted that condensers act as deaerators and reduce the oxygen content in condensate to a very low level, for the use of independent deaerator is unnecessary for many feed systems. However, in order to ensure that the amount of dissolved oxygen in a system is at a minimum, it is usual to install a deaerator to supplement the condenser. The solubility of air in water decreases with a rise in temperature and a reduction in pressure. Therefore, a deaerator is based on the principle that the amount of gas in solution in water is proportional to the pressure of that gas in the water on the surface of the water. Theoretically, if water is boiling, no gas can exist in the water on its surface and any gases in the water are driven out. It has been found in practice that heating the water to boiling point deaerates the water only if sufficient time is allowed to elapse. However, water can be deaerated if it is boiled rapidly and violently; in order to achieve this the water is heated under reduced pressure (vacuum condition). If the steam used to remove the gases in this way is taken from the intermediate stage of the turbine, the heat required to remove the gases will also help to increase the thermal efficiency of the system by imparting heat to the feed water.

Another important requirement of feed water system is to have a reservoir or storage tank of suitable deaerated water having sufficient capacity, about one-sixth of the MCR feed flow, situated above the feed pump suction to meet any sudden or instantaneous demands for feed water by the boiler. It is convenient to carry out the deaeration of the feed water in a deaerating heater which acts as a direct-contact feed heaing stage in the system, and to combine the deaerating heater with the storage tank. In other words, *a deaerator (a direct contact heater) which removes oxygen content from the feed water using bled steam, combined with the storage tank and arranged to an elevated position on the suction of feed pump ensures a positive suction head under all conditions of operation to feed deaerated water to the boiler.*

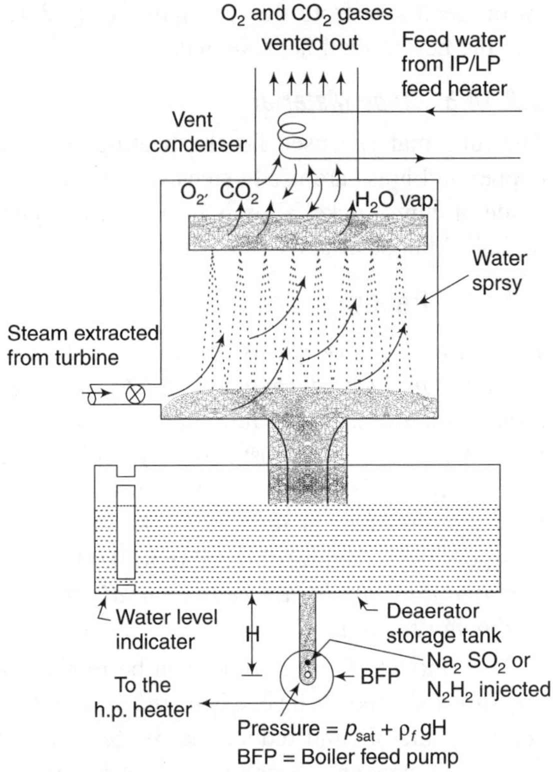

Figure 3.42 Deaerator and storage tank

Figure 3.42 shows a typical arrangement of a thermal deaerator and storage tank. The deaeration and heating is carried out in the two domes fitted to the storage tank. Bled steam is allowed to enter the storage tank above the water level from the top and flows along the steam space in the storage tank and up into the deaerating domes. Condensate passes through the surface type devaporizer and enters the deaerator dome at the top. There are essentially two stages in which the deaeration and heating are carried out in the deareating dome. There are a spray/stage and a tray stage. As the condensate enters, it passes through many small sprayers in parallel into the upper steam portion of the dome. The condensates are dispersed by the sprays into thin conical films which eventually coalesce to form small droplets which pass through the steam atmosphere into the trays below. The trays are arranged in series with the sprays on the condensate side and perform a similar duty so that a thin film of water is cascaded over the weir at the inner and outer edges and droplets cascade between trays through the perforations in them.

Approximately, 95% of the heating and deaeration takes place in spray stage to full load. But with decrease of load, the flow and pressure loss across the spray portion decrease resulting loss in efficiency in the atomization of the water and a reduced rate of heat transfer. With the decrease in water flow cascading over the trays, the water films become thinner resulting in efficient deaeration of that stage. Net result is a balance between two deaeration stages so giving an almost constant efficiency with decreasing load.

There is a problem of flashing at the feed pump suction following a load rejection. During a rapid load drop, the pressure in the heaters and the deaerator tends to drop. This causes flashing in the deaerator as the water is stored at boiling point corresponding to the pressure at full load. In order to avoid the flashing in the pump, the rate at which the water in the feed pump suction pipe gains static heads as it descends must be greater than the rate of pressure decay in the deareator. The suction pipe should be as near vertical as possible to avoid unnecessary head loss due to friction. Figure 3.43 shows the variation in pressure in a deareator and the suction of feed pump positioned beneath it following a load rejection. The curve of pump suction pressure is greater, by a constant figure equal to the static head of the deaerator on the pump than the curve of deaerator pressure. If the saturated pressure corresponding to the temperature of water arriving at the pump is greater than the total pressure (static head + deaerator pressure of that time-head loss due to friction) at the pump inlet, then flashing will take place in the pump.

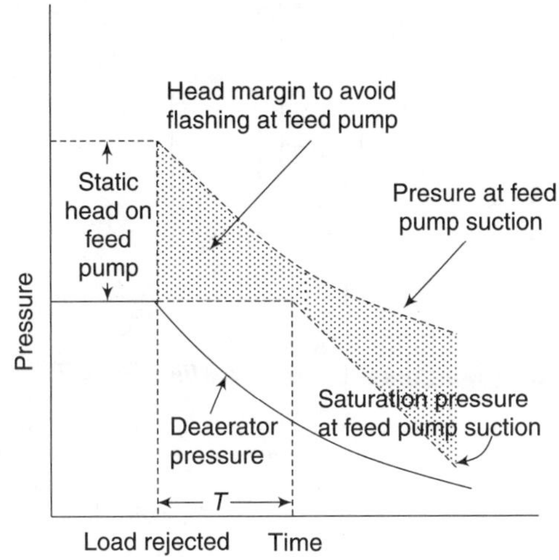

Figure 3.43 Pressure Variation in Dearator

In order to avoid this flashing, mixing baffles are installed in the tank to promote an adequate mixing of incoming water with the tank content to bring the total quantity quickly to the temperature appropriate to the unit loading. Even so, care has to be taken in the installation of these baffles, otherwise the increased friction loss caused by heavy flow rates during emergency conditions of sudden demand can cause a drastic reduction in the pressure at the feed pump suction

The deaerator is located at a sufficient height (H) from the pump location so that the pressure before suction is $(p_{sat} + \rho_f gh)$ to avoid cavitation.

The deaerator is generally placed near the middle of the feed water system so that the total pressure difference between the condenser and the boiler is shared equitably between the condensate pump and boiler feed pump. The feed water heaters before the deaerator are often termed high pressure (HP) heaters and those after deaerator are termed low-pressure (LP) heaters.

Example 3.3 A multi-stage steam turbine has to be provided with two stages of regenerative feed heating. The temperature rise of the feed has to be divided equally between the two heaters. Initial steam pressure is 27 bars with 300°C and exhaust pressure is 0.07 bar. Temperature of condensate is 35°C. The first tapping is taken at 2 bars. Determine the pressure for the second tapping point and the quantities to be tapped at each point. Calculate the gain in efficiency due to regenerative feed heating and increase in specific steam consumption rate. Use a stage efficiency of 70%.

Solution Refers to Fig. 3.44 and Fig. 3.45. As nothing is mentioned about drain arrangement the system shown in Fig. 3.37 has been assumed with ideal conditions.

From steam table, h_0 = 3000 kJ/kg, t_{f1} = 120.2°C (at 2 bar)

According to question

$$t_{f1} - t_{f3} = 2(t_{f1} - t_{f2}) = 2(t_{f2} - t_{f3})$$

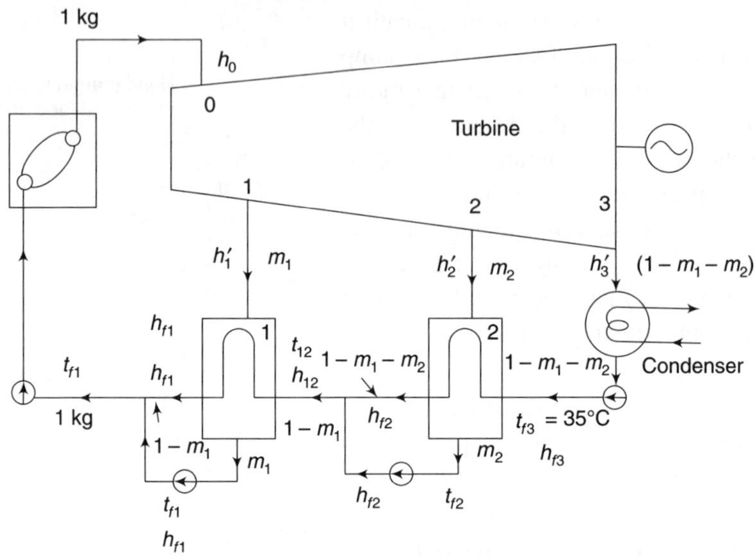

Figure 3.44 Example 3.3

$$\therefore \qquad t_{f2} = \frac{(t_{f1} - t_{f3})}{2} + t_{f3}$$

$$= \frac{(120.2 - 35)}{2} + 35 = 77.6°\text{C}$$

Corresponding to 77.6°C, the saturation pressure from Steam Table is 0.43 bar ***Ans.***

Now from chart; $h_0 - h_1 = 3000 - 2490 = 510$ kJ/kg

Since stage efficiency is 70%, so. $h_0 - h_1' = 0.7 \times 510 = 357$ kJ/kg

$$\therefore \qquad h_1' = 3000 - 357 = 2643 \text{ kJ/kg}$$

Thus, with enthalpy 2643 kJ/kg and pressure line 2 bars, the point 1′ is obtained.

Again $\qquad h_1' - h_2 = 2643 - 2400 = 243$ kJ/kg

$$\therefore \qquad h_1' - h_2' = 243 \times 0.7 = 170.1 \text{ kJ/kg or } h_2' = 2643 - 170.1 = 2472.9 \text{ kJ/kg}$$

Hence, the point 2′ can be located. Again, $h_2' - h_3 = 2472.9 - 2230 = 242.9$ kJ/kg

$$\therefore \qquad h_2' - h_3' = 242.9 \times 0.7 = 170.03 \text{ kJ/kg}$$

$$\therefore \qquad h_3' = 2472.9 - 170.03 = 2302.87 \text{ kJ/kg}$$

Hence, the point 3′ can be located on chart.

Since temperature at different stages are given, so

$$h_{f1} = 505 \text{ kJ/kg}, h_{f2} = 325 \text{ kJ/kg}, h_{f3} = 163 \text{ kJ/kg}$$

Using the heat balance equation:

At heater No. 1: $m_1 h_1' + (1 - m_1)\, h_{f2} = (1 - m_1)\, h_{f1} + m_1 h_{f1}$

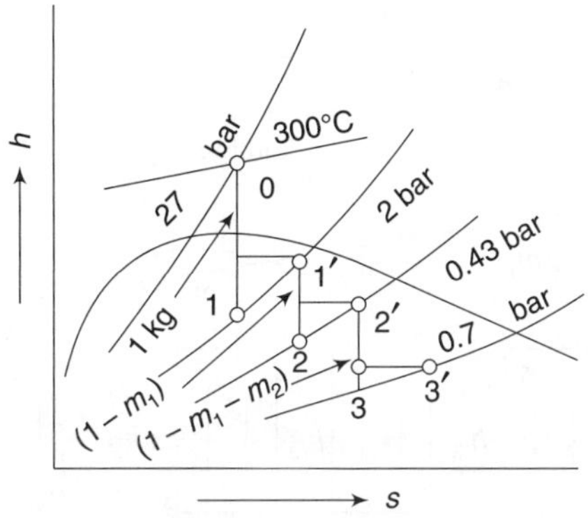

Figure 3.45 Example 3.3

$$\therefore \quad m_1 = \frac{h_{f1} - h_{f2}}{h_1' - h_{f2}} = \frac{505 - 325}{2643 - 325} = 0.0776 \text{ kJ/kg of entering steam} \quad \textbf{\textit{Ans.}}$$

At heater No. 2: $m_2 h_2' + (1 - m_1 - m_2)\, h_{f3} = (1 - m_1 - m_2)\, h_{f2} + m_2 h_{f2}$

$$\text{or} \quad m_2 = \frac{(325 - 163) - 0.0776\,(325 - 163)}{2472.9 - 163}$$

$$= 0.0647 \text{ kg/kg of entering steam} \quad \textbf{\textit{Ans.}}$$

$$\text{Work done} = w_t = (h_0 - h_1') + (1 - m_1)\,(h_1' - h_2') + (1 - m_1 - m_2)\,(h_2' - h_3').$$

$$q_R = (1 - m_1 - m_2)\,(h_3' - h_{f3}). = (1 - 0.0726 - 0.0647)(2302.87 - 163) = 1835.36 \text{ kJ/kg}$$

$$\text{Heat supplied} = q_A = h_0 - h_{f1} = 3000 - 505 = 2495 \text{ kJ/kg}$$

$$(\eta_{th})_{Reg} = \frac{q_A - q_R}{q_A} = \frac{2495 - 1835.36}{2495} = 26.43\% \quad \textbf{\textit{Ans.}}$$

Work done without regeneration $= h_0 - h_3 = 3000 - 2302.87 = 697.13$ kJ/kg
Heat supplied without regeneration $= h_0 - h_{f3} = 3000 - 163 = 2837$ kJ/kg

$\therefore$ Thermal efficiency without regeneration $(\eta_{th})_{rank} = \dfrac{697.13}{2837} = 24.57\%$

Percentage gain in thermal efficiency due to regenerative feed heating

$$= \frac{(\eta_{th})_{reg} - (\eta_{th})_{Rank}}{(\eta_{th})_{Rank}} = \frac{26.43 - 24.57}{24.57} \times 100 = 7.57\% \quad \textbf{\textit{Ans.}}$$

Now, specific steam consumption (scc) with feed heating $= \dfrac{3600}{w_{net}} = \dfrac{3600}{2495 - 1835.36}$

$$= 5.457 \text{ kg/kWh}$$

Again, scc without feed heating $= \dfrac{3600}{697.13} = 5.16$ kg/kWh

Hence, increase in scc rate due to feed heating $= 5.457 - 5.16 = 0.297$ kg/kWh **_Ans._**

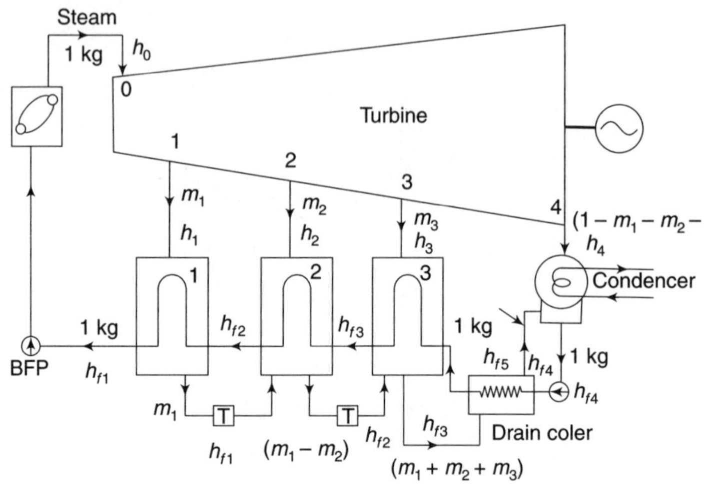

Figure 3.46 Example 3.4

Example 3.4 A steam turbine operates under the following conditions.

Initial pressure = 20 bars, Initial superheat = 93°C, Final pressure = 0.03 bar

Steam is bled off from turbine stages at pressures at 4.7, 1.3 and 0.25 bar for the purpose of heating the main condensate. The expansion in the turbine is isentropic. It may be assumed that condensate is heated in each heater up to the saturation temperature of the steam in that heater, and also that the drain water from each heater, is cascaded through a trap in the next heater on the low pressure side of it. The combined drain from the heater operating at 0.25 bar pressure are cooled in a drain cooler to condenser temperature. Calculate the following:

(a) Mass of steam extracted from each heater per kg of steam entering the turbine. (b) Thermal efficiency of regenerative cycle. (c) Thermal efficiency of Rankine cycle. (d) Theoretical gain due to regenerative feed heating. (e) Steam consumption of turbine in kg/kWh with and without regenerative feed heating. (f) Quantity of steam passing through the last stage nozzles of 50,000, kW turbine with and without regenerative feed heating. (g) Amount of heat carried away by the cooling water in the condenser with and without regenerative feed heating.

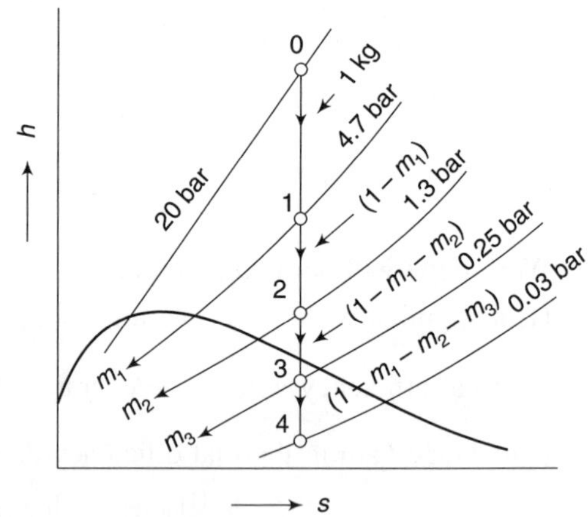

Figure 3.47 Example 3.4

Comment on the results obtained.

Solution Refer to Fig. 3.46 and Fig. 3.47. From Mollier chart,

$$h_0 = 3035 \text{ kJ/kg}, h_1 = 2715 \text{ kJ/kg}, h_2 = 2495 \text{ kJ/kg}, h_3 = 2260 \text{ kJ/kg}, h_4 = 2012 \text{ kJ/kg}$$

From Steam Table,

$h_{f1} = 629.8$ kJ/kg (at 4.7 bar), $h_{f2} = 449$ kJ/kg (at 1.2 bar)

$h_{f3} = 277$ kJ/kg (at 0.25 bar), $h_{f4} = 181$ kJ/kg (at 0.03 bar)

(a) Using heater No. 1 from energy balance: $m_1 h_1 + h_{f2} = m_1 h_{f1} + h_{f1}$

$$\therefore \quad \text{(a) } m_1 = \frac{h_{f1} - h_{f2}}{h_1 - h_{f1}} = \frac{629.8 - 449}{2715 - 629.8} = 0.0867 \text{ kJ/kg of entering steam} \qquad \textbf{\textit{Ans.}}$$

At heater No. 2: $m_2 h_2 + h_{f3} + m_1 h_{f1} = h_{f2} + (m_1 + m_2)\, h_{f2}$

$$\therefore \quad m_2 = \frac{(h_{f2} - h_{f3}) - m_1\,(h_{f1} - h_{f2})}{(h_2 - h_{f2})} = \frac{(449 - 272) - 0.0867(629.8 - 449)}{2495 - 449}$$

$$= 0.0788 \text{ kJ/kg of entering steam} \qquad \textbf{\textit{Ans.}}$$

At heater No. 3: $m_3 h_3 + h_{f5} + (m_1 + m_2)\, h_{f2} = h_{f3} + (m_1 + m_2 + m_3)\, h_{f3}$ ...(a)

At drain cooler: $(m_1 + m_2 + m_3)\, h_{f3} + h_{f4} = h_{f5} + (m_1 + m_2 + m_3)\, h_{f4}$

$$\therefore \quad h_{f5} = (m_1 + m_2 + m_3)\,(h_{f3} - h_{f4}) + h_{f4} \qquad \text{...(b)}$$

Putting the value of h_{f5} in Eq. (a), we gave

$$m_3 h_3 + (m_1 + m_2 + m_3)\,(h_{f3} - h_{f4}) + h_{f4} + (m_1 + m_2)\, h_{f2} = h_{f3} + (m_1 + m_2 + m_3)\, h_{f3}$$

$$\therefore \quad m_3 = \frac{(h_{f3} - h_{f4}) - (m_1 + m_2)\,(h_{f2} - h_{f4})}{h_3 - h_{f4}}$$

$$= \frac{(272 - 101) - (0.0867 + 0.0788)(449 - 101)}{2260 - 101}$$

$$= 0.0525 \text{ kJ/kg of entering steam} \qquad \textbf{\textit{Ans.}}$$

Neglecting pump work: Work done in the cycle = $w_T = q_A - q_R$

Also $w_T = (h_0 - h_1) + (1 - m_1)\,(h_1 - h_2) + (1 - m_1 - m_2)\,(h_2 - h_3)$
$+ (1 - m_1 - m_2 - m_3)\,(h_3 - h_4)$

$$q_R = (1 - m_1 + m_2 + m_3)\,(h_4 - h_{f4})$$

$$= (1 - 0.0867 - 0.0788 - 0.0525)(2012 - 101) = 1494.4 \text{ kJ/kg}$$

Heat supplied = $q_A = h_0 - h_{f1} = 3035 - 629.6 = 2405.2$ kJ/kg

$$\therefore \text{(b) } (\eta_{th})_{Reg} = \frac{q_A - q_R}{q_A} = \frac{2405.2 - 1494.9}{2405.2} = \frac{910.3}{2405.2} = 37.84\% \qquad \textbf{\textit{Ans.}}$$

(c) Work done without regeneration, = $h_0 - h_4 = 3035 - 2012 = 1023$ kJ/kg

Heat supplied = $h_0 - h_{f4} = 3035 - 101 = 2934$ kJ/kg

$$\therefore \text{Thermal efficiency of Rankine cycle } (\eta_{th})_{Rank} = \frac{1023}{2934} = 34.86\% \qquad \textbf{\textit{Ans.}}$$

(d) Theoretical gain of efficiency due to regenerative feed heating

$$\frac{(\eta_{th})_{reg} - (\eta_{th})_{Rank}}{(\eta_{th})_{Rank}} = \frac{37.84 - 34.86}{34.86} = 8.54\% \qquad \textbf{\textit{Ans.}}$$

∴ (e) $(ssc)_{Reg}$ with regenerative feed heating

$$= \frac{3600}{w_{net}} = \frac{3600}{2405.2 - 1494.9} = 3.954 \text{ kg/kWh} \qquad \textbf{\textit{Ans.}}$$

(f) With regeneration, steam entering the turbine $= \dfrac{50000}{910.3} = 54.92$ kg/s

With regeneration steam entering the nozzles of last stage $= (1 - m_1 - m_2 - m_3)54.92$

$$= 0.781 \times 54.92 = 42.92 = 0.782 \times 54.92 = 42.947 \text{ kg/s} \qquad \textbf{\textit{Ans.}}$$

Without regeneration steam entering the turbine $= \dfrac{50000}{123} = 48.875$ kg/s **_Ans._**

(g) Amount of heat carried away by cooling water in condenser per kWh with regeneration

$$= (1 - m_1 - m_2 - m_3)\,(h_4 - h_{f1}) \times \text{Mass of steam/kWh}$$

$$= 0.782(2012 - 101) \times 3.954 = 5723.34 \text{ kJ/kWh} \qquad \textbf{\textit{Ans.}}$$

This example shows clearly that regenerative feed heating (i) increases the thermal efficiency, (ii) requires a larger boiler capacity for a given output; and (iii) requires a reduced condenser capacity.

Example 3.5 Steam is being bled off from two stages of a steam turbine for feed heating purposes which operates under the following conditions:

Initial steam pressure = 26.5 bars, initial steam temperature = 370°C, steam pressure in first heater = 5.3 bars, steam pressure in second heater = 1 bar, Exhaust pressure = 0.06 bar, Condensate temperature = 39°C, Feed temperature after second heater = 95°C, Feed temperature after first heater = 150°C.

The bled steam is condensed in the feed heater and there is no cooling of the condensate. The drains from the first heater are passed through a steam trap in the second heater and the combined drain from the second heater are pumped by a drain pump into the feed pipe after the second heater. If the internal power developed by the turbine is 12,000 kW, find the required capacity of heater drain pump. Assume the efficiency ratio of the turbine is 0.8 and the condition curve may be assumed to be a straight line.

Solution Refer to Fig. 3.48 and Fig. 3.49. Let 1 kg of steam flowing through the turbine. From the steam table; $h_0 = 3170$ kJ/kg (at 26.5 bar and 370°C), $h_2 = 2122$ kJ/kg.

Since the efficiency ratio is 0.8 given which is the ratio of actual coupling work per kg to the isentropic heat drop per kg, thus we have

$$h_0 - h_3' = \eta_r\,(h_0 - h_3) = 0.8\,(3170 - 2122) = 838.4 \text{ kJ/kg}$$

$$h_3' = 3170 - 838.4 = 2331.2 \text{ kJ/kg}$$

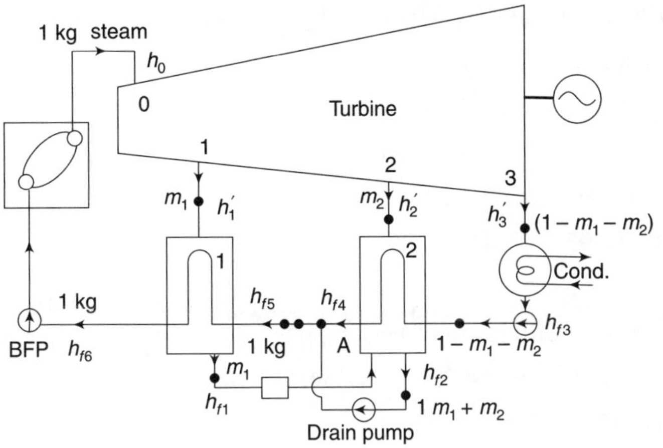

Figure 3.48 Example 3.5

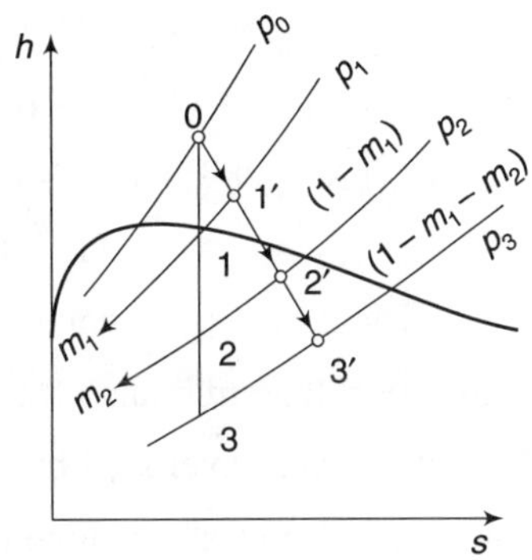

Figure 3.49 Example 3.5

Now, we can locate point 3′ corresponding to pressure 0.6 bar and enthalpy 2331.2 kJ/kg. By joining points 0 and 3′, we get the condition line. On this condition line point 1′ and 2′ are found where the condition line cuts 5.3 bars and 1 bar pressure line respectively, Thus, from chart, h'_1 = 2895 kJ/kg (5.3 bar), (h_2' = 2650 kJ/kg (1 bar)

From Steam Table : h_{f1} = 649.6 kJ/kg (at 5.3 bars), h_{f2} = 417 kJ/kg (at 1 bar), h_{f3} = 163 kJ/kg (at 39°C), h_{f4} = 398.17 kJ/kg (at 95°C), h_{f6} = 632.15 kJ/kg (at 150°C)

Using heat balance equation:

At heater No. 1: $m_1h'_1 + h_{f5} = h_{f6} + m_1h_{f1}$ ...(3.45)

At point *A*: $(1 - m_1 - m_2)\ h_{f4} + (m_1 + m_2)h_{f2} = h_{f5}$ ...(3.46)

Putting the value of h_{f5} in (3.45), we get

$$m_1h'_1 + (1 - m_1 - m_2)\ h_{f4} + (m_1 + m_2)\ h_{f2} = h_{f6} + m_1h_{f1}$$

or $$m_1(h'_1 + h_{f2} - h_{f1} - h_{f4}) + m_2(h_{f2} - h_{f4}) = h_{f6} - h_{f4} \quad ...(3.47)$$

At heater No. 2:

$$m_2h'_2 + (1 - m_1 - m_2)\ h_{f3} + m_1h_{f1} = (m_1 + m_2)\ h_{f2} + (1 - m_1 - m_2)\ h_{f4}$$

or $$m_1(h_{f1} - h_{f3} - h_{f2} - h_{f4}) + m_2(h'_2 - h_{f3} - h_{f2} + h_{f4}) = h_{f4} - h_{f3} \quad ...(3.48)$$

Now putting the value in Eqs. (3.47) and (3.48), we have

$$m_1(2895 + 417 - 699.6 - 398.17) + m_2(417 - 398.17) = 632.15 - 398.17$$

or $$2264.23m_1 + 18.83m_2 = 233.98 \quad ...(3.49)$$

and $$m_1(649.6163 - 417 - 163 - 398.17) + m_2(2650 - 163 - 417 + 398.17)$$
$$= 398.17 - 163$$

or $$793.77m_1 + 2794.17m_2 = 235.17 \quad ...(3.50)$$

Solving Eqs. (3.49) and (3.50), we get

m_1 = 0.102395 kg/kg of entering steam and m_2 = 0.08416 kg/kg of entering steam

∴ Total steam bled off for regeneration

$$= m_1 + m_2 = 0.102393 + 0.08416 = 0.18655 \text{ kg/kg of entering steam}$$

$$w_T = (h_0 - h'_1) + (1 - m_1)(h'_1 - h'_2) + (1 - m_1 - m_2)(h_2' - h_3')$$

$$= (3170 - 2895) + (1 - 0.102395)(2895 - 2650)$$

$$+ (1 - 0.102395 - 0.08416)(2650 - 2331.20)$$

$$= 754.23 \text{ kg/kg or entering steam}$$

This work done may also found from difference of heat supplied and heat rejected.

Internal power developed by the turbine = 754.23 × 1 = 754.23 kW/kg

The internal power developed = 12,000 kW

∴ Amount of steam to produce 2,000 kW $= \dfrac{12000}{754.23} = 15.19$ kg/s

Hence, total steam bled off = $(m_1 + m_2)$ 15.19 kg/s = 0.18055 × 15.19 = 2.968 kg/s

∴ Capacity of drain pump = 2.968 kg/s = 2.968 × 3600 = 1068.4 kg/h. ***Ans.***

3.4 REHEATING CYCLE

3.4.1 Effect of Flow of Wet Steam in Nozzles and Blades

Practically, in the L.P. stages of the turbine the flowing steam will no longer remain dry and saturated. The steam will be in wet condition, i.e., there will be suspended water particles in the steam. Until now, we have considered steam as a homogeneous fluid and that the steam and the suspended water particles are moving in the same direction and with the same velocity. But in actual practice this is not true. The direction and velocity of wet steam is different from that of dry steam due to difference in specific volumes.

The water particles produce a very detrimental effect on the nozzles and the blades and ultimately on the performance while passing through it.

Let us consider the expansion of wet steam through a nozzle. Now a large number of suspended water droplets of varying size will be carried by the flowing steam. There will be exceedingly minute water particles and also larger droplets due to joining of these smaller particles.

Let us consider two transverse section at distance δx part as shown in Fig. 3.50. Let p and $p - \delta p$ be uniform pressure at section 1-1 and 2-2 respectively. Let A and B having area δA represent the elementary mass of the steam and water particles respectively. B having a prismatic shape instead of natural spherical shape; also assume that all the particles have the same velocity at any moment at a section considered.

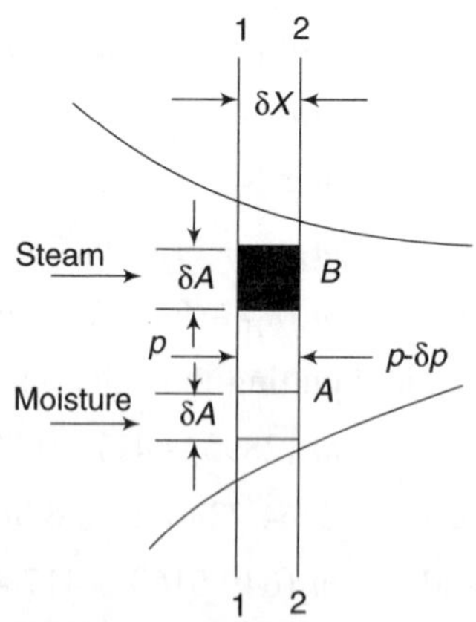

Figure 3.50 Effect of wet steam

Let ρ_s and ρ_w be the density of steam and water particles in kg/m^3 respectively.

Elementary mass of steam $A = \rho_s\ \delta A \cdot \delta x$.

Elementary mass of water particles, $B = \rho_w \cdot \delta A \cdot \delta x$ ...(3.51)

Accelerating force on the elementary steam mass, $A = \delta p \cdot \delta A$; N ...(3.52)

Accelerating force on the elementary water particles = $\delta p \cdot \delta A$; N

We see that accelerating force in both cases is equal but not the mass.

From Newton's second law of motion,

Accelerating force, F = Mass × Acceleration (a)

Thus accelerating force on elementary steam mass = $\delta p \cdot \delta A = \delta_s\ \delta A \cdot \delta x \cdot a_s$

or Acceleration of steam particles = $a_s = \dfrac{\delta \rho}{\delta x} \cdot \dfrac{1}{\rho_s}$; m/s^2 ...(3.53)

Acceleration of water particles = $a_w = \dfrac{\delta p}{\delta x} \cdot \dfrac{1}{\rho_w}$; m/s^2 ...(3.54)

Dividing (3.53) by (3.54), we get $\dfrac{a_s}{a_w} = \dfrac{\rho_w}{\rho_s} = \dfrac{v_s}{v_w}$ ...(3.55)

Let us compare this ratio at two different pressures.

At 0.7 bar, $\dfrac{a_s}{a_w} = \dfrac{2.364}{0.001036} = 2.2818 \times 10^3 = 2281.88$

At 0.05 bar, $\dfrac{a_s}{a_w} = \dfrac{28.2}{0.001006} = 28031.8$

It is thus clear that acceleration of steam is 2281.45 times that of water particles at 0.7 bar and 28031.8 times at pressure 0.05 bar thus a great difference exists at different pressure. This also shows that velocity of steam and water particles at any pressure is not the same. The water particle is accelerated at a much smaller rate than the steam, and soon lags behind. This is so because of the fact that the specific volume of water particles is very little than that of steam at the same pressure. Now, due to the difference in velocity of steam and water particles, there is a certain relative velocity of steam past the lagging particles which introduces a new accelerating force.

If C_s and C_m are the velocity of steam and moisture at any point, then $(C_s - C_m)$ will be the relative velocity.

The accelerating force, F on the water particles due to relative velocity $(C_s - C_m)$ only, is given by

$$F = y \cdot \rho_s\ (C_s - C_m)^2 \cdot \delta s\ ;\ \text{N} \qquad ...(3.56)$$

where δs = external surface of the droplet

y = a number, function of Reynolds number

Though, we have assumed the prismatic shape of the water droplets for simplicity, but if the droplets are of spherical type, then external surface area $\delta_s = \pi d^3$ where d is the diameter of sphere in metre and volume is $\pi d^3/6$.

Hence, the mass of water particles = $m_w\ \rho_w\ \pi d^3/6$

We can thus find the acceleration of the water particles caused by relative velocity, from Newton's second law of motion.

$$\text{Accelerating force} = \text{Mass} \times \text{Acceleration}$$

$$y \cdot \rho_s (C_s - C_m)^2\ \delta s = \frac{\rho_w \cdot \pi d^3}{6} \cdot a_w$$

or

$$a_w = \frac{6 \cdot y}{d}\left(\frac{\rho_s}{\rho_w}\right)(C_s - C_m)^2;\ \text{m/s}^2 \qquad \text{...(3.57)}$$

It is obvious from the above expression that the acceleration of the water particles will be least for larger droplets and also least in the low pressure stages where ρ_s is very small and ρ_w is very large.

It is also clear from the above analysis that in the stages of a turbine where the steam is wet, the velocity of water particles is much less than the expanding steam and also that this velocity difference is the maximum when the steam pressure is the lowest for the particles of larger diameter. The ratio of moisture velocity to steam velocity varies from 0.1 to 0.3. In addition to this; the moisture has a detrimental mechanical effect on the blades.

3.4.2 Velocity Diagram for Dry Steam and Water Particles

Figure 3.51 shows the velocity diagram for an impuslse turbine for dry steam and water particles.

Let C_{ml} = absolute velocity of moisture at exit from nozzle, m/s, represented by *AG*.

C_{sl} = absolute velocity of dry steam at exit from nozzle, m/s, represented by *AB*.

$$n_1 = \frac{C_{m1}}{C_{s1}} \qquad \text{...(3.58)}$$

C_{rm1} = Relative velocity of moisture at inlet to the blade, m/s, represented by *DG*

C_{rs1} = Relative velocity of moisture at inlet to blades, m/s, represented by *DB*

C_{rm2} = Relative velocity of moisture at outlet from blades m/s represented by *DH*

C_{rs2} = Relative velocity of steam at outlet from blades, m/s represented by *DF*

C_{m2} = Absolute velocity of moisture at exit from blade, m/s represented by *AH*

C_{s2} = Absolute velocity of steam at exit from blade, m/s represented by *AF*

C_{wm} = Velocity of whirl for moisture.

C_{ws} = Velocity of whirl for steam.

$$n_2 = \frac{C_{rm2}}{C_{rs2}} \qquad \text{...(3.59)}$$

The main assumption in drawing the velocity diagram are (a) that the water particles issue from the *nozzle* in the same direction as the steam and (b) that the relative velocity direction of the dry steam and water particles are the same at the blade outlet. *ABD* and *AGD* are the

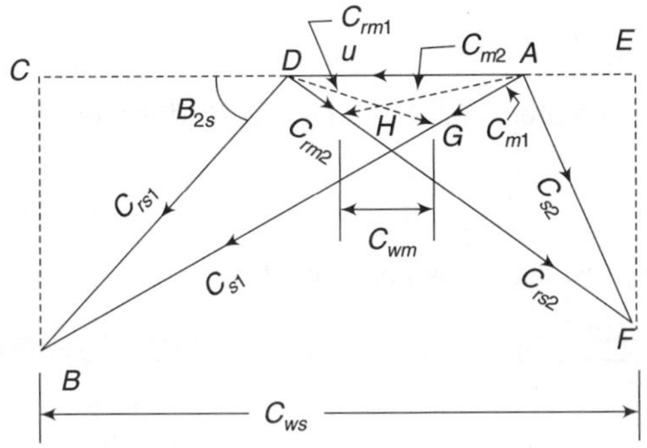

Figure 3.51 Velocity diagram for dry steam and water particles

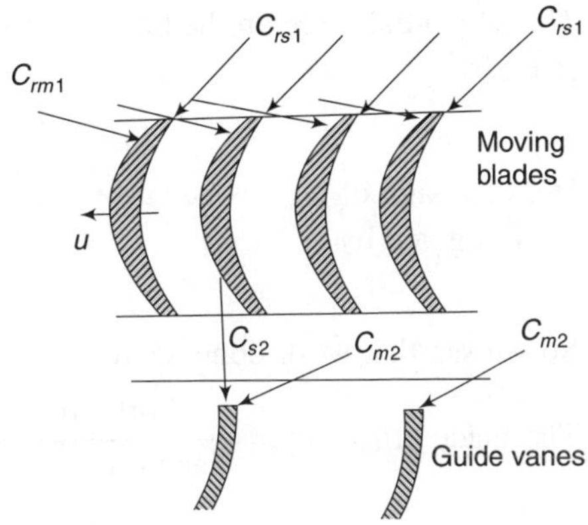

Figure 3.52 Diagram showing oblique impact of water drops on moving blades and nozzle guidevanes

inlet velocity diagrams for the steam and water particles respectively and *ADF* and *ADH* are the outlet velocity diagrams for the steam and water particles respectively. We see that the direction of relative velocity C_{rm1} of water particles is not the same as that of steam and thus it strikes the leading face of the moving blades instead of gliding off as shown in Fig. 3.52. Consequently, the spherical water droplets are flattened out and broken up into smaller droplets which are then carried away be the rapidly flowing steam. The relative velocity of water particles C_{rm2} at the blade outlet is shown by *DH* which is very low compared to that of steam C_{rs2} shown by *DF*. Thus, the direction C_{rs2} is not the same as that of C_{s2} and it strikes the fixed blades as shown in Fig. 3.52 causing impact.

Due to the striking of moisture at the inlet of the moving blades the momentum of the incoming moisture particles is practically destroyed at inlet and hence the momentum of dry steam is only available. To calculate the value of C_{rm2}, we assume that x is the dryness fraction of steam, then $x.\ C_{rs1}$ will be momentum per kg of the stuff, i.e., of the steam and moisture. The momentum at outlet, assuming homogeneous flow $K.x.C_{rs1}$ where K is C_{rs2}/C_{rs1}. If we equate the momentum of the steam and moisture at outlet from the blade channel, then we can find out C_{rm2} and C_{rs2}.

$$x \cdot C_{rs2} + (1 - x)\ C_{rm2} = x \cdot K \cdot C_{rs1}$$

or
$$C_{rs2}\ [x + (1 - x)n_2] = x \cdot K \cdot C_{rs1}$$

$$C_{rs2} = \frac{x \cdot K \cdot C_{rs1}}{x + (1 - x)n_2} \qquad ...(3.60)$$

and
$$C_{rm2} = \frac{n_2 \cdot x \cdot K \cdot C_{rs1}}{x + (1 - x)n_2} \qquad ...(3.61)$$

Equations (3.60) and (3.61) give the value of C_{rs2} and C_{rm2} and hence C_{rs2} and C_{rm2} can be drawn.

The tangential force on the blades per kg of the stuff (i.e., steam and moisture) per second is given by

$$F_t = x \cdot C_{ws} + (1 - x)C_{wm}\ ; \quad N$$

Here the sign of C_{wm} is negative because C_{m2} is greater than C_{m1}. The work done per kg of stuff is given by

$$w_T = u\ [x \cdot C_{ws} + (1 - x)C_{wm}]\ ; \quad W \qquad ...(3.62)$$

So we see that work done decreases due to the presence of water particles.

$$\text{The blade efficiency} = \frac{\text{Work done per kg of stuff}}{\text{Incoming energy per kg of stuff}}$$

$$\text{The incoming energy per kg of stuff} = \frac{x\,C_{sl}^2}{2} + \frac{(1 - x)C_{ml}^2}{2}$$

$$\text{Therefore, the blade efficiency} = \eta_b = \frac{2u[x\ C_{ws} + (1 - x)\ C_{wm}]}{C_{sl}^2\ [x + (1 - x)n_1^2]} \qquad ...(3.63)$$

Figure 3.53 shows the reduction in blade efficiency due to moisture of steam. We see that for a fixed value of n_1 blade efficiency decreases as the moisture content increases.

The stage efficiency of wet steam, given by Dr. K. Baumann is

$$\eta_s' = x \cdot \eta_s \qquad ...(3.64)$$

where x is the dryness fraction and η_s is the stage efficiency of dry steam.

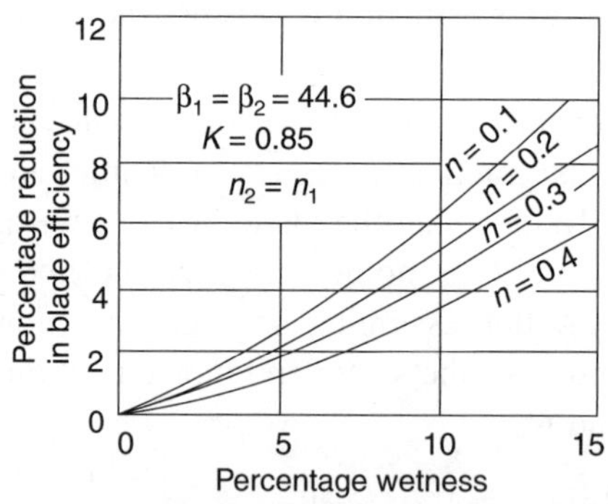

Figure 3.53 Reduction in blade efficiency due to water particles in steam

3.4.3 Correction to Condition Curve for Wetness

The condition curve of the wet steam would not be same as that of dry steam and so it needs correction. If Δh_{c1} and Δh_{c2} are the cumulative heat drop in the superheat part and wet part of expansion, then

$$\Delta h_{c1} = \frac{h_1 - h_c}{\eta_s} \quad \text{and} \quad \Delta h_{c2} = \frac{h_c - h_2'}{\eta_s}$$

where η_s is the constant stage efficiency.

The work done in a wet stage is x. Δh_{isen} where x is the dryness fraction of steam and Δh_{isen} is isentropic heat drop in that stage. Figure 3.54 shows the values of $\Sigma\Delta h_{isen}$ and x plotted against saturation temperature for wet stages. It is obvious that the relations are linear. Hence, from this relationship, internal work done in the turbine is given by

$$\Delta h_{i1} + \Delta h_{i2} = \eta_s (\Delta h_{c1} + x_m \Delta h_{c2})$$

where x_m is the mean dryness fraction in all wet stages and Δh_{i1} and Δh_{i2} (Fig. 3.54a) are the internal work done is dry stage respectively. In Fig. 3.54(b) the dotted line shows the actual condition line in the wet stage because the internal work done in wet stage is less than the dry stage. So modified condition curve will give, for a constant dry stage efficiency, slightly higher reheat factor and slightly greater specific volumes in L.P. stages.

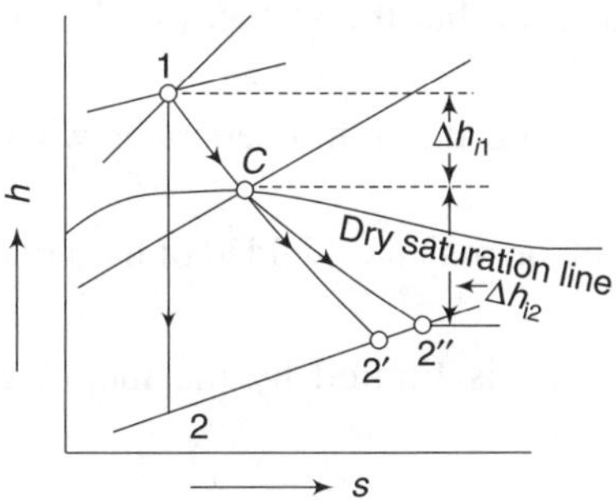

Figure 3.54(a) Correction to condition curve for wetness

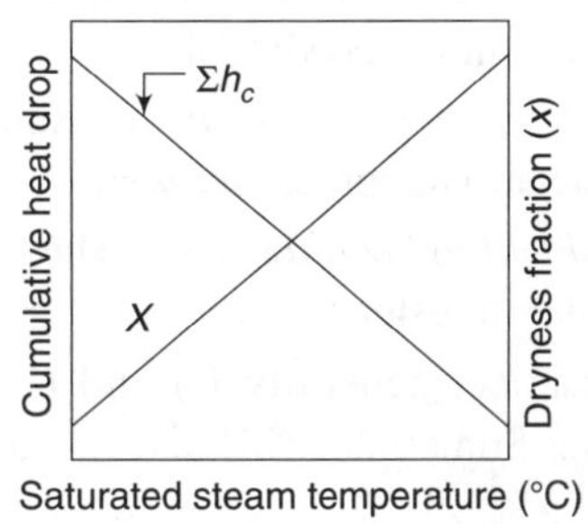

Figure 3.54(b) Relation of Δh_{isen} and x for wet steam

3.4.4 Erosion and Corrosion of Blades

The presence of water particles in steam not only reduces the nozzle and blade efficiency but also brings about a serious erosion and corrosion of blades. *Erosion of turbine blading is the wasting away or grooving of the back of the inlet edge of the blade, and only occurs where the moisture in steam exceeds about 10% and the blade speed is high.* We have seen that the water particles strike the leading surface of the blades instead of gliding smoothly to produce an impact on the blades, and if this impact is sufficiently heavy, severe local stresses develop in the blade material, causing the surface metal to fail and flake off. The water particles also corrode the metal surface. In the last one or two stages of the turbine where steam is the wettest, the erosion is more likely to occur. The erosion problem is mostly affected by the two factors (i) *wetness of steam* (ii) *peripheral velocity.*

(i) ***Wetness of steam.*** It has been found that a nominal wetness fraction exceeding 10 per cent produces severe erosion of the blades and the final wetness depends on the initial temperature and pressure of steam, the stage efficiency and the exhaust pressure.

(ii) ***Peripheral velocity.*** It has been found that the intensity of pressure produced by the impact of water droplets does not depend on the size of the drops but is directly

proportional to the velocity of the impact and this velocity of impact depends upon the blade speed. If the blade tip speed is very high, severe stresses are set up by the impact of water droplets. *The erosion is worst towards the tips of low pressure blades because centrifugal force tends to concentrate the water particles in the outer annulus, the tip speed is greater than the root speed.*

3.4.4.1 Prevention of Erosion and Corrosion

The following are the methods to combat the erosion and corrosion problems in steam turbines:

(i) *Raising initial temperature.* Raising the initial temperature of steam so that the wetness never exceeds 10%, even in the last stage of the low-pressure turbine.

(ii) *Reheating.* Remove the whole of the steam before the moisture content in the expanding steam approaches 10% and reheat it in a reheater so that the wetness at the exhaust does not exceed 10%.

(iii) *Water extractions.* By passing of whole of the steam through a separator where the steam is appreciably wet.

(iv) *Shielding leading edge.* The leading edge of the blade may be shielded of hard material like tungsten.

In practice, generally (i) and (ii) are preferred but (i) is limited by the metallurgical condition. Sometimes (iii) and (iv) are also used.

3.4.5 Reheating of Steam

By reheating of steam erosion and corrosion difficulties due to the presence of water particles in the steam are avoided because of the fact that the steam becomes dry after reheating. This is done by taking out the whole of the steam from the turbine at a suitable point, (i.e., where the steam tends to become wet or before reaching 10% wetness and it is reheated, i.e., a further supply of heat is given to it at constant pressure by the flue gases or live steam until it is superheated to near about the same initial temperature, after which the steam is re-admitted to the turbine and expanded to the condenser pressure. Generally, the expansion of steam is carried out in several stages and the steam is reheated by addition of heat between the stages at suitable points. Any power plant using reheating of steam also employs regenerative feed heating. The reheater may be incorporated in the walls of the main boiler or it may be a separately fired superheater. Double reheating may be employed.

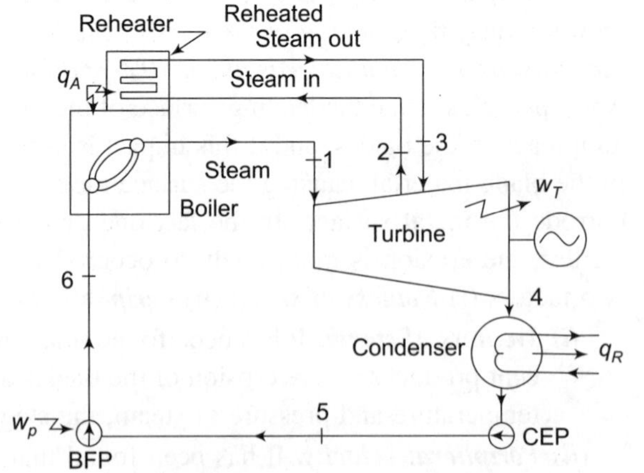

Figure 3.55 Reheat cycle

A schematic diagram of a theoretical single stage reheat cycle and its corresponding representation of ideal reheating process on *T-s* and *h-s* charts are shown in Figs. 3.55 and and 3.56. In Fig. 3.56(a), process 6–1 shows the formation of steam in the boiler. Steam at pressure p_1 and temperature t_1 represented as state point 1 enters the turbine and is expanded isentropically to a certain pressure p_2 where the state point is 2 and temperature t_2. At this point the whole of the steam is taken out and is reheated in a reheater to temperature $t_3 (t_3 \approx t_1)$ and is represented by state point 3. After this, the steam is re-admitted to the turbine in which it is expanded isentropically to condenser pressure p_3. Assuming 1 kg of steam flowing through the cycle let us compare the thermal efficiency with and without reheating.

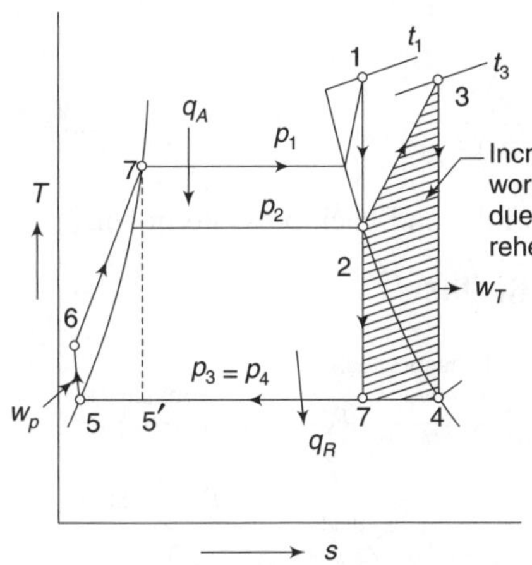

Figure 3.56(a) Reheated turbine process on *T-s* chart

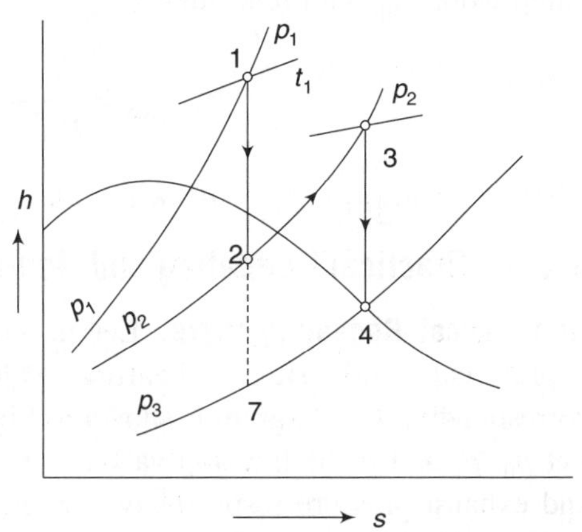

Figure 3.56(b) Reheated turbine process on *h-s* chart

With Reheating, Neglecting Pump work

Heat supplied = $(h_1 - h_{f4}) + (h_3 - h_2)$ kJ/kg.

where $h_{f4} = h_{f3} = h_{f5}$ = Enthalpy of water at exhaust pressure p_3 (bar)

Heat rejected = $q_R = h_4 - h_{f4}$ kJ/kg

Work done by the turbine = $q_A - q_R = (h_1 - h_{f4}) + (h_3 - h_2) - (h_4 - h_{f4})$

$$= (h_1 - h_2) + (h_3 - h_4) \quad \text{...(3.65)}$$

Also, $$w_{net} = w_T = (h_1 - h_2) + (h_3 - h_4) \quad \text{...(3.66)}$$

Hence, the theoretical thermal efficiency of the reheat cycle is

$$(\eta_{th})_{Reh} = \frac{(h_1 - h_2) + (h_3 - h_4)}{(h_1 - h_{f4}) + (h_3 - h_2)} \quad \text{...(3.67)}$$

Without reheating: $$w_{net} = w_T = (h_1 - h_7) \quad \text{...(3.68)}$$

$$\text{Heat supplied} = (h_1 - h_{f4}) \quad \text{...(3.69)}$$

So, the theoretical thermal efficiency without reheating, i.e., Rankine is

$$(\eta_{th})_{Rank} = \frac{(h_1 - h_7)}{(h_1 - hf_4)} \quad ...(3.70)$$

We see from equations (3.66) and (3.68) that the work done by the turbine is more in the cases of reheating as compared to without reheating and it is obvious from the *T-s* hatched diagram.

With pump work. If pump work has to be considered, then the result will slightly differ. Since pump work is very small (2 to 2.5% of turbine output), it is generally neglected. Considering pump work, η_{th} with reheating is

$$(\eta_{th})_{Reh} = \frac{(h_1 - h_2) + (h_3 - h_4) - w_p}{(h_1 - h_{f4}) + (h_3 - h_2) - w_p} \quad ...(3.71)$$

where $w_p = v_b\,(p_1 - p_b) \times 10^2$ kJ/kg and p_1 and p_b are inlet and back pressure in bar.

3.4.6 Practical Reheating and Non-reheating Cycles

(i) Practical Reheating Cycle. Let us consider a practical single stage reheating cycle. The corresponding *h–s* diagram is shown in Fig. 3.57. Let p_0, p_1 and p_3 be the stop valve, nozzles box and exhaust pressure respectively. Let the steam condition before stop valve be represented by 0 then process 0–1 shows the throttling of steam in the stop valve. Process 1–2′ shows the actual expansion line before the steam is taken out for reheating.

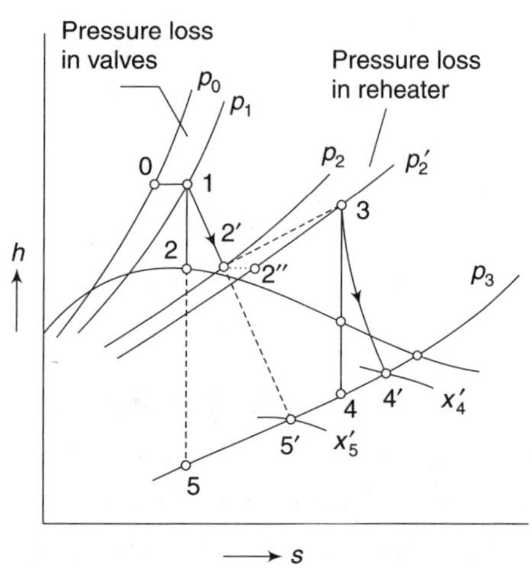

Figure 3.57 Practical *h-s* chart for single stage reheating

At pressure p_2 where the steam tends to become wet as represented by state point 2′ the whole of steam is extracted and reheated to temperature, t_3. Actually, due to resistance in the steam pipe and reheater tubes there is a pressure loss and the steam then re-enters the turbine at pressure p_2' and represented by the state point 3. This drop in pressure is assumed to occur at constant enthalpy and is represented by process 2′–2″. Therefore, 2′–3 shows the reheating of steam and 3–4′ the subsequent expansion of steam to the exhaust pressure, p_3. If there is no reheating of steam, then the actual expansion line is 1–5′ where the actual dryness fraction is x_5'. Neglecting the inlet and outlet kinetic energy of steam, we can find the total internal work in both cases (a) ignoring the effect of wetness and (b) allowing the effect of wetness.

(a) **Ignoring First the Effect of Wetness.** The total internal work done in the cycle is given by

$$\Delta h_1 = (h_1 - h_2') + (h_3 - h_4')$$

Heat supplied in the cycle = $(h_1 - h_{f4}) + (h_3 - h_2')$ where h_{f4} is the enthalpy of water at pressure p_3.

The actual thermal efficiency of reheating cycle is given by

$$(\eta_{th})_{Reh} = \frac{(h_1 - h_2') + (h_3 - h_4') - w_p}{(h_1 - h_{f4}) + (h_3 - h_2') - w_p} \quad \text{...(3.72)}$$

If pump work is neglected, $w_p \approx 0$

(b) **Allowing the Effect of Wetness.** Let us assume before reheating.

$\Delta h_{c1} \cdot \Delta h_{c2}$ = cumulative heat drop down to and below saturation line

x_2' = final dryness fraction.

Let after reheating:

Δh_{c3}, Δh_{c4} = cumulative heat drop down to and below saturation line.

Let η_s be the dry stage efficiency and it is uniform for all stages then the total internal work is given by

$$\Delta h_i = \eta_s \left[\Delta h_{c1} + \left(\frac{1 + x_2'}{2}\right)\Delta h_{c2} + \Delta h_{c3} + \left(\frac{1 + x_4'}{2}\right)\Delta h_{c4}\right]$$

$$\therefore \quad (\eta_{th})_{Reh} = \frac{\Delta h_i}{(h_1 - h_{f4}) + (h_3 - h_2')} \quad \text{...(3.73)}$$

(ii) Non-reheating cycle. When reheating is not done, then this is simple Rankine cycle. Process 1–5 and 1–5′ show the isentropic and actual expansion line respectively. The actual final dryness fraction in this case is x'_5.

(a) **First Ignoring the Effect of Wetness in the L.P. Stages.** The internal work done is given by $(h_1 - h_5')$ and the internal efficiency is

$$(\eta_{th})_{Rank} = \frac{h_1 - h_5'}{h_1 - h_{f5}} = \frac{h_1 - h_5'}{h_1 - h_{f4}} \quad \text{...(3.74)}$$

(b) **Allowing the Effect of Wetness**

Let Δh_{c1} = cumulative heat drop down to saturation line.

Δh_{c2} = cumulative heat drop from saturation line down to exhaust pressure.

x_5' = actual dryness fraction at p_3.

$$\text{So the actual work done} = \Delta h_i = \eta_5 \left[\Delta h_{c1} + \left(\frac{1 + x_5'}{2}\right)\Delta h_{c2}\right] \quad \text{...(3.75)}$$

$$\text{And the thermal efficiency} = (\eta_{th})_{Rank} = \frac{\Delta h_i}{(h_1 - h_{f4})} \quad \text{...(3.76)}$$

3.4.7 Advantages of Reheating

(i) Erosion and corrosion problems in the steam turbine are avoided.

(ii) The output of the turbine is increased.

(iii) The thermal efficiency is increased. A gain of 4 to 7% of thermal efficiency over an equivalent non-reheat cycle is possible. Figure 3.58 shows the gain in thermal efficiency due to single stage reheating.

(iv) Final dryness fraction is increased.

(v) Nozzle and blade efficiencies are increased.

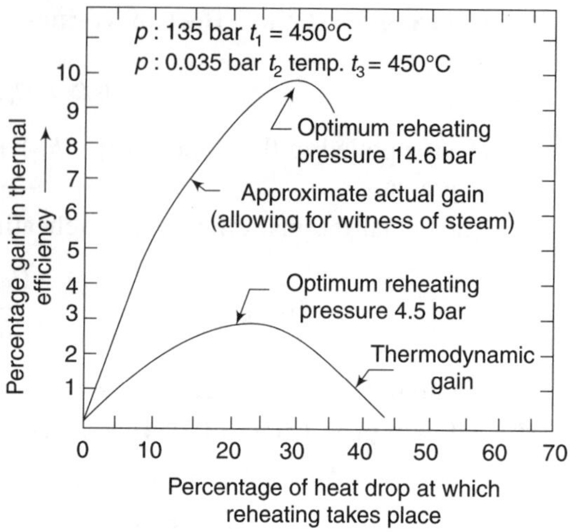

Figure 3.58 Curves of the gain in thermal efficiency due to single stage reheating at various pressures

3.4.8 Disadvantages of Reheating

(i) Relative to the expenditure incurred in reheating, the thermal efficiency does not increase appreciably.

(ii) Maintenance is more.

Note: It is to be noted that there is a particular pressure in the expansion at which it is most economical to reheat the steam. If the steam is reheated early in its expansion then the additional quantity of heat supplied will be small, consequently the gain in thermal efficiency will be small. If reheating is done at a fairly low pressure, then, although a large amount of additional heat is supplied, the steam will have a high degree of superheat as is obvious from the Mollier diagram, thus, a large proportion of the heat supplied in the reheating process will be thrown to waste in the condenser. So reheating should be done at the optimum pressure.

In high pressure steam turbine, as is obvious from the Mollier diagram the initial enthalpy content at any superheat temperature is less, corresponding to moderate or low pressure at that temperature. During expansion, up to certain H.P. stages only, the steam will be wet, so it is essential to have reheating arrangement in H.P. turbine than the turbine working with moderate or low pressure.

The clue for the solution of problems on reheating is to know the properties of steam at various states represented on h–s chart by possible methods.

Example 3.6 Steam is supplied to a turbine equipped with two-stage reheating at pressure 200 bars and temperature 500°C. At pressure 35 bars and temperature 280°C, whole of the steam passes through the first reheater and comes out at a pressure 34 bars and temperature 480°C to expand in the next stages. Again, at pressure 3 bars and temperature 180°C whole of the steam is taken out and made to pass through the second reheater. The steam comes out from the second reheater at 2.5 bars and temperature 480°C and expands in the L.P. stage to pressure 0.04 bars and dry and saturated. Calculate: (a) Ideal and actual heat supplied to turbine, (b) Ideal and actual work done by turbine, (c) Ideal and actual efficiency of turbine, (d) Ratio of actual to ideal work of turbine, (e) Ideal cycle efficiency for the same state points.

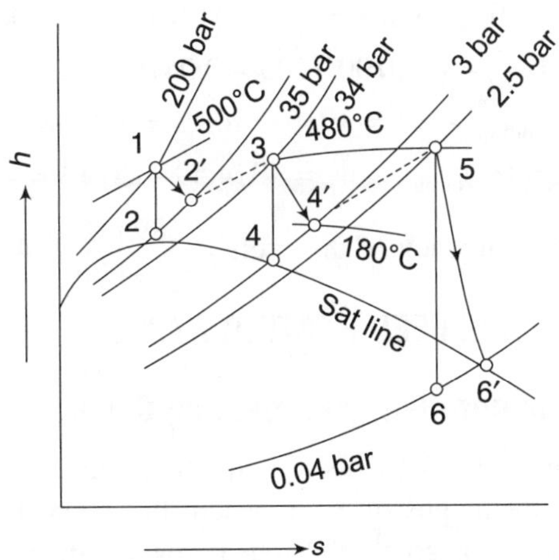

Figure 3.59 Example 3.6

Solution Refer Fig. 3.59, pressure and temperature give actual condition.

From Mollier chart and steam table: h_1 = 3245 kJ/kg, h_2 = 2884 kJ/kg, h_4 = 2778 kJ/kg, h_3 = 3405 kJ/kg, h_3' = 2925 kJ/kg, h_4 = 2778 kJ/kg, h_4' = 2820 kJ/kg, h_5 = 3448 kJ/kg, h_6 = 2520 kJ/kg, h_6' = 2550 kJ/kg, h_{f6} = 121.41 kJ/kg

(a) Ideal heat supplied will be obtained by neglecting the losses in turbine.

Ideal heat supplied = $(h_1 - hf_6) + (h_3 - h_2) + (h_5 - h_4)$

= (3245 – 121.41) + (3405 – 2884) + (3448 – 2778) = 4314.59 kJ/kg. ***Ans.***

Actual heat supplied = $(h_1 - hf_6) + (h_3 - h_2') + (h_5 - h_4')$

= (3245 – 121.41) + (3405 – 2925) + (3448 – 2820) = 4231.59 kJ/kg. ***Ans.***

(b) Ideal work done = Ideal heat supplied – Ideal heat rejected.

= 4314.59 – (2520 – 121.41) = 1916 kJ/kg ***Ans.***

Actual work done = 4231.59 – (2550 – 121.41) = 1803 kJ/kg ***Ans.***

(c) Ideal thermal efficiency = $\dfrac{1916}{4314.59}$ = 44.4% ***Ans.***

Actual thermal efficiency = $\dfrac{1803}{4231.59}$ = 42.60%

(d) Ratio of actual to ideal work $= \frac{1803}{1916} = 0.941.$ ***Ans.***

(e) It includes pump work also. Whenever the turbine efficiency and cycle efficiency are required separately, it means pump work has to be considered in the cycle efficiency.

$$w_p = v_b(p_1 - p_2) \times 10^2, \quad \text{where } v_b \text{ at 0.04 bar} = 0.001007 \text{ m}^3/\text{kg}.$$

Hence, pump work $= w_p \approx 0.001007(200 - 0.04)10^2 \approx 20.13$ kJ/kg

$$(w_{net})_{cycle} = (w_{net})_{turbine} - w_p = 1916 - 20.13 = 1895.87 \text{ kJ/kg}$$

Cycle heat supplied $= (q_2)_{turbine} - w_p = 4314.59 - 20.13 = 4294.46$ kJ/kg

Hence, ideal thermal efficiency of the cycle $= \frac{1895.87}{4294.46} = 44.14\%.$ ***Ans.***

3.5 REHEAT-REGENERATIVE FEED HEATING CYCLE

3.5.1 Ideal Reheat-Regenerative Feed Heating Cycle

All the modern steam turbine power plants are equipped with reheat and regenerative feed heating arrangement. The advantages of each system has been discussed separately in the previous section. The purposes of combined reheat–regenerative feed heating arrangement are (i) to increase thermal efficiency, (ii) to increase output, (iii) to avoid erosion, and (iv) to avoid corrosion. Figure 3.60 shows the diagrammatic arrangement of one stage reheating and two-stage regenerative feed heating. Its corresponding representations of ideal processes are shown on *T–s* and *h–s* diagram in Figs. 3.61 and 3.62.

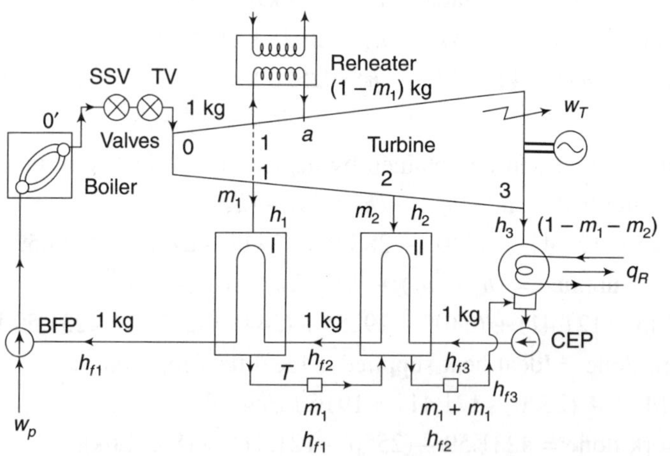

Figure 3.60 Reheat-regenerative feed heating cycle

Let 1 kg of steam enter the turbine. The state point 0 represents initial condition of steam entering the turbine, i.e., p_0 is the nozzle box pressure. At state point 1, m_1 kg of steam is extracted for regenerative feed heating and the remaining $(1 - m_1)$ kg passes through a reheater

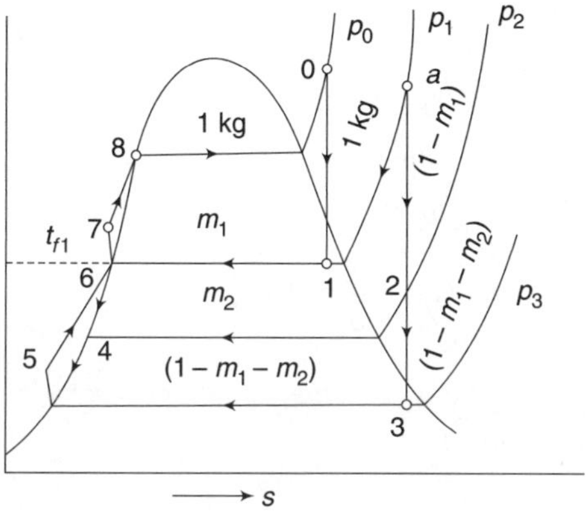

Figure 3.61 Representation of reheat-regenerative cycle on *T-s* chart

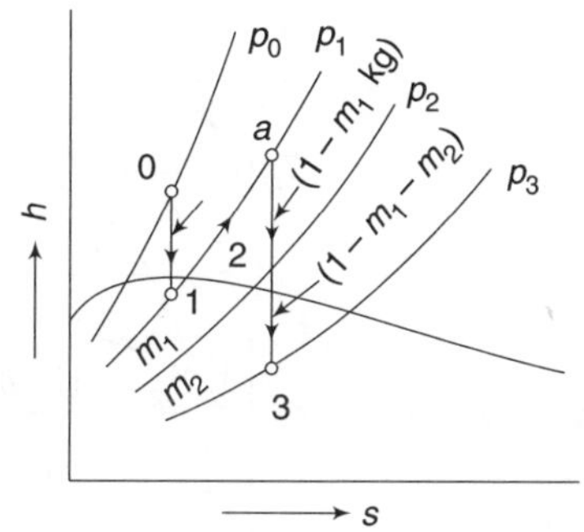

Figure 3.62 Representation of reheat-regenerative cycle on *h-s* chart

where the steam is reheated to state point a, near about the initial temperature. The reheating process occurs at a constant pressure p_1. At point a steam is re-admitted in the turbine and expands to pressure p_2. At state point 2, again m_2 kg of steam is blade-off for the second heater for feed heating and the remaining quantity of steam $(1 - m_1 - m_2)$ expands to the back pressure p_3.

Considering pump work:

Ideal heat supplied $= q_A = (h_0 - h_{f1}) + (h_a - h_1)(1 - m_1) - w_p$ kJ/kg ...(3.77)

Ideal work done in kJ/kg of entering steam.

$$w_T = (h_0 - h_1) + (1 - m_1)(h_a - h_2) + (1 - m_1 - m_2)(h_2 - h_3) \quad ...(3.78)$$

The value of m_1 and m_2 will be calculated by writing the enthalpy balance at heater No. 1 and 2 as done in previous section.

If pump work has to be considered then, the net work will be turbine work minus pump work, i.e.,

$$w_{net} = w_{turbine} - w_{pump}$$

But $$w_{pump} = v_b(p_1 - p_b) \times 10^2 \text{ kJ/kg}$$

where v_b = specific volume of water at back pressure, p_b (m^3/kg).

p_1 = initial pressure (bar), p_b = back pressure, bar

Hence, thermal efficiency $= \eta_{th} = \dfrac{\text{Net work done}}{\text{Heat supplied}} = \dfrac{w_{net}}{q_A}$ ...(3.79)

3.5.2 Actual Reheat-Regenerative Feed Heating Cycle

Figure 3.63(a) shows actual reheat-regenerative cycle on *T–s* and *h–s* diagrams of the system is shown in Fig. 3.63(b).

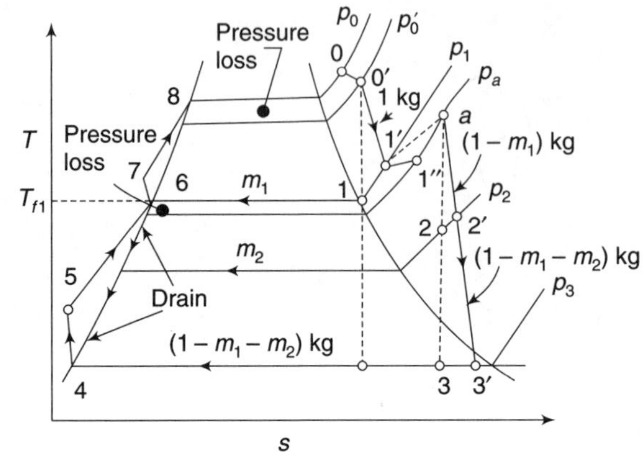

Figure 3.63(a) Actual *T-s* representation of reheat-regenerative cycle

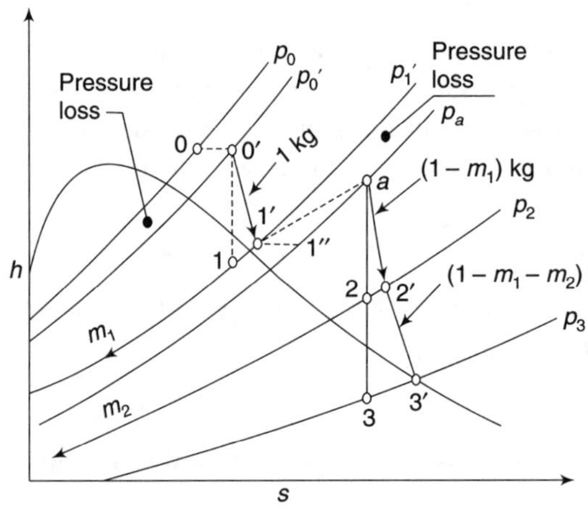

Figure 3.63(b) Actual *h-s* representation of reheat-regenerative cycle

In the actual cycle, process 0′–1′ shows the actual expansion of steam before extraction. In the reheater, there is pressure loss and so steam from the reheater enters the turbine at pressure p_a and state point a. Process a–2′ shows the actual expansion of steam before m_2 kg of steam extracted at pressure p_2 and process 2′–3′ shows the actual expansion in the L.P. stages.

Actual heat supplied = $q_{Aa} = (h_0 - h_{f1}) + (h_a - h_1')(1 - m_1) - w_{pa}$ kJ/kg ...(3.80)

Work done by turbine = $w_{Ta} = (h_0 - h_1') + (1 - m_1)(h_a - h_2') + (1 - m_1 - m_2)(h_2' - h_3')$ kJ/kg

where $h_0 - h_1' = \eta_t (h_0 - h_1)$, $h_a - h_2' = \eta_t(h_a - h_2)$ and $(h_2' - h_3') = \eta_t (h_2' - h_3)$

The values of m_1 and m_2 will be calculated by energy balance as discussed above. The actual thermal efficiency is given by

$$\eta_{tha} = \frac{w_{neta}}{q_{Aa}} \quad \text{where} \quad w_{neta} = w_{Ta} - w_{pa} \quad \text{and} \quad w_{pa} = \frac{w_p}{\eta_p} \qquad \text{...(3.81)}$$

Generally, η_p = overall pump efficiency = 85 to 90%. If mechanical, generator and boiler efficiencies are given, these have to be considered.

Example 3.7 Steam enters a turbine at 80 bars and temperature 400°C. At pressure 30 bars some quantity of steam is bled off for regenerative feed heating. The remaining steam expands to pressure 20 bars from where steam is passed through a reheater and comes out at 19 bars and 410°C. After that it expands to condenser pressure 0.04 bar. In actual turbine, the steam enters the reheater at 250°C, the feed water enters the boiler at 225°C. The combined steam rate is 4.5 kg/kWh. The generator efficiency is 0.95. The actual quantity of steam bled-off is 0.26 of the throttle flow, i.e., entering steam. Determine, for the ideal turbine—(a) Percentage of steam bled off (b) Thermal efficiency. For actual turbine: (c) Shaft work (i.e., brake work) (d) The brake and electrical thermal efficiency.

Solution Refer to Fig. 3.64(a) and 3.64(b). From Mollier chart, h_0 = 3142 kJ/kg, h_2 = 2814 kJ/kg, h_3 = 3274 kJ/kg, h_2' = 2908 kJ/kg, h_4 = 2176 kJ/kg.

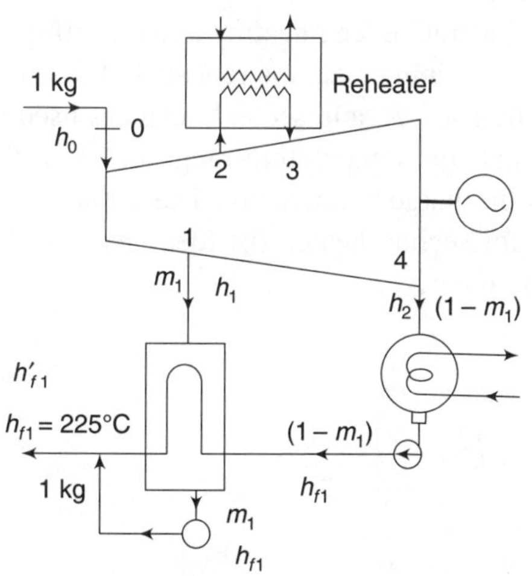

Figure 3.64 (a) Example 3.7

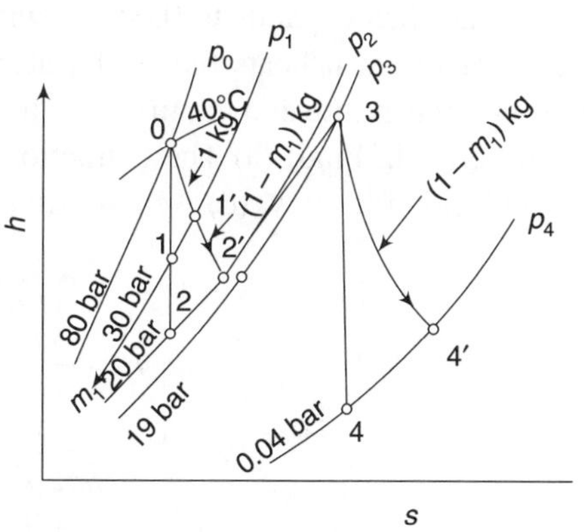

Figure 3.64 (b) Example 3.7

From steam tables:

h_{f4} = 121.41 kJ/kg, h'_{f1} = 966.89 kJ/kg at 250°C, h_{f1} = 1008.4 kJ/kg, m'_1 = 0.26 of throttle flow. Using energy balance,

Ideal turbine: (a) $m_1 h_1 + (1 - m_1) h_{f4} = h_{f1}$

or
$$m_1 = \frac{h_{f1} - h_{f4}}{h_1 - h_{f4}} = \frac{1008.4 - 121.41}{2908 - 121.41} = 31.83\% \qquad \textbf{\textit{Ans.}}$$

(b) Heat supplied = $q_A = (h_0 - h_{f1}) + (1 - m_1)(h_3 - h_2)$

$$= (3142 - 1008.4) + 0.6817(3274 - 2814) = 2447.18 \text{ kJ/kg}$$

Heat rejected = $q_R = (1 - m_1)(h_4 - h_{f4}) = 0.6817 \times (2176 - 121.41) = 1400.6$ kJ/kg

Work done by turbine = $q_A - q_R = 2447.18 - 1400.6 = 1046.58$ kJ/kg

Hence, thermal efficiency = $\dfrac{1046.58}{2447.18} = 42.76\%$ ***Ans.***

(c) Actual or brake work = $w_{Ta} = \dfrac{3600}{\text{Steam rate} \times \eta_{gen}} = \dfrac{3600}{4.5 \times 0.95} = 842.1$ kJ/kg ***Ans.***

(d) Actual heat supplied = $q_{Aa} = (h_0 - h_{f1}') + (h_3 - h_2)(1 - m_1')$

$= (3142 - 966.89) + (3274 - 2908) \times 0.74 = 2445.95$ kJ/kg

Hence, the brake thermal efficiency = $\eta_{brake} = \dfrac{w_{Ta}}{q_{Aa}} = \dfrac{842.1}{2445.95} = 34.42\%$ ***Ans.***

Therefore, the electrical thermal efficiency is given by

$= \eta_{brake} \times \eta_{gen} = 0.3442 \times 0.95 = 32.69\%$ ***Ans.***

Example 3.8 A 32 MW turbo-generator receives steam at 40 bars and 390°C. The steam while in expansion at 10 bars and 230°C is taken out for regenerative feed heating and reheating. A part of steam is taken to regenerative heater and the remaining steam is reheated at 9.5 bars to 360°C and then expands to 0.04 bar with 80% efficiency. Again, second heater is used in between, such that it heats the feed water through half the temperature range between the condenser temperature and the first feed heater water discharge temperature. The actual steam rate is 4 kg/kWh. Find: (a) The extraction pressure for second heater, (b) Ideal cycle steam rate, (c) Cycle efficiency, (d) Actual efficiency of the plant.

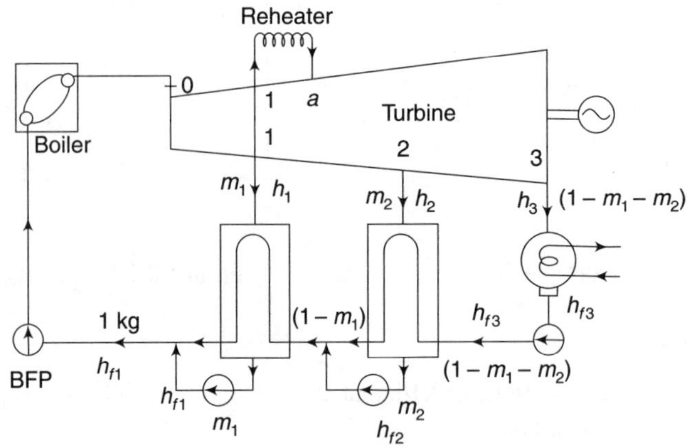

Figure 3.65 Example 3.8

Solution Refer to Fig. 3.65 and 3.66. From Mollier chart; $h_0 = 3190$ kJ/kg, $h_1' = 2866$ kJ/kg $= h_1''$, $h_a = 3184$ kJ/kg.

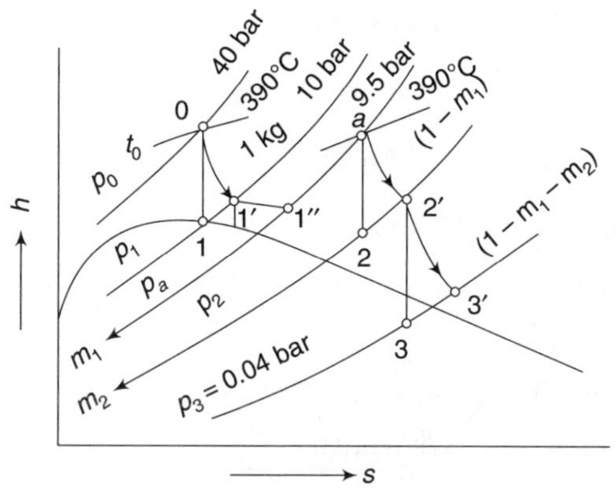

Figure 3.66 Example 3.8

In the absence of specified feed heating arrangement, let us assume the arrangement. shown in Fig. 3.65. Condenser temperature = 28.983°C at 0.04 bar

First heater discharge temperature = 179.88°C at 10 bars

∴ Temperature range = 179.88 – 28.983 = 150.897°C

Half of this temperature range = $\dfrac{150.897}{2}$ = 75.448°C

∴ Exit temperature of feed from second heaters = t_{f2} = 28.893 + 75.448 = 104.43°C

(a) Hence, pressure corresponding to 104.43°C = 1.1205 bars

(b) Now, from Mollier chart: h_2 = 2688 kJ/kg,

$$h_2' = h_a - (h_a - h_2)\,\eta_t = 3184 - (3184 - 2688) \times 0.8 = 2787.2 \text{ kJ/kg}$$

So, h_3 = 2301 kJ/kg

$$h_3' = h_2' - (h_2' - h_3')\,\eta_t = 2787.2 - (2787.2 - 2301) \times 0.8 = 2398.4 \text{ kJ/kg}$$

From steam table: h_{f3} = 121.41 kJ/kg at 0.04 bar, h_{f2} = 440 kJ/kg at 1.1205 bars.

h_{f1} = 762.61 kJ/kg at 10 bars

Writing enthalpy balance equation

At Heater No. 1: $m_1 h_1' + h_{f2} = h_{f1} + m_1 h_{f1}$

$$\therefore \quad m_1 = \frac{h_{f1} - h_{f2}}{h_1' - h_{f1}} = \frac{762.61 - 440}{2866 - 762.61} = 0.1533 \text{ kg/kg of entering steam}$$

At heater No. 2: $m_2 h_2' + (1 - m_1 - m_2) h_{f3} = (1 - m_1) h_{f2}$

$$\text{or} \quad m_2 = \frac{(1 - m_1)\,(h_{f2} - h_{f3})}{h_2' - h_{f3}} = \frac{(1 - 0.1533)\,(440 - 121.41)}{(2787.2 - 121.41)} = 0.1011 \text{ kg/kg}$$

Work done = $(h_0 - h_1') + (1 - m_1)\,(h_a - h_2') + (1 - m_1 - m_2)\,(h_2' - h_3')$

$= (3190 - 2866) + 0.8467(3184 - 2787.2) + 0.7456\ (2787.2 - 2398.24)$

$= 949.97$ kJ/kg

Hence, steam rate $= \dfrac{3600}{949.97} = 3.789$ kg/kWh ***Ans.***

(c) Heat supplied $= (h_0 - h_{f1}) + (1 - m_1)\ (h_a - h_1')$

$= (3190 - 762.61) + 0.8467(3184 - 2866) = 2696.64$ kJ/kg

Cycle efficiency $= \dfrac{951.187}{2696.64} = 35.27\%$ ***Ans.***

(d) Since actual steam rate = 4 kg/kWh, so

Heat supplied/kWh $= 4 \times 2696.64 = 10786.56$ kJ

Actual thermal efficiency $= \dfrac{3600}{10786.56} = 33.37\%$ ***Ans.***

Example 3.9 A steam turbine plant developing 120 MW of electrical output is equipped with reheating and regenerative feed heating arrangement consisting of two-feed heaters–one surface type on h.p. side and other direct contact types on 1.p. side. The steam conditions before the steam stop valve are 100 bar and 530°C. A pressure drop of 5 bars takes place due to throttling in valves.

Steam exhausts from the h.p. turbines at 25 bars. A small quantity of steam is bled off at 25 bars for h.p. surface heater for feed heating and the remaining is reheated in a reheater to 550°C and the steam enters at 22 bars in 1.p. turbine for further expansion. Another small quantity of steam is bled off at pressure 6 bars for the 1.p. heater and the rest of steam expands up to the back pressure of 0.05 bar. The drain from the h.p. heater is led to the 1.p. heater and the combined feed from the 1.p. heater is pumped to the high-pressure feed heater and finally to the boiler with the help of boiler feed pump.

The component efficiencies are: turbine efficiency 85%, overall pump efficiency 90%, generator efficiency 96%, boiler efficiency 90% and mechanical efficiency 95%. It may be assumed that the feed-water is heated up to the saturation temperature at the prevailing pressure feed heater. Work out the following :

(a) Sketch the feed heating system and show the process on *T-s* and *h-s* diagrams.
(b) Amounts of steam bled off.
(c) Overall thermal efficiency of turbo-alternator considering pump work.
(d) Specific steam consumption in kg/kWh and mass of steam in kg/h.
(e) Specific cooling water consumption circulating in condenser in condenser in kg/kWh if temperature rise is 10°C.
(f) Fuel cost per kWh generated in steam power plant, if $(LCV)_f = 29000$ kJ/kg and cost of coal is ₹1800.00 per tonne and 5% of steam generated is used to cover losses and auxiliary power.

Solution Refer to Fig. 3.67 and 3.68 for part (a). From Mollier diagram : $h_0 = 3500$ kJ/kg $= h'_0$, $h_1 = 3080$ kJ/kg, $h_2 = 3165$ kJ/kg, $h_3 = 2310$ kJ/kg, $h_a = 3575$ kJ/kg

$$h_1' = h_0' - \eta_t(h'_0 - h_1) = 3500 - 0.85(3500 - 3080) = 3143 \text{ kJ/kg}$$

$$h_2' = h_a' - \eta_t\,(h_0 - h_2) = 3575 - 0.85(3575 - 3165) = 3226.5 \text{ kJ/kg}$$

$$h_3' = h_a - \eta_t\,(h_a - h_3) = 3575 - 0.85(3575 - 2310) = 2499.75 \text{ kJ/kg}$$

From Steam Table : $h_{f1} = 961.8$ kJ/kg, $v_{f1} = 1.4521\ m^3$/kg, $h_{f2} = 670.5$ kJ/kg

$$h_{f3} = 137.83 \text{ kJ/kg},\ v_{f3} = 1.0053 \times 10^{-3} \text{ m}^3\text{/kg}$$

Energy balance at heater No: 1 $\quad m_1h_1' + h_{f2} = h_{f1} + m_1h_{f1}$

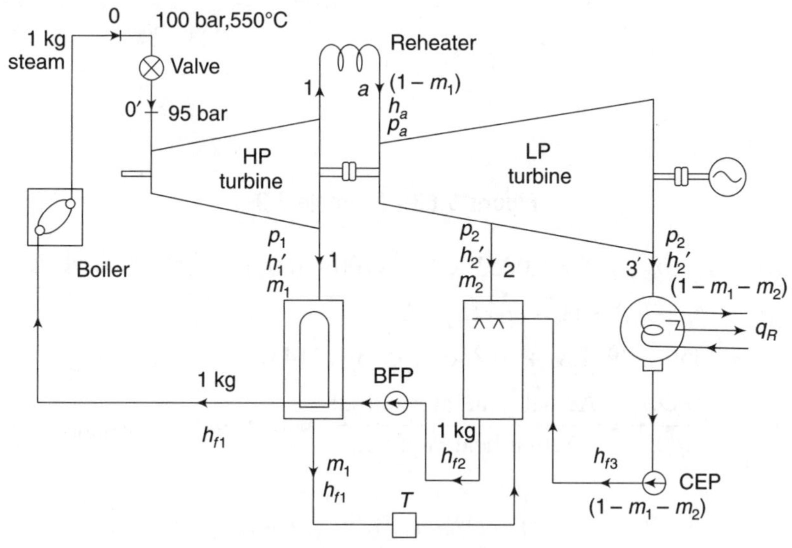

Figure 3.67 Example 3.9

(b) or $m_1 = \dfrac{h_{f1} - h_{f2}}{h_1' - h_{f1}} = \dfrac{961.8 - 670.5}{3143 - 961.8} = 0.1335$ kg/kg of entering steam ***Ans.***

Energy balance at heater No. 2: $m_2h_2' + (1 - m_1 - m_2)h_{f3} + m_1h_{f1} = h_{f2}$

or
$$m_2 = \frac{(h_{f2} - h_{f3}) - m_1\,(h_{f1} - h_{f3})}{h_2' - h_{f3}}$$

$$= \frac{(670.5 - 137.83) - 0.1335(961.8 - 137.83)}{3226.5 - 137.83} = 0.136 \text{ kg/kg of steam} \quad \textbf{\textit{Ans.}}$$

$$w_{Ta} = (h_0 - h_1') + (1 - m_1)\,(h_a - h_2') + (1 - m_1 - m_2)\,(h_2' - h_3')$$

$$= (3500 - 3143) + (1 - 0.1335)\,(3575 - 3226.5)$$

$$+ (1 - 0.1335 - 0.136)\,(3226 - 2499.75) = 1189.89 \text{ kJ/kg}$$

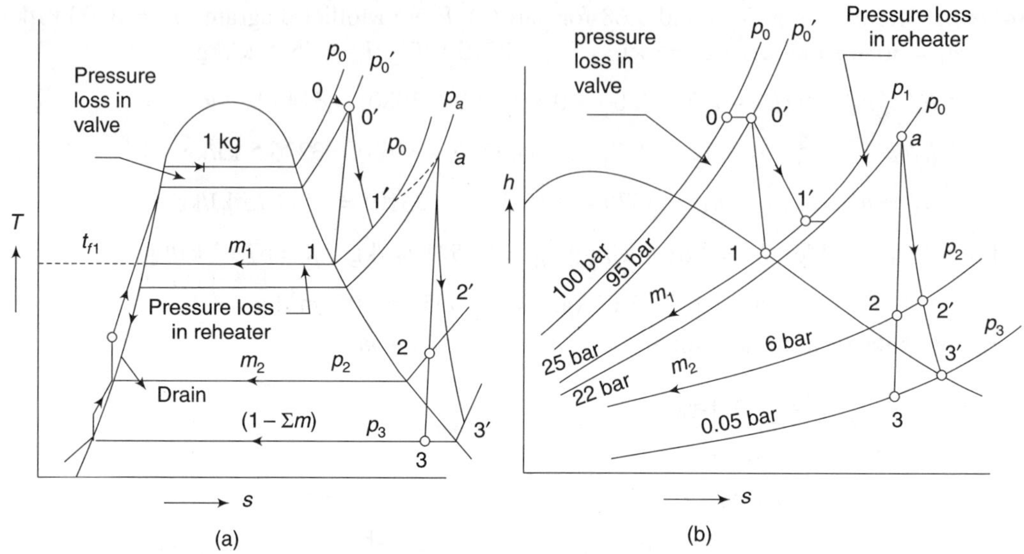

Figure 3.68 Example 3.9

$$\Sigma w_p \approx v_{f3}(p_0 - p_3) = 1.0053 \times 10^{-3}(100 - 0.05) \times 10^2 \approx 10.04 \text{ kJ/kg}$$

$$(q_A)_a = (h_0 - h_{f1}) + (1 - m_1)(h_a - h_1')$$

$$= (3500 - 961.8) + (0.8665)(3575 - 3143) = 2912.52 \text{ kJ/kg}$$

(c) $$\eta_{oth} = \frac{(G.O.)_a}{(q_A)_a} = \frac{\text{Actual generator output}}{\text{Actual heat supplied}} \quad \text{where } (q_A)_a = q_A/\eta_{boiler}$$

or $$\eta_{oth} = \frac{\left(w_{ta}\,\eta_m\,\eta_g - \dfrac{w_p}{\eta_p}\right)}{(q_A)_a} = \frac{\left(w_{ta}\,\eta_m\,\eta_g - \dfrac{w_p}{\eta_p}\right)\eta_{boiler}}{q_A}$$

$$= \frac{\left(1189.89 \times 0.95 \times 0.96 - \dfrac{10.05}{0.9}\right)}{2912.52} = \frac{1074.01}{2912.52} = 36.875\% \qquad \textit{Ans.}$$

(d) Specific steam consumption (ssc) based on G.O. is

$$\text{ssc} = \frac{3600}{w_{neta}\,\eta_g} = \frac{3600}{\left(w_{ta}\,\eta_m\,\eta_g - \dfrac{w_p}{\eta_p}\right)} = \frac{3600}{1074.01} = 3.351 \text{ kg/kWh}$$

Mass of steam = $\dot{m}_s = 3.351 \times 120 \times 10^3 = 402120$ kg/h ***Ans.***

(e) Specific cooling water consumption (scwc) circulating in condenser is

$$= \{(\text{scwc})(\text{kg/kWh}) \cdot c_{pw}\,(\text{kJ/kgK}) \cdot \Delta t_w(K) = (\text{ssc})\,(\text{kg/kWh}) \cdot q_R\,(\text{kJ/kg})$$

or $$\text{scwc} = \frac{(\text{ssc})\,(1 - m_1 - m_2)\,(h_3' - h_{f3})}{c_{pw} \cdot \Delta t_w}$$

$$= \frac{3.351 \times 0.7305(2499.75 - 137.83)}{4.18 \times 10} = 138.31 \text{ kg/kWh}$$ ***Ans.***

(f) $\dot{m}_f$(kg/kWh) × LCV (kJ/kg) × η_{boiler} = ssc (kg/kWh) × q_{Aa} (kJ/kg) × 1.05

or $\quad \dot{m}_f \times 29000 \times 0.9 = 3.351 \times 2912.52 \times 1.05$

or $\quad \dot{m}_f = 0.391$ kg/kWh

Fuel cost = 0.391 × 1800/1000 = ₹0.7038/kWh. ***Ans.***

3.6 REGENERATIVE WATER-EXTRACTION CYCLE

Erosion and corrosion problems in the L.P. stage of turbines are also avoided by the partial or complete extraction of suspended water particles from the steam during its flow through the wet stages. The extracted water particles are used to heat the feed-water hence the name, regenerative water-extraction cycle.

The larger droplets of water are concentrated in the outer half of the blades because of the fact that due to inertia, the water droplets move radially outwards during their passage through the moving blades. So, a certain quantity of moisture is flung off the blade, probably at the outlet. For the purpose of trapping this water, various types of water-catching devices are used. Here, water is cascaded from stage to stage; the advantage of this is that there is a partial recovery of the sensible heat.

Nowadays complete extraction of suspended water particles is done at a definite point. Arrangement is made such that only water is extracted. The cascaded water is driven by pump into the feed pipe to heat the feed water. Figure 3.69 shows such a regenerative two-point extraction system and its corresponding h–s diagram is shown in Fig. 3.70.

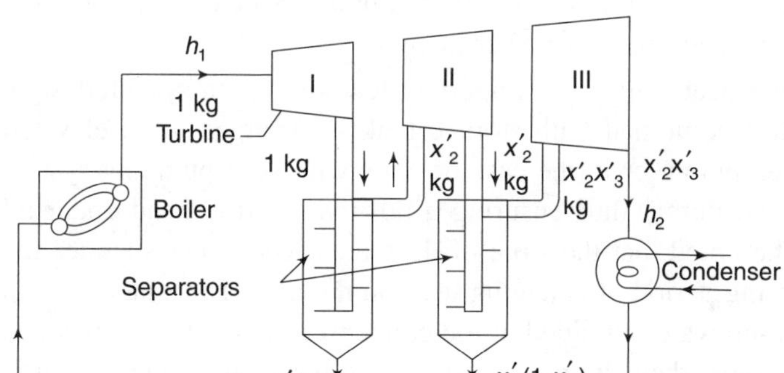

Figure 3.69 Diagrammatic arrangement of regenerative water-extraction cycle

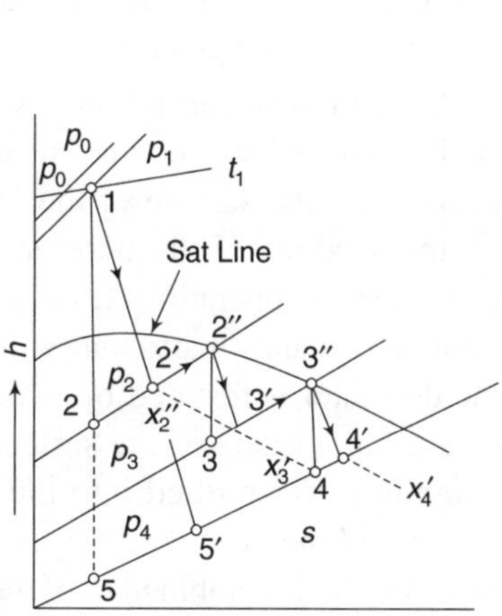

Figure 3.70 *h-s* diagram for water extraction cycle

Process 0–1 shows throttling in the stop valve. Let p_1 be the pressure in the nozzle box of first turbine. Process 1–2′ shows the actual expansion of steam before first point extraction of water particles. Complete extraction is assumed to take place at pressures p_2 and p_3. Let the dryness faction at 2′ be x_2', so the water extracted will be $(1 - x_2')$ kg per kg of steam entering the turbine and the steam passing through the turbine will be x_2' kg per kg of steam entering the turbine. The actual expansion line in the second stage turbine is 2″ 3′ and x_3' is the dryness fraction at state point 3′.

At the second point water extracted is $(1 - x_3')\, x_2'$ per kg of steam entering the turbine and $x_2'\, x_3'$ the second state of steam entering the third stage of the turbine. If we neglect the water extraction pumps work, then by equating the heat entering and the heat leaving at the junction points gives.

At junction 2 : $x_2'\, h_{f5} = x_2' \cdot x_3' \cdot h_{f4} + x_2'(1 - x_3')\, h_{f3}$

or $\quad h_{f5} = x_3'\, h_{f4} + (1 - x_3')\, h_{f3}$

At junction 1 : $\quad h_{f6} = x_2'(h_{f5}) + (1 - x_2')\, h_{f2}$

Putting the value of $h_{f5} \cdot h_{f6}$ can be calculated

Heat supplied $= q_A = h_0 - h_{f6} = h_1 - h_{f6}$

Internal work done $= (h_1 - h_2') + x_2'\,(h_2'' - h_3') + x_2'x_3'\,(h_3'' - h_4')$

So the internal thermal efficiency is

$$\eta_{ith} = \frac{(h_1 - h_2') + x_2'\,(h_2'' - h_3') + x_2'x_3'\,(h_3'' - h_4')}{h_0 - h_{f6}} \qquad \text{...(3.82)}$$

3.7 PRACTICAL FEED HEATING ARRANGEMENTS

There are many ways to arrange the various types of heater. Figures 3.71 to 3.72 show the feed heaters arrangement in various capacities of power plants.

As mentioned earlier, feed water deaeration is an important feature of feed system design. A deaerator (direct contact heater) combined with storage tank is arranged in an elevated position on the suction of the feed pump, or in the case of units with split pumping system on the suctions of the booster feed pump, thus ensuring a positive suction head under all conditions of operation. It is to be noted that the store of deaerated feed water supplies the boilers' demands, particularly during period of changing load on the unit. The water level in the deaerator is restored by make-up water admitted at the condenser. It is common practice to use h.p. heaters of the surface type, their drains cascaded from heater to heater and their combined drains lifted into the deaerator by the pressure differential between the first h.p. heater and the deaerator. So far the l.p. heaters are concerned, they are either of the surface type with their combined or of the direct contact type. This arrangement is a subject to variation only when the number of desuperheaters and drain coolers fitted are many. For example, the feed system shown in Fig. 3.71 has the maximum number of drain coolers, and drains from LP_3 are cooled by vapour condenser when the evaporator is not in service. Evaporators are used for make-up water needed in the boiler.

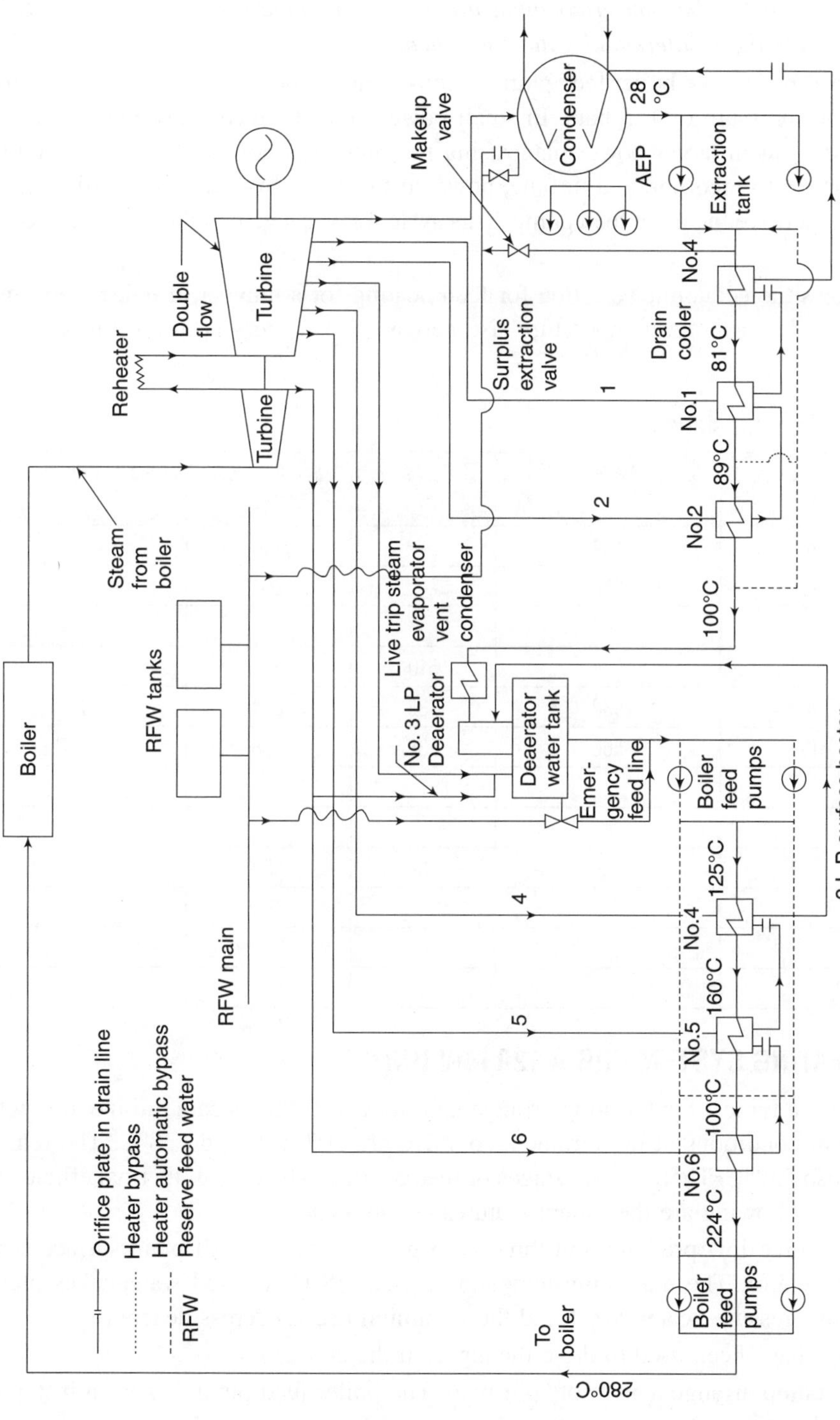

Figure 3.71 120 MW feed heating system

With larger units it has become a common practice to arrange the h.p. heaters in twin banks or increase the d.c. heaters and reduce h.p. heaters.

The power absorbed by the boiler feed pump is considerable about 2.5 to 3% of the turbine output for a steam pressure of 155 bars. In small power unit; the feed water pump may be electrically driven, but in larger power plant using reheater, it is now customary to modify the bled steam cycle to incorporate a turbine, powered by steam from the main turbine h.p. cylinder exhaust, to drive the main feed pump. This system gives a gain in thermal and overall efficiency.

Table 3.1 shows the economic selection for feed heating for a number of unit ratings and steam conditions. The average temperature rise across each heaters is in the range 28 to 38°C.

Table 3.1 Feed heating stages

Rating MW	*Turbine pressure bar*	*Stop valve temperature °C*	*Reheat temp °C*	*Final feed temp °C*	*Feed heating stages –*
30	41	454	–	166	4
60	103	566	–	204	5
		–			
100	103	524	510	224	6
120	103	538	538	224	6
200	162	566	538	238	6 or 7
275					
300					
350	158	566	566	252	7 or 8
500					
550					
660					

3.8 FEED HEATING SYSTEM FOR A 120 MW UNIT

Figure 3.71 shows a typical feed heating arrangement for a 120 MW steam turbine unit with reheat. The steam conditions at the turbine stop valve are 103 bars and 538°C. The reheat temperature is also 538°C. There are six stages of feed heating. The overall thermal efficiency obtained is 35%. Following are the salient features of this system:

(a) There are three low pressure and three high pressure heaters, all plain surface type except deaerator. The maximum temperature rise (38°C) in feed water takes place across low pressure heater No. 1 and the minimum (22°C) across deaerator.

(b) Air pumps have been used to drive the air from the condenser.

(c) The feed pump arrangement is of split type. The boiler feed pump has been installed after *h.p.* heaters and the booster feed pump has been arranged before *h.p.* heaters, and enough suction head for the boiler feed pump. Both types of pump are electrically driven and in duplicate.

(d) The high pressure cylinder is of double casing so bled steam cannot be taken along cylinder except at outlet.

(e) The cascade drains on the *h.p.* heaters return to the deaerator via an orifice plate.

(f) The feed water from all the heaters may be bypassed manually and if the water level in *h.p.* heaters gets too high, it may be automatically bypassed.

3.9 FEED HEATING SYSTEM FOR A 200 MW UNIT

Figure 3.72 shows a typical feed heating system for a 200 MW steam turbine unit with reheat. The steam conditions at the turbine stop valve are 162 bars and 566°C. The reheat temperature is 538°C. There are six stages of feed heating producing a final feed temperature of 243°C and giving an overall thermal efficiency of 36.9 per cent. Following are the salient features of this system:

(i) There are three low pressure and three high pressure heaters. Except deaerator, all heaters are of surface type. All *h.p.* heaters have separate desuperheater and drain cooler.

(ii) Number 5 desuperheater is hotter than No. 6 desuperheater and so it is the last heater in the system before the boiler feed pumps.

(iii) The cascade drains from all *h.p.* heaters pass through orifice plates and then into separate drain coolers where more heat is taken up by the feed water.

3.10 FEED HEATING SYSTEM FOR A 350 MW UNIT

Figure 3.73 shows a typical feed heating system of a 350 MW steam turbine unit with reheat. The steam condition at the turbine stop valve are 165 bars and 568°C. The reheat temperature is also 568°C. Seven stages of feed heating are employed to give a final feed water temperature of 252°C and an overall efficiency of 36.5 per cent.

Following are the salient features of this system:

- There are four low pressure and three high pressure heaters. Except deaerator, all heaters are of surface type. The three *l.p.* heaters are of plain surface type except the heater No. 1 which has a drain cooler. All the high pressure heaters are of surface type with integral desuperheater and drain-cooler.
- The condensate from the condenser passes through the hydrogen cooler, the generator stator cooler and the low pressure drain cooler for the recovery of the some of the waste heat.
- A unit evaporator is used in the system and the steam and water vapour from the evaporator outlet passes through a vapour condenser when it is condensed by the cooler feed water on its way to the deaerator. The distillate is taken out and is used as a make up water to the boiler. In preparing this distillate, the feed water gains little or more heat.

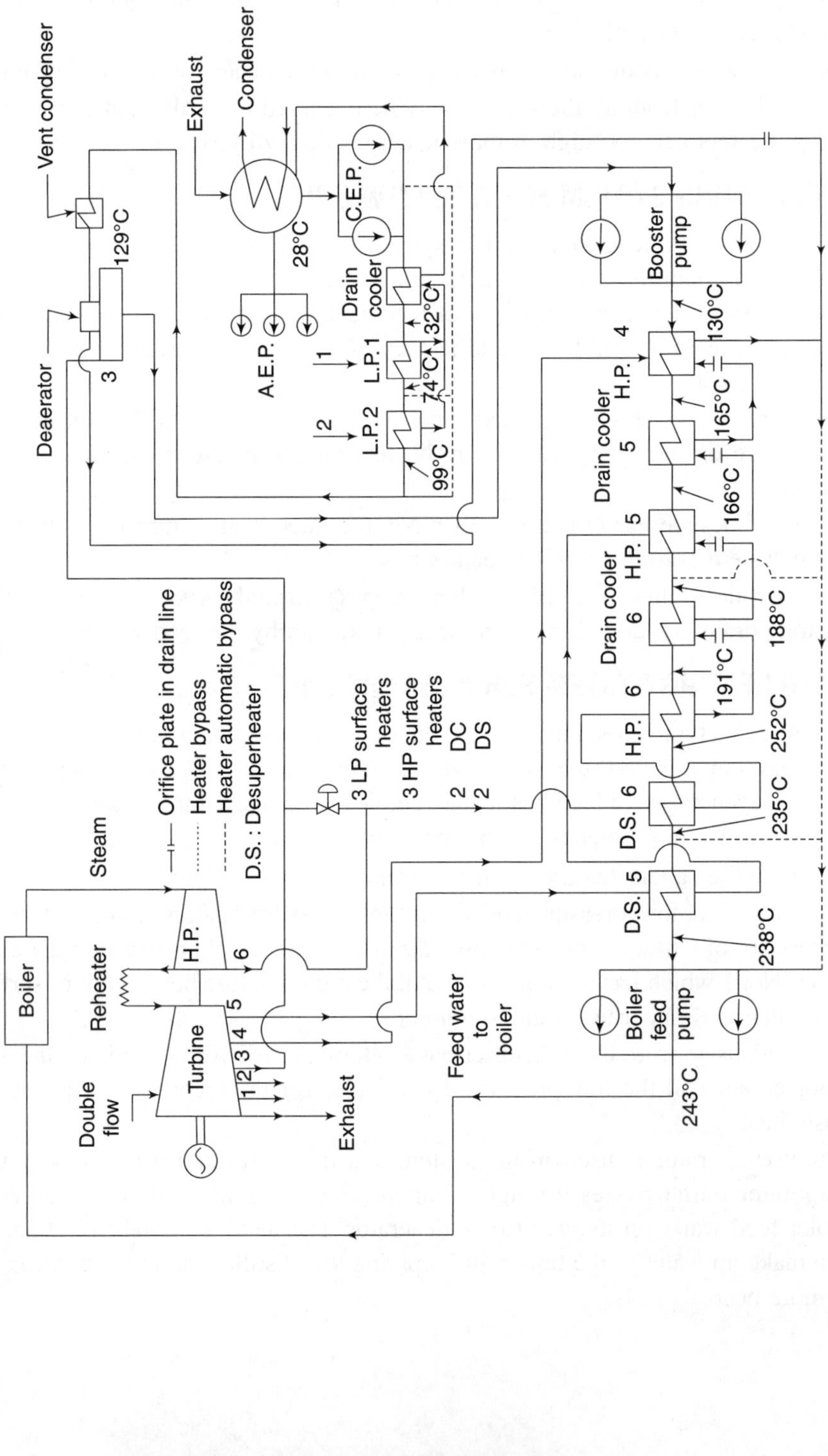

Figure 3.72 200 MW feed heating system

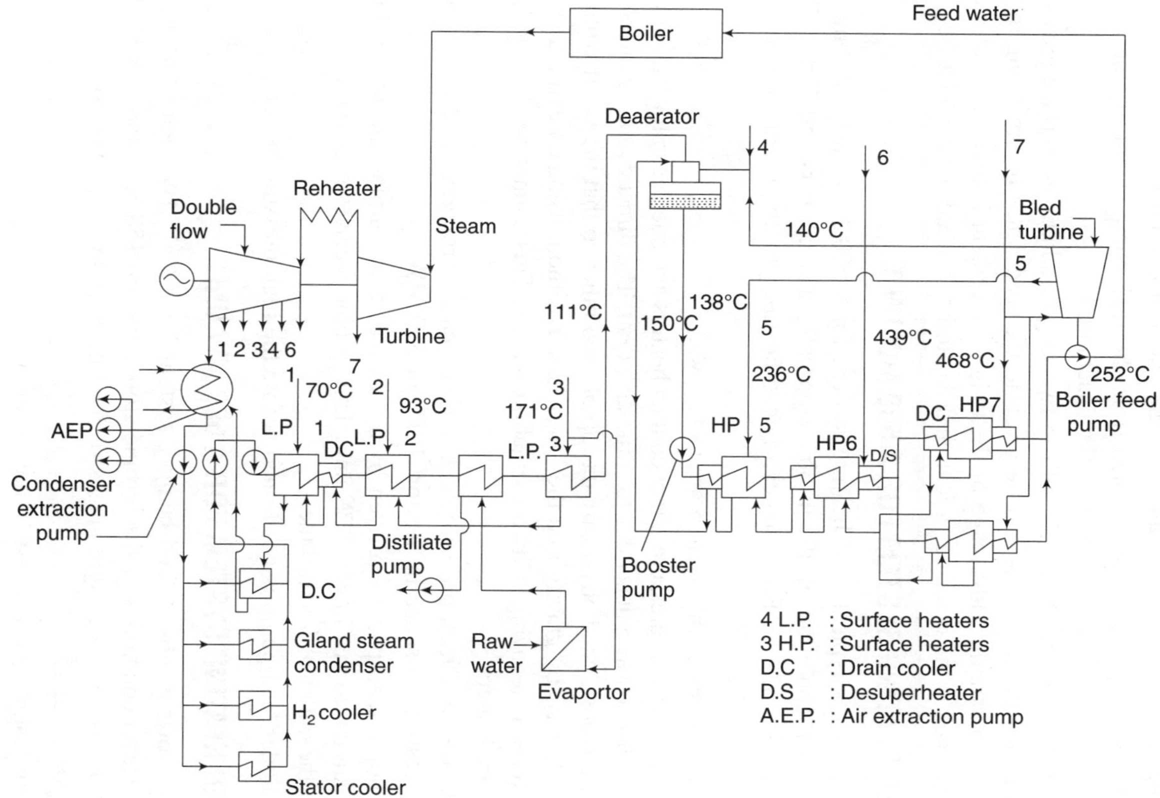

Figure 3.73 350 MW feed heating system

- The boiler feed pump is driven by a separate steam turbine meant for this purpose only. Due to larger size of the pump, steam driven pump is more economical. This turbine receives required steam from the outlet of *h.p.* turbine.

  For the heater No. 5, steam is bled from the feed pump steam turbine. The exhaust steam from this turbine returns to the deaerator.

(v) The final *h.p.* heater No. 7. is split in two to give additional surface area so that the maximum amount of heat can be given to the feed water before going to the boiler feed pump. Another advantage of the twin arrangement is that one heater may be in service even if the other has a leak and has to be isolated for maintenance.

3.11 FEED HEATING SYSTEM FOR A 500 MW UNIT

Figure 3.74 shows a typical feed heating system for a 500 MW steam turbine unit with reheat. The steam conditions at the turbine stop valve are 158 bars and 538°C. The reheat temperature is also 538°C. Eight stages of feed heating are employed to give a final feed water temperature of 254°C.

The salient features of this system are as follows:

- There are four low pressure direct contact heaters and one deaerator. The direct contact heaters also act as a deaerator to some an extent. The direct contact heaters 1 and 2, 3 and 4 are positioned physically on top of each other so that the water from the upper heater to pass by gravity into the lower heater without using an extraction pump. By this, there is a saving of two extraction pumps. This cannot be done with the high pressure heaters.
- There are three high pressure surface type heaters arranged in parallel banks for the same reasons as stated in Section 3.10, for 350 MW.
- The boiler feed pump is driven by a separate steam turbine meant for this purpose only similar to that in 350 MW unit. This turbine receives required quantity of steam from the outlet of *h.p.* turbine.
- The boiler feed pump is positioned before the high pressure heaters.

3.12 FEED HEATING SYSTEM FOR A 660 MW UNIT

Figure 3.75 shows a typical feed heating system for a 660 MW steam turbine unit with reheat. The steam conditions at the turbine stop valve are 158 bars and 538°C. The reheat temperature is also 538°C. Eight stages of feed heating are employed to give a final feed water temperature of 254°C.

The salient features of this system are as follows:

- There are five direct contact *l.p.* heaters in two cascades one deaerator and two *h.p.* surface type feed heaters.
- There is a single bank of *h.p.* heater instead of two.
- Similar to 350 and 500 MW, boiler feed pump is driven by a separate steam turbine meant for this purpose only.

For more details on feed heaters references [2, 6] may be consulted.

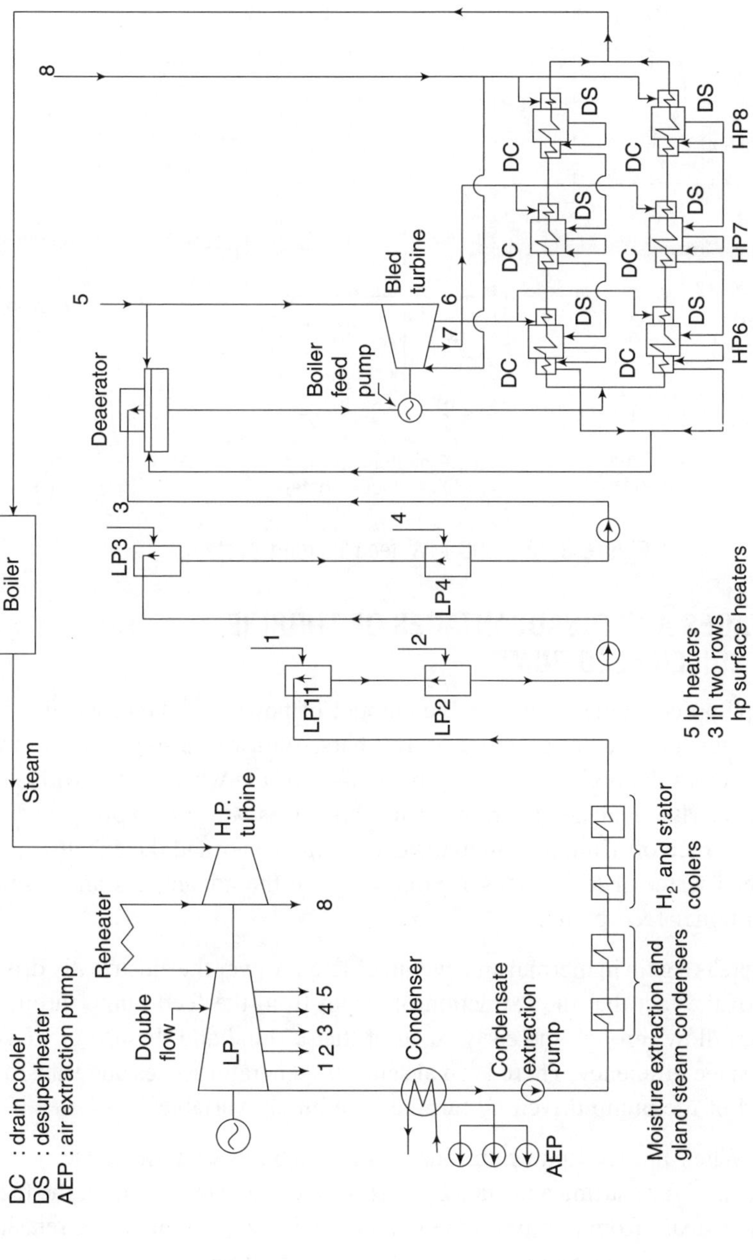

Figure 3.74 500 MW feed heating system

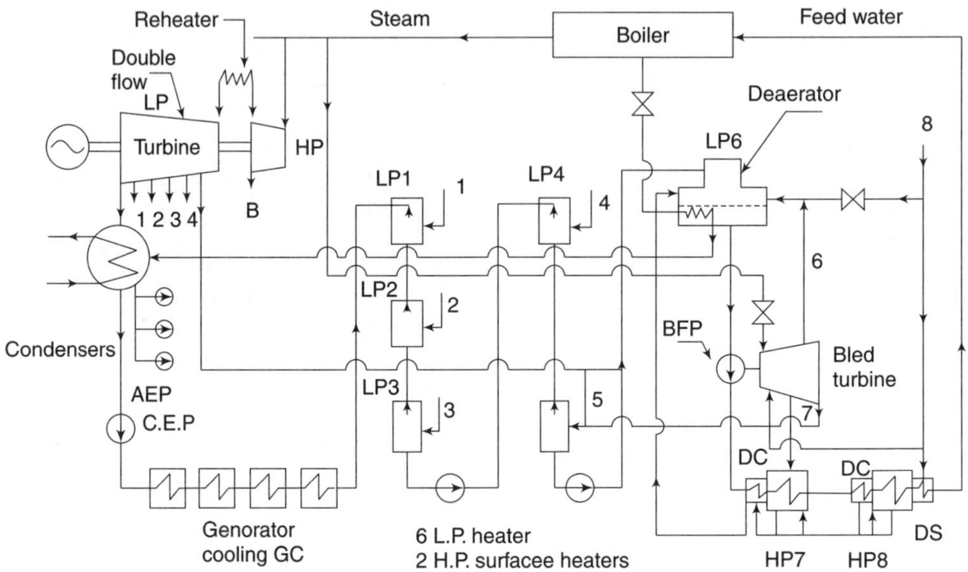

Figure 3.75 600 MW feed heating system

3.13 ADVANTAGES AND DISADVANTAGES OF TURBINE DRIVEN BOILER FEED PUMP

The boiler feed pump consumes considerable amount of power, of the order of 2.5% to 3% of the turbine output for a steam pressure of 160 bars. With reheat-regenerative cycle, it is now customary to drive the boiler feed pump by a *bled turbine* which is fed with steam from *h.p.* turbine exhaust. This bled steam turbine is also known as *boiler feed pump turbine*. Some quantity of steam is bled off from the pump turbine to *h.p.* heater and the exhaust steam either is led to some feed water or to condenser. Following are the advantages and disadvantages of turbine driven boiler feed pump.

Advantages: There is a gain in thermal and overall efficiency over the electrically driven pump. This is because of the fact that the extraction of steam from the feed pump turbine requires an increased steam flow through the early stage of that turbine and the use of longer blades results in higher stage efficiency. There is an absence of generator losses and transformer iron losses. The speed of the pump driven by turbine is infinitely variable.

Disadvantages: When the feed pump turbine is unavailable, two 50% capacity electrically driven pumps are used for starting and standby purpose. Under this condition, the feed heaters which were getting steam from pump turbine will not be in use and the boiler reheater circuit will be running at excessive capacity. Though, there are methods to deal with such situation but they are uneconomical.

3.14 HYDROGEN FUEL COMBUSTION TURBINES BASED POWER PLANT (UNDER DEVELOPMENT)

The rapid industrial revolution, transportation and energy consumption using fossil fuel have led to high level of air pollution, which is highly alarming for human civilization. The development of a comprehensive global energy system based on renewable energy using hydrogen as a fuel has become a necessary step towards realization of pollution free civilization. Realising this, the Japanese government through its New Energy and Industrial Development Organisation (NEIDO) has created the World Energy Network (WE-NET) Program, *a 28 years effort from 1993 to 2020*, directed as research and development of the technologies needed to develop a hydrogen based energy conversion system. Westinghouse, Toshiba and Mitsubishi heavy industries are working under the time frame to develop a hydrogen fuelled combustion turbine system designed to meet the goals set by WE-NET program which include 500 MW output and greater than 70.9 per cent low heating value thermal efficiency, a reliability-availability-maintainability (RAM) equivalent to current base loaded natural gas fired combined cycle and elimination of CO, CO_2, NO_x and SO_x during power generation.

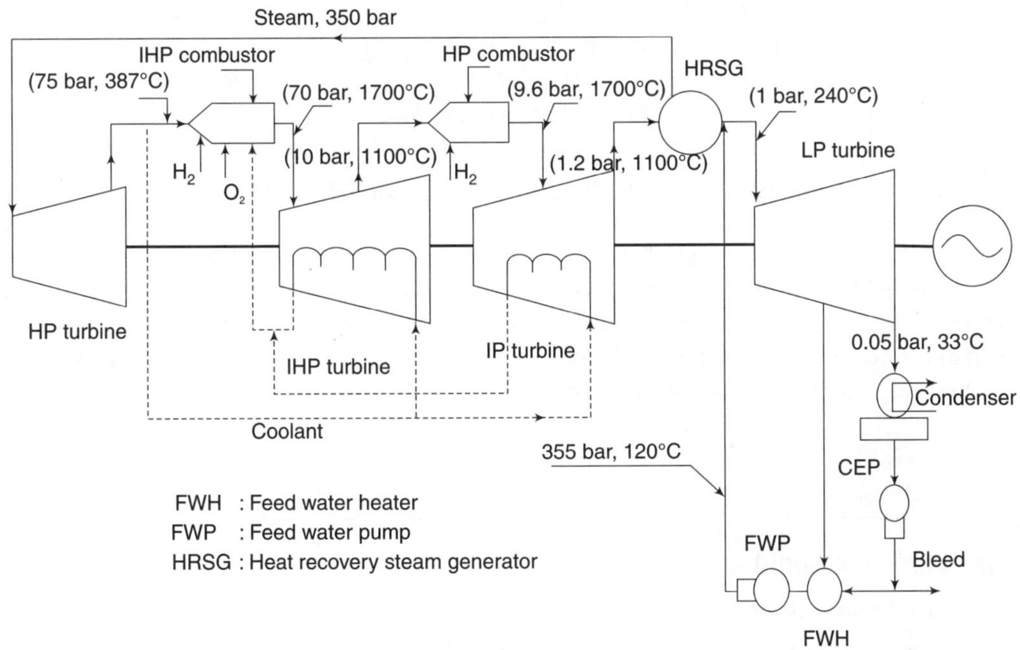

Figure 3.76 Schematic diagram of hydrogen-fuelled combustion turbine cycle [33, 34]

Figure 3.76 shows the schematic of hydrogen fuelled combustion turbine cycle, a long term plan [33, 34]. *This cycle works on Rankine cycle*. There are four turbines, namely, HP, IHP, IP and LP and two combustors IHP and IP. The steam with selected maximum pressure (p_{max}) is supplied to HP turbine while IHP and IP turbines receive steam at the selected maximum temperature. In both combustors, H_2 and O_2 are burnt resulting in the increase in steam temperature to a selected value of t_{max}. IPH and IP turbines are cooled by closed loop steam cooling drawn from HP turbine exhaust. The exhaust of IP turbine is used to heat the feed water to generate steam at the maximum cycle pressure in a heat recovery steam generator (HRSG) before expanding to L.P. turbine while the steam generated is led to the HP turbine for expansion. *The said plant is designed for maximum 350 bars and 1700°C steam pressure and temperature respectively.*

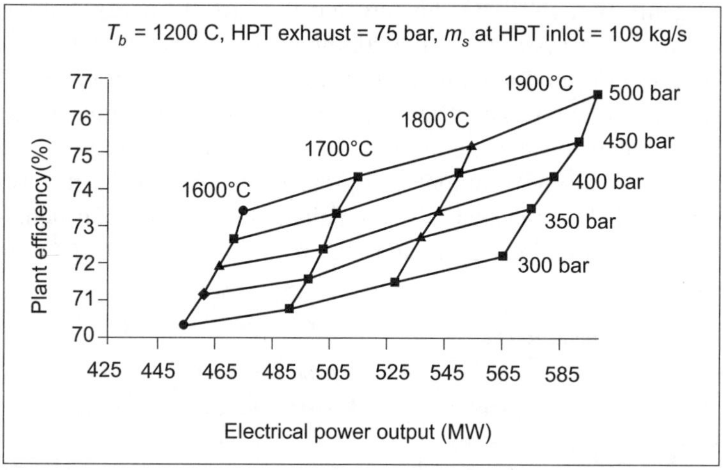

Figure 3.77 Plant efficiency vs. electrical output [33, 34]

Figure 3.77 shows variation of estimated plant efficiency versus plant work for various maximum steam pressures and temperatures in the cycle. The plant efficiency and work increase with maximum steam pressure and temperature. With the success of this plant at comparable cost there will be a revolution in electricity generation without using fossil fuels and the energy problem of the world may be nearly solved.

PROBLEMS

Things to Think About

3.1 List the main elements of a steam power plant?

3.2 How many laws are applicable in steam power plant?

3.3 What limits in the increasing of boiler steam pressure and temperature?

3.4 Can you think of a steam power plant without producing air pollution? If yes, the measures.

3.5 What measures are needed for further improvement in efficiency of steam power plant with reference to present value of 40%?

Rankine Cycle

Short and Long Theoretical Questions

3.1 What is the reason that the Carnot cycle is not a practical cycle for steam turbine plants even though its efficiency is higher?

3.2 What is the effect of maximum pressure, superheater and back pressure on the thermal efficiency of a steam turbine power plant?

3.3 Why is water used as working fluid in steam turbine plant?

3.4 Is it possible to operate the steam turbine with steam having temperature and pressure higher than critical?

3.5 What are the principles and methods for increasing the thermal efficiency of Rankine cycle?

3.6 What do you mean by the critical temperature and pressure of water?

3.7 What do you mean by binary-vapour cycle and in what case is it required?

3.8 What are the requirements of an ideal working fluid in steam turbine?

3.9 What are the advantages and disadvantages of mercury as working fluid?

Numerical Questions

3.1 Two steam turbines A and B operate under the following conditions: Initial pressure ; 17 bars, exhaust pressure, 0.046 bars. In turbine A, the steam is initially dry and saturated, whereas in turbine B it is initially superheated by 94.5%C. Calculate: (a) Theoretical steam consumption, kg/kWh, (b) Theoretical thermal efficiency of the two turbines, (c) Percentage reduction in steam consumption due to superheating, (d) Percentage increase in thermal efficiency due to superheating. [*Ans.* (a) 4.28, 3.73; (b) 31.9%; (c) 12%; (d) 3.18%]

3.2 Calculate the fuel cost per kWh generated in a steam power plant under the following conditions: Initial pressure = 20 bars, Initial superheat = 160°C, Exhaust pressure = 0.05 bar, Turbine efficiency ratio = 0.78, Boiler efficiency = 0.82, Alternator efficiency = 0.96, C.V. of coal = 29260 kJ/kg, Cost of coal = ₹1500 per tonne.

Assume: (a) that the feed temperature to the boiler is equal to the saturation temperature at 0.05; bar (b) that 6% of the steam generated is used to cover losses and auxiliary power.

3.3 A Rankine cycle has steam entering the turbine at 140 bars and 550°C. If the exhaust pressure is 0.06 bar and all processes are reversible, (a) find the cycle thermal efficiency, the work ratio and the steam rate, (b) if the efficiency of turbine is 80% and the efficiency of pump is 80% what is the thermal efficiency? (c) with the above value of turbine and pump efficiencies, if the boiler efficiency is 85%, find the power plant thermal efficiency.

[*Ans.* (a) 43.3%. 99.9%, 2.52 kg/kWh; (b) 34.3%; (c) 29.2%]

3.4 A power plant operating on the Rankine cycle has steam entering the turbine at 37 bars and 425°C. If the turbine output is equivalent to 10,000 kW, determine the efficiency of the cycle and the mass flow rate of steam for condenser pressure of (a) 0.07 bar and (b) 0.035 bar.

[*Ans.* (a) 36.6%, 31800 kg/h; (b) 39.1%, 29500 kg/h]

3.5 Steam at 56 bars enters a 20,000 kW condensing turbine and exhausts at 0.07 bar. The steam, loses 2220 kJ/kg to the condenser cooling water which increases 8°C in temperature while passing through the condenser. The Rankine cycle efficiency is 32.2%, the turbine isentropic efficiency is 82% and the actual steam rate is 6 kg/kWh. Assuming that the steam condensate leaves the condenser as a liquid subcooled 2°C, find the temperature of steam entering the turbine.

3.6 A steam turbine power cycle operates with turbine inlet conditions of 56 bars and 570°C. The condenser pressure is 0.07 bar. Calculate the change in thermal efficiency if the turbine has an efficiency of only 80% instead of operating isentropically. What is the change in moisture content leaving the turbine?

3.7 A steam generating unit has a steam consumption of 228 kg of steam per hour. This unit, running on the Rankine cycle, is equipped with superheater. The boiler receives 117×10^6 kJ/h of heat. Water enters the boiler at 30 bars and 95°C. After passing through the superheater, the steam goes to a turbine which produces net work. The steam is then condensed in the condenser where 1.3×10^6 kg/h of cooling water is circulated with a temperature rise of 8°C. The steam leaves the condenser at 0.1 bar and 43°C. Calculate (a) the enthalpy of steam leaving the boiler in kJ/kg; (b) enthalpy of steam entering the condenser; (c) from Mollier diagram, the quality of steam entering the condenser; (d) using Mollier diagram, enthalpy of steam entering the turbine assuming isentropic expansion from the turbine, (e) power output of the ideal surface.

[***Ans.:*** (a) 2732 kJ/kg; (b) 2175 kJ/kg; (c) 0.84; (d) 3284 kJ/kg; (e) 6688 kW]

3.8 In a nuclear power plant, liquid sodium is heated to 370°C in the reactor. The liquid sodium then flows through a heat exchanger where water is heated and vaporized to saturated steam at a pressure of 56 bars. The steam generated flows to a turbine, which has one extraction point at 4.2 bars where steam flows to an open feed water heater. If the turbine efficiency is 75%, and the condenser pressure 0.07 bar, calculate the heat transfer rate in the reactor from a turbine power output of 65,000 kW.

3.9 A simple steam power cycle uses solar energy for the heat input. Water in the cycle enters the pump as saturated liquid at 49°C and it is pumped to 2.1 bars. It then evaporates in the boiler at this pressure and enters the turbine as saturated vapour. At the turbine exhausts the conditions are 49°C and 6% moisture. The flow rate is 136 kg/h and the pump is rated at 0.37 kW. Calculate: (a) the net work output, (b) the energy conversion efficiency, (c) area of the solar collector needed if collector pick-up is 703.6 W/m^2.

3.10 A Rankine cycle operates with steam of 45 bars and 430°C and a condenser pressure of 5 cm of Hg. Assuming steady state and kinetic and potential energy changes are negligible, find: (a) the ideal work done by the turbine per kg of steam, (b) the amount of steam required per net kWh output (steam rate), (c) the thermal efficiency of the cycle, (d) the amount of heat required per net kWh output, (e) the unavailable energy with respect to the lowest temperature in the cycle, (f) Calculate the above for actual cycle where $\eta_t = 85\%$, $\eta_m = 96\%$. $\eta_b = 90\%$, $\eta_{op} = 85\%$, $\eta_{gen} = 96\%$.

Regenerative Feed Heating and Reheating Cycles

Short and Long Answer Questions

3.1 What is the thermodynamics principle of regenerative feed heating?

3.2 Is it theoretically possible to achieve the efficiency of Rankine cycle with that of Carnot? State the conditions for this.

3.3 What is the effect of the number of heaters on the thermal efficiency? Discuss the law of diminishing return.

3.4 How many heaters are used in practical regenerative cycle?

3.5 What are the advantages and Disadvantages of regenerative feed heating cycle over the simple Rankine cycle?

3.6 Explain how the regenerative cycle efficiency may approach the Carnot cycle efficiency.

3.7 Discuss the saving in heat rate from regenerative heating. How is the number of heaters related to the increase in feed water temperature for the maximum saving?

3.8 What are the effects of moisture content on the performance of steam turbines?

[***Ans.*** Erosion, corrosion, less blade efficiency, less nozzle efficiency, less thermal efficiency, less power developed]

3.9 Show the impacts of water droplets on blade surface by drawing velocity triangle.

3.10 What do you mean by reheating of steam?

3.11 Is it essential to reheat at a optimum pressure? If yes, why?

3.12 It is true that modern plants are equipped with reheat and regenerative arrangement? Why?

3.13 Discuss the advantages and disadvantages of reheating in steam power plants and explain why it becomes more necessary to adopt a reheat cycle with high steam pressure.

3.14 Draw a schematic diagram of a steam turbine plant equipped with reheat-regenerative arrangement. Draw the temperature distribution in various direct and indirect contact feed heaters. Give a feed heating arrangement of 120, 200, 500 and 600 MW turbines.

3.15 Derive the expressions for work done and blade efficiency when the steam contains water particles.

Numerical Questions

3.1 A steam turbine plant operating between pressure of 40 bars with 72°C superheat and 0.04 bar, employs regenerative feed heating. The feed-water leaving the hot well passes through two heaters in succession. The steam for the heaters is bled from the turbine at 6.0 bars and 1.4 bars, and the condensed steam leaving the heaters is pumped into the feed-pipe at points immediately after each heater. Assuming that the feed-water leaves each heater at saturation temperature, find the masses of steam bled and the thermal efficiency of the turbine.

[***Ans.*** $m_1 = 0.101$ kg, $m_2 = 0.123$ kg, $\eta = 30.23\%$]

3.2 An ideal reheat-regenerative cycle operates with throttle conditions at the turbine inlet of 140 bars and 600°C, and reheat at 7 bars to 540°C. A closed feed water heater operates at 14 bars, and the drain from the closed heater is trapped back into an open feed water which operates at 2 bars. The condenser pressure is 0.07 bar. Calculate (a) the efficiency of the cycle, and (b) the required mass flow rate through the steam generator for a turbine output of equivalent to 100,000 kW. [***Ans.*** (a) 47.9%]

3.3 A steam power generating station operates on the reheat-generative cycle, with part of the steam extracted for heating the feed water and the rest of the steam is reheated before being expanded to the back pressure in the low-pressure turbine. Steam enters the high pressure turbine at 70 bars, and the rest is superheated to 485°C before entering the low pressure turbine where it is expanded to 0.07 bar. Pump work in low pressure pump is 0.625 kJ/kg. Pump work in the high pressure pump is 8.3 kJ/kg. Assume isentropic expansion in each turbine. Calculate (a) mass of steam extracted for heating per kg of steam entering the turbine, and (b) thermal efficiency of the cycle.

3.4 Consider a combined reheat-regenerative steam turbine cycle where the inlet conditions to the high pressure turbine are 140 bar and 600°C and reheat it at 14 bars with a temperature at the low pressure turbine inlet 540°C. Closed feed water heaters operate at pressure of 14 and 3.5 bars with a trap outlet of 3.5 bars connected to the condenser drain. There is just one feed water pump located at the inlet to the 3.5 bars heater. Calculate (a) the enthalpy change of the feed water as it passes through the heater, (b) the ratio of the mass of steam bled at 140 bars to the mass of the steam entering the high pressure heater. [***Ans.*** (a) 242.6 kJ/kg; (b) 0.118]

3.5 In a 100 MW (generator output) commercial steam power plant operating on a reheat and regenerative cycle steam leaves the boiler superheater at 77 bars and 400°C and enters the high pressure turbine after a loss of 2 bars in the supply line. Steam exhausts from the high pressure turbine at 17 bars. It is passed through a reheater and then supplied to the low pressure turbine at 16.5 bars and 400°C. Steam exhausts from the low pressure turbine at 0.035 bar to condenser. 12% and 15% quantities of the flows to the high pressure and low pressure and low pressure turbines are expected at 26 and 5.2 bars respectively and supplied to open-type heaters for feed water heating. The condenser cooling water enters at 15°C and leaves at 23.3°C. Component efficiencies are : generator 94%, boiler (including superheater and reheater) 85%, high pressure turbine, 90%, low pressure turbine 85%. Mechanical efficiency, 95%.

Work out the following:

(a) Sketch the actual cycle on *T–s* coordinates, (b) Calculate the thermal efficiency of the actual cycle, (c) Calculate the overall plant efficiency, (d) Calculate the steam consumption rate in kg/kWh and kg/h, (e) Calculate the condenser cooling water flow rate in kg/kWh and in kg/h.

3.6 A steam turbine power plant developing 200 MW of electrical output is equipped with reheating and regenerative feed heating arrangement consisting of three feed heaters—one surface type on h.p. side and other two direct contact type on 1.p. side. The steam condition before the steam stop valve are 162 bars and 566°C. A pressure drop of 6 bars takes place due to throttling in valves. Steam exhaust from the h.p. turbine at 30 bars. A small quantity of steam is bled off at 30 bars for h.p. surface heaters for feed heating and the remaining is reheated in a reheater to 538°C. and the steam enters at 26 bars in 1.p. turbine expansion. Another small quantity of steam is bled off at 15 bars and 5 bars for first and second 1.p. feed heaters, and the rest of steam expands up to back pressure of 0.05 bar. The drain from h.p. heaters is led to the 1.p. heater and the combined feed from 1.p. heater pumped to the high pressure feed heater and finally to the boiler with the help of a boiler feed pump. The component efficiencies are: turbine efficiency 85%, overall pump efficiency 90% and mechanical efficiency 96%, boiler efficiency 90% and alternator efficiency 95%. It may be assumed that the feed water is heated up to the saturation temperature of the prevailing pressure in feed heater.

Workout the following:

(a) Sketch the feed heating system and show the process on *T–s* and *h–s* diagrams, (b) Amount of steam bled off, (c) Overall thermal efficiency of turbo-alternator considering pump work, (d) Specific steam consumption in kg/kWh, and amount of steam in kg/h. (e) Specific cooling water consumption in condenser in kg/kWh if temperature rise in 10°C. (f) Fuel cost per kWh if the lower calorific value of fuel (coal) is 19,000 kJ/kg and the price of fuel is ₹1800 per tonne, by assuming that 5% of steam generated is used to cover losses and auxiliary power.

Software Based Questions

3.1 Develop a software for the design of feed heating system for a steam power plant unit developing power from 120 MW to 1000 MW.

3.2 Develop a computer software for the analysis of steam power plant working on simple Rankine cycle.

3.3 Develop a computer software for the analysis of binary-vapour cycle.

3.4 Develop a computer software for the cost analysis per kWh supply including generation, transmission and distribution for steam power plant.

Design and Open Ended Question

3.1 Design a feed heating for the following power plant unit: (a) 120 MW, (b) 200 MW, (c) 350 MW, (d) 500 MW, (e) 660 MW, (f) 1000 MW. Assume suitable throttle and condenser conditions and other values to suit real conditions.

Multiple-Choice Questions

3.1 A steam power plants work on

(a) Carnot cycle (b) Brayton cycle (c) Joule cycle (d) Rankine cycle

3.2 The thermal efficiency of steam power plant ranges between

(a) 20% to 30% (b) 30% to 40% (c) 40% to 50% (d) 50% to 60%

3.3 The thermal efficiency of Rankine cycle

(a) increases with increase of inlet steam pressure.

(b) increases with increase of inlet steam temperature.

(c) increases with both inlet steam pressure and temperature.

(d) decreases with the decrease of both inlet steam pressure and temperature.

3.4 The number of feed heaters in regenerative feed heating cycle ranges from

(a) 3 to 10 (b) 5 to 15 (c) 10 to 18 (d) 15 to 20

3.5 The temperature rise per feed heaters ranges from

(a) 10°C to 20°C (b) 20°C to 30°C (c) 30°C to 50°C (d) 40°C to 50°C

3.6 With reheating of steam

(a) the work done increases

(b) the efficiency increases

(c) the work done and efficiency increase

(d) the work done and efficiency increase and also the problems of erosion and corrosion of blades increase

3.7 Terminal temperature difference in d.c. heater is around

(a) –4°C (b) 0°C (c) 4°C (d) 8°C

3.8 Terminal temperature difference in surface heater ranges from

(a) –6°C to –4°C (b) –4°C to 0°C (c) 0°C to 4°C (d) 4°C to 8°C

3.9 Terminal temperature difference in surface heater integral with dc and dc ranges from

(a) 4°C to 6°C (b) 6°C to 10°C (c) 10°C to 14°C (d) 14°C to 18°C

3.10 Corrosion and erosion problems in steam turbine bladings occur in

(a) hp stage (b) ip stage (c) 1p stage (d) all of them

3.11 The feed pump consumes about

(a) 1% (b) 2% (c) 3% (d) 4% of output energy

Answers to Multiple-Choice Questions

3.1 (d) 3.2 (b) 3.3 (c) 3.4 (a) 3.5 (b) 3.6 (a) 3.7 (b) 3.8 (a)
3.9 (a) 3.10 (c) 3.11 (b)

CHAPTER

4

Fuels and Combustion in Steam Power Plants

OVERVIEW

Steam power plants use fossil fuels. The choice of fuels, their preparation and feeding, and their methods of firing deserve special attention. The combustion of fuel gives heat energy which is utilised to generate steam in a steam generator or boiler and the steam so generated is expanded in steam turbine and is again fed back to the boiler after condensing it in a condenser. The flue gases coming out from the boiler are to be taken out through a chimney to safeguard the air pollution in the vicinity. Energy produced from fossile fuels is called black energy due to black colour of pollution.

The increasing worldwide demand for energy has focused attention on fuels, their availability and environmental effects. The world energy reserves is given in Table 4.1.

Table 4.1 World energy reserves

Fuel	*Type*	*Energy ($\times 10^{15}$ MJ)*
Fossil	Coal- oil and gas	32 6
Fissile	Uranium and thorium	600
Fusile	Deuterium	10^{10}

This chapter deals with fuels used in steam power plants, their combustion, draught needed for boiler system and the type of boiler furnace.

4.1 LAYOUT OF STEAM POWER PLANT

Figure 3.1 (in chapter 3) shows a *simple layout of steam power plant depicting all the systems such as coal handling, steam generation, ash handling, power conversion, circulating cooling water, etc. and other subsystems.* Combustion of fuels takes place in the combustion chamber segment of boiler and the combustion products called flue gases goes out of chimney to the atmosphere. *A sufficient draught is necessary to move the combustion gases from the furnace and out of the chimney.*

4.2 TYPES OF FUELS FOR STEAM GENERATORS

The primary fuels used in steam generators (boilers) are *coal, fuel oil and natural gas* that are preserved in the *sedimentary rocks.* Besides these, industrial wastes like *blast furnace gas, coke oven gas, refinery gas, sugar factory refuse (bagasse), saw mill wood dust, rice husk, etc. are also used as boiler fuels* as a boost to one of the primary fuels. The calorific value of industrial waste is low.

4.2.1 Coal

Coal is the primary energy source and available in India in large quantity. As per geological order of formation, coal may be of the following types: (i) peat, (ii) lignite, (iii) sub-bituminuos, (iv) bituminuos, (v) subanthracite, and (vi) anthracite, with increasing percentage of carbon. After anthracite, the formation of graphite takes place. Anthracite is often difficult to burn as it contains more than 86% fixed carbon (in amorphous form), and less volatile matter. Bituminuos coal is the largest group containing 46 to 86% of fixed carbon and 20 to 40% of volatile matter which helps in burning. Lignite is the lowest grade of coal containing moisture as high as 30% and high volatile matter. Peat contains about 90% of the moisture is not attractive at all as fuel for utility boilers.

The analysis of coal is done either by *proximate* or *ultimate analysis*. The properties of coal are described by swelling index (an index of caking of coal), grindability (power required to grind the coal to a specified particle size), weatherability (stock piling ability for a long period). Sulphur content (should be as minimum as possible), heating value (low and high), ash softening temperature, and spontaneous combustion. The heating value of coal is

$$\text{HHV} = 33.83\text{ C} + 144.45\left(\text{H} - \frac{0}{8}\right) + 9.385;\ \text{MJ/kg} \qquad ...(4.1)$$

$$\text{LHV} = \text{HHV} - 2.395\, m_w;\ \text{MJ/kg} \qquad ...(4.2)$$

where HHV and LHV are high and low heating values respectively.

4.2.2 Fuel Oil

Liquid fuels are an excellent energy source as it is easy to handle, store and burn. They are primarily a mixture of hydrocarbon compounds. They are obtained by refining the crude oil [C (83–87%), H_2 (11–16%), and N_2 (0–77%), S(0–4%)] which is believed to have been formed during past geological ages from decayed marine life, both vegetables and animals. In the refining process, it is distilled into a number of fractions–lighter and heavier. The lighter fraction (having lower boiling point) like gasoline, aviation fuel, kerosene, light diesel oil, heavy diesel oil, lubrication oil and so on are principally transportation and machine fuels. The heavier fractions (having higher boiling point) are used for boiler fuels and chemical production.

The physical properties of oil are described by specific gravity, viscosity, pour point, flash point, heating value, etc.

4.2.3 Natural and Petroleum Gas

The formation of natural gas took place millions of years ago from decaying vegetable matter generally alongwith petroleum. The oil is taken out by drilling oil wells into the geologic formation containing the gas and trapped oil. The best method of transporting the natural gas is through pipelines. At present in India, there is one HBJ (Hazira–Bijapur–Jagdishpur) pipeline. Many new pipelines will be coming up from Godavari basin. It is the cleanest fuel among all fossil fuels. The calorific value of fuel varies from 33.5 to 40 MJ/m^3. It is most suited for utility boilers, power gas turbine and combined cycle power plants.

Since the major constituent of all natural gas is methane (92 to 94%) whose critical temperature is –83°C, so cryogenic temperature is required to keep as a liquid at moderate pressure (–100°C at 36 bars). That is why liquid natural gas (LNG) is transported by special tankers and stored in spherical vessels to be used when needed.

For automobiles compressed natural gas (CNG) is gaining ground as an alternative fuel. Liquid petroleum gas (LPG) belongs to propane, propylene, butane, butylenes, etc. group is liquefied under moderate pressure at normal temperatures. At present, it is widely used as domestic fuel. It is also used as a supplement to natural gas.

4.2.4 Coal-Oil and Coal-Water Mixture

Liquid fuel using mixtures of fine coal in oil is known as *colloid fuel*. Nowadays, it is known as coal-oil mixtures (COM). It has been observed that combustion of colloidal fuels up to 40%, coal concentration with 90% of the coal powder having size 61 μm is comparable to coal or oil fuel. It is cheaper than oil and has less ash content than coal.

Coal-water mixture (CWM) is superior to COM. It is often preferred to replace oil. It contains 70 to 80% of coal mass. It is prepared in the form of slurry and fired like oil.

4.2.5 Synthetic Fuels

Synthetic fuels are gaining ground. They are called synfuels and are in the form of gaseous and liquid fuels produced largely from coal in an economically and environmentally acceptable manner either through coal gasification or coal liquedification. These are used as utility boilers but also as domestic, industrial and transportation purposes.

4.2.5.1 Coal Gasification

In the past, there has been widespread use of coal gas produced by destructing distillation of coal for illuminating and cooling purpose. If the coal is heated in the absence of air, it is carbonized to *coke* by removing its volatiles and using them as byproduct gas. In the past, this gas was distributed in urban areas as *town gas*. The coke so produced is burnt in big beds with less than the stoichiometric quantity of air to yield the *producer gas*.

$$\underset{\text{coke}}{2C} + \underset{\text{air}}{(O_2 + 3.76\ N_2)} \rightarrow \underset{\text{producer gas}}{2CO + 3.76\ N_2} \quad ...(4.3)$$

$$\underset{\text{coke}}{C} + \underset{\text{steam}}{H_2O} \rightarrow \underset{\text{water gas}}{CO + H_2} \quad ...(4.4)$$

Further, as above.

The modern gasification techniques are based on the producer and water gas reactions and are continuous and efficient process. Figure 4.1 shows the basic steps in coal gasification process. The coal is first ground into a sand-like powder and then preheated and dried (by air and steam) to reduce caking during conversion. For a caking coal a high temperature pretreatment is often used by steam to give the coal particles a thin coating of oxygen to prevent sticking.

Three gas mixtures, classified according to heating values, are produced by coal gasification namely low, medium and high heating gases. The process illustrated by the lower flow sheet (Fig. 4.1) yields a low heating value gas while the upper flow sheet a high heating value gas, called pipeline gas with properties close to natural gas ($CO + 3H_2 \rightarrow CH_4 + H_2O$) having heating value 38 MJ/m^3.

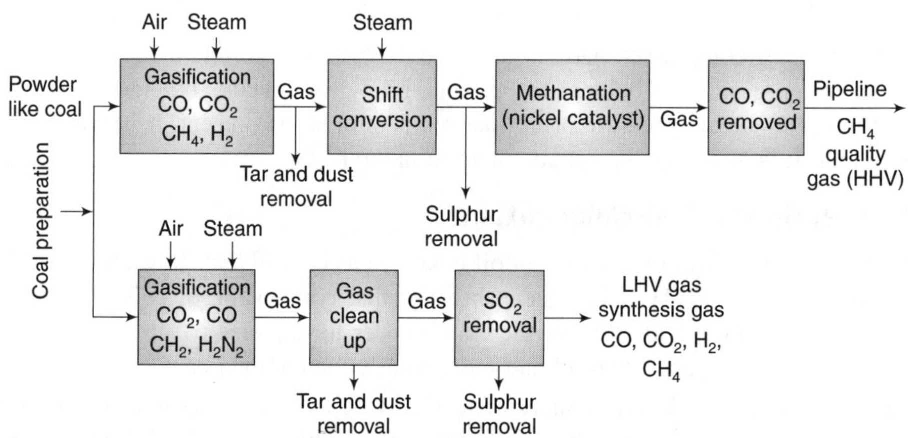

Figure 4.1 Coal gasification process

There is another method of gasification called hydrogasification in which the fluidized coal is gasified directly with hydrogen-rich steam to a methane-rich gas that requires very little additional shifting. The overall reaction is expressed as

$$CH_{0.8} + 0.55\ H_2O + 1.15\ H_2 \rightarrow 0.575\ CH_4 + 0.425\ CO_2 \quad ...(4.5)$$

The conversion efficiency of this method is higher than the above mentioned method. Now low heating value gas finds an attractive application as fuel in combined cycle power plant.

There are various commercial coal gasification processes, namely, Lurgi, Winklr, Koper-Totzek, Synthane process, etc. Figure 4.2 shows synthane coal gasification process developed by US Buensue of Mines. In this process, dry crushed coal is fed from a pressurized hopper into a preheater which is also fed with steam and oxygen. The reaction of coal with oxygen and steam takes place at 400°C, 7 MPa to devolatilize the coal and prevent caking. The coal mixture flows to the gasifier where it is partially gasified in a dense phase at 800°C and later

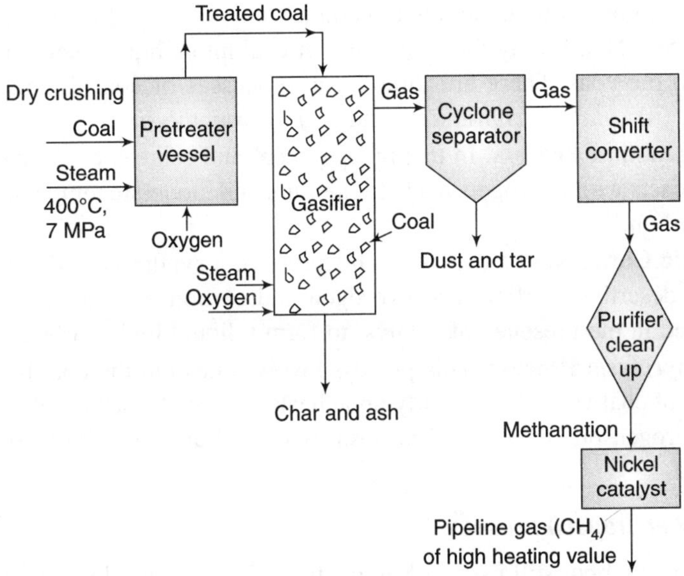

Figure 4.2 Synthane coal gasification process

at 1000°C, 7 MPa. Removal of unreacted char and ash is accomplished from the bottom of the vessel. The gas passes through a cyclone separator which removes dust and tar. For further cleaning the gas passes through shift converter, purifier clean-up vessel and nickel catalyst vessel and finally comes out gas pipeline gas (CH_4) of high heating value.

4.2.5.2 Underground Coal Gasification

Underground coal gasification can be carried out economically in those deposits of coals which could not be mined competitively by other processes. The gasification of coal is carried out *in situ* and the gas so produced is conveyed to the surface for utilisation in various requirements. Two deep drilled holes are made: one for supplying air and other for product gas (CO, CH_2, CH_4, H_2). With initial ignition, combustion takes place at the bottom of air hole and the combustion zone proceeds towards the product gas hole forming carbon dioxide ($C + O_2 \rightarrow CO_2$). While in reduction zone (ahead of combustion zone), carbon monoxide is formed ($CO_2 + C \rightarrow 2CO$). Due to the moisture in the coal, CO and H_2 are formed ($C + H_2O \rightarrow CO + H_2$). The heating values of product gas is low, about 5.5 MJ/m^3.

The *in situ* coal gasification is attractive due to (i) capacity to extract energy from inaccessible reserve of coals, (ii) reduction of mining personnel and equipment, (iii) safer and minimum occupational hazards, (iv) fuel production, and (v) easy to handle SO_2 formed.

4.2.5.3 Coal Liquefaction

Due to shortage of world petroleum supplied and rapidly increasing cost of oil, the interest of producing liquid fuel from coal has revived. In South Africa, major oil demand is met by coal liquefaction by *Fischer-Tropsch process.*

Coal has a ratio of hydrogen atoms to carbon atoms of only 0.8 : 1, while in petroleum, this ratio is 1.75:1. That is why the conversion of coal into a liquid fuel requires the addition of hydrogen to the coal. There are three basic processes of coal liquefaction, namely (i) *hydrogenation,* (ii) *catalytic conversion,* and (iii) *hydropyrolysis.*

(i) Hydrogenation Process. In this process, coal and catalyst are suspended as a slurry, which reacts with hydrogen at a high pressure and moderate temperature to form liquid hydrocarbon.

(ii) Catalytic Conversion Process. In this process, a synthesis gas is produced from the coal as described earlier. The hydrogen and carbon monoxide of the gas are then combined in the presence of catalyst to form a liquid hydrocarbon.

(iii) Hydropyrolysis Process. This process involves heating the coal beyond 450°C. The fraction of coal volatilized greatly upon heating exceeds the volatile matter in the coal. The hydrogen pulverized coal is flesh pyrolyzed and up to 50% of the coal can be liquefied.

4.2.5.4 Fischer-Tropsch Process

This is an old method but still used in South Africa. This process first produces a mixture of CO and H_2 from coal and steam followed by catalytic reactions at about 150°C and 150 bars which yields a range of hydrocarbon from gaseous methane to higher liquid hydrocarbons. These hydrocarbons are then separated with methane going as pipeline gas going to different liquid fuels.

4.2.6 Industrial Wastes and Byproducts

There are various types of industrial wastes which are combustible in nature and therefore are receiving increased attention to utilise them as fuels for steam generators serving two purposes: one is disposal and other is reduction in the use of oil. Such industrial wastes include *refinery gas, coke-oven gas, regenerator gas, blast furnace gas, liquid wastes, wood wastes, sugarcane (waste called bagasse), solid wastes, municipal wastes, refuse-derived fuels (RDF), etc.*

Refinery gas is a high heating value gas which is produced during the conversion of crude oil to gasoline and other refinery products. It is often blended with low heating value gas prior to combustion. Generation of *coke-oven gas* is accomplished during the production of coke from raw coal in a coke oven. It is a mixture of hydrogen (50%), methane (33%) and other gases and its heating value ranges from 6.2 to 21.3 MJ/m^3. *Regenerator gas* is produced in refinery by catalytic cracking process at about 540°C and has low heating value. It is burnt at the site. *Blast furnace* gas is produced during reduction process of iron oxide to iron in blast furnace. This gas contains about 30% CO and high dust loading and thus requires cleaning for burning.

There are large number of *liquid wastes* such as solvents, waste oils and oil sludges, phenols, tars, combustible chemicals, greases and fats. These liquid wastes have different heating values and other properties.

Solid wastes include wood waste and sugar factory refuse. Sugarcane waste (bagasse) is the residue of sugarcane after juice extraction. Its heating value is 8.4 to 9.77 MJ/kg. Boilers installed in sugar industry use bagasse to produce steam. Solid wastes generated by industrial and domestic processes in the form of garbage are often used as boiler fuel in incinerators, particularly in cities either to generate electricity or to use as process heat. Use of ***unserviceable types*** because of high heating value (32 MJ/kg) is becoming popular as a fuel of boiler. It is very common in USA.

4.2.7 Biomass

Biomass is an organic matter produced by plants both in land and water and includes forest crops, the crops grown in energy forms and animal manure. It is a renewable resource of energy as the plant life renews and adds to itself every year unlike fossil fuels which take millions of year to form and do not nonrenew. In biomass, it is the solar energy stored by way of photosynthesis as given below.

$$\underset{\text{Air}}{6\,CO_2} + \underset{\text{Soil}}{12\,H_2O} \xrightarrow[\text{Chlorophyll}]{} \underset{\substack{\text{glucose}\\\text{or starch}}}{C_6H_{12}O_6} + \underset{\substack{\text{ejected by air}\\\text{transpiration}}}{6H_2O + 6O_2} \quad ...(4.6)$$

In other words, *biomass (also called biofuels) is the products of photosynthesis.* These are bulky and contain large amount of water, so it is not economical to transport them over long distance rather it is utilised by direct burning in region close to the source. They can also be converted to a variety of gaseous, liquid or solid fuels.

The following are the various types of land crops.

(i) Sugar crops, such as sugar cane refuse or bagasse

(ii) Herbaceous crops, such as non-woody plants.

(iii) Forest crops, such as eucalyptus, sycamore, sweetgum, cultural hybrid poplar, etc.

Animal and human wastes are the indirect land crops from which methane and ethylene can be produced while their manure values are retained.

Aquatic crops which include seadwoods, marine algae and various kelps are grown in fresh, sea and brackish water.

There are three types of bioconversion process (i) Direct combustion such as wood waste and bagasse (ii) Thermochemical conversion, and (iii) Biochemical conversion

In ***thermochemical conversion process,*** there are two stages, gasification followed by liquefaction. Gasification is performed by heating with limited oxygen and liquefaction yields methanol and ethanol. In *biochemical conversion* process, there are two routes, namely, *anaerobic digestion* and *fermentation. Anaeration digestion* is the bacterial decomposition of organic matter (biomass) in the absence of air to produce gaseous mixture (biogas) of methane and carbon dioxide in an approximately 2 : 1 volume ratio CO_2 so produced can be converted to *synthetic natural gas* (SNG) and the sludge formed due to anaeration digestion serves as manure.

In ***fermentation process***, the breakdown of complex molecules in organic compounds takes place with the help of a ferment such as yeast, bacteria, enzymes, etc. Grains and sugar crops are converted by fermentation into ethanol which is mixed with gasoline to produce gasohol fuels for automobiles. In non-fertile areas, sycamore and eucalyptus trees are grown to produce energy by combustion. The heating values of some waste bioproducts is given in Table 4.2.

Table 4.2 Heating value of some waste bioproducts

Waste Bioproducts	*Heating value (MJ/kg dry)*
Wood chips	18.6-20.9
Bagasse	9.5
Cereal straw	16-17
Organic refuse	13.2

4.3 COMBUSTION REACTIONS

Combustion is defined as the high temperature oxidation of the combustible elements in coal and fuel oil having the presence of carbon, hydrogen and sulphur contents. The basic chemical equations for complete combustion of fuels are expressed as

$$C + O_2 \rightarrow CO_2; \qquad 2H_2O + O_2 \rightarrow 2H_2O; \qquad S + O_2 \rightarrow SO_2 \qquad ...(4.7)$$

In the case of insufficient oxygen, combustion will be incomplete forming CO. Thus,

$$2C + O_2 \rightarrow 2CO \qquad ...(4.8)$$

It is worth to note that for burning a fuel completely, *four basic conditions* must be fulfilled, namely (i) supply of enough air, (ii) creation of enough turbulence for thorough mixing of air and fuel, (iii) maintaining a furnace temperature sufficiently high enough to ignite the incoming mixture of air and fuel, and (iv) providing a furnace volume sufficiently large enough to allow time for complete combustion.

In other words, combustion is governed by a four letter word, **"MATT"** where M stands for sufficient mixture turbulence, A stands for proper air-fuel ratio, T stands for temperature and other T is for time-enough for combustion. The analysis of coal is performed either by *proximate (volume basis) analysis or by ultimate (mass basis) analysis.*

4.3.1 Stoichiometric Air-Fuel Ratio

The ultimate analysis of fuel (coal) shows the following components on mass basis : carbon (C), hydrogen (H), oxygen (O), nitrogen (N), sulphur (S), moisture (M) and ash (A). Therefore

$$C + H + O + N + S + M + A = 1.0 \qquad ...(4.9)$$

The mass of oxygen needed for oxidation processes are calculated as follows:

(i) C +	O_2	= CO_2	Similarly (ii)	$2H_2$	+ O_2	=	$2H_2O$
12 kg	32 kg	44 kg		4 kg	32 kg		36 kg
1 kg	2.67 kg	3.67 kg		1 kg	8 kg		9 kg
C kg	2.67C kg	3.67C kg		H kg	8H kg		9H kg

(iii) $S + O_2 = SO_2$

32 kg 32 kg 64 kg

S kg S kg 2 S kg

Therefore, mass of oxygen required for complete combustion of 1 kg of fuel:

$m_{02} = 2.67C + 8H + S - O$ where O is the oxygen in the fuel, if any.

As we know that air contains 23.2% oxygen by mass, therefore, theoretical air required for complete combustion of 1kg of fuel

$$\frac{m_{at}}{m_f} = \text{Air-fuel ratio} = \frac{m_{02}}{0.232} = \frac{2.67C}{0.232} + \frac{8}{0.232}\left(H - \frac{O}{8}\right) + \frac{S}{0.232}$$

or

$$A/F = \frac{m_{at}}{m_f} = 11.5C + 34.5\left(H - \frac{O}{8}\right) + 4.3S \qquad ...(4.10)$$

where C, H, O and S stand for mass fraction of carbon, hydrogen, oxygen and sulphur respectively.

It is a well-known fact that complete combustion of fuel cannot be achieved without supplying excess air than stoichiometric. The percentage of excess air is given by

$$\%\text{ excess air} = \frac{m_{aa} - m_{at}}{m_{aa}} \times 100 \qquad ...(4.11)$$

where m_{aa} is the actual air supplied for complete combustion of 1 kg of fuel. Due to excess air, a term *dilution coefficient* may be defined

$$d = \frac{m_{aa}}{m_{at}} \qquad ...(4.12)$$

For large utility boilers, the percentage of excess air varies from 15 to 30%.

Natural gas in which methane constitutes about 94% is also one of the fuels for utility boilers. The combustion of methane is expressed as

$$CH_4 + 2O_2 \rightarrow CO_2 + 2H_2O \qquad ...(4.13)$$

Since for each mole of oxygen taking part in combustion reaction, there are 0.79/0.21 or 3.76 moles of nitrogen, so for combustion of methane with 150% excess air

$$CH_4 + 2(1.5)O_2 + 2(3.76)(1.5)\ N_2 \rightarrow CO_2 + 2H_2O + O_2 + 7.5\ (1.5)N_2$$

The actual air required for 1 kg of fuel is given by

$$m_{aa} = \frac{3.04\ N_2\ C_{ab}}{CO_2 + CO} - \frac{N}{0.768} \qquad ...(4.14)$$

where C_{ab} is the fraction of carbon in fuel which has been burnt to CO_2 and CO.

and N is the nitrogen present in fuel (%) if any nitrogen in absent, $N = 0$

The analysis of combustion products is carried out by (i) Orsat analyser (ii) Haldane apparatus, (iii) infra-red gas analyser, and (iv) gas chromatograph.

4.3.2 Combustion Equation

In order to find the combustion equation based on the ultimate analysis of fuel and volumetric analysis of combustion products, let us consider the following example.

$$C = 62\%,\ H = 4\%,\ S = 3.0\%,\ O = 4\%,\ N = 2\%,\ H = 4\% \text{ and } A = 21\%$$

The exhaust gas has the following volumetric analysis,

$$CO_2 + SO_2 = 13\%,\ CO = 2\%,\ O_2 = 4\%, \text{ and } N_2 = 83\%$$

Let a moles of oxygen be supplied for 100 kg fuels thus combustion equation for above example may be written as

$$\frac{62}{12} C + \frac{4}{2} H_2 + \frac{3}{32} S + \frac{4}{32} O_2 + a\, O_2 + 3.76a\, N_2$$

$$\rightarrow\ b\, CO_2 + d\, CO + e\, SO_2 + f\, O_2 + g\, N_2 + H_2O \qquad ...(4.15)$$

By equating the coefficients of C, H_2, S, O_2 and N_2, the constants are evaluated.

For example, let us consider the combustion of propane gas with stoichiometric air

$$C_3H_8 + 5O_2 + 5(3.76)N_2 \rightarrow 3CO_2 + 4H_2O + 18.8\, N_2 \qquad ...(4.16)$$

With 80% theoretical air, the above combustion equation becomes with addition of formation of carbon monoxide due to incomplete combustion

$$C_3H_8 + 5(0.8)O_2 + 5(3.76)(0.8)N_2 \rightarrow a\, CO_2 + b\, CO_2 + 4H_2O + 15.04\, N_2 \qquad ...(4.17)$$

Carbon balance gives, $3 = a + b$

Oxygen balance gives, $8 = a + 2b + 4$

By solving : $a = 2$, $b = 1$, hence combustion equation is

$$C_3H_8 + 4O_2 + 15.04\, N_2 \rightarrow 2CO + CO_2 + 4H_2O + 15.04\, N_2 \qquad ...(4.17a)$$

4.3.3 Dew Point Temperature of Exhaust Gases in Boiler

The partial pressure of water vapour in the mixture of gases of combustion product having total pressure of 1 atm is given by

$$p_{H_2O} = x_{H_2O} \cdot p = x_{H_2O} \cdot (\text{atm}) = x_{H_2O} \text{ atm} \qquad ...(4.18)$$

where, x_{H_2O} = the mole fraction of water vapour formed

$$= \frac{n_{H_2O}}{N_{total}} = \frac{n_{H_2O}}{n_{CO_2} + n_{CO} + n_{O_2} + n_{SO_2} + n_{H_2} + n_{H_2O}} \qquad ...(4.19)$$

The saturation temperature at the partial pressure of water vapour is called the *dew point temperature (dpt)* as shown in Fig. 4.3. The flue gases coming out from superheaters/reheater are further cooled in economizer and air preheater in order to minimise the exhaust gas losses though chimney.

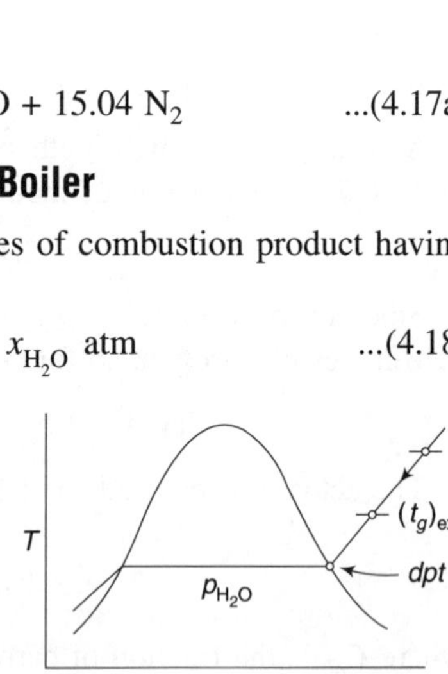

Figure 4.3 Dew point temperature of exhaust flue gases

It is essential that the exhaust gases should not be cooled below their dew point temperature. If it is cooled below the dpt, the water vapour in exhaust gases condenses into liquid droplets which combines with SO_2 or SO_3 to form acid (H_2SO_4). The acid so formed corrodes the metal surfaces of the duct or stack while flowing through it. That is why the temperature of flue gases leaving the preheater is kept around 160°C.

4.3.4 Heating Value of a Fuel (Coal)

The heating value of coal may be calculated based on the heat of reaction of its constituents which is given in Table 4.3.

Table 4.3 Heats of reaction of coal constituents

Constituents	*Formulae and state*	*Products of combustion and state*	*Heat of reaction (kJ/kg mole)*
Carbon (coke)	C (s)	CO_2 (g)	-407×10^3
Carbon	C (s)	CO_2 (g)	-397×10^3
Carbon monoxide	CO (s)	CO_2 (g)	-283×10^3
Hydrogen	H_2 (s)	H_2O (l)	-286×10^3
Sulphur	S (s)	SO_2 (g)	-291×10^3

The heat released by combustion of carbon (C) constituents in a 1 kg of coal is

$$\text{C (kg)} \times 407 \times 10^3 \left(\frac{\text{kJ}}{\text{kg.mole}}\right) \times \frac{\text{kg.mole}}{12 \text{ kg}} = 33917 \quad \text{kJ/kg}$$

Similarly, heat released by sulphur (S) $= \dfrac{291 \times 10^3}{32} = 90945 \quad \text{kJ/kg}$

It is worth to note that available hydrogen for combustion is the total hydrogen less than that required to combine with oxygen in the coal, $\left(H - \frac{O}{8}\right)$, and the heat released by it

$$= \left(H - \frac{O}{8}\right) \frac{286 \times 10^3}{2} = 143 \times 10^3 \left(H - \frac{O}{8}\right) \text{kJ/kg}$$

Thus, the higher heating value (HHV) for complete combustion of 1 kg of coal is expressed as

$$\text{HHV} = 33.917 \text{ C} + 143 \left(H - \frac{O}{8}\right) + 9.0945 \text{ S MJ/kg} \qquad ...(4.20)$$

The above equation is very close to Dulong's formula given as

$$\text{HHV} = 33.83 \text{ C} + 144.45 \left(H - \frac{O}{8}\right) + 9.38\text{S MJ/kg} \qquad ...(4.21)$$

$$\text{LHV} = \text{HHV} - 2.395\, m_w \text{ MJ/kg} \qquad ...(4.22)$$

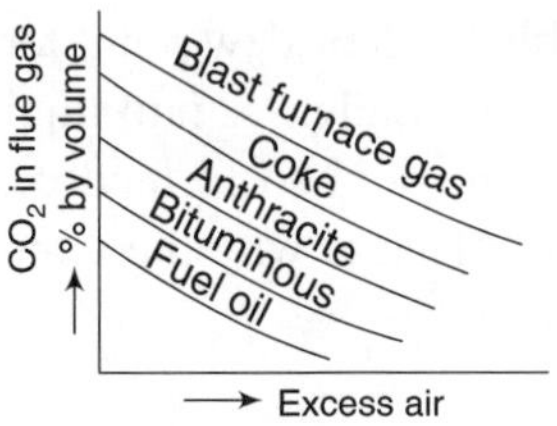

Figure 4.4 CO_2 variation in flue gases

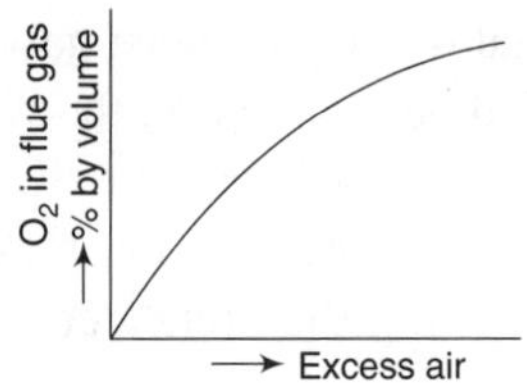

Figure 4.5 O_2 variation in flue gases

4.3.5 Optimum Excess Air in Boiler

In order to maintain optimum combustion efficiency in a boiler, proper control of right amount of excess air is essential. The amounts of CO_2 and O_2 in combustion gases are the indexes of excess air and which is a function of fuel. Figure 4.4 shows the variation of CO_2 in flue gases with fuels and excess air while the variation of O_2 in flue gases with excess air is given in Fig. 4.5. Boiler losses are estimated for different steam outputs and the excess air is then adjusted by controlling air supply to show the optimum value of CO_2 or O_2. By the method shown in Fig. 4.6, the optimum value of excess air for best combustion efficiency is determined and instruments are used to control this. The excess air is also calculated by equation proposed by Skrotzki and Vopat (1960).

$$\% \text{ excess air} = \frac{O_2 - 0.5 \text{ CO}}{0.264 \text{ N}_2 - (O_2 - 0.5 \text{ CO})} \times 100 \qquad ...(4.23)$$

where O_2, CO, N_2 are the volumetric percentages in the dry flue gases.

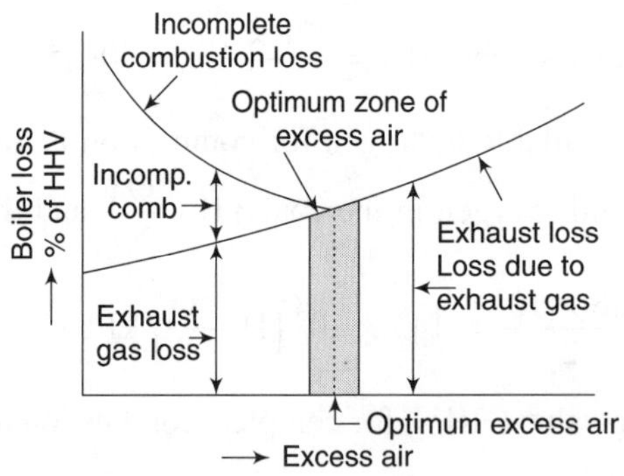

Figure 4.6 Optimum excess air for maximum combustion efficiency

4.4 MASS BALANCE OF A BOILER (STEAM GENERATOR) FURNACE

For the analysis of boiler, mass balance is needed. Figure 4.7 shows the mass entering and leaving the boiler furnace. 1 kg of coal with amount of air (m_a) needed for combustion are entering while dry flue gases ($\dot{m}_{dfg}$), steam ($\dot{m}_{H_2O}$) and refuse ($\dot{m}_{ref}$) are coming out.

Thus, $\dot{m}_f$ (1 kg coal/s) + $\dot{m}_a = \dot{m}_{dfg} + \dot{m}_{H_2O} + \dot{m}_{ref}$

or $\quad (C + H + O + S + N + M + A) + \dot{m}_a = \dot{m}_{dfg} + 9H + M + A + C - C_{ab}$

or
$$\dot{m}_a = \dot{m}_{dfg} + 8\left(H - \frac{O}{8}\right) - C_{ab} - N - S;\ \text{kg/kg fuel} \qquad ...(4.24)$$

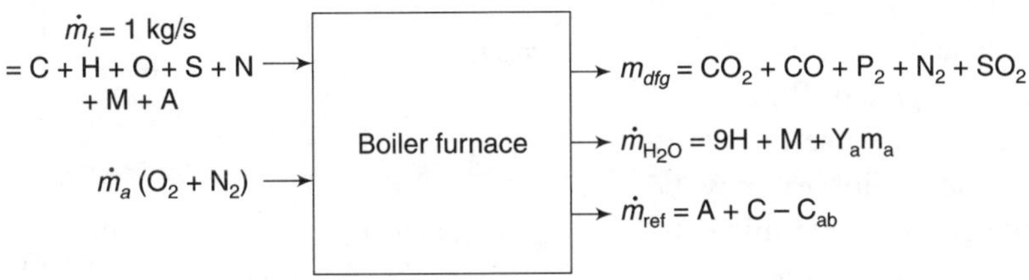

Figure 4.7 Mass balance of a boiler furnace

Mass of dry flue gases, $\dot{m}_{dfg}$ is expressed as

$$\dot{m}_{dfg} = \frac{C_{ab}\,(44\,CO_2 + 28\,CO + 28\,N_2 + 32\,O_2)}{12\,(CO_2 + CO)}$$

$$= \frac{C_{ab}[11\,CO_2 + 7\,CO + 8\,O_2 + 7(100 - CO_2 - CO - O_2)]}{3(CO_2 + CO)}$$

$$= \frac{C_{ab}[4CO_2 + O_2 + 700]}{3(CO_2 + CO)};\ \text{kg/kg fuel} \qquad ...(4.25)$$

Using perfect gas law, the volume of flue gases (wet) per kg coal is expressed as

$$V_{fg} = \left[\frac{\dot{m}_{dfg}}{M_{dfg}} + \frac{\dot{m}_{H_2O}}{18}\right] \times 22.4\,\frac{T_g}{273} \times \frac{101.325}{p_g};\ \text{m}^3\text{/kg} \qquad ...(4.26)$$

where p_g = gas pressure, (kPa) and M_{dfg} = molecule weight of dry flue gas

The mass of refuse (m_{ref}) per kg coal is expressed as

$$\dot{m}_{ref} = \frac{A}{A_{ref}}$$

where $\quad A$ = ash content in 1 kg of fuel = $A_{ref} + C_{ref}$

A_{ref} = mass fraction of ash in the refuse

Mass of unburnt carbon is refuse per kg coal = $C_{ref} \cdot \dot{m}_{ref} = C_{ref}\,\dfrac{A}{A_{ref}}$

The fraction of carbon burnt to CO_2 = $C_{ab} = C - C_{ref}\,\dfrac{A}{A_{ref}}$ ...(4.27)

4.5 ENERGY BALANCE IN A STEAM GENERATOR

The fuel and air are supplied to the steam generator and due to complete combustion, the fuel releases its heating value. The heat so released is absorbed by feed water supplied by the boiler feed pump and it is converted into steam while passing through economizer (ECO), evaporator (EVA) and superheater (SH) and reheater (RH). Figure 4.8 shows the control volume of a boiler furnace in which various masses and energies entering and leaving are depicted. Based on 1 kg of fuel (coal), the useful energy and various energy losses may be computed which will determine the efficiency of a steam generator.

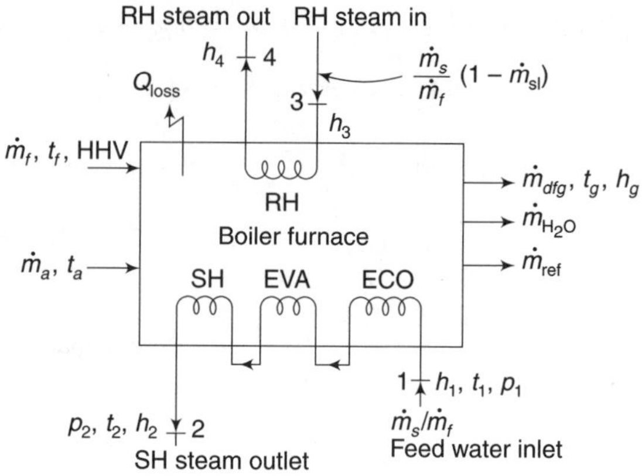

Figure 4.8 Energy balance in a steam generator

(i) Useful energy is the energy absorbed by feed water

$$= Q_{use} = Q_1 = \frac{\dot{m}_s}{\dot{m}_f}[(h_2 - h_1) + (1 - \dot{m}_{s1})(h_4 - h_3)] \text{ kJ/kg} \quad ...(4.28)$$

where $\dot{m}_f$ = fuel burning rate, (kg/s), $\dot{m}_s$ = steam generation rate, (kg/s), $\dot{m}_{s1}$ = mass fraction of inlet steam bled from *hp* turbine before reheating.

(ii) Energy loss due to *dry exhaust gases*

$$Q_{exh} = Q_2 = \dot{m}_{dfg}\,\bar{c}_{p_{dfg}}(t_g - t_a) \text{ kJ/kg} \quad ...(4.29)$$

where $\dot{m}_{dfg}$ = mass of *dfg* produced per kg of fuel, $\bar{c}_{p_{dfg}}$ = average specific heat of *dfg*. (kJ/kg), t_g = exhaust gas temperature °C, t_a = ambient temperature (°C)

(iii) Energy loss due to unburnt carbon

$$= Q_{uc} = Q_3 = (C - C_{ab})\frac{407 \times 10^3}{12} = 33917\,(C - C_{ab}) \text{ kJ/kg} \quad ...(4.30)$$

where C and C_{ab} are unburnout and burnout carbon in dry gas per kg of fuel

(iv) Energy loss due to incomplete combustion is caused by the formation of CO. The loss of energy per kg of C oxidized to CO

$$= Q_{co} = \frac{283 \times 10^3}{12} = 2358.3 \text{ kJ/kg of carbon}$$

Thus, energy loss due to incomplete combustion

$$Q_{incomb} = Q_4 = 2358.3\left(\frac{\text{kJ}}{\text{kg carbon}}\right) \times \dot{m}_{dfg}\left(\frac{\text{kg dfg}}{\text{kg fuel}}\right)$$

$$\times \left[\frac{28\text{CO}}{44\text{CO}_2 + 28\text{CO} + 32\text{O}_2 + 28\text{N}_2}\right]\left(\frac{\text{kg CO}}{\text{kg fuel}}\right) \times \frac{12\text{ (kg carbon)}}{28\text{ (kg CO)}}$$

$$= 10100\, \dot{m}_{dfg} \left[\frac{28CO}{44CO_2 + 28CO + 32CO_2 + 28N_2}\right] \frac{kJ}{kg} \quad ...(4.31)$$

(v) Energy loss due to moisture in fuel

$$= Q_{moist} = Q_5 = M[c_{pw}\,(100 - t_f) + h_{fg} + c_{pg}\,(t_g - 100)]\,\frac{kJ}{kg}$$

$$= M[4.18\,(100 - t_f) + 2256.8 + 2.09\,(t_g - 100)]\,\frac{kJ}{kg} \quad ...(4.32)$$

where t_f = temperature of fuel entering the fuel, (°C), t_g = temperature of exhaust gas (°C), M = mass of moisture per kg fuel

(vi) Energy loss due to hydrogen in fuel which converts into steam

$$Q_H = Q_6 = 9H\,[4.187\,(100 - t_f) + 2256.8 + 2.09\,(t_g - 100)]\,\frac{kJ}{kg} \quad ...(4.33)$$

(vii) Energy loss due to moisture coming with air supplied

$$Q_{am} = Q_7 = y_a\, \dot{m}_a\, \bar{c}_{p_s}\,(t_g - t_a)\,\frac{kJ}{kg} \quad ...(4.34)$$

where y_a = specific humidity of air, kg moisture/kg dry air

$\bar{c}_{ps}$ = average specific heat of superheated water vapour in air

(viii) Energy loss due to ash and slag (refuse)

$$Q_{ref} = Q_8 = (A + C + C_{ab})\,\bar{c}_{p_{ref}}\,(t_{furnace} - t_a)\ kJ/kg \quad ...(4.35)$$

where $c_{p_{ref}}$ average specific heat of refuse (ash)

(ix) Energy loss due to convection and radiation from the boiler surface

$$= Q_{cr} = Q_9 = (h_c + h_r)\,A_s\,(T_s - t_a)\,\dot{m}_f\ ;\ kJ/kg \quad ...(4.36)$$

where h_c = convective heat transfer coefficient (W/m^2 K), h_r = radiative heat transfer coefficient (W/m^2 K), A_s = total boiler surface area exposed to ambient air (m^2), t_s = temperature of wall surface of boiler (°C), t_a = ambient temperature (°C)

Energy released by complete combustion of 1 kg of fuel = high heating value (HHV)

Energy utilized in heating the feed water = HHV – Σ energy losses = Q_1; kJ/kg ...(4.37)

The efficiency of steam generator

$$\eta_{steam\ gas} = \eta_{boiler} = \frac{\text{Energy utilized}}{\text{Energy released}} = \frac{Q_1}{HHV} = \frac{HHV - \Sigma\ \text{energy losses}}{HHV}$$

$$= \left[\frac{\dot{m}_s\,(h_2 - h_1) + (1 - \dot{m}_{s1})\,(h_4 - h_3)}{\dot{m}_f \times HHV}\right] \quad ...(4.38)$$

Based on the useful energy and energy losses, heat balance sheet of boiler in bar chart or graph or table form may be prepared. A tabular form of heat balance sheet is shown in Table 4.4.

Table 4.4 Heat balance sheet per kg of dry fuel

Item	*Heat (kJ/kg fuel)*	*Heat (%)*
Heat : Heat supplied	–	100
Expenditure (i) Heat consumed to generate steam		
(ii)	–	–
(iv)	–	–
Total		100

4.6 HEAT OF COMBUSTION

The heat of combustion from reactive system is obtained by applying the steady flow energy equation considering the chemical energy which is ignored in non-reactive system.

4.6.1 Open System

Figure 4.9 shows a chemically reactive system. Applying steady flow energy equation ignoring kinetic and potential energy, we have

$$H_R + Q_{cv} = H_p + W_{cv} \quad ...(4.39)$$

where H_R and H_p are the enthalpies of reactants and products respectively evaluated for their constituents at their respective pressure and temperature. Q_{cv} is the net heat added to the control volume (which is usually negative, as the combustion heat is rejected to the surroundings) and W_{cv} is the net work done by the control volume.

Since the reactants and products are usually composed of several constituents, so the above equation may be expressed as mass basis

$$\sum_R (mh) + Q_{cv} = \sum_p (mh) + W_{cv} \quad ...(4.40)$$

For example, to define the enthalpies value, consider the combustion of ethane (C_2H_6) with oxygen

$$C_2H_6 + 3.5O_2 \rightarrow 2CO_2 + 3H_2O \quad ...(4.41)$$

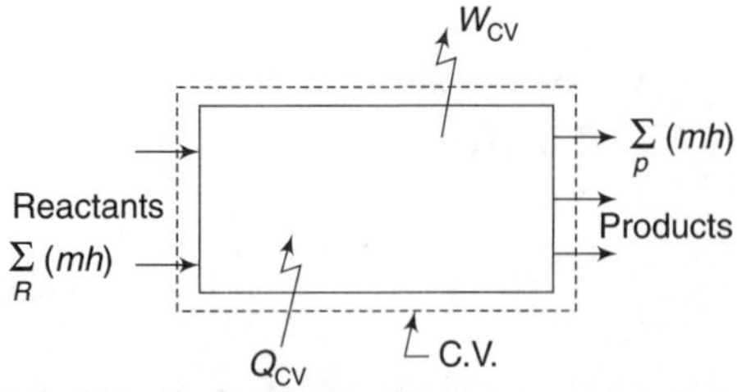

Figure 4.9 Natural (chimney) draught

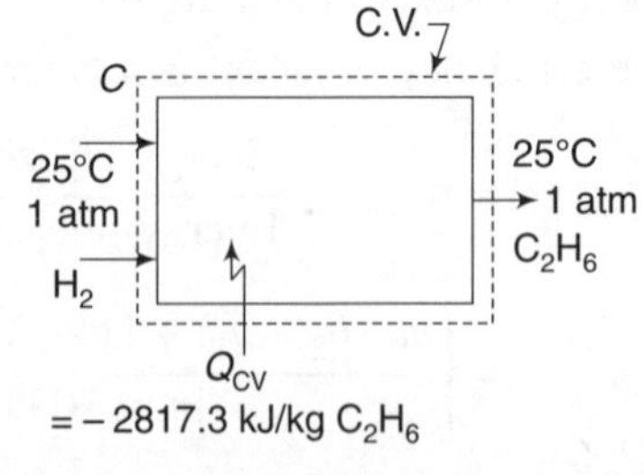

Figure 4.10 Dispersion of flue gases at stack exit

It is worth to note that the enthalpies of various reactants and products are those that start at the same datum of composition temperature and pressure which are arbitrarily chosen as the elemental substance at 25°C and 1 atm respectively. For example, C_2H_6 is formed from elemental carbon C and hydrogen H_2, CO_2 from C and O_2 on so on. These are exothermic reactions, that when they begin at 25°C and are cooled back to 25°C after the reaction takes place, yield 2817.2 kJ/kg and 8946.8 kJ/kg of product respectively (Fig. 4.10). In other words, in steady flow at 25°C, formation reactions are

$$2C + 3H_2 \rightarrow C_2H_6 + 2817.3 \text{ kJ/kg } C_2H_6 \quad \text{...(4.42)}$$

and

$$C + O_2 \rightarrow CO_2 + 8946.8 \text{ kJ/kg } CO_2 \quad \text{...(4.43)}$$

The quantities 2817.3 and 8946.8 kJ/kg leave the system and hence are negative. They are called enthalpies of formation *(h_f°)*. The value of h_f° for various substances at 25°C and 1 atm are given in *Appendix* in the end of this book.

In general, chemical equations such as Eq. (4.40) are balanced in terms of moles and not masses and since $m = nM$ where n is the number of moles and M is molecular weight, so Eq. (4.40) takes the form

$$\sum_R (n\,M\,h_f) + Q_{cv} = \sum_p (n\,Mh_f) + W_{cv} \quad \text{...(4.44)}$$

where h_f is the kJ/kg mole.

At any other temperature than 25°C and 1 atm

$$h_{T,p} = h_f^\circ + \Delta h \quad \text{...(4.45)}$$

4.6.2 Closed System

In many situations, burning takes place in closed system such as cylinder or bomb. The combustion equations for fuels burning in a closed system may be obtained by writing the energy balance for closed system. Thus,

$$U_R + Q = U_p + W \quad \text{...(4.46)}$$

where W is the non-flow work

For gases $H = U + pV = U + nR_0T$ where R_0 is universal constant hence

$$\sum_R (n\,M\,h_f - n\,R_0T) + Q = \sum_p (n\,M\,h_f - n\,R_0\,T) \quad \text{...(4.47)}$$

4.7 HEATING VALUES

There are two types of heating values—*higher heating value (HHV)* and *lower heating value (LHV)*. The former is the heat released when water vapour in the product due to combustion of hydrogen in the fuel condenses, the later when it stays in the vapour state. Based on the analysis of open system, the heating value (HV) is given by putting $W_{cv} = 0$ and replacing Q_{cv} by HV for heating value

$$\text{HV} = \left[\sum_R (n\,Mh_f) - \sum_p (n\,Mh_f)\right]_{T_1} \quad \text{...(4.48)}$$

where T_1 indicates the standard temperature. This heating value is sometimes called the *enthalpy of combustion* and depends on T_1 : whether the water vapour formed condenses giving HHV otherwise LHV. It also depends upon whether the reactant fuels were in liquid or vapour state, as it takes a certain amount of energy to vaporize a liquid fuel such as octane.

Based on the closed system, HV_{nf} is sometimes called the *internet energy of combustion*. The heating values of fuels are given in Appendix.

4.8 THEORETICAL (OR ADIABATIC) FLAME TEMPERATURE

A fuel burning with no heat exchange with the surroundings and no work done will result in the theoretical or adiabatic flame temperature. It is greater for stoichiometric than either a lean or rich mixture because, as above, a lean mixture has a dilution effect, whereas a rich mixture results in incomplete combustion. It will also be greater if the fuel is burnt in oxygen rather than air because of the dilution effect of nitrogen.

At the adiabatic flame temperature, $Q_{cv} = 0$ and $W_{cv} = 0$, so

$$\sum_{R} (n\ Mh_f) = \sum_{p} (n\ Mh_f) \quad \text{...(4.49)}$$

Adiabatic flame temperature so calculated is higher than actual due to dissociation of some of the products takes place at higher temperature. However, the effect of dissociation is lower for lean mixtures because the excess oxygen tends to drive the reaction towards completion, and for higher furnace pressure. That is why, the effect of dissociation is not significant in steam generator furnace combustion which occurs with excess air at about 1500°C.

4.9 FREE ENERGY OF FORMATION

Gibbs function is defined as $G = H - TS$ is called free energy. For equilibrium in a system at constant temperature and pressure, $\Delta G_{TP} < 0$. For a process at constant T and p, G always decreases and is minimum at the state of equilibrium. For driving force in a chemical reaction at constant temperature

$$\Delta G = \Delta H - T\,\Delta S \quad \text{...(4.50)}$$

The free energy of the chemical elements at 1 atm and 25°C is arbitrarily assigned zero value. Thus, g_f° is the free energy of formation of a compound when one mole of the compound is formed directly from its constituent elements, both the reactants and the products being at 25°C and partial pressure of 1 atm.

4.10 TYPES OF BOILER FURNACE

The factors such as temperature, concentration, preparation, distribution of reactants and mechanical turbulence greatly affect the rate of chemical reaction in the combustion process. In boiler, the firing system introduces the fuel and air for combustion, mixes these reactants to ignite the combustible mixture, and distributes, the flame envelop and the products of combustion.

There are two types of boiler furnaces as given below.

(i) **Grate fired furnaces** and (ii) **Chamber type or flame furnaces.**

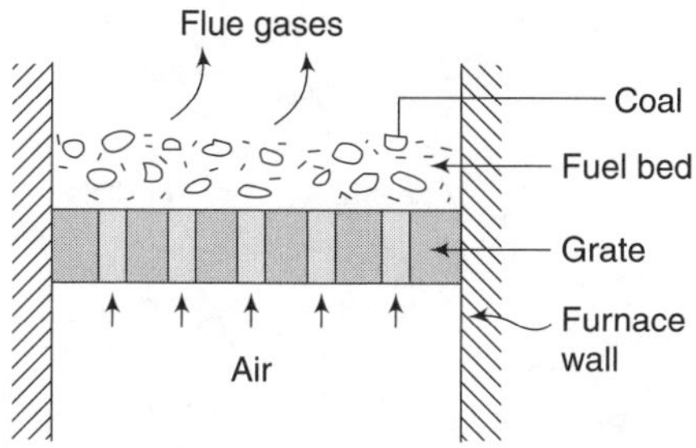

Figure 4.11 Balance draught system

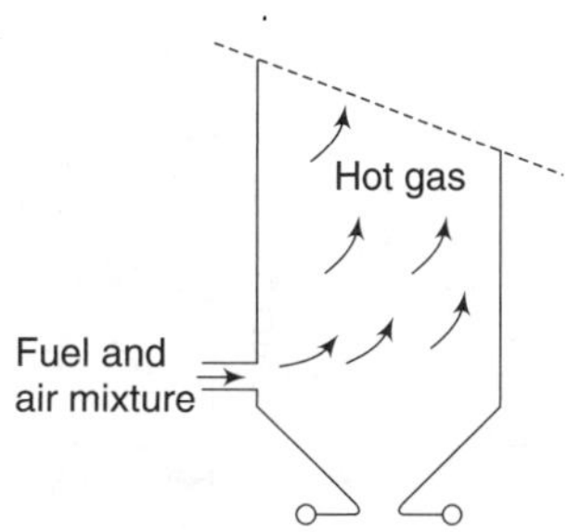

Figure 4.12 Chamber type or flame furnace

Figure 4.11. shows the grate-fired furnace. These are *suitable for having solid fuels* such as coal, wood, chips, bagasse, urban waste, etc. The *chamber type or flame furnace* is shown in Fig. 4.12 *which is used for firing pulverized coal, fuel oil and gas.* A mixture of fuel and air is delivered into the furnace chamber for combustion where it burns and the combustion gases go upwards as shown in Fig. 4.12.

4.11 KINETICS OF COMBUSTION REACTIONS

The phenomenon of combustion reaction is very complex. It is determined by physical and chemical factors. The physical factors consist of the proces of mixing of fuel and air, the size of fuel particles and the surface area exposed to reaction. The chemical factors include the temperature and concentration of reactants. The kinetics of chemical reactions are determined by the complex fields of velocities, temperaturers and concentration. The combustion reactions are accomplished with exothermic type of evolution of heat. The chemical reactions can proceed in either direction—forward or backward and they are called reversible reactions.

Reaction rates (r) are governed by the law of mass action and are expressed as

$$r = K\, c_A^m\, c_B^n = K\, p_A^m\, p_B^n \qquad ...(4.51)$$

where c_A and c_B are the concentration reactants of A and B. Here p_A and p_B are the partial pressures of components in gas mixture. The reaction rate constant (K) is expressed as

$$K = K_0\, e^{-E/R_0 T} \qquad ...(4.52)$$

where K_0 is constant, E is the activation energy, kJ/kg mole, R_0 is the universal gas constant (kJ/kg mole) and T is the absolute temperature.

From the nature of the curve, it is clear that combustion is possible only during a certain concentration limit.

4.12 MECHANISM OF SOLID FUEL COMBUSTION

Initiation of solid fuel (pulverized coal) combustion starts first with the stage of *thermal preparation* which includes the evaporation of residual moisture and *distillation of volatiles.* Volatiles are evolved in the temperature range of 400–600°C and this occurs within 0.2 to 0.5 seconds. A high yield of volatile produces enough heat to ignite coke particles. The final stage

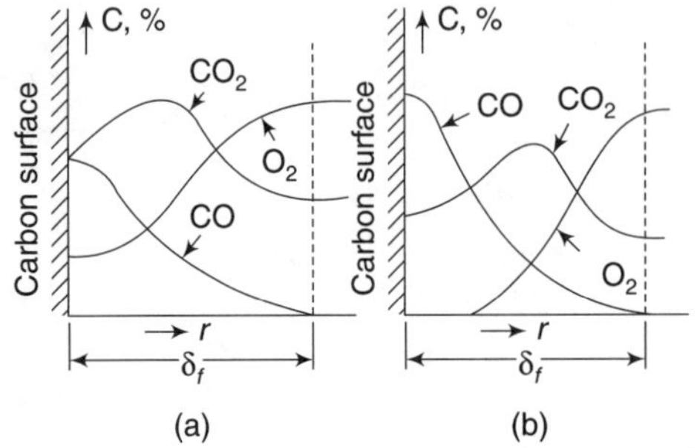

Figure 4.13 Effect of variation of concentration of gases near the burning of carbon surface (a) At low temperature (b) At high temperature

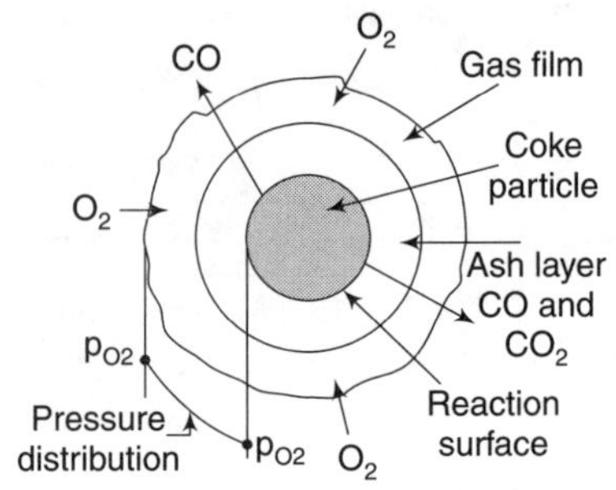

Figure 4.14 Burning of a coke particle

of combustion is the combustion of coke particles at a temperature above 800–1000°C. The burning of a coke particle takes up 1/2 to 2/3 of the total combustion time (1-2.5 seconds). The carbon-oxygen reactions at 1200°C and 1700°C are as follows

$$4C + 3O_2 = 2CO + 2CO_2 \text{ at } 1200°C \quad ...(4.53)$$

$$3C + 2O_2 = 2CO + CO_2 \text{ at } 1700°C \quad ...(4.54)$$

At 1700°C, the ratio of CO/CO_2 is around 2. The effect of variation of gases near the burning carbon surface at moderate and high temperature are shown in Fig. 4.13. The burning of a coal particle and its various zones around it are shown in Fig. 4.14.

4.13 DRAUGHT SYSTEM OF A BOILER

Draught (or draft) is defined as small pressure difference causing flow of air and gas (in and out) through the boiler. It is essential to supply a sufficient quantity of air to support combustion in boilers and to remove the products of combustion. *This is done by producing draught.* The functions of draught producing equipment are *(i) to create a differential pressure sufficient to allow the desired volume of air flow into the furnace at the required velocity, (ii) to overcome the resistance of passage in boiler, and (iii) to discharge gases, i.e., products of combustion at a sufficient height in order to avoid pollution to atmosphere.*

The amount of draught necessary for a certain boiler depends on the following:

(i) rate at which combustion takes place,
(ii) characteristics of fuel and its depth on the grate,
(iii) design of combustion chamber and the method of burning the fuel, and
(iv) resistance in flue gases circuits offered by baffles, tubes, superheaters, reheaters, evaporator, economizer, preheater, chimney, etc.

There are mainly two ways of producing draught: (a) **natural (chimney draught)** and (b) **artificial (mechanical) draft.** The artificial draught is produced by steam jet or by mechanical fans. Mechanical (fan) draught is of three types; (i) induced, (ii) balanced, and (iii) forced. In large output boiler furnace, forced and induced, draughts are in common use.

4.13.1 Natural Draught

The natural draught is produced by a chimney or stack and is cuased by the density difference between the atmospheric air and the hot gas in the stack. For a chimney of height H metres (Fig. 4.15), the draught or pressure difference, Δp produced is expressed as

$$\Delta p = gH\,(\rho_a - \bar{\rho}_g);\ \text{N/m}^2 \qquad ...(4.55)$$

where ρ_a = density of atmospheric (cold) air (kg/m^3), $\bar{\rho}_g$ = average density of hot gas in the chimney (kg/m^3), g = acceleration due to gravity = 9.81 m/s^2.

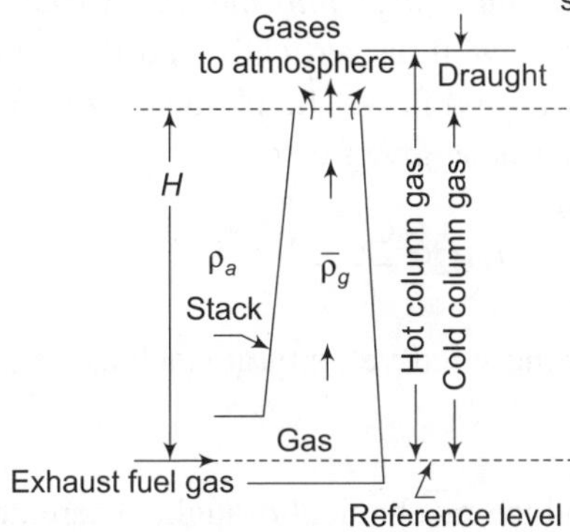

Figure 4.15 Natural (Chimney) draught

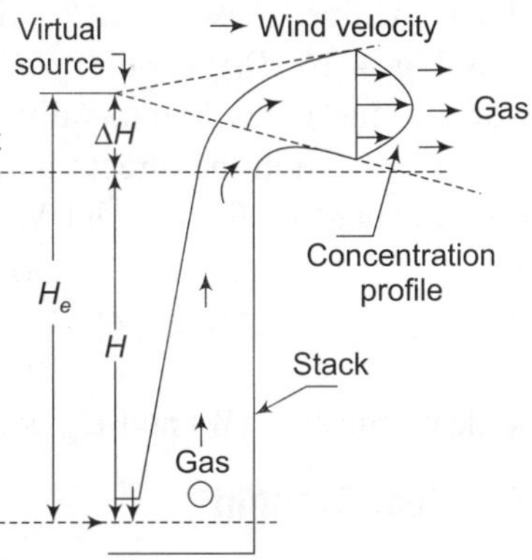

Figure 4.16 Dispersion of flue gases at stack exit

Using perfect gas laws, we have

$$\Delta p = gH\left[\frac{p_a}{R_a\,T_a} - \frac{p_g}{R_g\,T_g}\right] \qquad ...(4.56)$$

where $p_a = p_g = p_{atm}$ = 101.325 kN/m^2, $R_a = R_g$ = 0.287 kJ/kgK and $\Delta p = \rho_w \cdot gh$, N/m^2

where h = draught in water gauge.

T_a = ambient (cold) air temperature (K), T_g = average hot gas temperature in the chimney, in $K = \dfrac{T_0 + T_H}{2}$, T_0 = hot gas temperature at inlet to the chimney, K, T_H = hot gas temperature at exit from chimney, K

Thus,
$$\Delta p = \frac{gHp_a}{R_a}\left[\frac{1}{T_a} - \frac{1}{T_g}\right];\ \text{N/m}^2 \qquad ...(4.57)$$

From the above equation, it is evident that the natural draught produced is inversely proportional to the average hot gas temperature for a given head, H. The draught produced increases with the increase of T_g but the boiler efficiency goes down due to increase in stack loss. Therefore, an optimum value of T_g is selected. *The natural draught created by stack or chimney is not sufficient for modern boiler due to high fuel burning rate and large heat transfer surface areas, so mechanical draught is used.*

The natural draught produced, Δp given by above expression is the theoretical draught. Stack introduces pressure losses of their own caused by wall friction and the pressure equivalent to the kinetic energy of the gases leaving the stack. Thus, the actual draught produced is less than the theoretical value, Δp and is given by the following expression

$$\Delta p_{act} = \Delta p - \bar{\rho}_g \frac{C_s^2}{2}\left(1 + \frac{fH}{D}\right) \quad \text{N/m}^2 \qquad ...(4.58)$$

where f is the friction factor, H is the stack height (m), D is the stack inside diameter (m), C_s is the stack exit velocity (m/s), $\bar{\rho}_g$ is the average gas density in stack (kg/m^3).

The exit velocity of flue gases at stack exit results in a plume rise ΔH above the actual stack as shown is Fig. 4.16. *Dispersion of the flue gases into the atmosphere* is caused by the movement of the flue gases horizontally as well as vertically and their dilution by the atmosphere. The gases bend in the direction of wind flow. The plume height, ΔH is given by imperial equation prepared by Carron and Moses as given below

$$\Delta H = 2.62\,\frac{Q_c^{0.5}}{C_w} - 0.029\,\frac{C_s \cdot D}{C_w} \qquad ...(4.59)$$

where Q_c is the heat emission (W) and C_w is the wind velocity at stack exit (m/s).

4.13.2 Mechanical Draught

Centrifugal or axial flow fans are used to produce mechanical draught. There are three types of mechanical draught: *(i) induced draught produced by induced draught (1D) fan, (ii) forced draught produced by forced draught (FD) fan, and (iii) balance draught produced by the combination FD and ID fans.* When either one is used alone, it should overcome the total air and gas pressure losses within the steam generator.

4.13.2.1 Forced Draught

It is worth to note that forced draught fans installed at inlet to the air preheater and so they handle cold air. The power input is given by

$$\text{Power input} = \int \dot{m}v\,dp = \dot{m}_f\,(A/F) \times v \times \Delta p/\eta_{FD}\,;\ \text{kW} \qquad ...(4.60)$$

where $\dot{m}_f$ is the mass flow rate of fuel (kg/s), A/F is the air-fuel ratio, v is specific volume of inlet air (m^3/kg), Δp is the pressure developed by fan to overcome resistance in flow path (kN/m^2) and η_{FD} is the overall efficiency of FD fan.

Since FD fans handle cold air so they consume less power and have less maintenance problems. In general, in view of good reliability, two feed forced draught fans operate in parallel, each capable of undertaking 60% of full load air flow in the event of breakdown of one of them.

In many large steam generators and almost in all marine applications, FD fans alone are used and they maintain the entire system up to the stack under *positive gauge pressure*. As a result, the furnace has to be gas tight. The height of the stack for forced draught fan alone is shorter in height just to meet the pollution regulation.

4.13.2.2 Induced Draught

Induced draught fans are normally located at the foot of stack so they handle hot combustion gases. As a result, the power requirements are greater than that of FD fans. The maintenance problems due to handling of corrosive combustion gases and fly ash are increased. Induced draught fans are seldom used alone. They discharge gases essentially at atmospheric pressure and place the system under *negative gauge pressure* in upstream side. The power input is given by

$$\text{Power Input} = \int \dot{m}_g\, v_g\, \Delta p = \dot{m}_f\left[(1 + A/F)v_g \times \frac{\Delta p_{ID}}{\eta_{ID}}\right];\ \text{kW} \qquad ...(4.61)$$

where v_g is the specific volume of hot gas (m^3/kg), Δp_{ID} is the pressure head developed by *ID* fan (kN/m^2) and η_{ID} is the overall efficiency of *ID* fan.

4.13.2.3 Balance Draught

As the name implies, balance draught system uses both FD and ID fans. The FD fans located at the inlet of preheater push atmospheric air through the air preheater, dampers, various air ducts and burners into the furnace and ID fans are located at the foot of stack such that the flue gases through the heat transfer surfaces in the superheaters, reheaters, economiser, gas-side air preheater and dust collectors and discharge into the stack as shown in Fig. 4.17. The natural driving pressure is generated by the stack due to its height. When both FD and ID fans are used, the furnace is said to operate at approximately atmospheric pressure. In real situation, it is maintained at slightly negative gauge pressure ensuring leakage if any would be invariably inward. Due to the advantages of balance draught system, designers adopt this system for modern boilers.

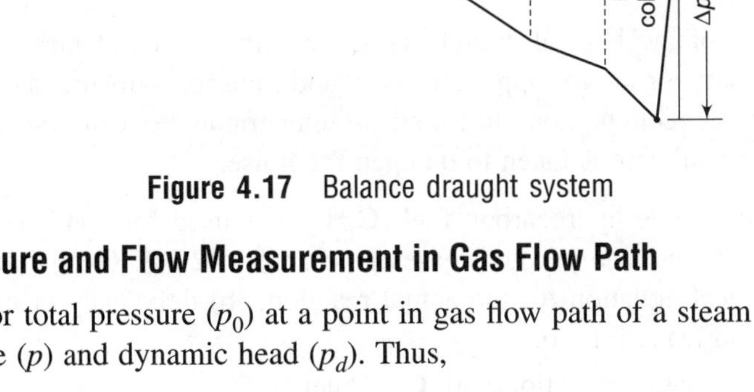

Figure 4.17 Balance draught system

4.13.3 Pressure and Flow Measurement in Gas Flow Path

The stagnation or total pressure (p_0) at a point in gas flow path of a steam generator is sum of static pressure (p) and dynamic head (p_d). Thus,

$$p_0 = p + p_d = p + \rho_g\,\overline{C}^2/2 \qquad ...(4.62)$$

The static pressure is measured by tapping wall connected to a water tube manometer while dynamic pressure is measured by a pilot tube connected to water tube manometer. Static and dynamic pressure may be simultaneously measured by a three hole probe.

Since the pressure in boiler gas flow path is slightly more than atmospheric or less than atmospheric, so it is expressed in mm of water.

One metre head of water = $p = \rho gh = 10^3 \times 9.81 \times 1 = 9810$ N/m^2

The dynamic pressure head in terms of mm of water, is expressed as

$$p_d = \frac{\rho_g C^2}{9810 \times 2} \times 10^3 = \frac{\rho_g C^2}{19.62} \text{ mm of } H_2O$$

If the area of the duct is A, the volume flow rate of gas in the duct is given by

$$V = A[19.62 p_d/\rho_g]^{1/2} = 4.43\, A\sqrt{p_d/\rho_g} \text{ m}^3\text{/s} \qquad ...(4.63)$$

where p_d is in mm of water and ρ_g is in m^3/kg.

4.13.4 ID, FD and other Fans

There are two types: of fans: centrifugal and axial. Centrifugal fans are cheaper and rugged than axial flow fans. The blades (or vanes) of centrifugal fans may be *backward, radial and forward swept.* Generally, *FD* fans are centrifugal type with backward swept blades. Since ID fans handle dust-ladden flue gases, so they are operated at low speeds and therefore, centrifugal fan with forward swept blades or radial blades may be used.

For controlling the output of fans, two methods are used. They are damper control and variable-speed control. Dampers are generally placed at the output of the fan. Variable-speed control has less power consumption and it is the most efficient method of fan control but the control equipment is costly. The effect of speed on the fan performance is as follows

Volume flow $\propto N$ where N is the speed in rpm.

Pressure head $\propto N^2$ and power input $\propto N^3$

The types of drives used for fan are: (i) Variable-speed steam turbine, (ii) hydraulic coupling, (iii) variable-speed dc motor, (iv) multi-speed ac motor, and (v) electronically adjustable motor drive.

Over and above of mechanical draught fans, two other types of fans are used in power plants. One is primary air fan to supply air to dry and transport pulverized coal to the furnace and the second is gas recirculation fan for steam temperature control system. Fans are noisy in operation, so special care is taken to dampen the noise.

Example 4.1 A unknown hydrocarbon fuel (C_xH_y) was used for combustion with air. The Orsat analysis of combustion products/gases give : CO_2 12.2%, O_2, 3.7%, and CO, 0.8%. Find (a) the chemical equation for the actual reaction, (b) deficiency or excess air used in the combustion, and (c) A/F ratio.

Solution The combustion equation with C_xH_y fuel is

$$C_xH_y + aO_2 + 3.76a\, N_2 \rightarrow 12.2\, CO_2 + 3.7 O_2 + 0.8\, CO + 83.3\, N_2 + bH_2O$$

Equating moles for N_2, O_2, C and H_2, we have

Nitrogen: $3.76\,a = 83.3$ or $a = 22.15$

Oxygen: $2 \times 22.15 = 2 \times 12.2 + 2 \times 3.7 + 0.8 + b$ or $b = 11.7$

Carbon: $x = 12.2 + 0.8 = 13.0$

Hydrogen: $y = 2b = 2 \times 11.7 = 23.4$

(a) The combustion equation becomes

$$C_{13}H_{23.4} + 22.15\ O_2 + 83.28\ N_2$$
$$\rightarrow 12.2\ CO_2 + 3.7\ O_2 + 0.8\ CO + 83.3\ N_2 + 11.7\ H_2O \qquad \textit{Ans.}$$

The composition of fuel is $C_{13}H_{23.4}$ ***Ans.***

The stoichiometric equation is given by

$$C_{13}H_{23.4} + 18.75\ O_2 + 3.76(18.75)\ N_2 \rightarrow 13\ CO_2 + 11.7\ H_2O + 18.75(3.76)\ N_2$$

(b) The percentage of excess air $= \dfrac{22.15 - 18.75}{18.75} \times 100 = 18.13\%$ ***Ans.***

The actual air supplied per mole of $C_{13}H_{23.4}$ is expressed as

$$= \frac{22.15 \times 32}{0.232} = 3055.17 \text{ kg}$$

(c) A/F ratio $= \dfrac{3055.17}{(12 \times 13 + 1 \times 23.4)} = 17.02$ ***Ans.***

Example 4.2 The following data relate to a steam generator—Coal analysis : carbon 62% hydrogen 3.0%, oxygen 3.0%, nitrogen 2.0%, sulphur 1.0%, moisture 4.0%, ash 2.5%, HHV of coal = 24.3 MJ/kg

Dry flue gas analysis: CO_2 12.5%; CO 1.5%: O_2 8.5%; N_2 77%

Unburnt carbon : nil, exhaust gas temperature = 180°C, unaccounted energy loss = 2.5% of HHV, steam generation rate = 200 t/h, steam condition at boiler outlet = 120 bar, 500°C, feed water inlet temperature = 170°C.

Heat of reaction for CO and CO_2 = 33.083 MJ/kg carbon and 9.5 MJ/kg carbon, c_p for dry flue gases = 1.05 kJ/kgK; ambient temperature = 30°C

Calculate: (a) the amount of dry flue gas produced per kg fuel; (b) the dry exhaust loss and the incomplete combustion loss per kg fuel, (c) the boiler efficiency; (d) the fuel burning rate; (e) the percentage of excess air used, and (f) prepare the heat balance sheet of steam generator.

Solution The mass of dry flue gas produced per kg fuel is expressed as

(a) $$m_{dfg} = \frac{C_{ab}\,(44CO_2 + 28CO + 32O_2 + 28N_2)}{12(CO_2 + CO)}$$

$$= \frac{0.62\ (44 \times 0.125 + 28 \times 0.015 + 32 \times 0.085 + 28 \times 0.77)}{12\ (0.125 + 0.015)}$$

$$= \frac{18.724}{1.68} = 11.14 \text{ kg}$$ ***Ans.***

(b) Energy loss due to dry exhaust gas is given by

$$= m_{dfg}\, c_{pg}\, (t_g - t_a) = 11.14 \times 1.05\ (180 - 30)$$

$$= 1754.55 \text{ kJ/kg} = 1.7545 \text{ MJ/kg}$$ ***Ans.***

Energy loss due to incomplete combustion in expressed as

$$= m_{dfg} \frac{28\ \text{CO}}{44\text{CO}_2 + 28\text{CO} + 32\text{O}_2 + 28\text{N}_2} (\text{H}_{\text{R.CO}} - \text{H}_{\text{R.CO}_2}) \frac{12}{28}$$

$$= 11.14 \times \frac{28 \times 0.015}{44 \times 0.125 + 28 \times 0.015 + 32 \times 0.085 + 28 \times 0.77} [33083 - 9500] \frac{12}{28}$$

$$= \frac{11.14 \times 28 \times 0.015 \times 10107}{30.2} = 1565.84 \text{ kJ/kg fuel} = 1.5658 \text{ MJ/kg fuel}$$ ***Ans.***

(c) Unaccounted energy loss = 0.025 × 24.3 = 0.6075 MJ/kg fuel

Total energy loss = 1.7545 + 1.5804 + 0.6075 = 3.9424 MJ/kg fuel

$$\text{Boiler efficiency} = \frac{\text{HHV} - \Sigma\ \text{Losses}}{\text{HHV}} \times 100 = \frac{24.3 - 3.9424}{24.3} \times 100 = 83.77\%$$ ***Ans.***

(d) The energy balance gives: $m_f \times \text{HHV} \times \eta_{\text{boiler}} = \dot{m}_s\,(h_1 - h_f)$

or
$$\dot{m}_f = \frac{m_s(h_1 - h_f)}{\text{HHV} \times \eta_{\text{boiler}}} = \frac{200 \times 10^3\ (3347 - 719.2)}{24.3 \times 10^3 \times 0.8377}$$

$$= 25818.29 \text{ kg/h} = 7.17 \text{ kg/s}$$ ***Ans.***

(e) Theoretical air required per kg of fuel

$$= m_{at} = 11.5\text{C} + 34.5 \left(\text{H} - \frac{\text{O}}{8}\right) + 4.3S$$

$$= 11.5 \times 0.62 + 34.5 \left(0.03 - \frac{0.03}{8}\right) + 4.3 \times 0.01 = 8.078 \text{ kg/kg of fuel}$$

Actual air supplied to the furnace when nitrogen is present in the fuel

$$m_{aa} = \frac{3.04\ \text{N}_2\ C_{ab}}{\text{CO}_2 + \text{CO}} - \frac{1}{0.768} \text{N}_2$$

$$= \frac{3.04 \times 0.77 \times 0.62}{(0.125 + 0.015)} - \frac{1}{0.768} \times 0.02 = 10.60 \text{ kg/kg of fuel}$$

Percentage of excess air used $= \frac{m_{aa} - m_{at}}{m_{at}} = \frac{10.60 - 8.078}{8.078} \times 100 = 31.22\%$ ***Ans.***

(f) The heat balance sheet is given by

	Item	*Heat (MJ/kg fuel)*	*Heat (%)*
Heat input	Heat supplied	24.3	100
Expenditure	(i) Heat consumed to generate steam	20.3576	83.77
	(ii) Energy loss in dry exhaust gas	1.7545	7.22
	(iii) Energy loss due to incomplete combustion	1.5658	6.443
	(iv) Unaccounted energy loss	0.6237	2.566
Total		24.3	100

Example 4.3 During a test of a boiler trial, the following data were obtained.

Coal analysis: C = 85.2%, H = 4.8%, ash = 8.2%, gross calorific value of dry coal = 35295 kJ/kg, moisture content = 1.8%, coal consumption = 18 t/h, boiler room temperature = 30°C, feed water temperature = 150°C, steam pressure = 140 bars, steam temperature = 500°C, steam raised = 200 t/h.

The analysis of dry flue gases by volume gave: CO_2 = 9.4%, O_2 = 11.1%, N_2 = 79.5%, the temperature of gas in stack = 180°C, mean specific heat of dry flue gas = 1.005 kJ/kgK and temperature of atmospheric air 30°C.

Prepare a complete heat balance sheet per kg of dry coal.

Solution Steam generated per kg of fuel

$$= \frac{200}{18} = 11.111 \text{ kg}$$

Heat consumed in steam generation $= m_s (h_1 - h_f)$

$$= 11.111\,(3321 - 632.2) = 29872.5 \text{ kJ/kg fuel}$$

Actual air supplied

$$= m_{aa} = \frac{3.04\, N_2\, C_{ab}}{CO_2 + CO} = \frac{3.04 \times 0.795 \times 0.852}{0.094 + 0} = 21.905 \text{ kg/kg fuel}$$

Theoretical (minimum) air required $= m_{at} = 11.5C + 34.5\left(H - \frac{O}{8}\right) + 4.3S$

$$= 11.5 \times 0.852 + 34.5\,(0.048 - 0) + 4.3 \times 0 = 11.454 \text{ kg/kg fuel}$$

Excess air = 21.905 – 11.454 = 10.451 kg/kg fuel

Total mass of flue gases = 21.905 + 1 = 22.905 kg/kg coal

Mass of water vapour in product = 0.018 + 0.048 × 9 = 0.45 kg/kg coal

Mass of dry products excluding excess air is

$$m_{dfg} = 22.905 - 0.45 - 10.451 = 12.004 \text{ kg/kg coal}$$

Heat loss in dry flue gas products $= m_{fgd}\, c_{pg}\, (t_g - t_a)$

$$= 12.004 \times 1.005\ (180 - 30) = 1809.603 \text{ kJ/kg fuel}$$

Heat loss due to moisture in fuel $= M\,[c_{pw}\,(100 - t_f) + h_g + c_{p\text{ steam}}\,(t_g - 100)]$

$$= 0.018\ [4.18\ (100 - 30) + 2256.8 + 2.09\ (180 - 100)] = 48.89 \text{ kJ/kg fuel}$$

Heat loss due to hydrogen in fuel which converts into steam

$$9\text{H}\ [c_{pw}\,(100 - t_f) + h_{fg} + c_{p\text{ steam}}\,(t_g - 100)]$$

$$= 9 \times 0.048\ [4.18\ (100 - 30) + 2556.8 + 2.09\ (180 - 100)] = 1299.97 \text{ kg/kg fuel}$$

The heat balance sheet is given below.

Heat balance sheet per kg of fuel

	Item	Heat (kJ/kg fuel)	Heat (%)
Heat input	Heat supplied	35295	100
Expenditure	(i) Heat consumed in steam generation	29872.5	84.63
	(ii) Heat loss in dry flue gas products	1809.603	5.127
	(iii) Heat loss in excess air	1409	0.1385
	(iv) Heat loss due to moisture in fuel	48.89	0.138
	(v) Heat loss due to hydrogen in fuel and converted into steam	1299.97	3.683
	Unaccounted heat loss (by difference)	855.037	2.42
Total		35295	100

Example 4.4 An oil fired steam generator uses fuel whose analysis by mass gives: carbon 85%, hydrogen 8.5%, sulphur 3%, oxygen 1.5%, and remainder uncombustible. The analysis if dry flue gases by volumes gives : combined (CO_2 + SO_2) 16.5%, O_2 2%, and there is no traces of CO or SO_3. Evaluate per kg of fuel (a) mass of air supplied, (b) percentage excess air supplied in kg mass of dry flue gas forced, and (c) mass of water vapour formed due to combustion.

Solution Since the fuel analysis is given on mass basis, so let us consider 100 kg of fuel. Let x moles of oxygen are supplied for combustion. The combustion equation on mole basis is given as

$$\frac{85}{12}\text{C} + \frac{8.5}{2}\text{H}_2 + \frac{3}{32}\text{S} + \frac{1.5}{32}\text{O}_2 + x\text{O}_2 + 3.76\,x\,\text{N}_2 \rightarrow$$

$$a\,\text{CO}_2 + b\,\text{SO}_2 + c\,\text{O}_2 + d\,\text{N}_2 + e\,\text{H}_2\text{O}$$

Equating coefficients from the equation

Carbon : $\frac{85}{12} = a$ or $a = 7.08$ Hydrogen : $\frac{8.5}{2} = e$ or $e = 4.25$

Oxygen : $\frac{1.5}{32} + x = a + b + c + e/2$ Sulphur : $\frac{3}{12} = b$ or $b = 0.25$

Nitrogen : $3.76\, x = d$

Based on volumetric analysis of dry flue gases

$$\frac{a + b}{a + b + c + d} = 0.165 \quad \text{(i)} \qquad \text{and} \quad \frac{c}{a + b + c + d} = 0.02 \quad \text{(ii)}$$

Dividing Eq. (i) by Eq. (ii), we have : $\frac{a + b}{c} = \frac{0.165}{0.02} = 8.25$

or $c = \frac{a + b}{8.25} = \frac{7.08 + 0.25}{8.25} = 0.888$

From oxygen balance, we have

$$x = 7.08 + 0.25 + 0.888 + 4.25/2 - 1.5/32 = 10.296 \text{ moles}$$

Thus, the mass of oxygen supplied of $10.296 \times 32 = 329.47$ kg

The mass of actual air supplied for 100 kg fuel $= \frac{329.47}{0.232} = 1420.12$ kg

(a) Therefore, the mass of actual air supplied per kg of fuel $= \frac{1420.12}{100} = 14.2012$ ***Ans.***

(b) Theoretical air required is $= m_{at} = 11.5\,\text{C} + 34.5\left(\text{H} - \frac{\text{O}}{8}\right) + 4.3\,\text{S}$

$$= 11.5 \times 0.85 + 34.5\left(0.09 - \frac{0.015}{8}\right) + 4.3 \times 0.03 = 12.944 \text{ kg}$$

Percentage of excess air $= \frac{m_{aa} - m_{at}}{m_{at}} = \frac{14.2012 - 12.944}{12.944} \times 100 = 9.7\%$ ***Ans.***

(c) The mass of dry flue gases formed from 100 kg of fuel

$$= a\, CO_2 + b\, SO_2 + c\, O_2 + d\, N_2$$

$$= 7.08 \times 44 + 0.25 \times 64 + 0.888 \times 32 + 3.76 \times 10.296 \times 28 = 1439.89 \text{ kg}$$

The mass of dry flue gas per kg of fuel $= e\, H_2O = 4.25 \times 18/100 = 0.765$ kg ***Ans.***

Example 4.5 A boiler uses a propane gas. The analysis of dry products of combustion gives CO_2 11.6%, O_2 2.8%, CO 0.6%. Calculate the percentage of excess air used in the combustion.

Solution Let x moles of propane react with a mole of oxygen supplied and thus the combustion equation becomes

$$x\, C_3H_8 + a\, O_2 + 3.76a\, N_2 \rightarrow 11.6\, CO_2 + 0.6\, CO + 2.8\, O_2 + 85\, N_2 + b\, H_2O$$

Equation moles, we have:

Nitrogen : $3.76a = 85$ or $a = 22.60$

Carbon : $3x = 11.6 + 0.6$ or $x = 4.066$

Oxygen : $22.60 = 11.6 + 0.3 + 2.8 + 2.8 + \frac{b}{2}$ or $b = 15.8$

Hydrogen : $8 \times 4.066 = 2 \times 15.8$ (Approx.)

Thus, the combustion equation becomes

$$4.006\ C_3H_8 + 22.6\ O_2 + 3.76 \times 22.6\ N_2 \rightarrow 11.6\ CO_2 + 0.6\ CO + 2.8\ O_2 + 85\ N_2 + 15.8\ H_2O$$

For one mole of C_3H_8, the above equation becomes

$$C_3H_8 + 5.558\ O_2 + 20.899\ N_2 \rightarrow 2.852\ CO_2 + 0.147\ CO + 0.688\ O_2 + 2.09 + 3.885$$

The stoichiometric propane combustion equation is given by

$$C_3H_8 + 50_2 + 5\ (3.76)\ N_2 \rightarrow 3CO_2 + 4H_2O + 18.8\ N_2$$

The excess air supplied $= \frac{(5.558 - 5)}{5} \times 100 = 11.16\%$ ***Ans.***

Example 4.6 A steam generator generates 100 t/h steam at 100 bars and 500°C. The calorific value of fuel used by steam generator is 41 MJ/kg with an overall efficiency of 80%. For efficient combustion 16 kg of air per kg of fuel is required for which draught of 22 mm of water gauge is required at the back of stack. The flue gases leave the steam generator at 310°C. The average temperature of the gases in the stack may be taken as 290°C and the atmospheric temperature is 30°C. Calculate the height of the stack and the diameter at its base. The velocity of the gases at the stack exit may be assumed negligible.

Solution The draught produced is expressed by

$$\Delta p = gH\,(\rho_a - \bar{\rho}_g) = gH\,\frac{p_a}{R_a}\left[\frac{1}{T_a} - \frac{1}{T_g}\right]$$

where Δp = draught in N/m^2 $= \rho_w \cdot g \cdot h$

or
$$22 \times 10^{-3}\,(m)\cdot g\left(\frac{m}{s^2}\right)10^3\left(\frac{\text{kg}}{\text{m}^3}\right) = gH\,\frac{1.013 \times 10^2}{0.287}\left[\frac{1}{303} - \frac{1}{563}\right]$$

or
$$H = \frac{22 \times 10^{-3} \times 10^3 \times 0.287 \times 303 \times 563}{1.013 \times 10^2\,(563 - 303)} = 40.89\ \text{m}$$ ***Ans.***

The overall efficiency of steam generator is given by

$$\eta_{\text{boiler}} = \frac{\dot{m}_s\,(h_1 - h_f)}{\dot{m}_f \times CV} = \frac{100 \times 10^3\,(3372 - 632.2)}{\dot{m}_f \times 41 \times 10^3} = 0.8$$

or
$$\dot{m}_f = 8353\ \text{kg/h} = 2.32\ \text{kg/s}$$

∴
$$\dot{m}_a = 16\dot{m}_f = 16 \times 2.32 = 37.12\ \text{kg/s}$$

The mass flow rate of flue gases $= \dot{m}_{fg} = (16 + 1) \times 2.32 = 39.44$ kg/s

From continuity equation applied at base of stack

$$\dot{m}_{fg} = \rho_{fg}\,A\,C_g = \frac{p_{fg}}{RT_g}\,\frac{\pi}{4}\,D^2\,(2gH)^{1/2}$$

or $$39.44 = \frac{1.013 \times 10^2}{0.287 \times 583} \cdot \frac{\pi}{4} D^2 (2 \times 9.81 \times 40.89)^{1/2}$$

or $$D = 1.71 \text{ m}$$ ***Ans.***

Example 4.7 A FD fan is used by a boiler to supply the air. Consumption of coal is 20 *t*/h. The coal analysis gives : C 77.0. H 3.0, O 2.5, S 0.8. The excess air used is 35%. The plenum chamber pressure is 200 mm water gauge. The mechanical efficiency of the fan is assumed to be 70%. The room temperature is 30°C. Calculate: (a) the amount of air to be handled by the fan, (b) the motor capacity of the fan. If an *ID* fan is used to produce draught to the same magnitude having gas temperature 179°C and fan efficiency 60%, then calculate (c) the fan motor capacity of *ID* fan.

Solution (a) The mass of theoretical air required is given by

$$m_{at} = 11.5 \text{ C} + 34.5\left(\text{H} - \frac{\text{O}}{8}\right) + 4.3 \text{ S}$$

$$= 11.5 \times 0.77 + 34.5\left(0.03 - \frac{0.025}{8}\right) + 4.3 \times 0.008$$

$$= 9.815 \text{ kg of air/kg of fuel}$$

Since the excess air 35% thus the actual air supplied given by

$$m_{aa} = 9.816 \times 1.35 = 13.25 \text{ kg of air/kg of fuel}$$ ***Ans.***

(b) The fan (FD) motor capacity is given by

$$P_{\text{motor}} = \frac{m_{aa} \cdot \dot{m}_f \, v_{\text{air}} \, \Delta p}{\eta_{\text{fan}}} \quad \text{where } \Delta p = \rho_w \, g.h., \text{ N/m}^2$$

But $$v_{\text{air}} = \frac{RT}{p} = \frac{0.287 \times 303}{1.0135 \times 10^2} = 0.858 \text{ m}^3\text{/kg}$$

$\therefore$ $$(P_{\text{motor}})_{FD} = \frac{13.25 \times 20 \times 10^3 \times 0.858 \times (200 \times 10^{-3} \times 9.81 \times 10^3)}{0.7 \times 3600}$$

$$= 177 \times 10^3 \text{ W} = 177 \text{ kW}$$ ***Ans.***

(c) The FD fan handles flue gas. Thus,

$$\dot{m}_{fg} = \dot{m}_f + \dot{m}_a = 20 \times 10^3 + 13.25 \times 20 \times 10^3 = 285 \times 10^3 \text{ kg/h,}$$

$$= 79.16 \text{ kg/s}$$

The specific volume of fuel gas at the inlet of FD fan is

$$v_{fg} = v_a \frac{t_g + 273}{t_a + 273} = 0.858 \frac{170 + 273}{30 + 273} = 1.254 \text{ m}^3\text{/kg}$$

Thus, FD fan motor capacity = $(P_{\text{motor}})_{FD} = \dfrac{\dot{m}_{fg} \, v_{fg} \, \Delta p}{\eta_{\text{fan}}} = 324.62 \times 10^3 \text{ W} = 324.62 \text{ kW}$

Ans.

Comment: The power input to FD fan is more than FD for producing the same draught.

Example 4.8 A steam generator uses a stack of 250 m high and 5 m diameter at its exit. The mass of the gases leaving the stack is 1200 kg/s at 120°C into the atmospheric air at 10°C. The prevailing wing velocity is 50 km/h and the atmospheric condition is in its neutral stability. Determine the height of plum (ΔH) based on correlation of Carson and Moses.

Solution The height of plum (ΔH) may be calculated by the correlation of Carson and Moses as given below:

$$\Delta H = 2.62 \frac{Q_e^{1/2}}{C_W} - 0.029 \frac{C_S D}{C_W}$$

where C_s = stack gas exit velocity (m/s), D = stack diameter at exit (m), C_W = wind velocity (m/s), Q_c = heat emission from plum (W).

Further, $$Q_c = m_g\, c_p\, (T_s - T_a) = 1200 \times 1.005\ (120 - 10) = 132660 \text{ kW}$$

$$C_w = 50 \times 10^3/3600 = 13.88 \text{ m/s}$$

$$C_s = \frac{m_g}{\rho_g \cdot A} = \frac{1200}{1.01385 \times (\pi/4) \times 5^2} = 60.28 \text{ m/s}$$

$$\therefore \quad \Delta H = 2.62 \times \frac{132.66^{1/2}}{13.88} - 0.029 \frac{60.28 \times 5}{13.88} = 1.544 \text{ m.}$$ ***Ans.***

PROBLEMS

Things to Think About

4.1 Is it possible to have a fuel with less carbon content, so that the emission could be slightly cleaner?

4.2 Is it wise preparation to have oil from coal?

4.3 Which is the best suited draught system in power plant and why?

4.4 Which is the best type of boiler furnace suited for modern boiler and why?

Short and Long Theoretical Questions

4.1 What are types of fuels for steam generators?

4.2 How was coal formed? What are the various types of coal?

4.3 How was natural gas and petroleum formed?

4.4 What are synthetic fuels?

4.5 How is coal gasification achieved ?

4.6 What do you mean by coal liquefaction?

4.7 How is lower heating value of fuel different from the higher heating value?

4.8 What is the main constituents of natural gas?

4.9 What are the major constituents of coal?

4.10 What is compressed natural gas (CNG)? Mention its applications.

4.11 What is LPG? Mention its constituents and applications.

4.12 What is town gas? Mention its applications.

4.13 What are the various forms of biomass? Mention various bioconversion processes.

4.14 What do you mean by industrial wastes and byproducts?

4.15 Derive the expression for stiochiometric air-fuel ratio.

4.16 Discuss the energy balance in a steam generator.

4.17 What do you understand by draught of a boiler?

4.18 Derive the expression for draught.

4.19 What are the various types of draught?

4.20 What are the various locations of ID and FD fans for producing draught.

4.21 What do you mean by balance draught?

4.22 What is the approximate height of stack in natural draught?

4.23 Mention the type of fans used in mechanical draught system.

4.24 What do you mean by heat of reactants and products?

4.25 What is the adiabatic temperature?

4.26 What do you mean by Gibbs free energy of formation?

4.27 Mention the types of boiler furnace.

Numerical Questions

4.1 A boiler consumes coal having the following composition : carbon 85.5%, hydrogen 5.0%. oxygen 4.0%, and incombustibles 5.5%. The ashes discharged from the furnace contained 0.015 kg of unburnt carbon per kg of coal supplied to the furnace. The analysis of dry flue gas gives : CO_2 14.5%, and CO 1.3% by volume. Determine the mass of air supplied and the complete volumetric composition of the flue gas.

[***Ans.*** 13.05 kg/kg-coal, by volume: CO_2 13.77%, CO 1.23%, O_2 3.0%, N_2 76.66%, H_2O 5.34%]

4.2 A boiler uses coal whose analysis by mass is as follows:

C 73%, ash 12%, moisture 15%. After burning the residue contain 18% C by mass. The analysis of dry flue gases by volume gives : CO_2 11.8%, CO 1.3%, O_2 5.5%. Determine the percentage of carbon in the coal which was really used in combustion and actual air used.

[***Ans.*** 97.05%, 13.46]

4.3 A steam generator employing a pulverized coal firing system relate to the following data : Evaporation rate = 200 *t*/h, boiler pressure = 160 bars, superheater outlet temperature = 570°C, economizer inlet temperature of feed water = 180°C, overall boiler efficiency = 90%, calorific value of coal = 23.5 MJ/k, heat release rate in furnace = 190 kW/m^3. The ratio of width : length : height of furnace = 1 : 2 : 3.

Calculate: (a) the dimension of the furnace, (b) the fuel burning rate, and (c) the percentage of each heating surface of boiler.

4.4 The following data relate to a boiler trial : coal analysis by mass = C 80%, H.4%, O 5%, N 1%, M (moisture) 4%, ash (A) 6%, HHV of coal = 26 MJ/kg, Refuse analysis : C 10%, A 90%.

Dry flue gas analysis : CO_2 13%, CO 1%, O_2 6%, N_2 80%, Flue gas temperature = 180°C. Air data : dry bulb temperature 20°C, wet bulb temperature 15°C. Prepare the heat energy balance sheet.

4.5 The following data relate to a boiler trial. Coal analysis (by mass) : C 76%, H 5%, O 3%, N 2%, S 4%, M 1%, A 9%. Refuse analysis : A 92%, C 8%. Air (ambient) has 1.8% moisture. Determine per kg of coal (a) actual air supplied, (b) theoretical air required, and (c) the percentage of excess air.

4.6 A fuel oil to be used in a boiler plant composed only of carbon, hydrogen, and sulphur. The volumetric flow of gas analysis on a dry basis is CO_2 11.7%, CO 0.44%, O 4002%, SO_2 0.176%, and N_2 83.682%, Calculate: (a) the fuel mass composition, (b) the A/F ratio by mass, (c) the excess air used, (d) the dew point of the flue gases if the pressure is 1.5 bars.

4.7 The following data relate to a boiler trial : Coal analysis : C 84%, H 5%, M 2.5% and A 8.5%. The coal consumption is 16 t/h, boiler room temperature = 30°C. Feed water temperature = 160°C, steam pressure = 150 bars, steam temperature 600°C, steam raised = 200 t/h, dry flue gas analysis : CO_2 10%, O_2 18%, N_2 78%. The temperature of the gas in the stack = 180°C. Mean specific heat of dry flue gases 1.005 kJ/kgK and temperature of atmospheric air = 30°C. Prepare a complete heat balance sheet per kg of fuel.

Software Based Question

4.1 Develop a software for the design of all types of boiler furnace.

Design and Open Ended Question

4.1 Design a modern boiler furnaces for 1000 MW coal based steam power plant

Multiple-Choice Questions

4.1 For steam power plant, the most suitable fuel form for coal based plant is

(a) big size coal lump (b) small size coal lump
(c) pulverized form of coal (d) all of the above

4.2 In modern steam power plant, the fan used for producing draught is

(a) only ID (b) only FD (c) both (d) None of them

4.3 In steam power plant, the chimney height should be

(a) very low (b) medium
(c) very high (d) high enough to suit regulation

Answers to Multiple-Choice Questions

4.1 (c) 4.2 (c) 4.3 (d)

CHAPTER

5

Combustion Equipment and Fuel Handling Systems for Steam Power Plant

OVERVIEW

Combustion equipment and fuel handling systems for steam power plant are one of the **main systems and their design and shape** depends upon the type of fuel used. As mentioned in chapter 4, that there are two *types of boiler furnace, namely, **grate fired furnace*** and ***chamber*** or ***flame furnace***. The former is suitable for *solid fuels* and the latter is for *pulverized coal.* This chapter deals with combustion equipment and fuel handling systems for steam power plant using different type of fuels.

5.1 COMBUSTION EQUIPMENT FOR COAL

There are two ways to feed coal into the furnace–(i) in ***lump pieces*** and (ii) in ***powder form.*** Based on this, the following furnaces may be used for the combustion of coal.

(i) Fuel bed furnaces (coarse particles): Grate fired

(ii) Pulverized coal furnaces (fine particles) : Flame or chamber type

(iii) Cyclone furnaces (crushed particles)

(iv) Fluidized bed furnaces (crushed small particles)

Construction and working principle of each system is discussed separately.

5.2 FUEL BED COMBUSTION FURNACE

In the case of fuel bed combustion furnace utilising coarse particles of coal, a ***grate*** is used at the furnace bottom to hold a bed of fuel as shown in Fig. 4.14 as mentioned in chapter 4. In general, there *are two ways to feed the coal on to the grate* as given below:

(i) Overfeeding and (ii) underfeeding.

5.2.1 Overfeed Fuel Bed

Figure 5.1 shows an overfeed fuel bed section. It receives fresh coal on its top surface and is characterized by the following fives zones from top to bottom.

(i) a layer of fresh or green coal—*fresh coal zone*

(ii) a layer of coal losing moisture—*drying zone.*

(iii) a coking layer of coal losing its volatile content—*distillation zone.*

(iv) a layer of incandescent coke where the fixed carbon is consumed, combustion zone. The ignition plane lies between green coal and incandescent coke.

(v) a layer of ash progressively getting closer—*ash zone.*

The heat is released due to combustion in the combustion zone and as a result the heat transfer occurs upward and downward by conduction, while the air flow from below will tend to carry away the heat upward by convection. The primary air supplied from the bottom gets warmed up as it flows through the ash layer and when it passes through the incandescent coke layer at about 1200°C, the exothermic reaction ($C + O_2 = CO_2$) taking place initially provides sufficient heat release for continuing the combustion process till all the oxygen is consumed. In the event of thick incandescent layer, the CO_2 may partly or fully reduced to CO ($CO_2 + C = 2CO$). Further, water gas reaction ($H_2O + C = H_2O + CO$) may also take place with the moisture from air. These reactions the are of endothermic nature resulting inconsiderable temperature drop of bed and gas stream.

As the hot gas stream passes through the distillation zone where the volatile matter is added and then through the drying zone where the moisture is picked up and finally comes out from the fuel bed containing N_2, CO_2, CO, H_2, volatile matter and water vapour. Some of the contents such as CO, volatile matter and hydrogen are not desirable, and if the combustion of these components is to be completed before the gases leave the furnace, the following requirements have to be met:

(i) A *hot ignition surface* in the range of 1000°C to 1300°C is to be provided by using *fire back-lined arch* which stores up the heat and remains at a high temperature.

(ii) Sufficient fresh air or secondary air or overfire air is to be supplied.

(iii) *Turbulence* is to be created in the secondary air by feeding at right angle about the fresh coal zone to the up flowing gas stream from the fuel bed for thorough mixing of air and gases (Fig. 5.1). It is not desirable to feed secondary air along with primary air as it will produce more carbon monoxide.

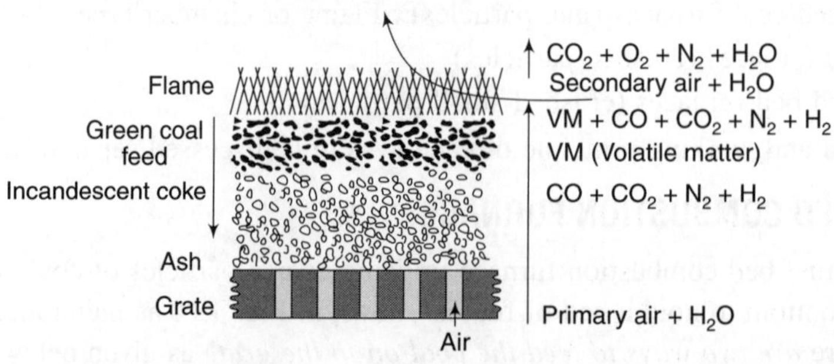

Figure 5.1 Overfeed fuel bed section

It is worth to note that the carbon contents of coal combined with hydrogen in the volatile matter leaves the fuel bed as part of the hydrocarbon gases, the molecules of which break down or disintegrates at high temperature of the furnace releasing some free carbon in the form of finally divided particles suspended in the gas stream. For burning these suspended carbon particles, it is essential that coal feeding apparatus must be designed properly and if the well designed furnace is not operated properly, it may appear at the chimney top as black smoke indicating energy wastage and inefficient combustion.

5.2.2 Underfeed Fuel Bed

Figure 5.2 shows the underfeed fuel bed showing the various zones. In this case, coal is fed from underneath the grate between two tyres by a screw conveyer or retort. The primary air passing through the holes in the tyres goes through the various zones. The ash zone is above the incandescent coke zone. The secondary air fed in the mid of flame zone which helps in burning the unburnt carbon and other coal constituents. Figure 5.3 *shows the single retort stoker showing contour of fuel bed and ash dumps.*

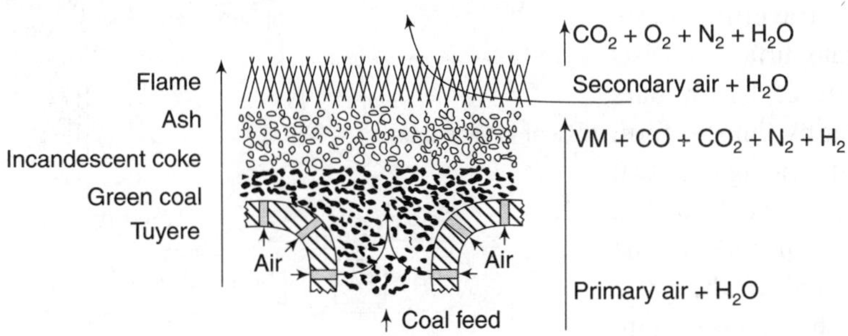

Figure 5.2 Underfeed fuel bed section

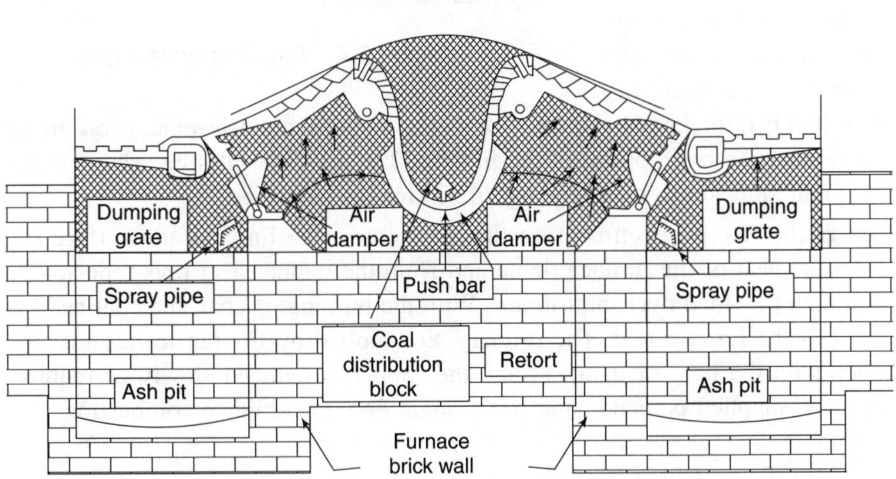

Figure 5.3 Contour of fuel-bed and ash dumps for single-retort stokers

5.3 MECHANICAL STOKERS FOR FUEL BED COMBUSTION

As mentioned earlier, *grate is used to hold the fuel in fuel bed combustion system.* In small steam generator, the **grate is stationary and coal is fed manually by shovels.** *But the bigger units demand more uniform operating conditions, higher burning rate and greater efficiencies and these are accomplished by moving grate or stokers.* There are various types of stokers as given below:

(i) Travelling grate stoker (ii) Chain grate stoker
(iii) Spreader stoker (iv) Vibrating grate stoker
(v) Underfeed stoker

Advantages. (i) No necessity of coal preparation, (ii) free from danger of explosion, (iii) capital and maintenance costs are less, (iv) suitable for medium capacity unit, and (iv) firing system reliable.

5.3.1 Travelling Grate Stoker

Figure 5.4 shows the schematic diagram of a travelling grate stoker. The grate surface consists of a series of cast-iron bars joined together by links to form an endless belt running over two sets of sprocket wheels with a wide surface as per need. The depth of the coal on the grate is regulated by a coal gate provided at the rear of the coal hopper which can be raised or lowered as per need. The burning rate of the coal is controlled by the simultaneous adjustment of grate speed, fuel bed thickness, and air flow so that nothing but ash remains on the grate by the time it reaches the rear portion of the furnace. The ash falls into the ash pit as the grate turns on the rear sprocket to make the return trip for the next feed.

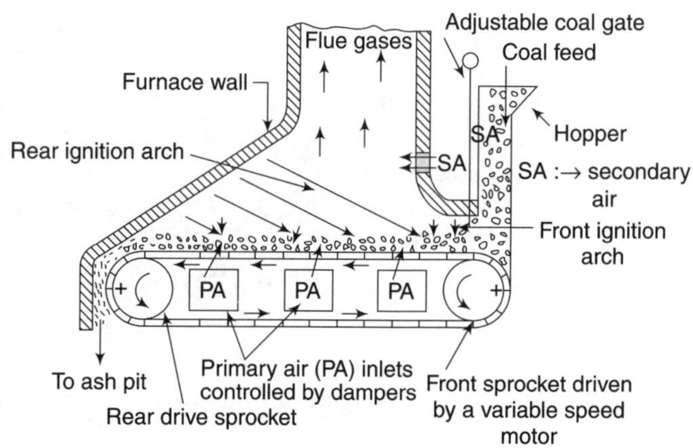

Figure 5.4 Travelling grate stoker

As soon as the raw or green coal on the grate enters the furnace, the surface coal gets ignited from the heat of the furnace flame and from the radiant heat rays reflected back by the ignition arch provided over the furnace. With the burning of coal, the coal bed becomes thinner towards the furnace rear. The primary air supplied by FD fan from under the grate is controlled with the help of dampers and the supply is reduced slowly as required. The secondary air is supplied perpendicular to the main flow to complete combustion.

5.3.2 Chain Grate Stoker

The chain grate stoker is similar to travelling grate stoker except that it is made of a series of cast iron (CI) links connected by bars or pins to form an endless chain. It can burn a wide range of coal. Overfeed stokers are suited for industrial power plants with steady demand. It is simple in construction and initial cost is low.

5.3.3 Spreader Stoker

Figure 5.5 shows the schematic of a spreader stoker. It consists of a rotating feeder (a drum fitted with short blades) and spreader installed at the bottom of hopper one over the other. Coal from the hopper is fed by a rotating feeder to the spreader or distributor below, which projects the coal particles in a continuous stream on to the grate holding an ignited fuel bed. The fine particles of coal burn in suspension where coarse particles burn on the grate. The speed of the feeder is a function of load (steam output) of the boiler. The secondary air is admitted above the fuel bed to promote turbulence and to complete combustion. The grate may be stationary or travelling as per the capacity of the boiler. High capacity boilers need travelling grate. The grate made of CI bars is connected to a lever through links underneath the grate. In order to allow the ash to fall below the ash pit, the lever is moved back and forth which makes the bar rock about the pivot. A variety of coal is burnt over this grate. However, it is necessary to size the fuel property for spreader stokers. Its size should be around 32 mm or less.

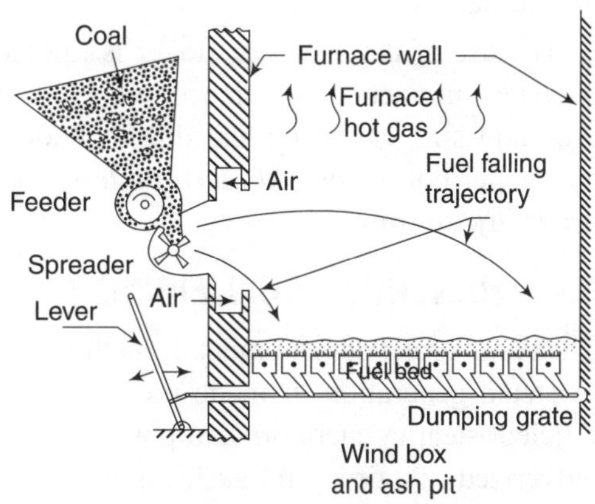

Figure 5.5 Spreader stoker

Spreader stoker is applied to a wide range of boiler sizes, 155 MW to 265 MW.

5.3.4 Vibrating Grate Stoker

In order to distribute the fuel bed properly, the stoker is vibrated which shakes the fuel bed intermittently. The frequency and amplitude of vibration depend on the load of boiler. The fuel bed is made inclined in order to move the fuel by gravity towards the rear of the boiler. In order to prevent slagging, the grate is water cooled.

5.3.5 Underfeed Stoker

Figure 5.6 shows a underfeed retort type stoker. This stoker consists of either single or multiple retort in the form of troughs into which coal is pushed by ram or screws. The primary air is fed by FD fan into the fuel bed through tyres. The volatile matter is distilled of the coal and burns above the incandescent fuel bed. It is worth to note that coal forms a thick sloping pile (15 to 25°) over the whole surface of the grate and the carbonization occurs above and proceeds downwards. Ram pushes the fuel bed forward till the refuse or ash is discharged to the ash pit. The fuel bed may be as thick as 0.6 m. The underfeed stoker gives higher thermal efficiency as compared to chain grate stoker. The part load efficiency is better with multiple retort. The grate is self cleaning. The initial cost is high.

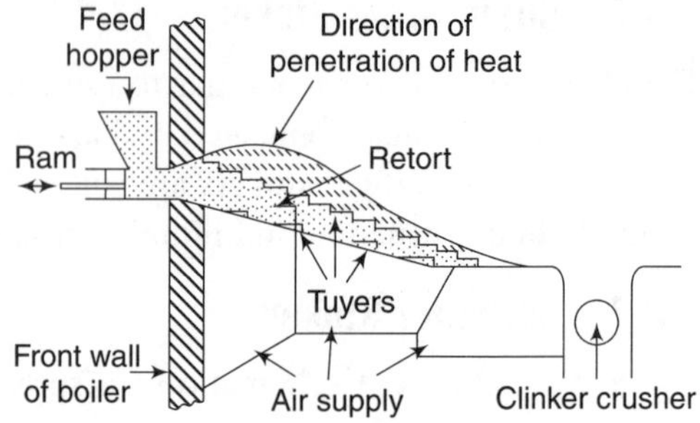

Figure 5.6 Retort type underfeed stoker

The use of any types of stoker is limited. It becomes impractical and uneconomical beyond a certain size of the boilers due to floor area and boiler house cost. ***Modern boilers use pulverized coal combustion system which grow vertically upward.***

5.4 PULVERIZED COAL FIRING SYSTEM

The commercial development of methods for firing coal is treated as a landmark in the history of steam generation. It made possible the construction and reliable operation of large and efficient steam generators and power plants. John Anderson and his associates used first the pulverized coal firing in Lakside Station, Wisconsin (USA) in 1920. Since then, it has become almost universal in central utility stations using coal as fuel.

In pulverized coal firing system, coal is first ground to dust-like size and powdered coal is then carried in a stream of primary air to be fed through burners into the furnace of boiler. With the heating of coal particles in high temperature flames in the furnace, the volatile matter is distilled off resulting in the formation of minute sponge-like masses of fixed carbon and ash. After the mixing of volatile gases with air, they get ignited and burn quickly releasing heat energy. Thus, a pulverized coal system consists of *pulverizing, delivery and burning equipment.*

In order to burn pulverized coal successfully in a furnace, the following requirements must be satisfied.

(i) The existence of large quantities of *very fine particles of coal*, usually those that would pass a 200 mesh screen, to ensure ready ignition because of their large surface to volume ratio.

(ii) The existence of a minimum quantity (0.9%) of coarser particles to ensure high combustion efficiency. These coarser particles should be able to be retained by a 50 mesh screen.

The following is a typical screen analysis for a high volatile bituminous coal sample pulverized to :

- 99.5% – 50 mesh (0.297 mm openings)
- 96.5% – 100 mesh
- 80% – 200 mesh (0.074 mm openings)

The coarser particles which would be retained over 50 mesh screen is just 0.5 per cent. The above size of pulverized coal represents a surface area of roughly *150,000 mm²/g with 97% of the surface in the 200 mesh portion.*

It is worth to note that the greater surface area per unit mass of coal provides faster combustion reaction as more carbon particles are exposed to oxygen and heat. This results in less requirement of excess air, reduction in dry exhaust loss and increase in steam generator efficiency.

The proper burning of fuel requires the supply of correct proportion of air, thorough mixing of fuel and air, high temperature and adequate time to complete combustion reaction. As a result of combustion, ash is formed. A part of ash falls to the furnace bottom and rest is carried in gas stream as ***fly ash,*** which is separated before exhausting through chimney.

5.4.1 Advantages and Disdvantages of Pulverized Coal Firing System

Advantages

(i) Greater surface area of coal per unit mass allows faster combustion reaction.
(ii) Excess air requirement is low.
(iii) The ability to use preheated air is high to reduce exhaust losses.
(iv) The boiler efficiency is high.
(v) It can burn wide varieties of coal.
(vi) The response to load changes is very fast enabling almost constant steam pressure under wide load conditions.
(vii) It can burn combination of gas and oil with ease.
(viii) It is able to release large amount of heat enabling to generate steam above 2000 t/h in one boiler.
(ix) Pressure losses are less and so draught.
(x) Fly ash emerging out has the ability to be used in brick preparation.
(xi) The system is free from clinker and slagging troubles.
(xii) The boiler can be started from cold very rapidly and efficiently which is an important factor during emergency.
(xiii) There is no moving parts in the furnace subjected to high temperature resulting in more system life.
(xiv) Practically there is no ash handling problem.
(xv) The furnace volume required is considerably less enabling minimum travel of flame length.

Disadvantages

(i) The capital cost of the pulverized system is considerably high.
(ii) Additional power is required for pulverizing coal.
(iii) Fly ash removal equipment such as electrostatic precipitator is costly.
(iv) Due to high flame temperature, the refractory line furnaces are inadequate, so water cooled walls are essential.

(v) The storage of powder coal requires special attention against fire hazards.
(vi) Special starting up equipment are needed.

5.4.2 Crushers

To prepare coal for pulverization, the following types of crushers are used:

(i) Ring crusher or granulator
(ii) Hammer mill crusher
(iii) Bradford breaker

Figures 5.7 and 5.8 show a ring and hammer type coal crusher. The coal is fed at the top and is crushed by the action of rings that pivot off centre on a rotor or by swinging *hammers* attached to it. Adjustable screen bars determine the maximum size of the discharge coal. Ring crushers and hammer mills are used off or on plant site. They discharge a large amount of fines suitable for further pulverized but not for cyclone furnace firing.

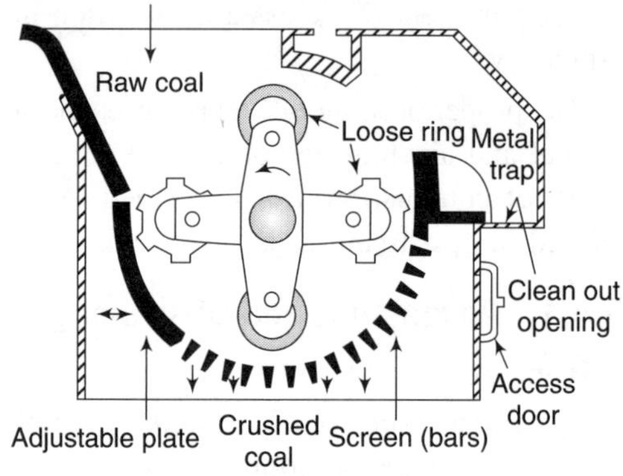

Figure 5.7 A ring coal crusher

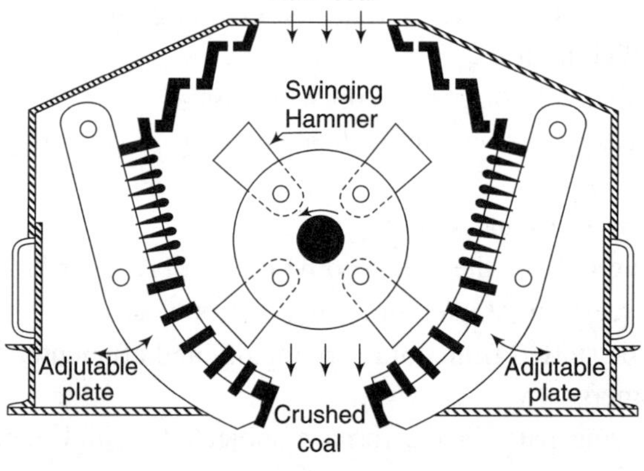

Figure 5.8 A hammer mill coal crusher

Bradford breaker is shown in Fig. 5.9. It is used for large capacity work. It comprises a large cylinder consisting of perforated steel or screen plates to which lifting shelves are attached on the inside. The cylinder rotates slowly at about 20 rpm and receives feed at one end. The coal is lifted by the shelves, and the breaking action is accomplished by the repeated dropping of the coal until its size permits it to discharge through perforation. The size of perforations determines the size of crushed coal. The crushing force due to gravity limits the quantity of fines. The main advantage of Bradford breakers is to reject foreign matter and to produce relatively uniform size coals.

5.4.3 Pulverizers

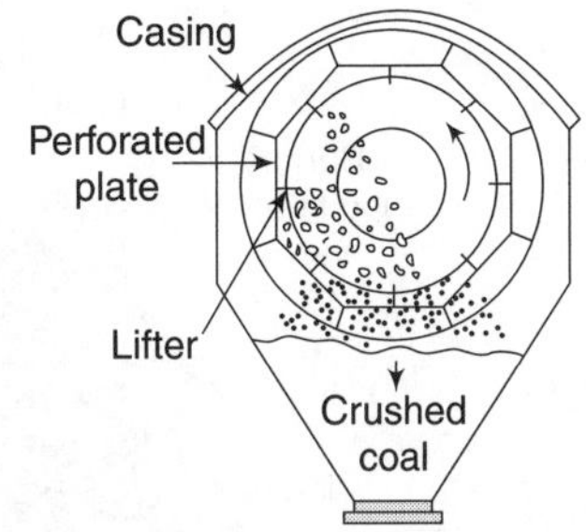

Figure 5.9 A Bradford Breaker

Pulverizing process consists of three stages, namely: (i) *feeding,* (ii) *drying and* (iii) *grinding.* The feeding system must automatically control the fuel-feed rate according to the boiler demand and the air rates required for drying and transporting pulverized fuel to the burner (primary air). For pulverization, coal has to be dry and dusty. Since coals have varying quantities of moisture and in order that lower rank coals can be used, *dryers* are an integral part of pulverizing equipment. A part of the air (primary air) from steam-generator air preheater is forced into the pulverizer at 350°C or more by primary air fan where it is mixed with the coal as it is circulated and ground. The third stage is the grinding and its equipment is called *pulverizer* or *grinding mill.* Grinding is performed by *impact, attrition, crushing or combination of these.*

There are several types of pulverizers, classified by speed. They are: (i) low speed (below 75 rpm): the ball-tube mill, (ii) medium speed (75 to 225 rpm) : the ball–and-race and roll-and race mill, (iii) high speed (above 225 rpm): the impact or hammer mill, and the attrition mill.

5.4.3.1 Ball-tube Mill

Figure 5.10 shows the schematic of a ball-tube mill. It is one of the oldest in the market. Basically, it is hollow cylinder with conical ends and heavy-cast wear resistance liners less than half-filled with forged steel balls of mixed size. Pulverized is accomplished by attrition and impact as the balls and coals ascend and fall with cylinder rotation. The primary air is supplied over the charge which carries the pulverized coal to classifiers. The system suffers from the drawback of more power consumption due to heavier construction and poor air circulation. Now efficient system is used.

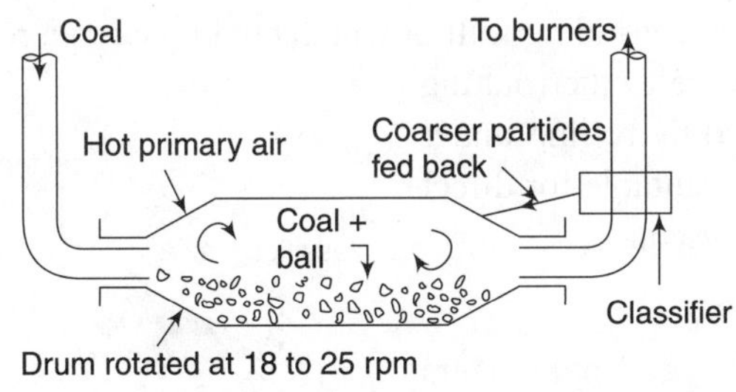

Figure 5.10 Ball-tube mill

5.4.3.2 Ball-and-Race Mill

Figure 5.11 shows a widely used ball-and-race type pulverizer. It operates on the principle of *crushing* and *attrition.* Pulverization takes place between two surfaces, one rolling on top of the other. The rolling elements may be balls or ring-shaped rolls that roll between two

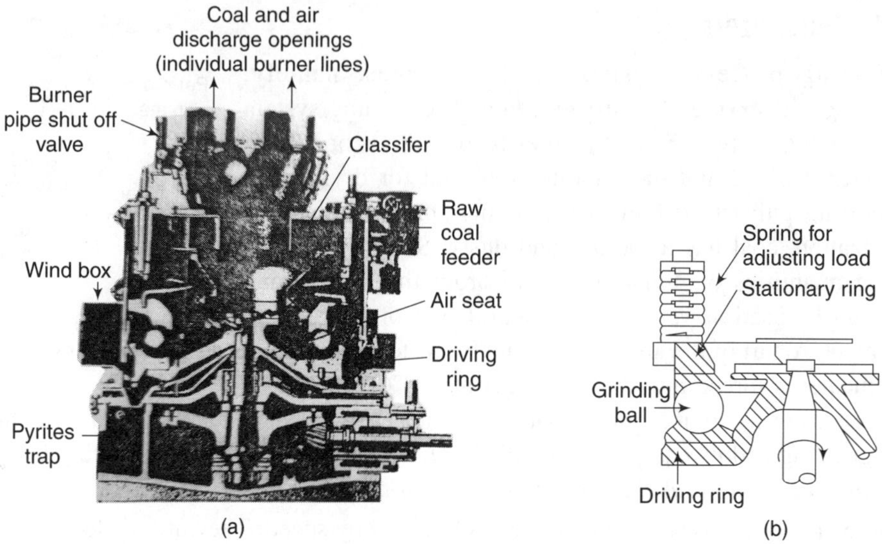

Figure 5.11 Ball-and-race coal pulverizer

races. The balls are between a top stationary race or ring and a rotating bottom ring, which is driven by vertical shaft of the pulverizer. The primary air supplied to the pulverizer causes coal feed to circulate between the grinding elements and when it becomes fine enough, it becomes suspended in the air and is carried to the classifier. In order to achieve most efficient grinding of various coals, the grinding pressure is varied by externally adjustable springs on top of the stationary rings.

The ball-and-race pulverizer has ball circle diameter ranging from 4.2 cm to 20 cm and capacities between 2 to 20 tonnes/h. It is suitable for direct-firing systems.

5.4.3.3 Bowl Mill

Figure 5.12 shows bowl type grinding mill in which coal is fed to a rotating bowl and thrown outward by centrifugal force to the rotating ring where it is pulverized by rolling under spring-loaded rollers. It is suitable for direct firing system and widely used.

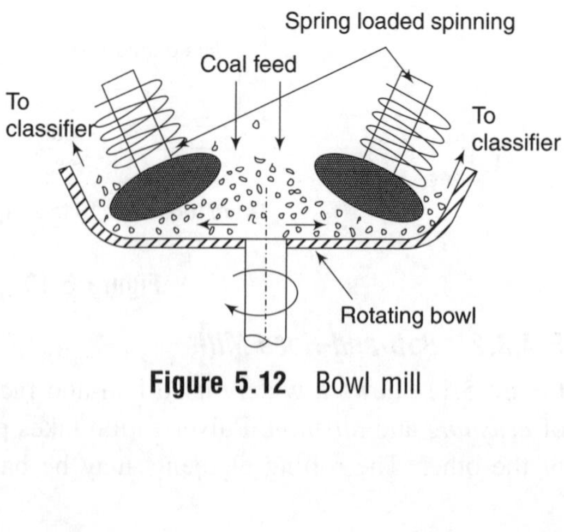

Figure 5.12 Bowl mill

5.4.3.4 Hammer Beaters

Hammer beaters are high speed pulverizers that revolve in a chamber equipped with high-wear resistant liners. They are suitable for low-rank coals with high moisture content and use fly gas for drying but they are not widely used for pulverized coal systems.

As mentioned above the classifier receives the pulvierzed coal. It is usually a cyclone with adjustable inlet vanes and is located at the pulverizer exit. *The function of the classifier is to separate the oversized coal and return it to the grinders to maintain proper fineness for the particular application and coal used.* It is the inlet vanes that is used for varying the gas suspension velocity in the classifier.

5.5 PERFORMANCE OF PULVERIZERS

The main factors affecting the performance of pulverizer are (i) the grindability of the coal, (ii) the surface moisture on the coal and (iii) the fineness of the grind needed. It is worth to note that the power consumption of the pulverizer should be low to minimize the operating cost.

The grindability of a coal is measured in terms of Hardgrove index. A low index number designates coal hard and difficult to grind, while a high index indicates soft coals which are easy to grind. The fineness of the pulverized coal needed for successful burning depends largely on the ratio of volatile matter (VM) to fixed carbon. Low VM coals should be ground finer as carbon burns slower than volatile matter. Moisture makes fine coal particles adhere to each other defeating the very purpose of grinding.

5.6 THE PULVERIZED COAL SYSTEM

A total pulverized coal system consists of *pulverizing, delivery and burning equipment.* The system must be capable of both continuous operation and rapid change as required by load demands. The following are the three main pulverized coal systems.

1. Bin or storage system, 2. Direct firing system, 3. Semidirect firing system.

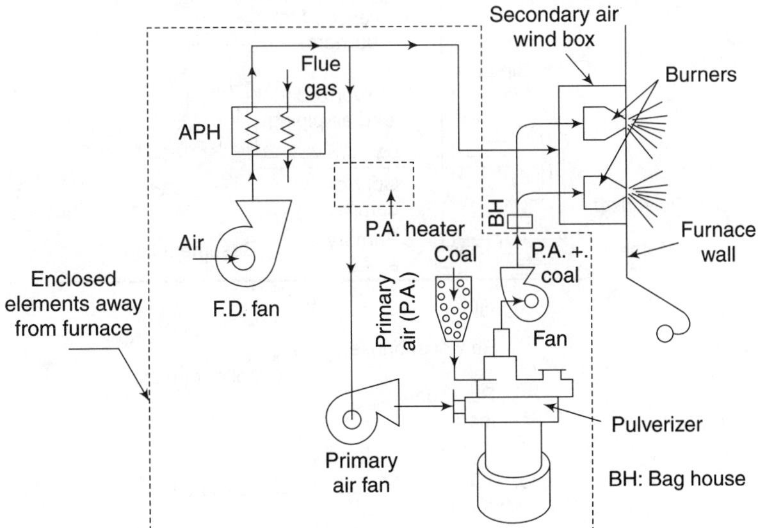

Figure 5.13 Bin or storage system

5.6.1 Bin or Storage System

It is essentially a batch system by which the pulverized coal is prepared away from the furnace. The resulting pulverized-coal-primary-air mixture goes to a cyclone separator and fabric bag filter that separate and exhaust the moisture laden air to the atmosphere and discharge the pulverized coal to storage bin as shown in Fig. 5.13. From there, the coal is pneumatically conveyed through pipelines to utilisation bins near the face to fed to burners. This system was widely used before the development of direct firing system. Since the drying, storing and transporting are carried out in many stages, so the bin system is prone to fire hazard. However, it is still in use in old plants.

5.6.2 Direct Firing System

Pulverized coal direct firing system (Fig. 5.14), is extensively *used in modern power plants.* In comparison to bin system, this system offers greater simplicity and hence greater safety, lower space requirements, lower capital and operating costs and greater plant cleanliness. It continually processes the coal from the storage receiving bunker through a feeder, pulverizer, and primary air fan to the furnace burners, The fuel flow suitable to load demand

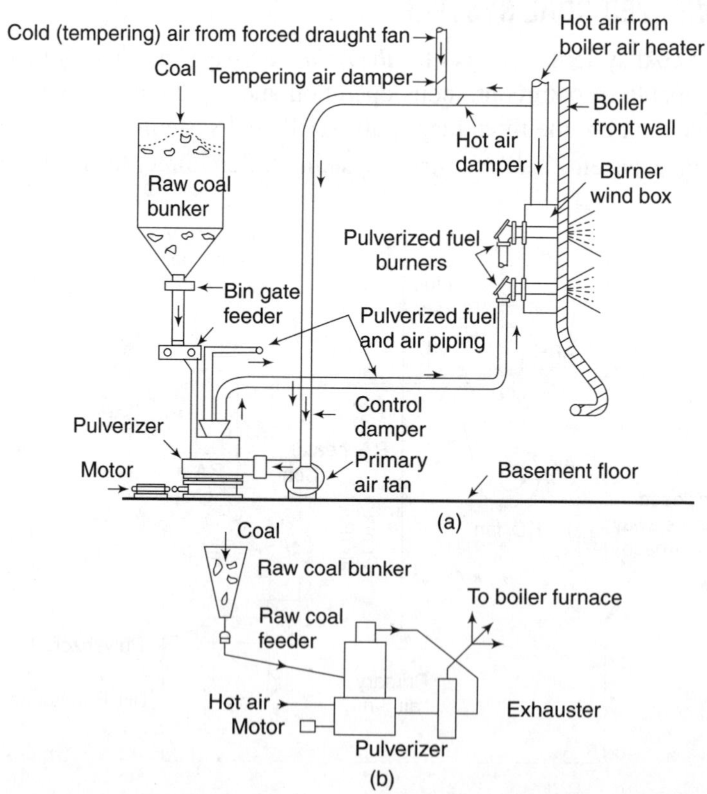

Figure 5.14 Pulverized coal direct firing system

is accomplished by a combination of controls on the feeder and on the primary air fan in order to give air-fuel ratio as per load. The control operating range is only about 3 to 1. It is worth to note that large steam generators are provided with more than one pulverizer system, each feeding a number of burners enabling wide control range.

5.6.3 Semi-Direct Firing System

Figure 5.15 shows a semi-direct firing system. A cyclone separator between the pulverizer and furnace separates the conveying medium from the coal. The coal particles fall on by gravity from the cyclone to feed the coal particles through the burners into the furnace.

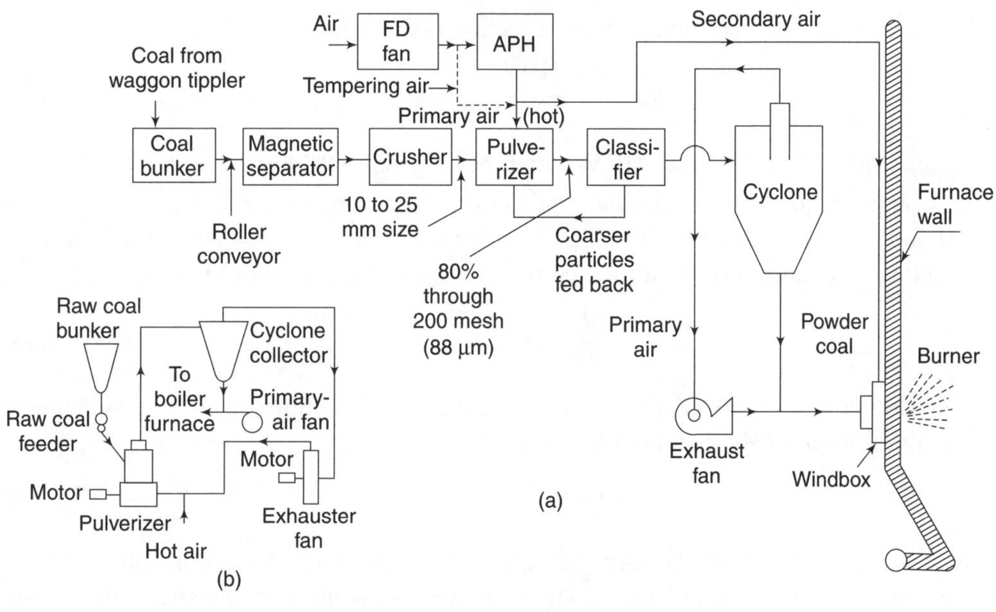

Figure 5.15 Semi-direct pulverized coal firing system

5.7 DESIGN ASPECT OF PULVERIZED COAL FIRED FURNACE

The shape, type and design of the furnace plays an important role in achieving the higher steam generator efficiency. Pulverized coal burns in suspension and the heat released by the combustion is transferred to the water wall tubes lined around the furnace by radiation mechanism. The temperature of the gas leaving the furnace is such which does not cause clinkering to the subsequent heating surfaces The water wall tubes receive water from the drum via downcomer tubes and discharge the water-tube mixture to the drum.

The shape of pulverized coal fired furnace is characterized geometrically by its linear dimensions such as front width (w), depth (d) and height (h_f) as shown in Fig. 5.16. The cross-sectional area of the furnace is $h_f = w\ x\ d$, through which hot furnace gas passes at a speed ranging from 7 to 12 m/s.

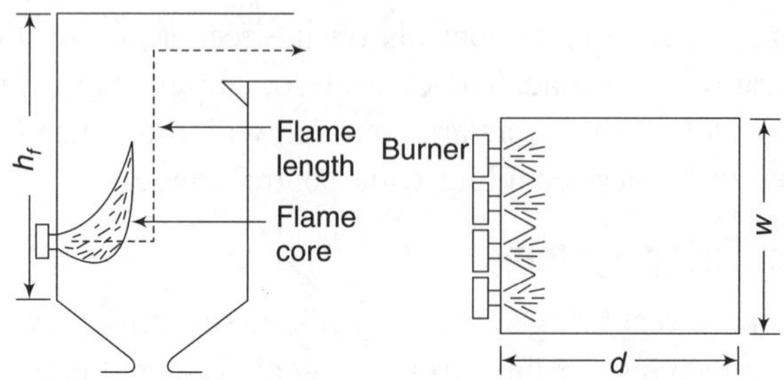

Figure 5.16 Main dimensions of a furnace

The heat release rate per unit bed cross section is expressed as

$$q_A = \frac{\dot{m}_f \cdot \text{HHV}}{a_f}, \text{ kW/m}^2 \qquad \text{...(5.1)}$$

where $\dot{m}_f$ = mass of fuel (kg/s) ; HHV = high heating value (kJ/kg)

q_A is supposed to be the most important characteristic of furnace and depends on the kind of fuel and method of burning. It ranges from 3500 kW/m^2 to 6400 kW/m^2 for brown coal. For meeting a certain power of a furnace (Q_f), cross-sectional area (a_f) is expressed as

$$a_f = \frac{Q_f}{q_A}, \text{ m}^2 \qquad \text{...(5.2)}$$

In general, w = 9.5 to 31 m, d = 6 to 10.5 m are taken. Sometimes, w is expressed in terms of steaming capacity ($\dot{M}_S$) is tonnes per hour as

$$w = 0.67\sqrt{\dot{M}_s} \qquad (5.3)$$

In choosing the height of the furnace (h_f) which ranges from 15 to 70 m, care is taken to ensure complete combustion of fuel along the flame length and to allow space for arranging the water tubes on the furnace for absorbing heat for its safety and to allow desired natural circulation. The furnace height (h_f) is expressed as

$$h_f = \overline{C}_g\,\tau \qquad \text{...(5.4)}$$

where $\overline{C}_g$ is the average velocity of the gas in furnace (m/s) and τ is resident time (s) a unit volume of gas present in the furnace.

In addition to q_A, energy release per unit volume (q_v) is also used to characterize the furnace. Thus,

$$q_v = \frac{Q_f}{V_f} = \frac{\dot{m}_f \cdot \text{HHV}}{V_f} \qquad \text{...(5.5)}$$

where V_f = volume of the furnace = $w \times d \times h_f$

The value of q_V ranges from 125 to 220 kW/m^3. Obviously, the volumetric heat release rate, q_v increases with decrease of resident time (τ). In general, τ is about 2 seconds for q_v = 200 kW/m^3.

It is worth to note that surface area provided inside the furnace consisting of water walls and secondary superheater absorbs heat by *radiation*, whereas the surface area of the primary superheater, reheater, economizer and air preheater absorbs heat by *convection*. In order to maintain the best ratio between the radiant and convective heat transfer in boiler surfaces, it has been found that the temperature of the gases at the exit of the furnace should range between 1100 to 1200°C and it is advisable to maintain this value.

Generally, pulverized coal furnaces are of two types. 1. Dry-bottom furnace and 2. Wet-bottom or slag type furnace.

5.7.1 Dry-bottom Furnace

Figure 5.17 shows a dry-bottom type furnace. In this case, ash or slag is removed in the dry state. This is possible by allowing the exit temperature of gases below the ash fusion temperature. About 85% of ash, called *fly ash* flows with the furnace and get collected in electrostatic precipitator (ESP) while remaining 15% falls through the furnace bottom to the clinker-grinder and then to the hydraulic sluice to which also falls the ash collected in dust collector ESP. A hopper is provided at the bottom of the furnace whose walls are inclined (50 to 60°) to allow gravity fall. The dry bottom furnace is cooler (about 700°C or less) to cool molten slag and the solid slag falls to a slag pit containing water. A screw mechanism is used to remove the granulated slag particles.

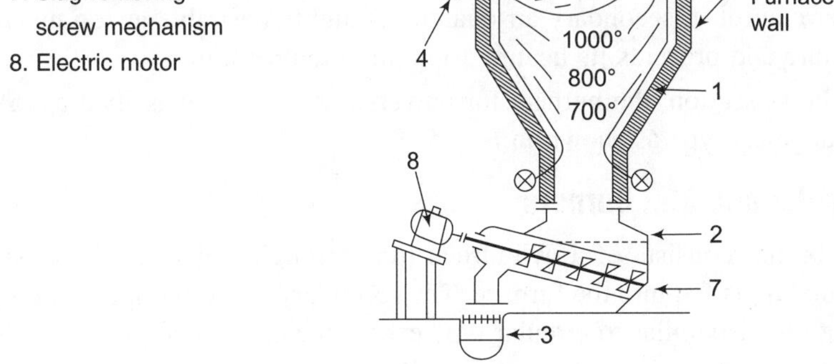

Figure 5.17 Dry-bottom furnace

5.7.2 Wet-Bottom Furnace

Figure 5.18 shows the schematic of a wet-bottom furnace. In this case, about 40 to 60% of fly ash moves with the furnace and rest falls through the furnace bottom in a continuous molten stream. This is possible by maintaining the temperature of lower portion of the furnace above the ash melting point. A slag tank containing water receives the molten slag where it is formed in granular form.

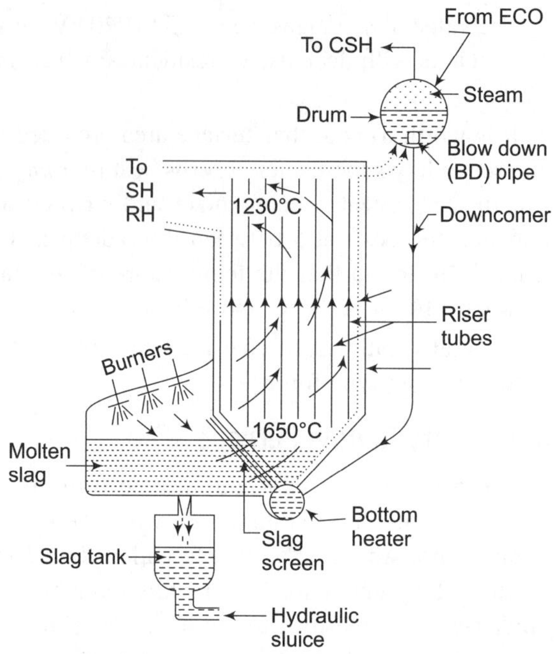

Figure 5.18 Wet-bottom furnance

5.8 PULVERIZED COAL BURNERS

The performance of steam generator largely depends upon the ability of burners to produce uniform mixing of coal and air and proper flame first. For efficient utilization of pulverized coal, the burners should meet the following requirements:

- It should prepare two individual flows, a coal dust primary air mixture and secondary air. This enables ignition and active burning in the furnace space and creates a proper turbulent environments for thorough mixing and sucking hot furnace gases for preheating.
- In addition to maintaining stable ignition of the fuel-air mixture, it should also control the flame shape and travel in the furnace as the rate of flame propagation control the combustion.
- The coal-air mixture must move away from the burners at a rate equal to the flame-front travel in order to prevent flash back into the burner.
- A proper control of secondary air quantity is highly desirable as too much air can cool the mixture and prevents its heating to ignition temperature.

Based on cross section, the burners for pulversized coal are classified as (a) circular and (b) slot (rectangular) type as shown in Fig. 5.19.

5.8.1 Circular and Slot Burners

The circular burner consists of a cylindrical tube through which the pulversized coal and primary air mixture is fed into the furnace. The secondary air is developed through a separate tube. The slot type is similar to circular type except the cross section.

Based on *kind of flow* of *the two streams,* the burners are classified as:

1. Straight-flow burner (external mixing of air-fuel mixture and secondary air in the furnace space).

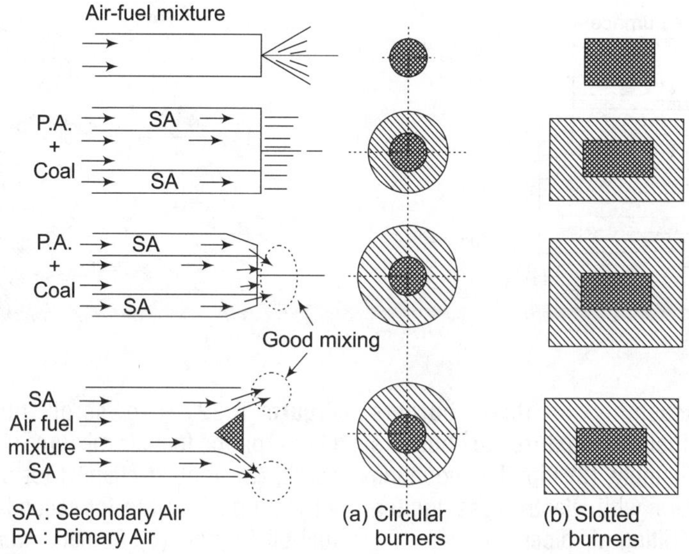

Figure 5.19 Pulverized coal burner: (a) Circular, and (b) Slot

2. Turbulent or vortex burner (internal mixing of air-fuel mixture and secondary air before entering the furnace).

5.8.2 Straight Flow Burners

Figure 5.20 shows straight flow burners for pulverized coal in which, air fuel mixture and secondary air are blown in as parallel jets and the inter mixing takes place in the furnace space, not in the burners, by properly arranging burners on the furnace walls. In high capacity boiler, burner assembly consisting of many burners are used (Fig. 5.21). The range of the primary air velocity at the burner outlet varies from 20 to 28 m/s, while the range of secondary air velocity is 35 to 45 m/s.

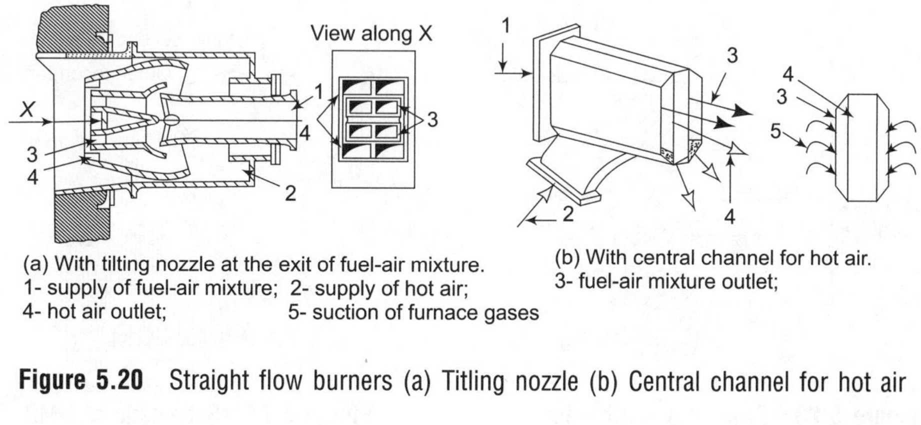

Figure 5.20 Straight flow burners (a) Titling nozzle (b) Central channel for hot air

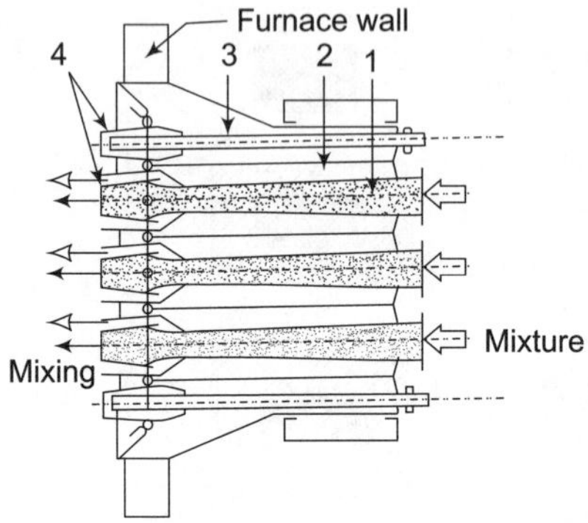

Figure 5.21 Assembly of three burners (1) Supply of fuel-air mixture to burners (2) Secondary air supply, (3) Pipe for mounting the fuel-oil burners with electric gas igniter, (4) Tilting air pipes

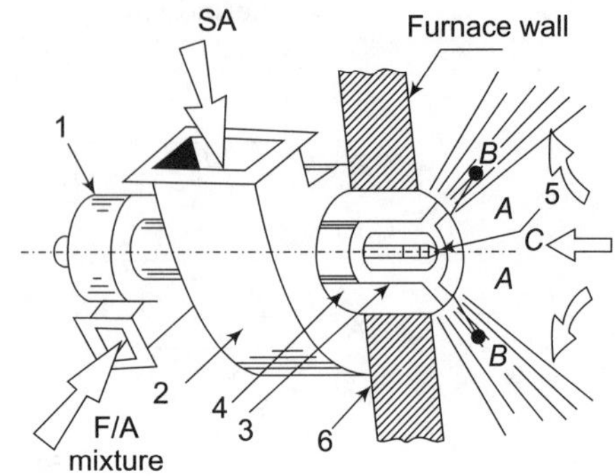

Figure 5.22 Two scroll turbulent burner (1) Scroll for fuel-air mixture, (2) Secondary air scroll, (3) Annular channel for discharging fuel-air mixture, (4) Ditto-for secondary air, (5) Main fuel-oil burner, (6) Furnace wall, AB-boundary of fuel-air mixture ignition, C-suction of gases to flame root

5.8.3 Turbulent or Vortex Burners

Figure 5.22 shows a turbulent or vortex burner. In such burners, a whirling rotary motion is imparted to the coal-air mixture by entering it tangentially to a central nozzle and also to the secondary air that flows through the annular space surrounding the nozzle. The mixing of two streams takes place as soon as they enter the furnace. Sometimes, it is called *two-scroll burner*, where two scrolls are provided for whirling the dust-air mixtures and secondary air.

Figure 5.23 depicts a straight scroll types turbulent (vortex) burner. In this case, the dust air mixture is supplied through a central straight flow nozzle and spread by a deflector and the secondary air is whirled in a scroll. Vane type vortex burner is also sometimes used in which

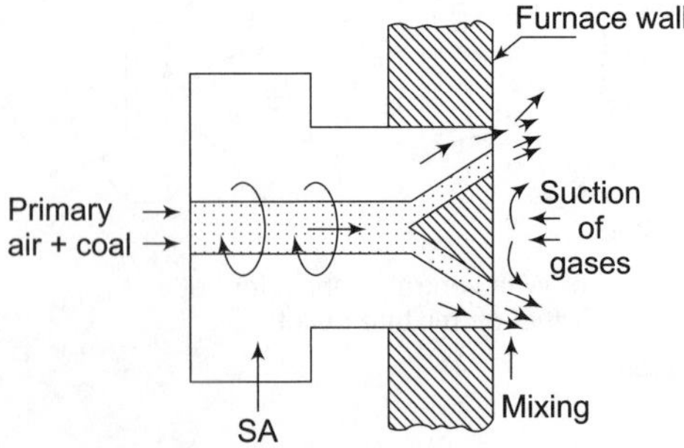

Figure 5.23 Straight scroll burner

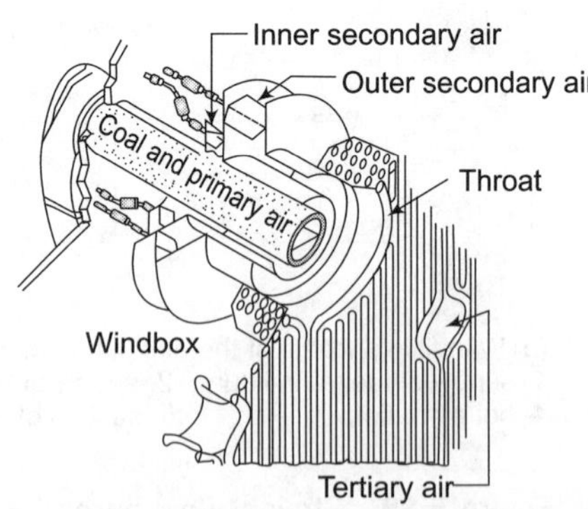

Figure 5.24 Schematic of DMB

the whirling of the dust-air mixture and the secondary air is effected by axial and tangential vanes. Such burner generally uses horizontally firing of coal.

5.8.4 Distributed Mixing Burner (DMB)

Figures 5.24 and 5.25 show the distributed mixing burner (DMB) and its mixing concept. Here, the total air supplied is distributed in such a way that complete combustion of fuel is achieved with minimum excess air. Basically, the total input is divided in three zones. In the first zone, stoichiometric air-fuel ratio is maintained at a very rich condition when devolatization starts. In the second zone, supplementary secondary air is then added to the rich product so as to delay the volatilized fuel-nitrogen compounds. The third zone is the tertiary air zone which provides the oxygen necessary for complete combustion. The third zone is also called the oxidizing zone around the rich primary zone prevents corrosion and limits the interactions between adjacent burners. DMB burner is supposed to be very efficient one.

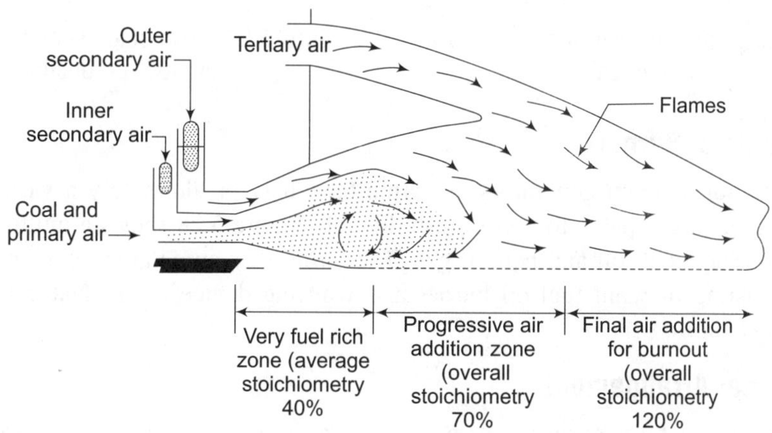

Figure 5.25 Mixing concept of DMB

5.8.5 Low NO_x Burners

It is the oxidation of both atmospheric nitrogen and the nitrogen chemically combined with the fuel which produces NO_x. The mechanism of NO_x formation suggests that the nitrogen available in primary combustion process is to be controlled by fuel-rich operation through reduction of primary air and subsequently to control the overall gas temperature, the combustion is completed with fuel-lean mixture. Figure 5.26 shows the effect of varying quantities of over-fire on NO_x production for various values of percentages of excess air.

Figure 5.27 depicts a *two-stage venturi furnace for reducing NO_x emissions. Two separate furnaces are used for burning coal.* In the primary furnace, coal is burnt at sub-stoichiometric conditions resulting in incomplete combustion. As a result, NO_x production is less. The secondary air is added in secondary furnace to complete combustion. The combined effect will be the reduction of NO_x formation. *The other method of NO_x reduction is the re-circulation of flue gases which reduces the bulk flame temperature and flame oxygen concentration.*

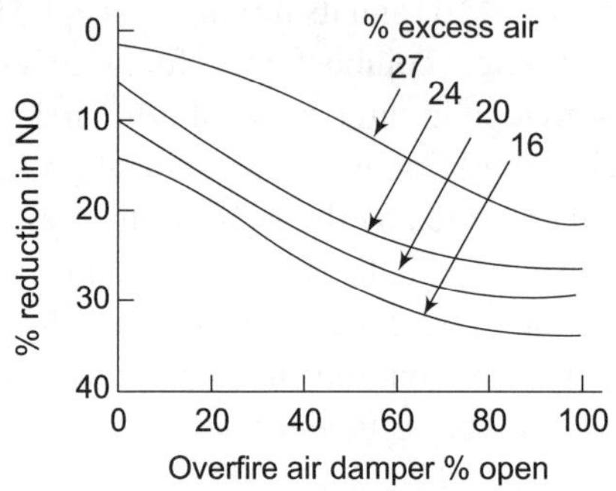

Figure 5.26 Effect of over-fire air on NO_x production

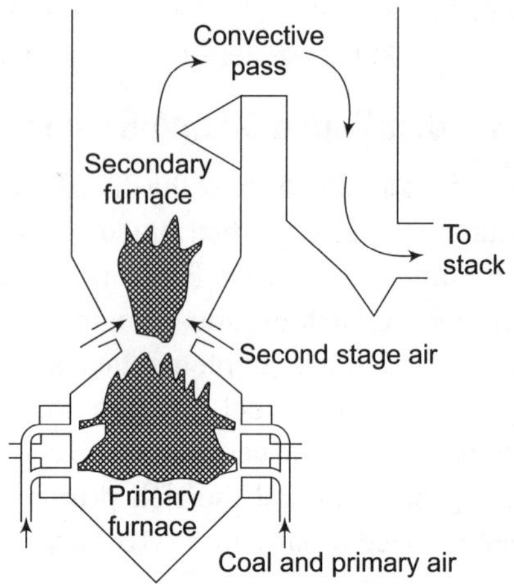

Figure 5.27 Two-stage venturi furnace for low NO_x production

5.8.6 Multifuel Burners

In many situations, steam generators use simultaneously or alternately a variety of fuels, multifuel burners are capable to accomplish this job. Figure 5.28 shows a multi-fuel burners to burn pulverized coal, oil and natural gas. The burner has an enlarged diameter in central channel consisting of main fuel oil burner and whirling device of air. Natural gas is fired through feeder tubes.

5.8.7 Burner Arrangement

The arrangement of burners on the furnace plays an important role on the performance of combustion as it ensures complete combustion in the flame that cares itself. Based on the type of burners and their characteristics, burners are arranged optimally on the furnace.

(1) Turbulent Burner Arrangement. The arrangement of turbulent burners on the furnace is as follows (Fig. 5.29):

(a) Front firing (b) Double front or opposed firing
(c) Side firing (d) Double-side firing.

The diameter of the port (D_p) is taken as the basis in the arrangement. Burners are arranged (2.2 to 3) D_p from the side walls of the furnace. This arrangement helps in preventing the flames from interacting prematurely and touching the walls. In the case of single wall firing, the opposite water wall has a high heat absorption (10 to 20% above the average value) which may clinker the wall. To prevent clinkering, the surface depth in dry bottom furnace must not be less than d = (6 to 7) D_p. In high capacity boilers, arrangement of all required number of burners is not possible in single wall, so opposite arrangement is employed ensuring more uniform heat absorption by water walls.

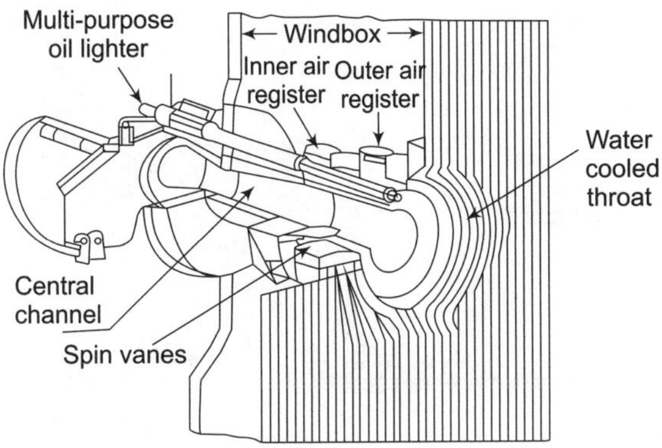

Figure 5.28 Multi-fuel burners

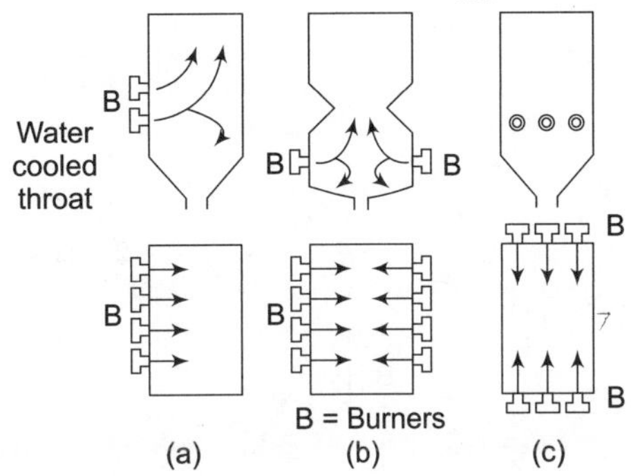

Figure 5.29 Arrangement of turbulent burners (a) Front firing, (b) Opposed firing, (c) Side firing (double), (d) Double-side fring

2. Straight Flow Burner Arrangement. They are arranged on the furnace wall as follows (Fig. 5.30).

(a) Opposite displaced (offset) firing
(b) Corner firing with encountering flame jets
(c) Corner firing with tangential jets
(d) Vertical firing

(a) The opposite displaced (offset) firing arrangement [Fig. 5.30(a)] produced, intensive turbulization of flames in the main combustion zone as a result of large velocity gradients between adjacent jets moving in opposite direction. This is suitable for peat and brown coal.

(b) The corner firing with encountering flames jets arrangement [Fig. 5.30(b)] produces a highly turbulized flame core but if the flame moves away from the centre to side walls there is a possibility of clinkering of the walls.

(c) Corner firing with tangential jets arrangement [Fig. 5.30(c)] is widely used in pulverized coal fired boilers. It directs the mixture of fuel and air tangentially to an imaginary circle 1–2.5 m in diameter in the furnace centre. Both fuel and air are projected horizontal from the corners of the furnace along lines tangent to a vertical cylinder at the centre of the furnace. At the meeting point of two streams, intensive mixing occurs which is desirable for complete combustion. A rotating motion, similar to that of a cyclone, is imparted to the flame body, which spreads out and fills the furnace area. The need of the air for the each compartment is controlled by dampers. The furnace employing tangential burner arrangement may be employed in a furnace having w/d ratio equal to 1 to 1.2.

(d) Vertical firing arrangements [Fig. 5.30 (d)] are suitable to fire solid fuels which are difficult to ignite, like anthracite and low volatile coals. Pulverized coal and primary air mixture is fed through vertical nozzle which are heated by heated combustion air. High pressure jets supply in the furnace. The turbulence is provided by supplying tertiary air from the front and rear rows. The travel of the flame is long which helps in combustion.

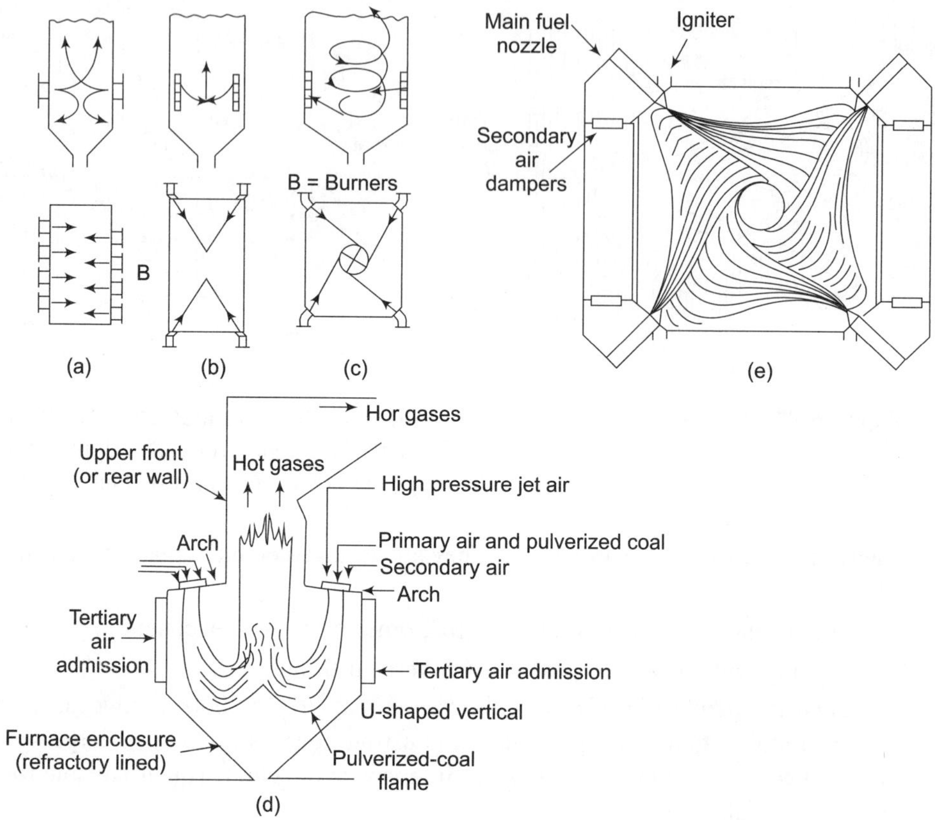

Figure 5.30 Arrangement of straight flow burners (a) Opposite displaced, (b) Corner firing with encountering jets, (c) Corner firing with tangential jets, (d) Vertical firing arrangement

Table 5.1 depicts the excess air required by some fuel system. The excess air requirement is more in spreader stoker.

Table 5.1 Excess air required by some fuel systems

Fuel	*System*	*Excess air (%)*
Coal	Pulverized, completely water-cooled furnace	15–20
	Pulverized, partially water-cooled furnace	15–40
	Spreader stoker	30–60
	Chain grate and travelling stoker	15–40
	Crushed, cyclone furnace	10–15
Fuel oil	Oil burners	5–10
	Multi-fuel burners	10–20
Gas	Gas burners	5–10
	Multi-fuel burners	7–12

5.9 CYCLONE FURNACE

Cyclone furnace is widely used to burn poor grades of coal having high ash and moisture content. Further, it is also suitable to burn biofuels like rice husks.

Figure 5.31 shows a cyclone furnace. It is essentially a water-cooled horizontal cylinder located outside the main boiler furnace in which the crushed coal is fed and fired with a high rate of heat release. The completion of combustion is accomplished before resulting hot gases enter the boiler furnace. The feeding of the crushed coal into the cyclone burner is accomplished at left along with primary air which is about 20% of combustion or secondary air. The primary air enters the burner tangentially, thus imparting a centrifugal motion to the coal. So far the secondary air is concerned, it is also admitted tangentially at the top of cyclone at high speed imparting further centrifugal motion. Tertiary air which is a small quantity of air is admitted at the centre.

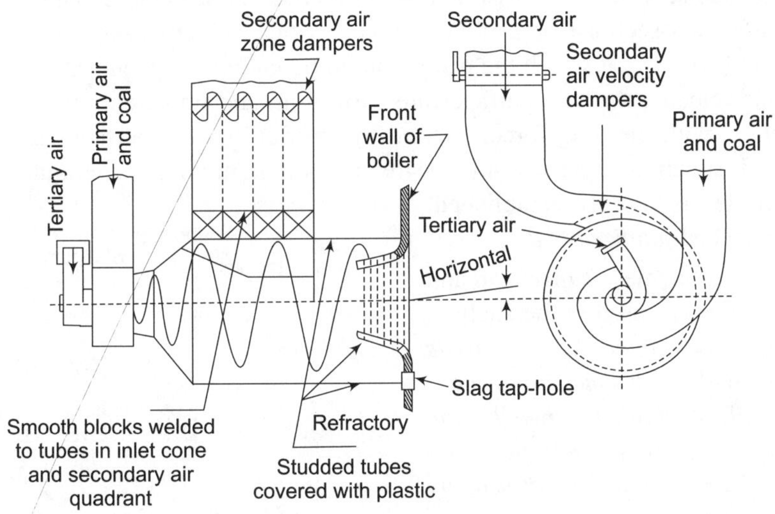

Figure 5.31 Cyclone furnace

It is worth to note that the whirling motion of air and coal results in large reheat release rate volumetric densities (about 4700 to 8300 kW/m^3) and high combustion temperature more than 1650°C. As a result of this high temperature, ash is melted into a liquid slag that covers the surface of the cyclone and eventually drains through the slag-tap opening to a slag tank at the bottom of boiler furnace, where it is solidified and after breaking it is removed. The resulting slag layer provides insulation against too much heat loss through the walls and contribute in enhancing the efficiency of cyclone firing. The *main disadvantage* is the large production of NO_x due to high temperature. The initial ignition is done by small retractable oil or gas burners in the secondary air ports.

It may use bin, or storage or direct firing system. The size and number of cyclones per boiler depends upon the boiler size and load. The burners of cyclone furnace need frequent maintenance as a result of erosion due to high velocity of the coal.

5.10 FLUIDIZED BED COMBUSTION

Fluidized bed combustion is of the concurrent type and has been under development since 1950s, with the aim of perfecting the process. It differs from the cyclone furnace in that sulphur is removed during the combustion process. In addition, fluidized bed combustion occurs at lower temperature which results in lower production of NO_x as well as the avoidance of slagging problems with some coals.

A fluidized bed is defined as the bed that contains solid particles which are in intimate contact with a fluid (air) passing through at a velocity sufficiently high to cause the particles to separate and become freely supported by the fluid. A fixed bed, on the other hand, is one in which the velocity of fluid (air) is too slow to cause fluidization. With the passing of air through a fixed or packed bed of particles, air supply percolates through the interstitial gaps between the particles. As the air flow rate through the bed is steadily increased, a point is eventually attained at which the pressure drop across the bed becomes equal to weight of the particle per unit cross-sectional area of the bed. This critical velocity is known as the *minimum fluidization velocity,* U_{mf}, at which the bed is said to be *incipiently fluidized.* With the further increase in air velocity, the coal particles are buoyed up and imparted a violently turbulent *fluidlike motion*, with the drag forces exerted by the fluid on the particles exceeding their weight. In such situation, there is a high degree of particle mixing and equilibrium between gas and particles is rapidly established. Such a bed is called *fluidized bed.*

Two more terms, namely, ***superficial*** and ***terminal velocities*** are used in the analysis of fluidzed bed. *The superficial velocity is obtained by dividing the measured mass flow rate of air through the bed by the product of the bed cross-sectional area and density of air. The superficial velocity of air at which particles entrainment starts is called terminal velocity* (U_l). Figure 5.32 shows the bed pressure drop with superficial velocity. The state of bed between U_{mf} (i.e., point *B*) and U_f (i.e., point *C*) is the fluidized bed regime, the bed becomes empty at point *D* due to increased superficial velocity more than U_t.

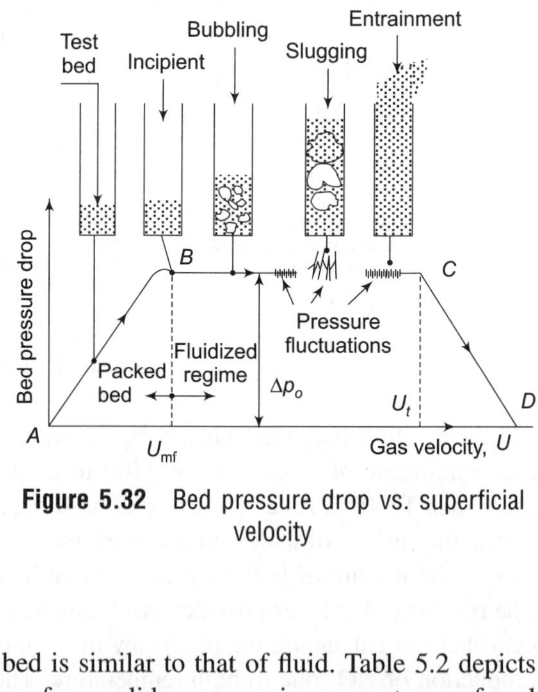

Figure 5.32 Bed pressure drop vs. superficial velocity

5.10.1 Regimes of Fluidization

The characteristics displayed by a fluidized bed is similar to that of fluid. Table 5.2 depicts a comparison of some characteristical features of gas-solid processes in vapour types of coal fired boilers given by Basu and Fraser [89].

Table 5.2 A comparison of characteristics of gas-solid processes [89]

Properties	*Packed bed*	*Fluidized bed*	*Fast bed*	*Pneumatic transport*
Boiler application	Stoker fired	Bubbling fluidized	Circulating fluidized	Pulverized coal fired
Mean particle diameter (mm)	< 300	0.03–3	0.05–0.5	0.02–0.08
Gas velocity through combustor zone (m/s)	1–3	0.5–3	3–12	15–30
Typical U/U_t	0.01	0.3	2	40
Gas motion	Upward	Upward	Upward	Upward
Gas mixing	Near plug	Complex	Dispersed	Near plug
Solid motion	Static	Up and down	Mostly upward	Upward
Solid-solid mixing	Negligible	Near perfect mixing	Near perfect mixing	Near plug flow
Overall voidage	0.4–0.5	0.5–0.85	0.85–0.99	0.98–0.998
Temperature gradient	Large	Very small	Small	Significant
Typical bed-to-surface heat transfer coefficient (W/m^2 K)	50–150	200–550	100–250	50–100
Attrition	Little	Some	Some	Appreciable
Agglomeration	Appreciable	Some	Without problem	Without problem

Figure 5.33 shows the gas-solid flow regimes for different types of combustion bed. The highest solid volume fractions is with static bed (*E*) and this value goes on decreasing with bubbling bed (*A*), turbulent bed (*B*), fast fluidized bed (*C*).

1. Fixed Bed. The pressure drop per unit height of a fixed bed [Fig. 5.34] (a)] of uniformly sized particles is given by Ergun's equation.

$$\frac{\Delta p}{L} = \frac{(1-\varepsilon)^2}{\varepsilon^3} \cdot \frac{\mu U}{(\phi d^2{}_p)} + 1.75 \left(\frac{1-\varepsilon}{\varepsilon^3}\right) \frac{\rho_g U^2}{\phi d_p} \quad \text{...(5.6)}$$

where U stands for superficial velocity, ε stands for void fraction in the bed, d_p and ϕ stand for the diameter and sphericity of particles respectively, μ and ρ_g stand for viscosity and density of the gas respectively.

The *voidage* (ε) is defined as

$$\varepsilon = \frac{\text{Volume of voids or gas}}{\text{Total volume of bed}} = \frac{\text{Volume of bed} - \text{Volume of solids}}{\text{Volume of bed}}$$

$$= 1 - \frac{\text{Volume of solids}}{\text{Volume of bed}} = 1 - \frac{m_s}{\rho_s} \cdot \frac{\rho_b}{m_b}$$

where m_s, m_b and ρ_b are the mass and density of steam and bed respectively. Since $m_s = m_b$, so

$$\varepsilon = 1 - \frac{\rho_b}{\rho_s} \quad \text{...(5.7)}$$

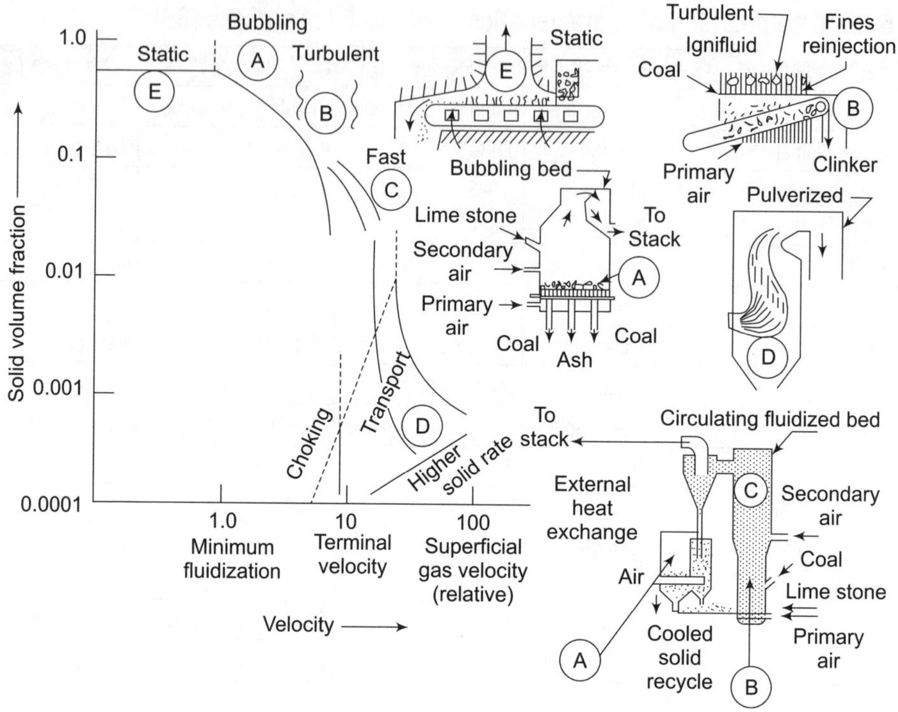

Figure 5.33 Gas-solid flow regimes (Source: 89)

The expression of voidage is valid for all types of bed.

The *sphericity* (ϕ) of a coal particle is defined as

$$\phi = \frac{\text{Surface area of a sphere of the same volume as the particle}}{\text{Surface area of the particle}} \quad ...(5.8)$$

2. Bubbling Fluidized Bed. As mentioned earlier, at the minimum fluidization velocity U_{mf}, the fuel bed starts behaving as a fluid [Fig. 5.34(b)] and under this condition the pressure drop across the bed is equal to the weight of the bed. Thus, the fluid drag, F_0 is expressed as.

$$F_0 = \Delta p A = AL\,(1 - \varepsilon_{mf})\,(\rho_s - \rho_g)g \quad ...(5.9)$$

where A and L = the cross-sectional area and height of bed respectively. ρ_s and ρ_g are the densities of solid fuel and gas respectively.

or
$$\frac{\Delta p}{L} = (1 - \rho_{mf})\,(\rho_s - \rho_g)g \quad ...(5.10)$$

The minimum superficial velocity, U_{mf} is obtained by solving Eqs. 5.11 and 5.16 in terms of Reynolds number.

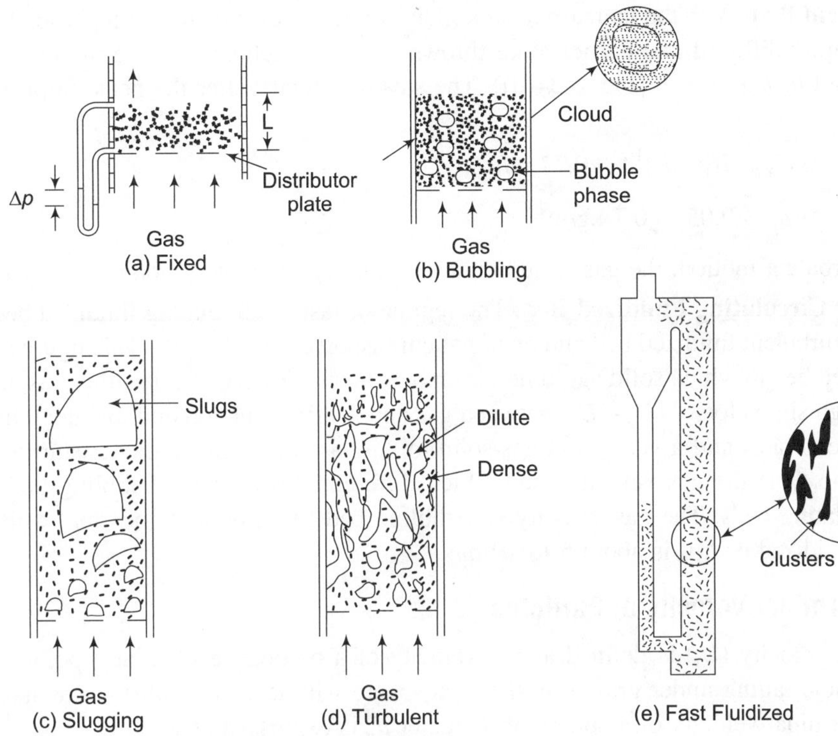

Figure 5.34 Gas-solid contacting mode for various beds

$$\text{Re}_{mf} = \frac{\rho_g\, d_p\, U_{mf}}{\mu} = [C_1^2 + C_2\, A_f]^{0.5} - C_1 \qquad \text{...(5.11)}$$

where A_r = Archimedes number

$$= \frac{\rho_g(\rho_s - \rho_g)gd_p^3}{\mu_g^2},\ C_1 = 27.2 \text{ and } C_2 = 0.0408$$

With the increase of superficial velocity beyond U_{mf}, the extra gas flow in the form of bubbles. The section of the bed outside the bubbles zone is known as *emulsion flow*.

In a bubbling bed, the bubble rise velocity (U_b)

$$U_b = K(U - U_{mf}) + 0.71\ \{gD_b\}^{1/2} \qquad \text{...(5.12)}$$

where $K \approx 1$ and D_b is the bubble diameter

3. Slugging. In the event of small bed cross-sectional area and deep bed, the bubble size may increase up to the diameter or width of the bed and thus the bubble then passes through the bed as a slug as shown in Fig. 5.34 (c). This phenomenon is known as slagging. The slagging produces unstable large fluctuation pressure drop across the bed.

4. Turbulent Bed. With the increase in superficial velocity beyond emulsion phase, the bed surface is highly diffused and particles are thrown into the freeboard (above the bed), such a bed is called *turbulent bed* [Fig. 5.34(d)]. The steady velocity after the peak amplitude is expressed as

$$U_k = 7(\rho_p\, d_p)^{1/2} - 0.77 \text{ m/s} \qquad \text{...(5.13)}$$

where $\rho_p\, d_p \approx 0.05$ to 0.7 kg/m^2

Due to turbulent motion, the gas-solid contact in this regime is very good.

5. Fast or Circulating Fluidized Bed. The regime of fast or circulating fluidized bed lies between the turbulent fluidized bed and pneumatic transport [Fig. 5.34 (e)]. This figure shows a fast fluidized bed in which solid agglomerates move up and down in a very dilute dispersion of solids. High slip velocity ($U_g - U_s$) between gas and solid, formation and disintegration of particle agglomerates and a very good gas-solid mixing are the main characteristic features of this regime. It is the gas velocity used which differentiates between bubbling beds and fast or circulating beds. The gas velocity in bubbling bed lies around 1 to 3 m/s while for circulating fluidized bed, it is about 5 to 10 m/s.

5.10.2 Terminal Velocity of Particles

The terminal velocity (U_t) is defined as the velocity of a particle reached at a point where a single particle falling under gravity in the atmosphere will accelerate till the gravitational force. The terminal velocity of a sphere of diameter d_p is expressed as

$$(\rho_s - \rho_g)\frac{g\pi d_p^3}{6} = \frac{1}{2}\rho_g\, U_t^2\, S\, C_D$$

where S is the frontal area $[(\pi/4)\, d_p^2]$ of a sphere, C_D is the drag coefficient to be determined experimentally, ρ_s and ρ_g are the densities of solid and gas respectively. Thus,

$$U_s = \left[\frac{(\rho_s - \rho_g)\, gd_p}{3\rho_g\, C_D}\right]^{1/2} \qquad \text{...(5.14)}$$

where, $$C_D = \frac{24}{R_{ep}} \text{ for } R_{ep} < 0.4 \qquad \text{...(5.15)}$$

and $$C_D = \frac{10}{(R_{ep})^{1/2}} \text{ for } 0.4 < R_{ep} < 500 \quad \text{where} \quad R_{ep} = \rho_g\, U_1\, d_p/\mu_g \qquad \text{...(5.16)}$$

The concept of terminal velocity is important as it constitutes the maximum velocity of the gas at which bubbling fluidized should be operated. Many solid materials undergo attrition when fluidized or fragmentation due to thermal stresses when put into a hot fluidized bed. Attrition and fragmentation result in the production of fines which becomes entrained in the gas.

5.10.3 Fluidized Bed Combustion of Solid Fuels

Figure 5.35 shows the simplest arrangement of fluidized bed combustion of solid fuels. The bed consists of inert particles such as sand, through which fluidizing air can be passed via a distributor plate at the bottom. It is worth to note that the bed normally contains only a small percentage of burning fuel (about 1 to 2%), so that the inert particles carry the heat away from the burning particles, thus keeping their temperature below the melting point of ash. In general, the bed temperature is controlled around 850°C by removing the heat from the bed by installing cooling tubes. The variation of bed temperature with time from the starting time till it reaches to steady state is depicted in Fig. 5.36. During starting up to solid fuel ignition temperature (about 450° to 500°C) depending on the type of fuel by burning fuel gas or oil in the bed and this supply of gas or oil is stopped gradually when self sustaining combustion of solid fuel is attained.

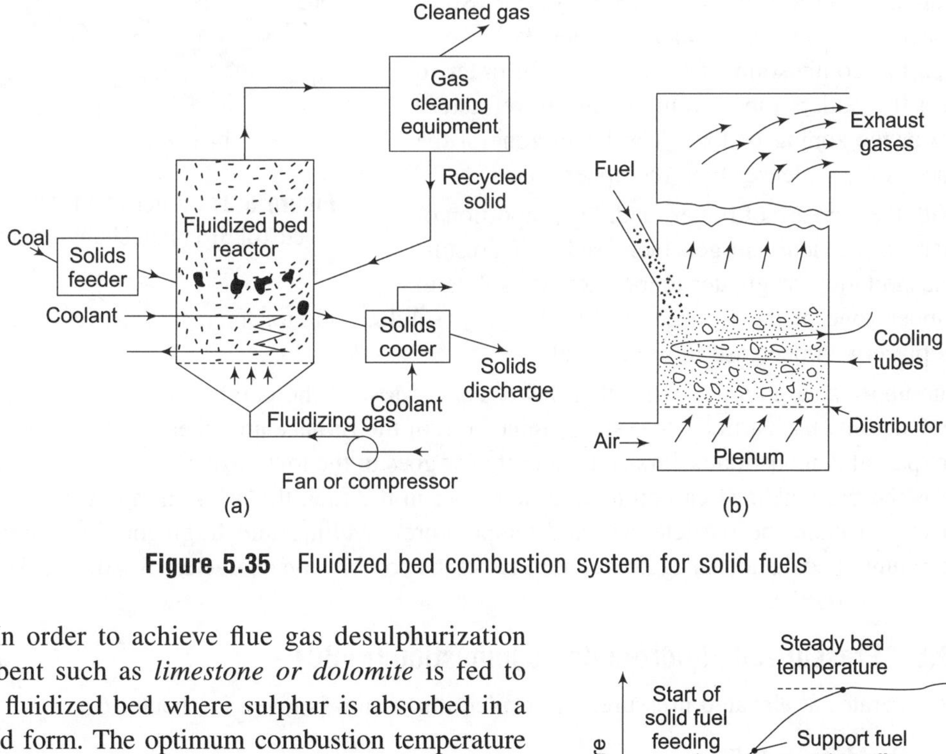

Figure 5.35 Fluidized bed combustion system for solid fuels

In order to achieve flue gas desulphurization sorbent such as *limestone or dolomite* is fed to the fluidized bed where sulphur is absorbed in a solid form. The optimum combustion temperature is around 850°C. It has been estimated that the amount of sorbent required to absorb about 90% of the sulphur in the fuel is about in the molar ratio of about 3 of calcium-sulphur. The variation of bed temperature is shown in Fig. 5.36.

There are various *advantages* of fluidized bed combustion and that are given in *chapter 6* dealing with fluidized bed combustion steam generator.

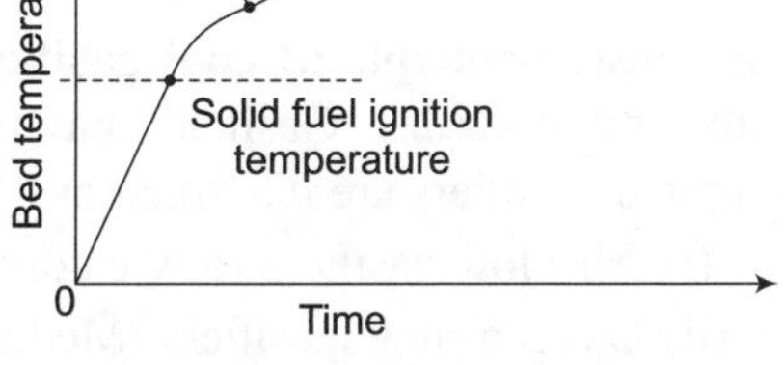

Figure 5.36 Transient fluidized bed combustor temperature variation

5.10.4 Circulating Fluidized Bed (CFB) Combustion System

As mentioned above, the circulating fluidized bed uses a much higher gas velocity (5 to 10 m/s) as compared to bubbling fluidized bed (1 to 3 m/s). Figure 5.37 shows a circulating fluidized bed combustion system which is divided into three distinct zones namely (i) lower zone (below the secondary air entry, (ii) upper zone (above the secondary air entry), (iii) hot gas/ solid separator.

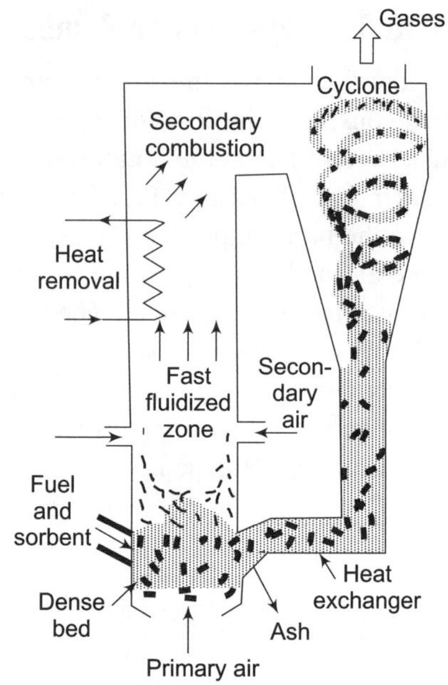

Figure 5.37 Circulating fluidized bed (CFB) combustion system

Primary combustion air constituting about 40 to 80% of the stoichiometric air (limestone or dolomite) as well as unburnt char from the hot cyclone. Since this section is oxygen deficient, so devolatilization and partial combustion of char occurs. In order to protect the boilers tubes from corrosion, refractory lining or the zoning is done. The lower zone of the furnace is much denser than the upper zone.

With the increase of boiler load, the proportion of primary to secondary air gets increased which results in transportation of greater amount of hot solids to the upper zone of the furnace and by this way, solid circulation increases. Reverse action takes place with decrease in boiler load.

The upper zone is much taller than lower zone. Most of the combustion occurs in upper zone. Unburnt char particles that are entrained out of the furnace are received by the cyclone where partial combustion occurs and unburnt char goes to the lower zone. Various parameters such as the ratio of fixed carbon to volatile matter in the fuel, the attrition rate, the intrinsic reactivity of char, the particle size and temperature, swelling and fragmentation, ignition temperature of coal and ash agglomeration affect the combustion characteristics of circulating fluidized bed (CFB).

5.10.5 Pressurized Fluidized Bed Combustion (PFBC)

PFBC operates at elevated pressure. Its detailed discussion is given in chapter 6.

5.11 COAL GASIFIER

The basic principle of coal gasification has been discussed in Previous Chapter 4 in Sub-section 4.5.2.1. Gasifiers may be classified in many ways. Based on the oxygen or air supplied, gasifiers are classified as

(i) Air-blow gasifiers (low calorific value; 5000-6000 kJ/kg)

(ii) Oxygen-blow gasifiers (Medium calorific value; 15,000 kJ/kg)

Based on the flow pattern in the gasifier, they are classified as

(i) Fixed gasifier using counter-current flow

(ii) Fluidized bed gasifier with bubbling or circulating bed

(iii) Entrained bed gasifiers with co-current flow.

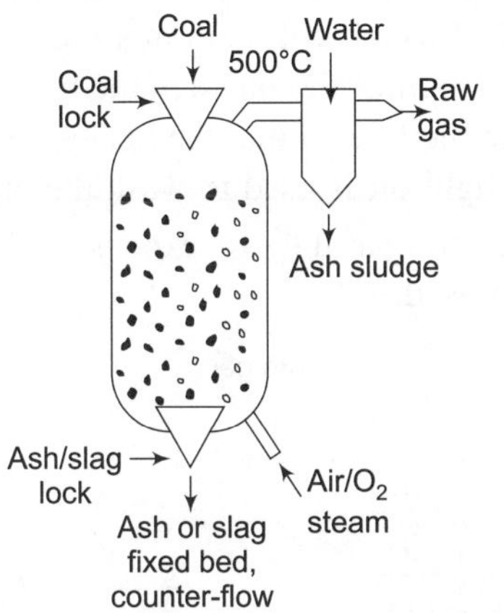

Figure 5.38 Fixed bed counter-flow gasifier

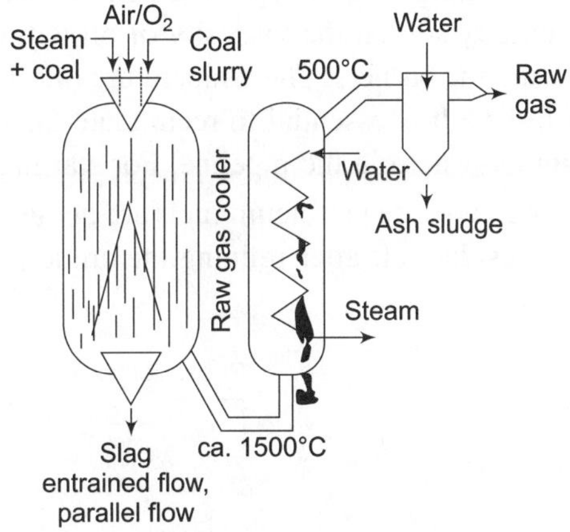

Figure 5.39 Entrained type

The commercial gasifiers are manufactured by British Gas/Lurgi is known as slagging gasifier and by Texaco, Shell, DOW, Prenflow as entrained type. Both gasifiers are oxygen-blown. Figures 5.38 and 5.39 show fixed bed and entrained type gasifier. Fluidized bed gasifier is shown in Fig. 5.40. Figure 5.41 shows the integrated gasification combined cycle (IGCC). The various steps are shown in block forms.

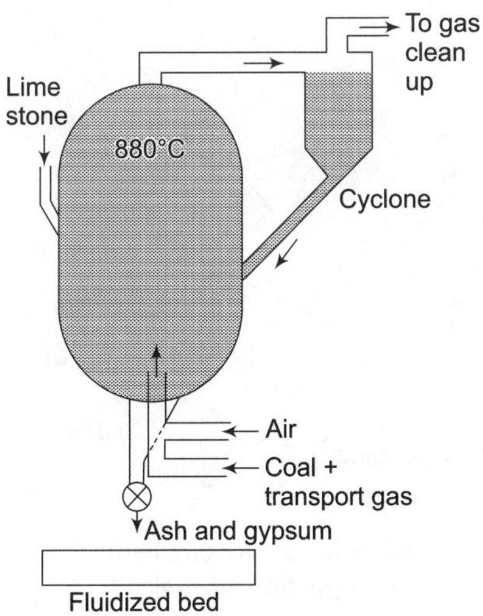

Figure 5.40 Fluidized bed gasifier

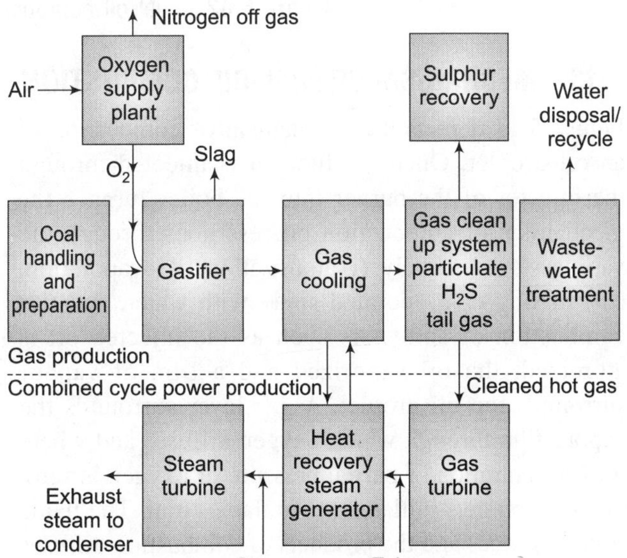

Figure 5.41 Basic IGCC power plant

5.12 FUEL OIL COMBUSTION SYSTEM

Figure 5.42 shows a fuel oil combustion system. It is essential to heat the fuel oil as its viscosity is high so that it can be easily pumped. This is accomplished by installing steam heating coils in the fuel tank. Between the heater pump strainer/filter is provided. The pump may be reciprocating or gear type. Further, heating is provided to the pumped oil either by steam or electricity so that the viscosity of oil is further reduced which will help in atomization after leaving the burners. The temperature of oil entering the burner should be in the range of 65°C to 115°C. It is essential to recirculate the oil to purge the line so that the heavy oil does not plug anywhere in the pipeline. For starting, initially light oil is used to wash the pipeline.

High volatile petroleum fuels are not used for boiler firing as they are expensive. The heavy oils, residue left after refining the crude petroleum is used.

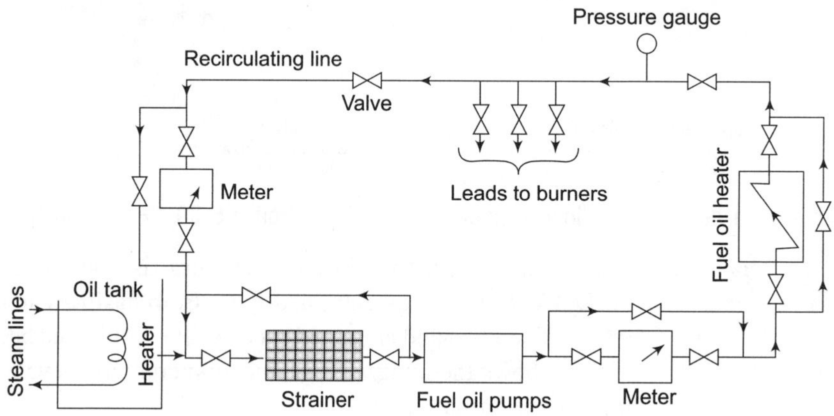

Figure 5.42 Fuel oil combustion system

5.13 MECHANISM OF FUEL OIL COMBUSTION

Figure 5.43 depicts the mechanism of combustion of an oil droplet. Once the fuel oil is injected through the injector of the burner into the boiler furnace the mechanism of combustion process goes through the various stages, namely: (i) atomization, (ii) evaporation, (iii) mixing of evaporated spray with water, and (iv) combustion of spray. As soon as the injected oil is atomized, the oil vaporized and a film of vapour surrounds the oil droplet. A gas layer surrounds the vapour film through which oxygen diffuses and where the concentration or partial pressure of oxygen is zero, the reaction gets initiated at the flame front. The flame front advances and the products of combustion diffuses out to the bulk fluid.

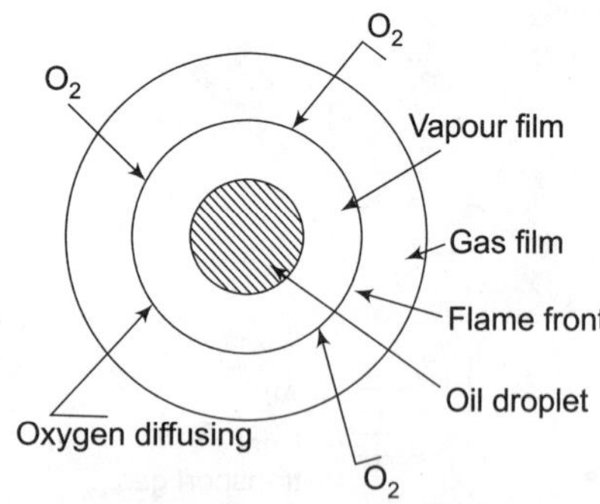

Figure 5.43 Mechanism of combustion of fuel oil

5.14 OIL BURNERS

Various types of oil burners are used in power plant practice. Atomization is the first requirement from an oil burner. It is accomplished mechanically or with steam, or air stream. Figure 5.44 shows the steam or air atomized oil burners. In Fig. 5.44(a) oil and steam or air passes through separate passage made in the burner. The shape at the exit of each passage is such that mixing starts immediately. Figure 5.44(b) uses an annulus, the inner passage is meant for oil. In Fig. 5.44(c), steam or air passes in straight passage while passage is inclined, Fig. 5.44(d) shows a converging passage for both. In Fig. 5.44(e), mixing starts before the burner exit, while arrangement shown in Fig. 5.44 (f) uses double channel for oil and through central passage, air or steam passes. The large size boiler uses pressure atomizer (6 to 20 bars) with tangential feed. The control of fuel oil in such burners is either done by oil recirculation or using piston which covers partly or fully the tangential shot as per load. (Fig. 5.45).

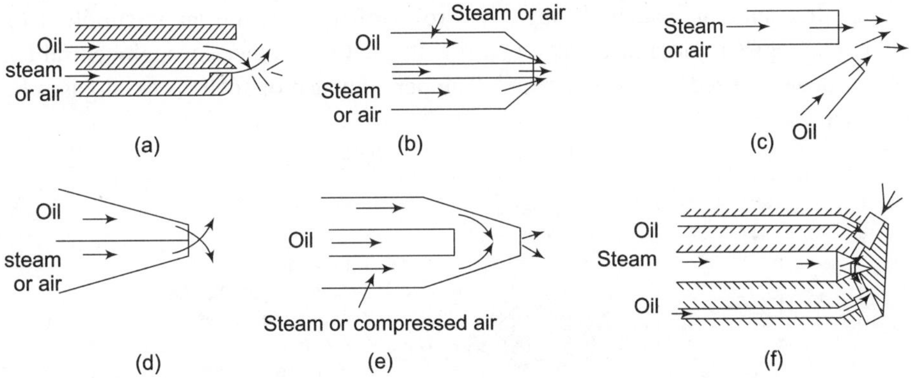

Figure 5.44 Air or steam atomized oil burners

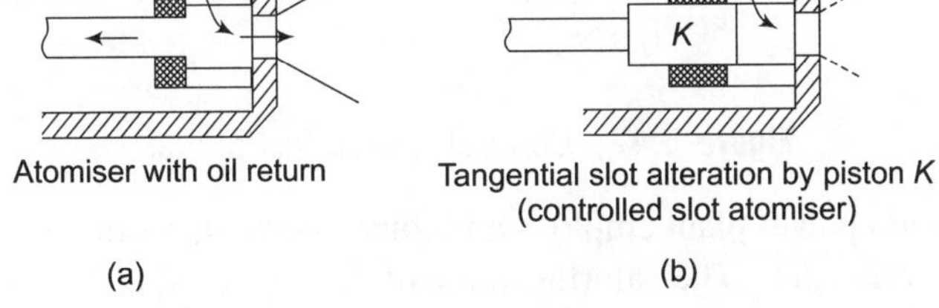

Figure 5.45 Output control of pressure atomizer

5.15 GAS BURNER

It is a well-known fact that the burning of gas is easy and clean. It does not require any atomization. Proper mixing of gas and air is made in the form of thin jets of high penetrability (Fig. 5.46).

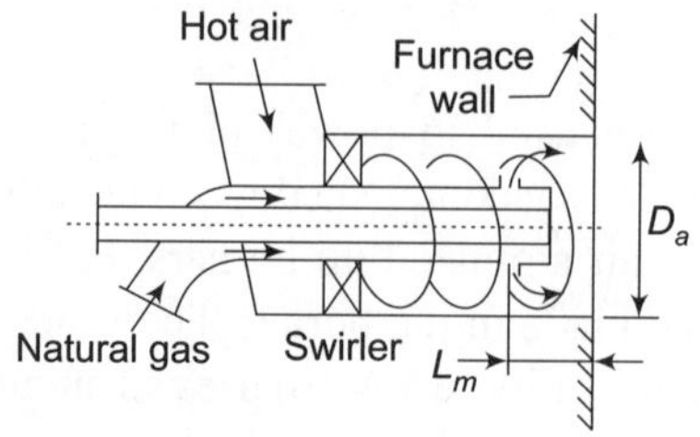

Figure 5.46 Gas burner with central air supply

5.16 COMBINED GAS-FUEL OIL BURNERS

In many situations, gas as well as oil is used for generation of steam in the boiler. This is taken care of by using single combined gas fuel oil burner as changing from one fuel to another is easy and it may be burnt with optimal conditions (Fig. 5.47). Natural gas flows from the central annular header through two rows of holes while air is supplied through tangential vane register and a disc valve controls it. Fuel is supplied through central channel. An electric means is used to ignite the fuel. The performance of such co-axial gas-oil burner is quite satisfactory.

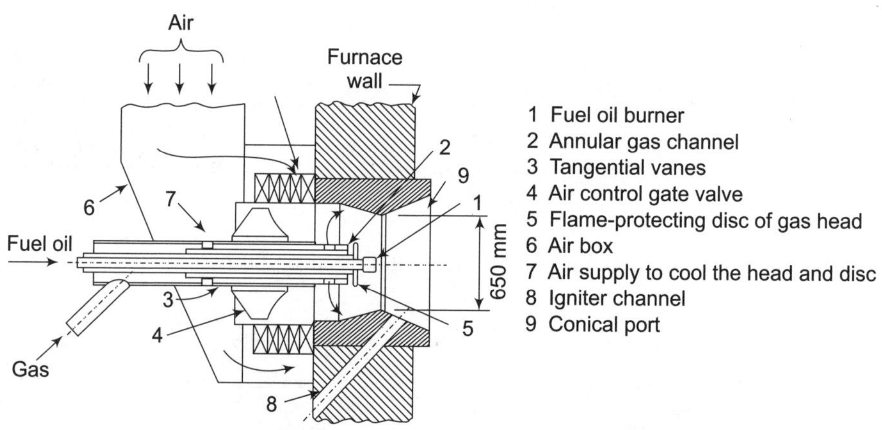

Figure 5.47 Co-axial gas-oil fuel burner

Example 5.1 A steam power plant employs a fluidized bed combustion system and its output is 150 MW at 90% efficiency. The calorific value of the coal used is 26 MJ/kg with a sulphur content of 3.8%. In order to limit SO_2 emission, limestone is fed to it at a calcium-sulphur ratio of 3.0. The limestone used contains 88% $CaCO_3$. Calculate: (a) the coal burning rate and (b) mass flow rate of limestone to be fed with coal.

Solution (a) The coal burning rate is expressed as

$$\dot{m}_f = \frac{150}{0.9 \times 26} = 6.41 \text{ kg/s} = 23076 \text{ kg/h} = 23.076 \; t\text{/h} \quad \textbf{\textit{Ans.}}$$

(b) The flow rate of sulphur with coal $= \dfrac{23076 \times 0.038}{32} = 27.4$ kmol/h

Calcium-sulphur ratio is 3:0. Thus, the flow rate of calcium is = 27.4 × 3.0 = 82.2 kmol/h. The mass flow rate of $CaCO_3$ required = 82.2 M = 82.2 × 100 = 8220 kg/h where molecular weight of $CaCO_3$ is 100.

Thus, the mass flow of limestone = 8220/0.88 = 9340.9 kg/h = 9.3409 *t*/h ***Ans.***

Example 5.2 A fluidized bed combustor uses high voltalile coal with calorific value 24 MJ/kg. The design of the bed and the combustion system allows the release of only 70% of calorific value in the bed and remainder in the above bed zone from which the combustion products leave at 850°C. The inlet temperature of air is 30°C, the bed temperature is 850°C and the air-fuel ratio by mass is 14 : 1. The specific heat of gas leaving the bed surface is 1.05 kg/kgK. If the burning rate of coal is 20 t/h, calculate the rate of heat removal (a) from the bed and (b) from the above-bed zone of the combustion system.

Solution (a) The heat removed from the fluidized bed is given by

$$= \text{Heat released} - \text{heat carried away by the products of gases}$$

$$= 0.7 \times 24000 - (14 + 1) \times 1.05 \times (850 - 30) = 3885 \text{ kJ/kg}$$

Thus, the rate of heat removal from the bed is given by

$$= \frac{3885 \times 20 \times 10^3}{3600} = 21583.3 \text{ kW} = 21.5833 \text{ MW.}$$ ***Ans.***

(b) The rate of heat removal from the above-bed zone is

$$= \frac{20 \times 10^3 \times 0.3 \times 24000}{3600} = 40 \times 10^3 \text{ kW} = 40.6 \text{ MW}$$ ***Ans.***

Example 5.3 A fluidized bed operates at atmospheric pressure having bed temperature of 850°C. The fuel used has a calorific value of 26 MJ/kg. The excess air supplied is 30%. The heat release rate is 120 MW. Assume density of air at 850°C is 0.32 kg/m^3. Calculate the platform area required for two cases, namely, if (a) the firing rate is 2 MW/m^2 and (b) the fluidized velocity is 3 m/s. Take A/F = 14.

Solution (a) The platform area $= \dfrac{120}{2} = 60 \text{ m}^2$ ***Ans.***

(b) The fuel burning rate $= \dot{m}_f = \dfrac{120 \times 10^3}{26 \times 10^3} = 4.615$ kg/s

The air flow rate $= \dot{m}_a = 14 \times 1.3 \times 4.651 = 84.648$ kg/s

The platform area $= \dot{m}_a/\rho_a\, U = 84.648/0.32 \times 3 = 88.175 \text{ m}^2$ ***Ans.***

Example 5.4 In a fluidized bed boiler, the fuel particles of mean size of 430 μm is used under the ambient conditions. The gas density is 1.22 kg/m^3 and its viscosity is 1.82 × 10^{-5} kg/ms. The density of the loosely packed bed is 1650 kg/m^3 while that of solid is 2800 kg/m^3. Calculate (a) voidage of the bed, and (b) the minimum fluidization velocity.

Solution The voidage at the minimum fluidization velocity is given by

$$\varepsilon_{mf} = 1 - \frac{\rho_b}{\rho_s} = 1 - \frac{1650}{2800} = 0.41 \qquad \textbf{\textit{Ans.}}$$

(b) The minimum fluidization velocity is expressed by Archimedes number (Ar);

$$Ar = \frac{\rho_g(\rho_s - \rho_g)gd_p^3}{\mu_g^2} = \frac{1.22(2800 - 1.22)\ 9.81 \times (430 \times 10^{-6})^3}{(1.82 \times 10^{-5})^2} = 8040$$

Ar is related to Re_{mf} as follows

$$Re_{mf} = [C_1^2 + C_2 \cdot A_r]^{0.5} - C_1 = [27.2^2 + 0.0408 \times 8040]^{0.5} - 27.2 = 5.478$$

But $$Re_{mf} = \frac{U_{mf}\,\rho_g \cdot d_p}{\mu_g} = 5.478 \text{ or } U_{mf} = \frac{5.478 \times 1.82 \times 10^{-5}}{1.22 \times 430 \times 10^{-6}} = 0.19 \text{ m/s}$$

Ans.

Example 5.5 A fluidized bed of coal particles has a mass of 3000 kg with a density of 2700 kg/m^3. The mean particle size is 850 μm having sphericity of particles is 0.84. Calculate the total surface area of the particles in the fluidized bed.

Solution The surface to volume ratio of bed is given by

$$S_v = \frac{A_s}{V_s} = \frac{N_p \pi d_p^3}{N_p\ (\pi/6) d_p^3} = \frac{6}{d_p} \text{ where } N_p \text{ is the number}$$

For no-spherical particles, the surface area of the particles is expressed as

$$A_s = \frac{6V}{\phi d_p} = \frac{6 \times m_s}{\phi d_p\,\rho_s} = \frac{6 \times 3000}{0.84 \times 850 \times 10^{-6} \times 2700} = 9337 \text{ m}^2 \qquad \textbf{\textit{Ans.}}$$

PROBLEMS

Things to Think About

5.1 What criteria should be adopted for selecting combustion equipment for coal?

5.2 What is the physics involved in combustion of coal as a fuel? How will be a complete combustion possible?

5.3 What criteria should be adopted in selecting fuel handling system for pulverized coal based steam power plant?

5.4 What are the basic principles of combustion of oil and gaseous fuels?

Short and Long Theoretical Questions

5.1 What are the various combustion equipment for coal?

5.2 Discuss the working principle of overfeed and underfeed fuel bed systems.

5.3 What are the various mechanical stoker for fuel bed combustion?

5.4 Discuss the working principle of travelling grate stoker, chain grate stoker, spreader stoker, vibrating grate stoker and underfeed stoker.

5.5 What do you mean by pulverized coal firing system?

5.6 List the advantages and disadvantages of pulverized coal firing system.

5.7 Discuss the working principle of various types of crushers.

5.8 What are the various types of pulverized coal system?

5.9 With the aid of simple sketch discuss the working principle of bin or storage system, direct firing system and semi-direct firing system.

5.10 Discuss the design aspects of pulverized coal fired furnace.

5.11 Mention the various types of pulverized coal furnace.

5.12 Discuss the working principle of dry-bottom and wet-bottom furnace.

5.13 Mention the various types of pulverized coal burners.

5.14 What are the requirements of pulverized coal burners?

5.15 With the help of sketch discuss the working principle of circular c/s, rectangular c/s, distributed mixing, burners, low NO_x burners and combined burners.

5.16 What are the various arrangement of burners?

5.17 What is tangential or corner firing? What are merits and demerits?

5.18 What is a cyclone furnace? Mention uses, advantages and disadvantages.

5.19 What is a fluidized bed? Define minimum fluidization velocity. What are the various types of fluidized bed?

5.20 Discuss the various regimes of fluidization.

5.21 Define voidage and sphericity of a particle.

5.22 Discuss the working principle of bubbling and circulating fluidized bed combustion.

5.23 What is the advantage of using limestone or dolomite as the bed material in a fluidized bed?

5.24 Mention and optimum combustion temperature and the approximate calcium sulphur molar ratio for maximum absorption of sulphur oxides.

5.25 What is pressurized fluidized bed combustion (PFBC)?

5.26 Explain the sequence of events of burning of a fuel particle in a fluidized bed.

5.27 With the help of sketch discuss, the working principle of fluidized bed combustion of solid fuels.

5.28 What are the various types of coal gasifiers?

5.29 Discuss the working principle of fixed bed counter flow and entrained parallel flow.

5.30 Discuss the working principle of integrated gasification combined cycle.

5.31 With the help of a sketch, discuss the working principle of fuel oil combustion system.

5.32 Discuss the mechanism of fuel oil combustion.

5.33 Sketch the various types of oil burners and describe their merits and demerits.

5.34 Sketch the gas burner and explain its working.

5.35 With the help of sketch discuss the combined gas-fuel oil burners.

Numerical Questions

5.1 A boiler used a fluidized bed combustion system having an output of 50 MW operating at 80% efficiency with a coal heating value of 26 MJ/kg with a sulphur content of 4%. The system requires a limestone to be bed at a calcium-sulphur ratio of 3.2 to limit SO_2 emission. The limestone used contains 86% of $CaCO_3$. Calculate the consumption rate of limestone.

5.2 A pulverized coal bed boiler uses coal of mean size 480 μm. The air density is 1.22 kg/m^3 and the viscosity is 1.88 × 10^5 kg/m.s. The bulk density of locally packed bed is 1700 kg/m^3. The density of the solid in 2800 kg/m^3. Determine (a) the voidage of the bed, (b) the minimum fluidization velocity, and (c) the sphericity of particles.

Software Based Question

5.1 Develop a software for the design of combustion equipment and fuel handling systems for a steam power plant.

Design and Open Ended Question

5.1 Write a comprehensive report with design features of all types of combustion equipment and fuel handling systems for all modern power plants.

Multiple-Choice Questions

5.1 Modern boilers used in steam power plant employ
(a) travelling grate stoker firing system (b) chain grate stoker firing system
(c) pulverised coal firing system (d) cyclone firing system

5.2 Fluid bed combustion system offers
(a) low NO_x (b) high NO_x
(c) low NO_x and low slagging (d) high NO_x and high slagging

5.3 In fluidized bed combustion, the mean particle diameter of coal ranges from
(a) 0.03 mm to 0.3 mm (b) 0.3 mm to 0.6 mm
(c) 0.6 mm to 1 mm (d) 1 mm to 2 mm

5.4 The gas velocity in combustion zone of fluidized bed combustion system ranges from
(a) 0.5 to 3 m/s (b) 2 m/s to 4 m/s (c) 3 m/s to 5 m/s (d) 4 to 6 m/s

5.5 Heat transfer coefficient in fluidized bed is
(a) less than packed bed (b) more than packed bed
(c) equal to packed bed (d) more than fast bed

5.6 Excess air required for coal is
(a) more than oil only (b) less than oil
(c) less than gas (d) more than oil and gas

Answers to Multiple-Choice Questions

5.1 (c) 5.2 (c) 5.3 (a) 5.4 (a) 5.5 (b) 5.6 (d)

CHAPTER

6

Steam Generators, Ash Handling Systems and Feed Water Treatment

OVERVIEW

The function of a steam generator (i.e., boiler) is to generate steam at a desired pressure and temperature by transferring heat produced by burning of fuel in a furnace to water to change it into steam. According to ASME Code, a steam generating unit is defined as follows:

"A combination of apparatus for producing, furnishing or recovering heat together with the apparatus for transforming the heat so made available to water which could be heated, vaporized and superheated to steam form".

Steam generators are used in both fossil fuel and nuclear fuel electric generating stations. In this chapter, only fossil fuel steam generators are discussed and nuclear fuel steam generators will be discussed in the chapter of "Nuclear Power Plant".

A steam generator is very complex, integration of furnace, drum, a evaporator, superheater, reheater, economiser and air preheater along with various mountings and auxiliaries such as safety valves, pulverizers, burners, fans, stokers, dust collectors and precipitators, ash-handling system, chimney (or stack), etc. Superheaters, reheaters, economizers and preheaters are called accessories of boiler. The water is supplied to the boiler by a feed pump (multi-stage centrifugal pump) at a pressure slightly higher than the pressure at which steam is generated. *Heating takes place at constant pressure and there is only a conversion of liquid phase of water into steam in the boiler.* This chapter deals with steam generator and their ash handling systems and feed water treatment.

6.1 CLASSIFICATION OF STEAM GENERATOR

Steam generators are classified on the basis of many parameters *such as tube contents, method of firing, pressure of steam, method of circulation of water, nature of service, position and number of drums, gas passage, heat source, once through boiler, fluid used, boiler shell material,* etc.

(a) Based on the contents of tubes, steam generators or boilers are classified as: (i) fire-tube boilers, and (ii) water-tube boilers.

(i) Fire-tube Boilers. *In the fire-tube boilers, the hot gases (i.e., flue gases) pass through tubes and water surrounds them.* Figure 6.1 shows the principle of operation of fire-tube boilers. The products of combustion (hot gases) leaving the furnace pass through fire tubes which are surrounded by water. Heat from hot flue gases is transferred to water which is converted into steam. The spent flue gases are then discharged to atmosphere through the chimney (stack). The examples of fire-tube boilers are Cochran, Lancashire, Cornish and Locomotive boilers.

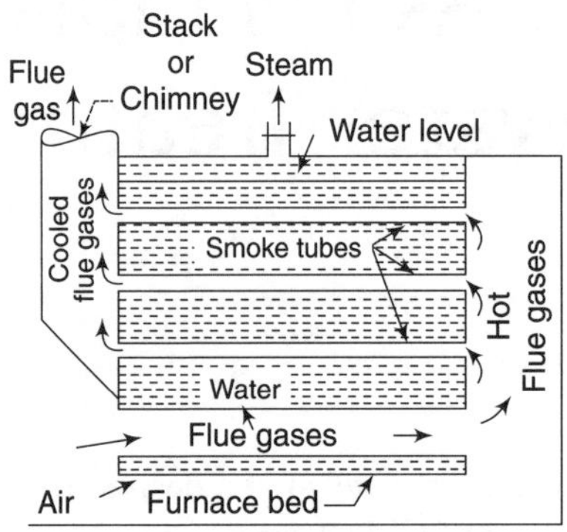

Figure 6.1 Fire-tube boiler

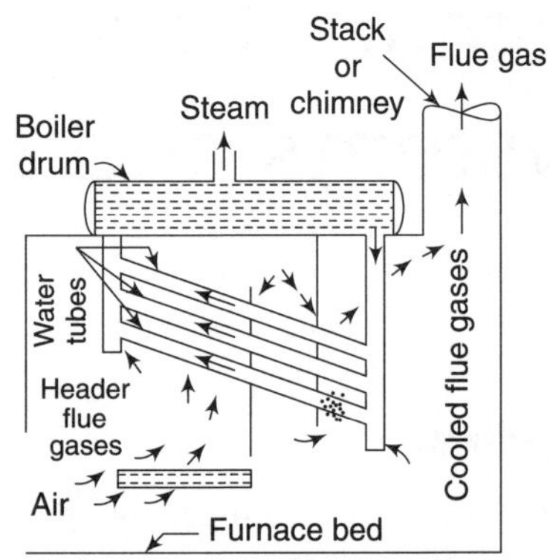

Figure 6.2 Water-tube boiler

(ii) Water-tube Boilers. *In water-tube boilers, water flows inside the tubes and the hot flue gases flow outside the tubes.* Figure 6.2 shows the principle of operation of water tube boilers. A bank of water tubes containing water is connected with a steam-water drum by means of two sets of headers. The hot flue gases from the furnace pass over the tubes and are discharged through the chimney. The water thus absorbs heat from the hot gases and evaporates in the form of steam. The steam thus formed gets accumulated in the steam space of drum from here it may be taken to superheater to superheat steam.

The examples of the water-tube boilers are Babcock and Wilcox boiler, Stirling boiler, Power boilers for utility, etc.

(b) Based on *nature of service,* steam generators or boilers are classified as follows:

(i) Utility boilers (ii) Industrial boilers

(iii) Marine boilers (iv) Locomotive boilers (now obsolete)

(i) Utility Boilers. These boilers or steam generators are used to utilities for electric power generating plants. Based on whether the pressure of steam is below or above the critical pressure (p_c = 221.2 bars), they can be either *subcritical* or *supercritical boilers.* The subcritical boilers are invariably water tube-drum type and they usually operate at between

130 bars and 180 bars and steam is superheated between 540°C and 560°C with one or two stages reheating, whereas supercritical boilers are drumless once-through type and operate at 240 bars or higher and steam is superheated between 570°C to 600°C with one or no stage reheating. The steam capacities of modern utility boilers ranges from 120 to 1300 MW with power plant unit output ranging from to 120 to 1300 MW.

So far the fuel is concerned, pulverized coal is mainly used. However, LNG and natural gas are also used. Now *circulating fluidized bed* steam generator are also gaining popularity.

(ii) Industrial Boilers. These boilers are used in process industries like sugar, carpet, paper, jute and so on, and institutions like hospitals, commercial and residential buildings. The sizes of industrial boilers are small. In general, the fuel used is coal or oil and it may be pulverized or fluidized or fired with the help of stoker. They can be heat recovery type using the waste of industries. Now, oil-fired boiler is gaining popularity. The pressure generated varies from 3 to 100 bars with steam capacities vary up to 110 kg/s. They supply dry and saturated steam or wet steam or even hot water. Superheated steam is not common with industrial boilers.

(iii) Marine Boilers. They are used by marine ships and ocean liners and in general prime mover is steam turbine. They are usually oil-fired. They operate at pressure between 50 to 70 bars and steam is superheated up to 540°C.

(iv) Locomotive Boilers. They were fire-tube type and used in locomotive which **are now obsolete**. At normal load, boiler had an evaporating capacity of 8500 kg/h at 14 bars and 370°C temperature and the rate of firing coal was 1585 kg/h.

6.2 CONSTRUCTION AND WORKING OF FIRE-TUBE BOILERS

Fire-tube boilers are not used in utility power plants but they are used in industrial boilers, locomotives (now obsolete) and river launches (now obsolete). They operate at pressure equal or less than 18 bars with steam capacity upto 6.5 kg/s.

Fire-tube boilers are suitable for small steam requirements. The advantages are low first cost, reliability in operation, quick response to load changes, need of only unskilled labour, less draught required, relatively inexpensive, etc.

A fire-tube boilers may be externally fired (e.g., locomotive boiler, Lancashire boiler, horizontally return tubular (HRT) boilers, etc.) or it may internally fired (Scotch marine boilers, package boilers, etc.).

The major shortcomings of the fire-tube boilers is the definite size and pressure limitation inherent in its basic design. In other words, the maximum size of the unit and the maximum operating pressure are limited. The tensile stress on the drum wall is a function of the drum diameter, and the internal pressure. The thickness of the drum is given by the well-known stress formula for this cylinder, $t = pd/2\sigma$ where p is the gauge pressure (N/m^2), σ = tensile stress (N/m^2), d = internal diameter of the shell (m), and t = thickness of wall (m). Thus, with high steam pressures and capacities, the thickness would become very large. Further, they are also subjected to large scale deposit and susceptible to boiler explosions, and become very costly.

6.2.1 Horizontally-Return Tubular Boiler

The horizontal-return tubular (HRT) boilers are very common in USA. Figure 6.3 shows a HRT boiler. Its simplicity and cheapness are the main advantages. They produce pressures up to approximately 17.5 bars with capacities from 4500 to 7000 kg/h. They are not suitable for large power plants because of their small capacities, limited steam pressure and low rates of steam capacity.

Construction. The boilers consist of a horizontal cylindrical shell closed at each end by a flat tube sheet or head. A large number of fire tubes which are usually 75 to 100 mm in diameter are supported between the two end covers are expanded into the cover at each end, thus serving not only the flues through which the hot combustion gases flow but also act as tie rods to hold the flat and covers in place against the steam pressure in the boiler.

The flat surface of the heads above the tubes are braced to the boiler shell by diagonal braces. The normal water level is such that all the fire tubes remain immersed in water. The water level in the boiler is indicated by a water column which is connected to the boiler by two pipes, one above and the other below the water level.

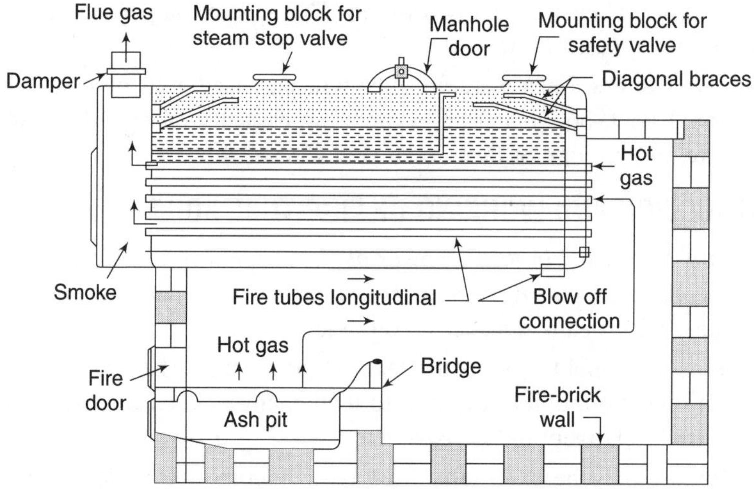

Figure 6.3 Horizontal-return tubular (HRT) boiler

The inside surface of the furnace wall is lined with fire bricks. A brick wall bridge is provided to deflect the hot combustion gases, A fire door is provided along with a single retort under feed stoker. It is an externally fired boiler.

A blow-off line is connected to the bottom of the drum at the rear. Valves in this line are opened periodicallly and some of the boiler water is blown down to a sewer, thus cleaning out the impurities carried to the boiler by the feed water. In small boilers of this type, chemicals are added to the feed water to prevent scale forming impurities in the feed water from precipitating on the heating surfaces as an adherent scale. If the boiler produced dry steam all these impurities remain in the boiler and have to be periodically removed by blowing down.

The boiler shell is provided with suitable openings for the attachment of spring loaded safety valves, feed water inlet, steam stop valve, manhole, etc., as boiler mountings.

Working. Since the furnace grate is below the front portion of the water shell, the hot gases of combustion heat the bottom of the shell and are deflected by the brick wall bridges to travel its full length as shown by the arrow in Fig. 6.3 to the rear of the boiler. From this position, the hot combustion gases pass into the horizontal tubes and return to the front of the boiler where they pass through the smoke box and are finally discharged to the chimney. The damper provided in the smoke box regulates the quantity of hot combustion gases. Thus, the hot gases complete two passes, one to the back and often from back to the front.

The water circulation in the boiler is natural circulation caused by convection currents. In one type of arrangement shown in Fig. 6.4(a), the fire tubes are staggered and the circulation of water takes place down the shell walls and then up; as shown by the arrows. Figure 6.4(b) shows of other types of arrangement in which internal baffles are provided to help circulation of water. The arrows show the circulation of water.

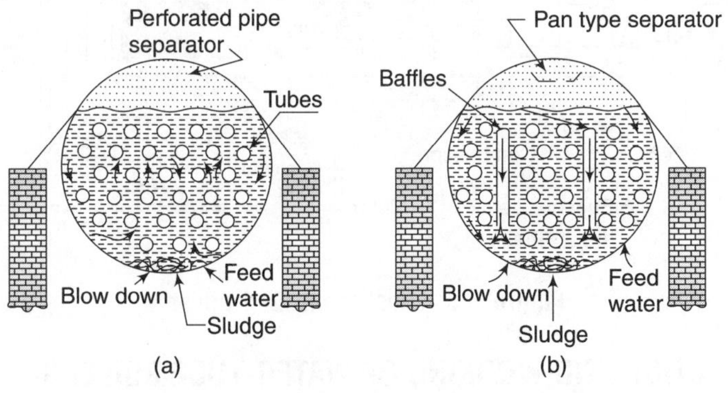

Figure 6.4 Water circulation in HRT boiler

6.2.2 Oil-fired Package (Nestler) Boiler

It is an oil fired, fire tube horizontal boiler having working pressure, temperature and steam capacity equal to 11.8 bars gauge, 238°C and 545 kg/h respectively. Water is fed to the boiler through a feed check valve with the help of a multi-stage centrifugal pump run by an electric motor. When the boiler is in operation, a duplex feed pump (reciprocating type) run by a steam engine using superheated steam feeds the water to the boiler to save the first class energy. The light diesel oil from the tank is pumped by a fuel pump (gear type) and it passes through an oil heater to decrease its viscosity for better atomization and is finally injected in the combustion chamber of boiler by an injector. The air is supplied to the combustion chamber with the help of centrifugal blower through swirler for rapid and proper combustion.

The hot flue gases formed as a result of combustion make three passes as shown in Fig. 6.5 and finally escape through the chimney. The number of fire tubes employed is 88. The wet steam formed as a result of heat taken by water from flue gases is allowed to pass through the superheater tubes and finally becomes superheated. Various mountings such as steam stop valve, spring loaded safety valve, water level indicators, pressure gauge, manhole, fusible plug, etc., are installed over the boiler for its safe and reliable operations.

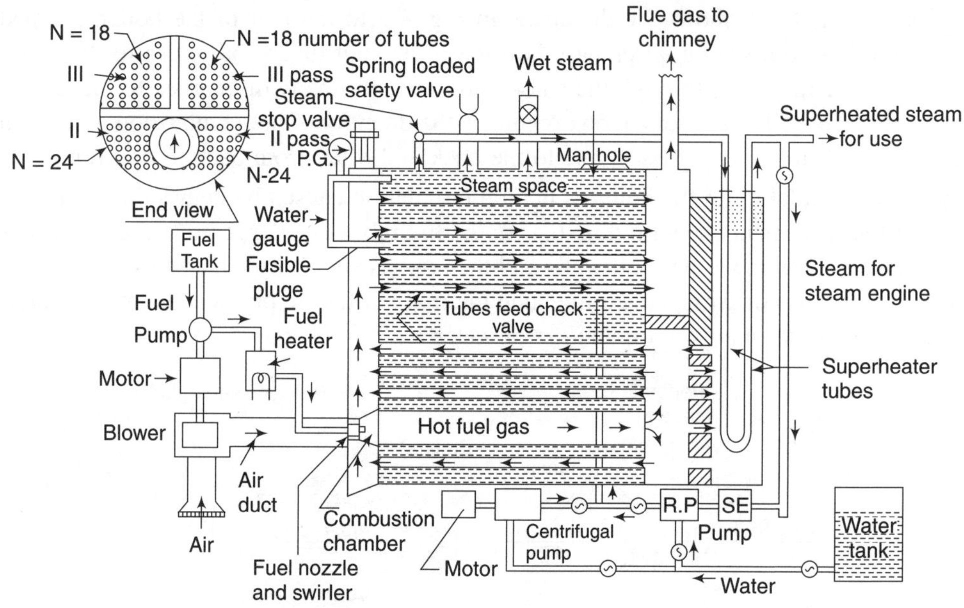

Figure 6.5 Nestler oil fired boiler

6.3 CONSTRUCTION AND WORKING OF WATER-TUBE BOILERS

The limitation of fired-tube boilers forced to develop water-tube boilers to permit the increase in boiler pressure and capacity with reasonable stress. Water-tube boilers are classified on the basis of circulation of water—natural or forced circulation, tube layout—horizontal straight tube or bent tube, or vertical tube, layout of drum—longitudinal drum or cross drum, number of drums—one, two, three, four or drumless and pressure of steam– subcritical or supercritical boilers. They are invariably used in utility plants.

6.3.1 Babcock and Wilcox Water-Tube Boiler

It is a longitudinal drum, externally fired, water tube, natural circulation type of stationary boiler. Since the boiler is externally fired, it is suitable for all types of fuels and for hand and stoker firing. Evaporating capacity in such boilers ranges from 20,000 to 40,000 kg/h of steam and operating pressure from 11.5. to 17.5 bars are quite common. But the operating of pressure may be as high as 42 bars. It is suitable for small size thermal power plants and other industrial works.

Construction. Figure 6.6 shows the details of Babcock-Wilcox water-tube boiler. It consists of a high pressure drum mounted at the top. The drum of the boiler is made of welded steel or single course joined by longitudinal butt strap. The heads of the drum are forged by hydraulic press and dished to a radius equal to the diameter of the drum. From each end of the drum, connections are made with the *upper header and downtake headers*. A large number of *water tube* connect the uptake and downtake headers. The water tubes are inclined 5 to 15 degrees to promote water circulation. The water tubes are straight, solid drawn steel tubes about 10 cm in diameter and are expanded into the bored holes of the headers to ensure proper fixing. The headers have a *serpentine* (sinusoidal) form. This serpentine form of headers arranges the water tubes such that they are staggered and this exposes the complete heating surface to flue gases. The heating surface of the unit is the outer surface of the tubes and half of the cylindrical surface of the water drum which is exposed to flue gases.

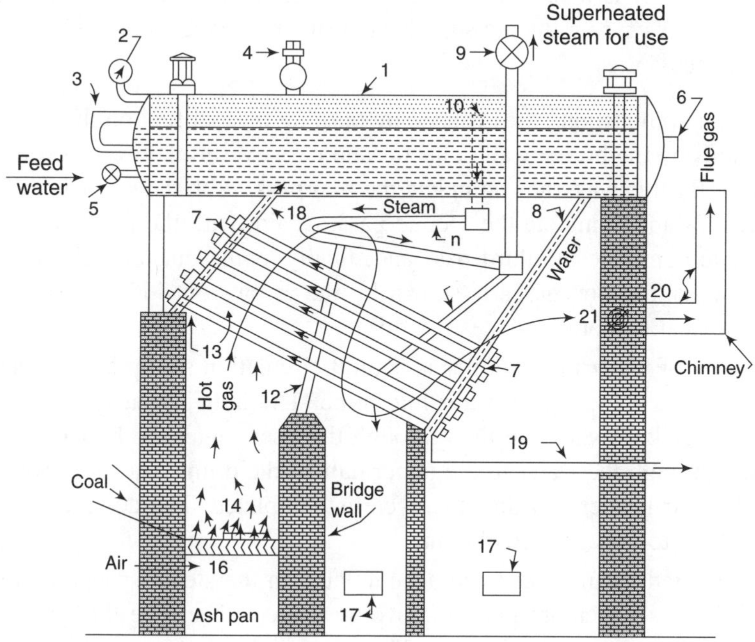

Figure 6.6 Babcock and Wilcox water-tube boiler. (1) Drum, (2) Pressure gauge, (3) Water gauge, (4) Safety valve, (5) Feed water inlet, (6) Manhole, (7) Headers, (8) Down comer, (9) None-return valve, (10) Anti-priming pipe, (11) Superheater, (12) Baffles, (13) Water tube, (14) Fire grate, (15) Fire door, (16) Ash pot, (17) Clean outdoors, (18) Riser, (19) Blow-off pipe, (20) Chimney, (21) Damper

A mud box is attached to the bottom of the rear header (i.e., downtake header). Any foreign matter held in suspension in the water gets collected in the mud box due to gravity and it can be blown off from time to time. A hand hole is also provided for cleaning the mud box. To clean the outside of the water tubes and to remove the soot, door is provided so that access to the interior of boiler can be made.

The whole of assembly of water tubes is hung along with the drum from a steel girder frame by steel rods called slings in a room made of masonry work, lined with fire bricks. Slings pass around each of the drum and is thus entirely independent of the brickwork of the setting. Boiler can expand or compress freely without straining the brickwork setting.

Below the uptake header, the furnace of the boiler is arranged. The coal is fed to the chain grate stoker through the fire door. The chain speed is so adjusted that by the time the coal reaches the other end of the grate, its combustion is complete. There is bridge wall deflector which deflects the combustion gases upwards. Baffles are arranged across the water tubes to act as deflectors for the flue gases and to provide them the gas passes. Here two baffles are arranged which provide three passes of the flue gases. A chimney is provided for the exit of the gases. A damper is placed at the inlet of the chimney to regulate the draught. There are superheating tubes for producing superheated steam. Connections are provided for other mountings and accessories.

Working. The hot combustion gases caused by burning of coal fuel on the grate rise and are deflected upwards by the bridge wall deflector and pass over to the front portion of the water tubes and drum. By this way, they complete the first pass. With the provision of baffles, they deflect downwards and complete the second pass. Again, with the provision of baffles they rise upwards and complete the third pass and, finally, come out through the chimney. The flow path of the combustion gases is shown by the arrows outside the tubes. During their travel, they give heat to water and steam is formed.

The circulation of water in the boiler is natural circulation set up by convective currents. Feed water is supplied by a feed water inlet pipe. The hottest water and steam rise from the tubes to the uptake header and then through the risers enter the boiler drum. The steam vapours escape through the water to the upper half of the drum. The cold water flows down the drum to the rear header and thus the cycle is completed. The down-comers are located outside the furnace to enhance circulation.

To get superheated steam, the steam accumulated in the steam space is allowed to enter the superheated tubes which are placed above the water tubes. The flue gases passing over the flue tubes produce superheated steam. The steam thus superheated is finally supplied to the user through a steam and stop valve. It is to be noted that when the steam has to be raised from the cold boiler, the superheater tubes are filled with water to the drum water level. The superheaters remain flooded until the steam reaches the working pressure. The superheater is then drained and steam is allowed to enter it for superheating purposes.

6.3.2 Stirling Bent Water-Tube Boilers

In 1888, Allan Stirling invented the bent water-tube boiler which has the following features:

(a) It consists of steam-water drums, more than two in number. This improves the evaporation rates of water and the unit can meet the demand of load variations with less pressure drops.

(b) The tubes connecting the drums are bent instead of being straight. Tubes are bent for the followings reasons:

(i) The bent tube allows free expansion and contraction and thus prevents the undue thermal stresses which are often met with in straight tubes.

(ii) Bent tubes enter the drum in an approximately radial direction which permits easy connection with drums.

(iii) It permits great flexibility in design, particularly in regards to drum arrangement.

(iv) Tube replacement becomes easier.

Construction and Working. Figure 6.7 shows the details of a stirling bent water tube boiler. It consists of four drums. The drums 1, 2 and 3 are arranged in the upper part of the setting called steam drums containing water and steam and the drum 4 which is arranged in the lower part is called mud or water drum. The drums 1, 2 and 3 are connected with drum 4 by a series of bent tubes, arranged to give ample space for superheater tubes and with alternate wide and narrow space to permit removal of tubes without difficulty. The ends of the steam drums are dished and the drum has a manhole in one end. The water and steam spaces of the steam drums are connected by tubes. The tubes which connect the water space of adjacent steam drums are called *circulating tubes* and these help in the equalization of the water level in the drum. The tubes which connect the steam space of the steam drums are called *equalizing tubes* and these serve to equalize the steam pressure. The middle drum is fitted with a water gauge to show the water level in the steam drums. The steam drums are usually suspended by a steel structure and the mud drum is supported by tubes to permit free expansion and contraction of tubes.

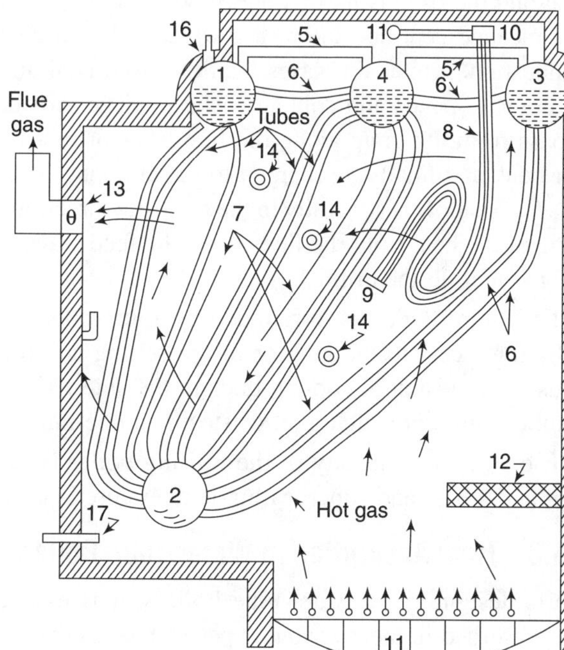

Figure 6.7 Stirling Bent Tube Boiler. 1, 2, 3. Steam drums, 4. Mud drum, 5. Equalizing tubes, 6. Water header, 7. Baffles, 8. Superheat tubes, 9. Inlet super heater, 10. Outlet superheater tubes, 11. Grate, 12. Fire brick arch, 13. Fire door, 14. Cleaning doors, 15. Steam stop valve, 16. Safety valve, 17. Blow-off cock

The feed water enters the rear drum 1 which is farthest from the fire and passes downward to the rear bank of tubes into the mud drum. Any foreign matter suspended in water settles down in the mud drum which is removed by blow off cock at regular intervals and only pure water passes upwards through the bank of the bent tubes to the front drum 3. The major portion of evaporation takes place in the front tubes and drums because the furnace is below them. The water from the front drum 3 passes through the water circulating tubes into the middle drum 2 and then downward through the middle bank of tubes to the mud drum and then retraces its course. The rear steam drum 3 is provided with safety valves to the main steam outlet and a dry pipe to ensure dry steam at the boiler outlet. The middle drum thus acts as a dryer and there is little moisture in the steam leaving the middle drum, and this is effectively removed by the baffles arrangement in the rear drum from which it passes to the dry pipe into superheater tubes.

The steam from the drum is led into the superheater tubes which are bent and zig-zag and placed in the front portion of the boiler. The steam passing through the superheater tubes gets further heat due to the flow of hot gases over it. The steam thus superheated passes to the pipe leading to the steam stop valve from where it passes to its destination.

The furnace is provided below the front portion of the drum. A cocking arch is sprung over the front of the grate with the grate directly under the arch when the boiler is fired by hand stocking using solid fuels. The large triangular space above this arch (between the boiler front and the front bank of water tubes) serves as a combustion chamber. The fire brick arch gets incandescent hot due to fire which helps in combustion by heating the air required for combustion, ignites the gases distilled from coal and prevents chilling of the furnace when the door is opened and cold air rushes in. The furnace gases are directed by fire brick baffles to pass from the grate along the first bank of tubes; then down the middle bank and up the rear bank and finally discharged to the chimney. Thus, we see that the flue gases make three passes. Gas flow is generally in the opposite direction of the feed water. With a view to recover some of the heat from the flue gases, the feed water is first circulated through an *economiser* before entering the drum 1.

The entire boiler is surrounded by four walls. Door are provided in front side and rear walls for access to the interior of the setting and for inspection cleaning and repairs. There is also a provision of sooting door in the side wall and opposite to the middle of each bank of tubes, through which a jet of steam can be introduced to clean the outside of the tubes.

The steaming capacity of the Stirling boiler is as high as 50,000 kg/h and pressure is as high as 60 bars and superheating temperature is up to 150°C.

6.3.3 Heat Absorption in Water-Tube Boilers

While designing the water-tube boilers, it is essential to estimate the percentage of total heat absorbed in their various types of heat exchangers. In a water-tube boiler, feed water is heated in four types of heat exchangers, namely, economizer, evaporator (downcomer-riser circuit), superheater and reheaters (Fig. 6.8). It is to be noted that theoretically heating takes place at constant pressure.

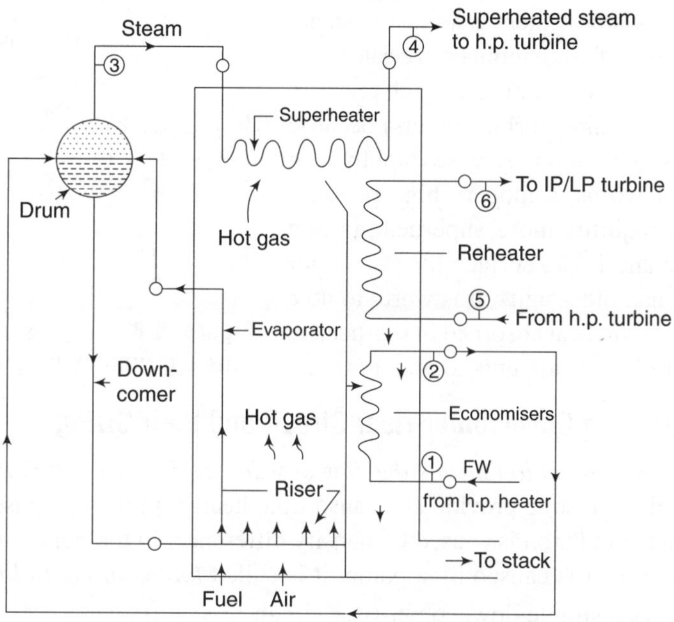

Figure 6.8 (a) Heat absorption in a water-tube boiler

For each kg of water, the heat absorbed by different heat exchangers.

$$q_{eco} = h_2 - h_1 \text{ (sensible heating)}$$

$$q_{eva} = h_3 - h_2 \text{ (evaporation)}$$

$$q_{sup} = h_4 - h_3 \text{ (superheating)}$$

$$q_{reh} = h_6 - h_5 \text{ (reheating)}$$

The percentage of total heat absorbed in the economizer, evaporator, superheater and reheater are given by

$$\% \text{ economizer} = \frac{q_{eco} \times 100}{q_{eco} + q_{eva} + q_{sh} + q_{reh}} = \frac{(h_2 - h_1) \times 100}{[(h_4 - h_1) + (h_6 - h_5)]} \quad ...(6.1)$$

$$\% \text{ evaporator} = \frac{q_{eva} \times 100}{q_{eco} + q_{eva} + q_{sh} + q_{reh}} = \frac{(h_3 - h_2) \times 100}{[(h_4 - h_1) + (h_6 - h_5)]} \quad ...(6.2)$$

$$\% \text{ superheater} = \frac{q_{sup} \times 100}{q_{eco} + q_{eva} + q_{sh} + q_{reh}} = \frac{(h_4 - h_3) \times 100}{[(h_4 - h_1) + (h_6 - h_5)]} \quad ...(6.3)$$

$$\% \text{ reheat} = \frac{q_{reh} \times 100}{q_{eco} + q_{eva} + q_{sh} + q_{reh}} = \frac{(h_6 - h_5) \times 100}{[(h_4 - h_1) + (h_6 - h_5)]} \quad ...(6.4)$$

The use of feed water heaters is a must in steam power plant. The use of large number of heaters (up to ten) means a smaller economiser, a high pressure means a small evaporation surface (risers) because the latent heat of vaporization decreases rapidly with pressure. In other words, a modern high pressure steam generator requires more superheating and reheating surface and lesser surface for evaporator and economiser than older units. It is worth to note that the fraction of total heat absorbed in suerheaters alone in a modern boiler amounts to about 60%.

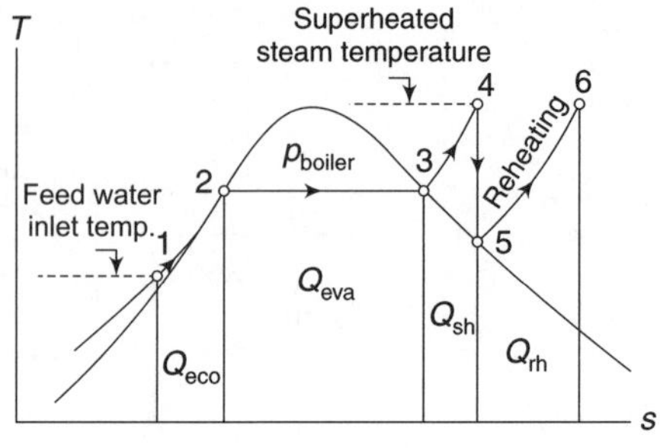

Figure 6.8 (b) *T-s* representation of heating process in a water-tube boiler

6.3.4 Circulation in Downcomer-riser Circuit and their Sizing

In water-tube boiler, *circulation means the flow of water and steam within the boiler circuit.* Adequate circulation must be provided to transfer the heat from the flue gases of the furnace to the water. If the circulation is caused by density difference, in the boiler circuit, it is called *natural circulation.* If it is caused by a pump, it is called *forced* or *controlled circulation.*

Figure 6.9 shows a simple downcomer-riser circuit connecting a drum and a header. The downcomer is outside the furnace, so it is insulated, while riser is inside the furnace. The number of downcomer tubes is less as compared to the riser while the size of downcomer tubes in larger than that of riser. From the drum nearly saturated water falls by gravity through the downcomer into the bottom header and from this bottom header water flows along the riser where it partially boils with the formation of bubbles and then back into boiler drum. In other words, the flow is in two phase (steam-water mixture) in the riser. The density of steam-water mixture in the riser is less than that of saturated water in the downcomer and as result of this density difference, a circulation current is set up within the downcomer-riser circuit. The steam so collected in the drum is taken out of the drum to the superheater where it is superheated. The feed water from the economiser enters the drum.

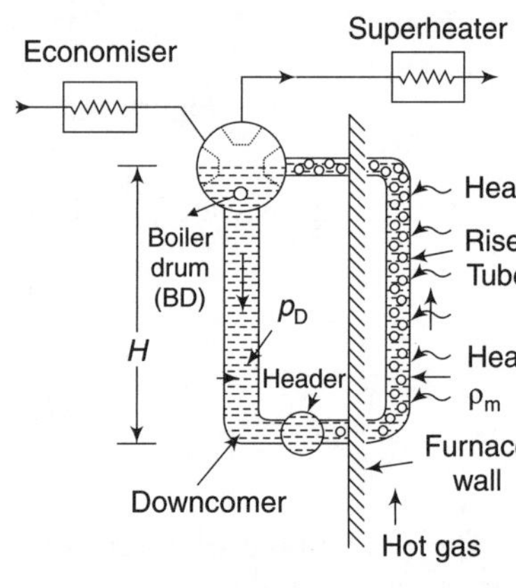

Figure 6.9 Natural circulation system

The available pressure head for natural circulation is expressed as

$$\Delta p = gH\,(\rho_D - \rho_m) \qquad \text{...(6.5)}$$

where H is the height of risers (i.e., furnace), ρ_D is the density of saturated water in the **downcomer** and ρ_m is the **mean density** of steam-water mixture in the riser. Though, the density of water-steam mixture (two-phase) varies along the height of riser, however, for simplicity, the mean density, ρ_m may be taken as the arithmetic mean of the densities at the bottom and top of the riser. Thus,

$$\rho_m = \frac{\rho_{\text{bottom}} + \rho_{\text{top}}}{2} \quad \text{...(6.6)}$$

Here, $\rho_{\text{bottom}} = \rho_D$ and $\rho_{\text{top}} = \dfrac{1}{v_{\text{top}}}$.

But $v_{\text{top}} = v_f + x_{\text{top}} \cdot v_{fg}$

where x_{top} is the quality of mixture (dryness fraction) at the top of the riser.

The burning of fuel and the heat release rate decides the maximum heat of the furnace (H). From Eq. (6.5), for a certain height H of the furnace, we have

$$\Delta p \propto (\rho_D - \rho_m) \quad \text{or} \quad \Delta p \propto \left(\frac{1}{v_D} - \frac{1}{v_m}\right) \quad \text{...(6.7)}$$

Equation (6.7) suggests that higher is the density difference, more will be the pressure head available for natural circulation. But the density difference decreases with the increase of pressure of steam as obvious from p-v and ρ-p diagrams shown in Fig. 6.10. At critical pressure, $\rho_g = \rho_f$ and thus there will be no circulation.

The heat transfer takes place from the gas produced by burning fuel in the furnace to the water flowing through the risers which are installed (embedded) all around four walls of the furnace and thus act as a *cooling tubes or water walls.* It is essential that the rate at which heat is released must be absorbed at the same rate and for this reason, adequate circulation must be provided in the circuit.

In the event of inadequate circulation, the rate at which heat is carried away will be less than the rate of heat release and thus the difference will be stored in the metal of the riser tubes leading to their overheating and rupturing when the tube temperature exceeds the metal melting point.

The location of downcomers depends on the boiler pressure. As a general practice, if the boiler pressure is more than 30 bars, downcomers are placed outside the furnace in order to obtain more density difference for more natural circulation. If the boiler pressure is less than 30 bars, both downcomers and risers are placed inside the furnace with the risers installed in the hotter zone. The tubes which are heated less will act as downcomers while the tubes which are heated more will act as risers.

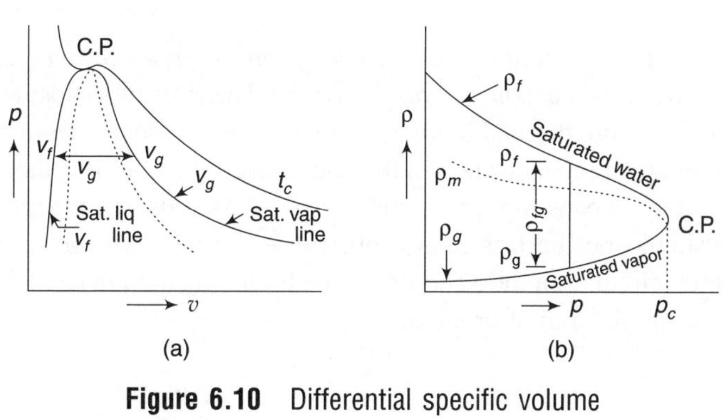

Figure 6.10 Differential specific volume

In the event of boiler pressure exceeding 180 bars, the density difference ($\rho_D - \rho_m$) becomes very small, as obvious from Fig. 6.10, which means poor and unreliable natural circulation and thus forced circulation becomes essential. The forced circulation is achieved by using a pump installed in the downcomers which circulates saturated water through all the risers installed inside the furnace as shown in Fig. 6.11.

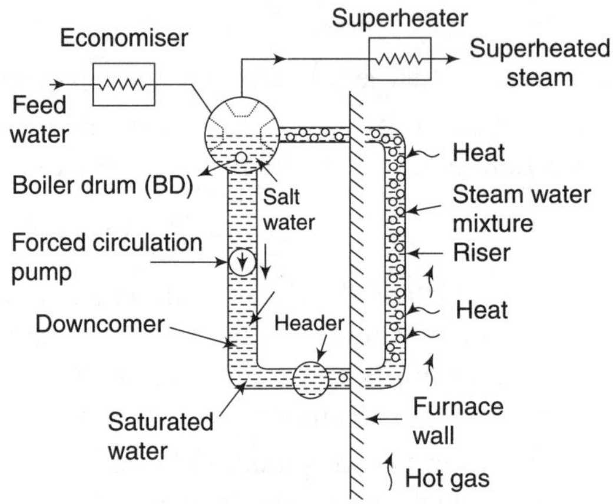

Figure 6.11 Forced circulation system

For the analysis of process of steam generation, a term called "circulation ratio" (CR) is used which is defined as

$$\text{CR} = \frac{\text{Flow rate of saturated water in downcomers}}{\text{Flow rate of steam released from the drum}}$$

$$= \frac{\dot{m}}{\dot{m}_g} = \frac{\dot{m}_f + \dot{m}_g}{\dot{m}_g} = \frac{1}{\dot{m}_g/(\dot{m}_f + \dot{m}_g)} = \frac{1}{x_{\text{top}}} \quad ...(6.8)$$

where $\dot{m}$ is the mass of saturated water flowing through the downcomer-riser circuit during a certain time, $\dot{m}_g$ is the mass of steam released during the same time, $\dot{m}_f$ is the mass of the saturated liquid water (fluid) at the riser exit and x_{top} is the equality of mixture at the riser top.

It is worth to note that circulation ratio varies from one riser to another due to varying heat absorption owing to placement position of risers. The riser tubes located opposite to the burner are subjected to more thermal loading and thus generate more steam which means less circulation. While the riser tubes located in the corner are subject to less thermal loading (i.e., relatively cooler) which means higher circulation ratio. As a design practice, the circulation ratio in any tube should not be less than 6 otherwise the tube will get overheated and fail prematurely. Further, circulation ratio should not be higher than 25 in any tube for its efficient utilisation in the steam generation process.

It has been observed that too much steaming (i.e., steam formation) is not desirable in the riser tube and the circulation ratio has to be maintained above 6 in any riser tube. From the point of view of heat transfer consideration, a wetted surface is always desired and nucleate boiling should always persist in riser tubes. During boiling heat transfer, the initiation of bubbles start on the heated surface of the riser tubes. With the increase in heat transfer to the risers, the formation of bubbles may increase and finally it may form a stable film. The thermal conductivity of a vapour film is much less than that of a liquid film, so it offers a large thermal resistance, almost blanketing the surface where it forms and as a result the heat absorbed and carried away (Q_2) will be much less than the heat transferred to the wall (Q_1) (Fig. 6.12). The difference of heat ($Q_1 - Q_2$) will be stored in the metal of the riser tubes which will enhance the temperature of metal. If the temperature of metal exceeds its melting points, the tube will rupture and the leakage of water and steam will start.

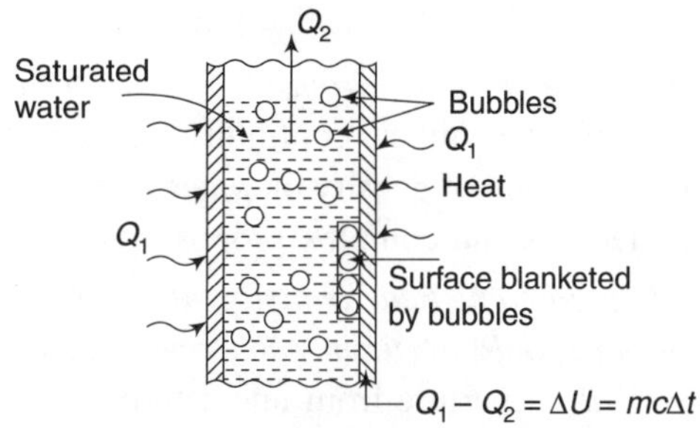

Figure 6.12 Nucleate and film boiling in a riser tube

6.3.4.1 Sizing of Downcomers

From the point of view of design considerations, downcomers are fewer in number and bigger in diameter, whereas risers are more in number and smaller in diameter. In downcomers, water falls by gravity. Bigger the diameter, less the pressure drop due to friction since pressure drop is inversely proportional to tube diameter which is expressed as

$$\Delta p = f \frac{L}{D} \rho \frac{C^2}{2} \quad \text{...(6.9)}$$

Therefore, the downcomers are made bigger in diameter which may vary from 150 mm to 200 mm or even higher. For estimating the number of downcomers, continuity equation is used which is given as

$$\dot{m}_f = A \rho_f C = \left(n \frac{\pi}{4} D^2\right) \rho_f C \quad \text{...(6.10)}$$

where $\dot{m}_f$ = mass flow rate of saturated water in the drum (kg/s), n = number of downcomer tubes, D = diameter of downcomer tubes (m), ρ_f = density of saturated water (kg/m^3), C = average circulation velocity of water in downcomers (m/s).

The value of circulation velocity in downcomer (C) depends on the circulation ratio in tubes and varies from 0.4 to 1.4 m/s depending on the boiler capacity. Thus, the number of tubes estimated from Eq. (6.10) can be 6, 8 or even high as 20 depending upon the boiler capacity.

6.3.4.2 Sizing of Risers

Since risers absorb heat from the furnace, so for the same total cross-sectional area, the smaller the diameter, the larger the surface area exposed to hot gas for heat transfer. *From this point of view, risers are of small diameter and as practice, the diameter varies from 62.5 to 76.5 mm.* For estimating the number of tubes, continuity equation is used where the determination of density of water-steam mixture (two-phase) is a tedious and complex proposition.

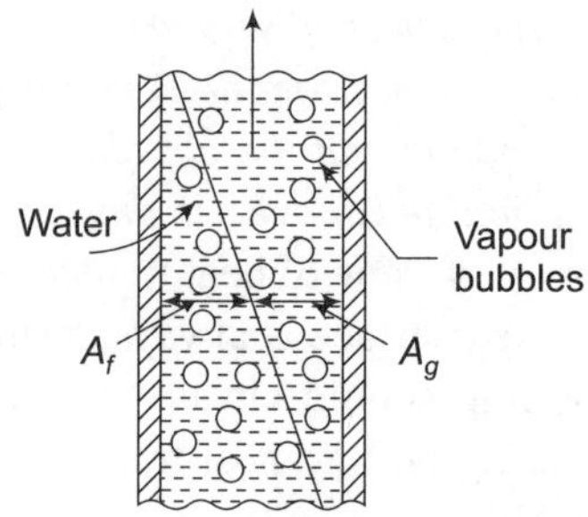

Figure 6.13 Two-phase flow in a riser tube

For efficient heat transfer, it is essential that nucleate boiling should occur in each risers tube and film boiling is avoided. Departure from nucleate boiling and on-set of film boiling occur due to too much steaming and less circulation in risers. The recent trend is to provide *internal twisters and springs* in riser tubes placed in high absorbing areas of the furnace which break the vapour film and retard the onset of film boiling. Tubes ribbed helically on their inside surface are often used now as ribbing creates centrifugal action which washes away the vapour film formed.

A term **"void fraction** (α)" or volumetric quality used to analyse the two-phase flow which is defined as

$$\alpha = \frac{\text{Volume of vapour}}{\text{Volume of liquid + Volume of vapour}}$$

The specific volume of a two phase mixture is given by

$$v = (1 - x)v_f + xv_g \approx v_f + xv_g$$

The density of a two-phase mixture with void fraction α is given by

$$\rho = (1 - x)\, \rho_f + \alpha\rho_g \approx \rho_f + \alpha\rho_g$$

Consider a two-phase flow in a riser tube where A_f is the cross section occupied by saturated liquid and A_g is the cross section occupied by vapour bubbles (Fig. 6.13). Thus, the void fraction for such a flow is expressed as

$$\alpha = \frac{A_g}{A_f + A_g} \quad \text{or} \quad \frac{A_f}{A_g} = \frac{1 - \alpha}{\alpha} \qquad ...(6.11)$$

In the movement of a two-phase mixture, the vapour, because of buoyancy has a tendency to slip past the liquid, i.e., the velocity of vapour is higher than the liquid. To account this, a slip ratio (S) is defined as

$$S = \frac{C_g}{C_f} = \frac{\text{Velocity of vapour}}{\text{Velocity of liquid}} \qquad ...(6.12)$$

The value of S at any cross section of the riser is more than unity.

$$\text{Since quality} = x = \frac{\dot{m}_g}{\dot{m}} = \frac{\rho_g A_g C_g}{\dot{m}} = \frac{A_g C_g}{\dot{m} v_g}$$

or

$$C_g = \frac{\dot{m} \cdot v_g \cdot x}{A_g} \qquad ...(6.13)$$

Similarly, $$C_f = \frac{(1-x)\dot{m}v_f}{A_f} \quad ...(6.14)$$

The ratio of C_g and C_f is expressed from Eqs. (6.13) and (6.14)

$$\frac{C_g}{C_f} = \frac{xv_g \cdot A_f}{A_g(1-x)v_f} \quad ...(6.15)$$

$$\therefore \quad S = \left(\frac{x}{1-x}\right)\left(\frac{A_f}{A_g}\right)\left(\frac{v_g}{v_f}\right)$$

From Eq. (6.11), we have, $S = \left(\frac{x}{1-x}\right)\left(\frac{1-\alpha}{\alpha}\right)\left(\frac{v_g}{v_f}\right)$ ...(6.16)

From Eq. (6.16), it is obvious that as the boiler pressure increases, the slip ratio decreases. Experimentally, it has been observed that the slip ratio varies between 1 and 10. Value of $S = 1$ occurs at high pressure where liquid and vapour densities approach each other.

After rearranging the Eq. (6.16), we have

$$\alpha = \frac{1}{1+\left(\frac{1-x}{x}\right)\cdot\frac{v_f}{v_g}\cdot S} = \frac{1}{1+\left(\frac{1-x}{x}\right)\phi} \quad ...(6.17)$$

where $$\phi = \frac{v_f}{v_g}\cdot S \quad ...(6.18)$$

Similarly, $$x = \frac{1}{1+\left(\frac{1-\alpha}{\alpha}\right)\cdot\frac{1}{\phi}} \quad ...(6.19)$$

The quality distribution along the height is obtained by assuming a reasonable value of S (say between 1 and 2).

The mixture density distribution may be obtained from

$$\rho = (1-\alpha)\,\rho_f + \alpha\rho_g \quad ...(6.20)$$

The average mixture density in the risers of height H is given by

$$\rho_m = \frac{1}{H}\int_0^H \rho(z)dz \quad ...(6.21)$$

where z stands for axial distance from the bottom of the riser.

For uniform axial heating, El Wakil (87) has suggested the following distribution for the mean density in the riser

$$\rho_m = \rho_f - \frac{\rho_f - \rho_g}{1-\phi}\left\{1 - \left[\frac{1}{\alpha_e - \phi} - 1\right]\ln\frac{1}{1-\alpha_e(1-\phi)}\right\} \quad ...(6.22)$$

where α_e = void fraction at the riser exit

It is worth to mention that the total pressure head produced due to density difference (given by Eq. 6.5) is balanced by the pressure losses in the downcomer-riser circuit.

$$\Delta p = gH\,(\rho_D - \rho_m) = \Delta p_D + \Delta p_R + \Delta p_{\text{bends}} + \Delta p_{\text{header}} \quad ...(6.23)$$

where, suffixes D, R, bends, header placed on Δp indicates the pressure losses in downcomers, risers, bends and header respectively.

From the heat transfer analysis, we have

$$Q = U\,A_s\,\Delta T_m = U(\pi\,DN)\,\Delta T_m \quad ...(6.24)$$

where U is the overall heat transfer coefficient and ΔT_m is the log-man temperature difference. By knowing the value of U, D and ΔT_m, the number of riser tubes (N) can be estimated from Eq. (6.24). The value of U depends on many factors such as heat transfer coefficients (inside and outside), thermophysical properties, geometry of tubes, flow behaviour inside and outside tubes, etc.

The circulation circuit in a natural circulation boiler showing for simplicity three downcomers, a lower header from which water enters the riser tubes, nine heated risers, an upper collecting drums and four pipes for delivery of the steam is shown in Fig. 6.14. As mentioned, the circulation rate depends upon the thermal loading of the circuit concerned. The steam formation rate will depend upon configurations and dimensions, density of water and steam, and thermal loading of the circuit.

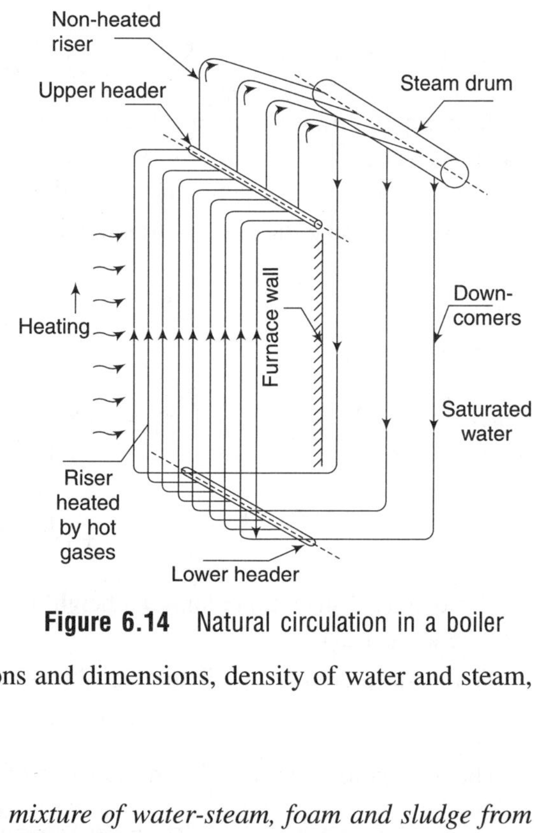

Figure 6.14 Natural circulation in a boiler

6.3.5 Steam Drum and Its Internals

The purpose of steam drum is to receive the mixture of water-steam, foam and sludge from the risers and to separate the steam from the mixture before the steam leaves the drum for superheaters.

Any moisture carried with steam to the superhater tubes contains dissolved salts. During superhating process water evaporates and the salts remain deposited on the inside surface of the tubes in the form of scale which is difficult to remove. The scale so formed reduces the rate of heat absorption, which in turn, leads to overheating of the tubes and finally failure through ruptures. The superheater tubes are subjected to the highest stream pressure and temperature on the inside and the maximum gas temperature on the outside, and so they are made of the costiliest materials which can withstand high temperature and pressure. This suggests that utmost care should be taken so that no damage is done to superheaters due to moisture carried away with steam. Further, some of the impurities in the steam may vaporize silica which may cause turbine blade deposits resulting in poor turbine performance due to disturbed flow.

The another requirement of steam drum is that no vapour bubble should flow along with saturated water from the drum to the downcomers as it will reduce the density difference and the pressure head for natural circulation. Further, the bubble due to its own characteristics tends to flow upward and thus it will impede the flow in the downcomers resulting in reduction in circulation. This suggests that *the steam drum has to ensure moisture-free steam going to the superheater and bubble-free water going to the downcomer as shown in* Fig. 6.15.

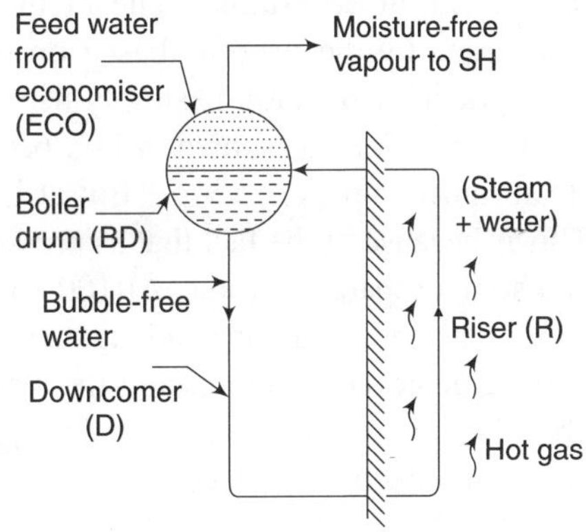

Figure 6.15 Separation of two-phase mixture in steam drum

Though, adequate treatment of feed water along with make-up water is done prior to feeding but still it carries impurities in the form of total dissolved solids (TDS) which are generally expressed in ppm (parts per million) by mass. One ppm means one part of impurity in millions parts of water by mass. Since whole circulation and feeding process are continuous, so solid content of water goes on increasing. In order to maintain a certain ppm in the drum, continuous or intermittent blowdown is essential. Continuous blowdown is more common in utility boilers. Hot pressurized water is used to blowdown solids precipitated at the bottom by opening the blowdown valve at the drum bottom. The ppm of drum water depends on pressure and varies from 1000 to 2000.

In order to help the precipitation process, trisodium phosphate or tannin is injected periodically in the drum. For continuous blowdown, the total solids entering will be equal to the total solids leaving. Thus,

$$\dot{m}_{fw} \times ppm_{fw} = BD(ppm)_{\text{drum}} + (1 - x)\, \dot{m}_s\, (ppm)_{\text{carry over}} \tag{6.24a}$$

where $\dot{m}_{fw}$ is the rate of feedwater entering the drum. BD is the necessary blowdown rate, $\dot{m}_s$ is the rate of steam flow from the drum and x is the quality of steam leaving the drum.

From the above discussion, it follows that **steam drum has to perform many functions** as given below:

(i) It stores water and steam in sufficient quantity to meet the varying load requirements.

(ii) It helps in circulation of water.

(iii) It separates steam from water-steam mixture discharged by risers and sends moisture free steam to superheaters.

(iv) It supplies bubble-free water to downcomers.

(v) It provides enough surface area for liquid vapour disengagement.

(vi) It permits the blowdown to remove precipitated dissolved solids.

(vii) It maintains a certain desired ppm of dissolved solids in the drum by phosphate injection and blowdown.

Mechanism of Separation. The method of separation of steam from steam-water mixture discharged by the risers is based on the boiler pressure. If the boiler pressure is below 20 bars, gravity separation is simply used with the condition that sufficient disengaging surface must be provided as shown in Fig. 6.16(a). Water being heavier separates out by gravity and steam moves up as $\rho_f >> \rho_g$. But at high pressure, separating force due to gravity is small. This is because of the fact that as pressure increases, the density difference $(\rho_f - \rho_g)$ decreases and so the separating force. At 100 bars, the separating force is about one-third of at 20 bars so it is difficult to achieve adequate separation. Though, an improvement in separating force may be achieved by increasing the size of the drum but the construction of drum is a costly affair, so it is not economical. This suggests that a positive mechanical means of steam-water separation becomes a necessicity.

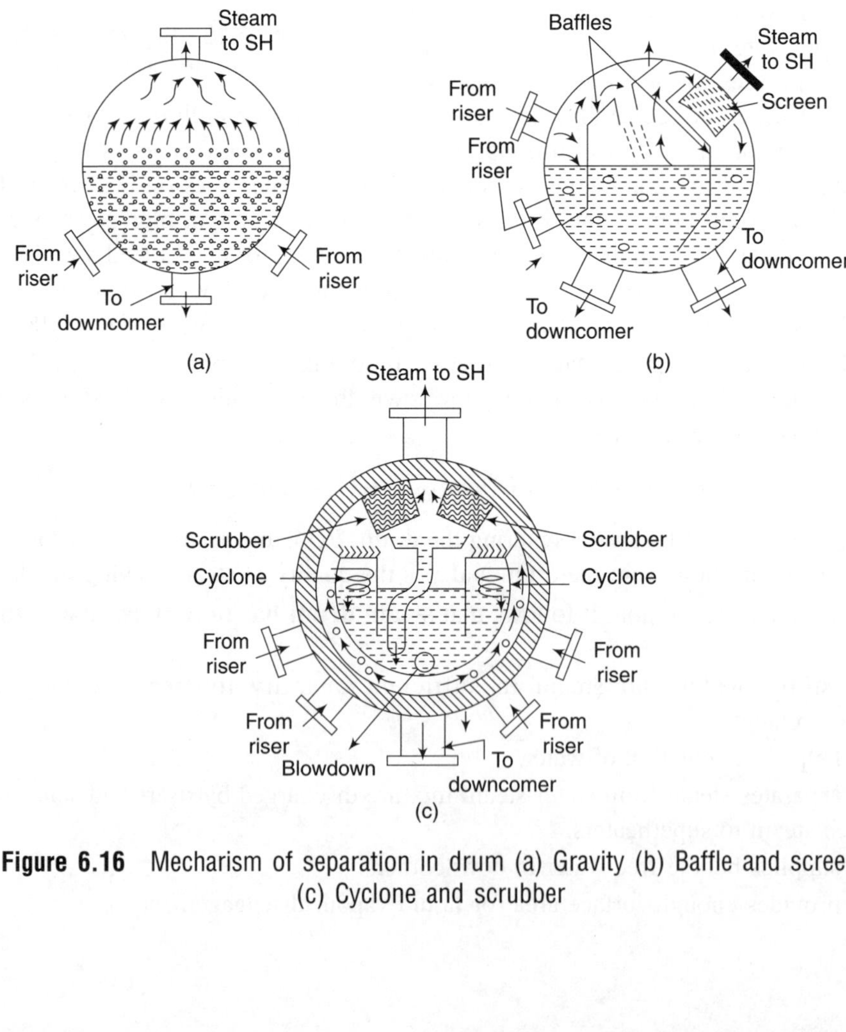

Figure 6.16 Mecharism of separation in drum (a) Gravity (b) Baffle and screen, (c) Cyclone and scrubber

There are various types of *mechanical separators such as baffles, screens and cyclones* which are housed inside the drum and are known as *drum internals*. Figure 6.16(b) shows the baffle plates which act as primary separators. These baffle plates change or reverse the steam flow direction and act as impact plates that cause water to drain off. In other words, baffle plates assist in gravity separation. Screens [Fig. 6.16(b)] made of wire mesh acts as secondary separators where individual wires attract and intercept the fine droplets, just as fabric filters attract dust from gases. The water accumulated on the screen wires drops and then falls by gravity back to the main body of water. In this way, we can say that screen also assists in gravity separation. Figure 6.16(c) shows the cyclone separator which utilizes the effect of centrifugal force for separation of two-phase mixture. The mixture is entered tangentially to direct the water downward and to make the steam flow upward.

After coming out from cyclone, the steam then goes through the zig-zag path in corrugated plates, called the *scrubber or dryer* positioned on the upper portion of the drum which removes even the last traces of moisture from steam. Finally, the perforated plates or screen under the drum exit perform the final drying action of steam. Figure 6.17 shows the drum internals of a controlled (forced) circulation boiler.

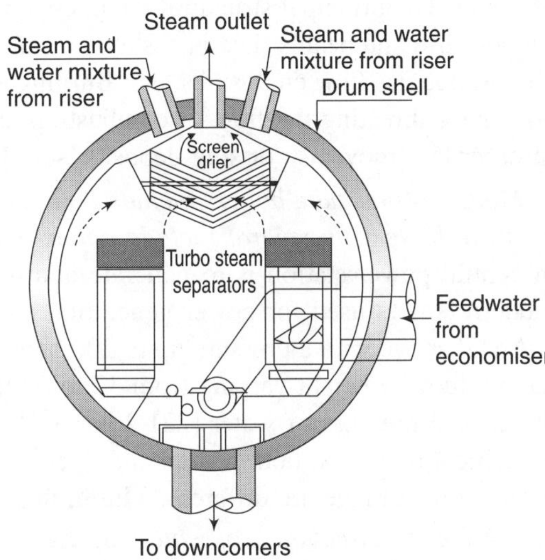

Figure 6.17 Drum internals of a controlled (forced) circulation boiler

In many cases, the separation process also involves a *washing* of the steam by passing it though a rain of the relatively pure feed water having very low ppm before passing through the scrubber which reduces the ppm of moisture carry-over. The other advantage of washing of the steam lies in the reduction of silica concentration in steam as silica vapour is absorbed in the feed water. If there is too much moisture carry-over because of a high water level or a high steam rate, such a boiler drum is said to be *priming*. *Priming* phenomenon is induced by erratic feedwater control and rapid changes in steaming rate. This suggests that in order to reduce priming, a certain water level in the drum should be maintained to give adequate steam space.

***Foaming** is another phenomenon encountered in the delivery of impure steam resulting from the formation of bubbles on the water surface.* This is due to the presence of saponification agents in the water drum such as oil, certain dissolved salts and high alkalinity. Some of the foam can be drawn off the water surface and wasted by blowing down through connection made on the drum at the normal water level.

The design of the shell of steam drum is based on the theory of pressure vessel. The size of the steam drum in the utility boilers having diameter as high as 5 m, length more than

30 m and weight a few hundred tonnes largely depends upon the space requirements of all the internals, streaming rate and minimum water level. The steam drum being a pressure vessel is made in cylindrical section, called courses, which are welded together, and two nearly hemispherical heads which are welded to the ends. It has to be safeguarded against the explosion.

6.4 MODERN AND HIGH PRESSURE WATER-TUBE BOILERS

Modern pulverized coal-fired boilers stand up vertically integrated with a furnace, a steam drum, evaporators, (downcomers and riser circuits) superheaters, reheaters, economisers, air preheater, tube mills, forced draught and induced draught fans, coal feeders and burners, wind box, various control systems, duct work and all accessories and mounting for safe operation of boiler. The principal structural members consist of vertical structural steel columns, a top support grid and beams (i.e., backstays) meant to provide stiffness to the furnace combustion air and flue gas flow enclosures, etc. Burning of the fuel is initiated by igniters and warm up burner for directing the flow of combustion air and flue gases, a duct work system is used. In order to remove ash or slug deposits, soot blowers are used.

Modern water tube boilers include both natural and forced circulation boilers while high pressure boilers in general include boiler only with forced circulation. The modern trend in central power station is to generate and use steam at high pressure and efficiency. The modern boilers used for power generation are for steam capacities ranging from 4 to 1300 kg/s and above with a pressure up to 200 bars and maximum steam temperature about 600°C having furnace height varying from 32 to 62 m. Now, it is possible to generate steam above the critical pressure of water (221.2 bars). But these supercritical boilers are different from subcritical boilers in both design and operation because of the fact that properties of steam in the critical range are different. **The high pressure boilers are** *characterised as–*

(i) *Forced circulation of water.* In high pressure boilers, circulation of water is forced instead of natural. Therefore, a pump is used to force the water in the boiler.

(ii) *Arrangement of drums and tubing.* In order to avoid any large resistance to flow of water, these boilers have a parallel set of tubes arrangement. They have a very small steam separating drum or may be entirely free of drum.

(iii) *Improved method of heating.* The following methods are used to improve heating:

  (a) Superheated steam is used to heat the water.

  (b) Saving of latent heat by evaporation of water at pressure above critical.

  (c) The heat flow through the tube walls may be increased with use of hot gases travelling with supersonic velocity.

(a) Advantages of High Pressure Boilers. The following are the advantages of high pressure boilers.

(i) Smaller bore and therefore lighter tubes make the unit more compact. The space and weight requirements are minimized which reduces the erection time and cost.

(ii) Reduction in the number of drums.

(iii) There is greater freedom for disposing of the heating surface and hence greater evaporation for a given size.

(iv) Lighter structure for a given output.

(v) The boiler is capable of meeting rapid changes of load without the use of complicated or delicate control devices.

(vi) All the parts are heated uniformly which eliminate the danger of overheating and setting up thermal stresses.

(vii) Due to uniform temperature of parts the differential expansion are minimized, this reduces the leakage of gas and air.

(viii) There is complete elimination of high head which is needed for natural circulation.

(ix) Due to high circulation velocity the tendency to form scales is eliminated to a large extent.

(x) If a external supply of power is available, very rapid start from cold state is possible. Hence the boiler is suitable for carrying peak loads. It is also useful for standby purposes in hydraulic station.

The system is slightly complicated and certain percetange of the power is consumed by the circulating pump. *Examples of high pressure boilers are* ***Lamont, Benson, Vedox Loeffier, Monotube, Radiant type (Natural circulation), etc.***

Three design concepts of water tube boilers are shown in Fig. 6.18. Type A is a boiler with **natural** circulation used up to 180 bars. Heat transfer to the water tubes around the walls is mostly by radiation from the fuel flame and less by convection from the flue gases. Type B is the forced circulation by a special pump and suitable up to pressure of 200 bars. Type C is once-through boiler, there is no drum, so no circulation, and no separation of water and steam and they operate above critical pressure (221.2 bars).

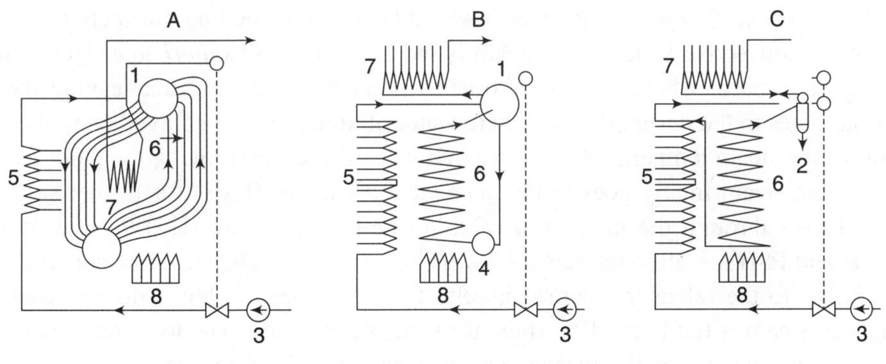

Figure 6.18 Boiler Concept based on Circulation. (A) Natural circulation, (B) Forced circulation, (C) Once-through boiler. 1. Boiler drum, 2. Steam separator, 3. Boiler feed pump, 4. Circulation pump, 5. Economiser, 6. Evaporator, 7. Superheated, 7. Superheated, 8. Furnace

6.4.1 Radiant Type Power Boiler with Natural Circulation

Most of the steam power plants are of the radiant type, and have moderately rated, water cooled (water wall) combustion chamber in which the heat is transferred to the evaporating surface by direct radiation as shown in Fig. 6.19. These boilers employ natural circulation and may use pulverized coal or employ cyclone furnace or fuel oil burning equipment. Some radiant boilers are stoker fired. The boiler shown in Fig. 6.19 employs a pulverizing mill. The pulverized fuel is fed into the combustion chambers with the help of burners. There are three zones of heat transfer (Fig. 6.19). In the zone 1, heat transfer is predominantly by radiation as the flame in this zone is diffused yellow-flame which radiates much more than the premixed flue flame. With the upward movement of burnt gases and the addition of secondary air, the effect of radiation gets reduced and the convection becomes predominantly as the flame (hot gases) changes from diffused to premixed. In the upper part of zone I, the heat transfer takes place by radiation as well as convection provided suitable heat transfer surface is introduced into the path. This space is marked by (R + C). The heat transfer in the zones II and III takes place mainly by convection and they are identified as high and low temperature zones respectively. The height of furnace varies from 42 m to 70 m.

Water at about 170 to 260°C depending on boiler pressure from high pressure feed heater enters the economizer and leaves saturated liquid or as a two-phase mixture of low quality and then enters the steam drum at midpoint along lengthwise. The downcomers are insulated and are situated outside the furnace through which water flows and feeds to the bottom headers through which riser tubes situated inside the furnace and lined to the furnace walls are connected.

The water in the riser tubes receives heat mainly by radiation (zone I). From the combustion gases and boils. *The circulation is set up due to the difference in densities between fluid in the risers and downcomers.* Steam is separated from the bubbling water in the drum and goes to the *convective or primary superheater* (CSH) where heat is absorbed by convective mode (zone II). From the convective superheater steam is fed to the *radiant superheater* (RSH) installed at the top of furnace, where heat is absorbed by radiation. Steam after leaving the radiant superheater goes to the desuperheater where water of high purity is sprayed on to the steam to bring down the steam temperature to its desired value if it exceeds the latter. After leaving the desuperheater, steam finally goes to the *pendent superheater* (PSH) for further superheating before it leaves through the main stop valve to the high pressure (HP) turbine. In general, both RSH and PSH are often termed *secondary superheaters*. The steam exhausting from HP turbine is fed to the reheater where it is reheated to the desired temperature. After leaving the reheater, steam is fed to the IP turbine for further expansion. The location of superheaters may vary as per design of the furnace and flue gas path. In many cases, desuperheater are not provided.

Forced draught **(FD)** fan supplies the air for combustion which is preheated in a heat exchanger called air preheater **(APH)** by the outgoing flue gases. Air is heated to the desired temperature in APH which varies from 250°C to 400°C depending upon type of coal burnt, is split into two streams, *primary* and *secondary. Hot primary air* constituting about 15 to

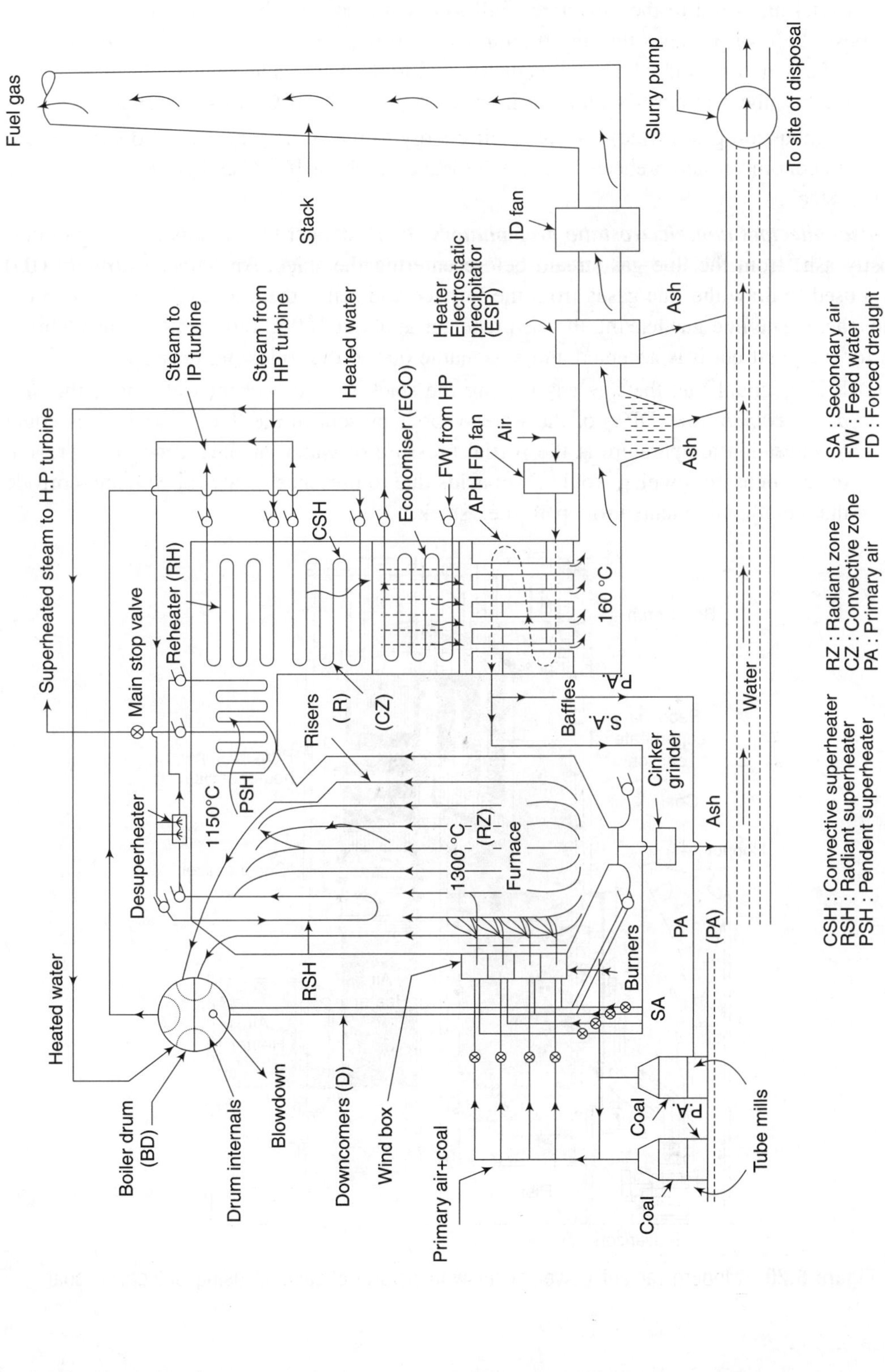

Figure 6.19 Modern radiant power boiler with natural circulation

30% of total air is fed to the pulverizer mill to dry the coal for better grinding and also to help push the coal particles through the burners into the furnace. The *hot secondary air* is fed into the wind box and where it gets distributed to the burners to help combustion. Based on the quality, the fuel burns with great intensity to some 1350°C or even higher.

The combustion gases (flue) transfer their energy to the risers tubes, then to superheater, reheater, economizer and preheater and finally leaves at about 160°C and goes to atmosphere through stack.

***Dust collectors and electrostatic precipitators* (ESP)** collect the entrained solid particles (mostly ash) from the flue gas stream before entering the stack. An *induced draught* (ID) fan is used to draw the flue gases from the furnace and sends them through the stack to the atmosphere. The flue gas leaving the air preheater at about 160°C represents the availability loss for the plant but it is accepted and reasonable due to the following reasons:

(i) It is essential that the flue gas leaving the stack should be kept well above the *dew point temperature* (DPT) of the water vapour present in the flue gas (which is equal to saturation temperature at the partial pressure of water vapour) in order to prevent the condensation which would form acids due to presence of SO_2 or SO_3 and corrode the metal components in its path, i.e., stack.

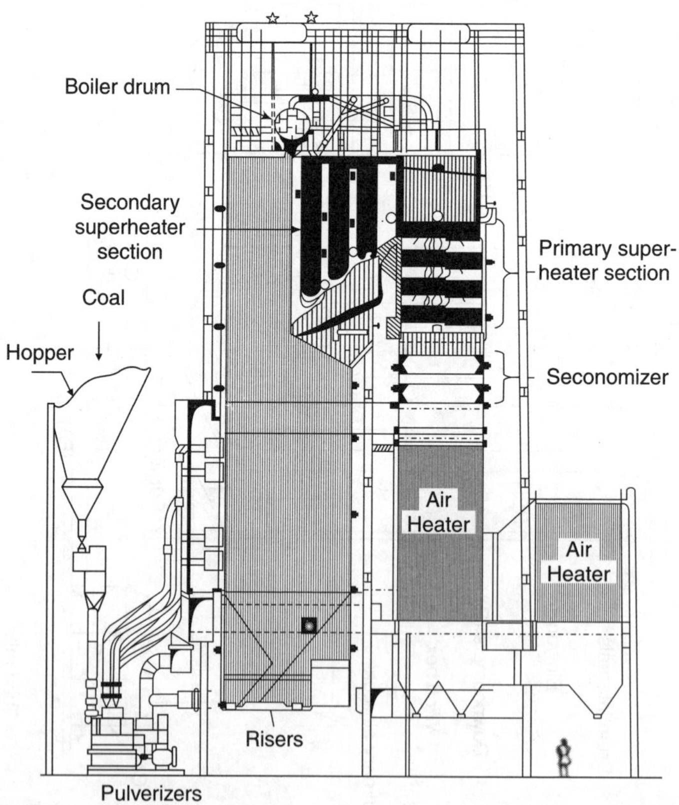

Figure 6.20 Modern radiant power boiler with natural circulation using pulverized coal

(ii) Enough buoyancy must be present in the flue gases in order to rise in high plume above the stack for proper atmospheric dispersion.

The clinkers from the dry bottom furnace fall to the clinker grinder which are broken into pieces and then carried by water in a *hydraulic sluice wherein also falls the ash from the dust collectors and the precipitator and finally it is led to the site of disposal with the help of slurry pump.*

An **another modern radiant power boiler with natural circulation** and single drum using pulverized coal is shown in Fig. 6.20.

Figure 6.21 shows the **cross section of 660 MW CEGB (UK) steam** generator and the flow path of flue gases water steam through the various elements of boilers. The flow path is shown in Fig. 6.22.

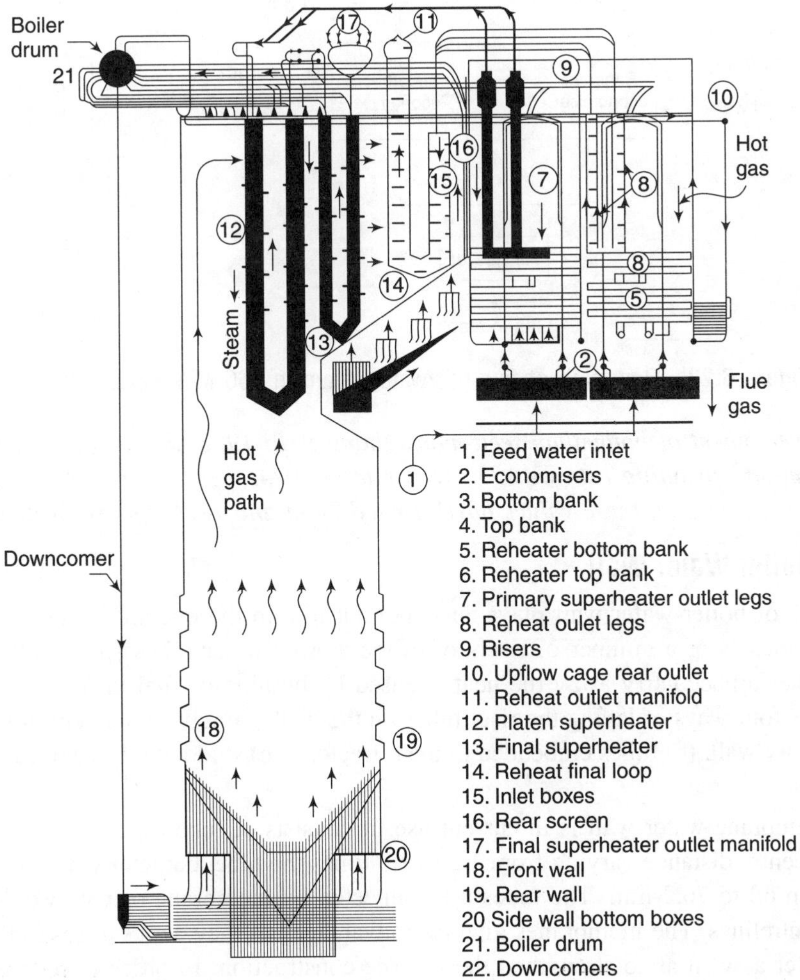

Figure 6.21 A 600 MW CEGB (UK) steam generator

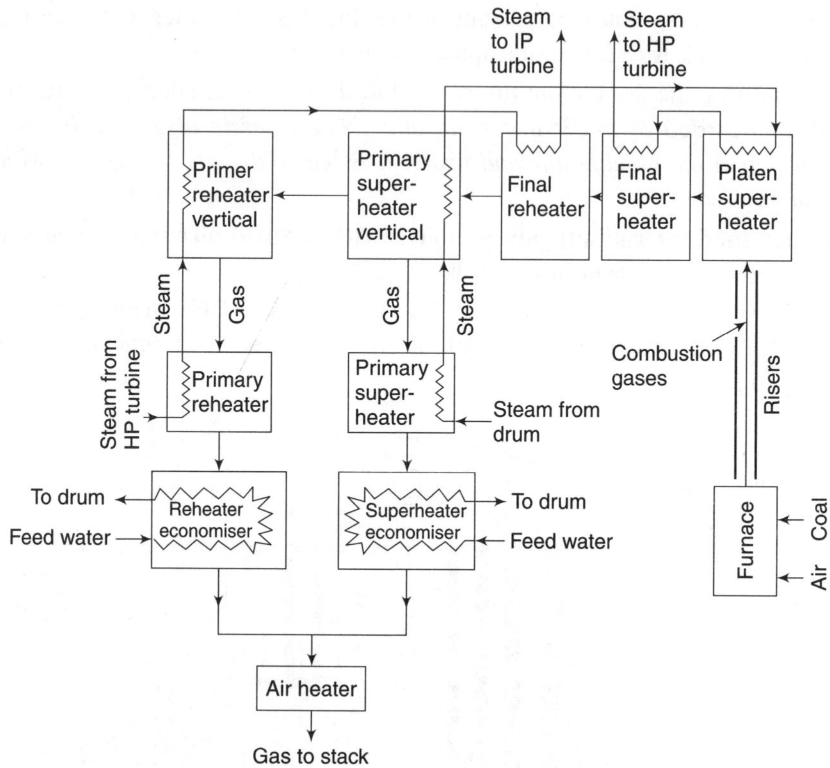

Figure 6.22 Flow path of flue gas/water/steam in 660 MW CEGB (UK) boiler

The development of boiler has seen many changes. In 1952, the ***height of furnace*** *was about 32 m and* ***steaming rate*** *was 35 kg/s while in 1974, these were 61.4 m and 560 kg/s respectively. At present, these values have crossed* ***70 m and 1300 kg/s respectively.***

6.4.2 Boiler Water Wall

The design of boiler water wall plays an important role in the overall design of water tube boiler. It consists of a number of risers of evaporator or water tubes spaced all around the walls of the furnace carry away the heat released by burning of fuel in the furnace. Figure 6.23 shows four ways of laying the riser tubes on the wall, namely (a) tangent tubes touching the refractory wall, (b) tubes embedded in the refractory, (c) studded tubes and (d) membrane tubes.

The membrane water wall is in current use. It consists of tubes spaced on centre having centre-to-centre distance varying from 1.2 to 1.3 times the tube diameter while tube diameter varies from 68 to 76.2 mm. The tubes are connected by membrane or bars welded to them at their centrelines. The membranes give two advantages as they act as a fins to increase the heat transfer as well as to strengthen the furnace construction. In order to reduce heat loss, insulation to provided on the outer side of the wall and then metal logging is provided to protect the insulation.

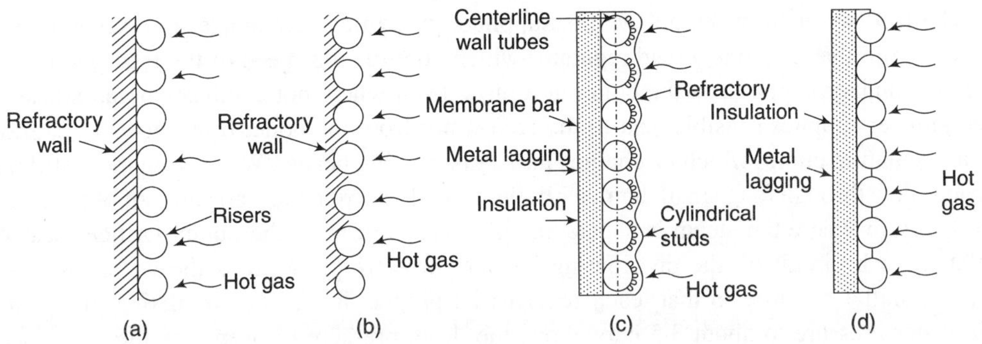

Figure 6.23 Types of water wall

Since the riser tubes are located on the four sides of furnace wall, so the water wall absorbs mainly by radiation. The radiant energy is expressed as

$$Q_R = \sigma A\, \varepsilon F\, (T_g^4 - T_w^4)$$

where T_g and T_w are the gas and wall temperatures respectively, σ is Stefan Boltzmann constant, F is the shape factor and ε is the emissivity which depends upon the constituents of combustion gases (CO_2, H_2O, SO_2, etc.) and their partial pressures.

6.4.3 LaMont Boiler

This is a high pressure, forced circulation, water tube type boiler invented by LaMont in 1925. Figure 6.24 shows the details of a LaMont high pressure boiler.

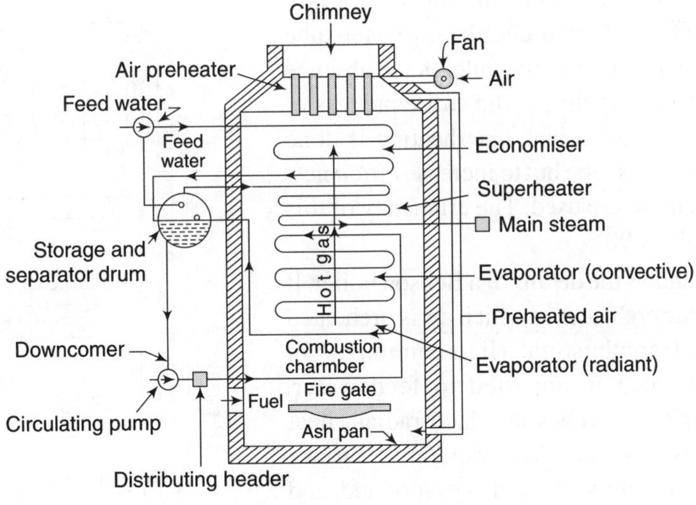

Figure 6.24 LaMont boiler

The feed water from the hot-well is supplied through an economizer to a separating and strong drum that contains a feed regulator which controls the speed of the feed pump. Since the economizer is placed in the boiler at a place from where hot combustion gases pass, the economizer supplies sensible heat to the feed water from the boiler drum flows by a gravity to a circulating pump, which discharges into a distributing header. Water from the distributing header flow through long small diameter boiler tubes located in the wall and roof of the furnace to the drum where the steam is separated and water returns to the pump. Orifice located at inlet to each circuit on the distributing header correctly proportions the water among the many parallel circuits, so that, each receives its proper share. The circulating pump raises the water pressure to about 3.5 bars above the drum pressure to overcome the resistance to the flow-controlling orifices and the long circuit of small diameter tubing. At normal load, the quantity of water circulated in each tube is about 3.8 times the steam evaporated in the long circuit of small tubes. This avoids the overheating of the tubes.

Since three circuit tubes are placed in the combustion chamber through which hot gases are passing upwards, the steam is generated in them. The mixture of water and steam from these tubes enter the boiler drum where the moisture is separated from the drum. Now the steam is led to the superheater tubes in which the steam gets superheated. The superheated steam goes to ultimate destination through the steam stop valve.

It is essential to maintain a constant level of water in the drum. This is possible by supplying feed water equivalent to the steam quantity blow-off continuously. The pump consumes about 0.5 to 0.6% of the boiler output which is supplied by the power unit using the steam from boiler. For economic combustion, forced air preheated by the flue gases is used.

6.4.4 Benson Boiler

The Benson high pressure boiler introduced by Mark Benson in 1923 is a forced circulation water tube boiler. The main feature of this boiler is the absence of the steam separating drum. The entire process of heating, steam formation and superheating is done in a single continuous tube but to increase efficiency, many parallel circuit are used. The efficiency of this boiler is as high as 90%.

Figure 6.25 shows the details of a Benson boiler. It consists of (i) Air preheater in which air is preheated for economical combustion, (ii) economizer in which sensible heat is imparted to feed water, (iii) radiant surface of tubes in which radiant heat is supplied by combustion, (iv) evaporative surface where the major quantity of water is evaporated, and (v) superheater tubes in which steam is superheated and led to the steam turbine.

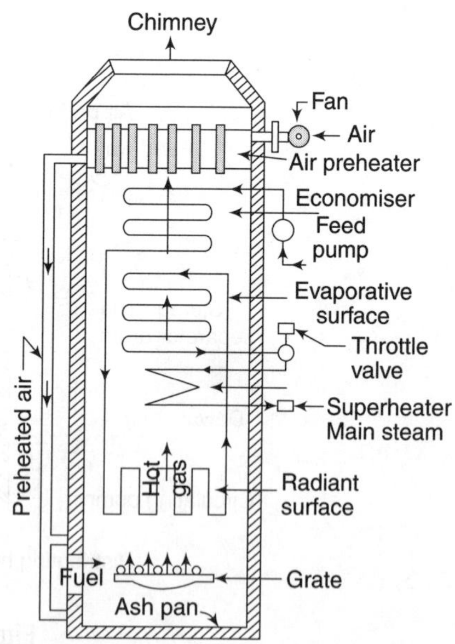

Figure 6.25 Benson boiler

It is an established fact that the rate of heat transmission from flue gas to water is seriously impaired by the presence of steam bubbles in contact with the tube, and that the release of these bubbles causes the water circulation to pulsate. It was pointed out by Mark Benson that if the boiler pressure was raised to the critical pressure, the steam and water would have the same density and therefore no bubbles would be formed, thus, eliminating the trouble. But to raise the pressure above critical, a large amount of power is consumed by the feed pump which lowers the efficiency of the plant. Hence, by using pressure slightly lower than critical improved efficiency is possible. At normal load, this boiler has an operating pressure of 210 bars with 13500 kg/h evaporating capacity at 405°C superheating temperature. Benson boiler is very near to supercritical boiler.

The Benson boiler has the following:

Advantages:

(i) There is no pressure limitations and it may be as high as supercritical.
(ii) Absence of circulating pump and downcomers.
(iii) Absence of drum hence the cost is less.
(iv) Lighter in weight, higher specific output with high safety factor.
(v) Evaporation is quick.
(vi) Self-contained unit and can be easily erected.

Disadvantages:

(i) Evaporation process is accompanied by formation of salt and solids in the tubes. Special arrangements are required to remove this.
(ii) On evaporative surface, there is change of corrosion of the tubes.
(iii) Overheating of the tubes in case of insufficient water supply.
(iv) Since the storage capacity is small, it requires close coordination among steam, feed water and fuel input.

6.4.5 Loeffier Boiler

This boiler is a forced circulation, indirect heating type of boiler. In this, water is evaporated solely by means of superheated steam, i.e., steam is used as the heat carrying and heat absorbing media.

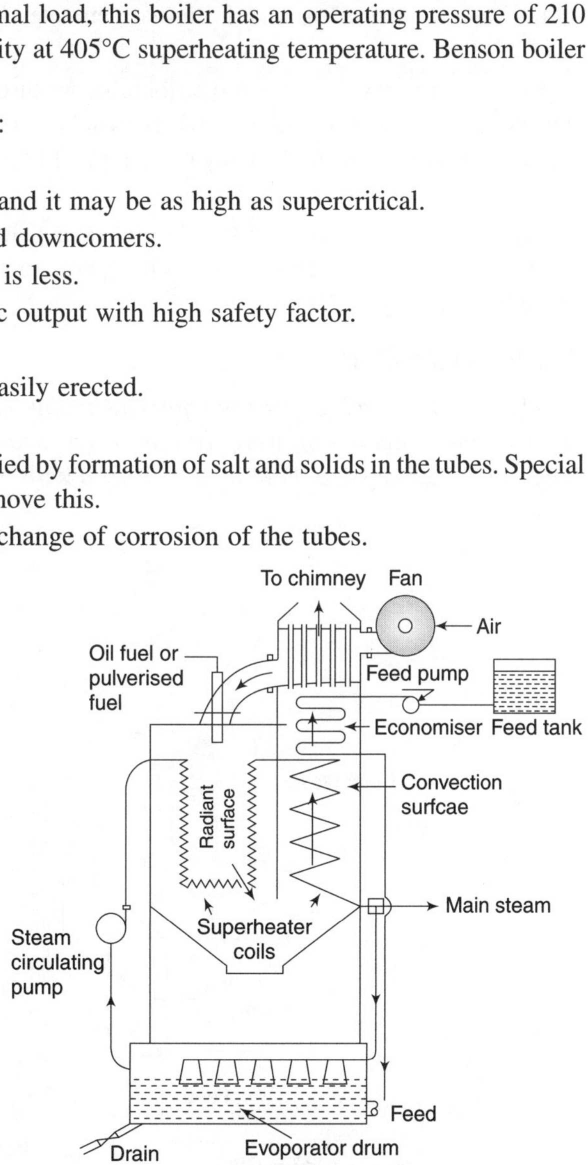

Figure 6.26 Loeffier boiler

Figure 6.26 shows the details of a Loeffier Boiler. It consists of an evaporator drum which may be placed at any convenient point outside the furnace setting. The feed water pump is placed where the feed water passes through as economizer on the way to the drum. Since the economizer is placed in the path of the outgoing hot gases, the economizer gives sensible heat to water. The steam circulating pump extracts steam from the evaporator drum and forces it to pass through the radiant and convective surface of the superheater tubes placed in the path of the hot combustion gases. From the superheater tube, a big portion (about 3/4) of the superheated steam is trapped off for external cause while the remainder (1/4) passes on to the evaporator drum where it gives its superheat to the water contained in this drum that generates an amount of steam equal to that trapped off.

Nozzles are provided which distribute steam throughout the water contained in the drum. The nozzles are of special design which avoid priming and noise so that the boiler can carry a higher salt concentration than any other type of boiler. This makes it very suitable for marine transport and for power generation.

The air, preheated through the air preheater placed in the path of the outgoing hot gases is used for economical combustion. The operating pressure of this boiler is 140 bars and the steaming capacity is 95000 kg/h.

6.4.6 Velox Boiler

The Velox boiler is a high pressure, forced circulation, pressurized or forced combustion boiler but with the limitation of firing with oil or gas under pressure with little or no ash problem. The evaporator capacity is limited to 10,200 kg/h of steam.

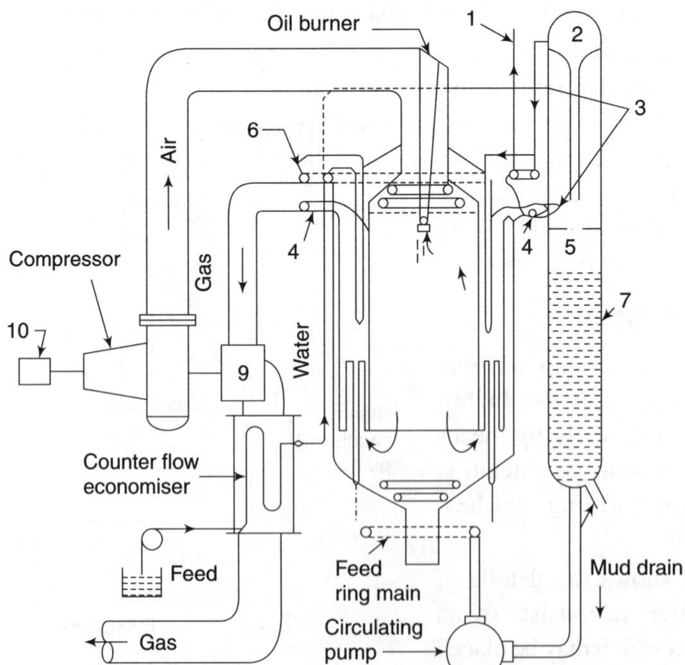

Figure 6.27 Velox boiler

Figure 6.27 shows the details of a Velox boiler. Air compressed to about 2.5 bars in an axial compressor (which, at the time of start is driven by a motor) before being supplied to an oil fuel furnace. The purpose of this compression of air is to secure a high velocity of gas and also at the same time release of a great amount of heat. The fuel and the compressed air are injected downwards into a vertical combustion chamber which is surrounded by hollow evaporator tubes on reaching the bottom of the combustion chamber the products of combustion are deflected upward into the evaporator tubes by means of a spiral water coil. The evaporator tubes consist of an outer annulus through which 10 to 20 times the water evaporated is circulates at a high velocity. The core of the lower half of the evaporator tube or element is occupied by a central pipe which supplied water to the outer annulus, while the upper half is occupied by a central pipe which supplies water to the outer annulus, while the upper hill is occupied by U-type superheater tubes. In the space between the inner pipes and outer annulus, the flue gas rushes at a speed of about 250 m/s.

There is provision of a ring main (4) which collects the steam and water, and discharges it tangentially into the separating chamber (5). This forms a forced vortex, which, by centrifugal loading on the water particles, allows steam release, without priming about two hundred times as great as in boilers of normal design. The dry steam then passes up the central tube (2) the superheater ring main, which distributes it to the various superheat elements. The mud drum collects the separated water which is extracted by means of a circulation pump. This circulating pump also creates a high water velocity through the evaporator tubes.

The flue gases, after passing through the superheater enter and exhaust gas turbines that drive the compressor. The exhaust from the turbine passes through the counter-flow feed heater where the feed water is preheated and which is discharged tangentially into the separating drum.

This boiler is a compact steam generating plant of great flexibility. It is capable of quick starting and its thermal efficiency is about 90 to 95%.

6.4.7 Once-through or Monotube Boiler (Supercritical Pressure Boiler)

The once-through or monotube boiler as the name implies has only one tube almost about 2 km long and there is no steam drum. It is the only type suited to supercritical pressure operation, i.e., above 221.2 bars as there is no latent heat of vaporization involved, and water on being heated at constant pressure (theoretically) suddenly flashes into vapour and steam is further heated to desired temperature in the superheater. Water is fed to the boiler by boiler feed pump at inlet through a number of parallel paths and comes out at outlet as superheated steam. LaMont may be called a monotube boiler. Subcritical boilers may also be monotube boilers.

Many supercritical pressure once-through boilers were built either with single or double reheating having steam pressure 240–345 bars and steam temperature 540 to 650°C. A simplified view of CE Sulzer once-through boiler with triflux reheater where steam is reheated partly by steam and partly by the gas is shown in Fig. 6.28.

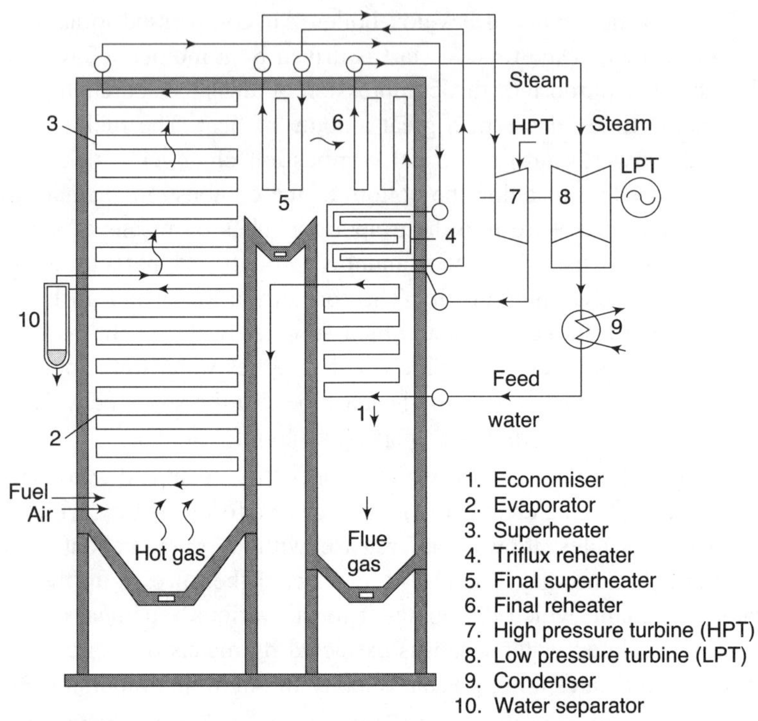

Figure 6.28 Sulzer once-through boiler (steam generator)

The main advantages of once-through boiler are the absence of drum, and best suited to quick starts rapid load changes. The main disadvantage is the poor quality boiler water.

The **following are the** *advantages* **of supercritical boilers over sub-critical**

Advantages:

(i) Large heat transfer coefficient.
(ii) Stable pressure level,
(iii) Higher thermal efficiency.
(iv) Minimum erosion and corrosion
(v) Easy to generate peak load
(vi) Ease in operation

6.5 FLUIDIZED BED BOILERS

The efforts to minimize the pollution from conventional steam power plants and to have efficient use of low grade fuels led to the discovery of fluidized bed boiler in which steam is generated by utilising the technique of fluidized bed combustion. The detailed analysis of fluidized bed combustion is discussed in this chapter.

Solid (coal) particles when subjected to a stream of air may be either (i) unaffected if the air velocity is below a certain value or (ii) air borne if the air velocity is above a certain value. Between the limits of velocity, a bed of small solid particles of fairly uniform size will not become fully air-borne but will have their reduced effective density and will move about in the bed as if bed is fluid. By proper regulation of air velocity the solid particles can be burnt in the combustion chamber almost as easily as if they are fluid. The *critical velocity* is called the *minimum fluidization velocity* at which the bed is said to be *incipiently fluidized.* With the further increase in velocity, the particles are buoyed up and imparted a violently turbulent fluid like motion. The mixing is rapid and equilibrium between gas and particles is rapidly established which is called a fluidized bed. There are two types of *fluidized bed* boilers. They are:

(i) Bubbling fluidized bed (BFB) boilers

(ii) Circulating fluidized bed (CFB) boilers.

Both boilers may be operating at ***atmospheric pressure or pressurized*** **condition.**

6.5.1 Bubbling Fluidized Bed Boiler (BFB)

Figure 6.29 shows a schematic of a bubbling fluidized bed boiler **working under atmospheric condition**. *Crushed coal having the size 6 to 20 mm is injected into the fluidized bed along with limestone through the feed hopper just above an air-distribution grid at the bottom of the bed.* In some cases, coal and limestone are subjected into the lower section of the furnace from bottom. The air flows upwards through the grid from the air plenum into the bed, which now becomes the furnace where combustion of the swirling mixture takes place. The combustion products leaving the bed contain a large proportion of unburned particles that are collected in a cyclone separator, which separates these particles from the gas by imparting centrifugal acceleration into the mixture. These separated unburnt particles return back to the fluidized bed to complete their combustion. The boiler water tubes are located in the furnace. Finer solid residues (ash and spent sorbents) generated during combustion and desulphurization

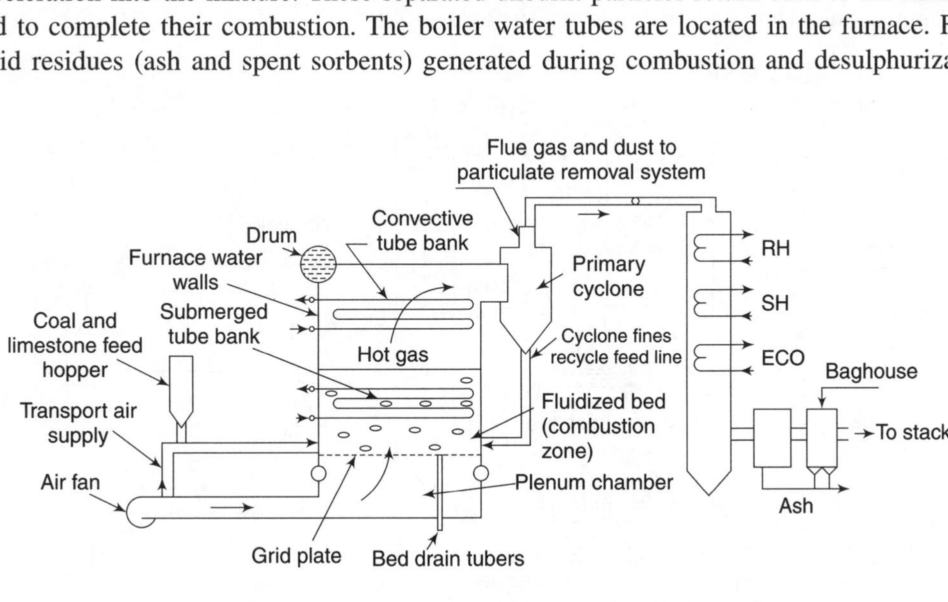

Figure 6.29 Bubbling fluidized bed (BFB) boiler

leave the furnace escaping through the cyclones, but they are collected by a baghouse or electrostatic precipitator. The most important advantage of fluidized bed combustion lies in the concurrent removal of the sulphur dioxide that results normally from the combustion of sulphur content of the coal. Desulphurization takes place by the addition of limestone directly to the bed together with the crushed coal. The combustion temperature is low (800–900°C). Due to low combustion temperature, inferior grades of coal can be used *without slagging problem* and there is no *formation of* NO_x. The other advantage of low combustion temperature is the use of cheaper alloy materials which results in economical construction. Since there is no pulverizer, so further economy is achieved. Further, a fluidized-bed combustor can be designed to incorporate the boiler within the bed, resulting in volumetric heat transfer rates that are 10 to 15 times higher and surface heat transfer rates that are 2 to 3 times higher than a conventional boiler. All these make the boiler more compact.

Economiser, superheater, reheaters, preheaters, baghouse or electrostatic precipitator, ID fan, stack, etc. are used like conventional boiler.

6.5.2 Circulating Fluidized Bed (CFB) Boilers Working under Atmospheric Pressure

Figure 6.30 shows a second generation CFB boiler working on atmospheric pressure. The whole system is divided into two sections, namely, solid fuel circulation loop and back-pass loop. The first section comprises furnace or fast fluidized bed, cyclone (gas- solid separator), solid recycle device, (loop seal or L-valve) and external heat exchanger (optional). *Similar to pulverized coal-fired (PC) boiler, the furnace enclosure is generally made of water tubes.* In the by-pass section, the remaining heat from the flue gases is absorbed by the superheaters, reheaters, economizers and air-preheater surfaces.

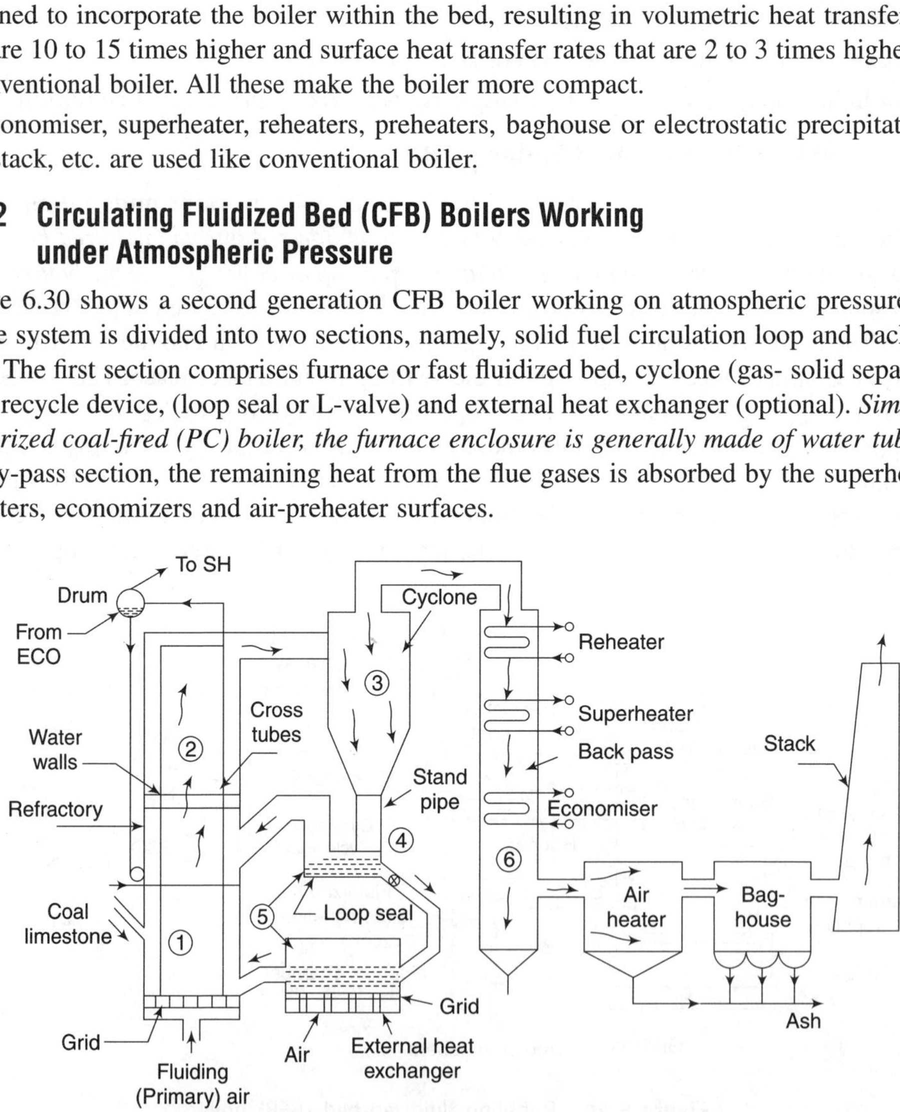

Figure 6.30 Circulating fluidized bed (CFB) boiler

In the first section, the furnace walls lined with refractory up to the level of secondary air entry and beyond this, furnace walls are generally cooled by evaporative superheater or reheater surface. The cyclone dust collector and the non-mechanical valves are also lined with refractory. Coal/limestone are injected into the lower section of the furnace through hopper system. There happens to be a loop furnace in which the recycled unburnt coal particles are burnt. The fluidizing (primary) air is fed to the furnace through an air distributor or grate and the secondary air is admitted at some height above the grate to complete combustion. With the injection of primary and secondary air, the bed is fluidized, combustion takes place resulting in lower tempetaure of the gas (800 to 900°C). The ash from the bed falls down. The cyclone separator, the coarse sorbent (limestone) and coal particles which are recycled back near the base of the furnace while the finer solid residue (ash and spent sorbents) after passing over superhaters, reheaters, economiser, air preheaters get captured by Baghouse or electronic precipitator and go finllay to atmosphere via stack. *The flow regimes are (i) turbulent fluidized bed, (ii) fast fluidized bed, (iii) swirl flow, (iv) moving packed bed, (v) bubbling fluidized bed, (v) pneumatic transport.*

The first commercial CFB boiler went into service at Pihlava, Finland in 1979. After that, many CFB boilers were commissioned. *The recent commissioned CFB boilers are 165 MWe Point Aconi Power Station at Cape Breton in Nova Scotia, Canada and 250 MWe Power unit at Lyon in France.*

6.5.3 Cyclone Separators

As the name implies, *the function of cyclone separator is to separate solid particles from the gas by utilizing the centrifugal force created by a spinning gas stream.* Figure 6.31 shows a standard tangential inlet vertical cyclone separator in which the gas flow is forced to follow the curved geometry of the cyclone, while the inertia of the particles in the flow causes them to move towards the outer wall, where they colide with wall, fall down along the wall by gravity and are collected at the bottom end.

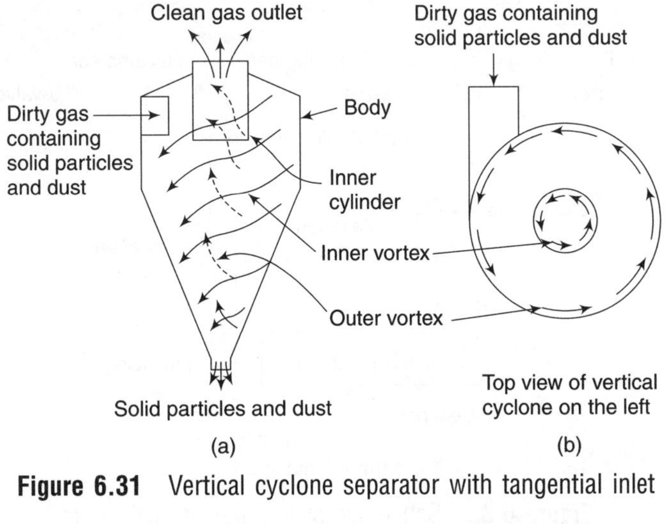

Figure 6.31 Vertical cyclone separator with tangential inlet

The centrifugal force (F_c) on a particle of mass $\dot{m}_p = \dot{m}_p C_0^2/r$ while gravity force is given by, $F_g = 9.81\ \dot{m}_p$, thus the centrifugal force is about 20 times the gravity force.

As a result of strong centrifugal force, the particles in the spinning gas stream move progressively closer to the outer wall as they flow through the cyclone and the gas stream executes several complete turns.

The particles laden gas comes out from the furnace and enters tangentially to the cyclone. The centrifugal force forces the particles to the wall where they fall down along the wall by gravity force. It is worth to note that at the bottom of the cyclone, the gas flow reverses to form the inner core that leaves to the top of the unit.

The performance of cyclone separator is described by a parameter called *collection efficiency which is defined as the ratio of the amount of particles separated to the amount of gas-solid mixture.* The collection efficiency increases with increasing particle size, particle density, inlet gas velocity, cyclone body length, number of gas revolutions and smoothness of cyclone wall surface. Further, the collection efficiency goes on increasing by decreasing cyclone diameter, gas outlet duct diameter and gas inlet area. *Cyclones used in CFB boilers are either steam or water cooled, due to relatively high temperature.*

6.5.4 Non-mechanical Valves

Non-mechanical valves, extensively used in FBB are devices that allow the flow solids between return leg (stand pipe) and the furnace without any external mechanical force.

There are various types of non-mechanical valves (Fig. 6.32) such as (i) L-valve, (ii) J-valve, (iii) reverse seal valve, (iv) loop valve and (v) seal pot.

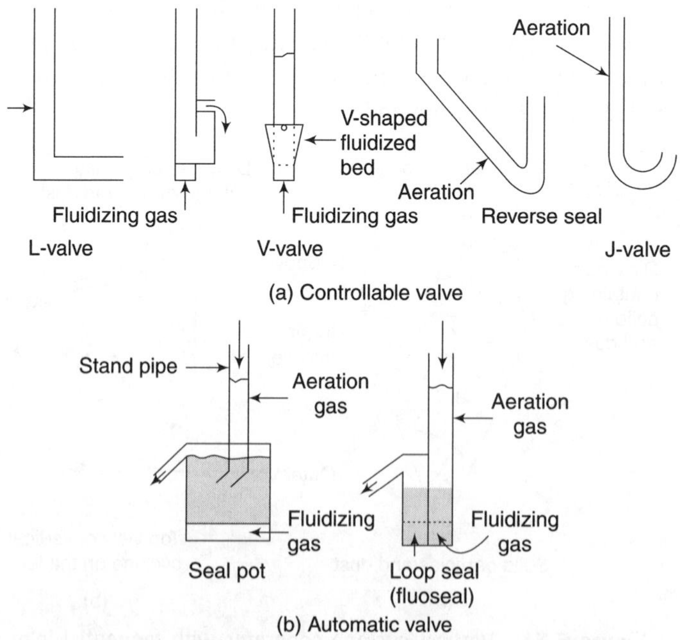

Figure 6.32 Schematic of non-mechanical valves

The movement of solids in these valves takes place by air which is fed at a short height above the exit of valve. The L-valve comprises a right-angled bent L-shaped pipe connecting the two vessels between which the solids are to be transferred. It is designed from pressure balance across the CFB loop. There is a difference in the density of bed. Even a small amount of air injected into the dense log moves solid to the diluted bed. The working principle of J-valve is similar to L-valve. It is the variation in the aeration rate in J-valve that controls the solid flow rate. A seal pot comprises of a bubbling fluidized bed with an overflow pipe and the standpipe of the cyclone. It is the higher pressure in the seal pot that prevents the furnace air from entering the standpipe and by this way it serves as valve. There is minor difference between the loop seal and the seal pot. The loop seal is smaller in size and here the solid movement is horizontally around a vertical partition separating the standpipe and overflow part of the seal pot and thus requires less air. The L-valve, V-valve and J-valves are controllable while seal pot and loop seal automatic.

6.5.5 Pressurized Fluidized Bed Combustion (PFBC) Boilers

In pressurized *fluidized bed combustion (PFBC) boiler, combustion takes place in a pressurized environment created by supplying the compressed air by compressor of gas turbine unit*

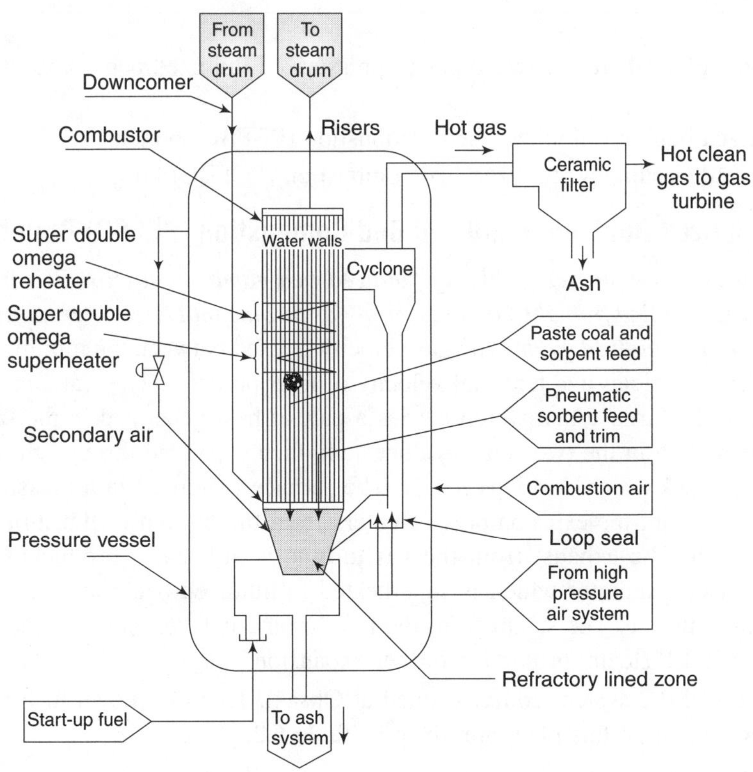

Figure 6.33 PFBC hot loop

in the furnace instead of atmospheric environment resulting in a compact furnace and improved combustion efficiency. Like atmospheric fluidized bed combustion (AFBC), the other advantage of PFBC is the capturing of sulphur and low NO_x generation due to lesser operating temperature (900°C). Since the pressure of the gas is high (about 15 to 20 bars), gas turbine may be driven to develop power by using inexpensive coal. The furnace along with cyclones is housed in a pressure vessel and the compressed air from the compressor enters the bottom of furnace via pressure vessel. The hot loop of PFBC system is shown in Fig. 6.33. The system may be used in turbocharged, combined and hybrid advance cycles.

Table 6.1 depicts the values of important parameters in the major classes of FBCS.

Table 6.1 Values of important parameters in major classes of FBCS

	Atmospheric		*Pressurized (17 bars)*	
	BFBC	*CFBC*	*BFBC*	*CFBC*
(i) Gas velocity (m/s)	1.4–3.6	4–10.5	1.2	5
(ii) Bed height (m)	1.1–1.4	36	3.5–4.2	40
(iii) Coal feed points/m^2	0.5–1.1	0.006–0.00	0.26–6	0.2–0.6
(iv) Bed residence time (s)	0.5	5	4	8

Similarly, to AFBC, there are two types of pressurized fluidized bed combustion boilers. These are:

(i) Pressurized bubbling fluidized bed combustion (PBFBC) boilers
(ii) Pressurized circulating fluidized bed combustion (PCFBC) boilers

6.5.6 Pressurized Bubbling Fluidized Bed Combustion (PBFBC) Boiler

Figure 6.34 shows a *pressurized bubbling fluidized bed combustion (PBFBC) boiler where compressed air is supplied from the compressor of gas turbine unit and the granular solids are fluidized in bubbling fluidized mode.* The gas velocity is fixed between the minimum bubbling velocity of coarser particles and terminal velocity of finer particles. The water carrying tubes are embedded in the furnace bed which carries a part of heat generated in the furnace. The dirty gas is cleaned up in the two stage cyclones and then expanded through the gas turbine as shown in Fig. 6.34 which develops power. The furnace is housed in a pressure vessel in which primary air from the exit compressor enters from the bottom of the furnace through the pressure vessel. The exhaust from the gas turbine is utilised in producing steam in a heat recovery steam generator which is superheated in fluidized bed combustor and finally expands in steam turbine. The steam from the h.p. turbine is taken out for reheating which is readmitted in IP/LP steam turbine for further expansion.

A 250 MWe PBFBC system commissioned at Ohsaki, Japan is shown in Fig. 6.35. The principle specifications of this plant are given in Table 6.2.

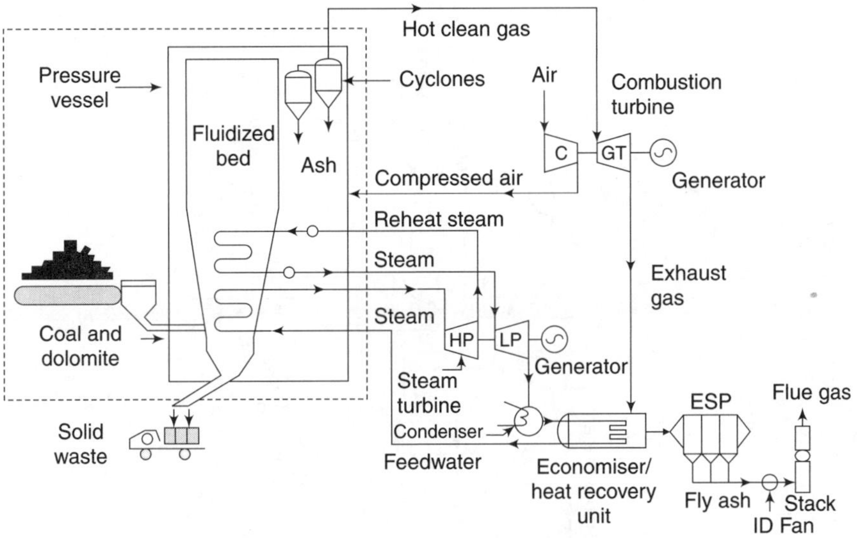

Figure 6.34 Pressurized bubbling fluidized bed boiler serving gas/steam turbine

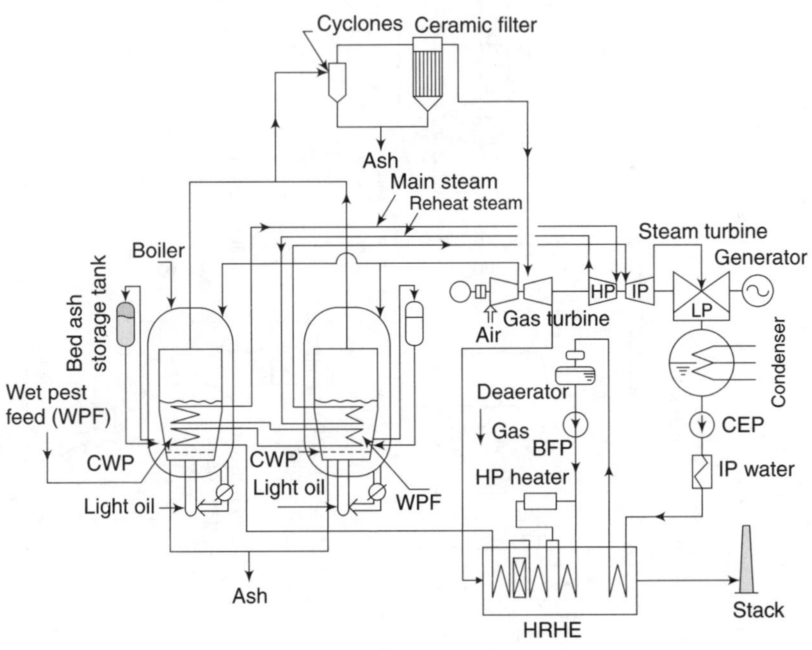

Figure 6.35 Schematic of a 250 MWe PFBC system at Ohsaki, Japan

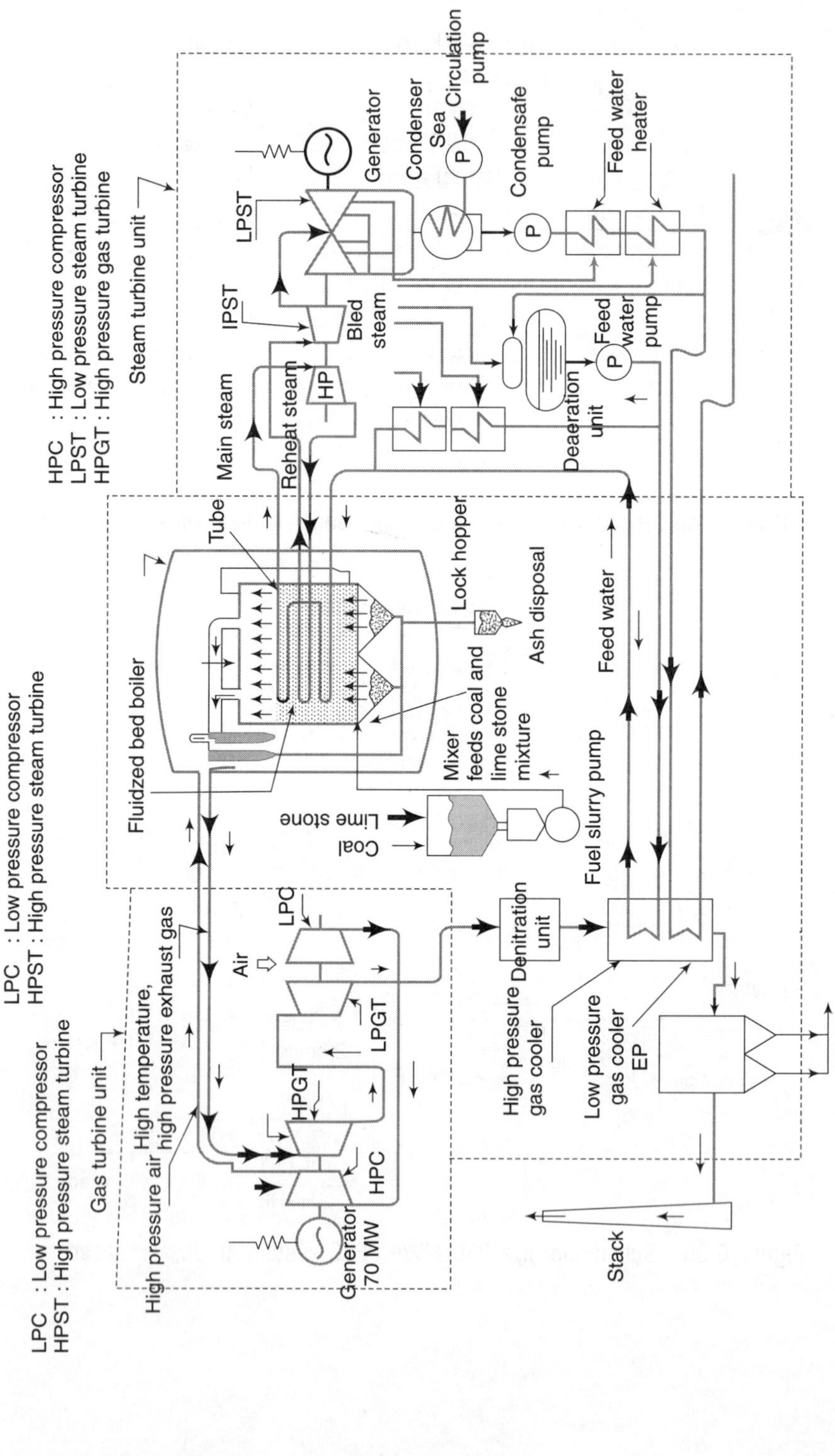

Figure 6.36 350 MWe PFBC plant at Kartia, Japan

Table 6.2 Specifications of PBFBC plant (250 MWe) at Ohsaki Japan

Gross electrical output	*250 MWe x2* *(ST : 210 MWe. GT : 40 MWe) x2*
Boiler	Pressurized bubbling fluidized bed type Combustion regenerative once-through type Steaming capacity : (520t/h) x 2 Bed temperature : 865°C Combustion pressure : 10 bars Cool and limestones feed system : Wet paste feed
Gas turbine	Single shaft, simple open cycle
Steam turbine	Tandem compound regenerative condensing type Steam condition : 166 bars, 566/593°C
Emission control	SO_2 : In-site limestone desulfurization NO_x : NH3 injection SNCR and SCR Particulate : cyclone and ceramic filters
Stack	Height : 200 m

Figure 6.36 shows a 350 MW PFBC plant commissioned at Kartia, Japan which is the biggest plant at present in the world.

6.5.7 Pressurized Circulating Fluidized Bed Combustion (PCFBC) Boiler

Figure 6.37 shows a *pressurized circulating fluidized bed combustion (PCFBC) boiler in which furnace is enclosed in a pressure vessel.* In this case, the solid particles of coal and limestone in wet form is injected into the furnace and are kept in fast fluidized condition, they are mixed through the combustor, but there is very little mixing of gas in the axial direction. The gas flows upward with negligible back mixing. The water tubes are lined along the furnace of the wall vertically. It is worth to note that primary air, less than stoichiometric amount, enters the combustor through the grid at the bottom of the combustor through pressure vessel. The admission of secondary air takes place at some distance above the grid. On account of absence of back mixing of the air, substoichiometric condition prevails in the lower section of the bed and above this boiler tubes are located. Due to staging of air, NO_x emission gets reduced and the chance of corrosion diminishes. Since air is added in stages, so desulphurization is better. The circulation system of the boilers may be either once-through, natural circulation and forced circulation.

In order to expand the high temperature and high pressure gas through the turbine two-stage cyclones are used. The gas velocities through the cyclones at inlet are 30 to 42 m/s respectively. The gas is further cleaned by passing through a ceramic filter before admission to gas turbine.

6.5.8 Advantages and Disadvantages of Fluidized Bed Combustion (FBC) Boilers

Advantages

(i) **Fuel flexibility.** The most attractive features of FBC boilers is the use of any cheap fuels without the support of any auxiliary fuel due to fluidized bed combustion.

(ii) **High combustion efficiency.** As a result of superior mixing of solid fuels with air, availability of reaction space up to the top of furnace (even up to 40 m and further beyond into the hot cyclone) and long residence time of combustion, the combustion efficiency is very high (upto 99%).

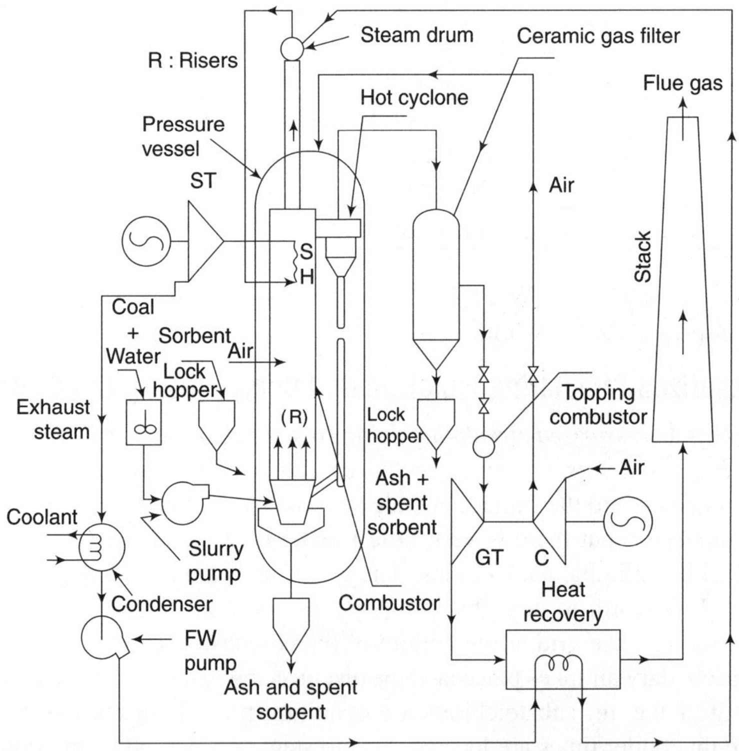

Figure 6.37 Pressurized circulating fluidized bed combustion (PCFBC) boiler

(iii) **Concurrent removal of SO_2.** The pressure of sulphur content in the coal produces SO_2. Desulphurization is accomplished by the addition of limestone or dolomite which acts as sorbent directly to the bed together with the crushed coal. Limestone ($CaCO_3$) absorbs the sulphur dioxide with the help of some oxygen from the excess air according to

$$CaCO_3 + SO_2 + 1/2O_2 \rightarrow CaSO_4 + CO_2$$

Thus, $CaSO_4$ retains the sulphur in the bed in the solid form rather than allowing it to eascape as gaseous SO_2. It is worth to note that the rate of formation of $CaSO_4$ is maxmized at about 850°C and after 900°C, the sulphur capture gets reduced. That is why the temperature of FBC furnace is not allowed to go beyond 900°C. The effect of pressure on sulphur to retention is appreciable.

(iv) Low NO_x emission. As a result of low combustion temperature (around 900°C) and the staged combustion process, NO_x emission in FBC boiler is very low in the range of 50–150 ppm.

(v) Simple fuel handling and feed system. Since the requirement of coal size in FBC boiler is coarse so there is no need of any conventional pulverizing mill and thus feed preparation is simple. The high degree of lateral solid mixing in turbulent zone at the bottom ensures uniform feed distribution within the bed. Further, the large height of the furnace prevents maldistribution of fuel.

(vi) Compact furnace. One of the main advantages of FBC boiler is the compactness of furnace due to high heat release per unit furnace cross section (about 5 MW/m^2). The high heat release and heat dispersion in the bed is possible due to intense gas-solid mixing and velocity (4 to 7 m/s).

(vii) Good turndown and load following capability. As a result of relative high fluidizing gas velocity and staged combustion, ensure a fairly good turndown ratio (about 5:1) simple reducing the proportionate amount of fuel and air.

(viii) High availability. FBC boilers offer high availability (above 90%).

Disadvantages

(i) Erosion of reactor walls, **(ii)** Attrition of fuel particles, **(iii)** Difficulty in immersing internals due to possible erosion and **(iv)** complexities of hydrodynamics in furnace.

6.5.9 Advantages of Pressurized Bubbling Fluidized Bed Combustion (PBFBC) Boiler

(i) Reduced capital cost. The specific power output gets increased, which in turn, reduces the capital cost. At a typical fluidizing velocity of 2 m/s, the furnace bed area of PBFSC boiler at 1 bar is 2 m^2/MW whereas it is 0.2 m^2/MW at 10 bars in non-fluidized.

(ii) Increase in overall efficiency. The overall efficiency of power generation gets increased from 35% for conventional power plants to more than 42% in the case of PBFBC plants.

(iii) NO_x emission reduces enormously.

(iv) The combustion efficiency of PBFBC is higher than that of AFBC boiler.

(v) Due to low fluidizing velocity (say 1 m/s) in PBFBC boiler, the gas residence time is high (about 5 seconds), whereas it is about 0.5 seconds in AFBC which results in better capturing of sulphate content.

6.5.10 Advantages of Pressurized Circulating Fluidized Bed Combustion (PCFBC) Boilers

(i) **Less space requirements.** The pressure vessel enclosing the furnace is smaller in diameter due to high fluidzing velocity, higher heat release per unit area and fewer components inside the pressure vessel.

(ii) **Less erosion and more accessibility.** The erosion of the furnace is less due to low particle density fluidized bed with fine particles distributed throughout the furnace. The heat transfer in PBFBC is submerged whereas in PCFBC boiler, it is not submerged which results in less possibility of erosion. This also helps in better accessibility for inspection and maintenance of plate superheaters.

(iii) **Vigorous mixing.** PCFBC boiler offers vigorous mixing of solid particles with air in the furnace which results in fewer feed points, thus a simpler and reliable furnace.

(iv) **Absence of material handling system.** There is no need of bed material handling system to control the bed temperature.

(v) **Higher exit gas temperature.** The exit gas temperature is higher which is very favourable for gas turbine performance.

(vi) **Lower NO_x emissions.** NO_x emission is lower due to low temperature of combustion and enhanced staged combustion.

(vii) **Better load following.** The following of load as per demand is better by varying the fuel and air flow and the proposition of primary and secondary air.

(viii) **Less bed inventory, less total metal and refractory.** Due to less bed inventory, less total metal and refractory, there is a shorter start-up and cool-down period and a quicker response to load changes.

(ix) **Use of dolomite.** At higher pressure, the partial pressure of carbon dioxide in a bubbling bed is higher which requires higher temperature for calcinations. Thus dolomite is more used in PBFBC boiler.

6.6 BOILER MOUNTINGS AND ACCESSORIES

For safe and efficient operations, boilers are equipped with two categories of components such as (a) Boiler mountings, and (b) boiler accessories

(a) Boiler Mountings. *Boiler mountings are those machine components which are mounted over the body of the boiler itself of the safety of boiler and for complete control of the process of steam generation.* These mountings form an integral part of the boiler. According to the Indian boiler regulations, the following mountings should be fitted on the boiler.

(i) Two safety valves, (ii) Two water level indicators, (iii) Pressure gauge, (iv) Fusible plug, (v) Steam stop valve, (vi) Feed check valve, (vii) Blow-off cock, (viii) Inspector test gauge, (ix) Man and mud hole, etc. Among the above mountings, safety valve, water level indicator and fusible plug are called ***safety fittings*** and the remaining are the **control fittings**.

(b) Boiler Accessories. *These are those machine components which are installed either inside or outside the boiler to increase the efficiency of the plant or to help in the proper working of the plant.* The following accessories are generally used in the boiler.

(i) Air preheater, (ii) Economiser (iii) Superheater, (iv) Feed pump (v) Steam trap, (vi) Steam separator, (vii) Pressure reducing valve, etc.

Only some of the important accessories will be discussed here.

6.6.1 Economiser

The function of economiser is to heat the feed water coming out form high pressure feed heater up to nearly the saturation temperature corresponding to the boiler pressure by utilizing the energy of hot flue gases leaving the superheater or reheater which is generally at a temperature varying from 370°C to 570°C. It is a heat exchanger, so its design will be based on the analysis of heat exchanger. The word economiser has been used to indicate that the energy leaving the superheater/reheater which was going as a waste is utilised to heat feed water.

Economisers are generally placed between the last superheater-reheater, and the air preheaters. They have been built with plain or extended surface tubes. The size of the economiser tubes are commonly 45–70 mm in outside diameter. They are built in vertical coils (sections) of continuous tubes connected between inlet and outlet headers with each section formed into several horizontal paths connected by 180° vertical at a pitch of 45 to 50 mm spacing which depends upon the type of fuel and ash characteristics.

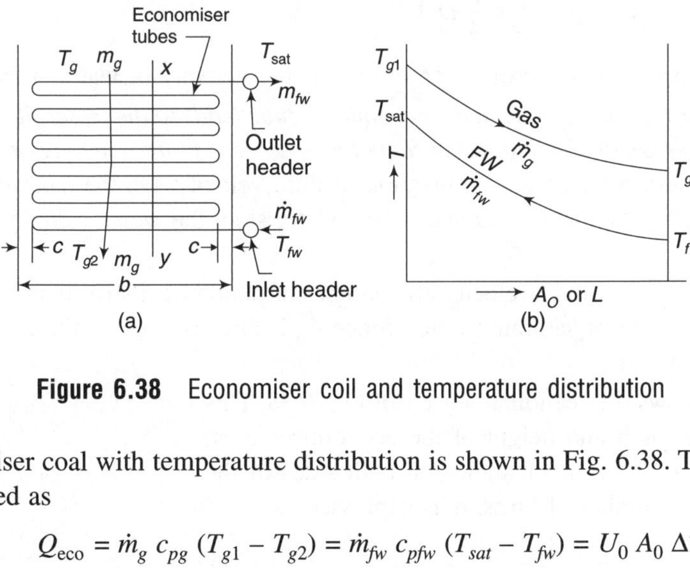

Figure 6.38 Economiser coil and temperature distribution

An economiser coal with temperature distribution is shown in Fig. 6.38. The heat transfer rate is expressed as

$$Q_{\text{eco}} = \dot{m}_g\, c_{pg}\, (T_{g1} - T_{g2}) = \dot{m}_{fw}\, c_{pfw}\, (T_{sat} - T_{fw}) = U_0\, A_0\, \Delta T_m \qquad ...(6.25)$$

where ΔT_m = log-mean temperature difference $= \dfrac{\Delta T_i - \Delta T_e}{\ln (\Delta T_i/\Delta T_e)}$ ...(6.26)

$\Delta T_i = T_{g1} - T_{\text{sat}}$, and $\Delta T_e = T_{g2} - T_{fw}$

and
$$\frac{1}{U_0} = \frac{1}{h_{sc}(A_i/A_0)} + \frac{1}{h_i(A_i/A_0)} + \frac{x_w}{k_w(A_{1m}/A_0)} + \frac{1}{h_{f0}} + \frac{1}{h_0} \qquad ...(6.27)$$

Here U_0 = overall heat transfer coefficient, h_{sc}, inside h_{f0} and h_0 are heat transfer coefficients due to scale formed on the side tubes, inside of water side, due to solids (ash and soots) on the outer surface of tubes and the outside film (gas side) respectively. $A_i = n\,\pi\,D_i\,l$, $A_0 = n\pi D_0 l$ where l is the length of one coil, n = number of coils, A_{1m} = log-mean area = $(A_0 - A_i)/\ln(A_0/A_i)$, x_w = thickness of tube, k_w = thermal conductivity of wall material. If h_{sc} and h_{f0} are negligible, then,

$$\frac{1}{U_0} = \frac{1}{h_i} + \frac{x_w}{k_w} + \frac{1}{h_0} \qquad ...(6.28)$$

The heat transfer coefficients, h_i and h_0 may be calculated from the following standard equation

$$Nu = 0.023\ \text{Re}^{0.8}\ Pr^n \qquad ...(6.29)$$

where n = 0.4 for heating, n = 0.3 for cooling, $Nu = hD/k$ = Nussett number, $\text{Re} = \frac{CD}{\mu_f}$; $Pr = (\mu\ c_p/k)$ = Prandtle number. From Eqs. (6.25) and (6.28), U_0 may be estimated and finally A_0 may be found out where $A_0 = \pi\ D_0\ n\ L$.

Using continuity equation at the exit of economiser where the water is saturated, we have

$$\dot{m}_{fw} \cdot v_f = AC = \left(n\,\frac{\pi}{4}\,D_i^2\right) C_{fw} \qquad ...(6.30)$$

where v_f is the specific volume of water and C_{fw} is the velocity of water at exit.

The high value of fluid velocity in the economiser tube leads to the lesser film thickness, i.e., *more heat transfer coefficient but at the same times it needs more pumping power.* These two conflicting requirements lead to optimization of fluid velocity. On the gas-side, the velocity of *flue gases is about 10 to 12 m/s* and on the water side, the *water velocity is limited to 1 to 1.5 m/s.*

Thus, for the selected fluid velocity, the number of parallel coils (n) may be estimated for a given mass flow rate of saturated water. Since A_0 is already known, the length of one coil may be computed.

Figure 6.39 shows the section along x-y of Fig. 6.38. This figure shows the pitch and height of the economiser duct. Let b be the width, c is the clearance on both sides of the gas duct, then the number of turns, n_t is expressed as

$$n_t = \frac{l}{b - 2c} \qquad ...(6.31)$$

The height of duct occupied by the economiser is expressed as

$$H_{eco} = n_t \cdot p \qquad ...(6.32)$$

where p is the vertical pitch on the centre-to-centre distance of the consecutive horizontal tubes forming the coil.

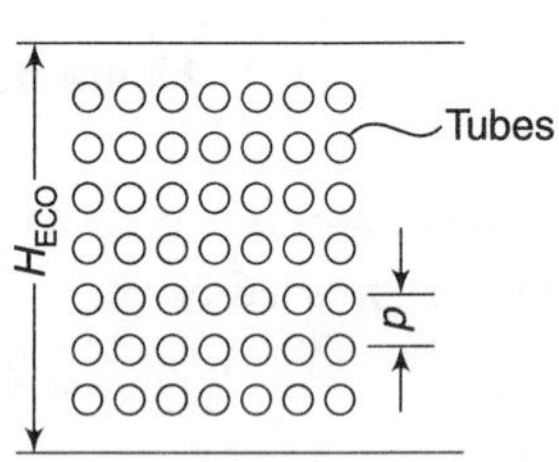

Figure 6.39 Section through economiser-coil (Fig. 6.38) showing height of duct

6.6.2 Superheaters and Reheaters

The function of a superheater is to superheat the saturated steam up to the desired temperature. It is essentially a surface heat exchanger located in the passage of a hot flue gases. In some cases, superheaters may be placed in an independent fire furnace. With the use of superheater in the steam power plant, thermal efficiency increases as well as the mostiure content reduces in the last stage of turbine. About 40% of heat generated in modern steam generator is utilised by superheater. This indicates that a large surface area is needed by superheaters. Modern superheaters and reheaters operating at about 540°C, however are usually made of special high strength steels chosen for both strength and corrosion resistant. Below 460°C, carbon steel is sufficient.

There are *three types of superheaters, namely,* **convective (CSH), radiant (RSH) and platen (PSH)** depending upon how heat is transferred from the gases to steam. As the name implies, convective superheaters are located in the convective zone, while radiant is in radiation zone and combined in a convective-radiation (pendent type) zone respectively. The radiant and combined superheaters are sometimes called *secondary superheaters.* Figure 6.40 shows heat absorption in three types of heaters. The heat transfer rates from the gas to steam in these three superheaters are

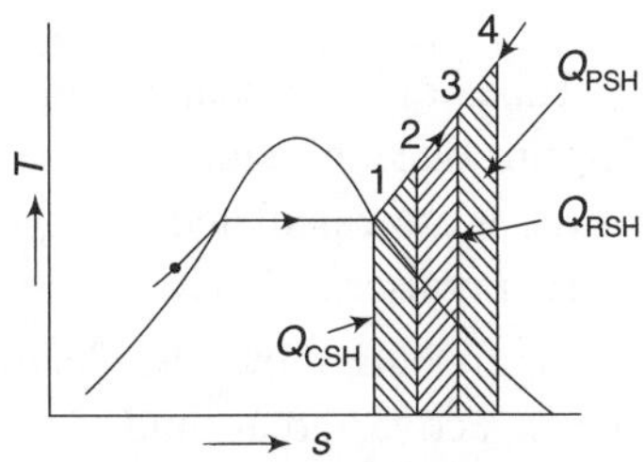

Figure 6.40 Heat absorption in superheaters

$$Q_{CSH} = \dot{m}_g\, c_{pg}\, (T_{g1} - T_{g2}) = \dot{m}_s\, (h_2 - h_1) = U_0\, A_0\, (\Delta T)_m \quad ...(6.33)$$

$$Q_{RSH} = \sigma A_T\, F_{f-w}\, (T_f^4 - T_w^4) = \dot{m}_s\, (h_3 - h_2) = U_0\, A_0\, (\Delta T)_m \quad ...(6.34)$$

$$Q_{PSH} = \gamma\, \dot{m}_g\, c_{pg}\, (T_{g3} - T_{g4}) + (1 - \gamma)\, \sigma\, A_T\, F_{fw}\, (T_f^4 - T_w^4)$$

$$= \dot{m}_s (h_4 - h_3) = U_0\, A_0\, (\Delta T)_m \quad ...(6.35)$$

where T_f and T_w are the flame and tube wall temperature, F_{f-w} is the shape factor with respect to flame and wall and γ is the fraction of energy shared by convection (say 50%).

Figure 6.41 shows the convective superheater coil and its temperature profiles. For radiant, it will be identical. The mass flow rate passing through the superheater is given by

$$\dot{m}_s \cdot v_g = \left(n\, \frac{\pi}{4}\, D_i^2\right) C_s \quad ...(6.36)$$

where C_s is the average velocity of steam in the superheater. For high intermediate and low pressure, C_s are selected as 10, 12 and 22 m/s respectively.

The erosion of superheaters tube is proportional to the square of gas velocity. This suggests that the greater the ash content of the fuel, the less will be the allowable gas velocity in the system. But higher gas velocity yields higher heat transfer coefficient. So as a compromise, the gas velocity for low-ash coals and high-ash coal should not exceed 12 and 8 m/s respectively.

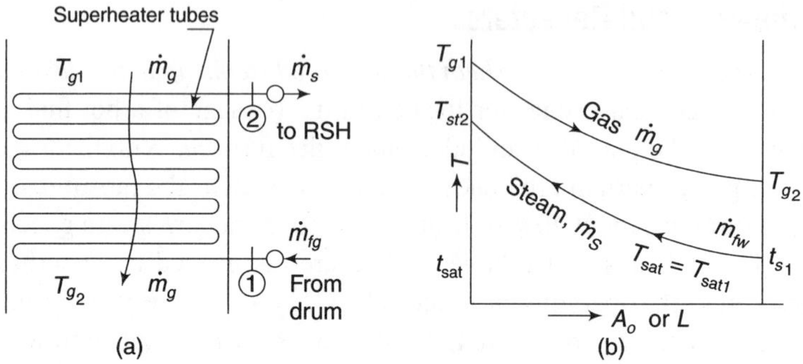

Figure 6.41 Convective superheaters and temperature distribution

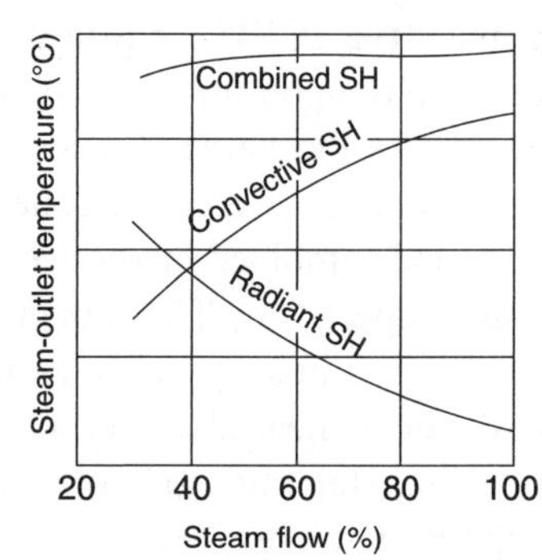

Figure 6.42 Steam-outlet temperature response of superheaters

Tubes of superheaters and reheaters are made of 50 to 75 mm outside diameter. The number of coils and length of tubes are calculated using Eqs. (6.34) to (6.36), similar to economiser.

Figure 6.42 shows the steam outlet temperature response of convective, radiant and combined (in series) superheaters. In convective superheater, the steam receives greater heat transfer per unit mass flow and its temperature increases with load. ΔT_m is calculated similar to economiser.

The radiation heat transfer to radiant superheater is proportional $(T_f^4 - T_w^4)$ and since T_f is much greater than T_w, thus the heat transfer is essentially proportional to T_f^4. Because T_f is not strongly dependent on load, the heat transfer per unit mass flow of steam decreases as the steam flow increases. Thus, an increase in steam flow due to an increased load demand would result in reduction in exit steam temperature, the opposite effect to convective superheater (Fig. 6.42). Convective superheaters alone are used with low temperature steam generators. Radiant and convection superheaters and reheaters are used for high temperature service. The radiant units are arranged in plant panels or platen sections with wide spacings.

Based on mechanical construction, there are **three kinds of superheaters, namely, *pendent, inverted and horizontal*** as shown in Fig. 6.43.

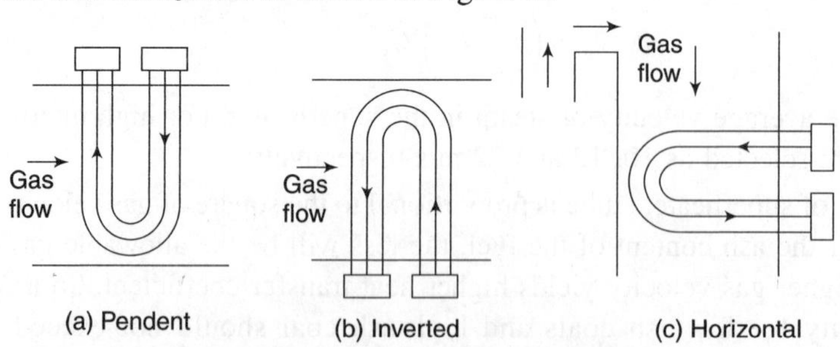

Figure 6.43 Kinds of superheaters

Pendent type superheaters and reheaters are those that are hung from above. They have the advantages of firm structural support but the disadvantages of flow blockage by condensed steam after a cold shutdown, which necessitates slow restart to purge the water that accumulates in the bottom. Pendent type is generally used in radiant superheater. As the name implies, ***inverted type*** units are supported from below, so they have proper drainage of the condensed steam but the structural rigidity is poor and that is why they are not commonly used. ***Horizontal type units*** are usually supported horizontally in the vertical gas ducts parallel to the main furnace and receive the hot gases after a U-turn at the top. They don't view the flame directly and hence are mainly of the convection type. They have both proper drainage and good structural rigidity.

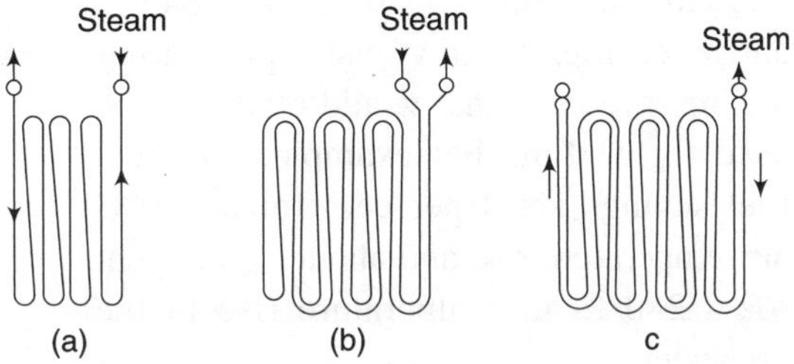

Figure 6.44 Types of superheat coils, (a) Single, (b) Double with one tube, (c) Double with two tubes

As mentioned earlier, radiant and **pendent superheaters** may be made in the form of coils or platens. The coils may be *single, double with one tube and double coil with two tubes* as shown in Fig. 6.44. Platens may be *rectangular* and rectangular type as shown in Fig. 6.45.

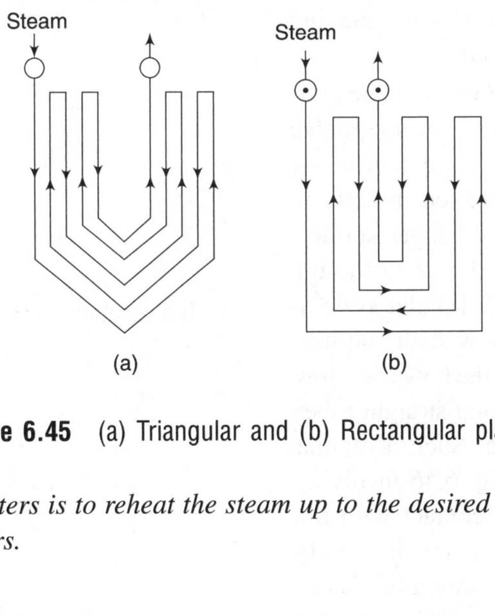

Figure 6.45 (a) Triangular and (b) Rectangular platens

The function of reheaters is to reheat the steam up to the desired temperature. The design is similar to superheaters.

6.6.3 Air Preheater

The function of air preheaters is to preheat the air before entering the furnace by utilisng some of the energy left in the flue gases before exhausting them to the atmosphere. They receive gases at 320 to 430°C. As mentioned earlier, the flue gases are cooled not below 160°C to avoid gas condensation and corrosion problems and to allow for proper dispersion in the atmosphere. The air is heated from forced draught fan outlet temperature (slightly more than atmospheric) to 260°C to 345°C and even higher. Obviously, preheating the air saves fuel that would otherwise be used for heating. For example, a typical fuel savings are 4 per cent for a 110°C air temperature rise and about 11 per cent for a 280°C air temperature rise in the preheater.

There is an additional requirement of the preheated air for the operation of pulverized coal furnace. *Air in the range of 150 to 420°C is needed for drying the fuel. Air is also used for transporting it to the furnace and burning it there.* Though, small stoker fire units don't require preheated air, but large stoker-fired bituminous coal steam generators have beneficial effect from preheated air, but only up to 180°C to prevent damage to the stoker's moving parts due to high temperature.

There are two general types of air preheaters namely, ***recuperative*** and ***regenerative***. In *recuperative* air preheaters, heat is transferred directly to the air across the heat exchanger surface. They are commonly tubular, although some plates are still used. Tubular systems are essentially counter-flow shell and tube heat-exchanger in which the hot gases flow inside vertical or horizontal straight tubes and the air flow outside. Such a tubular preheater is shown in Fig. 6.46 involving baffles to maximize air contact with the hot tubes. The tubes are mechanically expanded into top and bottom tube sheets. A hollow-type expansion unit shown at the bottom sheet is used to provide the thermal expansion. A hopper provided below the tubes collects root and dust that deposit on the inside tube surfaces. There should not

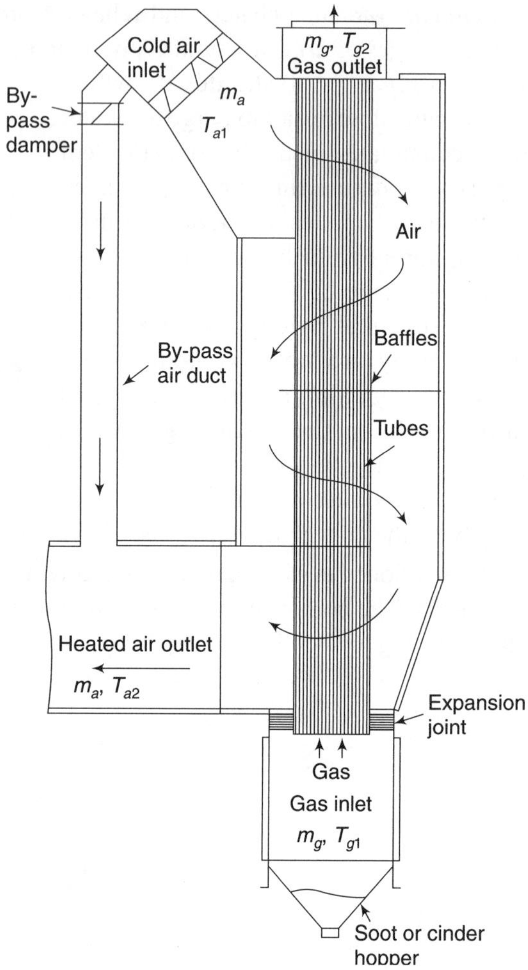

(a) Tubular counter-flow recuperative preheater

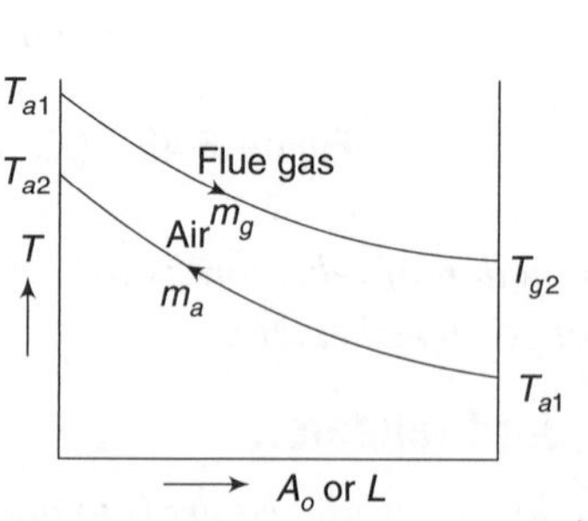

(b) Temperature profiles in APH

Figure 6.46

be any leakage from air (which is at high pressure) to the gas otherwise the power consumption of both forced and induced draft fan will increase along with reduction in effectiveness of preheater.

In order to suit particular steam generator spaces and duct layout, tubular preheaters are built in a variety of designs. They may be used for one or more passes for both and air gases in counter- or cross-flow, in vertical or horizontal arrangements. The size of the tubes varies from 50 to 65 mm OD. Cost, cleaning requirements and fuel type determine the diameter to be used in a given situation.

The design of the preheater is based on the heat exchanger analysis. The rate of heat transfer from gas to air is given by

$$Q_{APH} = \dot{m}_g\, c_{pg}\, (T_{g1} - T_{g2})$$
$$= \dot{m}_a\, c_{pa}\, (T_{a2} - T_{a1}) = U_0\, A_0\, \Delta T_m \qquad ...(6.37)$$

The value of U_0 is calculated similar to that of economiser.

Regenerative air preheaters are storage type heat exchangers in which heat is transferred from the hot flue gases, first to an intermediate heat storage medium called the matrix, then to the air. The most common type is the *rotary air preheater*, known as the ***Lyungstrom preheater***. It consists of a rotor driven by an electric motor at a low speed of 1 to 3 rpm depending on diameter, through reduction gearing arrangement.

The rotor consists of 12 to 24 radial members that form sectors as shown in Fig. 6.47. The matrix consisting of flat, corrugated, notched or undulated ribbing is filled in the sector. A stationary seal covers the equivalent of two opposite sectors and half of the remaining sectors are exposed at any instant to the hot gases. Rotary air preheaters are designed with vertical or horizontal shaft depending upon the layout and ducting. The main advantage of rotary preheater is its compactnesss with a large heat transfer surface accommodate in a small volume. They are as popular as tubular preheater. The main disadvantages include (i) plugging of flow openings by fly ash, (ii) sealing of gas to air path and vice-versa, and (iii) large pressure drop.

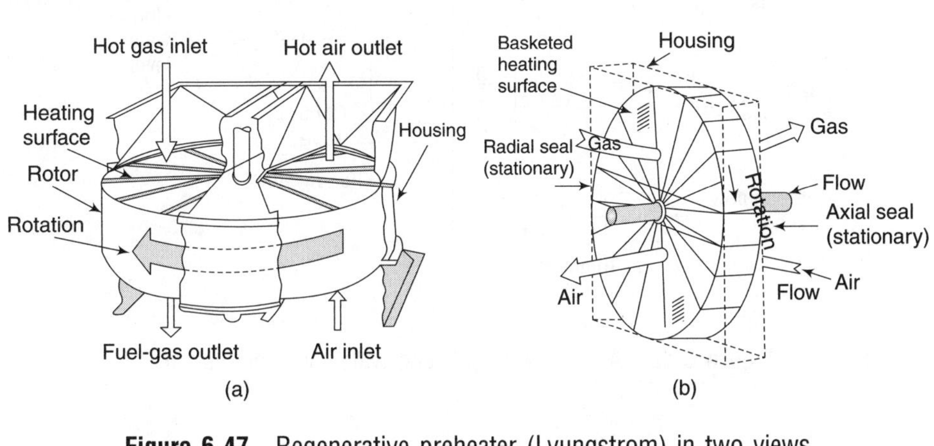

Figure 6.47 Regenerative preheater (Lyungstrom) in two views

6.7 STEAM GENERATOR CONTROL

The purpose of steam generator control is to supply the steam flow required by the turbine at design pressure and temperature and according to the load.

Steam generator, and indeed total power plant control is rather a broad subject that includes every important aspects of plant such as instrumentation, data processing and controls for combustion, steam flow, temperature, and pressure, drum level, burner sequencing, desulphurization precipitators, ash handling, system integration, start-up and shut-down and automation. This wide subject is not possible to be covered. The control system must act on the measurement of these parameters to control the plant.

The current practice in control system is to employ closed loop control. The actual output of the system is measured and compared to some demand signal (set point). The error signal (i.e., difference in output and demand) is used to control the quantity of steam. Proportional control is either directly or inversely proportional to the error signal.

Since the subject is very wide, only a few basic control systems that apply to steam generator such as feed water and drum control, steam pressure control and steam-temperature control will be discussed here.

6.7.1 Feed Water and Drum-level Control

The aim of feed water and drum-level control is to meet load demand by the turbine and at the same time maintain the level of water in the steam drum, within relatively narrow limits. It is common to maintain the normal drum level half-full, in reference to sight glass tube used outside the drum. In a particular situation of high steam consumption by the turbine, combined with low water supply would lower the water level in the drum. Figure 6.48 shows a three-element automatic control system, of which drum level is one element.

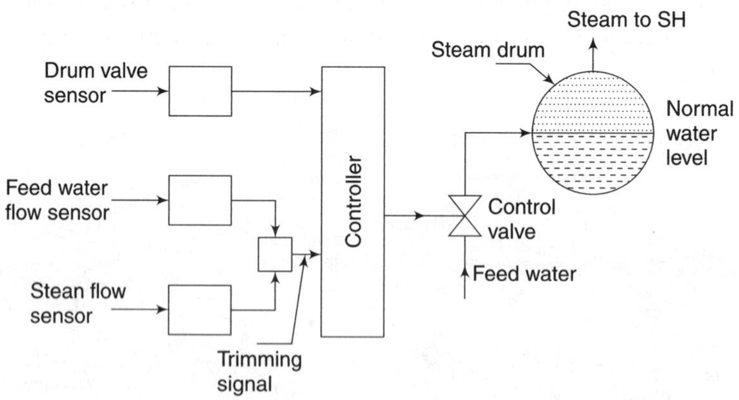

Figure 6.48 A Three-element feed water control system

The drum level sensor senses the level of water and responds to the error signal between actual drum and its set point, such as in the case of high steam consumption and low feed water supply, and acts on the controller to increase the feed water valve opening to meet the steam flow demand. This action may be too slow and is supplemented by sensor for feed water and steam flow to meet the demand as soon as possible. The difference between the signals from these two sensors anticipates changes in drum level and sends a signal to the controller to actuate the valve in the proper direction. The zero difference in these two signals mean constant water level in the drum as per demand.

6.7.2 Steam Pressure Control

The purpose of steam pressure control system sometimes called the boiler master is to maintain steam pressure by adjusting fuel and combustion air flow to meet the desired pressure. When pressure drops, the flows are increased. Three sensors, namely, steam flow, air flow, and steam pressure are incorporated in the control system. The steam pressure sensor senses the steam pressure and acts directly on the fuel flow and air flow control such as the pulverized coal power drives and forced drought fan to affect the desired changes as per demand (Fig. 6.49). A trimming signal obtained from the processing of fuel flow and air flow maintains the proper fuel-air ratio. Because it is often difficult to obtain accurate fuel flows, a steam flow sensor is sometimes substituted for the fuel-flow sensor.

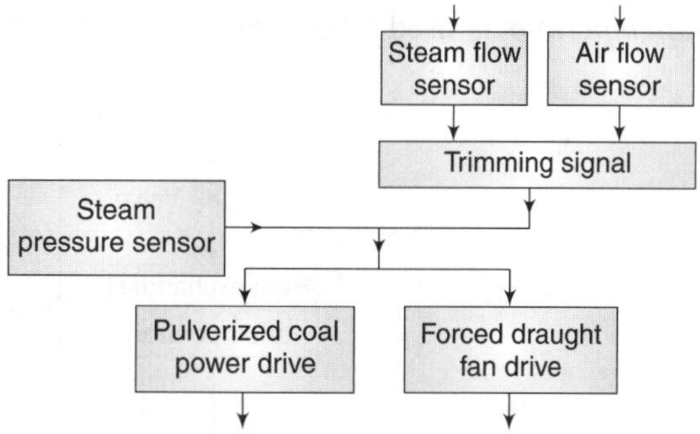

Figure 6.49 Steam Pressure Control System

In general, about 5 seconds delay is permitted when changing coal flow and air flow to ensure prevention of a momentary rich mixture and thus assume smoke-free combustion in the furnace.

6.7.3 Steam Temperature Control

For efficient power plant operation, it is necessary to control accurately the steam superheat temperature. The principal parameters affecting superheat temperature are furnace temperature, cleanliness of radiant and pendent superheater, temperature of gases entering the convective superwater, cleanliness of convective superheater, mass flow rate of gases through the convective superheaters and variation of load on the unit which is the most important among all. The interrelationship among these variables is complex.

A reduction in superheat temperature from the rated (design) value results in loss of plant efficiency. For example, a reduction of 20°C of superheat temperature results in above 1% increase in heat rate, whereas, a rise in steam temperature above the rated value may result in overheating and failure of superheater, reheater tubes and turbine blades.

As we know that steam leaving the drum is approximately saturated and its temperature corresponds to the boiler pressure and remains constant if the steam pressure controller is effective. The superheater and reheater temperature change as per load and so it needs control. The following are the various methods to control the superheater/reheater temperature.

(i) Combined radiant-convecting superheaters. In this method, radiant and convective superheaters are arranged in series to yield a relatively flat final steam temperature over a wide load range (Fig. 6.42). This is possible as they have opposite steam temperature responses to load variation.

(ii) Attemperation. *Attemperation is the reduction of steam temperature.* It is also called desuperheating. **Attemperation is achieved by two methods.** In the first method, a ***surface attemperator*** **is used** which removes heat from the steam in a heat exchangers.

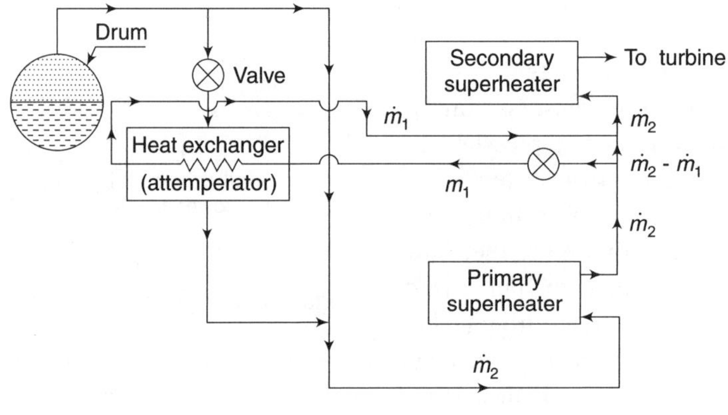

Figure 6.50 Principle of surface attemperator

There are two types of surface attemperator, namely, *shell type* and *drum type*. The *shell type* has a portion of the steam ($\dot{m}_1$) taken out through tubes from between the primary and secondary superheater by an automatic valve and diverts to a shell-and-tube heat exchanger (Fig. 6.50) containing some of the boiler water. The steam gives up some of its heat to that water and then remixes with the primary steam ($\dot{m}_2$) upon entering the secondary superheater. Temperature control is affected by controlling the amount of diverted steam. In *drum type*, heat exchange takes place between the diverted steam and boiler water within the main steam drum which is made longer to accommodate the attemperator tubes.

In the second method, a ***spray or direct contact attemperator*** is used as shown in Fig. 6.51 which reduces the steam temperature by spraying low temperature water from the boiler or economiser exit into the line between the primary and secondary superheaters (Fig. 6.52). The spray nozzle injects water into throat of a mixing venture, where the water mixes with high-velocity steam in the throat, vaporizes and cools the steam. Further, the venturi and a thermal sleeve made of high chrome steel protect the main steam pipe from thermal shock caused by any unvaporized water droplets that might otherwise impact the pipe leading to

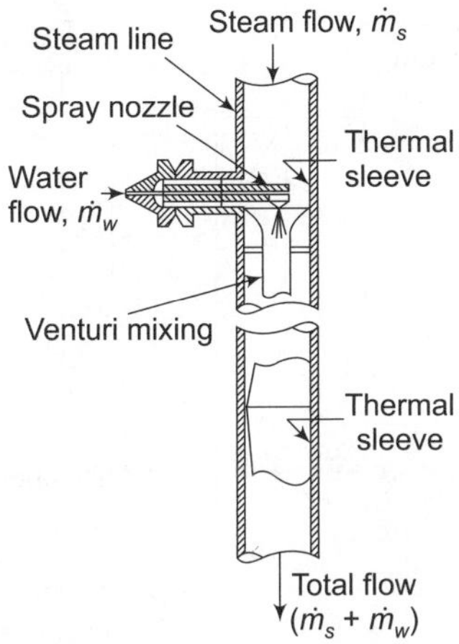

Figure 6.51 Spray attemperator for steam temperature control

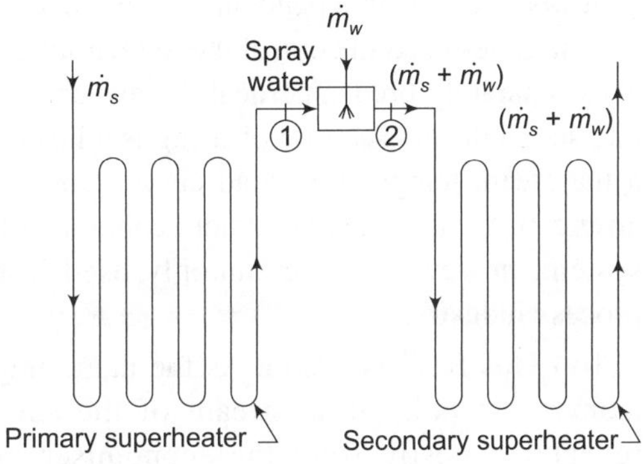

Figure 6.52 Location of attemperator

its failure. It is worth to note that water used must be of high purity to avoid adding deposits on the superheat tubes, pipes and turbine blades leading to drop in performance and failure. It has been found operating satisfactorily all over the world. *The main advantage of spray attemperator is to provide a rapid and sensitive means of temperature control.* Figure 6.53 shows the performance of a spray attemperator with primary and secondary superheaters. Steam temperature is controlled by regulating the amount of spray water to produce a flat temperature curve beyond the point '*a*'. A simple energy balance illustrates this principle of attemperation.

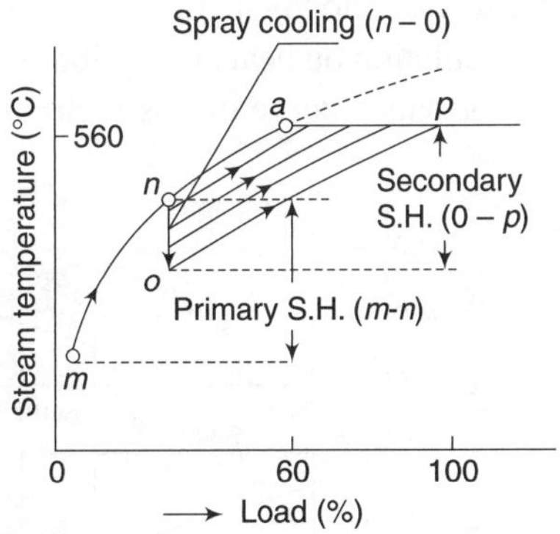

Figure 6.53 Attemperation with oversized primary and secondary superheaters

$$\dot{m}_s h_{s1} + \dot{m}_w h_w = (\dot{m}_s + \dot{m}_w) h_{s2} \qquad ...(6.38)$$

where $\dot{m}_s$ and $\dot{m}_w$ = Mass flow rate of steam and water respectively, kg/s

h_s and h_w = Specific enthalpies of steam and water respectively, kJ/kg

Subscripts 1 and 2 refer to steam inlet and exit respectively.

The location of attemperator of either type plays an important part. If it is located before the two superheaters, then it will cause condensation of the saturated steam from the boiler before it enters the superheater. If it is located after the superheater the steam temperature exceeds the final desired steam temperature before attemperation. The mid-location between the primary and secondary sections of the preheater is therefore preferred.

(iii) Separately Fired Superheater. A superheater with its own burner, fans, combustion chamber, controls, etc., all independent of the steam generator is sometimes used to control the temperature of steam. This separately fired superheater may serve more than one steam drum. The rate of firing is adjusted to yield a flat steam temperature load curve. Due to separate arrangement, this method is not economical for utility system. however, it is commonly used in chemical process industry.

(iv) Gas Recirculation. As the name implies, gas from some point downstream of the superheater-reheaters, mostly from the economiser outlet but sometimes from the air preheater outlet, is recirculated back to the furnace by means of a gas circulation fan to control the temperature of steam. This method of control is shown in Fig. 6.54 and the effect of gas recirculation on heat absorption is shown in Fig. 6.55 which may form the basis for design. The performance curve of gas recirculation and attemperator in series is shown in Fig. 6.56.

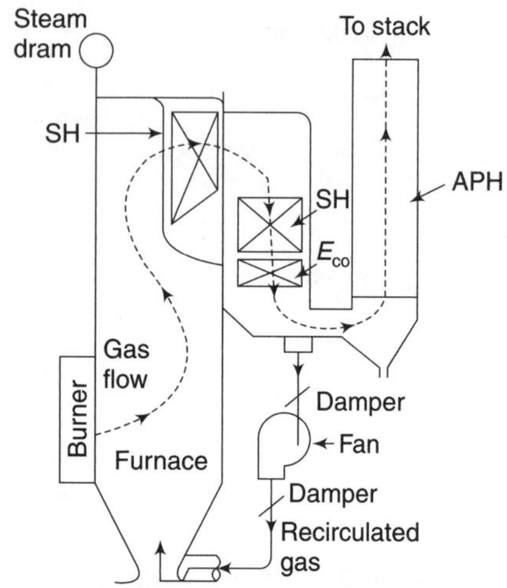

Figure 6.54 Gas recirculation for steam temperature control

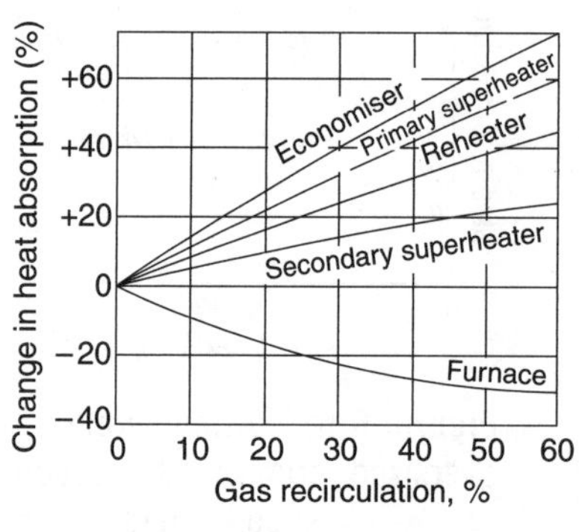

Figure 6.55 Effect of gas recirculation on the heat absorption

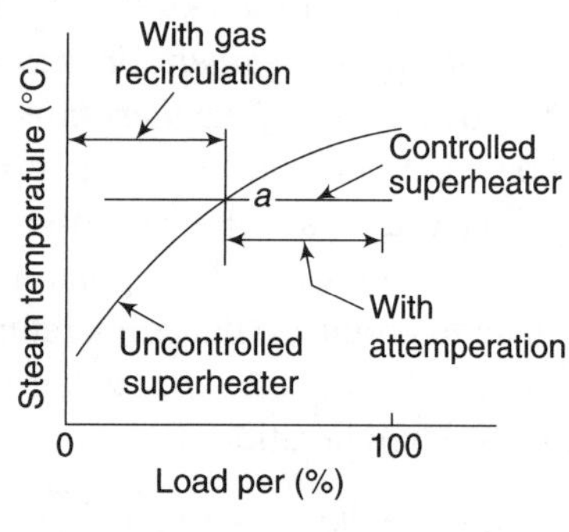

Figure 6.56 Performance of gas recirculation attemperation in series

The term recirculation is restricted to the case when recirculated gas is fed before the burning zone, while if admitted downstream of the burner it is called *gas tempering*.

(v) Gas Bypass or Damper Control. The temperature of steam is controlled by using a a damper control placed before the convective superheater to bypass the gas passing over it. This is achieved by creating desired energy exchange between the hot gas and steam.

(vi) Excess Air. By admitting excess combustion air, the furnace temperature gets reduced. The heat transfer to the furnace wall is by radiation which is a function of difference in the fourth power temperature of hot gases and wall. As a result, the gas temperature increases so the heat transfer by the radiation to the steam gets increased, which in turn, increases the steam temperature. But the increased mass of the gas will lower boiler efficiency.

(vii) Tilting Burners. Tilting or vertically adjustable burners are designed in such a way that they change the gas temperature entering the superheater section by changing the elevation of the fireball within the furnace, which in turn, will change the steam temperature to the desired value. For example, at low loads (steam flow), the burners are tilted upward in the range of ±20°C so that the lower portion of the furnace becomes less effective in absorbing energy. As a result, the gas enters the superheater at a higher temperature than if the burners were fixed, and the steam temperature leaving the convection superheater can be maintained constant from less than one-half load to full load. This method of temperature control is found satisfactory as well as economical.

(viii) Burner Selection. In utility steam generators, several burners are used in horizontal rows. At partial loads, only some burners are needed. So the upper burners are not fired which reduce the effective heat absorption by the furnace, which in turn, maintain the correct gas temperature level entering the convective economiser. By this way, steam temperature is controlled.

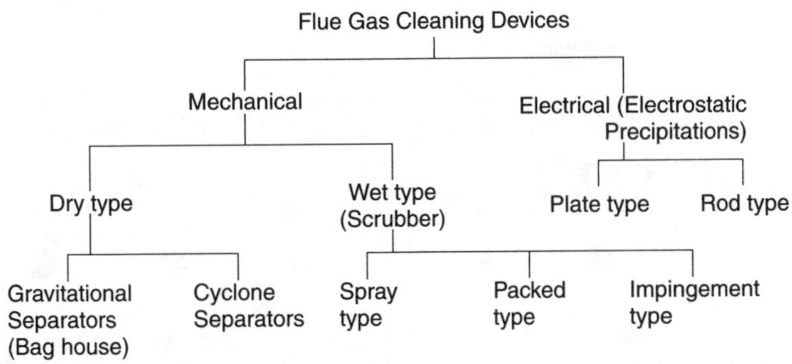

6.8 ASH COLLECTION AND ITS DISPOSAL

The subject of flue gas cleaning has become a problem of prime importance in the design of large thermal power stations using coal as fuel as the atmospheric pollution is highly restricted as per worldwide regulations. The major emissions from thermal power station are fly ash, carbon ash (known as cinder), smoke, dust and irritating vapours like CO, SO_2 and NO_x. It is necessary to clean the gas streams contaminated with particulate matter before the gases are discharged to the atmosphere. The classification of flue gas cleaning devices given above are discussed below.

6.8.1 Electrostatic Precipitator

The function of electrostatic precipitator (ESP) is to collect the fly ash particles from the flue gases by electrostatic principle. The first commercial electrostatic precipitators developed by Professor Cottrel of Physical Chemistry at the University of California in 1903. It is now extensively used in power plants, smelters, steel plants, paper mills, etc.

Figure 6.57 shows the basic elements of an electrostatic precipitator (ESP). It consists of two sets of electrodes insulated from each other. The first set is comprised of rows of *electrically grounded vertical parallel plates* called the *collection electrodes,* between which the dust-laden gas flows. The second set of electrodes consists of wires called the *discharge or emitting electrodes* which are centrally located between each pair of parallel plates. An external d.c. source supplies a negatively charged high voltage (20 to 100 kV) current through the wires which generates unidirectional, non-uniform electrical field whose magnitude is the greatest near the discharge electrodes as shown in Fig. 6.58. Due to high voltage, a blue luminous glow, called a *crona*, is produced around them and the electrical forces in the crona accelerate the

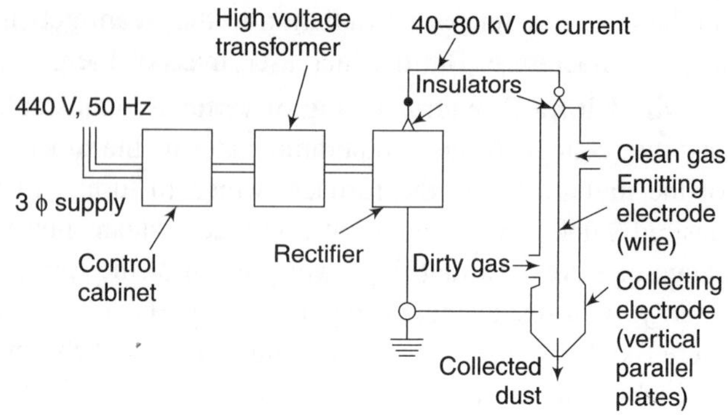

Figure 6.57 An electrostatic precipitator (ESP) for fly ash particles

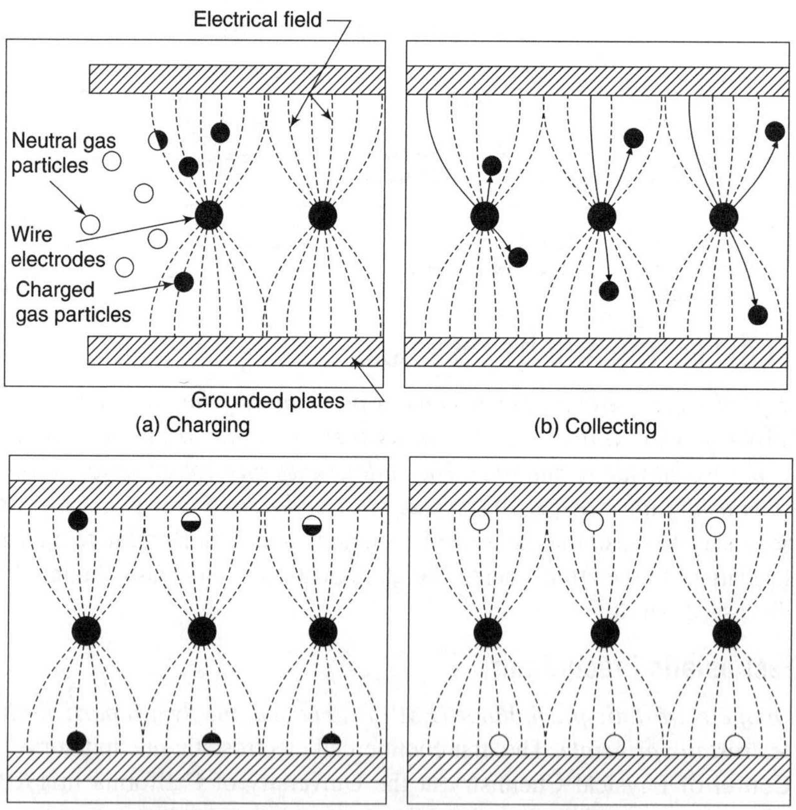

Figure 6.58 Four basic operations of ESP

free electronics present in the gas so that they ionize the gas molecules, thus forming more electrons and positive gas ions. By this way, a chain reaction of formation of free electrons and positive gas ions is produced.

Due to their characteristics, the positive ions travel to the negatively charged wire electrodes. The electrons follow the electrical field towards the grounded electrodes with a velocity, whose value diminishes forward the ground plates. Gas molecules capture the low velocity electrons and become negative ions. These negative ions, as they migrate to the collecting electrodes, collide with fly ash particles in the gas stream and give them negative charge which are driven to the collecting plates by an electrical force which is proportional to the product of the strength of the electrical field produced and this negative charge.

With the collection of ash particles on the grounded (collecting) plates, they lose their charge on the ground. The electrical resistivity of the particles, however causes only partial discharging and the retained charge tends to hold the particles to the plates. Ash particulates collected must be removed from the collecting plates on a regular schedule to ensure efficient collector operation. A mechanical hammer scrapping (rapping) system (a slow rotating camshaft) knocks the plates and the ash particles get removed from the plates due to vibration and fall into the hopper at the bottom of precipitator as shown in Fig. 6.58.

The **performance of the electrostatic precipitator** is described by a term called **overall collection efficiency** (η_0) which is expressed as

$$\eta_0 = \frac{\text{Mass of particles retained by collector}}{\text{Mass of all particles entering collectors}}$$

Dentsch has given a simple expression to calculate η_0 as given below

$$\eta_0 = 1 - \exp\left(-A\ C_{m0}/Q\right) \qquad \text{...(6.39)}$$

where A = area of the collector plate (m^2), Q = flue gas volume flow rate for each plate, m^3/s, C_{m0} = effective migration velocity of particles, (m/s)

$$= \frac{2.95 \times 10^{12}\ K\ (E/s)^2\ d}{\mu_g} \qquad \text{...(6.40)}$$

where $K = 2.0$

E = applied voltage (V), s = distance between charging and collecting electrodes (m), d = particle diameter (m), μ_g = gas viscosity, (kg/ms)

The performance of electrostatic precipitator is governed by corona characteristics, resistivity of the particles, rapping behaviour, gas velocity, particles size, and field strength.

Electrostatic precipitator is most effective to remove very small particles like smoke, mist and fly ash. Its range of dust removal is sufficiently large (0.01 μ to 1.00 μ). The small dust particles below 10μ cannot be removed with the help of mechanical separation and wet scrubber cannot be used if sufficient water is not available. Its efficiency is as high as 99.5%. The draught loss is also the least.

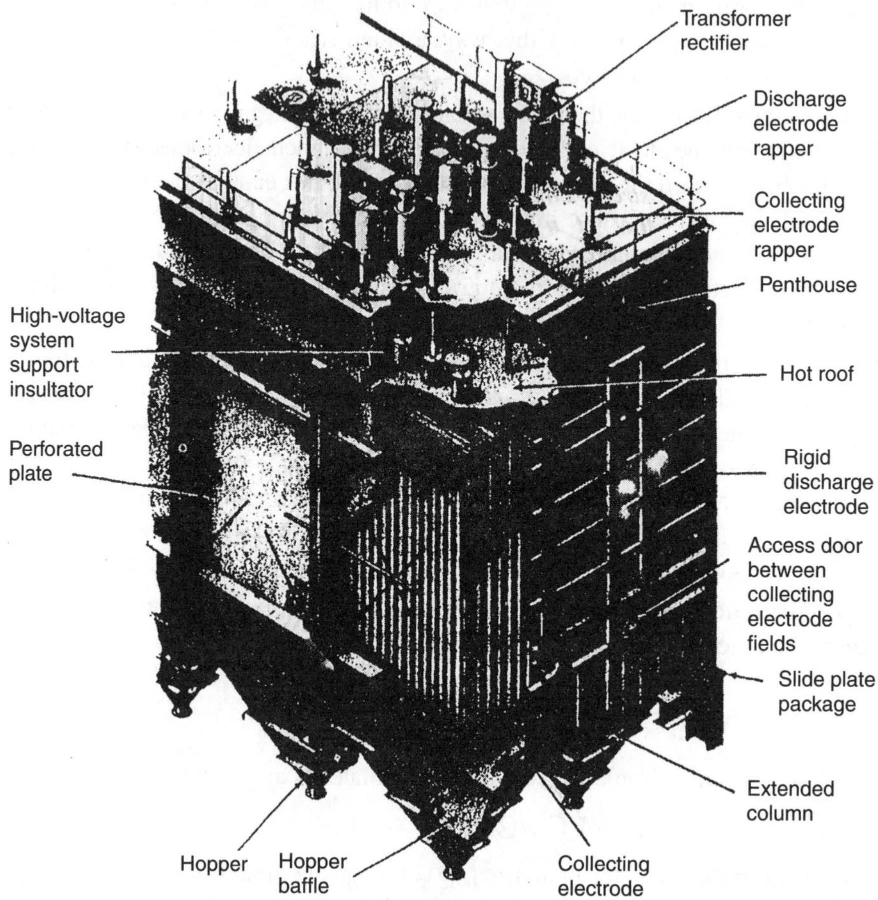

Figure 6.59 A view of an electrostatic precipitator

6.8.2 Fabric Filters and Baghouses

Fabric filters fall under the category of mechanical filters and are used in power plants to remove dust particles from a gas stream on the principle similar to that of a household vacuum cleaner except, of course, that the size of fabric filters is far greater. These fabric filters are made of porous material that retains particulate matter as the carrier gas passes through the voids. The fabric filter system usually known as *baghouse* is normally placed after the preheater.

In general, a fabric filter element is usually made in the form of long, hollow cylindrical tube that provides a large surface per unit of gas volumetric flow rate. *The filtering or air-to-cloth,* ratio is the inverse of this parameter, i.e., volumetric flow rate per unit surface area which is equal to the superficial gas velocity. It is based on the surface area rather than the void flow area within the fabric and ranges between 0.5 to 4 cm/s.

A fabric filter system usually contains a large number of fabric filter elements arranged in parallel rows as shown in Fig. 6.60. A power plant Baghouse may contain several thousands vertical hollow cylinders each ranging in diameter from 125 to 350 mm and upto 12.5 m in height.

Commonly, the elements have an open bottom and closed top and rest on a tube sheet above a dirty-air plenum (Fig. 6.60). The sheet distributes the gas evenly to the bags, allowing it to enter the elements at bottom, deposit its particulate matter on the inside of the tubes, and pass laterally through the fabric.

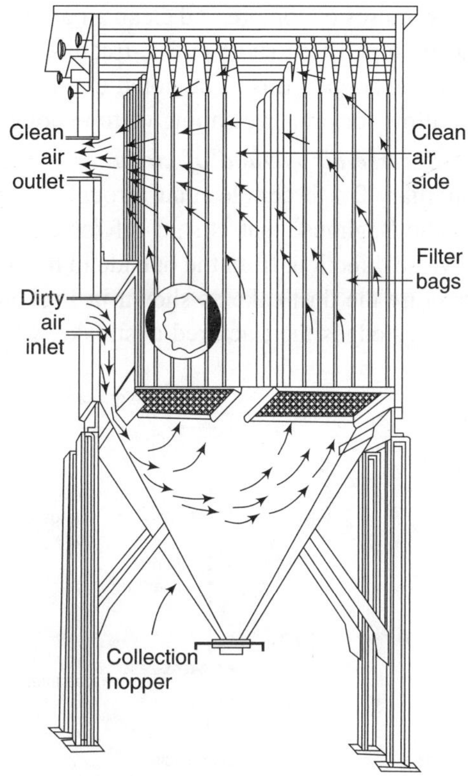

Figure 6.60 A typical baghouse with mechanical shakers

Below the tubes sheets, collection hoppers are provided for collecting removed particulate matters. This matter usually collects on the inside of the elements in the form of dust cake that must be made to fall down to the hopper by some means. There are three types of each means, namely, *mechanical shaking, pulse jet, and reverse air flow.*

The accomplishment of *mechanical shaking* is done by oscillating or vibrating rods attached to the top of the filter elements. Due to oscillations, the bag surfaces are rippled and dust cakes are broken and as a result the dust is discharged into the hopper below.

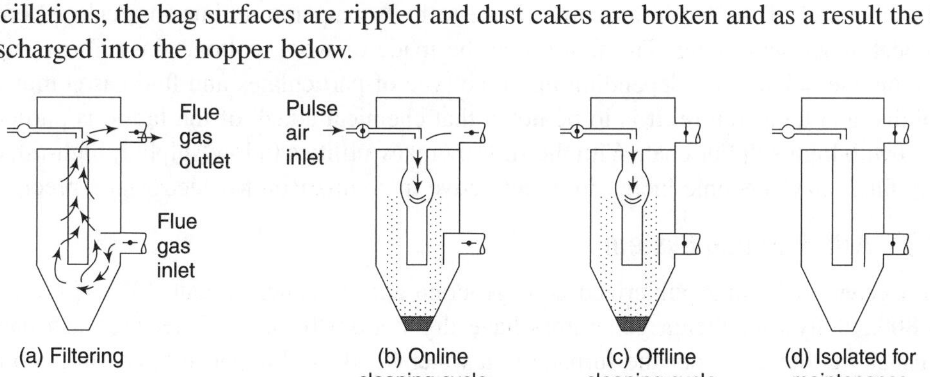

Figure 6.61 Pulse-jet cleaning

In ***pulse-jet cleaning*** (Fig. 6.61), the bags are hung from the top, normal flue gas flow is from the outside in, and the dust collects on the outside of the tubes. For cleaning operation, short pulses of compressed clean air are forced down into the bags through venturies provided at the top. These short pulses fracture and dislodge the dust resulting in falling of the dust into the hopper. It is to be noted that the bags in *pulse jet* system is shorter (less than 4.5 m) to allow for proper cleaning of tube bottoms periodically.

Reverse-air method of cleaning is the most popular method in power plants. Here, an auxiliary fan forces clean air through a flapper valve into the clean air plenum of the portion of the Baghouse that is shut-off for cleaning. This air passes through the filter elements in reverse direction, from the outside in resulting a "backwash action" that collapses the bag and fracture the dust cake as shown in Fig. 6.62. By bringing the bag back on line, it reinflates and dislodges the fractured dust cake, which falls into the hopper.

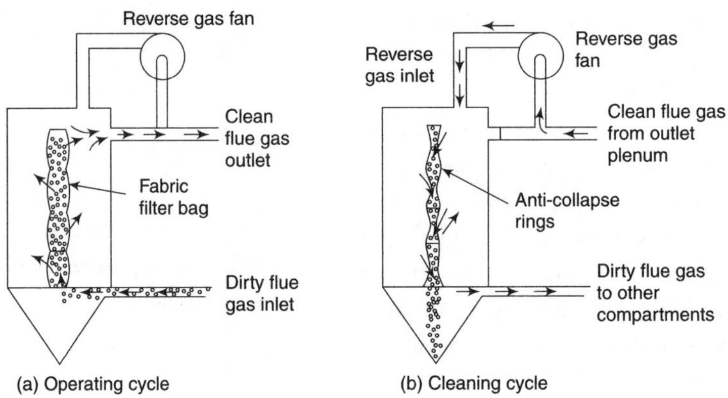

Figure 6.62 Reverse-air fabric filter cleaning

The material of the bag filter can fail due to high temperature, burning, caking, erosion, chemical attack and aging. The fabric may be made of wool, cotton, glass fibre, polyester and aromatic polyamides depending upon the type of particulates and flue gas composition, humidity, and temperature. It is to be noted that chemical attack of the fabric is particularly served with high-sulphur coal. With the shift by many utilities to low-sulphur, high-ash coals, fabric filters have become increasingly attractive in comparison to electrostatic precipitator.

6.8.3 Ash Handling System

Steam generators using pulverized coal generate approximately about 20% of bottom ash and 80% of fly ash. Steam generators have dry bottom furnaces. The large ash particles are usually collected under the furnace in a water filled ash hopper and fly ash is collected in dust collectors with either an electrostatic precipitator or a baghouse. Pyrites, which are rejected from the puvlerizers are collected and disposed with the bottom ash system (Fig. 6.63). There are three major factors while choosing and designing the ash disposal system, namely (i) plant size, (ii) fuel source, and (iii) environmental regulations. For many ash handling systems, water and land requirement are important considerations. It is the fuel source that decides the possible ash quality while ash storage and disposal sites are guided by environmental regulations.

For *handling bottom ash, the sluice conveyer system is widely used as shown* in Fig. 6.64 while for handling fly ash hydraulic vacuum conveyer system is the most frequently used as shown in Fig. 6.65.

Though, this ash is the waste item, however now it is in use. Bottom ash and slag are used as filling *material for road construction while bricks are made out of fly ash.*

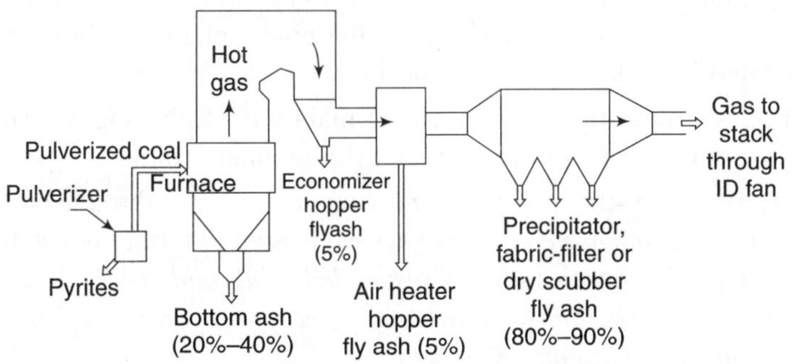

Figure 6.63 Ash collection and transportation from pulverized coal fuelled steam generator

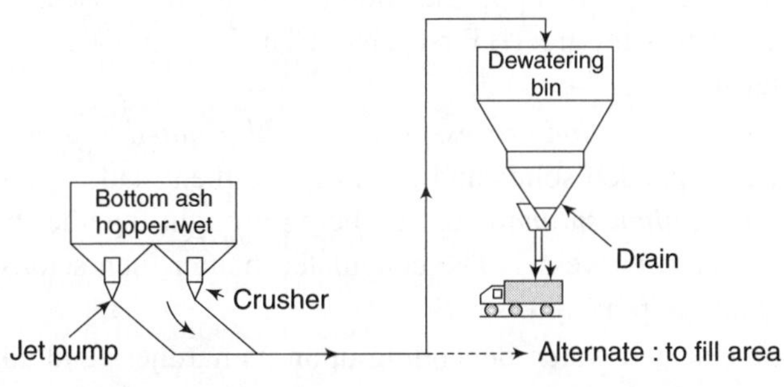

Figure 6.64 Bottom ash sluice conveyer system

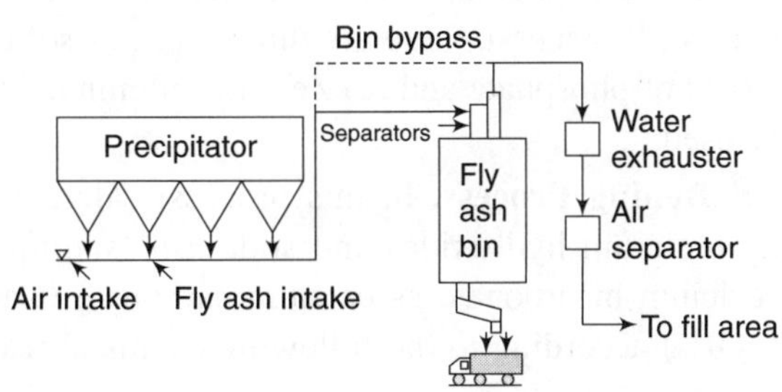

Figure 6.65 Fly ash hydraulic vacuum conveyer system

6.9 BOILER MAKEUP AND FEED WATER TREATMENT

Steam power plants lose a fraction of their water and steam during operation owing to the leakage from glands fittings and bearings, escape with noncondensable gases in deaeration processes, boiler blowdown and other causes. This loss amounts to be about 0.5 to 1.5 per cent of total flow. Since this loss of water and steam is a continuous process and so this loss must be made up continuously during operation. *It is highly essential that this make-up water must be treated well in water treatment plants to maintain water and steam purity in order to prevent (i) deposition of suspended solids and scale on boiler surfaces, (ii) silica deposition on turbine bladings, and (iii) corrosion damage to turbine blades and condensate-feed water system.* Further, in the beginning of first starting the power plant, whole water requirement will have to be treated in such water treatment plants.

The treatment of raw make-up water is done in mainly the following two stages one after other: (a) Pretreatment or external treatment, and (b) demineralization

For plants with stringent water quality requirements such as once through boiler and boiling water nuclear reactors, a *condensate polishing system* is used to further polish the water. Raw water contains a variety of impurities such as *suspended solids and turbidity, organic hardness (calcium and magnesium), alkalinity (bicarbonates, carbon hydrates) other dissolved ions (sodium, sulphate, chloride, etc.) and dissolved gases (O_2, CO_2).*

6.9.1 Pretreatment

The extent of pretreatment depends upon the source of raw water used. There are generally three stages of pretreatment. They are: (i) First stage: Clarification (ii) Second stage: Softening (iii) Third stage: Filtering

(i) Clarification. *In clarification process, water is chlorinated to prevent befouling of the equipment.* The suspended solids and turbidity are then made to coagulate by special chemical like *aluminium sulphate* and by being brought together by a *slow agitation* in the middle of clarifier vessel. The coagulated matter then settles by gravity in the clarifier vessel and is removed.

(ii) Softening. The clarified water, depending upon its hardness and alkalinity undergoes softening process. Hardness which is supposed to be the chief source of scale in heat exchangers, boilers and pipeline, is caused by the presence of *calcium and magnesium salts containing Ca and Mg. Alkalinity is mostly bicarbonate* HCO_3^-, but is also carbonate and hydrate. In general, there are three types of softening process, namely, (a) limesoda ash (b) hot-phosphate and (c) Zeolite softening. Among these, lime-soda process is widely used.

(a) Lime-soda Ash Softening Process. In this process, calcium and magnesium salts are removed using lime (calcium hydroxide) and soda ash (sodium carbonate). Calcium hydroxide precipitates calcium bicarbonate as calcium carbonate ($CaCO_3$) and magnesium salts as magnesium ($MgOH_2$) according to the following chemical reactions.

$$Ca(OH)_2 + Ca^{2+}(HCO_3^-)_2 \longrightarrow 2CaCO_3 \downarrow + 2H_2O \quad ...(6.41)$$

$$2\,Ca\,(OH)_2 + Mg^{2+}(HCO_3^-)_2 \longrightarrow 2CaCO_3 \downarrow + Mg(OH)_2 \downarrow + 2\,H_2O \quad ...(6.42)$$

$$(CaOH)_2 + Mg^{2+}(SO_4^-)_2 \longrightarrow Mg\,(OH_2) \downarrow + CaSO_4 \quad ...(6.43)$$

The soda ash, sodium carbonate (Na_2CO_3), is added to react with calcium chloride and calcium sulphate to form calcium carbonate.

$$Na_2CO_3 + CaSO_4 \longrightarrow CaCO_3 \downarrow + Na_2SO_4 \quad ...(6.44)$$

The products, calcium carbonate ($CaCO_3$) and magnesium hydroxide are insoluble in water and settle to the bottom of the vessel. During softening, calcium, magnesium and carbonate alkalinity are reduced to a few, 10 ppm each. Here, the problem of sludge removal arises. As per environmental regulations, sludge is not supposed to be discharged. Instead, it is "dewatered" either in a settling basis or thickeners and centrifuges. The water is then discharged to its source or recycle. When lime-soda process is carried out at normal raw water temperature, it is called a *cold process* and when carried at or near the boiling point, it is called hot *process.* The hot process is mostly preferred for *boiler water treatment.*

(b) Hot Phosphate Softening Process. In this process, calcium and magnesium hardness is removed using phosphate and caustic soda as due to reaction tricalcium phosphate [$Ca_3(PO_4)_2$] and magnesium hydroxide are precipitated. This process is carried out at a temperature of 100°C or above. The reaction is given below:

$$3Ca^{2+}(HCO_3^-)_2 + 6NaOH \longrightarrow 2CaCO_3 + 2Na_2CO_2 + 6H_2O \quad ...(6.45)$$

$$CaCO_3 + 2Na_3PO_4 \longrightarrow Ca_3(PO_4)_2 \downarrow + 2Na_2CO_3 \quad ...(6.46)$$

$$Mg(HCO_3)_2 + 4NaOH \longrightarrow Mg(OH)_2 + 2Na_2CO_3 + 2H_2O \quad ...(6.47)$$

$$3CaSO_4 + 2Na_3PO_4 \longrightarrow Ca_3(PO_4)_2 + 3Na_2SO_4 \quad ...(6.48)$$

This process requires costly chemicals as compared to lime-soda process, so it is used where the initial hardness is 60 ppm or less. Where hardness is greater than this, the lime-soda ash process may be used first, followed by phosphate softening.

(c) Sodium Zeolite Softening. In sodium zeolite softening process, raw water is passed through a bed of sodium zeolite, which may be natural compounds of sodium, aluminium, silicate with the caution of calcium and magnesium removed in the process. The reaction is as follows:

$$CaCO_3 + Na_2Z \longrightarrow CaZ + Na_2CO_3 \quad ...(6.49)$$

$$MgSO_4 + Na_2Z \longrightarrow MgZ + Na_2SO_4 \quad ...(6.50)$$

$$Ca(HCO_3)_2 + Na_2Z \longrightarrow CaZ + Na_2(HCO_3)_2 \quad ...(6.51)$$

The drawback of zeolite softening is deterioration of softening capacity in course of time. However, this can be regenerated by flushing it with brine (NaCI).

Zeolite softening is not ideal due to delerious effect of zeolite, reduction in efficiency, no reduction in alkalinity, silica gain in water, etc. Hydrogen zeolite softening may also be used.

(iii) Filteration. The third stage of pretreatment is filtration which further removes residual suspended solids and turbidity. Filtration can be done under gravity or pressure, although the latter is preferred. Various filter media are used, sand being the most common. The pressure difference across the solid medium is an indication of soild accumulation which are removed. Charcoal may be used to remove residual chlorine.

6.9.2 Demineralisation

Demineralisation is defined as the process of removing dissolved solids by an ion exchange. Two types of resins, namely, ***cation*** and ***anion*** are commonly used. *Cation resin* has a **positively charged hydrogen** ion attached to a negatively **charged polymer**. The hydrogen ion is exchanged for the cations calcium, magnesium and sodium. Anion resin is similar to cation except that it has negatively charged hydroxide ion (OH^-), attached to a positively charged polymer structure. The hydroxide ion is exchanged for the anions, sulphates, chlorides and bicarbonate. It is worth to note that both ion exchange processes are reversible, and the resins are restored to their original form by regeneration. The need of regeneration arises when the resins are about to be exhausted and traces of dissolved solids start appearing at the exit of demineralizer. Regeneration is achieved by passing a concentrated acid through the cation resin and caustic or sodium hydroxide through the anion resins.

Figure 6.66 shows a typical demineralizing train which consists of a series of demineralizers that contains weak-acid cation, strong-acid cation, decarbonators, weak-base anion, strong-base anion and mixed bed units. In decarbonators, air is passed and CO_2 is evolved which is taken out which reduces alkalinity. It is to be noted that demineralizing and regeneration equipment including tanks, pumps, etc., are programmed to operate automatically.

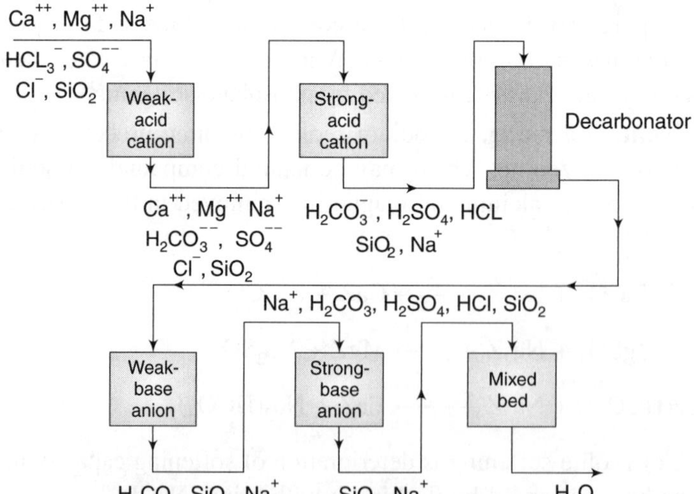

Figure 6.66 A typical demineralizing train

A latest demineralization method, called **reverse osmosis,** *is gaining popularity for boiler make-up water. Osmosis is the diffusion process of a solvent (in this case water) through a semipermeable membrane from a region of no or low solid concentration to a region of high solid concentration.* The membrane blocks the passage of the solutes (in this case the dissolved solids). Obviously, the motion will be in the direction of high partial pressure of pure water to the low partial pressure of less pure water due to the response of **osmotic presence.** If an external pressure greater than osmotic pressure is applied, the flow will be reversed. In boiler make-up water, pure water flows from the region of high concentration to the region of pure water. This is called *reverse osmosis.* Since the membrane are costly, so they are protected using first coagulation, settling, and filteration.

6.9.3 Condensate Polishing

Although, water treatment system comprising pretratement and demineralizing processes discussed above produce high-quality make-up water for the plant, the water in its journey through the cycle can pick up metallic ions, such as iron and copper, from pipelines, etc., as well as impurities due to condenser leakage from the circulating water. But once through boilers and boiling water nuclear reactors require continued high quality water which is possible by *condensate polishing system.* This polishing is achieved by passing the condensate through large demineralizing vessel characteristics of resins, they remove dissolved salts as well as act as filter for impurities.

6.9.4 Deaeration

Deaeration, also known as degasification, is one of the most important steps in boiler water treatment. It is based on the decrease in solubility of dissolved gases, notably O_2 and CO_2 as the water temperature is increased. Deaeration is carried out in a direct-contact feed heaters which has been already discussed in chapter 3.

6.9.5 Internal Treatment

There is a need of internal treatment of boiler water to prevent scale deposition and corrosion. As we know that dissociation and recombination of water go on till equilibrium is reached which is expressed as

$$HOH = H^+ (OH)^- \quad ...(6.52)$$

The *presence of excess hydrogen ions makes the water acidic and corrosive.* The pH value of water is logarithm of reciprocal of hydrogen ion concentration and decreases with increase of temperature. At 15°C, the pH of neutral water is 7. If pH > 7, water is alkaline and if pH < 7, it becomes acidic. In order to increase alkalinity in water, trisodium phosphate (Na_3PO_4) is injected into water, while to reduce alkalinity, mono-sodium phosphate (NaH_2PO_4) is injected into water. For knowing pH value of water, pH meter is used. It is the presence of hydrogen ions (low pH) and dissolved oxygen produce corrosive effects. Scale formation and corrosion are reciprocal phenomena. *A pH value of 10.5 is usually maintained for boiler water in order to avoid corrosion but it will cause scale formation.* There are two treatments of scale formation, periodic blowdown and pretreatment.

Internal treatment to prevent scale deposition is usually carried out as discussed by injecting some form of sodium phosphate such as trisodium, disodium or monosodium and reacting with calcium salts it forms tricalicum phosphate which precipitates. Magnesium salts also get precipitated. In order to prevent the precipitates adhering to the boiler surfaces, sometimes tannins or starches are injected into boiler water.

6.9.6 Boiler Blowdown

Drum type boilers utilise blowdown to control water purity by removing solid precipitates. Blowdown lines periodically remove a portion of water from the drum where concentration increases due to steam flashing and this water is replaced by pure feed water. The blowdown is then cooled and treated for reuse. The common treatment of blowdown in the use of cartridge filters to remove suspended solids, especially iron and copper oxides. The blowdown in expressed as

$$\% \text{ blowdown} = \frac{\text{Quantity of water blow down}}{\text{Quantity of feed water}} \times 100$$

6.9.7 Evaporators

One of the old methods of water treatment for boiler make-up water is the evaporation of pretreated water by steam blade from turbine. Now it is obsolete but still used in ships. Evaporation is performed by *flash type distillating units or evaporators.* It could be of one stage or single-effect type, or multi-stage multi-effect type.

Figure 6.67 shows a single-effect evaporator. Raw water ($\dot{m}_w$) pretreated as above, is introduced in the evaporator motive steam in the form of bled steam from turbine ($\dot{m}_s$) is introduced in the steam chest, flows through tubes and discharged as condensate which is returned to the plant. The pretreated water boils and pure saturated vapour is extracted from the evaporator as make-up water.

Figure 6.67 Single-effect submerged evaporator

Example 6.1 In a power plant employing natural circulation steam generator, the furnace wall riser, 30 m long, 76 mm OD and 6 mm thick receives saturated water at 100 bars. The flow velocity is 2 m/s. If the circulation ratio is 10 and a slip ratio is 1.2, calculate: (a) the quality of steam at top of riser; (b) the pressure head developed, (c) the void fraction at riser exit, and (d) the heat transfer rate per unit projected area of the riser tube.

Solution Refer to Fig. 6.68. The quality of the steam at the top of riser is given by

$$x_{top} = TDF = \frac{1}{CR} = \frac{1}{10} = 0.1 \qquad \textbf{\textit{Ans.}}$$

(a) Thus, the quality of steam at the top of riser = 10%.
From Steam Table at 100 bars; $v_f = 0.00148$ m^3/kg, $v_g = 0.018026$ m^3/kg, $h_f = 1407.56$ kJ/kg, $h_{fg} = 1317.1$ kJ/kg, $h_g = 2724.7$ kJ/kg, $v_{fg} = 0.016546$ m^3/kg

$$\therefore \qquad v_{\text{top}} = v_f + x_{\text{top}} \cdot v_{fg} = 0.00148 + 0.1 \times 0.016546 = 3.1346 \times 10^3 \text{ m}^3/\text{kg}$$

$$\rho_{\text{top}} = \frac{1}{v_{\text{top}}} = \frac{1}{3.1346 \times 10^{-3}} = 319.01 \text{ kg/m}^3$$

and
$$\rho_f = \frac{1}{v_f} = \frac{1}{0.00148} = 675.67 \text{ kg/m}^3$$

$$\rho_m = \frac{(\rho_f + \rho_{\text{top}})}{2} = \frac{319.01 + 675.67}{2} = 497.3 \text{ kg/m}^3$$

(b) The pressure head developed due to natural circulation is expressed as

$$\Delta p = H_g\,(\rho_f - \rho_w) = 30 \text{ (m)} \times 9.81 \text{ (m/s)}^2 \;(675.67 - 497.34) \text{ (kg/m}^3)$$

$$= 52482.5 \text{ Pa} = 52.4825 \text{ kPa} \qquad \textbf{\textit{Ans.}}$$

(c) Void fraction, α is expressed as

$$\alpha = \frac{1}{1 + \left(\frac{1-x}{x}\right)\phi} = \frac{1}{1 + \left(\frac{0.9}{0.1}\right)\phi}$$

where
$$\phi = \frac{v_f}{v_g} S = \frac{0.00148}{0.018026} \times 1.2 = 0.0985$$

$$\therefore \qquad \alpha = \frac{1}{1 + \frac{0.9}{0.1} \times 0.0985} = 0.53 \qquad \textbf{\textit{Ans.}}$$

(d) The heat transfer rate per unit projected area = $\ddot{Q} = \dfrac{\dot{m}_s h_{fg}}{D_0 \cdot H}$

The mass flow rate of saturated water in each riser is given by

$$\dot{m}_w = \rho_f A_i\, C_f = 675.67 \times \frac{\pi}{4}(0.064)^2 \times 1.2 = 2.608 \text{ kg/s}$$

The rate of formation of steam in each riser tube is given by

$$\dot{m}_{rs} = x\,m_w = 0.1 \times 2.608 = 0.2608 \text{ kg/s}$$

$$\ddot{Q} = \frac{0.2608 \times 1317.1}{0.076 \times 30} = 150.65 \text{ kW/m}^2 \qquad \textbf{\textit{Ans.}}$$

Example 6.2 A downcomer-riser circuit having 20 m height operates at 150 bars. The riser receives uniform heat flux and saturated water. The quality of steam at the riser-exit is 0.6. Assuming slip ratio of 1.2, calculate the present head developed due to natural circulation.

Solution Refer to Fig. 6.68. For uniform axial heating, the mean density of water-steam mixing is given by

$$\rho_m = \rho_f - \left(\frac{\rho_f - \rho_g}{1 - \phi}\right)\left[1 - \left\{\frac{1}{\alpha_e (1 - \phi)} - 1\right\} \ln \frac{1}{1 - \alpha_e(1 - \phi)}\right]$$

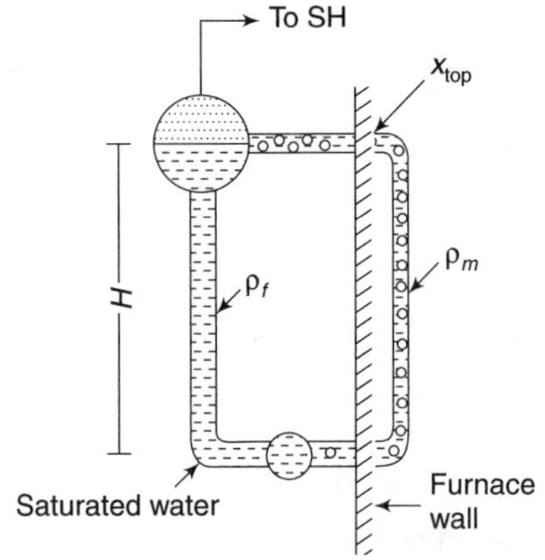

Figure 6.68 Example 6.1 and 6.2

Here, $$\phi = \frac{v_f}{v_g} S = \frac{0.001658}{0.010377} \times 1.2 = 0.1917$$

$$\alpha_e = \frac{1}{1 + \dfrac{1 - x_e}{x_e}\phi} = \frac{1}{1 + \dfrac{0.4}{0.6} \times 0.1917} = 0.886$$

$$\rho_m = 603.13 - \left(\frac{603.13 - 96.36}{1 - 0.1917}\right)\left[1 - \left\{\frac{1}{0.886(1 - 0.1917)} - 1\right\} \ln \frac{1}{1 - 0.886(1 - 0.1917)}\right]$$

$= 130.06 \text{ kg/m}^3$

Pressure developed due to natural circulation is expressed as

$\Delta p = g\,(\rho_f - \rho_m)\,H = 9.81\,(603.13 - 130.06) \times 20 = 92816.33 \text{ Pa} = 92.8163 \text{ kPa}$ ***Ans.***

Problem 6.3 In order to control the temperature of steam in steam generator, a spray type desuperheater is used, It is fed with water at 60°C. Stream at 90 bar and 300°C with flow rate 300 tonnes/h enters the desuperheater. Calculate the amount of water that must be sprayed per hour to maintain steam at 456°C. Neglect any pressure drop due to mixing and heat loss to the surroundings.

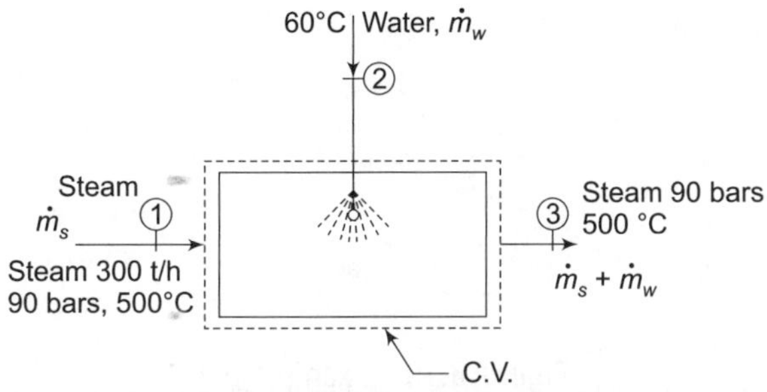

Figure 6.69 Example 6.3

Solution Refer to Fig. 6.69. Let $\dot{m}_w$ be the rate of water required to bring the steam temperature from 500°C to 450°C at pressure 90 bars. Applying steady flow energy balance across control volume, we have

$$\dot{m}_s\, h_1 + \dot{m}_w\, h_2 = (\dot{m}_s + \dot{m}_w)\, h_3$$

or
$$\dot{m}_w = \frac{\dot{m}_s\, (h_1 - h_3)}{(h_3 - h_2)}$$

From Steam Table, h_1 (at 90 bars and 500°C) = 3386.1 kJ/kg

h_2 (at 60°C) = 251.11 kJ/kg, h_3 (90 bar and 450°C) = 3256.6 kJ/kg

$$\dot{m}_w = \frac{300\,(3386.1 - 3256.6)}{(3256.6 - 251.11)} = 12.92 \text{ ton/h} = 3.59 \text{ kg/s}$$ ***Ans.***

Example 6.4 In steam power plant generating 250 MW of electrical power, boiler supplies steam at a pressure 150 bars and temperature 550°C. The condenser pressure is 0.05 bars. The turbine, mechanical and generator efficiencies are 88, 98 and 99%, respectively. The fuel used is coal having calorific value 25.7 MJ/kg. The boiler efficiency is 90%. The feed water enters the boiler at 160°C. The risers of the furnace are 50 m high while the downcomers are outside the furnace producing natural circulation. The quality of the steam at the top of risers is 10% and a minimum exit velocity of mixture leaving the riser and entering the drum is 1.5 m/s. The dimensions of the riser tubes is 60 mm OD and 3 mm wall thickness while the dimensions of downcomers are 150 mm and 6 mm thick. Assuming no pressure drop and heat loss in the risers, calculate: (a) the generation of rate of steam, (b) the rate of fuel flow required, (c) the evaporation factor, (d) the pressure head developed due to natural circulation, (e) the circulation ratio, (f) the number of risers, (g) the number of downcomers, and (h) the heat absorption rate per unit projected area of the risers.

Solution Refer to Fig. 6.70. From Steam Tables: h_1 = 3448.6 kJ/kg, s_1 = 6.5199 kJ/kgK = s_2 = 0.476 + 7.919 x_2, x_2 = 0.763, h_2 = 137.8 + 0.763 × 2423.7 = 1987 kJ/kg, h_{f3} = 137.8 kJ/kg, h_{f5} = 657.5 kJ/kg.

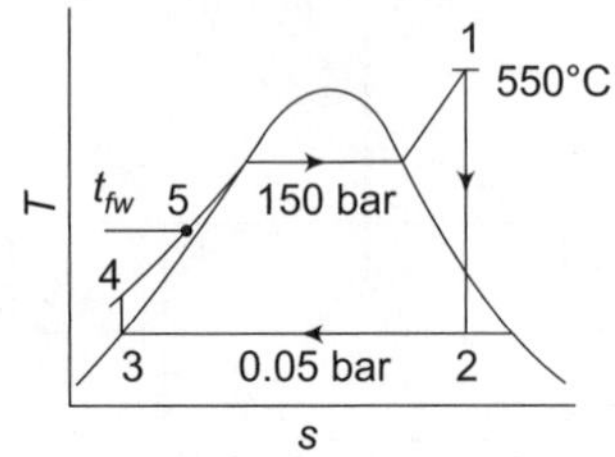

Figure 6.70 Example 6.4

Assuming isentropic expansion in turbine and other losses.

$$\dot{m}_s (h_1 - h_2)\, \eta_t \times \eta_m \times \eta_g = 250 \times 10^3$$

(a) $$\dot{m}_s = \frac{250 \times 10^3}{(3448.6 - 1987) \times 0.88 \times 0.98 \times 0.99} = 200.28 \text{ kg/s}$$ ***Ans.***

(b) $\dot{m}_f \times (C.V.)_f\, \eta_b = \dot{m}_s (h_1 - h_{f5})$ where $\dot{m}_f$ = mass of fuel burning rate

or $$\dot{m}_f = \frac{\dot{m}_s(h_1 - h_5)}{(CV)_f \times \eta_b} = \frac{200.28\,(3448.6 - 657.5)}{25.7 \times 10^3 \times 0.9} = 24 \text{ kg/s}$$ ***Ans.***

(c) Evaporation factor $= \dfrac{\dot{m}_s}{\dot{m}_f} = \dfrac{200.28}{24} = 8.345$ ***Ans.***

(d) Pressure head available due to natural circulation is given by

$$\Delta p = gH\,(\rho_f - \rho_m)$$

At 150 bars, $v_{\text{top}} = 0.001658 + 0.1 \times 8.719 \times 10^{-3} = 2.5299 \times 10^{-3}$ m^3/kg

$\therefore \quad \rho_{\text{top}} = 395.27$ kg/m^3, $\rho_f = 603.13$ kg/m^3

$$\therefore \quad \rho_m = \frac{\rho_f + \rho_{\text{top}}}{2} = \frac{603.13 + 395.27}{2} = 499.2 \text{ kg/m}^3$$

$\therefore \quad \Delta p = 9.81 \times 50(603.13 - 499.2) = 50977.66$ Pa = 50.97 kPa ***Ans.***

(e) The circulation ratio, CR $= \dfrac{1}{x_{\text{top}}} = \dfrac{1}{0.1} = 10$ ***Ans.***

(f) Rate of steam formation in one riser is given by

$$\dot{m}_{rs} = A_i\, \rho_{\text{top}}\, C_i\, x_{\text{top}} = \frac{\pi}{4}(0.054)^2 \times 395.27 \times 1.5 \times 0.1 = 0.1357 \text{ kg/s}$$

Number of risers $= \dfrac{\dot{m}_s}{\dot{m}_{rs}} = \dfrac{200.28}{0.1357} = 1457.9 \approx 1476$ ***Ans.***

(g) Mass flow rate of feed water through one downcomer,

$$\dot{m}_{dw} = A_i\, \rho_f\, C_i = \frac{\pi}{4}(0.148)^2 \times 603.13 \times 1.5 = 15.56 \text{ kg/s}$$

$$\text{Number of downcomers} = \frac{\dot{m}_s}{\dot{m}_{dw}} = \frac{200.28}{15.56} = 12.87 \approx 13 \qquad \textit{Ans.}$$

(h) The heat absorption rate per unit projected area of the riser

$$= \frac{\dot{m}_{rs}\, h_{fg}}{D_0 H} = \frac{0.1357 \times 1000}{0.06 \times 50} = 45.23 \text{ kW/m}^2 \qquad \textit{Ans.}$$

Example 6.5 In a steam generator, the feed water from the economiser enters the drum at the rate of 98 kg/s and has 3 ppm solid impurities. Steam leaves the drum at the rate of 100 kg/s at 120 bars with 3% moisture. The make-up water at the rate of 2 kg/s is fed into the drum with 50 ppm solid impurities. Effective steam washing is used which reduces the solid in the moisture leaving the drum to 5 ppm. If the solid concentration in the drum water is to be maintained at 1000 ppm, calculate (a) the blowdown (BD) required in kg/s, (b) heat loss in blowdown as a percentage of total heat released in the furnace, if blowdown recovery is not used, when the fuel burning rate is 8 kg/s, the calorific value of fuel is 23 MJ/kg, and the room temperature is 30°C, and (c) the possible deposition of scale in superheater tube in kg/day. Comment on the result.

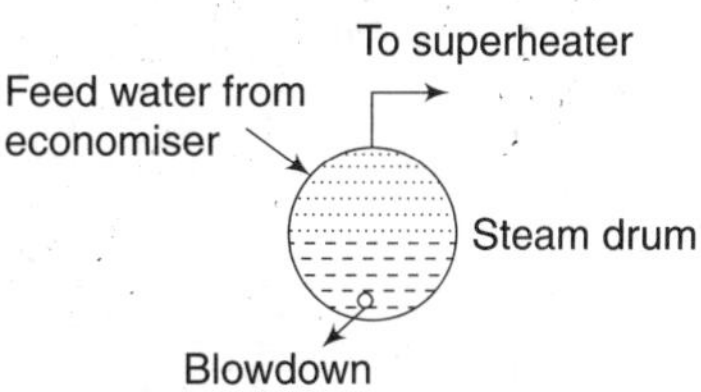

Figure 6.71 Example 6.5

Solution Refer to Fig. 6.71. Using an impurity balance of the drum at steady state, we have:

(a) Impurities entering the feed water + Impurities entering with make-up water = Impurities leaving with blowdown + Impurities leaving with moisture carry-over with steam.

Thus, $98 \times 3 \times 10^{-6} + 2 \times 60 \times 10^{-6} = \text{BD} \times 1000 \times 10^{-6} + 0.03 \times 100 \times 5 \times 10^{-6}$

$$\text{or} \quad \text{BD} = \frac{294 + 120 - 15}{1000} = 0.399 \text{ kg/s} \qquad \textit{Ans.}$$

(b) Heat loss in blowdown as a percentage of total heat released

$$= \frac{\text{BD}(h_{f,\,100\text{ bar}} - h_{\text{ambient}})}{\dot{m}_f \times (CV)_F} = \frac{0.399(1393 - 125.79)}{8 \times 23 \times 10^3} \times 100 = 0.274\% \qquad \textit{Ans.}$$

(c) Scale deposition in superheater tubes

$$= 0.03 \times 100 \times 5 \times 10^{-6} \times 3600 \times 24 = 1.296 \text{ kg/day} \qquad \textit{Ans.}$$

Scale deposition in superheater tubes is not desirable and so moisture content in steam has to be avoided by better method.

Example 6.6 An economiser of a modern steam generator is to be designed which receives feed water at its inlet header from the high pressure feed heater at the rate of 650 kg/s and at 150 bars and 180°C. The hot flue gases pass over the economiser and as a result of heating the feed water becomes saturated liquid at that pressure and leaves the outlet header to flow into the drum. The mass flow rate of flue gases passing over the economiser coils is 1350 kg/s. The flue gases leave the economiser at 450°C. In order to restrict the erosion rate by fly ash, its velocity is restricted to 12 m/s while the optimum velocity of water leaving the economiser coils is 1.2 m/s. The tubes are of 70 mm OD and 60 mm ID. The overall heat transfer coefficient may be taken as 70 W/m^{2}K. The specific heat of the gases may be taken as 1.2 kJ/kgK. The vertical pitch of the coil is 80 cm and the clearance on the two sides of the duct of width 4.8 m is 5 mm and pitch is 8 cm. Calculate the following: (a) The number of coils needed in the economiser and the length of one coil, and (b) The vertical height of economiser coils.

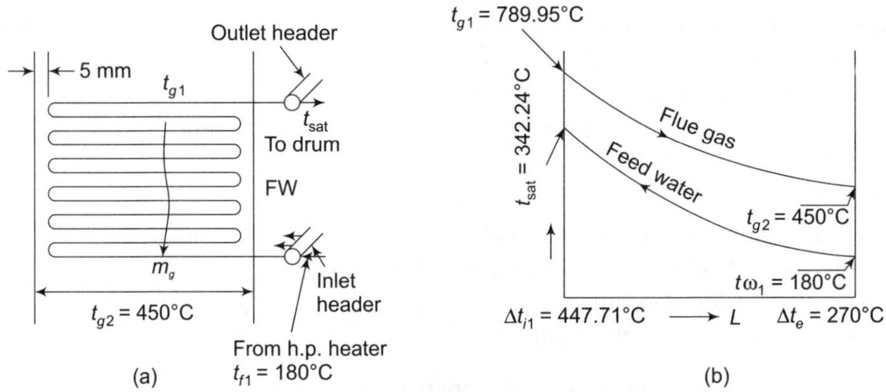

Figure 6.72 Example 6.6

Solution Refer to Fig. 6.72

From Steam Tables: At 150 bars, h_{f180} = 1610.5 kJ/kg, h_{f180} = 763.22 kJ/kg, $v_f = 0.001127$ m^3/kg, $t_s = 342.24$°C

The rate of heat transfer in the economiser is expressed as

$$Q = \dot{m}_s\,(h_f - h_{f180}) = \dot{m}_g\, c_{pg}\,(T_{g1} - T_{g2}) = U_0\, A_0\, \Delta T_m$$

$$= 650\,(1610.5 - 763.22)$$

or $\quad T_{g1} = 789.95$°C and Q = 550732 kW.

$$\text{LMTD} = \Delta T_m = \frac{\Delta T_i - \Delta T_e}{\ln \dfrac{\Delta T_i}{\Delta T_e}} = \frac{(789.95 - 342.24) - (450 - 180)}{\ln \dfrac{(789.95 - 342.24)}{(450 - 180)}} = 351.39°\text{C}$$

The outside heat transfer surface area is given by

$$A_0 = \frac{Q}{U_0\, \Delta T_m} = \frac{550732}{70 \times 10^{-3} \times 351.39} = 22.38 \times 10^3 \text{ m}^2$$

$$\dot{m}_s = 650 = A_i\, C_{wi}\, \rho_f = \left(n\,\frac{\pi}{4}\, D_1^2\right)\frac{C_{wi}}{v_f} = n\,\frac{\pi}{4} \times (60 \times 10^{-3})^2 \times \frac{1.2}{0.001127}$$

(a) Therefore, number of coils = n = 215.9 ≈ 216 ***Ans.***

$$A_0 = n\,\pi\,D_0\,l = 22.38 \times 10^3 = 216 \times \pi \times 70 \times 10^{-3}\,l$$

or $$l = 471.14 \text{ m}$$ ***Ans.***

(b) The number of turns of one coil,

$$n_i = \frac{1}{b - 2c} = \frac{471.14}{4.8 - 2 \times 0.005} = 98.35 \approx 98$$ ***Ans.***

The vertical height of the duct occupied by the economiser coils is given by

$$= n_i \times p = 98 \times 0.08 = 7.84 \text{ m}$$ ***Ans.***

Example 6.7 A superheater is to be designed in which steam is superheated to 600°C at pressure 150 bars. The mass flow rate of steam is 600 kg/s and steam flows through tubes at a velocity of 10 m/s. Due to material constraints, the heat flux in the superheater coils is limited to 150 kW/m^2. The outer tubes diameter are 50 mm having 5 mm thickness. Calculate the number and length of superheater coils.

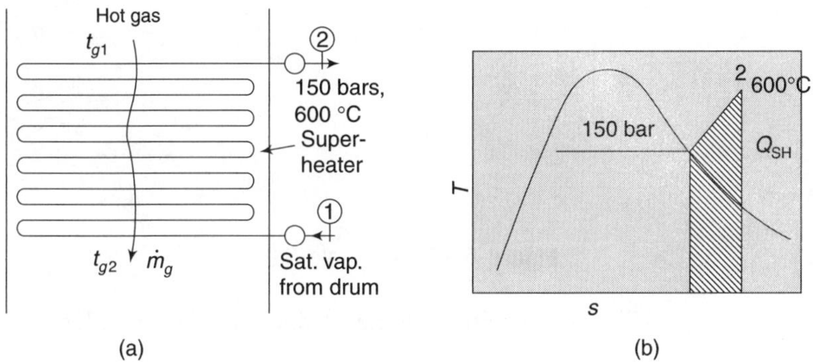

Figure 6.73 Example 6.7

Solution Refer to Fig. 6.73

From Steam Tables, $h_1 = h_g = 2610.5$ kJ/kg, $h_2 = 3582.3$ kJ/kg, $v_2 = 0.02491$ m^3/kg

The heat absorption rate in superheater coils is given by

$$Q = \dot{m}_s\,(h_2 - h_1) = 600\,(3582.3 - 2610.5) = 583080 \text{ kW}$$

Surface area required, $A_0 = \dfrac{Q}{\ddot{Q}} = \dfrac{583080}{150} = 3887.2 \text{ m}^2$

From continuity equation, we have

$$\dot{m} = \rho_1\,A_1\,C_1 = \rho_2\,A_2\,C_2 = \left(n\,\frac{\pi}{4}\,D_i^2\right)\frac{C_2}{v_2} = 600$$

or $$n = \frac{600 \times 4 \times 0.02491}{\pi(40 \times 10^{-3})^2 \times 10} = 1189.36 \simeq 1190$$ ***Ans.***

The outside heat transfer surface area, $A_0 = 3887.2 = n\,\pi\,D_0\,l$

$\therefore$ Length of one coil, $l = \dfrac{3887.2}{1190 \times \pi \times 0.05} = 20.79$ m **_Ans._**

Example 6.8 A tubular type air preheater is to be designed for a modern boiler where the hot flue gases of mass flow rate 1350 kg/s enters the preheater at 450°C and leaves the preheater at 160°C. The air enters the preheater at 35°C and flows over the tube at the rate of 1200 kg/s. The inlet velocity of flue gases is 12 m/s and the tube are of 60/65 mm in size. If the overall heat transfer coefficient is 30 W/m^2 K, calculate the number of tubes and length of tubes. Assume for flue gases $c_p = 1.1$ kJ/kg K and $R = 0.287$ kJ/kg K. For air, $c_p = 1.005$ kJ/kgK.

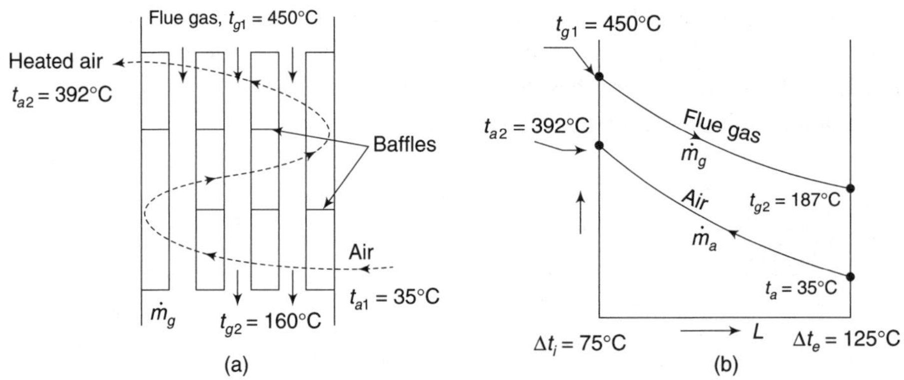

Figure 6.74 Example 6.8

Solution Refer to Fig. 6.74. The specific volume of flue gases at the inlet of air preheater is given by

$$v_{g1} = \frac{RT_{g1}}{p} = \frac{0.287 \times (450 + 273)}{1.0135 \times 10^2} = 2.0473 \text{ m}^3/\text{kg}$$

From the heat transfer analysis, we have

$$Q = \dot{m}_g\, c_{pg}\, (T_{g1} - T_{g2}) = \dot{m}_a\, c_{pa}\, (T_{a2} - T_{a1}) = U_0\, A_0\, \Delta T_m$$

or $\quad Q = 1350 \times 1.1\ (450 - 160) = 430650 = 1200 \times 1.005\ (t_{a2} - 35)$

or $\quad t_{a2} = 392$°C

$$\Delta T_m = \frac{\Delta T_1 - \Delta T_2}{\ln\left(\dfrac{\Delta T_1}{\Delta T_2}\right)} = \frac{58 - 152}{\ln\left(\dfrac{58}{152}\right)} = 122.55\text{°C}$$

$$Q = 430650 = 30 \times 10^{-3} \times A_0 \times 122.55$$

or $$A_0 = \frac{430650}{30 \times 10^{-3} \times 122.55} = 117.135 \times 10^3 \text{ m}^2$$

$$\dot{m}_g = \left(n \frac{\pi}{4} D_i^2\right) \frac{C_{g1}}{v_{g1}} = 1350$$

or $$n = \frac{1350 \times 4 \times 2.0473}{\pi \times (60 \times 10^{-3})^2 \times 12} = 81459.47 \approx 81460$$ ***Ans.***

$$A_0 = n\,\pi\,D_0\,l = 81460 \times \pi \times 65 \times 10^{-3} \times l = 132.67 \times 10^3$$

or $$l = \frac{177.135 \times 10^3}{81460\,\pi \times 65 \times 10^{-3}} = 7.041 \text{ m}$$ ***Ans.***

PROBLEMS

Things to Think About

6.1 In the boiler, which type of heating takes place in the conversion of water into steam?

6.2 It's essential to make superheater moisture free. What is the reason behind?

6.3 Why there is a need of boiler feed pump in boiler system?

6.4 Why there is a need of boiler mountings are accessories in boiler system?

Short and Long Theoretical Questions

6.1 Define steam generator. Enlist its main components.

6.2 Differentiate between utility boilers and industrial boilers.

6.3 Difference between fire-tube and water-tube boilers. Discuss the working principle of a fired tube and water tube boilers.

6.4 List the advantages and disadvantages of fire-tube and water-tube boilers.

6.5 What is a Stirling boiler? What are its main features? List the shortcomings of straight tube boilers?

6.6 What is circulation? Differentiate between natural and forced circulation boilers. What is the pressure head developed by natural circulation boilers?

6.7 Under what circumstances downcomers are placed outside and inside the furnace?

6.8 Define and mention the range of circulation ratio.

6.9 Why is too much steaming not desired in a riser tube and when it becomes vulnerable to rupture?

6.10 Why are downcomers fewer in number and bigger in diameter, while risers are more in number and smaller in diameter?

6.11 What is departure from nucleate boiling (DNB)?

6.12 Define void fraction and slip ratio. How these are related to quality?

6.13 Why should no moisture flow along with steam from the drum to the superheater and why is no vapour bubble desired to flow along with saturated water from the drum to the dwoncomers?

6.14 Why is blowdown needed?

6.15 What are drum internals and why are they required?

6.16 Explain the method of removing moisture from steam in the steam drum so that no mositrue goes to the superheater.

6.17 What do you mean by priming and foaming?

6.18 What are water wall and membrane water wall?

6.19 Discuss the advantage of high pressure boiler.

6.20 Discuss the working principle of natural circulation modern power boiler.

6.21 Discuss the working principle high pressure boilers, namely, LaMount, Benson, Velox and Loeffler.

6.22 Discuss the working principle of supercritical boiler. What are the advantages of supercritical boilers?

6.23 Discuss the working principle of fluidized bed boilers and pressurized fluidized bed boiler.

6.24 What are boiler mountings and accessories? List them.

6.25 What are the functions of preheaters, economisers and superheaters? Where is reheater located?

6.26 What are primary, secondary and pendent superheaters?

6.27 How is the erosion rate of fly ash related to gas velocity?

6.28 Explain the fact that the steam temperature from the exit of convective superheater increases with load.

6.29 Why does the exit steam temperature decrease with the increase in load in a radiant superheater?

6.30 Discuss the various methods of steam temperature control. What is attemperation?

6.31 What are titling burners?

6.32 Explain the working principle of bubbling fluidized bed boiler and circulating fluidized bed boiler.

6.33 Discuss the working of various non-mechanical valves?

6.34 What are cyclone separators? Explain principles of gas-solid separation in a cyclone.

6.35 Define the collection efficiency of a cyclone. On what factors does the collection efficiency of a cyclone depend?

6.36 Discuss the advantages and disadvantages of CFB boiler.

6.37 What do you mean by acid rain? How is sulphur removed in a fluidized bed combustor? At what temperature is sulphur capture maximized? What should be the Ca/S ratio for high sulphur capture?

6.38 Explain the working principle of an electrostatic precipitator

6.39 Explain the working principle of fabric filter and Baghouse.

6.40 Explain the working principle of ash handling system.

6.41 Why is there a need of make-up feed water treatment? Discuss the treatment of feed water in brief?

6.42 Discuss the working principle of demineralization plant.

6.43 What do you mean by clarification?

6.44 Discuss the basic process of water softening

6.45 Discuss the lime-soda ash and hot phosphate softening processes.

6.46 Explain the working principle of radium zeolite softening.

6.47 What do you understand by ion exchanger?

6.48 What is reverse osmosis? Explain condensate polishing.

6.49 Explain internal treatment of feed water.

Numerical Questions

6.1 A 220 MWe steam power plant has steam conditions, at boiler outlet as 160 bars, 600°C and the condenser pressure is 0.05 bar. The calorific value of fuel used is 25 MJ/kg and the boiler efficiency is 90%. The feed water temperature at the boiler inlet is 170°C. The height of the unheated downcomer and heated risers of 50 m ensures natural circulation and produces circulation ratio equal to 15. A minimum exit velocity of water-steam mixture leaving a riser and entering the drum is required to be 1.8 m/s. The inner and outer diameter of riser tubes are 54 mm and 60 mm respectively. Assume generator and turbine efficiencies as 98 and 92 per cent respectively and neglect any pressure drop and heat loss. Calculate (a) the steam generation rate (b) the fuel burning rate, (c) the evaporation factor, (d) the pressure head developed due to natural circulation, (e) the quality of the steam at the top of riser, (f) the void fraction of steam at the riser exit, if slip ratio is 1.2, (g) the number of riser tubes required, (h) the heat absorption rate per unit projected area of riser, (i) the overall plant efficiency of mechanical efficiency is 98%, overall pump efficiency is 85% and 3 per cent of total power is consumed in driving auxiliaries.

6.2 An economiser of a steam power plant producing 220 MWe working at boiler pressure 160 bars, 600°C and condenser pressure of 0.05 bar allows steam to enter the economiser inlet header at 160 bars, 170°C and leaves it at the saturated vapour state. The economiser consists of tubes 70 mm outer diameter and 60 mm inner diameter made in the form of vertical coils. The flue gas velocity is limited to 12 m/s. The velocity of water leaving the outlet header is 1.5 m/s. The ash content of coal is 30 per cent by mass and 10 kg of air is supplied per kg of coal. The flue gases leave the economiser coils at 450°C. If the overall heat transfer coefficient is 80 W/m^{2}K, calculate the number of coils need in the economiser and length of a coil assuming boiler, turbine and generator efficiency as 90%, 92% and 98% respectively. If the vertical pitch of the coils is 80 mm and the clearance on the two sides of the duct having width 4.8 m, and 5 mm, calculate vertical height of economiser.

6.3 A tubular type preheater is to be designed for a steam generator which supplies steam at 160 bars, 600°C to a steam turbine where the condenser pressure is 0.05 bar. The boiler, turbine generator and pump efficiencies are 90%, 92%, 98% and 85% respectively. The flue gas enters the preheater at 450°C and leaves at 160°C. Air enters the preheater at 35°C. The outlet velocity of the flue gases is 12 m/s and the tubes are of the air preheater and the length of tubes. Assume c_p for flue gas as 1.1 kJ/kg K and 0.286 kJ/kg K.

6.4 A steam generator comprises evaporator (boiler), superheater, economiser and air preheater. The feed water enters the economiser at 50°C. Air enters the preheater at 30°C and leaves at 150°C. The steam is generated in the steam drum at 100 bars, 0.99 dry and leaves the superheater at 500°C. The fuel used is oil, having calorific value 42 MJ/kg. Evaporation rate is 10 kg steam per kg of fuel and air-fuel ratio is 20:1 by mass. Neglecting heat losses and pressure drops, calculate the heat transfer per kg of fuel in each component and the steam generator plant efficiency.

6.5 A boiler drum has a concentration of 1300 ppm. Calculate the percentage of water in the boiler must be wasted through the blowdown to reduce the concentration to 1000 ppm when the feed water concentration is 12 ppm. For the desired concentration of 1 ppm in steam flow from the boiler having a concentration of 1000 ppm, calculate the maximum moisture in the steam.

6.6 A drum type steam generator operates at 180 bars. Water from the drum, subcooled by 7°C flows down the downcomer at the rate of 1350 kg/s. The height of downcomers and risers are each 15 m. The average density of water-steam mixture in the risers is 350 kg/m^3. The pressure losses in the downcomer and risers are 0.5 bar. Determine the power needed to drive a forced circulation pump of 80% efficiency.

6.7 A steam generator is to generate steam at the rate of 600 kg/s at 550°C and 120 bars. The condition of steam after the primary superheater is expected to be 540°C and 122 bars. It is necessary to reduce the temperature to 450°C prior to entering the secondary superheater if the desired final temperature is to be achieved. Calculate the quantity of desuperheater water at 130 bars, 180°C must be supplied per second to the desuperheater spray nozzles.

6.8 A marine boiler operates under the following conditions: feed water at 60 bars and 130°C is supplied to the boiler at the rate of 30 kg/s. Of this 25 kg/s leaves the sueprheater outlet at 45 bars and 450°C and the remainder leaves the steam drum as dry and saturated steam at 48 bars for auxiliary purpose. The fuel used is oil having calorific value of 40,414 MJ/m^3 and supplied at the rate of 7.5 m^3/h. Calculate the efficiency of marine boiler.

6.9 A boiler plant incorporating an economiser and air preheater generates steam at 50 bars, 400°C with the fuel of calorific value 33 MJ/kg burnt at the rate of 0.12 kg/s. In the economiser, the feed water is heated from 45°C to 135°C by hot flue gas which is cooled from 470°C to 250°C. The flue gases enter the air preheater where air is heated to 80°C. A forced drought fan delivers air at 1.02 bars and 20°C and produces a pressure head of 200 mm of water. The power input to the fan is 5 kW with a mechanical efficiency of 0.8. Calculate (a) mass flow rate of air, (b) temperature of flue gases leaving the plant, (c) the mass flow rate of steam, and (d) the efficiency of boiler plant. Neglect heat losses and assume c_p for gases equal to 1.01 kJ/kg.

6.10 Design a natural circulation and forced circulation steam generator for the following data: Steam pressure = 160 bars, Steam temperature = 600°C, Mass flow rate of steam = 1250 kg/s. Design a supercritical steam boiler for the following data: Steam pressure = 350 bars, Steam temperature = 650°C. Mass flow rate of steam = 1200 kg/s.

Software Based Questions

6.1 Develop a computer software for the design of steam generator for steam power plant. Validate the results with the existing one.

6.2 Develop a computer software for the design of ash handling system for coal based steam power plant.

6.3 Develop a computer software for the design of air preheater.

Design and Open Ended Question

6.1 Give a design concept of a modern boiler which is highly efficient and less polluting and is also capable of using any type of fuel.

Multiple-Choice Questions

6.1 Downcomers of a water-tube boiler are

(a) more in number (b) less in diameter
(c) more in number and less in diameter (d) less in number and more in diameter

6.2 Risers of a water-tube boiler are

(a) less in number (b) more in diameter
(c) less in number and more in diameter (d) more in number and less in diameter

6.3 Circulation ratio in any tube should be in the range of

(a) 3 to 5 (b) 6 to 25 (c) 26 to 50 (d) 5 to 100

6.4 Forced circulation in the risers of a water tube is achieved when the boiler pressure is exceeds

(a) 30 bars (b) 60 bars (c) 120 bars (d) 180 bars

6.5 The slip ratio

(a) increases with boiler pressure (b) decreases with boiler pressure
(c) remains constant (d) increases and decreases with boiler pressure

6.6 High pressure boilers have

(a) small bore tube (b) lighter structure
(c) lower number of drums (d) all of the above

6.7 The temperature of flue gases passing through chimney is around

(a) 60°C (b) 90°C (c) 120°C (d) 160°C

6.8 The quantity of water in each riser is about

(a) 1.8 times the steam evaporated (b) 3.8 times the steam evaporated
(c) 5.8 times the steam evaporated (d) 7.8 times the steam evaporated

Answers to Multiple-Choice Questions

6.1 (d) 6.2 (d) 6.3 (b) 6.4 (d) 6.5 (b) 6.6 (d) 6.7 (c) 6.8 (b)

CHAPTER

7

Steam Turbines

OVERVIEW

The **first steam turbine** historically recorded was ***"Hero turbine"*** **around 150 BC** by Hero of Alexandria which worked on reaction principle. In the year 1890, deLaval, Swedish engineer dealt a single disc steam turbine of 3.5 kW capacity working on the principle of **impulse**. After this, steam turbines were commercialised and many improvements were made which has now become a proven technology.

Steam turbine is one of the most important prime movers for generating electricity. This falls under the category of power producing turbomachines. *In the turbine, the energy level of the working fluid goes on decreasing along the flow steam.* Single unit of steam turbine can develop power ranging from 1 MW to 1000 MW. In general, 1 MW, 2.5 MW, 5 MW, 10 MW, 30 MW, 120 MW, 210 MW, 250 MW, 350 MW, 500 MW, 660 MW, 1000 MW, 1800 MW are in common use. The thermal efficiency of modern steam power plant above 120 MW is as high as 38% to 40%.

The purpose of turbine technology is to extract the maximum quantity of energy from the working fluid, to convert it into useful work with maximum efficiency, by means of a plant having maximum reliability, minimum cost, minimum supervision and minimum starting time. This chapter deals with the types and working of various types of steam turbines. The construction details are given in chapter 8.

7.1 PRINCIPLE OF OPERATION OF STEAM TURBINE

The principle of operation of steam turbine is entirely different from the steam engine. In reciprocating steam engine, the pressure energy of steam is used to overcome external resistance and the dynamic action of steam is negligibly small. But the steam turbine depends completely upon the dynamic action of the steam. According to Newton's second law of motion, the force is proportional to the rate of change of momentum (mass × velocity). If the rate of change of momentum is caused in the steam by allowing a high velocity jet of steam to pass over curved blade, the steam will impart a force to the blade. If the blade is free, it will move off (rotate) in the direction of force (Fig. 7.1(a)]. In other words, *the motive power in a steam turbine is obtained by the rate of change in moment of momentum of a high velocity jet of steam impinging on a curved blade which is free to rotate.* The steam from the boiler is expanded in a passage or nozzle where due to fall in pressure of steam, thermal energy of steam is converted into kinetic energy of steam, resulting in the emission of a high velocity jet of steam which impinges on the moving vanes or blades attached on a rotor which is mounted on a shaft supported on bearings, and here steam undergoes a change in direction of motion due to curvature of blades which gives rise to a change in momentum and therefore a force. This constitutes the driving force of the turbine. This arrangement is shown in Fig. 7.1(b). It should be realized that the blade obtains no motive force from the static pressure of the steam or from any impact of the jet, because the blade in designed such that the steam jet will glide on and off the blade without any tendency to strike it.

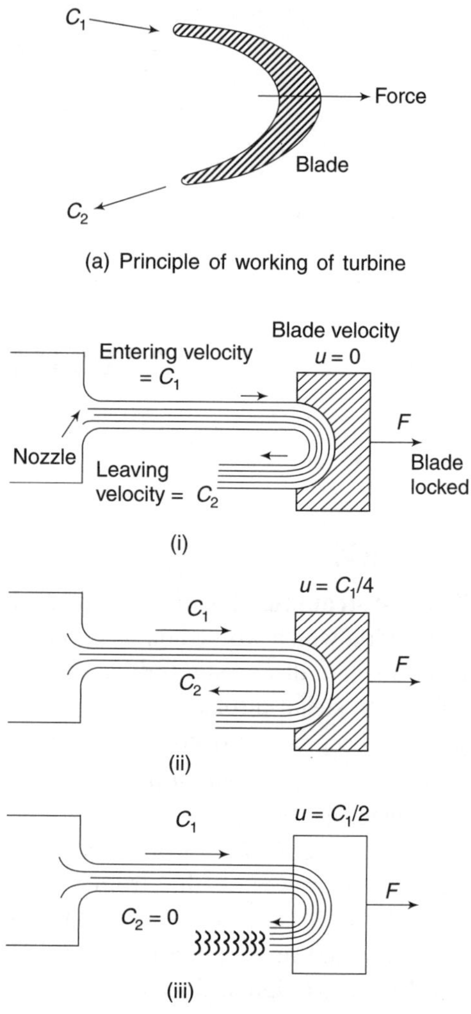

(a) Principle of working of turbine

(b) Effect of blade velocity

Figure 7.1

As shown in Fig. 7.1(b), and Fig. 7.2, when the blade is locked the jet enters and leaves with equal velocity, and thus develops maximum force if we neglect friction in the blades. Since the blade velocity is zero, no mechanical work is done. As the blade is allowed to speed up, the leaving velocity of jet from the blade reduces, which reduces the force. Due to blade velocity the work will be done and the maximum work is done when the blade speed is just half of the steam speed. In this case, the steam velocity from the blade is near about zero, i.e., it is trial of inert steam since all the kinetic energy of steam is converted into work. The force and work done become zero when the blade speed is equal to the steam speed.

From the above discussion, it follows that a steam turbine should have a row of nozzles, a row of moving blades fixed to the rotor, and the casing (cylinder). *A row of nozzles and a row of moving blades constitutes a stage of turbine.*

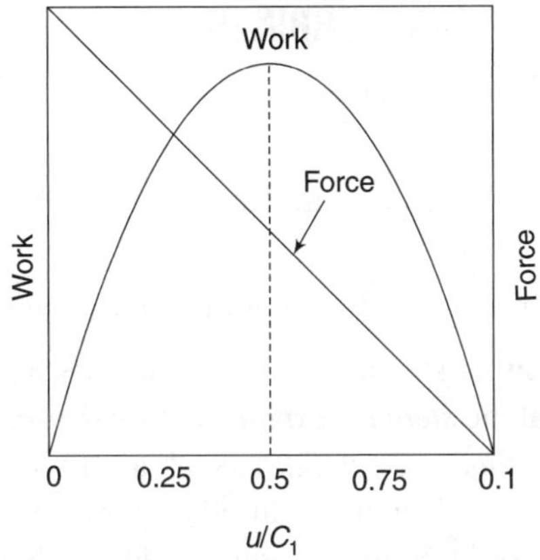

Figure 7.2 Theoretical relationship among work, force and blade velocity

7.2 COMPARISON OF STEAM ENGINE AND TURBINES

The following differences between the steam engine and steam turbine will show that the steam engine is almost obsolete compared to the steam turbine.

Steam Engine (Now Obsolete)	*Steam Turbine*
(i) Presence of reciprocating masses makes the machine *poorly balanced* and hence vibration problems are greater.	(i) Absence of reciprocating masses makes the machine *better balanced.*
(ii) Presence of numerous sliding or bearing members *lowers its mechanical efficiency.*	(ii) Absence of numerous sliding or bearing members *improves its mechanical efficiency.*
(iii) Higher speed above 300 r.p.m. is not advisable. *Higher range of speed is not possible.*	(iii) Speed up to 40,000 r.p.m. may be used and hence power output per unit volume of the working substance is increased. *Far greater* range of speed is *possible.*
(iv) *Much internal lubrication* is required, so the steam gets mixed with liquid and consequently the condensate is not suitable for being used as feed water.	(iv) *No internal lubrication* is required, hence steam does not get mixed with oil and condensate is suitable for feed water.
(v) Thermal efficiency is *very poor* (say 15%).	(v) High thermal efficiency is possible (say 40%).
(vi) Due to reciprocating motions and valve operations *wear, tear* and maintenance costs per unit of power are *very high.*	(vi) Wear, tear and maintenance costs are very low.
(vii) A back pressure lower than atmospheric is not recommended at all to increase the useful work.	(vii) Maximum useful work is possible by reducing the condenser pressure at any low vacuum.
(viii) It is difficult to design, build and operate a large power unit.	(viii) Large power unit is possible. Up to 1000 megawatts capacity units are already in use.
(ix) Electrical generators are difficult to be coupled and driven with engines.	(ix) Steam turbines are most suitable to drive generators.
(x) Steam consumption of the engine increase with years of service.	(x) Steam consumption does not increase with years of service.

7.3 CLASSIFICATION OF STEAM TURBINES

(A) On the basis of Principle of Operation

(i) Impulse turbine

(a) Simple, (b) Velocity stage, (c) Pressure stage, (d) Combination of (b) and (c).

(ii) Impulse-reaction turbine

(a) 50% (Parson's) reaction, (b) Combination of impulse and reaction.

(i) Impulse Turbine. If the flow of steam through the nozzles and moving blades of a turbine takes place in such manner that the *steam is expanded only in nozzles and pressure at the outlet sides of the blades is equal to that at inlet side"*, such a turbine is known as impulse turbine because it works on the principle of impulse. In other words, in impulse turbine, *the drop in pressure of steam takes place only in nozzles and not in moving blades*. This is obtained by making the **blade passage of constant cross-sectional area** (Fig. 7.12).

As a general statement, it may be stated that *energy transformation takes place only in nozzles, and moving blades (rotor) only cause energy transfer.* Since the rotor blade passages do not cause any acceleration of fluid, hence chances of flow separation are greater which results in lower stage efficiency.

(ii) Impulse-Reaction Turbine. In this turbine, *the drop in* pressure of steam takes place in *fixed (nozzles) as well as moving blades.* The pressure drop suffered by steam while passing through the moving blades causes a further generation of kinetic energy within the moving blades, giving rise to reaction and adds to the propelling force which is applied through the rotor to the turbine shaft. Since this turbine works on the principle of impulse and reaction both, so it is called impulse-reaction turbine **(Fig. 7.17. This is achieved by making the blade passage of varying cross-sectional area** *(converging type).*

In general, it may be stated that *energy transformation occurs in both fixed and moving blades.* The rotor blades cause both energy transfer and transformation. Since there is an acceleration of flow in moving blade passage, hence chances of separation of flow is less which results in higher stage efficiency.

(B) On the basis of "Direction of Flow

(i) Axial flow turbine, (ii) Radial flow turbine, (iii) Tangential flow turbine.

(i) Axial Flow turbine. In axial flow turbine, the steam flows along the axis of the shaft (Fig. 7.17). It is the most suitable turbine for large turbo-generators and that is why it is used in all modern steam power plants.

(ii) Radial Flow Turbine. In this turbine, the steam flows in the radial direction. It incorporates two shafts, end to end, each driving a separator generator. A disc is fixed to each shaft. Rings of 50% reaction radial-flow bladings are fixed to each disk. The two sets of bladings rotate counter to each other (Fig. 7.3). In this way, a relative speed of twice the running speed is achieved and every blade row is made to work. The final stage may be of axial flow design in order to achieve a larger area of flow. Since this type of turbine can be warmed and started

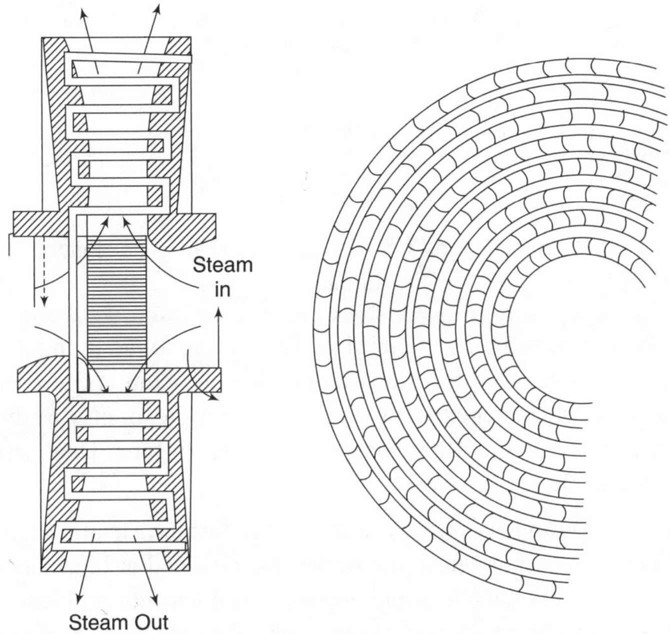

Figure 7.3 Ljugstrom turbine

quickly, so it is very suitable for use at times of peak load. Though, this type of turbine is very successful in the smaller sizes but formidable design difficulties have hindered the development of large turbines of this type. In Sweden, however, composite radial/axial flow turbines have been built of outputs upto 275 MW. Sometimes, this type of turbine is also known as Ljungstorm turbine after the name of its inventor B and F. Ljungstorm of Sweden.

(iii) Tangential Flow Turbine. In this type, the steam flows in the tangential direction. This turbine is very robust but not particularly efficient, sometimes used for driving power station auxiliaries. In this turbine, nozzle directs steam tangentially into buckets milled in the periphery of a single wheel, and on exit the steam turns through a reversing chamber, re-entering bucket further round the periphery. This process is repeated several times, the steam flowing a helical path. Several nozzles with reversing chambers may be used around the wheel periphery.

(C) On the basis of Means of Heat supply

(i) Single pressure turbine,

(ii) Mixed or dual pressure turbine.

(iii) Reheated turbine. (a) Single (b) Double

(i) Single Pressure Turbine. In this type of turbine, there is single source of steam supply (Fig. 7.4).

(ii) Mixed or Dual Pressure Turbine. This type of turbines use two sources of steam at different pressures. The dual pressure turbine is found in nuclear power stations where it uses both

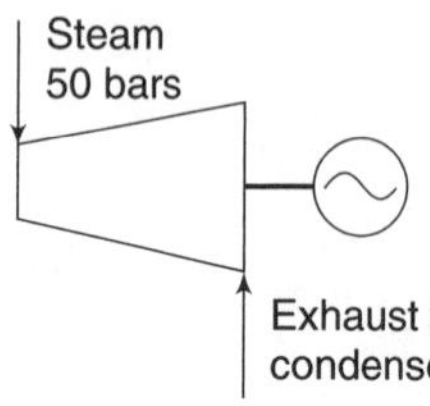

Figure 7.4 Single pressure turbine

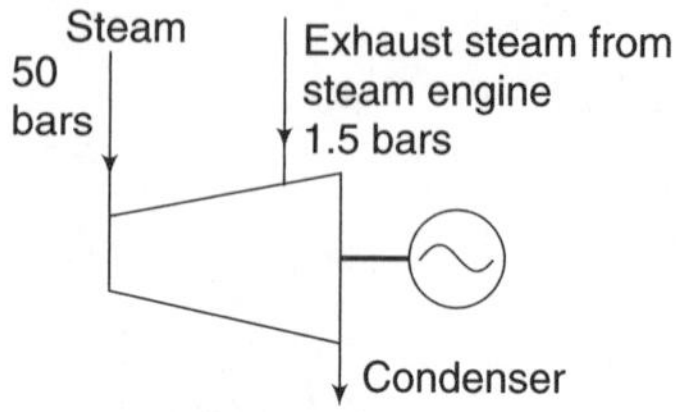

Figure 7.5 Mixed pressure turbine

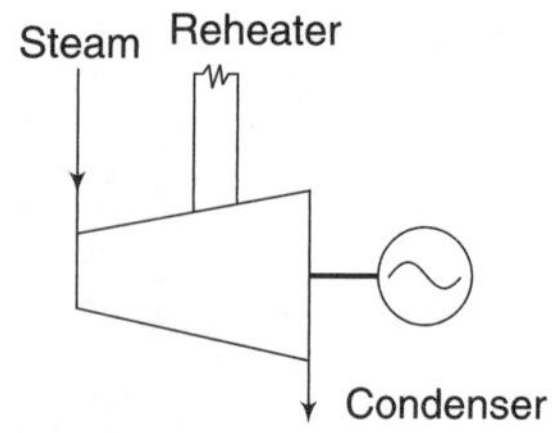

Figure 7.6 Reheated turbine

sources continuously. The mixed pressure turbine (Fig. 7.5) is found in industrial plants (e.g., rolling mill, colliery, etc.) where there are two supplies of steam and use of one supply is more economical than the other, for example, the economical steam may be the exhaust steam for engine which can be utilised in the L.P. stages of steam turbine. Dual pressure system is also used in combined cycle.

(iii) Reheated Turbine. During its passage through the turbine steam may be taken out to be reheated in a reheater incorporated in the boiler and returned at higher temperature to be expanded in (Fig. 7.6). This is done to avoid erosion and corrosion problems in the bladings and to improve the power output and efficiency. The reheating may be single or double or triple.

(D) On the Basis of Means of Heat Rejection

(i) Pass-out or extraction turbine, (ii) Regenerative turbine, (iii) Condensing turbine, (iv) Non-condensing turbine, (v) Back pressure or topping turbines.

(i) Pass-out Turbine. In this turbine (Fig. 7.7) a considerable proportion of the steam is extracted from some suitable point in the turbine where the pressure is sufficient for use in process heating; the remainder continuing through the turbine. The latter is controlled by separate valve-gear to meet the difference between the pass-out steam and electrical load requirements. This type of turbine is suitable where there is dual demand of steam–one for power and the other for industrial heating, for example, sugar industries. Double pass-out turbines are sometimes used.

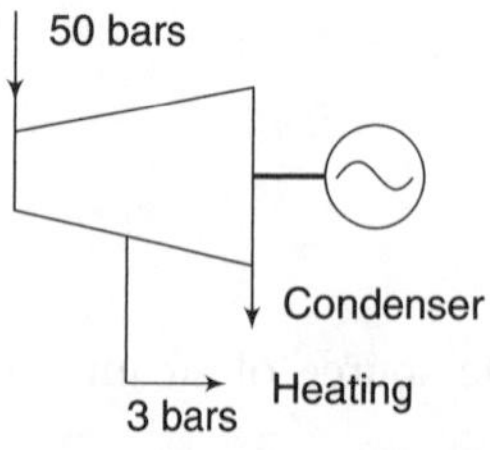

Figure 7.7 Pass-out turbine

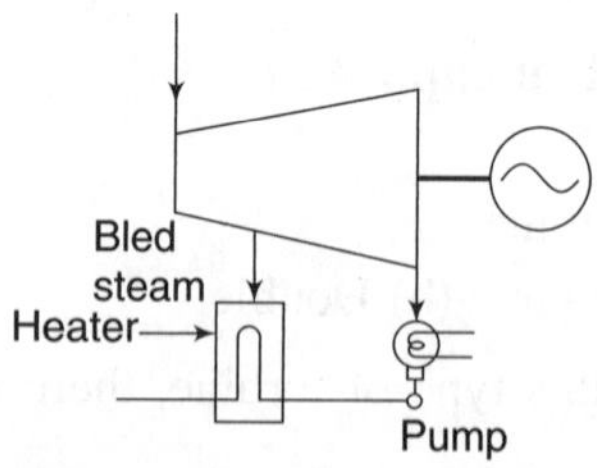

Figure 7.8 Regenerative turbine

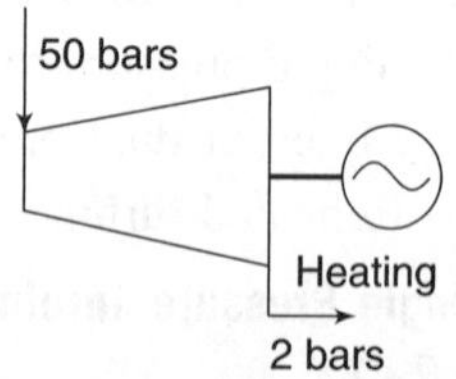

Figure 7.9 Back pressure turbine

(ii) Regenerative Turbine. This turbine incorporates a number of extraction branches, through which small proportions of the steam are commonly extracted for the purpose of heating the boiler feed water in a feed heater in order to increase the thermal efficiency of the plant (Fig. 7.8). Nowadays, all steam power plants are equipped with reheating and regenerative arrangement.

(iii) Condensing Turbine. In this turbine, the exhaust steam is condensed in a condenser and the condensate is used as feed water in the boiler. By this way, the condensing turbine allows the steam to expand to the lowest possible pressure before being condensed. All steam power plants use this type of turbine.

(iv) Non-Condensing Turbine. When the exhaust steam coming out from the turbine is not condensed but exhausted in the atmosphere is called non-condensing turbine. The exhaust steam is not recovered for feed water in the boiler.

(v) Back Pressure or Topping Turbine. This type of turbine rejects the steam after expansion to the lowest suitable possible pressure at which it is used for heating purpose. Thus, back pressure turbine supplies power as well as heat energy.

The back pressure turbine (Fig. 7.9) generally used in sugar industries provides low pressure steam for heating apparatus, whereas a topping turbine exhausts into a turbine designed for lower steam conditions.

(E) On the Basis of Number of Cylinder

Turbine may be classified as

(i) Single cylinder and (ii) Multi-cylinder.

(i) Single Cylinder. When all stages of turbines are housed in one casing, then it is called single cylinder. Such a single cylinder turbines uses one shaft.

(ii) Multi-Cylinder. In large output turbines, the number of the stages becomes so high that additional bearings are required to support the shaft. Under this circumstances, multi-cylinders are used.

(F) On the Basis of Arrangement of Cylinder Based on General Flow of Steam

(i) Single flow, (ii) Double flow and (iii) Reversed flow.

(i) Single Flow. In a single flow turbines, the steam enters at one end, flows once through the bladings in a direction approximately parallel to this axis, emerges at the other and (Fig. 7.10a). High pressure cylinder uses single flow. This is also common in small turbines.

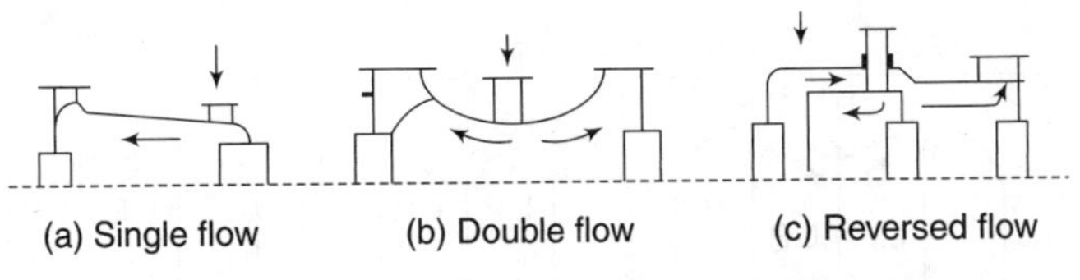

Figure 7.10 Cylinder flow arrangement

(ii) Double Flow. In this type of turbines, the steam enters at the centre and divides, the two portions passing axially away from other through separate sets of blading on the same rotor (Fig. 7.10b). The low pressure cylinder normally uses double flow. This type of unit is completely balanced against the end thrust and gives large area of flow through two sets of bladings. This also helps in reducing the blade height as mass flow rate becomes half as compared to single flow for the same conditions.

(iii) Reversed Flow. Reversed flow arrangement is sometimes used in h.p. cylinder where higher temperature steam is used on the larger sets in order to minimise differential expansion, i.e., unequal expansion of rotor and casing (Fig. 7.10c).

Typical Cylinder Layouts: Figure 7.11 shows the various **typical cylinder layouts**. This illustrates the way in which the modern turbine has evolved. The use of single, double and reversed flow is shown in the layout.

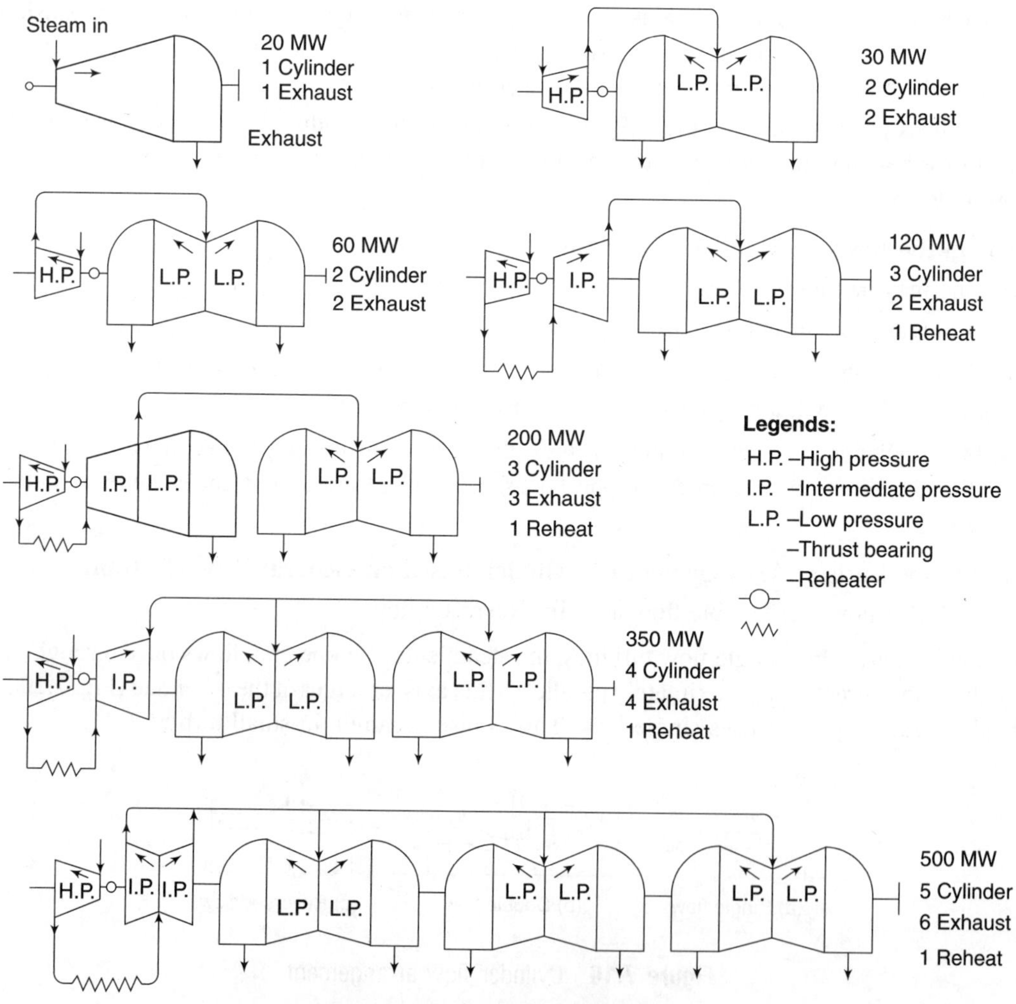

Figure 7.11 Typical cylinder layout

(G) On the Basis of Number of Shaft

(i) Tandem compound, (ii) Cross compound

(i) Tandem Compound. Most multi-cylinder turbines drive a single shaft and single generator. Such turbines are known as tandem compound turbines.

(ii) Cross Compound. In this type, two shafts are used driving separate generator. The reason may be one of turbine house arrangement, limited generator size, or a desire to run the *h.p.* shafting at half speed. The latter choice is sometimes preferred so that for the same centrifugal stress, longer blades may be used, giving a larger leaving area, a smaller velocity and hence a small leaving loss.

(H) On the Basis of Rotational Speed

(i) Constant speed turbines, (a) for 60 cycle/s generators, (b) 50 cycles/generator, (c) for 25 cycles/s generator,

(ii) variable speed turbines

(i) Constant Speed Turbines. Requirements of rotational speed are extremely rigid in turbines which are directly connected to electric generators as these must be *a-c* unit except in the smallest sizes and must therefore run at speeds corresponding to the standard number of cycles per second and governed by the following equation:

$$N = \frac{120 \times \text{Number of cycles per second}}{\text{Number of poles}} = \frac{120f}{p}$$

The minimum number of poles, in a generator is two and correspondingly the maximum possible speed for 60 - cycle is 3600 rpm; for 50 c/s of frequency, the speeds would be 3000. 1500 and 750 rpm for 2, 4 and 8 poles machines respectively.

(ii) Variable Speed Turbines. These turbines have geared units and may have practically any speed ratio between the turbine and the driven machine so that the turbine may be designed or its own most efficient speed. Such turbines are used to drive ships, compressor, blowers and variable frequency generators.

Table 7.1 depicts a typical cylinder layout of General Electric Co., USA showing cylinder layout, single or double flow, number of shafts, speed and output range [87].

7.4 SIMPLE IMPULSE TURBINE

This type of turbine works on the principle of impulse and is shown diagrammatically in Fig. 7.12. It mainly consists of a nozzle or a set of nozzles, a rotor mounted on a shaft, one set of moving blades attached to the rotor and a casing. The uppermost portion of the diagram shows a longitudinal section through the upper half of the turbine. The middle portion shows the development of the nozzles and bladings, i.e., the actual shape of the nozzle and blading, and the bottom portion shows the variation of absolute velocity and absolute pressure during flow of steam through passage of nozzles and blades. The example of this type of turbine is the de-Laval turbine.

Table 7.1 Typical cylinder layout with generator of GE Co., USA [87]

Fossil	*Fossil*	*Nuclear*
TC-2F LSB 26, 30 and 33.5 in Two casings, 3600 r/min HP-IP LP G 125–400 MW	TC-6F LSB 26, 30 and 33.5 in Five casings, 3600 r/min HP IP LP LP LP G 550–1000 MW	TC-4F LSB 38 and 43 in Three casings, 1800 r/min HP LP LP G 450–1000 MW
TC-2F LSB 26, 30 and 33. 5 in Three casings, 3600 r/min HP-IP LP LP G 250–650 MW	TC-6F LSB 30 and 33.5 in Five casings, 3600 r/min (double reheat) HP-IP IP LP LP LP G 450–725 MW	TC-6F LSB 38 and 43 in Four casings 1800 r/min HP LP LP LP G 600–1100 MW
TC-4F LSB 26, 30 and 33.5 in Four casings 3600 r/min HP IP LP LP G 550-850 MW LP LP G	CC-4F LSB 38 and 43 in Four casings 3600/1800 r/min HP IP G 3600 r/min LP LP G 1800 rpm 600-1250 MW	Legend : TC= Tandem Compound, CC = Cross-Compound, F = number of flow Ducts to condenser, LSB = last-stage blade HP : High pressure IP : Intermediate pressure LP : Low pressure G : Generator

It is obvious from the figure that the complete expansion of steam from the steam chest pressure to the exhaust pressure or condenser pressure takes place only in one set of nozzles, i.e., the pressure drop takes place only in nozzles. It is assumed that the pressure in the recess between nozzles and blades remains the same. The steam at condenser pressure or exhaust pressure enters the blade and comes out at the same pressure, i.e., the pressure of steam in blade passages remains approximately constant and equal to the condenser pressure.

Generally, converging-diverging nozzles are used. Due to the relatively large ratio of expansion of steam in the nozzles the steam leaves the nozzles at a very high velocity (supersonic), of about 1100 m/s. It is assumed that the velocity remains constant in the recess

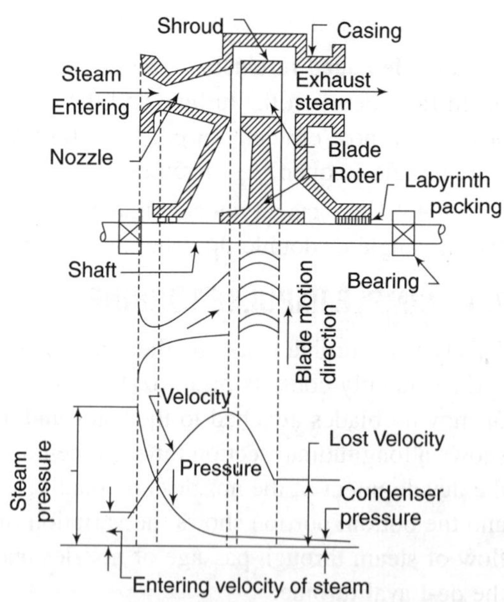

Figure 7.12 Diagrammatic arrangement of a simple (de-Laval) turbine

between the nozzles and the blades. The steam at such a high velocity enters the blades and reduces along passage of blades and comes out with an appreciable amount of velocity.

As it has been already shown, that for the good economy or maximum work, the blade speed should be one half of the steam speed, so blade velocity is of about 500 m/s which is very high. This results in a very high rotational speed, reaching 30,000 r.p.m. Such high rotational speeds can only be utilised to drive generators or machines with large reduction gearing arrangements. In this turbine, the leaving velocity of steam is also quite appreciable resulting in an energy loss, called "carry over loss" or "leaving velocity loss". This leaving loss is so high that it may amount to about 11% of the initial kinetic energy. This type of turbine is generally employed where relatively small power is needed and where the rotor diameter is kept fairly small. Construction aspects of turbine is given in chapter 8.

7.5 COMPOUNDING OF IMPULSE TURBINE

Compounding is a method for reducing the rational speed of the impulse turbine to practical limits. As we have seen, if the high velocity of steam is allowed to flow through one row of moving blades, it produces a rotor speed of about 30,000 r.p.m. which is too high for practical use. Not only this, the leaving loss is also very high. It is, therefore, essential to incorporate some improvements in the simple impulse turbine for practical use and also to achieve high performance. This is possible by making use of more than one set of nozzle blades, rotors, in a series, keyed to a common shaft, so that either the steam pressure or the jet velocity is absorbed by the turbine in stages. The leaving loss also will then be less. This process is called compounding of steam turbines. There are three main types.

(a) Pressure-compounded impulse turbine.

(b) Velocity-compounded impulse turbine.

(c) Pressure and velocity compounded impulse turbine.

7.6 PRESSURE COMPOUNDED IMPULSE TURBINE

In this type of turbine, the *compounded is done for pressure of steam only, i.e., to reduce the high rational speed of turbine, the whole expansion of steam is arranged in a number of steps by employing a number of simple turbines in a series keyed on the same shaft as shown* in Fig. 7.13. Each of these simple impulse turbine consisting of one set of nozzles and one row of moving blades is known as a stage of the turbine and thus this turbine consists of several stages. The exhaust from each row of moving blades enters the succeeding set of nozzles. Thus, we can say that *this arrangement is nothing but splitting up the wheel pressure drop from the steam chest pressure to the condenser pressure into a series of smaller pressure drops across several stages of impulse turbine and hence this turbine is called pressure compound impulse turbine.*

Pressure and velocity variation are also shown in Fig. 7.13. Nozzles are fitted into a diaphragm which is locked in the casing. This diaphgram separates one wheel chamber from another. All rotors are mounted on the same shaft and the blades are attached on the rotor. Rotor (i.e., disc) may be keyed to the shaft or it may be integral part of shaft.

The expansion of steam only takes place in the nozzles while pressure remains constant in the moving blades because each stages is a simple impulse turbine. So it is obvious from the pressure curve that the space between any two consecutive diaphragms is filled with steam at constant pressure and the pressure on either side of the diaphragm is different. Since the diaphragm is a stationary part, there must be clearance between the rotating shaft and the diaphragm. The steam tends to leak through this clearance for which devices like labyrinth packings, etc. are used.

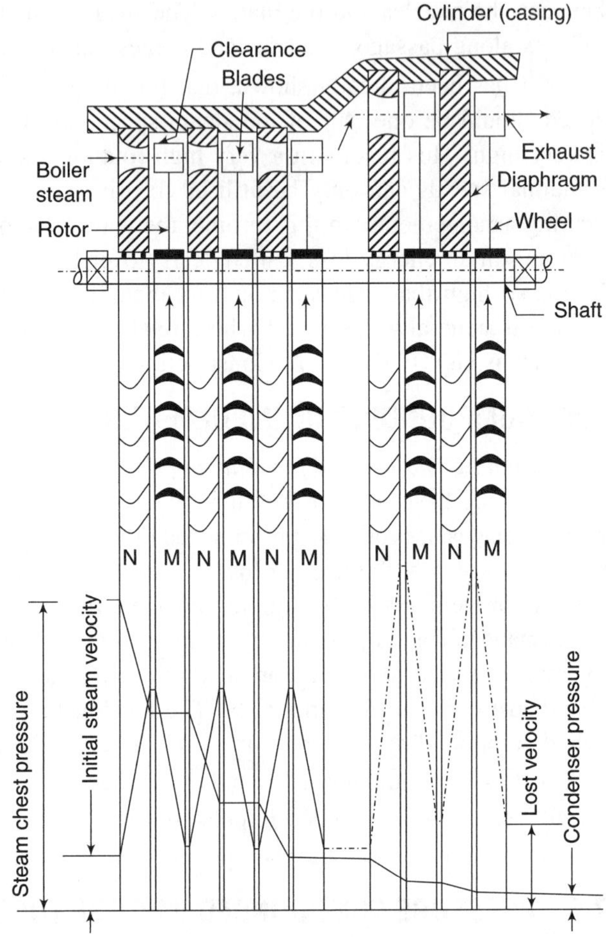

Figure 7.13 Diagrammatic arrangement of pressure compound impulse turbine

Since the drop in pressure of steam per stage is reduced, so the steam velocity leaving the nozzles and entering the moving blades is reduced which reduces the blade velocity. Hence, for good economy or maximum work, shaft speed is significantly reduced by increasing the number of stages according to onces need. The leaving velocity of the last stage of the turbine is much less compared to the deLaval turbine and and the leaving loss amounts to about 1 to 2 per cent of the initial total available energy. This turbine was invented by the late Prof. L. Rateau and so it is also known as Rateau Turbine.

7.7 SIMPLE VELOCITY-COMPOUNDED IMPULSE TURBINE

In this type of turbine, *the compounding is done for velocity of steam only, i.e., drop in velocity is arranged in many small drops through many moving rows of blades instead of a single row of moving blades.* It consists of nozzle or a set of nozzles and rows of moving blades attached to the rotor or wheel and rows of fixed blades attached to casing as shown in Fig. 7.14.

The fixed blades are guide blades which guide the steam to succeeding rows of moving blades suitably arranged between the moving blades and set in a reversed manner. In this turbine, three rows or rings of moving blades are fixed on a single wheel or rotor and this type of wheel is termed **three-row wheel**. There are two rows of guide blades or fixed blades placed between first and second and second and third rows of moving blades respectively.

The whole expansion of steam from the steam chest pressure down to the exhaust pressure takes place in the nozzles only. There is no drop in pressure either in the moving blades or the fixed blades, i.e., the pressure remains constant in the blades as in a simple impulse turbine. The steam velocity from the exit of the nozzle is very high as in a simple impulse turbine. Steam with this high velocity enters the first row of moving blades and on passing through these blades the velocity slightly reduces, i.e., the steam gives up a part of its kinetic energy and reissues from this row of blades with a fairly high velocity. It then enters the first row of guide blades which directs the steam to the second row of moving blades. Actually, there is a slight drop in velocity in the fixed on guide blades due to friction. On passing through the second row of moving blades, some drop in velocity again occurs, i.e., steam gives up another portion of its kinetic energy to the rotor. After this, it is redirected again by the second row of guide blades to the third row of moving blades where again some drop in velocity occurs and finally the steam leaves the wheel with a certain velocity in more or less axial direction. Compared to the simple impulse turbine, the leaving velocity is small and it is about 2 per cent of initial total available energy of team.

So we can say that *this arrangement is nothing but splitting up the velocity gained from the exit of the nozzles into many drops through several row of moving blades and hence the name velocity-compounded.*

This type of turbine is also termed *Curtis turbine*. Due to its low efficiency the

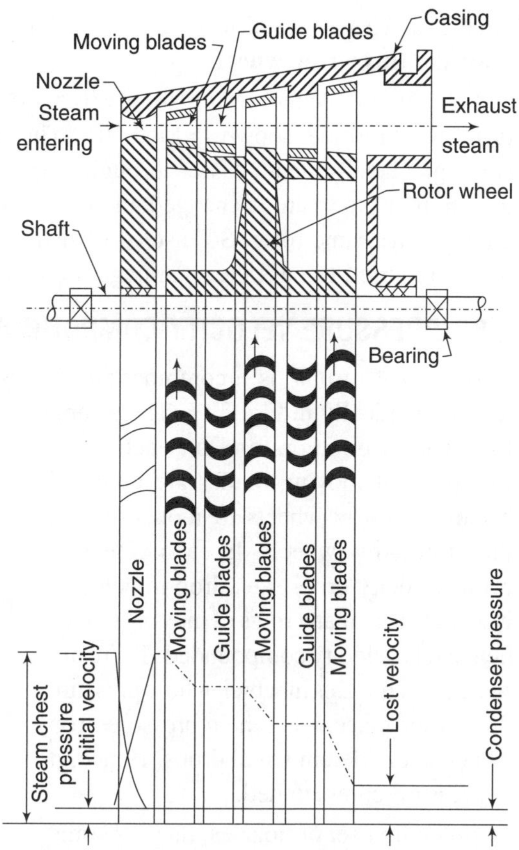

Figure 7.14 Diagrammatic arrangement of velocity-compounded impulse turbine

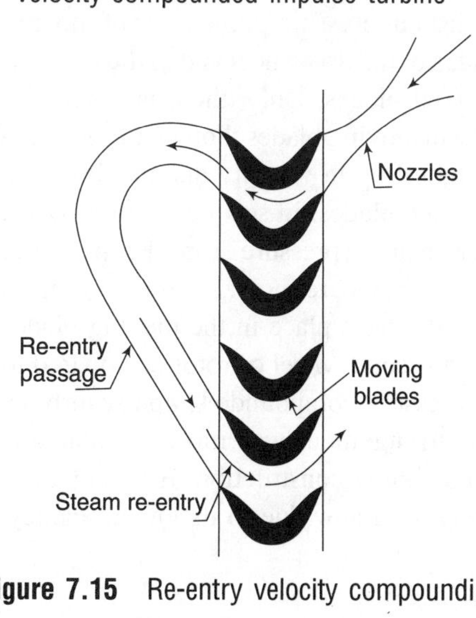

Figure 7.15 Re-entry velocity compounding

three-row wheel is used *for driving small machines.* **The two-row wheel is more efficient than the three-row wheel**.

Velocity compounding is also possible with only one row of moving blades as in the case of re-entry turbine, shown in Fig. 7.15. The whole pressure drop takes place in the nozzles and high velocity steam passes through the moving blades into a reversing chamber where the direction of the steam is changed and the same steam is arranged to pass through the moving blade of the same rotor. So instead of using two or three rows of moving blades, only one row is required to pass the steam again and again; thus in each pass, velocity decreases.

7.8 PRESSURE VELOCITY-COMPOUNDED IMPULSE TURBINE

This type of turbine is a combination of pressure and velocity compounding and is shown diagrammatically in Fig. 7.16. The arrangement shown here is only for two rotors. There are two wheels or rotors and on each, only two rows of moving blades are attached because two-row wheels are more efficient than three-row wheels. In each wheel or rotor, velocity drops, i.e., drop in velocity is achieved by many rows of moving blades hence it is velocity compounded. There are two sets of nozzles in which whole pressure drop takes place, i.e., whole pressure drop has been divided in small drops. Hence, it is pressure-compounded.

In the first set of nozzles, there is some decrease in pressure which gives some kinetic energy to the steam and there is no drop in pressure in the rows of moving blades of the first wheel and in the first row of fixed blades. Only, there is a velocity drop in moving blades though, there is also a slight drop in velocity due to friction in the fixed blades. In second set of nozzles, the remaining pressure drop takes place but the velocity here increase and the drop in velocity takes place in the moving blades at the second wheel or rotor. Compared to the pressure-compounded impulse turbine, this arrangement was more popular due to its simple construction. It is, however, rarely used now due to its low efficiency.

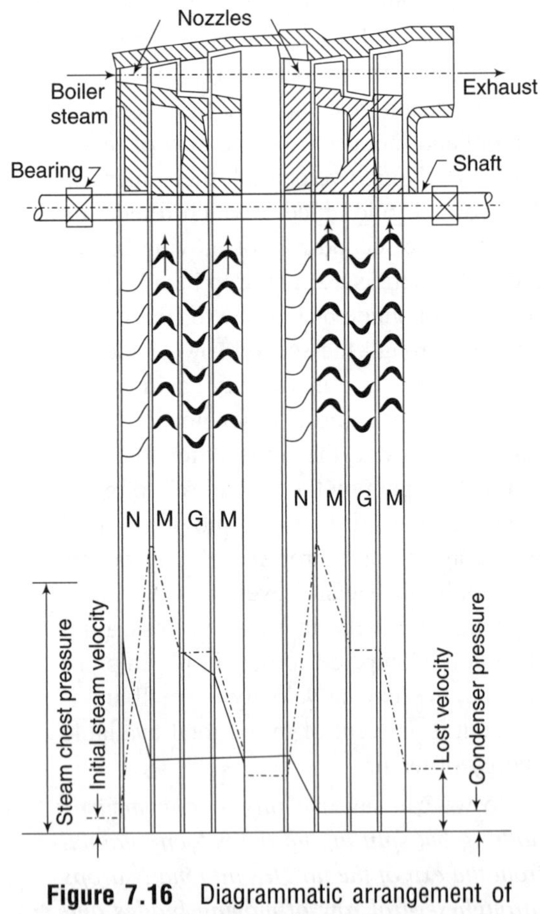

Figure 7.16 Diagrammatic arrangement of pressure-velocity compounded impulse turbine

7.9 IMPULSE-REACTION TURBINE

As the name implies, this type of turbine utilizes the principle of impulse and reaction both. Such a type of turbine is diagrammatically shown in Fig. 7.17. There are a number of rows of moving blades attached to the rotor and an equal number of fixed blades attached to the casing. In this type of turbine, the fixed blades which are set in a reversed manner compared to the moving blades, corresponds to nozzles mentioned in connection with the impulse turbine, Due to the row for fixed blades at the entrance, instead of the nozzles, steam is admitted for the whole circumference and hence there is all-round or complete admission. In passing through the first row of fixed blades, the steam undergoes small drop in pressure and hence its velocity somewhat increases. After this, it enters the first row of moving blades and just as in the impulse turbine, it suffers change in direction and therefore in momentum, This momentum gives rise to an impulse on the blades.

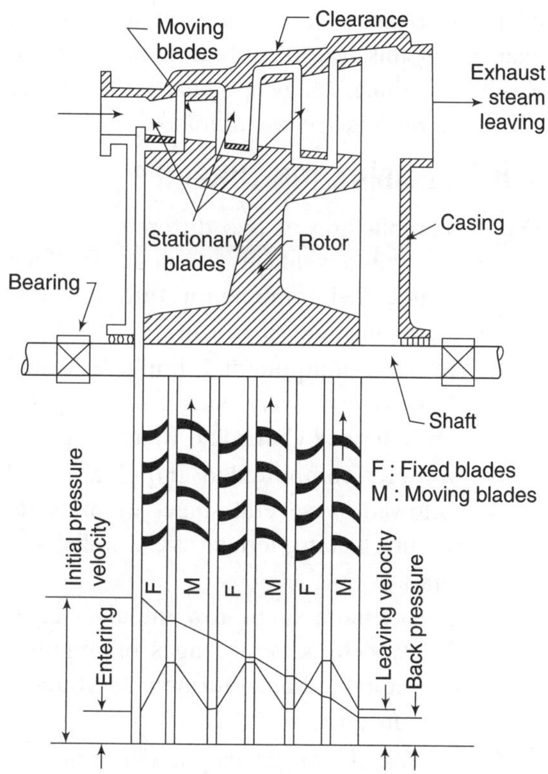

Figure 7.17 Diagrammatic arrangement of impulse-reaction turbine

But in this type of turbine, the passage of the moving blades is so designed (*converging*) that there is a small drop in pressure of steam in the moving blades which results in an increase in kinetic energy of steam. This kinetic energy gives rise to reaction in the direction opposite to that of added velocity. Thus, the rows propelling force on driving force is the vector sum of impulse and reaction forces. Commonly, this type of turbine is called **Reaction turbine**. The pressure and velocity variations are shown in Fig. 7.17. It is obvious from the figure that there is a gradual drop in pressure in both moving blades and fixed blades. The blades of impulse-reaction are of **aerofoil type**.

As the pressure falls, the specific volume increases and hence in practice, the height of blades is increased in steps, i.e., say upto 4 stages it remains constant, then it increases and remains constant for the next two stages.

In this type of turbine, the steam velocities are comparatively moderate and its maximum value is about equal to blade velocity. In general practic, to reduce the number of stages, the steam velocity is arranged greater than the blade velocity. In this case, the leaving loss is about 1 to 2 per cent of the total initial available energy. This type of turbine is used mostly in all power plants where it is great success. An example of this type of turbine is the Parsons-Reaction Turbine. *The power plants 30 MW and above are all impulse-reaction type.*

7.10 COMBINATION TURBINES

Due to combination of two different types of turbines certain advantages are possible. Such a turbine is called combination turbine. Most common combinations are as under:

(a) Velocity-compounded impulse wheel followed by several single-row impulse wheel termed Curtis and Rateau type.

(b) Velocity-compounded impulse wheel followed by several stages of impulse-reaction blading, termed as Curtis-Parsons types.

(c) One or more single-row impulse stages followed by several stages of impulse-reaction bladings, termed as Rataeu-Parsons type.

In practice, the larger output steam turbine which employs impulse-reaction turbine uses first or second stage impulse turbine to reduce the number of stages as expansion ratio in impulse turbine is more.

7.10.1 Overview of Turbine

Figure 7.18 shows a 3600 rpm tandem compound steam turbine with one high pressure (HP), one intermediate-pressure (IP), and two low-pressure (LP) sections.

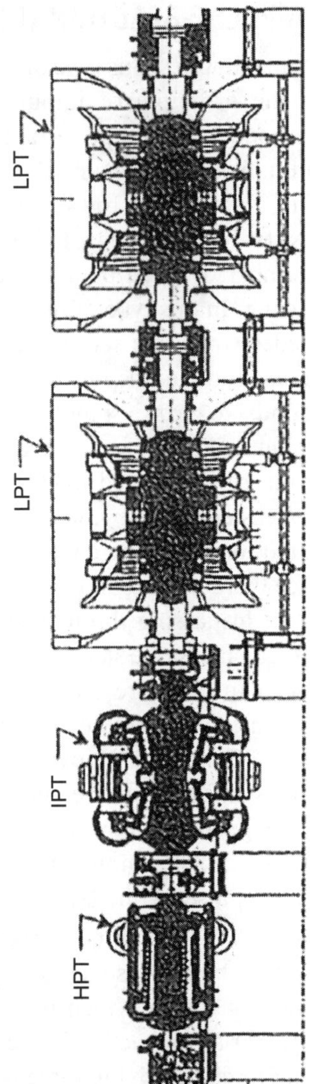

Figure 7.18 A tandem compound steam turbine, one HP turbine, one IP and two LP sections

7.11 COMPARISON BETWEEN IMPULSE AND REACTION TURBINES

Impulse Turbine	*Reaction Turbine*
(i) Pressure drops only in nozzles and not in moving blade channels	(i) Pressure drops in fixed blades (nozzles) as well as moving blade channels.
(ii) Constant blade channels area.	(ii) Varying blade channels area (converging types).

Contd.

Contd.

(iii) Profile type blades	(iii) Aerofoil type blades
(iv) Not all round or complete admission of steam	(iv) All round or complete admission
(v) Diaphragm contains nozzles	(v) Fixed blades similar to moving blades attached to casing serve as nozzles and guide the steam
(vi) Not much power can be developed	(vi) Much power can be developed
(vii) Occupies less space for same power	(vii) Occupies more space for same power
(viii) Lesser blading efficiency	(viii) Higher blading efficiency
(ix) Suitable for small power requirements	(ix) Suitable for medium and higher power requirements
(x) Blade manufacturing is not difficult and thus not costly	(x) Blade manufacturing process is difficult compared to impulse and hence costly
(xi) Velocity of steam is slightly higher	(xi) Reduced velocity of steam in stages

7.12 STEAM FLOW THROUGH NOZZLES

A nozzle is a passage of smoothly varying cross section by means of which the pressure energy of working fluid is converted into kinetic energy. The chief use of nozzles is to produce a jet of working fluid of high velocity for the purpose of driving turbines. There are three types of nozzles.

- convergent, • divergent, • convergent-divergent

In steam turbine, generally convergent nozzles are used.

Consider the steady flow of 1 kg of steam through a nozzle (Fig. 7.19a) and its h-s representation (Fig. 7.19b). Applying SFEE equation, we get

$$h_0 + \frac{C_0^2}{2 \times 10^3} + \frac{gz_0}{10^3} + q = h_1 + \frac{C_1^2}{2 \times 10^3} + \frac{gz_1}{10^3} + w$$

If $q = 0$ and obviously $w = 0$ and neglecting changes in PE, we get

$$h_0 + \frac{C_0^2}{2 \times 10^3} = h_1 + \frac{C_1^2}{2 \times 10^3}$$

or
$$\frac{C_1^2 - C_0^2}{2 \times 10^3} = (h_0 - h_1)$$

or
$$C_1 = \sqrt{2 \times 10^3 (h_0 - h_1) + C_0^2}$$

Neglecting C_0 in comparison to C_1, we get

$$C_1 = \sqrt{2 \times 10^3 (h_0 - h_1)} = 44.72 \sqrt{h_0 - h_1}$$

If there is a friction, then the actual velocity (Fig. 7.19)

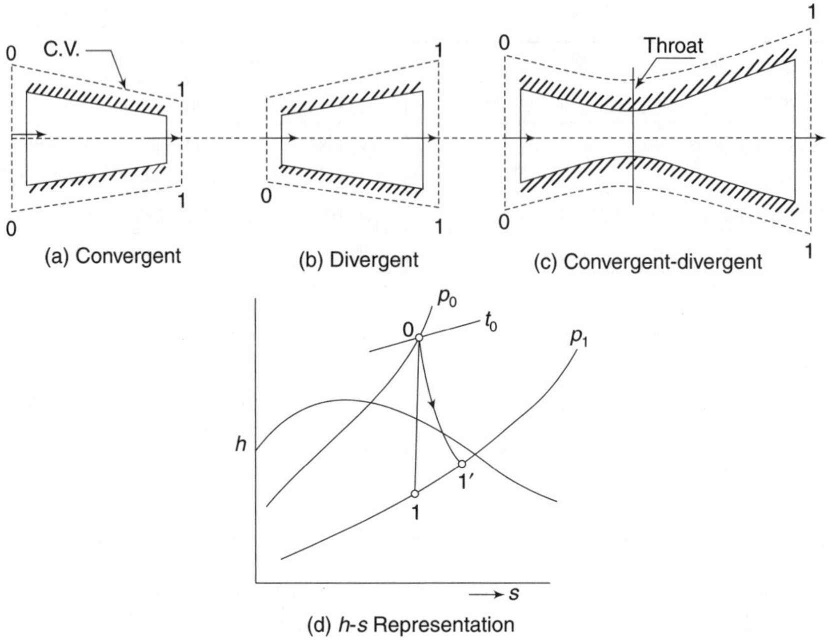

Figure 7.19 Flow through steam nozzles

$$C_1' = \sqrt{2 \times 10^3 (h_0 - h_1')} = 44.72\sqrt{(h_0 - h_1')} = 44.72\sqrt{(h_0 - h_1)} \times \eta_n$$

where η_n = nozzle efficiency = $(h_0 - h_1')/(h_0 - h_1)$

The smallest section of nozzle is known as **throat**, where the discharge is the maximum and the velocity is **sonic**.

Critical pressure ratio at throat = $r_c = p_c/p_0 = (2/n + 1)^{n/n - 1}$

where n = 1.3 for superheated steam. If the back pressure is critical pressure, it is called **choked flow** otherwise unchoked. A nozzle is designed for **choked flow**.

7.13 VELOCITY DIAGRAMS FOR IMPULSE TURBINES

It is a established fact that velocity is a vector quantity. It has (a) *magnitude* (b) *direction* and (c) *sense of direction*. Thus, we can represent velocity by a straight line and indicate (a) its magnitude by the length of that straight line to a suitable scale, (b) its direction by the direction of the said straight line with reference to some fixed direction, and (c) its sense of direction by an arrow placed on the straight line. *A row of nozzles and a row of moving blades constitutes a stage of turbine.*

Suffixes 1 and 2 refer to inlet and outlet conditions of moving blades as shown in Fig. 7.20(a) and Fig. 7.20(b). Let, C_1 = absolute velocity of steam at inlet to the moving blade, i.e., exit velocity of nozzle.

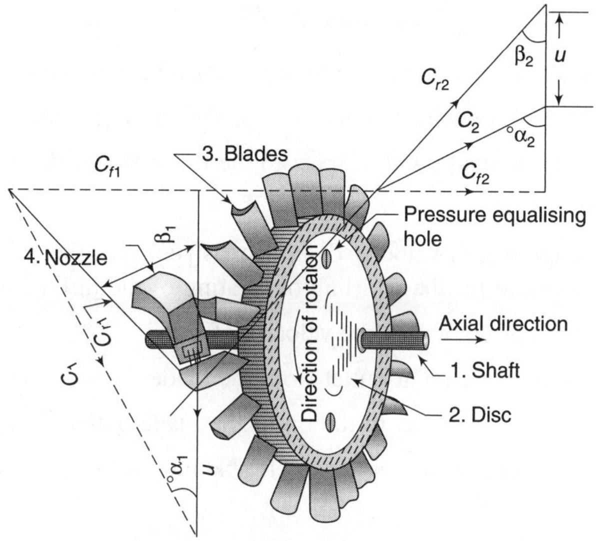

Figure 7.20 (a) Rotor and approaching flow

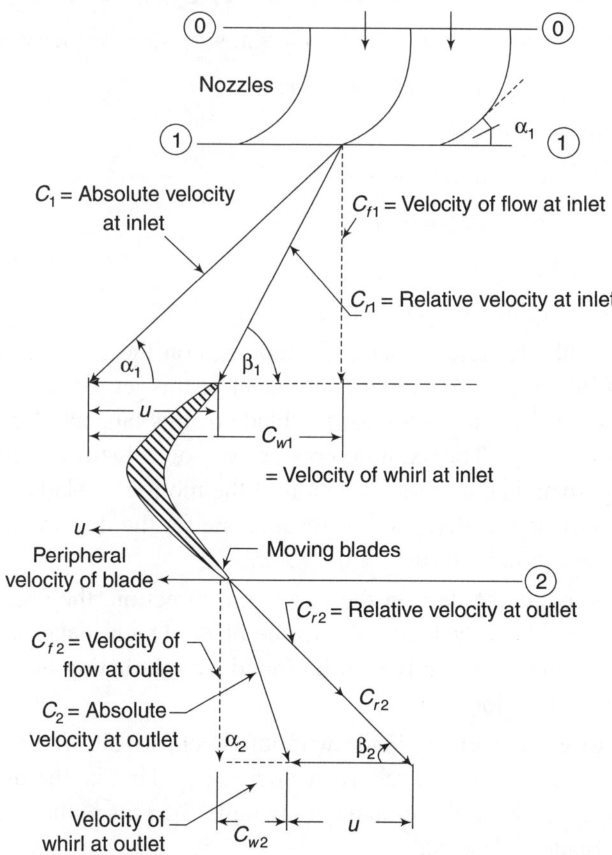

Figure 7.20 (b) Velocity diagram for impulse turbine

$= \sqrt{2 \times 10^3 (h_0 - h_1) + C_0^2}$; m/s (from SFEE)

α_1 = angle of the nozzle or angle made by the entering steam having velocity C_1 with tangent of the wheel at entrance of the moving blade.

C_{r1} = relative velocity of steam with respect to the tip of the blade at inlet.

β_1 = angle of blade at inlet.

C_{w1} = tangential component of velocity C_1 known as *velocity of whirl* or the component of velocity responsible for the whirling or rotating of the turbine rotor.

C_{f1} = axial component of C_1, called the velocity of flow.

C_2 = absolute velocity of steam at outlet from the blade.

C_{r2} = relative velocity with respect to the tip of the blade outlet.

α_2 = angle to the tangent of wheel at which absolute velocity of steam C_2 leaves the blades.

β_2 = outlet angle of the blade.

C_{f2} = outlet axial velocity or the axial component of C_2 or the velocity of flow at outlet.

C_{w2} = tangential component of the velocity C_2, known as the velocity of whirl at outlet.

u = peripheral velocity of blade (mean value)

K = blade velocity coefficient = C_{r2}/C_{r1}.

$C_{r1} = C_{r2}$ for the case of without friction.

$KC_{r1} = C_{r2}$ for the case of with friction.

$\dot{m}$ = mass flow rate kg/s.

All the velocities are in metres per second.

The jet of steam with absolute velocity C_1 impinges on the moving blade at an angle of α_1 to the tangent of blade. The tangential component of this jet works on the blade, because it is in the same direction as the motion of the blade u; this tangential component is known as the velocity of whirl, C_{w1}. The axial component C_{f1} of velocity C_1 does no work on the blade because it is perpendicular to the direction of the motion of blade. This component C_{f1} is known as the *velocity of flow* because it is responsible for the flow through the turbine. The velocity of flow causes an axial thrust on the rotor.

Due to the motion of the blade u in the peripheral direction, the entering steam jet with velocity C_1 will have a relative velocity, C_{r1} to the blade which makes an angle of β_1 to the tangent of blade. This relative velocity may be found by subtracting the vectors of the steam velocity C_1 and the blade velocity, u.

As the steam is to glide over the blade without shock, it follows that the surface of the blade at inlet must be parallel to the relative velocity, C_{r1}. That is, the moving blade at inlet must be inclined to the tangent of the blade at an angle β_1. This is the principle of shockless entry on which the blade it designed.

A similar vector diagram is shown at the outlet tip of the moving blade. The steam glides off the blade and comes out with a relative velocity, C_{r2} inclined at angle β_2 to the tangent. Addition of vector of blade velocity u and relative velocity, C_{r2} gives the absolute velocity of the leaving steam, C_2 inclined at angle α_2 to the tangent. After obtaining the vector C_2, its tangential component, C_{w2} and axial components, C_{f2} may be drawn.

The effect of angle of divergence of the nozzles on the fluid friction is more pronounced than the angle of convergence. If the angle of divergence of a converging-divergent nozzle is made too large, there is a chance that fluid may break away from the walls resulting in increased friction losses. **The included angle of the divergent portion is usually kept below about 12°, (generally 6 to 12° is kept)**. From this practice, it is concluded that for a given pressure ratio across a convergent-divergent nozzle, the divergent portion must be long compared to the convergent portion. The length of the divergent portion of the nozzle is determined by the equation (for circular cross section)

$$l = \frac{D_{\text{exit}} - D_{\text{throat}}}{2\tan(\theta/2)}$$

where θ = angle of divergence, 6 to 12°.

Since the divergent portion of the nozzle is comparatively longer and diverging flow is more susceptible to losses and the velocities in this portion are higher, it follows that bulk of the friction losses occurs in the divergent portion.

The angle of nozzle position made by the centreline of the nozzle passage at exit with the plane of rotation of blading is determined mainly as compromise between conflicting conditions. Very small angle of α_1 may lead theoretically to higher efficiency since the blading work is larger and exit velocity smaller. However, the small exit angle means a large angle of flow deflection in the nozzle with increase in friction losses. If α_1 is made large then the loss is in the blading becomes large. In practice, α_1 is chosen between 10 to 30°, 12 to 20° being more common.

7.14 COMBINATION OF VECTOR DIAGRAMS

The industrial method of constructing velocity diagrams is to combine both velocity diagrams at inlet as well as at outlet of the blade into a single diagram as shown in Fig. 7.21(a). In this, both velocity diagrams are drawn on a common base representing the blade velocity, u. When there is no friction ($C_{r1} = C_{r2}$) and $\beta_1 = \beta_2$, then $C_{f1} = C_{f2}$.

7.15 EFFECT OF BLADE FRICTION ON VELOCITY DIAGRAM

In practice, there is frictional resistance to the flow of steam jet over the blade, the effect of which is to cause a slowing down of the relatively velocity, C_{r2} which is in the range of 85 to 90 per cent of C_{r1}. Thus, due to frictional resistance [Fig. 7.21(b)].

$$C_{r2} = K \cdot C_{r1}$$

where K is called the blade velocity coefficient which takes into account the blade losses due to friction.

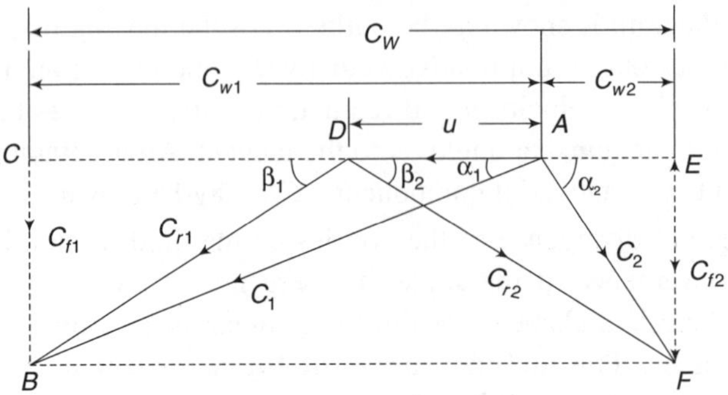

(a) Combined velocity diagram

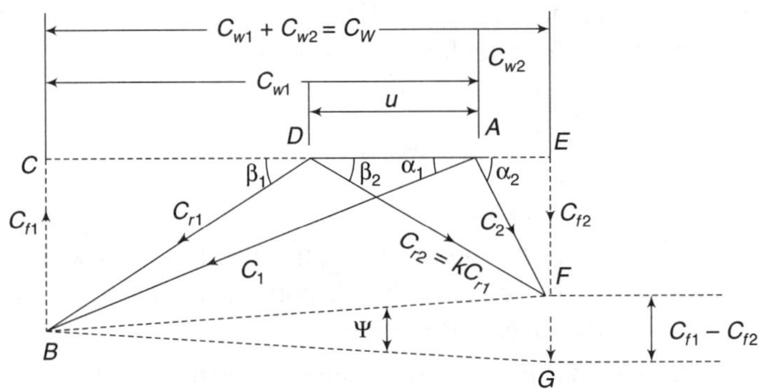

(b) Velocity diagram showing effect of friction

The resultant force parallel to BF is inclined at an angle Ψ to the plane in which the wheel rotates. The resultant force may be resolved into two components as shown in Fig. 7.21(c).

(1) A useful thrust or tangential force acting in the direction of blade motion, F_t.

(2) An idle thrust or axial force; perpendicular to the wheel plane, i.e., parallel to the turbine axis, F_a.

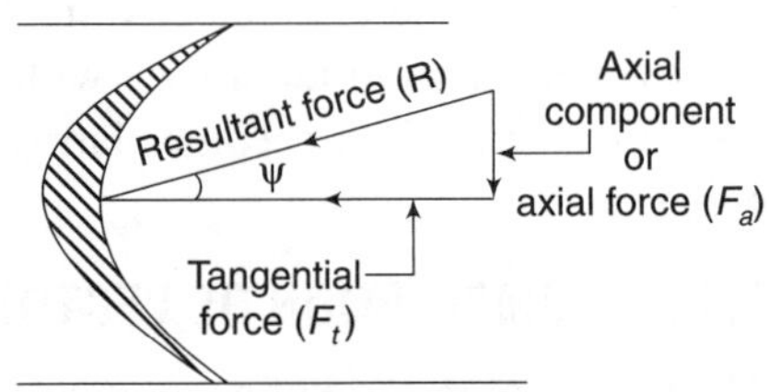

(c) Force diagram of impulse bladings

Figure 7.21

7.16 FORCES ON THE BLADES AND WORK DONE BY BLADES

7.16.1 Force

From Newton's second law of motion, tangential force on the blade,

$$F_t = \text{Mass} \times \text{Acceleration in the tangential direction}$$

$$= \text{Mass per second} \times \text{Change in velocity in tangential direction}$$

or

$$F_t = \dot{m}\,(C_{w1} - C_{w2})$$

where $\dot{m}$ = Mass flow rate of steam in kg per sec.

It's obvious from the velocity diagram that C_{w2} is opposite in direction to the blade motion. By giving the same direction of C_{w2}, the force becomes

$$F_t = \dot{m}\,(C_{w1} + C_{w2});\quad N \qquad ...(7.1)$$

or $$F_t = \dot{m}\,C_w \quad \text{where} \quad C_w = C_{w1} + C_{w2}$$

Thus, in general way it can be written as $F_t = \dot{m}\,(C_{w1} \pm C_{w2})$

An easier way to get C_{w1} and C_{w2} is to draw perpendiculars from B and F on the horizontal line, such that $CA = C_{w1}$ and $AE = C_{w2}$, then

$$F_t = \dot{m}\,(CA + AE)$$

or $$F_t = \dot{m}\,(CE) \qquad ...(7.2)$$

where CE is measured from the diagram and multiplied by the scale of the diagram.

7.16.2 Work and Power

Work done by blade per second is given by

w = Tangential force × Distance moved in unit time in the direction of force

= Tangential force × Blade velocity in the direction of force

$$= F_t \cdot u = \dot{m}\,(C_{w1} + C_{w2})\,u;\quad W \qquad ...(7.3)$$

Power developed is given by

$$\text{Power, } P = \frac{\dot{m}\cdot(C_{w1} + C_{w2})\cdot u}{1000} = \frac{\dot{m}\cdot C_w \cdot u}{1000};\quad \text{kW} \qquad ...(7.4)$$

or $$\text{Power, } P = \dot{m}\cdot\frac{(CE)\,(AD)}{1000},\quad \text{kW}$$

Note that this power is known as ***diagram* or *the rim power* (r.p.)** to distinguish it from the ***shaft power***.

It is to be noted that power developed given by Eq. (7.4) is by one stage. If n is the number of identical stages then the power developed is given by

$$\text{Power developed, } P = \frac{n\,\dot{m}u\,C_w}{1000},\quad \text{kW} \qquad ...(7.5)$$

The number of stages, n for a constant stage expansion pressure ratio, $(r_p)_{\text{stage}}$ is given below.

$$(r_p)_{\text{stage}} = \frac{p_1}{p_2} = \frac{p_2}{p_3} = \ldots\ldots = \frac{p_1}{p_{i+1}} = \frac{p_n}{p_{n+1}}$$

The overall expansion pressure ratio (r_p) of the entire machine,

$$(r_p) = (r_p)^n_{\text{stage}} = \frac{p_1}{p_{n+1}} = \frac{p_1}{p_{\text{cond}}}$$

Taking logarithm, the number of stages

$$n = \frac{\ln r_p}{\ln (r_p)_{stage}} = \frac{\ln(p_1/p_{cond})}{\ln(r_p)_{stage}} \quad ...(7.6)$$

Generally, the expansion ratio per stage varies from 1.5 to 2.5 in impulse turbines.

The number of blade channels varies from 50 to 150 depending upon the pitch selected and rotor diameter.

7.17 BLADE OR DIAGRAM EFFICIENCY

Blade or diagram efficiency is defined as the ratio of the work done by blade per second to the energy entering the blade per second.

Energy of the steam entering the blade per sec. $\frac{\dot{m} C_1^2}{2}$, W ...(7.7)

Therefore, the blade efficiency

$$\eta_b = \frac{\dot{m} u\, C_w}{\dot{m}\, C_1^2/2} = \frac{2u\, C_w}{C_1^2} \quad ...(7.8)$$

7.18 AXIAL THRUST OR END THRUST ON ROTOR

Axial force, F_a = Mass × Acceleration in axial direction

or F_a = Mass per second × Change in velocity in axial direction

∴ Axial force, $F_a = \dot{m}\,(C_{f1} - C_{f2})$; N ...(7.9)

Due to the reduced axial velocity (C_{f2}) at the outlet from blade of an impulse turbine, the force producing acceleration of the steam does not act wholly in the plain of the wheel, hence the force component ($\dot{m}$) ($C_{f1} - C_{f2}$) tends to push the shaft axially (Fig. 7.21(c). The axial thrust on the wheel must be taken by a thrust bearing. When there is no blade friction, i.e., $C_{r1} = C_{r2}$ and if $\beta_1 = \beta_2$, then $C_{f1} = C_{f2}$, hence axial thrust is zero. *It should be noted that this axial thrust is also augmented by the steam pressure acting on each side of the wheel or disc.*

7.19 GROSS STAGE EFFICIENCY

Gross stage efficiency (η_{gs}) is defined as the ratio of the work done by the blades per kg of steam flowing though the stage (i.e., nozzles and moving blades or a set of fixed and moving blades constitute a stage) to the isentropic enthalpy drop per kg of steam in that stage. Thus,

$$\eta_{gs} = \frac{\text{Work done per kg in the stage}}{\text{Isentropic enthalpy drop in the stage}}$$

Let, η_{gs} = Gross stage efficiency

Δh_{isen} = Isentropic enthalpy drop per kg in nozzles only in the impulse turbines

Now, work done on the blades per kg of steam flow is $u \cdot C_w$, J

$$\therefore \qquad \eta_{gs} = u \cdot C_w / \Delta h_{\text{isen}} = \left(\frac{2u \cdot C_w}{C_1^2}\right)\left(\frac{C_1^2}{2\,\Delta h_{\text{isen}}}\right)$$

or $$\eta_{gs} = \eta_b \times \eta_n \qquad \text{...(7.10)}$$

Thus, the gross stage efficiency is the product of blade efficiency and the nozzle efficiency. For any given nozzle angle, therefore, since the nozzle efficiency (η_n) would remain constant, the gross stage efficiency will be maximum when η_b is the maximum. It will be noticed that the stage efficiency is the same as the blade efficiency except when there is loss in the nozzles.

7.20 ENERGY CONVERTED TO HEAT BY BLADE FRICTION

$$Q_{\text{friction}} = \dot{m}\left(\frac{C_{r1}^2 - C_{r2}^2}{2}\right), \qquad W \qquad \text{...(7.11)}$$

When there is no friction, $C_{r1} = C_{r2}$, so Q_{friction} is zero.

Example 7.1 Steam issues from the nozzle of deLaval turbine with a velocity of 1200 m/s. The nozzle angle is 20°, the mean blade velocity is 400 m/s, and the inlet and outlet angles of blade are equal. The mass of steam flowing through turbine per hour is 900 kg. Calculate: (a) The blade angles, (b) The relative velocity of steam entering the blades, (c) The tangential force on the blades, (d) The power developed, and (e) The blade efficiency. Assume that $K = 0.8$.

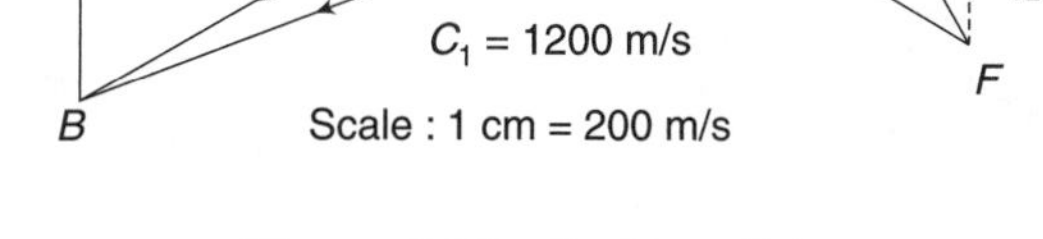

Figure 7.22 Problem 7.1

Solution Refer to Fig. 7.22.

deLaval turbine is a single stage impulse turbine. The velocity diagram is shown in Fig. 7.22. Scale used is 1 cm = 200 m/s. Given: C_1 = 1200 m/s, α_1 = 20°, u = 400 m/s, $\beta_1 = \beta_2$ and $\dot{m}$ = 900 kg/h. = 900/3600 kg/s. With this data the inlet velocity triangle ABD is drawn and the blade angle at inlet β_1 is measured which gives β_1 = 30°.

Since the blades are symmetrical so

(a) $\beta_1 = \beta_2 = 30°$ ***Ans.***

(b) C_{r1} = Relative velocity at inlet = 4.2 × 200 = 840 m/s ***Ans.***

The relative velocity at outlet = $C_{r2} = DF = K\,C_{r1} = 0.8 \times 840 = 672$ m/s

and $\beta_2 = 30°$, so the outlet triangle *ADF* can be drawn.

∴ $C_w = CE = 1320$ m/s

(c) Tangential force on the blades $= \dot{m}\,(C_{w1} + C_{w2}) = \dot{m}C_w = m\,(CE)$

$$= \frac{900}{3600} \times 1320 = 330 \text{ N}$$ ***Ans.***

(d) Power developed $= \dot{m} \cdot \dfrac{u \cdot C_w}{1000} = \dfrac{900}{3600} \times \dfrac{400 \times 1320}{1000} = 132$ kW ***Ans.***

(e) The blade efficiency $= \eta_b = \dfrac{2 \cdot u \cdot C_w}{C_1^2} = \dfrac{2 \times 400 \times 1320}{1200^2} = 0.733$ ***Ans.***

Note that this problem may also be solved analytically.

7.21 INFLUENCE OF RATIO OF BLADE VELOCITY TO STEAM VELOCITY ON BLADE EFFICIENCY IN A SINGLE STAGE IMPULSE TURBINE

Let us consider the case of single stage impulse turbine. Here the effect of friction in the blade channel is taken into account and accordingly the velocity diagram is shown in Fig. 7.23. Due to friction C_{r2} is less than C_{r1} and $C_{r2}/C_{r1} = K$ where *K* is blade velocity coefficient, which is always less than unity. Generally, in practice the angles β_1 and β_2 are equal, in some cases β_2 is somewhat smaller than β_1 but in pure impulse turbines the difference between the angles is always small.

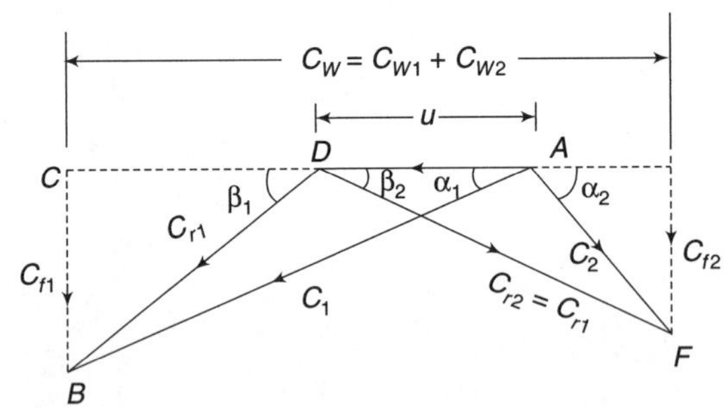

Figure 7.23 Velocity diagram for single stage impulse turbine

We know that the blade efficiency,

$\eta_b = \dfrac{2u\,(C_{r1} + C_{w2})}{C_1^2} = \dfrac{2u \cdot C_w}{C_1^2}$. From the velocity diagram find the value of $u \cdot C_w$

$$C_w = CD + DE = C_{r1}\cos\beta_1 + C_{r2}\cos\beta_2 = C_{r1}\cos\beta_1\left(1 + \frac{C_{r2}\cos\beta_2}{C_{r1}\cos\beta_1}\right)$$

$$= C_{r1}\cos\beta_1\,(1 + KB) \qquad \text{where} \qquad B = \frac{\cos\beta_2}{\cos\beta_1}$$

Since β_1 and β_2 are usually almost equal, so for this analysis we assume that B is a constant. We have,

$$C_{r1}\cos\beta_1 = C_1\cos\alpha_1 - u \qquad \text{So} \qquad C_w = (C_1\cos\alpha_1 - u)\,(1 + KB)$$

Therefore, $2u\,.\,C_w = 2u\,(C_1\cos\alpha_1 - u)\,(1 + KB)$

The blade velocity ratio, $\rho = u/C_1$ is an important factor in the design of steam turbines. The efficiency of the turbine depends upon this factor.

Now $\quad u = \rho C_1 \quad$ and $\quad 2u\,.\,C_w = 2\rho\,.\,C_1\,(C_1\cos\alpha_1 - \rho C_1)\,(1 + KB)$

$$2u\,.\,C_w = 2C_1^2\,(1 + KB)\,(\rho\cos a_1 - \rho^2)$$

Putting the velocity of $2uC_w$ in Eq. (7.8), we have,

$$\eta_b = \frac{2u\,.\,C_w}{C_1^2} = 2(1 + KB)\,(\rho\cos\alpha_1 - \rho^2) \qquad ...(7.12)$$

Equation (7.12) is the equation of the parabola which is shown in Fig. 7.24 (a) and (b).

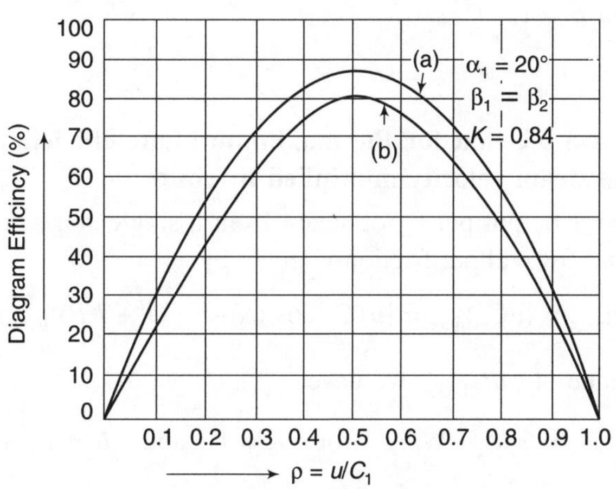

Figure 7.24 Diagram efficiency (a) without losses; (b) with losses

From the graph, it is obvious that the blade efficiency increases as ρ is increased until ρ reaches a value which is approximately one-half, after which blade efficiency decreases.

If α_1, K and B are assumed to be constant; then the blade efficiency depends on the value of ρ. For maximum blade efficiency, the first differential with respect to ρ should be zero and the second differential with respect to ρ should be negative. In this first differential

$$\frac{d\eta_b}{d\rho} = 0 = 2(1 + KB)(\cos\alpha_1 - 2\rho)$$

and in the second differential $\dfrac{d^2\eta_b}{d\rho^2} = -4(1 + KB) = -\text{ ve}$

So for the maximum blade efficiency, $2(1 + KB)(\cos\alpha_1 - 2\rho) = 0$

or $$(\cos\alpha_1 - 2\rho) = 0$$

Therefore, optimum value of ρ is

$$\rho_{\text{opt}} = \left(\frac{u}{C_1}\right)_{\text{opt}} = \frac{\cos\alpha_1}{2} \qquad ...(7.13)$$

The maximum value of the blade efficiency will be given by putting the value of ρ_{opt} for ρ in the Eq. 7.12. Thus, we have

$$(\eta_b)_{\max} = 2(1 + KB)\left(\frac{\cos\alpha_1 . \cos\alpha_1}{2} - \frac{\cos^2\alpha_1}{4}\right) = (1 + KB)\frac{\cos^2\alpha_1}{2} \qquad ...(7.14)$$

If $K = 1$ and $B = 1$,

$$(\eta_b)_{\max} = \cos^2\alpha_1 \qquad ...(7.15)$$

From Eq. (7.15), we see that for the maximum blade efficiency, the blade velocity should be half of the steam velocity multiplied by cos α_1.

The maximum power output per kg of steam from a single stage impulse turbine having equiangular blades and for without friction is given by

$$(\text{Power})_{\max} = (uC_w)_{\max} = [u(C_1\cos\alpha_1 - u)(1 + KB)]_{max}\ ; \quad \text{W/kg} \qquad ...(7.16)$$

Substituting the value of $(u/C_1)_{\text{opt}}$, we have

$$(\text{Power})_{\max} = 2\,u^2, \quad \text{W/kg} \quad \text{for} \quad K = 1 \quad \text{and} \quad B = 1 \qquad ...(7.17)$$

Example 7.2 The mean diameter of the blades of an impulse turbine with a single row wheel is 105 cm and the speed is 3000 r.p.m. The nozzle angle is 18°, the ratio of blade velocity to steam velocity is 0.42. and the ratio of the relative velocity at outlet from the blades to that at inlet is 0.84. The outlet angle of the blade is to be made 3° less than the inlet angle. The steam flow is 8 kg/s. Draw the velocity diagram for the blades and calculate the following. (a) Resultant thrust on the blades, (b) Tangential thrust on the blades, (c) Axial thrust on the blades. (d) Power developed by the blades, and (e) Blading efficiency.

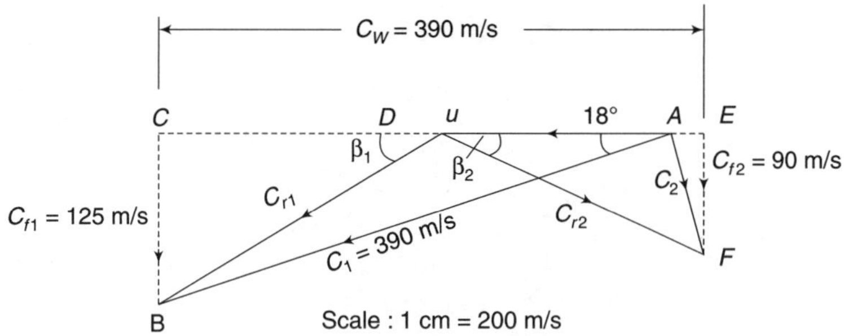

Figure 7.25 Example 7.2

Solution Refer to Fig. 7.25. Given: $D = 105$ cm, $\rho = u/C_1 = 0.42$, $\dot{m} = 8$ kg/s, $n = 3000$ rpm. $K = 0.48$, $\alpha_1 = 18°$, $\beta_2 = \beta_1 - 3°$

$$u = \text{the mean speed of blades} = \frac{\pi DN}{60} = \frac{\pi \times 105 \times 3000}{100 \times 60} = 164.93 \text{ m/s}$$

$$\rho = 0.42 = \frac{u}{C_1} \quad \text{or} \quad C_1 = \frac{164.93}{0.42} = 391.69 \text{ m/s}$$

(a) Tangential thrust on the blades, $F_t = \dot{m}C_w = 8 \times 390 = 3120$ N **Ans.**

(b) Axial thrust, $\dot{m}(C_{f1} - C_{f2}) = 8(125 - 90) = 280$ N **Ans.**

(c) Power developed, $\dfrac{m.u.C_w}{1000} = \dfrac{8 \times 164.93 \times 390}{1000} = 514.58$ kW **Ans.**

(d) Blade efficiency, $\dfrac{2u \cdot C_w}{C_1^2} = \dfrac{2 \times 164.93}{391.69^2} = 83.88\%$ **Ans.**

(e) Resultant thrust, $m(BF) = 8 \times 393.25 = 3280$ N **Ans.**

Resultant thrust may be also found out as follows

Resultant thrust, $\sqrt{F_t^2 + F_a^2} = \sqrt{(3133.52^2 + 280^2)} = 3280$ N **Ans.**

Example 7.3 The isentropic heat drop in given stage of a multi-stage impulse turbine is 33.5 kJ/kg of steam. The nozzle outlet angle is 20°. The efficiency of the nozzles, defined as the ratio of the actual gain in kinetic energy in the nozzle to the adiabatic (isentropic) heat drop is 92 per cent. The mean diameter of the blades is 95.5 cm and the revolutions per minute 3000. The carry-over factor is 0.88. Blades are equiangular with a velocity coefficient of 0.87. Calculate steam velocity at the outlet of the nozzle, blade angles and gross stage efficiency.

Solution Refer to Fig. 7.26. Given: $\Delta h_{\text{isen}} = 33.5$ kJ/kg $D = 95.5$ cm, $\alpha_1 = 20°$, $\phi = 0.88$, $\eta_n = 0.92$, $K = 0.87$, $\beta_1 = \beta_2$

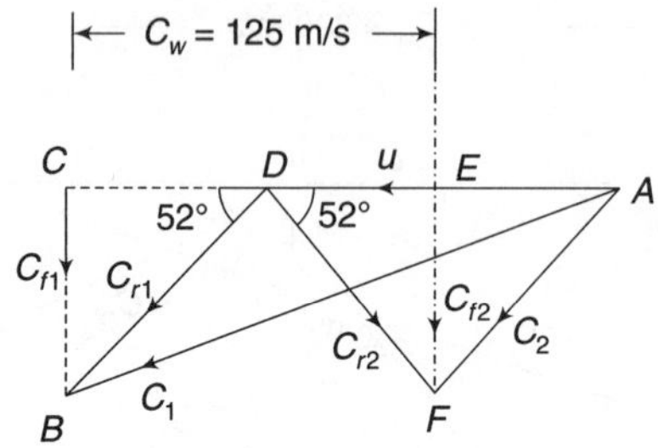

Figure 7.26 Example 7.3

The peripheral speed, $u = \dfrac{\pi DN}{60} = \dfrac{\pi \times 95.5 \times 3000}{100 \times 60} = 153$ m/s

We know that the nozzles efficiency in a stage of multi-stage turbine is given by

$$\eta_n = \frac{C_1^2 - \phi C_2^2}{2\,\Delta h_{\text{isen}}} \quad \text{where } \phi = \text{carry-over coefficient}$$

From the velocity diagram

$$C_2^2 = C_{r2}^2 + u^2 - 2u\,.\,C_{r2}\cos\beta_2 = K^2C_{r1}^2 + u^2 - 2u\,KC_{r1}\cos\beta_2$$
$$= K^2(C_1^2 - u^2 - 2u\,C_{r1}\cos\beta_1) + u^2 - 2K\,u\,.\,C_{r1}\cos\beta_2$$
$$= K^2\,C_1^2 + u^2\,(1 - K^2) - 2u\,C_{r1}\cos\beta_1\,(K^2 + K\cos\beta_2/\cos\beta_1)$$

Since $\beta_1 = \beta_2$ and $C_{r1}\cos\beta_1 = C_1\cos\alpha_1 - u$

$\therefore$ $$C_2^2 = K^2C_1^2 + u^2\,(1 - K)^2 - 2u\,(C_1\cos\alpha_1 - u)\,(K^2 + K)$$

Substituting the values of K, u and α_1, we have

$$C_2^2 = (0.87)^2\,C_1^2 + (150)^2\,(1 - 0.87)^2 - 2 \times 150\,(C_1\cos 20° - 150) \times (0.87^2 + 0.87)$$
$$= 0.757\,C_1^2 - 458C_1 + 78700$$

Now putting the value of C_2^2 in the expression of nozzle efficiency

$$0.92 = \frac{C_1^2 - 0.88\,[0.757\,C_1^2 - 458\,C_1 + 78700]}{2 \times 1000 \times 33.5}$$

or $$C_1^2 + 1225\,C_1 - 33400 = 0$$

or $$C_1 = 232.5 \text{ m/s}$$ ***Ans.***

With the help of u, C_1 and α_1, the inlet velocity triangle *ABD* can be drawn.

This gives $\beta_1 = \beta_2 = 37°$ ***Ans.***

Now $\dfrac{C_{r2}}{C_{r1}} = 0.87$, the outlet velocity triangle can be drawn.

$$\text{The blade efficiency, } \eta_b = \frac{2u \cdot C_w}{C_1^2} = \frac{2 \times 150 \times 125}{232.5 \times 232.5} = 0.693$$

The gross stage efficiency = $\eta_{gs} = \eta_n \times \eta_b = 0.92 \times 0.693 = 63.75\%$ ***Ans.***

7.22 VELOCITY DIAGRAM FOR THREE-ROW VELOCITY-COMPOUNDED WHEEL

We know that in the velocity-compounded impulse turbine, the whole pressure drop up to the back pressure is in only one set of nozzles and the velocity drop is brought about in so many moving blade rows by mounting sets of moving blades and set of guide blades. The purpose of guide blades is to simply direct the steam for succeeding set of blades, though the flow of steam will be accompanied with some losses due to friction.

Figure 7.27 shows the velocity diagram for a three-row wheel, i.e., three sets of moving blades. The method employed in drawing the velocity diagram is nothing but the extension of the method used for a single-row wheel. Let prefixes 1, 2 and 3 indicates the first row, second row and third row of blades respectively.

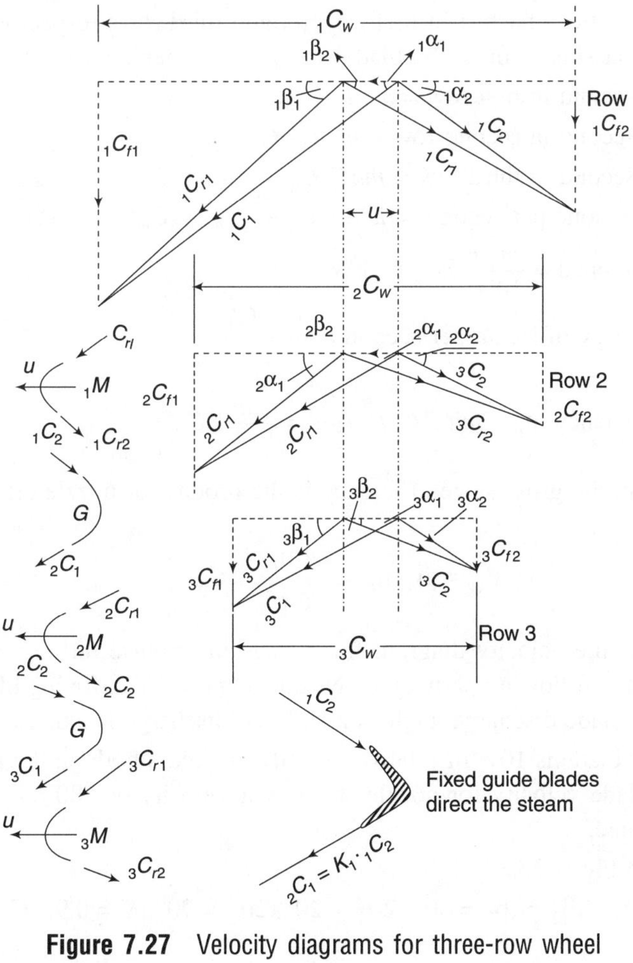

Figure 7.27 Velocity diagrams for three-row wheel

For the first row, let

$_1C_1$ = Absolute velocity of steam entering blades

$_1Cr_1$ = Relative velocity of steam entering blades

$_1Cr_2$ = Relative velocity of steam leaving blades

$_1C_2$ = Absolute velocity of steam leaving blades

$_1\alpha_1$ = Nozzle angle, $_1\beta_1$ = Blade inlet angle, $_1\beta_2$ = Blade outlet angle.

We see that $_1C_2$ = is the velocity with which the steam enters the first row of fixed blades and for the minimum shock loss the angle $_1\alpha_2$ is the inlet angle of the fixed blades.

For the second row of blades, $_2C_1$ is the absolute velocity of the steam issuing from the first row of fixed guide blades. Due to friction in these blades, $_2C_1$ will be less than $_1C_2$ and so it may be written as

$$_2C_1 = K.{}_1C_2$$

where K is the blade velocity coefficient. Similar will be the case for other rows. Let $_1C_w$; $_2C_w$ and $_3C_w$ be the velocities of whirl in first, second and third rows respectively and $\dot{m}$ be the mass of the steam passing through the blades per second, then

Work done per second in first row = $\dot{m}.u.{}_1C_w$

Work done per second in second row = $\dot{m}.u.{}_2C_w$

Work done per second in third row = $\dot{m}u.{}_3C_w$

So the total work done per second = $\dot{m}.u\,({}_1C_w + {}_2C_w + {}_3C_w) = \dot{m}.u\Sigma C_w$ ...(7.18)

Total power developed = $\dfrac{\dot{m}.u}{1000}\Sigma C_w$; kW ...(7.19)

Initial kinetic energy of steam per second = $\dfrac{\dot{m}.{}_1C_1^2}{2}$

So the blade efficiency, $\eta_b = \dot{m}u\,\Sigma C_w / \dfrac{\dot{m}.{}_1C_1^2}{2} = \dfrac{2u}{{}_1C_1^2}\Sigma C_w$ ...(7.20)

As we have seen, the gross stage efficiency is the product of nozzle efficiency and blade efficiency, so

$$\eta_{gs} = \eta_n.\eta_b = \frac{2u\,\eta_n}{{}_1C_1^2}.\Sigma C_w \quad ...(7.21)$$

Example 7.4 Draw the velocity diagram for a stage in impulse turbine with two rings of moving blades for the following particulars. Nozzle angle = 15°. Moving blade tip discharge angle = 30°. Fixed blade discharge angle = 20°. Final discharge is axial.

Friction loss per blade is 10% of relative velocity. It occurs both on the fixed and moving blades. Find the blade velocity for nozzle steam exit velocity of 500 m/s and the diagram efficiency of the stage.

Solution Refer to Fig. 7.28.

Given: $_1\alpha_1$ = 15°, $_1\beta_2 = {}_2\beta_2$ = 30°, $2\alpha_1$ = 20°, $2\alpha_2$ = 90°, K = 0.9, $_1C_1$ = 500 m/s.

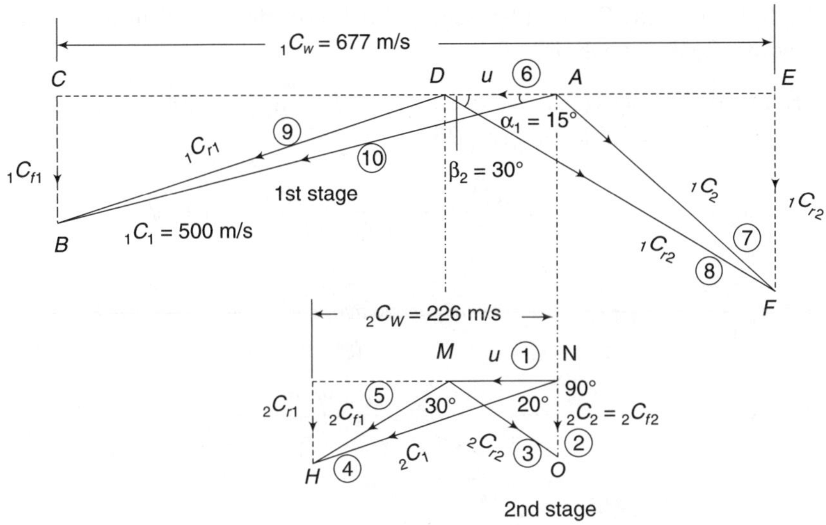

Figure 7.28 Problem 7.4

According to data given, the velocity diagram for the first stage cannot be drawn. It is only possible to draw the second stage velocity diagram. Let the scale of blade velocity $u = 2.5$ cm. Since the final discharge is axial so line NO can be drawn on MN. With angle $2\beta_2 = 30°$, $_2Cr_2$ can be drawn. We also have

$$\frac{_2C_{r2}}{_2C_{r1}} = 0.9, \frac{_1C_{r2}}{_1C_{r1}} = 0.9 \text{ and } \frac{_1C_2}{_2C_1} = 0.9$$

All steps followed are given in Fig. 7.28 in a serial order. In this way, $_1C_1$ gives the scale of the diagram. $_1C_1$ measures 11.6 cm. So the scale is

$$1 \text{ cm} = \frac{500}{11.6} = 43.1$$

Velocity of blade, $u = 43.1 \times 2.5 = 107$ m/s **Ans.**

$$\text{Blade efficiency} = \frac{2u(_1C_w + {}_2C_w)}{_1C_1^2} = \frac{2 \times 107.5 \times (677 + 266)}{500 \times 500} = 81\%$$ **Ans.**

Example. 7.5 A velocity-compounded impulse wheel has two rows of moving blades with a mean diameter of 70 cm. The speed of rotation is 3000 r.p.m., the nozzle angle is 16°, and the estimated steam velocity at the nozzle outlet is 610 m/s. The mass of steam passing through the blades per second is 6.5 kg.

Assuming that the energy loss in each row of blades (moving and fixed) is 24% of the kinetic energy of the steam entering the blades and referred to as the relative velocity, and that the outlet angles of the blades are: (1) first row of moving blades 18°, (2) intermediate guide blades 22°, and (3) second row of moving blades 38°. Draw the velocity diagrams and calculate the following: (a) Blade inlet angles, (b) Power developed in each row of blades, (c) Efficiency of wheel as a whole.

Solution Refer to 7.29. Given : ${}_1\alpha_1$ = 16°, D = 70 cm, N = 3000 rpm, ${}_1C_1$ = 610 m/s, ${}_1\beta_2$ = 18°, ${}_2\alpha_1$ = 22°, $\dot{m}$ = 5.5 kg/s, ${}_2\beta_2$ = 38°.

$$u = \frac{\pi DN}{60} = \frac{\pi \times 70 \times 3000}{100 \times 60} = 110 \text{ m/s}$$

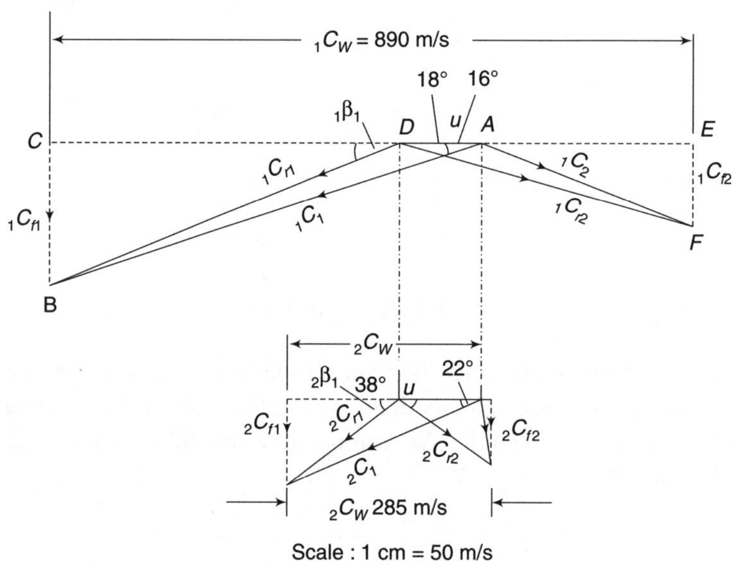

Figure 7.29 Problem 7.5

With these data, the inlet velocity triangle ABD for the first stage can be drawn.

K = blade velocity coefficient = C_{r2}/C_{r1}

Here $\eta_n = 0.76 = K^2 = \dfrac{{}_1C_{r2}^2}{{}_1C_{r1}^2}$

$\therefore$ $K = \sqrt{(0.76)} = 0.872$ So $0.872 = \dfrac{{}_1C_{r2}}{{}_1C_{r1}}$

$0.872 = \dfrac{{}_2C_{r2}}{{}_2C_{r1}} =$ and $0.872 = \dfrac{{}_2C_1}{{}_1C_2}$

After calculating required velocities, the velocity diagrams for the first and second stage are completed. These give

(a) ${}_1\beta_1$ = 20° and ${}_2\beta_1$ = 37° ***Ans.***

(b) Power developed in first row $= \dfrac{\dot{m} \cdot {}_1C_w \cdot u}{1000} = \dfrac{5.5 \times 890 \times 110}{1000}$ = 538.54 kW ***Ans.***

$$\text{Power developed in second row} = \frac{\dot{m}\,._2C_w\,.\,u}{1000} = \frac{5.5 \times 285 \times 110}{1000} = 172.42 \text{ kW} \quad \textbf{\textit{Ans.}}$$

$$\text{(c) Blade efficiency, } \eta_b = \frac{2u({}_1C_w + {}_2C_w)}{{}_1C_1^2} = \frac{2 \times 110\,(890 + 285)}{610 \times 610} = 69.5\% \quad \textbf{\textit{Ans.}}$$

7.23 BLADE HEIGHTS IN VELOCITY-COMPOUNDED AND PRESSURE-COMPOUNDED IMPULSE TURBINE

In designing the blades, blade height plays an important role. Assume $\dot{m}$ kg of steam per second flowing through two rows of moving blades of velocity-compounded impulse turbine. Let l_n, l_1, l_f and l_2 be the radial heights at the outlet side of the nozzle, first row of moving blades, fixed row and second row of moving blades respectively as shown in Fig. 7.30(a). For simplification, the specific volume (v) is assumed constant throughout the blade system, though, due to friction, losses in the blade channels, the steam will be reheated and the specific volume (v) will be slightly increased (the increase is small).

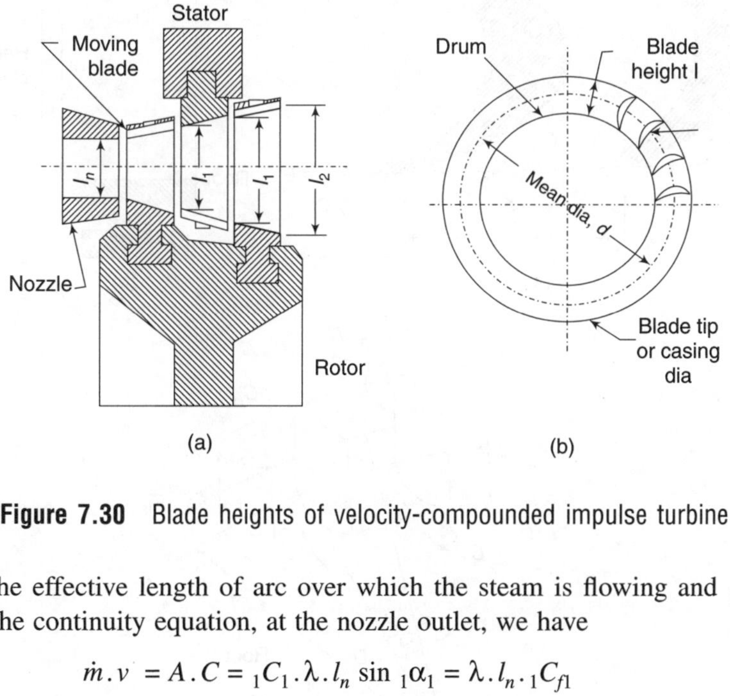

Figure 7.30 Blade heights of velocity-compounded impulse turbine

Let λ be the effective length of arc over which the steam is flowing and ${}_1\alpha_1$ = Nozzle angle. From the continuity equation, at the nozzle outlet, we have

$$\dot{m}\,.\,v = A\,.\,C = {}_1C_1\,.\,\lambda\,.\,l_n \sin {}_1\alpha_1 = \lambda\,.\,l_n\,.\,{}_1C_{f1} \quad \text{...(7.22)}$$

If l_n is the nozzle thickness, so the pitch of blades and for full admission, $\lambda = \pi D$, and the number of nozzles $(z_n) = \lambda / S_n = \pi D/S_n$ then from continuity equation

$$\dot{m}\,.\,v = {}_1C_1 \frac{\lambda l_n}{S_n} (S_n \sin {}_1\alpha_1 - t_n) \quad \text{...(7.23)}$$

Let, S_1 be the mean circumferential pitch of blades in first row, S_f = Mean circumferential pitch of blades in fixed row, S_2 = Mean circumferential pitch of blades in second row.

Let us assume that the arc covered by the steam as it passes through the blade system is constant. In this case, the number of the blade channel through which steam is flowing in the first row is λ/S_1 and total area through the blade channels at outlet is

$$A_1 = \frac{\lambda}{S_1}.l_1 \,(S_1 \sin {}_1\beta_2 - t_1) \quad ...(7.24)$$

where t_1 is the thickness of the outlet edge of the blade in the first row.

Therefore, continuity equation at the outlet of the first row of blade gives

$$\dot{m}.v = A.C = {}_1C_{r2} \frac{\lambda.l_1}{S_1} (S_1.\sin {}_1\beta_2 - t_1) \quad ...(7.\ 25)$$

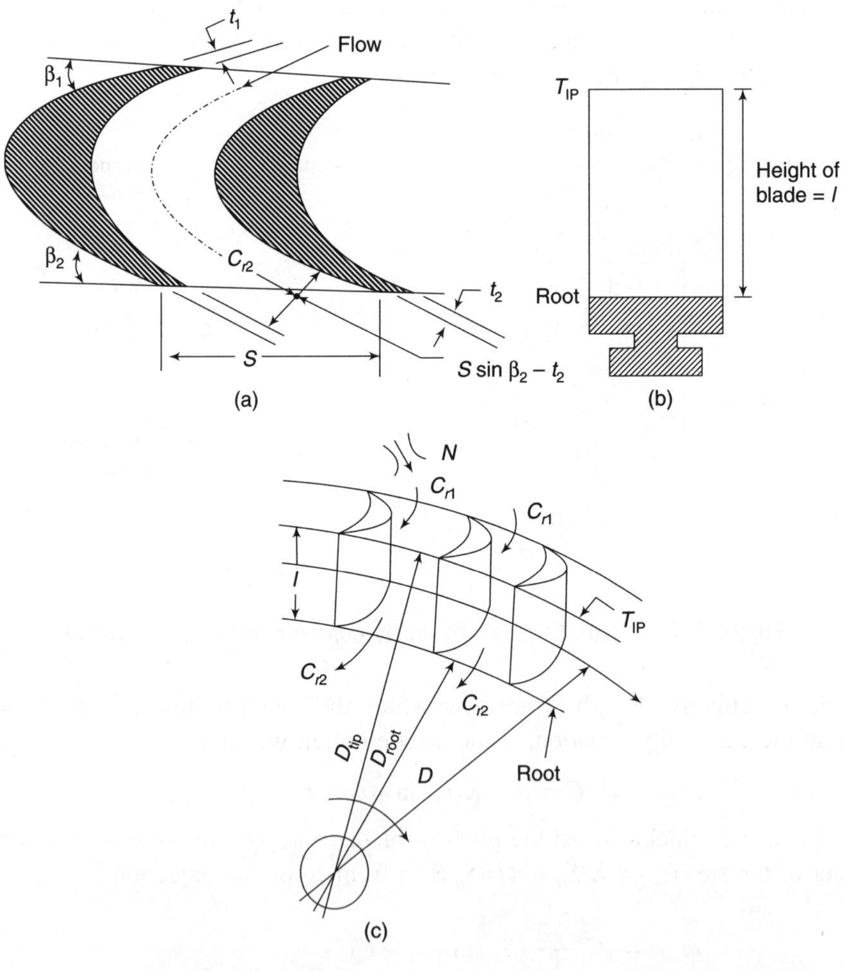

Figure 7.31 A general view of moving blade channel

From Eq. (7.26) it is clear that angle ${}_1\beta_2$ may be calculated for a given value of l_1 and if ${}_1\beta_2$ is given, then l_1 can be calculated.

In some cases, ${}_1\beta_2 = {}_1\beta_1$ but in majority of cases ${}_1\beta_2$ is less than ${}_1\beta_1$.

For the fixed blade it may be shown that

$$\dot{m}.v = A.C = {}_2C_1.\frac{\lambda}{S_1}.l_f\,(S_f.\sin {}_2\alpha_1 - t_f) \qquad ...(7.26)$$

where, t_f is the thickness of the outlet edge of fixed or guide blades.

Similarly, for the second row of moving blades, continuity equation gives (see Fig. 7.31).

$$\dot{m}.v = A.C = {}_1Cr_2.\frac{\lambda}{S_2}.l_2\,(S_2.\sin {}_2\beta_2 - t_2) \qquad ...(7.27)$$

where t_2 is the thickness of the outlet edge of the second row of moving blades. Eq. (7.28) is also applicable for *pressure compounded impulse turbine.*

Generally, the ratio of the blade height in the last moving row to the nozzle height, has the ranges—two-row wheels, l_2/l_n = 1.8 to 2.5 and three-row wheels, l_3/l_n = 2.5 to 3.5.

Thus, in general, the blade height of simple, velocity-compounded or pressure-compounded impulse turbine will be calculated at the exit condition by using the continuity equation.

$$\dot{m}v = A.C. = C_{r2} \times n.l(S.\sin\beta_2 - t) \qquad ...(7.28)$$

But $\quad n.S = \pi D$

and thus neglecting the blade thickness, in general (Fig. 7.30b)

$$\dot{m}v = \pi DlC_{r2}\sin\beta_2 = \pi Dl\,C_{f2} \qquad ...(7.29)$$

where D = Mean diameter (m), l = Blade height (m), C_{r2} = Relative velocity at outlet (m/s), $\dot{m}$ = Mass flow rate, (kg/s), v = Specific volume (m^3/kg).

All the above equations are also valid for pressure compounded.

7.24 ADVANTAGES OF VELOCITY-COMPOUNDED IMPULSE TURBINE

- Velocity-compounded impulse turbine requires a comparatively small number of stages due to relatively large heat drop per stage. *This is the reason that in all modern large impulse reaction turbines, the initial stage is two-row velocity stage or simple stage impulse turbine.*
- Due to small number of stages the initial cost is less.
- The maximum stage efficiency and the optimum value of ρ decreases as the number of rows of moving blades in a wheel increases as shown in Fig. 7.32.
- In a two- or three-row wheel, the steam temperature is sufficiently low, hence a cast iron cylinder may be used, thus saving in material cost. Generally, two-row wheel is more common.

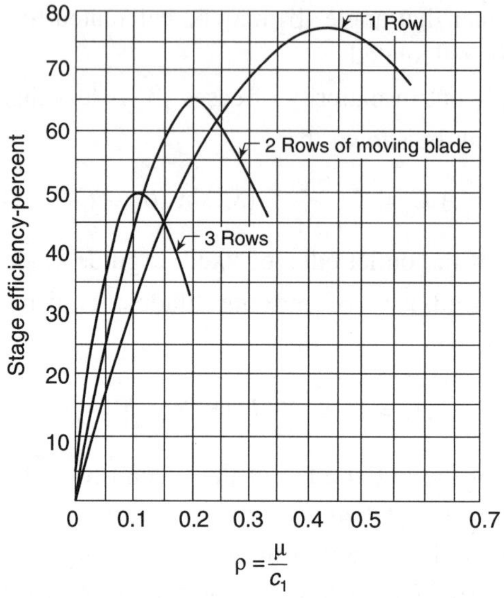

Figure 7.32 Stage efficiency vs. ρ

7.25 DISADVANTAGES OF VELOCITY-COMPOUNDED IMPULSE TURBINE

- Velocity-compounded impulse turbine has low efficiency and high steam consumption.
- In a single-row wheel, the steam temperature is high so a cast iron cylinder cannot be used due to phenomenon of growth. Cast-steel cylinder is generally used which is coastlier than cast iron.

Example 7.6 The nozzles in a stage of an impulse turbine are convergent type and formed by casting. The radial height of the nozzles is 15 cm and the mean diameter of the wheel is 1150 cm. The nozzles angle is 18°. The vanes are of uniform thickness amounting to 3 mm. The number of nozzles in the complete ring is 72. The specific volume of the steam at the nozzle outlet is 18 m^3/kg and the velocity of steam at nozzle outlet is 400 m/s. Calculate the mass flow rate of steam flowing through the nozzles.

Calculate the power developed in the blades of the given stage assuming that there is a single row of impulse blades having equal inlet and outlet angles, that the ratio of the relative outlet velocity to relative inlet velocity is 0.86, and that the rotor turns at 3000 rpm.

Solution Refer to Fig. 7.33, Given : $l_n = 15$ cm, $n = 72$, $D = 115$ cm, $v_1 = 18$ m^3/kg. $\alpha_1 = 18°$, $C_1 = 400$ m/s, $t = 3$ mm, $N = 3000$ rpm, $\beta_1 = \beta_2$, $m = ?$ and Power = ? $K = 0.86$.

From continuity equation, $\dot{m} \times v = A \times C$

The effective area at nozzles outlet = $n(S \,.\sin \alpha_1 - t)l_n$

where S = Mean circumferential pitch of nozzles and t = vane thickness at outlet

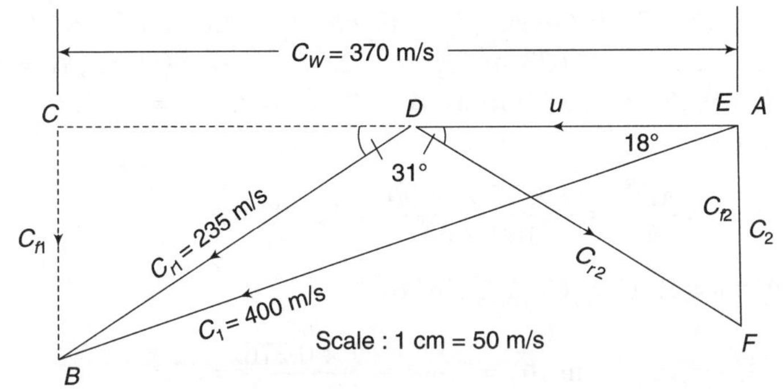

Figure 7.33 Example 7.6

$$S = \frac{\pi D}{n} = \frac{\pi \times 115}{72} = 5.017 \text{ cm}$$

At nozzle outlet, $\dot{m}v_1 = n . l_n(S \sin \alpha_1 - t) . C_1$

$$\dot{m} \times 18 = 72 \times \frac{15}{100}\left(\frac{5.017}{100} \times \sin 18° - \frac{0.3}{100}\right) 400$$

$$\dot{m} = 2.988 \text{ kg/s}$$ ***Ans.***

The peripheral speed of blades, $u = \frac{\pi DN}{60} = \frac{\pi \times 115 \times 3000}{100 \times 60} = 180.641$ m/s

With the known value of u, C_1 and α_1 the inlet velocity triangle *ABD* can be drawn. This gives : $\beta_1 = 13° = \beta_2$ and $C_{r1} = 235$ m/s.

$$\therefore \qquad C_{r2} = 0.86 \times 235 = 202 \text{ m/s}$$

With the values of β_2, u, C_{r2} the outlet velocity triangle *DAF* can be drawn.

$$\text{Power developed} = \frac{m . u . C_w}{1000} = \frac{2.988 \times 180.641 \times 370}{1000} = 199.7 \text{ kW}$$ ***Ans.***

Example 7.7 The nozzles in a two-row velocity-compounded impulse stage are 19 cm in height and extend over an arc of 76 cm. The effective angle at outlet is 16° and the specific volume of steam is 0.623 m^3/kg. The mean diameter of blades is 72 cm and the rotational speed 3000 rpm. The steam velocity at outlet from the nozzles is 558 m/s. The height of the second row of moving blades is 2.5 times radial height of the nozzle, i.e., 47.5 cm and the blade heights increase in arithmetic progression, i.e., $l_1 = 28.5$ cm, $l_f = 38$ cm and $l_2 = 47.5$ cm. Find the blade inlet and outlet angles, assuming that the blades have an axial width of 2.54 cm; that the thickness of the outlet edge 0.0508 cm and that blade velocity coefficient, $K = 0.86$. The mean circumferential pitch of the blades may be calculated by means of Briling's formula, $S = b/2 \sin 2\beta$.

Solution Refer to Fig. 7.34. Given : $l_n = 19$ cm, $l_1 = 28.5$ cm, $\lambda = 76$ cm, $l_f = 38$ cm, ${}_1\alpha_1 = 16°$, $l_2 = 47.5$ cm, $v = 0.623$ m^3/kg, $b = 2.54$ cm, $D = 72$ cm, $t_1 = t_2 = 0.0508$ cm, ${}_1C_1 = 558$ m/s, $K = 0.86$, $N = 3{,}000$ rpm, ${}_1\beta_1 = ?$, ${}_1\beta_2 = ?$, ${}_2\beta_1 = ?$, ${}_2\beta_2 = ?$

First moving row:

$$u = \frac{\pi DN}{60} = \frac{\pi \times 72 \times 3000}{100 \times 60} = 113.09 \text{ m/s}$$

We know that, $\dot{m}v = A.C = {}_1C_1.\lambda\, l_n \sin {}_1\alpha_1$

or
$$\frac{\dot{m}v}{\lambda} = {}_1C_1\, l_n \sin {}_1\alpha_1 = \frac{558 \times 19 \times 0.276}{100} = 29.3$$

With the value of u, ${}_1\alpha_1$ and ${}_1C_1$ the inlet velocity triangle can be drawn. This gives ${}_1\beta_1 = 20°$ and ${}_1C_{r1} = 452$ m/s.

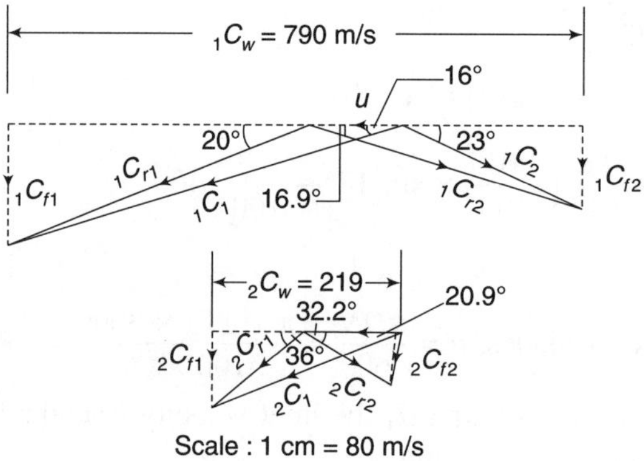

Figure 7.34 Example 7.7

∴ $${}_1C_{r2} = 0.86 \times 452 = 388.72 \text{ m/s}$$

Mean pitch of blades, $S_1 = \dfrac{b}{2 \sin 2\, {}_1\beta_1} = \dfrac{2.54}{2 \sin 40°} = \dfrac{2.54}{2 \times 0.645} = 1.975$ cm

We have $S_1 \sin {}_1\beta_2 = \dfrac{\dot{m}v}{\lambda}\dfrac{S_1}{l_1\, {}_1C_{r2}} + t_1$

or
$$\sin {}_1\beta_2 = \left(\frac{\dot{m}v}{\lambda}\right)\frac{1}{l_1\, {}_1C_{r2}} + \frac{t_1}{S_1} = 29.3 \times \frac{100}{28.5 \times 388.72} + \frac{0.0508}{1.975}$$

or $${}_1\beta_2 = 16.9°$$ ***Ans.***

Now the outlet velocity triangle for first row can be computed which gives

$${}_1\alpha_2 = 23° \quad \text{and} \quad {}_1C_2 = 272 \text{ m/s}$$

Fixed row: $$S_f = \frac{2.54}{2 \sin 46°} = \frac{2.54}{2 \times 0.72} = 1.765 \text{ cm}$$

$$\sin {}_2\alpha_1 = \left(\dot{m}\frac{v}{\lambda}\right)\frac{1}{l_f \cdot {}_2C_1} + \frac{t_f}{S_f} = (29.3)\,\frac{100}{38 \times 234} + \frac{0.0508}{1.765}$$

or $${}_2\alpha_1 = 21° \qquad \textit{Ans.}$$

Second row of moving blades: ${}_2C_{r1} = 145$ m/s and ${}_2\beta_1 = 36°$

$$S_2 = \frac{2.54}{2 \sin 72°} = \frac{2.54}{2 \times 0.95} = 1.34 \text{ cm}$$

$${}_2C_{r2} = 0.86 \times 145 = 124.5 \text{ m/s}$$

$$\sin {}_2\beta_2 = \left(\frac{\dot{m}v}{\lambda}\right)\frac{1}{l_2 \cdot {}_2C_{r2}} + \frac{t_2}{S_2} = 29.3 \times \frac{100}{47.5 \times 124.5} + \frac{0.0508}{1.34} = 0.5328$$

or $${}_2\beta_2 = 32.2° \qquad \textit{Ans.}$$

Now the velocity diagram is complete for second stage.

Let $\eta_n = 0.9$ be the nozzle efficiency, then

$$\eta_s = \frac{2 \,.\, \eta_n u\,({}_1C_w + {}_2C_w)}{{}_1C_1^2} = \frac{2 \times 0.9 \times 113.09 \times 1009}{558^2} = 65.9\% \qquad \textit{Ans.}$$

7.26 TWISTED BLADING

We have seen that at part load, the approaching flow angle to the moving blade is larger than the inlet angle of blades. If the blades are parallel having uniform sectional area and angles throughout its active length, the shock losses would occur over the whole length of the blade at inlet. This loss, may be minimized by the use of twisted bladings having angles that vary from root to tip. In this case, it is also possible to diminish the cross section from the root outward. This decreases the stress at the root or if desired allows the use of a longer blade with the same stress. Thus, the length of tapered blade may be 25 to 35 per cent of the blade ring mean diameter.

In the larger size impulse turbine, the blading of l.p. turbines for handling larger volume are longer. As a result, the peripheral velocity at the blade root will be much smaller than at the tip. Hence, for shockless entry and better performance, the blade angles should vary along the blade height as shown in Fig. 7.46. This results in twisting of the bladings. This is also true for impulse-reaction turbines since the peripheral velocity (u) is directly proportional to radius, so the blade angles vary accordingly. The velocity diagram at the root is impulse type while at the tip it is reaction type.

7.27 VELOCITY DIAGRAM AND WORK DONE FOR IMPULSE-REACTION TURBINE

It has been already discussed that in the impulse-reaction turbine, the pressure of steam at outlet from the moving blade is less than that of the inlet side, i.e., steam expands while passing over the moving blades,. Such a fall in steam pressure in the moving blades must be accompanied by an increase in the specific volume v_2 and hence and increase in C_{r2}. Increase in C_{r2} and specific volume v_2 are brought about by making the moving blades outlet angle actually smaller than the angle given by following equation.

$$\sin \beta_2 = \frac{1}{S}\left\{\frac{10^4 . \dot{m}.v^2}{n.t_2.C_{r2}} + 1\right\}$$ in order that this equation is satisfied.

It is also obvious from this relation that if β_2 is reduced, keeping S, $\dot{m}$, n and t constant, the value of v_2 and C_{r2} must be varied to satisfy the equation. When pressure drops in the moving blade channels, C_{r2} increases at a much faster rate than v_2 and the equilibrium point is reached at which the equation of continuity is satisfied. *This turbine blade section is similar to aerofoil section.*

The stage and velocity diagram of a impulse-reaction turbine is shown in Fig. 7.35 (a) and (b). The force diagram is shown is Fig. 7.35 (c) and (d).

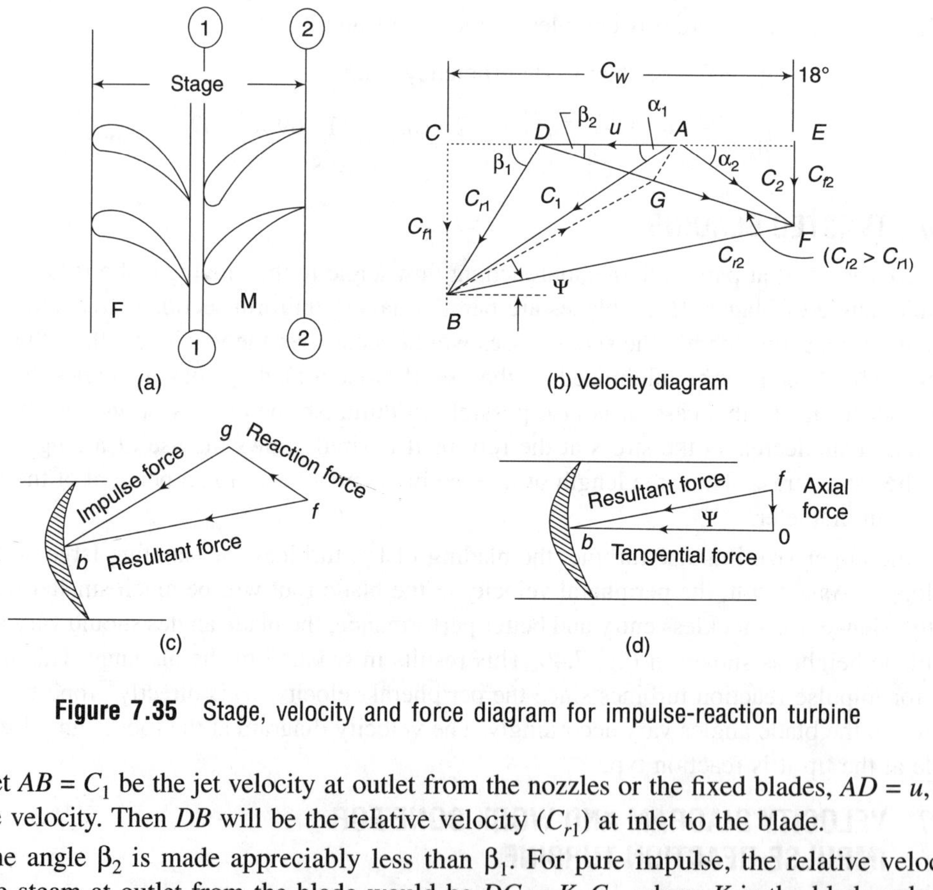

Figure 7.35 Stage, velocity and force diagram for impulse-reaction turbine

Let $AB = C_1$ be the jet velocity at outlet from the nozzles or the fixed blades, $AD = u$, the blade velocity. Then DB will be the relative velocity (C_{r1}) at inlet to the blade.

The angle β_2 is made appreciably less than β_1. For pure impulse, the relative velocity of the steam at outlet from the blade would be $DG = K.C_{r1}$ where K is the blade velocity coefficient. So the final absolute velocity would be AG. The change in velocity suffered by the steam would be represented by BG. And this change in velocity would give rise to the impulse force gb, which is equal to $\dot{m}$ (GB) and parallel to GB.

Due to the expansion of steam in the moving blades, the steam velocity is increased from DG to GF. The increase in outlet relative velocity gives rise to a reaction force fg which is equal to m (FB), and parallel to FG. By combining the impulse force (gb) and reaction force, fg. We get the resultant force (fb) which is equal to $\dot{m}$ (FB). Component of the resultant force in tangential and axial direction gives the tangential force and axial force respectively.

Like the impulse turbine, the tangential force,

$$F_t = \dot{m}\, C_w \quad \text{where } C_w = \text{whirl velocity} \qquad ...(7.30)$$

$$\text{Work done} = \dot{m}\, u.\, C_w$$

$$\text{Power developed} = \frac{\dot{m}\,.\,u\,.\,C_w}{1000}\,;\ \text{kW} \qquad ...(7.31)$$

The same definition will be used for blade efficiency and gross stage efficiency as in the impulse turbine but the expressions are different.

In reaction turbine, the velocity of flow per reaction pair is sensibly constant, so that the main end thrust is due to a difference in area and pressure on the rotor; this thrust is usually balanced by allowing low pressure steam to act on a *dummy piston* but thrust bearing is also provided.

7.28 DEGREE OF REACTION

The degree of reaction, (R) is defined as the ratio of isentropic heat drop in the moving blades to the sum of the isentropic heat drops in the fixed and the moving blades, i.e., in a stage. Mathematically, the degree of reaction is

$$R = \frac{(\Delta h_{\text{isen}})_b}{(\Delta h_{\text{isen}})_n + (\Delta h_{\text{isen}})_b} \qquad ...(7.32)$$

where $(\Delta h_{\text{isen}})_b$ and $(\Delta h_{\text{isen}})_n$ are the isentropic heat drops in moving blades and fixed blades respectively. To calculate the value of degree of reaction R, the value of $(\Delta h_{\text{isen}})_n$ and $(\Delta h_{\text{isen}})_b$ should be known. *In pure impulse turbine, degree of reaction is zero. In pure reaction turbine,* $R = 1$.

Let us consider a number of stages of an impulse-reaction turbine and let the mean diameter of all the stages be the same as shown in Fig. 7.36. Assume that the sections of the blades in each row are identical and the velocity ratios in each row are the same. Due to uniform blade velocity, the absolute velocities, C_1 and C_2, and the relative velocity, C_{r1} and C_{r2} will be the same for each pair of blades, i.e., for each stage.

Note that C_{r2} is always greater than C_{r1} in reaction turbine. As steam expands in each row of blades, its specific volume increases and since there is a constant rate of steam flow through the blading the blade heights must be increased progressively as shown in Fig. 7.36.

Kinetic energy $C_2^2/2$ in J per kg of steam comes out from each row of the moving blades and the steam with this energy is directed to pass over the fixed blades. Due to the losses at inlet, the available energy steam for utilization in each row of fixed blade will be $\phi\, C_2^2/2$ where ϕ is the carry-over coefficient.

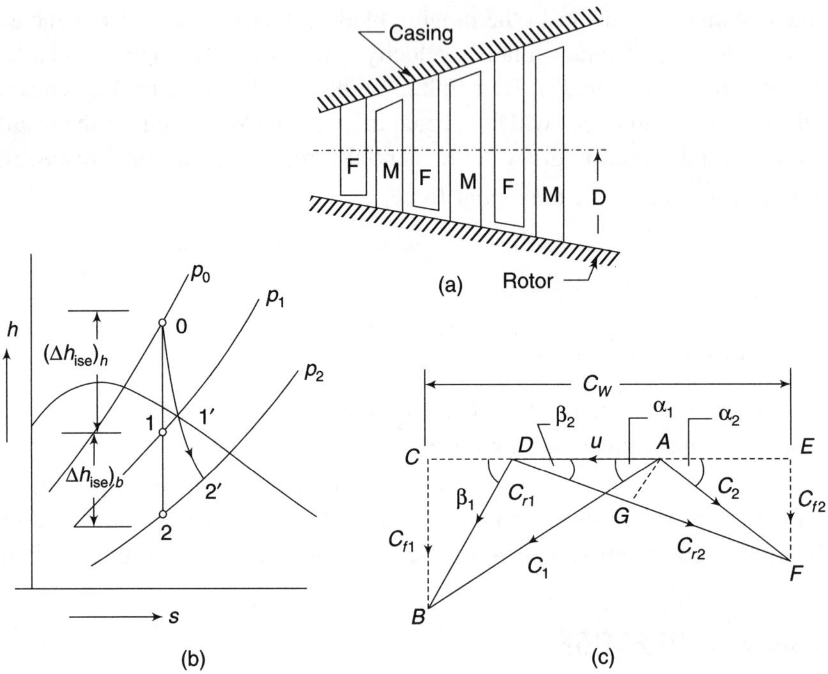

Figure 7.36 Impulse-reaction turbine

If η_n is the efficiency of the fixed blades when regarded as nozzles, then

$$\eta_n = \frac{C_1^2 - \phi C_2^2}{2(\Delta h_{\text{isen}})_n} \quad \text{or} \quad (\Delta h_{\text{isen}})_n = \frac{C_1^2 - \phi C_2^2}{2\eta_n} \text{ kJ/kg} \qquad ...(7.33)$$

From the velocity triangle (Fig. 7.36)

$$C_2^2 = u^2 + C_{r2}^2 - 2u\, C_{r2} \cos \beta_2 = C_{r2}^2 (1 + \rho_2^2 - 2\rho_2 . \cos \beta_2)$$

where $\rho_2 = \dfrac{u}{C_{r2}}$. Putting the value of C_2^2 in the Eq. (7.33), we get

$$(\Delta h_{\text{isen}})_n = \frac{\{C_1^2 - \phi\, C_{r2}^2 (1 + \rho_2^2 - 2\rho_2 \cos \beta_2)\}}{2\eta_n}$$

The available energy of each row of moving blades at inlet is $\phi\, C_{r1}^2/2$. If we assume η_n to be the efficiency of moving blades also for the sake of simplicity, then we have

$$\eta_n = \frac{C_{r2}^2 - \phi\, C_{r1}^2}{2(\Delta h_{\text{isen}})_b} \quad \text{or} \quad (\Delta h_{\text{isen}})_b = \frac{C_{r2}^2 - \phi\, C_{r1}^2}{2\eta_n} \qquad ...(7.34)$$

From the velocity diagram, $C_{r1}^2 = C_1^2 + u^2 - 2u\, C_1 \cos \alpha_1$

or $$C_{r1}^2 = C_1^2\,(1 + \rho - 2\rho\cos\alpha_1), \quad \text{where} \quad \rho = \frac{u}{C_1}$$

Putting the value of C_{r1}^2 in the expression for $(\Delta h_{\text{isen}})_b$ in the Eq. 7.1, we get

$$R = \frac{C_{r2}^2 - \phi\, C_1^2\,(1 + \rho - 2\rho\cos\alpha_1)}{C_1^2 - \phi C_{r2}^2\,(1 + \rho_2 - 2\rho_2\cos\beta_2) + C_{r2}^2 - \phi C_{r1}^2\,(1 + \rho - 2\rho\cos\alpha_1)} \quad \text{...(7.35)}$$

Work done per kg of steam = $u\,.\,C_w$

The velocity of whirl, $C_w = CD + DE = C_1\cos\alpha_1 - u + C_{r2}\cos\beta_2$

By putting the value of C_w in the above equation of work done per kg of steam, it becomes

$$w = u\,(C_1\cos\alpha_1 - u + C_{r2}\cos\beta_2) = \{C_1^2\,(\rho\cos\alpha_1 - \rho^2) + \rho_2\,.\,C_{r2}^2.\cos\beta_2\} \quad \text{...(7.36)}$$

$$\text{The gross stage efficiency, } \eta_{gs} = \frac{\text{Work done per kg of steam}}{\text{Isentropic heat drop in a stage}}$$

$$\eta_{gs} = \frac{C_1^2\,(\rho\cos\alpha_1 - \rho^2) + \rho_2\, C_{r2}^2\cos\beta_2}{\{(\Delta h_{\text{isen}})_n + (\Delta h_{\text{isen}})_b\}} \quad \text{...(7.37)}$$

Due to the leakage of steam through the clearances between the fixed blades and the rotor and between the moving blades and the cylinder, the actual stage efficiency would be less than calculated from the Eq. (7.37).

7.29 IMPULSE-REACTION TURBINE WITH SIMILAR BLADE SECTION AND HALF DEGREE REACTION (PARSONS TURBINE)

Consider particular case of reaction turbine, i.e., Parsons reaction turbine in which the degree of reaction is half and the section of blades is the same in both fixed and moving rows of blades (Fig. 7.37).

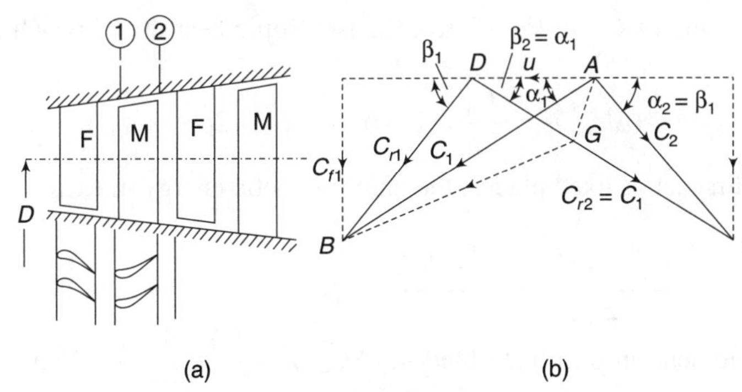

Figure 7.37 Parsons reaction turbine

In most of the cases, the blades of the turbine are divided into groups, each group contains several rows of blades with the same mean diameter and the same blade height; the results are, (i) increase in specific volume as the pressure falls, (ii) increase in the heat drop per stage compared to the preceding stage, and (iii) there is variation in the blade speed ratio from row to row, thus the degree of reaction is more than one-half.

But in the Parsons reaction turbine, *it is assumed that the blade section and the mean diameter of fixed as well as the moving blades are the same and the blade height is progressively so increased such that the velocity of steam at exit from each row of blades is uniform throughout the stage thus, the velocity triangle at the inlet and outlet of moving blades will be similar.*

The angle $\alpha_1 = \beta_2$, $\alpha_2 = \beta_1$ and velocity $C_{r2} = C_1$ and $C_2 = C_{r1}$ (Fig. 7.37).

7.29.1 Degree of Reaction

For Parsons turbine we can prove that the degree of reaction is half. For the pure impulse working condition *DG* represents the velocity of steam from the moving blades which is equal to $K.\ C_{r1}$. *Due to the expansion of steam in the moving blades, the relative velocity is increased to the value of* $C_{r2} = C_1$ represented by *DF*. Gain in the kinetic energy in the moving blades,

$$\frac{C_{r2}^2 - K^2 C_{r1}^2}{2} = \frac{C_1^2 - \phi C_{r1}^2}{2} \text{ where } \phi = K^2 = \text{carry-over coefficient}$$

If η_n is the efficiency of the moving blades as well as fixed blades, then the isentropic heat drop in each row of moving blades,

$$(\Delta h_{\text{isen}})_b = \frac{C_1^2 - \phi C_{r1}^2}{2\eta_n} \text{ kJ/kg} \qquad \text{...(7.38)}$$

From the velocity triangle, we know $C_{r1}^2 = C_1^2 + u^2 - 2u\, C_1 \cos \alpha_1$

Since $\rho = \frac{u}{C_1}$, then $C_{r1}^2 = C_1^2 (1 + \rho^2 - 2\rho \cos \alpha_1)$

By putting the value of C_{r1}^2 in Eq. (7.38), the isentropic heat drop in each row of moving blades becomes

$$(\Delta h_{\text{isen}})_b = \frac{C_1^2}{2\eta_n} \{1 - \phi (1 + \rho^2 - 2\rho \cos \alpha_1)\} \qquad \text{...(7.39)}$$

Except for the first set of fixed blades, the gain in kinetic energy in each row of fixed blade is given by

$$\frac{C_1^2 - \phi\, C_2^2}{2} = \frac{C_1^2 - \phi\, C_{r1}^2}{2}$$

So the isentropic heat drop in fixed blades, $(\Delta h_{\text{isen}})_n = \dfrac{C_1^2 - \phi\, C_{r1}^2}{2\eta_n}$ kJ/kg

Thus, the isentropic heat drop in the fixed and the moving blades is the same, i.e..

$$(\Delta h_{\text{isen}})_n = (\Delta h_{\text{isen}})_h, \text{ hence } R = 0.5$$

7.29.2 Gross Stage Efficiency and Optimum Value of ρ

The isentropic heat drop in a stage, i.e., in a pair of blades consisting of fixed and moving is

$$(\Delta h_{\text{isen}})_n + (\Delta h_{\text{isen}})_b = \frac{C_1^2}{\eta_n}\left|1 - \phi\,(1 - \rho^2 - 2\rho\cos\alpha_1)\right| \qquad \text{...(7.40)}$$

$$\text{Gross stage efficiency, } \eta_{gs} = \frac{\text{Work done per kg}}{\text{Isentropic heat drop in a stage per kg}}$$

Work done per kg of steam $= u\,.\,C_w$; kJ/kg

$$C_w = C_1\cos\alpha_1 - u + C_{r2}\cos\beta_2 = C_1\cos\alpha_1 - u + C_1\cos\alpha_1$$

So work done per kg of steam $= u\,.\,C_w = u\,(C_1\cos\alpha_1 - u + C_1\cos\alpha_1)$

$$= C_1^2\,(2\rho\cos\alpha_1 - \rho^2) \qquad \text{...(7.41)}$$

$$\eta_{gs} = \frac{(C_1^2)\,(2\rho\cos\alpha_1 - \rho^2)}{(C_1^2/\eta_n)\left\{1 - \phi(1 + \rho^2 - 2\rho\cos\alpha_1)\right\}} = \frac{\eta_n(2\rho\,.\cos\alpha_1 - \rho^2)}{1 - \phi(1 + \rho^2 - 2\rho\cos\alpha_1)} \qquad \text{...(7.42)}$$

$$\text{It may be put in another form also, } \eta_{gs} = \frac{\eta_n}{\left\{(1 - \phi/(2\rho\cos\alpha_1 - \rho^2)\right\} + \phi} \qquad \text{...(7.43)}$$

The above equation shows that the grow stage efficiency is maximum when $2\,.\,\rho\cos\alpha_1 - \rho^2$ is a maximum where η_n and ϕ are constant. For maximum value of gross stage efficiency

$$\frac{d(\eta_{gs})}{d\rho} = 0, \qquad \text{which gives} \qquad \frac{d(\eta_{gs})}{d\rho} = 0 = 2\cos\alpha - 2\rho$$

or $$\rho = \rho_{\text{opt}} = \cos\alpha_1 \qquad \text{...(7.44)}$$

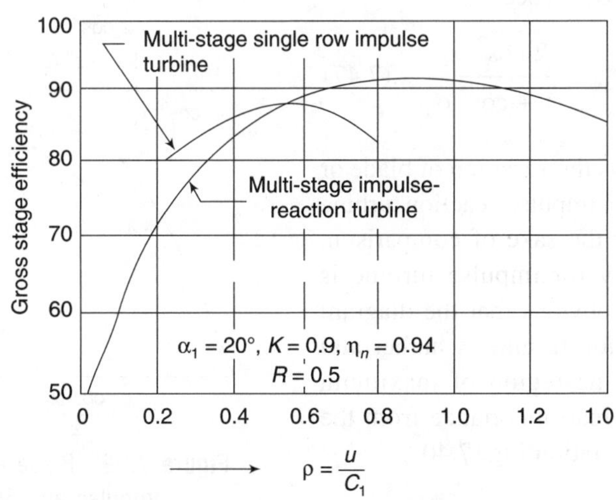

Figure 7.38 Gross stage efficiency vs. ρ

So we see that for maximum gross stage efficiency ρ_{opt} is equal to cos α_1 which is greater than ρ_{opt} of multi-stage turbine.

Putting the value $\rho = \cos \alpha$, in Eq. (7.42), we get

$$(\eta_{gs})_{max} = \frac{\eta_n . \cos^2 \alpha_1}{1 - \phi (1 - \cos^2 \alpha_1)} \quad ...(7.45)$$

Figure 7.38 shows the variation of η_{gs} with ρ. The curve is flat and the optimum value of ρ is 0.9 for α_1 = 20 degres.

7.29.3 Blade Efficiency

The diagram or blade efficiency, $\eta_b = \dfrac{\text{Work done per sec}}{\text{Energy input per sec}}$

Work done per kg, $u C_w = C_1^2 (2\rho \cos \alpha_1 - \rho^2)$ and energy input to fixed blades = $C_1^2/2$

Energy input to moving blades = $\dfrac{C_{r2}^2 - C_{r1}^2}{2}$

Total energy input = $\dfrac{C_1^2}{2} + \dfrac{C_{r2}^2 - C_{r1}^2}{2} = \dfrac{C_1^2}{2} + \dfrac{C_1^2 - C_{r1}^2}{2}$ for Parsons turbine

$$= C_1^2 - \frac{C_{r1}^2}{2} = C_1^2 - \frac{C_1^2}{2}(1 + \rho^2 - 2\rho \cos \alpha_1) = \frac{C_1^2}{2}[1 - \rho^2 + 2\rho \cos \alpha_1]$$

$$\eta_b = \frac{C_1^2 (2\rho \cos \alpha_1 - \rho^2)}{(C_1^2/2)[1 - \rho^2 + 2\rho \cos \alpha_1]} = \frac{2(2\rho \cos \alpha_1 - \rho^2)}{(1 - \rho^2 + 2\rho \cos \alpha_1)} \quad ...(7.46)$$

Since $\rho_{opt} = \cos \alpha_1$, hence

$$(\eta_b)_{max} = \frac{2 \cos^2 \alpha_1}{1 + \cos^2 \alpha_1} \quad ...(7.47)$$

Figure 7.39 shows the variation of blade or diagram efficiency of impulse-reaction turbine with $\rho = u/C_1$. For the sake of comparison, the blade efficiency for impulse turbine is also depicted. It is obvious that the diagram efficiency for reaction turbine is higher and reasonably flat in the region of maximum value which is of great importance from the point of view of operation Fig. 7.40.

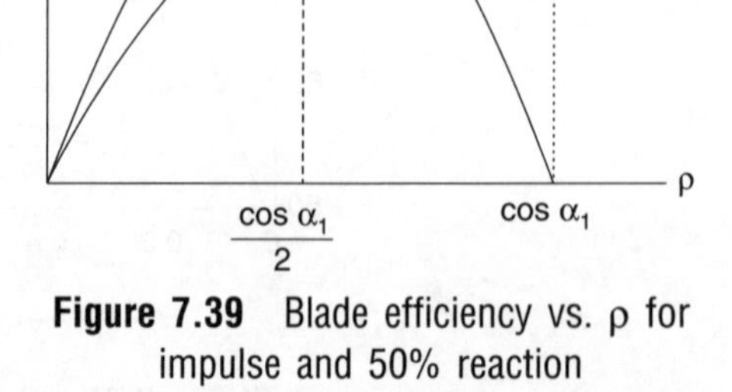

Figure 7.39 Blade efficiency vs. ρ for impulse and 50% reaction

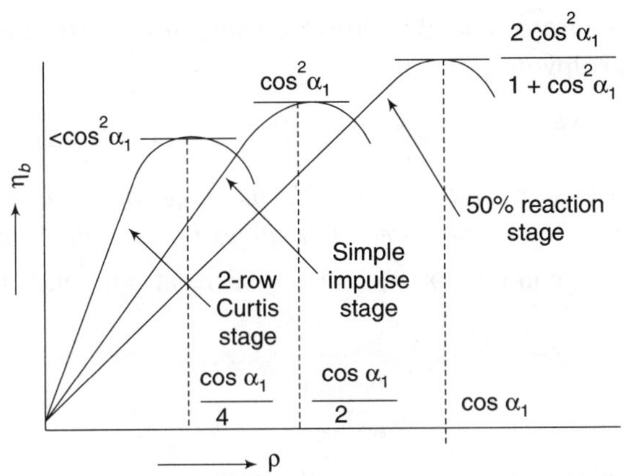

Figure 7.40 Blade efficiency vs. ρ for three turbines

7.30 COMPARISON OF ENTHALPY DROPS IN VARIOUS STAGE IMPULSE-REACTION TURBINES AND IMPULSE TURBINES

The optimum ratio of blade velocity to steam velocity for 50% impulse-reaction turbine is given by (Fig. 7.41).

$$\rho_{opt} = \frac{u}{C_1} = \cos\alpha_1$$

But C_1 is the velocity of steam coming out from the nozzle or fixed fixed blade neglecting initial kinetic energy is given by

$$C_1 = \frac{u}{\cos\alpha_1} = 44.72\sqrt{(\Delta h_{isen})_{nozzle}}$$

or
$$(\Delta h_{isen})_{nozzle} = \left(\frac{u}{44.72\cos\alpha_1}\right)^2$$

Thus,
$$(\Delta h_{isen})_{50\%\text{ reaction stage}} = 2\left(\frac{u}{44.72\cos\alpha_1}\right)^2 \text{ due to } F \text{ and } M \text{ rows} \quad ...(7.48)$$

Equation (7.48) gives isentropic enthalpy drop in a single stage of 50% impulse-reaction turbine.

The isentropic enthalpy drop in a single stage of impulse turbine and in a two-row Curtis stage impulse turbine for maximum efficiency are given as

$$(\Delta h_{isen})_{impulse\text{-}stage} = 4\left(\frac{u}{44.72\cos\alpha_1}\right)^2 \quad ...(7.49)$$

and
$$(\Delta h_{isen})_{\text{Two-row Curtis stage}} = 16\left(\frac{u}{44.72\cos\alpha_1}\right)^2 \quad ...(7.50)$$

For comparing the isentropic enthalpy drop the value of u and α_1 are kept the same, hence the relative ratio of enthalpy drop becomes

$$(\Delta h_{isen})_{50\%\ reaction\ stage} : (\Delta h_{isen})_{simple\ impulse\text{-}stage} : (\Delta h_{isen})_{2\text{-}row\ Curtis\ stage} = 1 : 2 : 8 \qquad ...(7.51)$$

The above comparison show that a 2-row Curtis stage ideally equivalent to four simple impulse stages and eight 50% impulse-reaction stages for enthatpy drop.

The number of stages required for a certain total isentropic enthalpy drop is expressed as

$$n = \frac{(\Delta h_{isen})_{total}}{(\Delta h_{isen})_{stage}} \qquad ...(7.52)$$

Equation (7.52) suggests that *the lower is the enthalpy drop per stage, the higher is the number of stage required for the same output.*

For u = 350 m/s and α_1 = 18°, we have

$$(\Delta h_{isen})_{50\%\ impulse\text{-}reaction} = 135.44 \text{ kJ/kg}$$
$$(\Delta h_{isen})_{impulse\text{-}stage} = 270.88 \text{ kJ/kg}$$
$$(\Delta h_{isen})_{2\text{-}row\ Curtis\ stage} = 1083.52 \text{ kJ/kg}$$

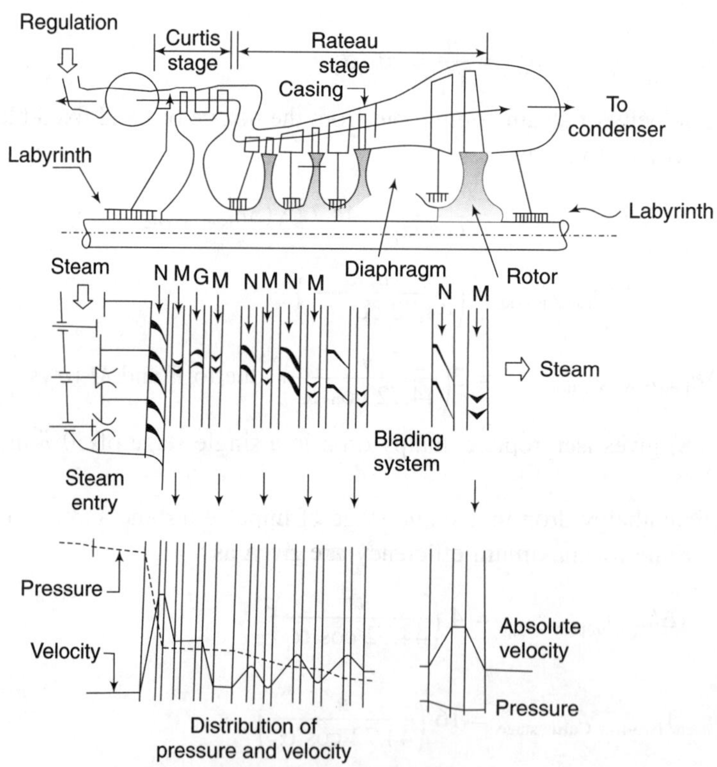

(a) Impulse turbine

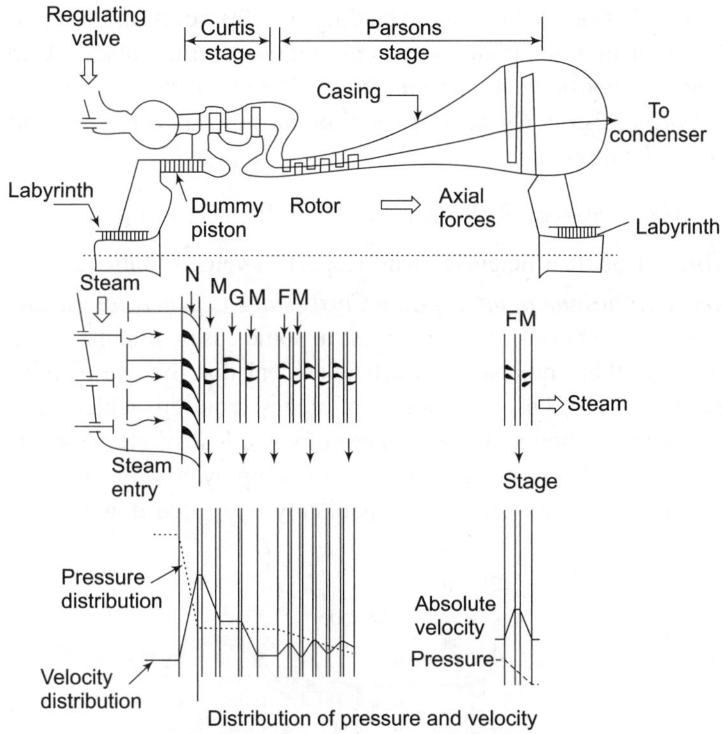

(b) Impulse-reaction turbine

Figure 7.41 (a) and (b) Simple layouts of impulse and impulse-reaction turbine

The above results illustrate that for the same output of a turbine, the number of 50% impulse-reaction stages will be about twice the number of simple impulse (i.e., Rataeu) stages and about eight times the number of 2-row Curtis stages. Obviously, it means that reaction turbine are costlier than impulse turbine for the same output.

Further, for the same blade (peripheral) velocity, the velocity at inlet to moving blades for maximum efficiency is

$C_1 = u/\cos\alpha_1$ for 50% reaction stage

$C_1 = 2u/\cos\alpha_1$ for simple impulse stage

$C_1 = 4u/\cos\alpha_1$ for a 2-row Curtis stage

We know that energy is due to friction is proportional to the square of the velocity of fluid (steam). Since the fluid velocity is the highest in 2-row Curtis stage and the lowest in 50% reaction stage, so the energy loss due to friction in the 50% reaction stage is the least while in Curtis stage, it is the highest. The energy loss in impulse stage will lie between these two values.

Further, the cross section of the moving bladings of 50% reaction turbine is of converging type, so the change of flow separation is less as compared to impulse or Curtis stage where cross-sectional area is constant. *As a result, the efficiency of the reaction stage will be the highest and that of the 2-row Curtis stage will be the lowest, while the impulse stage will exhibit efficiency in between, i.e.,*

$$\eta_{50\%\text{ reaction stage}} > \eta_{\text{simple impulse stage}} > \eta_{2\text{-row Curtis stage}} \quad ...(7.53)$$

The comparison of blade efficiencies with respect to velocity ratio is shown in Fig. 7.40.

In impulse-reaction turbine plant, a 2-row Curtis stage is generally used as the first stage, followed by a series of impulse-reaction stages. In some cases, impulse stages are also used after the Curtis stage, then impulse stages. It is evident that the use of initial Curtis stage, reduces the length of rotor, i.e., the number of stages required, which in turn, reduce the cost of rotor significantly, though at the expense of some loss of efficiency. It is often called the control stage as speed governing mechanism is employed with this stage where steam at the high pressure and temperature is immediately expanded down to a region of more

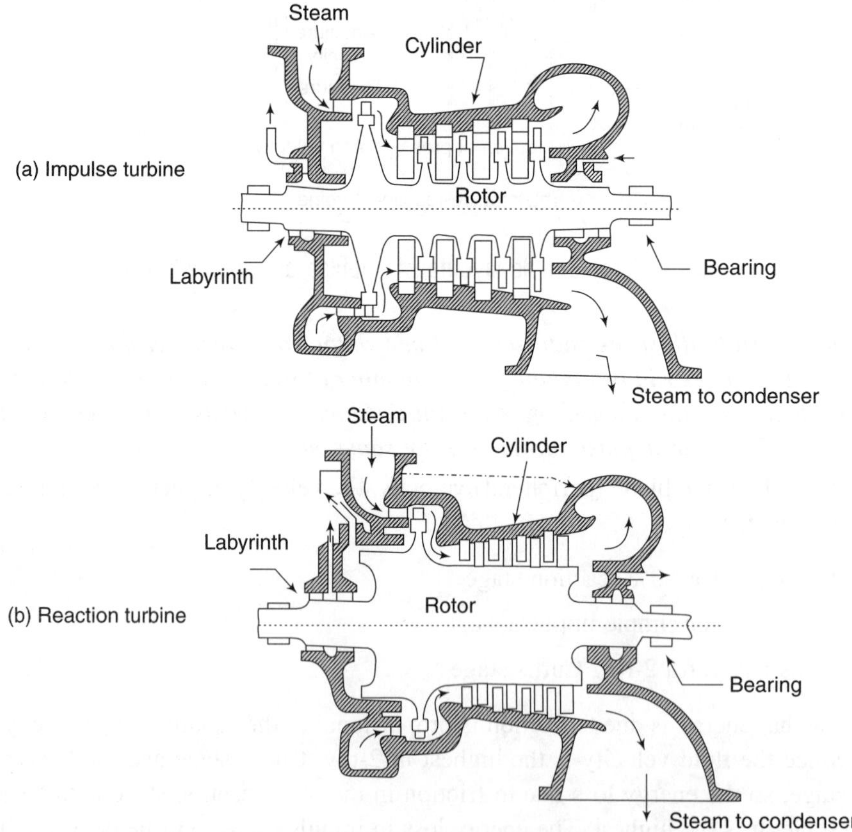

Figure 7.42 Typical design of impulse and impulse-reaction turbine

moderate pressures and temperature resulting in the reduction in the leakage of steam between stages.

The number of stages after first 2-row Curtis stage in impulse or reaction stage are given by

$$n_{\text{impulse or reaction}} = \frac{(\Delta h_{\text{isen}})_{\text{total}} - (\Delta_{\text{isen}})_{\text{2-row Curtis stage}}}{(\Delta_{\text{isen}})_{\text{impulse or reaction stage}}} \quad ...(7.54)$$

For a 20 stages impulse-reaction steam turbine, the distribution of stages will be:

2-row Curtis stage, 1(R), 2(R), 3(R), ... 20(R)

(N, MB, GB, MB) (FB, MB), (FB, MB) (FB, MB, etc.)

where N stands for nozzle, MB for moving blades, GB for guide blades, FB for fixed blades and R for reaction (50% or other) stage.

However, in large output turbine, all impulse-reaction stages are employed using throttle control governing.

Figure 7.41 shows simple layouts of an impulse and impulse reaction turbine showing the regulating valve, Curtis stage and followed by Rataeu stages or impulse-reaction reaction stage as the case may be with pressure and velocity distribution. Typical designs of impulse and impulse-reaction turbines are shown in Fig. 7.42.

Example 7.8 Steam flows into the nozzles of a impulse-reaction turbine stage from the blades of the preceding stage with a velocity of 100 m/s and issues from the nozzles with a velocity of 325 m/s at an angle of 20° to the wheel plane. Calculate the gross stage efficiency for the following data: Mean blade velocity = 180 m/s. Expansion efficiency for nozzles and blades. = 0.9, Carry-over factor for nozzles and blades = 0.9, Degree of reaction = 0.26, Blade outlet angle = 28°.

Solution Refer to Fig. 7.43

With the value of u, C_1, and α_1, the inlet velocity triangle ABD can be drawn.

Isentropic heat drop in nozzles

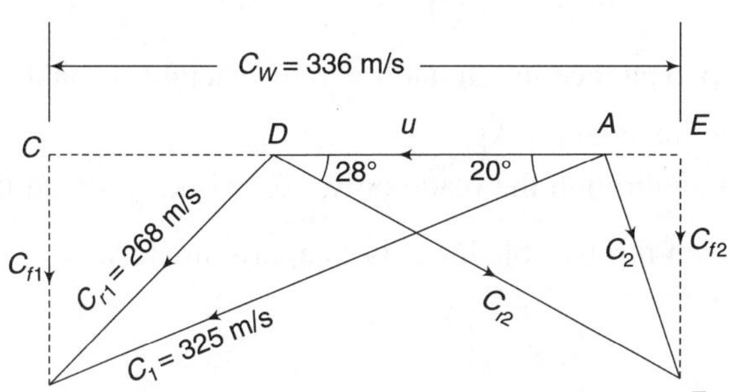

Scale : 1 cm = 40 m/s

Figure 7.43 Example 7.8

$$(\Delta h_{\text{isen}})_n = \frac{C_1^2 - \phi\, C_2^2}{2\eta_n \times 1000} = \frac{325^2 - 0.75 \times 100^2}{2 \times 0.9 \times 1000} = 54.51 \text{ kJ/kg.}$$

$$R = \frac{(\Delta h_{\text{isen}})_b}{(\Delta h_{\text{isen}})_n + (\Delta h_{\text{isen}})_b}$$

or $$(\Delta h_{\text{isen}})_b = \frac{R}{1 - R}(\Delta h_{\text{isen}})_n = \frac{0.26}{0.74} \times 54.51 = 19.15 \text{ kJ/kg}$$

We have, $$(\Delta h_{\text{isen}})_b = \frac{C_{r2}^2 - \phi\, C_{r1}^2}{2.\eta_b}$$

or $$C_{r2}^2 = 2 \times 1000 \times \eta_b\, (\Delta h_{\text{isen}})_b + \phi\, C_{r1}^2$$

$$= 2 \times 1000 \times 0.9 \times 19.15 + 0.9 \times 168^2 = 59871.6$$

or $$C_{r2} = \sqrt{(59871.6)} = 244.68 \text{ m/s}$$

With this value of C_{r2}, velocity diagram is completed. From diagram, $C_w = 336$ m/s

$$\eta_{gs} = \frac{C_w.u}{[(\Delta h_{\text{isen}})_n + (\Delta h_{\text{isen}})_b]} = \frac{336 \times 180}{1000 \times (54.51 + 19.15)} = 82.1\%$$ ***Ans.***

7.31 HEIGHT OF IMPULSE-REACTION TURBINE BLADING, PARALLEL EXHAUST AND CASING

7.31.1 General

Since the blade is the heart of the turbine, so its dimensions must be calculated accurately. In calculating the height of the turbine blading, the leakage of steam over the blade tips must be taken into account where unshrouded blading is used. Generally, the quantity of steam passing over the blade tip is given by the expression, (approximately),

$$\dot{m}_{bet} = 3\left(\frac{\epsilon}{l}\right) \times \text{full steam flow}$$

where ϵ = radial tip clearance in cm and l = radial height of blade in cm

Let, $\dot{m}$ = total flow of steam in kg/s.

$\dot{m}_1$ = steam flow through the blade only, i.e., working steam flow in kg/s.

Then the flow of steam through blades only is approximately

$$\dot{m}_1 = \dot{m}\left(\frac{1 - 3\epsilon}{l}\right) \qquad ...(7.55)$$

When there is no leakage over the blade tips, $\dot{m} = \dot{m}_1$.

In order to calculate the blade (fixed or moving) height, the total area required at outlet is equated to the total area for flow at outlet.

To know the true outlet angle of blades (say fixed blades), we draw a tangent to the leading and trailing surface at outlet to make angles α' and α'' respectively with the tangent of the wheel. Arithmetic mean of α' and α'' gives the true angle of blade at outlet, if the difference between these two angles is less.

It has already been discussed that for 50% reaction steam turbine working at constant ρ and constant outlet blade angle, the lengths of the blade should be increased progressively to accommodate the increasing volume of steam. But in practice, a gradual increase is not provided. The total number of stages in multi-stage have been divided into a number of groups which vary from twelve to thirty. In the high pressure part of the turbine, where the mean diameter is comparatively small, such a group might consist of five to seven pairs of blades and in the low pressure part of the turbine, a group consists only of one pair, i.e., one fixed and one moving blades rows.

The blade height is maintained constant in each group which contains more than one pair of blades. In such cases, two possible conditions arise: (i) constant outlet blade angle α_1 and (ii) constant value of ρ. Generally, the first case, i.e., constant α_1 is adopted.

On the assumption of condition (i), i.e., a uniform blade angle, the necessary blade height may be based on the steam conditions at the middle of the expansion. (Fig. 7.44)

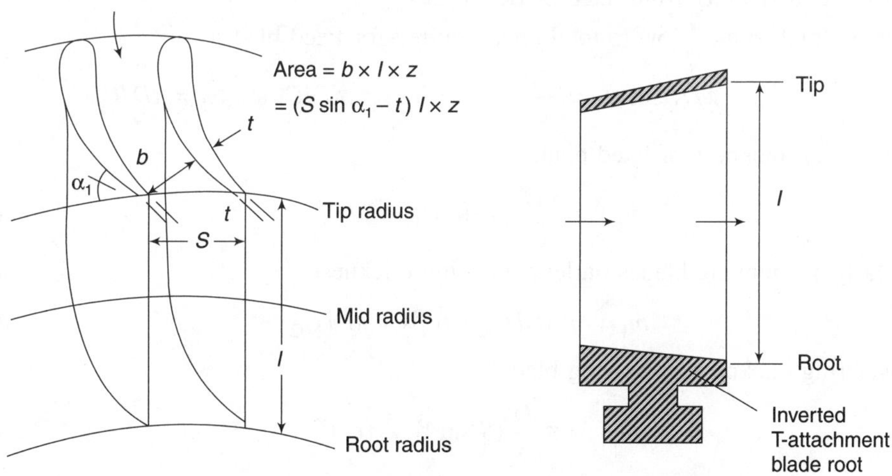

Figure 7.44 Impulse-reaction turbine blade passage and height

If b = with of blade passage

l = blade height in meter

N = r.p.m.

S = pitch of blades

z = number of blade passage

v = actual specific volume at the middle of expansion, m^3/kg

D = mean diameter of the blade ring, m.

t = thickness of blades

The number of blades per ring, $z = \pi\ D/S$

Area of one blade passage (Fig. 7.44) = $b \times l$

Total area of flow, $A_b = b \times l \times z = (S \sin \alpha_1 - t)\ l\ \pi D/S$ ...(7.56)

or $$A_b = S \sin \alpha_1 \left(1 - \frac{t}{S \sin \alpha_1}\right) \times l\ \pi\ D/S$$

or $$A_b = \pi Dl\ F_1 \sin \alpha_1 \quad \text{...(7.57)}$$

where F_t = nozzle/blade thickness factor = $\left(1 - \dfrac{t}{S \sin \alpha_1}\right)$

= 1 for negligible thickness

Neglecting thickness of blade.

Total area of flow, $A_b = \dfrac{\pi D}{S} Sl . \sin \alpha_1 = \pi Dl \sin \alpha_1$; m^2 ...(7.58)

Total area required on the basis of working steam, $A = \dfrac{\dot{m}_1 . v .}{C_1}$

where C_1 is exit velocity from fixed blades in m/s.

Equating total area of flow to total area required for fixed blades

$$\pi Dl \sin \alpha_1 = \frac{\dot{m}_1 v}{C_1} \quad \text{or} \quad \dot{m}_1 . v = \pi DlC_1 \sin \alpha_1 = \pi DlC_{f1}$$

Considering thickness of fixed blade

$$\dot{m}_1\ v = \frac{\pi D}{S} (S \sin \alpha_1 - t)\ lC_1 \quad \text{...(7.59)}$$

Similarly, for moving blades outlet neglecting thickness ...(7.60)

$$\dot{m}_1 . v = \pi Dl C_{r2} \sin \beta_2 = \pi Dl C_{f2} \quad \text{...(7.61)}$$

Considering thickness of moving blade

$$\dot{m}_1 v = \pi \frac{D}{S} (S \sin \beta_2 - t)\ l C_{r2} \quad \text{...(7.62)}$$

For Parsons turbine $\alpha_1 = \beta_2$, $C_1 = C_{r2}$ and v is assumed constant, so blade height for fixed as well as moving blades is the same in a stage.

But we know that

$$C_1 = \frac{u}{\rho} = \frac{\pi DN}{60 \times \rho}$$

$$\therefore \quad \pi Dl \sin \alpha_1 = \frac{\dot{m}_1 . v}{\pi DN} \times 60\ \rho$$

or $$D^2\ l = 6.08 \times \frac{\dot{m}_1 v . \rho}{N \sin \alpha_1} \quad \text{...(7.63)}$$

In any group if the values of $\dot{m}_1$, v. ρ. N and $\sin \alpha_1$ are known we can calculate the value of $D^2 l$, and if D is known then l can be found out.

Generally, D is given as

$$D = D_r + l$$

where D_r = diameter of drum

Then, $$D^2.l = (D_r + l)^2.l \quad ...(7.64)$$

When $D^2 l$ and D are known the value of l must be found by trial and error. When the values of D and D_r are known, the blade height can be easily calculated.

7.31.2 First Stage Nozzle and Blades

From the continuity equation, the nozzle area required for a given flow rate ($\dot{m}$) is expressed as

$$A_n = \frac{\dot{m}C_1}{v_1}$$

At high pressure of steam, specific volume (v_1) is small, $C_1 = u/\cos \alpha_1$ for 50% reaction and $C_1 = 2u/\cos \alpha_1$ for impulse bladings if the blades are assumed to run at maximum efficiency. Thus, at initial stage of turbine, the flow area of the nozzle (A_n) required is small. From Eq. (7.57).

$$A_n = \pi D l_n F_I \sin \alpha_1$$

where D is estimated from $u = \pi D N/60$. Since A_n is small and for a given value of F_t and α_1, the estimated value of nozzle height, l_n is often found too small to manufacture. Even with conical nozzle, the diameter is very less. It is an established fact that small height/diameter nozzles is less efficient.

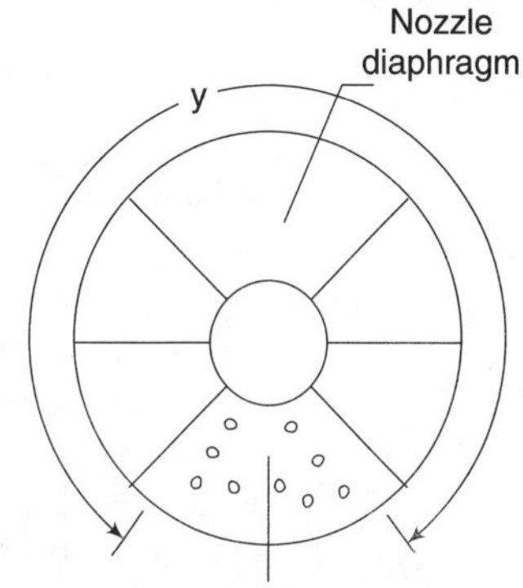

Figure 7.45 Partial admission of steam in the initial stages

Similarly, the blade following the nozzle in first stage are very short and difficult to manufacture. Under this situation, in general, the minimum height of blades of initial stage is taken as 20 mm. Thus,

$$(l_n)_{\min} = (l_n)_{\min} = 20 \text{ mm for initial stages} \quad ...(7.65)$$

But the required height, l_n for a given flow is much less than 20 mm, and so to provide the same flow area as required, some nozzle in the arc is blocked and thus the nozzle area required is expressed as

$$A_n = y\pi D (l_n)_{\min} F_{in} \sin \alpha_1 \quad ...(7.66)$$

where y is the fraction of total arc of nozzle which is open for steam flow and such a situation is known as *partial* admission of steam. *This is managed by blocking some of the nozzles by steel plates* (Fig. 7.45). By this way, the required flow area is maintained with minimum nozzle/blade height. *It is worth to note that the performance of the turbine deteriorates with partial admission.*

7.31.3 Last Stage Blade Height

The design of last stages of turbines poses several problems. Due to expansion of steam, the pressure in the last stage is very less and consequently the specific volume is very large. To accommodate the large volume of steam, the flow area needed is large and thus l_b and D increase with decrease in pressure and attain the maximum value in the last stage.

The blades are subjected to bending and centrifugal stresses and with increase in the height of blades, both stresses increase and put a limit to the height of mean diameter of blades. The blade velocity (u_{max}) is also limited depending upon the material of blades which is about 350 to 400 m/s. Taking

u_{max} = 350 m/s, we have

$$u_{max} = 350 = \frac{\pi D_{max} N}{60}$$

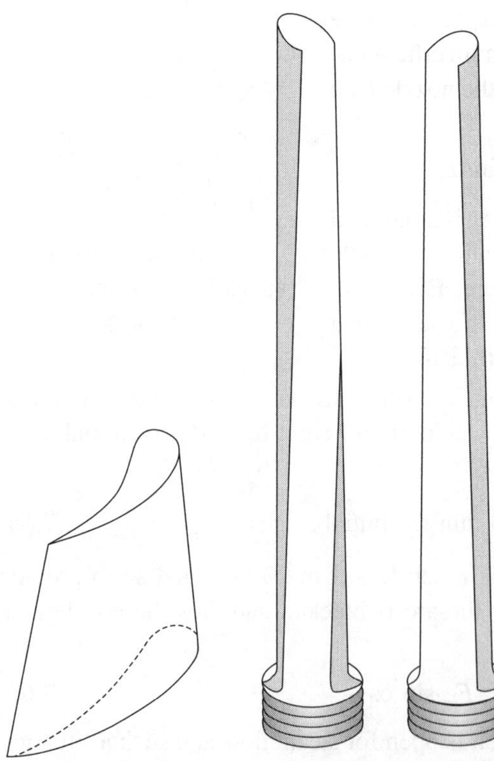

(a) Tapered blade (b) Twisted blade

Figure 7.46 Tapered and twisted blades

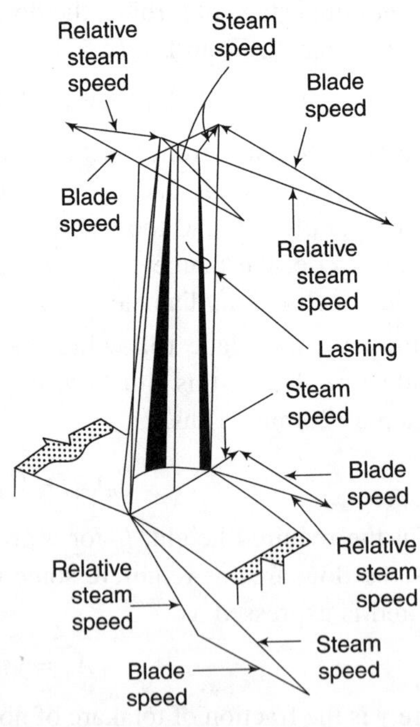

Figure 7.47 A twisted blade with impulse flow at root and reaction at tip

For a 2-pole, 50 Hz alternator, $N = \frac{120f}{p} = \frac{120 \times 50}{2} = 3000$ rpm, we have

$$D_{max} = \frac{350 \times 60}{\pi \times 3000} = 2.23 \text{ m}$$

As a common practice, for straight blades, the maximum blade height is about 20% of the mean ring blade diameter. Thus,

$$(l_b/D)_{max} = 0.2 \quad \text{...(7.67a)}$$

But in the last stage, this ratio gets exceeded because of flow requirement, or for shockless entry the blades may be tapered or twisted as shown in Figs. 7.46 and 7.47 to reduce both bending and centrifugal stresses. For twisted blading the blade height may approach about 30% of mean blade ring diameter. Thus,

$$(l_b/D)_{max} = 0.3 \text{ for twisted blades} \quad \text{...(7.67b)}$$

or $$(l_b)_{max} = 2.23 \times 0.3 = 0.67 \text{ m} = 67 \text{ cm}$$

The velocity diagram at the root and tip of twisted blading is shown in Fig. 7.47. The root is designed for impulse flow while tip is designed for pure reaction flow.

7.31.4 Parallel Exhaust: Number of Last Stages

The maximum flow area provided in the last stage of blade at the inlet is given by

$$(A_b)_{max} = \pi D (l_b)_{max} F_{tb} \sin \beta_1 \quad \text{...(7.68)}$$

Thus, the maximum volume flow that the last stage can accommodate, is expressed by

$$V_{max} = \pi D (l_b)_{max} F_{tb} . C_{r1} \sin \beta_1$$

or $$V_{max} = \pi D (l_b)_{max} F_{tb} C_{f1} \quad \text{...(7.69)}$$

For reaction, $\frac{u}{C_1} = \cos \alpha_1$ or $C_1 = \frac{350}{\cos \alpha_1}$

If $\alpha_1 = 20°$, and $F_{tb} = 90$,

$$V_{max} = \pi \times 2.23 \times 0.67 \times \frac{350}{\cos 20°} \times 0.9 = 534.5 \text{ m}^3/\text{s}$$

For impulse bladings, $V_{max} = 1075 \text{ m}^3/\text{s}$

Thus, the maximum mass flow of steam that the last stage can accommodate can be given by

$$(\dot{m})_{max} = 534.5/v_2 \text{ for 50\% reaction turbine} \quad \text{...(7.70)}$$

$$= \frac{1075}{v_2} \text{ for impulse turbine} \quad \text{...(7.71)}$$

For the turbine exhaust condition at pressure 0.1 bar and 0.9 dry,

$$v_2 = v_{f2} + x_2 v_{g2} = 1.01043 \times 10^{-3} + 0.9 \times 14.68 = 13.223 \text{ m}^3/\text{kg}$$

Thus, $$(\dot{m})_{max} = \frac{534.5}{13.223} = 32.859 \text{ kg/s for 50\% reaction turbine}$$

$$= \frac{1075}{13.233} = 65.718 \text{ kg/s for impulse turbine}$$

Form the above, the *number of parallel exhaust* (n) required to pass a mass of ($\dot{m}$) kg/s of steam is

$$n = \frac{\dot{m}}{(\dot{m})_{max}} = \frac{\dot{m}}{32.859} \text{ for 50\% reaction turbine} \quad ...(7.72)$$

$$= \frac{\dot{m}}{65.718} \text{ for impulse turbine} \quad ...(7.73)$$

7.31.5 Casing Arrangement

If the number of parallel exhausts estimated to be 4 (i.e., $n = 4$), the casing arrangement could be 1 HP, 1 IP and two parallel double flow LP turbine. The four LP turbine exhaust will enter into condenser. The typical casing arrangement is shown earlier in Fig. 7.11.

7.32 EFFECT OF WORKING STEAM ON THE STAGE EFFICIENCY OF PARSONS TURBINE

Let $\dot{m}_1$ = steam flow through the blade only, i.e., working steam flow in kg/s.

$\dot{m}$ = total steam flow per second.

Then, the work done $= u.C_w = \dot{m}_1\, C_1^2\, (2\rho \cos \alpha_1 - \rho^2)$

Total isentropic heat drop per second $= \dot{m}\,\{(\Delta h_{isen})_n + (\Delta h_{isen})_b\}$

The above expression consists of $\dot{m}$ *not* $\dot{m}_1$ *because the heat drop takes place in total steam flow.*

Putting the values of $(\Delta h_{isen})_h$ and (Δh_{isen}), we get the value of total isentropic heat drop per second as

$$= \frac{\dot{m}.C_1^2}{\eta_n} \left|1 - \phi(1 + \rho^2 - 2\rho \cos \alpha_1)\right|$$

Then the stage efficiency, $\eta_{gs} = \dfrac{\dot{m}_1}{\dot{m}} \cdot \dfrac{\eta_n\,(2\rho \cos \alpha_1 - \rho^2)}{1 - \phi(1 + \rho^2 - 2\rho \cos \alpha_1)}$ ...(7.74)

When ordinary unshrouded blades are used, then $\dfrac{\dot{m}_1}{\dot{m}} = 1 - \dfrac{3\epsilon}{l}$

And in this case, the stage efficiency,

$$\eta_{gs} = \left(1 - \frac{3\epsilon}{l}\right) \frac{\eta_n\,(2\rho \cos \alpha_1 - \rho^2)}{1 - \phi\,(1 + \rho^2 - 2\rho \cos \alpha_1)} \quad ...(7.75)$$

Since $\dot{m}_1$ is less then m, Eq. (7.74) shows that gross stage efficiency decreases by the consideration of working steam.

Example 7.9 A 50% reaction turbine running at 500 rpm consumes 13.6 kg/s of steam. If the adiabatic heat drop in turbine is 460 kJ/kg and turbine internal efficiency is estimated to be 0.89, what is the internal power developed by the turbine?

At a certain stage of this turbine, the mean steam conditions are 1.1 bars and 0.85 dry. The discharge blade tip angle is 20° and the blade to steam velocity is 0.7. If the blade height is 1/12 of the mean ring diameter, determine the drum diameter and height of blade.

Solution Refer to Fig. 7.44.

Internal power developed = $\dot{m} \times \Delta h_{\text{isen}} \times \eta_{gs}$ = 13.6 × 460 × 0.89 = 5567.84 kW ***Ans.***

Here $$\frac{u}{C_1} = 0.7 \quad \text{so} \quad C_1 = \frac{u}{0.7} \text{ m/s}$$

Specific volume of steam at 1.1. bars and 0.85 dry = $x.v_s$ = 0.85 × 1.549 = 1.3166 m^3/kg.

Assuming that specific volume is constant in the considered stage, thus from the continuity equation

$$\dot{m}v = \pi D l C_1 \sin \alpha_1$$

or $$\dot{m}v = \pi D.\frac{D}{12}.\frac{u}{0.7} \sin \alpha_1 = \frac{\pi D^3 \times \pi N \times \sin \alpha_1}{12 \times 0.7 \times 60}$$

or $$13.6 \times 1.3166 = \frac{\pi \times D^3 \times \pi \times 500 \times \sin 20°}{12 \times 0.7 \times 60}$$

or $$D = \left[\frac{13.6 \times 1.3166 \times 12 \times 0.7 \times 60}{\pi \times \pi \times 500 \times 0.342}\right]^{1/3} = 1.7486 \text{ m}$$

$$\text{Blade height, } l = \frac{D}{12} = \frac{1.7486}{12} = 0.1457 \text{ m} = 14.57 \text{ cm}$$ ***Ans.***

$$\text{Drum dia, } D_r = D - l = 1.7486 - 0.1457 = 1.6029 \text{ m}$$ ***Ans.***

Example 7.10 The following particulars refer to one stage of an impulse-reaction turbine: Mean diameter = 96 cm, rpm = 3,000, Nozzles outlet angle = 20°, Nozzle height = 12 cm, Blade height = 12 cm, Specific volume of steam: (a) at nozzle outlet = 4.4 m^3/kg, (b) at blade outlet = 4.8 m^3/kg, Steam velocity at nozzle outlet = 275 m/s, *Power developed by blade* = 265 kW.

Calculate the heat drop in the stage, the degree of reaction, the outlet angle of moving blades and the gross stage efficiency. Assume that the expansion efficiency is 0.94 and that the carry-over coefficient is 0.81. Neglect the effect of nozzles and blade thickness.

Solution Refer to Fig. 7.48.

$$\text{Peripheral velocity of blades, } u = \frac{\pi DN}{60} = \frac{\pi \times 0.96 \times 3000}{60} = 150.796 \text{ m/s}$$

Applying continuity equation at nozzle outlet : $\pi D l_n C_1 \sin \alpha_1 = \dot{m}v$

or $$\dot{m} = \frac{\pi \times 0.96 \times 0.12 \times 275 \times 0.342}{4.4} = 7.735 \text{ kg/s}$$

Applying continuity equation at moving blades outlet

$$\pi D l_b\, C_{r2} \sin \beta_2 = \dot{m} v$$

$$C_{r2} \sin \beta_2 = C_{f2} = \frac{7.735 \times 4.8}{\pi \times 0.96 \times 0.12} = 102.588 \text{ m/s}$$

$$\text{Power} = \frac{\dot{m}.C_w.u}{1000}$$

or

$$C_w = \frac{\text{Power} \times 1000}{\dot{m}.u} = \frac{265 \times 1000}{7.735 \times 150.796} = 227.193 \text{ m/s}$$

With the value of u, C_1 and α_1, the inlet velocity triangle *ABD* is drawn. With the help of value of C_w and C_{f2} the outlet velocity triangle *ADF* is completed.

From the diagram, C_2 =105 m/s, C_{r2} = 165 m/s, C_{r1} = 140 m/s

Now,

$$(\Delta h_{\text{isen}})_n = \frac{C_1^2 - \phi C_2^2}{2\,\eta_n \times 1000} = \frac{(275)^2 - 0.81\,(105)^2}{2 \times 1000 \times 0.94} = 35.475 \text{ kJ/kg}$$

$$(\Delta h_{\text{isen}})_b = \frac{C_{r2}^2 - \phi C_{r1}^2}{2 \times 1000 \times \eta_b} = \frac{(165)^2 - 0.81 \times (140)^2}{2 \times 1000 \times 0.94} = 6.036 \text{ kJ/kg}$$

Heat drop in the stage = $(\Delta h_{\text{isen}})_n + (\Delta h_{\text{isen}})_b$ = 35.457 + 6.036 = 41.51 kJ/kg.

= 41.51 × 7.735 = 321.07 kJ ***Ans.***

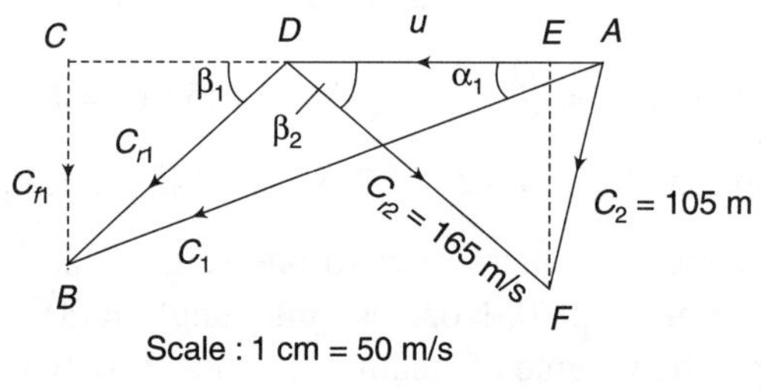

Figure 7.48 Example 7.10

The outlet angle of moving blades from the diagram = β_2 = 40 degrees ***Ans.***

$$\text{Degree of reaction} = \frac{(\Delta h_{\text{isen}})_b}{(\Delta h_{\text{isen}})_n + (\Delta h_{\text{isen}})_b} = \frac{6.036}{41.51} = 14.51\% \quad \textit{Ans.}$$

$$\eta_{gs} = \frac{u.C_w.}{1000\,\{\Delta h_{\text{isen}})_n + (\Delta h_{\text{isen}})_b\}} = \frac{150.796 \times 227.19}{1000 \times 41.51} = 82.53\% \quad \textit{Ans.}$$

Example 7.11 The steam consumption in a Parsons reaction turbine running at 400 rpm is 5 kg/s. The pressure of the steam at a certain pair is 2 bars, its dryness is 0.96 and the power developed by the pair is 4.4 kW. The discharging blade tip angle is 20° for both fixed and moving blades and the axial velocity of flow is 0.72 of the blade velocity. Calculate the drum diameter and the blade height, assuming the tip leakage as 5 per cent and neglecting the blade thickness.

Solution Refer to Fig. 7.49.

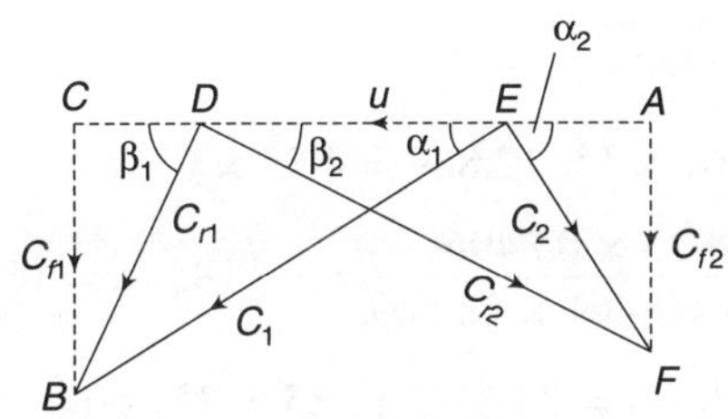

Figure 7.49 Example 7.11

The peripheral velocity of blades, $u = \dfrac{\pi DN}{60} = \dfrac{\pi D.400}{60} = 20.94\ D$ m/s

Actual mass of steam flowing through the blades for producing work

$$\dot{m}_1 = 5(1 - 0.05) = 4.75 \text{ kg/s}$$

$$\text{Power developed} = \frac{\dot{m}_1 . C_w . u}{1000} = 4.4 = \frac{4.75 \times C_w \times 20.94\ D}{1000}$$

or $$C_w = \frac{4.4 \times 1000}{4.75 \times 20.94\ D} = \frac{44.236}{D} \text{ m/s}$$

Assuming Parsons reaction turbine and solving the problem analytically, we have from Fig. 7.49.

$$C_f = C_{f1} = C_{f2} = C_1 \sin \alpha_1 = CB = 0.72u$$

or $$C_1 = \frac{0.72 \times 20.94\ D}{\sin 20^\circ} = \frac{0.72 \times 20.94\ D}{0.342} = 44\ D \text{ m/s}$$

Also, $$C_w = CD + DA = (C_1 \cos \alpha_1 - u) + C_{r2} \cos \beta_2$$

$$= (C_1 \cos \alpha_1 - u) + C_1 \cos \alpha_1 = 2 . C_1 \cos \alpha_1 - u$$

or $$\frac{44.236}{D} = 2 \times 44\ D \times 0.94 - 20.94\ D$$

$$= D\ (2 \times 44 \times 0.94 - 20.94\ D)$$

or $$\frac{44.236}{D} = D \times 61.78$$

or $$D = 0.8461 \text{ m} = \text{mean diameter} = 84.61 \text{ cm}$$

From the Steam Tables, specific volume at 2 bars and 0.96 dry

$$= x . v_s = 0.96 \times 0.8851 = 0.8496 \text{ m}^3\text{/kg}$$

Using the continually equation, we have, $\pi Dl . C_f = \dot{m}_1 . v$

But $C_f = 0.72\ u = 0.72 \times 20.94 \times 0.8496 = 12.809$ m/s

Putting the value of C_f, we have

$$\pi \times 0.8461 \times l \times 12.809 = 4.75 \times 0.8501$$

or $$l = \frac{4.75 \times 0.8496}{\pi \times 0.8461 \times 12.809} = 0.1185 \text{ m} = 11.85 \text{ cm.}$$ ***Ans.***

The drum diameter, $D_r = D - l = 84.61 - 11.85 = 72.76$ cm ***Ans.***

Example 7.12 A Parsons turbine develops 1140 kW at 400 rpm and consumes 7.89 kg of steam per kWh. Steam is supplied at 11 bars and 260°C and the isentropic efficiency of expansion is 85%. Blade angles are 35° and 20° at inlet and outlet tips respectively. Find the drum diameter and the blade height at a stage where the pressure is 1.4 bars. Height to drum diameter ratio of 1/12 is recommended. Find the power developed at that stage.

Solution Refer to Fig. 7.50.

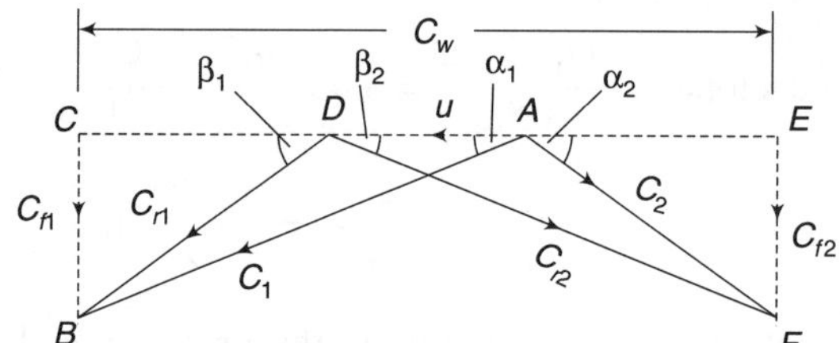

Figure 7.50 Example 7.12

Isentropic heat drop up to the given stage = $h_1 - h_2 = 2962 - 2563 = 399$ kJ/kg

∴ Actual heat drop = $0.85 \times 399 = 339.15$ kJ/kg

Actual enthalpy at the end of considered stage = $2962 - 339.15 = 2622.85$ kJ/kg

Actual dryness fraction = $x = 0.972$

At 1.4 bars and 0.972 dryness, the specific volume

$$v = 0.972 \times 1.231 = 1.196 \text{ m}^3\text{/kg}$$

Mass flow rate, $$\dot{m} = \frac{(\text{SSC}) \times \text{Power}}{3600} = \frac{7.89 \times 1140}{3600} = 2.498 \text{ kg/s}$$

Using the continuity equation,

$$\pi(D_r + l).l.C_f = \dot{m}.v.$$

or $$\pi(D_r + D_r/12)(D_r/12)\ C_f = \dot{m}.v. \quad \text{and} \quad u = \frac{\pi(D_r + D_r/12)\ N.}{60}$$

Draw the velocity diagram to an unknown scale, say $u = 3$ cm, then

$$C_f = C_{f1} = C_{f2} = 2.4 \text{ cm}$$

and $$\frac{C_f}{u} = \frac{2.4}{3} = 0.8$$

$\therefore$ $$C_f = 0.8\ u$$

Hence, putting value of

$$C_f\ \pi(D_r + D_r/12)\ (D_r/12) \times 0.8 \times \frac{\pi(D_r + D_r/12)\ N}{60} = \dot{m}.v$$

or $$\frac{D_r^3.\pi\ (1 + 1/12) \times 0.8 \times \pi\ (1 + 1/12) \times 400}{12 \times 60} = 2.498 \times 1.196$$

$$D_r = \left[\frac{12 \times 60 \times 2.498 \times 1.196 \times 144}{\pi \times \pi \times 0.8 \times 13 \times 13 \times 400}\right]^{1/3} = 0.8341 = 83.41 \text{ cm}$$ ***Ans.***

$$D_r = 0.8341 \text{ m} = 83.41 \text{ cm}$$ ***Ans.***

Hence, l = blade height $= \dfrac{D_r}{12} = \dfrac{0.8341}{12} = 0.0695$ m $= 6.95$ cm ***Ans.***

The blade velocity, $u = \dfrac{\pi DN}{60} = \dfrac{\pi(0.8341 + 0.0695) \times 400}{60} = 18.92$ m/s

Scale of the diagram, 1 cm $= \dfrac{18.92}{3} = 6.308$ m/s

Hence, $C_w = 9.78 \times 6.308 = 61.69$ m/s

Power $= \dfrac{\dot{m}.u.C_w}{1000} = \dfrac{2.498 \times 18.92 \times 61.69}{1000} = 2.915$ kW ***Ans.***

The student should solve this problem analytically.

Example 7.13 A 50% impulse-reaction turbine runs at 3000 rpm. The angles at exit of fixed bladings and inlet of moving bladings are 20° respectively. At a particular stage, the mean ring diameter is 0.7 m and the steam condition is 1.5 bars and 0.96 dry. Calculate (a) the required height of bladings to pass 50 kg/s of steam, and (b) the power developed by the stage. Solve the problem analytically.

Solution Refer to Fig. 7.51. Due to Parsons reaction turbine

$$\alpha_1 = 20° = \beta_2,\ \beta_1 = 30° = \alpha_2,\ C_1 = C_{r2},\ C_{r1} = C_2$$

$$u = \frac{\pi DN}{60} = \frac{\pi \times 0.7 \times 3000}{60} = 109.95 \text{ m/s}$$

By sin law, $\dfrac{C_1}{\sin 150°} = \dfrac{u}{\sin 10°} = \dfrac{C_{r1}}{\sin 20°}$

or $$C_1 = \frac{u \sin 150°}{\sin 10°} = \frac{109.95 \times 0.5}{0.1736} = 316.67 \text{ m/s}$$

$$C_{r1} = \frac{u \sin 20°}{\sin 10°} = \frac{109.95 \times 0.342}{0.1736} = 216.6 \text{ m/s}$$

$$C_w = C_{w1} + C_{w2} = C_1 \cos \alpha_1 + C_2 \cos \alpha_2$$

$$= 316.67 \cos 20° + 216.6 \text{ cost } 30° = 297.57 + 187.06 = 484.63 \text{ m/s}$$

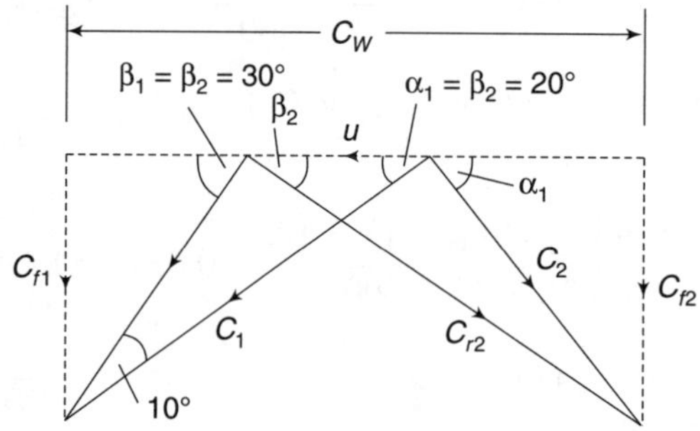

Figure 7.51 Example 7.13

The specific volume in the stage

$$v_1 = v_{f1} + x_1 v_g = 0.001052 + 0.96 \times 1.15937 = 1.114 \text{ m}^3\text{/kg}$$

From continuity equation: $\dot{m} v_1 = \pi D l \, C_{f1}$

(a) or $l = \dfrac{\dot{m} v_1}{\pi D C_1 \sin \alpha_1} = \dfrac{50 \times 1.114}{\pi \times 0.7 \times 316.67 \sin 20} = 0.2338 \text{ m} = 23.38 \text{ cm}$ ***Ans.***

(b) Power developed by the stage $= \dfrac{\dot{m} u C_w}{1000} = \dfrac{50 \times 109.95 \times 484.63}{1000} = 2664.25$ kW ***Ans.***

Example 7.14 At a certain point in low pressure (LP) stage of a 50% impulse reaction turbine, the steam leaves a moving blade row at 1.5 bars and 0.92 dry. The LP stage is double flow and the mass flow rate entering the LP turbine in 80 kg/s. The speed of the turbine is 3000 rpm. At entry to the moving blade row, the axial velocity of flow is 0.7 times and at exit from the row 0.75 times the mean blade velocity. The exit angles of both fixed and moving blades are 20°, measured from the plane of rotation. The height of moving bladings at exit is 1/10 of the mean diameter. Calculate the height of moving bladings at exit and power developed by the blade row. Solve the problem by analytical method.

Solution Refer to Fig. 7.52 (a) and (b).

Due to 50% reaction, $\alpha_1 = \beta_2 = 20°$ and $\beta_1 = \alpha_2$, $C_{f1} = 0.7\, u$ and $C_{f2} = 0.75\, u$

The mass flow rate passing through the stage in double flow is $\dot{m} = \dfrac{\dot{m}_1}{2} = \dfrac{80}{2} = 40$ kg/s

$$v = v_f + x\, v_g = 0.001052 + 0.92 \times 1.15937 = 1.06767 \text{ m}^3\text{/kg}$$

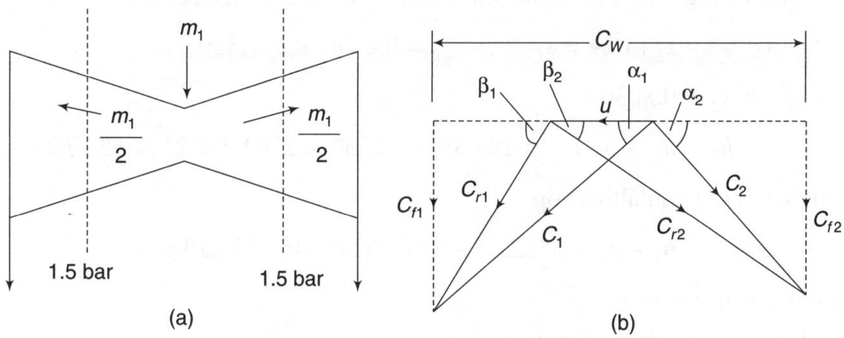

Figure 7.52 Example 7.14

From continuity equation at exit of moving blading

$$\dot{m} v = \pi D l C_{f2} = \pi D (0.1 D) \times 0.75\, \pi D \times 3000/60$$

or $$40 \times 1.06767 = \pi^2 D^3 \times 0.1 \times 0.75 \times 3000/60$$

or $$D = \left[\frac{40 \times 1.06767 \times 60}{\pi^2 \times 0.1 \times 0.75 \times 3000}\right]^{1/3} = 1.0488 \text{ m}$$

$\therefore$ $$l = \frac{D}{10} = \frac{1.0488}{10} = 0.10488 \text{ m} = 10.488 \text{ cm}$$ ***Ans.***

$$u = \pi D N/60 = \pi \times 1.0488 \times 3000/60 = 164.74 \text{ m/s}$$

$$C_w = 2\, C_1 \cos \alpha_1 - u = \frac{2\, C_{f1}}{\sin \alpha_1} \cos \alpha_1 - u$$

$$= \frac{2 \times 0.7\, u \cos \alpha_1}{\sin \alpha_1} - u = \left(\frac{1.4}{\tan \alpha_1} - 1\right) u = \left(\frac{1.4}{\tan 20°} - 1\right) \times 164.74$$

$$= 468.92 \text{ m/s}$$

Power developed by the stage

$$= \frac{\dot{m} C_w u}{1000} = \frac{40 \times 468.92 \times 164.74}{1000} = 3089.99 \text{ kW}$$ ***Ans.***

Example 7.15 A steam turbine plant works between the limit of 150 bars, 600°C and 0.1 bar. The mean blade velocity is 200 m/s. The average nozzle efficiency is 0.92. The nozzle (fixed blade) is 20°. All stages operate at the conditions of maximum efficiency. Determine the number of stages required for the following cases—(a) All simple impulse stages, (b) All 50% impulse-reaction stages, (c) A two-row Curtis stage followed by simple impulse stages, (d) A two-row Curtis stage followed by 50% impulse-reaction stages.

Solution Refer to Fig. 7.53. From steam table: h_1 = 3582.33 kJ/kg, h_{f2} = 191.85

$$s_1 = s_2 \text{ kJ/kg} = s_{f2} + x_2\, s_{g2} = 0.6490 + x_2\, 7.5009$$

or $$x_2 = 0.8036$$

∴ $$h_2 = h_{f2} + x_2\, h_{g2} = 191.85 + 0.8036 \times 2392.7 = 2114.62 \text{ kJ/kg}$$

The total isentropic enthalpy drop

$$\Delta h_{isen} = h_1 - h_2 = 3582.35 - 2114.62 = 1467.73 \text{ kJ/kg}$$

(a) All simple impulse stages

We have for the maximum efficiency

$$\frac{u}{C_1} = \frac{\cos \alpha_1}{2} = \frac{\cos 20°}{2} = 0.4698$$

or $$C_1 = \frac{u}{0.4698} = \frac{200}{0.4698} = 425.71 \text{ m/s}$$

Neglecting initial K.E. at the nozzle inlet, the nozzle velocity at exit in given by

$$C_1 = 44.72\, [(\Delta h_{isen})_{stage} \times \eta_n]^{1/2} = 425.71$$

or $$(\Delta h_{isen})_{stage} = \left(\frac{425.71}{44.72}\right)^2 \times \frac{1}{0.92} = 98.5 \text{ kJ/kg}$$

The number of simple impulse stages required is

$$n = \frac{\Delta h_{isen}}{(\Delta h_{isen})_{stage}} = \frac{1466.73}{98.5} = 14.89 \approx 15 \text{ stages}$$ ***Ans.***

(b) All 50% impulse-reaction stages

For maximum efficiency, we have

$$\frac{u}{C_1} = \cos \alpha_1$$

∴ $$C_1 = \frac{u}{\cos \alpha_1} = \frac{200}{\cos 20} = 212.83 \text{ m/s}$$

In 50% reaction, there is an equal heat drop in fixed and moving blades. Thus,

$$C_1 = 44.72 \left[\frac{(\Delta h_{isen})_{stage}}{2 \times \eta_n}\right]^{1/2} = 212.83$$

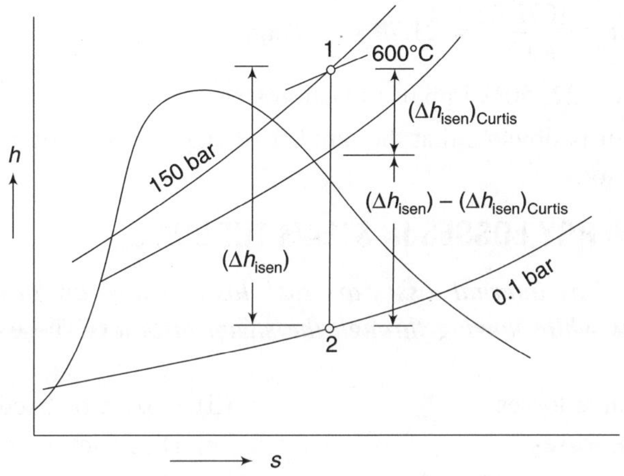

Figure 7.53 Example 7.15

or $$(\Delta h_{\text{isen}})_{\text{stage}} = \left(\frac{212.83}{44.72}\right)^2 \times \frac{2}{0.92} = 49.23 \text{ kJ/kg}$$

The number of 50% reaction stages required is

$$n = \frac{\Delta h_{\text{isen}}}{(\Delta h_{\text{isen}})_{\text{stage}}} = \frac{1466.73}{49.23} = 29.79 \approx 30 \text{ stage} \qquad \textit{Ans.}$$

(c) A 2-row Curtis stage followed by simple impulse stage. For maximum efficiency,

we have $$\frac{u}{C_1} = \frac{\cos \alpha_1}{4}$$

$\therefore$ $$C_1 = \frac{4u}{\cos \alpha_1} = \frac{4 \times 200}{\cos 20} = 851.31 \text{ m/s}$$

$$C_1 = 44.72 \, [(\Delta h_{\text{isen}})_{\text{stage}} \times \eta_n]^{1/2} = 851.31$$

or $$(\Delta h_{\text{isen}})_{\text{stage}} = \left(\frac{851.31}{44.72}\right)^2 \times \frac{1}{0.92} = 393.89 \text{ kJ/kg}$$

$$(\Delta h_{\text{isen}})_{\text{impulse}} = 1466.71 - 393.89 = 1072.82 \text{ kJ/kg}$$

Number of impulse stages required $= \dfrac{1072.82}{98.5} = 10.89 \approx 11$

1, two-row Curtis + 11 simple impulse stages *Ans.*

(d) A 2-row Curtis stage followed by 50% impulse-reaction stages

$$(\Delta h_{\text{isen}})_{\text{impulse}} = 1072.82 \text{ kJ/kg}$$

Number of 50% impulse-reaction stage required

$$n = \frac{1072.82}{49.23} = 21.79 \approx 22 \text{ stages.}$$

1, two-row Curtis + 22, 50% impulse reaction stages ***Ans.***

From the results, it is obvious that the number of stages has reduced from 30 to 22, thus saving the cost of rotor.

7.33 LIST OF ENERGY LOSSES IN STEAM TURBINES

(A) Internal Losses. *The internal losses are such losses which are directly connected with the steam conditions while flowing through the steam turbines.* These may be enumerated as follows:

(i) Regulating valve losses (ii) Nozzle or fixed blade losses
(iii) Moving blade losses (iv) Disc friction losses
(v) Blade windage or partially admission losses (vi) clearance losses
(vii) Losses due to wetness of steam (viii) Carry-over losses
(ix) Exhaust piping losses or residual or leaving velocity losses.
(x) Radiation and conduction losses

(B) External Losses. *External losses are those losses which do not influence the steam conditions. They are:*

(i) Mechanical losses and (ii) Leakage losses from the end seats.

The detailed discussions are out of the scope of the book. See Author's *"Steam Gas Turbines and Power Plant Engineering"*.

Example 7.16 The first stage wheel running at 1500 rpm of a 20,000 kW turbine is a single row wheel having a mean diameter 2.1 m. The conditions of steam in the first stage is as follows: Pressure = 17.5 bars, Superheat = 110°C, Specific volume = 0.1503 m^3/kg.

The blades are 3.2 cm long and the active nozzles cover 40% of total circumference at full load : Calculate: (a) Power absorbed by disc friction, (b) Power absorbed by blade windage and disc friction.

Solution From the Steam Tables, saturation temperature at 17.5 bars = 205.715°C

$\therefore$ Steam temperature = 205.715 + 110 = 315.715°C

Coefficient of viscosity, $\mu = (0.173t + 40.68) \times 10^{-7}$

$$= (0.173 \times 315.715 + 40.68) \times 10^{-7} = 95.298 \times 10^{-7} \text{ kg/m-s}$$

Again, $$\log_{10}(C/0.735) = 1.277 - 0.2 \log_{10}\left(\frac{uD\rho}{\mu}\right)$$

Here $$u = \frac{\pi DN}{60} = \frac{\pi \times 2.1 \times 1500}{60} = 164.93 \text{ m/s}$$

$\therefore$ $$\log_{10}(C/0.735) = 1.277 - 0.2 \log_{10}\left(\frac{2.1 \times 164.93 \times 10^{7}}{95.298} \times \frac{1}{0.1503}\right)$$

$$= 1.277 - 1.67676 = -0.39976$$

or $\quad \dfrac{C}{0.735} = 0.398 \quad$ or $\quad C = 0.398 \times 0.735 = 0.29253$

(a) Now power absorbed in disc friction is

$$(P_{\text{loss}})_{df} = CD^2 \left(\frac{u}{1000}\right)^3 \times \rho$$

$$= 0.29253(2.1)^2 \times \left(\frac{164.93}{100}\right)^3 \times \frac{1}{0.1503} = 38.50 \text{ kW} \qquad \textbf{\textit{Ans.}}$$

(b) Power absorbed by blade windage and disc friction is given by

$$(P_{\text{loss}})_{bw,\ df} = \lambda\ [1.07D + 0.61Z\ (1 - \varepsilon)l^{1.5}]\ D\left(\frac{u}{100}\right)^3 \rho$$

Here $\lambda = 1.15$; since steam in superheated $Z = 1$, $\varepsilon = 0.4$, $l = 3.2$ cm, $D = 2.1$ m

$$\therefore \quad (P_{\text{loss}})_{bw,\ df} = 1.15\ [1.07 \times 2.1 + 0.61 \times 1(1 - 0.4)\ (3.2)^{1.5}] \times 2.1 \times \left(\frac{164.93}{100}\right)^3 \times \frac{1}{0.1503}$$

$$= 189.962 \text{ kW} \qquad \textbf{\textit{Ans.}}$$

7.34 STAGE EFFICIENCY OF IMPULSE TURBINES

Stage efficiency is defined as the ratio of shaft power to the isentropic power.

$$\eta_s = \frac{s.p.}{is.p.} = \frac{\text{Shaft power}}{\text{Isentropic power}} = \frac{\text{Actual heat drop}}{\text{Isentropic heat drop}} \qquad ...(7.76)$$

where *s.p.* = shaft power developed in the stage, i.e., the actual power transmitted by the wheel to the rotor spindle.

is. p. = isentropic power of the stage, i.e., the power based on isentropic heat drop and total internal steam flow.

To find the value of *s.p.* and *is.p.* let us assume

$\dot{m}_1$ = total internal flow of steam kg per sec

$\dot{m}_n$ = steam flow through nozzles, kg per sec

$\dot{m}_g$ = steam flow through diaphragm gland, kg per sec

Then, $\dot{m}_1 = \dot{m}_s + \dot{m}_g \qquad ...(7.77)$

It is to be noted that in heat drop, $\dot{m}_1$ will be taken into account but if work done is calculated from velocity diagram, then $(\dot{m}_1 - \dot{m}_g)$, i.e., actual flow through blading $(\dot{m}_n)$ will be used. If $(\Delta h_{\text{isen}})_n$ be the isentropic heat drop in the nozzles of a pure impulse turbine, then the isentropic power by $\dot{m}_i$ kg of steam,

$$is.p. = \dot{m}_1\ (\Delta h_{\text{isen}})_n,\ \text{kW} \qquad ...(7.78)$$

In the case of impulse reaction turbine, the total isentropic heat drop in the heat drop in fixed blades $(\Delta h_{\text{isen}})_n$ and moving blades $(\Delta h_{\text{isen}})_b$ in a stage.

$$is.p. = \dot{m}_i\,[(\Delta h_{isen})_n + (\Delta h_{isen})_b];\ \text{kW} \quad ...(7.79)$$

Here, it is worth to note that the shaft power is less than blade or rim power (r.p.) by amount of power absorbed $(P_{\text{loss}})_{wf}$ in overcoming wheel friction, i.e., sum of power absorbed $(P_{\text{loss}})_{df}$ in disc friction and power absorbed $(P_{\text{loss}})_{bw}$ due to partial admission.

$$s.p. = r.p. - (P_{\text{loss}})_{wf} \quad \text{where r.p. is rim power}$$

or

$$s.p. = r.p. - (P_{\text{loss}})_{df} - (P_{\text{loss}})_{bw} \quad ...(7.80)$$

7.35 STATE POINT LOCUS OF AN IMPULSE TURBINE

Now, consider a stage of impulse turbine. Let A represent the initial condition of a steam on the h-s diagram shown in Fig. 7.54 and suppose that expansion of steam takes place from pressure p_1 to pressure p_2. AB shows the isentropic expansion of steam in the nozzle and the actual path of the steam is AC due to losses in the nozzle. The steam with the condition at C now enters the moving blades where the steam suffers a change in direction and where the rim power is developed.

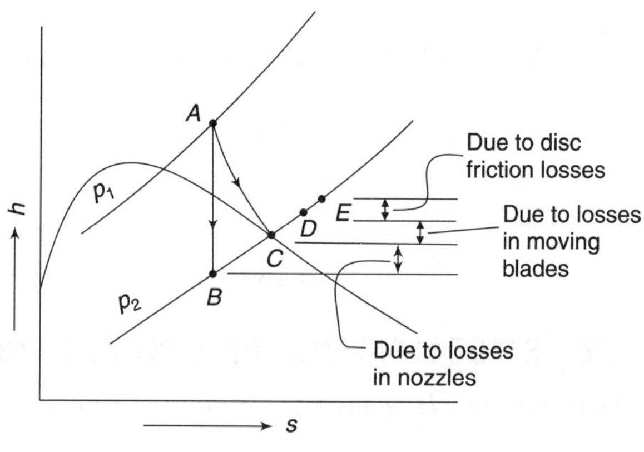

Figure 7.54 State point locus of an impulse turbine

The part CD of p_2 line shows the reheating of steam due to blade friction while passing over the blades, taking place, assuming constant pressure, although pressure in the blade channel is not constant. So, the $r.p.$ developed is equal to $\dot{m}_n\, C_w/1000$ and also

$$r.p. = \dot{m}_1\,(h_A - h_D)$$

or

$$h_A - h_D = \frac{r.p.}{\dot{m}_i} \quad ...(7.81)$$

From the above equation h_D is known, so we can locate the point D on p_2 line.

Part DE shows the further reheating of steam due to wheel friction, i.e., due to disc friction and blade windage. If $(P_{\text{loss}})_{wf}$ is the amount of power used to overcome wheel friction, then

$$h_E - h_D = \frac{(P_{\text{loss}})_{wf}}{\dot{m}_i} \quad ...(7.82)$$

By this method, we can locate the point E on p_2 line. It is also possible to locate the point E from the shaft power of the stage. The point E has been obtained after considering all types of losses in a stage of impulse turbine, so $(h_A - h_E)$ is the actual heat drop in the stage.

$$s.p. = \dot{m}_i(h_A - h_E) \qquad ...(7.83)$$

From Eq. (7.83), *is.p* is given by

$$i.sp. = \dot{m}_1\ (h_A - h_g)$$

$$\eta_s = \frac{s.p.}{is.p.} = \frac{h_A - h_E}{h_A - h_B} \qquad ...(7.84)$$

If η_s is known, the point E may be located on the p_2 line. The point E is the initial state point for the succeeding stage of turbine. *The point E is called the state point locus of a single stage impulse turbine because it gives the actual condition of steam* at the end of the stage.

7.36 STATE POINT LOCUS FOR MULTI-STAGE STEAM TURBINE

Figure 7.55 shows the expansion of steam through a multi-stage turbine. Assuming six stages only for simplicity. The initial pressure is p_1 and the black pressure is p_7. If the stage efficiency is known then the locus of the state point may be drawn on the *h-s* diagram.

State 1 is the initial point set down according to the initial condition of steam. 1–2 shows the isentropic expansion of steam in the first stage. But owing to losses the actual state point of the steam after expansion in the first stage is 2′ and not 2. Thus, the actual heat drop is 1 – 2′ and not 1–2 so, we have

$$h_1 - h_2' = \eta_s\ (h_1 - h_2)$$

and point 2′ is thus set down with the h_2' value on the p_2 line.

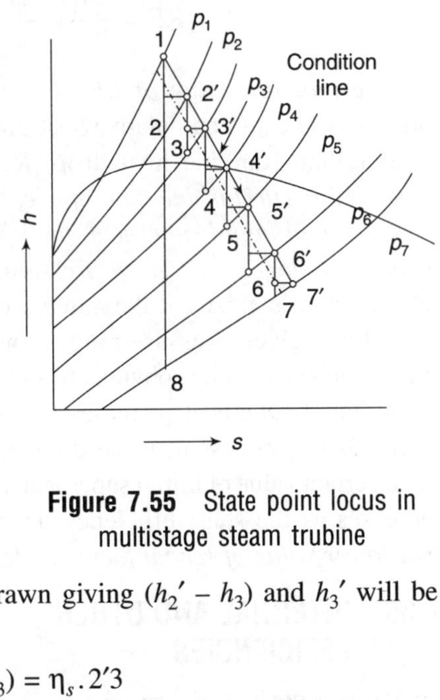

Figure 7.55 State point locus in multistage steam trubine

When point 2′ is known isentropic expansion is drawn giving $(h_2' - h_3)$ and h_3' will be calculated from the relation

$$(h_2' - h_3') = 2'3' = \eta_s\ (h_2' - h_3) = \eta_s . 2'3$$

and the point 3′ is set down. Similarly, other points in the succeeding stages are set down. The isentropic and actual heat drops in the succeeding stages are represented by 3′ – 4 and 3′ – 4′, 4′ – 5 and 4′ – 5′, 5′ – 6, and 5′ – 6, 6′ – 7, and 6′ – 7′. The points, 1, 2′, 3′, 4′, 5′, 6′, 7′, represent the condition of steam in the wheel chambers of the six stages, i.e., at the blade outlets. 1–8 is the Rankine heat drop for the initial and final pressure. *The curve joining the points 1, 2,′ 3,′ 4,′ 5,′ etc., is called the* ***condition line*** *of the expansion of steam because the steam probably expands along some such paths.* The heat drops 1 – 2,′ 2, 2′ – 3,′ 3′ – 4′, etc, which are actually converted into work may be called useful heat drops in the corresponding stages. *The dotted line shows the* ***condition of steam*** *at outlet from the nozzle.*

The sum of the isentropic heat drops 1 – 2, 2′ – 3, 3′ – 4, etc., called *cumulative heat drop,* is represented by Δh_c. Since the constant pressure line diverges as we move from left to right on the *h-s diagram the cumulative heat drop is always greater than direct isentropic or Rankine heat drop* 1 – 8.

7.37 REHEAT FACTOR

Reheat Factor (RF) is defined as the ratio of cumulative heat drop, Δh_c *to the direct isentropic or Rankine heat drop,* Δh_{isen} *(i.e., 18)*

$$\text{RF} = \frac{\Delta h_c}{\Delta h_{\text{isen}}} = \frac{\text{Cumulative heat drop}}{\text{Rankine heat drop}} \quad \text{...(7.85)}$$

Referring to Fig. 7.55,

$$\text{RF} = \frac{\Delta h_{12} + \Delta h_{2'-3} + \Delta h_{3'-4} + \ldots}{\Delta h_{\text{isen}}} = \frac{12 + 2'3 + 3'4 + \ldots}{18} \quad \text{...(7.86)}$$

Note that the value of RF is always greater than unity because cumulative heat drop is always greater than Rankine heat drop. *Reheat factor depends upon the turbine stage efficiency, the initial pressure, and initial superheat, final pressure and number of stages in a given pressure range.* The reheat factor is greater if the number of stages are large for a given pressure ranges and lower the stage efficiency. As it is obvious from Fig. 7.56 that if the initial superheat increases the reheat factor increases for a constant value of stage efficiency. For a certain value of initial superheat, reheat factor increases as the stage efficiency decreases. *Note that lower value of reheat factor is desirable.*

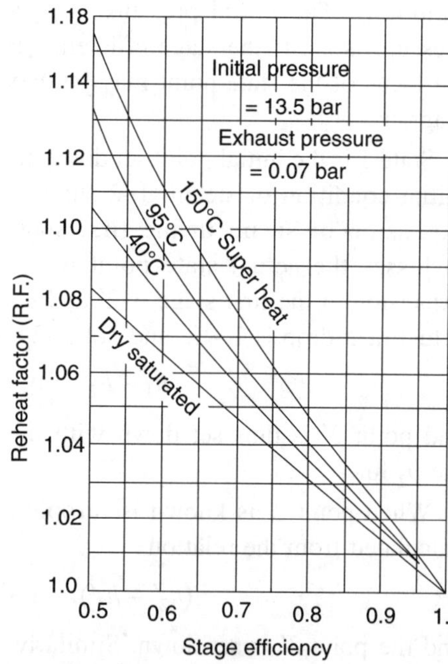

Figure 7.56 Effect of stage efficiency and superheat on reheat factor

7.38 INTERNAL AND OTHER EFFICIENCIES

(a) Internal Efficiency. *The internal efficiency of turbine* (η_i) *is the ratio of the internal work done in the turbine to the direct isentropic or Rankine work done.*

$$\eta_i = \frac{\Delta h_i}{\Delta h_{\text{isen}}} = \frac{\Delta h_{12}' + \Delta h_{2'3}' + \Delta h_{3'4}' + \ldots}{\Delta h_{\text{isen}}} = \frac{12' + 2'3' + 3'4' + \ldots}{18} \quad \text{...(7.87)}$$

If the stage efficiency η_s is the same for all stages or if η_s is the mean stage efficiency, then we have

$$\eta_s = \frac{\Delta h_{12}'}{\Delta h_{12}} = \frac{\Delta h_{2'3}'}{\Delta h_{2'3}} = \frac{\Delta h_{3'4}'}{\Delta h_{3'4}}, \text{ etc.} = \frac{\Delta h_{12}' + \Delta h_{2'3}' + \Delta h_{3'4}' + \ldots}{\Delta h_{12} + \Delta h_{2'3}' + \Delta h_{3'4} + \ldots} = \frac{\Delta h_i}{\Delta h_c} \quad ...(7.88)$$

Hence, $\Delta h_i = \eta_s \, \Delta h_c$

But $\quad RF = \dfrac{\Delta h_c}{\Delta h_{\text{isen}}}$

$\therefore \quad \Delta h_i = (RF) . \eta_s . \Delta h_{\text{isen}}$

Putting this value of Δh_i in Eq. (7.87)

$$\eta_i = (RF) . \eta_s \quad ...(7.89)$$

The internal power of a turbine,

$$ip = \dot{m}_1 . \Delta h_i ; \text{ kW} \quad ...(7.90)$$

(b) Isentropic Efficiency. Isentropic efficiency of turbine is the ratio of actual head drop to isentropic heat drop in whole turbine. Thus,

$$\eta_{\text{isen}} = \frac{(\Delta_h)_{\text{act}}}{(\Delta_h)_{\text{isen}}}$$

$$\eta_{\text{isen}} = \eta_s = \eta_s \quad \text{if} \quad RF = 1$$

$$i.p. = \dot{m}_i \, . \, \Delta h_i, \quad \text{kW} \quad ...(7.91)$$

(c) Overall Thermal Efficiency of Turbo-alternator (η_{oth}) Overall thermal efficiency of steam power plant is expressed as by considering all losses.

$$\eta_{\text{oth}} = \frac{\text{Generator output}}{\text{Actual heat supplied}} = \frac{\text{G.O.}}{q_{Aa}} = \frac{\left((w_t)_{\text{isen}} \times \eta_i \times \eta_m \times \eta_g - \dfrac{w_p}{\eta_p}\right) \eta_{\text{boiler}}}{q_A}$$

Neglecting pump work, $\eta_{\text{oth}} = (\eta_{\text{th}})_{\text{isen}} \times \eta_{\text{boiler}} \times \eta_i \times \eta_m \times \eta_g$ ...(7.92)

(d) Mechanical Efficiency (η_m) Mechanical efficiency is defined as

$$\eta_m = \frac{\text{Coupling power or Generator input}}{\text{Internal power}}$$

$$= \frac{i.p. - (p_{\text{loss}})_m}{i.p.} = \frac{\Delta h_i - (\Delta h_{\text{loss}})_m}{\Delta h_i} = 1 - \frac{(\Delta h_{\text{isen}})_m}{\Delta h_i} \quad ...(7.93)$$

(e) Relative Efficiency (η_{rel}): Relative efficiency is expressed as

$$\eta_{\text{rel}} = \frac{\text{Overall thermal efficiency based on coupling work}}{\text{Isentropic or Rankine efficiency}}$$

$$= \frac{\Delta h_{coup}}{q_A} \times \frac{q_A}{\Delta h_{isen}} = \frac{\Delta h_{coup}}{\Delta h_{isen}} = \eta_i \times \eta_m \quad ...(7.94)$$

(f) Generator Efficiency (η_g)

Generator efficiency is defined as $\eta_g = \frac{\text{Generator output (G.O.)}}{\text{Generator input (G.I.)}}$ ...(7.95)

(g) Electrical Efficiency (η_e) Electrical efficiency of a turbo-generator is given by

$$\eta_e = \eta_i \, \eta_m \, \eta_g \quad ...(7.96)$$

(h) Mass Flow Rate Mass flow rate of steam in kg/h is expressed as

$$\dot{M} = \frac{3600 \times \text{Generator output}}{\eta_i \, \eta_m \times \Delta h_{isen}} \quad ...(7.97)$$

Overall specific mass flow rate in kg/kWh is given as

$$SSC = \frac{3600}{\eta_i \, \eta_g \, \eta_m \, \eta_g \times \Delta h_{isen}} \quad ...(7.98)$$

Note. If we have to find out the generator output, it will be calculated as follows:

The losses in steam turbine are depicted in Fig. 7.57.

$s.p. = r.p. - (P_{loss})_{df} - (P_{loss})_{bw}$ for single stage.

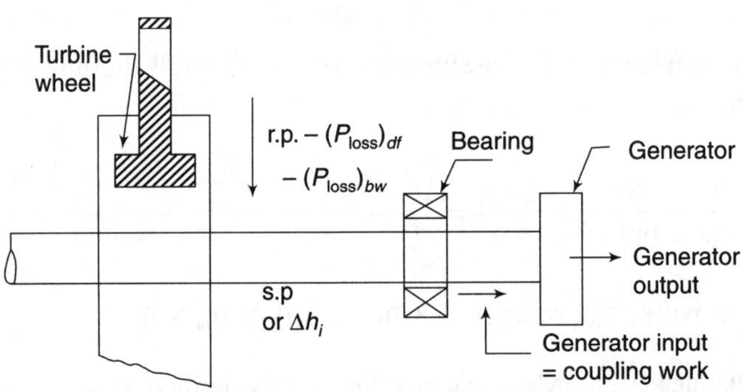

Figure 7.57 Losses in steam turbine

If internal efficiency is given, then

Internal work done = $\Delta h_i = \eta_i \times \Delta h_{isen}$

Generator input = *s.p.* – mechanical loss in bearing, etc., if any = $(s.p.)\,\eta_{mech}$

Also, generator input = Internal work done – Mechanical loss = $(\Delta h_i)\eta_m$

Generator output = (Generator input) η_g

If efficiency ratio is given, then

Turbine coupling work = Generator input = $\eta_r \times \Delta h_{isen}$

Note that for single stage turbine, stage efficiency and internal efficiency are the same.

Example 7.17 In a four-stage pressure-compounded impulse turbine, steam is at a pressure of 23 bars and superheated to a temperature of 345°C. The exhaust pressure is 0.07 bar and the overall turbine efficiency 0.72. Assuming that the work is shared equally between the stages and the condition line is straight, estimate the stage pressures, the efficiency of each stage and the reheat factor.

Solution Refer to Fig. 7.58. The initial condition of steam is represented by point 1. Here, the term overall turbine efficiency is used in the sense of internal efficiency of turbine.

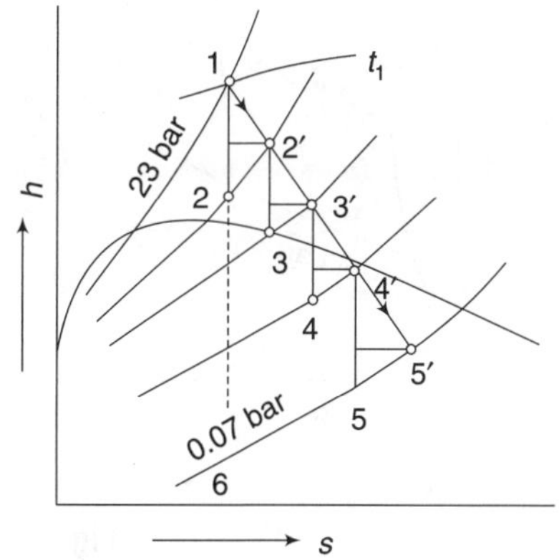

Figure 7.58 Example 7.17

Isentropic heat drop, Δh_{16} = 982.3 kJ/kg

Hence, actual heat drop

$$= 982.3 \times 0.72 = 707.25 \text{ kJ/kg}$$

This gives the point 5′ on h-s chart, so the condition, line can be drawn assuming a straight line.

Since the work is shared equally, so the heat drop, i.e., work done is each stage

$$= \frac{707.25}{4} = 176.81 \text{ kJ/kg}$$

Now $\qquad h_2' = h_1 - 176.81 = 3123 - 176.81 = 2946.19$ kJ/kg

$$h_3' = 2769.25 \text{ kJ/kg}, \; h_4' = 2592.64 \text{ kJ/kg}$$

With the help of line 1 5′ and h_2', h_3', h_4' the pressures are 7.25 bars, 1.9 bars, 0.39 bar. The efficiencies of the stage are

$$\eta_{s1} = \frac{\Delta h_{12}'}{\Delta h_{12}} = \frac{176.605}{286.33} = 61.677\% \quad \text{and} \quad \eta_{s2} = \frac{\Delta h_2'{}_3'}{\Delta h_2'{}_3} = \frac{176.605}{259.16} = 68.14\%$$

$$\eta_{s3} = \frac{\Delta h_3'{}_4'}{\Delta h_3'{}_4} = \frac{176.605}{250.8} = 70.41\% \quad \text{and} \quad \eta_{s4} = \frac{\Delta h_4'{}_5'}{\Delta h_4'{}_5} = \frac{176.605}{242.44} = 72.84\% \quad \textbf{\textit{Ans.}}$$

Cumulative heat drop, $\Delta h_c = \Delta h_{12} + \Delta h_2'{}_3 + \Delta h_3'{}_4 + \Delta h_4'{}_5$

$$= 286.33 + 259.16 + 250.8 + 242.44 = 1038.73 \text{ kJ/kg}$$

$$\therefore \qquad \text{RF} = \frac{\Delta h_c}{\Delta h_{\text{isen}}} = \frac{1038.73}{982.3} = 1.0574 \qquad \textbf{\textit{Ans.}}$$

Example 7.18 A 50% impulse-reaction turbine having 16 stages develops an electrical power of 14 MW. The steam enters the turbine at 15 bars, 350°C and exhaust to the condenser at 0.14 bar. The stage efficiency is 80% for each stage. The reheat factor is 1.04. At a certain stage, the steam is at 1 bar, dry and saturated. The exit angle of the blades is 20° and the ratio of blade velocity to steam velocity is 0.7. The blade height may be taken as 1/12 of the mean blade diameter. Work out the following: (a) The steam flow rate required by the turbine if the mechanical and generator efficiencies are each 98%, (b) the mean blade diameter, (c) the blade height at the stage considered, (d) The speed of the rotor.

Solution Refer to Fig. 7.59 (a) and (b).

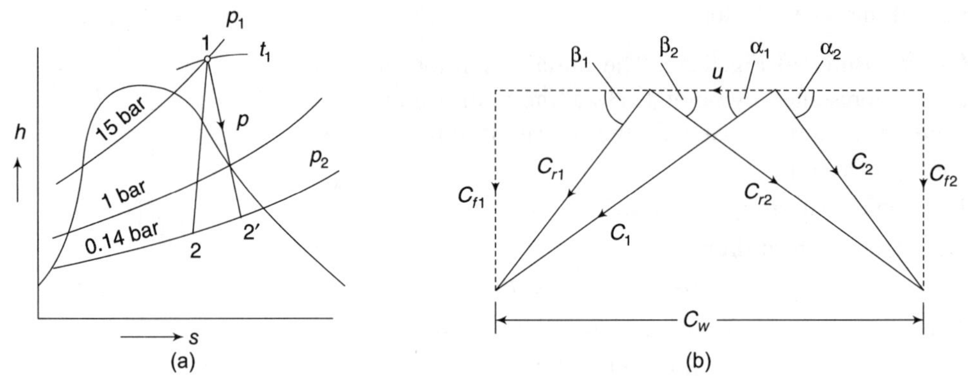

Figure 7.59 (a) and (b) Example 7.18

The internal efficiency is given by $\eta_i = \eta_s \times \text{R.F.} = 0.8 \times 1.04 = 0.832$

From Mollier chart, $\Delta h_{\text{isen}} = 855$ kJ/kg

$$\therefore \quad \Delta h_{\text{act}} = \eta_i \times \Delta h_{\text{isen}} = 0.832 \times 855 = 711.36 \text{ kJ/kg}$$

$$\text{Diagram power} = \frac{\text{Electrical power}}{\eta_m \cdot \eta_{\text{gen}}} = \frac{14 \times 10^3}{0.98 \times 0.98} = \dot{m}_s \times \Delta h_{\text{act}}$$

or
$$\dot{m}_s = \frac{14 \times 10^3}{0.98 \times 0.98 \times 711.36} = 20.49 \text{ kg/s}$$ ***Ans.***

Work done per kg, $w = u\,(2C_1 \cos \alpha_1 - u)$

But $$\frac{u}{C_1} = 0.7$$

$$\therefore \quad w = u\left(\frac{2u}{0.7}\cos 20° - u\right) = 1.6848\, u^2 \text{ J/kg}$$

$$\text{Actual enthalpy drop per stage} = \frac{711.36}{16} = 44.46 \text{ kJ/kg}$$

Thus, $$1.6848\, u^2 = 44.46 \times 10^3$$

or
$$u = \left(\frac{44.46 \times 10^3}{1.6848}\right)^{1/2} = 162.44 \text{ m/s}$$

$$C_{f1} = C_1 \sin \alpha_1 = \frac{u}{0.7}\sin \alpha_1 = \frac{162.44}{0.7}\sin 20° = 79.36 \text{ m/s}$$

Specific volume at 1 bar dry and saturated $v = 1.694\ m^3/kg$. From continuity equation

$$\dot{m}v = \pi D l\, C_{f1} = \pi\, D \times \frac{D}{12} \times 79.36$$

$$D = \left(\frac{\dot{m}v \times 12}{\pi \times 79.36}\right)^{1/2} = \left(\frac{20.49 \times 12}{\pi \times 79.36}\right)^{1/2} = 0.993 \text{ m} \qquad \textit{Ans.}$$

$$u = \frac{\pi D N}{60}$$

or

$$N = \frac{162.44 \times 60}{\pi \times 0.993} = 3124.2 \text{ rpm} \qquad \textit{Ans.}$$

7.39 NEED OF GOVERNING

Figure 7.60 shows the output-speed characteristics of ungoverned steam turbine that is not suitable for electrical generation purposes. The power output increases first with speed, reaches a maximum (rated) value at rated speed and after this decreases. The turbine speed (i.e., electrical frequency) is a function of electrical load (i.e., consumer's demand); the lower the load, higher the speed due to accelerating torque. The load, may not remain the same, e.g., sometimes full, sometimes half as per consumer's demand and under this condition, the electrical frequency will vary with the load. This is not at all desirable as all the electrical machines and appliances are designed for a nominal electrical frequency. This suggests that some additional form of control must be applied to ensure that the changes in the consumer's demand for electrical power are followed automatically without large deviations from the nominal electrical frequency. In addition to this, an excessive increase in speed may endanger the rotating parts of the turbine due to increased centrifugal stresses and therefore it needs some form of control device. For these reasons, all turbines are fitted with speed governors. *The governor is arranged to control the turbine so as to give its characteristics of falling output power with rising shaft speed above nominal value* (Fig. 7.60).

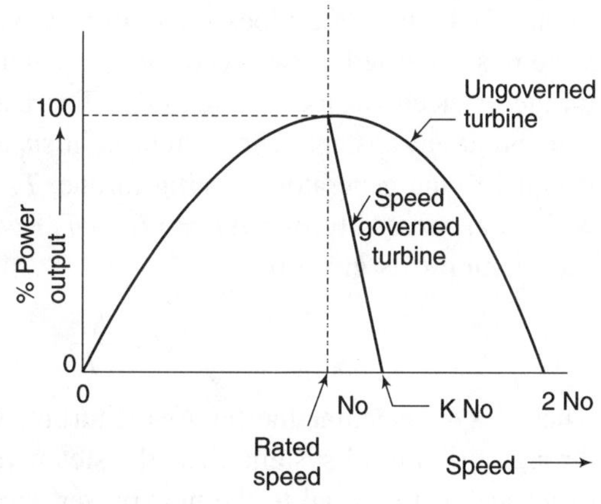

Figure 7.60 Output-speed characteristics of steam turbine

The purpose of governing is to maintain the speed of the turbine fairly constant irrespective of load. The change in speed from no load to full load, divided by the full load (or rated) speed is termed *regulation* (k) or the droop of the turbine. It is normally adopted within the range of ± 4%.

The turbine output power is controlled by varying the steam flow (power output which is a function of steam flow rate) by means of valves interposed between the boiler and the turbine. Based on the methods of varying steam flow rate, main methods of governing are:

(a) Throttle governing
(b) Nozzle control governing
(c) By-pass governing
(d) Combination of number (a) and number (b) or number (a) and number (c).

Over and above of forming the required load-speed characteristics of the turbine, the governing gear also provides the following:

(i) *Controlling the turbine speed at no load to permit the unit to be synchronised to the grid.*
(ii) *Controlling the turbine load running in parallel with other generating sets.*
(iii) *Providing protective gear for safe operation of the turbine.*

7.40 HYDROMECHANICAL SPEED GOVERNING LOOP

Figure 7.61 shows the block diagram of a typical hydrochemical speed governing loop. The system is subjected to two types of torque-turbine driving torque (T_T) and generator resisting torque or electrical load torque (T_E). The latter, T_E depends upon the consumers electricity demand at any instant. The system is in speed equilibrium only when the turbine torque T_T is equal to the generator resisting torque, T_E. If there is sudden fall in power demand, there will be an excess torque $\Delta T = (T_T - T_E)$ which will accelerate the machine. The rate of acceleration α is given by

$$\alpha = \frac{\Delta T}{I}$$

where I is the machine inertia. As the turbine speed rises, the speed governor will respond and through the control system close the steam valve partially or fully until the turbine response of torque (T_T) is equal to the new power demand. A glimpse of the response of turbine to a sudden change in load and the resultant change in speed may be obtained from Fig. 7.62(a).

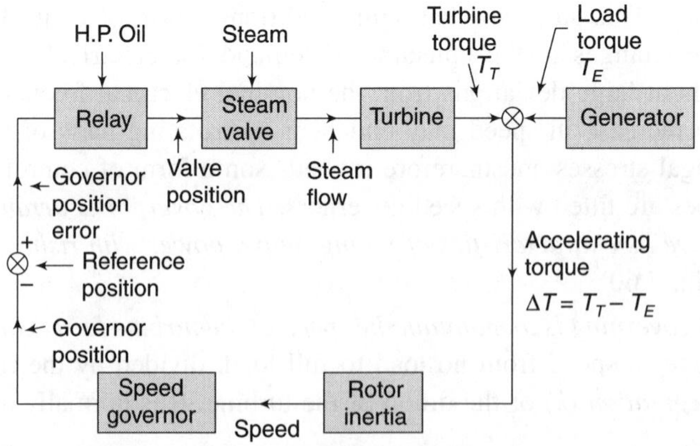

Figure 7.61 Hydromechanical speed governing loop

An idea of the change in load and speed (frequency) on the regulation diagram may be derived from Fig. 7.62(b). These figures show a small error (ΔT_e) in the actual power delivered to the consumer, as the electrical load is frequency sensitive.

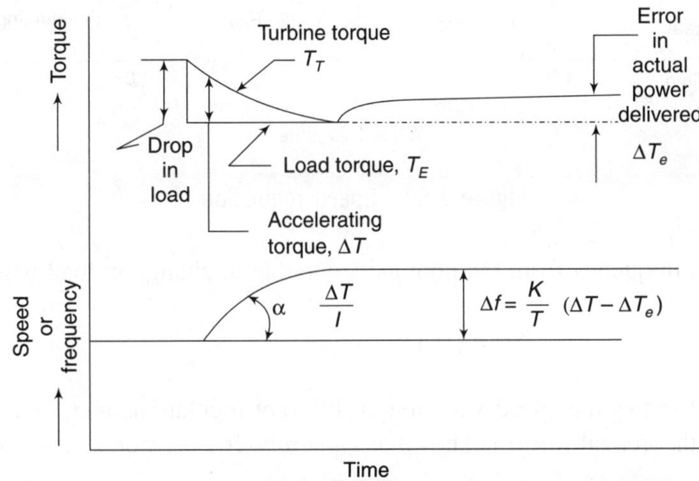

(a) Torque and speed vs. time

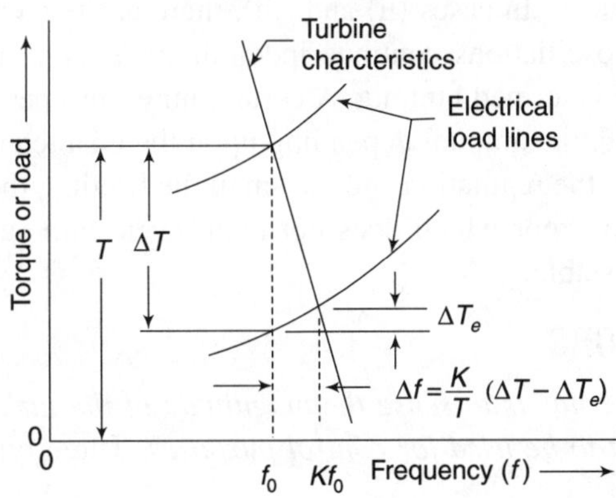

(b) Torque vs. frequency

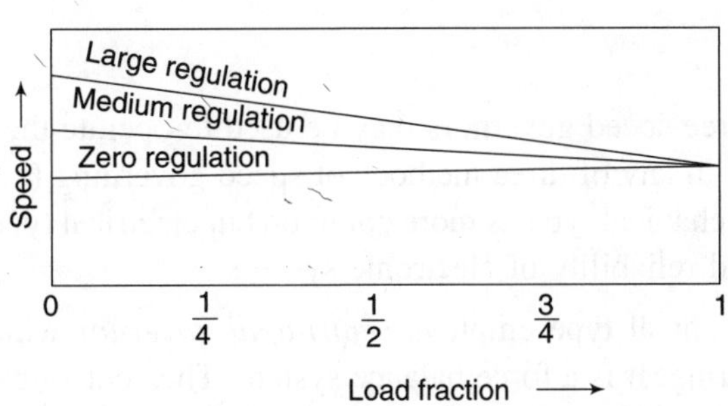

(c) Speed vs. load

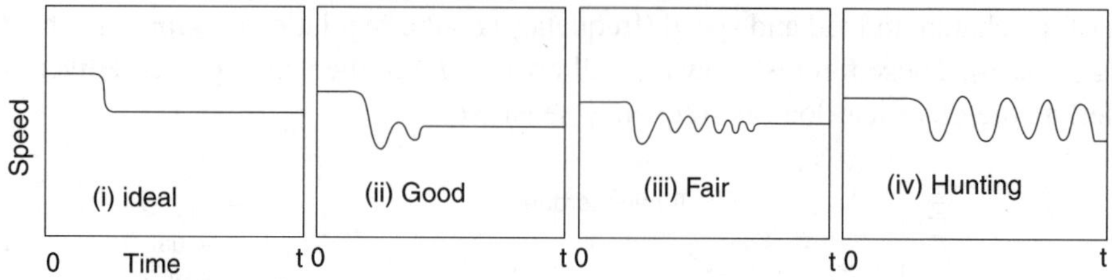

(d) Speed vs. time

Figure 7.62 Speed regulation

The change in frequency from the nominal value due to change in load is given by

$$\Delta f = \frac{K}{T}(\Delta T - \Delta T_e) \qquad ...(7.99)$$

Figure 7.62(c) shows the speed variation at different regulations and load fractions. The zero regulation (theoretical form) is known as *isochronous* governor.

Figure 7.62(d) shows the variations of rpm with time, before and after sudden addition of a load. In the case (i), the speed changes at once to the new value and continues there; this is ideal but difficult to achieve. In cases (ii) and (iii), there are few oscillations but diminish slowly while in case (iv) oscillations continue indefinitely. This oscillation of speed above and below the mean speed is termed hunting. A certain minimum percentage of regulation is essential to prevent *hunting*, the amount depending upon the characteristics of both governor and load variation. The less the regulation and the small the friction, the greater the sensitivity but less the *stability*. A governor which does not hunt, even with sudden and considerable change of load, is termed stable.

7.41 SPEED GOVERNORS

The function of speed governor is to sense the magnitude of the turbine speed and respond with a proportional output to be used for control purposes. Three types of speed governors are in common use.

- Mechanical
- Hydraulic
- Electrical

Any of the above three speed governors may be used to operate the valves proportionally through control system in any of three methods of speed governing (i.e., throttle, nozzle and by-pass). At present mechanical type, is more common but electrical type is gaining popularity due to its efficiency and reliability of electronic system.

(a) Mechanical. Mechanical type employs *centrifugal governor* which is driven from the main shaft through gearing. It is a force balance system. The centrifugal force of the rotating weights is balanced by the elastic force of spring and the radial position taken up by the weights depends upon the speed of rotation.

Figure 7.63(a) shows the principal part of a centrifugal governor which is coupled to the turbine shaft through a toothed gearing to sense the speed of turbine, As the speed of turbine increases, the weights (m) fly apart under the influence of centrifugal force changing their position along the arc of radius (AO) through a distance, Z_1. The position of the governor sleeve gets displaced from its original position B to new position B_1, i.e., through a distance, Z. If the speed decreases, the weight (m) would come closer and the sleeve would be pushed down along the governor spindle. The displacement of governor sleeve (i.e., radial position of weights with variation of shaft speed is utilised for effecting required change in the steam supply, both qualitatively and quantitatively. The sleeve displacement of the governor may be transmitted to the control through hydraulic relay.

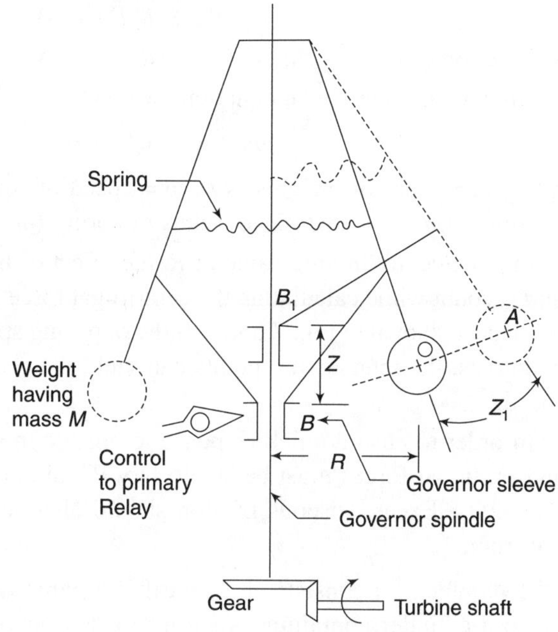

(a) Centrifugal governor

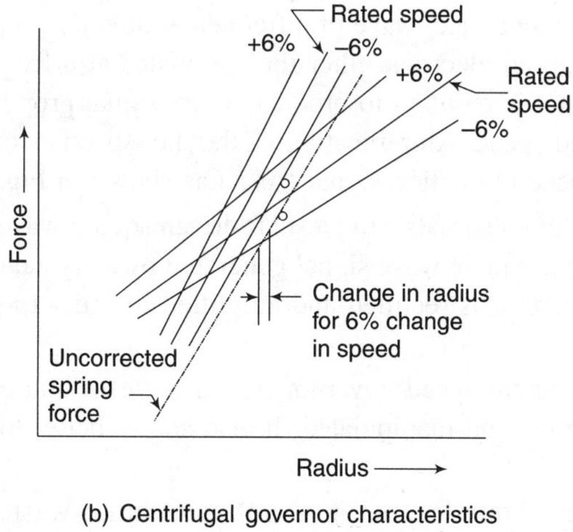

(b) Centrifugal governor characteristics

Figure 7.63

The radial position of the weight may be calculated by using the force balance principle. The centrifugal force of the weight for a simplified case of a single weight and spring is given by

$$F_w = m\omega^2 R, \quad N$$

and the force of the spring = $F_s = S(R - a), \quad N$

On equating the above equations, we get

$$\therefore \qquad m\omega^2 R = S(R - a) \qquad \text{...(7.100)}$$

where m = mass of weights (kg), ω = speed of rotation (rad/s), R = radial position of, e.g., of weight (m), a = unstrained radius of spring (m).

In practice, to find the value of R, the effect of linkage and the mass of the spring is taken into account while calculating the centrifugal force and a plot of speed, force and position can be made as shown in Fig. 7.63(b). If the opposing spring characteristics are then superimposed on curves, the intersection points can yield the relationship between the radial position and speed.

In order to ensure quick response to change in speed, it is essential that the radial inertia and frictional forces must be small. Dry (Coulomb) friction must be avoided as it results in sensitivity. Excess viscous friction should also be avoided to avoid oscillatory response to governor.

(b) Hydraulic. It consists of a centrifugal pump driven from the main shaft which supplies oil to a cylinder containing a spring loaded piston or bellows. The pressure developed in the oil is proportional to the square of the turbine speed; this means that the position of the piston or bellow is a function of the speed. The position of the piston may be used to operate steam valves through some control arrangement. This type of governor is used in some of the power plants of U.S.A.

(c) Electrical. One type of electrical speed governor consists of a shaft driven permanent magnet generator producing an a.c. wave of a frequency directly proportional to shaft speed. This a.c. wave is fed into an electrical filter that resonates at a frequency corresponding to rated speed and the output is rectified to produce a d.c. signal proportional to the difference between actual and rated speed. Amplification of the d.c. speed error signal is then made to produce the required speed correction signal ($\Delta f/K$) as shown in Fig. 7.62(b).

Another type uses a digital signals from a shaft driven square wave generator, the output is compared with a reference square wave signal generated by a crystal or tuning fork oscillator. The output of the comparator is rectified and amplified in order to get the required control signal.

The advantage of electrical speed governor lies in the ease and accuracy with which the speed signal can be formed and manipulated. It also shows better long-term stability than a mechanical governor.

In many large power plants located in U.S.A. and other western countries, electrical governors have been employed. India is also shifting for this.

7.42 RELAYS

Relays are devices by which a weak current or other small power (e.g., force and displacement) is used to control the strong one.

The output of the speed governor is not sufficient to actuate the steam valves and therefore requires amplification of both force and displacement to provide sufficient power to do the same. Governor relays (or servos) are employed to perform this duty. In practice, more than one relay is usually employed due to the relative positions of the governors and steam valves to provide the required force amplification. The hydraulic relay is the best as a force amplifier for steam valve actuation. Mechanical or electrical amplifier cannot actuate the steam valves of modern turbine and that is why the electrical signal is converted into a mechanical displacement so that hydraulic relay may be used.

Various forms of hydraulic relays are available. The relay used in steam turbine consists of a power cylinder, the piston of which is positioned by a flow of oil controlled by a spool pilot valve.

A power relay of the single-acting spring return type for operating the turbine governor steam valves is shown in Fig. 7.64(a). The function of the spring is to provide a reserve of energy which, in the event of a failure of the oil pressure, will close the valves automatically. Out of two bobbins on the pilot spool, only one is used as a valve, the other is meant to balance the pressure force.

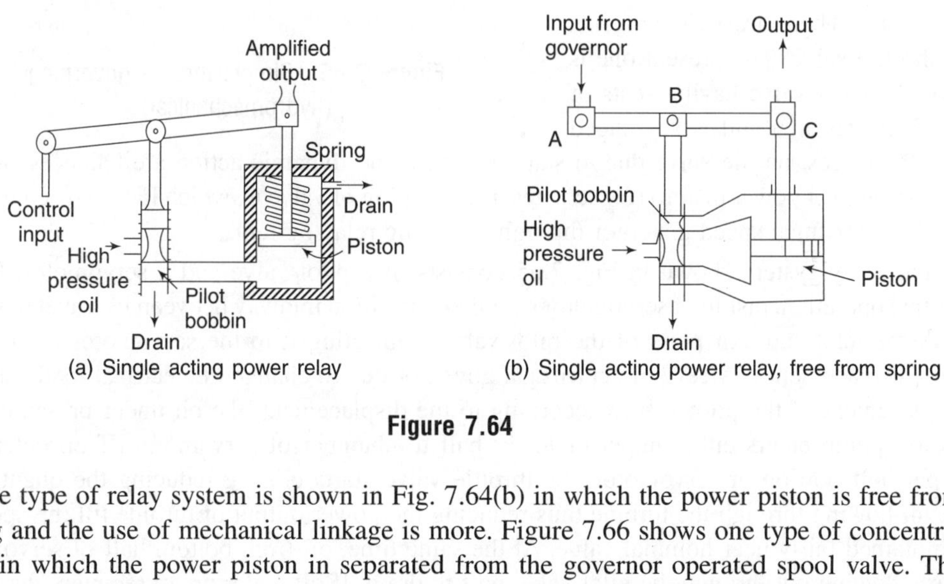

(a) Single acting power relay
(b) Single acting power relay, free from spring

Figure 7.64

One type of relay system is shown in Fig. 7.64(b) in which the power piston is free from spring and the use of mechanical linkage is more. Figure 7.66 shows one type of concentric relay in which the power piston in separated from the governor operated spool valve. The advantage of this relay lies in the fact that the mechanical linkage between relays can be dispensed with by using a hydraulic transmission for the governor signals.

An electrohydraulic relay system which is used in the case of electrical governing system is shown in Fig. 7.71.

7.43 THROTTLE GOVERNING EMPLOYING MECHANICAL SPEED GOVERNOR

The object of throttle governing is to throttle the steam to a suitable pressure and reduce the steam flow (i.e., to allow required quantity of steam to flow) through the turbine bladings whenever there is a reduction of load compared to economic or design load for maintaining the speed of the turbine.

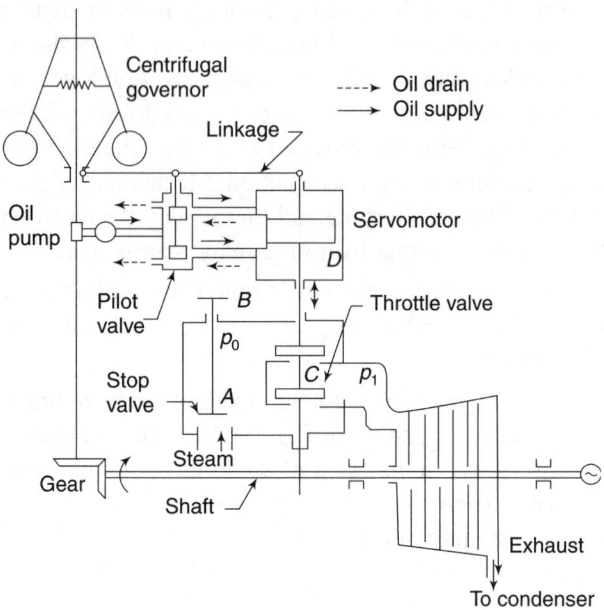

Figure 7.65 Simple throttle governing (hydromechanical)

Figures 7.65 and 7.66 show two schematic diagrams, one simple and other advance type of throttle governing employing mechanical (centrifugal) speed governor and hydraulic relay. '*A*' is the turbine steam stop wave which is used to start and stop the turbine by opening or closing it by means of a hand wheel or automatic control. When the turbine is on full load, the valve '*A*' is fully open. The governor valve or throttle valve '*C*' actually throttles, the steam. There are various forms of throttle valve. The present one is a double-beat valve having seats of nearly equal area and is so shaped that the forces, on the valve due to static pressure and dynamic action are balanced. An oil servomotor (or power piston) (*D*) actuates the operating of valve *C*, which in turn, is controlled by the centrifugal speed governor through hydraulic relay system.

The relay system shown in Fig. 7.65 consists of a pilot valve and a servomotor. Under normal operation, piston of servomotor occupies a position midway between its travel in which both the inlet and exit ports of the pilot valve connecting it to the servomotor are closed. Any displacement of sleeve of centrifugal governor due to change in electrical load causes a displacement of the pilot valve. According to the displacement, the oil under pressure from the oil pump enters either upper or lower half to chamber of servomotor. If oil enters the upper half portion of servomotor, the throttle valve starts closing reducing the quantity of steam flowing through the turbine thus reducing the power output of turbine till the speed is maintained fairly near nominal value. At the same time, oil from bottom half of servomotor starts flowing out through the pilot valve port to drain. If oil under pressure enters the lower half chamber of servomotor, an exactly opposite process is affected opening the throttle valve till the speed is maintained.

The relay system shown in Fig. 7.66 consists of primary and secondary relays. The spool valve (*E*) and the port in the sleeve (*F*) form an orifice whose area can be varied by either the speed governor through displacement of *E* or by the speeder gear that displaces the sleeve, *F*. This variation in the area of the orifice causes a change in the oil flow (*Q*) to drain which will change the oil pressure drop (Δp) through the fixed orifice, as Δp is proportional to the

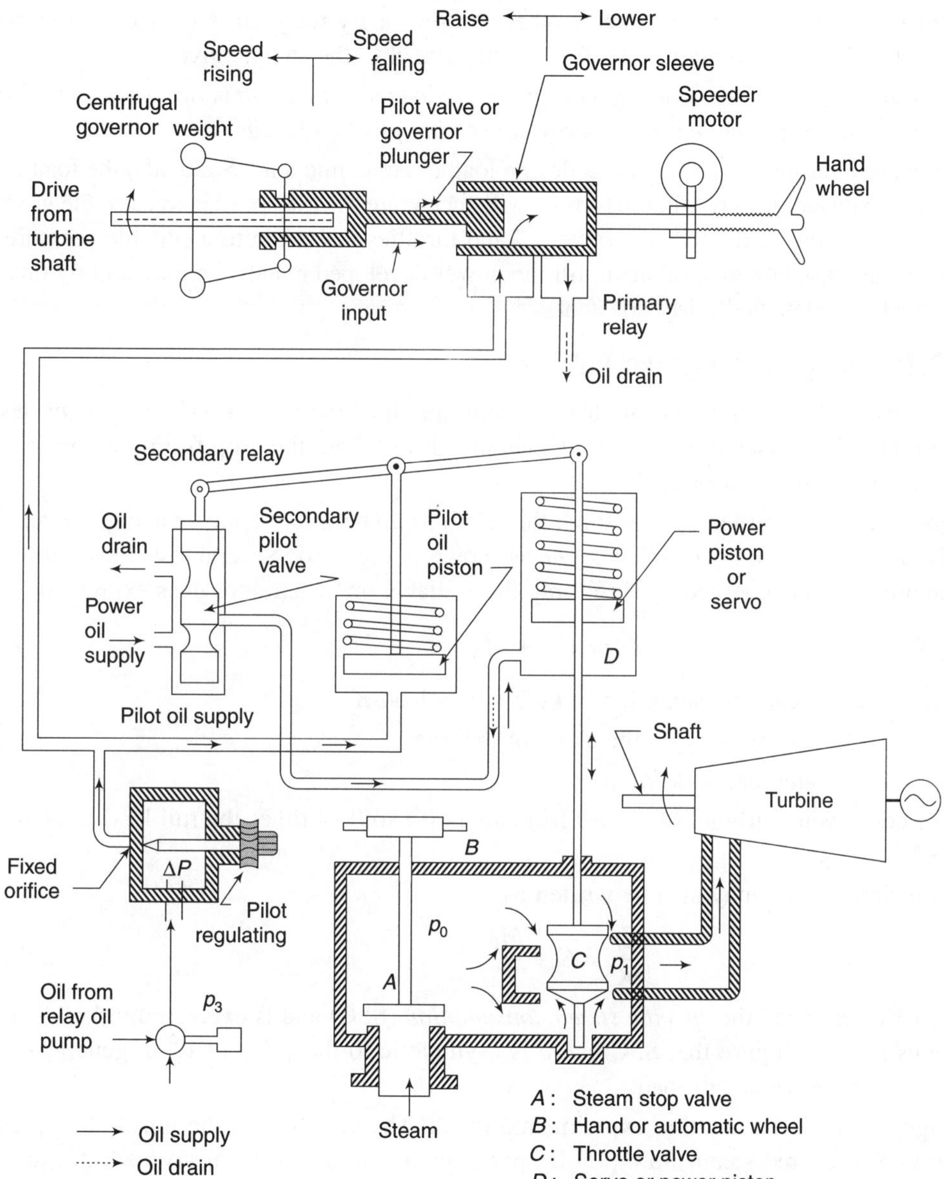

Figure 7.66 Throttle governing with concentric relay (hydromechanical)

oil flow Q squared. This means the pilot oil pressure transmitted to the steam valve relays is equal to the stabilised oil pressure (p_s) less the drop through the fixed orifice. In other words, the speed governor by positioning the spool (E) regulates the pilot oil pressure ($p_p = p_s - \Delta p$) and consequently the turbine inlet valve (throttle valve) with the help of secondary relay which controls the mass flow rate of steam. When the secondary relay oil is fully open to drain, it means that the governors gives the full closing signal to the steam valve.

Throttling of incoming steam affects the work delivered to the shaft in two ways—a reduction of rate of flow and a decrease of energy conversion per kg of steam.

Suppose a turbine is running at its design load or economic load. Suddenly, the load on the turbine is reduced, As a result, the turbine will accelerate. As the speed increases, the governor alters the opening of the throttle valve '*C*' and throttles the steam to a suitable pressure and reduces the mass flow rate of steam till the power developed equals the load and by this way the speed is maintained fairly constant.

7.43.1 Analysis of Impulse Turbine

Let p_0 and p_1 be the pressure of steam before and after the throttle valve; p_1 being less p_0 due to throttling losses in the valves. Generally, at full load, the drop from pressure p_0 to p_1 amounts 2 to 6 per cent of p_0.

For a turbine governed by throttling, the relationship between steam consumption and load is given by the well-known *Willan's line* as shown in Fig. 7.67. Several tests have shown that when a turbine is governed by throttling, the Willan's line is straight. It is expressed as

$$\dot{M} = CK + \dot{M}_0 \qquad ...(7.101)$$

where $\dot{M}$ = steam consumption in kg/h at any load K

C = a constant, i.e., the slope of the line

K = any load in kW.

For condensing turbine, $\dot{M}_0$ varies from about 0.1 to 0.14 times the full load consumption ($\dot{M}_1$).

Equation (7.101) may also be written as

$$\frac{\dot{M}}{K} = C + \frac{\dot{M}_0}{K} \qquad ...(7.102)$$

where $\dot{M}/K$ is called the *specific steam consumption* (SSC) and is expressed in kg/kWh. It is obvious from the figure that SSC curve is asymptotic to the value of *C*. In general, SSC is around 3.5 kg/kWh at full load.

Figure 7.68 shows another experimental result. The variation of the nozzle box pressure (p_1) with measured steam flow ($\dot{M}$) kg per hour for the range from no load, $\dot{M}_0$ to $\dot{M}_1$ is a straight line. In most of the cases, the line when produced passes through the origin. By combining the graph of Figs. 7.67 and 7.68 as shown in Fig. 7.69, we can get the value of nozzle box pressure at any load, *K*. It shows that with simple throttle governing, there is a straight line relationship between nozzle box pressure and load.

The throttle governing may be explained in a better way on the *h-s* chart as shown in Fig. 7.70. There are two assumptions here:

(i) There is a negligible loss in the throttle valve.

(ii) The vacuum in the condenser remains constant.

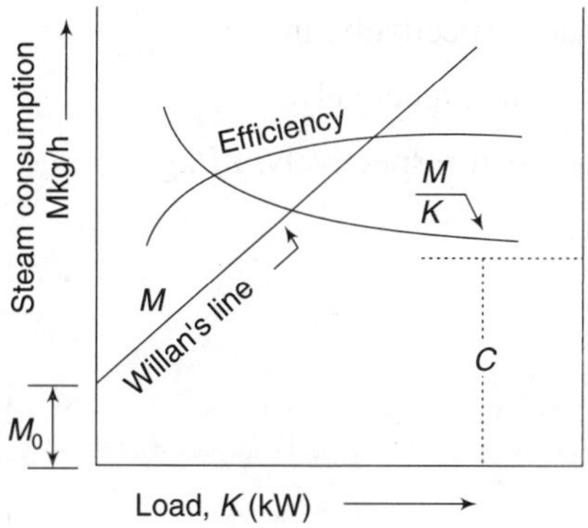

Figure 7.67 Throttle governing characteristics

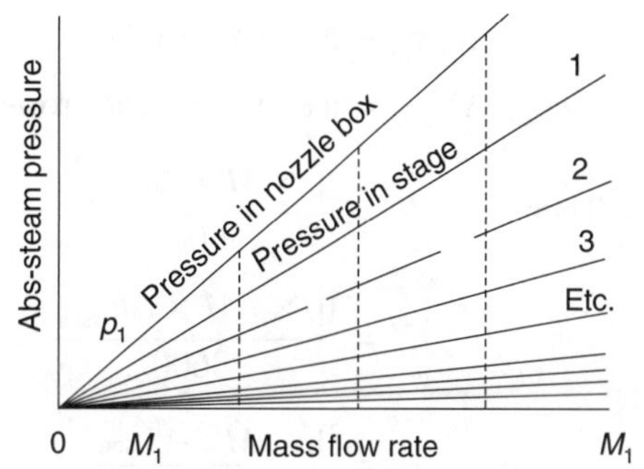

Figure 7.68 Variation of nozzle box pressure

The state point 0 represents the condition of steam before the stop valve where the pressure of steam is p_0 and its temperature is t_0. The state point 1 represents the condition of steam in the nozzle box where the pressure of the steam is p_1 and its temperature is t_1 at full load condition. The isentropic expansion of steam from p_1 to the condenser pressure (p_2) is shown by 1–2 and the actual expansion is shown by 1–2′.

If the turbine works at part load, the steam is throttled and let the nozzle box pressure be p_1' and the steam consumption be $\dot{M}'$, so according to Fig. 7.69,

$$p_1' = \frac{\dot{M}'}{\dot{M}} p_1 \qquad \text{...(7.103)}$$

We know that throttling process occurs at constant enthalpy and this is represented by 1–1′. Here, 1′–3 and 1′–3′ represent the isentropic and actual expansion of steam. *It is obvious from the figure that the effect of throttling apart from reducing the steam flow, is to reduce the isentropic heat drop so that the power developed is equal to the load.*

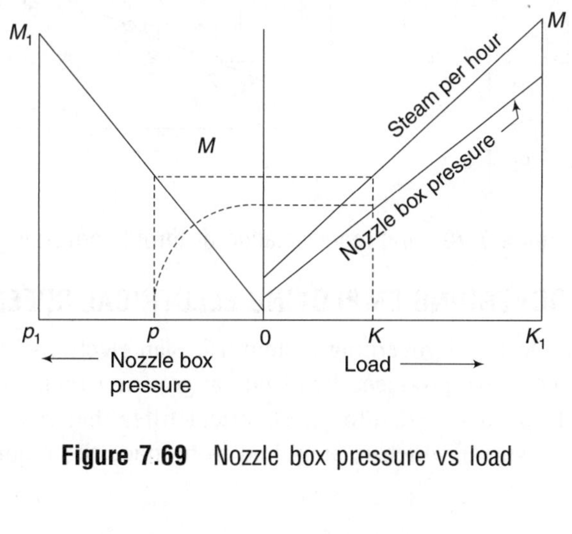

Figure 7.69 Nozzle box pressure vs load

Let K_i, K_i' = internal power at fully and part load respectively, in kW

η_i, η'_i = internal efficiency at full and part load respectively

$\Delta h_{\text{isen}}, \Delta h'_{\text{isen}}$ = isentropic heat drop at full and part load respectively, kJ/kg

Then,
$$K_i = \frac{\eta_i \times \dot{M} \times \Delta h_{\text{isen}}}{3600} \quad \text{...(7.104)}$$

$$K_i' = \frac{\eta_i \times . \dot{M} \times \Delta h'_{\text{isen}}}{3600} \quad \text{...(7.105)}$$

$$\frac{K_i'}{K_i} = \frac{\eta_i'}{\eta_i} \cdot \frac{\dot{M}'}{\dot{M}} \cdot \frac{\Delta h'_{\text{isen}}}{\Delta h_{\text{isen}}} \quad \text{...(7.106)}$$

The side effects of throttling are as follows:

(i) The effect of throttling is to increase the initial superheat and to delay the dew point.

(ii) The greatest variations in steam velocity occur in the stage nearest to the exhaust.

(iii) Even at no load, when the steam consumption is only one-tenth of the full load consumption, the steam velocities in the first half of stage are practically unaffected.

(iv) At partial loads, there is a reduction in stage efficiency due to reduction in the wetness in the later stages.

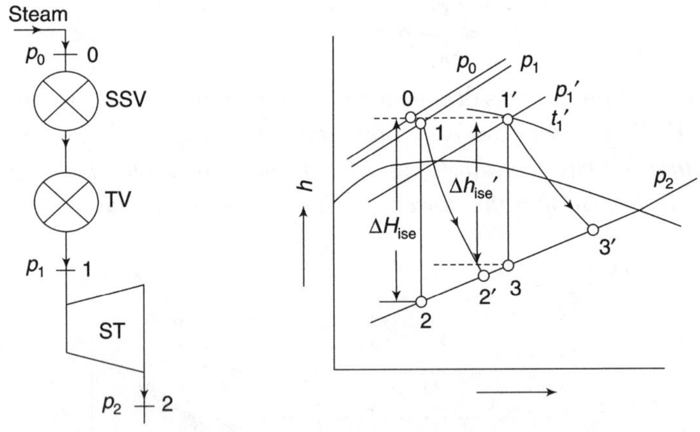

Figure 7.70 *h-s* representation of throttle governing

7.44 THROTTLE GOVERNING EMPLOYING ELECTRICAL SPEED GOVERNOR

Figure 7.71 shows an electrical governing system using an electrohydraulic relay to operate a throttle valve. The a.c. wave produced by electrical governor (generator) which is directly proportional to shaft speed is fed into an electrical filter that resonates at a frequency corresponding to a d.c. signal which is amplified to produce the required speed correction signal, $\Delta f / K$.

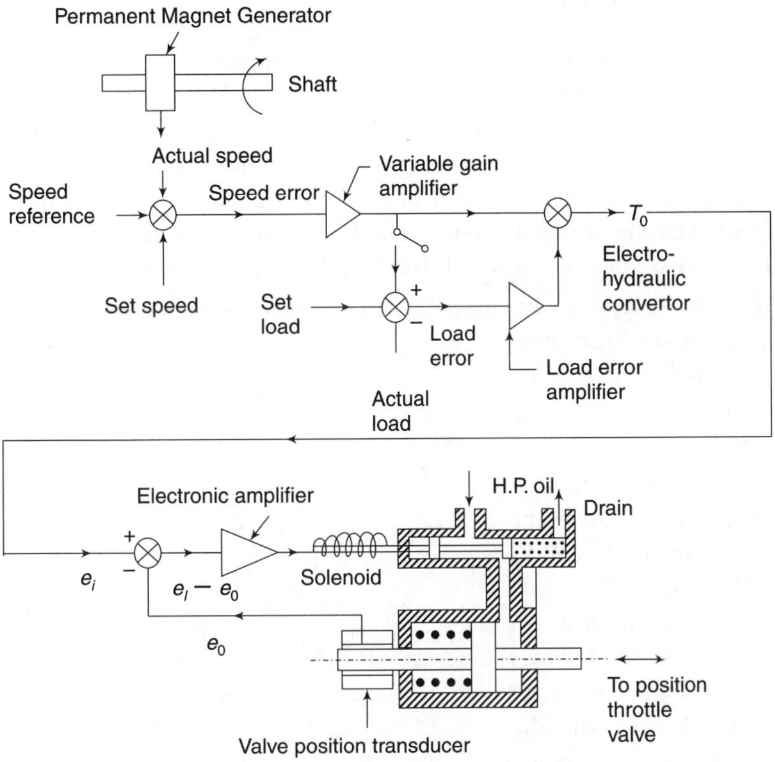

Figure 7.71 Electrical speed governor

An actual load signal from the generator transducer is subtracted and the load error signal is amplified and added to the speed correction signal to feed to the electrohydraulic convertor to actuate steam valve.

The signal from the set speed controller is used for running the machine and for synchronizing while the set load controller is used to control the load when the machine is synchronised.

In this type of electrohydraulic relay, the spool valve of the final relay is electrically operated by a solenoid or an electrical torque motor. The input signal from the electrical speed governor to electrohydraulic relay is a voltage (e_i) that represents the required valve position. The feed back single (e_0) of a transducer located on the steam valve rod represents the existing valve position. This feedback signal (e_0) is subtracted from the input signal (e_i) and the error ($e_i - e_0$) is applied to the electronic amplifier to drive a pilot valve through solenoid. The pilot valve controls the h.p. oil to the spring loaded steam valve which finally sets the required valve position.

The relations given by Eqs. 7.101 to 7.105 obtained with mechanical governor are valid in this case also.

7.45 NOZZLE CONTROL GOVERNING

In the nozzle control governing, the nozzles of the turbines are grouped in two, three or more groups and each group of nozzles is fed with steam controlled by valves. Various arrangements of valves and group of nozzles are employed. Whatever arrangement is employed, the nozzle control is necessarily restricted to the first stage of the turbine, the nozzle areas in the other stages remain constant. If the condition of the steam at inlet to the second stage is not materially affected by the changed condition in the first stage, the absolute pressure of steam in front of the second stage nozzles will be directly proportional to the rate of steam flow through the turbine.

Figure 7.72(a) shows an arrangement often employed in large steam turbines and turbines using high pressure steam.

Here, the nozzles are divided into three groups N_1, N_2 and N_3 under the control of three valves V_1, V_2 and V_3 respectively. The number of nozzle groups may vary from three to five or more. Figure 7.72(b) shows another type of arrangement and it differs from Fig. 7.72(a) in that the nozzle control valves are arranged in the forming part of casing and containing passages leading to the individual nozzle groups. Here, four groups of nozzles and four valves are shown. In this case, the nozzles are confined to the upper half of the casing and so the arc of admission is limited to 180° or less. The number of nozzle groups may vary from 4 to 12. A third arrangement is shown in Fig. 7.72 (c), In this case also, four groups of nozzles in which N_1 is under the control of the valve V_1, through which all the steam entering the turbines passes and further admission of steam is through the valves V_2, V_3, etc., in turn.

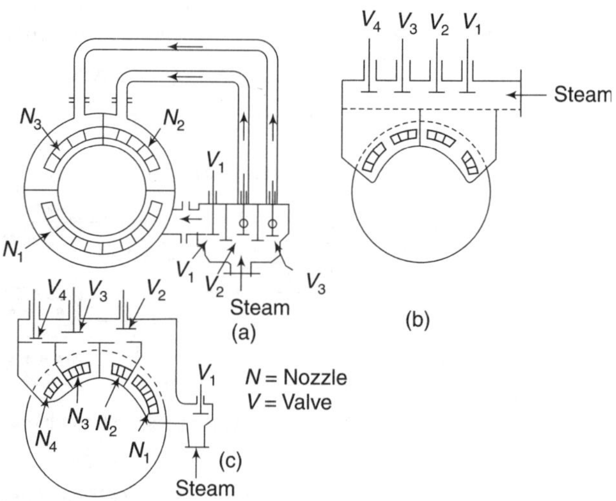

Figure 7.72 Diagram of nozzle control governing

7.46 CONDITION CURVE WITH NOZZLE CONTROL GOVERNING

Theoretically, it may be claimed that there is no throttling loss in the nozzle control governing but practically there will be throttling loss in the valves which will be partially opened. Let us consider the flow of $\dot{M}$ kg per hour of steam and let p_{n3} be the nozzle-box pressure for group No. 3 and p_3 be the pressure in the first stage. Let us also assume that $\dot{M}_1$, $\dot{M}_2$ and $\dot{M}_3$ are the steam flows through groups 1, 2 and 3 respectively. We assume here that the quantity $(\dot{M}_1 + \dot{M}_2)$ expands from p_1 to p_3. The locus of this expansion is 123 as shown in Fig. 7.73. In the nozzle valve V_3, throttling process occurs, so the quantity M_3 is throttled by the nozzle valve V_3 down to the pressure p_{n3} and so 1′ represents the initial condition of steam at inlet to the nozzles, 1′ 4 shows the expansion of steam in the nozzle and 4–5 show the reheating of steam due to blade friction. So by this way, some of steam issuing from the first stage blade has the condition represented by point 3, whereas the small quantity of steam has the condition 5 due to throttling losses.

Assuming perfect mixing in the second stage nozzles, though, it is not practically true, 6 will be representing the condition of the steam at inlet to the second stage nozzles, Thus,

$$(M_1 + M_2 + M_3)\, h_6 = (M_1 + M_2)\, h_3 + M_3 h_5 \quad ...(7.107)$$

Equation (7.107) gives the value of h_6.

Onward expansion of steam starts from 6 instead of 3 and so the specific volume from 3 to 6 is usually small; consequently the changes in pressure distribution in the remaining stages is also small which corrects our statement that the mass flow rate ($\dot{M}$) is directly proportional to pressure.

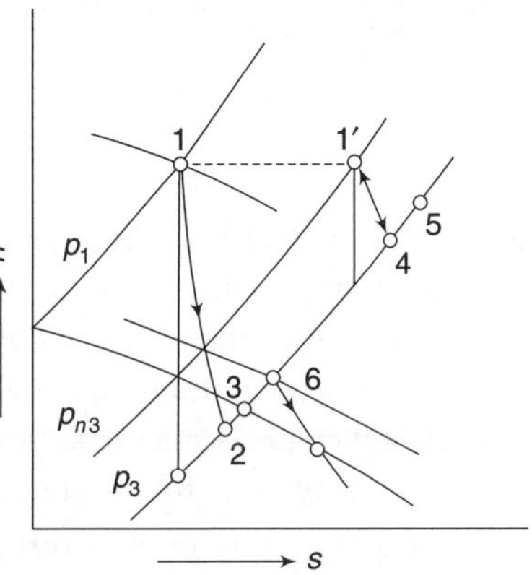

Figure 7.73 Condition curve with nozzle control governing

7.47 COMPARISON BETWEEN THROTTLE AND NOZZLE CONTROL GOVERNING

The following table shows the comparison between throttle and nozzle governing.

Throttle Control	*Nozzle Control*
(i) Severe throttling losses	(i) No throttling losses. (Actually there is slight throttling losses in nozzle valves which are partially open.)
(ii) Employed in impulse and reaction turbine both	(ii) Employed in impulse and also in reaction (if initial stage impulse)
(iii) Lesser partial admission losses	(iii) Much partial admission losses
(iv) Lesser heat drop available	(iv) Large heat drop available
(v) Suitable for small, medium and large turbines.	(v) Suitable for medium and large steam turbine if initial stage is impulse

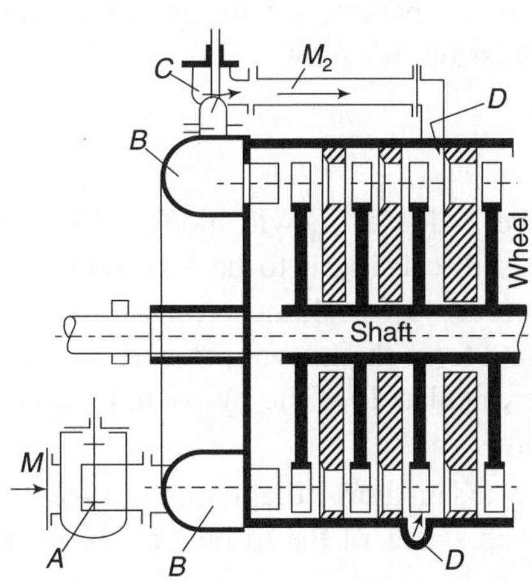

Figure 7.74 Diagram of by-bass governing

7.48 BY-PASS GOVERNING

The principle of by-pass governing is to by-pass some extra quantity of steam to the far down stream stages when the load is more than economic load. The modern high pressure impulse turbines consist of a number of stages of relatively small mean diameter in H.P. side and are designed to run at economic load at which the thermodynamic efficiency ratio of turbine is the maximum. Owing to small heat drop in the first stage of the high pressure turbine, it is not available to employ nozzle control governing. Full admission of steam in the high pressure stages at the economic load is desirable to eliminate the partial admission losses. Further, in case of higher loads, the extra steam required cannot be admitted through additional nozzles in the first stage due to various reasons.

By employing by-pass governing, these difficulties of regulation are overcome.

An arrangement of by-pass governing is shown in Fig. 7.74. The total amount of steam entering the turbine passes through the valve *A* which in under the control of speed governor *B* is the nozzle box or the steam chest. For all loads greater than the economic load, a by-pass valve *C* is opened, allowing steam to pass from the first-stage nozzle box into the steam belt (*D*) and so into the nozzles of the downstream stages, say 4. The valve *C* is designed such that it is not opened until the lift of the valve *A* exceeds a certain value. The valve *C* closes first as the load diminishes. The by-pass valve *C* remains under the control of a speed governor for all loads within its range. More than 1 by-pass valve may be used.

Let $\dot{m}_e$ = consumption of steam at the economical load with no by-pass, kg/s

$\dot{m}$ = consumption of steam at any load above the economical load; by-pass open, kg/s

$\dot{m}_1$ = quantity of steam passing through the stage which are by-passed, kg/s

$\dot{m}_2$ = quantity of steam which enters the by-pass stage, kg/s

p_n = pressure in by-pass stage at the economic load, bar

p_n' = pressure in by-pass stage at the high load, bar.

By neglecting the effect of by-passing on the steam condition at inlet to the nozzles following supplementary admission, we have

$$p_n' = p_n \cdot \frac{\dot{m}}{\dot{m}_e} \qquad \text{...(7.108)}$$

It is obvious that the higher value of p_n' will modify the mass flow through the nozzles of the stages that are by-passed, causing it to be less than it was at the economical load. We should proceed by the process of trial and error to find the mass flow, $\dot{m}_1$. Generally, trial values of $\dot{m}_1$ are taken and from the known pressure p_n and the Σh_{isen} and v_n curves the pressure distribution may be calculated for the by-passed stages. After finding correct value of $\dot{m}_1$, we can get value of $\dot{m}_2$.

This process can be represented on the *h-s* diagram. Let $1 - 2' - 3' - 4'$ represent the condtion of steam in the three by-passed stages of the turbine as shown in Fig. 7.75. The state point

1′ show the condtion of the steam passing through the by-pass throttle and thus by neglecting heat losses $h_1' = h_1$. After mixing $\dot{m}_1$ kg of steam having enthalpy h_4' with $\dot{m}_2$ kg of steam having enthalpy h_1', the state point is 5. h_5 will be calculate by

$$\dot{m}h_5 = \dot{m}_1 h_4' + \dot{m}_2 h_1' \qquad \text{...(7.109)}$$

A second approximation will be better to find the pressure distribution in the succeeding stages.

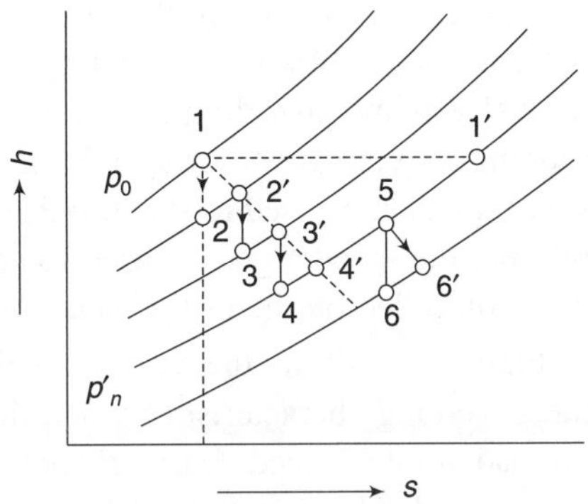

Figure 7.75 *h-s* representation of process for by-pass governing

7.49 BY-PASS GOVERNING OF REACTION TURBINES

Full admission of steam is essential in reaction turbine because of pressure drop in moving blades. So this rules out the possibility of employing any form of nozzle control governing. Therefore, the alternatives left are simply throttle governing and by-pass governing. In most of Parsons turbine, a single by-pass valve is used.

7.50 SPEEDER GEAR

The function of speeder gear is to adjust the speed of turbine within the limit of ± 6% of rated speed controlled by speed governor. It acts in parallel to the speed governor as shown in Fig. 7.76 and may be operated by hand or remotely by electrical control of a motor. Depending upon the design of speed governor, the adjustment may be affected.

In one common type, the speeder motor actuates the ported sleeve of the governor primary relay as shown in Fig. 7.76. In other type, a worm gear drive followed by a screw is used to adjust the spring compression on the governor.

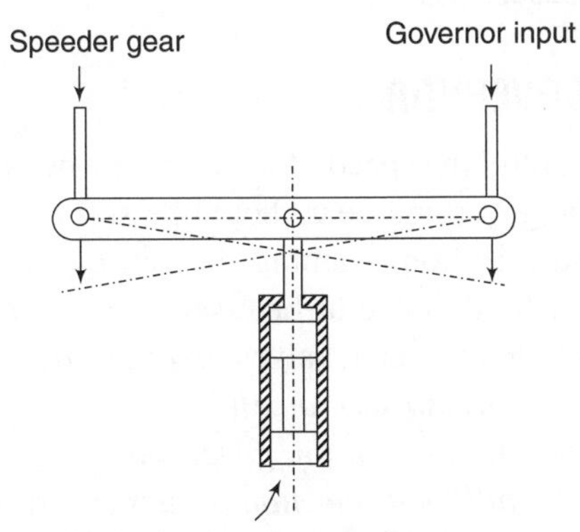

Figure 7.76 Speeder gear

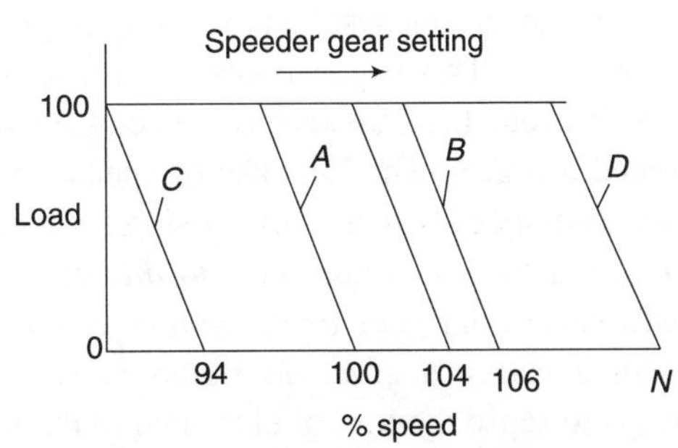

Figure 7.77 Regulation diagram due to speeder gear settings

At the time of starting the turbine, the governor weights are against their inner stop and the governor valves are opened by using a starting handwheel to move the pilot bobbin of the valve relay, providing a coarse adjustment in the linkage. As soon as the turbine starts running into the governor speed range, the governor weights begin to open and the speeder gear may then be used to synchronise the machine with grid. After the synchronization of the turbine, the speeder gear is used to control the power output, as the speed of the machine is now controlled by the grid frequency.

Figure 7.77 shows the regulation diagram due to speeder gear settings. Curve *A* represents the governing characteristics when the speeder gear is in the synchronizing position, i.e., at no load at rated speed. Curve *B* shows the characteristics when the machine is at full load at rated speed. The curves *C* and *D* represent the characteristics under the normal range (i.e., ± 6% of rated speed) provided. The speeder gear is equipped with a device to limit its setting if it is overwound, say to position *E* (Fig. 7.77) which means that the machine would still deliver full load at rated speed but would overspeed the machine to speed *N* if full load was rejected.

The speeder gear used in the electrical governing system is very simple. It consists of a motor driven potentiometer and the speeder gear setting is represented by a voltage that is added to the electrical speed governor signal as shown in Fig. 7.78.

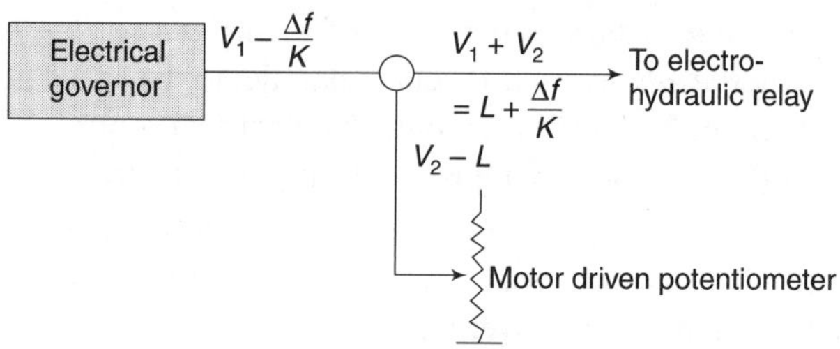

Figure 7.78 Electrical speeder gear

7.51 ANTICIPATORY GEAR OR SECONDARY GOVERNOR

The turbine governing system is severely required to control the speed of the turbine when full load is rejected. This suggests that the response of the governing gear should be sufficiently rapid to prevent the transient overspeed on full load rejection reaching the setting of the overspeed trip gear. The lone use of speed governor will allow the turbine speed to reach at least 4% overspeed before a full closing signal is applied to the steam valves. *In order to avoid waiting until the speed has risen to this figure, (i.e., 4% overspeed), auxilliary gear usually termed anticipatory gear or secondary governor is incorporated in the governing system to assist the governor to close the steam valves rapidly if a full load rejection occurs.* There are two ways to sense the loss of electrical load, one directly from the generator terminal and the other from the resultant acceleration of the machine.

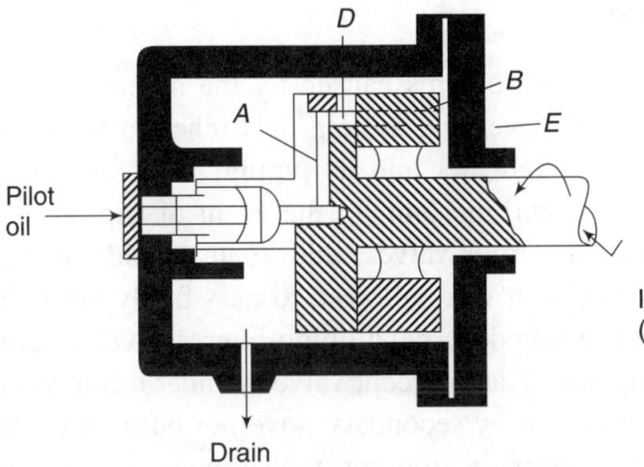

Figure 7.79 Mechanical anticipatory gear

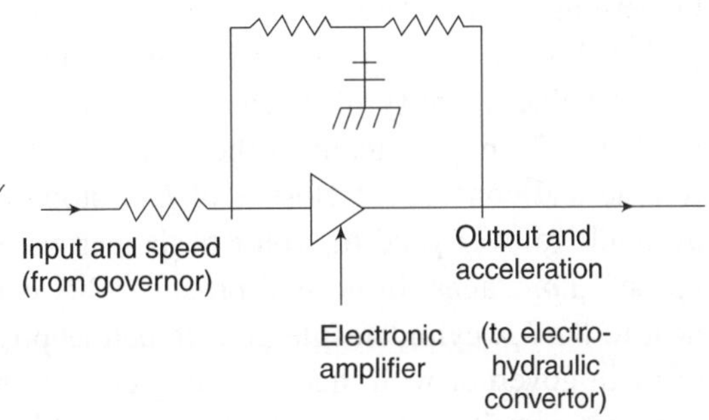

Figure 7.80 Electrical anticipatory gear

Figure 7.79 shows one mechanical type of anticipatory gear. It consists of drilled shaft driven by the turbine directly or via gearing. Pilot oil pressure is fed through the shaft seal arrangement to port *D* in the face of integral chamber, *A*. The flexible link (*C*) connects the flywheel (*B*) to the shaft. The system has a drain port (*E*) which is angularly displayed relative to port, *D*. The port (*D*) is fully covered by the flywheel when the turbine speed is uniform. As soon as the turbine accelerates, the flywheel will lag behind the shaft due to its inertia and mounting flexibility, and by an arrangement of mechanical stops, the two ports are brought into line drawing the pilot oil, which in turn, will close the steam valves.

Figure 7.80 shows an electrical type of anticipatory gear. The electrical circular employed may easily generate a signal proportional to acceleration, and may be fed to the electrohydraulic relay.

Figure 7.81 shows the **response of the governor and turbine to a full load rejection**. From Fig. 7.81, it is obvious that the turbine continues to accelerate after the steam valves are closed. This is attributed to the fact that the steam entrained between the steam valves and the turbine expands.

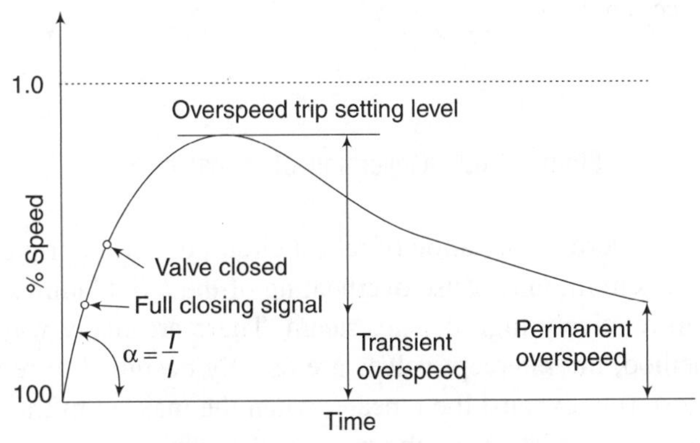

Figure 7.81 Full load rejection

7.52 GOVERNING OF REHEAT TURBINES

Governing of reheat turbines is complicated due to problems caused by the reheat and its piping. Figure 7.82 shows a schematic diagram of governing system of a reheat turbine. A considerable quantity of steam is entrained in the reheater and its pipings. This amounts to about 220 m^3 of steam in the case of 500 MW turbine unit. In the event of trip on the turbine and consequent closure of *h.p.* stop and governor valves, this steam would still be available to overspeed the rotor to dangerous level if it were allowed to pass freely through *i.p.* and *l.p.* stages. In normal practice, this is prevented by positioning intercept valves are near to the *i.p.* cylinder. Inlet nozzle belt as possible. The intercept valve is under the control of main governor with the overriding control provided by secondary governor but this valve does not contribute to the governing of turbine. The main stop and reheat emergency valve are under the control of overspeed trip gear. In normal case, the intercept valves are timed to commence closing at approximately 2% overspeed and are fully closed at 5% overpseed to avoid the acceleration of the rotor in the event of trip on the turbine.

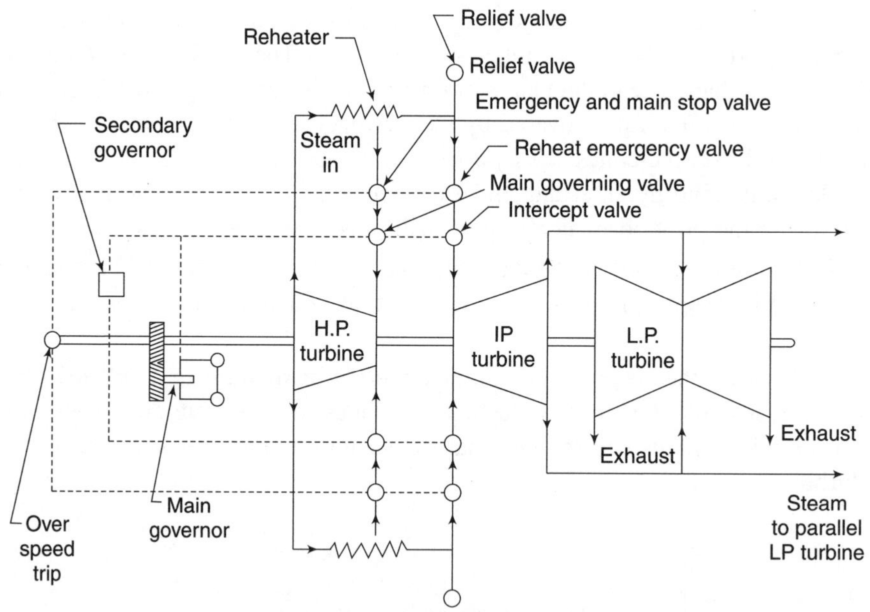

Figure 7.82 Governing of reheat turbine

There is a chance of overpressurisation of reheat circuit under the trip condition. Failure to depressurize the reheat circuit may cause overheating of the *h.p.* bladings due to windmilling action of the rotor in relatively high density steam. There are many ways to overcome this difficulty. In one method, the intercept valves are usually designed to reopen slowly (whilst the valves remain closed) to exhaust the reheater when the maximum transient speed rise has been allowed to pass following a trip and the speed of the machine has dropped to near normal.

The alternative method is to incorporate release valves controlled by the turbine governing system in the reheat circuit. Following a trip, these valves open releasing the trapped steam in the reheater.

7.53 DIRECT DIGITAL CONTROL (DDC)

Figure 7.83 shows the direct digital control system suggested by Mr. W. Wang [43]. One of the main boons of this technique is the increased reliability which results from electronic systems in place of mechanical systems thus avoiding wear and obtaining greater accuracy. Another advantage is that the engineer can record data on a routine basis automatically instead of manually. Modern systems also allow to pre-empt vibration monitoring techniques, which basically measure the extent and rate of deterioration of defects. By analysing other parameters, engineers can anticipate the formation of those defects which show up in vibration monitoring. For example, deposits on blades will lead to a fall-off in performance which can be detected long before the blades get out of balance.

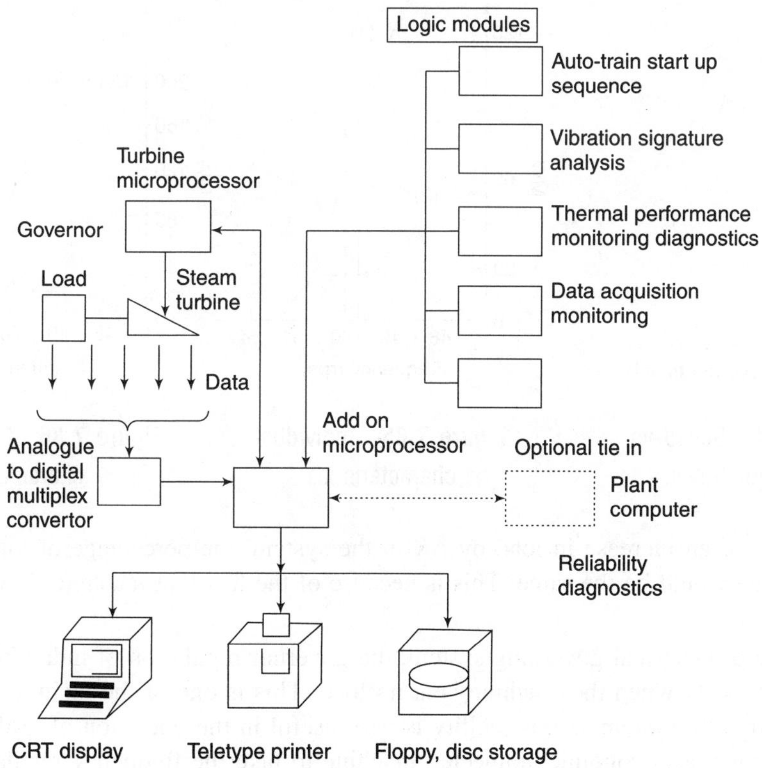

Figure 7.83 Direct Digital Control System

All the modern turbines are being equipped with this direct digital control system or improved version of it.

7.54 GOVERNING CHARACTERISTICS

Figure 7.60 shows the required characteristics of a steam turbine driving an a.c. generator. It is a linear relationship between load and speed, the load falling with the increase in speed. Mechanical governor characteristics is shown in Fig. 7.84. The relationship between speed and governor travel is more or less linear. In the case of throttled governed turbine, the relationship between steam consumption rate and the load is linear (Fig. 7.67). The variation of turbine efficiency and steam consumption per kWh with load are also depicted in this Fig. 7.67.

Sometimes several machines are connected to a common electrical load. In such case, the load-frequency characteristics of the system will be formed by cumulative governing characteristics of all connected machines. Figures 7.85 and 7.86 show the individual and combined characteristics of the two following cases:

(i) A 60 MW set generating 30 MW with 4% droop, i.e., 30 MW/c/s.

(ii) A 30 MW set generating 20 MW with 4% droop, i.e., 15 MW/c/s.

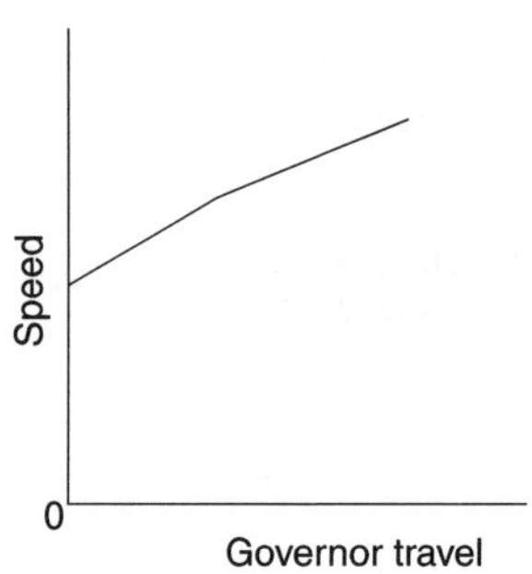

Figure 7.84 Speed vs. Governor travel

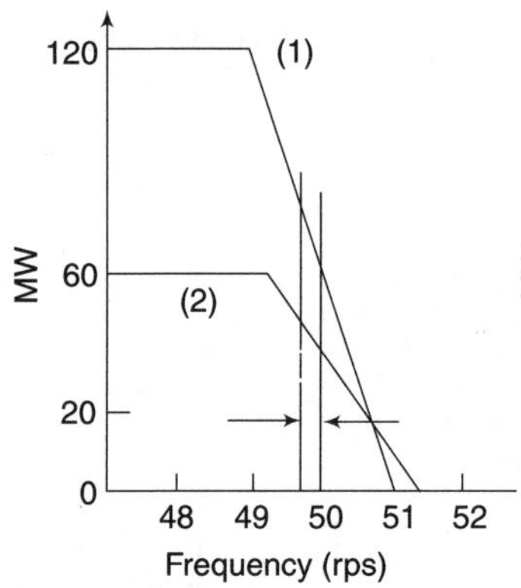

Figure 7.85 Individual characteristics

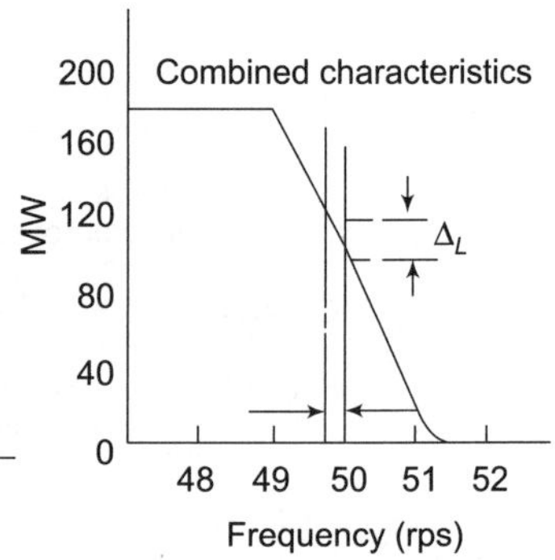

Figure 7.86 Combined characteristics

In the case of an increase in load by ΔK in the system, the percentage of load taken up by the machine would be the same. This is because of the fact that machines have the same regulation.

In the case of electrical governing system, the governor regulation of individual machine can be varied easily when the machine is at no load. This is one of the main advantages of electrical governing system as this facility is very useful in the operation of grid system by arranging or the least economic connected machine to take the frequency regulating duties and setting a low droop on their governor droop or inhibiting speed governor action together. While the most economic machine can be base loaded by the use of high governor droop or by inhibiting speed governor action together. Figure 7.87 shows the variable droop regulation diagram of a electrical governing system.

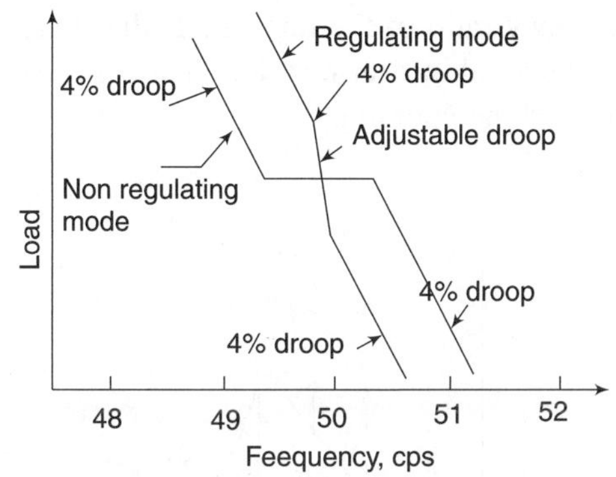

Figure 7.87 Variable droop regulation

7.55 HYDRAULIC SYSTEM AND FLUIDS

If the turbine lubricating oil is used to operate, the governor and valve relays, then the governor hydraulic system would be simplified as one single pump would serve the purpose. The oil pressure generated by the single pump is in the range of 10 to 13.5 bars to suit the relays gear. The lower pressure required for the bearing oil is bled off from a reducing valve. The high pressure oil is filtered before the complete flow is fed to the emergency trip valve, the major portion of the flow is then directed to the valve to the governor primary relay, to be converted into a speed sensitive pressure signal.

There is a limitation of oil pressure in the relay system of the order of 13.5 bars owing to fire risk in the event of oil leak. On using higher pressure cycle or large steam valves, this limitation on operating pressure necessitates the use of large diameter relays which are difficult to accommodate and maintain. In order to reduce the size of relays, a higher relay pressure must be obtained. The use of higher oil pressure also has the advantage of improving the steam valve response and thus reduces the transient speed rise on load rejection.

In modern steam turbines, higher relay pressures are used for which fire resisting fluids are adopted. The most common fire resisting fluids in use today are based on phosphate esters with various additives in order to improve their viscosity change with temperature and their hydraulic stability. With the adoption of fire resisting fluids, the governing system is entirely separate from the lubricating oil system except that a.c. motor driven positive displacement pumps (gear, piston or screw) are used rather than centrifugal pumps as the specific speed of the pump is low.

Example 7.19 Form the test of a steam turbine operating at fractional load, the following data were obtained:

Steam pressure before governor valve = 27 bars, Steam temperature before governor valve = 371°C, Steam pressure after governor valve = 8 bars, Steam pressure in exhaust branch = 0.07 bar, Steam temperature in exhaust branch = 49°C.

What is the internal efficiency ratio of the turbine? If the steam flow through the turbine is 818 kg/h, and the power absorbed in bearing friction and governor drive is 7.35 kW, what is approximate power at the turbine coupling?

Solution Refer to Fig. 7.88.

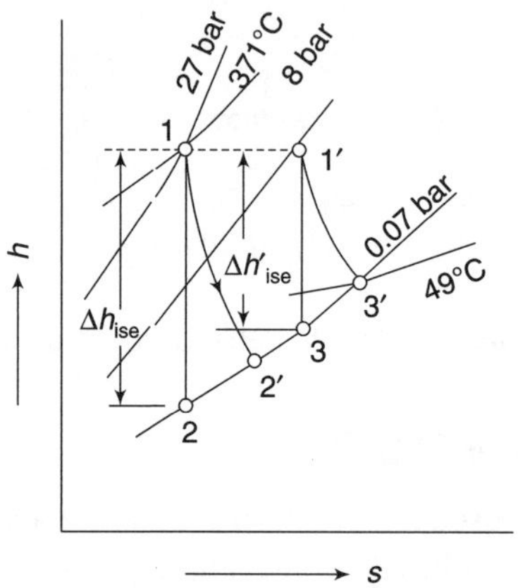

Figure 7.88 Example 7.19

Given: p_1 = 27 bars, p_2 = 0.07 bar, t_1 = 371°C, t_3' = 49°C, p_1' = 8 bar, $\dot{M}'$ = 1818 kg/h
Power absorbed in friction = 7.35 kW

From Mollier chart: $h_1 = h_1' = 3170$ kJ/kg

$h_1' - h_3 = 841$ kJ/kg = isentropic heat drop and $h_1 - h_3' = 573$ kJ/kg = actual heat drop

$\therefore$ Internal efficiency, $\eta_i = \dfrac{573}{841} = 68.13\%$

Internal power developed at part load,

$$K_i' = \frac{\eta_i . \dot{M}' \, \Delta h'_{isen}}{3600} = \frac{0.6813 \times 1818 \times 841}{3600} = 289.35 \text{ kW}$$

$\therefore$ Power at turbine coupling = 289.35 – 7.35 = 282 kW ***Ans.***

Example 7.20 The following results were obtained on a full load use of a turbo-alternator. Output at generator terminals = 10,000 kW, Steam pressure before stop valve = 30 bars, Steam temperature before stop valve = 371°C, Steam pressure in nozzle box = 27 bars, Steam pressure in exhaust branch = 0.07 bar, steam consumption = 45818 kg/h.

The power loss due to bearing friction, governor, and oil pump drive, etc., amounts to 110 kW. Assuming that the alternator efficiency is 95.5%, find the true internal efficiency ratio of the turbine at full load. What is the output of the generator when the steam consumption is $\frac{1}{4}$ and $\frac{1}{2}$ of that at full load? Assume that the turbine internal efficiency ratio, the mechanical losses in kW, the exhaust pressure and the alternator efficiency are the same as at full load, and that the turbine is governed by throttling.

Solution Refer to Fig. 7.89. Input to alternator = Internal output of turbine–Losses

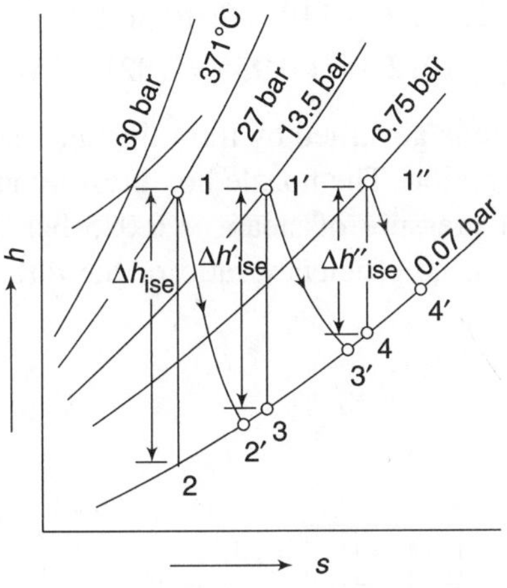

Figure 7.89 Problem 7.20

Input to alternator = $\dfrac{10000}{0.955}$ = 10471.2 kW

K_i = internal output of turbine = 10471.2 + 110 = 10581.2 kW

Here Δh_{isen} = 1037 kJ/kg, $\dot{M}$ = 45818 kg/h

$$\therefore \quad 10581.2 = \frac{\eta_i \times 45818 \times 1037}{3600} \quad \text{or} \quad \eta_i = \frac{10581.2 \times 3600}{45818 \times 1037} = 0.8017 \qquad \textbf{\textit{Ans.}}$$

At $\dfrac{1}{2}$ load the nozzle box pressure, $p_1' = \dfrac{27}{2}$ = 13.5 bars

At $\dfrac{1}{4}$ load, the nozzle box pressure, $p_1'' = \dfrac{27}{4}$ = 6.75 bars

From Mollier diagram, At $\dfrac{1}{2}$ load, $\Delta h'_{\text{isen}}$ = 936 kJ/kg and at $\dfrac{1}{4}$ load, $\Delta h'_{\text{isen}}$ = 832 kJ/kg

We know that $\dfrac{K_i'}{K_i} = \dfrac{\eta_i'}{\eta_i} . \dfrac{\dot{M}_i'}{\dot{M}} \dfrac{\Delta h'_{\text{isen}}}{\Delta'_{\text{isen}}}$

At $\dfrac{1}{2}$ load $\quad \therefore \quad K_i = 1 \times \dfrac{1}{2} \times \dfrac{936}{1037} \times 10581 = 4775.3$ kW

Output from turbine = 4775.3 – 110 = 4665.3 kW

∴ Output from generator = 0.955 × 4665.3 = 4455.3 kW ***Ans.***

At $\frac{1}{4}$ load

$$K_i' = 1 \times \frac{1}{4} \times \frac{832}{1037} \times 10581.2 = 2122.36 \text{ kW}$$

Output from the turbine = 2122.36 – 110 = 2012.36 kW

Output from the generator = 2012.36 × 0.955 = 1921.8 kW **Ans.**

Example 7.21 A steam turbine is governed by throttling and the no-load steam consumption is 12% of the full load consumption. The nozzle box pressure at full load is 17 bar the initial superheat 60°C. The exhaust pressure of steam is 0.035 bar. If the steam consumption at full load is 5 kg/kWh, what is the efficiency ratio of the turbine at quarter half, and three quarters load?

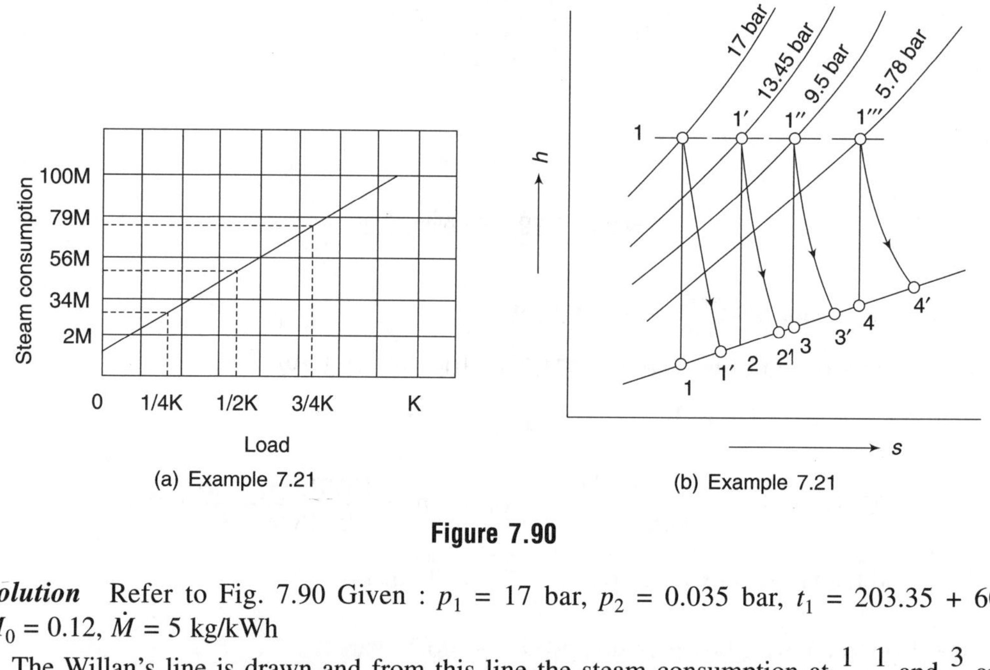

(a) Example 7.21 (b) Example 7.21

Figure 7.90

Solution Refer to Fig. 7.90 Given : p_1 = 17 bar, p_2 = 0.035 bar, t_1 = 203.35 + 60, $\dot{M}_0 = 0.12$, $\dot{M}$ = 5 kg/kWh

The Willan's line is drawn and from this line the steam consumption at $\frac{1}{4}, \frac{1}{2}$ and $\frac{3}{4}$ are 34%, 56% and 79% respectively of full load steam consumption.

We know that at part load, $p_1' = \frac{\dot{M}'}{\dot{M}} p_1$

Hence, the nozzle box pressure at different part loads are:

1/4 load = 0.34 × 17 = 5.78 bars, 1/2 load = 0.56 × 17 = 9.55 bars and 3/4 load = 0.79 × 17 = 13.45 bars

From the Mollier chart, isentropic heat drops are:

$$(\Delta h_{isen})_{\text{full load}} = 933 \text{ kJ/kg}, (\Delta h_{isen})'_{1/2 \text{ load}} = 848 \text{ kJ/kg}$$

$$(\Delta h_{isen})''_{1/4 \text{ load}} = 790 \text{ kJ/kg}, (\Delta h_{isen})'''_{3/4 \text{ load}} = 900 \text{ kJ/kg}$$

We know that at full load, $K_i = \dfrac{\eta_i \dot{M}\,\Delta h_{isen}}{3600}$

$$\therefore \quad 1 = \frac{\eta_i \times 5 \times 933}{3600}$$

$$\therefore \quad \eta_i = 0.771 \text{ at full load}$$

Also $$\frac{\eta_i'}{\eta_i} = \frac{K_i'}{K_i} \cdot \frac{\dot{M}'_i}{\dot{M}} \cdot \frac{\Delta h'_{isen}{}'}{\Delta h_{isen}}$$

$$\therefore \quad \eta'_i \text{ at } \frac{1}{4} \text{ load} = \frac{1}{4} \times \frac{1}{0.34} \times \frac{790}{933} \times 0.771 = 48.0\%$$

$$\eta'_i \text{ at } \frac{1}{2} \text{ load} = \frac{1}{2} \times \frac{1}{0.56} \times \frac{848}{933} \times 0.771 = 62.56\% \qquad \textit{Ans.}$$

$$\eta'_i \text{ at } \frac{3}{4} \text{ load} = \frac{3}{4} \times \frac{1}{0.79} \times \frac{900}{933} \times 0.771 = 70.6\%$$

Example 7.22 The full load output of a steam turbine, measured at the coupling is 2,000 kW and the losses due to bearing friction to the governor and oil pump drive, etc. is 60 kW. The steam is supplied at a pressure of 14 bars and 282°C. The exhaust pressure of steam is 0.07 bar. The internal efficiency ratio at full load is 74%. Estimating the coupling power of the turbine when the steam flow through the turbine is 20 per cent of that at full load. Assume that the turbine is governed by throttling, external losses are the same as that at full load, exhaust pressure is the same and the internal efficiency ratio is reduced to 68% at this load.

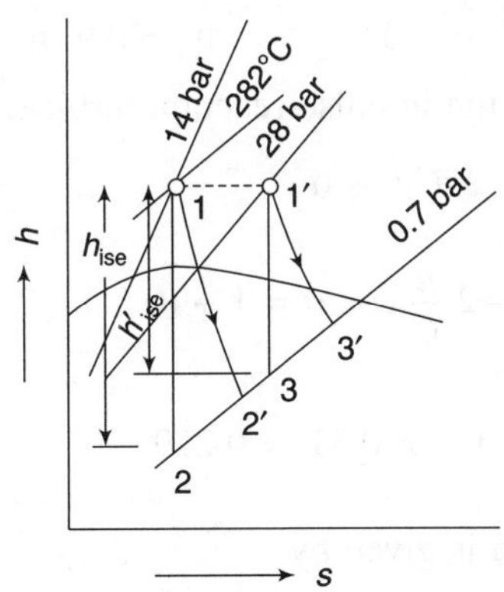

Figure 7.91 Example 7.22

Solution Refer to Fig. 7.91

Given: $p_1 = 14$ bars, $p_2 = 0.07$ bar, $\eta_i = 0.74$, $t_1 = 282°C$, $K_i = 2{,}000$ kW, losses = 60 kW, $\eta_i' = 0.68$, $\dot{M}' = M/5$

Total internal power developed by the turbine at full load, $K_i = 2000 + 60 = 2060$ kW.

The pressure after throttling or at part load, $p_1' = \frac{\dot{M}'}{\dot{M}} p_1 = \frac{1 \times 14}{5} = 2.8$ bars

From the h-s chart, $\Delta h_{\text{isen}} = h_1 - h_2 = 849$ kJ/kg

$$\Delta h_{\text{isen}}' = h_1' - h_3 = 614 \text{ kJ/kg}$$

We know that $\frac{K_i'}{K_i} = \frac{\eta_i'}{\eta_i} \cdot \frac{\dot{M}'}{\dot{M}} \cdot \frac{\Delta h_{\text{isen}}'}{\Delta h_{\text{isen}}}$

$$\therefore \quad K_i' = \frac{0.68}{0.74} \times \frac{1}{5} \times \frac{614}{849} \times 2060 = 273.8 \text{ kW}$$

$\therefore$ Power at turbine cooling = 273.8 – 60 = 213.8 kW ***Ans.***

Example 7.23 The following particulars refer to three successive "expansions" in the I.P. section of an impulse-reaction turbine.

Group No.	Mean diameter, cm	Blade height, cm	Blade outlet angle	Total no. of rows (fixed and moving)
1.	88	1.2	20°	8
2.	88.5	1.7	20°	8
3.	90	2.5	20°	8

The blade speed ratio is 0.5, $\phi = 0.9$, $\eta_n = 0.9$, the index n for the isentropic expansion is 1.135 and the stage efficiency is 0.72. Estimate the mass flow rate through the turbine blades when the condition at inlet to the first group is 6 bars dry and saturated and the pressure at outlet from the third group is 1.8 bars.

Solution Given $p_1 = 6$ bars $= 6 \times 10^5$ N/m², $\quad p_n = 1.8$ bars $= 1.8 \times 10^5$ N/m²,

$\rho = u/C = 0.5, \phi = 0.9 \quad \eta_n = 0.9, \eta_s = 0.72, n = 1.135$

From the velocity triangle of the impulse-reaction turbine, we have

$$C_r^2 = C^2 + u^2 - 2uC \cos \alpha$$

$$\therefore \quad \left(\frac{C_r}{C}\right)^2 = 1 + \left(\frac{u}{C}\right) - 2\frac{u}{C} \cos \alpha = 1 + (0.5^2) - 2 \times 0.5 \times 0.94 = 0.31.$$

Also $$y = \phi \left(\frac{C_r}{C}\right)^2 = 0.9 \times 0.31. = 0.279$$

The index of actual expansion is given by

$$n' = \frac{n}{n - \eta_n (n-1)} = \frac{1.135}{1.135 - 0.9 \times 0.135} = 1.1198$$

$$v_1 = 0.3155 \text{ m}^3\text{/kg (corresponding to 6 bar)}$$

$$p_1 v_1^{n'} = p_n v_n^{n'}$$

$$\therefore \quad v_n' = v_1 \left(\frac{p_1}{p_2}\right)^{1/n'} = 0.3245 \left(\frac{6}{1.8}\right)^{1/1.1198} = 1 \text{ m}^3\text{/kg}$$

The effective area at the blade outlet

$$A = \pi\, D l \sin \beta_2$$

$$\frac{1}{A_1} = \frac{1}{\pi \times 0.88 \times 0.012 \times 0.342} = 88.2\ (1/\text{m}^2)$$

$$\frac{1}{A_2} = \frac{1}{\pi \times 0.885 \times 0.017 \times 0.342} = 88.2\ (1/\text{m}^2)$$

$$\frac{1}{A_3} = \frac{1}{\pi \times 0.9 \times 0.025 \times 0.342} = 41.4\ (1/\text{m}^2)$$

The area of all rows in a group is the same, hence for 8 rows

$$\sum^{8}\left(\frac{1}{A}\right)^2 = 8\,\Sigma\left(\frac{1}{A}\right)^2$$

Since there are three groups, so area $= 8\left[\left(\frac{1}{A_1}\right)^2 + \left(\frac{1}{A_2}\right)^2 + \left(\frac{1}{A_3}\right)^2\right]$

We know that for impulse reaction turbine governing

$$\dot{m}^2 = \frac{2\eta_n}{1-y}\left(\frac{n'}{n'+1}\right)\frac{p_1/v_1 - p_n/v_n}{\sum_{1}^{n}\left(\frac{1}{A}\right)^2}$$

Putting all the values, we have

$$\dot{m}^2 = \frac{2 \times 0.9}{0.721}\left(\frac{1.092}{2.092}\right)\frac{[(6/0.3245) - (1.8/1.0)]10^5}{8[88.2)^2 + (61.6)^2 + 41.4^2]}$$

$$\therefore \qquad \dot{m} = 4.46 \text{ kg/s.}$$

7.56 SUMMARY OF FORMULAE

(1) $C_1 = 44.72\sqrt{h_0 - h_1}$

(2) $C_1' = 44.72\sqrt{(h_0 - h_1)\eta_n}$

(3) $P = \dfrac{\dot{m}.u.C_w}{1000}$; kW

(4) $\eta_b = \dfrac{2uC_w}{C_1^2}$

(5) $F_a = \dot{m}(C_{f1} - C_{f2})$

(6) $\eta_{gs} = \eta_b - \eta_n$

(7) $\rho_{\text{opt}} = \cos\alpha_1$

(8) $\eta_{b,\,\max} = \cos^2\alpha_1$

(9) $\eta_n = \dfrac{C_1^2 - \phi\, C_2^2}{2\,\Delta h_{\text{isen}}}$

(10) $\eta_{gs} = \dfrac{2u.\eta_n}{{}_1C_1^2}\cdot\Sigma C_w$

(11) $\dot{m}v = \pi D l C_{f2}$

(12) $R = \dfrac{(\Delta h_{\text{isen}})_b}{(\Delta h_{\text{isen}})_n + (\Delta h_{\text{isen}})_b}$

(13) $(\eta_{b,\,\max})_{\text{react}} = \dfrac{2\cos^2\alpha_1}{1 + \cos^2\alpha_1}$

(14) $\eta_s = \dfrac{sp}{is.p}$

(15) $RF = \dfrac{\Delta h_c}{\Delta h_{isen}}$ (16) $\eta_i = (RF)\eta_s$

(17) $\eta_{oth} = \left[(w_t)_{isen} \times \eta_t \times \eta_m . \eta_g - \dfrac{w_p}{\eta_p}\right]\eta_{boiler}$ (18) $\dfrac{M}{K} = C + \dfrac{M_0}{K}$

(19) $\dfrac{K_i'}{K_i} = \dfrac{\eta_i'}{\eta_i} . \dfrac{M'}{M} . \dfrac{\Delta h'_{isen}}{\Delta h_{isen}}$

PROBLEMS

Things to Think About

7.1 How will you apply Newton's 2nd law of motion to get continuous mechanical and electrical power using steam as working fluid?

7.2 What is the basic philosophy for increasing blade efficiency?

7.3 What is the basic philosophy in governing of steam turbines?

7.4 How will you reduce the number of stages without no appreciable efficiency drop?

7.5 What is the importance of cylinder layout?

Short and Long Theoretical Questions

7.1 What is the principle of operation of steam turbine?

7.2 What is the purpose of compounding of impulse turbine? Mention the various types of compounding.

7.3 Which factors decide the choice of impulse or reaction turbine for a particular need?

7.4 Mention the drawbacks of simple impulse turbine.

7.5 Give a cylinder layout of steam turbine ranging from 30 MW to 500 MW unit.

7.6 Under what conditions, single flow, double flow and reverse flow are used.

7.7 What is the reason that impulse-reaction turbine blading is more efficient than that of impulse type?

7.8 Find the condition of the maximum blade efficiency in a single stage impulse turbine.

7.9 Find the optimum ratio of blade velocity to steam velocity for a two-row velocity compounded impulse wheel turbine.

7.10 Define degree of reaction.

7.11 What is the condition of maximum gross stage efficiency in Parsons reaction turbine?

7.12 Comment on the statement:

Reaction turbine stages are preferred to impulse stages even through the reaction turbines occupy more space and run slower.

[***Ans.*** More power developed, normal speed, full admission, better efficiency are the possible causes for preference.]

7.13 On what factors does the reheat factor depend?

7.14 What is the importance of reheat factors and conditions curve is steam turbine design? Explain clearly.

7.15 What do you mean by governing of steam turbine? What are the chief methods of governing? Mention the various types of speed governors and their relative merits and demerits. What do you mean by direct digital control?

7.16 What is the difference between throttle and nozzle controlled governing? Discuss the hydromechanical control loop with regulation diagram.

7.17 Write notes on the following:

(i) Nozzle controlled governing

(ii) Throttle controlled governing

(iii) By-pass governing

(iv) Governing of reheated turbine

(v) Speeder and anticipator gear.

Numerical Questions

7.1 In a simple impulse turbine, the nozzle angle is 16° and the blade outlet angle is 25°. Inlet velocity of steam as it issues from nozzles is 720 m/s and the blade velocity is 180 m/s. The blade velocity coefficient may be taken as 0.75. The steam flow rate is 1 kg/s. Find: (a) Energy dissipated in blades due to friction, (b) Power developed, and (c) Diagram efficiency.

Ans. (a) 65.13 kJ/s (b) 158.098 kW (c) 61%

7.2 In a two-row velocity-compounded stage for an impulse turbine the initial velocity of steam from the nozzle is 550 m/s. The velocity of blade is 107 m/s. The nozzles have discharge angle 18° and the discharge angles of the three rows of blades in order are 21.5°. 28° and 45°. Assuming loss due to friction to be 15% in each row of blades, find the work done by the steam per kg and the efficiency of the blading. Neglecting any increase of volume of steam, find the height required for each row blades, taking the height of nozzles as 17.75 cm.

(*Hints.* Draw velocity diagram and find ${}_1C_{f1}$, ${}_1C_{f2}$, ${}_2C_{f1}$, ${}_2C_{f2}$)

[***Ans.*** 1st moving blade height = $17.75 \times \dfrac{{}_1C_{f1}}{{}_1C_f} = 21.7$ cm

Fixed ,, ,, = $17.75 \times \dfrac{{}_1C_{f1}}{{}_2C_{f1}} = 26.3$ cm.

Second moving ,, ,, = $17.75 \times \dfrac{{}_1C_{f1}}{{}_2C_{f2}} = 31.7$ cm.

Work done = 103005 J/kg, $\eta_b = 69.9\%$]

7.3 (a) The steam velocity at inlet to the first stage of a turbine having two-row velocity compounded impulse wheel is 600 m/s and the mean blade velocity is 120 m/s. The nozzle angle is 16° and the exit angles for the first row of moving blades, the fixed blades, and the second row of moving blades are 18°, 21° and 35° respectively. Find the blade inlet angles for each row. Find also for each row of moving blades, the driving force and the axial thrust on the wheel for a mass flow rate of 1 kg/s of steam. Find the diagram efficiency for the wheel and the diagram power. What is the maximum possible diagram efficiency for a given inlet velocity and nozzle angle? Take $K = 0.9$ for all blades.

[*Ans.* ${}_1\beta_1 = 20°$, ${}_1\alpha_1 = 24.5°$, ${}_2\beta_1 = 34.5°$, 874 N, 292.2 N, ${}_1F_a = 32$ N, ${}_2F_a = 9$ N, 77.9%, 140 kW; 92.3%]

(b) In the above problem, the steam flow rate is 5 kg/s and the nozzle height is 2.5 cm. Neglecting the wall thickness between the nozzles, calculate the length of nozzle arc. Take specific volume 0.375 m^3/kg. If all the blades have a pitch of 25 mm, and exit tip thickness of 0.5 mm, calculate the blade height at exit from each row.

[*Ans.* 0.454 m, $l_1 = 32.7$ mm, $l_f = 41.5$ mm, $l_2 = 44.2$ mm]

7.4 A set of steam nozzles in an impulse turbine stage is supplied with steam at 20 bars and 230°C. The mass flow rate of steam is 60 kg/s. The steam is expanded from rest to a back pressure of 14 bars with an efficiency of 98%. The mean diameter of the nozzle disc is 800 mm and the nozzle angle is 22°.

Assuming that steam is admitted all around the periphery of the nozzle disc, determine:

(i) whether a convergent or a convergent-divergent nozzle is used,

(ii) the velocity of flow of steam at the exit of the nozzle,

(iii) the area of the flow normal to the axis of the stage, and

(iv) the height of the nozzles. Assume adiabatic index for superheated steam to be 1.3. What would be the mass flow rate when the back pressure is reduced to 9 bars keeping upstream condition same.

7.5 A group of reacting blading consists of three fixed and three moving rings, all of the same height, and the mean blade speed of the moving rings is 76 m/s. For the mean moving ring the inlet absolute and relative velocities are 120.5 m/s and 44.8 m/s respectively and the specific volume 0.15 m^3/kg. Determine for a flow of 2.27 kg/s:

(a) the required area of blade annulus,

(b) the power developed by the groups, and

(c) the required heat drop for the group if the steam expands with an efficiency of 0.8.

[*Ans.* (a) $A = 0.1235$ m^2 (b) 40.8 kW (c) 22.15 kJ/kg]

7.6 In a certain stage in an axial-flow impulse-reaction turbine running at 3000 rpm, the fixed and moving blades are of identical shape and have an outlet angle of 20°. The blade length is 6.35 cm, the mean blade ring diameter is 71.2 cm and the steam is dry and saturated at 5.44 bars. If the mean blade speed is twice the axial velocity of flow of the steam, calculate the correct inlet-blade angle, the power developed in the stage, and the efficiency of the blading. What would be the rotational speed for maximum blading efficiency, other conditions being unaltered, and the maximum efficiency at that speed?

[*Ans.* $\beta_1 = 53.4°$, power = 499.8 kW, $\eta_b = 90\%$, (N) = 4130 rpm when $\eta_{\max b} = 93.8\%$]

7.7 A reaction turbine runs at 300 rpm and its steam consumption is 15560 kg/h. The pressure of the steam at a certain pair is 1.9 bars, its dryness is 0.93 and the power developed by the pair is 4.55 W. The discharging blade tip angle is 20° for both fixed and moving blades and the axial velocity of flow is 0.72 of the blade velocity. Calculate the drum diameter and the blade height if the tip leakage of steam is 8% and negligible blade thickness.

[*Ans.* $D_r = 101$ cm, $l = 8.26$ cm]

7.8 The fixed blades of a stage of a Parson's turbine bladings delivers dry and saturated steam at 2.7 bar at 90 m/s. The mean blade height is 4 cm and the moving blade exit angles is 20°. The axial velocity of the steam is 3/4 of the velocity at the mean radius. Steam is supplied to the stage at the rate of 9000 kg/h. Calculate the following: (a) the rpm, (b) the diagram power, (c) the diagram efficiency, and (d) the enthalpy drop in the stage.

[***Ans.*** (a) 1824 rpm, (b) 13.14 kW, (c) 78.7%, (d) 5.26 kJ/kg]

7.9 Steam enters in a stage of impulse-reaction turbine, with a speed of 280 m/s at an angle of 22° in the direction of blade motion. The mean diameter of the rotor, which rotates at 3200 rpm is 1 m. The blade height is 10 cm. The specific volume of steam at nozzle outlet and blade output are 3.6 m^3/kg and 4.1 m^3/kg respectively. The turbine develops 550 kW. Find (i) the mass of steam used per sec, (ii) blade angles, (iii) enthalpy drop in each stage, (iv) degree of reaction, (v) stage efficiency. Assume combined efficiency of nozzles and blades are 90% and carry-over coefficient as 0.8.

7.10 34×10^3 kg/h of steam expands in a impulse-reaction turbine from 20 bars, 400°C to a pressure of 0.2 bar. The turbine is designed such that the leaving steam is just dry and saturated. The reheat factor is 1.05 and the isentropic efficiency of each stage is the same throughout. There are fourteen stages. The enthalpy drop is the same in each stage. As the blades have an exit angle of 22° and the mean value of blade speed ratio is 0.82, the turbine speed is 2440 rpm, calculate the following:

(a) the stage efficiency, (b) the diagram power, (c) the drum diameter, (d) the blade height, and (e) the pressure at the entry to the last stage.

[***Ans.*** (a) 67.7%, (b) 6025 kW, (c) 1.34, (d) 175 mm (e) 0.32 bar]

7.11 The high-pressure (H.P.) supply to a turbine of 1158 kW output is 13.6 bars and superheated 37.8°C, the low pressure (L.P.) supply being dry saturated at 1.36 bars and the condenser pressure 0.07 bar.

The Willans lines for both H.P. and L.P. supplies are straight, and in each case the total steam used per hour at zero load is .012 of that required at full load. Assuming that the relative efficiency at full load is 0.8, whether H.P. or L.P. steam alone is supplied, find how much H.P. steam will be required for a load of 926 kW. 6370 kg/h of L.P. steam are available.

[***Ans.*** 1635 kg/h]

7.12 The full load steam consumption of a 11,000 kW turbo-alternator is measured to be 49,500 kg/h with steam conditions of 100 bars and temperature 500°C and exhaust pressure of 0.07 bar. Assuming an alternator efficiency of 94% and mechanical losses due to bearing friction governor, etc., of 100 kW, estimate the internal efficiency of the turbine.

During a part load operation of the turbine, the steam consumption was measured to be 34,000 kg/h with the same conditions at the stop valve and at exhaust. Assuming the same alternator efficiency and mechanical losses as obtained at full load, determine the alternator output under this condition. The turbine is governed by throttling. [***Ans.*** 81%, 7562 kW]

7.13 Following particulars relate to a steam turbine at full load conditions governed by throttling: Inlet conditions of steam = 20 bar, 350°C, Back pressure = 0.035 bar, Specific steam consumption = 5 kg/kWh. If the turbine at no load consumes 10% of steam consumption at full load, calculate the internal efficiency of the turbine. [**Hint:** Assume full load as K]

7.14 The steam consumption at full load of 93 MW turbine is 166×10^2 kg/h. At half load the steam consumption is 9100 kg/h. Applying Willans's line, estimate the steam consumption at no load and the consumption at a load of 25 MW. [***Ans.*** 16×10^2 kg/h, 14×10^2 kg/h]

Software Based Questions

7.1 Develop a software for the design of impulse and impulse-reaction steam turbines and validate the results with GE models.

7.2 Develop a software for predicting the performance of steam turbine.

7.3 Develop a software for governing system of a steam turbine. Assume suitable data wherever necessary.

Design and Open Ended Question

7.1 Develop a comprehensive report on various models of GE and Siemens. Based on these develop modern design procedure for a steam turbine.

Multiple-Choice Questions

7.1 In a simple steam turbine, the exit leaving velocity loss is around
(a) 5% (b) 10% (c) 15% (d) 20% of inlet energy

7.2 Which one is true for velocity compounded steam turbine?
(a) Velocity of steam remains constant in rotors.
(b) Velocity of steam goes on increasing in rotors.
(c) Pressure of steam goes on decreasing in rotors.
(d) Velocity drop takes place in stages in rotors.

7.3 Which one is true for pressure-compounded steam turbine?
(a) Pressure drop takes place in stages in various stages of nozzles.
(b) Pressure remains constant throughout the stages.
(c) Velocity goes on decreasing throughout the stages.

7.4 In impulse-reaction turbine, the two-row velocity compounded impulse turbine is installed
(a) in the beginning (b) at the end of *hp* stage
(c) at the end of *ip* stage (d) at the end of *lp* stage

7.5 The blading efficiency of impulse reaction is
(a) less than pure impulse (b) more than velocity compounded
(c) less than two-row velocity compounded (d) less than pressure compounded

7.6 The reheat factor of steam turbine is around
(a) 0.8 (b) 1.0 (c) 1.05 (d) 1.2

7.7 The speed regulation in steam turbine is around
(a) ±4% (b) ±6% (c) ±8% (d) ±10%

7.8 The isentropic efficiency of steam turbine ranges from
(a) 50 to 60% (b) 60 to 70% (c) 70 to 80% (d) 80 to 90%

7.9 Which one is true with reference to cylinder layout?
(a) Double flow in *hp* stage.
(b) Reverse flow in *hp* stage.
(c) Single flow in *lp* stage.
(d) Double flow in *lp* stage.

7.10 In cylinder layout of steam turbine, double flow is used because
(a) power output will be more.
(b) efficiency will be more.
(c) height of blading will be less.
(d) more balancing and less size of blading.

Answers to Multiple-Choice Questions

7.1 (b) 7.2 (d) 7.3 (a) 7.4 (a) 7.5 (b) 7.6 (c) 7.7 (a) 7.8 (d)
7.9 (d) 7.10 (d)

CHAPTER

8

Auxiliary Systems, Construction, Operation and Maintenance of Steam Turbines

OVERVIEW

This chapter deals with auxiliary systems, construction, operation and maintenance of steam turbines, It is desirable that at prefinal/final year stage, students of Mechanical Engineering, should know some aspects of these.

AUXILIARY SYSTEMS

In order to have reliable operation of steam turbines, various auxiliary systems are incorporated. The following are included in the steam turbine auxiliary systems:

(i) *Governing system* (already discussed in chapter 7)

(ii) Turbine protective devices (or turbine mountings)

(iii) Lubrication system

(iv) Turbine greasing

(v) Glands and sealing systems

(vi) Flange heating systems

Each system will be discussed separately in the subsequent pages in brief. Further, construction, operation and maintenance of steam turbines will be discussed.

8.1 TURBINE PROTECTIVE DEVICES

Turbine protective devices protect the turbine from damage during abnormal working conditions or malfunctions. Before discussing these devices, it is worth to first list the possible

* For more details, see the Author's book, *Steam & Gas Turbines, Turbomachines and Power Plant Engineering.*

hazards likely to be encountered and only after this it would be possible to think of protective devices to take care of these hazards.

8.1.1 Possible Hazards

During the operation of steam turbines, many dangers are likely to be encountered. These may be summarised as follows:

(a) Overspeeding
(b) Motoring
(c) Oil failure
(d) Thrust bearing failure
(e) Vacuum failure
(f) Boiler priming
(g) Excessive vibration
(h) Excessive temperature differentials
(i) Excessive eccentricity.

Among all the hazards, the first six, (i.e., a to f) items are catered by a series of automatic devices known as *(i)* ***trip gear*** *and (ii)* ***unloading gear***. The remainder last three (i.e., g to i) items are monitored by means of *(iii)* ***supervisory gear***.

(i) Trip gear. *The function of trip gear is to cause the immediate shut down of the turbo-generator by releasing the pressure from h.p. oil system (i.e., stopping the supply of h.p. oil to relay system and opening the servo oil lines to drain) resulting in the closing of emergency and main steam stop valve and reheat emergency valve.*

(ii) Unloading gear. *The function of unloading gear is to exert on overriding control on the turbine governing gear, and limit the output of the turbine when inlet and exhaust conditions are abnormal.*

It is to be noted that all the tripping devices operate on the fail-safe principle, in that the operation of any one of them causes complete collapse of relay oil pressure and thus immediate shut down of turbo-generator. Figure 8.1 depicts an example of a complete hydraulic control system for a modern turbo-alternator system.

8.1.2 Emergency Steam Stop Valve Operating Gear

Since steam enters the turbine chest only through the stop valve passage, so during emergency it will be wise to shutdown this valve, This suggests that the operating gear for the emergency valves must be very reliable and possess fail-safe characteristics. These valves are held open by high pressure relay oil against powerful spring and out of balance steam force hence they will automatically close when the oil supply to the valve is blocked and the oil in the operating relay is allowed to drain with the help of tripping devices.

An automatic steam stop and emergency valve control gear is shown in Fig. 8.2. The pilot valve which controls the admission and drainage of H.P. oil into and from the relay cylinder is spring biased. The bias is overcome only when high pressure oil is applied to the relay thereby admitting H.P. oil to the relay cylinder which raises the piston and opens the stop valve. When the relay connects oil to drain (e.g., to overspeed trip connects the H.P. oil to the drain) the pilot valve experiences low oil pressure causing it to go up and connect the oil of relay cylinder to drain resulting in the closing of stop valve. Another type of relay gear

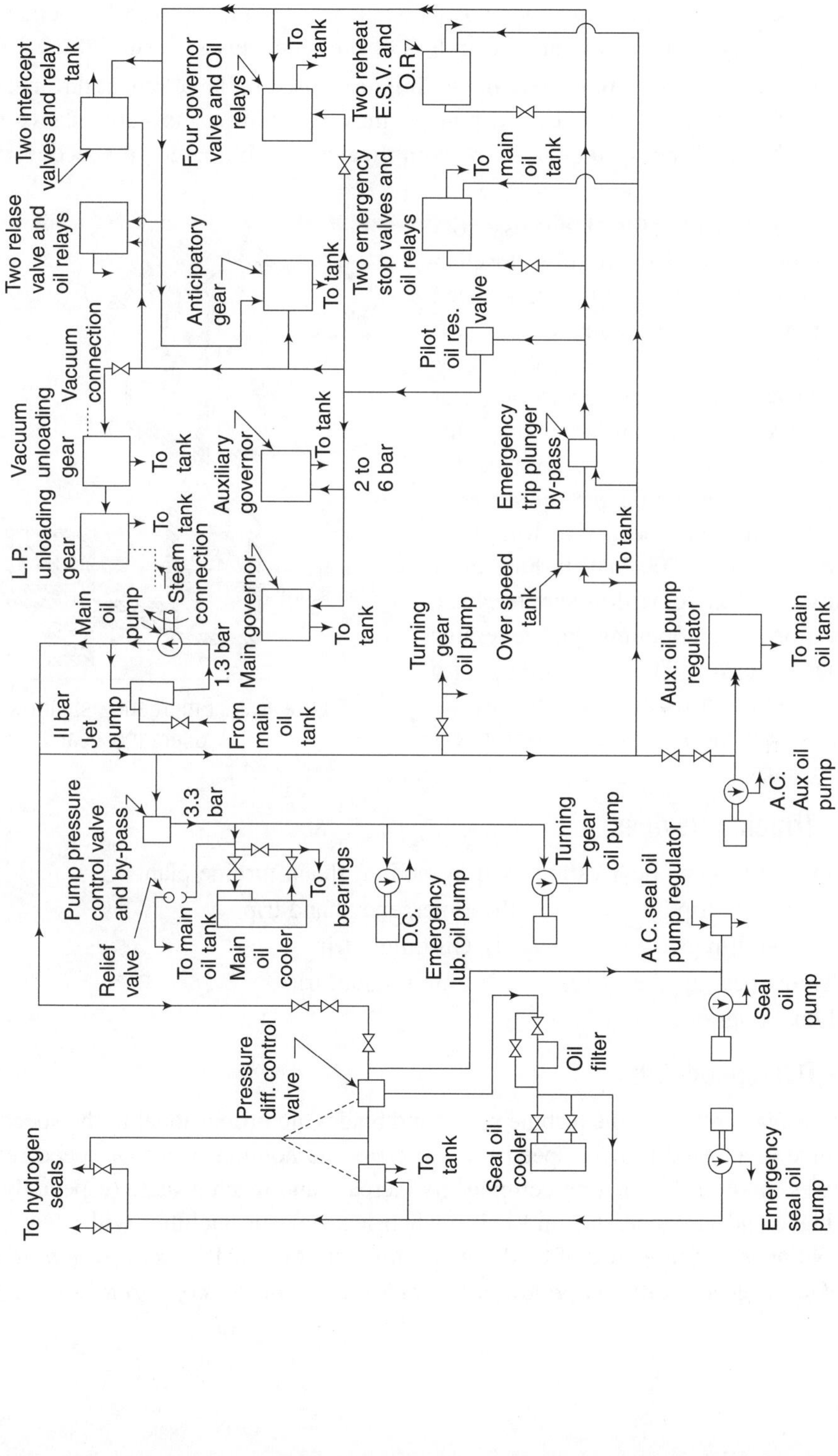

Figure 8.1 Hydraulic control system

consists of a second solenoid operated pilot valve arranged to dump the oil from the relay, when a load rejection signal is received by an electrical secondary governor.

Other type of automatic stop valve, not shown here, uses a hydraulically operated latch to release the energy of the spring and again loss of pressure causes closure of the valve.

Steam stop and emergency valve operating gear, used on large reheat turbine, basically consists of a hydraulically unbalanced plunger moving within a moveable sleeve and fixed bush (not shown). The system is more complicated as compared to the one discussed above.

When there is a loss of load, it is essential that the turbine stop valves are fully closed in time to prevent the resulting transient speed rise from exceeding the permitted 12.5%. It is the ratio of steam input energy to rotor inertia which determines the acceleration of turbine following a trip. In larger unit, this ratio is larger which suggests the need for more rapid valve operation gear, In order to prevent the transient speed rise following full load rejection, a 500 MW turbine must have all valves fully closed in about 0.4 sec after the instant of opening the generator main field breaker. This time takes care of the acceleration of machine due to the steam present in the loop pipes and cylinders following a trip.

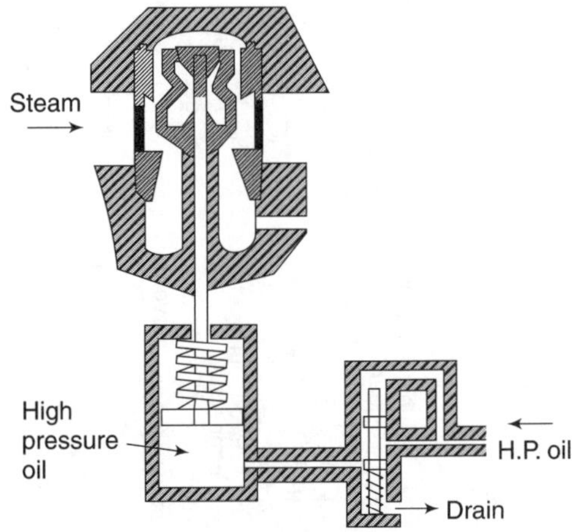

Figure 8.2 Emergency steam stop valve operating gear

8.1.3 Tripping Devices

The following tripping devices are used in modern steam turbine plants.

(a) Overspeed trip
(b) Emergency hand trip
(c) Remote trip gear
(d) Oil failure trip
(e) Thrust-bearing wear trip
(f) Low vacuum trip
(g) Low boiler pressure trip

8.1.4 Overspeed Trip

The centrifugal stresses in the turbine rotors and blades are proportional to the speed squared. If a turbine is allowed to over speed (i.e., to cross the nominal value of speed regulation), the stresses in all of the rotating components increase and reach a stage (especially in larger rotors, blade and the generator and bells) when local plastic yielding will occur and if the speed continues to rise failure of blades and rotors will eventually occur. *Due to this fact the speed of turbo-generators is not allowed to reach 12.5% in excess of normal speed.* This limit is only exempted for one work-proof test after manufactures, when the rotors are usually run for a few minutes at 20% over speed in a test pit.

There is an already speed governing system which should limit the speed to a level well before 110%, even then a safety device known an *over speed trip* is *invariably employed and usually duplicated to close the emergency stop valves when the speed rises above this figure.* The constructional details of an overspeed trip is shown in Fig. 8.3. The device comprises a spring loaded bolt or ring, incorporated into the shaft at the governor end, having its centre of gravity, a short distance away from the axis of rotation. This arrangement makes the bolt or ring unbalanced on the shaft and causes an unbalance force. Below 10% overspeed, the unbalanced force is compensated by the spring compression so that the bolt or ring should not knock the trip lever. At 10% overspeed and more, the out of balance force is arranged to overcome the spring compression, causing the bolt or ring to knock the trip lever, which releases a spring operated trip valve that blocks the supply of H.P. oil and opens all the servo oil lines to drain. By this way, there will be an immediate shutdown of the turbo-generator by closing of steam stop emergency and other valves. To adjust the tripping speed, thin shims under the spring are placed.

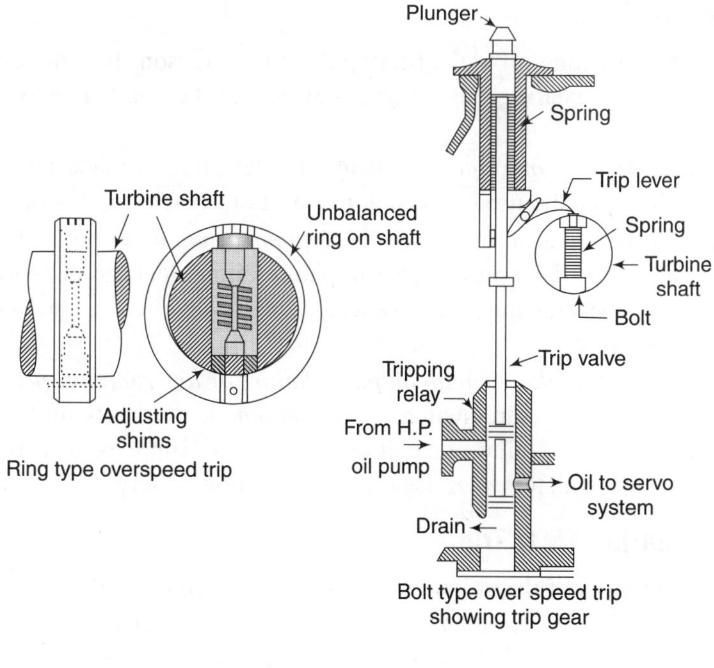

Figure 8.3 Overspeed trip

It is possible to test the overspeed trip without actually over speeding or taking the set off the load. Oil is injected into a small cavity in the unbalanced weight, increasing the out-of speed balance force so that it trips at normal.

Electrical over speed trip is more reliable and quick in response. It consists of a normally energised solenoid holding off the spring loaded trip valve (or the trip lever). The current supply to the solenoid is broken by a signal from an inductive shaft speed pick-up device resulting in the release of trip valve that blocks the supply of H.P. oil and consequently closure of emergency steam stop and other valves.

8.1.5 Emergency Hand Trip

As the name implies, this is an emergency trip operated by hand and may be used when the automatic trip gear fails. It consists of a lever on the governor pedestal that operates the trip valve of overspeed trip gear, thus closing all valves by releasing the H.P. oil pressure to drain.

8.1.6 Remote Trip Gear

In order to trip the turbine remotely from the control room, the emergency hand trip lever can be actuated by means of a solenoid or a torque motor. When the control room switch is closed, the station battery is used to energise the actuators.

8.1.7 Oil Failure Trips

If the supply of the lubricating oil to the bearing due to any reason stops there will be wiping of the journal and thrust bearing, resulting in damage to the shaft, It may caused by the following:

(i) *Low oil level in the sump due to leakage.* Its detection is made possible by a float operated switch may be used to sound an alarm or operate the remote trip gear.

(ii) *Low H.P. oil pressure generated from the oil pump common for both lubricating system and governing gear.* This may be due to pump failure, leakage or blockage. There is no need of any special tripping device as the steam valves close automatically on loss of relay oil pressure.

(iii) *Low H.P. oil pressure due to main oil pump failure or low running speed.* It is possible to detect this by a pressure switch, which automatically starts up the a.c. auxiliary oil pump. In case of a failure of unit and station a.c. supplies, a relay starts the d.c. emergency pump which is energised from the station battery.

8.1.8 Thrust-bearing Wear Trip

There is always a variation of thrust load on the rotor. On account of this variation, the thrust bearing is more vulnerable to wear than the journal bearing, This wear may damage the blades and glands. *In order to prevent this damage, a tripping device is sometimes incorporated on the turbine known as thrust-bearing wear trip.*

In this tripping device, a fine jet of H.P. oil issued from a nozzle plays on the face of a flange on the shaft. The nozzle position should be very near to the flange. Under normal condition, a small quantity of oil flows, since the clearance (wear) is small and the pressure behind the nozzles is maintained. Due to axial movement of shaft, the clearance (wear) increases, the oil leaks more resulting in fall of pressure and tripping of stop valve. Other simple type of thrust bearing wear trip is a "squealer" ring which rubs and gives audible warning of thrust bearing failure. This also provides a temporary face during run down.

8.1.9 Low Vacuum Trip

In the event of *failure of the vacuum unloading gear (see section 8.1.12), a vacuum trip device is incorporated independently that either operates in the same electrical circuit as the remote trip gear, or actuates a separate H.P. oil pump valve.*

This trip gear is actually in the form of pressure switch. The position of the switch contacts is controlled by a spring loaded bellows unit which is displaced by the vacuum that is applied to one side of the bellows.

Incorporation of vacuum trip depends by the ability of the last stage turbine blading to withstand the high temperature conditions resulting due to operation at poor vacuum.

8.1.10 Low Boiler Pressure Trip

In the event of falling in boiler pressure up to 80% of the design pressure, a low boiler pressure trip is installed which will trip out the turbine to prevent damage to it. This trip is in the form of pressure which operates in the remote trip gear circuit Fig. 8.3.

8.1.11 Unloading Gears

Unloading gears are of two types:

1. Low vacuum unloading gear
2. Low steam pressure unloading gear

8.1.12 Low Vacuum Unloading Gear

If the vacuum falls considerably, the temperature of the exhaust steam will increase and may possibly damage the last rows of blading and condenser tubes (tubes get buckled due to excessive temperature under the influence of high temperature). In order to avoid this, a devices known as low vacuum unloading gear is used which below a certain vacuum, progressively decreases the steam flow as the exhaust pressure rises, thus tending to restore the vacuum. However, it does not close the governing valve completely, as this would cause *motoring* which is an another hazard. The sensing of vacuum device consists of a spring loaded bellows which controls the pilot valve of a relay by measuring the vacuum. The output of the bellows unit is

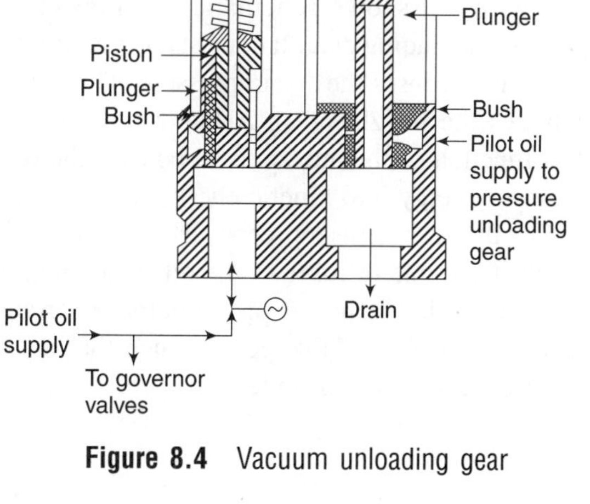

Figure 8.4 Vacuum unloading gear

arranged to open a valve in the governed oil pressure line and reduce the pressure to a value corresponding to the actual vacuum existing in the condenser.

Figure 8.4 shows a vacuum unloading gear. It has a vacuum connection to be made to the condenser. The condenser vacuum operates on a bellows connected to a plunger which has parts cut in it. The plunger is arranged such that it slides in a ported bush surrounded by a chamber to which governor pilot oil normally has access. If the vacuum is correct, the difference in pressure across the bellows compresses the spring and holds the plunger up against a stop. Under this condition, there is no line up of the ports in the plunger and the ports in the bush and hence the influence of unloading gear on the pilot oil pressure is nullified. In the event of falling condenser vacuum, the plunger will move downwards under the action of the spring and at a predetermined vacuum, the plunger ports will uncover and thus there will be a creation of a leak-off flow that will reduce pilot oil pressure and hence the lift of the governor valves. If the vacuum continues to fall, the gear will continue to operate, unloading the turbine, however, anti-motoring device will prevent the complete unloading.

8.1.13 Low Steam Pressure Unloading Gear

The requirement for the correct functioning of the turbine control gear is the correct maintenance of design steam conditions at the stop valve. In the event of boiler failure to maintain the steam pressure, the turbine would demand more steam which would cause still further fall in steam pressure, In addition to this, if the fall in steam pressure were sudden owing to any reason such as lots of ignition, sudden demand for more steam might cause water to be carried into the turbine, causing serious damage. This damage may be prevented by reducing the load on the turbine, should the inlet pressure fall below 90% of normal superheater outlet pressure of boiler.

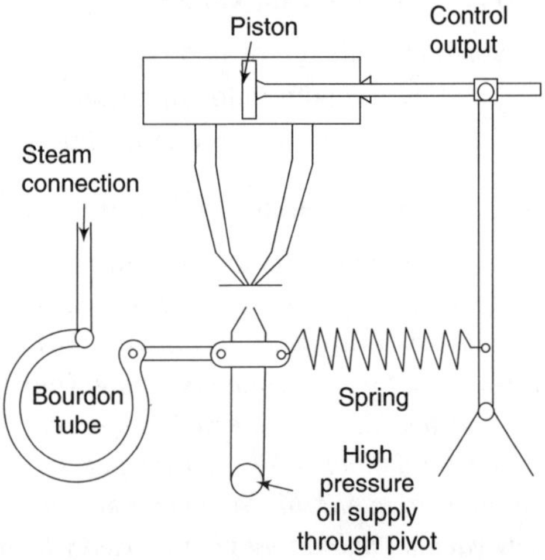

Figure 8.5 Low steam pressure unloading gear

Figure 8.5 shows schematic diagram of a pressure unloading gear. It consists of a bourdon tube which senses the steam pressure. This tube actuates the nozzle of an Askania type relay. The function of this nozzle is to convert the oil pressure energy into kinetic energy in a jet and one of the two diffusers reconverts back this energy to pressure energy. Thus, by deflecting the jet from one diffuser to the other, the pressures above and below the power piston may be controlled, The power piston output is arranged to override that of the governor primary relay. The design of this gear is such that it cannot automatically reload the turbine in the event of restoration of boiler pressure.

8.2 LUBRICATING SYSTEM

Lubricating system is one of the most important systems of a steam turbine. It is the life line of the bearings. Two purposes are served by lubricating oil. These are as follows.

(i) *Oil supports the shaft in the bearings by forming a continuous oil wedge on which the shaft revolves. Only a small quantity of oil is required for this purposes.*

(ii) *Oil cools the bearings and maintains them at a desired temperature level by removing heat produced considerably due to shaft conductivity, surface friction and turbulence set up in the oil. A large quantity of oil is needed to remove this heat.*

It is the general practice to supply oil to the bearings at a pressure of 1.35 to 2.75 bars to ensure that the pressure in the upper part of the bearing does not fall below atmospheric otherwise there will be a formation of discontinuities in the oil film. If the oil pressures are too high, the oil will be sprayed out from the ends of the bearings at a high velocity and will become finally atomized and may easily escape from the bearing housing.

It is essential to maintain the temperature of oil within certain limits otherwise the two purposes mentioned above will not be fulfilled. In the case of too low oil temperature entering

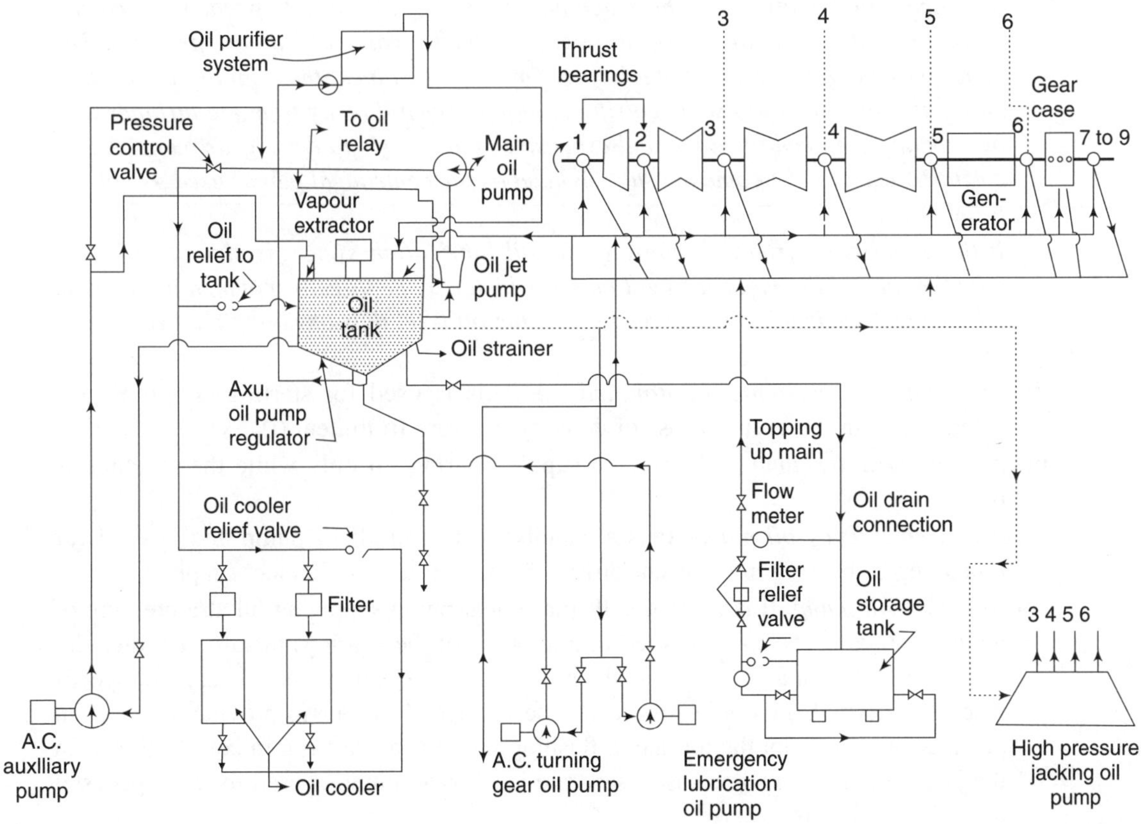

Figure 8.6 Lubricating system

the bearing, the bearing lubrication will be poor due to high viscosity of oil, whereas if the oil temperature leaving the bearing is too high, the oil property will be deteriorating due to high rates of oxidation. *The temperature of the oil leaving the bearing is limited to 74°C so that the maximum temperature within the bearing is not more than* 71°C. The temperature of the oil entering the bearing should be in the range of 38 to 43°C and the required oil leaving temperature is maintained by adjusting the oil supply to each bearing with the use of generous pipe diameters, removable orifice plate and control valve. The main elements of lubricating system are:

(a) Oil pumps (b) Oil relief valves (c) Oil coolers
(d) Oil tanks, strainers and filters (e) Oil

Figure 8.6 shows a simple sketch of lubricating system used on large turbo-generator set. The jet pump is generally submerged in the oil tank. For clarity it is shown outside.

8.2.1 Oil Pumps

There are **five types** of pump used in lubrication system.

(i) *A full duty shaft driven main oil pump. It supplies oil at a pressure 4 to 14 bars to bearing and relay when the machine is running at normal speed. For bearing lubrication, the oil is drawn off through a reducing valve at suitable pressure. The pump may be gear or centrifugal type. On large turbines, the quantity of oil used makes it economical to use a centrifugal pump driven directly from the turbines thus overcoming the need for large and expensive gear wheels to drive gear pump. An oil ejector is used to overcome the suction head as the centrifugal pumps used as not self priming type.*

If the governing system is separate, the oil lubrication system is simplified by the adoption of a.c. driven centrifugal pumps which may be directly mounted in the main oil tank with their inlets submerged below the oil level thus obviating the need for an oil ejector.

(ii) *An auxiliary, i.e., pump (motor driven).* This is used for starting, stopping and emergencies in the event of loss of delivery pressure from item (i).

(iii) *An a.c. flushing pump.* It is used to supply bearing oil only while the machine is barring.

(iv) *A d.c. emergency oil pump.* It is a standby to the auxiliary pump and is used for protecting of bearing in the event of loss of the unit and station a.c. supplies.

(v) *A jacking oil pump.* It is a positive displacement pump supplying oil at a pressure of around 136 bars. Jacking oil is fed into the base of the bearings in sufficient quantity to establish in each an oil film on which the heavy rotors may float at very low speeds when the normal oil wedge would not be formed, In order to provide a sufficient quantity of oil to cool the journal, a flushing pump is always run in conjunction with the jacking pump. In some cases, separate jacking oil pump is used for each pedestal to avoid the fire risk.

8.2.2 Oil Relief Valve

Since the pumps used in the lubricating system are of constant speed type, so the quantity of oil delivered is uniform. It is essential to bleed continuously the surplus oil. This is achieved by spring loaded relief valves. *The function of relief valve is to maintain a correct amount of oil flow to the bearing at desired pressure by allowing the extra oil to return to the sump.* Relief valves those fitted to both the main and auxiliary pumps limit the delivery pressure and each pump is fitted with a non-return valve in the discharge branch which prevents oil flowing back through the idle pump units.

If a single pump is used to provide oil for lubrication and relays, the pump delivers oil at relay pressure and a greater portion is then throttled through an adjustable reducing valve down to lubrication pressure.

Relief valve fitted in the oil cooler inlet manifold upstream of the coolers and associated filters units allows oil temporarily to by-pass the coolers in the event of excessive pressure drop occurring across the cooler units indicating blockage somewhere in the system. If a standby cooler is available then it is brought in service.

8.2.3 Oil Coolers

The oil in the lubricating system gets heated by two ways. In one way, by the losses of centrifugal pump as the efficiency of a centrifugal pump is of the order of 85% and the remaining 15% of the pump input energy generates heat in the oil. In another way, it gets heated by the bearing friction and shaft conduction. In order to remove this heat, the oil is passed through coolers in which it is cooled to a temperature required for the bearings.

Oil coolers are shell and tube type, counter flow heat exchangers to ensure the minimum formation of sludge in the oil (Fig. 8.7). These are designed to have the maximum heat transfer area with a small pressure drop between entrance and exit. In order to have easy removal of the tube nests, they are mounted vertically. Water enters at bottom and passes upwards through the tubes ensuring complete flooding at all times. Oil enters at the top and comes out at the bottom through zig-zag way. The tube nest consists of 70/30 brass tubes ensuring high thermal conductivity.

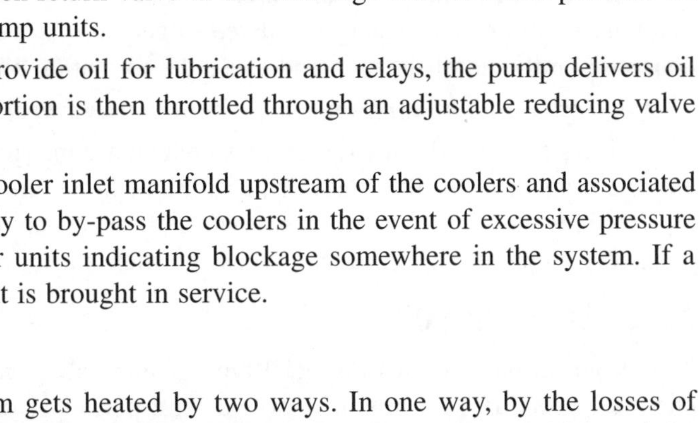

Figure 8.7 Oil cooler

In general practice, two 100% duty or three 50% duty oil coolers are provided so that one may be taken out of service for cleaning purposes without shutting down the plant. This is made possible by the use of parallel oil circuits and special combination valves.

8.2.4 Tanks, Strainers and Filters

The purpose of the oil tank is to store the lubricating oil for continuous use. *It is essential to change the oil many times per hour in order to liberate entrained air. The number of changes per hour may be 5 to 10 and it is this number which determines the capacity of the lubrication system, i.e., the size of the drain tank, which is generally situated below the floor level near the turbine.*

In order to disperse froth, the interior of the tank is divided by baffles. It is essential that tank should be adequately ventilated to get rid of water vapour which would otherwise condense on the lid of the tank and cause the formation of the rust.

It is essential to drain off the sludge periodically which settles in the bottom of the tank.

Strainers are the cylindrical wire-mesh which are interposed between the returning drains and the drain tank to filter the oil. In addition, filters are interposed in the pump delivery pipe to filter the foreign particles so that the bearings are relay systems are protected from damage. Sometimes, a centrifugal bypass purifier is also included in the system.

8.2.5 Lubricating Oil

The steam turbine requires a special type of lubricating oil to ensure that the physical and chemical properties are retained over a long period of operation. Although, mineral oils usually possess good resistance to oxidation, foaming and emulsifying and have good anti-rust properties. The extreme requirements for the modern steam turbine lubricating oil are usually met by the addition of inhibitors to a high quality base oil that is chosen for its high degree of chemical stability coupled with a capacity to absorb additives.

8.2.6 Generator Hydrogen Seals

The current practice is to cool generator by hydrogen gas. This means gas seals must be used on the shaft ends to prevent gas leakage into the atmosphere, otherwise there will be a formation of explosive mixture with air. Two types of seals are in use (a) radial and (b) axial seal.

8.2.6.1 Radial Seal

Figure 8.8 shows a simple sketch of radial type hydrogen seal. It consists of two series of sealing fins with a small radial clearance and is constrained by a relatively weak circumferential greater spring to limit the damage caused by an accidental rub. A seal oil pump supplies oil to the centre of the seal at a pressure slightly higher than the gas pressure. The oil flows through the restrictions in both directions into the air and hydrogen side

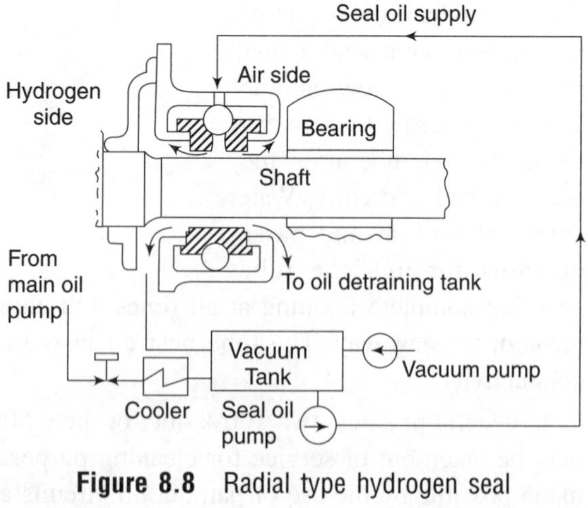

Figure 8.8 Radial type hydrogen seal

of the seal, This continuous liquid barrier between the shaft and the seal effectively prevents leakage of the hydrogen gas into the atmosphere.

Oil wiper rings are used to prevent oil entering the generator casing. A drainage chamber is formed by the oil wiper rings and the hydrogen seal from where the oil leaking into the hydrogen side of the seal can be collected in a vacuum tank where the hydrogen gas entrained in the oil is separated out and returned to the stator casing. The seal oil pump is backed up by the main turbine oil pump.

8.2.6.2 Axial Seal

Axial type of hydrogen seal is more efficient and used for the large generators. It is an axial or thrust type seal and normally used when high hydrogen gas pressures (3 bars or above) are used in the generator to improve the cooling. The main advantage of this seal is the small quantity of oil required to maintain an effective gas seal.

8.3 TURBINE GREASING

During start-up or shutdown, all sliding surfaces are supplied with grease to ensure smooth continuous movement of the turbine cylinders on bearing pedestals. Sticking at one pedestal could lead to 'crabwise' movement with the possibility of contact between fixed and moving parts. In order to ensure adequate movements at the bearing surface, efficient greasing must be carried out.

8.4 GLANDS AND SEALING SYSTEMS

8.4.1 Essence of Glands and Sealing System

Rotating and stationary components (i.e., shaft and cylinder or diaphragm) cannot be combined together. A small radial clearance space has to be provided to separate them. Due to pressure difference existing across them, steam or air will leak through this clearance. There are two zones of leakage, namely (a) recess made by the cylinder (casing) and the shaft at both external ends, (b) recess made by the diaphragm and shaft in impulse turbine. Where the cylinder pressure higher than atmospheric (e.g., h.p. end), there will be general steam leakage outwards, whilst where the cylinder contains steam below atmospheric pressure, (e.g., 1.p. end), there will be a leakage of air inwards. This suggests that a sealing system must be used to prevent this leakage of steam or air.

Glands and seals are the devices used on turbines to prevent or reduce the leakage of steam or air between rotating and stationary components that have a pressure difference across them.

Since steam leakage from glands does not pass through the turbine stages, a loss of power output results in and every effort is made to reduce this power loss by an efficient arrangement of seals and glands. There are three types of glands and seals in common use on steam turbines, namely *(a) labyrinth gland, (b) carbon ring gland and (c) water-sealed gland.* The first two glands restrict the steam and air leakage while the water-sealed gland prevents completely the steam and air leakage. Labyrinth gland is superior to carbon ring gland because it can withstand higher steam temperature.

8.4.2 Labyrinth Gland and its Forms

The labyrinth gland consists of a ring with a series of machined fins that form a number of fine annular restrictions, every restriction is followed by an expansion chamber. Figure 8.9 shows the various forms of labyrinth gland.

(i) *Plain Labyrinth (Fig. 8.9a).* This is a simple form of labyrinth seal in which the shaft is plain and the fins are made on the ring. The radial clearance and the number of restrictions (throttling) control the leakage.

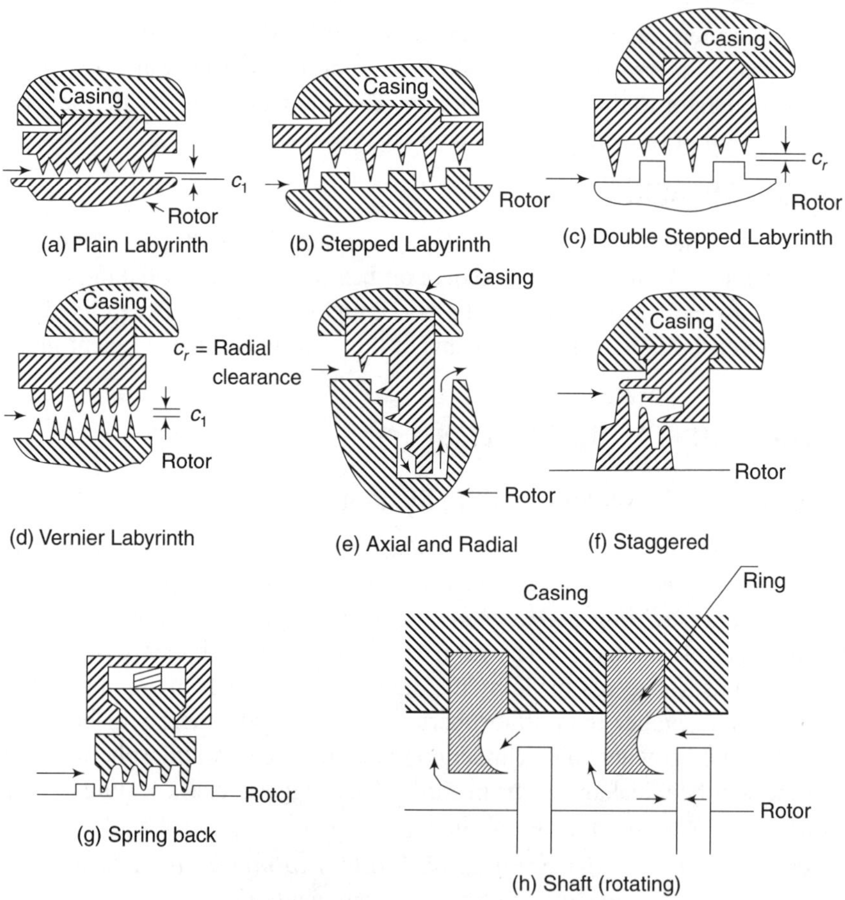

Figure 8.9 Labyrinth glands

(ii) *Stepped and Double Stepped Labyrinth (Fig. 8.9b, c):* In order to ensure that the maximum kinetic energy is converted into heat in the expansion chambers, the finned ring and shaft are usually stepped, Stepped labyrinth gland can only be used where axial differential expansion between the rotor and casing are small while the double stepped labyrinth gland can accept a large axial differential expansion, For effective restriction, two of the three fins per pitch are sufficient.

(iii) *Vernier Labyrinth (Fig. 8.9d).* In this type, the shaft and the seal ring are finned. The pitch of the fins is slightly different on the two seal components.

(iv) *Axial and Radial Labyrinth (Fig. 8.9e).* This type of gland is suitable where close control of axial differential expansion is possible, e.g., near the turbine thrust bearing. It is obvious from the figure that the steam is forced to flow radially inwards against a centrifugal pumping action resulting in a reduction in leakage of steam or air.

(v) *Staggered Labyrinth (Fig. 8.9f). In order to ensure that no kinetic energy is carried over from the preceding restriction, staggered labyrinth is used. The stationary fins are made axial and if an accidental rub occurs, the heat generated causes the stationary fins to expand increasing the clearance resulting in minimal damage.*

(vi) *Spring back Labyrinth (Fig. 8.9g).* In order to minimize the effect of a rub with close clearance glands, the gland rings are often spring loaded and are made up of four or more segments.

The flow through a labyrinth gland is function of the initial pressure and temperature, the final pressure and the clearance area under the restrictions.

(vii) *Axial Clearance Gland (Fig. 8.9h).* In this case, rings of non-ferrous metal are caulked into the gland casing. The packing may be regarded as axial or facial type. Shallow collars about 0.31 cm wide are turned on the shaft and the leakage is controlled by the axial clearance, c_0. This type is suitable at the H.P. end adjacent to the adjusting block.

8.4.3 Requirements of Glands

The following are the requirements of glands:

(i) The radial clearance and diameter of labyrinth gland should be made as small as practicable as the leakage through gland is directly proportional to the clearance. The minimum radial clearance adopted is 0.5 mm.

(ii) The pressure drop through throttling should not be large.

(iii) The number of throttling should be large.

(iv) The gland strip should be arranged to destroy as far as possible the kinetic energy acquired by the previous opening. This is only possible by staggering the openings.

(v) It should not add materially to the length of the bearing centres.

(vi) The tip thickness of glands should be made as thin as practicable so that if an accidental rub occurs between the shaft and glands, the fins will rub away with little heating of the shaft. In the case of heavy rube due to thick tip, the heat generated may bend the shaft and unbalance it.

8.4.4 Gland Sealing System

Figure 8.10 shows a typical gland sealing system used on a steam turbine. In order to achieve better sealing and to promote a gradual temperature gradients, many gland elements (sections) say about 20 are used. Each gland element is represented by a dot. The gland elements *A*, *B*, *C* and *D*, *E*, *F* marked in figure are the elements used as final glands on H.P. and L.P. cylinders respectively (Figs. 8.11 and 8.12).

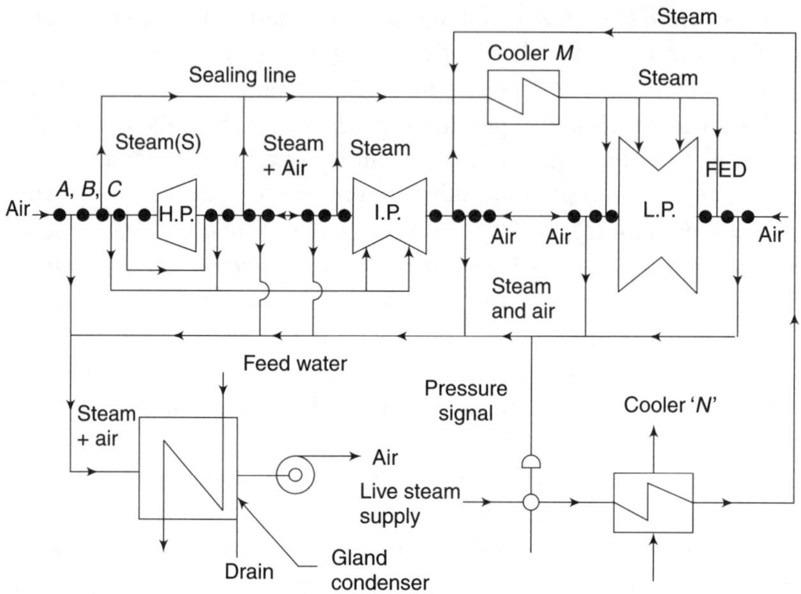

Figure 8.10 Gland sealing system

In order to prevent the leakage of steam from labyrinth glands into the turbine house where it would condense on the walls and plant, all final gland elements such as *A*, *B* and *D*, *E*, etc., are maintained at approximately 0.034 bar below atmospheric by the use of a gland condenser vented to atmosphere via a blower. This small vacuum is sufficient to draw air into the final gland section (Figs. 8.11 and 8.12) where it is mixed with steam leaking from the cylinder and by this way, a mixture of steam and air enters the gland condenser. The air is separated in the gland condenser and passed back to the atmosphere.

From the H.P. and L.P. cylinder gland, where the pressure at the gland leak off points a few kN/m^2 above atmospheric pressure (e.g., between *B* and *C*, etc.), is bled off to seal the gland of L.P. cylinder through cooler, *M*, thus the outflowing steam prevents the ingress of air into the cylinder and condenser. The steam flow H.P. and I.P. gland may be sufficient to seal the L.P. glands at high loads but at low loads, a separate supply of sealing steam has to be provided. This separate supply is usually taken from the main turbine supply and throttled through a reducing valve and cooled in a cooler, *N*. Coolers are required to reduce the steam temperature acceptable to turbine shaft. These

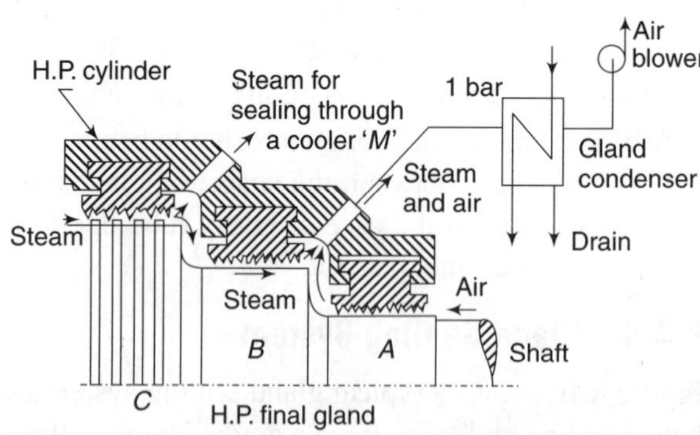

Figure 8.11 H.P. glands

coolers may be either water spray type or tube and shell type heat exchangers.

In order to reduce the power in external glands at the ends of the H.P. and I.P. cylinders, the leaking steam is bled off and taken to an appropriate stage in the turbine or to a feed heater after each section. In other words, the leaking steam returns heat to the cycle, a pressure controlled reducing valve is usually used to shut down automatically the sealing system.

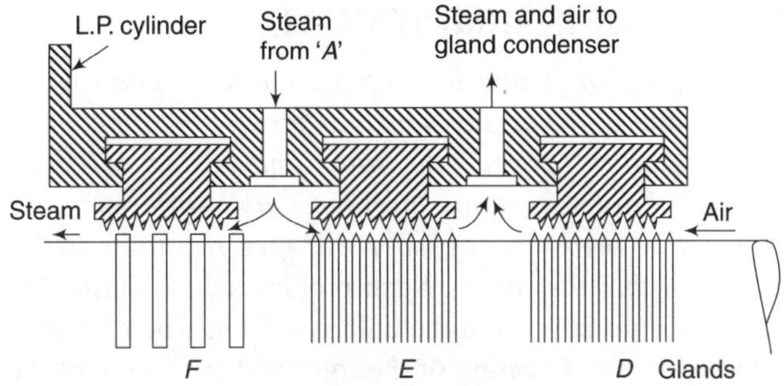

Figure 8.12 L.P. glands

8.4.5 Carbon Ring Gland

A compact and efficient form of gland may be made from carbon rings which are about 2.5 cm square in section. It consists of a number of segmental rings of graphite carbon and are enclosed in a suitable housing. It is self lubricating. Such a gland is shown in Fig. 8.13. The rings are held close to the shaft by a spring known as garter spring. If there is rubbing of ring to the shaft, the tension of the spring should be such that rings should exert only a slight pressure. The rotation of the ring gland is prevented by keys. Carbon ring glands are not used on large modern steam turbines as their maximum operating temperature and shaft speed is limited to about 120°C and 5 m/s rubbing speed.

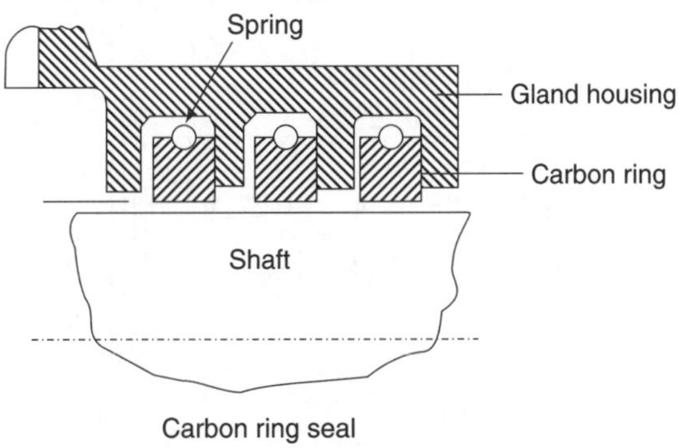

Figure 8.13 Carban ring gland

8.4.6 Water Sealed Glands

As stated earlier, water sealed gland completely prevents leakage. Figure 8.14 shows a typical water sealed gland. It consists of a shaft mounted impeller with a series of vanes or pockets machined in both faces which is enclosed within an annular chamber. When water enters the chamber, the impeller vanes force the water to rotate at a speed approximately equal to the impeller speed.

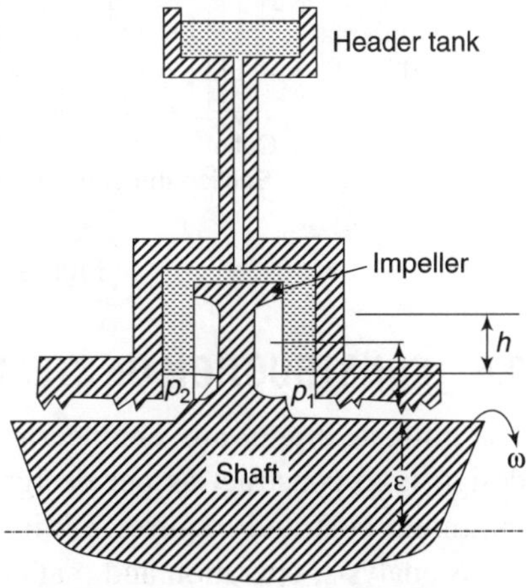

Figure 8.14 Water sealed gland

8.5 FLANGE HEATING SYSTEMS

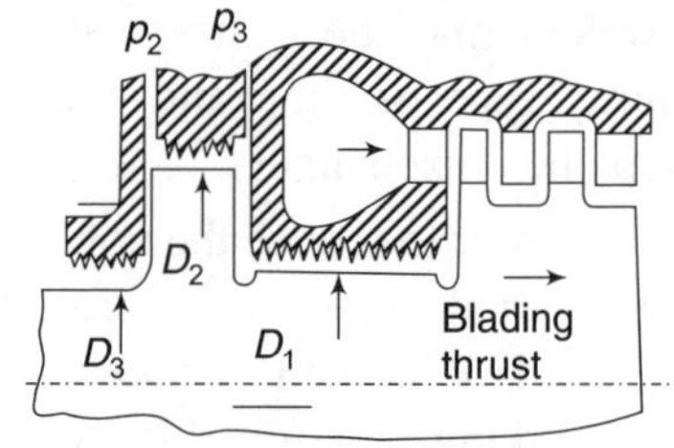

Figure 8.15 Dummy piston

The purpose of flange heating system is to control the heating rate of the casing flange, otherwise the differences in temperature between the rotor and casing will lead to differential expansion problems and differences in temperature across the cylinder will give rise to excessive stresses within the casing. Such problems arise during the warming especially from the cold condition where there is a different rate of heating of the rotor and the casing of the high and intermediate pressure cylinders; and inner and outer surface of cylinder particularly the flanges. In the flange heating system, steam is passed through passages within the flanges. On leaving the flanges, the steam is exhausted to a convenient point in the system (Fig. 8.16). Heating system may be from an external source such as the gland steam supply or it may be trapped from within the associated cylinder.

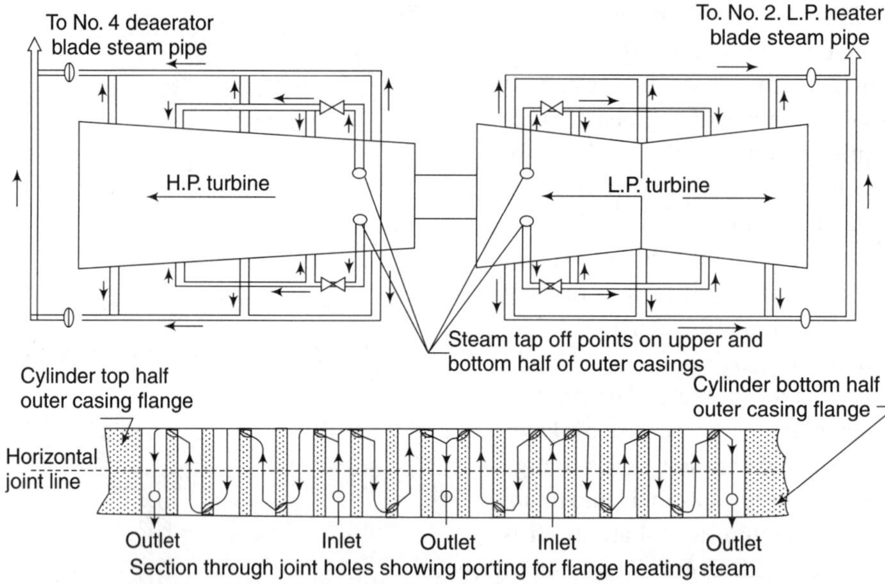

Figure 8.16 Flange heating system

8.6 CONSTRUCTION OF STEAM TURBINE COMPONENTS

The ultimate aim of any engineering science theory is to design and the ultimate aim of any design is to construct as per design so that the product maintains its high quality with the minimum cost and requires minimum maintenance. This chapter deals with construction, stress analysis, operation and maintenance of steam turbine components. *The construction of rotor will also be applicable for gas turbines.*

8.6.1 Construction of Nozzles and Diaphragm

8.6.1.1 Design Requirements of Nozzles

Basic requirements for a nozzle are:

(a) Design of nozzle should be such as to permit easy manufacturing and finishing and allow accurate channel sections to be obtained at high pressure end of turbine.

(b) The inlet of the nozzle should be so designed as to utilise the carry-over energy from the previous stage to the largest possible extent.

(c) Sudden change in direction of flowing steam should not occur at high velocity.

(d) The shape and finish of nozzle should be designed so that conversion of thermal energy into kinetic energy should take place with greatest possible efficiency.

(e) In order to reduce friction, especially where the steam velocity is high, the wall surface should be as smooth as possible.

8.6.1.2 Construction of Nozzles

Figures 8.17 to 8.18 show various constructional shapes of steam nozzles.

8.6.2 Construction of Turbine Blades

8.6.2.1 Production of Blades

Blades may be considered to be the heart of a turbine, and all other members exist for the sake of the blades. Without blading, there would be no power and the slightest fault in blading would mean a reduction in efficiency or lengthy and costly repairs.

The following are some of the methods adopted for the production of blades.

(i) *Rolling.* Sections are rolled to the finished size and used in conjunction with packing pieces. Blades manufactured by this method do not fall under combined bending and centrifugal force.

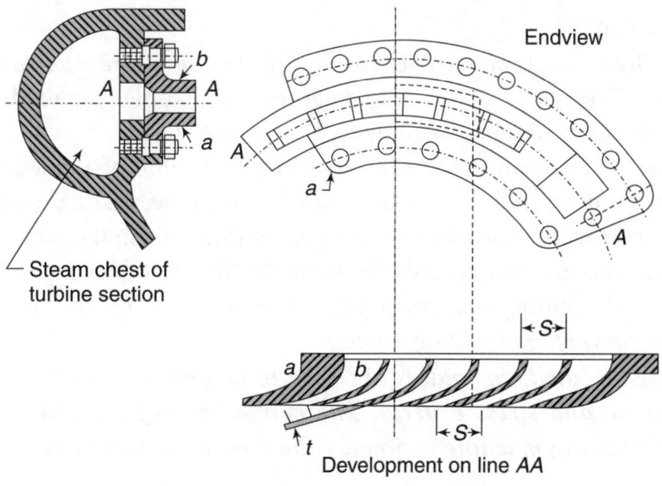

Figure 8.17 A segment of convergent nozzle

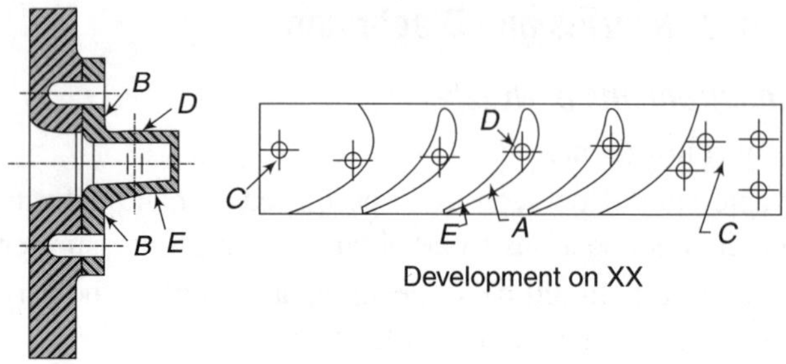

Figure 8.18 Built-up nozzle

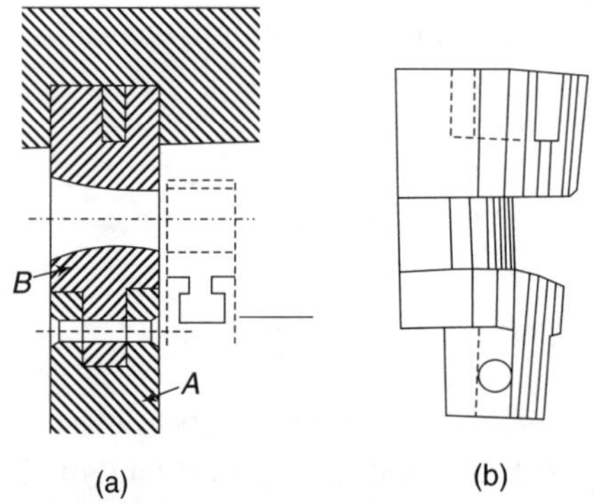

Figure 8.19 Erste brunner type of built-up nozzle

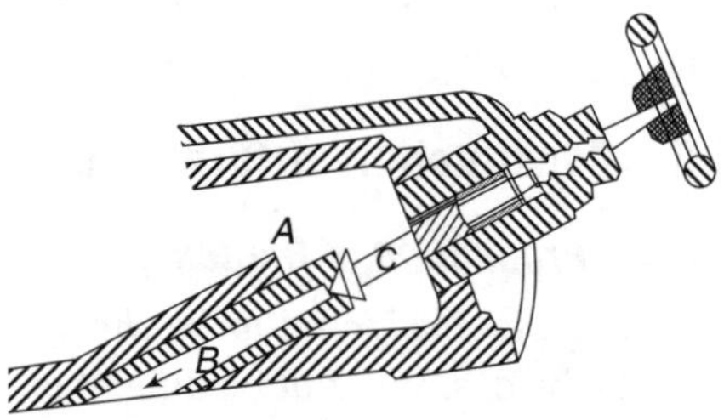

Figure 8.20 Convergent-divergent de-Leval nozzle

(ii) *Machining.* Blades are also machined from rectangular bars. This method has more or less the same advantage as that of the first. Impulse bladings are manufactured by this technique.

(iii) *Forging. Blade and vane surface having airfoil sections are manufactured by specialist techniques.* The simplest way is to determine the profiles required at the hub and tip, and join them by straight, ruled lines. For more accuracy, a profile, at middle to each end separately is obtained. Once the geometry of the family of ruled lines is established, they may be machined in turn by a milling machine, rest carefully for each line to generate the shape required in a master block from which the forging *die maybe copy-machined.* This method ensures the accurate forging of blades to their finished size, requiring only fettling and polishing. *The machining may be used in some parts to avoid the conventional cutting process.*

In advance method, computers are used to determine the blade shape required by aerodynamic and stress criteria. The computer may then instruct a numerically controlled milling machine to prepare the dies. Now, more advance system has come up.

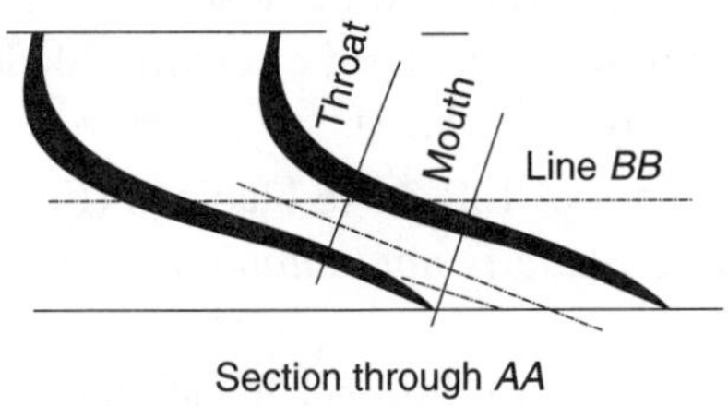

Figure 8.21 Cast-in-type convergent-divergent nozzle

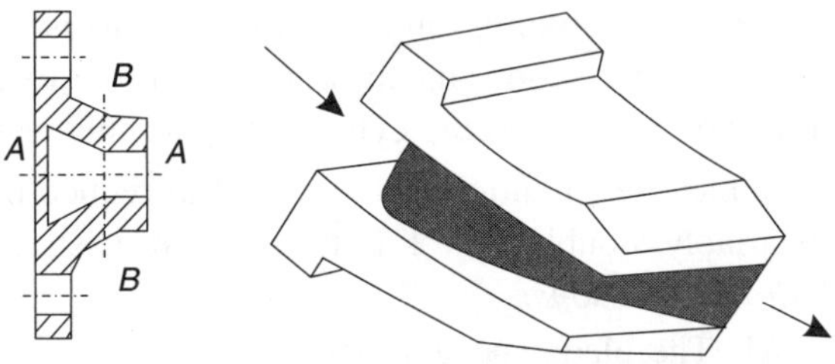

Figure 8.22 Built-up type convergent-divergent nozzle

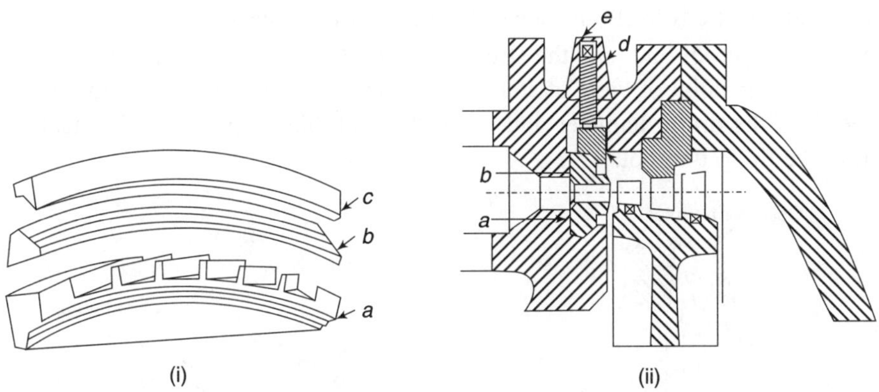

Figure 8.23 Three segment built-up nozzle

(iv) *Extrusion.* Blades are sometimes extruded and the roots are left on for subsequent machining. This method is not rolled section, because of narrow limits imposed on the composition of the blade material.

(v) *Cold drawing.* Blade are also cold drawn.

8.6.2.2 Long Blade (L.P. Stage Blading)

The blades of low-pressure stage must be long to cater for the greatly increasing specific volume of stream at the lower pressures. Irrespective of the design of previous stages, the final stage of I.P. turbine employs little or no reaction at the root and up to about 65% reaction at the tip. This design allows the steam velocity to match the peripheral blade velocity of all radii. In order to dampen vibration, long blades may be laced together in batches. It is to be noted that the lacing holes are source of weakness and disturb the flow path, so they should be avoided as far as possible by better designing against vibration. Sometimes, an arched cover band may be used to brace the blades instead of lacing wires.

L.P. stage bladings face an another problem of erosion of leading edges due to the condensation droplets. In order to avoid this, satellite protection stripe (extremely hard alloy of cobalt, chromium, tungsten, and carbon) is sometimes brazed to the leading edges. Due to centrifugal action, much moisture can be extracted after leaving the moving blade, and provision is made in the cylinder to lead this water away. (Fig. 8.24).

Sometimes, multi-exhaust is used to reduce the length of L.P. stage bladings (Fig. 8.25). Invariably, double flow L.P. turbines are used. *A long blade is not suitable because of the following reasons:*

(i) The blade speed varies from root to tip, thus there are different blade angles and if the steam is to flow on the blade without shock, the blade must be twisted. Generally, the discharge is axial. They are mainly impulse form at the roots and reaction form at the tips. The inlet angles of blades are varied to allow for change in blade speed.

(ii) The space between the adjacent blades may increase so much from the root to the tip as could effect adversely the steam flow through the blades.

(iii) At the tip sped of 330 m/s, the stress at the root of the blade is great. For this reason and from the point of view of stability, low pressure blades are not made longer than one-third the drum diameter and event then the blade section is frequently tapered from the tip to the root.

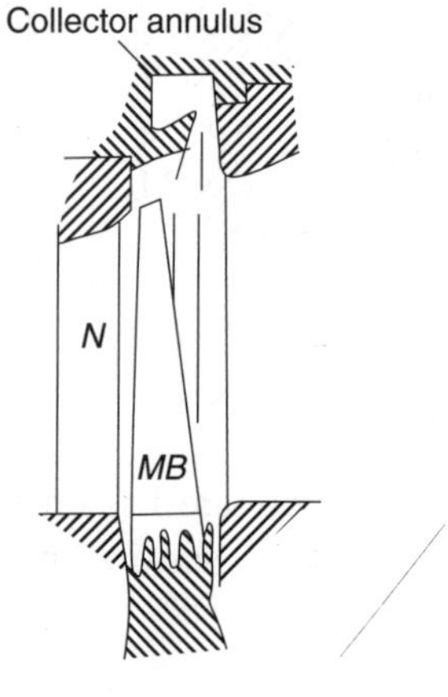

Figure 8.24 Moisture Drainage Bladings

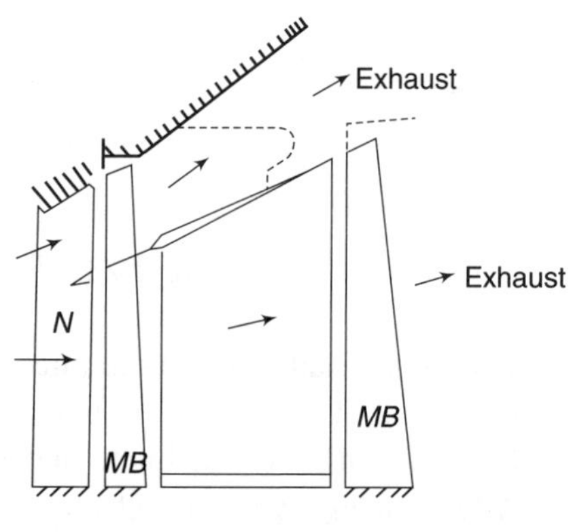

Figure 8.25 Multi-exhaust System

8.6.2.3 Hollow Blades

Hollow blades satisfy the condition for ideal blades, i.e., they give the most efficient control to the steam and are at the same time uniformly stressed. The hollow blades do not impose severe stresses in the rotor, and for that reason increased speed, leading to increased output is possible.

8.6.2.4 Shrouding

Stiffness against vibration and correct guidance to the steam is essential. To meet these conditions, the outer ends of the blades are usually tied together by a perforated ribbon of metal known as shroud. In case, where stress consideration is of primary importance for example, the last row of the low-pressure blades, the shroud is omitted. In longer blades of L.P. turbine, lacing or binding wires are also silver soldered to connect bundles of blades together at various radii.

8.6.2.5 Blade Materials

Proper selection of blade material plays an important role in the blade design. The factors that influence the selection of blade materials are:

(a) Method of manufacture

(i) Ease of machining.
(ii) The ability to produce the blade section free from flaws.
(iii) Ductility to allow of rolling to shape.
(iv) The capacity for being brazed or welded.
(v) Ease of forging.

(b) Conditions of operation

(i) Suitable tensile strength of material at high temperature.
(ii) Resistance to creep.
(iii) Resistance to corrosion and erosion specially in L.P. stages.

(c) Cost should be less

The commonly used blade materials are:

Brass. Brass (70 to 72% Cu and 28 to 30% Zn) is suitable for temperature up to 230°C. It is resistant to corrosion and it may be cold drawn. *This is rarely used nowadays.*

Copper Nickel. This is an alloy containing about 80% Cu, 19% Ni and fraction of iron and manganese, etc.

Nickel Brass. It is suitable for temperature range up to 230°C and 50% Cu, 40% Zn and 10% Ni. It may be cold drawn.

Manganese Copper. Its composition is 95 to 96% Cu, 4 to 5% Mn and small amounts of iron, carbon and lead. It is not suitable for high stress and temperature above 310°C. It may be cold drawn and cold rolled.

Phosphor Bronze. This is copper-tin alloy with a small amount of phosphorous. Its composition is 86% Cu, 14% tin and 1% phosphorous for hard bronze, and 92% Cu, 8% tin and traces of phosphorus for soft bronze. It is resistant to corrosion but not suitable for high temperature. It may be forged, rolled or cold drawn.

Monel Metal. The composition of monel metal is 67% Ni and 28% Cu with a small amount of iron, carbon manganese, etc. It is resistant to corrosion and is suitable for high temperature. It is used for marine work.

Mild Steel. In the past, it was used in many turbines *but is not used now*. It corrodes very soon with wet steam but it is very inexpensive.

Nickel Steel. This alloy is more resistant to corrosion than in mild steel. It may be forged and machined but not welded. Generally, steel with 3 to 5% Ni used. It is used in many turbines.

Stainless Steel. It is an alloy of iron, chromium and carbon, containing 12 to 14% chromium and normal percentage of carbon. It is very resistant to corrosion and erosion. It is also very hard. Stainless steel is also very suitable for bladings materials containing 0.1% carbon. It may be rolled, easily machined and welded.

In some cases, FV 520 B are used. *Nowadays, it is used in many steam turbines.*

ATV Steel. This is an alloy containing 36% Ni, 12% Cr, 0.25% C and the rest iron. It is highly resistant to corrosion and erosion. Its other qualities are that it can withstand high temperature. This steel is commonly used nowadays in many turbines.

CMV Steel. Most bladings are made of 12% chromium-molybdenum-vanadium steel which has good strength and creep properties. Niobium may be added to improve the creep properties. It is commonly used.

8.6.2.6 Turbine Blade Attachment to the Rotor

The attachment of the turbine blades to the rotor is the most critical aspect of steam turbine design. All the forces are transmitted through the attachment to the rotor. Especially, at the low pressure end of turbines of large output, the attachment has to bear relatively large forces due to high speed, the centrifugal force on the blade is many times it mass. Therefore, it becomes necessary to estimate the stresses in the attachment but sometimes it is difficult to get the exact value. There is always the possibility of stress concentration at the sharp corners. Therefore, selection of material is very important which can safeguard from this stress concentration and that is why the calculated stress is kept reasonably low.

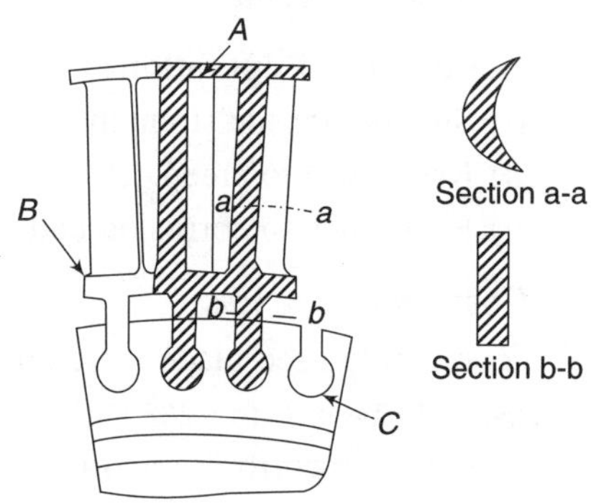

Figure 8.26 de Laval blades attachment

A careful study of the forms of attachment is also necessary because occasionally it influences the shape of the wheel rim and stresses in the disc. The form of the attachment should be such that the centrifugal force on the blade is transmitted to disc in the simplest and most direct manner and it should give the security of attachment. The various forms are given below.

(a) de Lavel blade root attachment, (Fig. 8.26), (b) Inverted-T-attachment, (Fig. 8.27), (c) Serrated blade (Annular Fir-tree) root attachment, (Fig. 8.28), (d) Straddle (single or multiple fork) attachment, (Fig. 8.29), (e) Modified straddle attachment, (Fig. 8.30), (f) Side entery blades attachment, (Fig. 8.31), (g) Shrouding strip attachment, (Fig. 8.32), (h) Parson's end-tightened blading. (Fig. 8.33), (i) Parson's integral blades (Fig. 8.34).

Most of the above attachments are also used in gas turbine bladings. However, Annular Fir-tree or its modified versions are most common in gas turbines.

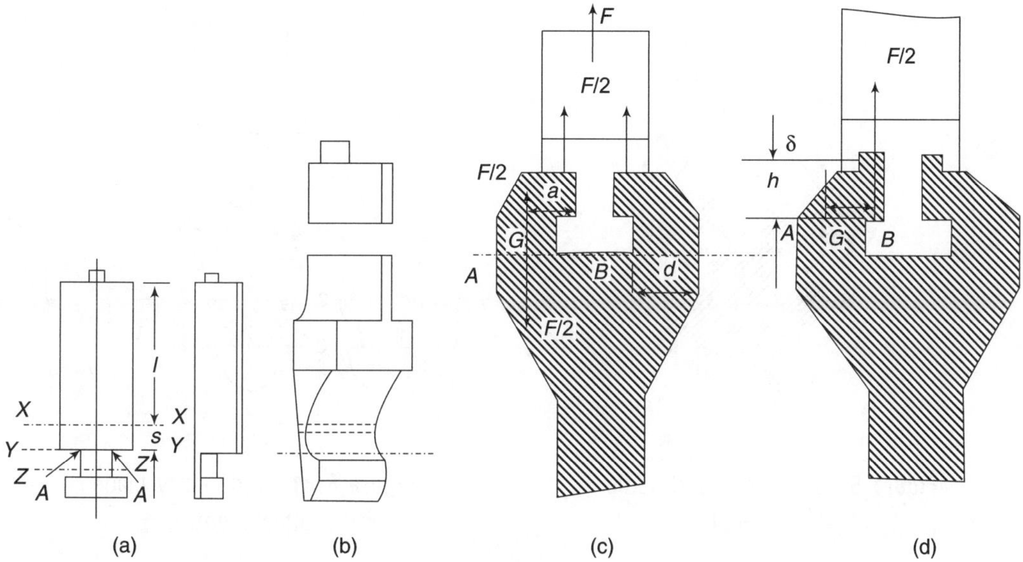

Figure 8.27 Various forms of inverted-T attachment

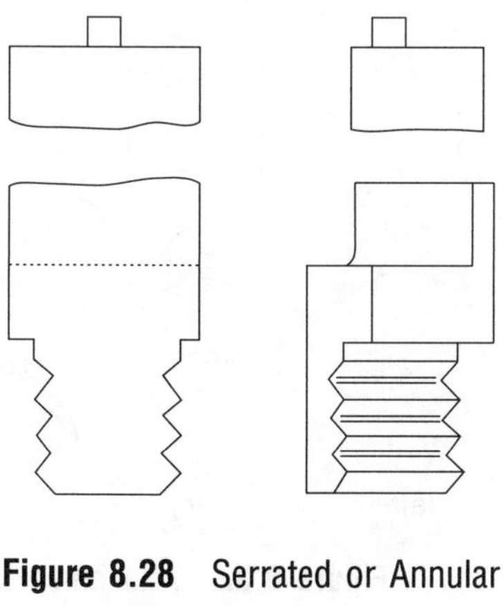

Figure 8.28 Serrated or Annular

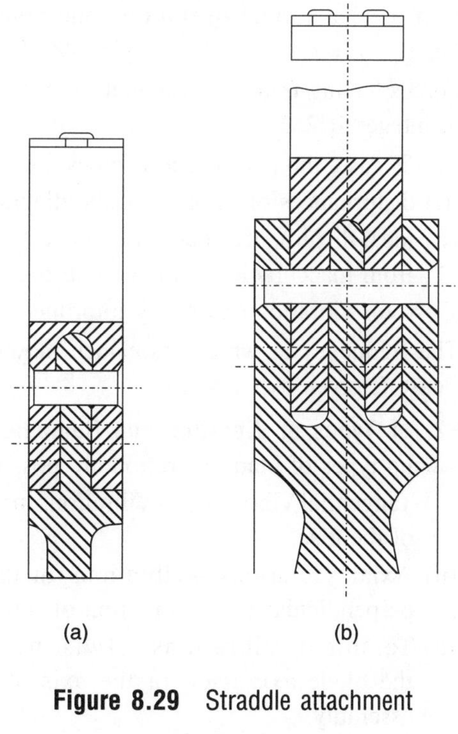

Figure 8.29 Straddle attachment

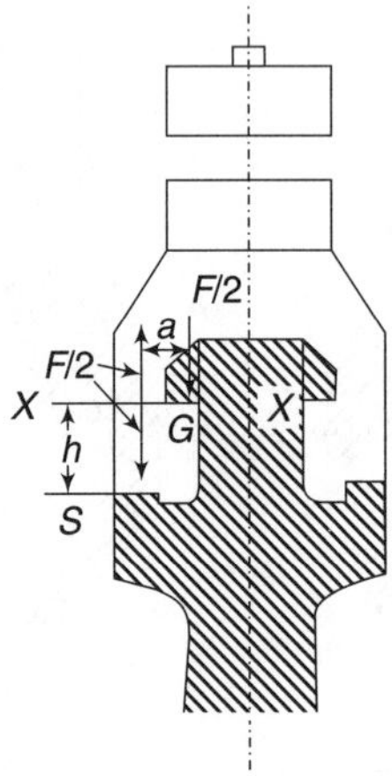

Figure 8.30 Modified straddle attachement

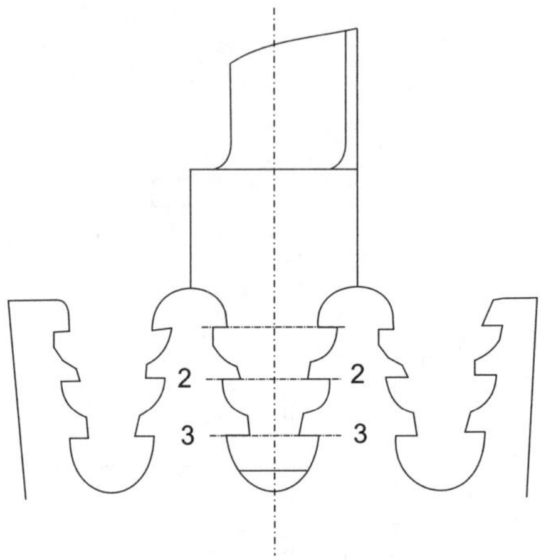

Figure 8.31 Side entry blade attachment

8.6.2.7 Vibration of Blades

The vibrations of turbine blades are caused by the periodic nature of steam flow (i.e., the irregularities of steam flowing out from nozzles or guide blades behaving as a periodic external forces). Under normal operation conditions, only *forced vibrations* are possible in steam turbines.

$$f = NZ \quad ...(8.1)$$

Forces having frequencies equal to the rotor rps or its multiple, i.e., equal to KN where K is an integer 1, 2, 3, etc.

In this case, the perturbation forces are caused by the following:

(i) disturbances of the flow at the diaphragm joints,

(ii) local disturbances caused by the differences in the guide passage dimensions resulting from inaccuracies in manufacturing, and

(iii) rotor vibration caused by improper balancing.

The frequency at which resonance may occur in this case is given by $f = K.N$ ...(8.2)

8.6.2.7.1 Types of Vibration and Remedies: The following types of blade vibration are possible:

(i) Tangential Vibrations—Vibrations in the plane of the disc.

(ii) Axial Vibrations—Vibrations in the plane perpendicular to the direction of rotation.

(iii) Torsional Vibrations—Twisting around the blade axis itself or the axis of a blade assembly.

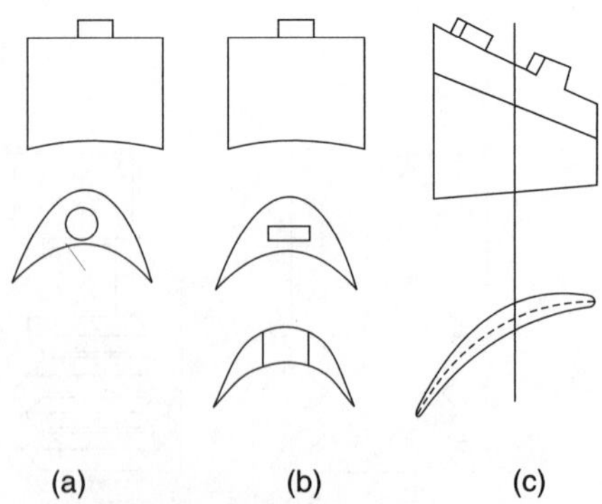

Figure 8.32 Attachment of Shrouding strip

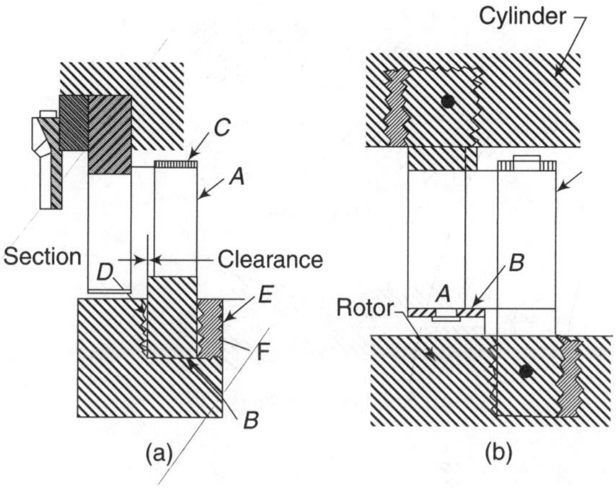

Figure 8.33 Early Parson's end-tightened bladings

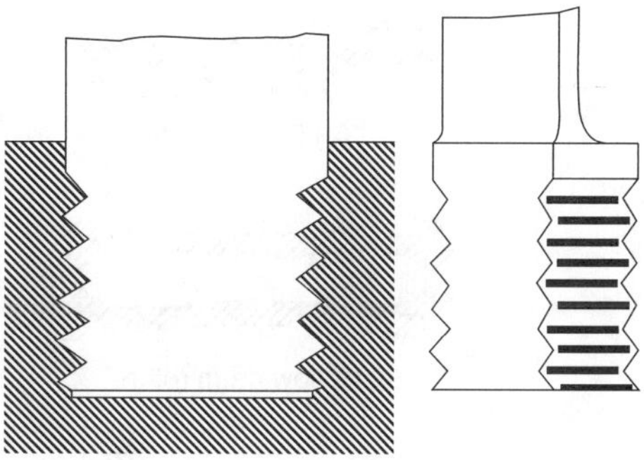

Figure 8.34 Parsons integral blades

Remedies. Following steps are taken to reduce blade vibrations:

(i) By changing the number of blades in a stage or increasing the spacing between blade rows. Use of different prime number of blade in each stage is also helpful.

(ii) By altering the thickness of the blade airfoil all over or selectively.

(iii) Introducing shrouds to provide a damping contact between adjacent blades. Wire lacing of blades has the same effect.

(iv) Selecting appropriate material combinations.

8.6.3 Rotor Construction

8.6.3.1 Types of Rotors

There are five types of steam turbine rotor in common use. These are as follows (Fig. 8.35).

(a) The built up rotor (b) The integral rotor (c) The hollow drum rotor
(d) The solid drum rotor (e) The welded disc rotor

Rotors (a) and (b) are in use in large turbines which make use of bladings of the *impulse type*. In large turbines using 50% *reaction* bladings, rotor (c), (d), (e) and (a) are used. The built-up and welded disc rotors are mainly used for L.P. rotors of reaction turbines.

(a) ***The built up rotor*** (Fig. 8.35a). It consists of a forged steel shaft on which separate forged steel discs are shrunk and keyed.

(b) ***The integral rotor*** (Fig. 8.35b). The shaft and wheels of this type of rotor are formed from one solid forging. Integral rotors are expensive and difficult to forge, and there is a high incidence of rejects. Over and above, a large amount of machining time and waste material are involved. Nevertheless, their advantages are such that they are invariably used for the H.P. rotors on modern reheat turbines, and sometimes for the I.P. and L.P. rotors as well. Following are the advantages of integral rotors:

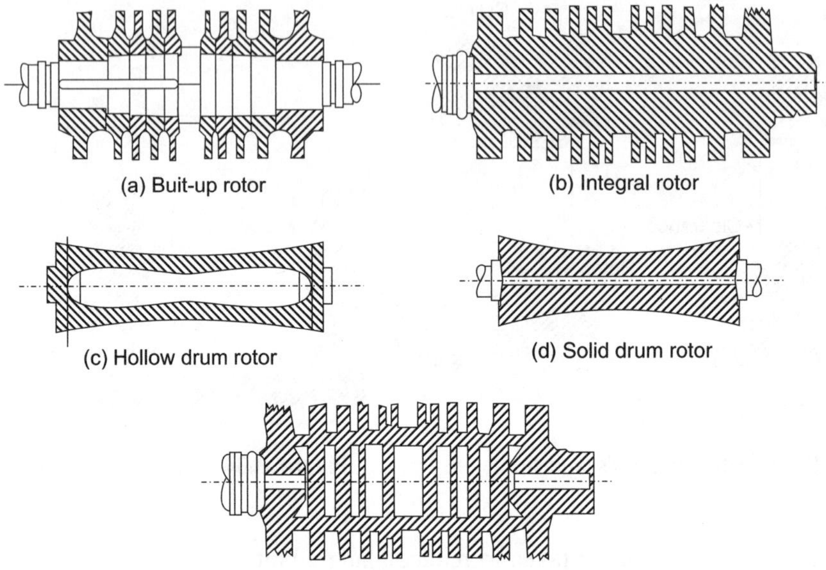

Figure 8.35 Turbine rotors

(i) There is no change of discs to become loose, particularly at the high temperature end where at times the wheels may be hot and the shaft cool as found in the built up rotor.

(ii) This rotor is also free from the effect of creep which may cause the shrink fit of built up rotor to disappear after a large number of running hours.

(iii) The hoop stress is of lower magnitude as it contains a small hole meant for inspecting the forging.

(iv) There is a saving in axial length and reduction in spindle diameter over the built-up type.

(c) ***The hollow drum rotor*** (Fig. 8.35c). This type of rotor promotes even temperature distribution because it is designed with the same thickness of materials as the casing.

(d) ***The solid drum rotor*** (Fig. 8.35d). This type of rotors are suitable for cylinders where there are lower temperature but large diameters, as in I.P. cylinders without reheat.

(e) ***The welded disc rotor*** (Fig. 8.35e). The last stage disc is the most heavily stressed part of the turbine and this one of the main problems of L.P. rotors.

These five types of rotors are also used in gas turbines.

8.6.3.2 Rotor Materials

The selection of rotor materials depends mainly on the estimated stress in the rotor and the temperature to which the rotor is subjected.

For rotors subjected to moderate stresses and temperature, plain carbon steel is used, It may be in two categories, steel No. 1 and steel No. 2.

Steel No. 1 has the following composition:

C 0.25; Mr 0.58; Si 0.11; Ni 0.13 and safe working stress = 10800 N/cm^2

Steel No. 2 has the following composition:

C 0.33; Mr 0.58; Sr 0.21; Ni 0.2 and safe working stress = 1220 N/cm^2

For rotors subjected to high stresses and temperature, H.P. and I.P. rotors are usually made of chromium-molybdenum-vanadium steel, a ferritic material suitable for wheelcase steam temperature up to 540°C. L.P. rotors are made either of 3% chromium molybdenum steel or 2.25% nickel chromium molybdenum steel.

The English Electric 200 MW turbine has the following composition of rotor material

H.P. and L.P. shafts

C 0.25 max; Si 0.35 max; Mr 0.4 to 0.7; S 0.04 max; P 0.04 max; Mo 05 to 0.85; Va 0.25; Ni 0.5 max; Cr 0.6 max.

H.P. Disc

C 0.35 to 0.4; Si 0.2 to 0.25; Ni 0.6 to 0.9; S 0.04 max; P 0.04 max; Ni 0.3 max.

L.P. Disc

C 0.3 to 0.4; Si 0.35 max; Mr 0.6 to 0.9; S 0.04 max; P 0.04 max; Mo 05 to 0.7 max; Ni 0.5 max; Cr 3.0 to 3.5 max.

In most of the impulse turbines, the rotor material is forged, heat treated with 3% Cr. Mo alloy steel to 13.5. En 29.

8.6.3.3 Thermal Stability of Rotors

A great care is taken during manufacture of rotor to ensure that the forging is stable. A stable rotor is one whose physical properties do not change in service. There are three types of instability:

(a) permanent, (b) temporary and (c) transient.

(a) ***Permanent instability:*** This is due to assymetrical coefficients of expansion across a diameter, A close metallurgical control at the ingot stage may avoid it.

(b) ***Temporary instability:*** It is due to locked-up stresses in the rotor. It is relieved by heating before and after machining in a special furnace in which the shaft is rotated.

(c) ***Transient instability:*** It is due to differences in conductivity and emissivity. The use of turning gear overcomes this difficulty.

8.6.3.4 Balancing of Rotors

From the theoretical point of view, a turbine (gas/steam/water) rotor is a balanced body but in actual practice, errors of balance are introduced by various causes such as:

(i) Lack of homogeneity of the material.

(ii) Slight error in machining.

(iii) Differences in the pitch of the blades and also in their individual masses.

Therefore, it is essential to test the balance of a completed turbine rotor and make any adjustments necessary to ensure that the balance is as good as possible.

The purpose of balancing of a rotor is to reduce the amplitude of vibration on a tolerable level which can be taken to be about 0.0254 mm at the bearing pedestals of a 300 rpm machine. There are two types of balancing ***static** and **dynamic***. Each rotor assembled with blades is balanced both statistically and dynamically.

8.6.3.4.1 Static Balance

It means that the weight of the rotor is evenly disposed around the axis of the shaft, it is checked by rolling the rotor on horizontal knife-edge supports.

8.6.3.4.2 Dynamic Balance

It means that the moments of the out-of-balance weights along the axis about either bearing add up to zero. It is checked by spinning the rotor on resilient bearings, detecting the vibration and adding or subtracting weights until the vibration is negligible. Normally, rotors are balanced at low speed say 400 rpm.

8.6.3.5 Critical Speeds of Rotors

Even with the utmost care taken up in the construction and balancing of turbine shaft and discs, i.e., rotor, due to some reason or another, the mass centre of the rotor does not coincide with the geometrical axis of the shaft. This means that after assembly the rotor has a certain amount of imbalance. The distance between the mass-centre and geometrical axis of the shaft is known as the *eccentricity of the rotor.* If the shaft is rotated, even a small eccentricity will give rise to a transverse force tending to deflect the shaft. The deflection of the shaft increases with the speed of the shaft. At a particular speed, the deflection may be so great that the shaft becomes permanently bent. *The speed at which the deflection of the shaft theoretically tends to infinity is known as the critical speed, In other words, the speed which numerically coincides with the natural frequency of transverse vibrations of the shaft is known as critical speed.*

This suggests that the operation of turbines at the critical speed should be avoided otherwise resonance may occur leading to failure, From experience, *it has been found that if the critical speed differs from the normal speed by 15 to 20%, safe working of the turbine is completely stored. Most of the manufacturers use normal speed either higher or lower than the critical by about 30 to 40%.*

Shafts having critical speeds less than their normal operating speed are known as *flexible shafts.* If the critical speeds are higher than the normal speeds, they are known as *rigid shafts. Rotors of impulse turbines are made of both these type of shaft, flexible and rigid. Care has to be taken in the case of critical speed less than operating speed while starting the turbine so that the critical speed must be passed over quickly allowing hardly any time* for the deflection to grow which if allowed could result in a bent shaft and damaged bearing.

8.6.4 Construction of Cylinders (Casing)

8.6.4.1 Cylinders and Diaphragms

A steam turbine cylinder has to withstand the high pressure and temperature of steam; therefore, it falls essentially under the category of a pressure vessel. Its weight is being supported at each end. Following are the design requirements of a steam turbine cylinder:

(i) *It should withstand hoop stresses in the transverse plane.*

(ii) *It should be very stiff in a longitudinal direction in order to maintain accurate clearance between the diaphragm or fixed blades and the rotor.*

A turbine cylinder has a very complicated to shape to accommodate rotor and blades of different sizes. Therefore, *with all practical purposes, casing is horizontally split to facilitate assembly and dismantling.*

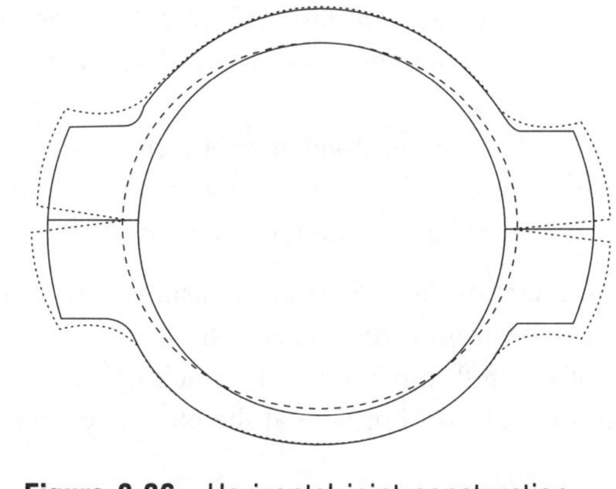

Figure 8.36 Horizontal joint construction

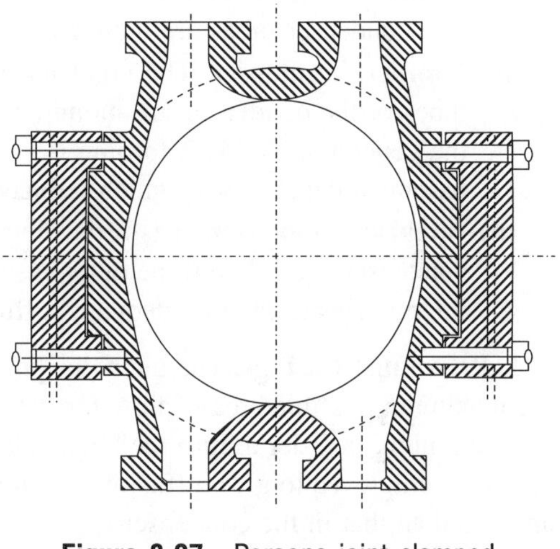

Figure 8.37 Parsons joint clamped

Figure 8.36 shows the horizontal joint construction. Figure 8.37 shows Parsons joint clamped construction for high pressure. For very high pressure double shell shown is used (Fig. 8.38). Figure 8.39 shows the force acting on the flange faces. Figure 8.40 shows cylinder support pedestal.

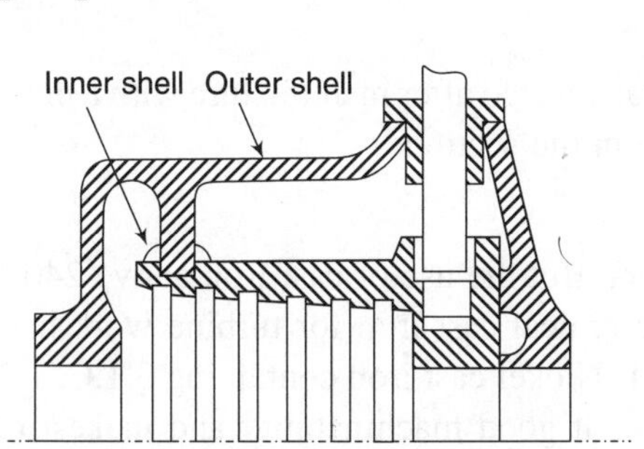

Figure 8.38 Double shell cylinder

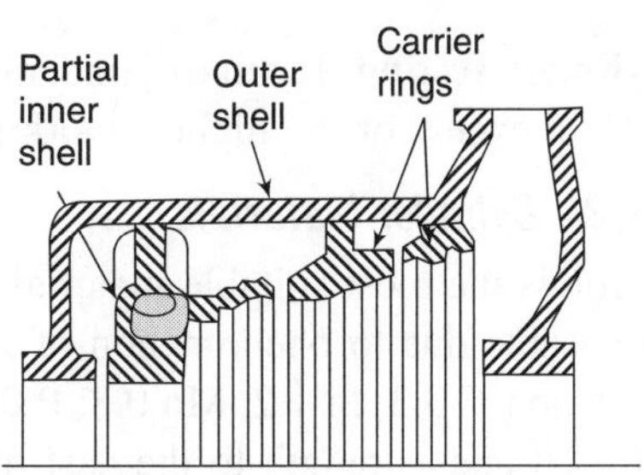

Figure 8.39 Force on cylinder

In the case of large modern sets with three double-flow l.p. cylinders, four condenser arrangements have been adopted in order to provide adequate rigidity of support for the turbine bearings. These are:

(a) axial condenser, (b) bridge condenser (c) pannier condenser, (d) integral condenser.

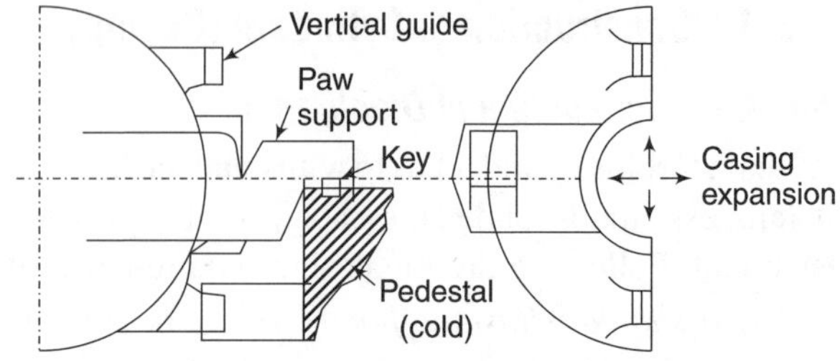

Figure 8.40 Cylinder support pedestal

- **(a) Axial Condenser.** In this case, the L.P. turbines are supported on two parallel concrete walls. The generator also rests on the extension of these two walls. Enough space has to be allowed beneath the generator for tube withdrawal.
- **(b) Bridge Condenser.** It bridges the gap between the i.p. turbines block and the generator, The condenser tubes are mounted transversely.
- **(c) Pannier Condenser.** The condensers are arranged parallel to the L.P. turbines. The tubes of the condenser are mounted axially but withdrawal takes place either side of the generator. The L.P. turbine is supported on a continuous concrete plinth and the basement depth is very much reduced.
- **(d) Integral Condenser.** It is a development of the pannier condenser. In this case, the condenser and L.P. turbine outer casings are combined in a single structure containing internal ribs to give it adequate stiffness. It also results a smaller basement depth.

L.P. Exhaust Casings. The main design requirement of the L.P. exhaust casings is to give the maximum possible area of flow for the exhaust steam in order to keep the pressure drop to a minimum. In order to turn the huge volume of steam through without 'bunching', curved vanes are employed to give diffusion action resulting in lower pressure at the blading exhaust annulus than that in the condenser.

Insulation of Cylinder. It is essential to insulate the hot external surface of the cylinders in order to obtain nearly adiabatic expansion of steam and to protect the cylinder walls from severe temperature gradients which would cause cracking.

The insulation is covered by outer covering known as ***cladding*** to reflect a certain amount of heat and improves the appearance of the turbine like shape of a cylinder. It consists of planished steel, enamelled steel or aluminium sheet.

Cylinder Testing. Turbine cylinder is hydraulically tested after manufacture with a pressure of 50% in excess of the highest working pressure in the section.

8.6.4.2 Cylinder Materials

Cast iron is the most suitable material for cylinders. But the use of cast iron above 240°C is not advisable due to phenomenon of growth. The typical cast iron for turbine work has the composition-C 3.3, Si 1.2, Mn 0.8; P 0.2 and S 0.1. Nickel cast iron containing 3.1.C, 1.2 Si, and 1.5 Ni adds strength to the cast iron and gives it good machinability and makes it less porous. Cast steel is suitable for moderate steam pressure and temperature range between 240°C and 433°C. The H.P. cylinder may be of cast steel and L.P. may be of cast iron.

For steam temperature more than 420°C, alloy steel is used. This alloy steel contains 0.5 to 0.75 molybdenum and small amounts of nickel and chromium. In special case, cylinders have been machined out of solid blocks of forged steel.

The following is the composition of H.P. and I.P. outer casings of the English Electric 200 MW turbine: C 0.15 max; Si 0.2 to 0.5; Mn 0.4 to 0.7; Ni 0.3 max; Cr 0.5 max; Mo 0.5 to 0.7; S.05 max; P 0.05 max; Va 0.20 min. The inner casing which is subjected to highest temperature has following composition: C 0.16 max; Si 0.4 max; Mn 0.4 to 0.6; Ni 0.5 max; Cr 2.75 to 3.25; Mo 0.5 to 0.7; S 0.05 max; P 0.05 max; Va 0.7 to 0.9; W 0.4 to 0.6.

8.6.5 Steam Chests and Valves

8.6.5.1 Steam Chests

Steam chest is fed with steam from the superheater outlet via the high pressure steam main. It normally houses the steam strainer, emergency stop valve and governing valves. It is steel casted or solid forged.

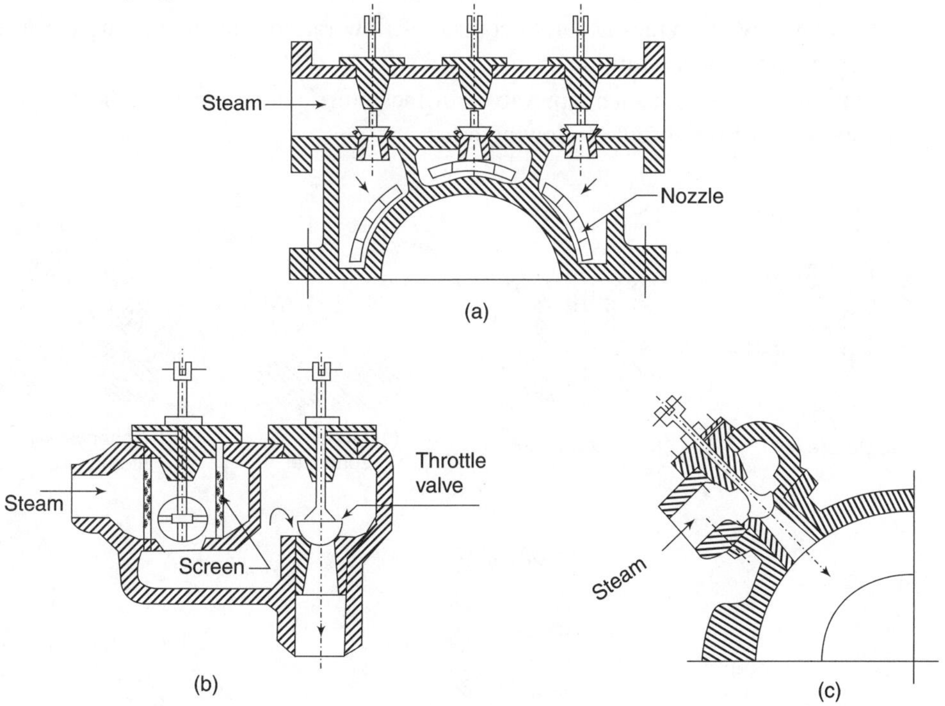

Figure 8.41 Steam chest

Figure 8.41 (a) shows a steam chest suitable for nozzle control governing. The portion containing the governing valves embodied in the H.P. cylinder casting so that the channels from the governing valves to the nozzle groups may be cored. This type of casting is very difficult and therefore sometimes casted in two parts and welded together. Multi-valve integral steam chest is not suitable for temperatures above 480°C due to development of cracks caused by high local stress as a result of uneven temperature distribution. Figure 8.41(b) shows another design *of steam chest suitable for high steam condition. It is very simple in construction and flexible enough to allow the turbine to expand freely without undue stress on flange.* In order to prevent solid particles of foreign matter from being carried into the turbine by the incoming steam and causing damage to the blading, the steam passes through a cylindrical strainer which surrounds the stop valve (Fig. 8.41b).

Figure 8.41(c) shows the third type of steam chest which provides better way of controlling overspeed by mountings a single governing valve chest on the cylinder. The disadvantage offered by this arrangement is the difficulty in accommodating satisfactorily the relay gear.

8.6.5.2 Steam Valves

A steam turbine is equipped with the following valves:

1. Emergency stop valves cut-off the steam supply during periods of shutdown and emergency. Figure 8.42 shows various steam valves.
2. Governing valve provides accurate control of flow rate of steam entering the turbine to control speed of the turbine.
3. Reheat emergency and intercept valves in the return path from the reheater prevent overspeeding of the turbine following a trip.

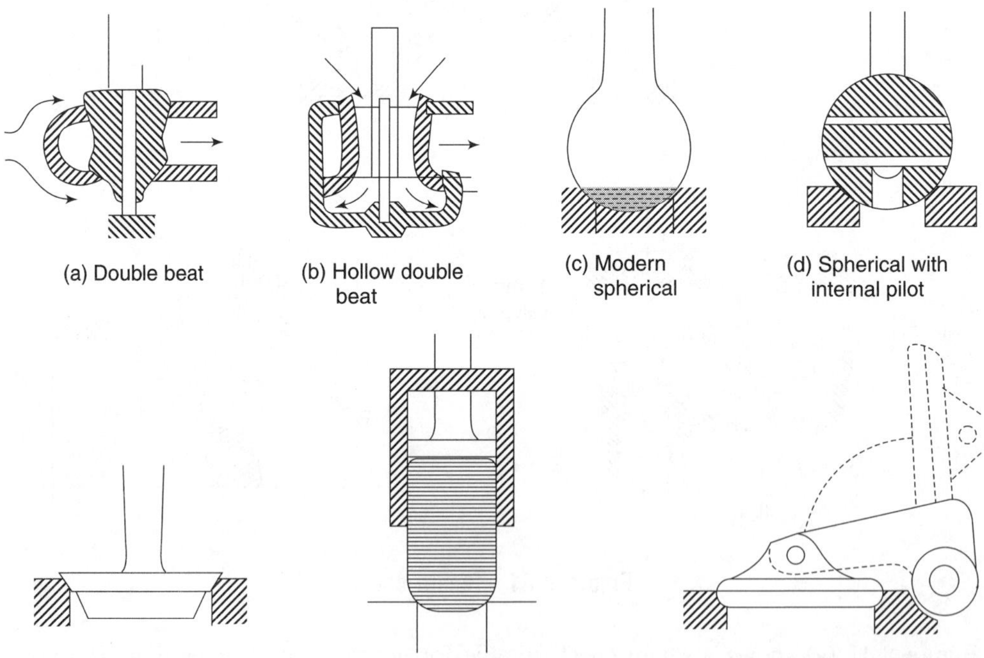

(a) Double beat (b) Hollow double beat (c) Modern spherical (d) Spherical with internal pilot

(e) Mushroom (f) Cylindrical (g) Flap valve

Figure 8.42 Steam valves

One of the main parameters in the design of the valve is the selection of steam velocity through valve. The diameter of valve opening are generally calculated to give maximum steam velocities of about 60 m/s for emergency valve and about 120 m/s for governing valves.

In order to resist wear the mating annular faces of valves and their seats are nitrided or faced with satellite. Such wear is more due to erosion than the mechanical impact. The seating upon which any such valve closes is invariably part of a removable sleeve which is replaceable when worn.

8.6.6 Couplings

These are needed to couple two shafts for the two main reasons:

(i) The limiting length of the shaft which is possible to forge in one piece.

(ii) Use of different material for the various rotors in view of the various conditions of temperature and stress.

The functions of coupling are the following:

(a) To transmit torque from one shaft to another

(b) To allow relative angular malalignment

(c) To transmit axial thrust from one shaft to another

(d) To ensure axial location or to allow relative axial movement.

There are three types of coupling. They are

• Flexible • Semi-flexible and • Rigid.

(i) Flexible coupling. *It absorbs small amounts of angular malalignment as well as axial movement. Double flexible couplings can also accommodate eccentricity.*

Figure 8.43 shows some designs of flexible couplings. The claw coupling is robust and slides easily when transmitting light load but on heavy load, it becomes axially rigid due to friction. For medium size load, Bibby couplings are suitable. The multitooth couplings transmit torque by external and internal gear teeth of involute form.

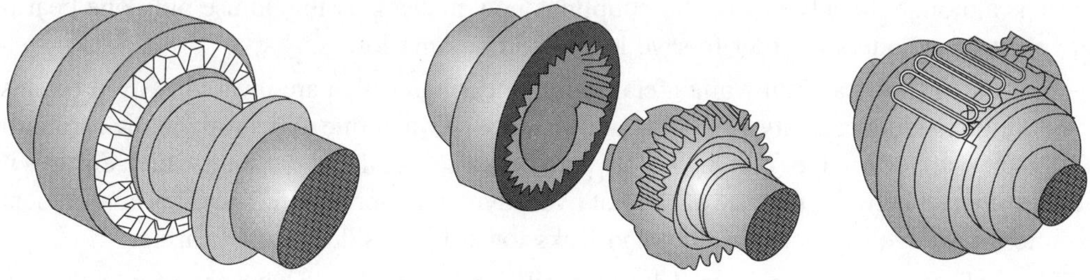

(a) Claw coupling (b) Multi-tooth coupling (c) Biby coupling

Figure 8.43 Couplings

Flexible couplings *require continuous lubrication.* A jet of oil feeding into an annular recess is led centrifugally to the coupling teeth through drilled passage ways.

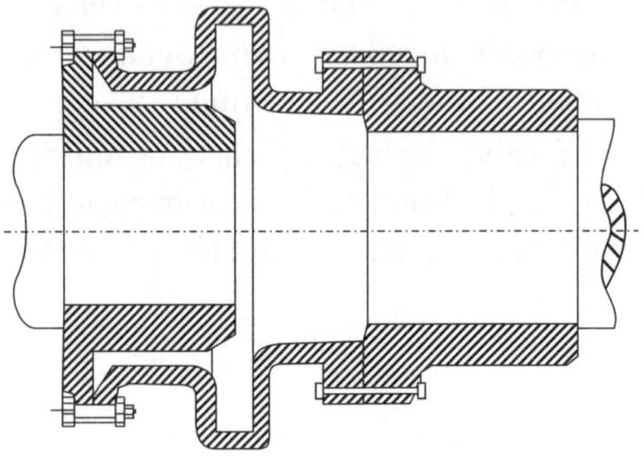

Figure 8.44 Semi-flexible coupling

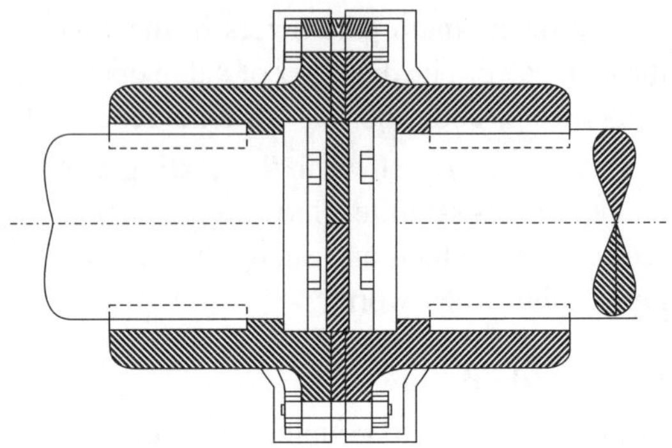

Figure 8.45 Rigid coupling

(ii) Semi-flexible Coupling. It allows angular bending only and requires no lubrication. It consists of a bellows-piece having one or more convolutions as shown in Fig. 8.44. It is interposed between the turbine and generator.

(iii) Rigid Coupling. The use of flexible couplings becomes impractical on the large turbines because high torque is to be transmitted. This requirement is met by the use of rigid couplings between the turbine cylinder, so that the turbine shaft behaves as one continuous rotor. Figure 8.45 shows such a rigid coupling in which a spigot locates the two half couplings and the flanged are joined by the numbered bolts.

8.6.7 Bearings

Two types of bearing are employed in the steam turbine. They are–

(a) Journal bearings and (b) Thrust bearings

(a) Journal Bearings. *The function of the journal bearings is to support the shaft with the help of formation of oil wedge.* Each section of the shaft is normally supported on two bearings although the advent of solid couplings now makes possible to use only one bearing between two cylinders in order to save length and bearing losses.

Figure 8.46 shows a plain white metal journal bearings which are invariably used because of their high loading capacity, reliability and absence of water due to hydrodynamic formation of high pressure oil wedge (Fig. 8.47) between the white metal and the shaft which ensures no metallic contact between them as a result of hydrodynamic lubrication. The oil is continuously fed into the wedge by frictional drag and leaks towards the sides of the bearing.

The shell of the bearing is split horizontally and dowelled. There are four pads fitted with shims on which the shell rests. A small adjustment to the alignment may be made by changing these shims. The pads may be spherical or cylindrical but the former is meant for self-alignment.

The white metal lining may be thick and thin. Dovetails are used to attach the thick white metal lining to the shell.

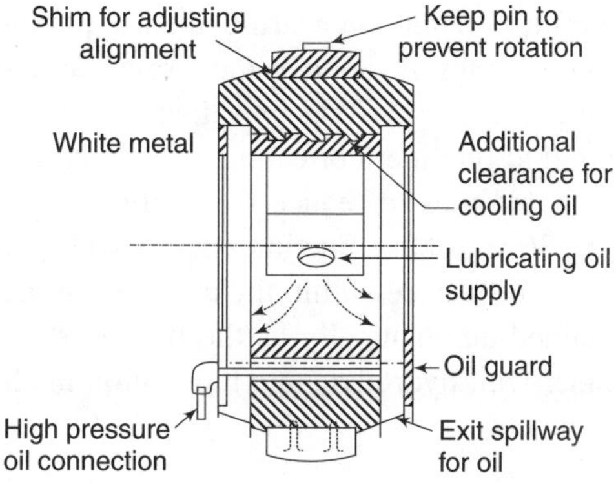

Figure 8.46 Journal bearing

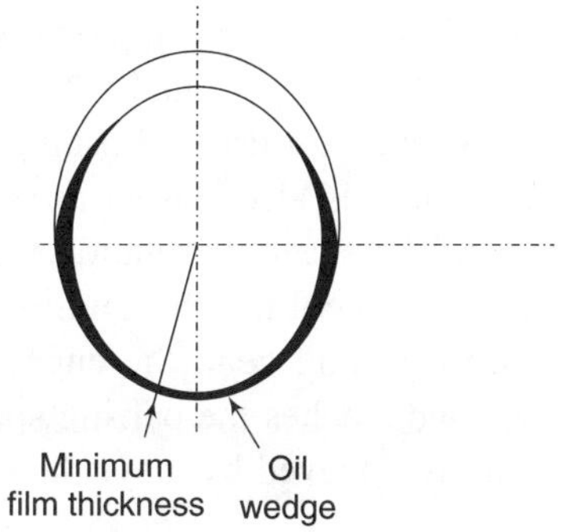

Figure 8.47 Formation of high pressure oil wedge

(b) Thrust Bearings. *The function of thrust bearing is to provide a positive location for the rotor relative to the casings and withstand thrust due to blade reaction and steam pressure acting on unbalanced areas.* Two types of thrust bearings are in common use, namely, *multi-collar and Michell tilting-pad thrust bearings.* The former is suitable for small turbines while the latter for large turbines, The Michell tilting pad bearing uses pads whichs are white metalled and loosely fastened to removable half-rings and automatically tilt to the appropriate angle to build up a wedge of oil (Fig. 8.48). In the case of rigidly coupled shafts, only one thrust bearing of non-adjustable type is required whereas flexible coupled shafts need one bearing to locate each shaft.

8.6.8 Turning Gear

There is an improper cooling of the rotor when the turbine comes to rest after running. This is because of the fact that the cooler steam being more dense tends to collect in the lower half of the rotor cools more quickly than the upper half. This improper cooling of the rotor causes the shaft to "hog" if special precautions are not taken. *It is the use of turning gear which overcomes this problem.*

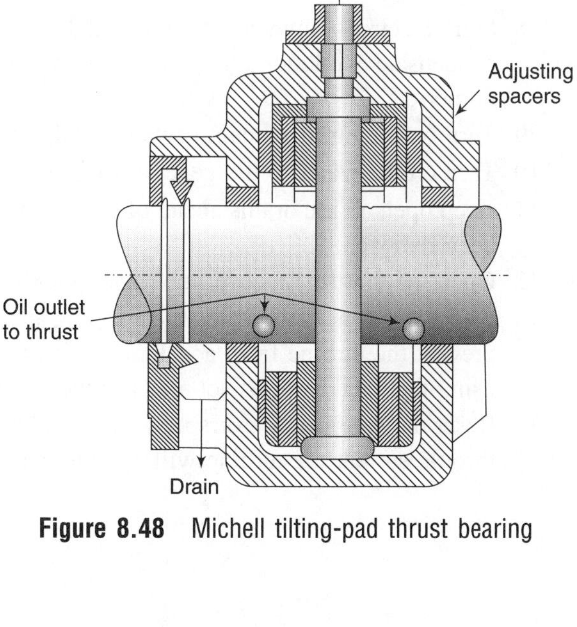

Figure 8.48 Michell tilting-pad thrust bearing

The *purpose of turning gear is to keep the shaft turning slowly until cool when the turbine comes to rest.* There are various types of turning gear. One type consists of an electric motor driving train of reduction gearing, the last pinion of which meshes with a gear ring on one of the shaft couplings. As soon as the turbine comes to rest, a gear lever is operated which slides the pinion to mesh along its splined shaft. At the same time, a motor driven flushing pump is started to provide lubricating oil to bearings and jacking pump is also started to support the shaft and thus to reduce the starting toque of the motor. The speed of rotation varies from 3 to 30 rpm but sometimes up to 100 rpm is also selected to cool the shaft uniformly by creating turbulence within the cylinder. In the recent type of turning gear, the engagement is obtained automatically as the turbine shaft slows down and reaches the barring speed. This renders full hydrodynamic lubrication in all bearings and is achieved by a synchro-self-shifting clutch.

8.7 STEAM TURBINE OPERATION

The following are the sequences of turbine operation.

A. Starting Sequences:

1. Application of control power illuminates all of the malfunction lights. This provides a check of the malfunction lights before starting the turbine.
2. Reset malfunction circuit by operating a reset switch. Malfunction lights go off and all control devices assume the condition for starting.
3. Inspect the governor mechanism, till all grease cups and oil where necessary.
4. Open boiler stop valve to permit heating of the line and avoid condensation in the line.
5. Open header, separator, throttle and turbine casing drains.
6. Start auxiliary oil pump. This has to be stopped when the main oil pump starts delivering oil at normal pressure.
7. Adjust needle valve to secure required oil pressure (1.3 to 3 bars) for the bearings.
8. Start the circulating water pumps and dry vacuum pumps of the condenser. Operate the condensate extraction pumps as found necessary to remove water during the warming up period.
9. Turn on the turbine steam or water seal.
10. Turn on the water to the generator oil cooler and other water requiring parts,
11. Keep open all the drains ahead of the throttle valve until all water of condensation has been removed.
12. Open the throttle or governor valve quickly to set the rotor in motion.
13. In order to check whether the tripping mechanism operates properly or not and to prevent the turbine from accelerating too rapidly, operate the overspeed trip valve by using the hand lever as soon as turbine starts rolling.
14. Reset the emergency overspeed valve and before the turbine comes to rest, adjust the throttle so that the turbine will operate between 200 and 300 rpm.

15. While the rotor is in slow motion, observe any rubbing or mechanical difficulty by using a metal rod or listening device.
16. As soon as the temperature of the oil leaving the bearing reaches about 48–50°C, start circulating water through the oil coolers to maintain bearing oil temperature.
17. Increase the speed gradually. Follow manufacturer's instruction.
18. Adjust the water seal on the turbine and the atmospheric relief valve.
19. Once the machine comes under the control of governor, test the emergency governor by opening the valve in the oil line to it. See that all valves controlled by this tripping mechanism close promptly. Reset open throttle valve and restore speed to normal.
20. Close all drains.
21. Open leak-off from H.P. side gland in order to flow any excess steam to the feed water heater or to one of the lower stage of the turbine.
22. Synchronize the generator and tie it in the line.
23. The speed is now under the control of governor. The turbine is now ready for load and is regulated from the turbine control panel.

B. Sequence when the Turbine is in Motion

1. Apply the load gradually.
2. Check up the oil pressure going to the bearings and control gear.
3. Observe the oil bearing temperature.
4. Observe turbine for any noise, vibration by watching the vibration and other indicators.

C. Sequence when Shutting the Turbine Down

1. Gradually reduce the load to zero.
2. Start the auxiliary oil pump and make sure that oil will be supplied to bearings at proper pressure while the turbine is coming to a stop.
3. Trip the emergency valve.
4. Close the leak off from the H.P. glands.
5. Stop the supply of cooling water to the condenser.
6. Shutdown condensing equipment and open drains on turbine pipings and casings.
7. Continue auxiliary oil pump in operation until the turbine rotor has stopped.
8. Operate turning gear to rotate rotor at about 3 to 30 rpm for some time.

During operation, it is good practice to keep a log sheet and record the hourly readings of the instruments. Some of the readings which might prove valuable are the following:

Load on the generator throttle, steam pressure and temperature, exhaust pressure, temperature of cooling water entering and leaving the cooler, bearing oil pressure and temperature, the throttle steam flow rate, speed, frequency, and vibration level.

The main requirements of a steam turbine while in operation are the proper application of oil to the bearings and a continuous flow of cooling water.

8.8 MAINTENANCE OF STEAM TURBINES

8.8.1 Aims and Objectives of Maintenance

The purpose of maintenance is to maintain the steam turbine in order to achieve as high a plant availability as possible at the minimum cost. The maintenance of large modern unit must be organised to minimise outages which affect availability. If the outages are unavoidable, the maintenance work should be planned so as to return plant to service in the minimum space of time. This may require round-the-clock working on some activities. Plant efficiency is also a major aspect of maintenance team and it is the duty of maintenance department to keep a close liaison with the planning and operation departments in observing the plant performance. These observations will help in arranging the maintenance on plant short of its rated efficiency.

It is essential to keep a close watch for faults which are inherent in the design of any equipment. These are generally known as 'type faults.' After careful study, its design may be improved with the help of manufacture. Maintenance of steam turbines used in power station may be classified broadly in three categories. These are as following.

(i) Rectification of defects. The defects may or may not be urgent. If immediate repairs are needed, these may be carried out on running units or the units should be shut down. If repair is not urgent, it may be carried out in planned programme of future work.

(ii) Planned preventive maintenance or running units.

(iii) Planned Preventive Maintenance on Shut-down Units, e.g., statutory surveys and major overhauls.

Each category needs a different planning approaches and the maintenance is carried out daily, weekly, monthly and yearly by maintenance department. For rectifying the defects in a proper way, a **"Defect Card"** system is used which is in the form of standard pattern. The defects are posted in the defect cards and reported to the maintenance department for their rectification.

(i) Planned Preventive Maintenance on Running Units. A system of routine maintenance is adopted to prevent the defects and breakdowns. The planning of a such system need careful consideration in order to arrive at the optimum levels of maintenance. The actual work content of each maintenance routine can only be determined after experience with the plant over a period of time. After carrying out the preventive maintenance, it is recorded on a 'Record Card' with full particulars of the items attended.

(ii) Planned Preventive Maintenance on Shutdown Units. In order to minimise the maintenance costs of shut-down units, it is necessary to plan the work to be done carefully in advance so that the duration of the plant shutdown is reduced to a minimum. This planning process should start some weeks or months before the shutdown takes place. A work list is prepared and based on this it is possible to programme the course of the overhaul; and the technique for critical path analysis is now generally used for this purpose. A record is maintained after doing the maintenance for future use.

8.8.2 Steam Turbine Overhaul

The following are the predictable factors which influence the frequency of plant outage.

(i) Progressive deterioration in output on account of blade deposits. (ii) Erosion of L.P. bladings. (iii) Retightening of nuts and bolts. (iv) Statutory inspection of associated pressure vessels and boiler plant and planned outage of the transmission system. (v) Availability of manpower to carry out the planned overhaul within the specific time.

It is to be noted that the unscheduled turbine outage is the most serious aspect of maintenance due to extra cost of operating less efficient plant. When mechanical defects such as excessive shaft eccentricity, bearing vibration or excessive bearing temperature are noticed, unscheduled outage must follow. A common reason for this type of outage is rotor unbalance due to displacement and erosion of blade shrouding, and lacing wire and blade erosion. Deposits on the moving blades produces the same effect. Another common source of outage is the malalignment of shaft due to changes in the levels of bearings on account of foundation settlement or changes in pedestal support length.

*Following is the **sequence of events** of a typical **turbine overhaul:***

1. Remove pipe works mounted on turbine and all top half cylinder and covers.
2. Remove coupling bolts and bearing covers leaving thrust bearing assembled.
3. Measure all relevant blade and gland clearances on the horizontal joint, with feeler or tapered gauges.
4. Remove the thrust bearings and then the turbine rotors to have a detailed examination of fixed and moving blades and diaphragms gland segments, casing, bearings, bolts, etc., concerning any type of irregularities and damages.
5. Reform all gland and baffle segment knife edges and restore all radial clearance. Examine the bedding of all bearings to journals and measure oil clearance. Check the adhesion of white metal of journal bearings.
6. After cleaning the blade deposits by water washing, chemical washing or blasting process, refit the rotors and measure clearance on the bottom points of glands and blades with lead or plastic strip. Measure similar clearances on the horizontal joint and compare for eccentricity of the shaft in the casing.
7. After cleaning the bearings, check the wear-down of bearing with the help of appropriate bridge gauge and feeler gauges and compare the figures with those taken on the erection or previous overhaul.
8. Check the alignment of shafts by taking readings on all couplings and also record the level of all journals.
9. Refit top half cylinders and before bolting-up take a further set of top blade and gland clearance to confirm the concentricity of the rotor within the cylinder.
10. Remark the horizontal casing joints and refit all heavy parts. After this, take the readings for final coupling alignment and adjust it if necessary.
11. After refitting the coupling bolts, refit the thrust bearings and all bearing covers.
12. Inspect turbine governor gear with stop, throttle and intercept valves.

13. Dismantle other accessories and mountings if necessary and inspect service and restall these.
14. Clean or replace the oil filters and other items of lubricating system. Replace the lubricating oil if necessary.
15. Inspect all measuring instruments installed on the turbine and replace if necessary.

PROBLEMS

Things to Think About

8.1 What is the purpose of auxiliary system? List them.

8.2 Why there is need of protective devices?

8.3 Why there is a need of automation of turbine protective devices?

Short and Long Theoretical Questions

8.1 What do you mean by turbine protective devices? Discuss each of them in brief.

8.2 Mention the possible hazards in steam turbines.

8.3 List the various tripping devices. Discuss each of them in brief.

8.4 What do you mean by unloading gears? Discuss the working principle of various unloading gears.

8.5 What are the main elements of lubricating system? Discuss the working principle of such a system used in turbine.

8.6 Discuss the working principle of a gland sealing system. Sketch various types of labyrinth glands and discuss their working.

8.7 Discuss the working of flange heating systems.

8.8 Mention the various types of pumps used in lubricating system.

8.9 What is the type of nozzle that is generally used in impulse turbine?

8.10 What is the method of manufacturing of the built-up nozzle?

8.11 What are the methods of manufacturing of turbine blades?

8.12 What is the reason that the blades of the L.P. stages are twisted?

8.13 Why are the long blades objectionable?

8.14 What factors influence the selection of blade materials?

8.15 Mention the suitable blade materials.

8.16 Mention the method of attachment of turbine bladings to the rotor.

8.17 Which type of attachment is generally used in impulse turbine for moderate speed and small and medium power?

8.18 What are the methods of manufacturing the steam turbine rotors? Discuss type of rotors.

8.19 Mention the rotor materials.

8.20 What type of steam turbine casing is generally used in practice? Discuss them.

8.21 Mention the cylinder materials.

8.22 Discuss Inverted T-attachment, serrated blade root attachment, Parsons integral blades and fir-tree.

8.23 What is the function of couplings? Mention their types?

8.24 What is the function of bearings? Mention the types of bearings used in steam turbine.

8.25 State the function of turning gear.

8.26 What do you mean by balancing of rotors?

8.27 Discuss the critical speed of rotors.

8.28 Discuss the causes of vibration of turbine blades.

8.29 Discuss the sequence of steam turbine operation.

8.30 Write an essay on maintenance and operation of steam turbine.

8.31 Discuss the various type of couplings used in steam turbine.

8.32 Discuss the various type of steam valves used.

8.33 Discuss the requirements of steam chest.

8.34 Derive the expression for stresses in rotor and bladings.

Software Based Questions

8.1 Develop a software for the design of various protective devices installed on steam turbines.

8.2 Develop a software for the design of tripping devices.

8.3 Develop a software for the design of lubricating system.

8.4 Develop a software for the design of hydraulic control system for a modern turbo-alternator system.

8.5 Develop a software for the design of gland and sealing system for steam turbines.

8.6 Develop a software for the design of rotor and bladings.

Design and Open Ended Question

8.1 Prepare a comprehensive report of constructional aspect of auxiliary systems, and various parts of steam turbines.

Multiple-Choice Questions

8.1 Overall speed control works when the overspeed is about

(a) 4% (b) 8% (c) 12% (d) 16%

8.2 The cylinder of steam turbine is

(a) single piece (b) vertically split (c) horizontally split (d) all of them

8.3 Excessive vibration and excessive eccentricity in the turbine are monitored by

(a) governing system (b) trip gear

(c) unloading gear (d) supervisory gear

8.4 Which one is true with case of steam turbine blade root attachment?

(a) Blades are attached with rotor through welding.

(b) Blades are attached with rotor through bolt and nut.

(c) Blades are attached with rotor through its blade root like inverted T-attachment or fir tree
(d) None of them

8.5 Shrounding of blading is done for
(a) providing stiffness against vibration
(b) giving correct guidance to the steam
(c) moisture removing
(d) for (a) and (b)

8.6 The heavy duty lubrication system meant for large system turbine employ.
(a) a full duty shaft driven main oil pump
(b) item (a) and auxiliary pump and d.c. emergency oil pump
(c) (a) and (b) and a.c. flushing pump
(d) (a), (b), (c) and jacking oil pump

Answers to Multiple-Choice Questions

8.1 (c) 8.2 (c) 8.3 (d) 8.4 (c) 8.5 (d) 8.6 (d)

CHAPTER

9

Condensers and Cooling Towers

OVERVIEW

Condenser is one of the main elements of a steam power plant. *About 50 to 60% of the energy associated with steam is lost in condenser.* This suggests that a comprehensive thermodynamic study of condenser is essential which will help in its designing.

Cooling tower is one of the main elements of cooling system where the cooling water is to be circulated for cooling the steam in condenser. This chapter deals with the thermodynamic aspects of condensers, cooling system and cooling towers.

9.1 FUNCTIONS OF A CONDENSER

The work done and efficiency of a steam turbine plant are increased if the back or exhaust pressure of the turbine is reduced. This is so because of the fact that the average temperature at which heat is rejected in the cycle is reduced. It can be made possible by employing a condenser in which steam exhausts and get condensed.

Figure 9.1 shows the effect of reducing the back pressure from p_b (atmospheric pressure) to p'_b. The work done is increased by hatched area due to reduction in back pressure. The back pressure can easily be reduced and maintained below atmospheric pressure by condensing it in a closed vessel as after condensing the steam volume gets reduced tremendously and vacuum is maintained and only latent heat is absorbed by cooling water. For example, one kg of steam at 0.1 bar occupies 19.9 m^3 whereas after condensation it occupies only 0.001016 m^3 of liquid. Further, little work is needed to discharge the condensate. From above discussion, it follows that the minimum condenser pressure is limited by the atmospheric condition.

Thus *condenser is defined as a closed vessel in which exhaust steam from steam turbine is condensed by cooling water and vacuum is maintained, resulting in an increase in work done and efficiency of the steam power plant and use of condensate as the feed water to the boiler.*

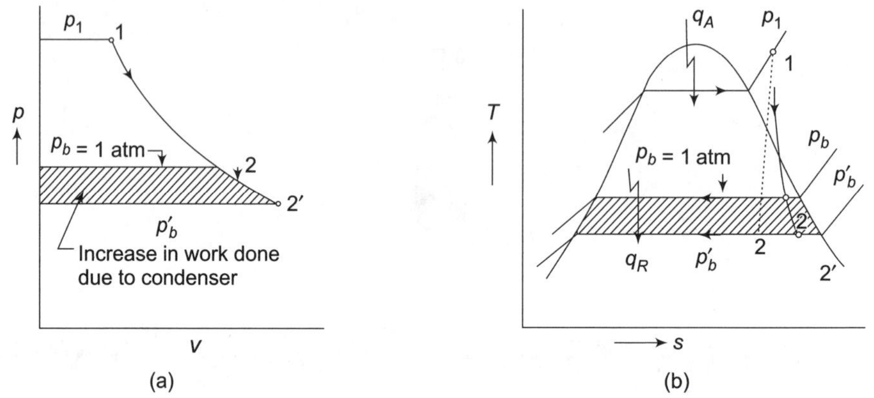

Figure 9.1 Effect of condenser pressure on *p-v* and *T-s* diagrams

Based on the above discussion, the **advantages of a condenser incorporated in steam power plant are the following:**

(i) Improved work done and efficiency due to low pressure (vacuum) of condenser.

(ii) Recovery of the condensate to be fed to the boiler as a high quality feed water for reuse.

(iii) Reduced steam consumption for the same power output due to increased work done.

(iv) Reduced thermal stresses due to high temperature of feed water entering to boiler.

(v) Economy in water softening plant as only make-up water is to be treated instead of full feed water.

9.2 COOLING SYSTEM

The cooling system of condenser is one of the most important system of power plant. The cooling medium may be water or air. The cooling water system may be open or closed. The availability of water plays a major role in deciding the plant size.

About 50 to 60% of the heat supplied in the cycle is rejected in the condenser. The cooling water requirement in open system is about 40 to 50 times the flow of steam in the condenser. Even with closed system using cooling towers, the requirement of circulating cooling water is considerably large, about 120 to 200 kg/kWh. This means that 2000 MW station will need at least 1.7×10^4 tonnes of water per day and out of this about 5% (2 to 10%) is *evaporated in cooling tower*, so make-up cooling water needed will be 8.5×10^2 tons per day.

Since the cooling water takes the latent heat of steam, hence the temperature of water may increase up to 10°C and the condensate temperature is the saturation temperature of steam at condenser pressure. In a closed system, it is necessary to precool this in cooling tower. In open system (suitable where large reservoir is available), the water gets cooled in long pipe or open channel before entering the reservoir (lake, river, etc.). The circulating cooling water system and its type are discussed in subsequent chapter.

9.3 ELEMENTS OF A WATER COOLED CONDENSING AND COOLING SYSTEM

Figure 9.2 shows main elements of a steam condensing unit (closed). These are:

(a) A closed vessel, condenser in which steam condenses. This is a heat exchanger.

(b) A dry air extraction pump (AEP) whose purpose is to remove air (leaked into condenser through joints) and other non-condensable gases from the condenser in order to maintain vacuum constant.

(c) A condensate extraction pump (CEP) which extracts the condensed steam collected in hot-well of the condenser and pumps it to the fed water line. Sometimes, a wet air pump serves the purpose of CEP and AEP.

(d) A circulating cooling water pump (CCWP) to circulate cooling water in condenser.

(e) A boiler feed water pump (BFWP) to pump the condensate to the boiler at a pressure slightly above than boiler pressure and is located below the deaerator.

(f) A cooling tower or spray pond to recool the circulating water in the condenser which is heated due to condensation of steam in a closed system. Cooling tower is essential where there is a scarcity of water. Cold and hot water flow in open channel with flow gates in open system.

(g) A relieve valve to relieve the steam from the condenser when condenser is not in working order. Using this, the plant becomes non-condensing.

(h) A makeup cooling water pump with screen at its inlet to supply make-up cooling water to the condenser as 2 to 5% of cooling water gets lost due to evaporation in cooling tower.

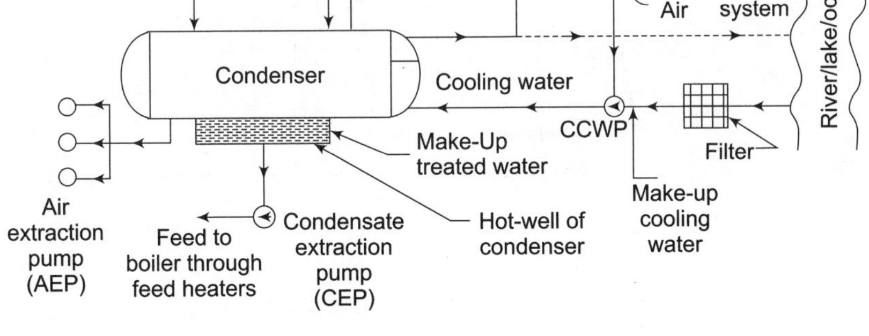

Figure 9.2 Main elements of a condensing unit

9.4 TYPES OF CONDENSERS

Broadly, condensers are classified in two categories.

(i) *Direct contact type,* where the cooling water and steam directly meet and come out as a single stream.

(ii) *Surface condensers* (Indirect type) where there is no mixing of cooling water and steam. It is a shell and tube heat exchanger. The heat released upon condensation is transferred to circulating cooling water through the walls of the tubes.

9.5 DIRECT CONTACT CONDENSERS

They are classified into three categories:

(a) Spray condenser, (b) Barometric condenser, and (c) Jet condenser.

9.5.1 Spray Condenser

Figure 9.3. shows the schematic of a spray condenser. The cooling water is sprayed into the steam and due to direct mixing, the steam gets condensed and mixed with cooling water. The exhaust steam from the turbine at state 2 mixes with cooling water at state 5 to produce saturated water at state 3, which is pumped to state 4. A part of the condensate (m_4), equal to the turbine exhaust flow (m_1) is sent back to the plant, The remaining condensate is cooled in a dry cooling tower to state 5 which is finally sprayed in the condenser on entering exhaust steam. The main drawback of direct contact spray condenser is the requirement of very high quality cooling water since a part of the condensate is used as feed water to the boiler. The mass and energy balances yield.

$$\frac{m_5}{m_2} = \frac{h_2 - h_1}{h_3 - h_5} \quad \text{as} \quad m_2 = m_4 \qquad ...(9.1)$$

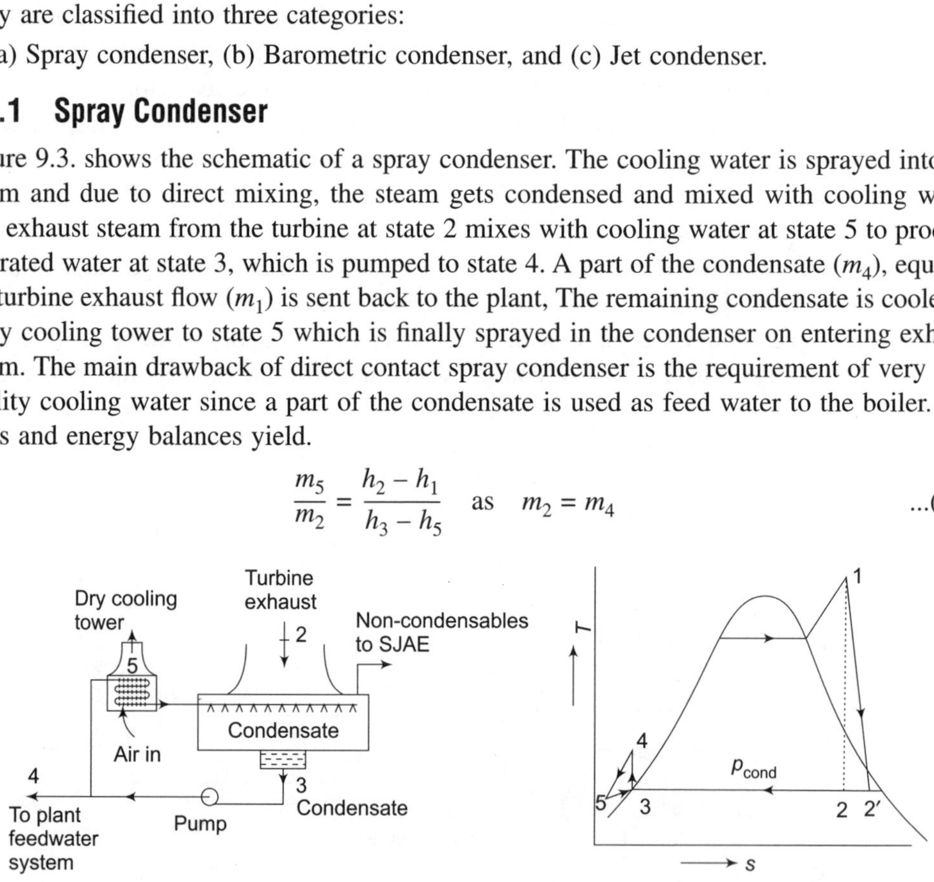

Figure 9.3 Spray condenser

9.5.2 Barometric Condenser

The schematic of barometric condenser is shown in Fig. 9.4(a). In this case, the cooling water is made to fall in a series of baffles to expose large surface area for the steam fed from below to come in direct contact. Condensation of steam takes place and the mixture falls in a tail pipe to the hot well below. As a consequence of its static head, the tail pipe compresses the mixture to atmosphere pressure. Thus, the height of tail pipe (H) is expressed as

$$p_{atm} - p_{cond} + \Delta p_f = \rho g H$$

where, ρ is the density of mixture and Δp_f is the pressure drop due to friction. For low value of Δp_f. H is around 9.5 m. Barometric condenser also suffers from the same drawback of spray condenser. Mixture is split and cooled in the same manner of spray condenser.

9.5.3 Jet Condenser

Figure 9.4(b) shows the schematic diagram of a jet condenser. Baffles of barometric type are replaced by cascade and to reduce the height of tail pipe, diffuser is provided. The diffuser acts on the same principle as the diverging section of a convergent-nozzle in subsonic flow, In this case also, the mixture is split and cooled in the same manner as in spray condenser.

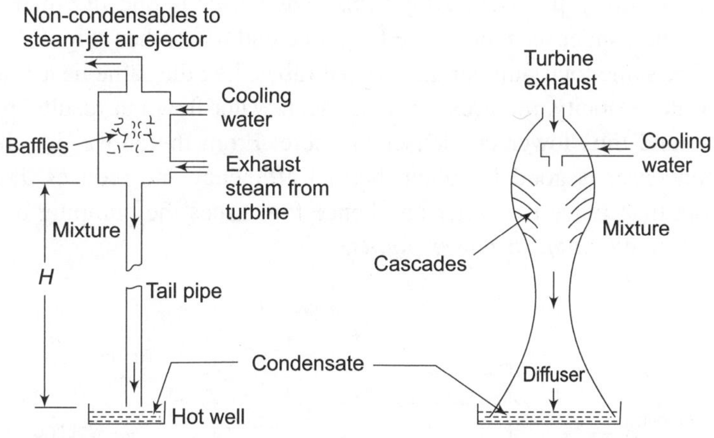

(a) Barometric condenser (b) Jet condenser

Figure 9.4

9.6 SURFACE CONDENSERS

Invariably, almost all the steam power plants employ surface condensers. They are essentially shell and tube type of heat exchangers where cooling water flows through tubes and exhaust steam fed into the shell and surrounds the tubes. As a result of heat transfer from the steam to tube wall and then to cooling water, steam condenses outside the tubes.

Figure 9.5 shows the schematic of a surface condenser with two passes on the water side. The shell is made of steel with water boxes on each side. The right water box is divided to allow for two water passes. Tube sheets are provided at each end into which the water tubes are rolled and also prevent leakage of circulating water into the steam. An expansion joint is provided which allows different rate of expansion between the tubes and shell. In order to provide support to the long tubes and to prevent tube vibration, vertical palates at intermediate points between the tube sheets are provided. The hot well provided at the bottom acts as a reservoir of the condensate with a capacity equal to the total condensate flow for a certain period of time (say five minutes) from where the condensate is extracted by a extraction pump and fed to the feed line.

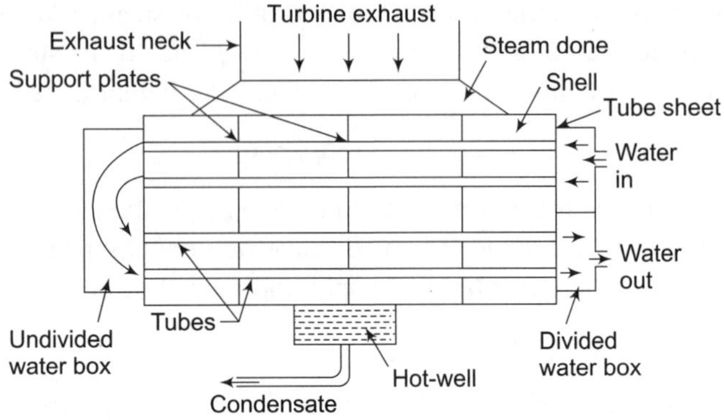

Figure 9.5 Schematic of a two-pass surface condenser

Condensers are normally designed with one, two or four cooling-water passes. It is the number of passes that determines the size and effectiveness of a condenser. Four passes are seldom used in utility plants. A *single pass condenser* is one in which cooling water flows through all the condenser tubes once from one end to another (Fig. 9.6). A single pass condenser with the same total number and size of tubes, i.e., the same heat-transfer area, and with the same water velocity, requires twice as much water flow but results in half the water temperature rise and thus lower condenser pressure. From the above, it follows that such a single-pass condenser is good for plant thermal efficiency and reduces thermal pollution but requires more than twice the water and hence four times the pumping power. ***Two-pass condenser are more common in power plants.***

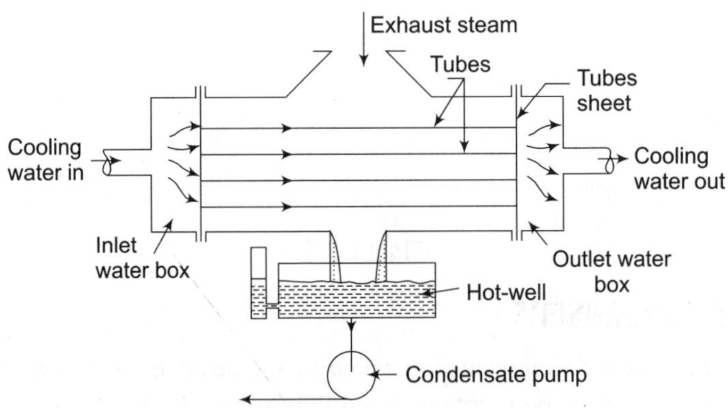

Figure 9.6 Single pass surface condenser

In order to facilitate the cleaning or repairing of tubes during operation in section of condenser, water boxes are often divided beyond the divisions required by the number of passes. A divided water box single pass condenser, for example, may have a partition in both inlet and outlet water at opposite end of the condenser. Similarly, in the case of divided two-pass condensers, the water boxes are divided into four quarters. Divided water boxes have duplicate inlet and outlet connections, each with its own circulating water circuit and the valves provided in the division plants permit backwashing by reversing water flow for cleaning purposes.

The current design philosophy of condenser is to have a tube layout in the shape of funnel with most tubes and the largest tube passage area, where the steam enters from the turbine. With the condensation of steam and consequent decrease in volume, there are fewer tubes and smaller passage areas. Steam is made to enter near the tube bundle or bundles from all sides towards central air cooler for deaeration (below). This results in a low and balanced (to avoid cross-flow) pressure drop. An expansion joint is usually provided between the turbine exhaust and condense steam inlet. *The tube layout may have triangular square or irregular arrangement.*

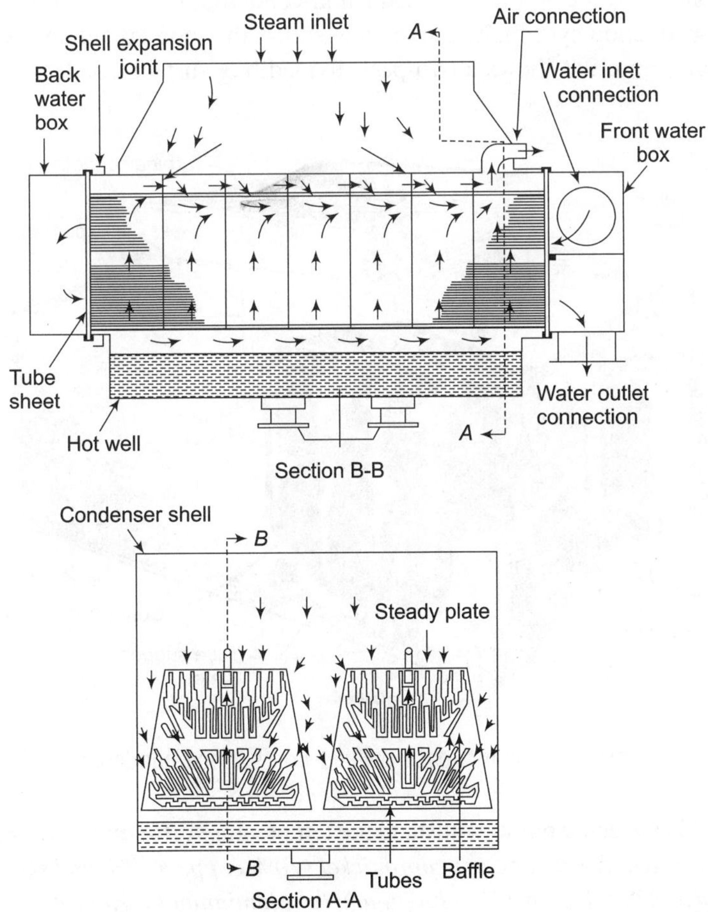

Figure 9.7 A typical two-pass surface condenser for large steam power plant

Large power plants, usually have two or more low pressure turbine sections in tandem. The condenser may be divided into corresponding sections or shells, situated below the low pressure turbine sections.

In the event of same exhaust pressure in all section, i.e., when the turbine exhaust ducts are not isolated from each other, a *single pressure* condenser is used. If the turbine exhaust ducts are isolated from each other, these individual condenser shell pressures will increase as it flows from shell to shell. Under this circumstances, *multipressure* surface condensers are used. It is worth to note that a multi-pressure results in efficiency improvement because the average turbine back pressure is less compared with of a single-pressure condenser. Multi-pressure condensers are more commonly used in nuclear power plants.

A typical *modern two-pass surface condenser for a large steam power plant* is shown in Fig. 9.7. Steam enters the tube bundles in two separate sections from the top, sides and bottom, and flows towards the centre of the tube nest in each section. At that point, most of steam has condensed leaving air and other non-condensable gases which are cooled and removed by air extraction system. The side view shows the tube layout and baffles provided in the condenser. Figure 9.8 shows a two-pass divided box surface condenser.

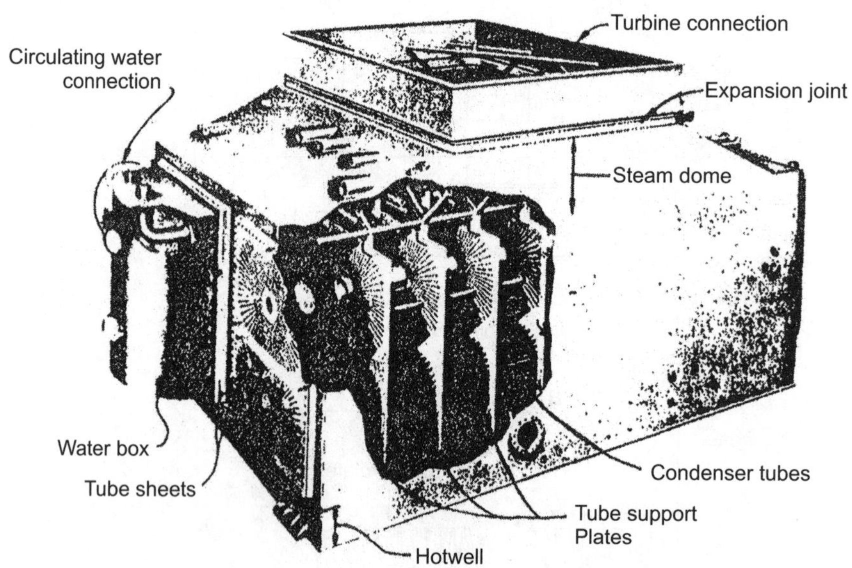

Figure 9.8 A two-pass divided box surface condenser

The tube material selected has to withstand the temperature difference, pressure differential and load. The tube material can be *(a) cupronickel (70% copper, 30% nickel), (b) aluminium brass (76% copper, 22% zinc and 2% aluminium, (c) aluminium bronze (95% copper and 5% aluminium, (d) muntz metal (60% copper and 40% zinc),* (e) admiratly alloy (71% copper, 28% zinc and 1% tin) and (f) stainless, etc. The outside diameter of the tubes is either 22 mm, 23 mm or 25.4 mm. The tube length for single pressure condenser varies from *about 9 to 15 m while for multi-pressure condenser, these ranges from about 21 to 27 m.* Table 9.1 gives typical condenser dimensions for 100, 200 and 500 MW capacity.

Table 9.1 Typical condenser dimensions (t_{w1} = 26.7°C, vacuum = 5.08 cm)

Capacity (MW)	*Steam flow (kg/h)*	*Surface area (m^2)*	*Number of passes*	*Tube length (m)*	*Tube OD (mm)*	*Turbine exhaust width (m)*	*Turbine exhaust length (m)*
100	22.04×10^4	6967	2	8.5344	22	4.267	4.267
200	41.2×10^4	13935	2	9.144	22	5.33	5.943
500	97×10^4	31586	2	9.144	22	5.486	6.096

The following parameters indicate the performance of surface condenser.

(i) Condenser efficiency $= \dfrac{\text{Actual temperature rise of water}}{\text{Maximum possible temperature rise}}$

(ii) Vacuum efficiency $= \dfrac{\text{Actual Volume}}{\text{Ideal Volume}}$

(iii) Undercooling of condensate $= t_s - t_{w2} = t_s - t_{c2}$

9.7 DESIGN ASPECTS OF SURFACE CONDENSERS

Calculations of heat transfer for determining the tubes and total surface area required by a surface condenser are rather complex. They require the knowledge of the total heat load on the condenser, the heat transfer mechanism and coefficients in various parts of the condenser. Generally, $(t_{w2} - t_{w1})$ varies from 8 to 10°C.

When the condenser is new, the outside surfaces are usually clean but quickly develop an oily film that changes condensation from *dropwise to filmwise condensation.* Thus, the shell side heat transfer coefficients are conservatively based on the *lower filmwise condensation mechanism.* The shell side heat transfer coefficient depends upon the difference between the steam saturation temperature and tube-wall temperature, the relative position of the tubes, steam velocity and turbulence, the extent of noncondensables, and the existence of superheat steam if any. The circulating water side heat transfer coefficient depends upon its velocity, temperature and cleanliness of the inside surface. It is worth to note that the circulating water is obtained from natural body of water, so heat transfer rate, Q is given below.

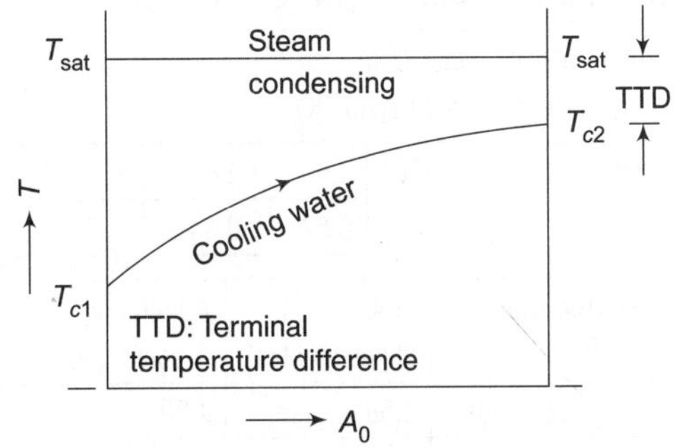

Figure 9.9 Temperature distribution in a condenser

$$Q = U_o A_o \Delta T_m \quad \text{...(9.2)}$$

where Q = heat load on condenser (W), U_o = overall condenser heat transfer coefficient based on outside tube area (W/m^2.K), A_o = total outside tube surface area (m^2), ΔT_m = log-mean temperature difference in condenser (°C). ΔT_m is expressed as (Fig. 9.9)

$$\Delta T_m = \frac{\Delta T_i - \Delta T_e}{\ln(\Delta T_i/\Delta T_e)} \quad \text{...(9.3)}$$

where ΔT_i = difference between saturation-steam temperature and inlet circulating water (°C) = $(T_{sat} - T_{c1})$.

ΔT_e = difference between saturation-steam temperature and outlet circulating water temperature (°C), also called *terminal temperature difference (TTD)* = $(T_{sat} - T_{c2})$.

9.7.1 HEI Method

The overall heat transfer coefficient (U_o) is expressed empirically by

$$U_o = C_1.C_2.C_3.C_4 \sqrt{C_w} \quad \text{...(9.4)}$$

where C_W = circulating cold water velocity in tubes at inlet (m/s), C_1 = dimensionless factor depending upon the tube's outer diameter, considering heat transfer coefficient, C_2 = dimensionless correction factor for circulating water inlet temperature, C_3 = dimensionless correction factor for the tube material and gauge, C_4 = values of these factors are given in Table 9.2. By knowing the values of ΔT_m and U_o, A_o can be calculated for a known value of heat load, Q.

Table 9.2 Factors (C_1, C_2, C_3, C_4)

Tube outer diameter (mm)					19		22		25.4	
C_1 (C_w in m/s, U_o in W/m^2 K)					2777		2705		2582	
Water temperature (°C)	1.66	4.44	7.22	10	12.27	15.55	21.11	26.66	32.22	37.7
C2	0.57	0.64	0.72	0.79	0.86	0.92	1.00	1.04	1.08	1.10

Tube material		304 stainless steel	Admirality Arsenic copper	Aluminium brass, muntz metal	Aluminium bronze 290–10 Cu-Ni	70-30 Cu-Ni
C_3	18 gauge	0.58	1.00	0.96	0.9	0.83
	17 gauge	0.56	0.98	0.94	0.87	0.80
	16 gauge	0.54	0.96	0.91	0.89	0.76
C_4	0.85	For clean tubes, less for algae covered or sludged tubes				

9.7.2 Conventional Method

In conventional method, the usual head transfer equations are used to calculate U_o. For filmwise condensation, the average heat transfer coefficients on steam side for a horizontal tube is given by Nusselt.

$$(h_{av})_{ss} = h_o = 0.725 \left[\frac{k_f^3\, \rho_f^2\, g\, h_{fg}}{N\, \mu_f\, d_o\, \theta}\right]^{1/4} \qquad ...(9.5)$$

where N = number of horizontal tubes in a vertical tier, $\theta = T_{sat} - T_{wall}$, h_{fg} = latent heat of condensation of steam, μ_f = viscosity of condensate (fluid), ρ_f = density of condensate (fluid), k_f = thermal conductivity of condensate (fluid), d_0 = outside tube diameter.

It should be noted that the Nusselt's equation for h_0 gives a conservative value for the condensing film coefficient for heat transfer. However, this value will also be influenced by many factors such as superheat, vapour velocity, turbulence, and the inside released gases and air leaked.

The inside heat transfer coefficient on the water side is given by Mc Adams equation

$$\text{Nu} = \frac{h_i d_i}{k} = 0.023\ \text{Re}^{0.8}\ \text{Pr}^{0.4} \qquad ...(9.6)$$

where Re = Reynolds number due to flow of circulating water through tubes = Cd_i/μ

Pr = Prandtle number = $c_p\mu/k$

The overall heat transfer coefficient for a condenser is given by

$$\frac{1}{U_0 A_0} = \frac{1}{h_i A_i} + \frac{1}{h_{scale} \cdot A_i} + \frac{x_{wall}}{k_{wall}\, A_{im}} + \frac{1}{h_0 A_0} \qquad ...(9.7)$$

where h_{scale} = heat transfer coefficient of scale formed

x_{scale} = wall thickness

k_{wall} = thermal conductivity of wall

A_{im} = mean inside area including scale formed

For simplicity, the tube wall resistance due to thin tube and good thermal conductivity may be neglected. Hence,

$$\frac{1}{U_0} = \frac{1}{h_i} + \frac{1}{h_{scale}} + \frac{1}{h_0} \qquad ...(9.8)$$

It is to be noted that h_o is much larger than h_i and U_0 mainly depends on water velocity as $h_i \propto C_w^{0.8}$

$$\frac{1}{U_0} = A + B \cdot \frac{1}{C_w^{0.8}} \qquad ...(9.9)$$

where

$$A = \frac{1}{h_0} + \frac{1}{h_{scale}} \qquad ...(9.10)$$

and

$$B = \frac{1}{0.023 \dfrac{k_f \rho_r^{0.4}}{d_i^{0.2}\, v_f^{0.8}}} \qquad ...(9.11)$$

The rate of heat transfer from the condensing vapour to the cooling water is expressed as

$$Q = \dot{m}_s\,(h_{\text{steam}} - h_{\text{condensate}}) = \dot{m}_c\,c_{pc}\,(T_{c2} - T_{c1}) = U_0 A_0\,\Delta T_m \quad ...(9.12)$$

where $\dot{m}_s$ = mass flow rate of steam entering to condenser, $\dot{m}_c$ is the mass flow rate of coolant and T_{c1} and T_{c2} are inlet and outlet temperature of cooling water. Generally, $(T_{c2} - T_{c1})$ is around 10°C, ΔT_i around 11 to 17°C and ΔT_e = TTD should not be less than 3°C.

In practice, due to losses in condenser, the undercooling of the condensate $(T_{\text{sat}} - T_{\text{condensate}})$ is around 4°C.

From Eq. (9.12), the mass flow rate of coolant, i.e., water ($\dot{m}_c$) is given by

$$\dot{m}_c = \frac{\dot{m}_s(h_{\text{steam}} - h_{\text{condensate}})}{c_{pc}\,(T_{c2} - T_{c1})} \quad ...(9.13)$$

The outside surface area A_0 is thus calculated from Eq. (9.12)

$$A_0 = \frac{\dot{m}_s\,(h_{\text{steam}} - h_{\text{condensate}})}{U_0\,\Delta T_m} = n\pi\,d_0\,l \quad ...(9.14)$$

where n = number of tubes, l = length of one tube (for a single pass condenser).

Further,

$$\dot{m}_c = \left(\frac{n\pi}{4}\,d_i^2\right)\rho_w\,C_w \quad ...(9.15)$$

where ρ_w = density of water (10^3 kg/m^3), and C_w is the velocity of water which varies from 1.8 to 2.5 m/s. Therefore, the length and number of tubes can be calculated from Eqs. (9.14) and (9.15). Generally, tube length and diameter are selected so the estimation is made for number of tubes. In the case of large size power plant, the ***number of tubes may be as high as 50,000 or even more.***

The pressure drop in the condenser is composed of (i) the pressure drop in the water box and (ii) the friction pressure drop in the tubes. The pressure drop in terms of head is

$$\Delta p = \rho\,gH$$

9.8 NON-CONDENSABLE GASES (AIR) REMOVAL OR DEAERATION

It is important to remove the non-condensable gases that get accumulated in the condenser. The non-condensables are mostly air that leaks through the joints into the condenser as condenser operates much below atmospheric pressure. It also includes other gases caused by the decomposition of water into oxygen and hydrogen by thermal action and by chemical reactions between water and materials of construction. *The performance of condenser due to presence of noncondensables gets affected badly because of the following reasons:*

(i) Increase in Condenser Pressure. The total pressure of the condenser gets increased as it is the sum of the partial pressure of its constituents, i.e., steam and air ($p = p_s + p_a$). An increase in condenser pressure lower plant efficiency.

(ii) Reduction in Heat Transfer Coefficient. They blanket the heat transfer surface such as outside the surface of condenser tubes which reduces the condensing heat transfer coefficient and thus condenser effectiveness.

(iii) Possibility of Corroding Tubes. The presence of noncondensables results in various chemical activities such as corrosion especially by oxygen in steam-generator. The maximum acceptable oxygen concentration in the condensate leaving the condenser is 0.003% by volume.

The process of removing noncondensables is called *deaeration*. Good deaeration within a condenser requires time, turbulence, and good venting equipment. The cold condensate falling from the lower tubes must have sufficient falling height and scrumbling steam for reheat and deaeration. The scrambling steam is provided by allowing some of the incoming steam to pass through an open flow area directly to the bottom tubes to reheat the condensate. This is done keeping in view that noncondensables are more easily released from a hotter than a colder liquid.

Once the noncondensables are released, they are cooled to reduce their volume before being pumped out of the condenser. For the convenience of noncondensable gases (air) removal, *an air cooler section* is provided in the condenser shell. The air cooler section may be provided either at the side below the centre position (Fig. 9.10) or in the centre (Fig. 9.11). In Fig. 9.10, most of the condensation is carried out on the main bank of the tubes and the air is drawn over another smaller bank (about 5 to 6%) which is shielded from the main bank by a baffle and is called *air-cooler*. Here, further condensation takes place at a lower temperature, and thus there is a saving in feed water as well as in air injection load. In the central air cooler section (Fig. 9.11), the air cooling tubes (about 5 to 6%) are in the centre of the condenser and the air is removed from this section. The incoming steam passes all round the bank of tubes and some is drawn upward to the centre. This air cooler section in baffled to separate the noncondensable from the main steam flow. The noncondensables flow toward the cold end of the condenser, where they connect to a vent duct that leads to the venting equipment.

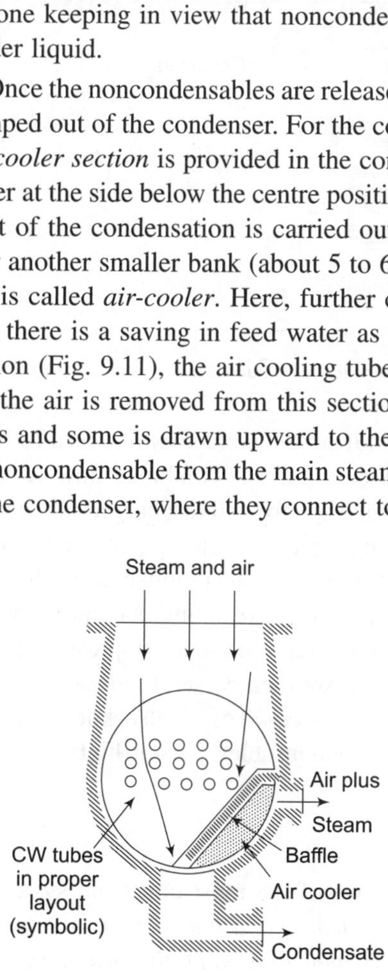

Figure 9.10 Air cooler section (side)

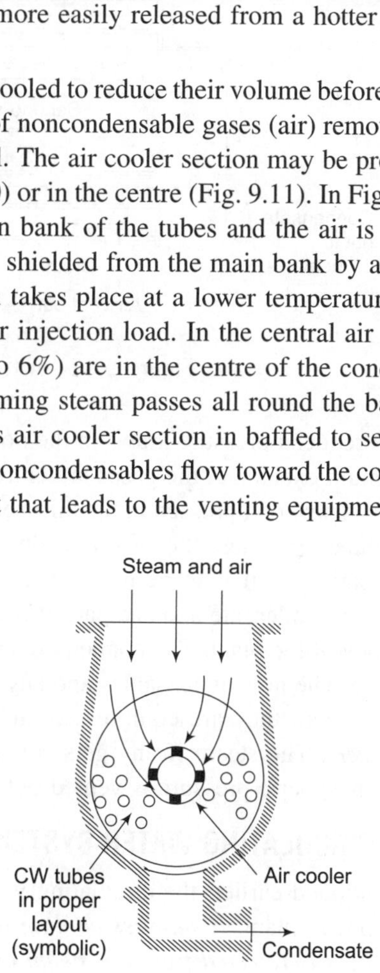

Figure 9.11 Air cooler section (central)

The venting equipment includes reciprocating compressor (called dry vacuum pump) and jet pump, called *steam-jet air ejector* (SJAE), the latter is now invariably used. SJAE uses steam jet as their motive or driving flow. They are usually multistage units, usually two or three. Figure 9.12 shows a two-stage SJAE. It uses main steam at a reduced pressure that enters a driving-flow nozzle in the first-stage ejector, from which it exits with high velocity and momentum but at reduced pressure.

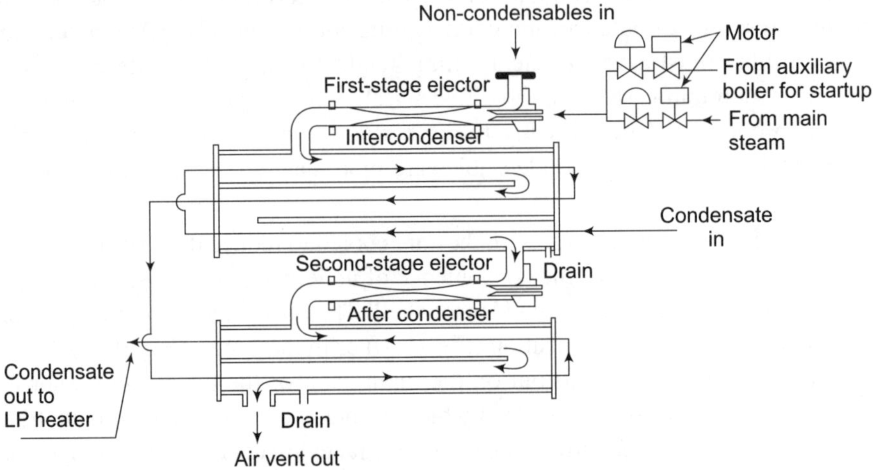

Figure 9.12 A two-steam jet air ejector (SJAE)

This reduced pressure draws in the noncondensables from the condenser and as a result of momentum exchange, the gases are entrained by the steam jet. The combined flow of steam and gas is now compressed in the diffuser of the first stage ejector and discharged into a small intercondenser, where the steam is condensed by passing across cooling pipes similar to the main condenser. It is to be noted that cooling here, however, is accomplished by the main condenser condensate and is a part of feed water heating system, resulting in enhancement in efficiency of the plant. The condensed steam is drained and returned to a low pressure part of the cycle. The noncondensables and any remaining steam are then passed to the second stage ejector, where they are compressed and passed to an *after condenser* (also known as *vent condenser*). The steam if any gets condensed and noncondensables (air) at higher pressure than atmospheric pressure is vented out.

9.9 CIRCULATING WATER SYSTEM

As mentioned earlier, the circulating water system is one of the most important systems of steam power plant. *It supplies cooling water to the condenser and thus acts as a vehicle by which heat is rejected from the steam cycle to the environment. Circulating water system is broadly classified as (a) once-through, (b) closed loop and (c) combination system.*

9.9.1 Once-through Cooling System

Figure 9.13 shows once-through system. In this case, water is taken from a natural body of water such as a lake, river or ocean and pumped through the condenser where it is heated and then discharged back to the source. There are mainly three methods of discharge namely (i) surface discharge, (ii) submerged discharge and (iii) diffuser discharge.

Once-through cooling system is thermodynamically, the most efficient means of heat rejection. Due to scarcity of water or other environmental regulation, closed loop system is universally used.

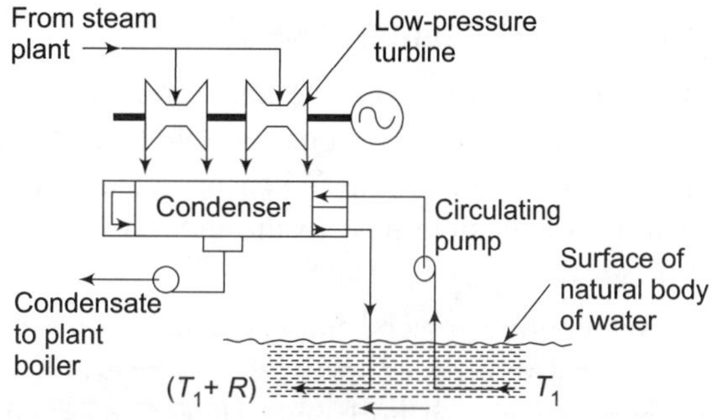

Figure 9.13 Once-through cooling water system

9.9.2 Closed-loop Cooling System

Closed-loop water cooling is shown in Fig. 9.14. In this case, hot water coming out from the condenser is passed through a cooling device (such as cooling towers, spray ponds, spray canals and cooling lakes) and is returned to the condenser with the help of a pump. A nearby natural body of water is still necessary to supply make-up water to replace the lost one by evaporation during the cooling process (say in cooling tower) and to receive blowdown from it.

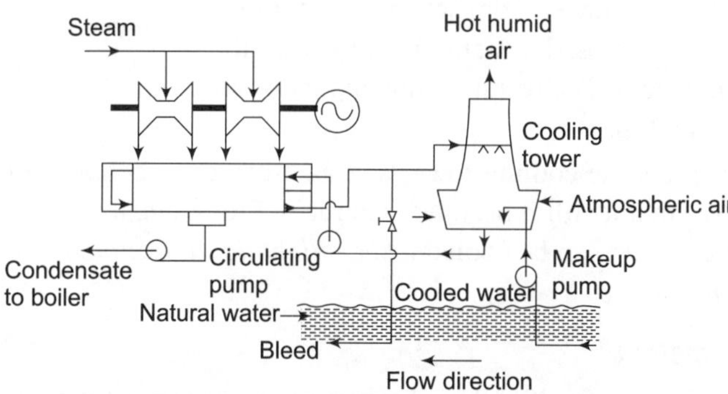

Figure 9.14 Closed loop cooling system

9.9.3 Combination Cooling System

A combination cooling system is such in which wet cooling tower is used and as per requirement either the cooling water coming out from cooling tower is led to the open mode or closed mode as per availability of water or scarcity of water.

9.10 COOLING TOWERS

The purpose of cooling tower *is to cool the warmed water discharged from the condenser and feed the cooled water back to the condenser.* By this way, the cooling water requirement gets reduced to make-up water supply only. The cooling tower may be *wet or dry type.*

9.10.1 Wet Cooling Towers

Wet cooling towers cool the hot water by dissipating heat to the environment through the mechanism of (i) addition of sensible heat to the air and (ii) evaporation of a portion of the recirculation water itself. When operated in the open mode, there is third mechanism (iii) addition of sensible heat to the natural body of water as a result of terminal temperature difference.

Wet cooling towers employ a *hot-water distribution system* that showers or sprays the water evenly over lattice work of closely set horizontal slats or bars called *fill*, or *packing* (Fig. 9.15). Since the water splashes down from one fill level to the next by gravity, there is a through mixing of falling water with air moving through the fill. Outside air enters the tower via louvers in the form of horizontal slats on the side of the tower. The slats are arranged in such way that they usually stop downward to keep the water in. The intimate mix between water and air results in the enhancement of heat and mass transfer (evaporation) which cools the water. The cold water gets collected in a concrete basin at the bottom of the tower from where it is pumped back to the condenser in closed system or returned to the water body in open system. The resulting hot and moist air leaves the tower at the top.

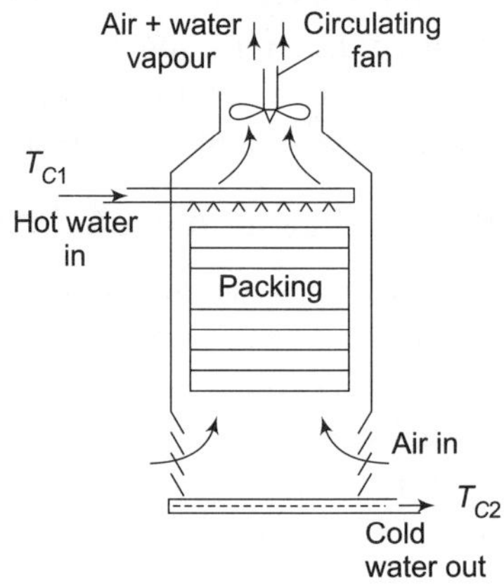

Figure 9.15 Packing or fill in a wet cooling tower

Unsaturated air enters the cooling tower and as it comes in contact with the water spray. Water continues to evaporate till it becomes saturated. From this, it follows that the minimum temperature to which water can be cooled is the *adiabatic saturation or wet bulb temperature* (T_{wb}) *of the ambient air.*

Performance Parameters

A cooling tower is characterized by three performance parameters namely (a) **approach**, (b) **range** and (c) **cooling efficiency**. *The approach temperature (A) is defined as the difference between the exit temperature of cooling water and the wet bulb temperature of the ambient air.* Thus,

$$A = T_{c2} - T_{wb} \qquad ...(9.16)$$

where T_{wb} = the wet bulb temperature of air

T_{c2} = cooling water exit temperature from cooling tower

This approach temperature (*A*) varies from 6 to 8°C.

The cooling range (R) is defined as the difference in temperature of the incoming warm water (T_{c1}) *and exiting cooler water* (T_{c2}). *Thus,*

$$R = T_{c1} - T_{c2} \qquad ...(9.17)$$

This range *R* varies from 8 to 10°C.

The cooling efficiency is defined as the ratio of actual cooling of water to the maximum cooling possible. Thus

$$\eta_{\text{cooling}} = \frac{\text{Actual cooling}}{\text{Maximum cooling possible}} = \frac{T_{c1} - T_{c2}}{T_{c1} - T_{wb}} \qquad ...(9.18)$$

Classification of Wet Cooling Tower. Wet cooling towers are classified as either (i) *mechanical* draught or (ii) natural draught cooling tower. Each of these may be *counter flow* or cross-flow.

9.10.1.1 Mechanical Draught Cooling Tower

Figure 9.16 shows the schematic of mechanical draught cooling tower. In this case, the air is moved by one or more mechanically driven fans. The fan could be of the forced-draught (FD) type or induced draught (ID) type. The FD fan (Fig. 9.16b) is mounted on the lower sides to force air into the two while ID fan (Fig. 9.16a) is located on top of the tower. Though, FD fan is thermodynamically superior as it handles cold air but it has shown some disadvantages because of air distribution problems, leakage, recirculation of the hot and moist exit air back to the tower and frost accumulation at fan inlets during winter operation. As a result, the majority of mechanical-draught cooling towers for utility application are therefore of the induced-draught (ID) type. In this case, air enters the sides of the tower through large onening at low velocity and passes through the fill. The fan located at the top of the tower exhaust the hot, humid air to the atmosphere. Induced-draught cooling towers are usually in the form of multi-cell with a number of fan stacks on the top and built in various arrangement like rectangular, octagonal, circular, etc. The fans are propeller type and driven by electric motor. The blades of fans are usually made of cast aluminium, stainless steel or fibres glass to safeguard against corrosion.

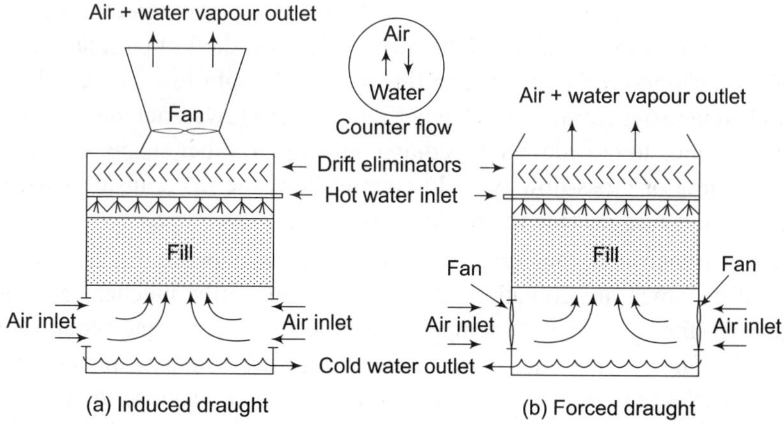

Figure 9.16 Mechanical draught (induced and forced) counter-flow cooling tower

The advantages of mechanical (forced) draught cooling towers include the assurance of moving the required quantity of air at all loads and climatic conditions, low initial capital costs, and low physical profile. The main disadvantages include power consumption, operating and maintenance costs and greater noise generated from fans.

9.10.1.2 Natural Draught Cooling Towers

There is no fan used by natural draught cooling tower. They depend for air flow upon the natural driving pressure caused by the difference in density between the cool outside air and the hot, humid air inside (Figs. 9.17 and 9.18). The driving pressure differential is expressed as

$$\Delta p_d = (\rho_o - \rho_i)\, gH \qquad \text{...(9.19)}$$

where H = height of tower above the fill (m)

ρ_o and ρ_i = density of outside and inside air respectively (m^3/kg)

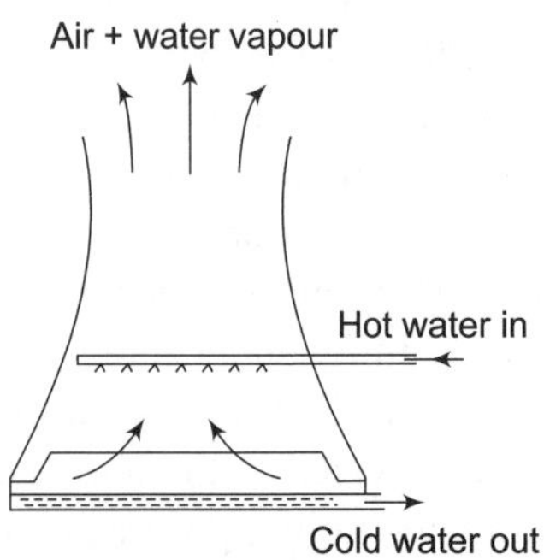

Figure 9.17 Natural draught cooling tower

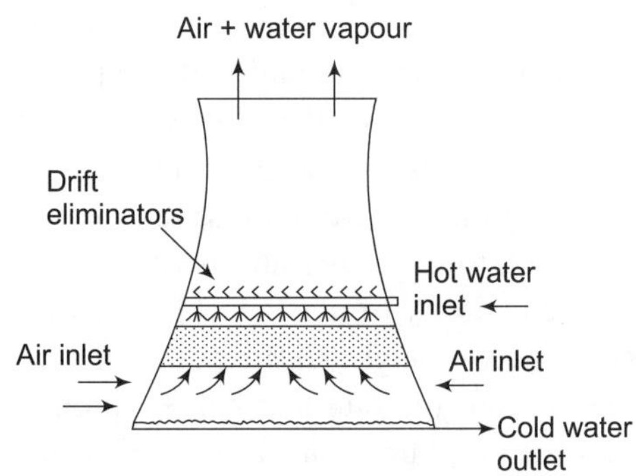

Figure 9.18 Counter-flow natural draught cooling tower

Since $(\rho_o - \rho_i)$ is relatively small, so H must be large to cause desired Δp_d and as a result natural draught cooling towers are therefore very tall. It is to be noted that the tower body, above the water distribution system and the fill, is an empty shell of circular cross section but with vertical *hyperbolic profile* and due to this natural draught towers are called *hyperbolic towers*. The advantage of natural draught hyperbolic cooling tower include its superior strength and greatest resistance to outside wind loading compared to other shape. The natural draught may be counter-flow or cross-flow type. In counter-flow, the fill is inside whereas in cross-flow, the fill sits in a ring outside the tower, outside the stilts.

The choice between different type of cooling towers depends upon many factors, the most important of which are climatic and economic. Generally, mechanical draught towers are selected when the approach (A) temperature is low and when the broad range of water flow is expected. The latter is met easily because they are usually built as multi-cell units with a variable airflow fan which offers versatility and good response to changes in cooling demands and parameters. Natural draught cooling towers are selected under the following conditions.

(i) In cool humid climates (low wet-bulb temperature and high relative humidity): (ii) when there is combination of low wet-bulb temperature and high condenser-water inlet and outlet temperature and (iii) in cases of heavy winter loads. In general, they are preferred for very large power plants where fewer and large towers can be built.

9.10.1.3 The Water Distribution System

It distributes hot condenser water evenly over the fill. There are various types of distribution system such as (i) gravity distribution, (ii) spray distribution, and (iii) rotary distribution. *Gravity distribution* system, mainly used in cross-flow towers, consists of vertical hot-wire risers that feed into an open concrete basin from which the water flows by gravity through orifices to the fill kept below. *Spray distribution* is used mainly on counter flow towers and has crosspiping with spray downward nozzles. *Rotary distribution* comprises two slotted distributor arms that rotate about a central hub through which water comes in under pressure. The slots are aimed downwards but slightly to one side which provides a curtain of water at an angle and a reaction force that rotate arms. There is always some water loss (about 2.5% or more) in the cooling tower due to (a) evaporation, (b) drift and (c) blowdown.

9.10.1.4 The Fill

The fill or packing is supposed to be the heart of cooling tower. It should provide good water-air contact for high rates of heat transfer and low resistance to air flow to minimise work input to fan. Further, it should also be strong, light and deterioration resistant. Basically, there are two types of fill namely (i) ***splash type*** and (ii) ***film or nonsplash type***. *Splash packing is made of bars may be of different shapes, narrow, square or grid and made of different materials such as redwood high impact polystyrene or polyethylene. Fill film* is generally made of vertical sheets that adheres to the vertical surfaces. As a result, this exposes the maximum water surface to the air without breaking into drops. They are made of redwood battens, cellulose, corrugated sheets, asbestos-cement sheet, waveform metal or plastic. Film type fill also provides less resistance to air flow and requires less area.

At present, the *trend in materials for wet cooling towers is for concrete structure with plastic fill, drift eliminators, fan stacks, fan blades and manifolds, valves and nozzles.*

9.10.1.5 Drift and Drift Eliminators

Drift is water fine droplets entrained by and carried with the air as an unevaporated drizzle. This is lost to the circulating water system and does not contribute to heat removal system. Drift is minimized by drift eliminators which are nothing but baffles placed in one, two or three rows. The function of the baffles is to change suddenly the direction of air (Fig. 9.19). Due to the momentum, the heavier drops get separated out from the air and impinges against the baffles, thus forming a thin film of liquid that falls into the tower. The materials of baffles may be wood, metal or plastic. The drift eliminators are situated at the air exit from the fill.

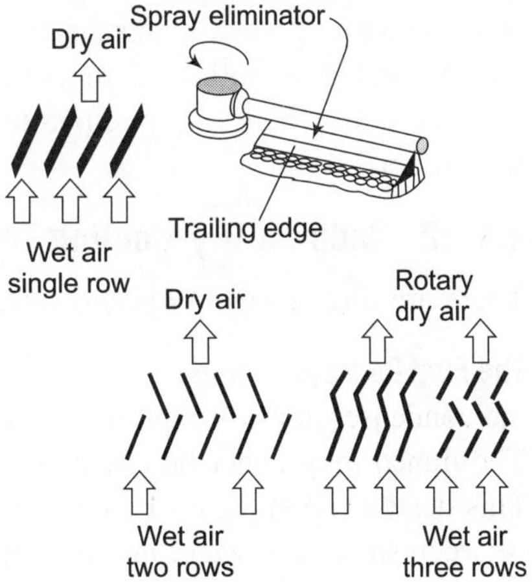

Figure 9.19 Types of drift eliminators

In order to maintain a certain solid concentration, blowdown is necessary from cold water basin of the bottom of tower. The total loss of water (evaporation about 1 to 1.5 per cent drift loss 0.03 per cent and blowdown 1–1.5%) is about 2 to 2.5% or even up to 5% which is met by make up-water.

9.11 DRY-COOLING TOWERS

As the name implies, a dry-cooling tower is one in which the circulating water is passed through finned tubes over which the cooling air is passed. As a result, all the heat rejected from the circulating water is thus in the form of sensible heat to the cooling air. It can also use mechanical or natural draught. The operation of all dry cooling towers is in the closed mode. In recent years, dry cooling towers have attracted much attention to their use. They are very suitable where there is scarcity of water (not even make-up water). The plant could be located, on fuel source site to avoid transportation cost. They are less expensive to maintain, free from thermal (moist air) and blowdown pollution. The main disadvantage is less efficient type than wet type, works at high back pressure which lowers plant output and efficiency.

Dry cooling towers are of two types—*direct* and *indirect*.

9.11.1 Direct Dry-Cooling Towers

Figure 9.20 shows a direct dry-cooling tower. In this case, turbine exhaust steam is admitted to a steam header through large ducts to minimize pressure drop and is condensed as it flows downward through a large number of finned tubes or coils arranged in parallel which are cooled by atmospheric air flowing in a natural draught cooling tower or by forced draught fan. A system is provided to prevent freezing in cold weather. The condensate flows by gravity and gets collected in condenser receiver from where it is pumped to plant feed water system.

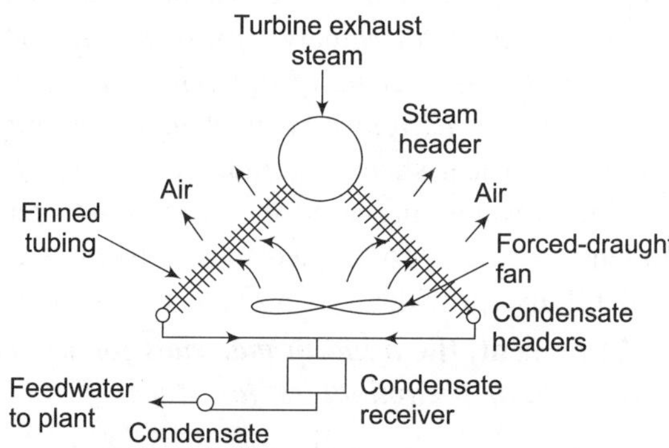

Figure 9.20 A direct dry-cooling tower

9.11.2 Indirect Dry-Cooling Tower

There are *three design concepts* available for indirect dry-cooling tower.

The First Concept. It uses a conventional surface condenser in which circulating water leaving the condenser goes through finned tubing cooled by atmospheric air in the tower (Fig. 9.21). The finned may either be cooled by air through natural draught or induced draught system. This design encompasses two heat exchangers in series and thus two temperature drops, one between steam and water and one between water and air. This double irreversibility imposes a server penalty on turbine back pressure, thus necessitating operation at condenser pressure of about 0.17 to 0.27 bar compared with 0.034 to 0.069 bar for once through system. As a result, there is increased heat rejection and low plant efficiency.

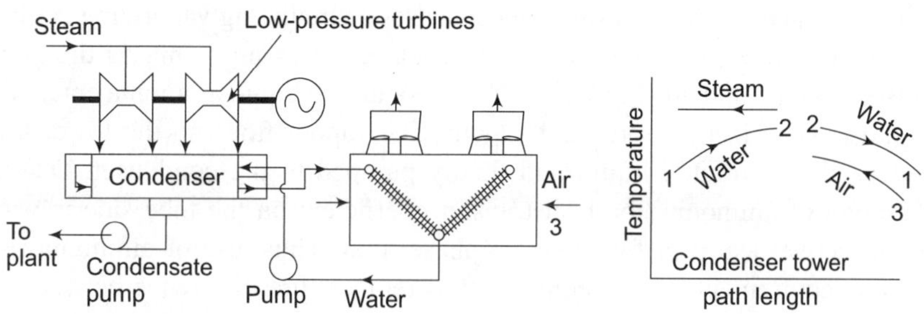

Figure 9.21 An indirect dry-cooling tower with conventional surface condenser

The Second Design Concept. It eliminates the intermediate water loop of first design concept and uses an open or direct contact condenser (jet or spray condenser). Since the operation is in the closed mode and no atmospheric or surface water impurities enter the system through make-up, the circulating water can be mixed with the steam from the plant and hence use of open-type condenser is justified (Fig. 9.22). The exhaust steam from turbine enters the open condenser where the cold circulating water is sprayed into the steam for intimate mixing. The condensate (including cooling water) falls to the bottom of the condenser, from which most of it is pumped by recirculation pump under positive pressure to finned tubing or coils in the lower. This part of the condensate is cooled and is returned to the condenser sprays. The balanced of the condensate, equal to the steam flow, is pumped to the plant feed water system by the condensate pump. In this case also, the cooling tower may be natural or forced draught. Note that the ratio of circulation to feed water is large. In some cases, instead of two pumps, single pump with control valve may serve the purpose. In order to maintain the

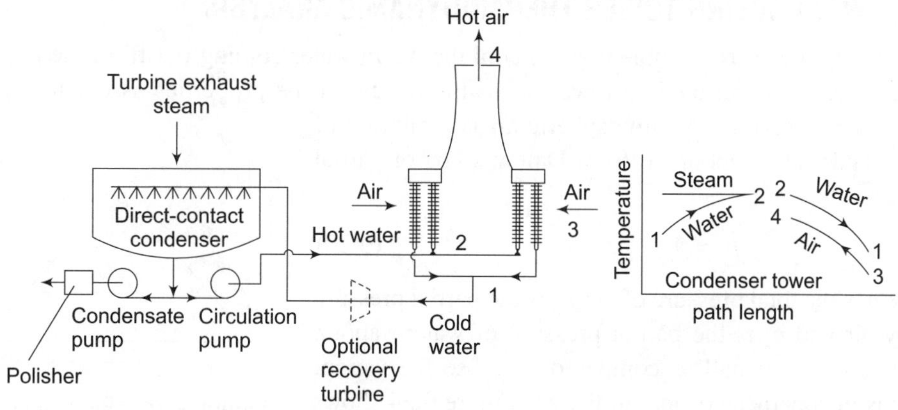

Figure 9.22 An indirect dry-cooling tower with an open-type condenser

quality of circulating water, condensate polishers may be used. A water recovery turbine (optional) may be used to recover some of the work of pump. This design concept of indirect dry cooling tower is efficient, more economical and more feasible for large plants.

The Third Design Concept. This design concepts uses a circulating vaporizing coolant instead of water. Currently, one plant that has been developed uses ammonia as the heat transfer medium between steam and air (Fig. 9.23). Nearly saturated liquid ammonia enters the surface condenser and is vaporized to saturated vapour. The vapour flows to the lower finned coils and is condensed to saturated liquid and finally pumped to the condenser. Due to boiling and condensation of ammonia, the heat transfer coefficient on the tube side is very high as compared convective heat transfer in single phase fluid. Thus, use of ammonia reduces the size and power requirements of the equipment as temperature difference between steam and ammonia, and ammonia and air get reduced.

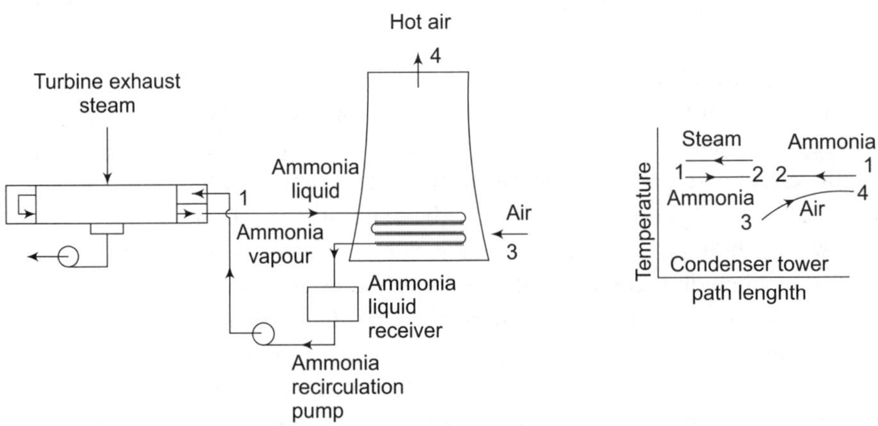

Figure 9.23 Indirect dry-cooling tower with a surface condenser and ammonia as coolant

9.12 WET COOLING TOWER THERMODYNAMIC ANALYSIS

Wet cooling tower uses ambient air to cool the warm water coming out from the condenser. Thus, analysis of wet cooling tower needs the review of the properties associated with air-water-vapour mixture. As atmospheric air is a mixture of dry air and water vapour, so from Dalton's law of partial pressure

$$p = p_a + p_v \qquad ...(9.20)$$

where p is the total pressure of air, p_a is the partial pressure of dry air and p_v is the partial pressure of water vapour. Since p_v is very small as compared to p_a, so the vapour remains in superheated state in the air. Figure 9.24 shows the representation p_a, p_v, t_{db}, t_{wb} and t_{dp} where suffixes

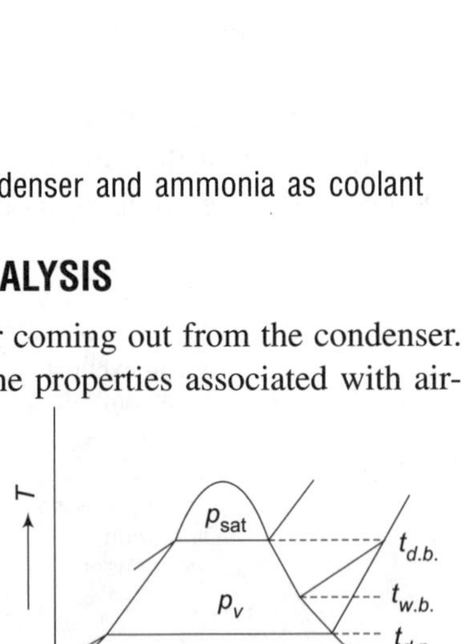

Figure 9.24 Representation of t_{db}, t_{wb} and t_{dp} of air

db, *wb* and *dp* represent dry, wet bulb and dew point temperature respectively. If the air at its t_{db} (unsaturated state) is cooled, it will attain its dew point temperature (i.e., saturation temperature of p_v). Thus, t_{wb} will be attained when the air will be cooled under adiabatic conditions.

Relative humidity (ϕ) is defined as the ratio of partial pressure of water vapour in air (p_v) to the saturation pressure (p_{sat}) at the air temperature,

i.e., $$\phi = \frac{p_v}{p_{sat}} \quad ...(9.21)$$

Specific humidity or humidity ratio (ω) is defined as the mass of water vapour per unit mass of dry air. Thus,

$$\omega = \frac{\dot{m}_v}{\dot{m}_a} = 0.622\frac{p_v}{p - p_v} \quad ...(9.22)$$

The degree of saturation (μ) is defined as the ratio of actual specific humidity to saturated specific humidity.

$$\mu = \frac{\omega}{\omega_{sat}} = \frac{p_v}{p_{sat}} \times \frac{p - p_{sat}}{p - p_v} \quad ...(9.23)$$

Figure 9.25 shows the states of water and air entering and leaving the wet cooling tower. The cooling tower utilises the phenomenon of evaporative cooling to cool the hot water below the t_{db}. Water never reaches t_{wb}. Water is cooled, the air is heated and humidified.

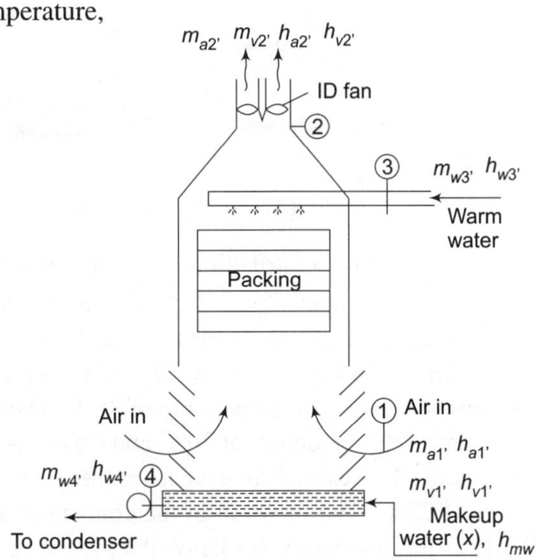

Figure 9.25 Analysis of a cooling tower

If x (kg/s) is the make-up water which is added to replenish the evaporation, thus

$$x = \dot{m}_a(\omega_2 - \omega_1) \quad ...(9.24)$$

where $\dot{m}_a$ is the mass flow rate of dry air (kg/s) and ω_1 and ω_2 are the specific humidities of air at inlet and outlet of cooling tower.

Using energy balance around the cooling tower, we have

$$(\dot{m}_{a1}\, h_{a1} + \dot{m}_{v1}\, h_{v1}) + \dot{m}_{w3}\, h_{w3} + x.h_{mw} = (\dot{m}_{a2}\, h_{a2} + \dot{m}_{v2}\, h_{v2}) + \dot{m}_{w4}\, h_{w4} \quad ...(9.25)$$

or $$h_{a1} + \omega_1 h_{v1} + \frac{\dot{m}_{w3}}{\dot{m}_{a1}} h_{w3} + (\omega_2 - \omega_1)\, h_{mw} = h_{a2} + \omega_2 h_{v2} + \frac{\dot{m}_{w4}}{\dot{m}_{a1}} h_{w4} \quad ...(9.26)$$

Here $\dot{m}_{a1} = \dot{m}_{a2} = \dot{m}_a$, $\dot{m}_{w3} = \dot{m}_{w4} = \dot{m}_w$ and $(h_{a2} - h_{a1}) = c_{pa}(T_{a2} - T_{a1})$

Thus, $$\dot{m}_w(h_{w3} - h_{w4}) = (h_{a2} - h_{a1}) + (\omega_2 h_{v2} - \omega_1 h_{v1}) - (\omega_2 - \omega_1)\, h_{mw} \quad ...(9.27)$$

Working with *moist air* directly, we have from energy balance

Here, $$\dot{m}_1 h_1 + \dot{m}_w h_{w3} + x\, h_{mw} = \dot{m}_2 h_2 + \dot{m}_w h_{w4} \quad ...(9.28)$$

Here, $\dot{m}_1 = \dot{m}_2$

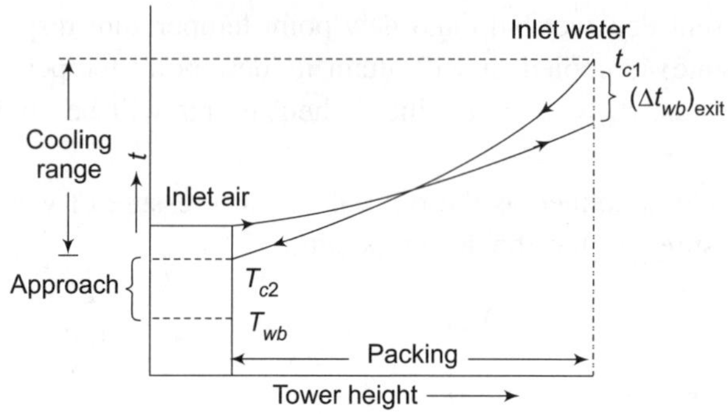

Figure 9.26 Temperature distribution in cooling tower

The temperature distribution of air and water in cooling tower is shown in Fig. 9.26.

Example 9.1 A steam power plant employs a surface condenser. The surface condenser receives steam from the turbine exhaust amounting 30 t/h with dryness fraction of 0.9 at 40°C. The air leakage to the condenser is estimated to be 0.3 kg per 1000 kg of steam. The condensate leaves at a temperature of 38°C. The make-up water is added at 30°C. The cooling water enters the condenser at 30°C and leaves at 38°C. A separate air extraction pump (ejector type) is used to extract air and from the air cooler section and air along with some steam leaves at 27°C. The pressure in the condenser for simplicity is assumed to remain constant. Workout the following: (a) Draw the system and show the process on *T-s* diagram, (b) The rate of saving of the condensate and the rate of saving in heat supplied in the boiler due to separate air extraction pump, (c) The percentages reduction in air ejector load due to this separate air extraction method, (d) The mass flow rate of cooling water, (e) The undercooling of condensate.

Solution Refer to Fig. 9.27 for the answer of part (a)

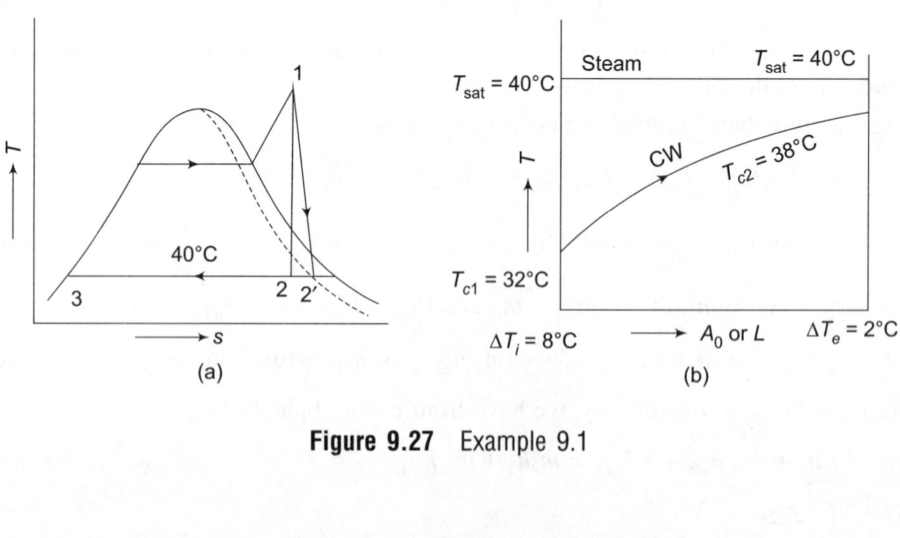

Figure 9.27 Example 9.1

At 40°C for steam, $p_{sat} = 0.07384$ bar and $v_g = 19.52$ m^3/kg, $\dot{m}_{s1} = 30\ t/h = \dot{m}_s$

Using perfect gas law $p_a\, v_a = \dot{m}_a\, R_a\, T_a$

Since air will occupy the condenser volume, thus

$$\dot{V}_a = \dot{V}_s = \dot{m}_s\, v_s = \dot{m}_s\, v_g$$

$\therefore\quad p_a\, \dot{m}_s\, v_g = \dot{m}_a\, R_a\, T_a \quad$ but $\quad \dot{m}_a = 0.3\, \dot{m}_s/1000$

or $$p_a = \frac{\dot{m}_a R_a T_a}{\dot{m}_s v_g} = \frac{0.3 \times 0.287\,(40 + 273) \times 10^3 \times 10^{-5}}{1000 \times 19.52} = 1.38 \times 10^{-5} \text{ bar (negligible)}$$

This shows that at entry, p_a is negligible and total pressure in the condenser is 0.07384 bar.

The condensate leaves at 38°C. At 38°C, $p_{sat} = 0.06624$ bar and $v_g = 21.63$ m^3/kg

The total pressure in the condenser $= p = p_{sat} + p_{air}$ or $0.07384 = 0.06624 + p_{air}$

or $$p_{air} = 0.07384 - 0.06624 = 7.6 \times 10^{-3} = 0.0076 \text{ bar}$$

The mass flow rate of air removed at 38°C,

$$\dot{m}_a = \dot{m}_s \times 0.3/10^3 = 30 \times 10^3 \times 0.3/10^3 = 9 \text{ kg/h}$$

Volume of air removed per hour at 38°C

$$\dot{V}_a = \frac{\dot{m}_a\, R_a\, T_a}{p_a} = \frac{9 \times 287 \times 311}{0.0076 \times 10^5} = 1057 \text{ m}^3/\text{h}$$

The mass flow rate of steam accompanying this air at 38°C is

$$\dot{m}_{s,\,38°C} = \frac{\dot{V}}{v_g} = \frac{1057}{21.63} = 48.867 \text{ kg/h}$$

Due to installation of separate air extraction in cooler side, air along with some steam leaves the air cooler section at 27°C.

At 27°C, the steam properties are: $p_{sat} = 0.03564$ bar and $v_g = 38.81$ m^3/kg

$$\therefore\quad p_{air} = 0.07384 - 0.03564 = 0.03820 \text{ bar}$$

The volume of air removed at 27°C from condenser air cooler section

$$\dot{V}_a = \frac{\dot{m}_a\, R_a\, T}{p_{air}} = \frac{9 \times 287 \times 300}{0.03820 \times 10^5} = 202.85 \text{ m}^3/\text{h} = \dot{V}$$

Mass of steam accompanying this air as $\dot{V}_a = \dot{V}_s$

$$\dot{m}_{s,\,27°C} = \frac{\dot{V}}{v_g} = \frac{202.85}{38.81} = 5.226 \text{ kg/h}$$

(b) The saving in condensate by using separate extraction pump in cooler steam of the condenser

$(\dot{m}_c)_{\text{saved}} = \dot{m}_{s,\,38°\text{C}} - \dot{m}_{s,\,27°\text{C}} = 48.867 - 5.226 = 43.64$ kg/h **Ans.**

Saving in heat supply in the boiler $= (\dot{m}_c)_{\text{saved}}\, c_{pe}\, (t_{c2} - t_{\text{makeup}})$

$= 43.64 \times 4.18 \times (38 - 30) = 1459.32$ kJ/h $= 0.4053$ kW **Ans.**

(c) Air ejector pump capacity without air cooler section = 1057 m^3/h

Air ejector pump capacity with air cooler section = 202.85 m^3/h

Thus, percentage reduction in air ejector load

$$= \frac{1057 - 202.85}{1057} \times 100 = 80.8\%$$ **Ans.**

Taking the energy balance around the control volume of condenser (Fig. 9.27) and using subscripts *s*, *a* and *c* representing steam, air and condensate respectively, we have

$$Q_{\text{rej}} + \dot{m}_c h_c + \dot{m}_{a2} h_{a2} + \dot{m}_{s2} h_{s2} = \dot{m}_{a1} h_{a1} + \dot{m}_{s1} h_{s1}$$

or $$Q_{\text{rej}} = \dot{m}_{a1} h_{a1} + \dot{m}_{s1} h_{s1} - \dot{m}_c h_c - \dot{m}_{a2} h_{a2} - \dot{m}_{s2} h_{s2}$$

Here, $\dot{m}_{a1} = \dot{m}_{a2} = 9$ kg/h and $\dot{m}_c = (30 \times 10^3 - 5.226)$ kg/h, $\dot{m}_{s2} = 5.226$ kg/h

$h_{s1} = h_g$ at 40°C = 2574.4 kJ/kg, $h_{s2} = h_g$ at 27°C = 2550.2 kJ/kg,

$h_c = h_f$ at 38°C = 159.3 kJ/kg

$$\therefore \quad Q_{\text{rej}} = 9 \times 1.005(40 - 27) + 30 \times 10^3 \times 2574.4 - (30 \times 10^3 - 5.226)\, 159.3$$
$$- 5.226 \times 2550.2 = 117.58 + 77232000 - 4778167.4 - 13327.2$$
$$= 72440622.9 \text{ kJ/h} = 20.122 \text{ MW}$$

But $$Q_{\text{rej}} = 20.122 \times 10^3 = \dot{m}_w\, c_{pw}\, (t_{w2} - t_{w1}) = \dot{m}_w \times 4.18 \times 8$$

or $$\dot{m}_w = \frac{20.122 \times 10^3}{4.18 \times 8} = 601.73 \text{ kg/s}$$ **Ans.**

(e) The undercooling of condensate = 40 – 38 = 2°C **Ans.**

Example. 9.2 In a condenser test, the following observations were made—Vacuum = 70 cm of Hg, Barometer = 76 cm of Hg, Mean temperature of condensation = 35°C, Hot-well temperature = 29°C, Mass of cooling water = 45500 kg/h, Inlet water temperature = 16.5°C, Outlet water temperature = 31°C, Mass of condensate = 1200 kg/h.

Find (a) the mass of air present per unit condenser volume, (b) the state of steam entering the condenser, (c) the vacuum efficiency, (d) the condensate undercooling and (e) condenser efficiency.

Solution Corresponding to 35°C, the partial pressure of steam from Steam Table = 0.05622 bar.

Absolute pressure of mixture in condenser $= (76 - 70) \times \dfrac{1.013}{76} = 0.07997$ bar

Since $$p = p_s + p_a$$

So $\qquad p_a = p - p_s = 0.07997 - 0.05622 = 0.023575$ bar

(a) Mass of air $= \dot{m}_a = \dfrac{p_a \times 10^5 \times v}{R_a T_a} = \dfrac{0.02375 \times 10^5 \times 1}{287\,(273 + 35)} = 0.02686$ kg/m^3 ***Ans.***

(b) Heat absorbed by circulating water $= \dot{m}_w\, c_{pw}\, (t_{w2} - t_{w1})$

$$= 45500 \times 4.18(31 - 16.5) = 2757755 \text{ kJ/h}$$

Heat rejected by steam in condensing at 35°C and is being undercooled from 35°C to 29°C $= m_c\,(h_{f1} + x h_{fg1} - h_c) = 1200\,(146.56 + 2418\,x - 121.48) = 1200(25.08 + 2418x)$

Equating the heat absorbed by heat rejected, we have the state of steam at

inlet to condenser $= x = \dfrac{2757757 - 25.08 \times 1200}{1200 \times 2418} = 0.94$ ***Ans.***

(c) Vacuum efficiency $= \dfrac{\text{Actual vacuum}}{\text{Ideal vacuum}}$

$$= \frac{70}{76 - 0.05622 \times (76/1.013)} \times 100 = 97.51\%$$ ***Ans.***

(d) Underloading of condensate = 35 – 29 = 6°C

(e) Condenser efficiency $\dfrac{\text{Actual temperature rise of water}}{\text{Maximum permissible temperature rise}}$

$$= \frac{(31 - 16.5)}{(35 - 16.5)} \times 100 = 78.37\%$$ ***Ans.***

Example 9.3 A surface condenser of a large steam power plant receives steam from turbine exhaust at the rate of 1500 t/h at 40°C having 88% dryness fraction. The circulating cooling water enters the condenser at 30°C and leaves at 38°C. The pressure inside the condenser is maintained at 0.078 bar. The velocity of circulating water is 2 m/s. The condenser tubes are of 25 mm *OD* and 1.25 mm thickness. Assume overall coefficient as 2600 W/m^2 K. By showing the processes on *T–s* diagram, calculate (a) the rate of flow of cooling water, and make-up water if evaporate rate is 2% (b) the rate of air leakage into the condenser shell, (c) number of tubes and (d) the length of tubes.

Solution Refer to Fig. 9.28. At 40°C, from Steam Table, $h_{fg} = 2407$ kJ/kg

$$h_1 = h_{f1} + x_1\, h_{fg1} \qquad \text{but} \qquad h_{f1} = h_{f1} = h_{f3} = h_c = 157.67 \text{ kJ/kg}$$

During condensation, only latent heat of steam is released which is consumed by circulating water

Thus, $\qquad \dot{m}_s\,(h_1 - h_c) = Q_{\text{rej}} = \dot{m}_w\, c_{pw}\, (t_{w2} - t_{w1})$

or $\qquad \dfrac{1500 \times 10^3}{3600}\,(h_{f1} + 0.88 \times 2407 - h_c) = \dot{m}_w \times c_{pw}\,(38 - 30)$

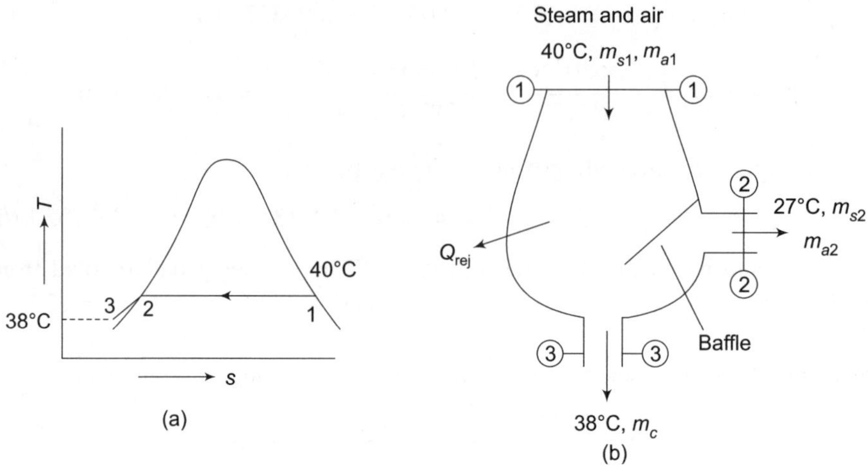

Figure 9.28 Example 9.3

(a) $\dot{m}_w$ = 26392.54 kg/s and make-up water = 0.02 × 26392.54 = 527.85 kg/s ***Ans.***

(b) At 40°C, from Steam Table, p_{sat} = 0.07375 bar

$$\therefore \qquad p_{air} = p - p_{sat} = 0.078 - 0.07375 \text{ bar} = 0.00425$$

$$v_1 = v_{f1} + x_1 v_{fg} = 0.001008 + 0.88 \times 19.544 = 17.2 \text{ m}^3/\text{kg}$$

From gas laws $p_a \dot{V}_a = \dot{m}_a R_a T_a$ but $\dot{V}_a = \dot{V}_s = \dot{m}_s v_s$

$$\text{or } \dot{m}_a = \frac{p_a \dot{m}_s v_s}{R_a T_a} = \frac{0.00425 \times 1500 \times 10^3 \times 17.2 \times 10^2}{3600 \times 0.28 \times 313} = 34.75 \text{ kg/s} \qquad \textit{Ans.}$$

(c) The log-mean temperature difference (ΔT_m) is expressed as

$$\Delta T_m = \frac{\Delta T_i - \Delta T_e}{\ln(\Delta T_i/\Delta T_e)} = \frac{(40-30)-(40-38)}{\ln(10/2)} = 4.97°\text{C}$$

The heat transfer rate = $Q = U_o A_o \Delta T_m = \dot{m}_s (h_1 - h_c)$

$$\text{or} \qquad 2600 \times 10^{-3} \times A_0 \times 4.97 = \frac{1500 \times 10^3}{3600} (0.88 \times 2407) = 882566.66$$

$$\text{or} \qquad A_0 = \frac{882566.66}{2600 \times 4.97 \times 10^{-3}} = 68.29 \times 10^3 \text{ m}^2$$

$$\dot{m}_w = \left(n \frac{\pi}{4} d_i^2\right) \rho\, C = 26392.54 \text{ kg/s}$$

or $$n.\frac{\pi}{4}(25-2.5)^2 \times 10^{-6} \times 10^3 \times 2 = 26392.54$$

or $$n = \frac{26392.54 \times 10^3 \times 4}{\pi(22.5)^2 \times 10^{-6} \times 10^3 \times 2} = 33{,}189 \text{ tubes}$$ ***Ans.***

(d) But $A_0 = n\pi D_0 l = 26392.54 \text{ m}^2$

$$\therefore \quad l = \frac{26392.54}{33{,}189\pi \times 25 \times 10^{-3}} = 31.8 \text{ m (Triple pass needed)}$$ ***Ans.***

Example 9.4 A steam power plant employs a wet type cooling tower which receives warm water at 30°C at the rate of 1.2 kg per kg of air. Air enters the tower at the dbt of 20°C and relative humidity of 60% and leaves it at a dbt of 28°C and 90% relative humidity. Makeup water is added at 20°C. Calculate (a) the temperature of water leaving the tower (b) the approach and range of the cooling tower and (c) the fraction of water evaporated.

Solution Refer to Fig. 9.25. Since relative humidity in air is given, so we can work directly for humid air. Using energy balance

$$m_1 h_1 + m_w h_{w3} + x h_{v1} = m_2 h_2 + m_w h_{w4}$$

Considering humid air, $m_1 = m_2 = m_{a1}$

or $$\frac{m_w}{m_{a1}}(h_{w3} - h_{w4}) = h_2 - h_1 - (\omega_2 - \omega_1) h_{v1}$$

From psyochrometric chart, for inlet and outlet air, we have

$$t_{wb1} = 15.2°\text{C}, t_{wb2} = 26.7°\text{C}, h_1 = 43 \text{ kJ/kg dry air}, h_2 = 83.5 \text{ kJ/kg dry air}$$

$\omega_1 = 0.0081$ kg water vapour/kg dry air and $\omega_2 = 0.02132$ kg water vapour/kg dry air. From Steam Table, $h_{v1} = 84$ kJ/kg.

Substituting the values in energy equation

$$1.2 \times 4.18 (30 - t_{w4}) = (83.5 - 43) - (0.02132 - 0.00881)84$$

(a) $t_{w4} = 22.13°\text{C}$

(b) Approach $= t_{w4} - t_{wb1} = 22.13 - 15.2 = 6.93°\text{C}$ ***Ans.***

Range $= t_{w3} - t_{w4} = 30 - 22.13 = 7.87°\text{C}$ ***Ans.***

(c) The fraction of water evaporated

$$x = m_{a1}(\omega_2 - \omega_1) = (0.02132 - 0.00881) = 0.01251 \text{ kg/kg dry air}$$ ***Ans.***

Example 9.5 It is required to cool 10^6 kg/h of water from 50°C to 40°C. Air enters the tower at 20°C dry bulb temperature and 15.5°C wet bulb temperature. It leaves the tower at 30°C and 90% relative humidity. Determine the fan capacity and loss of water.

Solution Refer to Fig. 9.25. The make-up water is not mentioned in the problem, so it is assumed that it is added outside the cooling tower. From energy balance.

$$m_{a1} h_{a1} + m_{v1} h_{v1} + m_{w3} h_{w3} = m_{a2} h_{a2} + m_{v2} h_{v2} + m_{w4} h_{w4}$$

Here, $m_{a1} = m_{a2} = m_a$ and $m_{w4} = m_{w3} - (\omega_2 - \omega_1)\, m_a$

or $$h_{a1} + \omega_1 h_{v1} + \frac{m_{w3}}{m_a} h_{w3} = h_{a2} + \omega_2\, h_{v2} + \left[\frac{m_{w3}}{m_a} - (\omega_2 - \omega_1)\right] h_{w4}$$

or $$(h_{a1} - h_{a2}) + \omega_1 h_{v1} + \frac{m_{w3}}{m_a} h_{w3} = \omega_2 h_{v2} + \left[\frac{m_{w3}}{m_a} - (\omega_2 - \omega_1)\right] h_{w4}$$

or $$c_{pa}\,(t_{a1} - t_{a2}) + w_1\, h_{v1} + \frac{m_{w3}}{m_a}(h_{w3} - h_{w4}) = w_2 h_{v2} - (\omega_2 - \omega_1)\, h_{w4}$$

From psychrometric chart, we have

$\omega_1 = 0.00925$ kg/kg dry air (at $t_{db} = 20°C$ and $t_{wb} = 15.5°C$)

$\omega_2 = 0.0244$ kg/kg dry air (at $t_{db2} = 30°C$ and $\phi_2 = 90\%$

From Steam Table, we have $h_{v1} = 2538.2$ kJ/kg at 20°C, $h_{v2} = 2556.4$ kJ/kg at 30°C

$h_{w3} = 209.3$ kJ/kg, $h_{w4} = 167.5$ kJ/kg

Substituting the values in energy equation, we have

$$1.005\,(20 - 30) + 0.00925 \times 2538.2 + \frac{m_{w3}}{m_a}(209.3 - 167.5)$$

$$= 0.0244 \times 2556.4 + (0.0244 - 0.00925) \times 167.5$$

or $$-10.05 + 23.47 + \frac{41.8\, m_{w3}}{m_a} = 62.478 + 2.537$$

or $$\frac{m_{w3}}{m_a} = 1.234 \text{ kg of water/kg of dry air}$$

Hence, mass flow rate of air entering the cooling tower

$$= \frac{10^6}{1.234} = 810372.7 \text{ kg/h}$$

The specific volume of air at section 2 = $v_{2a} = 0.8895$ m^3/kg

(a) Therefore, the induced draught fan capacity = $\dfrac{810372.7 \times 0.8895}{60} = 12013.7$ kg/min

Ans.

(b) The loss of water = $m_a(\omega_2 - \omega_1) = 810372.7 \times (0.0244 - 0.00925) = 12277$ kg/h

Ans.

PROBLEMS

Things to Think About

9.1 What is the reason that in general, the ends of condenser is either cylindrical or hemispherical?

9.2 What is the reason that there is an undercooling of condensate?

9.3 Is it possible to have a steam power plant without a condenser. Justify your answer with possible consequences?

9.4 Why there is a need of make-up water in feed heating system and cooling tower?

Short and Long Theoretical Questions

9.1 What are the functions of a condenser in steam power plant?

9.2 Mention the range of temperature rise in cooling water, why is it restricted in this range?

9.3 Classify condensers and mention their merits and demerits in the light of steam power plant.

9.4 What do you mean by passes in condenser?

9.5 What is the importance of tube arrangement in condenser shell? What are steam lines?

9.6 Why is it necessary to remove incondensable gases (air)?

9.7 What do you mean by undercooling of the condensate?

9.8 Mention the range of water velocity in condenser.

9.9 Discuss the effects of air leakage on the performance of condenser.

9.10 What are the causes of presence of air in condenser?

9.11 Discuss the various methods of air removal from condenser.

9.12 Discuss the method of designing a condenser.

9.13 Mention the sizes of tubes, and its length used in condenser.

9.14 What is the function of a cooling tower?

9.15 Differentiate between wet and dry type of cooling tower.

9.16 What are the advantages of induced and forced draught cooling tower? Which are preferred in general?

9.17 What is natural draught cooling tower? What are limitations?

9.18 How does water loss occurs in cooling water?

9.19 What are the various types of fills used in cooling tower?

9.20 What are the various types of water distribution system in cooling tower?

9.21 What is a drift? What are the functions of drift eliminators?

9.22 What do you mean by dry-cooling towers? With the help of sketches, discuss the various types of dry-cooling tower.

Numerical Questions

9.1 The following data relate to a single pass surface condenser:

Specific steam consumption = 3.8 kg/kWh, Plant output = 120 MW, Quality of steam at the turbine exhaust = 0.88, Pressure in the condenser = 0.05 bar, Temperature of inlet circulating water = 25°C, Temperature of outlet circulating water = 33°C, Velocity of circulating = 1.8 m/s, Tube outside diameter = 25 mm, Tube thickness = 1.25 mm, Overall heat transfer coefficient = 3000 W/m^{2}K.

By drawing the system and showing the process on *T-s* diagram, calculate (a) the volume flow rate of circulating water required, (b) heat transfer surface area, (c) the length of tubes and (d) the number of tubes.

9.2 A steam power plant employs a surface condenser. It receives steam 1500 t/h at 40°C with 10% moisture. The cooling water enters at 33°C and leaves at 40°C. The pressure inside the condenser is 0.078 bar. Assume subcooling of 2°C and overall heat transfer coefficient as 3000 W/m^{2}K, calculate (a) the condenser cooling water flow rate, (b) air ejector capacity in kg/s, (c) the heat transfer surface area of the condenser.

9.3 In a condenser, the air leakage is estimated to be 38 kg/h. The vacuum near the outlet to the air pump is 72.5 cm of Hg and the temperature at this point is 20°C. The barometer reads 76 cm of Hg. Calculate the minimum capacity of the air pump in m^3/min, and (b) the mass of vapour extracted with air per minute.

[***Ans.*** (a) 23.21 m^3/min, (b) 0.405 kg/min]

9.4 Surface condenser handles 20,000 kg of steam per hour and the air leakage is estimated at 0.3 kg per 100 kg of steam. The steam enters the condenser dry and saturated at 38°C. The condensate is extracted at the temperature of 36°C at the lower point of condenser. The condensate loss is made up with water at 7°C. Calculate the saving in condensate and the saving in heat supplied in the boiler, by fitting a separate air extraction pump which draws the air oven an air water. Assume that the air leaves the cooler at 27°C. If the temperature rise of cooling water is 5.5°C, calculate the mass of water required.

[***Ans.*** 28.11 kg/h 3411 kJ/h, 8.8 × 10^6 kg/h]

9.5 Dry and saturated steam at 39°C enters a surface condenser at the rate of 20,000 kg/h. The air leakage is estimated to be 0.3 kg per 1000 kg of steam. The temperature of the condensate is 30°C. There are separate air and condensate pumps and the air cooling section of tubes is to reduce the temperature of the air pump suction to 28°C. Calculate the following:

(a) Saving in condensate and saving in heat supply to the boiler by incorporating separate arrangement for the extraction of condensate and the air. The water available is 15°C is used to make up the loss of condensate.

(b) Percentage reduction in air pump capacity of separate air extraction pump is used.

[***Ans.*** (a) 16.816 kg/h, (b) 1482.5 kJ/h, (c) 68.27% (d) 48 × 10^4 kg/h]

9.6 A steam power plant employs a wet type cooling type. Cooling water enters a cooling tower the rate 20 t/h and 50°C. Water is pumped from the base of tower as make-up water at 25°C. Air enters the tower at 15°C, 50% RH. 1.013 bar and leaves the tower at 35°C and 1 bar. Find the flow rate of dry air in kg/h and the make-up water required in kg/h.

Software Based Question

9.1 Develop a software for the design of surface condenser for the power plant for the following data by assuming suitable data: Power output = 500 MW, Specific steam consumption = 3.5 kg/kWh, Dryness fraction of turbine exhaust steam = 0.88, Condenser pressure = 0.05 bar, Overall heat transfer coefficient = 2900 W/m^2 K.

9.2 Develop a software for the design of wet cooling tower, for the following data by assuming suitable data: Power output = 1000 MW, Specific steam consumption = 3 kg/kWh, Dryness fraction of steam at turbine exhaust = 0.9, Condenser pressure = 0.05 bar: Water inlet temperature = 30°C, Air inlet temperature = 30°C.

Design and Open Ended Questions

9.1 Prepare a comprehensive report about number and shape of condenser, their design and specification used in NTPC of India.

9.2 Develop a design procedure of very efficient modern surface condenser.

Multiple-Choice Questions

9.1 Which one is true in power plant steam surface condenser?

(a) Water enters the shell and flows along the shell.

(b) Steam flows through the tubes.

(c) Water and steam mixes and then steam condensers.

(d) Steam flows through the shell and water flows through tubes.

9.2 Undercooling of condensate in surface condenser is around

(a) 1°C (b) 4°C (c) 8°C (d) 10°C

9.3 Presence of air and gases in the condenser is due to

(a) presence of air in the condensing steam.

(b) leakage of air into the condenser.

(c) (b) and (a) release of gases from condensation of steam

(d) combination of (a), (b) and (c)

9.4 Make-up water needed in condenser is about

(a) 1.5% of inlet cooling water (b) 5% of inlet cooling water

(c) 10% of inlet cooling water (d) 15% of inlet cooling water

9.5 The make-up cooling water needed in the summer in cooling tower is about

(a) 10% of cooling water (b) 15% of cooling water

(c) 20% of cooling water (d) 25% of cooling water

9.6 The temperature rise in cooling water temperature in condenser is around

(a) 10°C (b) 20°C (c) 30°C (d) 40°C

Answers to Multiple-Choice Questions

9.1 (d) 9.2 (b) 9.3 (d) 9.4 (a) 9.5 (b) 9.6 (a)

CHAPTER 10

Nuclear Power Plant

OVERVIEW

Nuclear power plants are an economic necessity due to increasing price and depletion of fossil fuels. Now, the cost of electricity generated per kW from nuclear power plants is comparable with coal-fired power plants. But nuclear industry is facing many difficulties all over the world, primarily as a result of negative impact of the issues of nuclear safety (i.e., radiation), waste disposal, weapons proliferation and the economics on the public and government. However, after so many hurdles, electricity generated by nuclear plant takes the share around 30% in west and USA. In India, the nuclear power plants contribute only about 2.7% of total electricity generated.

The energy released in nuclear reaction is very large in comparison to the chemical reaction. The heat evolved by one kg of uranium is equivalent to the combustion of 1,13,350 quintals of coal or 30×10^6 litres of diesel oil.

The capital cost of nuclear power of larger capacity (say 250 MW) is nearly ₹2500 per kW installed in which the fuel cost is 22% and the capital cost is about 62%. The overall running cost of nuclear plant is about 85 paise per kWh. Now it has increased.

This chapter deals with the principles of nuclear energy and nuclear power plant systems.

10.1 THE ATOMIC STRUCTURE

In 1803, John Dalton postulated that all matter is composed of unit particles called *atoms.* This was simple but incomplete atomic hypothesis.

An atom consists of a relatively heavy, positively charged ***nucleus*** and a number of much lighter negatively charged ***electrons*** that exist in various orbits around the nucleus. The nucleus consists of subparticles, called ***nucleons***. Nucleons consist of ***neutrons*** which are electrically neutral, and the ***protons*** which are positively charged. Electric charge on the proton is equal in magnitude but opposite in sign to that on the electron. *Since the number of protons is equal to the number of electrons in orbit, so atom as a whole is electrically neutral.*

The most interesting aspect of this atomic structure is that one atom may be transferred into another by losing or acquiring some of the above subparticles. Such atomic reaction results in a change of mass (Δm) and therefore, release (or absorb) large quantity of energy (E) which is given by **famous Einstein's law.**

$$\Delta E = \Delta m\, C^2$$

where C is the velocity of light (3×10^8 m/s) and Δm = loss in mass.

Figure 10.1 depicts the structure of three light atoms. Hydrogen has a nucleus composed of one proton, no neutrons, and one orbital electron. It is worth to note that it is the only atom that has no neutrons. Deuterium has one proton and one neutron in its nucleus and one orbital electron while helium has two protons, two neutrons on nucleus and two orbital electrons.

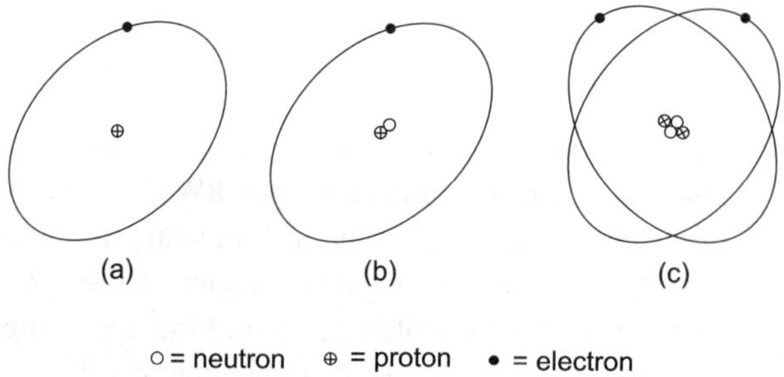

Figure 10.1 Structure of (a) hydrogen, (b) deuterum and (c) helium atoms

It is the nucleus that contains all the mass of the atom. The masses of the three primary atomic subparticles, which are supposed to be the building blocks of all atoms are given as follows:

Neutron mass = m_n = 1.008665 amu

Proton mass = m_p = 1.007277 amu

Electron mass = m_e = 0.0005486 amu

Where amu stands for *atomic mass unit* and 1 amu = 1.66×10^{-27} kg.

*Since the atoms contain these three primary atomic subparticles in varying numbers, so the atoms differ in their masses. The atoms with nuclei that have the same number of protons and have similar chemical and physical characteristics and differ in their masses are called **isotopes**.*

For example, **deuterium**, frequently called ***heavy hydrogen***, is an *isotope* of hydrogen. When combined with oxygen, ordinary hydrogen and deuterium form ***ordinary water*** and ***heavy water*** respectively.

The number of protons in the nucleus is called the ***atomic number (z)***, while the total number of nucleons in the nucleus is termed ***mass number (A)***. Thus, isotopes of the same element have the same atomic number. Nucleus symbols are generally written as

$$_{z}X^{A}$$

where *X* is the usual chemical symbol.

Thus, hydrogen nucleus is $_{1}H^{1}$, deuterium is $_{1}H^{2}$ and helium is $_{2}He^{4}$. For particles containing no protons, the subscript indicates the magnitude and sign of the electric charge. For example, an electron is $_{-1}e^{0}$ and (sometimes e^{-} or β^{-}) and neutrons is $_{0}n^{1}$.

There are many elements found in nature as mixture of isotopes of varying abundance such as **naturally occurring uranium**, called ***natural uranium*** is composed of **99.282 mass per cent U^{238}, 0.712 mass per cent U^{235}**, and **0.066 mass per cent U^{234}** where **the atomic number** if deleted which is **92 in all three cases**.

Over and above, **three above primary atomic subparticles**, the two other particles of importance are the ***positron and neutrino***. The *positron* is a positively charged electron having the symbol $_{+1}e^{0}$, e^{+} or β^{+} while *neutrino* is a tiny electrically neutral particle that is difficult to observe experimentally.

Further, there are many other atomic subparticles. For example, it is the ***mesons*** which are unstable, positive, negative or neutral particles where mass lies between electron and proton.

It is to be noted that electrons that orbit in the outermost shell of an atom are called ***Valence Electrons***, while the outermost shell is called the ***Valence Shell***.

10.2 COMPARISON BETWEEN CHEMICAL AND NUCLEAR EQUATIONS

There are some similarities and dissimilarities between chemical and nuclear equations. Chemical reactions involve the combination or separation of *whole atoms*. Consider the reaction between carbon and oxygen, which is

$$C + O_2 \rightarrow CO_2 \qquad ...(10.1)$$

This equation is accompanied by the release of about 4 *electron volts* (eV). An electron volt is a unit of energy in common use of nuclear engineering 1 eV = 1.6021×10^{-9} J = 4.44×10^{-26} kWh.

In chemical reactions, each atom participates as a whole and retains its identity. The molecules change but the nuclei are unaffected. There are as many atoms of each participating elements in the products as in the reactants.

It is to be noted that both chemical and nuclear reactions are *exothermic* or *endothermic.* In nuclear reactions, but reactant nuclei do not show up in the products, instead we may find either isotopes of the reactants or other nuclei. Further, in balancing nuclear equations, it is essential to observe that the same, or equivalent, nucleus show up in the products as entered the reactions. As an example, if *K, L, M* and *N* are chemical symbols, the corresponding nuclear equation becomes

$$ _{z1}K^{A1} + _{z2}L^{A2} \rightarrow _{z3}M^{A3} + _{z4}N^{A4} \quad ...(10.2)$$

For balancing the above equation, the following conditions must be satisfied

$$Z_1 + Z_2 = Z_3 + Z_4 \quad ...(10.3)$$

and

$$A_1 + A_2 = A_3 + A_4 \quad ...(10.4)$$

10.3 NUCLEAR BINDING ENERGY

There is a difference between theoretically calculated mass of nuclei (the sum of the masses of the protons and neutrons that comprise the nucleus) and the actual mass of the atomic nucleus. The difference in mass is called the **mass defect (Δm)**. This mass defect is found by adding up all the individual particles (neutron proton and electron) masses and subtracting the actual mass (m) of the atom. Thus,

$$\Delta m = n_n\, m_n + n_p\, m_p + n_e\, m_e - {}_Z m^A \quad ...(10.5)$$

where n refers to the number of particles, m the mass of particles and suffixes, n, p and e refer to neutron, proton and electron respectively. The mass defect is converted to energy in a nuclear reaction as given by Einstein's law

$$\Delta E = \Delta m\, C^2 \quad ...(10.6)$$

where E stands for energy (J), C stands for velocity of light = 3×10^8 m/s.

The energy associated with mass defect is known as the ***binding energy** of the nucleus.* The energy equivalent of 1 g of mass is

$$\Delta E = (1 \times 10^{-3}) \times (3 \times 10^8)^2 = 9 \times 10^{13} \text{ J} = 9 \times 10^7 \text{ MJ}$$

Similarly, energy equivalent of 1 amu of mass is

$$\Delta E = (1.66 \times 10^{-27}) \times (3 \times 10^8)^2 = 14.94 \times 10^{-11} \text{ J} = 9.3 \times 10^8 \text{ eV} = 931 \text{ MeV}$$

10.4 NUCLEAR FUSION AND FISSION

Nuclear reactions are of three types, namely, *fusion, fission* and *radioactivity*. In *fusion*, two or more light nuclei fuse to form a heavier nucleus while in *fission*, a heavy nucleus is split into two or more lighter nuclei. In both, there is a decrease in mass resulting in exothermic energy given by above Einstein's law.

10.4.1 Fusion

Energy is produced in the sun and stars by continuous fusion reactions. In this fusion process, four nuclei of hydrogen fuse in a series of reactions and culminates in one nucleus of helium and two positrons as given below:

$$4\ {}_1H^1 \rightarrow {}_2He^4 + 2\ {}_{+1}e^0 \quad ...(10.7)$$

This results in a decrease in mass of about 0.0276 amu and thus release of 25.7 MeV. The heat produced in these reactions maintains temperature of the order of several million degrees in their cores and serves to trigger and sustain succeeding reactions.

On the earth, it is difficult to have fusion reaction for meaningful use due to various reasons. In order to cause fusion, it is necessary to accelerate the positively charged nuclei to high kinetic energies, to overcome electrical repulsive forces, by raising their temperature to hundreds of millions of degree resulting in a *plasma formation*. Achieving such a high temperature is a difficult proposition. Further, plasma must be prevented from contacting the walls of the container and must be confined for a small period (about a second) at a minimum density. Fusion reactions are also called ***thermonuclear*** as very high temperatures are required to trigger and sustain them. Many problems have to be solved before artificially made ***fusion reactor*** becomes a reality.

10.4.2 Fission

Fission is a practical proposition and can be caused by the neutron, which being electrically neutral, can strike and fission the positively charged nucleus at high, moderate or low speeds without being repulsed and sustains chain reactions because two or three neutrons are usually released for each one absorbed in fission. These keep the reactions going. A typical fission reaction is depicted in Fig. 10.2.

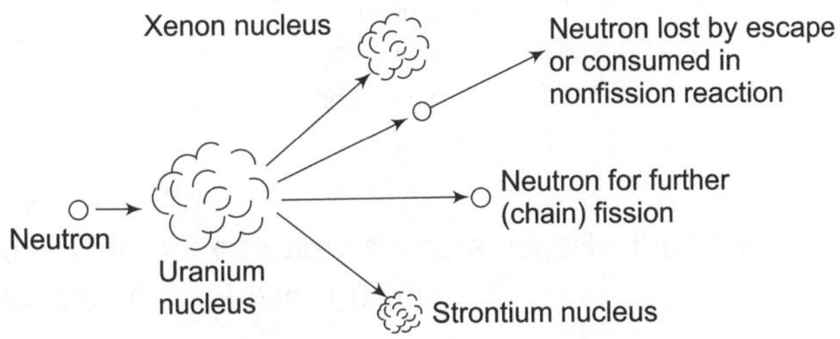

Figure 10.2 A typical fission reaction

It is worth to note that there are only a few fissionable isotopes. U^{235}, Pu^{239} and U^{233} are fissionable by neutrons of all energies while U^{238}, Th^{232} and Pu^{240} are fissionable by high energy neutrons only. The fission reaction with U^{235} is as follows

$$_{92}U^{235} + {_0n^1} \rightarrow {_{54}Xe^{140}} + {_{38}Sr^{94}} + 2\ {_0n^1} \quad ...(10.8)$$

The immediate (prompt) products of a fission reaction such as Xe^{140} and Sr^{94} are called *fission fragments* while they, and their decay products are called *fission products*. Figure 10.3 shows the fission product data for U^{235} by thermal and fast neutrons and for U^{235} and Pu^{239} for thermal neutrons with their mass number.

The two fission fragments in Eq. (10.8) are not equal in size and are radioactive. The fission fragments, therefore, undergo several stages of β decay (converting neutrons into protons) until a stable product is formed in each case

$$_{54}Xe^{140} \xrightarrow[16s]{\beta} {_{55}Cs^{140}} \xrightarrow[66s]{\beta} {_{56}Ba^{140}} \xrightarrow[12.8\ days]{\beta}$$

$$_{57}La^{140} \xrightarrow[40\ hrs]{\beta} {_{55}Ce^{140}}\ \text{(stable)}$$

and

$$_{38}Sr^{94} \xrightarrow[2\ min]{\beta} {_{39}Y^{94}} \xrightarrow[17\ min]{\beta} {_{40}Zr^{94}}\ \text{(stable)} \quad ...(10.9)$$

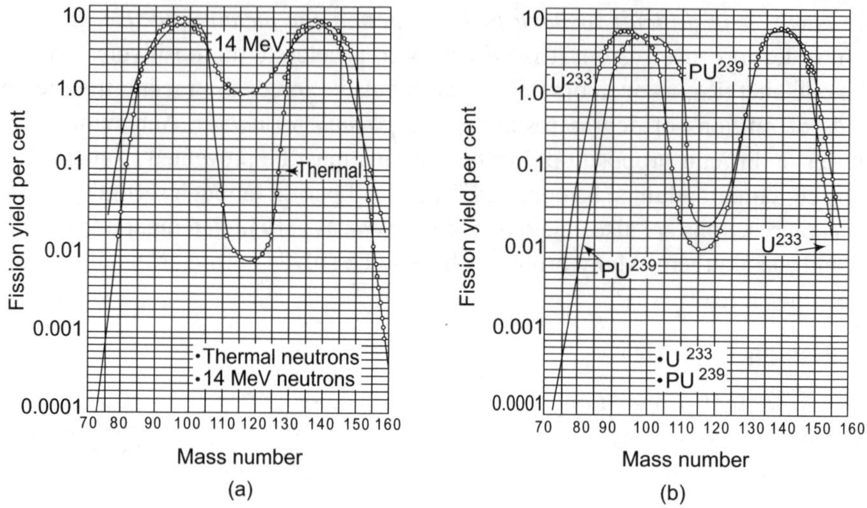

Figure 10.3 Fission products yield data for (a) U^{235} by thermal and 14 MeV neutrons and (b) U^{233} and Pu^{239} by thermal neutrons

The above series is known as *fission chain*. It is a well-known fact that β decay is usually accompanied by γ radiation, so suitable shielding against γ rays as well neutron must be provided in a reactor.

10.4.3 Energy from Fission and Fuel Burnup

Energy released by fission reaction depends upon the type of reaction. The reaction given by Eq. (10.8) yields 196 MeV. Consider the following reaction

$$_{92}U^{235} + {}_0n^1 \rightarrow {}_{56}Ba^{137} + {}_{36}Kr^{97} + 2\ {}_0n^1 \qquad ...(10.10)$$

Taking mass balance, we have

$$235.0439 + 1.00867 \rightarrow 136.9061 + 96.9212 + 2 \times 1.00867$$

or $$236.0526 \rightarrow 235.8446 \text{ amer}$$

$\therefore$ $$\Delta m = 235.8446 - 236.0526 = -0.2080 \text{ amu}$$

Thus, $$\Delta E = 931 \times (-0.2080) = -193.6 \text{ MeV}$$

It has been observed that on the *average* the fission of a U^{235} nucleus yields about 193 MeV. U^{233} and Pu^{239} also yield the similar value of energy release. It is to be noted that this amount in energy is *prompt*, i.e., released at the time of fission and more energy is produced because of (i) slow decay of fission fragments and (ii) the nonfission capture of excess neutrons in reaction that produce energy. **As a result, the *total energy* produced *per fission* reaction is about 200 MeV.**

The complete fission of **1 g of U^{235}** nuclei produces

$$\text{Energy released} = \frac{\text{Avogadro's number}}{U^{235}\text{ isotope mass}} \times 200\text{ MeV} = \frac{0.60225 \times 10^{24}}{235.0439} \times 200$$

$$= 0.513 \times 10^{24}\text{ MeV} = 2.276 \times 10^{24}\text{ kWh} = \mathbf{0.948\ MW\text{-}day}$$

In simple way, we can say that reactor burning 1 g of fissionable material generates nearly 1 MW-day of energy.

It is to be noted that nuclear fuel is defined as all uranium, plutonium and thorium isotopes. Further, depending upon fuel type and enrichment (mass per cent of fissionable fuel in all fuel), the energy released may vary from 10^3 to 10^5 MW-day/tonne.

Nuclear Fuels: In general, nuclear fuels used in nuclear power plants include $_{92}U^{235}$, $_{94}Pu^{239}$, **and** $_{92}U^{233}$.

10.5 RADIOACTIVITY

Radioactivity is an important aspect of nuclear science. It provides an important source of energy for small power devices and a source of radiation for use in research, industry, medicine and a wide variety of applications and as well as an environmental concern.

Most of the naturally occurring isotopes are *stable. The isotopes which are not stable are known as* ***radioactive***. The examples of radioactive are some naturally occurring isotopes of the heavy elements like thalium (Z = 81), lead (Z = 82) and bismuth (Z = 83) and all the isotopes of the heavier elements beginning with polonium (Z = 84). A few lower-mass naturally occurring isotopes are radioactive, such as K^{40}, Rb^{87} and In^{115}. In addition to the above, *several thousand artificially produced isotopes of all masses are radioactive.* It is to be noted that natural and artificial radioactive isotopes, also known as **radio isotopes,** have similar disintegration rate mechanisms.

A continuously undergoing spontaneous (i.e., without outside help) disintegration of radioactive isotopes is called ***radioactivity***. This is accompanied by the emission of one or more smaller particles from the *parent nucleus*. The resulting nucleus is known as ***daughter nucleus***. The parent nucleus is said to *decay* into the daughter nucleus. The daughter nucleus may or may not be stable, and several successive decays may occur until a stable isotope is formed. The following nuclear reaction shows an example of radioactivity.

$$_{49}In^{15} \rightarrow {}_{50}Sn^{115} + {}_{-1}e^{0} \qquad \text{...(10.11)}$$

where the parent, In^{115}, is a naturally radioisotope, and its daughter, Sn^{115}, is stable.

It is worth to note that radioactivity is always accompanied by a decrease in mass and thus is always exothermic. The energy liberated is in the form of kinetic energy and radiation.

The emission from naturally occurring radioisotopes include α, β or γ particles or radiations. The artificial isotopes, in addition to the above, emit or undergo the following particles or reactions – positron; orbital electron absorption, called K capture; and neutron. In addition, neutrino emission is accompanied with β emission of either sign.

(i) Alphs (α) decay. α particles are helium nuclei, each consisting of two protons and two neutrons and are commonly emitted by the heavier radioactive nuclei. The decay of Pu^{239} into fissionable U^{235} and α (He^4) particles, is an example of this.

$$_{94}Pu^{239} \xrightarrow{\alpha} {}_{92}U^{235} + {}_{2}He^4 \quad ...(10.12)$$

(ii) Beta (β) decay. It is commonly accompanied by the emission of neutrino (Δ) and γ radiation. An example of β decay is Eq. (10.13) and the following

$$_{82}Pu^{214} \xrightarrow{\beta} {}_{83}Bi^{214} + {}_{-1}e^0 + \Delta \quad ...(10.13)$$

The penetrating power of β particles is small compared to γ rays, however, it is larger than that of α particles.

(iii) Gamma (γ) radiation (decay). γ particles are electromagnetic radiation of extremely short wavelength and very high frequency resulting in high energy. γ rays originate from the nucleus while X rays from the atom. γ wavelengths are, on an average, about one-tenth those of x-rays, though energy ranges overlap somewhat. There is no alteration of atomic or mass number due to γ decay.

(iv) Positron decay. Positron decay is caused when the radioactive nucleus contains an excess of protons. An example of this is the decay of $_7Ni^{23}$ into $_6C^{13}$.

$$_7Ni^{13} \rightarrow {}_6C^{13} + {}_{+1}e^0 \quad ...(10.14)$$

Since the draught nucleus has one less proton than the parent, so one of the orbital electrons is released to maintain atom neutrality. As a result, there is combination of emitted and orbital electron resulting in γ radiation and known as *annihilation process*.

$$_{+1}e^0 + {}_{-1}e^0 \rightarrow \gamma \quad ...(10.15)$$

The energy emitted by γ radiation is equivalent to the sum of their rest masses, i.e., 1.02 MeV. The reverse of annihilation process is known as *pair production* (i.e., positron-electron pair), an endothermic process that converts energy into mass.

(v) K capture. K capture is a process which captures an orbital electron from the orbit or shell nearest to the nucleus which possesses an excess of protons but not the threshold of 1.02 MeV necessary to emit positron (Fig. 10.4). The vacancy in K shell in filled by another electron falling from the higher orbit resulting in x-ray emission from the atom. K capture also effectively changes a proton into neutron. The example of K capture is as follows

$$_{29}Cu^{64} + {}_{-1}e^0 \rightarrow {}_{28}Ni^{64} \quad ...(10.16)$$

Here, Ni^{64} is stable.

(vi) Neutron Emission. It occurs when a neutron possesses an extremely high excitation energy. If the excitation energy of nucleus is more than 8 MeV, it could decay by the emission of neutron. For example,

$$_{54}Xe^{137} \rightarrow {}_{54}Xe^{136} + {}_0n^1 \quad ...(10.17)$$

Here, the parent Xe^{137} is a fission product from the β decay of the fission fragment I^{137}, generally called precursor and daughter is an isotope of the parent. Generally, it happens rarely except nuclear reactor where it is the source of *delayed fission* neutrons. *In reactor, the control of this neutrons are of utmost importance.*

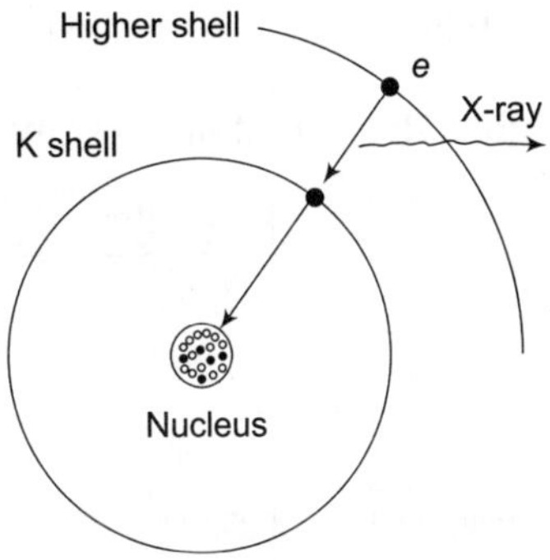

Figure 10.4 K. kapture phenomena

10.6 DECAY RATES AND HALF-LIVES

In the event of a very large number of radioactive nuclei of the same kind, there happens to be a definite statistical probability that a certain fraction will *decay* in a certain time. This decay rates increase exponentially with temperature and do not depend upon temperature, pressure or the physical and chemical state of matter.

If N is the number of radioactive nuclei of one species at any time θ, the rate of decay is directly proportional to N. Thus,

$$\frac{-dN}{d\theta} = \lambda N \qquad ...(10.18)$$

where, λ is a proportional factor called the *decay constant* and depends upon type of isotopes.

After integration of the above equation

$$\int_{N_0}^{N} \frac{dN}{N} = \lambda \int_{0}^{\theta} d\theta, \quad \text{where} \quad N = N_0 \text{ at } \theta = 0 \qquad ...(10.19)$$

Thus,

$$N = N_0\, e^{-\lambda\theta} \qquad ...(10.20)$$

The rate of decay ($-dN/d\theta$) is also called the **activity (*A*)** and commonly has dimension s^{-1},

In nuclear science, the common trend to represent decay rates is by using the concept of *half-life* (θ/2). This is used as the time during which one-half of a number of radioactive species decay or one-half of their activity cases. Thus,

$$N/N_0 = A/A_0 = 1/2 = e^{-\lambda\theta_{1/2}} \qquad ...(10.21)$$

and

$$\theta_{1/2} = \frac{\ln 2}{A} = \frac{0.6931}{A} \qquad ...(10.22)$$

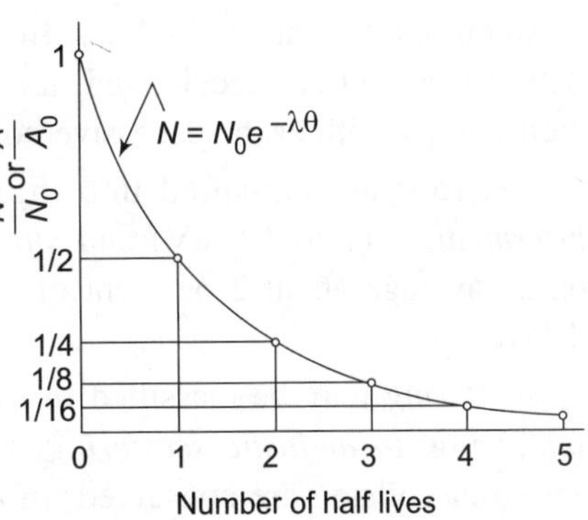

Figure 10.5 Radioactive decay rates vs. half-life

Equation (10.22) suggests that half-life is inversely proportional to the decay constant. Figure 10.5 shows the radioactive decay rates as a function of half-life.

Table 10.1 shows the half-life of some of the radio-isotopes.

Table 10.1 Half-life of some isotopes

Isotope	$\theta_{1/2}$	*Activity*	*Isotope*	$\theta_{1/2}$	*Activity*
Tritium (H^3)	12.26 yr	β	Thorium 232	1.41×10^{10} yr	α and γ
Carbon 14	5730 yr	β	Thorium 233	22.1 min	β
Krypton 87	76 min	β	Protactinium 233	27 days	β and γ
Strotium 90	28.1 yr	β	Uranium 233	1.65×10^5 yr	α and γ
Xenon 135	9.2 h	β and γ	Uranium 235	7.1×10^8 yr	α and γ
Barium 139	82.9 min	β and γ	Uranium 238	4.51×10^9 yr	α and γ
Radium 223	11.43 days	α and γ	Neputnium 239	2.35 days	β and γ
Radium 226	1600 yrs	x and γ	Plutonium 239	2.44×10^4 yr	α and γ

From Table 10.1, it is observed that the readily fissionable isotopes, U^{233}, U^{235} and Pu^{238} have extremely long half-lives, so they can be stored practically indefinitely. U^{235} is found in nature but U^{233} Pu^{239} are artificially produced from Th^{232} and U^{238} respectively.

10.7 NEUTRON ENERGIES AND SCATTERING

Since neutrons are essential in fission process, so it is essential to study neutron energies. The kinetic energy of a neutron KE_n or simply E_n is expressed as

$$E_n = m_n C^2/2 \text{ where } m_n \text{ is mass of neutron, } C \text{ is the speed of the neutron} \quad ...(10.23)$$

Since m_n = 1.008665 amu and if C is in cm/s, then

$$E_n = 1.008665\ C^2/(2 \times 0.965 \times 10^{18}) = 5.227 \times 10^{-19}\ C^2;\ \text{MeV}$$

As a result of fission reaction, the newly born fission neutrons have energies ranging between 0.075 to about 17 MeV. In the course of travelling through matter, neutrons collide with nuclei and are decelerated, mainly by the lighter nuclei resulting in losing of some of their energy with each successive collision. This process is called ***scattering***.

Neutrons are classified into three categories, namely, (i) *fast* (greater than 10^5 eV), *intermediate* (1 to 10^5 eV) and *slow* (less than 1 eV). Newly born fission neutrons carry on an average about 2 per cent of a reactor fission energy which may be either prompt or delayed.

Scattering may be classified in two types namely (i) *inelastic scattering* and (ii) *elastic scattering. In inelastic scattering* the momentum and total energy of the particles before and after collsion are conserved. In *elastic scattering* momentum and kinetic energy of the colliding particles are conserved. The number of collisions, n, required to slow down a neutron from initial energy E_{ni} to a final energy E_{nf} in elastic scattering is expressed by

$$n = \frac{\left[\dfrac{\ln E_{ni}}{E_{nf}}\right]}{\xi} \quad \text{where } \xi = \text{logarithmic energy decrement} \qquad ...(10.24)$$

For example, for U^{235} nucleus, the number of collision (n) between 2 MeV and 0.025 eV is about 4480.

10.8 THERMAL NEUTRONS

An effective scattering medium, called ***moderator*** is used to scatter or slow down the fission neutrons. The *moderator* has a small nuclei with high neutron-scattering and low neutron-absorption probabilities, such as *H* and *D* (in H_2O and D_2O), C (graphite) and Be or BeO. In the process of slowing down, the neutron achieve the lowest energy level that put them in *thermal equilibrium* with the molecules of the medium they are in. They become thermalized and are called *thermal* or *slow neutrons.*

The thermal neutrons possess a wide range of energies and corresponding speeds which are given by Maxwell distribution law. Most probable speed of thermal neutron is given by

$$C_m = \left(\frac{2kT}{m_n}\right)^{0.5} \qquad ...(10.25)$$

where k = Boltzmann constant = 8.617×10^{-11} MeV/K, T = absolute temperature (K)

m_n = mass of thermal neutron particle = 1.674×10^{-27} kg

The energy corresponding to the most probable speed is given by

$$E_m = \frac{1}{2} m_n C_m^2 = kT \qquad ...(10.26)$$

Equation (10.26) suggests that the energy of thermalized particle is independent of mass and a function only of absolute temperature of medium.

10.9 NUCLEAR CROSS SECTIONS

The concept of nuclear cross section arises when we are interested to know the beam intensity of neutrons. The actual cross-sectional area of a nucleus is obtained from its radius (r_c) as given by

$$r_c = r_0 A^{1/3} \qquad ...(10.27)$$

where r_0 is a constant and has average value of 1.4×10^{-23} cm and A is the mass number. The average cross-sectional area of nuclei is about 10^{-24} cm^2.

It is worth to note that the probability of neutrons colliding or interacting with nuclei is proportional to an effective, rather than actual cross-sectional area of the nuclei in question. This effective cross-sectional area is known as *microscopic cross section* or simply the *cross section* of the reaction and designated as σ. It depends upon nucleus, type of reaction and neutron energy.

In order to find the intensity of a neutron beam striking target area A (Fig. 10.6), consider a number of nuclei in volume $A\ \Delta x$ which is equal to $NA \cdot \Delta x$. If the beam intensities are I and $I\text{-}\Delta I$ at x and $x + \Delta x$ respectively, then in the limit.

$$\frac{-dI}{I} = \frac{\sigma(NAdx)}{A} = \sigma N dx$$

After integrating from I_0 (at $x = 0$) and I (at $x = x$), we have

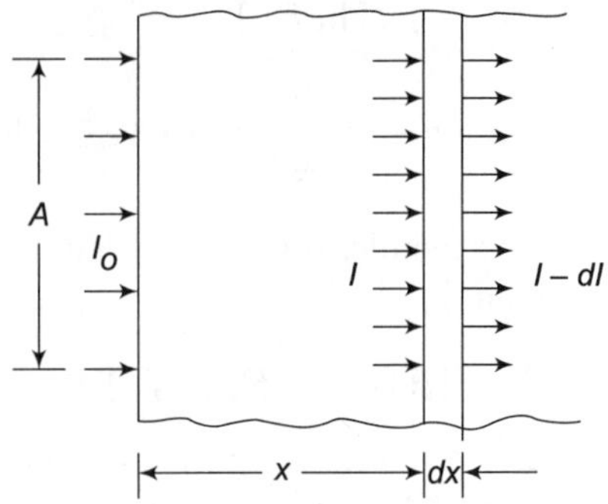

Figure 10.6 Neutron beam striking target area

$$-\int_{I_0}^{I} \frac{dI}{I} = \sigma N \int_{x=0}^{x} dx$$

or
$$I = I_0\, e^{-\sigma N x} \quad \text{...(10.28)}$$

Neutrons have as many cross sections as there are reactions. The total cross section, σ_t, is given by

$$\sigma_t = \sigma_s + \sigma_c + \sigma_f + \text{any other} \quad \text{...(10.29)}$$

where σ_s = microscopic cross section for scattering, σ_c = microscopic cross section for radiation capture and σ_f = microscopic cross section for fission.

The product $\sigma \cdot N$ is equal to the total cross sections of all the nuclei present in a unit volume and is called *macroscopic cross section* represented as $\Sigma = N\sigma$. Thus,

$$I = I_0\, e^{-\Sigma x} \quad \text{...(10.30)}$$

The value of N can be calculated for an element of atomic mass A_t, density ρ (g/cm^3)

$$N = \rho\, \frac{\text{Avogadro's number}}{A_t}\ ; \text{nuclei/cm}^3 \quad \text{...(10.31)}$$

10.10 NEUTRON FLUX AND REACTION RATES

If n is the neutron density (neutron/cm^3) and C is the neutron velocity (cm/s), the product nC is called the neutron flux (φ)

Thus,
$$\varphi = nC \quad \text{...(10.32)}$$

φ represents the number of neutrons crossing a unit area from all directions per unit. time. The reaction rate is expressed as

$$\text{Reaction rate} = n\, C\, N\, \sigma = \phi\Sigma\ ; \text{reaction/(s.cm}^3) \quad \text{...(10.33)}$$

where σ and Σ are the cross sections of the particular reaction in equation.

10.11 THE VARIATION OF NEUTRON CROSS SECTION WITH NEUTRON ENERGY

Figure 10.7 shows the variation of neutron cross section with neutron energy for U^{235}. The variations are represented in three regions, namely, (i) The $1/v$ region, (ii) the resonance region and (iii) the fast neutron region.

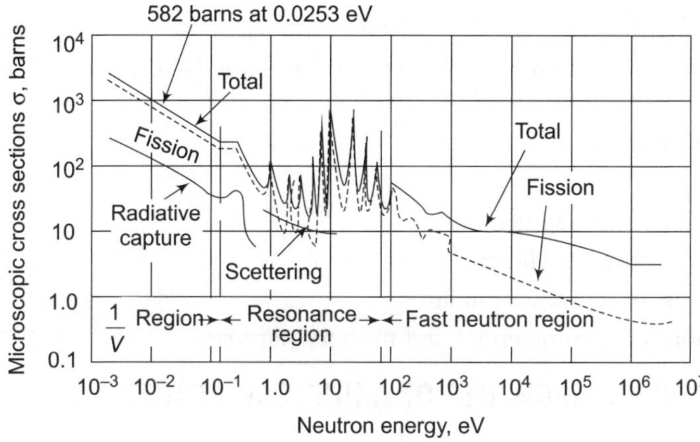

Figure 10.7 Neutron cross section for U^{-235}

(i) 1/v Region. It is the low energy range. The absorption cross section for many, but not all, nuclei are inversely proportioned to the square root of the neutron energy, E_n.

$$\sigma_a = \sigma_c + \sigma_f = C_1 \left(\frac{1}{E_n}\right)^{0.5} = \frac{C_2}{C} \quad ...(10.34)$$

where C_1 and C_2 are constant and C is the neutron velocity.

(ii) Resonance Region. In this region, most neutrons and absorbers exhibit one or more peaks occurring at definite neutron energies, called ***resonance peak*** **U^{235}** has many peaks. This affects the design of thermal reactor.

(iii) Fast-neutron Region. In the first-neutron region, the cross section usually undergoes a gradual decrease as neutron energies increase. The total cross section $\sigma_t = 0.125\ A^{2/3}$ barns (1 barn = 10^{-24} cm^2). In the very high neutron energy range, the total cross sections are rather low, usually less than 5 barns, each for the largest nuclei.

Table 10.2 gives the microscopic cross section of uranium fuels.

Table 10.2 Microscopic cross section for uranium fuels

Neutron	*Microscopic cross section, barns*	*Nucleus*		*Natural uranium*
		U^{235}	*U^{238}*	
Fast, 1 MeV	σ_s	5.30	6.6	6.6
	σ_c	0.093	0.14	0.19
	σ_f	1.2	0.018	0.026
Thermal, 0.0253 eV	σ_s	6.0	8.0	8.0
	σ_c	112	2.73	3.47
	σ_f	577	0	4.16

10.12 MAIN PRODUCTS OF THE REACTOR

For power production nuclear reaction is carried out in a *nuclear reactor*. The main products of the reactor are

(i) Thermal energy, i.e., heat
(ii) Neutrons
(iii) Radioactive fission fragments
(iv) Materials which are fissionable
(v) Gamma quantas and other radiation
(vi) Radio isotopes for commercial and military purposes.

10.13 PRINCIPAL COMPONENTS OF A NUCLEAR REACTOR

A nuclear reactor may be regarded as a substitute for the boiler fire box of steam power plant or combustion chamber of gas turbine plant. The heat produced in the nuclear reactor is by fission process whereas in steam and gas power plants, the heat is produced by combustion of fuel. The other cycle of operation and components required is the same either as steam plant if steam is generated by using the heat released due to fission or a gas turbine plant, if gas is heated by using the heat released due to fission. The steam or gas may be the working fluid in nuclear power plant.

Each reactor consists of the following principal parts as shown in Fig. 10.8.

1. Fuel Element. *The nuclear fuels which are generally used in reactor are* $_{92}U^{235}$, $_{94}Pu^{239}$ and $_{92}U^{233}$. Out of the three, the $_{92}U^{235}$ is only naturally available up to 0.7% in the uranium ore 239 and the remaining is $_{92}U^{238}$. The other two fuels, $_{94}Pu^{239}$ and $_{92}U^{233}$, are the byproducts and formed in the nuclear reactor during fissioning process from $_{92}U^{238}$ and $_{90}Th^{232}$ due to absorption of neutron without fission.

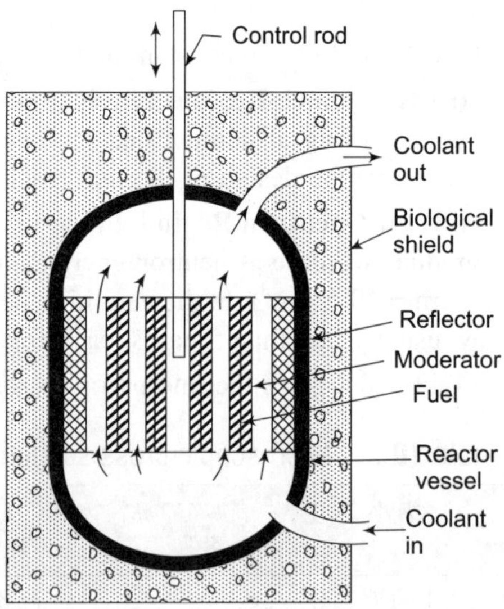

Figure 10.8 Principal parts of a nuclear reactor

The selection of the shape of the fuels and their locations in the reactors are made keeping in view of uniform heat production within the reactor. The fuel elements are designed taking into account the heat transfer, corrosion and structural strength. The cooling passage should be designed in such a way that there should not be any hot spot in the fuel element.

2. Moderator. It is a material used to slow down the neutrons from high kinetic energy (1 MeV or 13200 km/s) to slow neutron (0.25 eV or 2200 m/s) in a fraction of a second. Further, a moderator is to increase the probability of reaction and to maintain the chain reaction due to slow neutrons. The slowing down of the neutrons is effectively done by the light elements such as H_2, D_2, N_2, O_2, C and Be. The *graphite, heavy water* or *bergllium* can be used as a *moderator* with natural uranium. In one of the arrangements, the moderator (solid) surrounds the fuel.

3. Reflector. In order to keep the critical size of the reactor and hence the amount of fissionable material as small as possible, it is important to conserve neutrons. This is possible by surrounding the reactor core with a material which reflects escaping neutrons back into the core. This material is called *reflector*. The required properties of a good reflector are low absorption and high reflection for neutrons, high resistance to oxidation and irradiation as well as high radiation stability.

Many times, the material used as moderator is also used as reflector. The H_2O, D_2O and carbon are also used as reflector.

4. Coolant. The main purpose of the coolant in the reactor is to transfer the heat produced in the reactor and to keep the fuel assembly at a safe temperature to avoid their melting and destruction. The same heat carried by the coolant is used in the heat exchanger for further utilization in the power generation either generating steam or using hot gas.

The following are the requirements of a good coolant:

(i) It should not absorb the neutrons.

(ii) It should have high chemical and radiation stability.

(iii) It should be non-corrosive.

(iv) It should have high boiling point (if liquid) and low melting point (if solid).

(v) It should be non-oxidizing and non-toxic.

(vi) It should have high density, low viscosity, high conductivity, and high specific heat.

Water, heavy water, gas (He, CO_2), a metal in liquid form (Na) and organic liquids are used as coolant.

5. Control Rods. The control system controls the rate of energy generated. It starts, increases, decreases and stops the reaction. These rods may be shaped like the fuel rods themselves and are interspread throughout the core. Instead of containing fuel they contain *neutron absorber* such as boron, cadmium or indium. They have high neutron-absorption cross sections.

6. Biological Shield. The intensity of radiation fragments from the reactor core is too high for the human body to tolerate. Therefore, it is necessary to surround the reactor with shielding material to prevent damage to human body due to radiation.

7. Reactor Vessel. The reactor vessel encloses the reactor core, reflector and shield. It also provides coolant inlet and outlet passages. The reactor vessel has to withstand the pressure at 200 bars or above. The reactor core (fuel and moderator assembly) is generally placed at the bottom of the vessel.

In order to obtain *critically or steady power,* three methods are used (i) *building a heterogeneous reactor,* (ii) *enriching the fuel* or *both*. A heterogeneous reactor using natural uranium or low-enriched fuel, is one in which the fuel must be divided into separate fuels elements in the form of pins, rods, plates, hollow cylinders, pellets, spheres, etc. These are placed in the cores, with space between them filled with moderator (Fig. 10.8). The second method of attaining criticality is to enrich the fuel by artificially increasing the percentage of U^{235} in it. Ordinary water, often incorrectly called *light water*, may be used as moderator, resulting in thermal *heterogeneous reactor*.

10.14 CLASSIFICATION OF NUCLEAR REACTOR

Nuclear reactors are classified on the basis of the following:

(A) Coolant. On the basis of coolant used, the reactors are classified as:

(i) Gas cooled reactors, such as air, CO_2 or helium.

(ii Liquid cooled reactors such as boiling water, pressurized water, liquid metal cooled reactors.

The boiling water cooled reactor serves directly as a steam boiler. In the pressurized water cooled reactor, water is under pressure in excess of saturation pressure at the maximum water temperature in the system, so that it cannot boil in the reactor.

(B) Principal Product. On the basis of principal products, the reactors are classified as:

(i) Research reactors to produce neutrons.

(ii) Power reactor to produce heat.

(iii) Breeder reactor to produce fissionable materials.

(iv) Production reactor to produce isotopes.

(C) Neutron Energies. On the basis of neutron energies, reactors are classified as:

(i) Fast-fission reactors.

(ii) Intermediate or epithermal reactors.

(iii) Thermal (low energy) reactors.

(D) Fuel Moderator Assembly. On the basis of fuel moderator assembly, reactors are classified as:

(i) Homogeneous reactors, in which the fuel in the form of liquid such as uronyl sulphate dissolved in water.

(ii) Heterogeneous reactor in which the fuel is in the form of solid (bar, rod, etc.) surrounded by moderator. This is the most common reactor.

(E) Fuel State. On the basis of fuel state, reactors are classified as:

(i) Solid fuel reactors.

(ii) Liquid fuel reactors.

(iii) Gas fuel reactors.

(F) Fuel Materials. On the basis of fuel materials, reactors are classified as:

(i) Natural uranium with U^{235} reactors.

(ii) Enriched uranium with more than 0.71 of U^{235} reactors.

(iii) Pu^{239}, Pu^{241} or Pu^{239} (artificial) reactors.

(iv) U^{233} (artificial) reactors.

(v) Fertile materials are: U^{232} and Pu^{238}.

(G) Construction of Core. On the basis of construction of core, reactors are classified as (i) spherical, (ii) cylindrical, (iii) octagonal, (iv) cuboid, (v) slab, (vi) annulus reactor.

(H) Moderator. On the basis of moderator, reactors are classified as:

(i) Water, (ii) heavy water, (iii) graphite, (iv) beryllium (v) hydrocarbon reactors.

Table 10.3. gives the most common fission power reactor types in use commercially or under serious development world over.

Table 10.3 The most common fission-power reactor types

Neutron energy	*Reactor type*	*Coolant*	*Moderator*	*Fuel enrichment*
Thermal	(i) Pressurized water reactor (PWR)	H_2O	H_2O	UO_2, low
	(ii) Boiling water reactor (BWR)	H_2O	H_2O	UO_2, low
	(iii) Gas cooled reactor (GCR)	CO_2, He	Graphite	UC, low
	(iv) Heavy water reactor (HWR)	D_2O	D_2O	UO_2, natural
Fast	(i) Liquid-metal fast breader reactor (LMFBR)	Na	None	$Pu^{239}O_2 + U^{238}O_2$
	(ii) Gas cooled fast reactor (GCFR)	He	None	$Pu^{239}O_2 + U^{238}O_2$

Table 10.4 gives the characteristics of typical power reactors.

Table 10.4 Characteristics of typical power reactors

Parameters	*PWR*	*BWR*	*LMFBR*	*GCFR*
Electric power (MW_e)	1300	1050	1000	330
Thermal power (MW_{th})	3800	3000	2750	842
Specific power (kW_{th}/kg)	33	26	575	50
Power density (kW_{th}/m^3)	100	60	300	10
Core height (m)	4.25	3.75	1.5	50
Core diameter (m)	3.50	4.90	3.25	59
Coolant	H_2O	H_2O	NaI (liquid)	He
Pressure (bar)	155	72	8.0	48
Inlet temperature (°C)	280	275	330	400
Outlet temperature (°C)	310	285	500	770
Coolant flow rate (Mg/s)	20	12	11	0.45
Average linear heat rate	22.5	20	30	–

10.15 PRESSURIZED WATER REACTOR (PWR) POWER PLANT

PWR reactor is currently most extensively used type of power reactor in the world mainly due to excellent properties of water as a moderator and coolant. The fuel used is alloy of uranium or UO_2.

The most stringent limitation on a PWR is the critical temperature of water, 374°C. This is the maximum possible exit temperature of coolant in the reactor but in practice, its value taken is around 310°C or less. In order to avoid the boiling of coolant in the reactor, the coolant pressure must be far greater than the saturation pressure corresponding to 310°C, i.e., 98.7 bars. In practice, the pressure of the coolant is maintained at about 155 bars and by this way the chance of boiling of coolant is prevented in the reactor.

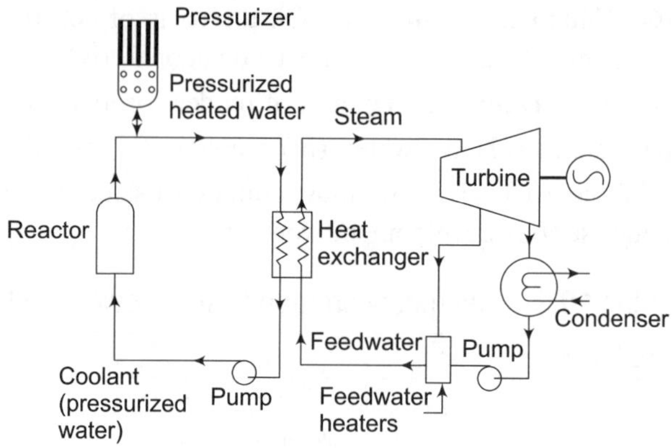

Figure 10.9 A schematic of PWR power plant

Figure 10.9 shows a PWR power plant. It consists of two loop in series, the *coolant loop*, called the *primary loop*, and the *water-steam* or *working fluid loop*. The coolant picks up the heat released due to fission reaction in reactor and transfers it to the working fluid in the steam generator. The steam so generated is then used in a Rankine type cycle to generate electricity.

10.15.1 Construction Details of a Reactor

Construction details of reactor are shown in Fig. 10.10. The reactor vessel is in the form of a cylinder closed at each end by a hemispherical head. The upper head is bolted to the barrel, so that, for an access to the anterior of reactor, it can be removed. The size of the reactor vessel may be 3 m in external diameter, 4.5 m to 11 m in height and from 18 cm to 23 cm thick. It is lined with stainless steel. The fuel assembly which contains the fuel elements are mounted in a cage in the centre of reactor vessel. The cage consists of a core barrel of cylindrical shape closed at the top and bottom by horizontal core support plates.

The length of fuel assembly is about 1.8 m to 2.8 m. It is mounted rigidly in vertical position between the core support plates and can be removed from the reactor vessel. Commonly, the fuel may be alloy of uranium which may be enriched by more than normal 0.7% of U-235 that is found in natural uranium or pillets of UO_2. The fuel is encased in a stainless steel tubes that are seal-welded at the ends to produce a corrosion-resistant gas tight cladding.

Baffle plates are used to keep these stainless steel tubes in accurate parallel alignment. The baffles plates, themselves are mounted in a housing of rectangular cross section that is constructed of some corrosion resistant alloy such as zircalloy 2. As the energy is released in the nuclear fuel within the small tubes, they are so spaced that cooling water flows around them in sufficient quantities so that they may be cooled properly and prevent any local boiling of water. Sometimes, fuel elements consist of flat bars of uranium alloy which are clad in stainless steel.

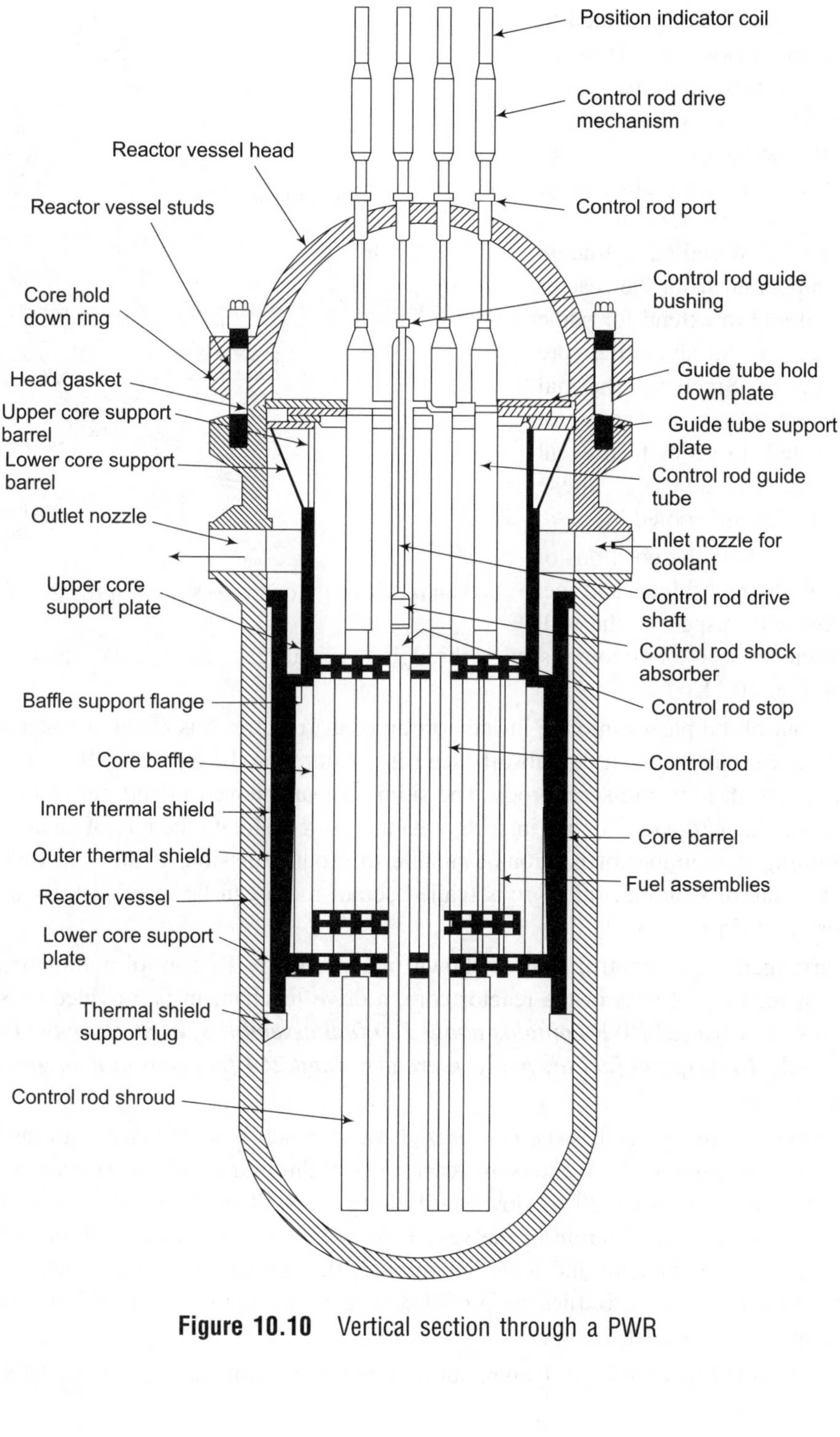

Figure 10.10 Vertical section through a PWR

Thermal shields are provided to absorb gamma rays and neutrons that are not reflected back into the core. Thus, the thermal shields protect the walls of the pressure vessel from excessive radioactivity. They are made of steel plates in the form of thin walled cylinders and are supported from the reactor vessel wall and to extend for some distance below and above the core support plates. Since the thermal shield absorbs gamma and neutrons, it gets heated. In order to prevent them from overheating, these thermal shields are cooled.

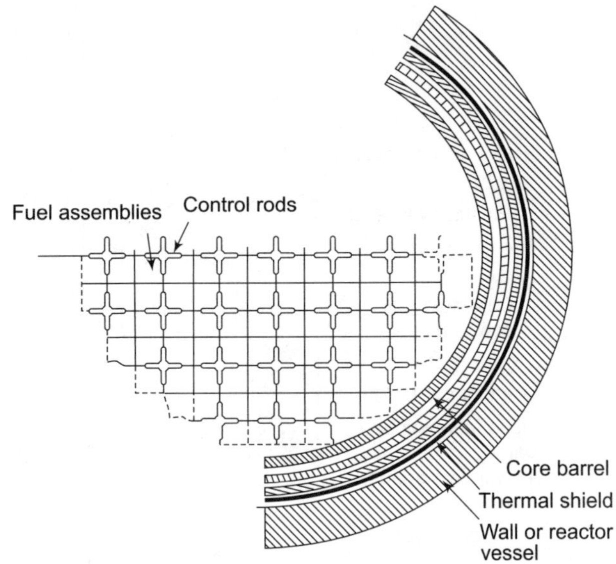

Figure 10.11 Horizontal section through a PWR

In order to show the positions of core barrel, thermal shields and fuel assemblies with respect to the wall of the reactor vessel, a horizontal section through a typical reactor may serve the purpose as shown in Fig. 10.11.

Since control rod passes into the inside portion of the reactor, it is made of materials such as boron or cadmium which can absorb neutrons. Control rods are generally classified as regulating or skin rods and scram rods. The skim rods are moved in and out of the reactor core under close automatic control in such a manner as to regulate the rate of energy release by controlling the number of neutrons available to produce fission. If the skin rods fail to control the rate of reaction, safety rods (called scram rods) can be inserted fully in a very short period of time.

The arrangement of control rods is shown in Fig. 10.11. To control automatically the position of the control rods in the reactor core, a drive mechanism is provided outside the reactor vessel. ***A typical PWR contains about 200 fuel assemblies, each assembly being an array of rods. In a typical fuel assembly, there are about 264 fuel rods and 24 guide tubes for control rods.***

The coolant in this reactor is the pressurized water which is maintained with the help of *pressurizer* (dealt separately). It enters the reactor vessel through an inlet nozzle located below the flanged head. The water flows downward along the wall of the reactor vessel through the thermal shields to the bottom of the vessel. After this, it flows upward through the fuel elements in the reactor core and goes out through the discharge nozzle which is located opposite to the inlet vessel. Baffles are provided so as to prevent the coolant flowing directly from the inlet to the outlet nozzle.

It is to be noted that the basic features of all types of reactors are similar to PWR.

10.15.2 Working Fluid Loop

The coolant leaving the reactor enters the steam generator which may be either shell and tube type with U-tube bundles or once through type, the former arrangement being more common. In the case of U-tube steam generator, the hot coolant enters at inlet channel head at the bottom, flows through the U-tube and reverses direction to an outlet at the bottom.

Since the coolant temperature is less, so it can generate only saturated steam at moderate pressure (i.e., around 40 bar). In the once-through arrangement, the primary coolant enters at top, flows downward through tubes and exits at the bottom of main pumps while feed water flows upward (counter-flow) and gets converted in water in the condenser and is again recirculated in the steam generator.

The first PWR power plant was built at shipping port USA in 1959 which generates 231 MW thermal output. The coolant pressure in the primary circuit is 14.1 bar and the coolant (water) outlet temperature is 282°C which generates dry and saturated steam in heat exchanger (i.e., steam generator) at 41 bar; 252°C. It develops 68 MW of electrical output at an thermal efficiency of 29.4%. The plant efficiency may be improved by using an oil-fired superheater.

10.15.3 Pressurizer

As mentioned above, the coolant (water) in the PWR primary loop is maintained at a pressure (about 155 bars) much greater than the saturation pressure corresponding to the maximum coolant temperature (about 310°C) in the reactor to prevent bulk boiling. Since liquids are incompressible, small changes in volume due to changes in coolant temperature caused either due to load variation or sudden nuclear activity insertions results in severe or oscillatory pressure changes which are not at all desirable. In the event of increase of coolant pressure, some water will flash into steam which will affect the reactor performance and may lead to possible burn-out of fuel element. In the event of decrease of coolant pressure, there may be a chance of cavitation. This suggests that a surge chamber often known as *pressurizer* may be incorporated in the primary loop that will accommodate the coolant volume changes while maintaining pressure within permissible limits. There are two

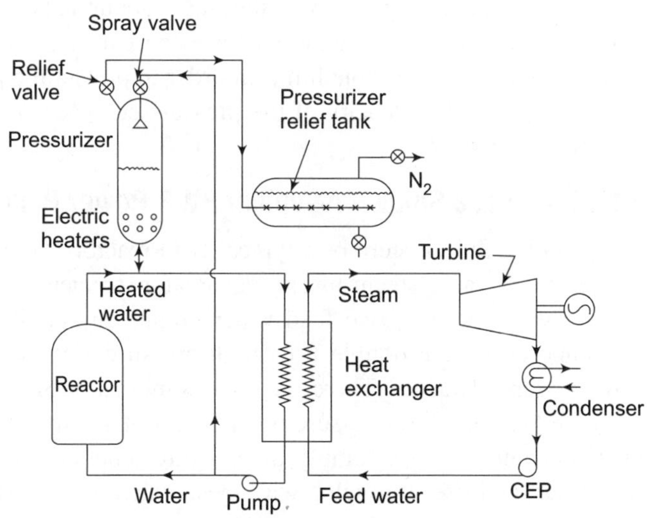

Figure 10.12 Flow diagram of a PWR primary loop with a vapour type pressurizer system

types of pressurizers in common use, namely, vapour pressurizers and gas pressurizer. PWR conveniently uses vapour pressurizers because their coolant, water, can be vapourized. Gas type pressurizers are commonly used in liquid metal cooled fast-breeder reactor.

A *vapour pressurizer* (Fig. 10.12) is essentially a small boiler. In this case, liquid, the same as the primary coolant, is maintained by controlled electrical heating at a constant temperature and consequently a constant vapour pressure above its full surface. It is to be noted that this vapour pressure has to be the same as that of the primary coolant at the junction between the pressurizer and the hot leg of the primary loop. Thus, the pressurizer temperature is higher than the primary coolant temperature because the latter is subcooled. The heaters are of the electric immersion type, located in the lower section of the pressurizer vessel. The bottom of the pressurizer is connected to the hot leg of the primary coolant system. A spray nozzle is located at the top of pressurizer which is connected via control valves to the cold leg of the primary coolant system after the pump. Under normal full power operation, the pressurizer is about half full of water and the top half is full of vapour.

In the case of positive surge, the volume of the primary coolant increases and the vapour in the top half is compressed. Entry of the cooler primary coolant into the pressurizer condenses some of the vapour thus limiting the pressure rise. Further, the spray valves are power-actuated and a cool spray (under pump pressure) enters the top, which helps condenser vapour at a rapid rate and limits the pressure rise. In order to prevent the excessive cooling of the spray piping and to maintain equal boron concentration in the primary coolant and pressurizer water, a small continuous spray is provided. Boron is used for chemical skim.

In the case of negative surge, the primary coolant volume gets decreased and the vapour in the pressurizer expands resulting in a momentary reduction in pressurizer pressure. The liquid in the pressurizer then partially flashes into vapour and assisted by further steam generation as a result of the automatic actuation in the electric heater. By this way, the pressure is maintained above a minimum allowable limit. In order to protect the pressurizer against pressure surges, that are beyond the capacity of the pressurizer, a power operated relief valve is attached to the top of pressurizer.

10.15.4 Case Study—A Typical PWR Power Plant

The working fluid system of a typical twin-reactor 100 MW (thermal) power plant consists of steam generator, steam turbine, generator, condenser, two condenser pumps, five stages of feed water heaters, two feed water pumps and auxiliary equipment. The steam turbine is composed of one double flow high-pressure element in tandem with two double flow, low pressure elements. There are four combination moisture separator-reheater assemblies between high and low pressure units. The wet steam from the exhaust of the high pressure elements enters each assembly of moisture-separator at one end, is distributed by internal manifolds, and rises through a wire mesh where the moisture is removed. After this, it flows over the tubes in the reheater where it is reheated by high pressure steam from the steam generators and after giving heat it gets condensed and is fed to the high pressure feed water heater. The reheated steam flows back to the low pressure turbine for further expansion. The generator coupled with steam turbine generates electricity.

10.15.5 Advantages

The following are the advantages of PWR:

(i) The water which is used as *coolant, moderator and reflector* is cheap in first cost and available in plenty.
(ii) Small number of control rods are needed.
(iii) The secondary circuit may be optimized for higher efficiency.
(iv) Fuel consumption is less.
(v) Safe and stable operation is possible.
(vi) The reactor responds quickly for more power demand.
(vii) The reactor is compact due to the use of enriched feed.

10.15.6 Disadvantages

The following are the disadvantages of PWR:

(i) High capital cost due to high primary circuit pressure.
(ii) The plant efficiency is low.
(iii) Commissioning problems are more severe.
(iv) Fuel suffers radiation damage.

10.16 BOILING WATER REACTOR (BWR) POWER PLANT

In BWR system, steam is produced directly in the reactor core and thus it directly serves the purpose of steam generator. Here, the coolant (water) is in direct contact with the heat-producing nuclear fuel and boils in the same compartment in which fuel is located. It boils because the reactor pressure is maintained at about 70 bars *less than half that in a PWR*, with the nuclear fuel temperature roughly comparable. Since water and vapour coexists in the core, the saturated steam is produced at about 283°C. ***The coolant in this case serves the triple functions of coolant, moderator and working fluid.***

Figure 10.13 shows a simple form of BWR power plant. It consists of a reactor, a steam turbine, a generator, a condenser, and associated equipment (air ejector, cooling system, etc.), and a feed pump.

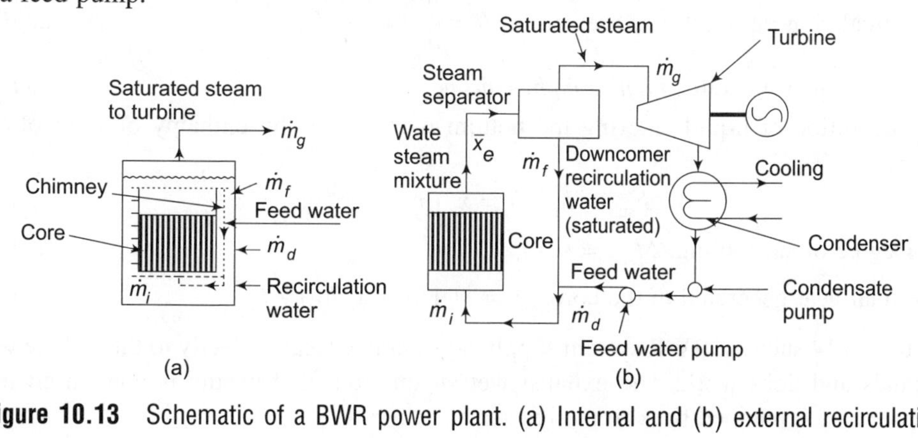

Figure 10.13 Schematic of a BWR power plant. (a) Internal and (b) external recirculation

The saturated liquid that separates from the vapour at the top of reactor or in a steam separator flows downward either *internally* within the reactor or *externally* outside the reactor via downcomer and mixes with the return condensate (Fig. 10.13). This recirculating coolant flows either naturally, caused by density differential between the liquid in the downcomer and the two-phase mixtures in the core or by the recirculating pump in the downcomer (not shown). ***The internal, forced recirculating type is most common in modern BWR.***

The performance of BWR is affected by *recirculating ratio* which is defined as the ratio of recirculating liquid to the saturated vapour produced. It is a function of the core average exit quality. Since boiling water exit qualities are low between 10 and 14 per cent, so recirculation ratio lies in the range of 6 to 10. This value of recirculation ratio is necessary to avoid large void fractions in the core which would materially lower the moderating powers of the coolant and result in low value of overall heat transfer coefficient or vapour blanketing and burnout of the fuel element.

10.16.1 Thermodynamic Analysis

For the analysis of a BWR power plant, let us assume that a slightly subcooled liquid enters the core bottom at a rate of $\dot{m}_i$ mass per unit. The liquid rises through the core and chimney if any. Here, the chimney is an unheated section above the core which enhances natural recirculation. The resulting vapour separates and proceeds to the power plant at a mass flow rate of $\dot{m}_g$. Let saturated recirculation liquid flow via the downcomer at a mass flow rate of $\dot{m}_f$. Here, the recirculated liquid mixes with the relatively cold return feed water ($\dot{m}_d$) from the power plant to form the slightly subcooled inlet liquid, $\dot{m}_i$ (Fig. 10.13).

The mass balance in the reactor core yields

$$\dot{m}_d = \dot{m}_g \quad ...(10.35)$$

and

$$\dot{m}_g + \dot{m}_f = \dot{m}_i \quad ...(10.36)$$

The average exit quality of the entire core ($\bar{x}_e$) is expressed as

$$\bar{x}_e = \frac{\dot{m}_g}{\dot{m}_g + \dot{m}_f} = \frac{\dot{m}_d}{\dot{m}_d + \dot{m}_f} = \frac{\dot{m}_d}{\dot{m}_i} \quad ...(10.37)$$

The rirculation ratio (R) is expressed as, $R = \dfrac{\dot{m}_f}{\dot{m}_g} = \dfrac{1 - \bar{x}_e}{\bar{x}_e}$ ...(10.38)

The heat balance results: $\dot{m}_i\, h_i = \dot{m}_f\, h_f + \dot{m}_d\, h_d$ ...(10.39)

The condition of liquid entering the bottom is expressed by enthalpy of subcooling as given by

$$\Delta h_{\text{sub}} = h_f - h_i = \bar{x}_e\,(h_f - h_d) \quad \text{or} \quad x_e = \frac{h_f - h_i}{h_f - h_d} \quad ...(10.40)$$

or Degree of subcooling, $\Delta t_{\text{sub}} = t_f - t_i$ ...(10.41)

The total heat generated in the core, $Q_1 = \dot{m}_i\left[(h_f + \bar{x}_e\, h_{fg}) - h_i\right]$ ...(10.42)

Figure 10.14 shows a BWR system supplying saturated steam directly to the turbine where it expands and does work. The exhaust wet steam from the turbine is condensed in the condenser and is pumped back to the reactor.

The advantages of the direct cycle are simplicity and it relatively low capital cost. Because of the direct loop arrangement, there are no heat exchangers. There is no need of any pressurizer, as there is large steam volume above water. A major drawback of the direct cycle is that the reactor is not load following.

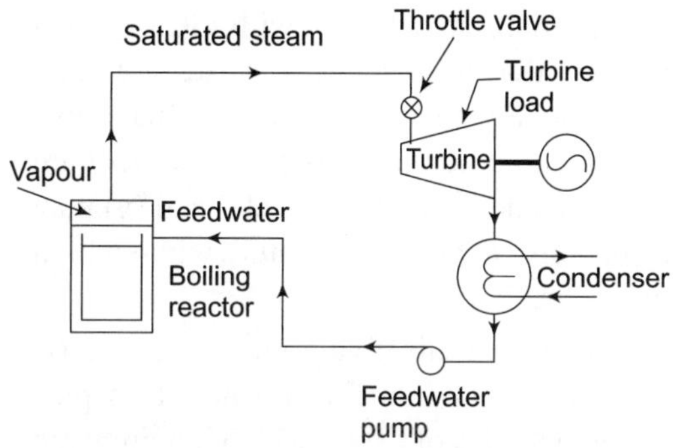

Figure 10.14 A direct cycle BWR power plant

10.16.2 The Current BWR System

The current BWR designs are of the direct cycle, forced internal-recirculation-type. The reactor is about 7 m ID and 24 m high for a 100 MW plant, the fuel rods are similar to those of a PWR. They are Zircalloy-clad, about 4 m high in active length and contain enriched UO_2 pellets. The fuel assemblies are either 7 × 7 or 8 × 8 arrays and are enclosed in a Zircalloy-4 channel. The control rods are cruciform in shape and occupy the space between four channels.

10.16.3. Case-Study—A Typical BWR Power Plant

A typical BWR power plant produces 780 MW from 0.4 per cent moisture steam at 66.7 bars at turbine inlet condensing to 0.067 bar. The turbine is a tandem compounded and double flow (with one HP section and two LP sections), four flows to the condenser and non-reheat. The cycle has a moisture separator between HP and LP turbines are five feed heaters. The throttle steam flow in 1210 kg/s. The thermal efficiency is 33.18 per cent.

10.16.4 Advantages

(i) The pressure vessel is less costly due to smaller pressure in the reactor vessel.
(ii) The reactor does not require steam generator, pressurizer, circulating pump and connecting pipes, thus further reduces the cost.
(iii) The metal surface temperature in the reactor is lower than PWR.
(iv) The reactor is more stable than PWR.
(v) The plant efficiency is higher than BWR.

10.16.5 Disadvantages

(i) It cannot meet sudden demand in load.
(ii) The steam leaving the reactor is slightly radioactive.
(iii) The power density is nearly 50% (33.6 kW/litre) of PWR, therefore the size of the vessel is large.
(iv) The possibility of burn-out of fuel is more.

10.17 GAS COOLED REACTOR (GCR) POWER PLANT

GCR power plants have received great attention in UK and to a lesser extent in Germany, France, USSR and USA. This type of reactor uses natural or enriched uranium fuel, CO_2 or helium as coolant and graphite or heavy water as a moderator.

The main advantages with GCR lies in the fact that in general, (i) gases are safe, (ii) relatively easy to handle, (iii) have low macroscopic neutron cross section, (iv) are plentiful and cheap (except He) and (v) may be operated at high temperature without high pressurization.

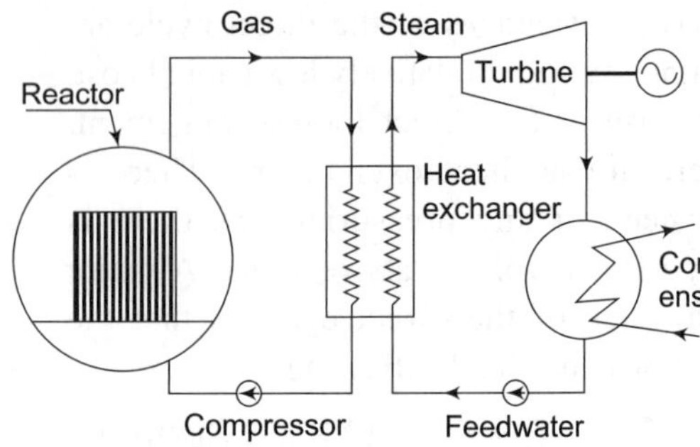

Figure 10.15 A GCR power plant

The major disadvantages include: (i) the lower heat transfer and heat-transport chara-cteristics of gases, which require large contact surface and flow passage within the reactor and heat exchangers, and (iii) their high pumping requirements (between 8 to 20 per cent of plant gross power), which offers challenging problems of fluid flow, pressure drop, etc.

Figure 10.15 shows a GCR power plant. The gas coolant absorbs the heat released due to nuclear reaction in the reactor and this heat is utilised to generate steam in a heat exchanger (i.e., generator) which expands in the steam turbine, get condensed in the condenser and fed back to steam generator.

10.17.1 Case Study–A Typical U.K. Advanced GCR Power Plant

An example of AGCR example is Hinkley Point B nuclear power station. It has twin reactors with a total output of 1250 MW. The reactor uses enriched ceramic fuels clad in stainless steel, graphite as moderator and CO_2 as coolant. The core is a 16 sided stack of graphite blocks, of which 308 form fuel channels and 81 are the control rods. The core is surrounded by an annular graphite reflector. The CO_2 coolant enters at bottom at 310°C and 42.6 bars and leaves at top at 655°C and 40.4 bars. It then flows down 12 superheater-reheat boilers surrounding the core, within the cavity of a prestressed concrete reactor vessel leaving at 276°C and 40.2 bars. The coolant is then pumped back to the reactor by eight constant speed, electrically driven (28 MW) centrifugal circulators with variable guide vane flow control. Then steam is generated at 166.8 bars and 541°C. A plant thermal efficiency of 41.7 per cent is obtained.

10.17.2 Case Study–A Typical-Temperature Gas-Cooled Reactor (HTGR)

In USA, PWR and BWR power plants constitute the major share among nuclear plants. However, HTGR has been conceptualized and installed (Fig 10.16). In this concept, graphite is used for fuel particles coating, fuel structural material, moderator and coolant channel walls, and fuel as U–235. In a 1160 MWe plant, the coolant is helium. The reactor (3000 MW thermal) is housed in a vertical prestressed concrete reactor vessel. It is 29.4 m in a diameter, 27.6 m high, and included the reactor core rods and their drives, and facilities. Helium is circulated at an average pressure of about 48 bars. It enters the reactor, core at the top at about 240°C, flows downward through the coolant holes in the graphite block and exits at 740°C. The steam generators are shell and tube heat exchangers in which helium flows on the shell side. Superheated steam from the six steam generators, a total about 1000

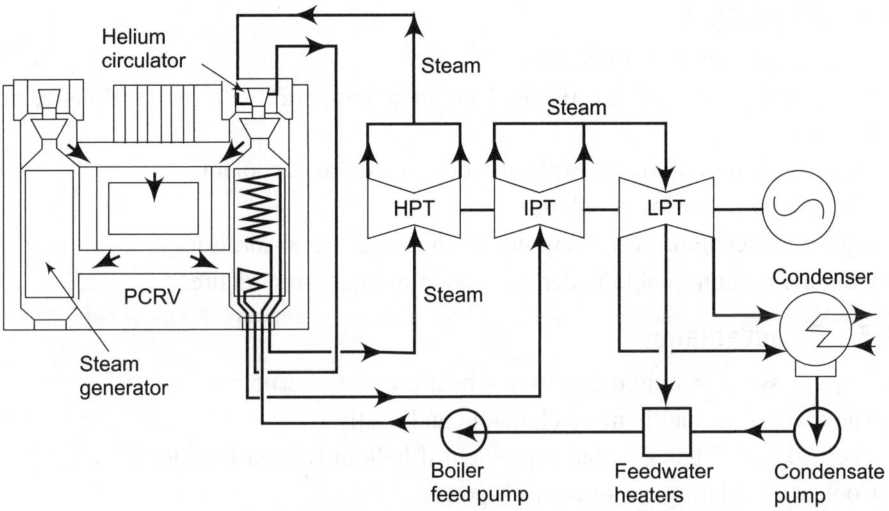

Figure 10.16 HTGR power plant diagram

kg/s, combines into a single heater and enters the double flow HP turbine at 166 bars and 510°C. The exhaust steam from HP turbine is reheated and it leaves the reheater at 38 bars and 538°C, recombines and enters the double flow HP turbine and then to double flow LP turbine. The exhaust steam gets condensed and sent back to the boiler.

10.17.3 HTGR Direct Cycle-Gas Turbine Plant

This type of plant uses gas turbine for power production and coolant as helium gas. It consists of a regenerator and multi-stage compression with intercooling (Fig. 10.17). The fuel used may be U-233 or Th–232.

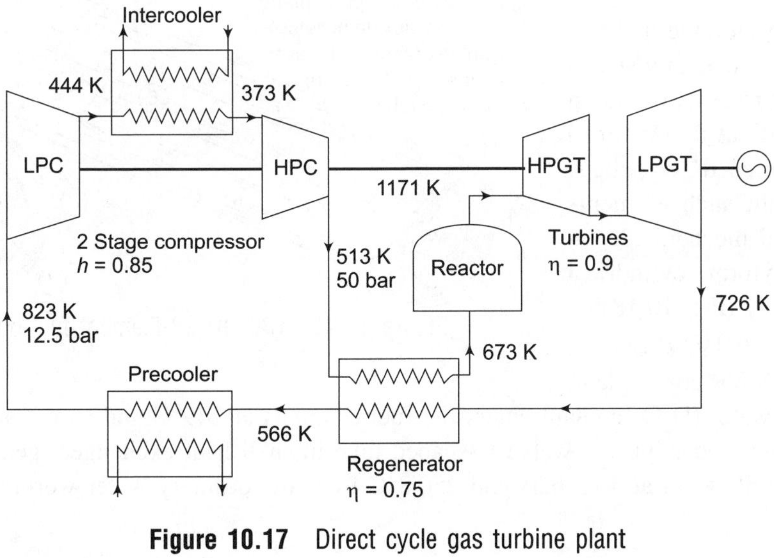

Figure 10.17 Direct cycle gas turbine plant

10.17.4 Advantages

(i) It needs simpler fuel processing.
(ii) The problem of limiting the fuel element temperature is not serious as in other reactors.
(iii) There is no possibility of explosion due to CO_2 as a coolant.
(iv) There is no corrosion problem.
(v) It gives better return economy due to low parasitic absorption.
(vi) Graphite remains stable under irradiation at high temperature.

10.17.5 Disadvantages

(i) Large vessel is required due to low heat transfer coefficient.
(ii) The loading of fuel is more elaborate and costly.
(iii) The leakage of gas is a major problem if helium is used instead of CO_2.
(iv) Coolant circulating requires much power.
(v) The control is more complicated.

10.18 HEAVY WATER REACTOR (HWR) POWER PLANT

The property of heavy water as very low neutron capturing cross section makes it an excellent moderator and coolant and permits the use of natural uranium (0.7% U-235) as fuel. The *advantages* of HWR include good neutron economy resulting in good breeding ratios, high fuel burn-up, and reduction in fuel cost. The *disadvantages* include the limitations imposed by critical temperature like water.

Due to economic incentive, Canada has developed this reactor for its maximum use as CANDU reactors.

In this reactor, the fuel elements are hermetically sealed Zircalloy-2 rods that contain compacted and sintered naturally enriched UO_2 pellets. Twenty-eight such elements are attached mechanically at their ends to form a cylindrical fuel assembly (Fig. 10.18).

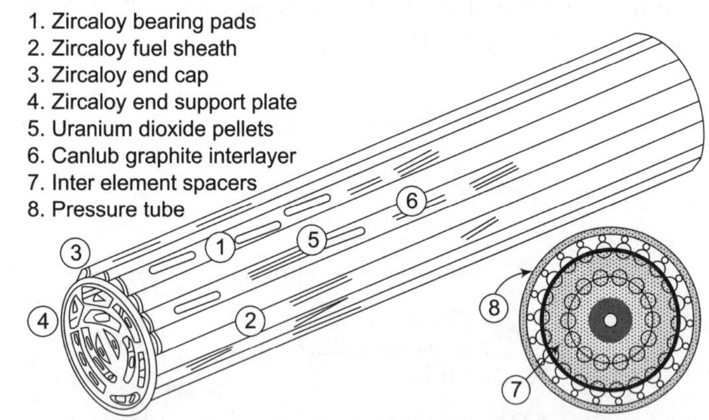

Figure 10.18 CANDU 28-Element fuel bundle

Figure 10.19 shows a diagram of HWR power plant. The heavy water (D_2O) coolant enters the fuel channels at 249°C and leaves at 293°C and 90.7 bars via feeder pipes. Twelve U-shaped tube in-shell heat exchangers generate a total of 814 kg/s of steam at 40.3 bars and 250.8°C from the ordinary water working fluid. The steam expands in the turbine, does work, gets condensed in the condenser and returns back to the steam generator.

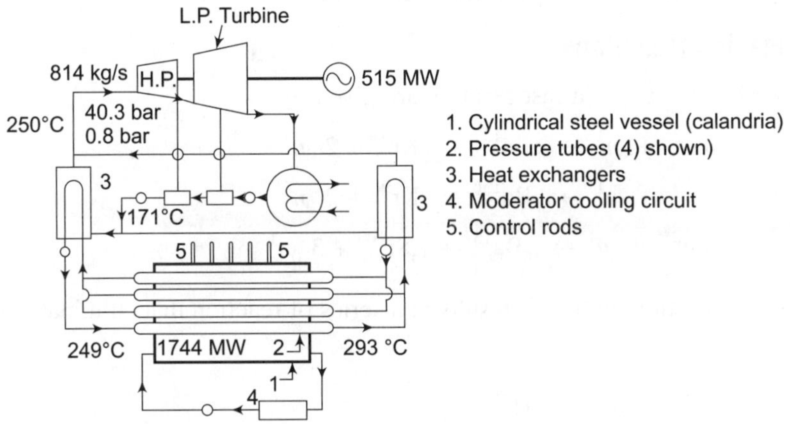

Figure 10.19 A schematic of HWR power plant, CANDU Reactor

10.18.1 Advantages

The following are the advantages of HWR plant:

(i) The fuel need not be enriched.

(ii) The reactor vessel may be built to withstand low pressure as compared to PWR and BWR.

(iii) Control is very easy and so less costly.

(iv) Good return economy is possible.

(v) The reactor has high multiplication factor and low fuel consumption as heavy water is a good moderator.

(vi) High-fuel burn-up is achieved.

(vii) A shorter period is required for plant construction.

10.18.2 Disadvantages

The following are the disadvantages:

(i) The cost of heavy water is high.

(ii) The leakage is a major problem.

(iii) Standard design, manufacture, inspection and maintenance are required.

(iv) The power density is considerably low (9.7 kW/litre) as compared to PWR and BWR.

10.19 LIQUID METAL FAST BREEDER REACTOR (LMFBR) POWER PLANT

Fast reactors are those whose neutrons do not slow down to thermal energies by a moderator. They are designed to create or breed new fissile material while producing useful electric power.

In fast-breeder reactor, the objective is to maintain a chain reaction with fast neutrons that have an average energy of about 1 MeV by fission in U^{235} and Pu^{239}. It also must provide additional fast neutrons sufficient to convert U^{238} to Pu^{239}.

10.19.1 Fission Reactions

The typical fission reactions in fast reactors are given by

$$_{92}U^{235} + {_0n^1} \rightarrow {_{56}Ba^{137}} + {_{36}Kr^{97}} + 2\ {_0n^1} \quad ...(10.43)$$

$$_{92}U^{233} + {_0n^1} \rightarrow {_{56}Ba^{136}} + {_{36}Kr^{96}} + 2\ {_0n^1} \quad ...(10.44)$$

$$_{92}Pu^{235} + {_0n^1} \rightarrow {_{56}Ba^{137}} + {_{36}Sr^{100}} + 3\ {_0n^1} \quad ...(10.45)$$

A fast neutron reaction with U^{238} results in a series of reaction that culminate in Pu^{239} and is given by

$$_{92}U^{238} + {_0n^1} \rightarrow {_{92}U^{239}} + \gamma \quad ...(10.46)$$

$$_{92}U^{239} \xrightarrow[\beta]{24\ min} {_{93}Np^{239}} + {_{-1}e^0} \quad ...(10.47)$$

$$_{92}Np^{239} \xrightarrow[\beta]{2\text{-}3\ days} {_{94}Pu^{239}} + {_{-1}e^0} \quad ...(10.48)$$

This is a *breeding* reaction that converts fertile U^{238} into fissionable Pu^{239}.

In the analysis of fast breeder reactor, the two following important fuel parameters are used

Δ = the average number of neutrons emitted per fission and depends on fissile isotopes.

η = the fission factor, equal to the average number of neutrons emitted per neutron absorbed. It is strong function of neutron energy.

The conversion or breeding ratio is given by $C = \eta - 1 - L$ ...(10.49)

where L represents the losses.

Based on the value of η and L, C can occupy three values which may be used to define a reactor.

If C is much less than unity, the reactor is called a *burner*. A reactor with low C is generally called a *converter*. One with high C but less than 1.0 is called an *advanced* converter.

Breeding gain is defined as the gain in fissionable nuclei per fissionable nucleus consumed.

Thus, $$G = C - 1 = \eta - 2 - L \quad ...(10.50)$$

Thus, $$G_{max} = C_{max} - 1 = \eta - 2 \quad ...(10.51)$$

On economic importance to a breeder reactor is its *doubling time*. It is defined as the time required to produce as many new fissionable nuclei as the total number of fissionable nuclei that are both normally contained in the core and tied up in the reactor fuel cycle. In general, doubling time ranges between 10 to 20 years, the shorter the better.

10.19.2 Working Principle

Figure 10.20 shows the schematic of a LMFBR power plant. Sodium and other liquid metals are used as a coolant but they suffer from high induced radioactive and are generally chemically active. This necessitates an intermediate cooling loop to be used between the primary radioactive coolant and the steam cycle. The intermediate coolant is usually also a liquid metal, often Na or NaK. This intermediate loop guards against reactions between the radioactive primary coolant and water. It also ensures against its high pressure water or hydrogen entering the reactor.

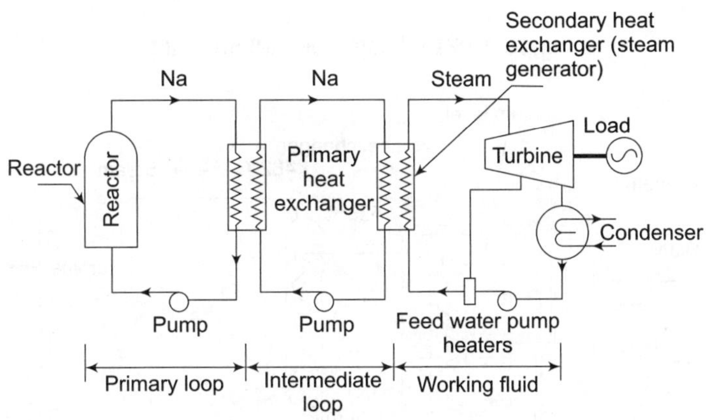

Figure 10.20 Schematic of a LMFBR power plant

There are two types of primary loop concept : (i) the loop or pipe type, (ii) the pool, tank or pot type.

The loop type is the most common in USA and is shown in Fig. 10.21. In this case, the reactor vessel, heat exchangers, liquid metal pumps, and other components of the primary system and their interconnecting piping are separated within a large building containing an inert atmosphere to preclude sodium fires in the case of sodium leak.

The *pool type* is more common in UK and is shown in Fig. 10.22. In this case, the entire primary system, including reactor, primary heat exchangers, and pumps, is submerged in a large tank filled with molten sodium. The tank is part of the primary coolant loops. The heat exchangers discharge coolant directly into the tank and the pumps receive coolant directly from it.

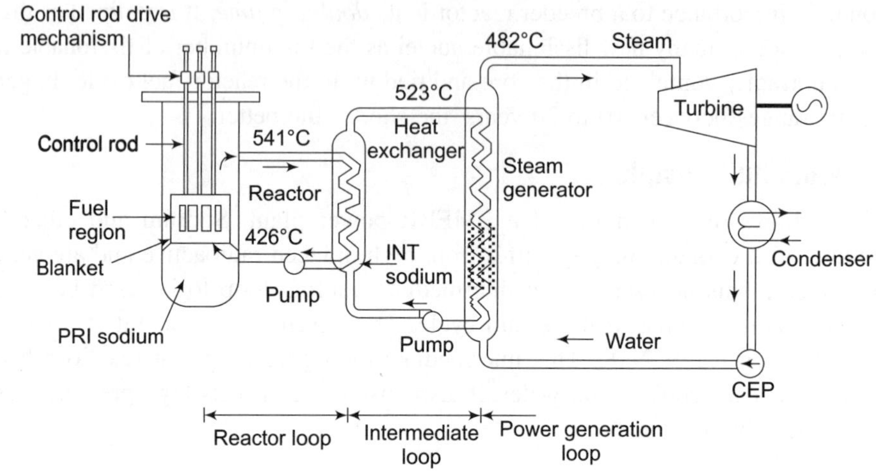

Figure 10.21 Loop type LMFBR plant

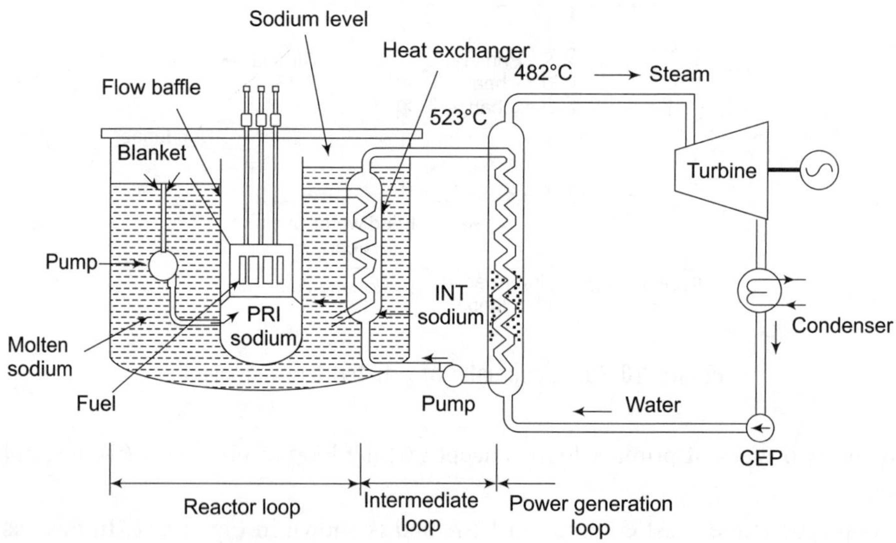

Figure 10.22 Pool type LMFBR power plant

It has been observed that the pool system has the edge in safety and economy while the *loop system* has the edge in the simple mechanical design.

10.19.3 Advantages

The following are the advantage of LMFBR

(i) High plant efficiency at low cost power is possible due to high temperature.

(ii) The sodium as a coolant need not be pressurized.

(iii) It is best suited to thermal reactor with slightly enriched fuel as the neutron absorption cross section of sodium is low.
(iv) The low cost graphite moderator can be used as it can retain its mechanical strength and purity at high temperature.
(v) The reactor is more stable due to liquid metal cooled.
(vi) The reactor size is comparatively small.

10.19.4 Disadvantages

The following are the disadvantages of LMFBR.

(i) The neutron economy is reduced.
(ii) It is necessary to shield the primary and secondary cooling systems.
(iii) The leak of sodium is very dangerous.

10.20 FUSION REACTOR POWER PLANT

As mentioned above the fusion reactor power plant is not yet commercialized due to very high temperature requirement resulting in plasma to initiate fusion process and to sustain it.

10.20.1 Fusion Reactions

Deuterium, a stable heavy isotope of hydrogen, present in natural water, is the main fuel for a fusion reactor. The following four reactions involving deuterium are given below:

Fusion Reaction	Energy per reaction	
(i) ${}_1H^2 + {}_1H^2 \rightarrow {}_2He^3 + {}_0n^1$	3.2 MeV	...(10.52)
$(D + D \rightarrow He^3 + n)$		
(ii) ${}_1H^2 + {}_1H^2 \rightarrow {}_1He^3 + {}_1p^1$	4.0 MeV	(10.53)
$(D + D \rightarrow T + p)$		
(iii) ${}_1H^2 + {}_1H^3 \rightarrow {}_2He^4 + {}_0n^1$	17.6 MeV	(10.54)
$(D + T \rightarrow {}_2He^4 + n)$		
(iv) ${}_1H^2 + {}_2He^3 \rightarrow {}_2He^4 + {}_1p^1$	18.3 MeV	(10.55)
$(D + He^3 \rightarrow He^4 + p)$		

The symbols n, p, D and T stand for neutron, proton, deuterium (${}_1H^2$) and tritium (${}_1H^3$) respectively. Third fusion reaction occurs most easily between deuterium (D) and tritium (T) which is self-sustaining at temperature of 50×10^6 K releasing 17.6 MeV per reactions. The fourth fusion reactions requires a temperature equivalent to 10^9 K.

10.20.2 Working of a Conceptual Fusion Reactor

A futuristic deuterium-tritium fusion reactor is shown in Fig. 10.23. The plasma is contained inside an evacuated tube of about 4 m. The surrounding vacuum wall through which 14 MeV neutrons from the plasma pass, is maintained at 750°C. There happens to be two concentric regions outside this wall, namely, the lithium breeding moderator and the magnetic shield.

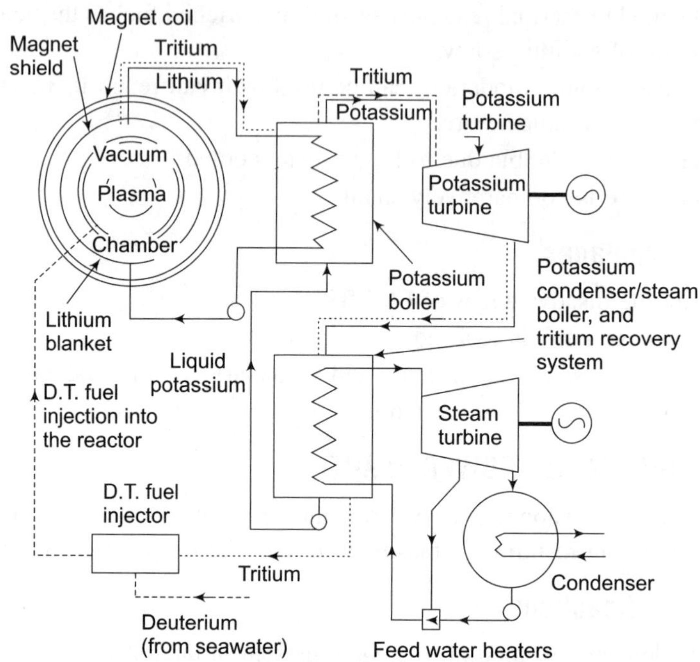

Figure 10.23 A conceptual design of fusion power plant

Tritium is manufactured in the lithium blanket. The magnetic shield is maintained by large cryogenic superconducting magnets of 7 to 8 m in diameter. The binary vapour cycle is used in which potassium acts as topping cycle and steam as bottoming cycle. There is a recovery system for tritium.

The following are the *advantages* of fusion power plant:

(i) The fuel deuterium needed is almost inexhaustible.

(ii) There is no production of radioactive waste.

(iii) It is very safe to operate.

(iv) The conversion efficiency is high as 60 per cent.

(v) The rejection of heat per kW generated is low.

If the large fusion power plant becomes successful, the energy crisis of the world will be over and a new area of high quality of standard of living of mankind will be achieved. Mankind will be waiting for that day.

10.21 INDIAN NUCLEAR POWER PLANTS

Among the various types of nuclear power plants available, India has selected pressurized heavy water reactor (PHWR) due to its several advantages. A PHWR plant uses natural uranium as fuel with heavy water as moderator and coolant. Natural uranium is available in India. Owing to its excellent neutron economy, a PHWR plant yields a greater amount of plutonium. Indian nuclear programme consists of three stages, namely:

First stage. Building of PHWR plant to produce electricity and plutonium fuel with 96% unused uranium (U-238).

Second stage. With the help of fast breeder reactor with plutonium as fuel, the electricity is generated and more plutonium from U-238 will be kept as blanket. It will also produce U-233 fuel from thorium, used as a blanket material.

Third stage. U-233 will be used as fuel and thorium as blanket.

Table 10.5 shows the existing and future nuclear power plant in India.

Table 10.5 Existing and future nuclear power plant in India

S.No.	*Units*	*Capacity (MWe)*	*Cumulative capacity (MWe)*	*Year of commissioning*
	Existing			
(i)	Tarapur (BWR) 1 & 2	2 × 210	420	1969
(ii)	Rajasthan 1 & 2	2 × 210	860	1973, 1981
(iii)	Chennai 1 & 2	2 × 235	1330	1983, 1985
(iv)	Narora 1 & 2	2 × 235	1800	1990, 1991
(v)	Kakrapur 1 & 2	2 × 235	2270	1991, 1992
(vi)	Kaiga 1 & 2	2 × 235	2740	1995, 1996
(vii)	Rajasthan 3 & 4	2 × 235	3210	1995, 1996
(viii)	Kaiga,3, 4, 5, & 6	4 × 235	940	1996, 1997
(xi)	Tarapur 3 & 4	2 × 500	1000	1997, 1998
(x)	Rajasthan 5, 6, 7, & 8	4 × 500	2000	1998, 2000
	Working and Planned			
(xi)	Kundankulum (PWR) 1 & 2	2 × 1000	2000	2016
(xii)	Kundankulam planned	2 × 1000	2000	2022

Still the nuclear power plants are very less in India and constitutes about 3 per cent of total capacity power plant generated by thermal, hydel and other.

10.22 ADVANTAGES AND DISADVANTAGES OF NUCLEAR POWER PLANT

10.22.1 Advantages

(i) Less space is required as compared to other conventional plants of same size.

(ii) It consumes small quantity of fuel, hence transportation cost is less and it needs small space to store.

(iii) Now, there is an increased reliability of operation.

(iv) The plant is not affected by *adverse weather conditions*.

(v) Plants are well suited to meet large power demands and give better performance of higher load factors.

(vi) The material expenditure on metal structures, piping and storage mechanism is less as compared to coal based steam power plant.
(vii) The size of the area requirement for the plant is about 1/3 or coal based plant of the same size.
(viii) The plant does not require large quantity of water.
(ix) The plant does not pollute the environment by not emitting CO_2, NO_x, CO, etc.

10.22.2 Disadvantages

(i) The initial cost of a nuclear power plant is higher as compared to hydro or steam power plant.
(ii) The plants are well suited for varying load conditions.
(iii) There is a possibility of radiation hazard.
(iv) Radioactive water if not disposed of carefully may produce radiation hazard.
(v) Handling of radioactive wastes and storing it for a long time pose a great problem and it is very costly affair.
(vi) Maintenance cost is high.
(vii) The plant needs trained personnel.

Example. 10.1 In a nuclear reactor, each fission of U-235 yields 190 MeV of useful energy. Assume that 85% of neutron absorbed by U-235 cause fission, the rest is absorbed by non-fission captures to produce an isotopes of U-236. Calculate the fuel consumption of U-235 per day to produce 5000 MW of thermal power.

Solution Energy produced by each fission = 190 MeV × 1.6 × 10^{-13} J/MeV

= 304 × 10^{-11} J of useful energy

Number of fission required to obtain energy in W.s.

$$= \frac{1}{3.04 \times 10^{-11}} = 3.3 \times 10^{10} \text{ ; W.s.}$$

In one day (i.e., 86,400 s) operation the number of nuclei burnt per MW of thermal power is given by

$$= \frac{(10^6 \text{ W})\ (3.3 \times 10^{10} \text{ fission/W.s})\ (86{,}400 \text{ s/day})}{0.85 \text{ fission/absorption}} = 3.35 \times 10^{21} \text{ absorption/day}$$

Mass of U-235 consumed to produce 1 MW power

$$= \frac{(3.35 \times 10^{21} \text{ day}^{-1})\ (235 \text{ g/g mole})}{6.023 \times 10^{23} \text{ (nuclei/g mole)}} = 1.3 \text{ g/day}$$

Thus, the fuel consumption to produce 5000 MW of thermal power per day

= 1.3 × 5000 g = 6.5 kg/day ***Ans.***

Example 10.2 Determine the mass defect and binding energy per nucleon of oxygen for the following masses of proton and electron.

$$m_p = 1.007277 \text{ amu}, m_n = 1.008665 \text{ amu} \quad \text{and} \quad m_e = 0.00055 \text{ amu}$$

atomic mass of oxygen = 16 = 15.99491 amu

Solution Molecule of O_2 has 8 proton, 8 electrons and 8 neutrons. The mass defect,

$$\Delta m = m_n n_n + m_p n_p + m_e n_e - {}_2m^A$$

$$= 1.00866 \times 8 + 1.007277 \times 8 + 0.00055 \times 8 - 15.99491 = 0.13701 \text{ amu}$$

The binding energy = $\Delta m \times 931 = 0.13701 \times 931 = 127.6$ MeV

$\therefore$ The binding energy per nucleon = $\dfrac{127.6}{16} = 7.97$ MeV ***Ans.***

Example 10.3 A certain nucleus has a cross section of 20 barns for 2200 m/s neutrons. Calculate the cross section if the kinetic energy of the neutrons increases to 0.1 to eV. The energies of two neutrons are within 1/V range of the nucleus.

Solution At 2200 m/s, $E_n = 0.02525$ eV

For 1/V region of energy

$$\sigma_a = C_1 \left(\frac{1}{E_n}\right)^{0.5}$$

$$\therefore \quad \frac{\sigma_{a1}}{\sigma_{a2}} = \left(\frac{E_{n1}}{E_{n2}}\right)^{0.5} = \left(\frac{0.1}{0.02525}\right)^{0.5} = 1.98$$

$$\therefore \quad \sigma_{a2} = \frac{\sigma_{a1}}{1.98} = \frac{20}{1.98} = 10.1010 \text{ barns.}$$ ***Ans.***

PROBLEMS

Things to Think About

10.1 On what scientific law, nuclear power plant has been based?

10.2 Why there is a need of absorbing some neutrons in a reactor?

10.3 What is the reason that nuclear fusion has not been commercialized?

Short and Long Theoretical Questions

10.1 Describe the structure of an atom.

10.2 What do you mean by amu?

10.3 What do you mean by isotope?

10.4 List the name of fuels which are used in nuclear power plants.

10.5 What do you mean by mass defect and binding energy?

10.6 What do you mean by radioactive decays and radioisotopes?

10.7 What do you mean by half-life?

10.8 Differentiate between fusion and fission processes.

10.9 How are neutrons are classified according to neutron energy?

10.10 What do you mean by scattering? What is logarithmic energy decrement?

10.11 With the help of a simple sketch describe the main elements of a nuclear reactor.

10.12 Discuss the functions of moderator and reflector.

10.13 With the help of a sketch describe the working principle of:
(a) PWR power plant, (b) BWR power plant, (c) GCR power plant, (d) HWR power plant, (e) LMFBR power plant, and (f) GCFR power plant

10.14 Describe the advantages and disadvantages of (a) PWR, (b) BWR, (c) GCR, (d) HWR, (e) LMFBR, and (f) GCFR power plants.

10.15 Explain the terms (a) breeding ratio, (b) burner, (c) converter, (d) breeder, and (e) doubling time.

10.16 Explain the working of CANDU type reactor.

10.17 Describe the futuristic fusion power plant.

Numerical Questions

10.1 A nuclear power plant is operated continually for one year developing 1500 MW. The reactor contained 200 tonnes of 3% enriched uranium dioxide fuel. If the plant efficiency is 34 per cent, determine (a) the mass of U-235 consumed in kg, and (b) the fuel burn-up in MWd/tonne.

10.2 Determine the power produced in MeV/cm^3 and kW/m^2 for a 3.5% enriched uranium dioxide fuel element in a thermal reactor of the effective fusion cross section is 400 burns and neutron flux per cm^3 s is 10^4. The density of UO_2 is 10.5 g/cm^3.

10.3 A reactor is fuelled with 200 tonnes of natural uranium (atomic mass 238.05) in which the average thermal neutron (220 m/s) flux is 10^{13} neutron/cm^2.s. The 2200 m/s cross section of U-235 (atomic mass is 235.04 are $\sigma_f = 579$ barns and $\sigma_c = 101$ barns. The energy release per fission in 200 MeV and 0.715% of natural uranium is U-235. Find (a) the rating of the reactor in MW/tonne, and (b) the rate of consumption of U-235 per day.

Software Based Question

10.1 Develop a software for the design of all types of nuclear power plants.

Design and Open Ended Question

10.1 Prepare a comprehensive report of different types of nuclear plant installed worldwide with special features which may be helpful for the selection.

Multiple-Choice Questions

10.1 Isotopes are those atoms which have

(a) the same number of protons
(b) similar chemical and physical characteristics
(c) different in their masses
(d) all of them

10.2 The natural uranium is composed of
(a) 69.282 mass per cent U^{238} (b) 79.282 mass per cent U^{238}
(c) 89.282 mass per cent U^{238} (d) 99.282 mass per cent U^{238}

10.3 The fission reaction with U^{235} is as follows:
(a) ${}_{92}U^{235} + {}_{\sigma}n' \rightarrow {}_{54}Xe^{140} + {}_{38}Sr^{94} + {}_{0}n'$ (b) $\rightarrow {}_{54}Xe^{140} + {}_{38}Sr^{94} + 2\ {}_{0}n'$
(c) $\rightarrow {}_{50}Xe^{100} + {}_{25}Sr^{50} + {}_{0}n'$ (d) $\rightarrow {}_{50}Xe^{100} + {}_{25}Sr^{50} + 2\ {}_{0}n'$

10.4 Which one is true with moderator of nuclear reactor?
(a) it slows down neutron energy (b) it increases the probability of reaction
(c) it needs to maintain chain reaction (d) all of them

10.5 Which one is true with heavy water reactor?
(a) the fuel is to be enriched (b) the fuel need not be enriched
(c) both case (a) and (b) (d) use (b) and no leakage problem.

Answers to Multiple-Choice Questions

10.1 (d) 10.2 (d) 10.3 (b) 10.4 (d) 10.5 (d)

CHAPTER

11

Hydroelectric Power Plant

OVERVIEW

The hydropower plays an important role in the development of a country as it provides power at cheaper rate being perpetual and renewable source of energy. *In hydroelectric power plant, the head energy of water (by erecting dam over the river or naturally available from fall) is utilised to drive the water turbine, which in turn, drives the generator to produce electricity.* It is a well known fact that the rain falling upon the earth's surface has potential energy relative to oceans towards which it flows. This head energy can be converted into shaft work by passing through hydraulic turbine (prime mover) and ultimately into electrical energy. *In hydraulic turbine, the rate of change of moment of momentum of water takes place which produces shaft work.* The shaft power (P) developed by hydraulic turbine is given by

$$P = g \rho Q H \; ; \quad W \quad \text{(ideal)} \qquad \text{...(11.1)}$$

where g is acceleration due to gravity (9.8 m/s^2), ρ is the water density (10^3 kg/m^3), Q is the discharge rate (m^3/s) and H is the head available of water (m).

If t is the operating time in hours (8760 h/year) and η is the overall hydropower plant efficiency, the electricity generated (E) in kWh is expressed as

$$E = 9.81 \, Q \, H \, \eta \, t/1000 \quad ; \quad \text{kWh} \qquad \text{...(11.2)}$$

Nearly, 20% of the total power generated in the world is met by hydroelectric power plants and so it comes next to thermal power plants. However, in some countries like Norway and Switzerland, the almost total electricity demand is met by hydropower.

In India, the first hydropower station was installed in 1987 near Darjeeling. After independence, large hydropower stations have been commissioned like Damodar Valley Corporation (DVC), Bhakra Nangal, Hirakud, Nagarjunsagar, Rihand, Koyna, Metture, Narmada, etc. Some more hydroelectric power plants will be coming in the 10th five year plan and beyond.

It is worth to note that most of the water resource projects are multi-purpose projects such as irrigation, navigation, flood control and generation of electric power.

11.1 ADVANTAGES AND DISADVANTAGES OF HYDROELECTRIC POWER PLANTS

11.1.1 Advantages

The following are the advantages of hydroelectric power plants.

(i) **No fuel requirement.** No fuel is required to generate electricity. Water source is perennially available.

(ii) **Low running cost.** The electricity generated per kWh is very cheap as compared to thermal or nuclear power.

(iii) **No problem of disposal of ash.** Since no fossil fuel is used, so there is no problem of disposal of ash.

(iv) **Pollution free electricity generation.** The electricity generated by hydropower does not produce any type of pollution.

(v) **Easily switched on and off in a short period.** The hydraulic turbine is switched on and off in a very short period unlike thermal or nuclear plants. The steam turbine of thermal or nuclear plant is put on turning gears for about 48 hours during start-up and shut-down.

(vi) **Simple in concept, self-contained and reliable in operation.** The design concept of hydraulic power plant is simple in concept and the operation is self-contained and reliable as compared to thermal or nuclear power plants.

(vii) **Greater life expectancy.** Modern hydraulic power plants have greater life expectancy (about 50 years) as compared to thermal or nuclear plant (about 30 years).

(viii) **Act as ideal spinning reserve.** Owing to greater ease of taking up and throwing off the load, the hydropower can be used as ideal spinning reserve in a system of mix of thermal, hydro and nuclear power stations.

(ix) **Higher plant efficiency.** The plant efficiency is high (about 85 to 94%) over a considerable range of load.

(x) **Ancillary benefits.** Hydroelectric power plants provide ancillary benefits like irrigation, flood control, afforestation, navigation and aqua-culture.

(xi) **Less skilled workers.** Owing to simple design and operation, hydroelectric power plants do not require highly skilled workers. Even the manpower requirements is much less as compared to thermal or nuclear power plants.

(xii) **Quick response to the change of load.** The outstanding features of hydroelectric plant is quick response to the change of load compared with thermal or nuclear. The rapid fluctuating loads are served most economically.

11.1.2 Disadvantages

The following are the disadvantages of hydroelectric power plants.

(i) **High capital cost.** Hydroelectric power plants are capital intensive with a low rate of return. The interest rate of this capital cost is a large part of annual cost of hydraulic power station.

(ii) Power dependent on quantity of water available. Power generated by the hydro plants is only dependent on the quantity of water available, which in turn, depends on the natural phenomenon of rain. The dry year is more serious for the hydroelectric project.

(iii) Site selection dependent on water availability. The site is selected on the criterion of water availability at economical load. Such sites are usually away from the load centres. Long transmission lines are needed to transmit power from station to consumers.

(iv) Long erection time. The completion of hydroelectric power plants takes a much longer period (about 10 years) as compared to thermal power plants (about 3 years).

(v) Disturbed ecology of the area. The large hydro plants disturb the ecology of the area by way of deforestation, destroying vegetation and uprooting people. There is a strong public opinion against large hydro plants. Now, emphasis is on small, mini- and micro-level hydro plants.

Presently, the government of India is planning to interconnect all the rivers of India. Since natural grid is now a reality so if rivers are interconnected it will not only help in power production but in irrigation and flood control also.

11.2 OPTIMIZATION OF HYDRO-THERMAL MIX LOAD SHARING

Hydroelectric power plants suffer from seasonal variation as power output is directly proportional to the availability of water. In order to meet the variable load demand, large amount of water has to be stored covering a large catchment area. When water flow is less, hydropower cannot meet the maximum load. Further, if the maximum capacity of the plant is based on the minimum flow of water in a particular season, then it will be highly uneconomical.

In order to overcome the above problems, hydroelectric plants are run in conjunction with thermal power in an interconnected system to achieve the minimum cost of electricity generation. In the pursuit of optimization of hydro-thermal mix load sharing, it has been estimated that 30% of hydro and 70% of thermal or 35% hydro and 65% of thermal may be used as sharing of the load.

Load sharing by hydropower is the maximum when the available flow of water is the maximum (say during monsoon season). As long as the sufficient water is available in the reservoir, the hydropower takes up the base load and the peak load is shared by thermal power plants. (Fig. 11.1). If the water availability is less in reservoir, hydro takes the peak load while thermal takes the base load. Optimisation of hydro-thermal mix load sharing leads to the following advantages:

(i) The necessary reserve capacity is reduced.

(ii) Construction programmes may be diversified.

(iii) Higher utilization factors for hydro plants are achieved.

(iv) Higher capacity factor for thermal plants may be achieved.

(v) Cost of electricity generation is reduced.

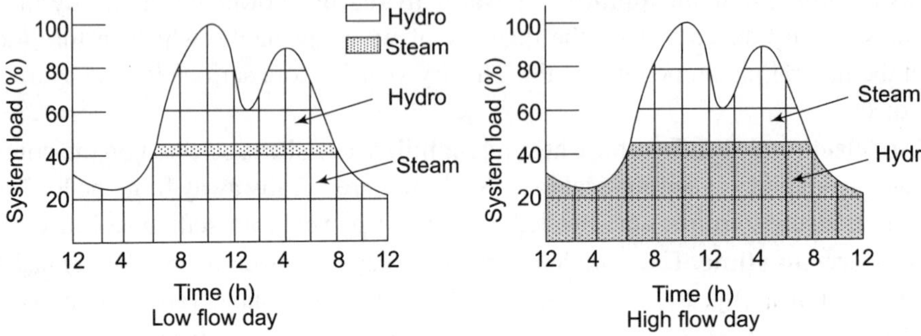

Figure 11.1 A typical load sharing by a hydro-steam mix system

11.3 SITE SELECTION FOR A HYDROELECTRIC PLANT

The site selection for a hydroelectric plant is not an easy job. It involves hydrological, topographical and geological investigations. The following factors should be considered for selecting the site.

1) Availability of water, 2) water storage capacity, 3) available water head, 4) accessibility of site, 5) distance from the load centre, 6) type of the land on site, 7) environmental aspects, and 8) sociological aspects.

1. *Availability of Water.* The availability of water plays an important role in deciding the design and capacity of the plant. The maximum and minimum rainfall in one year on the site and consequently the quantity of water available in a year decides the following aspects:

(a) Capacity of the plant, (b) size of the peak load plant such steam, diesel or gas turbine plant, and (c) adequate spillways or gate relief during floods.

2. *Water Storage Capacity.* Water storage capacity is a function of rainfall at round the year on the site and the size of reservoir including catchment area. The storage capacity is estimated with mass curve. This water storage capacity decides the minimum and maximum flow rate of water entering to the hydro turbine.

3. *Availability of Water Head.* The power output is a function of head available and quantity of water flow rate to turbine. If the water head is more, less quantity of water flow rate is needed to generate the same power. Further, the water head available depends upon the topography to the site.

4. *Accessibility of Site.* Accessibility of site by rail or road is essential otherwise the movement of the men and material will be difficult and a costly affair.

5. *Distance from the Load Centre.* Transmission and distribution (T & D) losses of electricity mainly depend up the distance from the generation site to the load centre. If the distance is more, obviously T & D losses will be more.

6. *Type of the Land on the Site.* The type of the land on the site should be conducive for the hydropower plants. The land should be cheap and rocky. The foundation rock for the masonry dam should be strong enough to withstand the stresses in the structure and the thrust

of water in the event of full reservoir. The dam to be constructed at the site should have large catchment area to store water at high head. The site should not be earthquake prone.

7. *Environment Aspects.* Environmental aspects of the site selection should go hand in hand with engineering aspects. The hydro plant located on a particular site should not disturb the ecology tremendously. The site selection should fulfill the following requirements in respects of environmental aspects.

(a) To assure safe, healthy, productive and culturally pleasing surroundings.

(b) To avoid the health hazards, or other undesirable and unintended consequences.

(c) To preserve important historic, cultural and natural aspects of the site.

8. *Sociological Aspects.* The catchment area requirements is very large and as a result many villages may submerge in the catchment area and a large number of people have to be displaced. This is a sociological aspects and has to be considered before site selection.

11.4 HYDROLOGY

The science which deals with rainfall and run-off is known as hydrology. The evaporation of the water from the surface of oceans and rivers and its precipitation (i.e., rainfall, on the earth is known as hydrological cycle. The knowledge is a great tool to design the irrigation and flood control works, hydropower projects, water supply schemes, navigation work, etc.

11.4.1 Hydrological Cycle

Water is evaporated from plants, rivers, oceans and lands and carried with air in the form of vapour which is known as cloud. When the vapour in the atmosphere is cooled below dew point temperature (dpt), it falls in the form of water droplets (rain) or snow depending upon temperature. This evaporation and precipitation is naturally a continuous process and therefore constitutes a perennial source of energy. The hydrological cycle is shown in Fig. 11.2.

The distribution of precipitation may be classified as (i) direct evaporation, (ii) absorption and transpiration by vegetation, (iii) seepage and storage, and (iv) direct surface run-off, eventually forming rivers. In others words, the water equation over a certain site during a given period is given by

$$\text{Run-off} + \text{Seepage} + \text{Evaporation} + \text{Transpiration} = \text{Precipitation} \pm \text{Change in storage} \quad ...(11.3)$$

11.4.2 Run-off

As the rain falls upon the drainage basin, a portion of it is evaporated directly by the sun, another large portion is taken up by vegetation and growing crops particularly in growing season and some percolates into the ground. The percolating water may partly be absorbed by the roots of vegetation or it may appear in course of time as spring water either inside or outside the catchment area. The remaining portion of rainfall flows through the catchment area on the surface of the earth and is known as *run-off.* In general, run-off (R) is expressed as

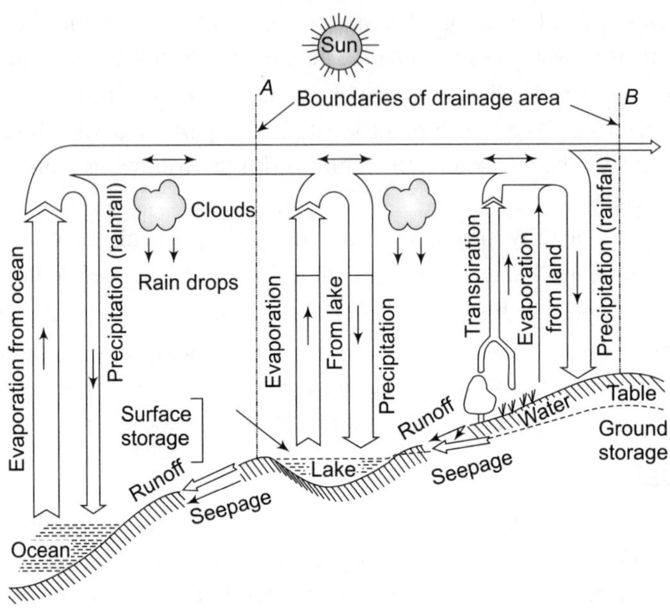

Figure 11.2 Hydrological cycle

$$R = P - L \qquad ...(11.4)$$

where P = precipitation by rainfall, L = all losses, $R = R_s + R_e$, where R_s = run-off over the surface, R_e = run-off reaching the catchment area through the porous earth.

The unit of run-off is cm^3/s for centimetre of water depth over the entire catchment area.

The suitability of run-off for power generation depends on the magnitude and time distribution of flow and head made available by the surrounding topography.

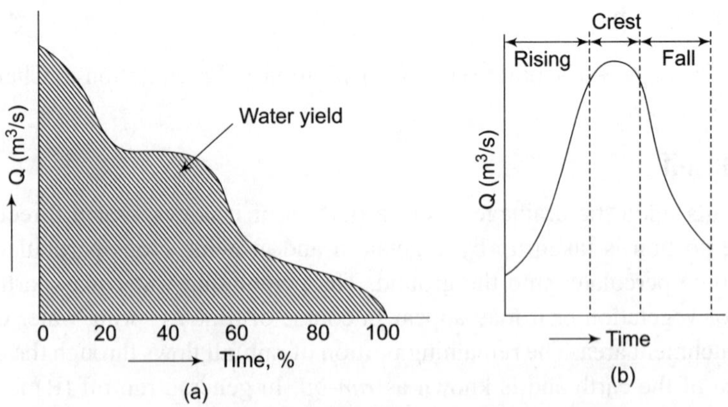

Figure 11.3 Flow duration curve

11.4.3 Hydrographs

Hydrograph is defined as a graph showing discharge (run-off) of flowing water (say. river) with respect to time for a specified time. The time period may be hour, day, week or month. The unit discharge may be m^3/s, km^2-cm/h or day-s-m.

Figure 11.3 shows a hydrograph. In Fig. 11.3 (a) the discharge is decreasing with time (expressed in percentage of time) while in Fig. 11.3 (b), the discharge is increasing with time (say onset of monsoon), reaches a maximum value and then decreases. The area under the curve shows the water yield. Similarly, power duration curve may be plotted by representing power (= $g\,\eta$ QH) on ordinate.

Figure 11.4 shows an another typical flow duration curve showing the minimum (Q_{min}) and maximum (Q_{max}) flow rate as a per cent of time. The power output available at Q_{min} (area *OABC*) is known as *primary power* and it will be available for all the times (i.e., 100% time). The additional output available at higher water flows is called *secondary power* (area *BCDE*). If the flow rate of Q_n is required for all the times (i.e., 100% times) as indicated by the area under the flow demand line *DEF*, then it would be possible to meet the uniform demand of the flow rate (or power) for all the times if storage equal to *BEF* is provided. Alternatively, a thermal power unit of *BF* capacity to work as a supplement to hydropower unit is to be installed.

The curve also shows the natural flow sufficient to meet the flow demand, Q_n is available for 53.5 per cent of times or 195 days in a year.

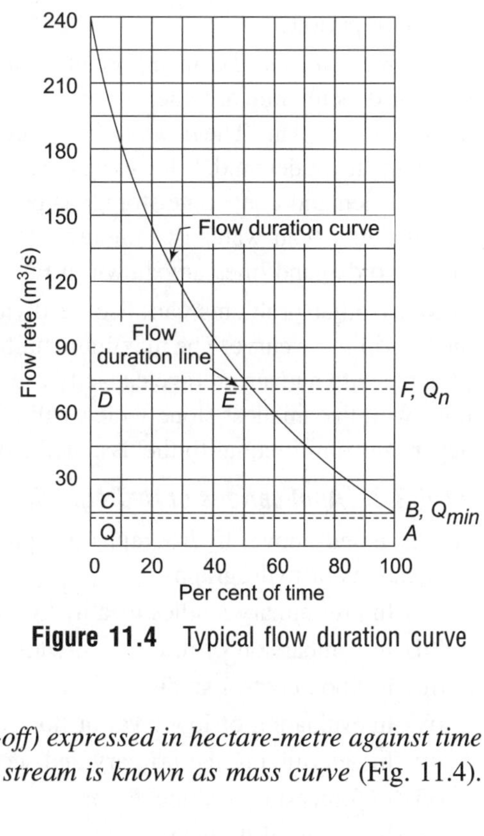

Figure 11.4 Typical flow duration curve

11.4.3.1 Mass Curve

A plot of cumulative value of water quantity (run-off) expressed in hectare-metre against time made from the records of mean monthly flows of stream is known as mass curve (Fig. 11.4). Mathematically, the mass curve is expressed as

$$V = \int_{t_1}^{t_2} Q(t)dt \qquad \text{...(11.5)}$$

where V is the volume run-off and $Q(t)$ is the discharge in m^3/s as a function of time.

A typical mass curve is shown in Fig. 11.5. The slope of the curve at any point indicates the rate of flow at that particular time. This means that if the curve is horizontal, the flow will be zero and if there is a high rate of flow, the curve will be steeper. The dry periods are indicated as concave depression on the mass curve.

The straight line AB is obtained by joining the end points of the mass curve (Fig. 11.5). The slope of this straight line AB ($\tan\theta = V/t$) represents the *average discharge* over the total period. Two straight lines $A'B'$ and $A''\,B''$ are drawn parallel to AB and tangent to the mass curve at the lowest tangent point C and the highest tangent point D respectively. The vertical intercept $A'A''$ represents the *storage volume* required to permit continuous release of water at this average discharge rate over the entire period.

The *rate of demand* is an important parameter and is defined as the rate at which water is required for use in a power plant. The *demand line* represents the constant rate of demand. If the *mass curve* is steeper than the demand line, some storage is possible and if it is flatter than water is withdrawn at a greater rate. Two demand lines are shown in Fig. 11.5.

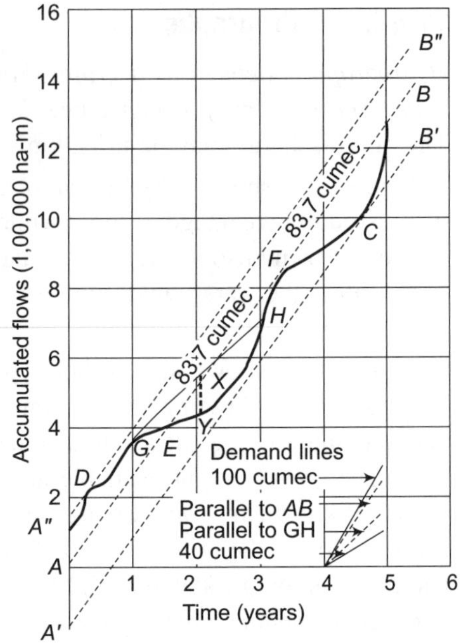

Figure 11.5 Mass curve

Many other problems can also be divided with the help of mass curve. The maximum uniform flow possible with a given storage capacity can be determined from the mass curve by locating the lines with the smallest slope which will be tangent to the mass curve and has the maximum departure from it equal to the assumed storage capacity.

11.4.3.2 Applications of Hydrographs

As discussed above, hydrographs are presented in different forms. The following are the applications of hydrographs.

(i) In preliminary studies of a hydro project.
(ii) In comparison of different streams.
(iii) In flood control studies.
(iv) In evaluation of low level of flows.
(v) In determining the primary and secondary powers.
(vi) In forecasting sedimentation.
(vii) In design of drainage system.
(viii) In forecasting average discharge and demand rate.

11.5 PONDAGE AND STORAGE

Since the rainfall and consequently run-off varies considerably over the years, so for harnessing water power to meet a certain power demand, it is essential that some arrangement in the form of pondage and storage of water in required.

***Pondage** is defined as a regulating body of water in the form of a relatively small pond or reservoir provided at the plant.* It is used to regulate the variable water flow to meet the power demand. Figure 11.6. shows the location of powerhouse using pondage at a suitable site. A weir is provided across the river and water is directed through water channel to feed to pondage. From pondage, water is fed to turbine through short penstocks.

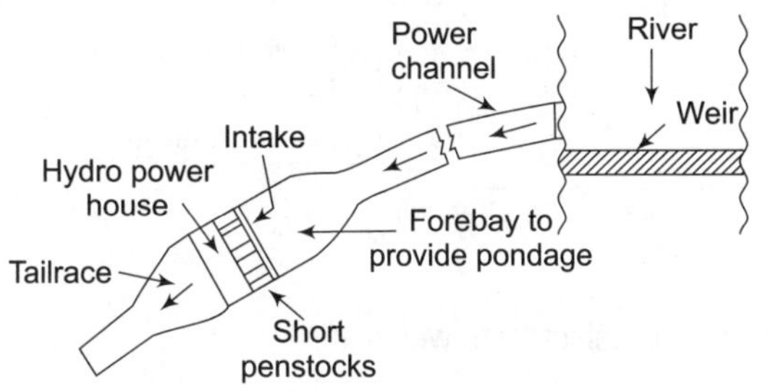

Figure 11.6 Powerhouse with pondage

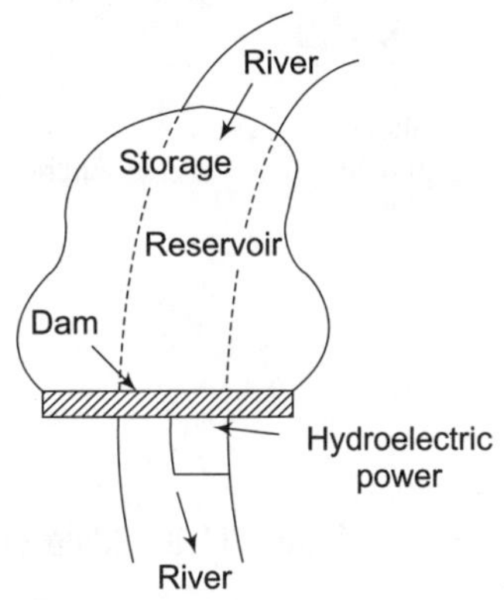

Figure 11.7 Powerhouse with storage

***Storage** is defined as impounding of a considerable amount of excess run-off during the season of surplus flow for use in dry seasons.* This is made possible by constructing a dam across the river stream at a suitable site and building a big storage reservoir on the upstream side of the dam. The water is taken to the turbine through the penstrock on the downstream side of the dam so that sufficient head is available. Figure 11.7 shows the location of a power house with storage.

It is worth to note that the pondage increases the capacity of a river for a short time such as week while storage increases the capacity of a river varying from 6 months to 2 years.

11.6 MAIN ELEMENTS OF A HYDROELECTRIC POWER PLANT

Figure 11.8 shows the main elements of a typical hydroelectric power plant. They are: (i) catchment area (ii) reservoir, (iii) dam, (iv) sluice gate or valve, (v) penstock or conduits, (vi) spillways, (vii) surge tanks, (viii) trash rock (ix) foreway, (x) powerhouse, (xi) draft tube, (xii) switch yard for transmission lines.

11.6.1 Catchment Area

A vast area upstream side of the river across which dam is constructed into which various streams of water in the form of run-off terminates into river is known as catchment area. The catchment area is characterised by many parameters such as size, shape, surface, orientation, altitude, topography, geological aspects, etc. The larger catchment area, steeper slope, and higher altitude result in greater total run-off water in the catchment area.

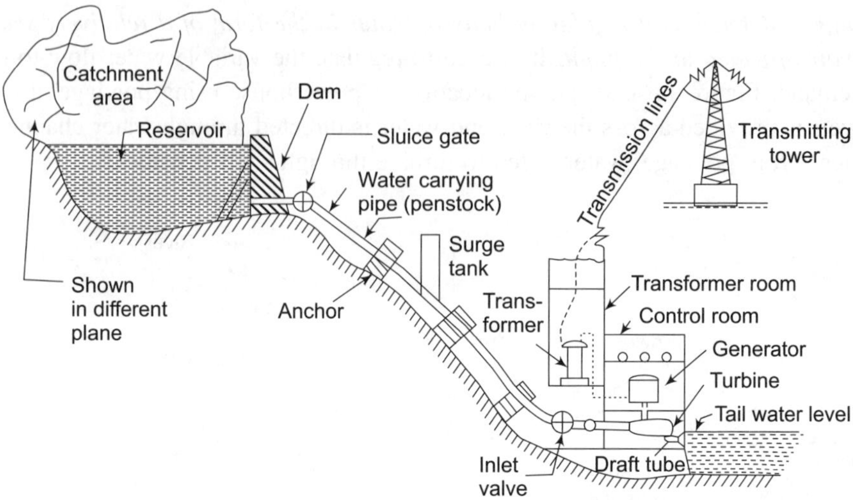

Figure 11.8 Main elements of hydroelectric power plant

11.6.2 Reservoir

A reservoir is a storage of water either natural like a lake on a mountain or artificially built by erecting a dam across a river. The efficient use of water resources lies in the storage of water during times of plenty for subsequent use in times of scarcity. Water stored in the reservoir is not only used for power generation, but also for irrigation, flood control, water supply and navigation. This suggests that efficient management is a must.

11.6.3 Dam

A dam is a thick wall made of RCC, earth, masonry, etc. constructed across a river to store water at a high head. The design of the dam is such that it takes care all types of possible stresses acting upon it.

11.6.3.1 Classification of Dam

Dams are classified on the basis of the following:

(i) **Function.** On the basis of functions, dams are classified as–(a) storage dam, (b) diversion dam and (c) detention dam.

  (a) *Storage dams* are constructed for storing water and using it subsequently as and when required for various uses such as hydropower, irrigation and water supply.

  (b) *Diversion dams* are constructed to raise the water level and to divert the river flow is another direction.

  (c) *Detention dams* are constructed primarily to store flood water.

(ii) **Shape.** Based on shape, the dams are classified as *trapezoidal* or *arch dams* to suit the structural functions.

(iii) **Construction material.** On the basis of construction materials, dams can be classified as *earth, rock pieces, stone masonry, concrete, RCC, etc. dams.* Concrete, RCC, earthen and rockfill dams are most popular.

(iv) **Hydraulic design.** On the basis of hydraulic design, dams are classified as *non-overflow type and overflow type.* In non-overflow type, water is not allowed to flow over the top of the dam while in overflow type, water flows over it.

(v) **Structural design.** On the basis of structural design, dams are classified as *gravity dam, arch dam* and *buttress dam* where water thrust is resisted by gravity, arch and buttresses action.

11.6.3.2 Selection of Site of Dam

Several factors should be taken into consideration in the selection of a site of dam. They are as follows:

(i) At neck formation of river (Fig. 11.9).
(ii) At confluence of two rivers (Fig. 11.9).
(iii) Geology of foundation.
(iv) Hydraulic factors and river diversion during construction.
(v) Availability of construction materials.
(vi) Accessibility of materials by transport.
(vii) Economics.
(viii) Availability of know-how.
(ix) Sociological factors resulting from submergence of area, displacing people and disturbing the ecology.
(x) Safety consideration as failure of dam is a big catastrophe. The chances of failure if any may be during flood and earthquake, and a dam has to be designed to withstand these shocks over and above the other stresses acting over the dam in normal situation.

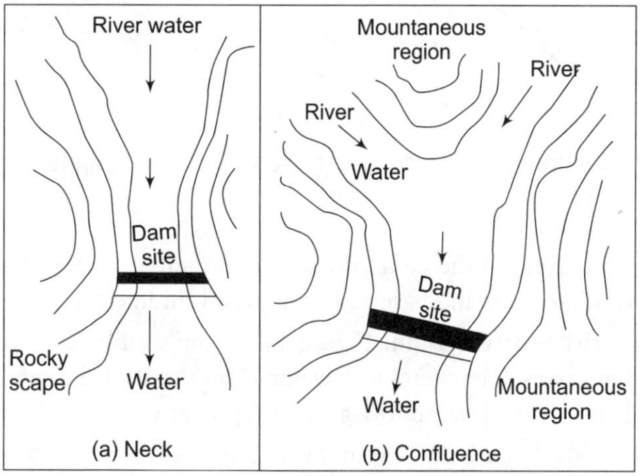

Figure 11.9 Neck and confluence

11.6.3.3 Choice of Dams

(i) Gravity Dams. In gravity dams, the water thrust is resisted by gravity. They may be constructed either in stone masonry or concrete or RCC. The concrete or RCC dam can be constructed faster with better quality control. The cross section of solid gravity dam is shown in Fig. 11.10. In a stable dam, the resultant of all the force acting on the dam will intersect the base at some point within the middle third of the base line. The water pressure (P) and

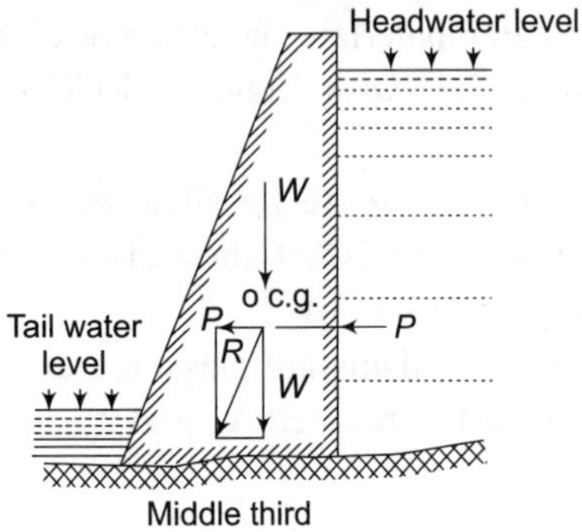

Figure 11.10 Cross section of solid gravity type masonry dam

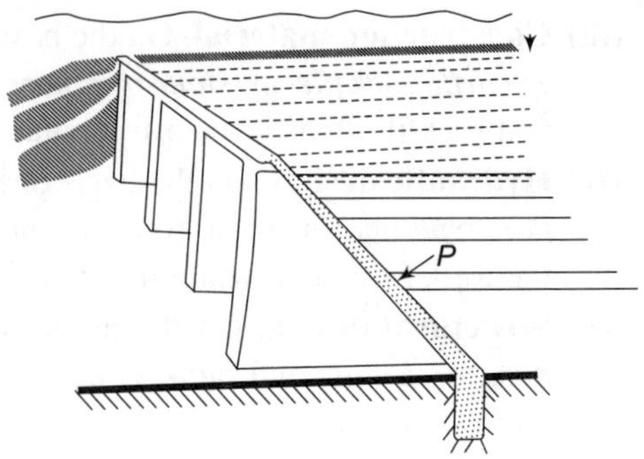

Figure 11.11 Buttress type masonry dam with flat deck

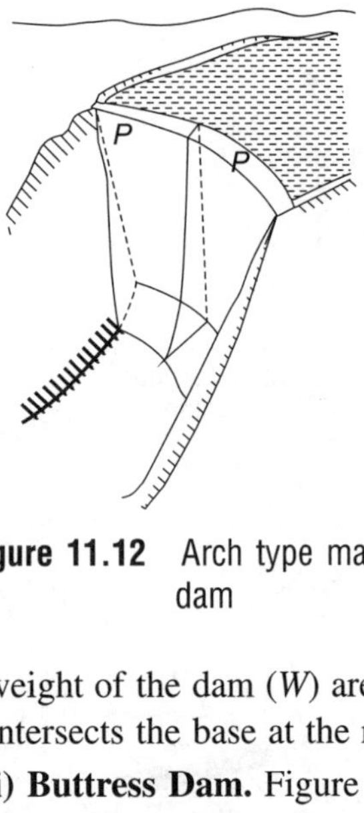

Figure 11.12 Arch type masonry dam

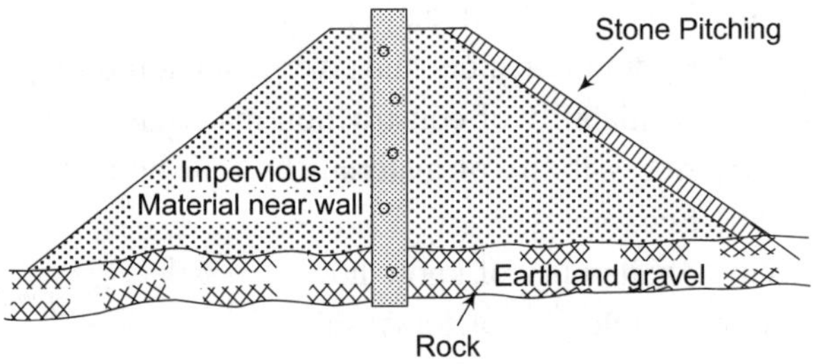

Figure 11.13 A cross section of earth dam

the weight of the dam (*W*) are the principal forces acting on the dam and the resultant force (*R*) intersects the base at the required middle third.

(ii) Buttress Dam. Figure 11.11 shows a buttress dam. It has a flat deck laid on sloping buttresses. The weight of water *P* on the deck is transmitted as a thrust to the buttresses. It is also called be hollow gravity type dam.

(iii) Arch Dam. Figure 11.12 shows an arch type masonry dam in which all forces are acting perpendicular to the surface so that they tend to compress the materials in the arch.

(iv) Earth Dam. Earth dam is constructed of earth fill with wide base. Generally, for small projects, up to 70 m in height, earth dam is constructed. A large volume of earth material is required and it should be available in the vicinity. The construction of dam varies with height and the side slopes are flatter. Figure 11.13 shows the cross section of a earth dam. Stone pitching is done on the slopes. Though, it is cheaper than masonry dam, but the seepage losses are more. The erosion by water overtapping the dam or seeping through it may cause serious damage to the dam.

(v) Rock-fill Dam. Rock-fill dams are constructed from loose rock of all sizes and has a trapezoidal shape with a wide base. It must have a water-tight section to reduce seepage as chances of seepage are more. It is suitable for mountainous regions where plenty of all sizes of rocks are available.

11.6.4 Trash Rock

The water intakes from the dam or from the forebay are provided with trash rock to prevent the entry of the debris which might damage the wicket gates or turbine runners or choking up the nozzles of impulse turbines. In severe winter, ice may be deposited on trash rock and to prevent the icing trouble, it is heated electrically. Sometimes, air bubbling system is provided to overcome icing trouble.

11.6.5 Forebay

The purpose of forebay is to serve as a regulating reservoir temporarily storing water when the load on the plant is reduced and provides water for initial increment of an increasing load while the water in the canal is being accelerated. In short, forebay is a naturally provided storage which is able to absorb the flow variations. The forebay is always provided with some type of outlet structure to direct water to the penstock depending upon local conditions.

11.6.6 Spillways

During the rainy season, the water level in the reservoir basin rises which may endanger the stability of the dam structure. In order to relieve the reservoir of this excess water, a structure is provided in the body of a dam or close to it which allows the excess water to spill from the reservoir so that water level does not cross high flood level (HFL). Such structure is known as spillway. There are various types of spillways as given below.

11.6.6.1 Overall (or Solid Gravity) Spillway

An overall spillway shown in Fig. 11.14. This provision is provided in concrete, RCC and masonry dam. Water spills and flows over the crest in the form of rolling sheet of water. The shape of the spillway is provided in such a way that it forms crest, upper nappe, lower nappe and finally a bucket. The excess energy is destroyed by the bucket.

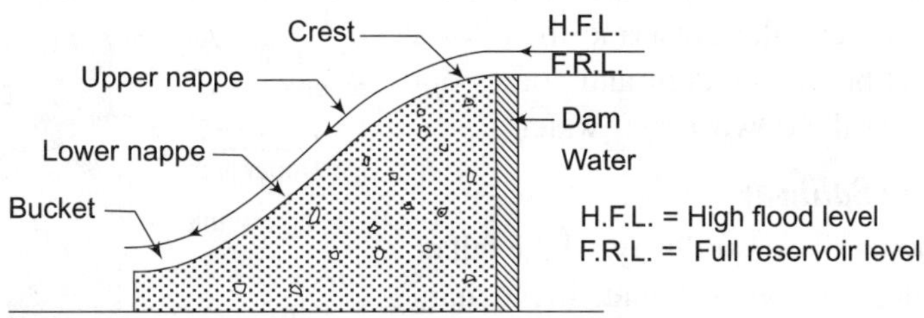

Figure 11.14 Overall spillway

11.6.6.2 Chute or Trough Spillway

It is similar to overall spillway except chute or trough is provided at the end. The excess water spill and flows over the crest and shoots down a channel or trough to meet the river downstream of the dam. Trough spillway is suitable to the situation when the valley is very much narrow which cannot accommodate the overall spillway.

11.6.6.3 Side Channel Spillway

In many situations, when the valley is too narrow, even the trough spillway cannot be accommodated and it is not desirable to allow the flow of water over the dam side channel spillway is provided (Fig. 11.15). After running parallel to downstream, the side channel either meets to the downstream or used as canal.

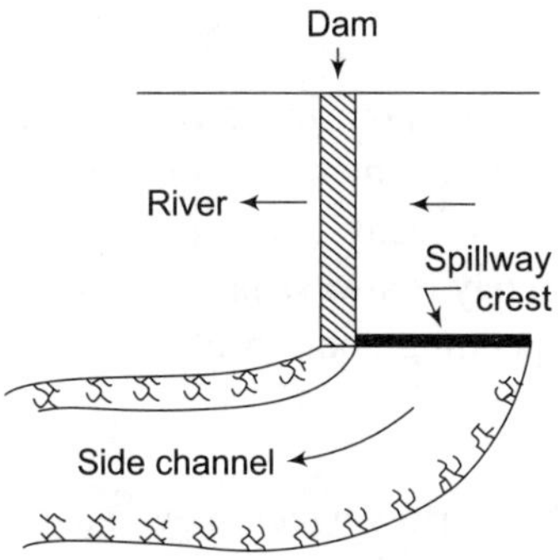

Figure 11.15 Side channel spillway

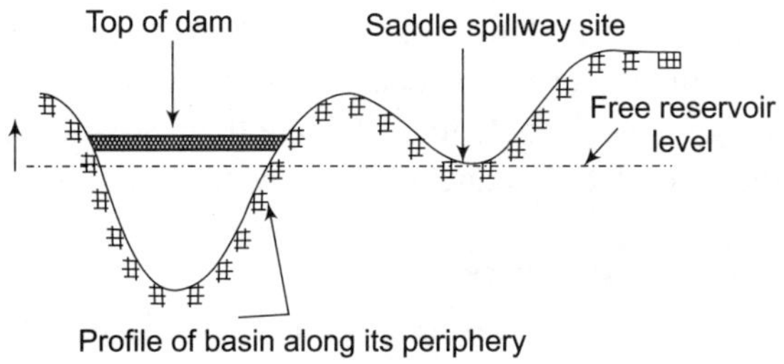

Figure 11.16 Saddle spillway

11.6.6.4 Saddle Spillway

A saddle spillway is preferred only when any of above types of spillway are not favourable. A saddle spillway is a natural depression type of spillway and is provided on the periphery of the reservoir basin away from the dam (Fig. 11.16).

11.6.6.5 Shaft Spillway

The shape of the shaft spillway is like a funnel constructed adjoining the dam, the lower end of which is turned at right angle and then taken out below the dam horizontally. The excess water spills over the circular crest and then flows through the funnel below the dam and comes out horizontally to meet the downstream water.

11.6.6.6 Siphon Spillway

Figure 11.17 shows a siphon spillway. At the fall of reservoir level (FRL), a siphon is made in the main body of the dam. When water level rises above FRL, water spills through the siphon. An air vent to provided to drive entrapped air.

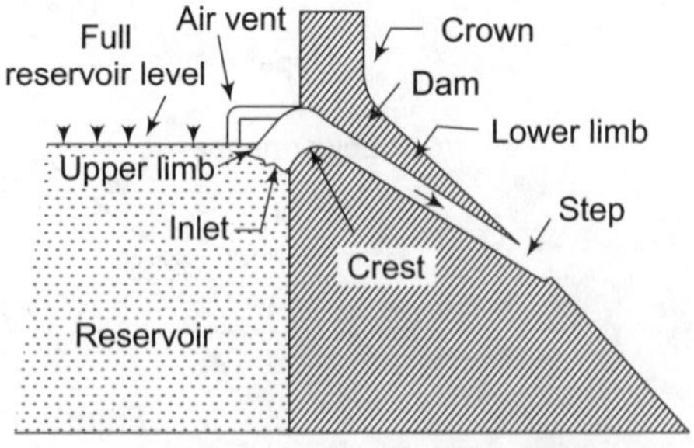

Figure 11.17 Siphon channel spillway

11.6.7 Conduit (Canal, Flume, Tunnel, Pipeline Penstock, etc.)

A conduit is an open or closed passage through which water flows from the head race to the turbine and water from the turbine goes to the tail race. There are various forms of conduits, namely, canals, flumes, tunnels, pipelines and penstocks. Canals and flumes are open, while tunnels, pipeline and penstock, and closed passages.

A combination of conduits is shown in Fig. 11.18. A *canal* is an open waterway excavated in natural ground following its contour while a *flame* is an open channel erected on a surface above the ground supported on a trestle. In many situations, there exists a ridge or mountain between the dam and powerhouse. To allow the water to enter the turbine, a closed channel called *tunnel* is excavated through the ridge or mountain from the reservoir to the power house or to another reservoir as the case may be. A *pipeline* is a closed conduit supported on or above the surface of the ground. *A penstock is a closed conduit for supplying water from reservoir/surge tank to the turbine.* Canals and flumes are non-pressure conduits while tunnel, pipe and penstock are pressure conduits.

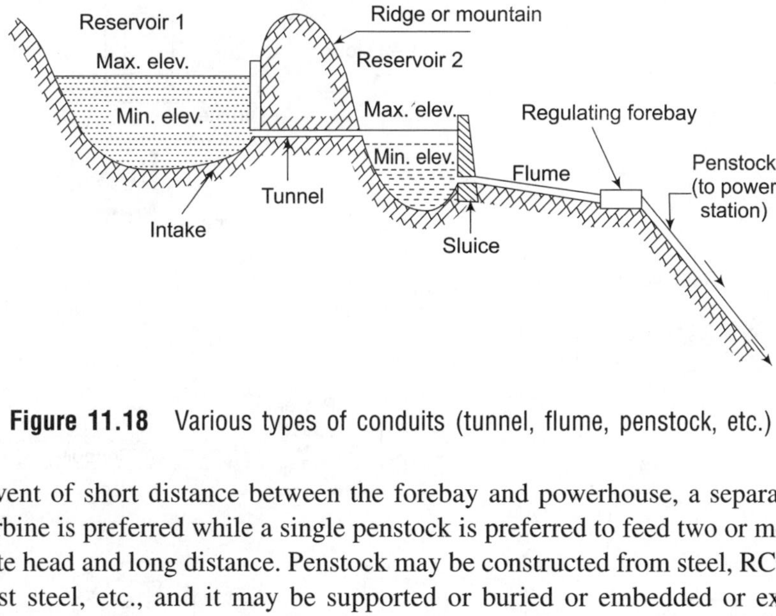

Figure 11.18 Various types of conduits (tunnel, flume, penstock, etc.)

In the event of short distance between the forebay and powerhouse, a separate penstock for each turbine is preferred while a single penstock is preferred to feed two or more turbines for moderate head and long distance. Penstock may be constructed from steel, RCC, asbestos, cement, cast steel, etc., and it may be supported or buried or embedded or exposed. The penstock may be interconnected either by flexible couplings or expansion joints with rigid pipes having anchorage, semi-rigid pipes or flexible pipes.

11.6.8 Surge Tank

When the load on the generator decreases, there is a sudden increase of pressure in the penstock due to sudden decrease in the rate of water flow to the turbine when the gates admitting water to the turbines are suddenly closed owing to the action of governor. *The sudden rise of pressure in the penstock above normal due to reduced load on generator is known as water hammer.*

When the load on the generator increases, the turbine gates open suddenly allowing more water to enter the turbine to meet the increased demand, water has to rush through the penstock and there is a tendency to cause vacuum in the water system. The penstock must withstand the positive hammer caused by sudden decrease in load and no vacuum should be produced in the system when the gates suddenly open under increased load condition.

A surge tank is a small reservoir introduced in the system between the dam and powerhouse, preferably on the high ground to provide better regulation of water pressure by reducing the pressure swing during variable load conditions so that they are not transmitted to the penstock. Figure 11.19(a) shows a surge tank placed in between the reservoir and powerhouse.

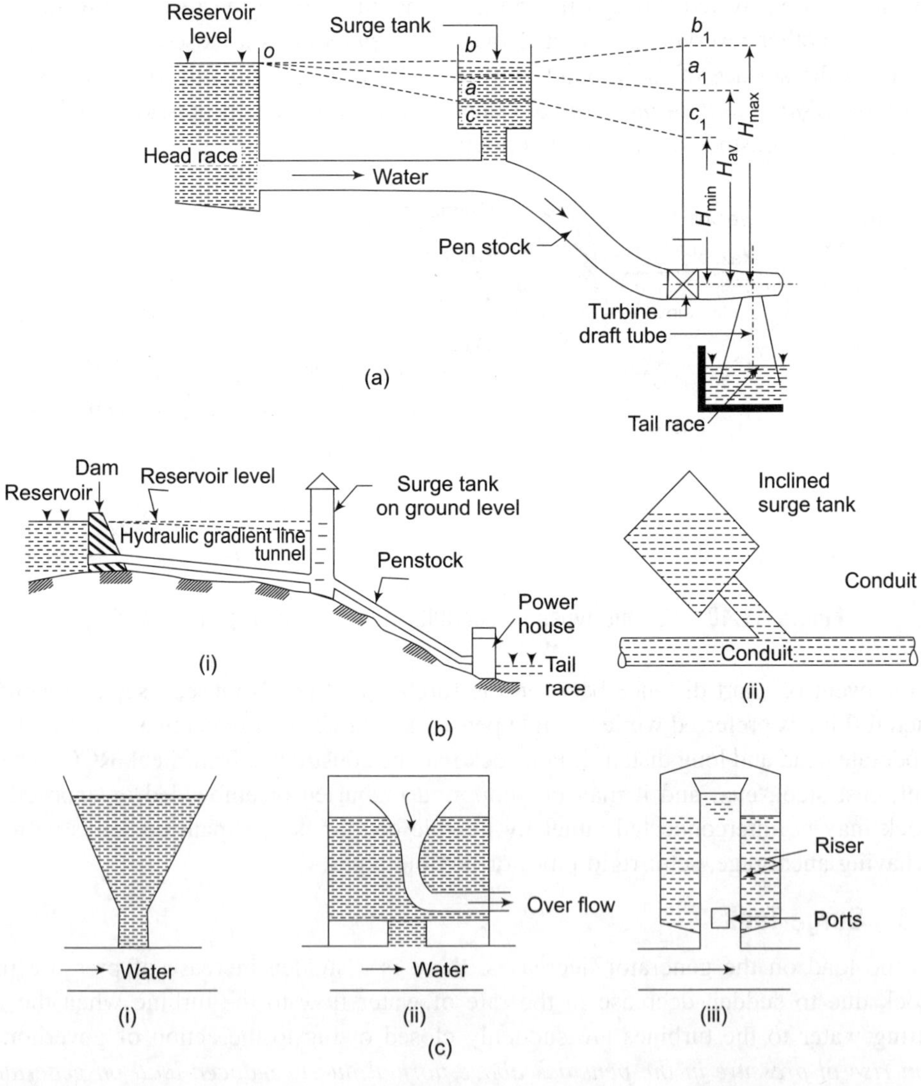

Figure 11.19 Surge tank and its types (a) simple surge tank (b) arrangement and (c) types

At low load water flows into the turbine gets reduced suddenly, so water level rises in the surge tank. This produces a retarding head and decreases the velocity of water in the penstock. When the velocity of water in the penstock is reduced to the value demanded by the turbine, the level of water in the surge tank starts falling and fluctuates up and down till its motion is damped out by friction. In the event of sudden rise in the load on the turbine, additional water is supplied from the surge tank. This lowers the water surface in the surge tank and thus producing an accelerating head which increases the flow of water in the penstock. When the discharge of water in the penstock corresponds to the turbine demand, the water surface in the penstock ceases to fall down. *By this way, surge tank helps in stabilizing the velocity and pressure in penstock and reduces the water hammer effect.* Figure 11.19(b) shows a typical arrangement of surge tank showing surge tank on ground level and inclined surge tank.

The various types of surge tank such as *conical, internal belt mouth spillway, and differential types* are shown in Fig. 11.19(c).

In the differential surge tank, there is a central riser pipe having small ports at its lower end. It has the advantage of providing both the retarding and accelerating heads with the decrease and increase of load. A surge tank is located near the powerhouse as is feasible to reduce the length of penstock thereby reducing the water hammer effect.

11.6.9 Draft Tubes

The draft tube is a diverging discharge passage connecting the water turbine outlet with tail race. *It is shaped as diverging passage to decelerate the flow with minimum loss so that the remaining kinetic energy of water coming out of runner which would otherwise a waste as an exit loss in efficiently converted into head energy thereby increasing the total pressure difference on the runner.* It also allows the inspection and maintenance of turbine. Figure 11.20 shows the various types of draft tubes. The conical type is mainly used for low head turbine while elbow type is more common as it allows the water to come out horizontally in tail race.

11.6.10 Powerhouse

A powerhouse comprises two main parts, a sub-structure to support the hydraulic and electrical equipment and a superstructure to house and protect these equipment.

The turbines (prime movers) which rotate on vertical axis are placed just below the floor level while those rotating horizontally are placed on the ground floor. The generating units and exciters are placed on the ground floor. The elevation of water turbine with respect to tail race is determined by the necessity of avoiding cavitation.

It is always advantageous to locate the powerhouse *underground under* certain topographical conditions where there is no convenient site for a conventional (on the surface) type. Many underground powerhouses are working world over, for example, Koyna powerhouse in India and Kemano powerhouse of Canada having generating capacity 1670 MW under a head of 785 m. The following equipment are installed in the powerhouse (Fig. 11.21).

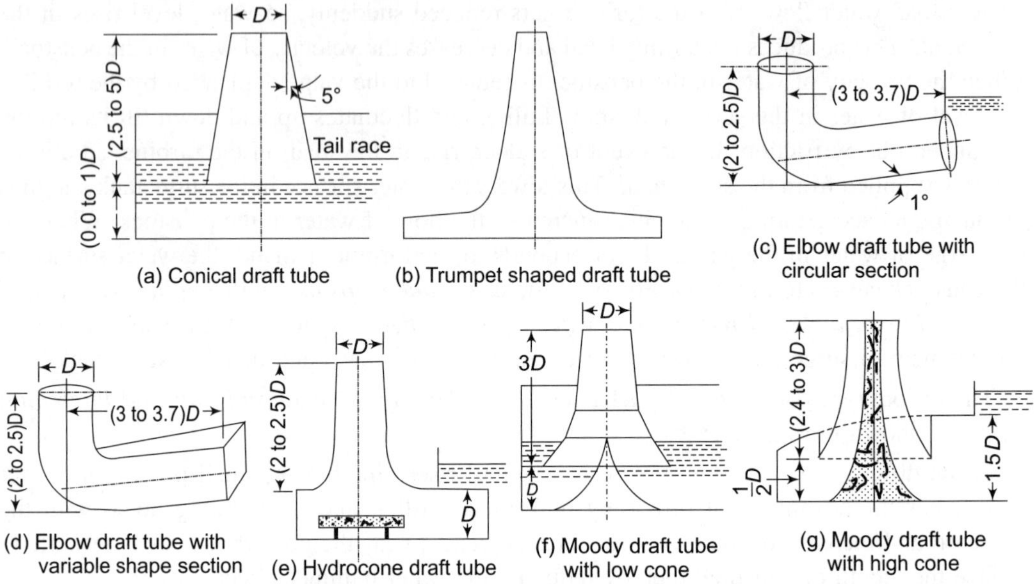

Figure 11.20 Various types of draft tubes

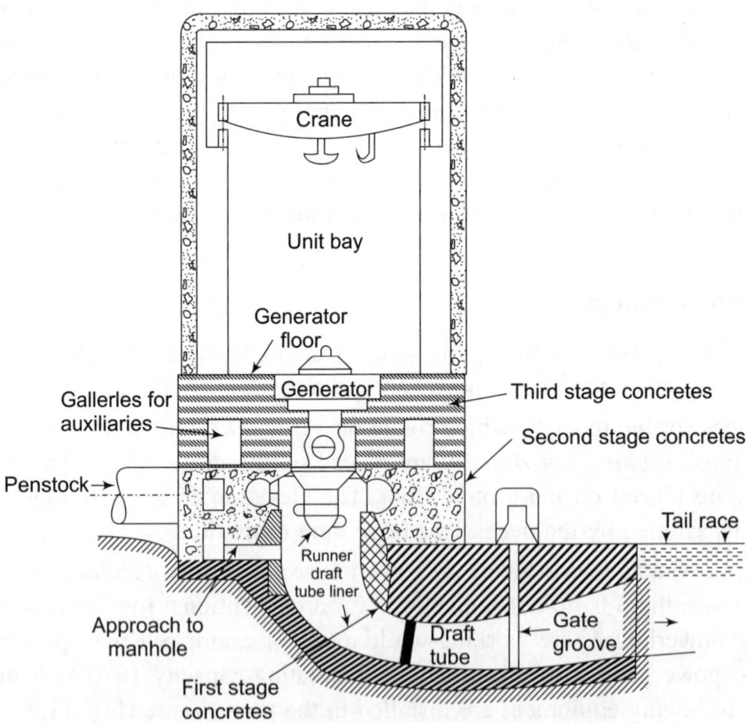

Figure 11.21 A sectional view of a powerhouse

(i) Hydraulic turbines	(ii) Electrical generators
(iii) Governors	(iv) Gate valves
(v) Relief valves	(vi) Water circulating pumps
(vii) Air duct	(viii) Switchboard and instruments
(ix) Control room	(x) Storage batteries
(xi) Cranes	

11.7 CLASSIFICATION OF HYDROELECTRICAL POWER PLANTS

There are various ways to classify the hydroelectrical power plants. They are as follows:

(a) Availability of head. Based on the availability of head, hydroelectric power plants are classified as: (i) high head power plants (ii) medium head power plants, (iii) low head power plants.

(b) Nature of loads. On the basis of nature of loads, power plants are classified as: (i) base load plants (ii) peak load plants.

(c) Availability of water. On the basis of quantity of water available power plants are classified as

(i) Hydel plant with storage reservoir
(ii) Run-off river hydel plant with and without pondage
(iii) Pump storage hydel plants
(iv) Mini and micro hydel plants

Various types of above hydel plants are discussed below in brief.

11.7.1 High Head Power Plants

The hydel power plants operating under a head of **100 m or above are called high head power plants**. Figure 11.22 shows a schematic of high head powei plants. In such plants, water is stored in lakes or high mountains during rainy season or when snow melts. Surplus water if any is discharged by a spillway. Tunnel is constructed through the mountain and a surge chamber is provided with the help of regulating valve, water is allowed to flow through the turbine. Due to large head, Pelton wheel is used as prime mover.

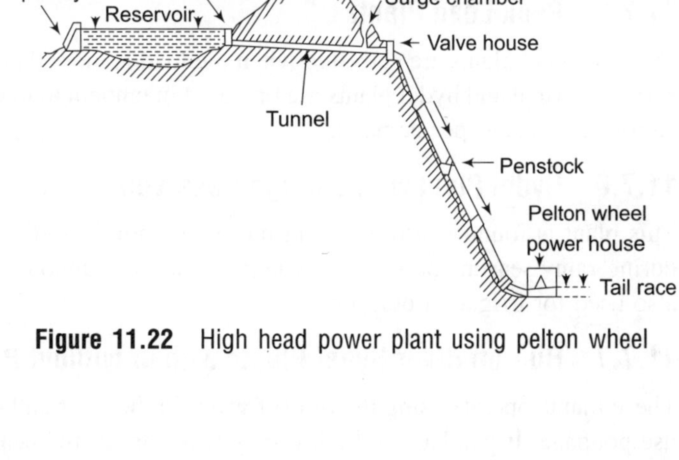

Figure 11.22 High head power plant using pelton wheel

11.7.2 Medium Head Power Plants

Such plants in large number are found world over. They operate under **heads varying from 30 m to 100 m**. Francis turbines are invariably used as prime mover Fig. 11.23 shows a schematic of such a plant. In such cases, forebay before the penstock acts as reservoir as well as surge tank. In the forebay, trash rock is provided to prevent the entry of the debris so that water gates or turbine runners are not damaged.

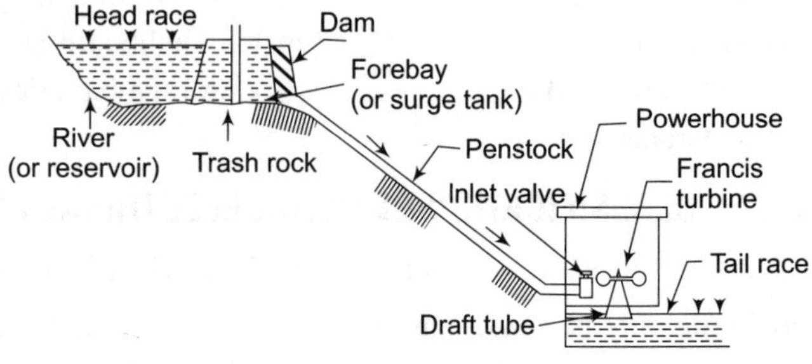

Figure 11.23 A sectional view of a powerhouse

11.7.3 Low Head Power Plants

Low head power plants are constructed where the head available is **below 30 m**. Under such a situation, a dam is constructed across a river and a sideway stream diverges from the river at the dam and a forebay is provided before the powerhouse. After the powerhouse, the water meets the original river. Since head is low, so Francis or Kaplan turbines are used as prime movers (Fig. 11.24) shows the management of such a plant.

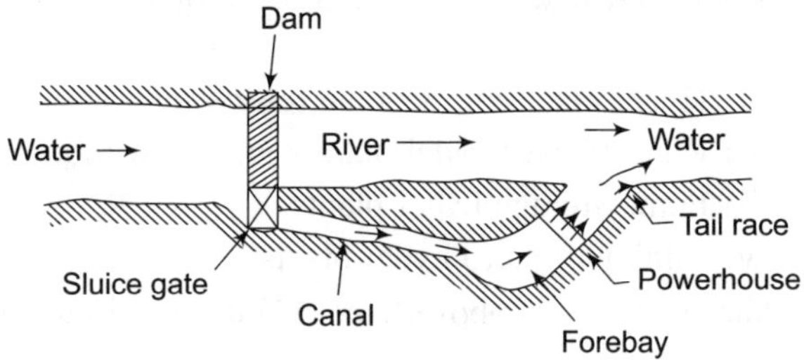

Figure 11.24 A low head power plant

11.7.4 Base Load Plants

Base load plants supply constant power to the grid without any interruption.

11.7.5 Peak Load Plants

The peak load plants are activated when the load is more than the nominal load. As discussed earlier, thermal and hydel plants are operated in tandem to meet the base and peak load during various reasons as per demand.

11.7.6 Hydel Plant with Storage Reservoir

This plant is found world over and more common in India. Water is stored in the reservoir during rainy season and electricity is generated throughout the year. In some cases, water is also used for irrigation purposes.

11.7.7 Run-off River Hydel Plants with or without Pondage

These plants operate using the run-off water in the river and operate daily. It may or may not use pondage. If pondage is used, it may take care of the peak load.

11.7.8 Pump Storage Hydel Plants

In such plants, a small dam is constructed across the tail race. Water after passing through the turbine is stored in the tail race reservoir. During low load (say night time), the water is pumped back from the tail race to the head water reservoir during excess electricity from the grid or from the nearby steam plant. The peak load is met by drawing excess water from the head reservoir and by this way about 70% of electricity consumed in pumping is recovered and at the same time peak load demand is met. Figure 11.25 shows the schematic of such a plant. The advantages of this plant lie in (a) substantial increase in peak load capacity at low cost (b) high operating efficiency, (c) higher load factor and (d) independence of stream flow conditions.

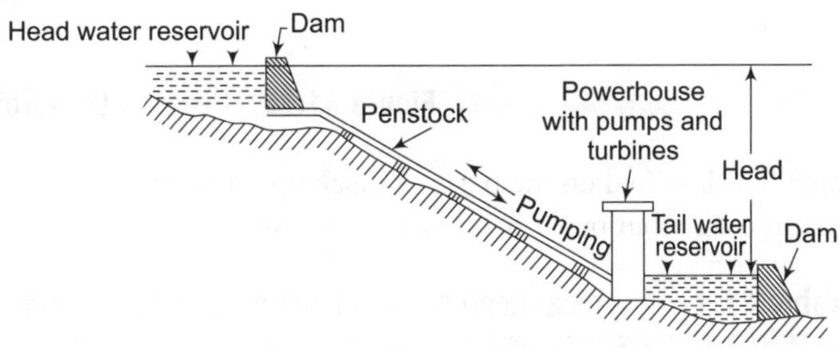

Figure 11.25 Pumped storage plant

11.7.9 Mini- and Micro-Hydel Plants

Due to high capital cost, large gestation period, sociological problems, the large hydel plants are now less matter of attention. Now, more attention is given to mini and and micro-hydel plants. The natural water source in hilly terrain may be easily utilised to generate electricity using low head turbogenerator sets. There is a negligible disturbance of ecology. The **mini** hydel plants operate under head of **9 to 20 m producing about 1 MW**. From the latest survey, it has been estimated that the potential energy from **mini and micro-plants** in India is about **22,000 MW**.

11.8 HYDRAULIC TURBINES AND THEIR CLASSIFICATION

Hydraulic turbines are the oldest prime movers serving the mankind. They convert the head energy of water into shaft work and finally electrical energy through electric generators when turbines and generators are coupled on the same shaft. The oldest turbine is waterwheel. However, the modern turbines have gone through a lot of technological advances and as a result, the size of the turbine is very large yielding high efficiency and large output.

Hydraulic turbines are classified on the basis of the following.

(a) Head and Discharge of Water Available. The head of water available is the most important parameter in the selection of turbine. The gross head (H) is the difference in the elevation of water surface between the headrace and tail race while the net head (H_{net}) is the difference between the gross head and head loss due to friction (h_f) (Fig. 11.26). The water turbine works under a wide range of heads varying from **2 to 2000 m**. Table 11.1 shows the classification of turbines with respect to head and the suitable turbine for that head. Turbines

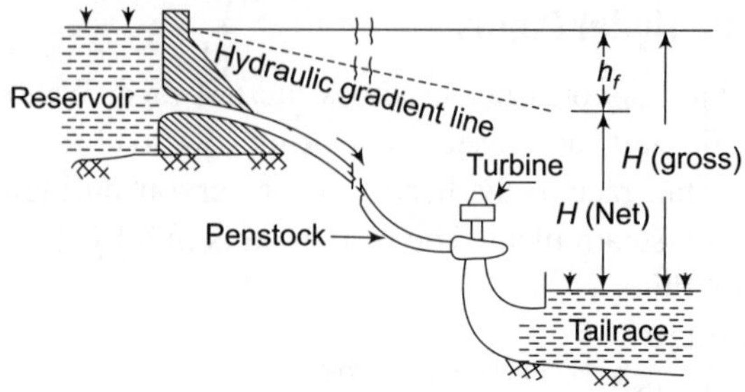

Figure 11.26 Head on the turbine

are also classified on the basis of discharge; low, medium and high discharge. It is to be noted that the distribution of head is typical one.

Table 11.1 A typical head based classification of turbines

Head base turbine	*Head (m)*	*Discharge*	*Turbine type*
Low head turbine	2–15 m	High discharge	Kaplan or Propeller
Medium head turbine	16–70 m	High or medium discharge	Kaplan or Francis
High head turbine	71–500 m	Medium or low discharge	Francis or Pelton
Very high head turbine	> 500 m	Low discharge	Pelton

(b) Inventor of Turbine. On the basis of name of the inventor, water turbines are named as follows:

(i) *Pelton turbine.* It is named after its inventor Lester Allen Pelton of USA. It is an impulse turbine and is suitable for high head and low discharge.

(ii) *Francis turbine.* It is named after James B. Francis who invented it. It is a reaction turbine and is suitable for medium head and medium discharge.

(iii) *Kaplan turbine.* It is named after its inventor Dr. Victor Kaplan. It is an reaction turbine and is used for low head and large discharge.

(iv) *Deriaz turbine.* It is named after its inventor Deriaz, a Swiss engineer. It is a reversible turbine-pump and suitable upto 200 m.

(c) Nature of Working on Blades. Turbines may be classified on the basis of energy conversion process when water impinges on blades. They may be *impulse type* or *reaction type*. The impulse type is one in which the whole head energy is converted into kinetic energy in nozzles and water comes out as a free jet and impinges on bucket, i.e., moving blades. During this process water is always in contact with air and thus theoretically there is no head or pressure drop in bucket. Such a turbine is Pelton turbine.

In reaction turbines, the entire flow takes place in a closed conduit. Part of head energy is converted into kinetic energy in nozzles and the runner converts both kinetic energy and head energy into mechanical energy. Many turbines such as Francis, Propeller, Kaplan and Deriaz are examples of reaction turbines.

(d) Direction of Flow. Based on the direction of flow with respect to wheel, the water turbines are classified as axial, radial (inward), tangential, mixed and diagonal flow turbine. In axial flow, water flows along the axis of shaft while in radial flow, the direction of flow is perpendicular to the axis of shaft. Mixed is the combination of axial and radial. Tangential flow is tangent to wheel while diagonal flow is neither axial or nor radial but angular with respect to axis. Table 11.2 gives the flow direction of various turbines used in practice.

(e) Orientation of Axis of Shaft. On the basis of orientation of axis of shaft, turbines are classified as *vertical* or *horizontal* turbine. Pelton turbines have horizontal shaft whereas Propeller, Kaplan, Francis and Deriaz turbines have vertical shaft.

Table 11.2 Flow directions in water turbines

Turbine types	*Flow direction*
Propeller and Kaplan turbines	Axial flow
Francis turbines	Radial inward flow or mixed flow
Pelton turbine	Tangential flow
Deriaz turbine	Diagonal flow

(f) Specific Speed. Specific speed of turbine is one of the most important parameter for the design of turbine. *It is defined as the speed of a geometrically similar turbine which produces 1 kW power under 1 m head. Mathematically, specific speed (N_s) is expressed as*

$$N_s = \frac{N\sqrt{P}}{H^{5/4}} \qquad \text{...(11.6)}$$

where N is speed in rpm, P is power developed in kW and H is head in metre. Although, technically, specific speed is not a dimensionless parameter but it exhibits the property of non-dimensional parameter. Table 11.3 shows a typical classification of water turbine based on specific speed. The specific speed varies from **4 to 1100 m** and the distribution has been made among slow, medium and fast speed for various types of turbines.

Table 11.3 A typical classification of turbines based on specific speed

Turbine	*Specific speed*		
	Slow	*Medium*	*Fast*
Pelton	4–20	21–35	36–70
Francis	60–150	151–250	251–400
Kaplan	300–450	451–700	701–1100

11.9 RANGE OF WATER TURBINE DESIGNS PARAMETERS

With the advancement of technology, water turbines have gone various changes with reference to their sizes, hydraulic design and efficiency. The present trend is towards bigger size. The turbine size and its numbers are chosen judiciously to obtain optimum working conditions as well as economy. Table 11.4 gives the ranges of water turbine design parameters.

Table 11.4 Ranges of water turbine design parameters

Types of turbine	*Maximum head (m)*	*Maximum power (MW) developed by the unit*	*Maximum wheel diameter of a runner (m)*	*Specific speed*
Pelton	300–2000	250	5.5	4–70
Francis	30–500	750	10.0	60–400
Kaplan	2–70	230	10.0	300–1100

11.10 THEORY OF PELTON TURBINES

11.10.1 Construction Details and Velocity Diagram

A Pelton turbine consists of one nozzle or a set of nozzles and a Pelton wheel. Figure 11.27(a) shows the construction details of a Pelton wheel. The runner consists of a large circular disc on the periphery of which a number of two-lobe ellipsoidal buckets are evenly mounted. Each bucket has a ridge or splitter in the middle which divides the jet into two equal stream as shown in Fig. 11.27(b). The symmetry of the bucket ensures zero momentum in the axial direction resulting in no axial thrust on the shaft bearing.

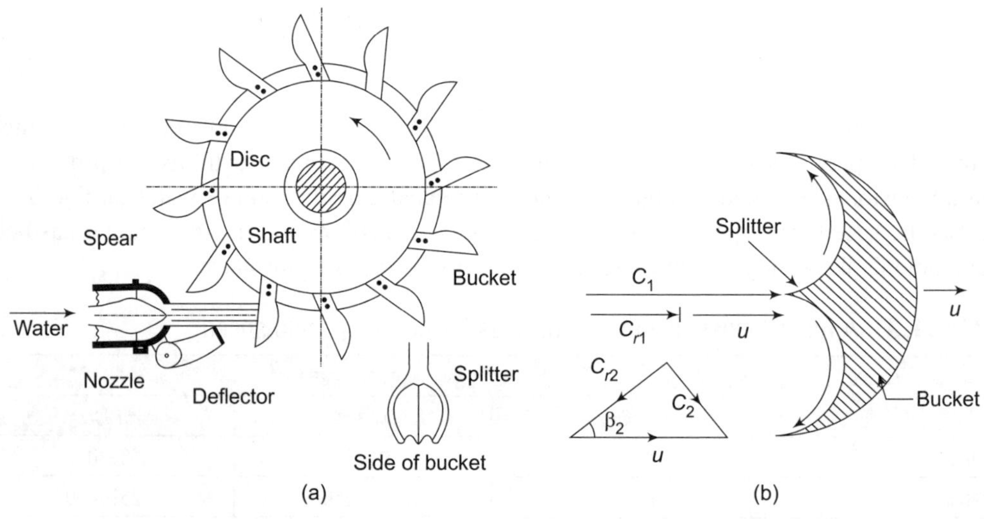

Figure 11.27 (a) Pelton wheel and nozzle (b) Shape of bucket and approaching flow

The whole head energy is converted into kinetic energy in the nozzle and comes out in the form of free jet. Pelton turbine is of impulse type.

The nozzle directs the flow on the wheel and the flow approaching to the bucket is governed with the help of a *spear valve* which is under the control of *governing system*. The water jet after impinging on the buckets is deflected through an angle of about 165° and flows axially in both directions thus avoiding the axial thrust.

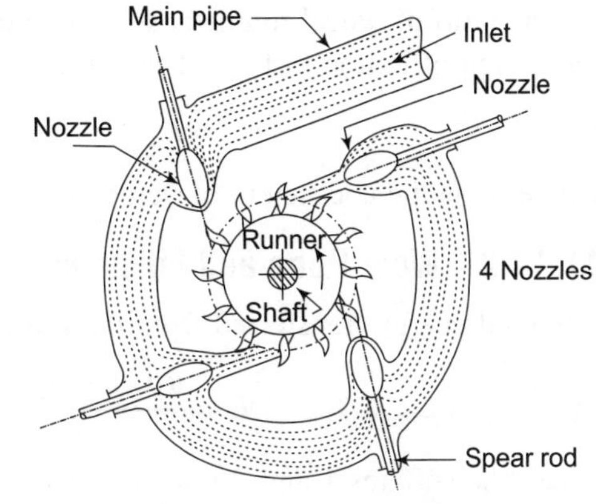

Figure 11.28 A four-jet Pelton wheel

In general, there is a single nozzle feeding water to the turbine, but if the discharge is more, up to six nozzles may be used, all symmetrically placed and causing rotation in the same direction. A four-jet Pelton wheel arrangement is shown in Fig. 11.28. Single jet Pelton wheel has invariably horizontal shaft but multi-jet Pelton wheel may have vertical shaft. The specific speed of the multi-jet Pelton wheel is given by

$$N_{s,mj} = \sqrt{n}\, N_s \qquad ...(11.7)$$

where n is the number of jets and N_s is the specific speed of a single jet.

The maximum number of nozzles used is 6 and the maximum specific speed of a single jet turbine wheel used is 6, so the maximum speed of multi-jet Pelton wheel will be about 70. In India, many Pelton turbines are in operation such as at Koyna (475 m head, 4 jets), Sharavathi (570 m head, 4 jets). The world's largest Pelton turbine is at Aurland-2, Brazil (840 m head, 6 jets) producing 243 MWe.

The **velocity diagram** for Pelton wheel is shown in Fig. 11.29. In Pelton wheel, $u_1 = u_2 = u$. The bucket deflection angle is around 160° or 165° and as a result the water after leaving the bucket falls freely to the tail race.

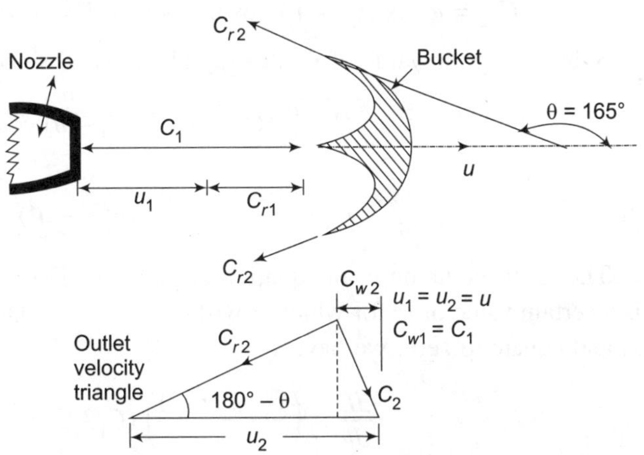

Figure 11.29 Velocity diagram for Pelton turbine

The peripheral velocity (u) of the bucket is given by

$$u = \omega \cdot r = \frac{\pi D N}{60} \qquad ...(11.8)$$

where r is the bucket circle radius, D is the wheel diameter, and N is the speed in rpm.

The ratio of wheel diameter to nozzle diameter (D/d) is an important size parameter for the design of Pelton turbine. The velocity of jet issuing from the nozzle is given by

$$C_1 = K_n \sqrt{(2gH)} \quad ...(11.9)$$

where K_n is the coefficient of velocity of nozzle and its value ranges from 0.97 to 0.99.

11.10.2 Work Done and Efficiency

The total energy transferred from the fluid to the wheel is given by Euler's equation. Thus,

$$E = \frac{u_1 C_{w1} - u_2 C_{w2}}{g} = \frac{u}{g}(C_{w1} - C_{w2}) \quad ...(11.10)$$

where the suffixes 1 and 2 refer to inlet and outlet conditions of runner, suffixes w, r, and f represent whirl, relative and flow velocities respectively.

From the outlet velocity diagram, we have

$$C_{w2} = u - C_{r2} \cos(180 - \theta) = u + C_{r2} \cos\theta \quad ...(11.11)$$

where θ is the bucket deflection angle (say $\theta = 165°$)

The relative velocity of water at the exit is reduced due to friction and thus given by

$$C_{r2} = kC_{r1} = k(C_1 - u) \text{ where } k = \text{blade velocity coefficient} \quad ...(11.12)$$

From Eqs. (11.11) and (11.12), we have

$$C_{w2} = u + k(C_1 - u)\cos\theta, \quad \text{where } k = \text{blade velocity coefficient} \quad ...(11.13)$$

Substituting the value of C_{w2} in Eq. (11.10) and since $C_{w1} = C_1$ we have

$$E = \frac{u}{g}[C_1 - u - k(C_1 - u)\cos\theta]$$

or

$$E = \left(\frac{1 - k\cos\theta}{g}\right)(uC_1 - u^2) \quad ...(11.14)$$

The above equation is an equation of parabola. For a given value of k, $\cos\theta$ and C_1, there is a certain value of u, for which E will be maximum. Differentiating above Eq. (11.14) w.r.t. u, and equate to zero, we have

$$\frac{dE}{du} = \left(\frac{1 - k\cos\theta}{g}\right)(C_1 - 2u) = 0$$

or

$$\frac{u}{C_1} = 0.5 \quad ...(11.15)$$

Substituting the optimum value of u in Eq., (11.14), we have

$$E_{max} = \left(\frac{1 - k\cos\theta}{g}\right)\frac{C_1^2}{4} \quad ...(11.16)$$

The blading or diagram efficiency ($\eta_b = \eta_d$) is defined as the ratio of work output by a stage of the turbine to inlet kinetic energy. Thus,

$$\eta_b = \frac{E}{C_1^2/2g} = \frac{(1 - k\cos\theta)}{g}(uC_1 - u^2)\frac{2g}{C_1^2}$$

or $$\eta_b = \eta_d = 2\,(1 - k\cos\theta(\rho - \rho^2) \qquad ...(11.17)$$

where $$\rho = u/C_1$$

The variation of η_b with u/C_1 is plotted and is shown in Fig. 11.30.

For maximum efficiency, η_b is differentiated with w.r.t. ρ and equated to zero.

$$\frac{d\eta_b}{d\rho} = 0,$$

Thus, $$2(1 - k\cos\theta)\,(1 - 2\rho) = 0$$

or $$\rho_{opt} = 1/2 \qquad ...(11.18)$$

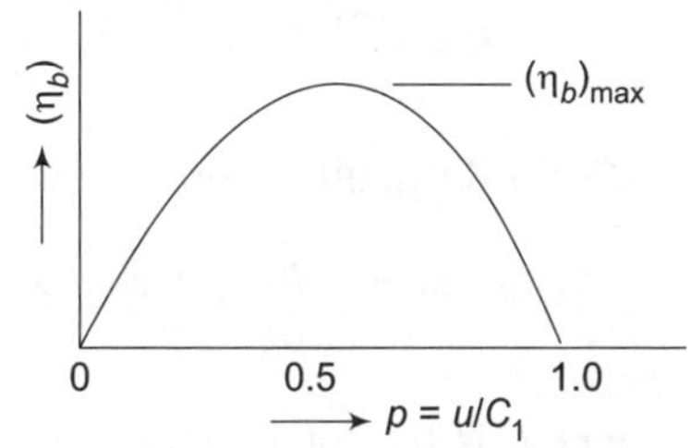

Figure 11.30 Variation of η_b with ρ

Thus, we find that for the maximum diagram or blade efficiency, the ratio of u/C_1 is just half.

The same expression has been found with impulse steam turbine.

Substituting the ρ_{opt} in Eq. (11.17), we get $(\eta_b)_{max} = (\eta_d)_{max}$

$$(\eta_b)_{max} = \frac{1 - k\cos\theta}{2}. \qquad ...(11.19)$$

The diagram of blading efficiency for the maximum work is given by

$$(\eta_b)_{max,\ work} = \frac{E_{max}}{C_1^2/2g} = \frac{1 - k\cos\theta}{2} = (\eta_b)_{max}$$

For no blade friction, $k = 1$, thus $(\eta_b)_{max} = \dfrac{1 - \cos\theta}{2}$

Generally, the value of k has been found to vary from 0.8 to 0.85.

If $$k = 0.8 \text{ and } \theta = 165°,$$

$$(\eta_b)_{max} = 0.886 \qquad ...(11.20)$$

Though, theoretically $\rho_{opt} = (u/C_1)_{opt} = 0.5$, but practically $\rho_{opt} = 0.46$ is used.

The minimum number of buckets (z) on the wheel is given by an empirical equation.

$$z = \frac{m}{2} + 15 \text{ where } m \text{ is the jet ratio } (D/d) \qquad ...(11.21)$$

11.10.3 Erosion and Cavitation

Erosion, and cavitation problems arise during operation of Pelton turbines. The erosion problems occur on the buckets due to erosive effect of flow and at the nozzle due to cavitation effect. Cavitation occurs at the exit of turbine or nozzles. In order to protect the buckets from wear and tear due to erosion and cavitation, chrome alloy steel or stainless steel is used as material.

11.11 DEGREE OF REACTION FOR WATER TURBINES

Degree of reaction is defined as the ratio of energy transfer due to static pressure drop in the rotor to the total energy transfer to the turbine.

$$R = \frac{\text{Energy transfer due to static pressure drop in the rotor}}{\text{Total energy transfer to the turbine}} \quad ...(11.22)$$

From Bernoulli's equation applied to the inlet of outlet of water turbine, we have

$$\frac{p_1}{\rho g} + \frac{C_1^2}{2g} = E + \frac{p_2}{\rho g} + \frac{C_2^2}{2g} \quad ...(11.23)$$

where E is the total energy transferred from the fluid to the turbine and ρ is the density of water. Thus,

$$E = \left(\frac{p_1 - p_2}{\rho g}\right) + \left(\frac{C_1^2 - C_2^2}{2g}\right) \quad ...(11.24)$$

In the above equation, the first term on the R.H.S. represents the energy transfer due to static pressure drop and the second term represents the energy transfer due to drop in velocity head.

For impulse turbine (i.e., Pelton wheel), $p_1 = p_2$, there is no static pressure drop, so for *impulse turbine*

$$(E)_{\text{impulse}} = \frac{C_1^2 - C_2^2}{2g} \quad ...(11.25a)$$

$$(E)_{\text{reaction}} = \frac{p_1 - p_2}{\rho g} \quad ...(11.25b)$$

If $C_1 = C_2$, then E is expressed by Eq. (11.25b). Equation (11.25b) is valid for *pure reaction turbine.*

So, the degree of reaction (R) is expressed as

$$R = \frac{(p_1 - p_2)/\rho g}{E} = \frac{(p_1 - p_2)/\rho g}{[(p_1 - p_2)/\rho g] + [C_1^2 - C_2^2)/2g]} \quad ...(11.26)$$

Thus, $R = 0$ for pure impulse and $R = 1$ for pure reaction.

Further, Eq. (11.26) is modified as below

$$R = 1 - \frac{(C_1^2 - C_2^2)}{2gE} \quad ...(11.27)$$

From Euler's equation, $E = (u_1 C_{w1} - u_2 C_{w2})/g$

Thus, $$R = 1 - \frac{C_1^2 - C_2^2}{2(u_1 C_{w1} - u_2 C_{w2})} \quad ...(11.28)$$

The reaction of a water turbine may be calculated either from Eq. (11.26) or (11.27) or (11.28).

11.12 THEORY OF FRANCIS WATER TURBINES

Francis turbines are widely used world over. It operates under the head varying from 30 to 500 m. The single unit may develop power as high as 750 MW. The specific speed ranges from 60 to 400. It is a reaction turbine. Formerly, its specific speed was limited to about 60 and it was *radial inward flow* type but at present they are *the mixed flow type with radial entry and axial exit.*

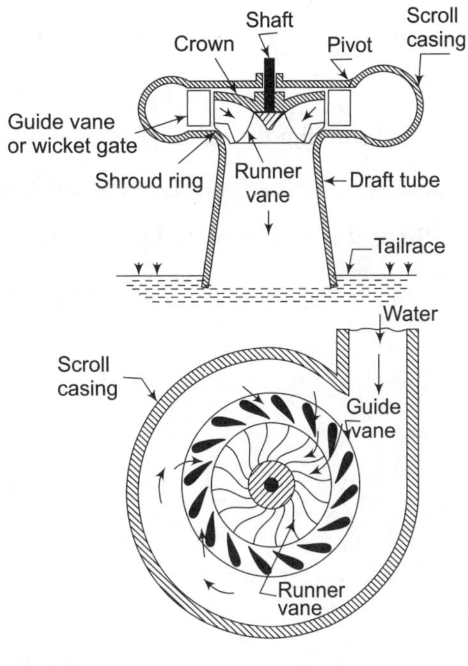

Figure 11.31 A Schematic of a Francis water turbine

11.12.1 Construction Detail

Figure 11.31 shows the schematic of a Francis turbine. It consists of a spiral casing, guide vanes (adjustable) and a runner. Water from the reservoir through the penstock enters the spiral or scroll casing which surrounds the runner. In order to keep the water velocity constant all along the flow path inside the spiral casing, its cross-sectional area diminishes uniformly along the circumference. The water enters the adjustable guide vanes or wicket gates which are pivoted and can be turned suitably to control the flow and output. The purpose of guide vanes is to impart a whirl component of velocity or angular momentum of the water before entering to the runner. The runner consists of number of curved blades numbering **12 to 22**. These blades are welded to the shroud of the runner. The blade shape is such that it changes the direction of water flow from radial to axial direction. After leaving the runner, water enters the draft tube (either straight or elbow type) which increases the head by reducing the kinetic energy and finally falls to the tail race.

It is worth to note that the pressure of water at the inlet of the runner is more than that at the outlet, so it is essential that the flow must take place in a closed conduit. The runner is always full of water and the draft tube is submerged into the tail race.

11.12.2 Velocity Diagram, Work Done and Efficiency

Figure 11.32 shows the velocity diagram at inlet and outlet of the runner. The water comes out from the guide vanes at velocity C_1 and with peripheral velocity u_1, the water impinges the curved blade with velocity C_{r1} at inlet and comes out with velocity, C_{r2}. The discharge is axial, i.e., $C_{w2} = 0$, $\alpha_2 = 90°$ and $C_2 = C_{f2}$.

From Euler's equation, the energy transfer to the turbine is

$$E = (u_1 C_{w1} - u_2 C_{w2})/g \quad ...(11.29)$$

Since $C_{w2} = 0$, so $E = u_1 C_{w1}/g$

But $C_{w1} = C_{f1} \cot \alpha_1$ where α_1 is exit angle of guide blades

$\therefore$ $u_1 = C_{f1} (\cot \alpha_1 - \cot \beta_1)$ where β_1 is inlet blade angle

$\therefore$ Substituting the value of u_1 and C_{w1} in Eq. (11.29), we have

$$E = C_{f1}^2 \cot \alpha_1 (\cot \alpha_1 - \cot \beta_1)/g \quad ...(11.30)$$

Since it is a reaction turbine, so the total energy input is

$$E_{\text{input}} = E + E_1 \quad \text{where} \quad E_1 = \text{loss of kinetic energy} = \frac{C_{r2}^2 - C_{r1}^2}{2g}$$

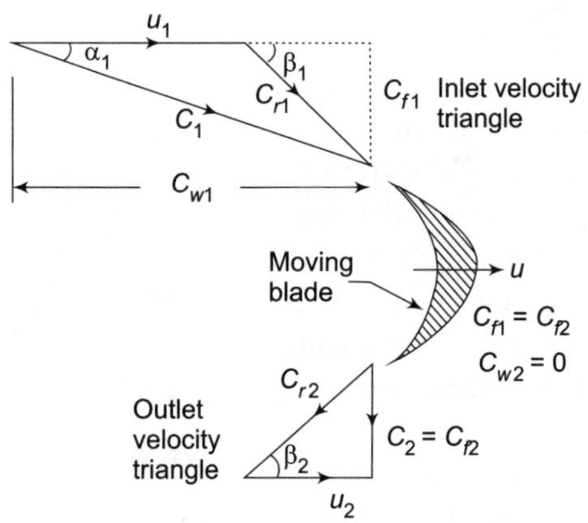

Figure 11.32 Velocity diagram for a francis turbine

Thus, the blading or diagram efficiency is expressed as

$$\eta_b = \eta_b = \frac{E}{E_{\text{input}}} = \frac{E}{E + E_1} = 1 - \frac{E_1}{E + E_1}$$

$$= 1 - \frac{1}{1 + 2 \cot \alpha_1 (\cot \alpha_1 - \cot \beta_1)} \quad ...(11.31)$$

Degree of reaction, $R = 1 - \dfrac{C_{f1}^2 \cot \alpha_1^2}{E} = 1 - \dfrac{C_{w1}^2}{E}$ ...(11.32)

Hydraulic efficiency is expressed as

$$\eta_h = \frac{E}{H} = \frac{u_1 C_{w1}}{gH} \quad ...(11.33)$$

The overall efficiency of the turbine (η_0) is expressed as

$$\eta_0 = \frac{P}{\rho Q g H} \qquad ...(11.34)$$

where P is total power output from turbine, ρ is the density of water (m^3/kg), Q is discharge (m^3/s) and H is head (m).

11.13 THEORY OF PROPELLER AND KAPLAN WATER TURBINES

Propeller turbines are axial flow and reaction types machines. They are suitable for low heads (2–70 m) and high specific speed (300–1100) capable of handling large volume of water.

11.13.1 Construction Details

Figure 11.33 shows a **propeller turbine** which consists of as axial flow runner usually with four to six blades of aerofoil shape. The spiral casing and guide blades are similar to those in Francis turbines. The runner blades are fixed and non-adjustable.

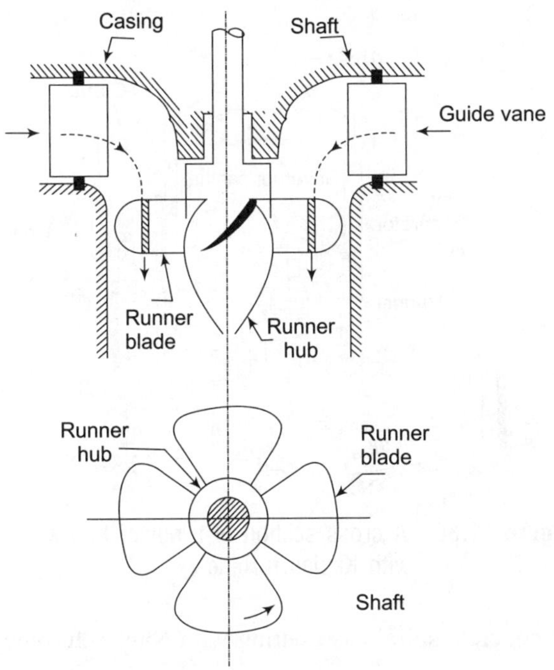

Figure 11.33 A schematic diagram of a propeller turbine

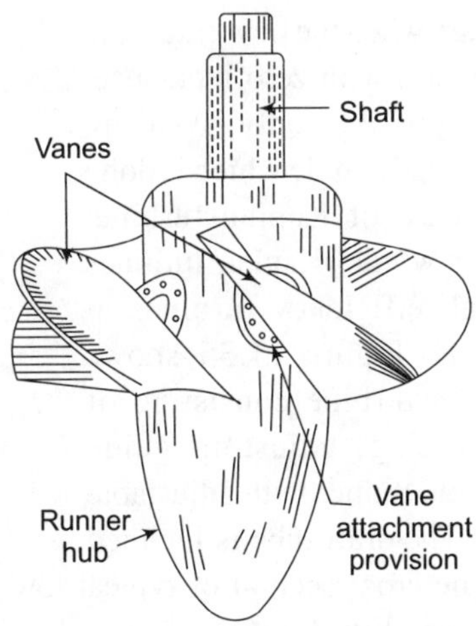

Figure 11.34 A schematic view of a Kaplan rotor with adjustable blades

Kaplan turbine is a *special type of a propeller* turbine in which the *individual runner blades are pivoted on the hub* so that their inclination may be adjusted during operation as per load (Fig. 11.34). The blades are adjusted automatically rotating about the pivots with the help

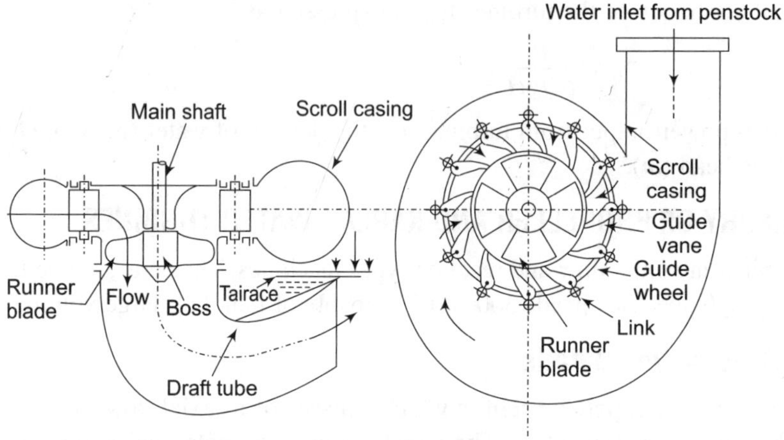

Figure 11.35 A schematic view of a Kaplan turbine

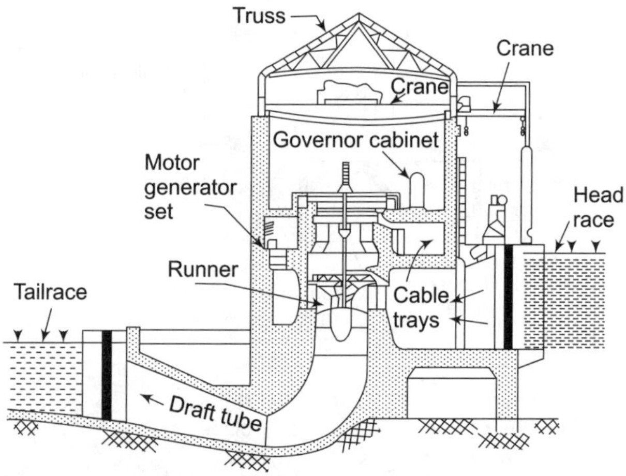

Figure 11.36 A cross section of a power house with Kaplan turbine

of a governor servo-mechanism. The varying demands mean varying flow rates which means varying approaching flow angle. For the maximum efficiency, it is necessary when the flow enters the axial blades with zero incidence. This is possible only by using varying blade angle which is done in the case of Kaplan turbine. In other words, Kaplan turbines are high efficiency turbines at all loads. Figure 11.35 shows a Kaplan turbine consisting of scroll casing, adjusting guide blades and runner with adjustable blades. The draft tube is inclined type. The cross section of typical low head concrete spiral case setting with Kaplan turbine is shown in Fig. 11.36.

11.13.2 Velocity Triangle, Work Done and Efficiency

Figure 11.37 shows the velocity triangle for propeller and Kaplan turbine. Since the peripheral velocity varies from root to rip, so the velocity triangles are different in symmetry. *The expressions for work done, and efficiency are the same as that of Francis turbine.*

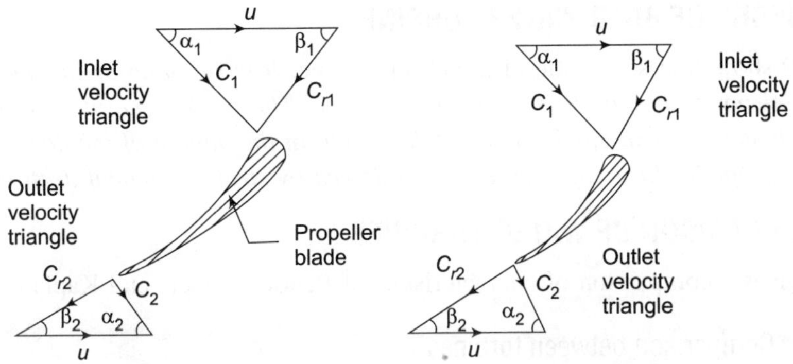

(a) Near the tip (b) Near the hub

Figure 11.37 Velocity triangles for propeller and Kaplan turbines

11.14 THEORY OF DERIAZ WATER TURBINE

A schematic diagram of Deriaz turbine is shown in Fig. 11.38. It is a diagonal turbine and the flow over the runner is at an angle of 45° to the axis. Like Kaplan turbine, its runner has adjustable blades.

The flow is diagonal or mixed. So far its suitability is concerned, it may be placed between two turbines, Kaplan and Francis, and is suitable for head up to 200 m. The number of blades on the runner varies from 10 to 12. The guide blades are inclined and adjustable.

This turbine is most suitable for reversible flow conditions when the turbine works as a pump in pump storage power plants. The draft tube is straight conical type.

The velocity triangles, the expression for work done, reaction and efficiency are the same as that of Kaplan turbine.

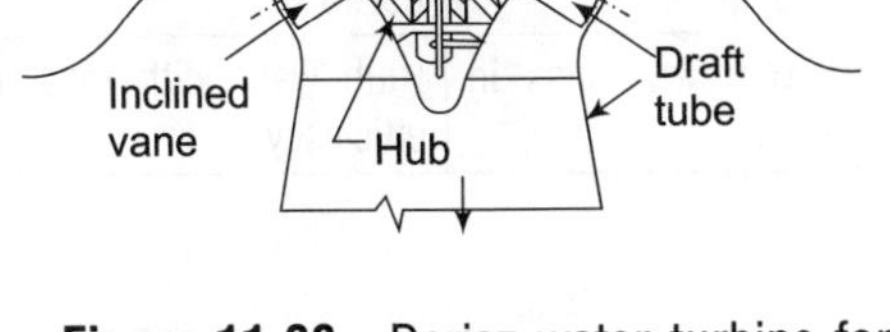

Figure 11.38 Deriaz water turbine for reversible operation

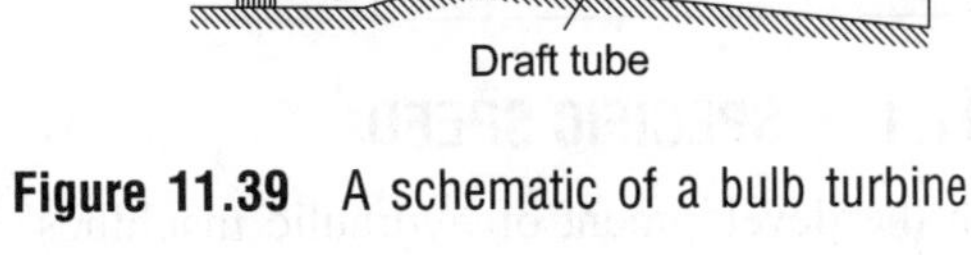

Figure 11.39 A schematic of a bulb turbine

11.15 THEORY OF BULB WATER TURBINE

Figure 11.39 shows the schematic of a bulb or turbine. *It is suitable for low head in micro, mini and tidal power plants. It is an axial flow type. The turbo-generator is housed in an enclosed bulb-shaped casing, which is installed right in the middle of the flow passage. The bulb and the propeller form an integral unit followed by straight conical draft tube.*

11.16 COMPARISON OF WATER TURBINES

Table 11.5 gives a comparison of characteristics of Pelton, Francis and Kaplan turbines

Table 11.5 Comparison between turbines

S. No.	*Parameters*	*Pelton turbine*	*Francis turbine*	*Kaplan/Propeller turbine*
(i)	Flow, stage and impulse or reaction	Tangential, single stage, impulse type	Inward radial or mixed flow, single stage, reaction type	Axial flow, single stage, reaction type
(ii)	Maximum generating capacity	250 MW	750 MW	230 MW
(iii)	Jets/guide blades	1 to 6 jets, 2 for horizontal and 6 for vertical	Adjustable guide blades	Adjustable guide blades
(iv)	Runner blades	Fixed	Fixed	Fixed for propeller and adjustable for Kaplan
(v)	Head	300-2000 m	30-500 m	2 to 70 m
(vi)	rpm	75-1000	93-1000	72 to 600
(vii)	Specific speed	4-70	60-400	300-1100
(viii)	Regulation mechanism	Spear nozzle and deflector plate	Guide vanes and relief valve	Guide vanes and runner vanes and relief valve better controlled
(ix)	Variation in operating head	Less controlled	Better controlled	Better controlled
(x)	Size of runner, generator and power house	Larger	Smaller	Smaller
(xi)	Mechanical efficiency	Decreases faster with time	Less decrease with time	Less decrease with time
(xii)	Hydraulic efficiency	85-90%	90-94%	85-93%
(xiii)	Uncleaned water	Less wear	High wear with loss in efficiency	High wear with loss in efficiency

11.17 SPECIFIC SPEED

In the development of hydraulic machines, first the model is prepared for hydraulic test to predict the performance of prototype machine. The concept of specific speed helps in this regard.

The specific speed of a turbine is defined as the speed of operation of a geometrically similar model of the turbine which produces 1 kW power when operating under 1 m head. The expression for specific speed is derived as follows.

$$\text{Power, } P = \rho QgH \propto (AC).H \qquad ...(11.35)$$

where A is the cross-sectional flow area and C is the velocity of water

or $$P \propto D^2(2gH)^{1/2}\, H \propto D^2 . H^{3/2} \qquad ...(11.36)$$

where D stands for wheel diameter through which water flows axially and H stands for net head of water.

The blade velocity = $u \propto C$

or $$DN \propto (2gH)^{1/2} \qquad ...(11.37)$$

or $$D \propto H^{1/2}/\text{N} \qquad ...(11.38)$$

From Eq. (11.36) and Eq. (11.38), we have

$$P \propto \frac{H}{N^2}.H^{3/2} \quad \text{or} \quad N^2 \propto \frac{H^{5/2}}{P} \quad \text{or} \quad N \propto \frac{H^{5/4}}{P^{1/2}} \quad \text{or} \quad N = K\frac{H^{5/4}}{P^{1/2}} \quad \text{or} \quad K = N\frac{P^{1/2}}{H^{5/4}}$$

If $P = 1$ kW, $H = 1$ m, $K = N = N_s$ = Specific speed

$\therefore$ $$N_s = \frac{NP^{1/2}}{H^{5/4}} \qquad ...(11.39)$$

It is to be noted that the specific speed is not a non-dimensional number but it serves its purpose.

The non-dimensional specific speed (N_s) is expressed as

N_s' is not used in practice.

The ranges of specific speed for various water turbines are already given earlier.

11.18 SCALE RATIO

The scale ratio is defined as the ratio of the diameters of the model turbine and the prototype turbine.

We have seen that $u \propto C$

or $$DN \propto \sqrt{H} \qquad ...(11.40)$$

Thus, the ratio of DN for model and prototype is given as

$$\frac{D_m N_m}{D_p N_p} = \sqrt{\frac{H_m}{H_p}}$$

or $$\frac{D_m}{D_p} = \sqrt{\frac{H_m}{H_p}}.\frac{N_p}{N_m} = \text{Scale ratio} \qquad ...(11.41)$$

where suffixes m and p stand for model and prototype.

11.19 UNIT SPEED, UNIT POWER AND UNIT DISCHARGE

The concept of unit speed (N_u), unit power (P_u) and unit discharge (Q_u) are frequently used to express the operational characteristics of hydraulic turbine. Using the concept, the variation of speed, power and discharged for various turbines are found out.

11.19.1 Unit Speed

The unit speed (N_u) is defined as the speed of a geometrically similar turbine working under a head of 1 m.

We have $u \propto C$

or $DN \propto \sqrt{H}$

or $N \propto \sqrt{H}$ for geometrically similar turbine

or $N = K\sqrt{H}$ where K = constant. If $H = 1$ m, $N = N_u = K$.

Thus,
$$N_u = \frac{N}{\sqrt{H}} \qquad \text{...(11.42)}$$

11.19.2 Unit Power (P_u)

It is defined as kW of power generated by a geometrically similar turbine working under a head of 1 m.

We have $P = \rho Q\, g\, H = \rho\,(AC)\, gH = \rho A\,\sqrt{(2gH)}\; gH$

or $P \propto H^{3/2} = KH^{3/2}$

If $H = 1$, $P = P_u = K$,

so,
$$P_u = P/H^{3/2} \qquad \text{...(11.43)}$$

11.19.3 Unit Discharge (Q_u)

It is the flow rate through the turbine working under a head of 1 m.

We have $Q = AC = A\sqrt{2gH} \propto \sqrt{H}$ or $Q = K\sqrt{H}$

If $H = 1$, $Q = Q_u = K$, thus
$$Q_u = \frac{Q}{\sqrt{H}} \qquad \text{...(11.44)}$$

11.20 CAVITATION PROBLEM IN WATER TURBINES

Cavitation is one of the undesirable problems in the operation of water turbines. *The formation of water vapour and air bubbles on the water surface due to reduction of pressure and sudden collapse are known as cavitation.* In any part of the turbine, if pressure drops below the vapour pressure at that temperature, some of the liquid flashes into vapour. The bubbles formed during vaporization and carried by the water stream to higher pressure zones, where the bubbles condense into liquid forming a cavity or vacuum. The surrounding liquid rushes towards the cavity giving rise to a very high level pressure as high as 7000 atm. The rapid formation and collapsing of the bubbles cause the pitting of the metallic surface accompanied with vibration of noise. This phenomenon is known as *cavitation.* It also reduces the hydraulic efficiency of turbine.

Since the chance of reduction of pressure below atmospheric occurs only *at the exit of runner blade of turbine or inlet of draft tube* (Eq. 11.45), so the cavitation occurs at the exit of runner. The energy balance at the exit of runner (*er*) and exit of draft tube (*ed*) yields

$$\frac{p_{er}}{\rho} = \frac{p_a}{\rho} - \left[h + \left(\frac{C_{er}^2 - C_{ed}^2}{2g}\right) - h_f\right] \quad \text{...(11.45)}$$

where h is height between turbines exit and tail race, h_f is the head loss between turbine and draft tube exit, p_a is the atmospheric pressure exerting on the tail race, p_{er} is the pressure of water at the exit of runner, C_{er} and C_{ed} are water velocities at the exit of runner and draft tube respectively.

11.20.1 Cavitation Factor

Prof. D. Thowe of Germany suggested a cavitation factor to determine the zone where the turbines can work without any danger of cavitation.

The critical value of cavitation factor is given by

$$\sigma_c = \frac{(H_a - H_v) - h}{H} \quad \text{...(11.46)}$$

where H_a = atmospheric pressure head (*m*), H_v = vapour pressure (*m*) at water temperature, H = working head of turbine (*m*), h = height of turbine outlet above tail race level (*m*).

The values of σ_c depend upon specific speed and turbine which are given in Table 11.6.

Table 11.6 Values of σ_c for various values of N_s and different turbines

Francis turbines		*Kaplan turbines*	
N_s	σ_c	N_s	σ_c
50	0.04	300–450	0.35–0.4
100	0.05	450–550	0.4–0.45
150	0.07	550–600	0.46–0.6
200	0.1	650–700	0.85
250	0.14	700–800	1.05
300	0.2		
350	0.27		

11.20.2 Methods to Avoid Cavitation

The following are the methods to avoid cavitation in water turbines.

(i) *Installation of turbine below tail race level.* The danger of cavitation increases in the case of low head and high speed propeller runner as the value of $(C_{er}^2 - C_{ed}^2)/2g$ is considerably large in Eq. 11.45. In order to keep the value of pressure at the exit of runner (p_{er}) within the cavitation limitation, the value of h is made negative keeping the runner below tail race level. For such installations, the turbines remain always under water. The main disadvantage of this method is the difficulty in inspecting and repairing the turbine.

(ii) *Cavitation free runner.* By designing properly the cavitation free runner, the problem of cavitation is avoided. This is possible by selecting the proper shape of blade, blade angle and thickness of blade, etc.

(iii) *Use of material.* The cavitation effect can be reduced by selecting a material which can resist the cavitation effect. Cast steel, stainless steel and alloy steel material for runner are in order of increasing resistant to cavitation. It has been observed that the welded part is more resistant to cavitation than the casted.

(iv) *Polishing.* By using polished surface, the cavitation effect is minimised. The cast steel runners are coated with stainless.

(v) *Selection of specific speed.* By selecting a runner of proper specific speed, it is possible to avoid cavitation.

11.21 GOVERNING OF HYDRAULIC TURBINES

The purpose of governing of hydraulic turbines is to maintain the speed of turbines fairly constant irrespective of the load. The generator coupled with hydraulic turbine is required to run at constant speed irrespective of the load so that the frequency of electricity generated remains fairly constant with a small permissible deviation (±2) otherwise electrical appliances will burn. The frequency or speed is expressed as

$$N = \frac{120 f}{p} \quad \text{...(11.47)}$$

where f is the frequency of electricity generated, N is the speed of generator known as synchronous speed and p is the number of poles on the generator.

When the load on the generator varies with the variation of consumer's demand and when input to the turbine remains the same, then the speed of the turbine or the frequency will vary due to the difference in power output and load which is not desired as all the electrical appliances and machines are designed to run on nominal frequency. Therefore, the speed of the turbine is always maintained fairly constant irrespective of the load.

This is made possible by using a **governing system** which regulates the quantity of water flowing through the turbine in proportion to the load so that the power output from the turbine is equal to the load and by this way the speed is maintained constant.

11.21.1 Governing System of Impulse Water Turbines

Figure 11.40 shows the governing system of impulse water (Pelton) turbines. It consists of a centrifugal governor mounted on shaft which is geared with turbine shaft relay valve, servomotor, spear, deflector plate, oil sump, oil pump and linkages. In this case, water flow is regulated by joint action of the spear and the deflector plate. When the load on the generator drops, the speed of the turbine runner increases due to accelerating torque caused by the difference between turbine output and generator load. As a result, the flyballs of centrifugal governor fly outward due to higher speed. The sleeve moves up, the portion of the lever to the right fulcrum moves down pushing the piston rod of relay or control valves downward. The relay consists of two valves – V_1 and V_2. With the downward movement of piston rod, valve V_1 closes and valve V_2 opens. The oil is supplied to the relay valve at high pressure by gear pump, the high pressure oil enters to the left of servomotor and pushes the piston rod which partly closes the flow area, and hence the required (less) amount of water enters to the turbine so that the turbine output is equal to the generator load and by this way speed is maintained constant.

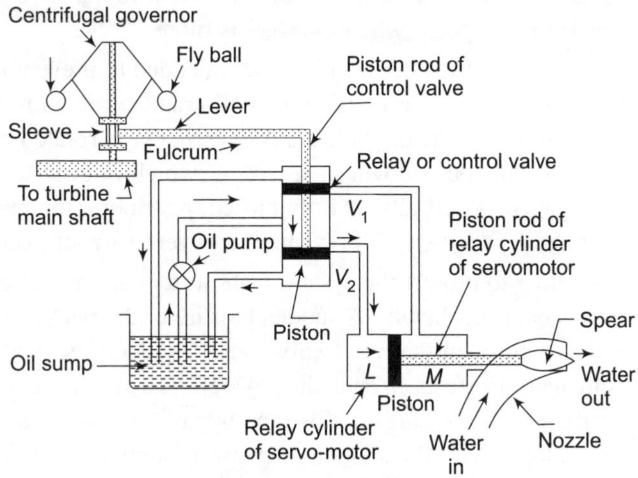

Figure 11.40 Governing system of impeller (Pelton) turbine

When the load on the generator increases, the speed decreases, the valve V_1 of relay opens while valve V_2 closes. The high pressure oils enters through valve V_1 opening to the right portion of servometer which pushes the piston to the left hand side and by this way the spear moves to the left and thus supply more water to the turbine resulting in turbine output equal to the load. By this way, the speed of the turbine in maintained fairly constant irrespective of the load.

It is worthy to note that the spear or needle valve is normally used for small load fluctuations. In the event of sudden fall of load, the spear moves very rapidly to close the nozzle which results in water hammer problem in penstock. In order to avoid the water hammer problem, a *deflector* is introduced in the system (not shown). The purpose of deflector is to deflect some water from the jet advancing to the turbine runner is the event of sudden drop in load. The deflected water without doing any work falls into the tail race. Thus, in the case of sudden rapid drop in load, the speed is maintained constant by the joint action of spear and deflector plate.

11.21.2 Governing System of Reaction Water Turbines

Figure 11.41 shows the governing system of reaction water turbines. This system is similar to that of impulse turbine except that the motion of piston in servomotor is used to partially close or open the guide vane gates through which water is supplied to the turbine instead of spear in the nozzle. When the load is less than design on the generator, the centrifugal governor senses the speed and valve V_1 closes while V_2 of relay valve opens which allows the high pressure oil to enter the left portion of servomotor. This pushes the piston forward and partially close the passage of guide blades through rock and pinion arrangement via linkages. Reverse takes place when the load is more.

A compensating device is, however, added to prevent the governor from overshooting. The compensating device consists of ball-crank lever with a pivot when the servomotor piston moves to the right, the bell-crank lever *EFG* is rotated downward about *F* and the arm *G* is lowered. This pulls down the pivot. *A*, which in turn, lowers the fulcrum *B*. By this way, the relay part *B* is partially or fully closed, restricting the piston motion of the servomotor to the right and by this way it prevents the governor from overshooting.

In order to protect the system from water hammer, a pressure *relief valve* is incorporated in the system (not shown). A sudden closure of the wicket gate of guide vanes due to sudden fall in load will open the relief valve due to sudden increase in pressure and protect the conduit from inertia effects of speeding water. The relief valve consists of a spear and is held by the fluid (oil) pressure to close the bypass of water from the spiral casing to the tail race at design load. With the sudden decrease in load, a bell-crank lever opens the pilot valve of the pressure chamber so that the pressure on the spear is reduced which permits the spear to be lifted up and allows a portion of water to flow directly from the spiral casing to the tail race through the bypass without passing through runner blades.

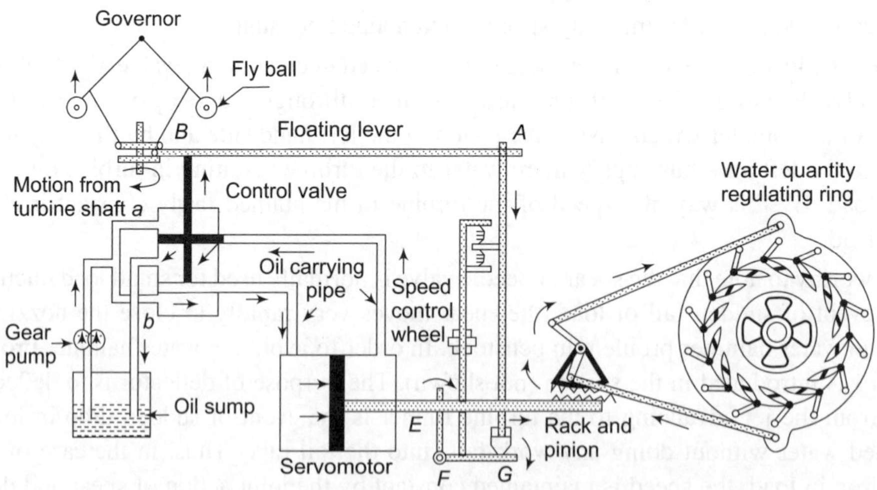

Figure 11.41 Governing of reaction (Francis) water turbine

Kaplan turbine is governed by the simultaneous control of adjustable guide vanes and runner vanes. Both are operated by separate servomotor and a control valve which are interconnected with those of guide vanes to proper inclination of runner vanes for a particular setting of guide vanes.

11.22 PERFORMANCE OF WATER TURBINES

In real situation, turbines are supposed to work also at off-design conditions. In order to know the off-design performance of water turbines, tests are conducted on prototype, a model machine and the results are plotted in graphical forms called *characteristic curves*. Generally, these curves are plotted in terms of unit quantities as given below:

(a) Constant head characteristics curves
(b) Constant speed characteristics curves
(c) Constant efficiency curves

11.22.1 Constant Head Characteristics Curves

In order to plot these curves, tests are conducted by maintaining a constant head and constant gate opening and the speed of the turbine is varied by varying the load on the turbine. By this way, a series of values of speed (N) are obtained. For each value of N_s, the discharge and output of the turbine are measured. Similar tests are conducted for different gate opening. The values of Q_u, P_u and N_u and overall efficiency (η_o) are computed for each gate opening. Such characteristics are shown in Fig. 11.42 for various types of turbines. Q_u is independent of N_u and it is only a function of gate opening as the discharge only varies with gate opening. For Kaplan turbine, Q_u increases with N_u linearly for all positions of gate opening. Obviously, the higher Q_u will be for full gate opening. But for Francis turbines, Q_u decreases with the increase of N_u.

The variation of P_u vs N_u and η_0 vs N_u are parabolic in nature for all types of turbines.

There is a maximum value of P_u and η_0 at a particular N_u for any gate opening, however, the optimum value of N_u is different for different gate openings.

11.22.2 Constant Speed Characteristics Curves

In order to obtain constant speed characteristics curves say efficiency vs load, the constant speed is attainted by regulating the gate opening, i.e., discharge as the load varies. The head may or may not remain constant. Figure 11.43 shows the overall efficiency vs load for different turbines. It is obvious from the results that overall efficiency increases with load and attains its maximum value at full load, however, Kaplan and Pelton turbines maintain the high efficiency over a longer of part load (i.e., about 40% to 100%) as compared to other turbines.

The variation of overall efficiency and power with discharge is shown in Fig. 11.44. The overall efficiency increases with discharge and it attains a maximum value while the power output increases with discharge linearly.

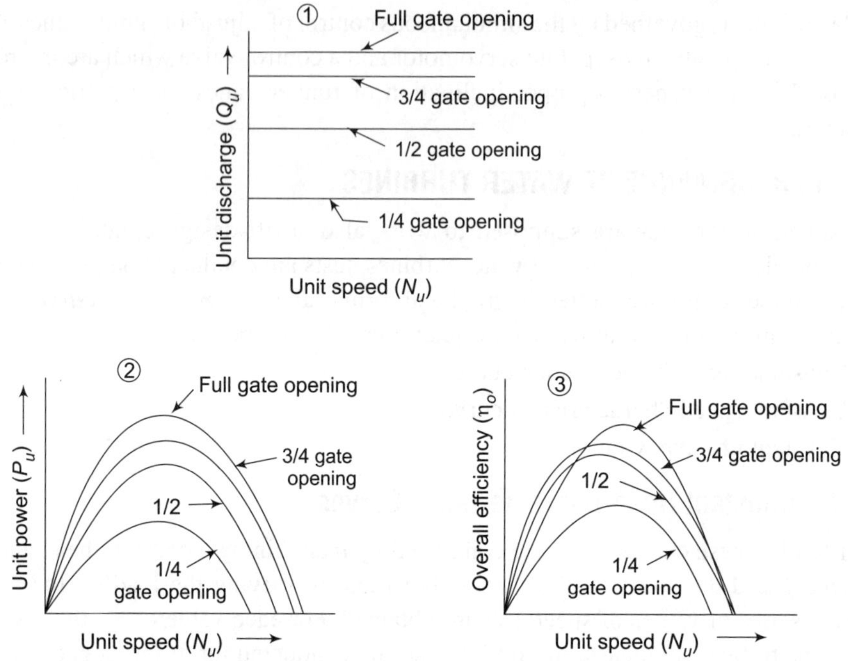

(a) For Pelton wheel

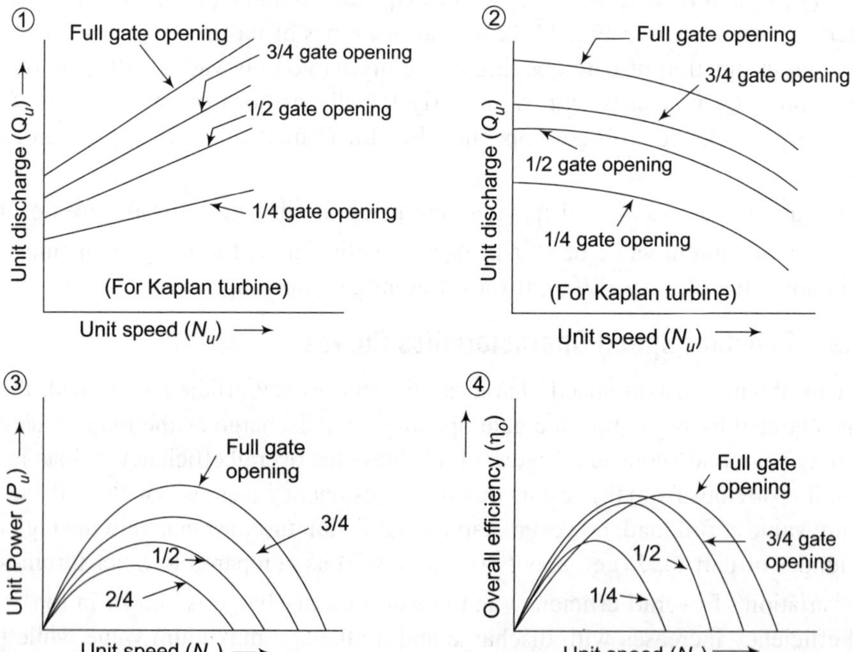

(b) For reaction turbine

Figure 11.42 Constant head characteristic curves for impulse and reaction turbines

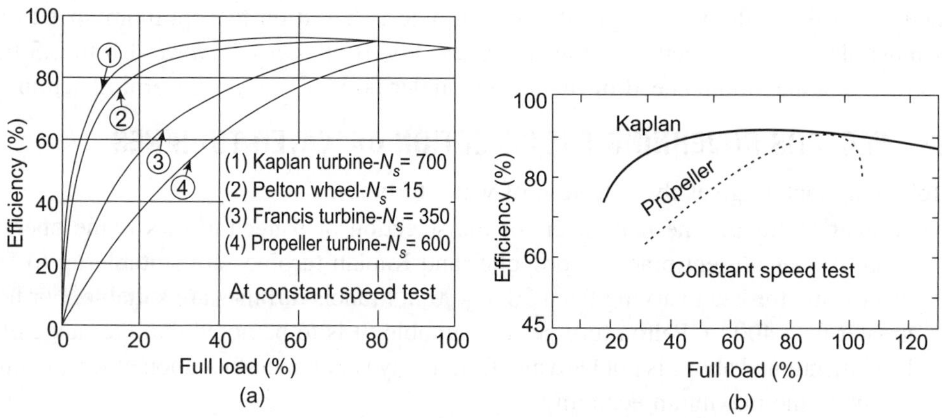

Figure 11.43 (a) and (b) Overall efficiency vs Load for various turbines at constant speed

11.22.3 Constant Efficiency Curves

Constant efficiency or universal characteristics curves are shown in Fig. 11.45. They represent for all conditions of running. Iso-efficiency, and best performance curves and plotted between N_u and Q_u for different gate opening. The innermost curve represents the highest efficiency while outer curves represents lower efficiency.

A vertical line drawn at a specific value of Q_u, it cuts the iso-efficiency curves at two points, giving two values of N_u for the same efficiency while it cuts at only one point at the best performance efficiency curves.

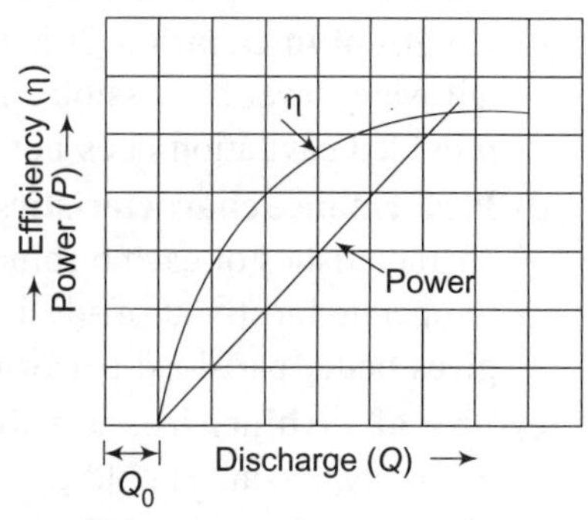

Figure 11.44 Variation of overall efficiency and power with discharge

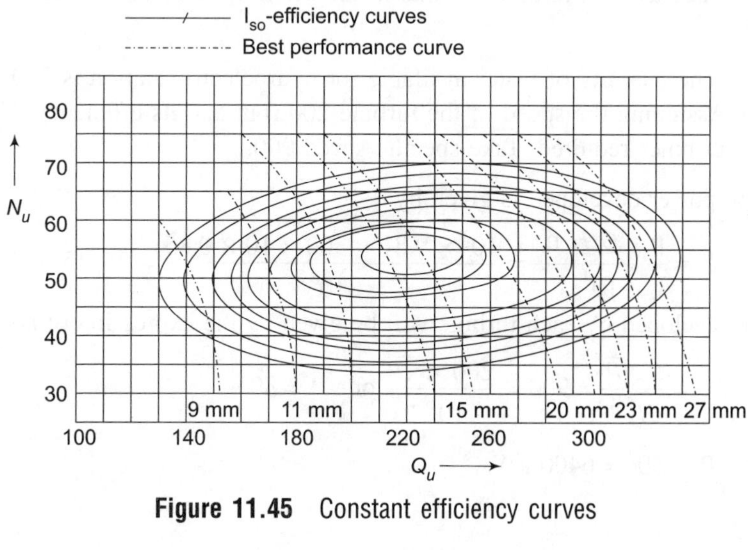

Figure 11.45 Constant efficiency curves

Though, all the hydraulic design of water turbine is based on the optimum speed (N_{opt}) but it must also satisfy structurally the safety conditions at *runway speed* (1.8 to 2.3 times N_{opt}) which is a maximum speed of the turbine under no load and no governing action.

11.23 FACTORS GOVERNING THE SELECTION OF WATER TURBINES

The following factors gives the selection of water turbines.

(a) **Operating head.** One parameter for the selection of water turbines is the operating head. As per present practice, propeller and Kaplan turbines are suitable up to 50 m head while for head varying from 50 to 400 m Francis turbines are suitable. For heads greater than 400 m, Pelton turbines are suitable. It is to be noted that the range of the head mentioned above is not hard and fast, it may vary depending upon other conditions to obtain the maximum economy.

(b) **Specific speed.** Specific speed plays an important role in the selection of turbines. It is always better to choose turbines of high specific speed as it means compact size of turbines, generator and power house, etc. The range of specific speeds of Pelton, Francis and Kaplan are 6-60, 50-400 and 250-1100 respectively.

(c) **Height of installation.** It is always advisable to install the turbines as high above the tail water level as possible. This concept saves excavation cost for placing draft tube provided cavitation does not occur.

(d) **Performance characteristics of turbines.** While selecting the turbine, the performance characteristics of each turbine is to be studied carefully. If the turbine to be installed is to operate mostly at off-design conditions, the turbines should be selected such which gives better part-load performance.

(e) **Size of turbine.** It is a well-known fact that large size of the turbines offer better economy, so one should go for large size turbine.

(f) **Number of turbines.** There is always a chance of breakdown, so one turbine does not serve the purpose for supplying constant electrical supply. It is always advisable to install at least two turbines so that in the event of breakdown, one may supply the power.

Example 11.1 The quantity of water available for hydroelectric station is 300 m^3/s under a head of 10 m. Assuming the speed of the turbine 200 rpm and its efficiency 85%, calculate the number of turbines required. Take specific speed 900.

Solution The power developed is given by

$$P = \frac{\rho\, g\, Q\, H\, \eta_o}{10^3} = \frac{10^3 \times 9.81 \times 300 \times 10 \times 0.85}{10^3} = 25015.5 \text{ kW}$$

The power developed by each turbine will be given by the expression of specific speed

$$N_s = \frac{N\sqrt{P}}{H^{5/4}} = 900 = \frac{200 \times \sqrt{P}}{10^{5/4}} \text{ or } \sqrt{P} = 80$$

or $$P = 80^2 = 6400 \text{ kW}$$

The number of Kaplan turbines required, $n = \frac{25015.5}{6400} = 3.9 \approx 4$ **Ans.**

Example 11.2 The following data relate to a hydroelectric power plant:

Available head = 30 m, Catchment area = 450 km^2, Rainfall = 150 cm/year, Percentage of total rainfall utilised = 70%, Penstock efficiency = 90%, Turbine efficiency = 85%, Generator efficiency = 90%, Load factor = 45%. Calculate the power developed by the turbine and suggest a suitable turbine for the plant.

Solution The quantity of water available per year in the reservoir

$$= (450 \times 10^6)\ m^2 \times 1.5\ (m) \times 0.7 = 472.5 \times 10^6 \text{ m}^3$$

Quantity of water available per second for utilisation

$$Q = \frac{472.5 \times 10^6}{365 \times 24 \times 3600} = 14.98 \text{ m}^3\text{/s}$$

$$\text{Power developed} = \frac{\eta_{\text{pen}} \times \eta_t \times \eta_g \times \rho g\, Q\, H}{10^3}$$

$$= \frac{0.95 \times 0.85 \times 0.9 \times 10^3 \times 9.81 \times 14.98 \times 30}{10^3} \text{ kW} = 3203.96 \text{ kW} \quad \textit{Ans.}$$

The load factor is expressed as $\frac{\text{Average load}}{\text{Peak load}} = 0.45$

$\therefore$ Peak load capacity $= \frac{3202.96}{0.45} = 7119.9$ kW

At least two turbines have to be provided.

The capacity of each unit $= \frac{7119.9}{2 \times 0.9} = 3955.5$ kW ***Ans.***

Example 11.3 A Pelton turbine is to be designed for the following specifications : Power to be developed = 120 MW, Net head available = 500 m, Speed = 200 rpm, Ratio of jet diameter to wheel diameter = 1/10, Hydraulic efficiency = 86%, Velocity coefficient = 0.98, Speed ratio (u/C_1) = 0.46. Calculate (a) volume flow rate of water required, (b) number of jets, (c) diameter of each jet, and (d) diameter of the wheel.

Solution The velocity at the exit of nozzle (jet) is

$$C_1 = K_n \sqrt{2gH} = 0.98 \sqrt{2 \times 9.81 \times 500} = 97.06 \text{ m/s}$$

The peripheral velocity, $u = \rho\, C_1 = 0.46 \times 97.06 = 44.64$ m/s

The overall efficiency, $\eta_0 = \frac{P}{\rho\, Q\, gH} = \frac{120 \times 10^6}{10^3 Q \times 9.81 \times 500} = 0.86$

(a) or $$Q = \frac{120 \times 10^6}{10^3 \times 0.86 \times 9.81 \times 500} = 28.44 \text{ m}^3\text{/s} \quad \textit{Ans.}$$

$$u = \frac{\pi D N}{60} = \frac{\pi D \times 200}{60} = 44.64$$

(b) Number of jets $= \dfrac{Q}{C_1 \frac{\pi}{4} d^2} = \dfrac{28.44 \times 4}{97.06 \times \pi\,(0.4344)^2} = 1.977 \approx 2$ **Ans.**

(c) d = diameter of each jet $= \dfrac{4.262}{10} = 0.4262$ m **Ans.**

Since, $d/D = 1/10$, thus,

(d) or $D = \dfrac{44.64 \times 60}{\pi \times 200} = 4.262$ m **Ans.**

Example 11.4 A Pelton turbine with single jet develops 20 MW while working under a head of 600 m. The specific speed of the turbine is 20, the overall efficiency is 85%, and the coefficient of velocity is 0.98. Assuming the ratio of bucket speed to jet speed as 0.46. Calculate (a) the volume flow rate. (b) the diameter of the wheel, and (c) jet diameter. Take blade velocity coefficient as 0.98.

Solution The specific speed, $N_s = \dfrac{N\sqrt{P}}{H^{5/4}}$

or $$N = \frac{N_s H^{5/4}}{\sqrt{P}} = \frac{20 \times 600^{5/4}}{\sqrt{20 \times 10^3}} = 419.95 \text{ rpm}$$

The velocity of jet, $C_1 = K_n \sqrt{2gH} = 0.98 \sqrt{2 \times 9.81 \times 600} = 106.32$ m/s

The peripheral speed, $u = 0.46 \times 106.32 = 48.90$ m/s

$$u = \frac{\pi D N}{60} = \frac{\pi \times D \times 419.95}{60} = 48.9$$

(a) or $D = \dfrac{48.9 \times 60}{\pi \times 419.95} = 2.223$ m **Ans.**

The overall efficiency, $\eta_0 = \dfrac{P}{\rho g Q H} = \dfrac{20 \times 10^6}{10^3 \times 9.81\, Q \times 600} = 0.85$

(b) or $Q = \dfrac{20 \times 10^6}{10^3 \times 9.81 \times 600 \times 0.85} = 3.997$ m^3/s

(c) $Q = A_1 C_1 = \dfrac{\pi}{4} d_1^2 \times 106.32 = 3.997$ where d_1 = jet diameter

or $d_1 = \left(\dfrac{3.997 \times 4}{\pi \times 106.32}\right)^{1/2} = 0.218$ m **Ans.**

Example 11.5 A Pelton turbine employs four jets of each 60 mm diameter. These jets strike the bucket and each jet deflects by an angle of 165°. The speed of the bucket wheel is 50 m/s. Calculate (a) the velocity of jet for the maximum efficiency, (b) the maximum power, and (c) the maximum blade efficiency. Take $K = 0.98$.

Solution (a) For the maximum efficiency, $C_1 = 2u = 2 \times 50 = 100$ m/s **Ans.**

Flow through each jet = $(\pi/4)\, d^2\, C_1 = (\pi/4)\,(0.06)^2 \times 100 = 0.2827\ \text{m}^3/\text{s}$

$$\text{Maximum power developed} = \left(\frac{1 - k \cos \theta}{g}\right) \frac{C_1^2}{4} . mg;\ W$$

$$(P_{max})_{\text{each jet}} = \left(\frac{1 - 0.98 \times \cos 165}{g}\right) \times \frac{100^2}{4} \times 0.2827 \times 10^3 \times g/10^3 = 1375.76 \text{ kW}$$

For four jets, the maximum power developed

(b) $P_{max} = 4 \times 1375.76 = 5503.04$ kW = 5.50304 MW **Ans.**

(c) The maximum blade efficiency, $(\eta_b)_{max} = \dfrac{1 - k \cos \theta}{2} = \dfrac{1 - 0.98 \cos 165}{2} = 97.33\%$

Ans.

Example 11.6 A Francis turbine operates under the following conditions : Outer diameter of runner = 2 m, Inner diameter of runner = 1 m, Head = 200 m, Specific speed = 130, Power developed = 20 MW. The inclination of water jet with tangential direction = 12°. The water leaves the blades radially with no velocity of whirl. Hydraulic efficiency = 92%. Calculate the inlet and outlet blade angles.

Solution Refer to Fig. 11.46. The specific speed is given by

$$N_s = \frac{N\sqrt{P}}{H^{5/4}} = 130$$

or

$$N = \frac{130 \times 200^{5/4}}{\sqrt{20 \times 10^3}} = 691.98 \text{ rpm}$$

$$\text{The peripheral velocity at inlet} = u_1 = \frac{\pi\, D\, N}{60} = \frac{\pi \times 2 \times 691.48}{60} = 72.41 \text{ m/s}$$

$$\text{The hydraulic efficiency, } \eta_b = \frac{u_1\, C_{w1} - u_2\, C_{w2}}{gH} = \frac{u_1 C_{w1}}{\text{gH}} \text{ as } C_{w2} = 0$$

or

$$C_{w1} = \frac{0.92 \times 9.81 \times 200}{72.41} = 24.92 \text{ m/s}$$

$$\text{From velocity triangle, } \tan \alpha_1 = \tan 12° = \frac{C_{f1}}{C_{w1}}$$

or

$$C_{f1} = C_{w1} \tan 12° = 24.92 \times \tan 12° = 5.296 \text{ m/s} = C_{f2}$$

$$\tan \beta_1 = \frac{C_{f1}}{u_1 - C_{w1}} = \frac{5.296}{72.41 - 24.92} \quad \text{or} \quad \beta_1 = 6.36°$$ **Ans.**

The peripheral velocity at outlet of runner, $u_2 = \dfrac{u_1}{2} = \dfrac{72.41}{2} = 36.205$ m/s

$$\tan \beta_2 = \frac{C_{f2}}{u_2 - C_{w2}} = \frac{5.296}{36.205} = 0.1462 \quad \text{or} \quad \beta_2 = 8.32° \qquad \textbf{\textit{Ans.}}$$

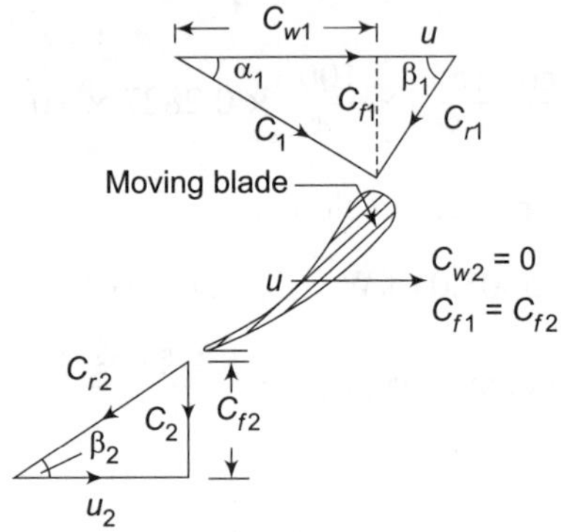

Figure 11.46 Example 11.6

Example 11.7 The following data relate to a Francis turbine. Shaft power = 500 kW, Net head = 80 m, Speed = 600 rpm, Overall efficiency = 85%, Hydraulic efficiency = 92%. Flow ratio $(C_{f1}/C_1) = 0.22$, Breadth ratio $(B/D) = 0.1$, Outer diameter = 2 × Inner diameter, C_f = constant. Outer discharge = radial. The thickness of vanes occupies 7% of circumferential area of the runner. Calculate (a) the volume flow rate, (b) the dia and width of wheel at inlet, (c) the guide vane angle, and (d) the runner vane inlet and outlet angles.

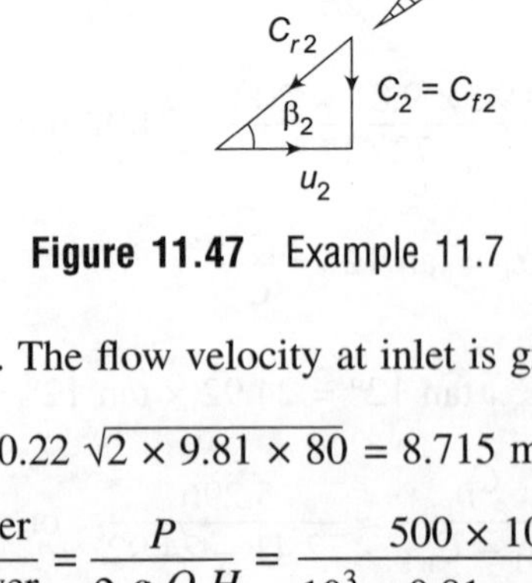

Figure 11.47 Example 11.7

Solution Refer to Fig. 11.47. The flow velocity at inlet is given by

$$C_{f1} = 0.22\, C_1 = 0.22 \sqrt{2 \times 9.81 \times 80} = 8.715 \text{ m/s}$$

$$\eta_0 = \frac{\text{Shaft power}}{\text{Water power}} = \frac{P}{\rho\, g\, Q\, H} = \frac{500 \times 10^3}{10^3 \times 9.81 \times Q \times 80} = 0.85$$

(a) or $$Q = \frac{500 \times 10^3}{10^3 \times 9.81 \times 80 \times 0.85} = 0.7495 \text{ m}^3/\text{s}$$ ***Ans.***

(b) $$Q = A\,C = 0.93\,\pi\,D_1\,B_1\,C_{f1}$$

or $$0.7495 = 0.93\,\pi \times D_1 \times 0.1\,D_1 \times 8.715$$

or $$D_1 = \left(\frac{0.7495}{0.93\,\pi \times 0.1 \times 8.715}\right)^{1/2} = 0.5425 \text{ m}$$ ***Ans.***

$$B_1 = 0.1 \times 0.5425 = 0.05425 \text{ m} = 5.425 \text{ cm}$$ ***Ans.***

(c) $$D_2 = \frac{D_1}{2} = \frac{0.5425}{2} = 0.227225 \text{ m}$$

$$u_1 = \frac{\pi\,D\,N}{60} = \frac{\pi \times 0.5425 \times 600}{60} = 17.043 \text{ m/s}$$

$$\eta_h = \frac{u_1 C_{w1} - u_2 C_{w2}}{gH} = \frac{u_1\,C_{w1}}{gH} = \frac{17.043 \times C_{w1}}{9.81 \times 80} = 0.92$$

or $$C_{w1} = \frac{0.92 \times 9.81 \times 80}{17.043} = 42.364 \text{ m/s}$$

From inlet velocity triangle, we have, $\tan \alpha_1 = \dfrac{C_{f1}}{C_{w1}} = \dfrac{8.715}{42.364} = 0.2057$

or $$\alpha_1 = 11.62°$$ ***Ans.***

(d) $$\tan \beta_1 = \frac{C_{f1}}{C_{w1} - u_1} = \frac{8.715}{42.364 - 17.043} = 0.3441$$

or $$\beta_1 = 18.992°$$ ***Ans.***

From outlet velocity triangle, we have, $\tan \beta_2 = \dfrac{C_{f2}}{u}$,

But $$u_2 = \frac{D_2 u_1}{D_1} = \frac{17.043}{2} = 8.521 \text{ m/s}$$

$\therefore$ $$\tan \beta_2 = \frac{8.715}{8.521} = 1.0227 \quad \text{or} \quad \beta_2 = 45.644°$$ ***Ans.***

Example 11.8 A hydropower plant employs an inward flow reaction turbine. The peripheral velocity of the runner of the turbine is 40 m/s. The velocity of whirl of inflowing water is 35 m/s and the radial velocity of flow is 5 m/s. If the flow is 2 m^3/s and the hydraulic efficiency is 90%, calculate (a) the head on the runner, (b) the power generated by the turbine and (c) the angles of the vanes. Assume radial discharge.

Solution Refer to Fig. 11.46. Since the discharge is radial, thus $C_{w2} = 0$. The hydraulic efficiency is expressed as

$$\eta_h = \frac{C_{w1}\, u_1}{gH} = 0.9 = \frac{35 \times 40}{9.81 \times H}$$

(a) or $H = \dfrac{35 \times 40}{9.81 \times 0.9} = 158.56$ m ***Ans.***

(b) Power generated, $P = \dfrac{\rho Q\, g\, H\, \eta_h}{10^3} = \dfrac{10^3 \times 2 \times 9.81 \times 158.56 \times 0.9}{10^3}$

$= 2799.85$ kW $= 2.79985$ MW ***Ans.***

(c) From velocity triangle, we have, $\tan \alpha_1 = \dfrac{C_{f1}}{C_{w1}} = \dfrac{5}{35} = 0.14285$

or $\alpha_1 = 8.13°$ and $\theta = 180° - 8.13° = 171.87°$ ***Ans.***

$$\tan \beta_1 = \frac{C_{f1}}{u_1 - C_{w1}} = \frac{5}{40 - 35} = 1 \text{ or } \beta_1 = 45°$$ ***Ans.***

Example 11.9 A hydroelectric power plant equipped with four units of vertical shaft Pelton turbine is coupled with 70,000 kVA, 3 phase, 50 hertz generators. The generators employ 10 pairs of poles. The gross design head is 550 m. The transmission efficiency of head race runner and penstock is 94 per cent. The power output from the four units is 280 MW with an efficiency of 92 per cent. The nozzle efficiency is 98% and each turbine has four jets. Take $u/C_1 = 0.48$, $K_n = 0.98$ and nozzle diameter is 28% bigger than jet diameter. The turning angle of buckets on wheel is 165°, the bucket friction factor is 0.98 and the discharge efficiency is 99.7%. Calculate: (a) the volume flow rate through turbine, (b) the jet diameter, (c) the nozzle tip diameter, (d) the pitch circle diameter of the wheel, (e) the specific speed, (f) the number of buckets on the wheel, and (g) the work done per kg of water on the wheel, the hydraulic efficiency.

Solution The synchronous speed of the generator is given by

$$N = \frac{120f}{p} = \frac{120 \times 50}{20} = 300 \text{ rpm}$$

Net head available $= \eta_{\text{penstock}} \times$ Gross head $= 0.94 \times 550 = 517$ m

Power developed per turbine $= \dfrac{\text{Total power}}{\text{Number of turbines}} = \dfrac{280}{4} = 70$ MW

Input water power for each turbine $= \dfrac{70 \times 10^3}{0.92} = \dfrac{\rho\, g\, QH}{10^3} = \dfrac{10^3 \times 9.81 \times Q \times 517}{10^3}$

(a) or Q = Discharge $= \dfrac{70 \times 10^3 \times 10^3}{0.92 \times 10^3 \times 9.81 \times 517} = 15\ \text{m}^3/\text{s}$ ***Ans.***

Discharge per jet $= \dfrac{15}{4} = 3.75\ \text{m}^3/\text{s}$

Velocity of jet = $C_1 = K_n \sqrt{2gH} = 0.98 \sqrt{2 \times 9.81 \times 517} = 98.7$ m/s, here $K_n \approx \eta_n$

Discharge per jet = $3.75 = (\pi/4)\, d^2\, C_1$

(b) or d = diameter of each jet = $\sqrt{\dfrac{4}{\pi} \times \dfrac{3.75}{98.7}} = 0.2199$ m **Ans.**

(c) The nozzle tip diameter = $1.20 \times 0.2199 = 0.26388$ m **Ans.**

The pitch circle diameter of Pelton wheel (D), $u = \dfrac{\pi D N}{60}$

But $u = 0.48\, C_1 = 0.48 \times 98.7 = 47.376$ m/s

or $u = 43.376 = \dfrac{\pi\, D \times 300}{60} \approx u_1$

(d) or $D = \dfrac{47.376 \times 60}{\pi \times 3000} = 3.016$ **Ans.**

(e) The specific speed, $N_s = \dfrac{N\sqrt{P}}{H^{5/4}} = \dfrac{300\sqrt{280 \times 10^3}}{(517)^{5/4}} = 64.392$ **Ans.**

The jet ratio = $\dfrac{D}{d} = \dfrac{3.016}{0.2199} = 13.715$

(f) The number of buckets, $z = \dfrac{D}{2d} + 15 = \dfrac{13.715}{2} + 15 = 21.85 \approx 22$ **Ans.**

The work done per kg, $E = \dfrac{(C_1 - u_1)\,(1 - k\cos\theta)\,u_1}{g}$; kg.m/kg

$$= \frac{(98.7 - 47.376)(1 - 0.98\cos 165) \times 47.376}{9.81} = 482.489$$

The hydraulic efficiency, $\eta_b = \eta_{\text{head}} \times \eta_{\text{discharge}} = \dfrac{\text{Work done per kg}}{H}\, \eta_{\text{discharge}}$

$$= \frac{482.489}{517} \times 0.997 = 93.04\%$$ **Ans.**

Example 11.10 An axial flow water turbine works under a gross head of 40 m. The mean diameter of the runner is 2 m and rotates at 150 rpm. Water leaves the guide vanes at 30° to the tangential direction. The outlet angle of the runner vanes is 55°. The head loss in the casing and guide vanes is 6 per cent and the blade velocity coefficient of the runner is 0.82. Calculate (a) the blade angle, (b) the work output per kg of water in metre head, and (c) the hydraulic efficiency.

Solution Refer to Fig. 11.46. The effective head = $H = 0.94 \times 40 = 37.6$ m

The velocity at the exit of guide vanes is given by

$$C_1 = \sqrt{2gH} = \sqrt{2 \times 9.81 \times 37.6} = 27.16 \text{ m/s}$$

The peripheral velocity, $u = \dfrac{\pi D N}{60} = \dfrac{\pi \times 2.0 \times 150}{60} = 15.7$ m/s

From inlet velocity triangle, we have

$$\tan \beta_1 = \frac{C_1 \sin \alpha_1}{C_1 \cos \alpha_1 - u} = \frac{27.16 \times \sin 30°}{27.16 \times \cos 30° - 15.7} = 1.736$$

(a) or $\beta_1 = 60°$ ***Ans.***

The relative velocities at inlet and outlet of runner are

$$C_{r1} = \frac{C_1 \sin \alpha_1}{\sin \beta_1} = \frac{27.16 \times \sin 30°}{\sin 60°} = 15.32 \text{ m/s}$$

and $C_{r2} = k\, C_{r1} = 0.82 \times 15.32 = 12.56$ m/s

$$C_{w1} = C_1 \cos \alpha_1 = 27.16 \cos 30° = 23.52 \text{ m/s}$$

$$C_{w2} = u - C_{r2} \cos \beta_2 = 15.7 - 12.56 \cos 55° = 3.14 \text{ m/s}$$

Work done per kg of water in metre head of water is given by

(b) $$E = \frac{u(C_{w1} - C_{w2})}{g} = \frac{15.7(23.52 - 3.14)}{9.81} = 32.62 \text{ m}$$ ***Ans.***

(c) The hydraulic efficiency, $\eta_h = \dfrac{E}{H} = \dfrac{32.62}{37.6} = 86.75\%$ ***Ans.***

Example 11.11 The following data relate to a Francis turbine having specific speed 300 and running at 200 rpm. An experimental model is to be made to operate at a flow rate of a 0.6 m^3/s and head of 5 m. Taking overall efficiency as 0.88, calculate (a) the power, (b) the speed (c) the scale ratio, and (d) the volume flow rate of protype. The turbine develops 60 MW.

Solution Power developed by the model turbine is

(a) $$P_m = \frac{\rho g Q H \eta_o}{10^3} = \frac{10^3 \times 9.81 \times 0.6 \times 5 \times 0.88}{10^3} = 25.898 \text{ kW}$$ ***Ans.***

The specific speed of the model and prototype has to be the same. Thus

$$N_s = \frac{N_m \sqrt{P_m}}{H_m^{5/4}} = 300$$

(b) or $$N_m = \frac{300 \times 5^{5/4}}{\sqrt{25.898}} = 440.75$$ ***Ans.***

Since $\quad DN \propto \sqrt{H}$, so $\dfrac{D_p N_p}{D_m N_m} = \sqrt{\dfrac{H_p}{H_m}}$

or $\quad$ Scale ratio $= \dfrac{D_p}{D_m} = \dfrac{N_m}{N_p}\sqrt{\dfrac{H_p}{H_m}}$

Here H_p is not known, so it has to be calculated

$$N_{sp} = \frac{N_p \sqrt{P_P}}{H^{5/4}} = \frac{200\sqrt{60 \times 10^3}}{H^{5/4}} = 300$$

or $$H_p^{5/4} = \frac{200\sqrt{60 \times 10^3}}{300} = 163.29$$

or $$H_p = (163.29)^{4/5} = 58.93 \text{ m}$$

(c) Now scale ratio $= \dfrac{D_p}{D_m} = \dfrac{440.75}{200}\sqrt{\dfrac{58.93}{5}} = 7.565$ ***Ans.***

The volume flow rate through prototype turbine is

(d) $Q_p = \dfrac{P_p}{\rho\, g\, H_p\, \eta_p} = \dfrac{60 \times 10^6}{10^3 \times 9.81 \times 58.93 \times 0.88} = 117.94 \text{ m}^3/\text{s}$ ***Ans.***

Example 11.12 A test has been conducted on one-sixth scale model of a Francis turbine working under a head of 2 m. The results indicated that it could develop 10 kW at 600 rpm. Calculate (a) the speed and (b) power of a prototype turbine while working under a head of 50 m.

Solution The scale ratio is expressed as

$$\frac{D_m}{D_p} = \frac{N_p}{N_m} \times \sqrt{\frac{H_m}{H_p}} = \frac{1}{6}$$

or $$\frac{1}{6} = \frac{N_p}{600}\sqrt{\frac{2}{50}}$$

(a) or $\quad N_p = \dfrac{600}{6}\sqrt{25} = 500$ rpm ***Ans.***

or $$N_s = \frac{N_m\sqrt{P_m}}{H_m^{5/4}} = \frac{600\sqrt{10}}{2^{5/4}} = 797.74$$

(b) $$N_s = \frac{N_p\sqrt{P_p}}{H_p^{5/4}} = 797.74$$

or $$P_p = \left[\frac{797.74 \times 50^{1.25}}{500}\right]^2$$

$$= 44999.50 \text{ kW}$$

Example 11.13 Table 11.7 below shows the run-off data of a river at a particular site:

Table 11.7 Run-off data

Month	*Mean discharge (Millions of cu.m/second)*	*Month*	*Mean discharge (Millions of cu.m/second)*
January	40	July	90
February	35	August	110
March	30	September	120
April	05	October	70
May	15	November	50
June	60	December	40

(a) Draw the hydrograph and calculate the mean flow.
(b) Draw the flow duration curve.
(c) Calculate the power developed for the available head of 200 m with overall efficiency of 90 per cent. Assume 30 days in a month for simplicity.

Solution The mean discharge from the river.

$$= \frac{40 + 35 + 30 + 05 + 15 + 60 + 90 + 110 + 120 + 70 + 50 + 40}{12} = 54.16 \text{ million m}^3\text{/s}$$

The hydrograph based on the above run-off data is given in Fig. 11.48.

In order to draw the flow duration curve, it is necessary to find the lengths of time during which certain flow is available as given in Table 11.8.

Table 11.8 Discharge-duration data

Discharge per month (Million m^3/s)	*Total number of months during which flows is available*	*Percentage time*
0	12	100
10	11	91.666
20	10	83.332
30	9	75.00
40	8	66.66
50	6	50.00
60	5	41.666

Contd.

Table 11.8 Contd...

70	4	33.333
80	3	29.00
90	3	25.00
100	2	16.66
110	2	16.66
120	1	8.33

Based on the Table 11.8, the flow duration curve is shown in Fig. 11.49.

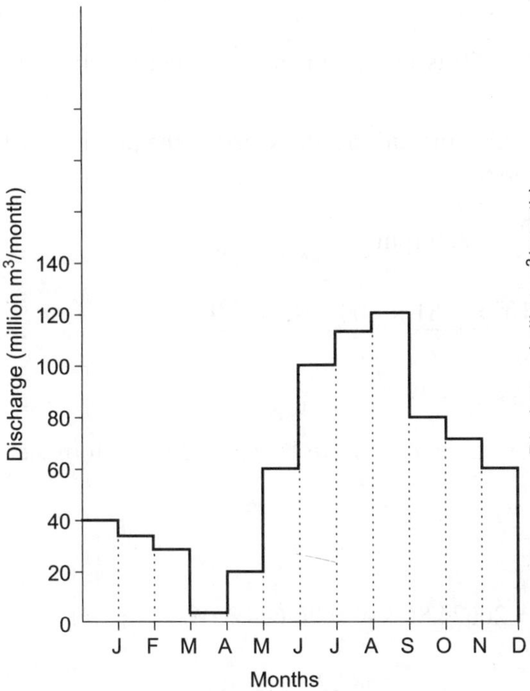

Figure 11.48 Hydrograph for Example 11.13

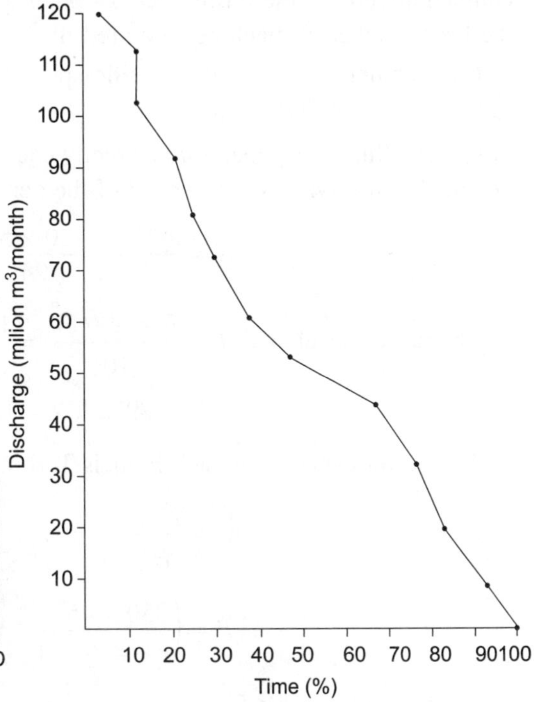

Figure 11.49 Flow duration curve for Example 11.13

Example 11.14 Calculate the maximum height of the Francis turbine above the tail race for the following data: Atmospheric pressure 1.013 bars, Pressure at the exit of runner = 0.5 bar absolute, Velocity of water leaving the runner = 6 m/s, Velocity of water leaving the draft tube = 2 m/s, Head of water lost passing through the draft tube due to friction = 0.25 m of water head.

Solution The maximum height of the turbine above tail race level is calculated by

$$h = \frac{p_a - p_{re}}{\rho} - \frac{C_{re}^2}{2g} + \left(\frac{C_{de}^2}{2g} + h_f\right)$$

$$= \frac{(1.013 - 0.5) \times 10^4}{10^3} - \frac{6^2}{2 \times 9.81} + \left(\frac{2^2}{2 \times 9.81} + 0.25\right)$$

$$= 5.13 - 1.834 + 0.2038 + 0.25 = 3.7498 \text{ m above tail race} \qquad \textit{Ans.}$$

Example 11.15 The available discharge and head at a proposed site of hydroelectric power plant is 400 m^3/s and 45 m respectively. The overall turbine efficiency is 90%. The generator coupled directly to the turbine has 24 poles and generates electricity at 50 cycles. Calculate the least number of machines required if:

(a) a Francis turbine with specific speed of 350 is used, and (b) a Kaplan turbine with specific speed of 700 is used.

Solution Since the generator is directly coupled to the turbine, the speed of the turbine must be equal to the synchronous speed of the generator.

$$N = \frac{120\,f}{p} = \frac{120 \times 50}{24} = 250 \text{ rpm}$$

The power developed, $P = \dfrac{\rho\, g\, Q\, H\, \eta_o}{10^3} = \dfrac{10^3 \times 9.81 \times 400 \times 45 \times 0.9}{10^3}$

$$= 158922 \text{ kW} = 158.22 \text{ MW}$$

The power capacity of each Francis Turbine can be calculated by the following formula

$$N_s = \frac{N\sqrt{P}}{H^{5/4}}$$

or

$$P = \left(\frac{350 \times 45^{5/4}}{250}\right)^2 = 26624.86 \text{ kW} = 26.624 \text{ MW}$$

(a) Therefore, number of Francis turbine required $= \dfrac{158.22}{26.62486} = 5.942 \approx 6$ ***Ans.***

The power capacity of each Kaplan turbine is calculated from N_s equation

$$N_s = \frac{N\sqrt{P}}{H^{5/4}} = 250\,\frac{\sqrt{P}}{45^{5/4}} = 700$$

or

$$P = \left(\frac{700 \times 45^{5/4}}{250}\right)^2 = 106499.4 \text{ kW} = 106.4994 \text{ MW}$$

(b) Number of Kaplan turbines required $= \dfrac{158.22}{106.4994} = 1.4856 \approx 2$ ***Ans.***

PROBLEMS

Things to Think About

11.1 Which type of conversion of energy takes place in hydroelectric power plant?

11.2 What is the reason that the plant efficiency of hydroelectric power plant is much higher than others?

11.3 Is it possible to generate power from flowing water of a river? Justify your answer.

11.4 Which type of water turbine can develop maximum power from a single unit ? Justify your answer.

Short and Long Theoretical Questions

11.1 What are the basic two parameters on which power output from water turbine depend?

11.2 Discuss the advantages and disadvantages of hydroelectric power plants.

11.3 Explain the optimization procedure of hydro-thermal mix to meet the power demand of a particular region.

11.4 Discuss the parameters on which the selection of site for a hydraulic power plant depend.

11.5 Discuss hydrology and hydrological cycle.

11.6 What do you mean by hydrographs?

11.7 What do you mean by mass curve? How is it plotted?

11.8 What are storage and pondage?

11.9 List the essential elements of a hydroelectric power plant.

11.10 What do you mean by catchment area, reservoir and dam?

11.11 Classify the dams and discuss each of them in brief.

11.12 What do you mean by gravity, buttress and arch dam?

11.13 Discuss the parameters for the selection of site for dam.

11.14 What do you mean by spillway? Discuss its various types?

11.15 What are the various types of conduits?

11.16 What do you mean by surge tank? What are the various types of surge tank?

11.17 Who do you mean by water hammer problem? How is it overcome?

11.18 What do you mean by draft tube? What are the various types of draft tubes?

11.19 Classify hydroelectric power plants and sketch each of them.

11.20 Explain the following: (i) headrace, (ii) tail race, (iii) canal, (iv) plume, (v) tunnel, (vi) pipeline, (vii) penstock.

11.21 Explain with neat sketch a pumped storage plant.

11.22 What type of turbine would you recommend for the following heads? (a) 1100 m, (b) 200 m, and (c) 20 m

11.23 Classify hydraulic turbines.

11.24 With a neat sketch discuss the construction in detail, velocity diagram, power out and efficiency of Pelton, Francis and Kaplan turbines.

11.25 On what basis you will select Pelton, Francis and Kaplan turbine for a particular site?

11.26 With neat sketch, discuss the constructional aspects of Deriatz and bulb turbines.

11.27 Define specific speed and derive the expression for it.

11.28 Define and derive the expressions for scale ratio, unit speed, unit power and unit discharge.

11.29 Give a comparison for turbines.

11.30 What is the purpose of governing of hydraulic turbines.

11.31 With a neat sketch discuss the governing impulse and reaction turbines.

11.32 Discuss the performance characteristics of water turbines.

11.33 What is runway speed? How does it affect the turbine design?

11.34 How is the type of turbine selected in a certain hydro unit?

11.35 What do you mean by cavitation? What are its effects and how can it be minimized?

11.36 Derive the optimum speed ratio for maximum power and maximum blade efficiency of water turbines.

11.37 Explain the characteristics features of Francis, Pelton and Kaplan turbines.

Numerical Questions

11.1 The Pelton wheel diameter is 2 m and the deflecting angle of the bucket is 162°. The diameter of jet is 165 mm. The pressure behind the nozzle is 700 kPa. The speed of the bucket is 320 rpm. Neglecting friction, calculate the power developed by the turbine and the hydraulic efficiency. [*Ans.* 352 kW, 0.62]

11.2 A Pelton wheel develops 4500 kW at 400 rpm, working under a head of 360 m. There are two equal jets and the bucket angle is 170°. The pitch circle diameter of the bucket is 1.82 m. If the blade velocity coefficient is 0.85, calculate (a) the blade efficiency and (b) the diameter of each jet. [*Ans.* (a) 0.91, (b) 0.1032 m]

11.3 A Pelton wheel is to be designed to produce 12 MW under a head of 300 m at a speed of 500 rpm. The diameter of the jet is 1/9 of wheel diameter, calculate the number and diameter of jets, diameter of the bucket wheel and the volume flows rate. Take overall efficiency = 88%, $K_n = 0.97$ and speed ratio = 0.45. [*Ans.* $D = 1.32$ m, $d = 0.146$ m, $n = 4$, $Q = 4.62$ m^3/s]

11.4 The runner of a Francis turbine works under a head of 150 m. The outer diameter is 2 m while the inner is 1 m. The specific speed is 200 and the turbine generates 15 MW. If the hydraulic efficiency is 90% and nozzle angle is 12° and leaves the blades with no whirl, calculate the outlet and inlet angles.

11.5 A hydraulic power plant develops 100 MW. The head available is 40 m and the speed is 70 rpm. Calculate the outlet and inlet angles.

11.6 A hydroelectric power station is to be designed for a catchment area of a 120 km^2, run-off 70%, average rain is 130 cm. The head available is 400 m. If the overall efficiency is 85% calculate the power developed.

11.7 The following Table shows the mean discharge of a river at a particular location.

Month	*Discharge m^3/s*	*Month*	*Discharge m^3/s*
January	150	July	1150
February	350	August	1350
March	450	September	1100
April	750	October	900
May	900	November	650
June	950	December	350

Draw (a) the hydrograph and (b) the flow duration curve.

11.8 The following data relate to a hydroelectric power plant:

Available head = 180 m. Catchment area = 2100 km^2, Annual average rainfall = 150 cm, Turbine efficiency = 85%, Generator efficiency = 95%, Percolation and evaporation losses = 15%. Calculate the power developed and suggest the type of turbine to be used if the runner speed is to be kept below 250 rpm.

Software Based Questions

11.1 Develop a computer software for design of a hydroelectric power plant using Pelton, Francis and Kaplan turbines.

11.2 Design a hydroelectric power plant for the following data:

(a) H = 1200 m, P = 250 MW, (b) H = 400 m, P = 700 MW, (c) H = 60 m, P = 200 MW

11.3 Develop a software for the data given in Question 11.8.

Design and Open Ended Question

11.1 Prepare a comprehensive report on a big size hydroelectric power plant involving its design features, specifications and plant efficiency and capacity. Prepare a case study of one of them which is highly exciting.

Multiple-Choice Questions

11.1 Hydrograph is defined as graph showing

(a) the measurement of depth and width of a river.

(b) the discharge of flowing water with respect to time for a specified period.

(c) the quantity of water in the catchment area.

(d) the quantity of rain fall in a year.

11.2 Which one is not true with storage dam of hydroelectric plant?

(a) To store water for hydropower.

(b) To store water for hydropower and irrigation.

(c) To store water for hydropower, irrigation and water supply.

(d) Case (a), (b), (c) and to divert river flow.

11.3 The surge tank is located

(a) at the beginning of dam.

(b) near the powerhouse.

(c) between dam and power house at very low ground.
(d) between dam and power house at a very elevated ground.

11.4 Cavitation takes place in water turbine
(a) in the beginning (b) in the middle (c) in the end (d) none of them

11.5 The plant efficiency of hydroelectric plant ranges from
(a) 59 to 60% (b) 60 to 70% (c) 70 to 80% (d) 70 to 90%

11.6 In Kaplan turbine, the flow
(a) radial inward (b) axial (c) tangential (d) diagonal

11.7 Kaplan is suitable for the head of
(a) low specific speed (b) medium specific speed
(c) high specific speed (d) all of them

11.8 Kaplan is suitable for the head of
(a) 2 to 70 m (b) 30 to 500 m (c) 200 to 1000 m (d) 300 to 2000 m

11.9 From the single unit, Francis turbine is capable to develop power around
(a) 230 MW (b) 250 MW (c) 550 MW (d) 750 MW

11.10 Francis turbine can operate under the head varying from
(a) 30 to 100 m (b) 30 to 300 m (c) 30 to 500 m (d) 30 to 700 m

11.11 Which one is true for the bulb water turbine
(a) suitable for low head
(b) suitable for medium head
(c) case (a) and installed right in the middle of flow passage
(d) case (b) and installed right in the middle of flow passage

Answers to Multiple-Choice Questions

11.1 (b) 11.2 (c) 11.3 (d) 11.4 (c) 11.5 (d) 11.6 (b) 11.7 (c) 11.8 (a)
11.9 (d) 11.10 (c) 11.11 (c)

CHAPTER

12

Diesel Engine Power Plants

OVERVIEW

Diesel electric power plants available in the range of 1 MW to 50 MW capacity are used sometimes as central station for meeting small requirements and universally employed as standby plant to supplement thermal power station or hydroelectric station, either for starting condition from cold or to meet emergency situations. Enough literature is available in I.C. engine books, here only systems and performance will be discussed. For more details, see the Author's book titled ***"I.C. Engine, Air Pollution and E-Vehicles."***

12.1 APPLICATIONS

The following are the applications of diesel electric power plants:

(i) *Peak load plant.* Diesel plants are suitable as peak load plants in combination with thermal or hydel plant. The peak load unit needs easy, starting and stopping and diesel plants serve this purpose.

(ii) *Standby unit.* There are many situations in which standby units are needed such as when main unit fails or cannot cope with the demand. As an example, due to less rainfall in a particular year, the hydro-plant cannot meet the demand. Thus diesel units are installed as standby unit to supply power in parallel to generate the shortfall of power.

(iii) *Central stations.* Due to ease in installation, starting and stopping, diesel electric plants can be used as central station where the capacity requirement is small.

(iv) *Starting station.* For starting large thermal power plant, auxiliaries such as FD and ID fans, BFP, circulating cooling water pump, etc., can be run by installing diesel electric power plant.

(v) *Energy unit.* In many developing countries like India, power failure/interruption is a common features for hours. Under this circumstance, the power to vital units such as hospitals or industrial plants can be supplied by installing diesel electric plants.

(vi) *Mobile unit.* In many remote areas, large civil construction work is carried out where there is no power. Diesel plants mounted on trailer are used to supply power to construction work.

(vii) *Nursery station.* In remote town where there is no supply of power by main grid, diesel electric plant can be installed. When the power becomes available by main grid, this plant can be shifted to other place. Such a diesel electric power plant is generally called nursery unit.

12.2 ADVANTAGES AND DISADVANTAGES

Advantages. The following are the advantages of diesel electric power plant: (i) Easy installation, (ii) Easily available in standard capacities, say 1 MW, 5 MW, (iii) Quick response for load changes, (iv) Less standby losses, (v) Ease in starting and stopping, (vi) Less space requirement, (vii) Less cooling water requirement per kW produced, (viii) Less capital cost, (ix) Less supervisory staff needed, (x) Reasonable thermal efficiency (about 32%), (xi) Better part load thermal efficiency, (xii) Less civil engineering work needed for installation, (xiii) No proper and specific site needed, (xiv) Fully capable of getting located at load centre, (xv) No ash handling problem, (xvi) Easy lubrication.

Disadvantages. The following are the disadvantages of diesel electric power plant: (i) High operating cost, (ii) High maintenance cost, (iii) High lubrication cost, (iv) Limited capacity, (v) Noisy, (vi) Inability to meet overload, (vii) Emissions not suitable to human life.

12.3 WORKING AND CLASSIFICATION OF DIESEL PLANTS

In a diesel (or compression ignition) engine, air is sucked through the opening of inlet valve and is first compressed to high pressure through a compression ratio varying from 13 to 22 such that the temperature of the compressed air is more than the self-ignition temperature of the diesel fuel. The diesel fuel is injected in the cylinder at the end of compression in a very fine atomized form and combustion starts automatically resulting in very high pressure and temperature which forces the piston down for power or expansion stroke till the exhaust valve opens for exhausting the combustion products out of cylinder. It is worth to note that there is no spark plug for installation of combustion. It is the high temperature of compressed air which initiates combustion and so the name is compression ignition engine also. The cycle works on Diesel cycle (constant volume).

Diesel engines may be classified as: (a) Four-stroke, (b) Two-stroke (c) Horizontal engine, (d) Vertical engine, (e) Single cylinder, (f) Multi-cylinder, (g) Indirect ignition, (h) Direct ignition, (i) Naturally aspirated, (j) Supercharged engine, (k) Air cooled, (l) Water cooled, etc. Four-stroke multicylinder engines are generally preferred for standing plant.

12.4 GENERAL LAYOUT OF A DIESEL ENGINE POWER PLANT

Figure 12.1 shows the general layout of a diesel engine power plant. The engine and its various auxiliaries are depicted with their proper positions. The flow path of air, fuel and gas are shown by arrows. The plant consists of the following.

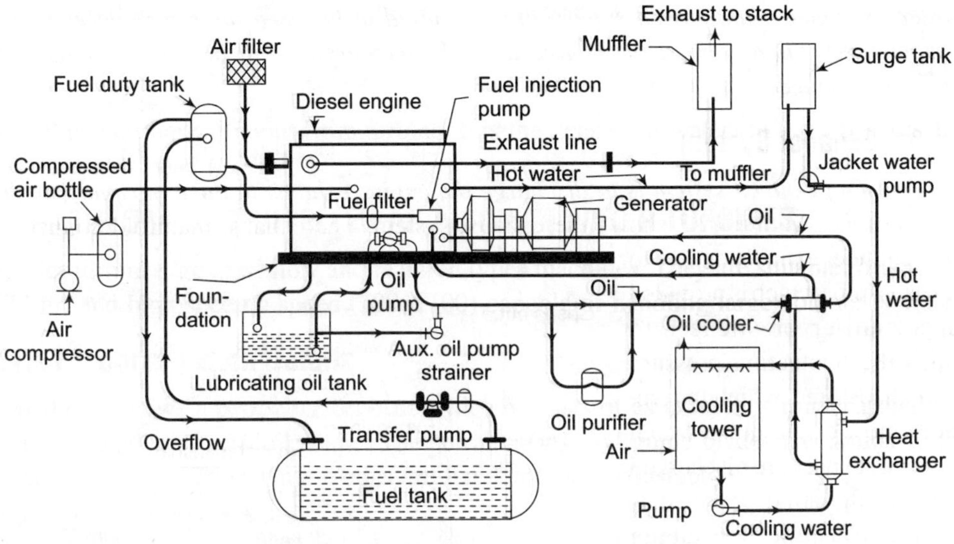

Figure 12.1 General layout of a diesel plant

(a) Engine (b) Air intake system, (c) Exhaust system, (d) Fuel system, (e) Fuel injection system, (f) Cooling system, (g) Lubrication system, (h) Starting system.

12.4.1 Principal Parts of a Diesel Engines

Figure 12.2 shows the cross section of an air cooled IC engine depicting the principal parts. Generally, for standby plant, water cooled engine is preferred but where there is scarcity of water or in mobile power plant, air cooled engine is preferred. The principal parts are: (a) Cylinder, (b) Cylinder head, (c) Piston, (d) Inlet valve, (e) Inlet port, (f) Exhaust valve, (g) Valve spring, (h) Cooling fins (i) Wrist pin, (j) Connecting rod, (k) Crankcase, (l) Crank-pin, (m) Crank, (n) Crankshaft, etc.

12.4.2 Air Intake System

The function of air intake system is to convey fresh air through louvers and air filter to the cylinder via intake manifold.

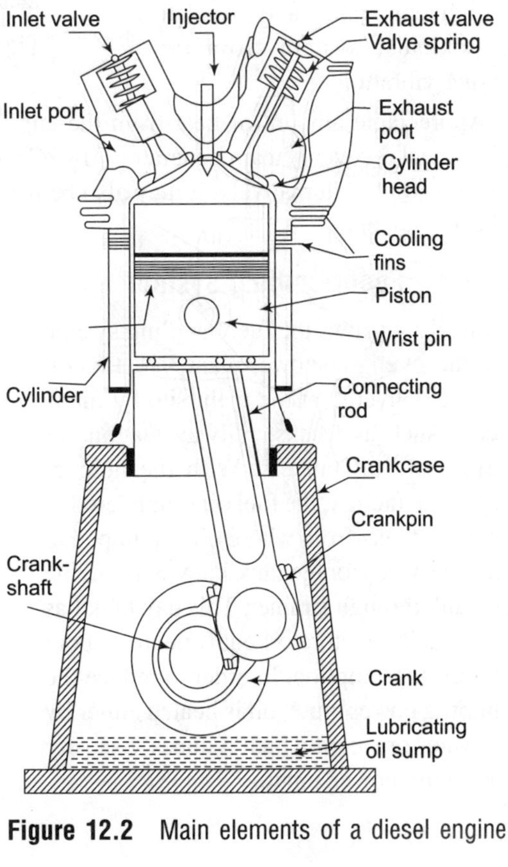

Figure 12.2 Main elements of a diesel engine

In order to augment the power, supercharger is fitted in between the filter and engine and the supercharger is driven by the engine itself. Supercharging will be discussed separately in subsequent pages.

12.4.3 Exhaust System

The purpose of exhaust system is to discharge the engine exhaust to the atmosphere with the minimum noise. Figure 12.3 shows the exhaust system. The exhaust manifold connects the engine cylinder exhaust outlets to the exhaust pipe which is provided with a muffler or silencer to dampen the fluctuating pressure of the exhaust line, which in turn, reduces most of the noise which may result if gases are discharged directly to the atmosphere. It is advisable to use flexible tubing system for exhaust pipe to take up the effects of expansion due to high temperature and to isolate the exhaust system from the engine vibration.

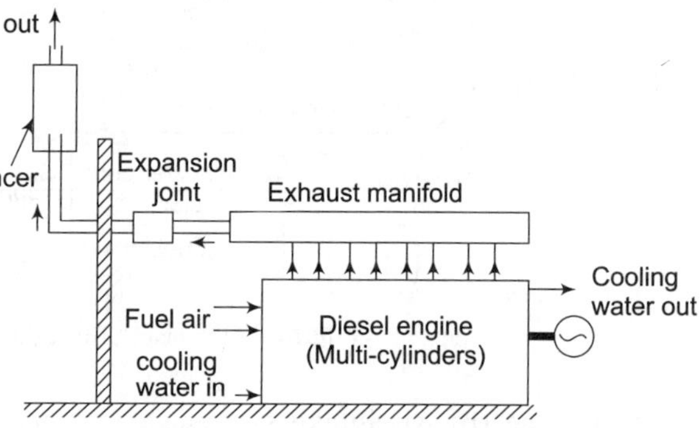

Figure 12.3 Diesel engine exhaust system

Appreciable amount of heat from the engine exhaust goes as a waste. In order to utilise this, a heat recovery steam generator (HRSG) may be used to generate low pressure steam for process work. Alternatively, it may also be used for air heating where exhaust passes through a heat exchanger.

12.4.4 Fuel Handling System

Figure 12.4 shows the fuel handling system of a diesel engine power plant. The fuel oil may be delivered at the plant site by many means such as trucks, railway wagons or barges and oil tankers. With the help of unloading facility, the fuel oil is delivered to the main tanks from where oil is pumped to small service storage tank known as engine day tank through strainer. This day tank has the capacity to store oil equivalent to about 8 hours consumption. In order to reduce the pumping power input, oil is heated either by hot water or steam which reduces viscosity and so the power input.

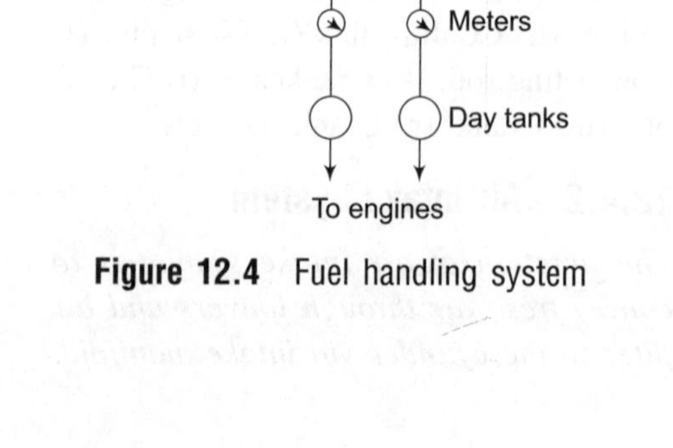

Figure 12.4 Fuel handling system

12.4.5 Fuel Injection System

It is supposed to be the heart of a diesel engine and its failure means stopping of the engine. The fuel injection system performs the following functions:

(i) It *filters* the fuel ensuring oil free from dirt.
(ii) It *meters* the correct quantity of the fuel to be injected in each cylinder.
(iii) It *times* the injection process in relation to the crankshaft revolution.
(iv) It *regulates* the fuel supply.
(v) It atomizes finely the fuel oil for better mixing with the hot air leading to efficient combustion.
(vi) It *distributes*, the atomized fuel properly in the combustion chamber.

There are two ways to atomize the fuel. In one case, air injection is used while in other method, pressure or mechanical or solid injection is used. Now, the air injection is obsolete and mechanical injection is invariably used. *In mechanical or solid injection system, the fuel oil is forced to flow through spray nozzles at a pressure above 100 bars.* There are three types of solid injection system, namely:

- Common rail injection system
- Individual pump injection system
- Distributor system

12.4.5.1 Common Rail Injection System

As the name implies, a single pump supplies fuel under high pressure to a fuel header or common rail as shown in Fig. 12.5 from where the fuel goes to each of the nozzle located in the cylinder. The timing of injection is maintained by a mechanically operated valve and the amount of fuel is controlled by the push rod stroke.

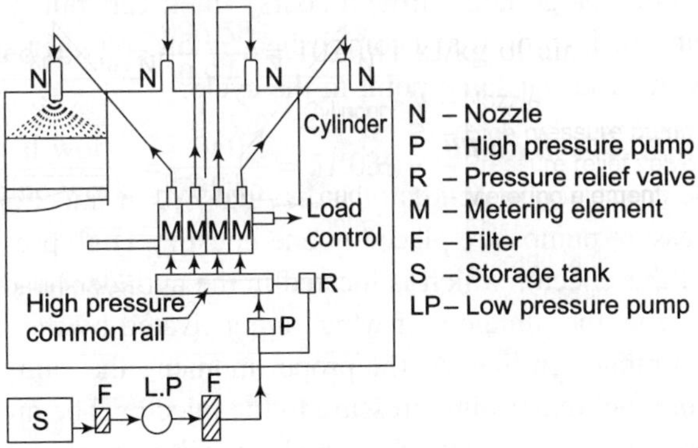

Figure 12.5 Common rail injection system

12.4.5.2 Individual Pump Injection System

As the name implies, the system has an independent high pressure pump for each cylinder which meters, pumps and controls the timing of fuel injection as shown in Fig. 12.6. Each cylinder is provided with one injector and the pump and injector may be integrated as one unit. The fuel is brought to the individual pump from storage tank through course filter, low pressure pump and fine filter. The high pressure pump is equipped with control mechanism and at the proper time, a rocker arm actuates the plunger and thus injects the fuel into the cylinder. The amount of fuel injected is regulated by the effective stroke of the plunger. It is the most popular fuel injection system in practice.

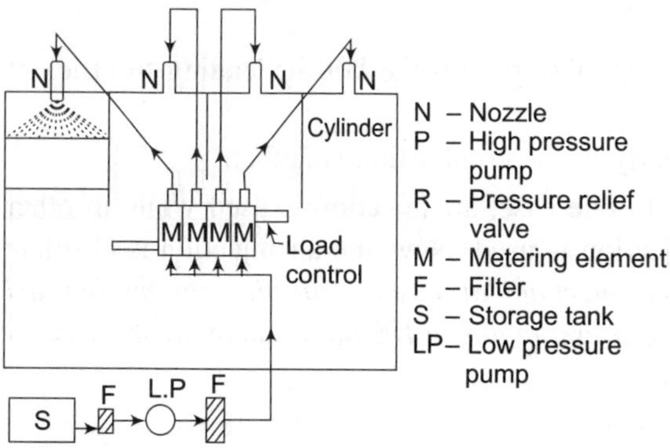

Figure 12.6 Individual pump injection system

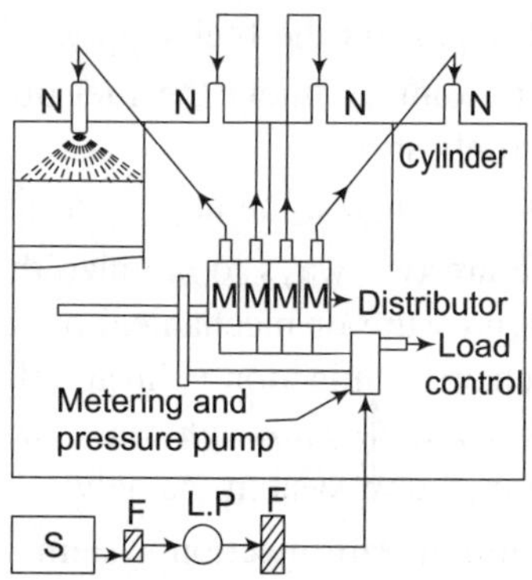

Figure 12.7 Distributor system

12.4.5.3 Distributor System

Figure 12.7 shows the arrangement of distributor system. In this system, a metering and high pressure pump is used to pump the metered quantity of fuel in the rotating distributor which distributes the fuel to the individual cylinder at the correct timing. The number of injection strokes per cycle for the pump is equal to the number of cylinders. The fuel is fed to the high pressure pump from storage tank through course filter, L.P. pump and fine filter. Since the metering and timing of injection is accomplished by one plunger, equal amount of fuel is supplied to each cylinder at the same point in the cycle.

12.4.5.4 Fuel Injection Bosch Pump

Figure 12.8 shows the construction details of individual Bosch injection pump. Here, all of the individual high pressure pumps are placed in one housing. High pressure lines start at the pump and terminate at the injector which is located in the cylinder head. The pump is of the positive displacement type, the plungers of which are activated by individual cam (one cam per pump) which are arranged in line. At the proper moment, the cam activates the plunger axially, which forces the fuel under high pressure to the injector. The amount of fuel injected is controlled by a horizontal movement of the rack, which in turn, rotates the plunger and thus varies the effective stroke.

12.4.5.5 Fuel Injector

The liquid fuel in the injection system enters the combustion chamber through the injector. Fuel injector employed in CI engine is of automatic type. It is mounted on the cylinder body

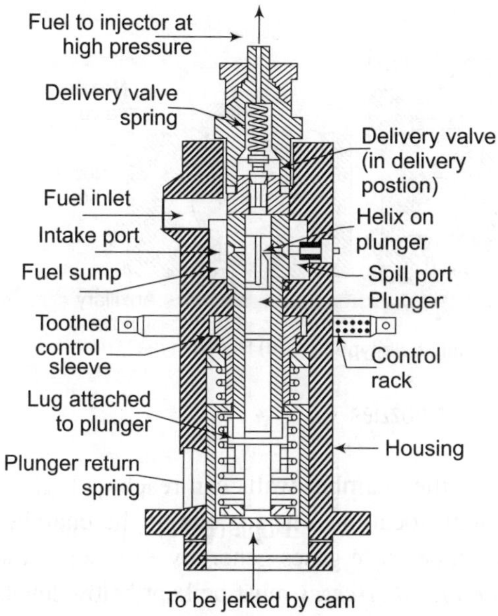

Figure 12.8 Construction details of bosch pump

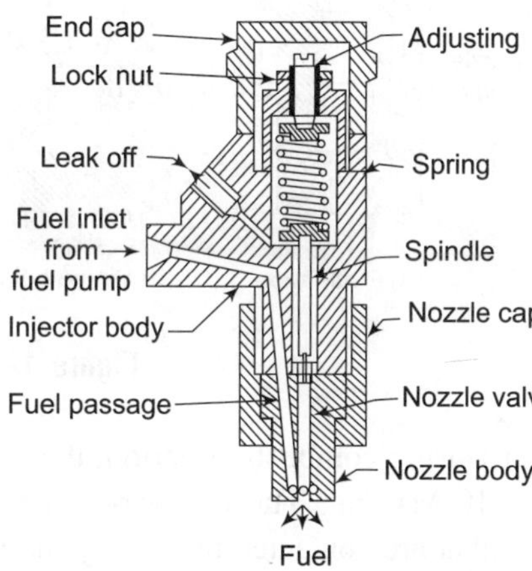

Figure 12.9 Bosch fuel injector

at such a location which yields better performance. Quick and complete combustion is ensured by a well designed fuel injector. The fuel injector assembly consists of the following:

(a) a needle or nozzle valve, (b) a compression ring, (c) a nozzle and (d) an injector body.

Figure 12.9 shows a cross-sectional view of a typical Bosch fuel injector. The fuel from the fuel pump is fed down to the nozzle mouth through long drilling passage. The fuel pressure acts on the differential area of the nozzle valve which lifts against the spring force, and thus allows the fuel to enter the combustion chamber via small orifice (holes) in the form of finally atomized spray. Once, the fuel from the delivery pump gets exhausted, the spring pressure pushes the nozzle valve back on its seat.

12.4.5.6 Types of Nozzles

The design of nozzle is mainly based on the types of combustion chamber used ensuring proper and efficient combustion of fuel. The types of nozzle used in diesel engines are (Fig. 12.10) (a) single hole, (b) multi-hole, (c) Pintle type, (d) Pintaux type.

12.4.6 Cooling System

During combustion process, the peak gas temperature in the cylinder of an internal combustion engine is of the order of 2500 K. The maximum metal temperature for the inside of the combustion chamber space is limited to much lower values than the gas temperature by a large number of considerations and thus cooling for the cylinder head, cylinder and piston must therefore be provided. Necessity of engine cooling arises due to the following facts.

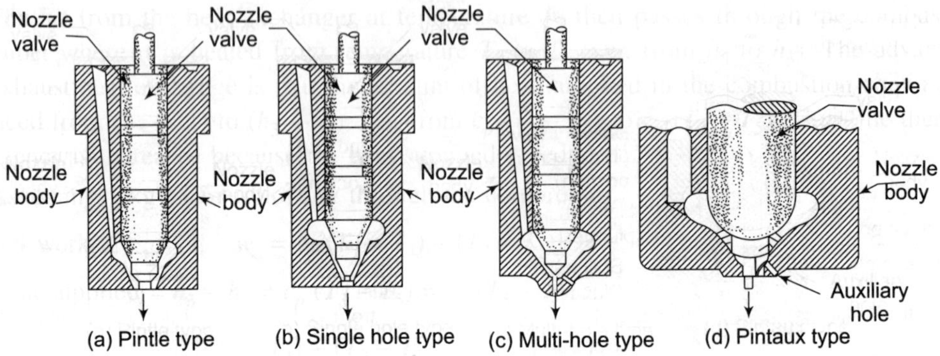

Figure 12.10 Types of nozzles

(i) During combustion period, the heat fluxes to the chamber walls can reach as high as 10 MW/m^2. The flux varies substantially with location. The regions of the chamber that are contacted by rapidly moving high temperature gases generally experience the highest fluxes. In region of high heat flux, *thermal stresses* must be kept below levels that would cause *fatigue cracking*. So temperature must be less than about 400°C for cast iron and 300°C for aluminium alloy for water cooled engines. For air cooled engines, these values are 270°C and 200°C respectively.

(ii) The gas side surface temperature of the cylinder wall is limited by the type of lubricating oil used and this temperature ranges from 160° to 180°C. Beyond these temperature, the properties of lubricating oil deteriorates rapidly and it might even evaporate and burn, damaging piston and cylinder surfaces. Piston seizure due to overheating resulting from the failure of lubrication is quite common.

(iii) The valves may be kept cool to avoid knock and preignition problems which result from overheated exhaust valve (true for S.I. engines).

(iv) The volumetric and thermal efficiency and power output of the engines decrease with an increase in cylinder and head temperature.

Based on cooling medium, two types of cooling systems are in general use. They are:

(a) Air as direct cooling system.

(b) Liquid or indirect cooling system.

Air cooling is used in small engines and portable engines by providing fins on the cylinder. Big diesel engines are always liquid (water/special liquid) cooled.

Liquid cooling system is further classified as:

(i) Open cooling system

(ii) Natural circulation (thermo system)

(iii) Forced circulation system

(iv) Evaporation cooling system

12.4.6.1 Open Cooling System

This system is applicable only where plenty of water is available. The water from the storage tank is directly supplied through an inlet valves to the engines cooling water jacket. The hot water coming out of the engine is not cooled for reuse but it is discharged.

12.4.6.2 Natural Circulation System

The system is closed one and designed so that the water may circulate naturally because of the difference in density of water at different temperature. Figure 12.11 shows a natural circulation cooling system. It consists of water jacket, radiator and a fan. When the water is heated, its density decreases and it tends to rise, while the colder molecules tend to sink. Circulation of water then is obtained as the water heated in the water jacket tends to rise and the water cooled in the radiator with the help of air passing over the radiator either by ram effect or by fan or jointly tends to sink. The direction of natural circulation which is slow is shown by arrows.

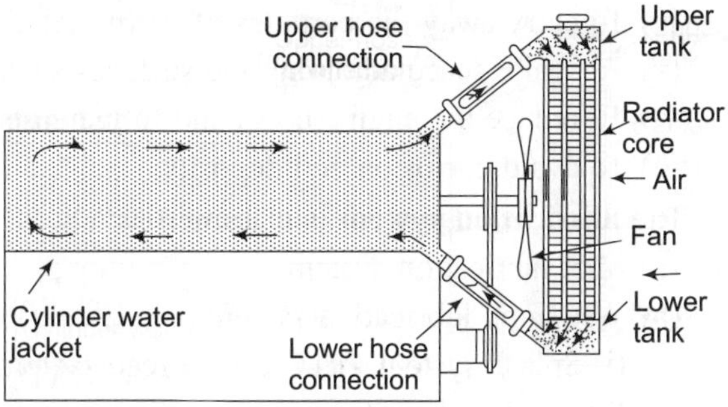

Figure 12.11 Natural circulation cooling system

12.4.6.3 Forced Circulation Cooling System

Figure 12.12 shows forced circulation cooling system which is closed one. The system consists of pump, water jacket in the cylinder, radiator, fan and a thermostat. The coolant (water or synthetic coolant) is circulated through the cylinder jacket with the help of a pump which is usually a centrifugal type, and driven by the engines. The function of thermostat which is fitted in the upper hose connection initially prevents the circulation of water below a certain temperature (usually up to 85°C) through the radiation so that water gets heated up quickly.

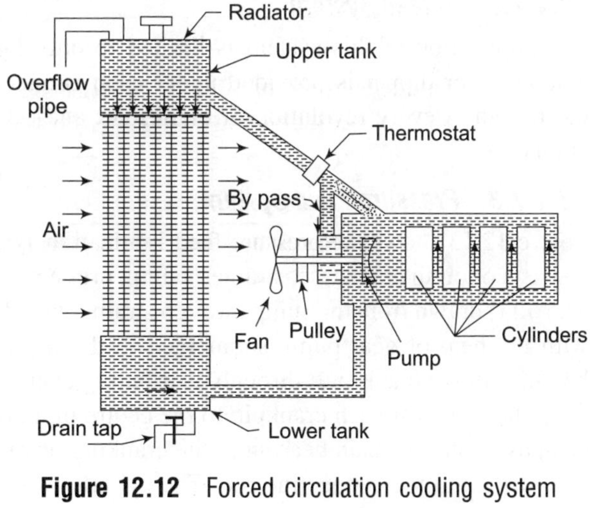

Figure 12.12 Forced circulation cooling system

Standby diesel power plants up to 200 kVA use this type of cooling. In this case of bigger plant, the hot water is cooled in a cooling tower and recirculated again. There is a need of small quantity of cooling make-up water.

12.4.7 Lubrication System

The purpose of lubrication system is to provide sufficient quantity of cool filtered oil to give positive and adequate lubrication to all the moving parts of the engine. The following are the functions of lubrication:

(i) To reduce friction which consequently reduces the power required to overcome the same.
(ii) To reduce wear and tear between the rubbing and wearing surface.
(iii) To cool the surfaces by carrying away heat generated by friction.
(iv) To carry away the particles of worn metal and carbon.
(v) To seal a space adjoining the surfaces such as piston moving in a cylinder.
(vi) To reduce the engine noise and to increase the engine life.
(vii) To avoid corrosion and deposits.

The **lubrication system** is classified as:

(a) Mist lubrication system.
(b) Wet sump lubrication system
    (i) Splash system, (ii) Pressure feed system, (iii) Splash and pressure feed system
(c) Dry sump lubrication

12.4.7.1 Mist Lubrication System

In mist lubrication system, small quantity of lubricating oil is mixed in the fuel tank. It is used in two-stroke engines.

12.4.7.2 Splash System

The application of this system is limited to only light duty engines. As the name implies, a splasher or dipper is provided under each connecting rod cap which dips into the oil in the trough at every revolution of crankshaft and oil is splashed all over the anterior of the crankcase.

12.4.7.3 Pressure Feed System

Figure 12.13 shows the pressure feed lubrication system. This main elements of this system consists of oil in crankcase, strainer, pump, pressure regulator, filter, breather and oil gallaries. The oil is drawn from the sump through strainer which prevents foreign particle and is pumped with the help of gear pump submerged in the oil and driven by crankshaft to all the main bearings of the crankshaft through distributing channel. An oil hole is drilled in the crankshaft from the centre of each crankpin to the centre of an adjacent main journal, through which oil can pass from the main bearing to the crankpin bearing. The piston pin receives oil through a hole drilled in the connecting rod. The cylinder walls, tapped rollers, cam, piston and piston rings are lubricated by oil spray from around piston pins and the main and connecting rod bearings. A pressure regulator fitted near the delivery point of the pump which opens when the pressure in the system attains a predetermined value in the case of filter clogging or an oil cold and excess oil is returned back to the sump.

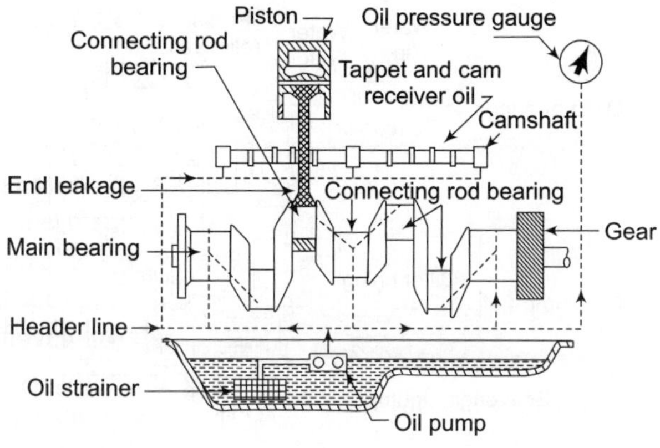

Figure 12.13 Pressure feed lubrication system

12.4.7.4 Splash and Pressure Feed Lubrication System

Figure 12.14 shows a splash and pressure feed lubrication system. In this case, lubricating oil is supplied under pressure to main and camshaft bearings. Splash is also used to lubricate crankpin bearings.

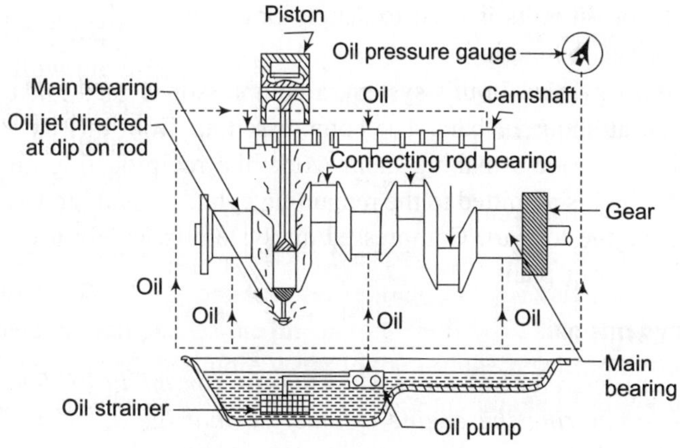

Figure 12.14 Splash and pressure feed lubrication system

12.4.7.5 Dry Sump Lubrication System

Figure 12.15 shows a dry sump lubrication system. In this dry sump, the supply of oil is carried in an exernal tank with the help of scavenging pump through strainer and filter. The scavenging pump is placed out of the sump. The capacity of the scavenging pump is always greater than oil feed pump. The supply tank is usually placed behind the radiator. The dry sump is generally used in large stationary marine engines.

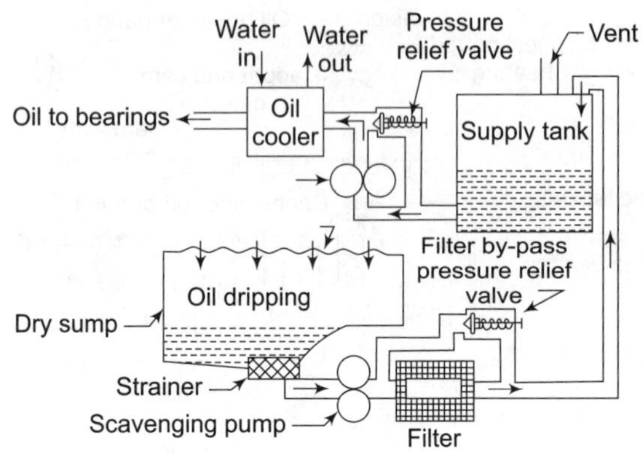

Figure 12.15 Dry sump lubrication system

12.4.8 Starting System

There are three methods to start the diesel engines. They are:

(a) *By an auxiliary engine.* An auxiliary engine is mounted close to the main engine and drives the diesel engine plant through clutch and gears.

(b) *Storage battery and electric motor.* By using an electric motor, in which a storage battery of 12 or 36 volts is used to supply power to an electric power to an electric motor that drives the engines.

(c) *Compressed air system.* In this system, a compressor is used to store compressed air in an air tank at about 17 bars. The compressed air from this air tank is admitted to a few engines cylinders making them work like reciprocating air motor to run the engine shaft. Fuel is admitted to the remaining cylinders and ignited in the natural way causing the engine to start. Compressed air system is widely used to start stationary diesel engine power plant.

12.4.9 Supercharging

The term supercharging refers to increasing the air (or mixture) density by increasing its pressure prior to entering the engine cylinder so that the amount of air inducted per cycle increases, which in turn, increases the power output of the engine. The objectives of supercharging may be one or more of the following:

(i) To increase the power output for a given weight and bulk of the engine (for aircraft, marine and automobiles).

(ii) To compensate for the loss of power due to altitude especially in aircraft engines where loss of power is about 1% at every 100 m altitude as air density decreases with increase in altitude.

(iii) To obtain more power from an existing engines especially rail and standby power plant.

(iv) To obtain better gas exchange process.

There are three basic methods of supercharging. They are as follows:

(a) *Mechanical supercharging.* In this supercharging, a separate pump or blower or compressor, usually driven by power taken from the engines, provides the compressed air.

(b) *Turbocharging.* Turbocharging employs a turbocharger, a compressor and a turbine on a single shaft, is used to boost the inlet air density. The exhaust gas from the engine cylinder expands through turbocharger turbine and the power developed from the turbine drives the turbocharger compressor.

(c) *Pressure wave supercharging.* This methods uses wave action in the intake and exhaust system to compress the intake mixture or air. This is rarely used.

Figure 12.16 shows the supercharging and turbocharging configuration, The widely accepted arrangement is mechanical supercharger (Fig. 12.16a) or turbocharger (Fig. 12.16b). The compressor may be positive displacement (roots blower, vane compressor, screw compressor), centrifugal compressor or axial flow compressor.

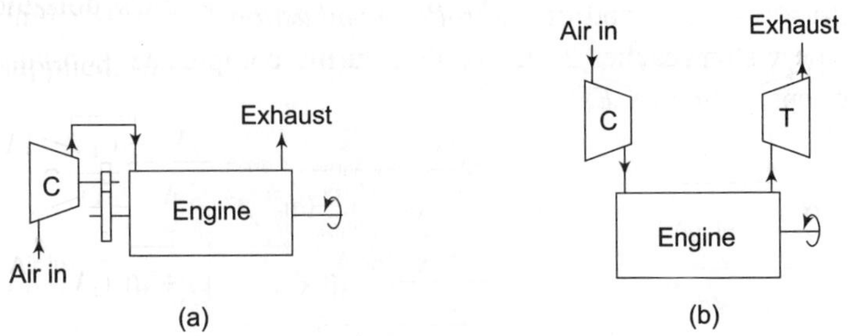

Figure 12.16 Supercharging and turbocharging configuration

Figure 12.17 shows the *p-v* diagram for unsupercharged and supercharged engines. From steady flow energy equation, the work input to compressor is given by

$$-W_c = \dot{m}_a \, c_{pa} \, (T_{0,\text{out}} - T_{0,\text{in}}) \quad ...(12.1)$$

$$W_T = \dot{m}_g \, c_{pg} \, (T_{0,\text{out}} - T_{0,\text{in}}) \quad ...(12.2)$$

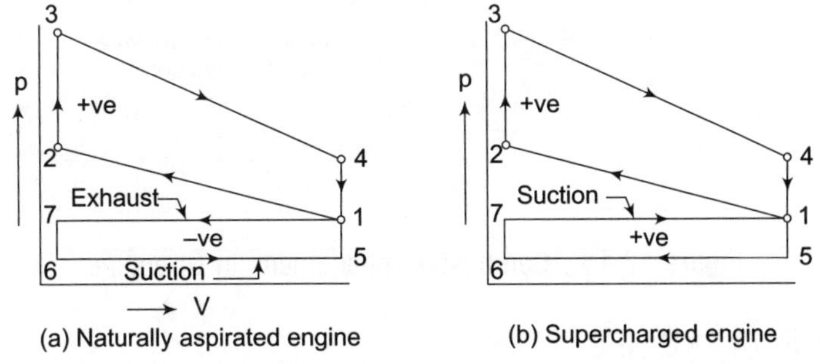

Figure 12.17 *p-v* diagram for naturally aspirated and supercharged engine

12.5 COMBUSTION IN DIESEL (CI) ENGINE

The combustion process in CI engine is divided into four stages, namely:

(i) ignition delay, (ii) rapid pressure rise or period of uncontrolled combustion, (iii) controlled or constant pressure rise, and (iv) burning on the expansion stroke.

Figure 12.18 shows the stages of combustion in CI engine. Point *A* represents the time at which injection starts. The fuel sprayed first attains self-ignition temperature and then undergoes preliminary chemical reactions before the actual ignition takes place. The time interval between the start of fuel injection (A) and the commencement of actual combustion (C) is called the *delay period or ignition delay.* At the end of the delay period, ignition takes place in some region or regions where the concentration of fuel is optimum. During the delay period, considerable amount of fuel accumulates in the combustion space and is properly atomized and vaporized. Consequently, when the actual ignition starts, the fuel burns at an extremely rapid rate causing a steep rise in cylinder pressure and temperature. The combustion phase is called the period of *uncontrolled combustion* and is represented by process *CD*. The process *DE* shows the period of *controlled combustion.* Since the combustion is never complete when the piston reaches *TDC* position during compression stroke, there is always an after burning represented by *EF*.

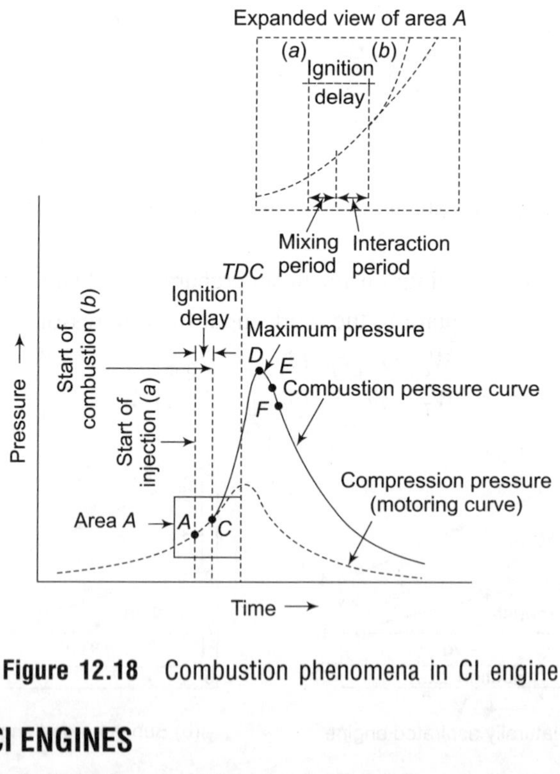

Figure 12.18 Combustion phenomena in CI engine

12.6 KNOCK IN CI ENGINES

If the ignition delay is longer, the actual burning of the first droplets is delayed and a greater quantity of fuel droplets are accumulated in the combustion chamber. When the actual burning starts, the ignition of large amount of accumulated fuel causes a violent and instantaneous rise in pressure as shown in Fig. 12.19. Under such conditions extreme pressure differentials are produced and violent gas vibration known as *detonation* or *knock occurs*. The properties of ignition is measured by *cetane number*.

To eliminate knocking in the CI engines, it is necessary to achieve auto-ignition as early as possible which may be possible from a fuel having a short delay period and low self-ignition temperature.

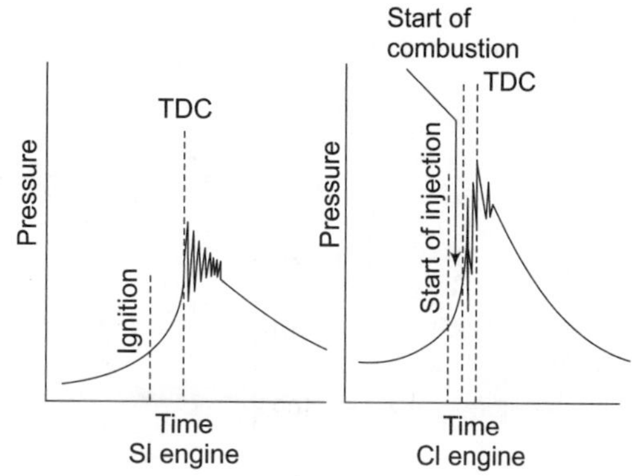

Figure 12.19 Knock in CI engines

12.7 PERFORMANCE CHARACTERISTICS

The following parameters are used to describe the performance characteristics.

(a) *Indicated mean effective pressure (p_{mi}).* It can be defined as theoretical constant pressure exerted (which can be imagined) during each power stroke of the engine to produce a work output equal to the indicated power.

$$p_{mi} = \frac{\text{Net work of cycle}}{\text{Stroke volume}} \quad ...(12.3)$$

(b) *Indicated power (ip).* It is the power developed inside engine cylinder and for 4-stroke engine, it is expressed as

$$ip = \frac{100\, p_{mi}\, LANx}{60 \times 2} \text{ kW} \quad ...(12.4)$$

where p_{mi} is in bar, L is in m, A is in m^2, x is the number of cylinders.

For 2-stroke engine, *ip* will be just double.

(c) *Brake power (bp).* The brake power is the useful power available at the crankshaft or clutch shaft. The brake power is less than indicated power.

$$bp = \frac{2\pi NT}{60 \times 10^3} = \frac{100\, p_{mb}\, \text{LAN}}{2 \times 60}\text{ ; kW} \quad ...(12.5)$$

and

$$bp = ip - fp \quad ...(12.6)$$

(d) *Indicated thermal efficiency (η_{ith}).* It is expressed as

$$\eta_{ith} = \frac{ip}{q_A} = \frac{ip}{\dot{m}_f \times LCV} \quad ...(12.7)$$

(e) *Brake thermal efficiency (η_{bth}).* It is expressed as

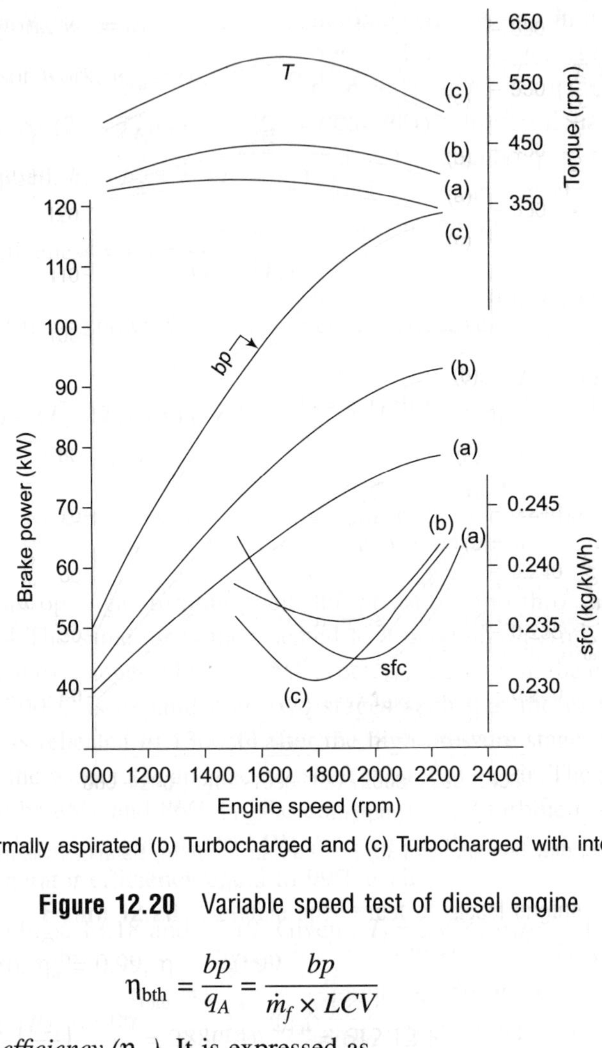

(a) Normally aspirated (b) Turbocharged and (c) Turbocharged with intercooling

Figure 12.20 Variable speed test of diesel engine

$$\eta_{\text{bth}} = \frac{bp}{q_A} = \frac{bp}{\dot{m}_f \times LCV} \qquad ...(12.8)$$

(f) *Mechanical efficiency* (η_m). It is expressed as

$$\eta_m = \frac{bp}{ip} = \eta_{bth}/\eta_{bth} \qquad ...(12.9)$$

(g) *Volumetric efficiency* (η_v). It is defined as

$$\eta_v = \frac{(\dot{m}_a)_{act}}{(\dot{m}_a)_{th}} = \frac{(\dot{m}_a)_{act}}{\rho_a\, n\, V_s} \text{ where } n = N/2 \text{ for 4-stroke} \qquad ...(12.10)$$

(h) *Scavenging efficiency* (η_s). It is expressed as

$$\eta_s = \frac{\text{Actual mass of air retained per minute}}{\text{Ideal mass of air retained per minute}} \qquad ...(12.11)$$

Figure 12.20 shows variable speed test curves of a naturally aspirated, turbocharged and turbocharged with intercooling direct injection CI engine.

Figure 12.21 shows the constant and variable speed curves of a 4-stroke CI engine.

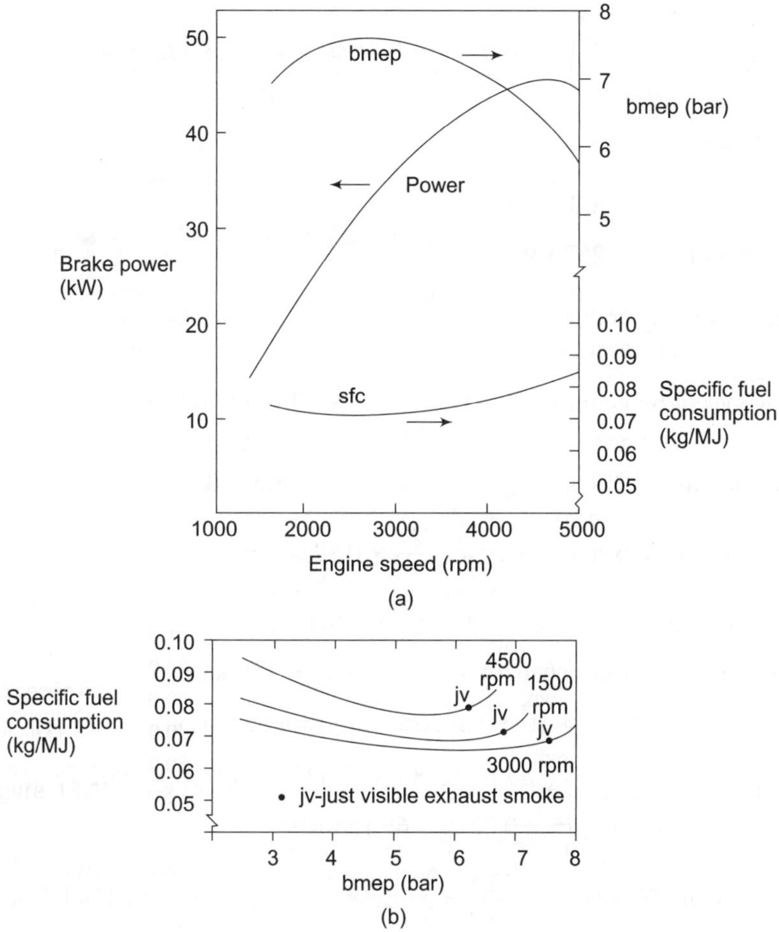

Figure 12.21 Variable and constant speed test of a diesel engine

Example 12.1 The data related to the testing of a four-cylinder, four-stroke diesel engine is –Bore = 36 cm, Stroke = 40 cm, Speed = 350 rpm, bp = 257 kW, mep = 7, Fuel consumption = 72 kg/h, CV of fuel = 43960 kJ/kg, Hydrogen content of fuel = 12%, Air consumption = 28.2 kg/min, Mass of jacket water = 8.6 kg/min, Rise in temperature of jacket cooling water = 41°C, Amount of piston cooling oil = 53 kg/min, Temperature rise of cooling oil = 23°C, Specific heat of cooling oil = 2.09 kJ/kgK, Room temperature = 20°C. Exhaust gas temperature = 325°C, c_p of dry exhaust gas = 1.045 kJ/kgK, Sp. Heat of superheated steam = 2kJ/kgK. Partial pressure of steam in gases = 0.035 bar.

Draw up the heat balance sheet and calculate the brake specific fuel consumption at half of the load.

Solution Heat supplied in the fuel $= \dfrac{72 \times 43960}{3600} = 879.2$ kW

$$ip = \frac{100\, p_{mi}\, LA\, (N/2)x}{60} = \frac{100 \times 7 \times 0.4\, (\pi/4)\, 36^2 \times (350/2) \times 4}{60 \times 10^4} = 332.5 \text{ kW}$$

$$fp = ip - bp = 332.5 - 257 = 75.5 \text{ kW}$$

Heat equivalent to ip = 332.5 kW

Heat equivalent to bp = 257 kW

Heat carried away by cooling water $= \dfrac{86}{60} \times 4.18 \times 41 = 245.64$ kW

Heat carried away by cooling oil $= 2.098 \times \dfrac{53}{60} \times 23 = 42.46$ kW

Total mass of exhaust gases $= \dfrac{72}{60} + 28.2 = 29.4$ kg/min = 0.49 kg/s

Mass of vapour with exhaust gases $= \dfrac{72}{3600} \times 0.12 \times 9 = 0.0216$ kg/s

Mass of dry exhaust gas = 0.49 – 0.0216 = 0.4684 kg/s

Heat in dry exhaust gases = 0.4684 × 325 × 1.045 = 159 kW

From steam tables, heat lost per kg of steam at given conditions

$$= (2550 - 83.86) + 2(325 - 26.694) = 3062.75 \text{ kJ/kg}$$

Heat lost in steam = 3062.75 × 0.0216 = 66.155 kW

Indicated thermal efficiency $= \dfrac{332.5}{879.2} = 37.81\%$

At half load, $bp = \dfrac{257}{2} = 128.5$ kW

Assuming fp to remain constant at half load, the ip at half load

$$= 128.5 + 75.5 = 204 \text{ kW}$$

Heat input at half load $= \dfrac{204}{\eta_{ith}} = \dfrac{204}{0.3781} = 539.53$ kW

Fuel consumption $= \dfrac{\text{Heat input}}{\text{C.V.}} = \dfrac{539.53 \times 3600}{43960} = 44.18$ kg/h

bsfc at half load $= \dfrac{44.18}{128.5} = 0.3438$ kg/kWh ***Ans.***

The heat balance sheet is given in the following Table

Item		*Heat in (kW)*	*Percentage (%)*
Input	Heat supplied in fuel	879.2	100
Expenditure	Heat equivalent to bp	257	29.23
	Heat in cooling water	245.64	27.93
	Heat in cooling oil	42.46	4.82
	Heat in dry exhaust gases	159	18.08
	Heat in steam in exhaust	66.155	7.52
	Unaccounted heat by difference	108.945	12.42
	Total	879.2	100

PROBLEMS

Things to Think About

12.1 What is the reason that diesel power plant is not used as base plant?

12.2 Why there is a need of auxiliary system in a diesel plant?

12.3 Which parameters limit the power output and efficiency of a diesel plant?

12.4 List the various auxiliaries of a diesel power plant.

Short and Long Theoretical Questions

12.1 List the applications of diesel electric power plant.

12.2 List the advantages and disadvantages of diesel electric power plant.

12.3 Mention the types of diesel engines.

12.4 Draw the layout of a diesel engine power plant showing flow path of various streams.

12.5 List the various auxiliaries systems of diesel electric power plant.

12.6 List the main parts of diesel engine.

12.7 Discuss the intake system of diesel engine.

12.8 Discuss the working of exhaust system of diesel engine.

12.9 Discuss the working of various methods of injection system of diesel engine.

12.10 Discuss the fuel handling system of a diesel electric power plant.

12.11 Discuss the working of cooling system of diesel engine.

12.12 Discuss the working of various lubrication system of diesel engine.

12.13 Discuss the various starting systems of diesel electric power plant.

12.14 What do you mean by supercharging?

12.15 Discuss the various methods of supercharging. Mention the type of compressors used in supercharging.

12.16 Discuss the performance curves of diesel engines.

Numerical Questions

12.1 The following data relate to a four-stroke, four-cylinder diesel engine : Cylinder diameter = 0.35 m, Stroke = 0.40 m, N = 320 rpm, imep = 7.5 bar, bp = 260 kW, $\dot{m}_f$ = 82 kg/h, CV = 44 MJ/kg, H_2 content in fuel =13% $\dot{m}_a$ = 32 kg/min, Cooling water circulated = 92 kg/min, Rise in cooling water temperature = 39°C. Exhaust gas temperature = 325°C, c_{pa} 1.005 kJ/kgK, c_{pa} = 1.05 kJ/kgK, T_a = 24°C, Specific heat of superheated steam = 2.093 kJ/kgK, partial pressure of steam in exhaust gases = 0.3 bar. Calculate: (a) Mechanical efficiency, (b) Indicated thermal efficiency, and (c) bsfc, also (d) Draw the heat balance sheet.

12.2 A turbocharged six-cylinder, 4-stroke diesel engine has a swept volume of 40 litres. The inlet manifold conditions are 2.0 bars and 54°C. The volumetric efficiency of the engine is 94%. The engine runs at 16.5 bars bmep at 1200 rpm with air-fuel ratio of 21.5. The power delivered to the compressor is 100 kW with entry conditions of 25°C and 0.95 dry. The fuel has a calorific value of 42 MJ/kg. Calculate: (a) the power output of the engine, (b) the brake efficiency of engine, (c) the compressed isentropic efficiency, and (d) the effectiveness of intercooler.

Software Based Question

12.1 Develop a software for the design of a diesel engine power plant.

Design and Open Ended Question

12.1 Develop a project involving design of a diesel power plant using CH_4 gas obtained from municipal and solid wastes.

Multiple-Choice Questions

12.1 In the fuel system used for diesel plant, which one is true for fuel heating system?
(a) It increase power output (b) It increases efficiency
(c) It is used for cold starting (d) Both (a) and (b)

12.2 The plant efficiency of a diesel power plant ranges from
(a) 10 to 25% (b) 25 to 30% (c) 30 to 35% (d) 35 to 40%

12.3 The cooling tower used in the diesel plant needs make-up water
(a) due to leakage in heat exchanger (b) due to leakage in pump
(c) due to evaporation of water (d) all of them

Answers to Multiple-Choice Questions

12.1 (c) 12.2 (c) 12.3 (d)

CHAPTER

13

Gas Turbine Power Plant

OVERVIEW*

The turbine in many respects is the most satisfactory means of producing mechanical power among the other various means. Further, gas turbine is very compact as compared to water and steam turbines. Serious development of gas turbine began not along before the Second World War with shaft power in mind, but the emphasis was soon diverted to the turbojet engine for aircraft propulsion.

The gas turbine obtains its power by utilising the energy of burnt gases and air which are at high temperature and pressure by expanding through the several rings of fixed and moving blades. It, thus, resembles to a steam turbine. To get a high pressure for the simple cycle (of the order of 4 to 40 bars) of working fluid which is essential for expansion, a compressor is required. *The quantity of the working fluid and speed required are more, so, generally, a centrifugal or an axial flow compressor is employed.* The compressor is driven by the turbine and so it is coupled to the turbine shaft. If after compression, the working fluid were to be expanded in a turbine, then assuming that there were no losses in either component the power developed by the turbine would be just equal to that absorbed by the compressor and the work done would be zero. But the power developed by the turbine can be increased by increasing the volume of the working fluid at constant pressure, or alternatively increasing the pressure at constant volume. Either of these may be done by adding heat so that the temperature of the working fluid is increased after the compression. To get a higher temperature of the working fluid, a combustion chamber is required where combustion of air and fuel takes place giving temperature rise to the working fluid theoretically at constant pressure, however, in actual practice, there is a small pressure loss.

Thus, a simple gas turbine cycle consists of *(i) a compressor, (ii) a combustion chamber and (iii) a turbine. Since the compressor is coupled with the turbine shaft, it absorbs some of the power produced by the turbine (about 50%) and hence lowers the efficiency.* The net work

* For more detail, see the the Author's book *"Steam & Gas Turbines, Turbomachines and Power Plant Engineering"*

is therefore the difference between the turbine work and work required by the compressor to drive it. Gas turbine power plant yielding efficiency from **35 to 47%** are working all over the world. In the present technology, gas turbines from **1 MW to 400 MW** are being built as simple open cycle gas turbine.

Gas turbines have been constructed to work on the following fuels; ***oil, natural gas, coal gas, producer gas, blast furnace and synthetic gas***. Various auxiliary systems are incorporated in gas turbine for their efficient and safe operations.

When gas turbine is used for electricity generation it is called ***power turbine*** and when it is used as propulsion unit in aircraft, it is called ***aviation or aircraft turbine***. In the case of power turbine, the expansion of high pressure and high temperature gas takes place fully up to the atmospheric pressure whereas in aviation turbine, the expansion takes place partly to drive compressor and other auxiliaries and rest expansion takes place in nozzle to produce a jet of gas resulting in production of thrust which propels the aircraft in opposite direction.

The applications of gas turbine are very wide. Gas turbines are extensively used in **base load electricity power generation**, standby power plants, industrial gas turbines, driving pumps/compressors in gas/oil pipe line, aircraft propulsion, rockets, combined cycle power plants, etc. At present, the cost of gas turbine plant including generator is about US dollar 350 per kWh.

Figure 13.1 shows a schematic of a simple gas turbine power plant. It consists of a compressor (axial flow), combustor, gas turbine and turbo-alternator and equipped with various auxiliaries systems. The auxiliaries systems include the following:

(i) Blade cooling system (ii) Lubrication system
(iii) Fuel system (iv) Speed control system
(v) Protective system (vi) Starting system
(vii) Ignition system

Blade coolings with air bled from compressor is essential due to safeguard against its failure.

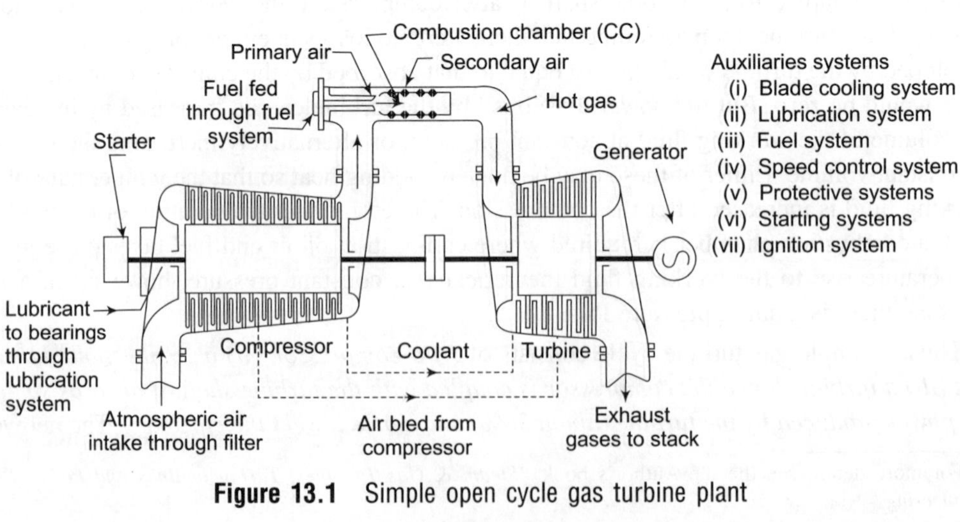

Figure 13.1 Simple open cycle gas turbine plant

13.1 CLASSIFICATION, APPLICATION AND MODERN DEVELOPMENTS

13.1.1 Classification

Gas turbines may be classified on the basis of the following:

(A) On the basis of *combustion process* the gas turbine is classified as follows:

(i) *Continuous combustion or constant pressure type. The cycle working on this principle is called Joule or Brayton cycle, and* (ii) *The explosion or constant volume type. The cycle working on this principle is called Atkinson cycle.*

(B) On the basis of the *action of expanding gases* similar to steam turbine, it is classified as

(i) *Impulse turbine, and* (ii) *Impulse-reaction turbine.*

(C) On the basis of *path of working substance,* the gas turbine is again classified as

(i) *Open cycle gas turbine* (working fluid enters from atmosphere and exhaust to atmosphere), (ii) *closed cycle gas turbine* (working fluid is confined within the plant), and (iii) semi-closed cycle (part of the working fluid is confined within the plant and another part flows like open system.

(D) On the basis of direction below:

(i) Axial flow and (ii) Radial flow

In power plants and aviation gas turbines, axial flow compressors and turbines are common. Radial flow is used in small plants.

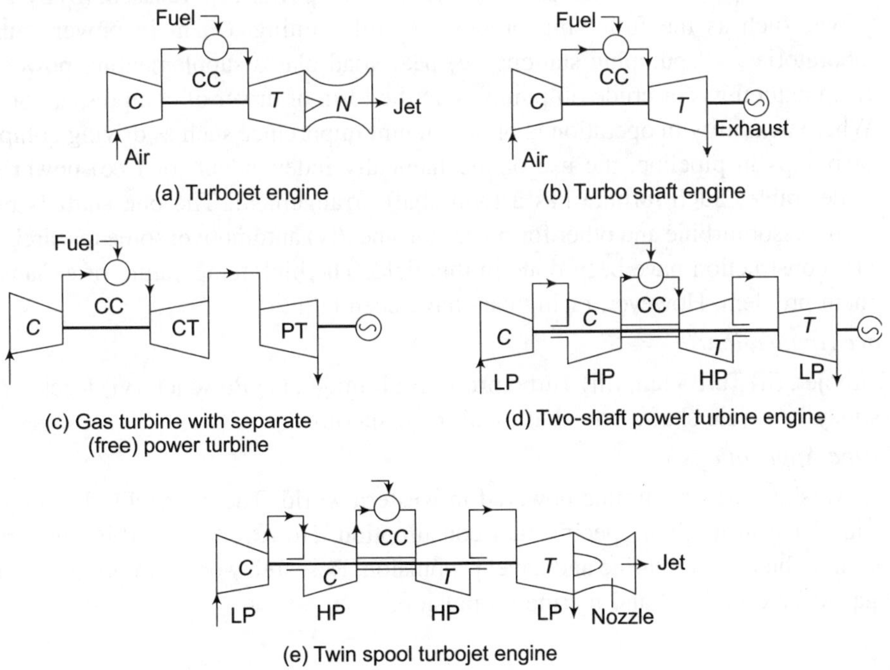

Figure 13.2 Gas turbine shaft configurations

(E) On the basis of *shaft configuration the gas turbine* may be classified as:

(i) Single shaft engine, e.g., turbojet engine (Fig. 13.1a) and turboshaft engine (Fig. 13.1b), and (ii) multi-shaft engine, e.g., industrial split shaft (Fig. 13c.), two-shaft engine with a power turbine (Fig. 13.1d) and twin-spool turbojet engine (Fig. 13.1e).

(F) On the basis of *thermodynamics cycle*, the gas turbine may be classified as

(i) Simple cycle (Fig. 13.1), (ii) Simple with regenerator (Fig. 13.10), (iii) Simple with reheat (Fig. 13.14), (iv) Simple with intercooler (Fig. 13.16), (v) Simple with intercooler reheater and regenerator. (Fig. 13.20)

(G) On the basis of *Application*

(i) Power or industrial turbine and (ii) aviation or aircraft turbine.

13.1.2 Applications

The following are the applications of gas turbine.

(A) *Load Applications:*

(i) Locomotive propulsion–still not in practice

(ii) Central power stations

(a) Units plants (b) Combined cycle and co-generation plants, (c) Standby plants for hydro installations (d) Fully automatic booster station at the end of transmission line, (e) Bomb-proof power plants, and (f) At location where water is not available.

(iii) *Industrial gas turbines*

(a) Standby plants for steam power plant or where power is purchased, (b) by-products power such as the following process for oil refining, (c) main power source for laboratories, (d) pumping stations, (e) peak load plants supplementing power plants, (f) in natural gas or crude oil pumping in the form of driver of compressors or pumps. Where flexibility in operation is of paramount importance such as driving compressors or pumps in pipeline, the use of mechanically independent (or free) power turbine is desirable. Such turbine has a twin shaft arrangement. The one shaft is used for compressor turbine and other for power turbine, (iv) automotive: some research, design and construction have been done in this field. The high temperature of exhaust is the main problem. However, racing cars have been built.

(B) *Space Application:*

(i) Turbojet, (ii) Turbo fan, (iii) Turbo prop, (iv) Ram jet, (v) Pulse jet, (vi) Rocket Aircraft engines may be single shaft, twin shaft, and three spool.

(C) *Marine Applications*

Many warships are gas turbine powered in western world. The major disadvantage of the gas turbine have in its poor specific fuel consumption. However, gas turbine has got little application in the field of marine and navel propulsion. Research work is in progress for better use of gas turbine in navel and marine propulsion.

13.1.3 Modern Developments

The development of the gas turbine has gone hand in hand with the development of aerodynamic science, and that of metallurgy, with the result that it is now possible to find advance turbines using pressure ratio above 35:1, component efficiency of 85–90 per cent, and turbine inlet temperature exceeding 1700 K.

The design evolution of industrial gas turbines over an extended period, shows how the power and efficiency are continuously improved by increasing pressure ratio, turbine inlet temperature and air flow. The evolution of Westinghouse. 501 gas turbine from 1968 to 1993 is depicted in the following Table 13.1 [58]. Table 13.1 depicts how the power increased from 42 MW to 160 MW. The pressure ratio has been raised from 7.5 to 14.6 and turbine inlet temperature was raised from 1153 K to 1533 K.

Table 13.1 Development of Westinghouse 501 gas turbine

Year	*1968*	*1973*	*1981*	*1993*
Power (MW)	42	80	107	160
Thermal efficiency (%)	27.1	30.5	33.2	35.6
Pressure ratio	7.5	11.2	14.0	14.6
Turbine inlet temp. (K)	1153	1266	1406	1533
Air flow (kg/s)	249	388	354	435
Exhaust gas temp. (K)	447	486	531	584
No. of compressor stages	17	17	19	16
No. of turbines stages	4	4	4	4
No. of cooled rows	1	1	4	6

Table 13.2 shows the specifications of some gas turbines built by various leading manufacturers, using natural gas as fuel.

Table 13.2 Specifications of latest gas turbines (*Courtesy.* GT World 2000)

S. No.	*Model*	*Year*	*ISO output rating (MW)*	*Pressure ratio*	*Mass flow rate (kg/s)*	*Turbine speed (rpm)*	*Turbine inlet temp (°C)*	*Turbine exhaust temp (°C)*	*ISO thermal efficiency (%)*
	A B B ALSTOM Power								
1.	GT 24 (Reheat)	1994	183.00	30.00	390.09	3600	1235	640.00	38.3
2.	GT 26 (Reheat)	1994	265.00	30.00	565.5	3000	–	640.00	38.51
	GE/Hitachis/BHEL								
3.	PG 9231 (Ec)	1997	169.2	14.2	498.5	3000	–	557.77	34.93
4.	9 HA.01	2000	448	–	–	3000	–	–	42.9

Contd.

Table 13.2 Contd...

	GE Ind. Aeroderivative								
5.	LM 6000 PC	1997	44.09	29.4	127.0	3600	837.7	450.0	41.95
6.	LM 250 + (PK)	1998	31.32	22.7	87.09	3600	835.55	515.0	39.5
	Flat Avio								
7.	TG 50 D55	1991	146.915	14.0	469.47	3000	–	522.0	34.23
8.	701 IF	1992	260.44	9.4	655.9	3000	–	578.0	37.28
	Mitsubishi HI								
9.	ASE 40	1996	2.6	8.4	12.29	15400	–	561.00	25.25
10.	ASE 50	1999	3.2	10.2	13.83	15400	–	556.11	27.67
11.	M 501G	1997	254.0	20.0	566.25	3600	–	596.11	38.74
12.	M 701G	1997	334.0	21.0	737.1	3000	–	596.2	39.54
	Siemens Westinghouse								
13.	V 643A	1996	68.0	16.2	190.96	5400	1315.5	588.88	34.72
14.	V 943A	1995	258.0	17.0	634.1	3000	1315.5	567.77	38.39
15.	V 942A	1997	189.0	14.0	520.27	3000	1176.66	585.0	35.29

Department of Energy (DOE) of USA had given a project to some leading manufacturers of gas turbine to develop simple gas turbine cycle using natural gas as fuel and yielding thermal efficiency of 47% by 2005 which has been achieved. This gas turbine is now in the market and will revolutionize the energy scene with its better version.

DOE of USA has given an another project to develop a hybrid system-fuel cell and gas turbine yielding about 70% efficiency. The fuel cell will be using natural gas and exhaust of fuel cell will be at about 1000°C which will be expanding in a gas turbine. The power output sharing will be about 80% by fuel cell and about 20% by gas turbine. The hybrid system will be free from combustion products. The prototype hybrid unit producing 40 MW had been completed by 2004. Many such plants have been commissioned in USA.

13.2 SIMPLE OPEN CYCLE GAS TURBINE (CONSTANT PRESSURE HEAT ADDITION) OR AIR STANDARD BRAYTON (OR JOULE) CYCLE

The Joule or Brayton cycle is the most idealized cycle for a simple gas turbine power plant shown in Fig. 13.3. Atmospheric air is compressed from p_1 to a high pressure p_2 in the compressor and delivered to the burner or combustion chamber (CC) where fuel is injected and burnt. The combustion process occurs nearly at constant pressure. Due to combustion heat is added to the working fluid in the burner (combustor) from T_2 to T_3. The products of combustion from the combustion chamber are expanded in the turbine from p_2 to atmosphere pressure p_1 and then discharged into the atmosphere. The turbine and compressor are mechanically coupled, so the net work is equal to the difference between the work done by the turbine and work consumed by the compressor. To first run a compressor, a starter is needed. When the turbine starts running, the starter is cut-off.

Representation of ideal Brayton cycle on *p-v* and *h-s* or (*T-s*) diagrams are shown in Fig. 13.3(b) and 13.3(c). All the assumptions are valid here for air standard cycle.

Process 1-2 is the isentropic compression in the compressor.

Process 2-3 is constant pressure heat addition in the combustion chamber.

Process 3-4 is isentropic expansion in the turbine.

Process 4.1 is constant pressure heat rejection.

Referring to Fig. 13.3, we may find the thermal efficiency on the basis of 1 kg of working fluid flow. Use the steady flow energy equation for analysis. Since the velocity of gas is high in gas turbine so the analysis should be based on stagnation values, but for simplicity, the

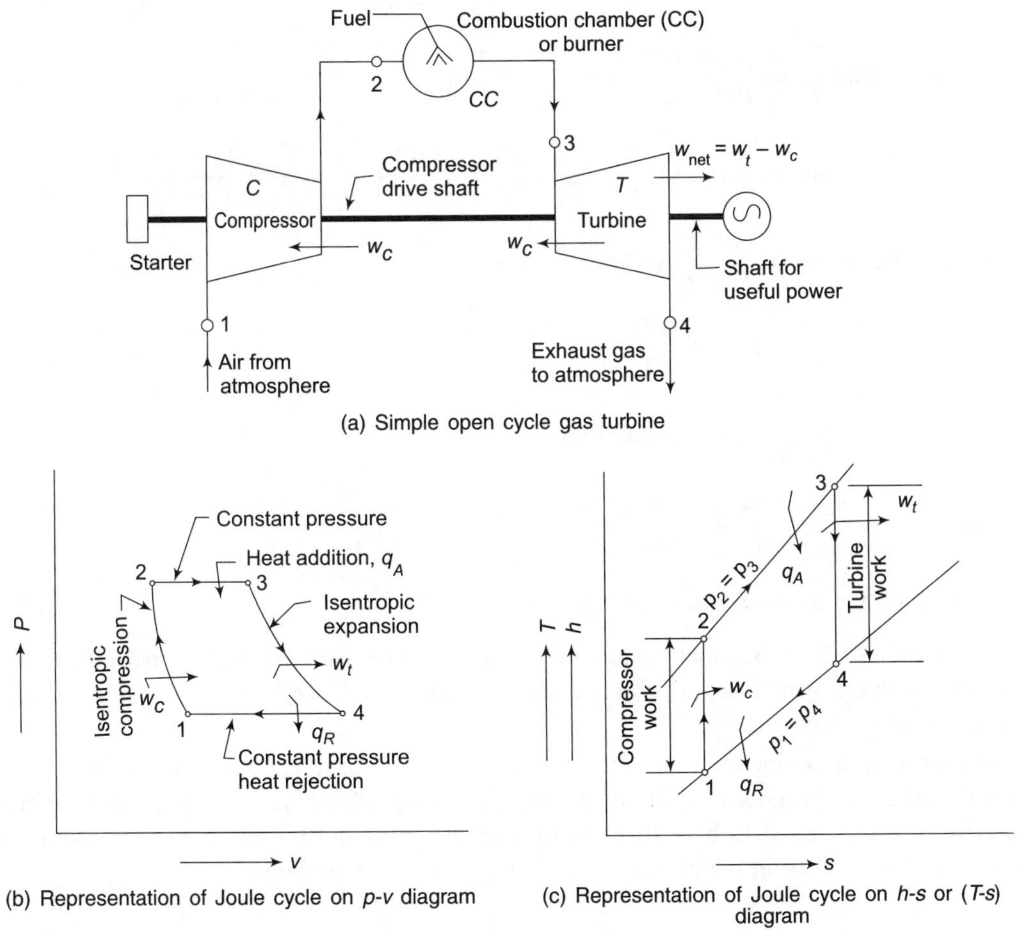

Figure 13.3

analysis is based on static values. *However, with the same analysis stagnation values may be put by placing suffix 0 in enthalpy, temperature and pressure terms.*

Heat supplied, $q_A = h_3 - h_2 = c_p (T_3 - T_2)$ if c_p is assumed constant in process 2-3.

Heat rejected, $q_R = h_4 - h_1 = c_p (T_4 - T_1)$ if c_p is assumed constant in process 4-1.

Net work, $w_{net} = q_A - q_R = c_p [(T_3 - T_2) - (T_4 - T_1)]$

This net work may also be found from turbine and compressor work.

Work done by turbine, $w_1 = h_3 - h_4 = c_p (T_3 - T_4)$

Work consumed by compressor = $w_c = h_2 - h_1 = c_p (T_2 - T_1)$

$$w_{net} = w_1 - w_c = c_p \{(T_3 - T_4) - (T_2 - T_1)\} = c_p \{(T_3 - T_2) - (T_2 - T_1)\}$$

The thermal efficiency, $\eta_{th} = \dfrac{w_{net}}{q_A} = \dfrac{c_p\{(T_3 - T_2) - (T_4 - T_1)\}}{c_0 (T_3 - T_2)} = 1 - \dfrac{T_4 - T_1}{T_3 - T_2}$ ...(13.1)

We know that $\dfrac{T_2}{T_1} = \left(\dfrac{p_2}{p_1}\right)^{(\gamma-1)/\gamma}$ and $\dfrac{T_3}{T_4} = \left(\dfrac{p_3}{p_4}\right)^{(\gamma-1)/\gamma}$

Since $p_1 = p_4$ and $p_2 = p_3$, so $\dfrac{T_2}{T_1} = \dfrac{T_3}{T_4} = \left(\dfrac{p_2}{p_1}\right)^{(\gamma-1)/\gamma}$ or $\dfrac{T_1}{T_2} = \dfrac{T_4}{T_3} = \dfrac{T_4 - T_1}{T_3 - T_2}$

Putting the value of $\dfrac{T_4 - T_1}{T_3 - T_2}$ in Eq. (13.1), we get

$$\eta_{th} = 1 - \frac{T_1}{T_2} = 1 - \frac{T_4}{T_3}$$

Let r_p = pressure ratio = $\dfrac{p_2}{p_1}$

So, $\dfrac{T_2}{T_1} = = \left(\dfrac{p_3}{p_4}\right)^{(\gamma-1)/\gamma} = (r_p)^{(\gamma-1/\gamma)}$

Therefore, the thermal efficiency, $\eta_{nth} = 1 - 1/r_p^{(\gamma-1)/\gamma}$ ...(13.2)

From Eq. (13.2), we see that the thermal efficiency of the Brayton cycle is the same as that of the Otto cycle. *Note that for turbine and compressor work, we have neglected the change in K.E. and P.E. energies.*

A plot of η_{th} as a function of r_p is the shown in Fig. 13.4(a). It is obvious from Fig. 13.4(a) that the efficiency progressively increases with increasing value of pressure ratio. The **turbine specific work** is found to be a function of pressure ratio and turbine inlet temperature as shown in Fig. 13.4(b) in which $T_3/T_1 = T_{max}/T_{min} = t$ and expressed as

$$\frac{w_{net}}{c_p T_1} = t\left[1 - \frac{1}{r_p(\gamma - 1/\gamma)}\right] - \left[r_p^{(\gamma-1/\gamma} - 1\right] \quad ...(13.3)$$

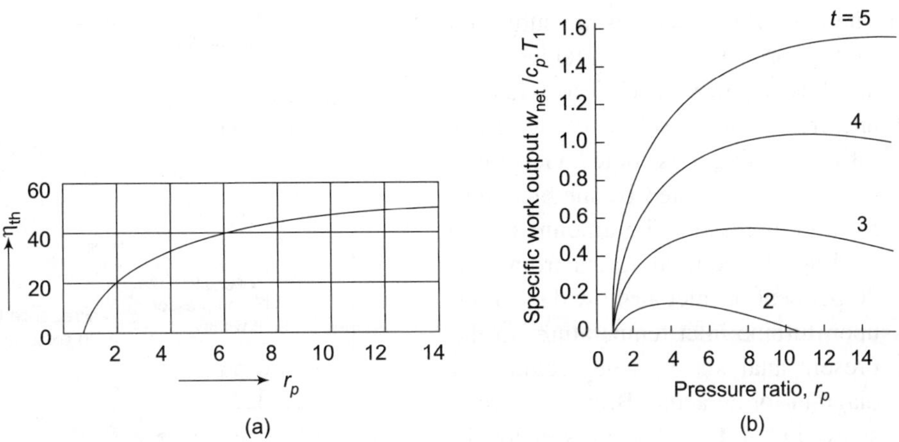

Figure 13.4 Thermal efficiency and specific work output for simple cycle

It is obvious that the specific work increases with t at any pressure ratio but there exists an optimum value of r_p for every value of t, i.e., $T_3/T_1 = T_{max}/T_{min}$.

13.3 ACTUAL BRAYTON CYCLE

The actual gas turbine cycle differs from theoretical cycle in the following main respects:

(1) The fluid velocities are high in turbo-machinery, hence the change in kinetic energy between inlet and outlet of each component cannot necessarily be ignored. This may be taken care by considering stagnation values of properties instead of static values.

(2) Due to *frictional losses* in the compressor and turbine, the compression and expansion processes are not frictionless and take place with some increase in entropy (i.e., the processes are irreversible adiabatic). In the ideal case, the compressor and turbine efficiencies were 100 per cent but actual turbine and compressor have efficiency less than 100 per cent. (Generally, $\eta_c = 87\%$ and $\eta_t = 88\%$).

(3) A *small pressure loss (about 2% of inlet pressure) occurs* in the combustion chamber. Combustion is never complete. This loss is so little that it can be neglected for the sake of simplification of the problem wherever necessary. Thus, $p_3 = p_2 - (\Delta p)_{loss,\ comb}$

(4) The mass of the gas flowing through the turbine is $(1 + F/A)$ times the mass of air flowing through the compressor where F/A represents the fuel-air ratio.

(5) The *specific heat of combustion gas is slightly higher than that of air* ($c_{pg} = 1.20$ kJ/kgK). This increase is so little that the specific heat of combustion gas may be taken as that of air for simplicity wherever necessary.

(6) There is a pressure loss in the exhaust hood of turbine and as a result, the expansion of gas will take place up to a pressure higher than atmospheric so that after pressure loss the exhaust pressure will be equal to the atmospheric pressure. Thus, $p_4 = p_1 + \Delta p_{loss}$. If $p_4 = p_1$, the exhaust will not go to atmospheric and there will be a back flow.

(7) The gas turbine bladings are subjected to high pressure and high temperature and thus they are vulnerable to oxidation and creep. For the safe working of gas turbine blading, it is cooled. Generally, the blade are cooled by air bleeding, from compressor. The amount of the cooling air required varies from 7 to 20 per cent of inlet air flow depending upon turbine inlet temperature. In the present analysis, it is neglected. *h–s* diagram for an actual Brayton cycle is shown in Fig. 13.5. The pressure loss in the combustion chamber is represented by $p_2 - p_3$ and the exhausthood, $p_4 - p_1$. In the cycle.

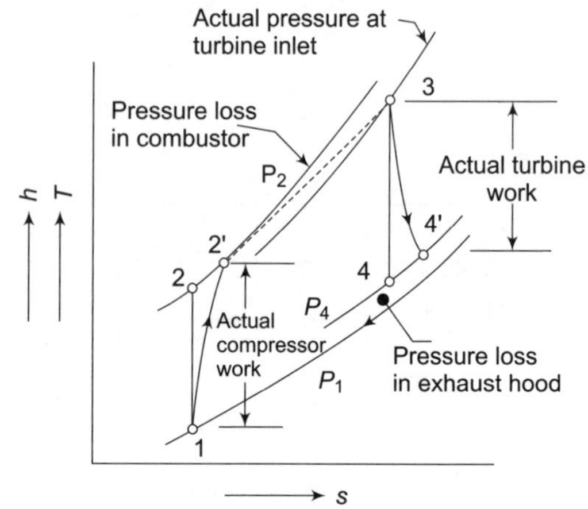

Figure 13.5 Actual gas turbine cycle representation on *h–s* chart

1 – 2 is isentropic compression. 1 – 2′ is actual compression

3 – 4 is isentropic expansion 3 – 4′ is actual expansion

The compressor efficiency, also known as, isentropic compressor efficiency, η_c is

$$\eta_c = \frac{\text{Isentropic compression work}}{\text{Actual compression work}} = \frac{w_c}{w_{ca}}$$

Based on static values, $\eta_c = \dfrac{h_2 - h_1}{h_2' - h_1} = \dfrac{c_p\,(T_2 - T_1)}{c_p\,(T_2' - T_1)} = \dfrac{T_2 - T_1}{T_2' - T_1}$ ...(13.4)

Based on total or stagnation values, $\eta_{oc} = \dfrac{h_{02} - h_{01}}{h_{02}' - h_1} = \dfrac{T_{02} - T_{01}}{T_{02}' - T_{01}}$ ...(13.4a)

Here suffix 0 denotes the stagnation values

The turbine efficiency, $\eta_t = \dfrac{\text{Actual turbine work}}{\text{Isentropic turbine work}} = \dfrac{w_{ta}}{w_t}$

Based on static values, we have:

Actual turbine work, $w_{ta} = h_3 - h_4' = c_{pg}\,(T_3 - T_4')$ kJ/kg

If specific heat of combustion gas c_{pg} may be taken as that of air, then

Actual turbine work, $w_{ta} = c_p\,(T_3 - T_4')$ kJ/kg

So, actual turbine efficiency, $\eta_{ta} = \dfrac{h_3 - h_4'}{h_3 - h_4} = \dfrac{c_p\,(T_3 - T_4')}{c_p(T_3 - T_4)} = \dfrac{T_3 - T_4'}{T_3 - T_4}$ ...(13.5)

Based on total or stagnation values, the turbine isentropic efficiency,

$$\eta_{ota} = \frac{h_{03} - h_{04}'}{h_{03} - h_{04}} = \frac{T_{03} - T_{04}'}{T_{03} - T_{04}} \quad \text{...(13.5a)}$$

The thermal efficiency of the actual cycle is computed as follows for static values:

Actual net work, $w_{neta} = w_{ta} - w_{ca} = [(1 + F/A)\,(h_3 - h_4')] - (h_2' - h_1)$ kJ/kg

where $F/A = \dot{m}_f/\dot{m}_a$ = fuel air ratio

Actual heat supplied, $q_{Aa} = (1 + F/A)\,h_3 - h_2'$ kJ/kg of air

So, actual thermal efficiency, $\eta_{tha} = \dfrac{(1 + F/A)\,(h_3 - h_4') - (h_2' - h_1)}{(1 + F/A)\,h_3 - h_2'}$

$$= \frac{(1 + F/A).c_{pg}\,(T_3 - T_4') - c_{pa}\,(T_2' - T_1)}{(1 + F/A).c_{pg}.T_3 - c_{pa}.T_2'} \qquad ...(13.6)$$

If we neglect the mass of fuel in comparison to the mass of air used, then

$$\eta_{tha} = \frac{(h_3 - h_4') - (h_2' - h_1)}{h_3 - h_2'} = \frac{c_{pg}(T_3 - T_4') - c_{pa}\,(T_2' - T_1)}{(c_{pg}\,T_3 - c_{pa}.T_2')}$$

If we assume that the specific heat of air and combustion gas be the same, then

$$\eta_{tha} = \frac{c_p(T_3 - T_4') - c_p(T_2' - T_1)}{c_p(h_3 - h_2')} = \frac{(T_3 - T_4') - (T_2' - T_1)}{(T_3 - T_2')} \qquad ...(13.7)$$

or
$$\eta_{tha} = \frac{\eta_t(T_3 - T_4) - (T_2 - T_1)/\eta_c}{T_3 - T_2'} = \frac{(T_3 - T_4')\,\eta_t.\eta_c - (T_2 - T_1)}{(T_3 - T_2')\,\eta_c}$$

Based on stagnation values, $\eta_{otha} = \dfrac{(T_{03} - T_{04})\,\eta_{0t}\,\eta_{0c} - (T_{02} - T_{01})}{(T_{03} - T_{02}')\,\eta_{0c}}$

Based on static values, we have $\eta_{tha} = \dfrac{(w_t)\,\eta_t\,\eta_c - w_c}{q_{Aa}.\eta_c}$ ...(13.8)

From Eq. (13.8), it is clear that the thermal efficiency is increased by improving either η_t or η_c or both.

It should be borne in mind that the extra work supplied to the compressor because of its inefficiency is not entirely wasteful because as compared to isentropic compression, the heat supplied in the combustion chamber is reduced. It is worth to find an expression for minimum temperature ratio, $t = t_{min} = T_{03}/T_{01}$ in the case of gas turbine. For stagnation values,

$$w_{ca} = \frac{c_p T_{01}}{\eta_{oc}}\left(r_p^{(\gamma - 1/\gamma} - 1\right) \quad \text{and} \quad w_{ta} = \eta_{ot}.c_p\,T_{03}\left[\frac{r_p^{(\gamma-1)/\gamma} - 1}{r_p^{(\gamma-1)/\gamma}}\right]$$

$$\therefore \quad w_{neta} = \eta_{ot}.c_p\,T_{03}\left[\frac{r_p^{(\gamma-1)/\gamma} - 1}{r_p^{(\gamma-1)/\gamma}}\right] - \frac{c_p\,T_{01}}{\eta_{oc}}\left(r_p^{(\gamma-1)/\gamma} - 1\right)$$

If $w_{ca} = w_{ta}$, the minimum temperature ratio $T_{03}/T_{01} = t = t_{min}$ can be calculated for which no net work is possible. This results in

or $$t = t_{min} = \frac{T_{03}}{T_{01}} = \frac{T_{max}}{T_{min}} = \frac{r_p^{(\gamma-1)/\gamma}}{\eta_{ot}\,\eta_{oc}}$$

If $$r_p = 5,\ \eta_{oc} = \eta_{ot} = 0.85 \text{ and } \gamma = 1.4,\ t = \boldsymbol{t_{min} = 2.19}$$

This means if $t_{01} = 15°C$, $t_{03} = 358°C$, then for *net work to be done, t_{03} has to be more than 358°C.*

The value of T_3, i.e., the **turbine inlet temperature (TIT)**, and hence the ratio t, can be used in practice, depends upon the maximum temperature which is highly stressed parts of the turbine can withstand for the required working life; it is often called the "metallurgical limit. In early gas turbine, t was selected between 3.5 and 4 but with the introduction of **air cooled turbine blades *t* is raised to 5 and 6**. Without cooling of gas turbine blades, with the air bled from compressor or by other method, the blade will fail due to fatigue and thermal stresses.

13.4 POLYTROPIC OR SMALL STAGE EFFICIENCY

The isentropic efficiency deals with the overall performance of a dynamic compressor (centrifugal and axial). Sometimes, it is desirable to establish the true aerodynamic efficiency of compression of a single stage or small stage within a compressor or turbine by considering a large number of small compression or expansion (Figs. 13.6 and 13.7).

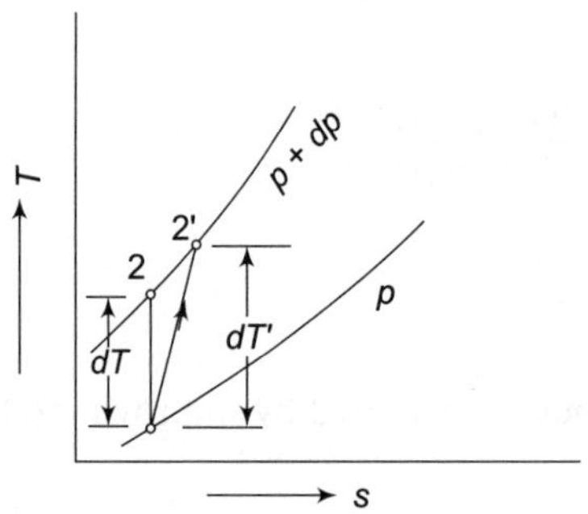

Figure 13.6 Polytropic compression

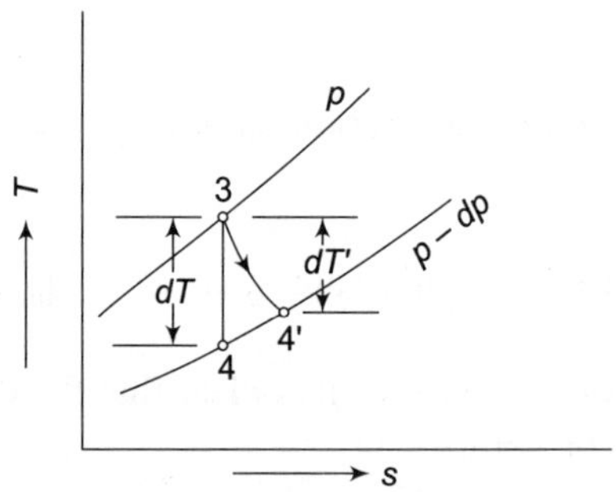

Figure 13.7 Polytropic expansion

Let dT = isentropic temperature rise, and dT' = actual temperature rise.

Thus, $$\eta_{pc} = \frac{dT}{dT'} \quad \text{and} \quad \eta_{pt} = \frac{dT'}{dT}$$

We know that, $$\frac{pv}{T'} = c_1 \quad \text{or} \quad \left(\frac{pv}{T'}\right)^{\gamma} = c_1^{\gamma} = c \qquad ...(13.9)$$

and $$pv = c_2 \qquad ...(13.10)$$

Dividing Eq. (13.9) by Eq. (13.10), $p^{\gamma-1} = K\,(T')\gamma$ where K is constant

or $$p = K\,(T')^{\gamma/(\gamma-1)} \qquad ...(13.11)$$

Differentiating, we have, $\dfrac{dp}{dT} = T\dfrac{\gamma}{\gamma-1}\,(T')^{(1/\gamma-1)}$ ...(13.12)

Note that here for a small difference, differentiation with respect to T in place of T' has been done.

Dividing Eq. (13.12) by Eq. (13.11) we get $\dfrac{dp}{p} = \dfrac{\gamma}{\gamma-1}\dfrac{dT}{T'}$

For compression process, $dT = \eta_{pc}\, dT'$

$$\therefore \qquad \frac{dp}{p} = \frac{\gamma}{\gamma-1}\,\eta_{pc}\,\frac{dT'}{T'}$$

Integrating between limits 1 and 2′, we get for compression process.

$$\frac{T_2'}{T_1} = \left(\frac{p_2}{p_1}\right)^{(\gamma-1)/\gamma.\eta_{pc}}$$

Note: If we write $(y - 1/\gamma\,\eta_{pc}$ as $(n-1)/n$ the above equation is the familiar relation between p and T for a polytropic process and thus the definition of η_p implies that non-isentropic process by is polytropic process. This is the origin of the term polytropic efficiency.

$$\therefore \qquad \eta_{pc} = \frac{\gamma-1}{\gamma}\,\frac{(\ln p_2/p_1)}{\ln (T_2/T_1)}$$

and $$\eta_{pc} = \frac{\gamma-1}{\gamma} \times \frac{n}{n-1} \quad \text{or} \quad \frac{n-1}{n} = \frac{\gamma-1}{\gamma.\eta_{pc}} \qquad ...(13.13a)$$

Similarly, for expansion from 3 to 4′

$$\frac{T_3}{T_4'} = \left(\frac{p_3}{p_4}\right)^{(\gamma-1)\eta_{pt}/\gamma}$$

Thus, $$\eta_{pt} = \left(\frac{\gamma}{\gamma-1}\right) \times \left(\frac{n-1}{n}\right) \quad \text{or} \quad \frac{n-1}{n} = \frac{(\gamma-1)\,\eta_{pt}}{\gamma} \qquad ...(13.3b)$$

Note: The clue for solving any problem on gas turbines is to know the values of temperature at all states in the cycle by possible methods.

13.5 THE CYCLE AIR RATE, WORK RATIO AND SPECIFIC FUEL CONSUMPTION

13.5.1 The Cycle Air Rate

In addition to the thermal efficiency which is a measure of fuel economy, the size of the plant is equally important in so many applications, particularly in the field of aviation. The size of the plant depends on the rate of flow of air in relationship to the useful power output.

Air rate, AR, is defined as the air flow required per kWh output.

Thus, $$AR = \frac{\dot{m}_a}{\text{kWh output}} = \frac{\text{mass of air required}}{\text{kWh output}}$$

$$AR = \frac{\dot{m}_a}{m_a\,(w_{net}/3600)} = \frac{3600}{w_{net}\text{ in kJ/kg}} = \frac{3600}{w_{net}}\ \frac{\text{kg of air}}{\text{kWh}} \quad ...(13.14)$$

The reciprocal of air rate is known as ***specific power****.* Air rate is actually the criterion of the size of the plant, i.e., the lower air rate, the smaller the plant.

13.5.2 The Cycle Work Ratio

A good design of gas turbine not only requires a high thermal efficiency and a low rate but it must also have a high work ratio.

The work ratio, WR is defined as the ratio of net work to the turbine work. Thus,

$$WR = \frac{w_{net}}{w_t} \quad ...(13.15)$$

From Eq. (13.15), it is also clear that for a good gas turbine, the work ratio should be high. The work ratio may be also used as a guide in the determination of the size of the gas turbine and as an indicator of the sensitivity of the plant to the decrease of the component efficiency.

13.5.3 Specific Fuel Consumption

It is one of the most important parameters and expressed in kg/kWh. Thus,

$$\text{sfc} = \frac{3600\,\dot{m}_f\text{ (kg/s)}}{W_{net}\text{ (kW)}}\text{ kg/kWh} = \frac{3600\,(\dot{m}_f/\dot{m}_a)}{w_{net}\text{ (kJ/kg)}}\text{ kg/kWh}$$

Example 13.1 An open gas turbine plant works between the fixed absolute temperature limits 300K and 1500K, the absolute pressure limits being 1 bar and 14 bars. The isentropic efficiency of compressor is 0.85 and that of turbine is 0.86. Estimate the actual thermal efficiency of the plant and the power developed. The calorific value of fuel is 42000 kJ/kg. Assume, $\eta_{comb} = 0.99$, $\eta_m = 0.98$ for whole assembly, $\eta_{gen} = 0.985$ and $\dot{m}_a = 500$ kg/s.

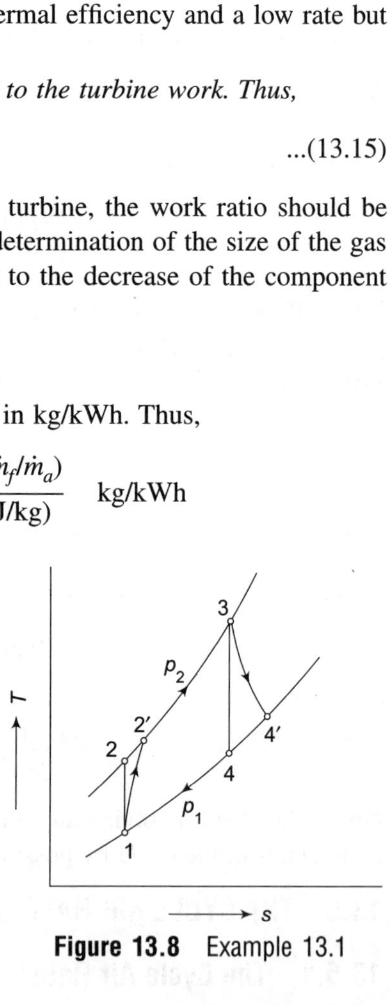

Figure 13.8 Example 13.1

Solution Refer to Fig. 13.3(a) and 13.8. Given T_1 = 300K, $\eta_m = 0.98$, $p_1 = 1$ bar, $\eta_{gen} = 0.985$, $p_2 = 14$ bar, $\eta_c = 0.85$ and $\eta_t = 0.86$, $T_3 = 1500$ K, $\dot{m}_a = 500$ kg/s, $\eta_{comb} = 0.99$, C.V. = 42000 kJ/kg.

$$T_2 = T_1\left(\frac{p_2}{p_1}\right)^{(\gamma-1)/\gamma} = 300\,(14)^{0.4/1.4} = 300 \times 2.125 = 637.5\text{ K}$$

The compressor efficiency, $\eta_c = 0.85 = \dfrac{T_2 - T_1}{T_2' - T_1} = \dfrac{637.5 - 300}{T_2' - 300}$

or $$T_2' = 300 + \frac{(637.5 - 300)}{0.85} = 697.05 \text{ K}$$

We have $$\frac{T_3}{T_4} = \left(\frac{p_3}{p_4}\right)^{(\gamma-1/\gamma)} = (14)^{0.4/14} = 2.125$$

$\therefore$ $$T_4 = \frac{T_3}{2.125} = \frac{1500}{2.125} = 705.88\text{K}$$

or $$T_3 - T_4' = \eta_t\,(T_3 - T_4) = 0.86\,(1500 - 705.88) = 682.94$$

or $$T_4' = T_3 - 682.84 = 1500 - 682.94 = 817.06\text{K}$$

Let $\dot{m}_a$ and $\dot{m}_f$ be the mass flow rates of air and fuel respectively.

$$(\dot{m}_a + \dot{m}_f)h_3 - \dot{m}_a h_2' = \dot{m}_f \times \text{C.V.} \times \eta_{\text{comb}}$$

or $$(\dot{m}_a + \dot{m}_f)\,c_{pg}\,T_3 - \dot{m}_a\,c_{pa}\,T_2' = \dot{m}_f \times \text{C.V.} \times \eta_{\text{comb}}$$

or $$\frac{\dot{m}_a}{\dot{m}_f} = \frac{\text{C.V.} \times \eta_{\text{comb}} - c_{pg}\,T_3}{c_{pg}\,T_3 - c_{pa}\,T_2'} = \frac{\text{C.V.} \times \eta_{\text{comb}} - c_p T_3}{c_p\,(T_3 - T_2')} \text{ if } c_{pg} = c_{pa} = c_p$$

or $$\frac{\dot{m}_a}{\dot{m}_f} = \frac{42000 \times 0.99 - 1.005 \times 1500}{1.005\,(1500 - 697.05)} = 49.65$$

Fuel-air ratio = F/A = 1 : 49.65

Actual work consumed by the compressor/kg of air

$$w_{ca} = c_p\,(T_2' - T_1)/\eta_m = 1.005(697.05 - 300)/0.98 = 407.17 \text{ kJ/kg}$$

Actual work done obtained by the turbine assuming c_p = 1.005 is given by

$$w_{ta} = c_p\left(1 + \frac{\dot{m}_f}{\dot{m}_a}\right)(T_3 - T_4') = 1.005\left(1 + \frac{1}{49.65}\right)(1500 - 817.06) = 700.15 \text{ kJ/kg}$$

Net work = w_{net} = 700.15 – 407.17 = 292.98 kJ/kg of air

Here, $\dot{m}_a$ = 500 kg/s and $\dot{m}_f$ = 500/49.65 = 10.07 kg/s

Total net power available = $\dot{m}_a \times w_{\text{net}} \times \eta_{\text{gen}}$

$$= 500 \times 292.98 \times 0.985/10^3 = 144.29 \text{ MW}$$ ***Ans.***

Thermal efficiency of the plant $= \dfrac{w_{\text{neta}}}{q_{Aa}} = \dfrac{w_{\text{neta}}}{\dot{m}_f \times \text{C.V.}} = \dfrac{144.29 \times 10^3}{10.07 \times 42000} = 34.11\%$

Note: In calculation of $\dot{m}_f$, η_{comb} has already been considered. **Ans.**

Problem 13.2 Air at temperature of 15°C enters a gas turbine plant working at pressure ratio of 15. Turbine inlet temperature is 1250°C. Polytropic efficiency (i.e., small stage efficiency) of compressor and turbine is 0.91. Assume $c_p = 1.005$ and 1.128 for air and gases respectively and calorific value of fuel used = 42000 kJ/kg of fuel, calculate (a) overall efficiency (b) specific output (c) fuel to air ratio and (d) specific fuel consumption.

Solution Refer to **Fig. 13.8**. Given: $p_2/p_1 = 15$, $T_1 = 288$ K, $T_3 = 1250 + 273 = 1523$ K, $c_{pg} = 1.128$ K, $c_{pa} = 1.005$, C.V. = 42000 kJ/kg, $\eta_{pt} = \eta_{pc} = 0.91$

$\therefore$ For polytropic compression process, $\left(\frac{n-1}{n}\right) = \left(\frac{\gamma-1}{\gamma}\right) \cdot \frac{1}{\eta_{pc}}$

Now, from gas laws, $\frac{T_2'}{T_1} = \left(\frac{p_2}{p_1}\right)^{(n-1)/n} = \left(\frac{p_2}{p_1}\right)^{(\gamma-1)/\gamma.\eta_{pc}}$

$\therefore$ $T_2' = 288(15)^{0.4/1.4 \times 0.91} = 673.98\text{K}$

Again, for expansion process $\frac{n-1}{n} = \frac{\gamma-1}{\gamma} \cdot \eta_{pt}$

Hence, $\frac{T_3'}{T_4} = \left(\frac{p_a}{p_1}\right)^{(\gamma-1)\eta_{pt}/\gamma} = (15)^{0.4 \times 0.91/1.4} = 2.022$

or $T_4' = \frac{1523}{2.022} = 753.21$ K

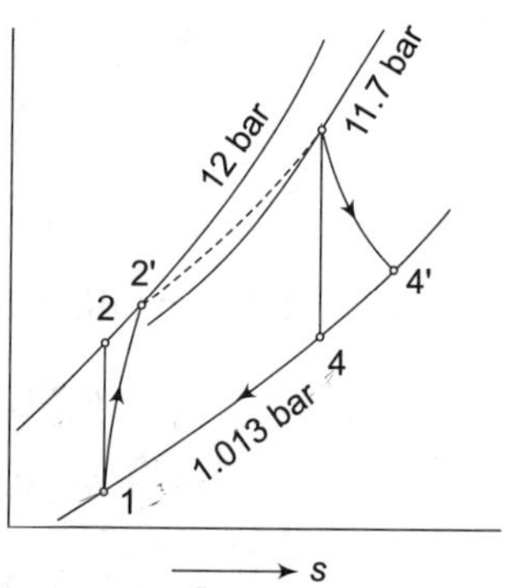

Figure 13.9 Example 13.3

Actual net work done = $w_{ta} - w_{ca} = c_{pg}(T_3 - T_4') - c_{pa}(T_2' - T_1)$

$= 1.128[(1523 - 753.2)] - 1.005\,(673 - 285) = 478.39$ kJ/kg of air

Heat supplied = $(c_{pg}\,T_3 - c_{pa}\,T_2') = 1.128 \times 1523 - 1.005 \times 673.98 = 1040.59$ kJ/kg of air

(a) Thermal efficiency $= \frac{\text{Net work}}{\text{Heat supplied}} = \frac{478.39}{1040.59} = 45.97\%$ ***Ans.***

(b) Specific output = 478.39 kJ/kg of air ***Ans.***

(c) Fuel-air ratio $= \frac{\text{External heat supplied in kJ/kg of air}}{\text{Calorific value of fuel in kJ/kg of fuel}} = \frac{1040.59}{42000} = 0.0247$ ***Ans.***

(d) Specific air consumption $= \frac{1}{\text{Specific output}} = \frac{1}{480.42}$ kg/s kW

Specific fuel consumption = Specific air consumption × (F/A)

$= \frac{3600}{478.39} \times 0.0247 = 0.1858$ kg/kWh ***Ans.***

Example 13.3 A simple open cycle gas turbine takes in air at atmospheric pressure and 15°C and compresses air in the compressor up to 12 bars. Then air enters the combustion chamber and is heated to a maximum temperature 1350°C, then it enters the turbine and expands to atmospheric pressure. If the isentropic efficiency of compressor and turbine is 0.86, combustion efficiency is 0.97, fall of pressure through the combustion system is 0.3 bar, c_p for both air and gas 1.005. $\gamma = 1.4$. Determine the flow of air and gas for net power of 200 MW to be developed. Calculate also the heat supplied per kg of air, work ratio, thermal efficiency and specific fuel consumption if C.V. of fuel is 42000 kJ/kg.

Solution Refer to Fig. 13.9. Given : $p_1 = 1.013$ bar, $T_1 = 288$ K, $T_3 = 1623$ K, $p_4 = 1.013$ bar; $p_2 = 12$ bar, $p_3 = 11.7$ bar, $\eta_c = \eta_t = 0.86$, $\eta_{comb} = 0.97$, Power = 200 MW

$$T_2 = T_1 \left(\frac{p_2}{p_1}\right)^{(\gamma-1)/\gamma} = 288 \times \left(\frac{12}{1.013}\right)^{(0.4)/1.4} = 288 \times 2.0264 = 583.61 \text{ K}$$

$$T_2' = T_1 + \frac{T_2 - T_1}{\eta_c} = 288 + \frac{583.61 - 288}{0.86} = 288 + 343.73 \text{ K} = 631.73 \text{ K}$$

$$T_4 = \frac{T_3}{(p_3/p_4)^{(\gamma-1)/\gamma}} = \frac{1623}{(11.7/1.013)^{(0.4)/1.4}} = 806.78 \text{ K}$$

and $$T_3 - T_4' = \eta_t (T_3 - T_4) = 0.86(1623 - 806.78) = 701.95 \text{ K}$$

Net work = Turbine work – compressor work = $c_p (T_3 - T_4') - c_p (T_2' - T_1)$

$= 1.005 (701.95 - 343.73) = 360$ kJ/kg of air

Power developed $= 200 \times 10^3 = \dot{m}_a \times 360$ where $\dot{m}_a$ is the rate of flow of air

or $\dot{m}_a = 555.55$ kg/s ***Ans.***

Effective heat supplied = $c_p (T_3 - T_2')$ kJ/kg = 1.005(1623 – 631.73) = 996.58 kJ/kg of air

Actual heat supplied = $q_{Aa} = \dfrac{996.58}{0.97} = 1027.41$ kJ/kg of air ***Ans.***

Work ratio $= \dfrac{\text{Actual work}}{\text{Turbine work}} = \dfrac{360}{705.45} = 51.03\%$ ***Ans.***

Actual thermal efficiency $= \dfrac{w_{neta}}{q_{Aa}} = \dfrac{360}{1027.41} = 35.03\%$ ***Ans.***

The specific fuel consumption, sfc $= \dfrac{3600\ \dot{m}_f\ (\text{kg/s})}{W_{net}\ (\text{kW})}$

But $\dot{m}_f = \dfrac{q_{Aa}}{\text{C.V.}} = \dfrac{1027.41}{42000} = 0.0244$ kg/kg of air, so, $\dot{m}_f = 0.0244 \times 555.55 = 13.55$ kg/s

$\therefore$ sfc $= \dfrac{3600 \times 13.55}{200 \times 10^3} = 0.2439$ kg/kWh ***Ans.***

13.6 OPTIMUM PRESSURE RATIO FOR MAXIMUM SPECIFIC OUTPUT IN ACTUAL SIMPLE GAS TURBINE CYCLE

Referring to Fig. 13.5 (without pressure loss), we have

$$\frac{T_2}{T_1} = \frac{T_3}{T_4} = \left(\frac{p_2}{p_1}\right)^{(\gamma-1)/\gamma} = \left(\frac{p_3}{p_4}\right)^{(\gamma-1)/\gamma} = (r_p)^{(\gamma-1)/\gamma} = x$$

Actual compressor work, $w_{ca} = (h_2' - h_1) = \dfrac{(h_2 - h_1)}{\eta_c} = \dfrac{c_p\,(T_2 - T_1)}{\eta_c}$ kJ/kg

Actual turbine work, $w_{ta} = (h_3 - h_4') = (h_2 - h_4)\eta_t = c_p(T_3 - T_4)\eta_t$ kJ/kg

$$w_{\text{neta}} = c_p\,(T_3 - T_4)\,\eta_t - \frac{c_p(T_2 - T_1)}{\eta_c} = c_p\,\eta_t\,T_3\left(1 - \frac{T_4}{T_3}\right) - \frac{c_p\,T_1}{\eta_c}\left(\frac{T_2}{T_1} - 1\right)$$

$$= c_p\,\eta_t\,T_3\left(1 - \frac{1}{x}\right) - \frac{c_p\,T_1}{\eta_c}(x - 1) \quad \text{...(13.16)}$$

For the maximum output, differentiating w_{neta} w.r.t. to x and equating to zero

$$\frac{dw_{\text{neta}}}{dx} = 0 = c_p\,\eta_t\cdot T_3\left(\frac{1}{x^2}\right) - \frac{c_p.T_1}{\eta_c}$$

So, $$x = \sqrt{\left(\eta_t\,\eta_c\,\frac{T_3}{T_1}\right)} \quad \text{or} \quad (r_p)^{(\gamma-1)/\gamma} = \left(\eta_t.\eta_c.\frac{T_3}{T_1}\right)^{1/2}$$

or $$(r_p)_{\text{opt}} = \left(\eta_t.\eta_c.\frac{T_3}{T_1}\right)^{\gamma/2(\gamma-1)} = \left(\eta_t.\eta_c.\frac{T_{\max}}{T_{\min}}\right)^{\gamma/2(\gamma-1)} \quad \text{...(13.17)}$$

For ideal simple cycle, $(r_p)_{\text{opt}} = \left(\dfrac{T_{\max}}{T_{\min}}\right)^{\gamma/2(\gamma-1)}$ ...(13.18)

13.7 MEANS OF IMPROVING EFFICIENCY AND SPECIFIC OUTPUT OF SIMPLE GAS TURBINE CYCLE

The improvement of a simple gas turbine cycle plant may be done by the following methods.

(1) *Regeneration* (recuperation). This is done by preheating the air with the turbine exhaust, thus saving the fuel consumption.

(2) *Improving turbine output.* This may be done by:

 (i) *Reheating.* The whole expansion in the turbine is achieved in two or more stages and reheating is done after each stage.

 (ii) Increasing the value of maximum cycle temperature, i.e., turbine inlet temperature (TIT). This requires:

(a) Better quality of fuel.
(b) New materials which can withstand high temperature and pressure.
(c) Better blade cooling methods and cooling medium.

(iii) *Improved turbine efficiency.* It depends on design improvement of each component.

(3) *Reducing compressor input.* This may be done by:

(i) *Intercooling.* Compressor work is reduced by intercooling the air between compressor stages.
(ii) *By lowering inlet temperature to compressor.* It is not practical because this will increase the pressure ratio.
(iii) *By increasing the compressor efficiency.* This depends upon the design improvement.
(v) *Water injection.* By injecting water at the inlet to compressor, the work-output and efficiency is increased due to the extra mass of water injected and cooling of air.

13.8 OPEN CYCLE GAS TURBINE WITH REGENERATION (RECUPERATION)

In this method, the air delivered by the compressor passes through a heat exchanger (regenerator or recuperator) utilizing the gases exhausted from the turbine; the heated air then passes into the combustion chamber and part of it is employed to burn the fuel. Since some heat is added to the air in the heat exchanger itself, so the same turbine gas inlet temperature is attained as that when no heat exchanger is employed; with a lower fuel consumption, hence, the thermal efficiency will accordingly be higher. The compressor, turbine and net work are not affected by the addition of the recuperator or regenerator to the cycle, only the heat required to be supplied in the combustion chamber is decreased which gives the gain in thermal efficiency.

13.8.1 Ideal Cycle

Process 1-2-3-4-6-1 shows the ideal open cycle gas turbine with regeneration as shown in Fig. 13.10. The high pressure air at state 2 is passed through a regenerator (or heat exchanger) where it is heated at constant pressure p_2 from T_2 to T_5 due to which the exhaust gas from the turbine at state 4 is cooled from T_4 to T_6. With perfect heat exchange in the heat exchanger, in the ideal case, the temperature of air leaving the heat exchanger will be equal to the temperature of exhaust gas entering the heat exchanger and the temperature of exhaust gas leaving heat exchanger will be equal to the temperature of air entering the heat exchanger, Thus, $T_5 = T_4$ and $T_6 = T_2$.

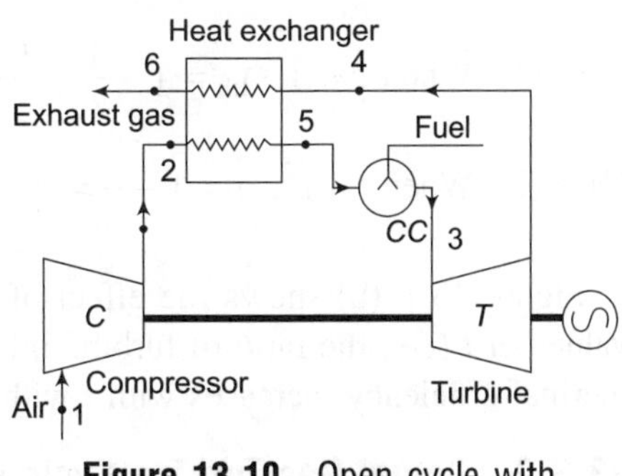

Figure 13.10 Open cycle with regeneration

The air from the heat exchanger at temperature T_5 then passes through the combustion chamber where it is heated from temperature T_5 to T_3 (i.e., from h_5 to h_3). The advantage of exhaust heat exchange is that the amount of heat supplied in the combustion chamber is reduced from $(h_3 - h_2)$ to $(h_3 - h_5)$, i.e., from $c_p\,(T_3 - T_2)$ to $c_p\,(T_3 - T_5)$. Thus, the thermal efficiency is increased because the heat supplied is reduced.

Assuming 1 kg of air following through the compressor.

Net work, $w_{net} = w_t - w_c = c_p\,(T_3 - T_4) - (T_2 - T_1)]$

Heat supplied $= h_3 - h_5 = c_p\,(T_3 - T_5) = c_p\,(T_3 - T_4)$

So, $$\eta_{th} = \frac{c_p[(T_3 - T_4) - (T_2 - T_1)]}{c_p\,(T_3 - T_4)} = 1 - \frac{T_2 - T_1}{T_3 - T_4} \qquad ...(13.19)$$

But we know that, $\dfrac{T_2}{T_1} = \left(\dfrac{p_2}{p_1}\right)^{(\gamma - 1)/\gamma} = (r_p)^{(\gamma - 1)/\gamma}$

and $$\frac{T_3}{T_4} = \left(\frac{p_3}{p_4}\right)^{(\gamma - 1)/\gamma} = \left(\frac{p_2}{p_1}\right)^{(\gamma - 1)/\gamma} = (r_p)^{(\gamma - 1)/\gamma}$$

So, $$\frac{T_2}{T_1} = \frac{T_3}{T_4} \quad \text{or} \quad \frac{T_2}{T_3} = \frac{T_1}{T_4} = \frac{T_2 - T_3}{T_1 - T_4}$$

By putting the value of $(T_2 - T_1)/T_3 - T_4)$ in Eq. (13.19), we get

$$\eta_{th} = 1 - \frac{T_2}{T_3} = 1 - \frac{T_2}{T_1}\cdot\frac{T_1}{T_3} = 1 - \frac{T_1}{T_3}\,(r_p)^{(\gamma - 1)/\gamma} \qquad ...(13.20)$$

Let $$t = \frac{T_3}{T_1} = \frac{T_{max}}{T_{min}} \quad \text{and} \quad c = \frac{T_2}{T_1} = \frac{T_3}{T_4} = (r_p)^{(\gamma - 1)/\gamma}$$

Then, the thermal efficiency $= \eta_{th} = 1 - \dfrac{c}{t}$ ...(13.21)

Case (1): When $c = 1$, $\eta_{th} = 1 - \dfrac{1}{t} = \dfrac{T_{max} - T_{min}}{T_{max}}$ = Carnot cycle efficiency

Case (2): When $t = c^2$, $\eta = 1 - \dfrac{1}{c} = 1 - \dfrac{1}{(r_p)^{(\gamma - 1)/\gamma}}$ = Simple open gas turbine efficiency

Figure 13.11(b) shows the effect of pressure ratio, r_p, on thermal efficiency for various values of t (i.e., the ratio of turbine inlet temperature to compressor inlet temperature). The thermal efficiency increases with t with a decrease in pressure ratio.

13.8.2 Actual Gas Turbine Cycle with Regeneration

1–2′–5′–3′–4′–6′–1 shows the actual cycle as shown in Fig. 13.11(a). In actual practice, the compression and expansion processes are not isentropic. Assuming 1 kg of air flowing through the compressor and if we neglect the mass of the fuel and assume the same value

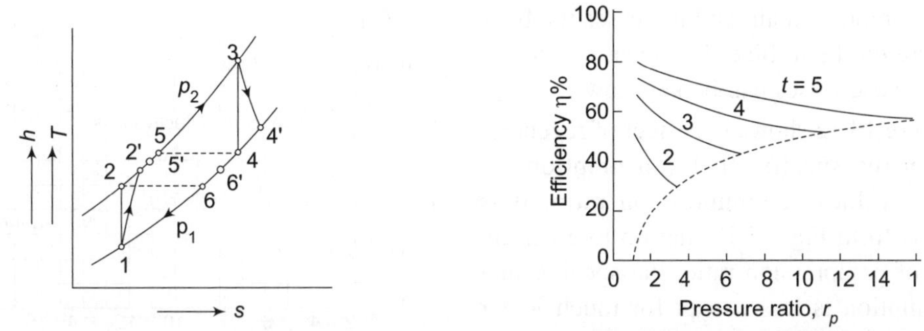

(a) *T-s* representation with regeneration (b) η_{th} vs r_p for various t

Figure 13.11

of c_p for air and exhaust gas, turbine and compressor work may be computed per kg of air as follows:

Actual net work, w_{neta}, $w_{ta} - w_{ca} = c_p\,[(T_3 - T_4') - (T_2' - T_1)]$

Actual heat supplied, $q_{Aa} = h_3 - h_5' = c_p\,(T_3 - T_5')$

$$\text{Actual thermal efficiency, } \eta_{\text{tha}} = \frac{w_{\text{neta}}}{q_{Aa}} = \frac{(T_3 - T_4') - (T_2' - T_1)}{(T_3 - T'_5)} \qquad ...(13.22)$$

or

$$\eta_{\text{tha}} = \frac{(T_3 - T_4)\,\eta_t - \dfrac{(T_2 - T_1)}{\eta_c}}{T_3 - T_5'} = \frac{(T_3 - T_4)\,\eta_t . \eta_c - (T_2 - T_1)}{\eta_c\,(T_3 - T_5')} \qquad ...(13.23)$$

In actual practice, perfect heat exchange is never possible in the regenerator. As a result, $h_5' < h_4'$ and $h_6' < h_2'$ (i.e., $T_5' < T_4'$ and $T_6' < T_2'$) in the actual cycle. The regenerator effectiveness $\in_r$ is defined as

$$\in_r = \frac{\text{Actual heat exchange}}{\text{Maximum theoretical heat exchange}} \qquad ...(13.24)$$

The actual heat exchange $(h_5' - h_2')$ or $(h_4' - h_6')$

The maximum theoretical heat exchange = $(h_4' - h_2')$

$$\text{Hence, } \in_r = \frac{h_5' - h_2'}{h_4' - h_2'} = \frac{h_4' - h_6'}{h_4' - h_2'} = \frac{T_5' - T_2'}{T_4' - T_2'} = \frac{T_4' - T_6'}{T_4' - T_2'} \qquad ...(13.25)$$

If $\in_r$ is known, T_5' can be found out and hence the thermal efficiency can be evaluated.

A drawback of such a heat exchanger, acting as a regenerator or recuperator, is the low overall heat transfer coefficient when both the heating and heated fluids are gases. The maximum effectiveness is about 75 per cent. This, in turn, requires a large heat exchange area. Another drawback is the resistance to flow of the gases, which causes a pressure drop in the compressed air, and increases the back pressure on the turbine. However, by careful design these losses can be kept low.

Figure 13.12 shows the effect of regeneration on the specific fuel consumption for various values of compression ratio. It is obvious from Fig. 13.12 that up to a certain value of compression ratio, the specific fuel consumption is less except for much lower value of regenerator effectiveness. After certain value of pressure ratio, i.e., when the temperature of compressed air is more than the exhaust temperature of gas, the specific fuel consumption increases. Thus, we can say that the use of regenerator is limited by pressure ratio. Note that the specific fuel consumption is the minimum at lower value of pressure ratio, i.e., a lower value of compressed air temperature.

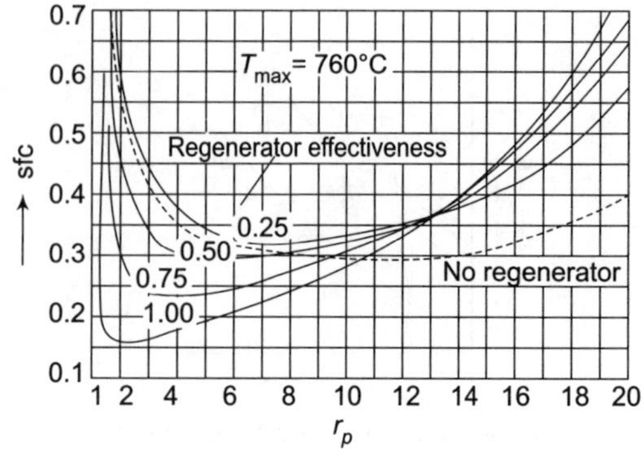

Figure 13.12 sfc vs. compressor pressure ratio

Example 13.4 In a gas turbine plant, the air at 10°C and 1 bar is compressed to 12 bars with compression efficiency of 80%. The air is heated in the regenerator and the combustion chamber till its temperature is raised to 1400°C, and during the process the pressure falls by 0.2 bar. The air then expands in the turbine and passes to regenerator which has 75% effectiveness, and causes a pressure drop of 0.2 bar. If the isentropic efficiency of the turbine is 85%. Determine the thermal efficiency of the plant.

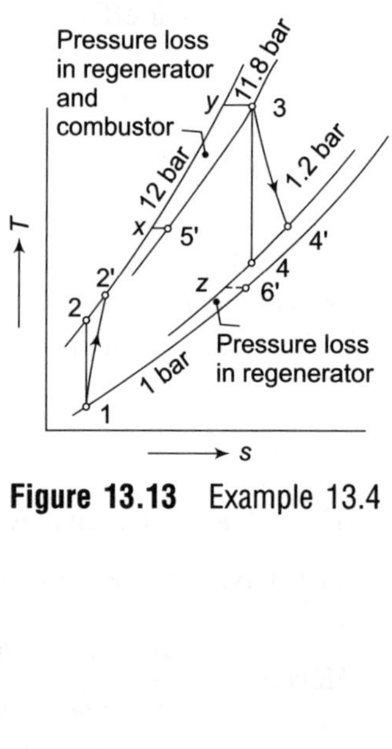

Figure 13.13 Example 13.4

Solution Refer to Figs. 13.10 and Fig. 13.13. Though, there is a fall in pressure in regenerator and combustion chamber but for simplicity, addition of heat will be assumed at constant pressure, i.e., the processes $2' - 3$ will be assumed as $2' - x - y - 3$ and process $4' - 6'$ will be assumed as $4' - z - 6'$.

Given: $T_1 = 273 + 10 = 283$ K

$$T_2 = T_1\left(\frac{p_2}{p_1}\right)^{(\gamma-1)/\gamma} = 283 \times (12)^{0.286} = 576 \text{ K}$$

Now $$T_2' = T_1 + \frac{(T_2 - T_1)}{\eta_c} = 283 + \frac{(576 - 283)}{0.8} = 649.25 \text{ K}$$

$$p_3 = 12 - 0.2 = 11.8 \text{ bar}, \quad p_4 = p_4' = 1 + 0.2 = 1.2 \text{ bar}$$

Again, $$\frac{T_3}{T_4} = \left(\frac{p_3}{p_4}\right)^{0.286} = \left(\frac{11.8}{1.2}\right)^{0.286} = 1.922$$

or $$T_4 = \frac{1673}{1.922} = 870.44 \text{ K}$$

and $$T_4' = T_3 - (T_3 - T_4)\,\eta_t = 1673 - (1673 - 870.4) \times 0.85 = 990.82 \text{ K}$$

$$w_{ca} = c_p(T_2' - T_1) = 1.005(649.25 - 283) = 368.08 \text{ kJ/kg}$$

$$w_{ta} = c_p(T_3 - T_4') = 1.005(1673 - 990.82) = 682.18 \text{ kJ/kg}$$

Net work $= w_{\text{neta}} = w_{ta} - w_{ca} = 682.18 - 368.08 = 314.1$ kJ/kg

The regenerator effectiveness, $\epsilon_r = \dfrac{T_5' - T_2'}{T_4' - T_2'} = 0.75$

or $$T_5' = 649.25 + (990.82 - 649.25) \times 0.75 = 905.42 \text{ K}$$

Heat supplied, $q_{Aa} = c_p\,(T_3 - T_5') = 1.005\,(1673 - 905.42) = 771.41$ kJ/kg

$$\eta_{\text{th}} = \frac{w_{\text{neta}}}{q_{Aa}} = \frac{325.58}{771.41} = 42.2\%.$$

13.9. OPEN GAS TURBINE CYCLE WITH REHEAT

By reheating or adding heat to the gases after they have passed through a part of the rows of the turbine blading (or a stage), a further increase in work done is obtained. In reheating, the gas temperature, which has dropped due to expansion is brought back to approximately the initial temperature for the expansion in the next stage. Since the working fluid contains about 85 per cent air, additional fuel can be burnt by injecting it into the gases without any additional air supply. GT24 and GT26 models are reheat gas turbines in operation.

13.9.1 Ideal Cycle

1–2–3–4–5–6–1 shows the ideal cycle gas turbine with reheat as shown in Fig. 13.14(a) and (b). The combustion gas from the burner (combustor) CC_1 at temperature T_3 is partially expanded in the H.P. turbine from pressure p_2 to intermediate pressure p_x. After this, it is then passed through combustion chamber CC_2 (reheater) where it is reheated or additional heat is

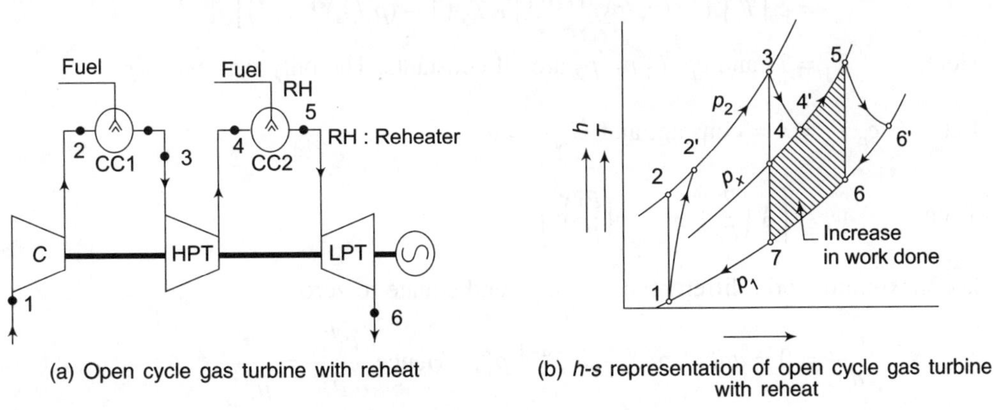

(a) Open cycle gas turbine with reheat

(b) *h-s* representation of open cycle gas turbine with reheat

Figure 13.14

added at constant pressure p_x so that the temperature of the gas is raised from T_4 to $T_5 = T_3$. After this, the gas is expanded in the L.P. turbine from pressure p_x to p_1.

The net work done is more with reheat compared to the ideal cycle (without reheat). The efficiency with which the heat is added to the gas in the combustion chamber CC_2 at lower pressure p_x is lower than the efficiency when this heat is added to the gas in the combustion chamber CC_1 at higher pressure p_2. So, efficiency decreases as a result of reheating, only the net work is more by hatched area 4-5-6-7 shown in Fig. 13.14(b). However, in real cooled cycle, it may increase efficiency at higher pressure ratio at a particular reheat pressure ratio.

Figure 13.15(a) and (b) show the variation of thermal efficiency and specific work with pressure ratio and both increase with the increase of *t*.

Net work = $w_{net} = (h_3 - h_4) + (h_5 - h_6) - (h_2 - h_1)$

Heat supplied = $q_A = (h_3 - h_2) + (h_5 - h_4)$

$$\eta_{th} = \frac{(h_3 - h_4) + (h_5 - h_6) - (h_2 - h_1)}{(h_3 - h_2) + (h_5 - h_4)} = \frac{(T_3 - T_4) + (T_5 - T_6) - (T_2 - T_1)}{(T_3 - T_2) + (T_5 - T_4)} \quad ...(13.26a)$$

Let $t = T_3/T_1$ and $c = r_p^{(\gamma - 1)/\gamma}$, then

$$\eta_{th} = \frac{2t - c + 1 - 2t/\sqrt{c}}{2t - c - t/\sqrt{c}} \quad ...(13.26b)$$

and

$$\frac{w_{net}}{c_p T_1} = 2t - c + 1 - \frac{2t}{\sqrt{c}} \quad ...(13.26c)$$

For the maximum output, there must be an optimum pressure at which reheating should be done. As we know, the compressor work w_c is not affected by reheating, so for the maximum output, we have to find the condition where turbine work w_t is the maximum.

$$w_t = c_p (T_3 - T_4) + (T_5 - T_6) = c_p \left\{T_3\left(1 - \frac{T_4}{T_4}\right) + T_5\left(1 - \frac{T_6}{T_5}\right)\right\}$$

$$= c_p\left[T_3\{1 - (p_x/p_2)^{(\gamma - 1)/\gamma}\} + T_5\left(1 - (p_1/p_x)^{(\gamma - 1)/\gamma}\right)\right]$$

Here, $T_3 = T_5$ and c_p, T_3 p_1, p_2 are all constants. The only variable is p_x

Let $c_p T_3 = A = \text{constant}$ and $\frac{\gamma - 1}{\gamma} = a$

Then $$w_t = \left\{A\left(\frac{p_x}{p_2}\right)^a + 2 - \left(\frac{p_1}{p_x}\right)^a\right\}$$

For maximum work, differentiate w.r.t. p_x and equate to zero.

$$\frac{dw_t}{dp_x} = 0 = ap_x^{a-1}\, p_2^{-a} - ap_x^{-a-1}\, p_1^a. \quad \text{Hence, } \frac{p_x^{a-1}}{p_x^a} = \frac{p_1^a}{p_x^{a+1}}$$

or $$p_x = \sqrt{(p_1 p_2)} \quad ...(13.27)$$

or $$\frac{p_x}{p_1} = \frac{p_2}{p_x} = \sqrt{\left(\frac{p_2}{p_1}\right)} \quad \text{or} \quad r_{p1} = r_{p2} = \sqrt{(r_p)}$$

For n stages, $r_{p1} = r_{p2} = r_{p3} = \sqrt[n]{(r_p)}$

So, we can say that the optimum intermediate pressure is the square root of initial and final pressure, i.e., the pressure ratio in each stage of turbine is equal to the square root of total pressure ratio. For finding out thermal efficiency for maximum output in this cycle, this value of intermediate pressure should be used.

13.9.2 Actual Cycle Gas Turbine with Reheat

Process 1–2′–3–4′–5–6′–1 shows the actual cycle with reheat as shown in Fig. 13.14(b).

Actual turbine work, $w_{ta} = (h_3 - h_4') + (h_5 - h_6') = c_p\,\{(T_3 - T_4') + (T_5 - T_6')\}$

Actual compression work, $w_{ca} = h_2' - h_1 = c_p\,(T_2' - T_1)$

Actual heat supplied, $(h_3 - h_2') + (h_5 - h_4') = c_p\,\{(T_3 - T_2') + (T_5 - T_4')\}$

$$\therefore \quad \eta_{tha} = \frac{(T_3 - T_4') + (T_5 - T_6') - (T_2' - T_1)}{(T_3 - T_2') + (T_5 - T_4')} \quad ...(13.28)$$

or $$\eta_{tha} = \frac{(T_3 - T_4)\,\eta_t + (T_5 - T_6)\,\eta_t - \dfrac{(T_2 - T_1)}{\eta_c}}{(T_3 - T_2') + (T_5 - T_4')} \quad ...(13.28a)$$

In this case also, the condition for maximum output will be the same.

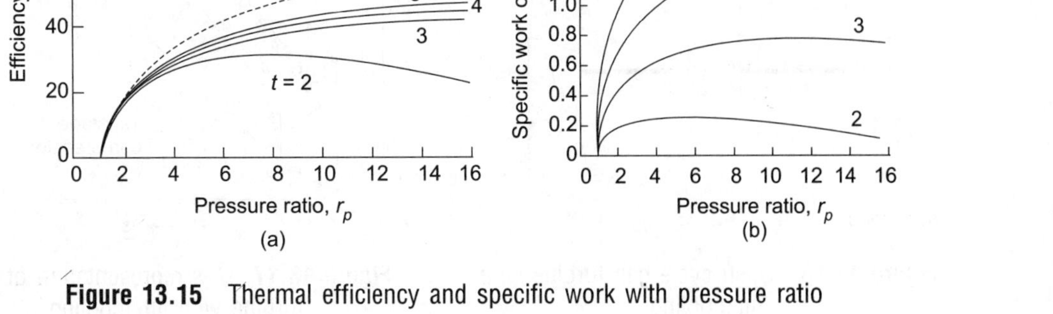

Figure 13.15 Thermal efficiency and specific work with pressure ratio

Note: With reheat system, two shafts system may be used; H.P. turbine may drive compressor called ***compressor turbine*** and L.P. turbine may produce power by driving generator called ***power turbine***. Thus,

$$w_c = (w_t)_{ct} \quad \text{or} \quad (h_2' - h_1) = (h_4' - h_3) \quad \text{and} \quad \text{power} = (w_t)_{pt}$$

and Power $= (h_5 - h_6')$, kJ/kg $= \dot{m}(h_5 - h_6')$, kW

With two shafts system, the conditions of the maximum output will not be applicable as it may be possible that the compressor turbine may not be able to drive compressor.

13.10 OPEN GAS TURBINE CYCLE WITH INTERCOOLING

The net wok of the gas turbine cycle may be increased by saving some compression work. This is done by using several stages of compression with intercooling of air between stages. The air from the low pressure stages compressor (LPC) is cooled in a intercooler approximately to initial temperature before entering the high pressure stage compressor (HPC). The effect of intercooling is to increase the net work and decrease the efficiency as compared to the simple ideal cycle without intercooling.

13.10.1 Ideal Cycle

Process 1–2–3–4–5–6–1 shows the ideal cycle gas turbine with intercooling as shown in Figs. 13.16 and 13.17. In L.P. stage compressor, atmospheric air is compressed from p_1 to p_x, it is then cooled from temperature T_2 to $T_3 = T_1$ in the intercooling at constant intermediate pressure p_x and finally compressed from p_x to p_2 in the H.P. stage of compressor. At constant pressure p_2, heat is added in the combustion chamber up to T_5 and after that it is expanded in the turbine from pressure p_2 and temperature T_5 to pressure p_1. It is evident from Fig. 13.17 that the turbine work is the same as for the simple cycle but the work supplied to the compressor is reduced because the constant pressure lines converge towards the left. The work supplied to the compressor is saved due to intercooling which may be shown as area 2–3–4–a–2, so this is the amount by which net work of the cycle is increased as compared

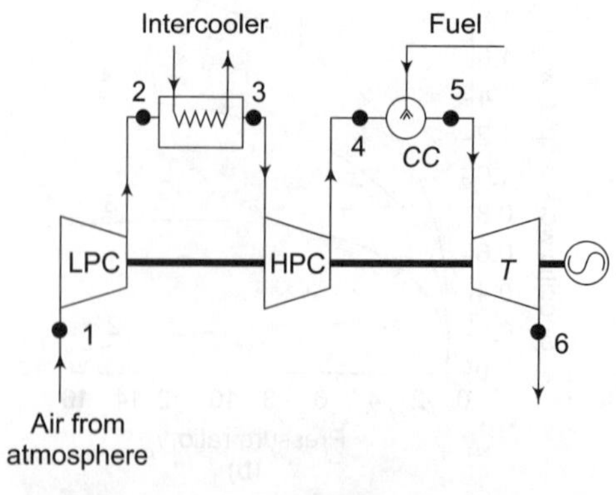

Figure 13.16 Open cycle gas turbine with intercooling

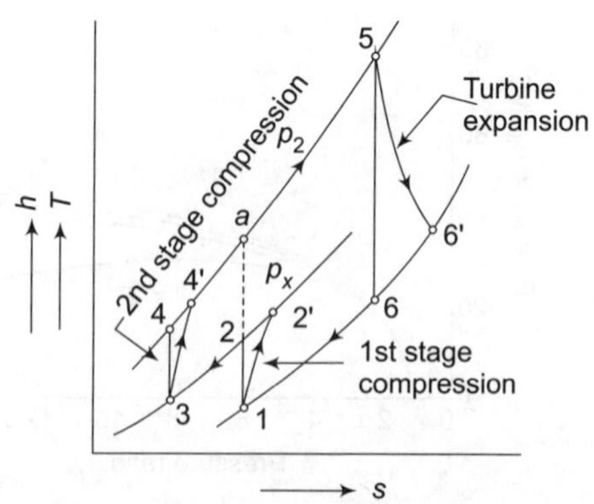

Figure 13.17 *h-s* representation of gas turbine with intercooling

to ideal cycle without intercooling. But the heat supplied is increased by amount $(h_a - h_4)$ because the compressed air delivered by second stage compressor is at lower temperature (T_4) than in simple cycle. So, the effect of the intercooling is to decrease the thermal efficiency of the cycle. This result is to be expected because the heat supplied in the intercooler is a complete loss, no part of it is converted into useful work.

Turbine work, $w_t = h_5 - h_6 = c_p\,(T_5 - T_6)$

Compressor work, $w_c = (h_2 - h_1) + (h_4 - h_3) = c_p\,\{(T_2 - T_1) + (T_4 - T_3)\}$

Heat supplied, $h_5 - h_4 = c_p\,(T_5 - T_4)$

So, the thermal efficiency $= \eta_{th} = \dfrac{(T_5 - T_6) - (T_4 - T_3) - (T_2 - T_1)}{T_5 - T_4}$ ...(13.29)

For the minimum compression work in the compressor or maximum work output in this cycle, only the compressor work is responsible. Here $T_3 = T_1$

$$w_c = c_p\,\{(T_2 - T_1) + (T_4 - T_3)\} = c_p\,T_1\left\{\frac{T_2}{T_1} - 1 + \frac{T_4}{T_1} - 1\right\}$$

$$= c_p\,T_1\left\{(p_x/p_1)^{(\gamma-1)/\gamma} - 1 + (p_2/px)^{(\gamma-1)/\gamma} - 1\right\}$$

Let $c_p\,T_1 = A$ = constant and $(\gamma - 1)/\gamma = a$

$$w_c = A\left[\left(\frac{p_x}{p_1}\right)^a - 1 + \left(\frac{p_2}{p_x}\right)^a - 1\right]$$

For the maximum work output or the minimum compression work, differentiate w_c w.r.t. to p_x and equate to zero.

Thus, $\dfrac{dw_c}{dp_x} = 0 = a\,.\,p_x^{a-1}\,p_1^{-a} - a\,.\,p_2^a\,.\,p_x^{-a-1}$ or $\dfrac{p_x^{a-1}}{p_1^a} = \dfrac{p_2^a}{p_x^{a+1}}$

or $p_x = \sqrt{(p_1 p_2)}$ ...(13.30)

or $\dfrac{p_x}{p_1} = \dfrac{p_2}{p_x} = \sqrt{\left(\dfrac{p_2}{p_1}\right)}$ or $r_{p1} = r_{p2} = \sqrt{(r_p)}$ ...(13.31)

For n number of stage, $r_{p1} = r_{p2} = r_{p3} \ldots \sqrt[n]{(r_p)}$

Thus, we see that for the minimum compression work, the optimum intermediate p_x is the square root of initial and final pressure. For the *maximum work output, this* value of intermediate pressure p_x is used.

13.10.2 Actual Cycle Gas Turbine with Intercooling

Process 1–2′–3–4′–5–6′–1 shows the actual cycle gas turbine with intercooling as shown in Fig. 13.17.

Actual turbine work, $w_{ta} = h_5 - h_6' = c_p(T_5 - T_6')$

Actual compressor work, $w_{ca} = (h'_2 - h_2) + (h'_4 - h_3) = c_p \{(T_2' - T_1) + (T_4' - T_3)\}$

Actual net work, $c_p\{(T_5 - T_6') - (T_4' - T_3) - (T_2 - T_1')\}$

Actual heat supplied, $h_5 - h_4' = c_p (T_5 - T_4')$

$$\text{Intercooler effectiveness} = \in_1 = \frac{T_2' - T_3}{(T_2' - T_1)}$$

T_3 may be found from known value of intercooler effectiveness

$$\eta_{tha} = \frac{(T_5 - T_6') - (T_4' - T_3) - (T_2' - T_1)}{(T_5 - T_4')} = \frac{(T_5 - T_6)\eta_t - \dfrac{(T_4 - T_3)}{\eta_c} - \dfrac{(T_2 - T_1)}{\eta_c}}{(T_5 - T_4')} \quad ...(13.32)$$

In this case also, we can show that for minimum work input to the compressor $r_{p1} = r_{p2} = \sqrt{r_p}$.

Example 13.5 In an open gas turbine plant, the pressure ratio through which air at 15°C is compressed is 14 The same air is then heated to a maximum permissible temperature of 1300°C first in a heat exchanger which is 75% efficient, and then in the combustion chamber. The same air at 1300°C is expanded in two stages such that the expansion work is the maximum. The air is reheated to 1300°C after the high pressure stage. Determine the cycle thermal efficiency, the work ratio and the net shaft work per kg air. The isentropic efficiency may be assumed to be 85% and 86% for the compressor and turbine respectively. Calculate the flow rate of air for an output of 240 MW. Assume mechanical efficiency for compressor and turbine and generator efficiency equal to 99% each.

Solution Refer to Figs. 13.18 and 13.19. Given : $T_1 = 288$K, $p_2/p_1 = 14$, $T_4 = T_6 = 1573$ K, $\eta_c = 0.85$, $\eta_t = 0.86$, $\eta_m = 0.99$, $\eta_{gen} = 0.99$

$$T_2 = T_1 \left(\frac{p_2}{p_1}\right)^{(\gamma - 1)/\gamma} = 288(14)^{0.4/1.4} = 612.12 \text{ K}$$

$$T_2' - T_1 = \frac{T_2 - T_1}{\eta_c} \quad \text{or} \quad T_2' = \frac{(612.12 - 288)}{0.85} + 288 = 669.31 \text{ K}$$

For the maximum work output, the reheating should be done at intermediate pressure

$$p_x = \sqrt{(p_1 p_2)} \quad \text{and} \quad \frac{p_x}{p_1} = \frac{p_2}{p_x} = \sqrt{r_p} = \sqrt{14} = 3.74$$

$$\frac{T_4}{T_5} = \left(\frac{p_2}{p_x}\right)^{(\gamma - 1)/\gamma} = (3.74)^{0.4/1.4} = 1.457 \quad \text{or} \quad T_5 = \frac{1573}{1.457} = 1079.6 \text{ K}$$

Now, $(T_4 - T_5') = \eta_t (T_4 - T_5) = 0.86\,(1573 - 1079.61) = 424.32$ K

$T_5' = 1573 - 424.32 = 1148.68$ K

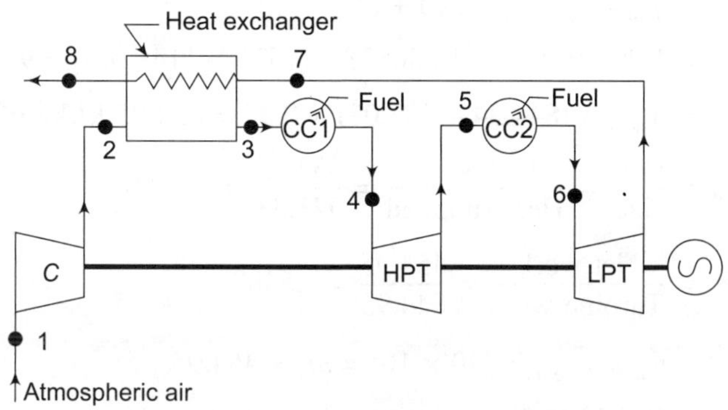

Figure 13.18 Gas turbine with reheat and regeneration

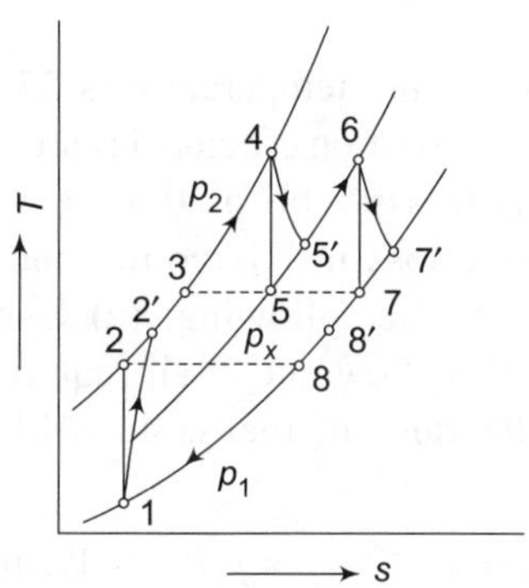

Figure 13.19 *T-s* representation of gas turbine with reheat and regeneration

Again, $\dfrac{T_6}{T_7} = \left(\dfrac{p_x}{p_1}\right)^{(\gamma - 1)/\gamma} = (3.47)^{0.4/1.4} = 1.457$

or $\qquad T_7 = \dfrac{T_6}{1.457} = \dfrac{1573}{1.457} = 1079.6 \text{ K}$

and $\qquad (T_6 - T_7') = \eta_t(T_6 - T_7) \quad$ or $\quad T_7' = 1573 - 0.86\ (1573 - 1079.6) = 1148.68 \text{ K}$

Alternatively, $T_7 = T_5 = 1079.61$ K by given data

Regenerator effectiveness, $\in_r = \dfrac{T_3' - T_2'}{T_7' - T_2'} = 0.75 = \dfrac{T_3' - 669.31}{1148.68 - 669.31}$

or $\qquad T_3' = 1028.83 \text{ K}$

$$w_{ca} = c_p\ (T_2' - T_1)/\eta_m = 1.005\ (669.31 - 288)/0.99 = 387.08 \text{ kJ/kg of air}$$

$$w_{ta} = c_p\ [(T_4 - T_5') + (T_6 - T_7')] \times \eta_m$$

$$= 1.005\ [(1573 - 1148.68) + (1573 - 1148.68)] \times 0.99 = 844.34 \text{ kJ/kg}$$

$$\text{Heat supplied} = q_{Aa} = c_p\,[(T_4 - T_3') + (T_6 - T_5')$$

$$= 1.005\,[(1573 - 1028.83) + (1573 - 1148.68)] = 973.33 \text{ kJ/kg}$$

$$\text{Net work} = w_{\text{neta}} = (844.34 - 387.08) \times 0.99 = 454.95 \text{ kJ/kg of air}$$

$$\eta_{\text{th}} = \frac{w_{\text{neta}}}{q_{Aa}} = \frac{\text{Net work}}{\text{Heat supplied}} = \frac{454.95}{973.33} = 46.74\%$$

$$\text{Work ratio} = \frac{\text{Net work}}{\text{Turbine work}} = \frac{454.95}{844.95} = 0.5388 \qquad \textbf{\textit{Ans.}}$$

$$\text{Power output} = \dot{m}_a \times w_{\text{neta}} = 240 \times 10^3 = \dot{m}_a \times 454.95 \qquad \textbf{\textit{Ans.}}$$

or $\dot{m}_a = 527.53$ kg/s

Example 13.6 The following data relate to an open cycle gas turbine plant with inter-cooling.

Pressure ratio is 49, compressor inlet temperature is 27°C, turbine inlet temperature is 1400°C, compressor, turbine and combustion efficiencies are 0.86, 0.87 and 0.99, respectively. Mechanical efficiency of the whole assembly is 0.97. Generator efficiency is 0.98. Inlet pressure is 1 bar. Neglect pressure loss in intercooler and combustion chamber. Assume $c_{pa} = c_{pg} = 1.005$ kJ/kg. Calculate the following: (a) Cycle rate, (b) Cycle work ratio, (c) Cycle thermal efficiency, (d) Mass flow rate of air required for power output of 250 MW, (e) Air-fuel ratio if lower calorific value of fuel is 44 MJ/kg of fuel, and (f) Comment on exhaust temperature

Solution Refer to Fig. 13.16 and 13.17. $r_p = p_2/p_1 = 49$, $\eta_c = 0.86$, $\eta_{ca} = 0.99$, $\eta_t = 0.87$, η_{mc} for whole assembly 0.97, $\eta_g = 0.98$, $p_1 = 1$ bar, $t_1 = 27$°C, $t_{\max} = t_5 = 1400$°C, $L_{CV} = 44$ MJ/kg of fuel, $T_5 = 273 + 1400 = 1673$ K.

The optimum pressure ratio for the minimum work of compressor

$$p_x = \sqrt{p_1 p_2} = \sqrt{49} = 7 \text{ bars}$$

$$T_2 = T_1\left(\frac{p_x}{p_1}\right)^{(\gamma-1)/\gamma} = 300\,(7)^{0.4/1.4} = 523.38 \text{ K}$$

$$T_2' = T_1 + \frac{(T_2 - T_1)}{\eta_c} = 300 + \frac{(523.38 - 300)}{0.86} = 559.74 \text{ K}$$

Work consumed by L.P. compressor is

$$w_{lpc} = c_p\,(T_2' - T_1) = 1.005\,(559.79 - 300)$$

$$= 261.03 \text{ kJ/kg}$$

For the perfect intercooling, and the minimum work to compressor, the work consumed by both stages will be the same. Hence, the total work done in both stages will be the sum of both and $T_4' = T_2' = 559.74$ K.

$$w_{ca} = (261.03 + 261.03)/0.97 = 538.2 \text{ kJ/kg}$$

From turbine expansion,

$$\frac{T_5}{T_6} = \left(\frac{p_5}{p_6}\right)^{(\gamma-1)/\gamma} = (49)^{0.4/1.4} = 3.043$$

or

$$T_6 = \frac{T_5}{3.043} = \frac{1673}{3.043} = 549.78 \text{ K}$$

Again,

$$\eta_t = \frac{T_5 - T'_6}{T_5 - T_6} = 0.87$$

or

$$T'_6 = T_5 - 0.87\,(T_5 - T_6)$$

$$= 1673 - 0.87\,(1673 - 549.78) = 695.8 \text{ K}$$

Actual turbine work is given by

$$w_{ta} = c_p(T_5 - T'_6) = 1.005\,(1673 - 695.8)$$

$$= 982.08 \text{ kJ/kg}$$

Net actual work, $w_{\text{neta}} = (w_{ta} - w_{ca})\eta_{\text{gen}}$

$$= (982.08 - 538.2)\ 0.98 = 435 \text{ kJ/kg}$$

Actual heat supplied, $q_{Aa} = c_p\,(T_5 - T'_4)/0.99$

$$= 1.005\,(1673 - 559.74)/0.99 = 1130.12 \text{ kJ/kg}$$

(a) Cycle rate, $\dfrac{3600}{w_{\text{neta}}} = \dfrac{3600}{435} = 8.27$ kJ/kWh **_Ans._**

(b) Cycle work ratio, $WR = \dfrac{w_{\text{neta}}}{w_{ta}} = \dfrac{435}{982.08} = 44.2\%$ **_Ans._**

(c) Cycle thermal efficiency is $\eta_{\text{tha}} = \dfrac{w_{\text{neta}}}{q_{Aa}} = \dfrac{435}{1130.12} = 38.49\%$

(d) Mass flow rate of air required for 250 MW output

$$\dot{m}_{er} = \frac{\text{Generator output}}{w_{\text{neta}}} = \frac{250 \times 10^3}{435} = 574.71 \text{ kg/s}$$ **_Ans._**

(e) Air-fuel ratio is given by

$$Q = LCV \times \dot{m}_f = \dot{m}_a \times q_{Aa}$$

or

$$\frac{\dot{m}_a}{\dot{m}_f} = \frac{44 \times 10^3}{1130.12} = 38.93 : 1$$ **_Ans._**

(f) Comment: Exhaust temperature is high, so it is most suitable for a combined cycle plant.

13.11 OPEN GAS TURBINE CYCLE (OGTC) WITH INTERCOOLING, REHEAT AND REGENERATION

The power output as well as the thermal efficiency of this cycle is much greater than that of the simple cycle. It is to be noted that the power output is increased due to reheat and intercooling and thermal efficiency is due to regeneration.

13.11.1 Ideal Cycle

Process 1-2-3-4-5-6-7-8-9-10-1 shows such a type of an ideal cycle as shown in Figs. 13.20 and 13.21. The following are the various processes.

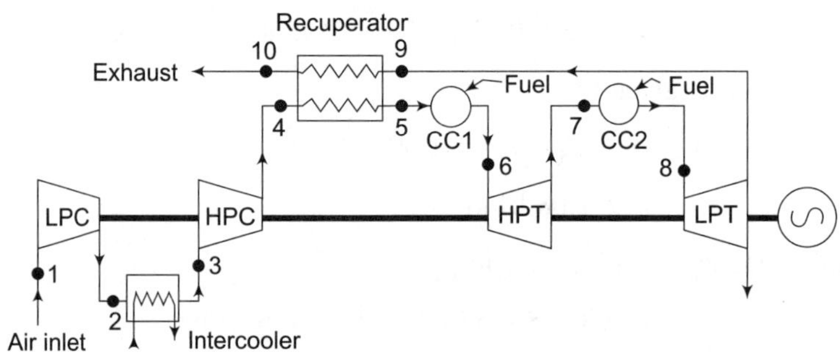

Figure 13.20 Gas turbine with intercooling, reheating and regeneration

Process 1-2. Isentropic compression of air in L.P. stage compressor (LPC). For the minimum compression work the pressure ratio r_{p1} of compressor (LPC) should be equal to $\sqrt{r_p}$.

Process 2-3. Cooling of air in the intercooler at constant pressure p_x from temperature T_2 to $T_3 = T_1$.

Process 3-4. Isentropic compression of air in H.P. stage compression (HPC). For the minimum compression work, the pressure ratio r_{p2} of (LPC) should be equal to $\sqrt{r_p}$.

Process 4-5. Heating of air in the regeneratior at constant pressure p_4 from temperature T_4 to T_5. For perfect heat exchange in the intercooler T_5 should be equal to T_7.

Process 5-6. Addition of heat at constant pressure p_2 in the combustion chamber (CC1) or from temperature T_5 to T_6.

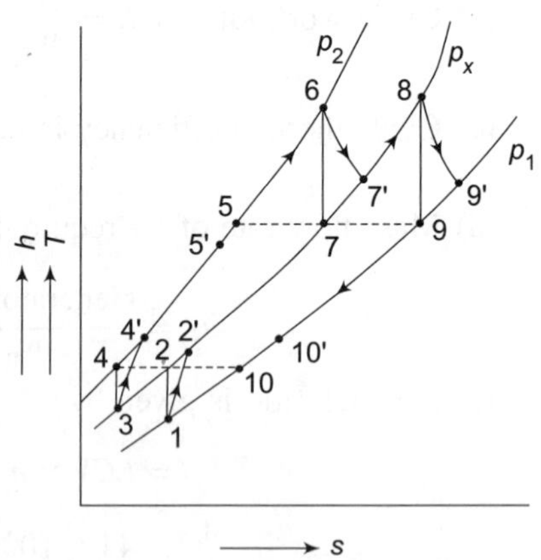

Figure 13.21 *T-s* representation of gas turbine cycle with intercooling, reheating and regenerative

Process 6-7. Isentropic expansion of gas from pressure p_2 to p_x in the first stage of turbine. For the maximum turbine work, pressure ratio, $p_2/p_x = \sqrt{r_p}$.

Process 7-8. Reheating of gas at constant pressure p_x in the combustion chamber (CC2) from temperature T_7 to $T_8 = T_6$.

Process 8-9. Isentropic expansion of gas from pressure p_x to p_1 in the second stage of turbine, for maximum turbine work pressure ratio $p_x/p_1 = \sqrt{r_p}$.

Process 9-10. Cooling of exhaust gas in the regenerator from temperature T_9 to T_{10}. If there is a perfect heat exchange in the regenerator $T_{10} = T_2 = T_4$.

Turbine work, $w_t = (h_6 - h_7) + (h_8 - h_9)$ and

Compressor work, $w_c = (h_2 - h_1) + (h_4 - h_3)$

Net work, $w_{net} = (h_6 - h_7) + (h_8 - h_9) - (h_2 - h_1) - (h_4 - h_3)$

Heat supplied, $q_A = (h_6 - h_5) + (h_8 - h_7)$

$$\eta_{th} = \frac{(h_6 - h_7) + (h_8 - h_9) - (h_2 - h_1) - (h_4 - h_3)}{(h_6 - h_5) + (h_8 - h_7)}$$

$$= \frac{(T_6 - T_7) + (T_8 - T_9) - (T_2 - T_1) - (T_4 - T_3)}{(T_6 - T_5) + (T_8 - T_7)} \quad ...(13.33)$$

13.11.2 Actual Gas Turbine Cycle with Regeneration Reheat and Intercooling

Process 1–2′–3–4–5′–6–7′–8–9–10′ shows such an actual cycle as shown in Fig. 13.21.

Actual turbine work, $w_{ta} = (h_6 - h_7') + (h_8 - h'_9)$

Actual compressor work, $w_{ca} = (h_2' - h_1) + (h_4' - h_3)$

Net work, $w_{neta} = (h_6 - h_7') + (h_8 - h_9') - (h_2' - h_1) - (h_4' - h_3)$

Heat supplied, $q_{Aa} = (h_6 - h'_5) + (h_8 - h_7')$

Actual thermal efficiency, η_{tha}

$$= \frac{(h_6 - h_7') + (h_8 - h'_9) - (h_2' - h_1) - (h'_4 - h_3)}{(h_6 - h'_5) + (h_8 - h'_7)}$$

$$= \frac{(T_6 - T_7') + (T_8 - T_9') - (T_2' - T_1) - (T_4' - T_3)}{(T_6 - T_5') + (T_8 - T_7')}$$

$$= \frac{(T_6 - T_7)\eta_t + (T_8 - T_9)\eta_t - (T_2 - T_1)/\eta_c - (T_4 - T_3)/\eta_c}{(T_6 - T_5') + (T_8 - T_7')}$$

$$\text{Regenerative effectiveness, } \in_r = \frac{h'_5 - h'_4}{h'_9 - h'_4} = \frac{h'_9 - h'_{10}}{h'_5 - h'_{10}} = \frac{T_5' - T_4'}{T_9' - T_4'} \quad ...(13.34)$$

If $\in_r$ is known then T_5' can be found out. Once T_5' is known, we can find out the thermal efficiency.

If the pressure loss in intercooler, combustion chambers, regenerator and exhaust hood are given, the *h-s* representation will be modified accordingly.

It is worth to note that in real cycle for the maximum efficiency, the value of p_x (intercooling pressure or reheating pressure) will not be the geometric mean.

13.12 EFFECT OF VARIOUS MODIFICATIONS

Effect of various modifications in the simple cycle on the work output and thermal efficiency are given here. The following are the results for a cycle having $r_p = 4$, $T_{max} = 864$ K, $T_{min} = 288$ K and $p_1 = 1$ bar.

Addition to simple cycle	Effect on efficiency	Effect on output
(a) Regeneration	+ 50%	Nil
(b) Intercooling	– 6.5%	+ 10.2%
(c) Reheat	– 10.4%	+ 24.5%
(d) Reheat + Regeneration	66.7%	+ 24.5%
(e) Intercooling + Regeneration	68%	+ 10.2%
(f) Reheat + Intercooling	– 18.2%	+ 34.7%
(g) Reheat + Intercooling + Regeneration	+ 80%	34.7%

So, we see that the thermal efficiency increases compared to simple cycle by addition of above in the simple cycle gas turbine as given in Table.

Example 13.7 The following data refer to an open cycle gas turbine using intercooling, regeneration and reheating arrangement.

Pressure ratio = 49, Compressor inlet temperature = 27°C, Turbine inlet temperature = 1300°C, Compressor efficiency = 0.86, Turbine efficiency = 0.87, Regenerative effectiveness = 0.8, Combustion efficiency = 0.99, Mechanical efficiency of whole assembly = 0.97, Generator efficiency = 0.98, Inlet pressure to compressor = 1 bar.

Neglect pressure losses in intercooler, combustion chamber and regenerator.

Determine (a) cycle specific work, (b) cycle thermal efficiency, (c) cycle work ratio, (d) cycle air rate, (e) for 220 MW power output, find the mass flow rate of air.

Solution Refer to Figs. 13.20 and 13.21. Given : $r_p = p_2/p_1 = 49$, $\eta_c = 0.86$, $\eta_t = 87$, $\in_r = 0.8$, $\eta_{comb} = 0.99$, $\eta_{gen} = 0.98$, $\eta_m = 0.97$, $T_1 = 300$ K, $p_1 = 1$ bar, $T_6 = T_8 = 1573$ K

The optimum pressure for the minimum work of compressor and the maximum work of turbine is given by $p_x = \sqrt{p_1 p_2} = \sqrt{49} = 7$ bars

$$T_2 = T_1 \left(\frac{p_x}{p_1}\right)^{(\gamma-1)/\gamma} = 300\ (7)^{0.4/1.4} = 523.38 \text{ K}$$

Now $$T_2' = T_1 + \frac{(T_2 - T_1)}{\eta_c} = 300 + \frac{523.38 - 300}{0.8} = 300 + 259.74 = 579.22 \text{ K}$$

Work consumed by L.P. compressor, w_{lpca}

$$= c_p (T_2' - T_1) = 1.005 (579.22 - 300) = 280.61 \text{ kJ/kg}.$$

For perfect intercooling and minimum work to compressor, the work consumed in both stages will be the same. Hence, total work done in two stages of compression including mechanical losses,

$$w_{ca} = (280.61 + 280.61)/0.97 = 578.57 \text{ kJ/kg}$$

$$T_7 = \frac{T_6}{(7)^{(\gamma - 1)/\gamma}} = \frac{1573}{1.744} = 901.94 \text{ K} = T_9$$

Again, $$T_7' = T_6 - \eta_t (T_6 - T_7) = 1573 - 0.87(1573 - 901.64) = 989.18 \text{ K} = T'_9$$

Work done by the L.P. stage of turbine, $c_p (T_6 - T_7') = 1.005 \times 583.82 = 586.73$ kJ/kg

Assume that the reheating is done at the optimum pressure up to in the initial inlet turbine.

Temperature, i.e., $T_6 = T_8$ and $T_7' = T_9'$ hence the work done by both stages will be the same, though in actual case, it is not true.

Total work done by the turbine, $w_{ta} = 586.73 + 586.73 = 1173.46$ kJ/kg ***Ans.***

Since mechanical efficiency as a whole assembly is given so in turbine it will not be considered. It is already considered in compressor work.

(a) Net work, $w_{\text{neta}} = (w_{ta} - w_{ca})\eta_{\text{gen}} = (1173.46 - 538.2) \times 0.98 = 622.55$ kJ/kg ***Ans.***

Regenerator effectiveness, $$\epsilon_r = \frac{T_5' - T_4'}{T_9' - T_4'} = \frac{T_9' - T_{10}'}{T_5' - T_{10}'}$$

or $T_5' = 579.22 + 0.8(989.18 - 579.22) = 327.96 + 579.22 = 907.18$ K

Heat supplied, $q_{Aa} = c_p [(T_6 - T_5') + (T_8 - T_7')]/\eta_{cc}$

$$= 1.005 [(1573 - 907.18) + (1573 - 989.18)]/0.99 = 1281.38 \text{ kJ/kg}$$

(b) Cycle thermal efficiency, $\dfrac{w_{ta} - w_{ca}}{q_{Aa}} = \dfrac{622.55}{1281.38} = 48.58\%$ ***Ans.***

(c) Cycle work ratio, $\dfrac{\text{Net work}}{\text{Turbine work}} = \dfrac{622.55}{1173.46} = 53.05\%$ ***Ans.***

(d) Cycle rate, $\dfrac{3600}{w_{\text{neta}}} = \dfrac{3600}{622.55} = 5.78$ kJ/kWh ***Ans.***

(e) $$\dot{m}_a = \frac{\text{Generator output}}{w_{\text{neta}}} = \frac{220 \times 10^3}{622.55} = 353.38 \text{ kg/s}$$

13.13 EFFECT OF REGENERATION, INTERCOOLING AND REHEATING ON EFFICIENCY

It is evident from the curves shown as Fig. 13.22 that higher thermal efficiency is possible from simple cycle with regeneration, intercooling and reheat. It is also obvious (from the curves) that due to addition of regeneration, intercooling and reheat, the peak value of thermal efficiency shifts towards slower pressure ratio compared to simple cycle. In all these cases, thermal efficiency first increases with pressure and reaches a maximum value and then decreases. The optimum pressure in each case is different and shift towards lower pressure ratio.

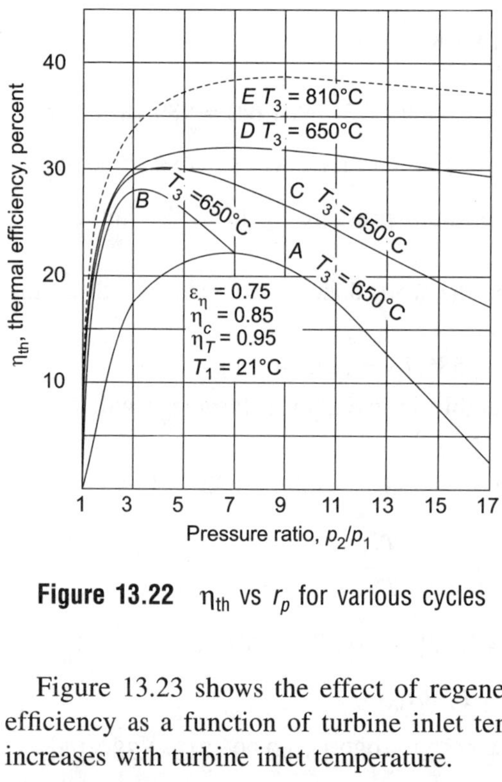

Figure 13.22 η_{th} vs r_p for various cycles

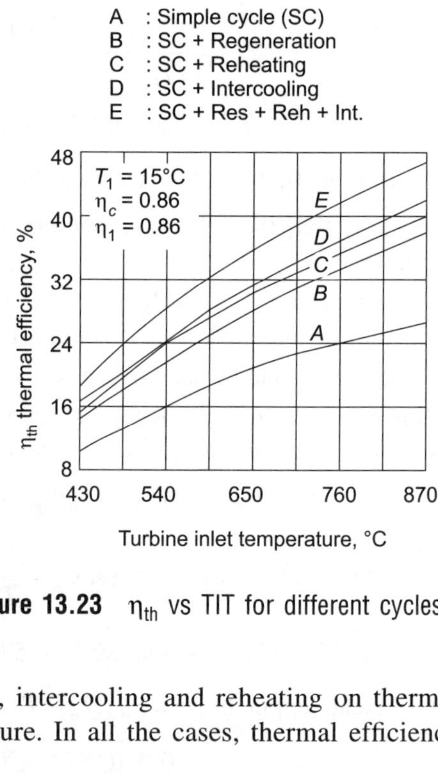

Figure 13.23 η_{th} vs TIT for different cycles

Figure 13.23 shows the effect of regeneration, intercooling and reheating on thermal efficiency as a function of turbine inlet temperature. In all the cases, thermal efficiency increases with turbine inlet temperature.

13.14 EFFECT OF OPERATING VARIABLES ON THERMAL EFFICIENCY

The thermal efficiency of an actual single cycle gas turbine depends upon the following operating variables:

(a) Pressure ratio, $r_p = p_2/p_1$.
(b) Turbine efficiency, η_t.
(c) Turbine inlet temperature, $T_3 = T_{max}$
(d) Compressor efficiency, η_c.
(e) Compressor inlet temperature, $T_1 = T_{min}$
(f) Specific fuel consumption.

(i) *Effect of Variation of Turbine Inlet Temperature and Pressure Ratio.* Figure 13.24 shows that as the turbine inlet temperature increases, the thermal efficiency increases with the other variables being held constant. For each value of turbine inlet temperature, there is an optimum pressure ratio for the maximum thermal efficiency. For lower value of turbine inlet temperature, the thermal efficiency first increases and after reaching a maximum value, it decreases rapidly. Similar is the trend on the effect of specific work. However, the optimum pressure ratio for the thermal efficiency and specific work is not the same. The optimum pressure ratio for the maximum work is at a lower value as compared to the maximum efficiency at the same of turbine inlet temperature. But for higher value of turbine inlet temperature, the peaks of the curves are flatter giving a greater range of optimum value of pressure ratio. The turbine inlet temperature is limited by turbine blading material.

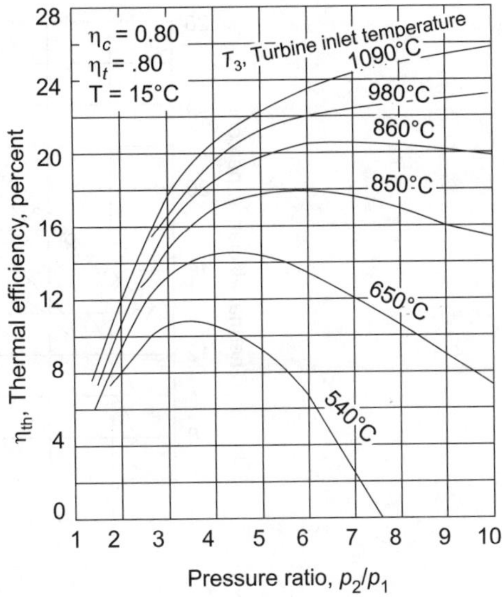

Figure 13.24 η_{th} vs r_p for various TIT (T_3)

(ii) *Effect of Variation of Turbine and Compressor Efficiency.* Figure 13.25 shows that the thermal efficiency is very much sensitive to the variations of turbine and compressor

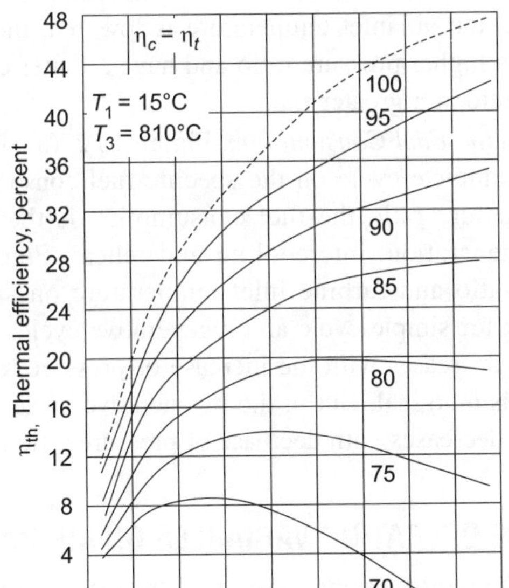

Figure 13.25 η_{th} vs r_p for various η_c and η_t

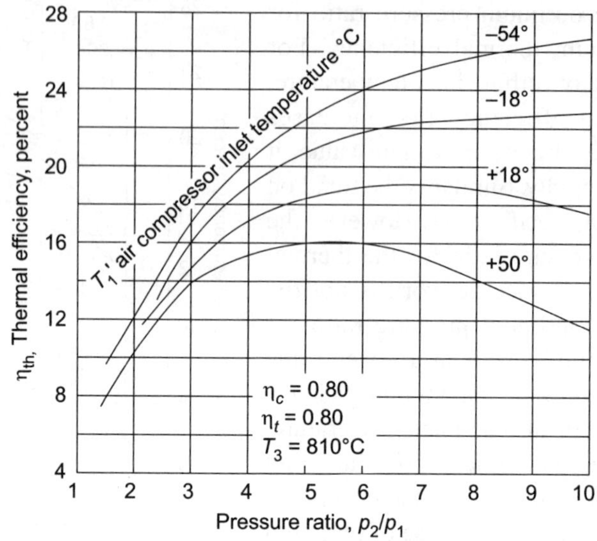

Figure 13.26 η_{th} vs r_p for various CITs (T_1)

efficiency. The dotted line shows the thermal efficiency for ideal simple open cycle. As the efficiency of turbine and compressor increases the thermal efficiency increases. There is an optimum pressure of each set of component efficiency.

(iii) *Effect of Variations of Compressor Inlet Temperature or Atmospheric Air.* Figure 13.26 shows that as the air inlet temperature is lowered, the thermal efficiency increases. Peaks are at a higher pressure ratio and have a flatter curve giving an optimum value of pressure ratio at a greater range.

(iv) *Effect on Specific Fuel Consumption.* Figure 13.27(a) shows the effect of regenerative, simple and complete cycle on the specific fuel consumption. It suggests that for all values of pressure ratio the fuel consumption is the minimum in complete cycle, i.e., with regeneration, intercooling and reheat. Figure 13.27(b) shows the effect of pressure ratio and turbine inlet temperature on specific fuel consumption and specific work for simple cycle and regenerative cycle. For simple cycle, specific fuel consumption decreases with the increase of pressure ratio rapidly while the effect of turbine inlet is marginal. But in the regenerative cycle story is different. Specific fuel consumption decreases with decrease of pressure ratio but decreases with turbine inlet temperature.

13.15 EFFECT OF OPERATING VARIABLES ON AIR RATE

(i) *Effect of Turbine Inlet Temperature and Pressure Ratio on the Air Rate.* Figure 13.28 shows that as the turbine inlet temperature increases, the air rate decreases, It is also clear from the figure that as the pressure ratio increases, the air rate decreases. to a minimum value first and after that it again increases.

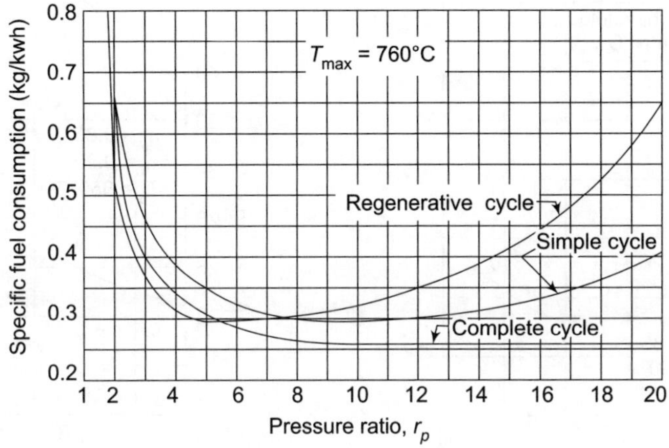

(a) sfc vs r_p for various cycles

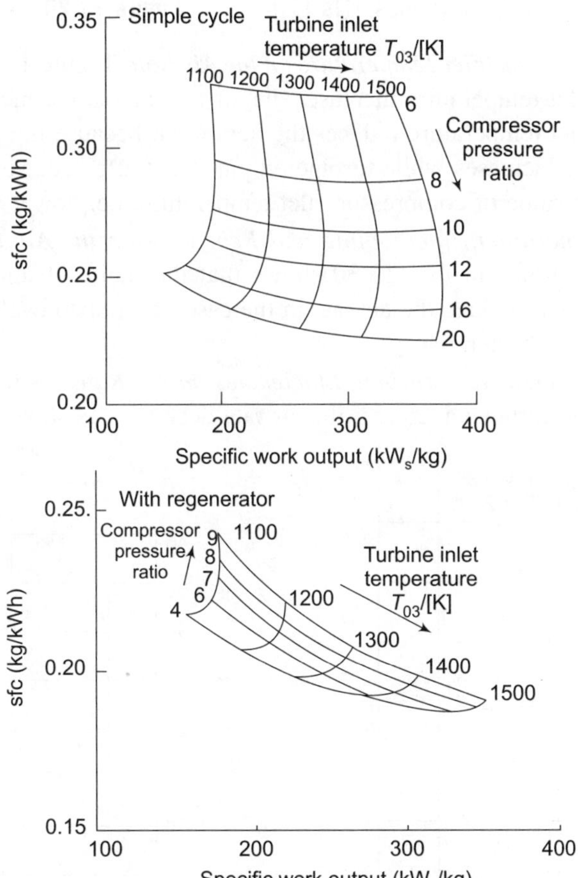

(b) sfc vs specific work for various r_p and TIT (T_3)

Figure 13.27

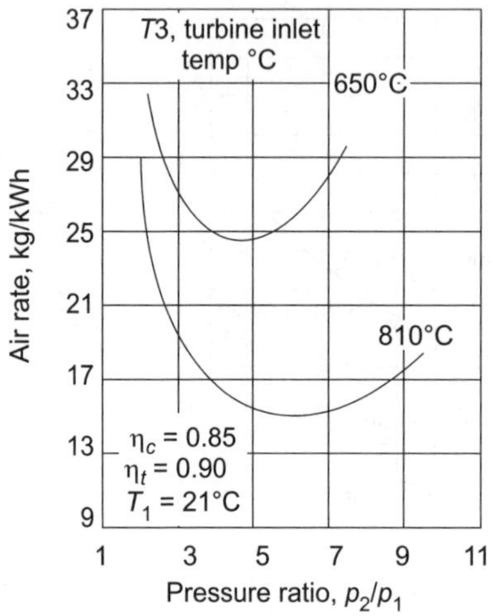

Figure 13.28 Air Rate vs r_p for various TITs (T_3)

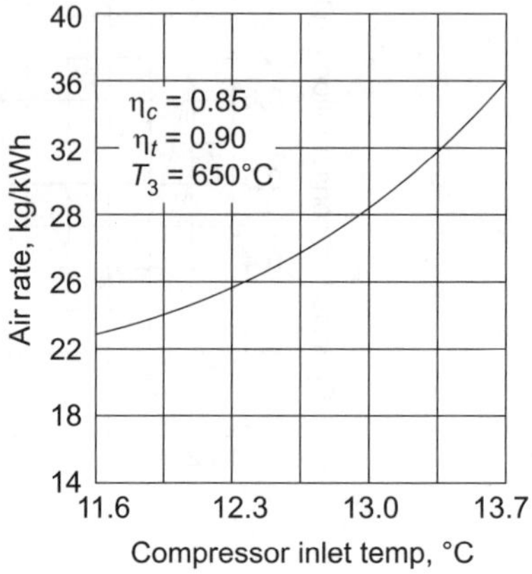

Figure 13.29 Air rate vs CIT (T_1)

(ii) *Effect of Compressor Inlet Temperature on the Air Rate.* Figure 13.29 shows that as the compressor inlet temperature increases the air rate increases. Increasing the value of compressor inlet temperature reduces the net output because the power consumed by the compressor increases while turbine output is not affected. So, a good gas turbine requires lower value of compressor inlet temperature, i.e., lower value of air rate.

(iii) *Effect of Regeneration, Intercooling and Reheating on the Air Rate.* A comparison of the curves shown in Fig. 13.30 shows that the main advantage of intercooling and reheating is to reduce the air rate in the case of an open cycle with regeneration, intercooler and reheater.

(iv) *Effect of Compressor and Turbine Efficiencies on Air Rate.* As the efficiencies of the compressor and turbines increase, the air rate decreases as shown in Fig. 13.31.

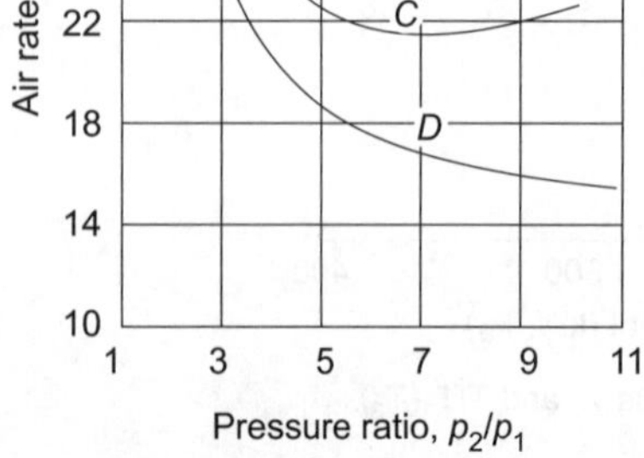

Figure 13.30 Air rate vs r_p for various cycles

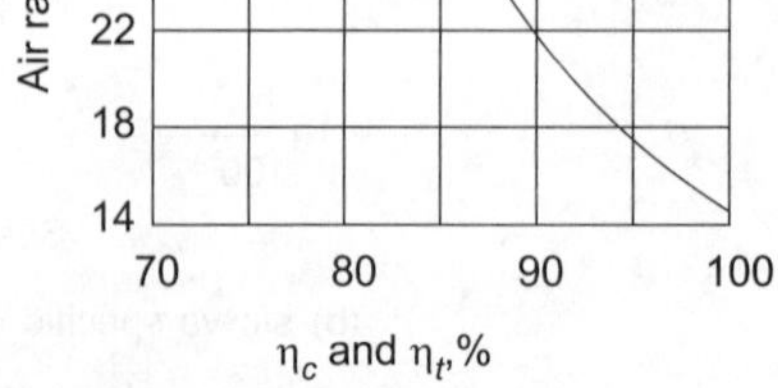

Figure 13.31 AR vs η_c & η_t

13.16 EFFECT OF OPERATING VARIABLES ON WORK RATIO

(i) *Effect of Compressor Inlet Temperature (CIT) on the Work Ratio (WR).* Figure 13.32 shows that as the compressor inlet temperature decreases the work ratio increases. This is due to the decrease in the power required by the compressor while the turbine output remains the same.

(ii) *Effect of Compressor and Turbine Efficiencies on the Work Ratio.* The work ratio is much affected by the variation in the efficiencies of turbine and compressor as shown in Fig. 13.32. It is obvious from the curve that as the turbine and compressor efficiencies increase, the work ratio increases.

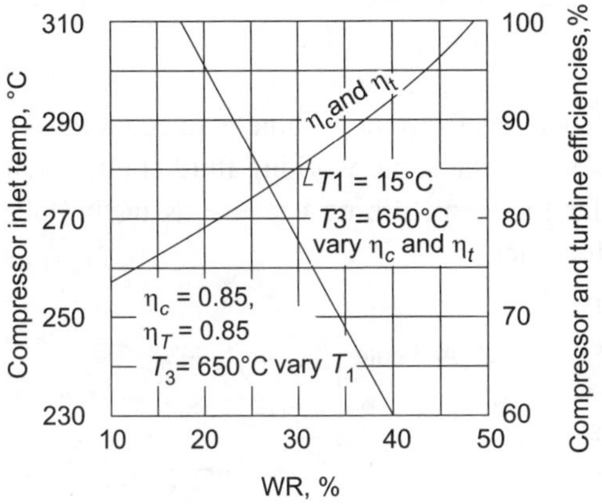

Figure 13.32 CIT vs. WR

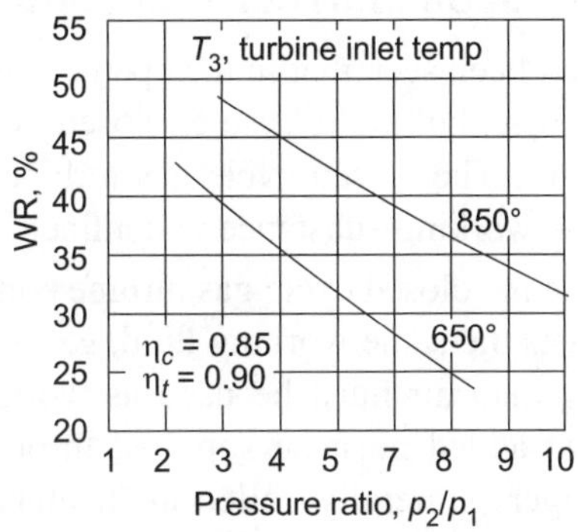

Figure 13.33 WR vs. r_p for various TITs (T_3)

(iii) *Effect of Turbine Inlet Temperature and Pressure Ratio on the Work Ratio.* Figure 13.33 shows that as the pressure ratio increases the work ratio decreases for a fixed value of turbine inlet temperature. It is also evident from the curve that as the turbine inlet temperature increases, the work ratio increases.

(iv) *Effect of Regeneration, Intercooling and Reheating on the Work Ratio.* Figure 13.34 shows that maximum work is possible in a simple cycle with reheat, intercooling and regeneration. Intercooler and reheater not only improve the work ratio but also decrease the slope of the curve to give a greater improvement in efficiency as the work ratio is raised.

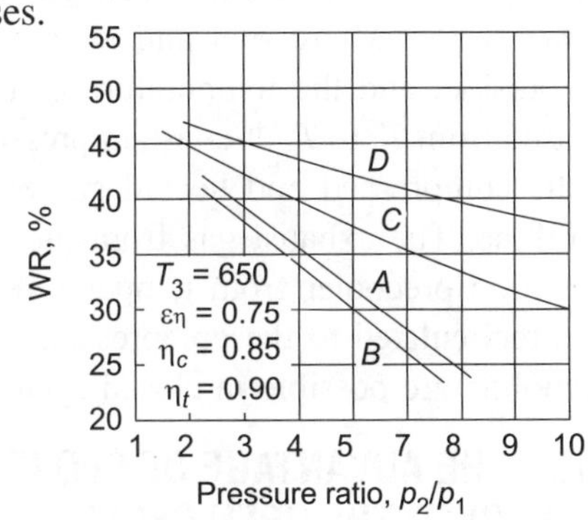

Figure 13.34 WR vs. r_p for various cycles

13.17 WATER INJECTION

Another method of improving the performance of the gas turbine is to inject water into the working air at the entrance to the compressor. This way the compressed air is cooled by the latent heat of vaporization of water. By injection of water, the total mass flow of the working medium is increased by the mass of the injected water and hence the power output of the cycle is increased, the work ratio is increased and the air rate is lowered.

Water injection system is commonly used as a power boost for take off and emergencies on jet propelled aircraft. It is possible for marine or land based gas turbines to use water injection for long duration or continuous injection. The water to be injected should be pure, otherwise, it may cause corrosion or deposits on the blades.

13.18 CLOSED CYCLE GAS TURBINE

As it has been seen that in the open cycle, the exhaust gas from the turbine is rejected into the atmosphere, but in the case of closed cycle gas turbine, the same working fluid continuously circulates. The closed cycle gas turbine is similar to steam turbine because as in the latter case the working substance is confined within the plant.

Since the closed cycle gas turbine continuously circulates the same working fluid, gases of higher density than air must be used as working fluid, the heat added must be supplied through a heat exchanger (generally called air heater) from an external source and the heat rejected from the system must be through another heat exchanger with a cooling medium generally called precooler.

Figure 13.35 shows a closed cycle gas turbine. The gas is compressed from pressure p_1 to p_2 in the compressor. The compressed gas is heated in a *air heater* where fuel and air are supplied from outside, and the temperature of the gas is increased from T_2 to T_3 at constant pressure. After this the compressed and heated gas expands in the turbine. The exhaust gas from the turbine is cooled in a precooler from temperature T_4 to T_1 and is recirculated to the compressor. Note that all the modifications such as regenerative, intercooling are possible in closed cycle.

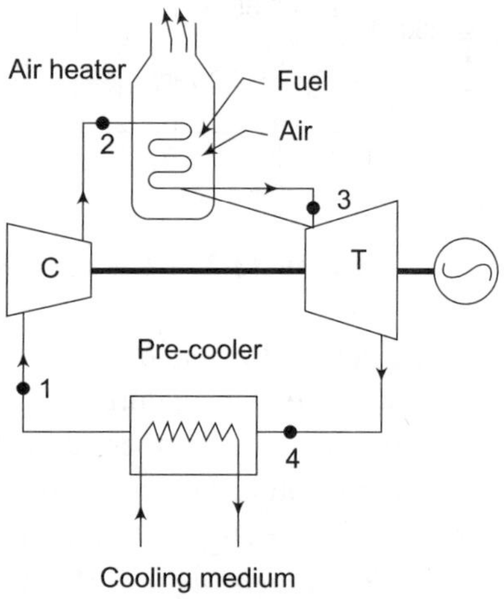

Figure 13.35 Closed gas turbine cycle

13.19 THE ADVANTAGE OF CLOSED CYCLE GAS TURBINE OVER THE OPEN CYCLE

(i) *Higher Thermal Efficiency.* Appreciably higher thermal efficiencies are obtainable for the same maximum and minimum, temperature limits as with open cycle gas turbine.

(ii) *Reduced Size.* The density of the working fluid is increased in the closed cycle by placing the system under an initial overall higher pressure.

Since the working medium is not required to support combustion, so it is possible to use a gas of higher density such as argon, xenon and mercury vapour as the working fluid. The increase in the density of the working fluid reduces the size of all components and ducts of the system for the same power output. It also permits the use of higher temperature for a given stress limit.

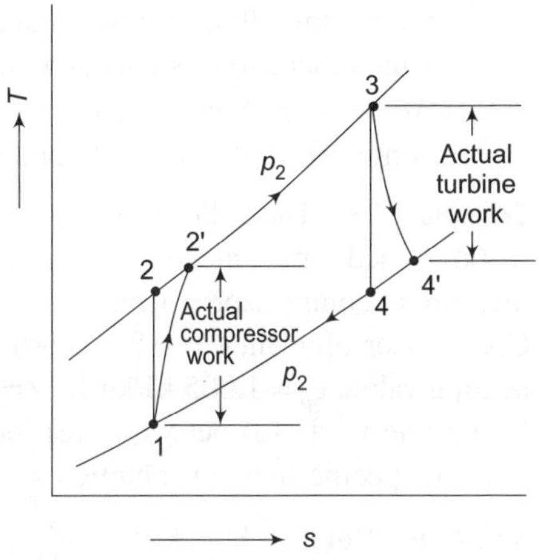

Figure 13.36 *h-s* representation of closed gas turbine cycle

(iii) *Improved Part Load Efficiency.* Part load efficiency of a closed cycle is improved as compared to the open cycle because the method of control of the output in the closed cycle does not vary the temperature and pressure ratio of the system, and also, the efficiency during the load change.

(iv) *Improved Heat Transmission.* The rate of heat transmission is improved.

(v) *Lesser Fluid Friction Loss.* The fluid friction loss is reduced because of higher Reynold's number.

(vi) *No Contamination.* No corrosive gases circulate through the closed system. As the working fluid does not contain the combustion gases, the turbine and regenerator are not subjected to carbon deposits and remain relatively clean. The compressor also remains free of ducts and other foreign deposits because the working medium is cleaned before being put into the system. This advantage plays an important part in increasing the life of the plant as well as its efficiency compared to the open cycle plant.

(vii) *Greater Output.* Much greater outputs are possible and these do not involve excessive weight.

(viii) *Inexpensive Fuel.* Because the closed cycle utilizes external heating, any inexpensive solid fuels such as coal can be used to heat the working fluid.

(ix) *No Loss of Working Medium.* Since the working circuit is closed, there is no loss of working medium to make good.

13.20 DISADVANTAGES OF CLOSED CYCLE AS COMPARED TO OPEN CYCLE

(i) *Dependent System.* The closed cycle is dependent on cooling water which must be provided for the precooler. This eliminates the use of this systems as an aeronautical engine. It is best suited for marine propulsion where the provision of cooling water is not a problem.

(ii) *Complexity. The load control of the closed system is complex and costly compared to open system. Since the initial pressure, i.e.*, the compressor inlet pressure is high and the working fluid is other than air, the system has to be gas tight against high pressure which increases its cost and also poses engineering problems.

(iii) *Air Heater.* A heavy and large air heater is required which is relatively inefficient compared to the internal combustion chambers of the open cycle.

Example 13.8 The following particulars relate to closed cycle gas turbine using air as the working medium–Atmospheric temperature = 26°C, Maximum temperature = 870°C, Initial pressure at compressor = 1 bar, Final pressure of compressor = 5, Turbine efficiency = 0.84, Compressor efficiency = 0.8, Calorific value of fuel = 41840 kJ/kg, Heater loss = 10% of heating value, c_p = 1.005 kJ/kg K, γ = 1.4. Calculate (a) compressor work, (b) heat supplied, (c) turbine work, (d) net work, (e) fuel-air ratio, (f) thermal efficiency (g) air-rate (h) work ratio (i) specific fuel consumption.

Solution Refer to Fig. 13.35.

Since the cycle is a closed one, so there is no chance of mixing of air and fuel.

$$T_2 = T_1(r_p)^{(\gamma-1)/\gamma} = 299 \times (5)^{0.286} = 473.56 \text{ K}$$

(a) $w_{ca} = \dfrac{c_p(T_2 - T_1)}{\eta_c} = \dfrac{1.005(473.56 - 299)}{0.8} = 219.29$ kJ/kg

(b) $w_{ca} = c_p\,(T_2' - T_1) = 219.29$ or $T_2' = \dfrac{219.29}{1.005} + 299 = 517.2$ K ***Ans.***

Heat supplied, $c_p\,(T_3 - T_2') = 1.005(1143 - 517.2) = 628.92$ kJ/kg ***Ans.***

(c) Now, $T_4 = \dfrac{T_3}{(r_p)^{(\gamma-1)/\gamma}} = \dfrac{1143}{1.5834} = 721.86$ K

Actual turbine work, $c_p(T_3 - T_4) \times \eta_t = 1.005\,(1143 - 721.86) \times 0.84$

$= 363.99$ kJ/kg ***Ans.***

(d) Net work, $w_{net} = 363.99 - 219.29 = 144.7$ kJ/kg ***Ans.***

(e) As 10% of heating value is lost, i.e., the efficiency of heater is 90% each kg of fuel contributors 0.9 × 41840 = 37656 kJ of fuel. Heat supplied = 628.92 kJ/kg of air

Therefore, Air-fuel ratio $= \dfrac{37656}{628.92} = 59.87:1$ ***Ans.***

(f) Thermal efficiency $= \dfrac{\text{Net work}}{\text{Heat supplied}} = \dfrac{144.7}{648.87} = 22.3\%$ ***Ans.***

(g) Air rate $= \dfrac{3600}{144.7} = 24.87$ kg/kWh ***Ans.***

(h) Work ratio, $\dfrac{\text{Net work}}{\text{Turbine work}} = \dfrac{144.7}{363.99} = 39.75\%$ ***Ans.***

(i) Specific fuel consumption, $\dfrac{\text{Air rate}}{\text{Air-fuel ratio}} = \dfrac{24.87}{59.87} = 0.412$ kg/kWh

Example 13.9 The following data relate to a simple cycle gas turbine with a free power turbine.

Ambient = 1 bar, 288 K, Total compressor pressure ratio = 13.5, Turbine inlet temperature = 1500 K, Total to total isentropic efficiency of compressor = 0.86, Total to total isentropic efficiency of each turbine = 0.89, Mechanical efficiency of each shaft = 0.98, Combustion efficiency = 0.98, Combustion chamber pressure loss = 5% of compressor delivery pressure, Exhaust pressure loss = 0.03 bar, Alternator efficiency = 97%, Calorific value of fuel = 42000 kJ/kg, c_{pa} = 1.005 kJ/kgK, c_{pg} = 1.15 kJ/kgK, Electrical power output = 200 MW. Calculate the specific work output, specific fuel consumption, mass flow rate of air, air fuel ratio, and cycle thermal efficiency. Comment on the result for the use of this gas turbine for combined cycle.

Solution Refer to Figs. 13.37 and 13.38. Assume ambient conditions also relate to total conditions, $p_1 = p_{01}$ = 1 bar and $T_1 = T_{01}$ = 288 K

$$T_{02} = T_{01}(r_p)^{(\gamma-1)/\gamma} = 288(13.5)^{0.4/1.4} = 605.82 \text{ K}$$

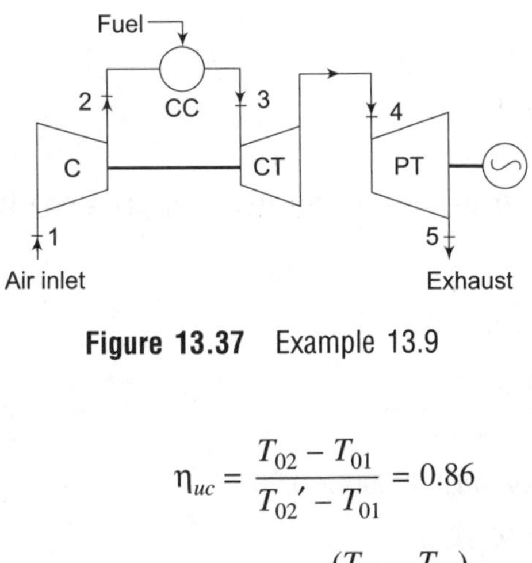

Figure 13.37 Example 13.9

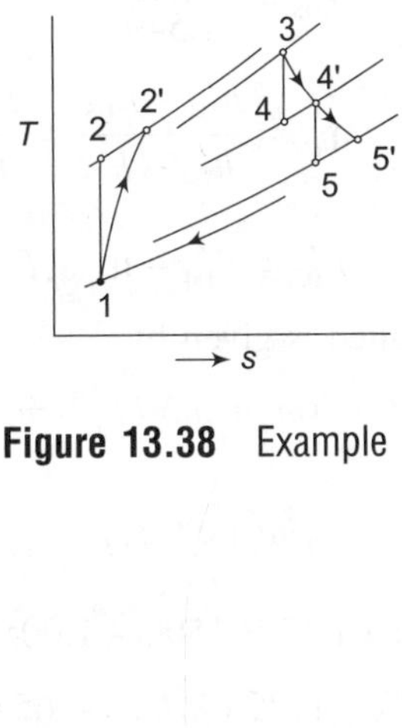

Figure 13.38 Example 13.9

$$\eta_{uc} = \frac{T_{02} - T_{01}}{T_{02}' - T_{01}} = 0.86$$

or $$T_{02}' = T_{01} + \frac{(T_{02} - T_{01})}{\eta_{uc}} = 288 + \frac{605.82 - 288}{0.86} = 657.55 \text{ K}$$

$$p_{03} = 0.95 \times 13.5 = 12.825 \text{ bars}$$

The compressor turbine exhaust pressure is not known. This can be evaluated by equating the work input to compressor equal to power developed by compressor turbine.

$$w_{ca} = \frac{h_{02}' - h_{01}}{\eta_m} = \frac{c_{pa}(T_{02}' - T_{01})}{\eta_m} = \frac{1.005(657.55 - 288)}{0.98} = 378.97 \text{ kJ/kg}$$

Since $$w_{ca} = w_{tca} = 378.97 = (h_{03} - h_{04}') = c_{pg}(T_{03} - T_{04}') = 1.15(1500 - T_{04}')$$

or $$T_{04}' = 1500 - \frac{378.97}{1.15} = 1170.49 \text{ K}$$

From the concept of total to total isentropic turbine efficiency, we have

$$\eta_{nct} = \frac{h_{03} - h'_{04}}{h_{03} - h_{04}} = \frac{T_{03} - T'_{04}}{T_{03} - T_{04}} = 0.89$$

or $$T_{04} = T_{03} - \frac{T_{03} - T'_{04}}{\eta_{nct}} = 1500 - \frac{1500 - 1170.97}{0.89} = 1130.33 \text{ K}$$

We have $$\frac{T_{03}}{T_{04}} = \left(\frac{p_{03}}{p_{04}}\right)^{(\gamma-1)/\gamma} = \frac{1500}{1130.33} = \left(\frac{12.825}{p_{04}}\right)^{0.4/1.4}$$

or $$p_{04} = \frac{12.825}{(1.327)^{3.5}} = 4.765 \text{ bar}$$

$$\frac{T'_{04}}{T_{05}} = \left(\frac{p_{04}}{p_{05}}\right)^{0.4/1.4} = \frac{1170.49}{T_{05}} = \left(\frac{4.765}{1.03}\right)^{0.4/1.4} = 1.549$$

or $$T_{05} = \frac{1170.49}{1.549} = 756.64 \text{ K}$$

$$\eta_{upt} = \frac{h'_{04} - h'_{05}}{h'_{04} - h_{05}} = \frac{T'_{04} - T'_{05}}{T'_{04} - T_{05}} = 0.89$$

or $$T'_{05} = T'_{04} - \eta_{upt}(T'_{04} - T_{05}) = 1170.49 - 0.89(1170.49 - 756.64) = 802 \text{ K}$$

Heat supplied is given by

or $$(\dot{m}_a + \dot{m}_f)\, h_{03} - \dot{m}_a\, h'_{02} = \dot{m}_f \times q_{LCV} \times \eta_{\text{comb}}$$

or $$(1 + \dot{m}_f/\dot{m}_a)\, c_{pg}\, T_{03} - c_{pa}\, h'_{02} = \frac{\dot{m}_f \times Q_{LCV} \times \eta_{\text{comb}}}{\dot{m}_a}$$

or $(1 + \dot{m}_f/\dot{m}_a)\ 1.15 \times 1500 - 1.005 \times 657.55 = (\dot{m}_f/\dot{m}_a)\ 42000 \times 0.98$

or $1725 - 1725\,(\dot{m}_f/\dot{m}_a) - 657.55 = 41160\,(\dot{m}_f/\dot{m}_a)$

or $(\dot{m}_f/\dot{m}_a) = 0.0259$ ***Ans.***

Specific turbine output, $w_{tpa} = w_{neta} = (1 + \dot{m}_f/\dot{m}_a)\,(h'_{04} - h_{05})\,\eta_m\,\eta_{al}$

$= (1 + \dot{m}_f/\dot{m}_a)\, c_{pg}\,(T'_{04} - T'_{05})\,\eta_m\,\eta_{al}$

$= (1 + 0.0259) \times 1.15(1170.49 - 802) \times 0.98 \times 0.97 = 413.26$ kJ/kg ***Ans.***

We have, $(\dot{m}_a + \dot{m}_f)\, w_{tpa} = 200 \times 10^3$ or $\dot{m}_a\,(1 + 0.0248) \times 413.26 = 200 \times 10^3$

or $\dot{m}_a = 472.24$ kg/s ***Ans.***

$$\text{sfc} = \frac{3600\, \dot{m}_f}{\text{output (kW)}} = 3600 \times 472.24 \times 0.0259/200 \times 10^3 = 0.22 \text{ kg/kWh}$$ ***Ans.***

Heat supplied = $(1 + \dot{m}_f/\dot{m}_a)\, c_{pg}\, T_3 - c_{pa}\, T'_{02}$

$$= 1.0259 \times 1.15 \times 1500 - 1.005 \times 657.55 = 1108.84 \text{ kJ/kg}$$

Thermal efficiency $= \dfrac{w_{\text{neta}}}{q_{Aa}} = \dfrac{w_{tpa} + w_{tca} - w_{ca}}{q_{Aa}} = \dfrac{w_{tpa}}{q_{Aa}} = \dfrac{413.26}{1108.84} = 37.30\%$ ***Ans.***

Comment. The gas turbine exhaust temperature is 801.89 K which is less for efficient steam cycle for combined cycle operation as high temperature steam will not be available. The exhaust temperature above 873 K is desirable. This is possible by using higher turbine inlet temperature say 1650 K or lower pressure ratio.

Example 13.10 The following data relate to high pressure ratio single shaft gas turbine cycle with reheat total ambient condition = 1.01 bars, 288 K, Total compressor ratio = 32, Polytropic efficiency = 0.9, for compressor and turbine both. Total turbine inlet temperature of both turbine = 1600 K, Pressures loss in first combustor = 2% of compressor delivery pressure, Pressure loss in second combustor = 4% of H.P. turbine, exhaust pressure. Total turbine exhaust pressure = 1.05 bars, Mechanical efficiency for assembly = 0.97, Combustion efficiency of both combustors = 0.96, Alternator efficiency = 0.97, Alternator power output = 240 MW, Lower calorific of fuel = 42000 kJ/kg, c_{pa} = 1.005 kJ/kgK, c_{pg} = 1.15 kJ/kgK.

Calculate the specific work output, air-fuel ratio, mass flow rate of air, specific fuel combustion and cycle thermal efficiency.

Solution Refer to Figs. 13.39 and 13.40. For polytropic compression and expansion, the value of $(n - 1)/n$ has to be evaluated and polytropic efficiency will help in evaluating this.

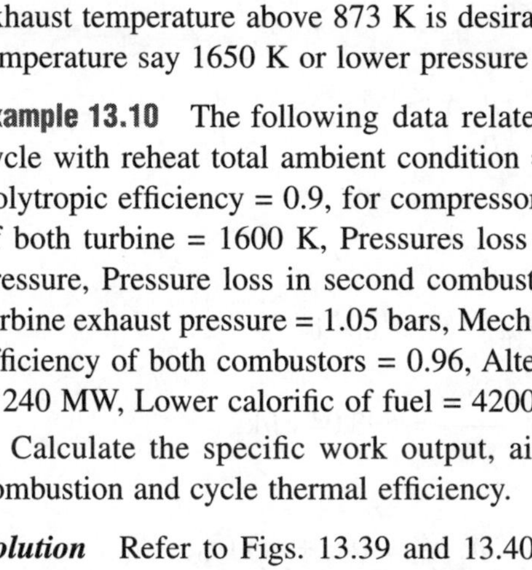

Figure 13.39 Example 13.10

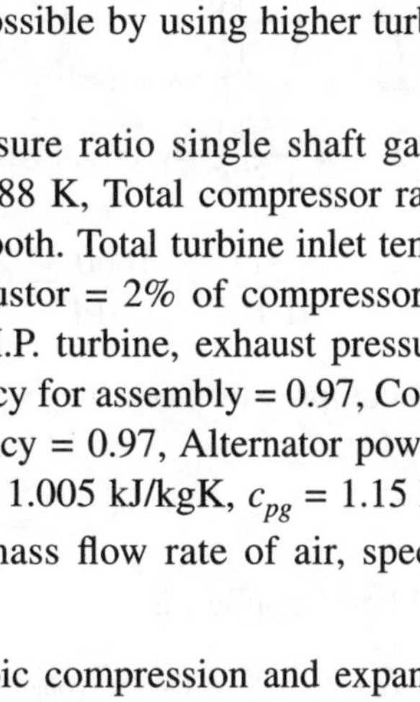

Figure 13.40 Example 13.10

For compression process, $\dfrac{n-1}{n} = \dfrac{\gamma - 1}{\gamma\, \eta_{pc}} = \dfrac{0.4}{1.4 \times 0.9} = 0.3174$

For expansion process, $\dfrac{n-1}{n} = \dfrac{(\gamma - 1) \times \eta_{pt}}{\gamma} = \dfrac{0.4 \times 0.9}{1.4} = 0.2571$

For actual compression process 1–2′: $T'_{02} = T_{01}\left(\dfrac{p_{02}}{p_{01}}\right)^{(n-1)/n} = 288(32)^{0.3174} = 865.22$ K

$$p_{03} = 0.98 \times 32 \times 1.01 = 31.673 \text{ bars and } p_{06} = 1.05 \text{ bars}$$

For optimum power output, the reheating pressure under ideal condition should be

$$p_4 = p_x = \sqrt{(r_p)_1} = \sqrt{\left(\frac{p_{03}}{p_{06}}\right)} = \sqrt{31.673/1.05} = 5.492 \text{ bar}$$

Since there is a pressure loss of 4% of H.P. turbine exhaust, the actual reheating pressure will be 0.96 × 5.492 = 5.27 bars

$$\frac{T_{03}}{T'_{04}} = \left(\frac{p_{03}}{p_{04}}\right)^{(n-1)/n} = \left(\frac{31.673}{5.27}\right)^{0.2571} = 1.585$$

or $$T'_{04} = \frac{1600}{1.585} = 1009.46 \text{ K} \quad \text{and} \quad \frac{T_{05}}{T'_{06}} = \left(\frac{p_{05}}{p_{06}}\right)^{(n-1)/n} = \left(\frac{5.242}{1.05}\right)^{0.2571} = 1.514$$

or $$T'_{06} = \frac{T_{05}}{1.154} = \frac{1600}{1.514} = 1056.8 \text{ K}$$

Actual compressor work, $$w_{ca} = \frac{(h'_{02} - h_{01})}{\eta_m} = c_{pa}\,(T'_{02} - T_{01})/\eta_m$$

$$= \frac{1.005\,(865.22 - 288)}{0.97} = 598.04 \text{ kJ/kg}$$

Actual heat supplied is

$$q_{Aa} = \dot{m}_f \times q_{LCV} \times \eta_{comb} = (\dot{m}_a + \dot{m}_f)\, h_{03} - \dot{m}_a\, h'_{02} + (\dot{m}_a + \dot{m}_f)\,(h_{05} - h'_{04})$$

or $$\frac{\dot{m}_f}{\dot{m}_a} q_{LCV} \times \eta_{comb} = [(1 + \dot{m}_f/\dot{m}_a)\, c_{pg}\, T_{03} - c_{pa}\, T'_{02}] + (1 + \dot{m}_f/\dot{m}_a)\,(h_{05} - h'_{04})$$

or $$\frac{\dot{m}_f}{\dot{m}_a} \times 42000 \times 0.96 = (1 + \dot{m}_f/\dot{m}_a) \times 1.15 \times 1600 - 1.005 \times 865.22$$
$$+ (1 + \dot{m}_f/\dot{m}_a) \times 1.15\,(1600 - 1019.75)$$

or $$40320\,(\dot{m}_f/\dot{m}_a) = 1840 + 1840\, \dot{m}_f/\dot{m}_a - 869.54 - 667.28 + (\dot{m}_f/\dot{m}_a)\, 667.28$$

or $$\dot{m}_f/\dot{m}_a = 0.0433$$

or $$\dot{m}_a/\dot{m}_f = 23.094$$ ***Ans.***

Actual turbine work, $w_{ta} = (1 + \dot{m}_f/\dot{m}_a)\,[(h_{03} - h'_{04}) + (h_{05} - h_{06})]\,\eta_{al}$

Since η_m is given for assembly, so it has already been considered in compressor work

$$= (1 + \dot{m}_f/\dot{m}_a)\, c_{pg}\, [(T_{03} - T'_{04}) + (T_{05} - T'_{06})].\eta_{al}$$

$$= 1.0433 \times 1.15\,[(1600 - 1009.46) + (1600 - 1056.8)] \times 0.97 = 1319.29 \text{ kJ/kg}$$

Plant net specific work, $w_{neta} = w_{ta} - w_{ca} = 1319.29 - 598.04 = 721.25$ kJ/kg ***Ans.***

The mass flow rate of air is given by, $\dot{m}_a \times 721.25 = 240 \times 10^3$

or $\dot{m}_a = 332.75$ kg/s ***Ans.***

$$\text{sfc} = \frac{3600 \times \dot{m}_f \text{ (kg/s)}}{\text{Power output (kW)}} = \frac{332.75 \times 0.0433 \times 3600}{240 \times 10^3} = 0.2161 \text{ kg/kWh}$$ ***Ans.***

Actual thermal efficiency is given by

$$\eta_{tha} = \frac{w_{neta}}{q_{Aa}} = \frac{721.25}{(\dot{m}_f/\dot{m}_a)q_{LCV} \times \eta_{comb}} = \frac{721.25}{0.0433 \times 42000 \times 0.96} = 41.31\% \qquad \textit{Ans.}$$

13.21 ADVANTAGES AND DISADVANTAGES OF GAS TURBINES OVER STEAM TURBINE POWER PLANTS

Advantages

(i) No feed water supply steam is required.

(ii) There is no need of a condensing plant.

(iii) No boiler is required.

(iv) There are less components to be designed.

(v) The maintenance cost is low because working fluid is usually air.

(vi) It is quick to start and there is no banking loss. This is important in aviation, locomotive and emergeous power plants.

(vii) It has lower specific weight and size, i.e., lower weight and size per kW output. This is advantageous in aviation.

(viii) It is an lower operating pressure.

(ix) It is an independent system except closed cycle plant. It is best suited for aviation purpose.

(x) Being comparatively small, it is suitable for mobile use.

(xi) The initial cost and operating costs are lower.

(xii) The installation period is very short. Now compact system installed on the frame are supplied to start generating power at the site.

Disadvantages

(i) The power developed is less than steam turbine because of the power consumed by the compressor. However, with all round development gas turbines have been built to develop 334 MW as a single unit yielding 39.54% efficiency.

(ii) Due to high temperature, the life of blading is short even after cooling.

(iii) Air rate is high, so that the open cycle gas turbine is a great disadvantage in marine applications.

(iv) The part-load efficiency is poor.

13.22 ADVANTAGES AND DISADVANTAGES OF GAS TURBINE OVER DIESEL OR PETROL ENGINES

Advantages

(i) Better balanced due to absence of reciprocating masses.

(ii) Better mechanical efficiency due to absence of numerous sliding or bearing members.

(iii) High power plant of greater capacity than diesel engines can be built.

(iv) The specific weight is lower and this size smaller. This is very important for aviation purpose.

(v) It can operate with cheaper fuels like any of the hydrocarbon fuels from natural gas, high octane gasoline down to heavy diesel oil, even solid fuels (pulverized coal).

(vi) It requires less lubrication and it is easy and inexpensive.

(vii) It has a much higher operating speed.

(viii) Smoke less exhaust.

(ix) Comparatively silent operation due to continuous exhaust.

(x) It can be used in a jet propulsion unit because of continuous exhaust at high temperature and pressure.

(xi) The system is flexible and thus reheating, regenerative and intercooling arrangements can be added to increase efficiency and output.

(xii) It has lower inlet temperature of air at high altitude which increases the thermal efficiency.

Disadvantages

(i) It has poor part-load efficiency.

(ii) It has a high air rate, so that open cycle is not suitable in marine applications.

(iii) It is sensitive to changes in compressor and turbine efficiencies.

(iv) It is not suitable for marine applications. Rarely closed cycle gas turbine is used for marine purpose.

13.23 ELEMENTS AND ACCESSORIES OF GAS TURBINE PLANT

The following are the main elements and accessories of a gas turbine plant

(1) Compressor (centrifugal and axial)
(2) Combustion chamber
(3) Gas turbine
(4) Fuel system
(5) Starting system
(6) Ignition system
(7) Speed control system
(8) Blade cooling system
(9) Protective system

13.24 CENTRIFUGAL COMPRESSOR*

A centrifugal compressor consists of (i) inlet part, (ii) impeller (single or double shrouded) keyed to a shaft supported on bearings, (iii) diffuser (vaneless or vaned or both) and (iv) outlet part (volute casing and outlet pipe). Figure 13.41(a) shows a schematic diagram of a rdial balded impeller with inducer ($\beta_{2b} = 90°$) of a centrifugal compressor. The blades of the impeller may also be backward [Fig. 13.41 (b), $\beta_{2b} < 90°$] or forward ($\beta_{2b} > 90°$) but these are not suitable for gas turbines because these need a larger size for the same pressure ratio and cannot stand the high stressed developed. The impeller may have single or double entry of air. In the case of radial blades, the inlet portion must have an inducer which may be supposed as a small size axial stage which turns the axial velocity into a radial velocity in a

* For more details see Author's book "*Steam & Turbines, Turbomachines and Power Plant Engineering*".

small passage. With the help of starting system, the impeller is rotated. Air enters through the inlet part due to suction created at the impeller eye and the impeller converts the mechanical energy imparted to the air by the rotation of the impeller into pressure and kinetic energy. Figure 13.41(c) shows a cut-view of industrial centrifugal compressor. The pressure and velocity distribution are shown in Fig. 13.41(d). The representation of flow process on h–s diagram is shown in Fig. 13.42. In the inlet part, the flow is accelerated accompanied by friction (process 0–1). The pressure rise in the impeller (process 1–2) is due to diffusion action (i.e., the relative velocity decreases from inlet to outlet due to diverging blade channel area equivalent to $(C_{r1}^2 - C_{r1}^2)/2$ and the centrifugal action (i.e., the air enters at a lower diameter and comes out at a higher diameter) equivalent to $(u_2^2 - u_1^2)/2$. The rest of kinetic energy available at the tip of the impeller is converted into energy in vaneless and vaned diffuser and volute casing. Due to friction and other losses there is an increase in entropy and the total pressure loss is $[(p_{eo})_{\text{without loss}} - p_{eo}]$.

The flow phenomena through the centrifugal compressor is very much complex and three dimensional. The flow coming out from the impeller is skewed having jets and wakes. The vaneless space converts some part of the kinetic energy into pressure energy and stabilizes the flow so that it enters the vaned diffuser without shock. In the vaneless diffuser, the flow is assumed to be logarithmic spiral. Generally, the width of diffuser is kept the same as that

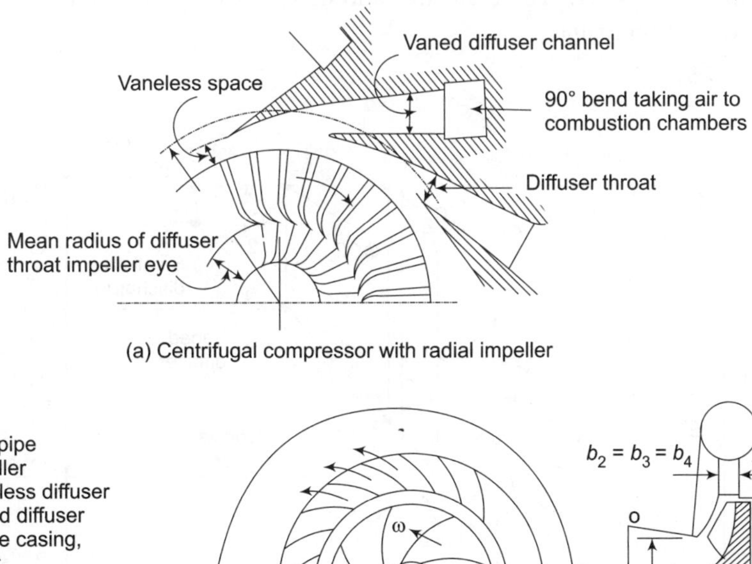

(a) Centrifugal compressor with radial impeller

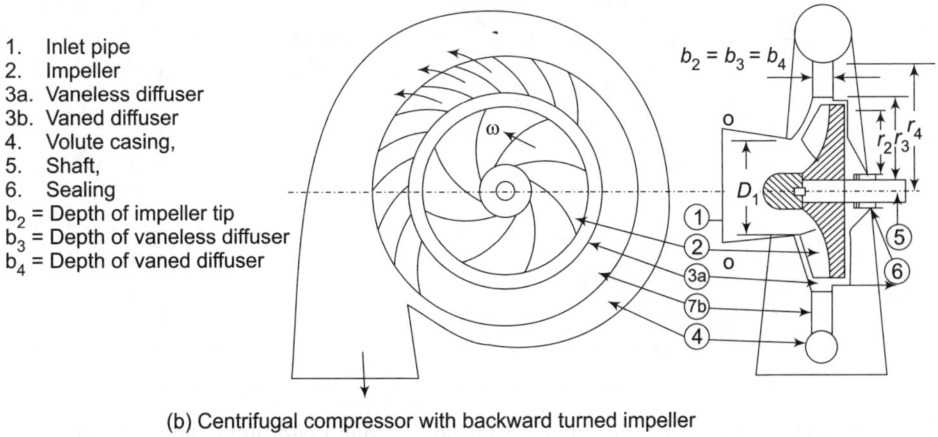

(b) Centrifugal compressor with backward turned impeller

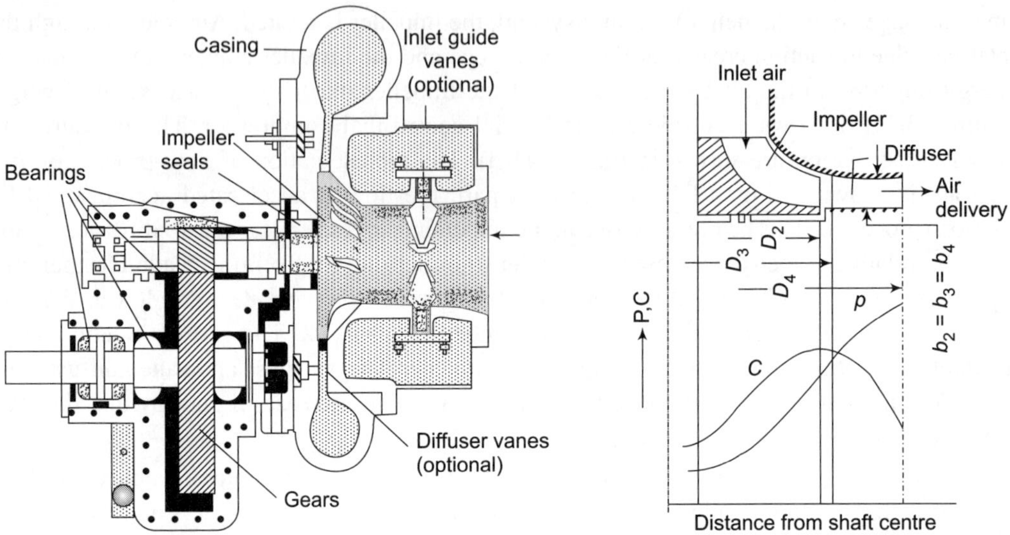

(c) An industrial centrifugal compressor (d) Pressure and velocity variation

Figure 13.41

of impeller tip (i.e., $b_2 = b_3$). The diameter ratio of vaneless space to impeller tip diameter, D_3/D_2 varies from 1.06 to 1.12.

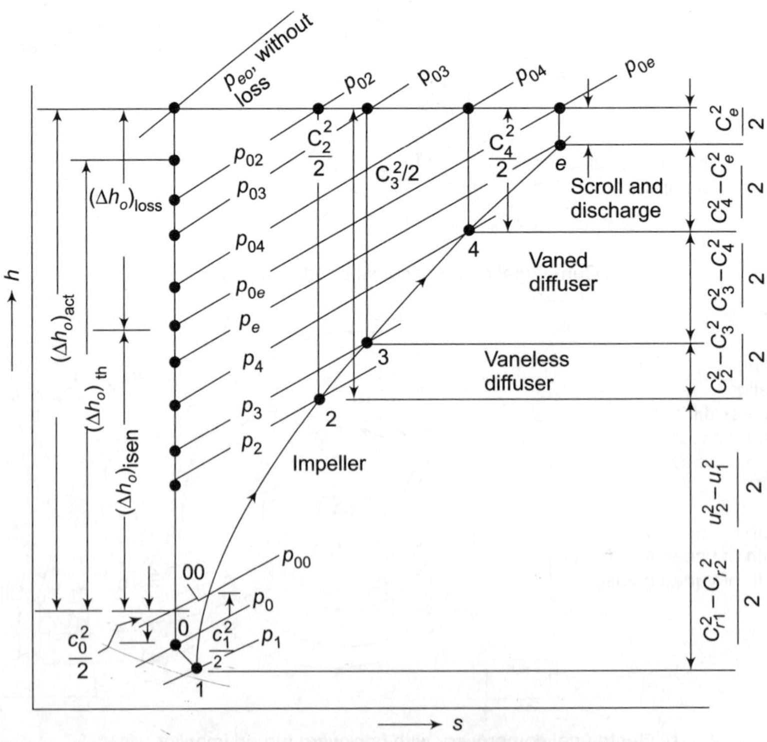

Figure 13.42 *h-s* representation on flow through centrifugal compressor

The vaned diffuser converts the major part of kinetic energy (about 40 to 50%) into pressure energy. As compared to the vaneless diffuser when used alone, the vaned diffuser is more compact and of small geometrical dimension for the same pressure recovery. The design efficiency of the vaned diffuser is more than that of the vaneless but its part load efficiency is poor as a result of shock and other aerodynamic losses due to presence of vanes. The vanes may be cambered, aerofoil, flat, circular arc, etc. The number of vanes may vary as that of impeller (Z_d = 10 to 30) but both numbers should not coincide to avoid resonance and interference effect. The diameter ratio, D_4/D_3 may vary from 1.25 to 1.6 and the maximum equivalent diffusion angle is around 10 degree.

From the vaned diffuser, the air is collected in the volute casing and comes out from the outlet pipe. The purpose of volute casing is mainly to collect the flow coming out from the impeller or diffuser and partly to convert the kinetic energy into pressure energy; this conversion is mainly achieved in the diffuser branch of the volute casing. *For gas turbine (instead of putting the volute casing), a 90° bend is provided to take air to the combustion chamber.*

One of the most critical regions of a high performance compressor is at the inlet. After a certain value of inlet relative velocity, some reduction in performance can be expected due to Mach number effects in compressor *For example, suppose that the entry velocity to the eye is at the inducer tip*. Now, consider two extreme situations as follows, in which the hub diameter and mass flow rate remain constant.

(a) In the case of larger eye tip diameter, the axial velocity is low but the blade speed is high.

(b) If the eye tip diameter is made small, the blade speed is also small but the axial velocity is high.

It is to be noted that both the situations give large relative velocity, and clearly, there must be some optimum eye tip diameter, where the relative velocity is the minimum.

A pressure ratio of 5 : 1 may be obtained at a reasonable isentropic efficiency in a single stage centrifugal compressor. By employing multi-staging, (up to 8 stages), the delivery pressure up to 400 atm may be obtained. The isentropic efficiency of centrifugal compressor is around 80%. *The impeller of centrifugal compressor is generally forged or die casted of low silicon aluminium alloy.*

Over and above of small size gas turbines, centrifugal compressors are mostly used in petrochemical, fertilizers, central air conditioning and other process industries.

13.25 VELOCITY DIAGRAM, EULER'S WORK AND EFFECT OF BLADE SHAPE

13.25.1 Velocity Diagrams

Figure 13.43 shows the velocity diagrams for impeller inlet, impeller outlet, diffuser inlet, and diffuser outlet.

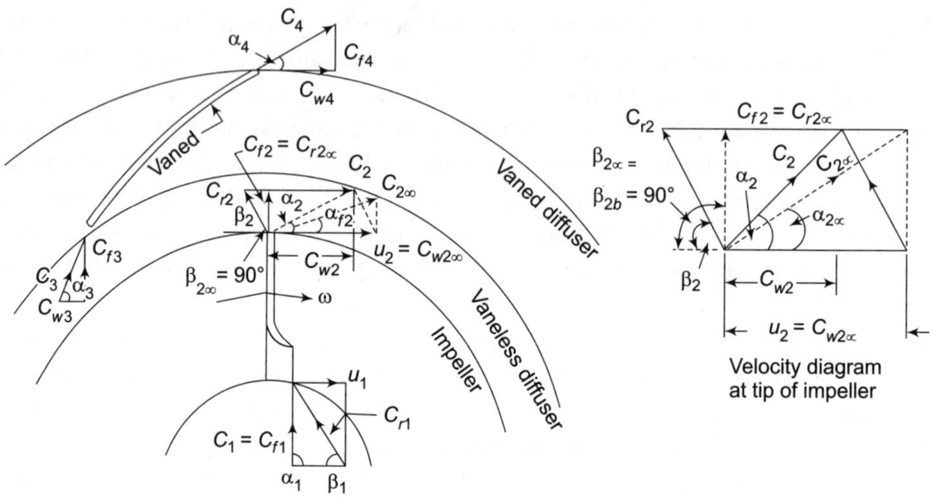

Figure 13.43 Velocity diagram of centrifugal

Let β_1, β_2 = Impeller inlet and outlet blade angles respectively.

α_1 and α_2 = Angles made by absolute velocity at inlet and outlet of impeller respectively.

α_3, α_4 = Vaned diffuser inlet and outlet angles respectively.

C_1, C_2 = Absolute velocity of air at impeller inlet and outlet respectively.

u_1 u_2 = Peripheral velocity at impeller inlet and outlet respectively.

C_{r1}, C_{r2} = Relative velocity of air with respect to blade at impeller inlet and outlet respectively.

C_{w1}, C_{w1} = Velocity of whirl or tangential components of absolute velocity at inlet and outlet respectively.

C_{f1}, C_{f2} = Flow or meridional or radial component of absolute velocity at inlet and outlet respectively.

$\dot{m}$ = Mass flow rate, kg/s.

Note that all velocities are in m/s.

13.25.2 Euler's Work

The theoretical torque which must be applied to the impeller will be equal to the rate of change of angular momentum experienced by the air. Consider a flow of 1 kg/s of air:

Theoretical torque = $(C_{w2} . r_2 - C_{w1} . r_1)$

If ω is the angular velocity in radian per second, the theoretical head developed or work input is given by

Theoretical work input, $w = (C_{w2} . r_2 - C_{w1} . r_1)\ \omega$

or

$$w = (u_2 . C_{w2} - u_1 . C_{w1}); \quad \text{watt/kg} \qquad ...(13.35)$$

Equation (13.41) is known as Euler's work.

Theoretical work input = $u_2\ C_{w2}$

Theoretical head developed = $u_2\ C_{w2}/g$ ...(13.36)

Hence, the maximum head or pressure rise per kg of air is given by u_2^2/g ...(13.36a)

Expressing C_{w1} and C_{w2} in terms of C_{r1}, $C_{r2}.C_1$ and C_1 from the relations of velocity triangles, the Euler'work is given by

$$w = \frac{C_{r1}^2 - C_{r2}^2}{2} + \frac{u_2^2 - u_1^2}{2} + \frac{C_2^2 - C_1^2}{2} \quad ...(13.37)$$

The first two terms of the above equation are responsible for the pressure rise in the impeller and the third term shows the amount of kinetic energy coming out from the impeller which has to be converted into pressure energy in the diffuser. The first term is called the *diffusion action* because the relative velocity decreases from inlet and outlet in the diverging blade passage of impeller. The second term is known as the *centrifugal action* as the fluid enters at a lower diameter and comes out at a higher diameter.

The driving torque applied to the rotor generates a circumferential pressure gradient across each passage, as each vane pushes the first layer of molecules, which pushes the next layer, etc. In a most simple form, we may say that the rotational flow and radial flow of impeller causes a *velocity gradient and pressure gradient across* an impeller passage (Fig. 13.44). The pressure gradient across adjacent passages causes a pressure difference across each vane, which exerts a tangential force in reaction to the applied force. The impeller puts work into the flow by accelerating each flowing element of the fluid in tangential direction until it reaches its final whirl velocity, C_{w2} at outlet. It is to be noted that radial blades are unable to apply any radial force caused by boundary layer drag. The tangential acceleration named after *Coriolis* is the inevitable result of rotating a flow vector about its source. The coriolis acceleration is the change in velocity divided by time and is equal to $2\omega C_r$. It acts in the direction if the flow is towards or against the direction of rotation if the flow is inwards.

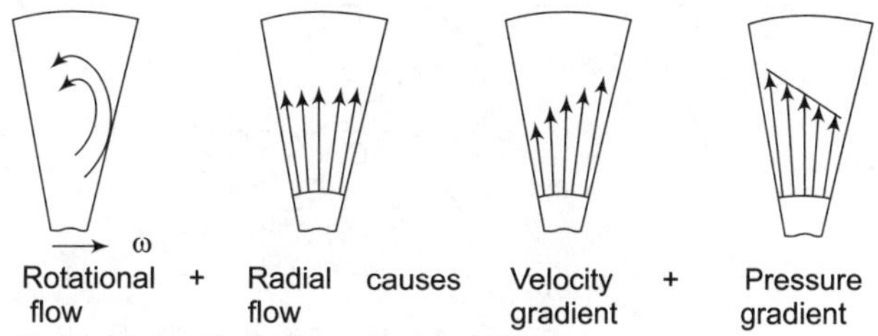

Figure 13.44 Velocity distribution in impeller

If b_1 and b_2 are widths of blades at inlet and outlet of impeller, then the mass flow rate neglecting the thickness of blades is given by

$$\dot{m} = \pi\ D_1\ b_1\ C_{f1}\ \rho_1 = \pi\ D_2\ b_2\ C_{f2}\ \rho_2 \quad ...(13.38)$$

If b_3 and b_4 are the widths of blades or passage at diffuser inlet and outlet, then

$$\dot{m} = \pi\ D_3\ b_3\ C_{f3}\ \rho_3 = \pi\ D_4\ b_4\ C_{f4}\ \rho_4 \quad ...(13.39)$$

13.25.3 Effect of Blade Shape

Figure 13.45 shows three shapes of blades, namely, backward radial and forward on the same impeller. The resulting velocity triangles are also depicted. The minimum work input needed by the compressor in the case which results in the minimum value of C_{w2}. It is the backward turned blades in which the C_{w2} is the minimum, thus it will be more efficient. Generally, β_2 = 40° to 50° for backward turned blade and β_{10pt} = 30°. Next number is that of radial and last is of the forward. However, radial blades are used for heavy duty work to withstand centrifugal stresses. [For more details, see the author's books: *Applied Thermodynamics, and Steam & Gas Turbines and Power Plant Engineering*].

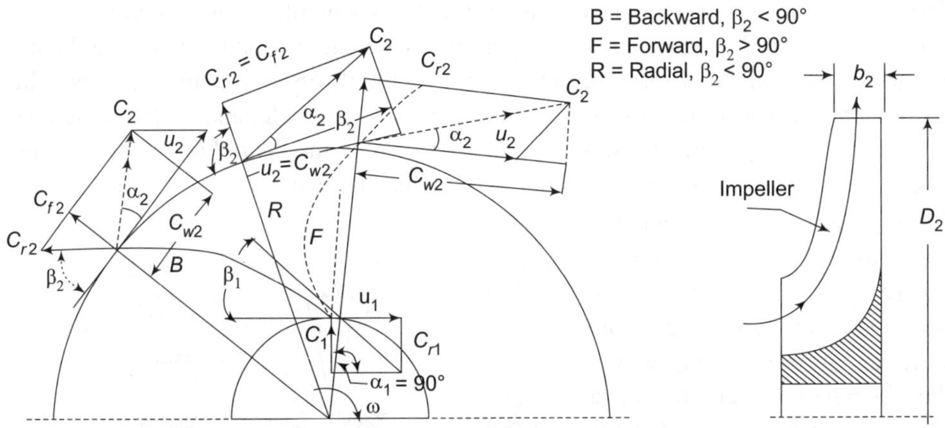

Figure 13.45 Effect of blade shape on velocity triangle

13.26 AXIAL FLOW COMPRESSOR*

As the name implies, the flow in the axial flow compressor is in the axial direction of the machine. Invariably, gas turbine power plant employs an axial flow compressor where provisions are made to bled compressed air at appropriate points for cooling of gas turbine bladings which are suspected to high pressure and temperature. It consists of a *rotor and a stator* as shown in Fig. 13.46. The rotor (i.e., moving component) consists of *rows of moving blades* and the stator (i.e., stationary component) consists of a rows of stationary blades. A row of moving blades with a succeeding set of stationary blades is called a ***stage* of an axial flow compressor**. It is usual practice to provide a row of so called inlet guide vanes (IGV) which are additional stator rows of blades to the first stage in the compressor. These serve to direct the axial approaching flow correctly the first row of rotating blades. Some part of the kinetic energy imparted to the air by the rotor is converted into pressure energy in the rotor due to diffusion action (i.e., decrease in relative velocity and the rest is converted into pressure in the stator due to diffusion action (i.e., decrease in absolute velocity). The stator blades also redirect the air into an angle suitable for entry to the succeeding set of moving blades. The rotor as well as the stator blade channels are of *diverging type* (i.e., area increases from inlet to outlet).

* For more details, see Author's book *"Steam & Gas Turbines, Turbomachines and Power Plant Engineering"*.

Generally, the maximum *pressure ratio achieved in a stage of axial flow compressor is about 1.12 to 1.3*. hence to obtain a pressure ratio of 15, attainable by an axial flow compressor, 18 to 20 stages are required and pressure ratio 30 may be obtained in 22 stages. The stage temperature rise $(\Delta T_0')_{\text{stage}}$ may vary widely in different compressor design, depending upon the application and the importance from *10K to 30K* for subsonic stages and may be 45K or higher in high performance transonic stages. In the present technology, **a multistage axial flow compressor is capable of achieving compressor pressure ratio upto 35 at 91% polytropic efficiency**.

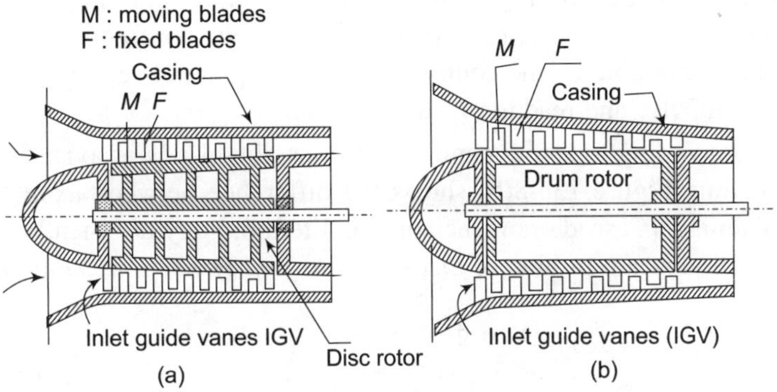

Figure 13.46 Axial flow compressor

Two forms of rotor have been used,, namely, the *drum and disc types* (Fig. 13.46). The disc type is used where consideration of low weight is more important than cost as in aircraft applications. The drum type is more suitable for static industrial applications. In some applications, combinations of both types have been used as shown in Fig. 13.47, however, it is not desirable. This is shown just for design features.

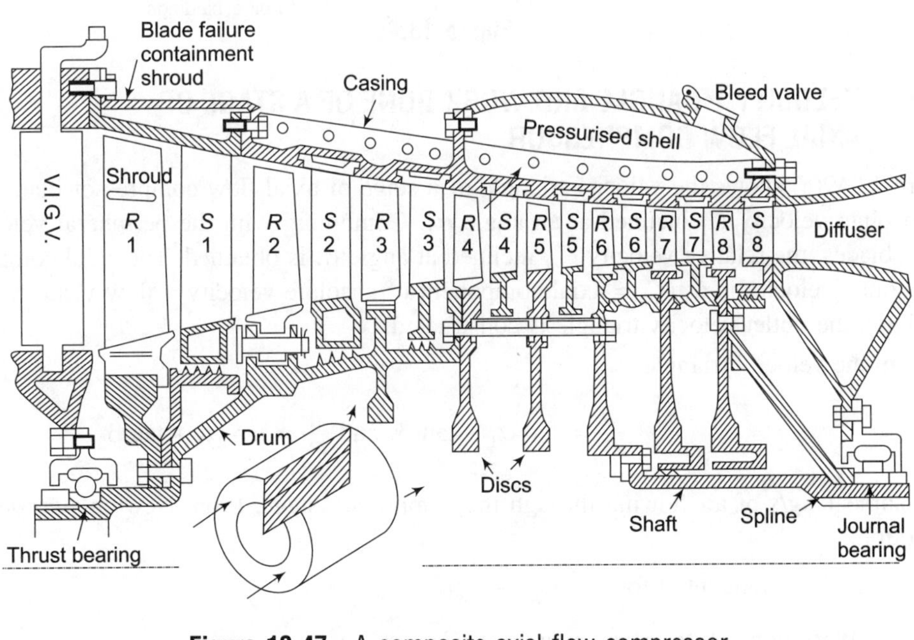

Figure 13.47 A composite axial flow compressor

The other important construction aspect is the *contraction of the flow annulus* from the low to the high pressure end of compressor. This practice is followed to maintain the axial velocity at a reasonably constant level throughout the length of the compressor despite the increasing density of the working fluid. *It is worth to note that most of the axial flow compressor are designed on the basis of constant axial velocity throughout the stage.* Contraction of annulus is also possible by the contracting the casing and diverging the rotor. Figure 13.48(a) shows the impulse and reaction bladings.

According to aerodynamic conditions, the blades of axial flow compressor are of aerofoil section. Figure 13.48(b) shows the difference between axial flow compressor and turbine bladings. [Cascade nomenclature and terminology are mentioned in the chapter on impulse reaction steam turbine of *Steam & Gas and Power Plant Engineering*].

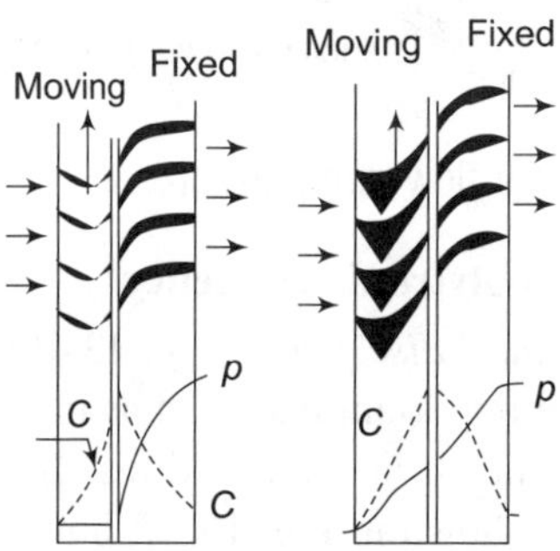

(a) Impulse and reaction blading

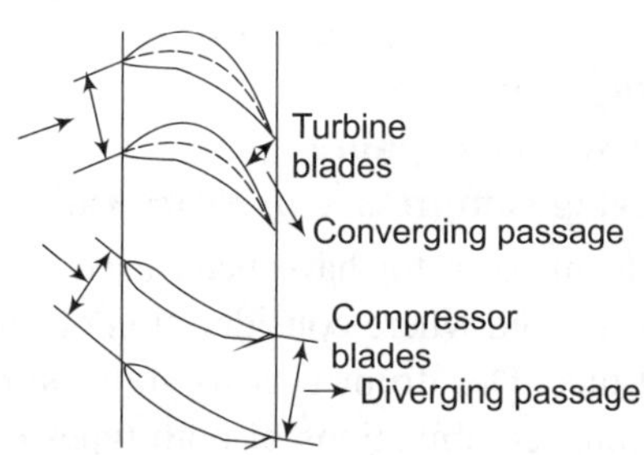

(b) Difference between axial compressor and turbine bladings

Figure 13.48

13.27 VELOCITY TRIANGLE AND WORK DONE OF A STAGE OF AXIAL FLOW COMPRESSOR

Figure 13.49(a) shows the velocity triangle of a stage of axial flow compressor blades with an absolute velocity C_1 inclined at an angle α_1. Combining with the peripheral velocity u of the blades, the relative velocity C_{r1} inclined at angle β_1 is obtained. The axial component of absolute velocity is C_f (C_a = axial component of absolute velocity = flow velocity = C_f). Similarly, the outlet velocity triangle is constructed.

From the velocity triangle,

$$\frac{u}{C_a} = \tan \alpha_1 + \tan \beta_1 \text{ and } \frac{u}{C_a} = \tan \alpha_2 + \tan \beta_2$$

Assume 1 kg/s of air flowing through the compressor stage. From Newton's second law of motion.

$$\text{Tangential force} = C_{w1} - C_{w1}$$

$$\text{Work done by the stage} = w_{st} = u\,(C_{w2} - C_{w1})$$

or $$w_{st} = u\, C_a (\tan \alpha_2 - \tan \alpha_1) \quad ...(13.40)$$

$$= uC_a (\tan \beta_1 - \tan \beta_2) \quad ...(13.41)$$

By the use of velocity triangle and cosine theorem, the above equation may be modified as

$$w_{st} = \frac{C_{r1}^2 - C_{r2}^2}{2} + \frac{C_2^2 - C_1^2}{2} \quad ...(13.42)$$

The first term on the right side of the above equation introduces the part of the work supplied by a rotating cascade which is inverted into pressure due to diffusion action in rotating cascade itself. The second term represents the increment of kinetic energy in rotating cascade that has to be converted with pressure energy in stationary cascade. *We find here the absence of centrifugal action and that is why the pressure ratio per stage in axial flow compressor is less than that of centrifugal.* Figure 13.49(b) shows the process in axial flow compressor represented on h–s chart.

From steady flow energy equation,

$$w_{st} = (h_{03}' - h_{01}) = c_p (T_{03}' - T_{01}) = c_p (T_{02}' - T_{01}) = c_p (\Delta T_0')_{\text{stage}}$$

$$= uC_a (\tan \beta_1 - \tan \beta_2)$$

or $$(\Delta T_0')_{\text{stage}} = \frac{u \,.\, C_a}{c_p} (\tan \beta_1 - \tan \beta_2) \quad ...(13.43)$$

As a result of adverse pressure gradient in compressors, the boundary layer along the annulus walls thicken as the flow progresses. This will reduce the area available for flow

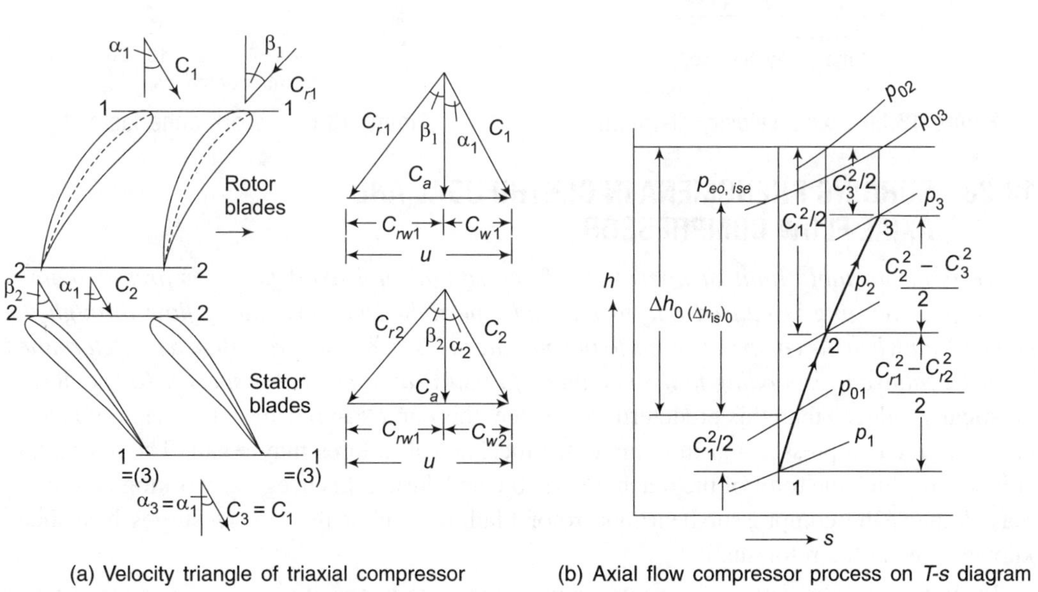

(a) Velocity triangle of triaxial compressor (b) Axial flow compressor process on *T-s* diagram

Figure 13.49

below the geometric area of the annulus. As a result, there will be a considerable effect on the axial velocity distribution through the compressor and must be considered in the design process. Since calculation of boundary layer growth is extremely difficult due to complex flow phenomena, hence empirical correlation factors based on experimental data are used.

In actual fact, the stage temperature rise $(\Delta T_0')_{stage}$ will be less than this owing to three dimensional effects. This is based on the fact that radial distributions of axial velocity is not constant across the annulus, but becomes increasingly peaky, as the flow proceeds, settling down to a fixed profile after fourth stage. Figure 13.50 shows the *axial velocity is affected due to axial velocity distribution in various stages.* An increased axial velocity will result in a decrease of W_{stage}. To take case of this aspect, $(\Delta T_0')_{stage}$ should be multiplied by a work done factor λ, its variation is shown in Fig. 13.51.

Thus, actual temperature rise

$$(\Delta T')_{stage} = (\Delta T_0')_{stage} = (T_{02}' - T_{01}) = \frac{\lambda}{c_p} u\,(\tan\beta_1 - \tan\beta_2) \quad ...(13.44)$$

and $$(r_p)_{st} = \frac{p_{03}}{p_{01}} = \left[1 + \eta_{st}\frac{(\Delta T_0')_{stage}}{T_{01}}\right]^{\gamma/(\gamma-1)} \quad ...(13.45)$$

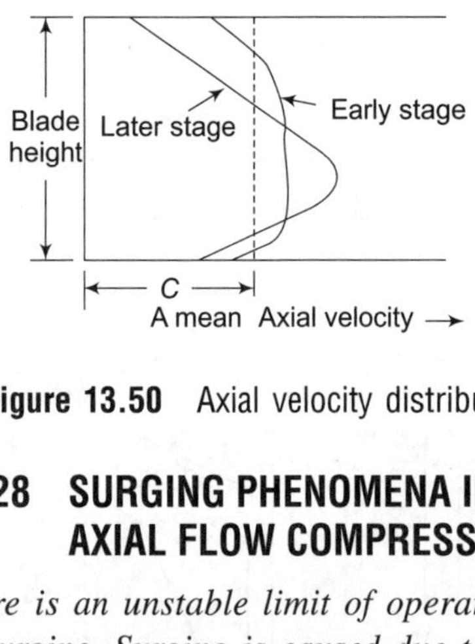

Figure 13.50 Axial velocity distribution

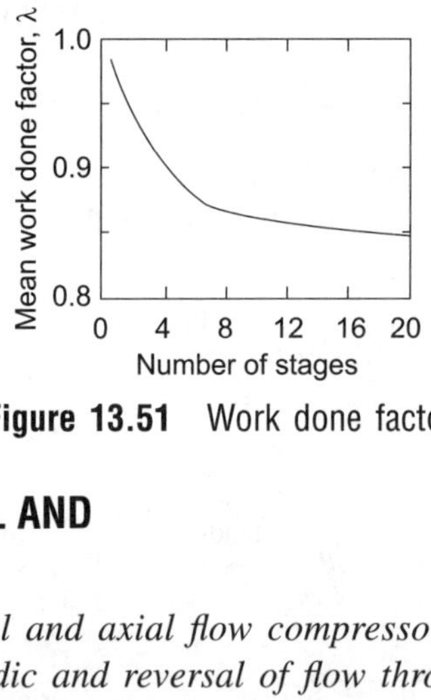

Figure 13.51 Work done factor

13.28 SURGING PHENOMENA IN CENTRIFUGAL AND AXIAL FLOW COMPRESSOR

There is an unstable limit of operation of centrifugal and axial flow compressor known as surging. Surging is caused due to unsteady, periodic and reversal of flow through the compressor when the compressor has to the operate at less mass flow rate than a predetermined value (a value corresponding to the maximum pressure, at a particular speed). As the flow is drastically reduced than this predetermined value, this surge can reach such a magnitude as to endanger the compressor and in many cases mechanical failures may result. The alternating stresses to which the rotor of the machine is subjected during this irregular working condition, may damage the compressor bearings, rotor bladings and seals. Severe surges have been known to bend the rotor shaft.

To understand the cause of surging, consider the head-capacity curve of a centrifugal or axial flow compressor as shown in Fig. 13.52. This curve consists of two portions, portion

AB having a positive slope and a portion *BC* having a negative slope. *A* and *C* are the two extreme points on the curve. Point *A* corresponds to the flow condition when the discharge valve is fully closed and point *C* corresponds to the flow condition when the discharge valve is fully open. In the case of fully closed discharge valve, the flow rate is zero and the static pressure developed is termed shut-off head. In the fully open discharge valve, the mass flow rate is maximum but the pressure developed is zero. This means that the compression efficiency is zero.

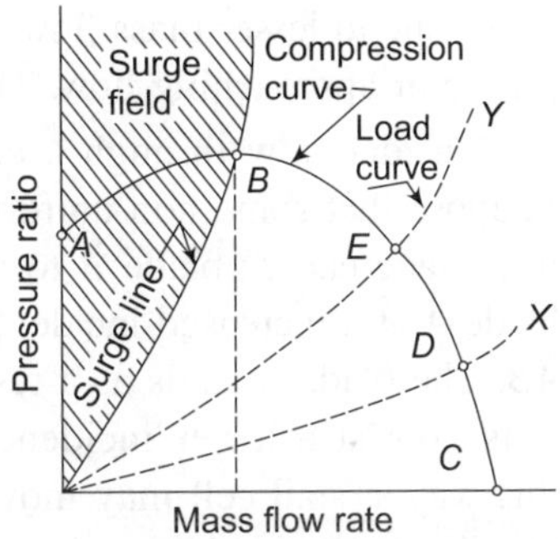

Figure 13.52 Surging phenomena

Now, consider the case when the discharge valve is opened slightly. The delivery of fluid will commence and the static pressure will rise due to contribution of diffuser towards pressure rise. As the discharge valve is further opened, the pressure goes on increasing till the maximum pressure rise represented by point *B* is attained. *This is the point of maximum efficiency for a given speed, inlet pressure and temperature,* with the further opening of the discharge valve beyond point *B*, the mass flow rate increases, but the pressure ratio approaches unity, the mass flow rate will be the maximum but the compression efficiency would be zero.

The opening and closing of the discharge valve can be treated on the external load on the compressor. This means that the compressor has to operate in conjunction with external load on the compressor. Thus, the point of intersection of the compressor curve will give the operation point. For example, *D* shows such an operating point. Now, suppose that the external load is increased due to the partial closing of the discharge valve. This will increase the back pressure in the discharge line. As a result, the point of operation will shift to point *E* which is the point of intersection of new load line *OY* and the compression curve. The new operating point *E* is possible and stable because the compressor will produce more head which will take care of increased back pressure in the discharge line. With the further increase in external load due to closing of the valve, the operating point shifts say between the region *AB*. Under this conditions of operation, the mass flow rate will be less than the predetermined value, a value corresponding to point *B*. The compressor produces less head and consequently the delivery will not be possible since high pressure would be existing in the discharge line. This will result momentarily in the reversal of flow. After a lapse of time, the fluid from the delivery pipe would leave and the back pressure would fall. The compressor would again start delivery fluid and the cycle would be repeated with instability occurring again. Thus, when the flow through the compressor is less than a ***predetermined value***, a surge or pulsation begins and the air surges fro and to through the whole compressor instead of giving a steady stream in one direction. This gives rises to ***surging***.

13.29 STALLING

Stalling of stage of an axial flow compressor is defined as the aerodynamic stall or the break-away of the flow from the suction side of the blade aerofoil. This break-away of the flow may occur due to lesser mass flow rate than designed value or due to non-uniformity in the blade profile or approaching flow. Thus, stalling is an ahead phenomenon of surging.

Consider flow through a cascade as shown in Fig. 13.53. Suppose that some non-uniformity in the approaching flow or in a blade causes blade *B* to stall. The air now flows on to a blade *A* at a increased incident due to the blockage of channel *AB*. The blade *A* stalls as a result but the flow on to the blade *C* is now at a lower incidence and blade *C* may unstall. In this way, a stall cell may move along the cascade opposite in the direction of blade motion or the direction of the lift of the blades. *In other words, the phenomenon of reduction in lift force at higher angles of incidence is known as stalling.* If the natural frequency of vibration of the blades coincides with that at which the stall cells pass a blade, the resonance and possible mechanical failure of the blades may occur.

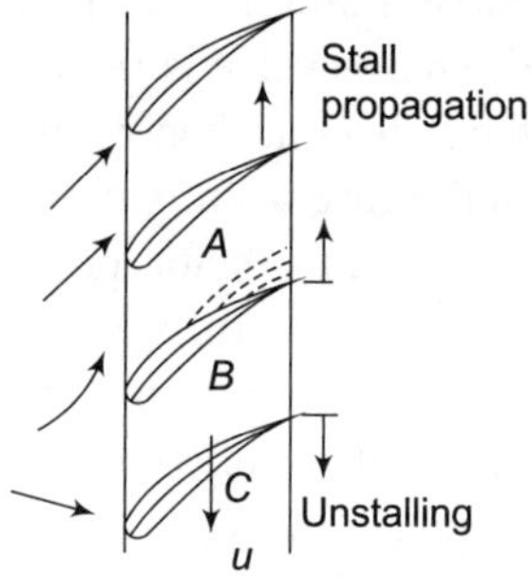

Figure 13.53 Stalling phenomena

13.30 CHOKING FLOW

Theoretically, the mass flow rate becomes the maximum when the pressure ratio is unity, i.e., there is no compression. This generally occurs when the Mach number corresponding to the relative velocity at inlet becomes sonic. Under this condition, the mass flow rate possible from a centrifugal or axial flow compressor becomes the maximum which is known as choking flow (Fig. 13.54). Choking means fixed mass flow rate regardless of pressure ratio, i.e., the characteristic becomes vertical (Fig. 13.54).

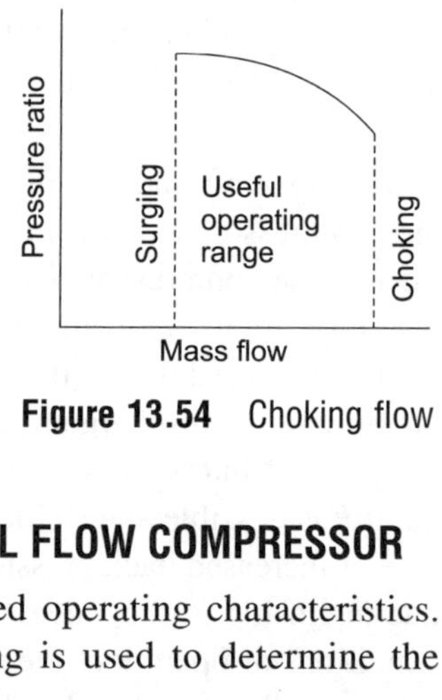

Figure 13.54 Choking flow

13.31 CHARACTERISTICS OF CENTRIFUGAL AND AXIAL FLOW COMPRESSOR

Both centrifugal and axial flow compressors have well defined operating characteristics. They are normally plotted in dimensionless form. Rig testing is used to determine the characteristics.

Figure 13.55 shows a typical constant speed characteristic for one stage of a compressor in terms of pressure rise and flow coefficient. The effect of positive and negative incidence on the stage pressure rise is also depicted.

Figure 13.56 shows the characteristics of a centrifugal compressor at various speeds in dimensionless from. The pressure curve is flatter in the case of centrifugal compressor. The maximum flow limit is set by choking at the eye, the critical flow limit causes the constant speed curves to turn down vertically. The minimum flow limit of the impeller is set by negative incidence stall of the inducer vanes. A typical working line may be located with 10 per cent surge margin, the mass flow being 10 per cent greater than at surge. The constant efficiency curves are also superimposed on the curve.

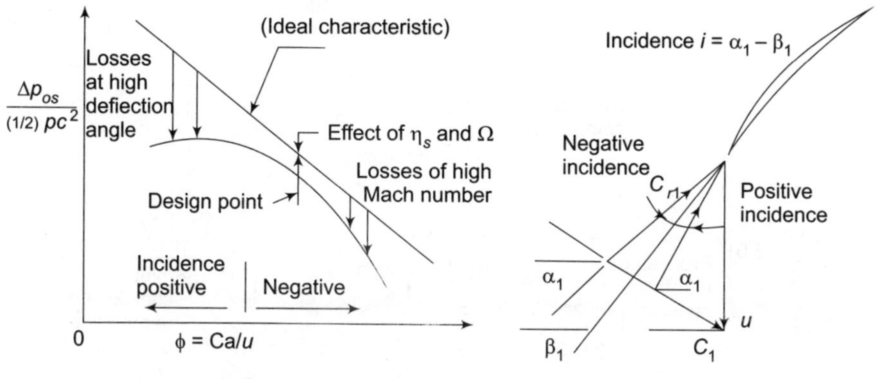

Figure 13.55 Characteristics of a compressor

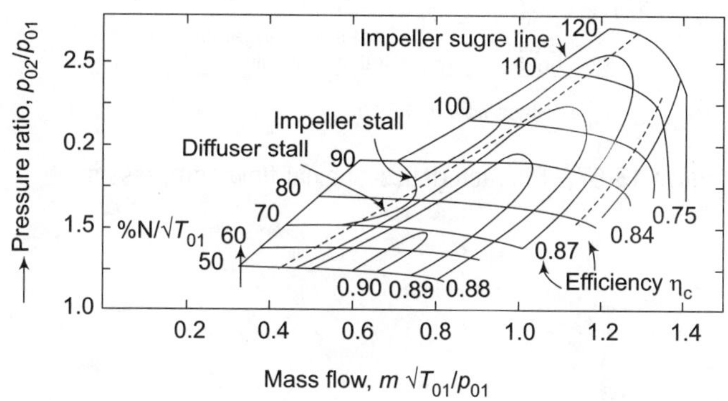

Figure 13.56 Characteristics of a centrifugal compressor

Figure 13.57 shows the characteristics of axial flow compressors. At design point A, $(N/\sqrt{T_{01}} = 100\%)$ the flow conditions are correct for all stages through the compressor with incidence angles being ideal, the efficiency at its maximum and usually constant axial velocity. This corresponds to the majority of engine operation. At the same speed, pressure ratio could rise as far as B or fall to C. At B, the axial velocity is so low that it causes surge. But at C, the axial velocity increases progressively until the last stage encounters negative shock stall—which may cause fatigue failure of the blades. Therefore, it is necessary to avoid both extreme conditions, B and C. At speed 110% and 120%, the characteristics become almost vertical below the working line and the inlet is chocked. At lower speed the characteristics is less steeper.

Figure 13.58 shows power and efficiency versus rate of flow for a particular speed. At a certain flow rate, the efficiency becomes the maximum. The power consumed increases with the increase of rate of flow. This is valid for both centrifugal and axial flow compressor.

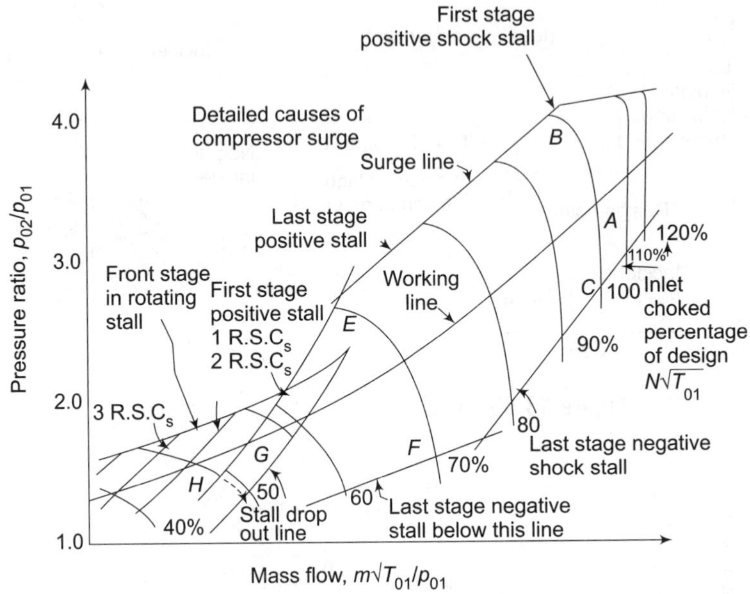

Figure 13.57 Characteristics of axial flow compressor

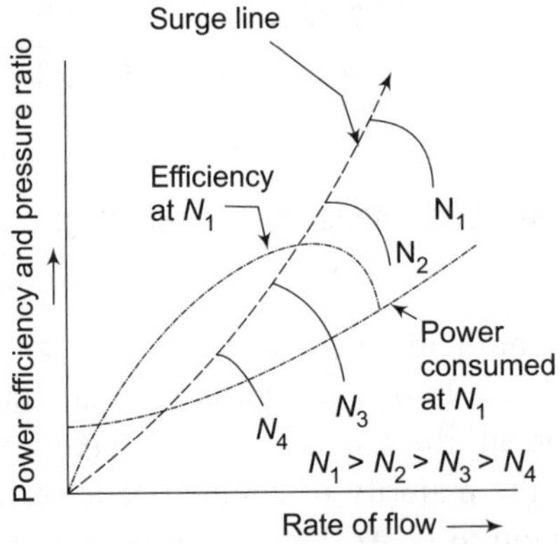

Figure 13.58 Power and efficiency vs rate of flow of axial flow compressor

13.32 COMPARISON BETWEEN CENTRIFUGAL AND AXIAL FLOW COMPRESSORS

The following main features give the comparison between centrifugal and axial flow compressor. The comparison helps us to form some criterion for the choice between these two compressors.

Centrifugal	*Axial*
(i) The flow is raidal	(i) The flow is axial i.e., parallel to the direction of the axis of the machine.
(ii) The pressure ratio per stage is high about 5:1. Thus, the unit is compact. In supersonic compressors, the pressure ratio per stage is about 10 but at the cost of efficiency. Operation is not so difficult and risky.	(ii) The pressure ratio per stage is low, about 1.2:1 due to absence of centrifugal action. To achieve the pressure ratio equal to that of per stage centrifugal, 10–12 stages are required. Thus, unit is less compact and less rugged. The pressure ratio per stage in supersonic compressor is about 10 but the efficiency drops rapidly. Compression is achieved by shock wave and the operation is risky.
(iii) Isentropic efficiency is about 82%.	(iii) The isentropic efficiency is about 88%.
(iv) Centrifugal compressors have a wide range of operation between surging and chocking limit. The head capacity curve is flat. The part load performance is better.	(iv) Axial compressor have a narrow range of operation between surging and chocking limit. The part load performance is poor.
(v) Centrifugal compressors have a larger frontal area and thus produces more drag force for the same mass flow rate and pressure ratio if used in aviation.	(v) Axial flow compressors have a small frontal area for the same mass flow rate and pressure ratio than that of centrifugal. This makes the axial flow compressor more suitable for jet engines. It is invariably used in aviation and power gas turbines.
(vi) When working with the contaminate fluids, the accumulation of deposits on the surface of flow passage do not adversely affect the performance.	(vi) The accumulation of deposits on the surface of flow passage effects adversely the performance of axial flow compressors.
(vii) It needs low starting torque.	(vii) It needs high starting torque.
(viii) Its construction is simple, rigid and relatively cheap. At high altitude, it is less sensitive to icing trouble	(viii) Its construction is complex and costly it is sensitive to icing troubles at high altitude.
(xi) Multistaging is slightly difficult and upto 400 bars delivery pressure is possible.	(xi) It is most suitable for multistaging and upto 35 bars delivery pressure in a single casing is possible.
(x) It is used in blowing engines in steel mills, low pressure refrigeration, big central air conditioning plants, fertiliser industry, supercharging gas pumping in long distance pipeline, petrochemical industries. Previously, it was used in jet engines, small aircraft and gas turbines.	(x) Due to higher efficiency and smaller frontal area axial flow compressors are mostly used in jet engines. In power plant gas turbines, axial compressor are invariably used.

13.33 COMBUSTION CHAMBERS

The combustion chamber has the difficult tasks of burning large quantities of fuel, supplied through the fuel burners with extensive volume of air supplied by the compressor, and releasing the heat in such a manner that the air is expanded and accelerated to give a smooth stream of uniformly heated gas at all conditions required by the turbine. Generally, the air-fuel ratio in an open gas turbines varies from 50 : 1 to 200 : 1 to get efficient combustion and to keep the turbine inlet temperature down to permissible limits.

Aircraft and ships must carry the fuel required for their mission which has resulted in the universal use of liquid fuels. However, hydrogen has been experimented but the commercial use seems to be unlikely. Aircraft gas turbines operate over a wide range of inlet pressure and temperature within the flight envelop of Mach number and altitude. A typical subsonic aircraft will operate at a cruise altitude of 11 km, where the ambient pressure and temperature for ISA conditions are 0.227 bar and 216.9 K compared with the sea level value of 1.013 bars and 288 K. Therefore, the combustor has to operate with a greatly reduced air density and mass flow at altitude, while using approximately, the same fuel-air ratio as at sea level to maintain an appropriate value of turbine inlet temperature. Atmospheric condition changes quite rapidly during climb and descent and combustor chamber has to meet all operational requirements. Industrial gas turbines have a wider scope of fuel. Natural gas is the most preferred fuel.

13.34 REQUIREMENTS OF COMBUSTION CHAMBER

The requirements of a combustion chamber are—(a) low pressure loss; (b) high combustion efficiency; (c) good flame stability; (d) low carbon deposit in combustion chamber, turbine and regenerator; (e) low weight and frontal area; (f) reliability and serviceability with reasonable life and (g) through mixing of cold air with the hot products of combustion. Combustion efficiency is defined as the ratio of actual heat released to the maximum theoretical possible.

13.35 TYPES OF COMBUSTION CHAMBERS

Types. In general types of combustion chamber used in open cycle are (a) tubular, (b) annular, (c) turbo-annular or cannular (d) industrial, (e) reversed and (f) cyclone. These are shown in Fig. 13.59.

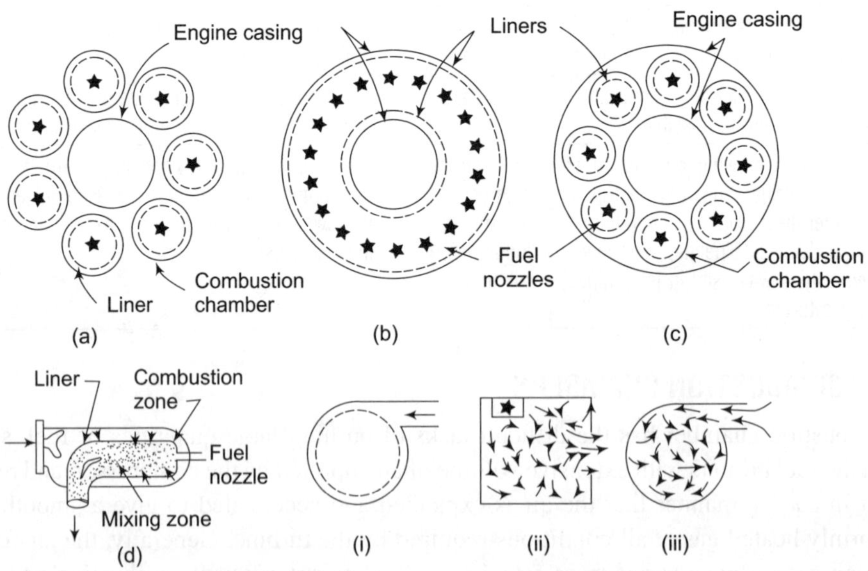

Figure 13.59 Forms of combustion chamber

13.36 THE COMBUSTION PROCESS

The combustion of a liquid fuel in the combustion chamber involves the following aspects:

(a) Breaking down of heavy hydrocarbons into lighter fractions.

(b) Intimate mixing of molecules of these hydro-carbons with oxygen molecules.

(c) Vaporization of droplets.

(d) Chemical reaction themselves.

Air from the compressor enters the combustion chamber at a velocity upto 170 m/s but because at this velocity, the air speed is far high for combustion; the first thing that the chamber must do is to diffuse it, i.e., decelerate it and raise its static pressure. In other words, a region of low axial velocity has to be created in the chamber, so that the flame will remain alight throughout the range of the engine operating condition. This is achieved with the help of a diffuser as shown in Fig. 13.60.

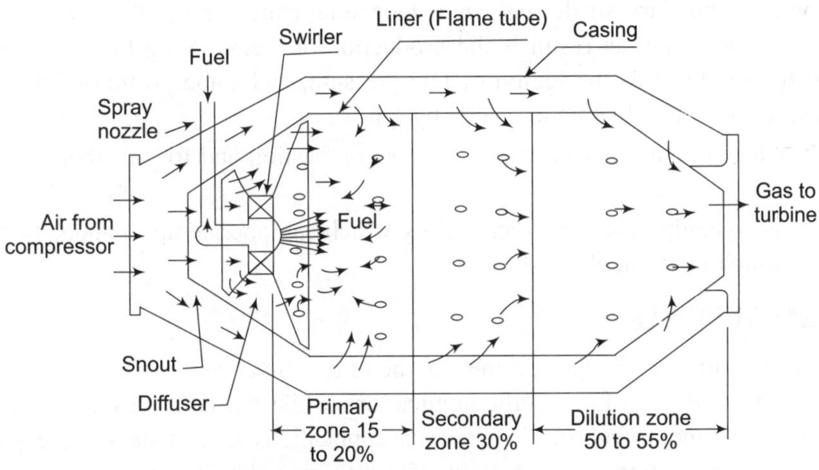

Figure 13.60 Construction of combustion chamber

Since the overall air-fuel ratio in the combustion chamber is in the region of 100 : 1 while the stoichiometric ratio is approximately 15:1, the first essential requirement is that the air should be introduced in stages. In other words, the fuel must be burnt with only a part of the air entering the chamber. This is achieved by means of a flame tube (combustion liner) that has various devices for metering the air flow distribution along the chamber.

There are three stages of introduction of air such as *–primary, secondary and tertiary* (dilution) as shown in Fig. 13.60. Around 15 to 20% of the air is introduced around the primary one to provide the necessary high temperature for rapid combustion. The air-fuel ratio in this zone is around 15:1. Some 30% of the total air is then introduced through the holes in the flame tube in the secondary zone to complete the combustion. In order to achieve high combustion efficiency, this air must be injected carefully at the right points in the process, to avoid chilling the flame locally and drastically reducing the reaction rate in the neighbourhood. Finally, in the dilution zone, the remaining air is mixed with the products of combustion to cool them down from about 2000°C to the temperature required at inlet to the turbine, around 1400°C. Sufficient turbulence must be promoted so that the hot and cold streams are thoroughly mixed to give the desired outlet temperature distribution.

It is essential to have a self-piloting flame (flame stabilization) in the air steam of combustion chamber liner. The zonal method of introducing the air does not provide flame stabilization. Therefore, the third essential feature is to produce a recirculating flow pattern (i.e., torodial vortex similar to a smoke ring) which directs some of the burning mixture in the primary zone back on to the incoming fuel and air. There are various method of flame stabilization.

Figure 13.60 shows one method of flame stabilization in which the fuel is injected in the same direction as the air stream and the primary air is introduced through twisted radial vanes, known as swirl vanes. This swirl vanes produces a vortex motion of the air which induces a region of low pressure along the axis of chamber. This vortex motion is sometimes enhanced by injecting the secondary air through short tangential chutes in the flame tubes, instead of through plain holes. The net result is the production of recirculating flow, i.e., the burning gases tend to flow towards the region of low pressure, and some portion of them is swept around towards the jet of fuel as indicated by arrows.

A spark initiated by a spark plug initiates the combustion and for this there is an *ignition system.*

The liner is typically made of a nickel alloy which can operate up to about 1400 K, or of cobalt alloy which may run above 1400 K.

13.37 GAS TURBINES

Similar to steam turbines, the gas turbine may be of the *radial or axial flow type*. The former is used only in small sizes for specific applications while the latter has a very wide utility ranging from jet engines to industrial and marine applications. Like steam turbine, gas turbines may be impulse or reaction types. A stage of gas turbine (like steam turbine) consists of a set of nozzles (called stator) and a set of moving blades (called rotor). Generally, nozzles are convergent type.

The basic requirements of gas turbines when used in aviation are different from those used in industrial applications. In aviation applications, a small turbine of low weight is essential. This is achieved by raising the allowable stress at the cost of working life and by designing with a high axial velocity through the turbine. Generally, one or at the most two stages are required because the rest of expansion has to take place in nozzle to produce jet thrust. Gas turbines when used in industrial or marine applications need a large number of stages up to five. This is required to reduce the carry-over losses and to reduce the load on the blades aiming to achieve a long working life of blades.

The life of industrial gas turbine between overhauls is 20,000 to 30,000 hours compared to 300 to 7000 hours for aero-engines. It is to be noted that all gas turbine blades are cooled either by air bled from compressor or by other means.

The disc stresses are very high which requires heavy hub to help the rim with stand the centrifugal blade loading and to achieve equal material stress at all radii. The rotor is located axially relative to the casing by a thrust bearing and its position must be set very accurately to ensure the minimum flow leakage between the blade tips and the seal segments. Since there is a considerable pressure drop across each row of blading to reduce number of stages, so it is necessary to seal between static and rotating assemblies between each blade row, and this is done by labyrinth seals which permit axial movement and adjustment. These seals permit small leakage of air flow which is used to cool the discs and shafts.

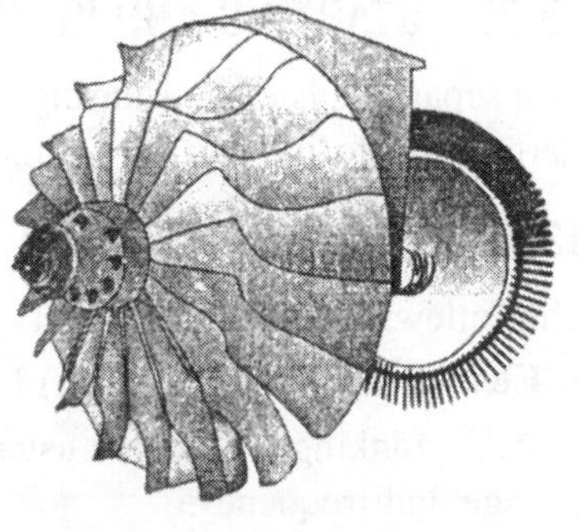

Figure 13.61 Gas turbine coupled to centrifugal compressor

Note: Velocity diagrams, analysis of stage efficiency vs u/C_1, degree of reaction are similar to steam turbine.

An axial turbine coupled with a centrifugal compressor is shown in Fig. 13.61. A cut-away view of Siemens Westinghouse 501 G gas turbine [Gas Turbine World 2002] is shown in Fig. 13.62.

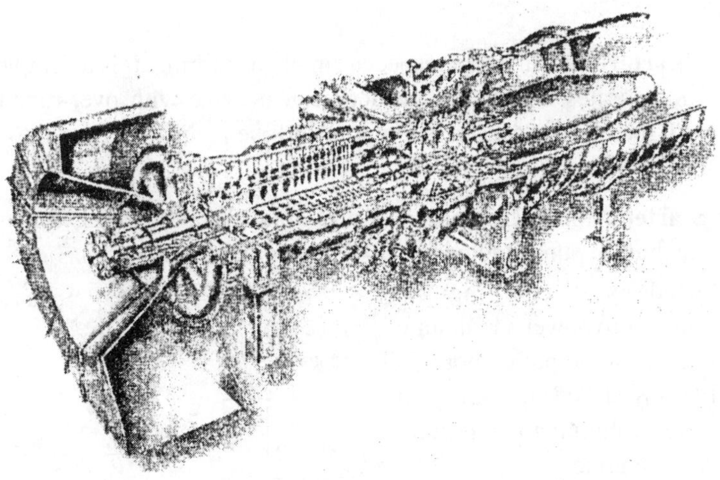

Figure 13.62 Cut-away view of Siemens Westinghouse 501G gas turbine [Gas Turbine World, 2002]

13.38 GAS TURBINE AUXILIARY SYSTEMS, OPERATION AND MAINTENANCE

Auxiliary systems are the backbone of the gas turbine plant. Without auxiliary system, the very existence of the gas turbine is impossible. It permits the safe working of the gas turbine. The auxiliary system includes *starting, ignition, lubrication and fuel system and control.*

13.39 STARTING AND IGNITION SYSTEMS

Two separate systems—starting and ignition—are required to ensure a gas turbine engine will start satisfactorily. During engine starting, the two systems must operate simultaneously.

13.40 TYPES OF STARTER

The following are the various types of gas turbine starter.

1. Electrical. (i) A.C. and (ii) D.C.

A.C. cranking motors are usually 3-phase induction types rated to operate on the available voltage and frequency.

D.C. starter motor takes the source of electrical energy from a bank of batteries of sufficient capacity to handle the starting load. Engine or disengaging clutch is used.

2. Pneumatic or Air Starter. Air starter is used mostly as it is light, simple and economical to operate. As air starter motor has a turbine rotor that transmits power through a reduction gear and clutch to the starter output shaft that is connected to the engine. The starter turbine is rotated by air pressure taken from an external ground supply, from an auxiliary power unit carried in the aircraft or from an engine that is running. An electrical control unit controls the air supply to the starter by opening the pressure reducing valve. When an engine starter is selected and is automatically closed at a predetermined starter speed, the clutch also automatically disengages as the engine accelerates. It is most suitable for natural gas pipeline gas turbine drive.

3. Combustion Starter. It is in every respect a small gas turbine. It is a completely integrated system which incorporates a planetary reduction gear drive with over-running clutch. The unit is started with the electric starter. The starter turbine is directly geared to the gas turbine shaft through a reduction gear.

4. Hydraulic Starter Motor. It consists of a hydraulic starter motor for main engine, an accumulator, a hydraulic pump motor for auxiliary power unit. (APU).

APU is started by discharging the hydraulic accumulator to power a hydraulic pump motor. The hydraulic pump motor is driven with APU (Fig. 13.63) to start main engine and recharge accumulator. It is more suitable for aircraft engine.

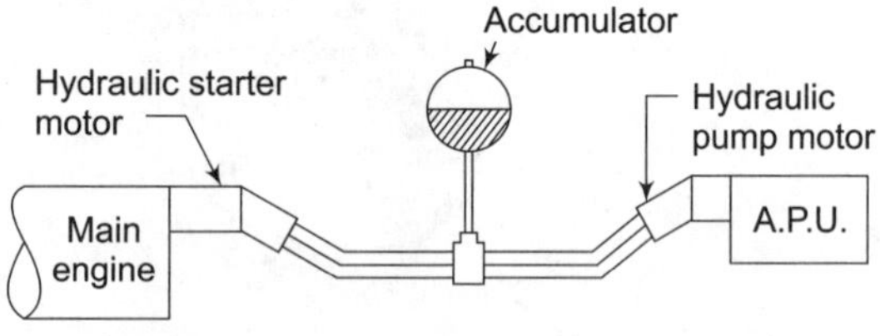

Figure 13.63 Hydraulic starter motor

13.41 IGNITION SYSTEM

Ignition system is utilised to initiate spark during the starting. Once it starts, the combustion is continuous and the working of ignition system is cut-off automatically. The following are the types of ignition system.

- Capacity discharge system.

(a) High tension system, and (b) low tension system

- Induction system
- A.C. power circuits.

Here A.C. power circuit ignition system is discussed. Figure 13.64 shows the arrangement of A.C. power circuit. It receives an alternating current that is passed through a transformer and rectifier to charge a capacitor. When the voltage in the capacitor is equal to the breakdown value of a sealed discharge gap, the capacitor discharges the energy across the face of the ignition plug. Safety and discharge resistors are fitted in the circuit.

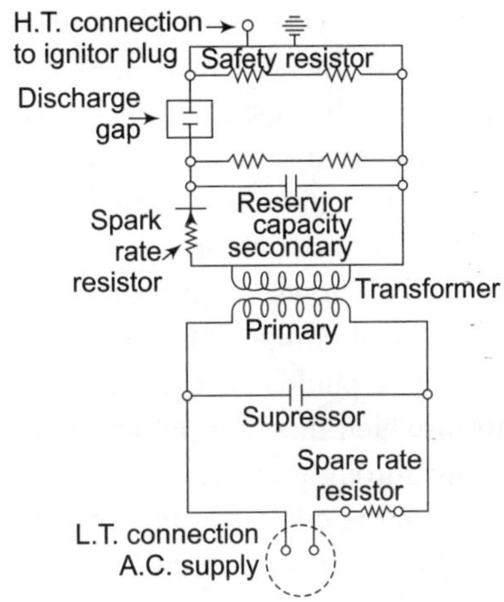

Figure 13.64 Ignition system

13.42 LUBRICATION SYSTEM

13.42.1 Elements of Lubrication System

The following are the elements of lubrication system of a gas turbine: (1) Oil tank, (2) Oil pump, (3) Filter and strainer, (4) Relief valve, (5) Oil cooler, (6) Oil and pipe line, (7) Magnetic drain plug, (8) By-pass, valve, (9) Warning devices.

13.42.2 Working Principle of a Lubrication System

Figure 13.65 shows the lubrication system used in a gas turbine plant. Sufficient oil is stored in an external tank to replace that which may be lost during the normal duration of run. Oil is drawn from the tank by a gear or rotor type lubrication pump. A relief valve is fitted to

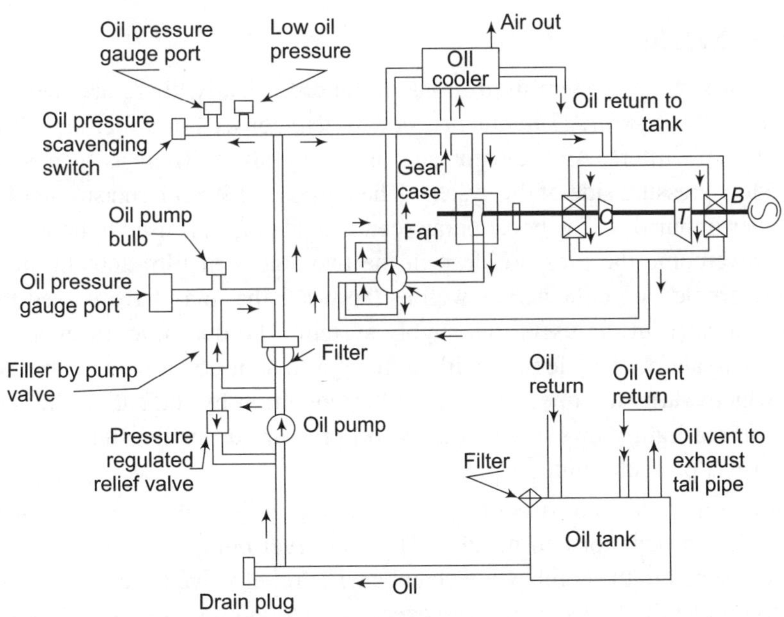

Figure 13.65 Lubrication system for a gas turbine

return the excess oil to the pump inlet than that required for lubrication. The oil flows through a paper or metal screen filter of 10 to 40 micron filtration capacity. In the case of clogged filter, the oil pressure drop across the filter increases which causes a by-pass valve to open and unfiltered oil to reach the bearings. This is undesirable. For such malfunctions, various, warnings, safety and indicating devices such as oil temperature indicators, bulb oil over temperature switch, low oil pressure switch and so on are installed in the filtered oil line. The filtered oil flows to the bearings and gear case of turbine and after lubricating and cooling the oil washed components, it is returned to the oil tank through a cooler by a scavenge pump. Scavenge pumps have a greater displacement than the quantity of oil flowing into the cavity from which they are pumping. This introduces a negative pressure which draws air into the sump through any leakage paths, precluding leakage of oil. The air-oil mixture returned through the cooler to the tank must be separated and for this it is recirculated in a separator and the air is permitted to escape. The separator vent usually connects to the tail pipe so that any remaining oil vapour will be burnt off by the hot gases flowing out of exhaust. This vent must be checked for obstruction regularly, otherwise the air may build high pressure in the tank and results in its failure. The separator is nothing but a simple baffle which causes the air-oil mixture to follow a devious path, the oil will tire to turning the corners and will find its way out through the vent into the tank, and the air will find its way out through the vent into the exhaust system. A drain plug of magnetic type is incorporated in the oil line before oil pump which can collect any metal particles circulating in the oil system. A thorough study is made of debris collected by the plug to decide whether it is metal chips or else.

13.43 FUEL SYSTEM AND CONTROLS

13.43.1 Fuel System

Figure 13.66 shows a fuel system used in gas turbine. Basically, there are *two sub-systems low pressure and high pressure. The low pressure system* consists of a fuel tank, boost pump, strainer, dual fine oil filters, Δp indicator and ordinary valves. There is a two-stages filter process on the low pressure side of the system. The purpose of 7.5 micron strainer is to collect the large particles impurities. The boost pump (centrifugal pump) is used to boost the pressure so that it may overcome the pressure drop in the two stages of filtration. In the fine filter, all small solid particles are collected as well as the water that may have passed through the water separator located in the users fuel supply system. The dual filter arrangement allows for switching to an alternative element without having to shut down for the change. It is the Δp indicator which warns the operator of the filter condition by indicating the red light for switch-over. The red light glows as soon as the oil pressure drops through the working fine filter crosses over a pre-determined value.

The *high pressure system* consists of main fuel pump, relief valve, strainer, fuel controls, shut-off valve, flow distributor and nozzles. The main fuel pump generates the oil pressure sufficient to be injected in the combustion chamber. The relief valve installed around the main fuel pump is to protect the fuel system from excessively high pressure created by the shut-off valve or clogging of the passage in the fuel system by transferring the excess fuel to the pump

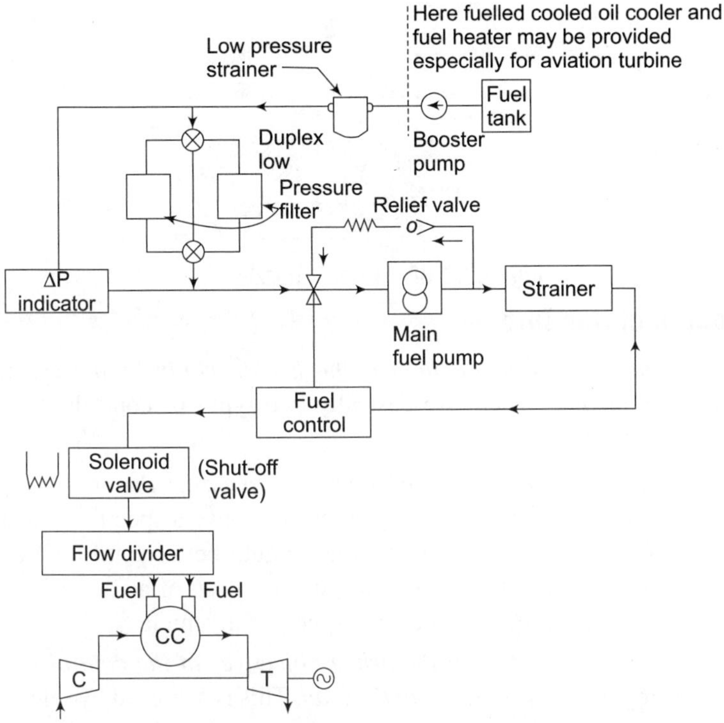

Figure 13.66 Fuel system for gas turbine

inlet. The function of shut-off valve is very simply to interrupt the supply of the fuel to the combustion chamber and affect a shut-down of the turbine automatically or manually. It is also used for interrupting the flow of other fluid driven pumps and the supply to auxiliary fuel lines such as the starting fuel injector. Generally, the shut-off valve is of solenoid type. The gas turbine requires that the fuel should be well distributed in the various fuel injectors. The work is accomplished by flow divider. It is a pressure sensitive valve which positions the valve opening determined by the pressure at inlet to the flow divider.

Injector. There are various types of injectors in current use. Figure 13.67 shows a splash plate type injector, In this case, the fuel at pressure about 40 bars is deflected into a cone by a splash plate, without spin. The conical fuel film thins with increasing radius and surface tension breaks the film into droplets of about 50 to 100 μm diameter. The included cone angle of the spray is typically about 80° to 100°. Figure 13.68 shows a duplex nozzle with single feed. It successfully atomises the fuel varying over a flow range of 50 to 1.

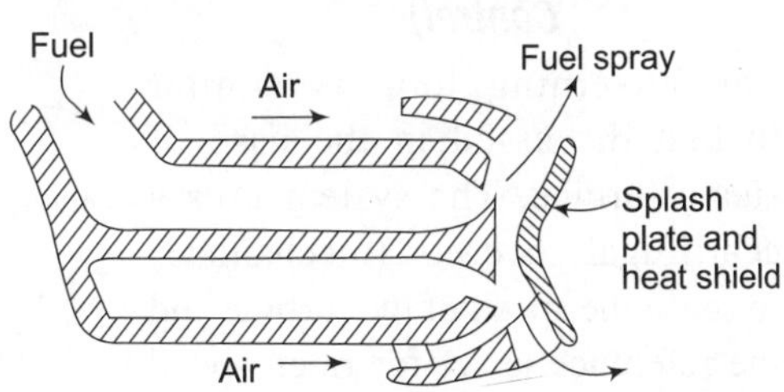

Figure 13.67 Splash plate type injector

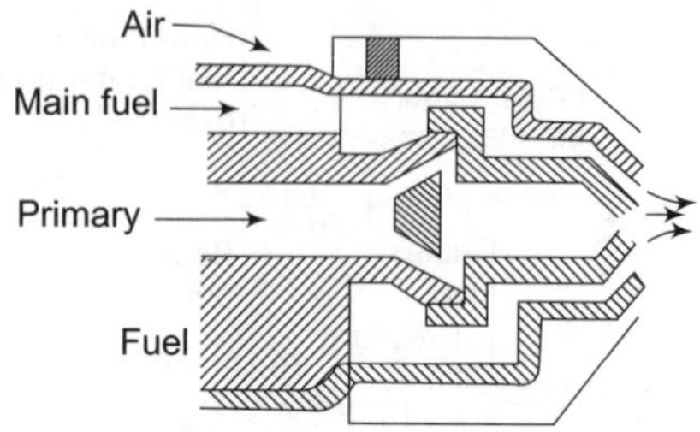

Figure 13.68 Duplex nozzle

13.43.2 Control of Gas Turbine

The purpose of gas turbine controls is to meet the specific control requirements of users and safe operation of the turbine. There are basically two types of controls: (a) Prime controls and (b) Protection control.

(a) Prime Control. *The objective of the prime control is to ensure the proper application of the turbine power to the load.* The users of the gas turbines have a specific control requirements according the use of gas turbines. The requirement might be to control: (i) the frequency of an a.c. generator, (ii) the speed of boat or ship, (iii) the speed of an aircraft, (iv) the capacity or head of a pump or compressor, (v) the road speed of a vehicle.

On a gas turbine a.c. generator set the prime objectives of the control is to maintain the constant electrical frequency irrespective of the load. This is achieved by selecting the primary controller as a speed governor (a speed sensing device) which maintains the constant electrical loads. On a gas turbine driven pipeline compressor, the prime objective of the control is to maintain a constant pipeline pressure downstream of the driven compressor. In this case, the pipe line pressure is sensed and turbine power is varied to maintain a constant pipeline pressure for varying flow conditions. There is relatively constant relationship between turbine power and fuel flow, *so in the prime control, the position of fuel valve is controlled.*

13.43.2.1 Hydro-mechanical Speed Governor System (Prime Control)

The governing loop is similar to that discussed in the case of steam turbine. The system uses a centrifugal governor (mechanical) to sense the speed of the turbine and the reference is set for rated speed of the turbine (Fig. 13.69). The governor senses the speed through

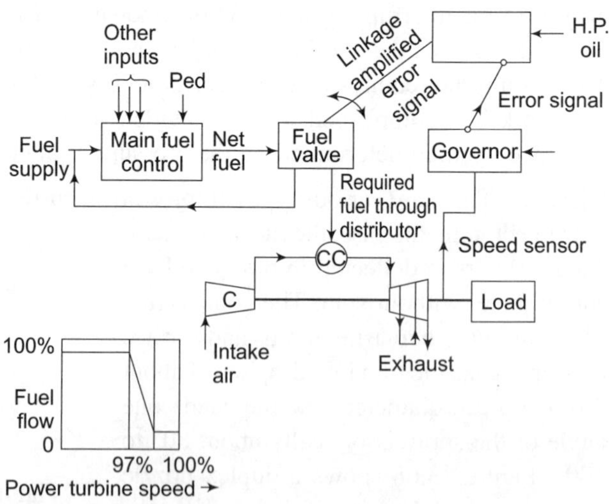

Figure 13.69 Prime control (hydro-mechanical)

the accessory drive shaft and flyweights. Internally, the governor hydraulically amplifies the error signal and provides an actuator output position, that can be linked to a fuel valve. The fuel valve either throttles or bypasses the metered fuel from the main fuel control. The speed sensing device may be electrical or hydraulic.

(b) Protective Controls. *The objective of the protective control is to ensure adequate protection for the turbine in preventing its operation under adverse condition.* Whenever, unsafe operating conditions are approached, the prime control is overtaken by the protective control to protect the turbine or driven equipment. Basically, the protective control are of two types:

(i) Shut down control and (ii) Modulating control

13.43.2.2 Shut Down Control

The shut down type control detects a condition which can cause a serious malfunction and actuate the shut-off valve to stop the turbine. The following are the various types of shut down controls.

(a) A turbine overtemperature, (b) Turbine overspeed, (c) Low lube oil pressure, (d) High lube oil temperature, (e) Excess vibration.

(a) Turbine Overtemperature. Figure 13.70 shows a turbine overtemperature control. Turbine temperature may be well sensed at the turbine inlet but the sensing device put at the turbine inlet goes wrong and therefore it is sensed at the turbine exhaust which is also an indication of turbine inlet temperature. The temperature sensor may be a thermocouple, bimetal or mercury vapour. As soon as the turbine inlet/exhaust temperature increases a predetermined value, the relay system acts upon the shut-off valve and shuts down the turbine by stopping the supply of fuel completely to the combustion chamber.

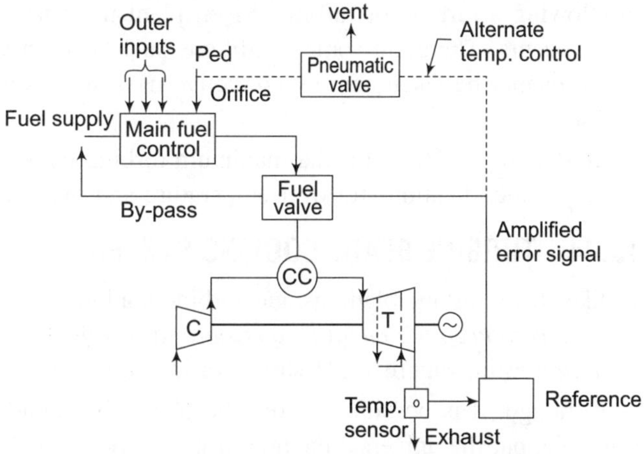

Figure 13.70 Turbine overtemperature control

(b) Turbine Overspeed. The turbine overspeed control is similar to the prime speed control as shown in Fig. 13.69. In this case, the reference speed is the maximum allowable speed instead of rated speed. The speed sensor could be a centrifugal governor, a tachometer generator or magnetic pick-up. If the speed increases a certain fixed value, the speed acts upon the relay system to shut-down the fuel valve.

(c) **Low Lube Oil Pressure.** It is essential to protect the turbine from low lube oil pressure to ensure proper lubrication and cooling of the turbine bearings. This is accomplished by the use of a pressure switch in the lube oil supply line. The switch operates an alarm signal or shut down fuel valve if the oil pressure drops a safe valve.

(d) **High Lube Oil Temperature.** High lube oil temperature in the lubrication system is a dangerous signal as it is an indication of low lube oil supply or failure in bearings. gears, etc. This protection is generally accomplished through the use of a temperature sensing device immersed in the lube oil. The sensing device may be a thermal switch which triggers an alarm or shuts down the turbine.

(e) **Excessive Vibration.** A slight increase in vibration is a cause of warning. Protection against vibration is accomplished by stalling one or more vibration pick-ups in control similar to Fig. 13.69. The output signal is fed to a monitoring device which may shut down the turbine if the vibration increases a certain value.

13.43.2.3 Modulating Controls

The purpose of modulating control is to sense an impending malfunction or a condition which could adversely affect turbine life and make some modifications to the operating condition of the turbine in order to alleviate the undesired conditions. An example of this control may be the maximum turbine inlet temperature and the maximum speed. The modulating control is more complex and more costly than the shut-down control. But it offers and advantage in allowing a turbine or turbine driven plant to continue operating when normally a shut-down occurs. There are some conditions such as high vibrations and low lube oil pressure which cannot be taken care by corrective measures as in this case shut-down of the turbine is essential.

Modulating controls for the maximum turbine inlet temperature and the maximum speed are slight modifications for overtemperature and overspeed controls.

13.44 TURBINE BLADE COOLING SYSTEM

Unlike steam turbine bladings, gas turbine bladings need cooling. *The objective of the blade cooling is to keep the metal temperature at a safe level, to ensure a long creep life and low oxidation rates.* Figure 13.71 shows various methods of blade cooling.

Although, it is possible to cool the blades by liquid using thermosyphon and heat pipe principle, but the universal method of blade cooling is by air bled from compressor or other working fluid flowing through internal passages in the blades. **The cooling technique may be *internal convection, film and transpiration cooling*.** The cooling medium may be air, water and steam. The cooling may be *open loop* or *closed loop*. In the open loop, the cooling fluid mixes with the working fluid (i.e., combustion products) while in closed loop, the cooling fluid does not mix with working fluid. In the film cooling, a film of cooling air is created on the both surfaces of the blade to reduce the heat transfer from gas to be turbine blade. The film cooling is more efficient than internal convection cooling and as a result the requirement cooling mass flow per kg of gas is much less. In the transpiration cooling, the cooling air is forced through a porous wall. This method is the most economical in cooling air, because

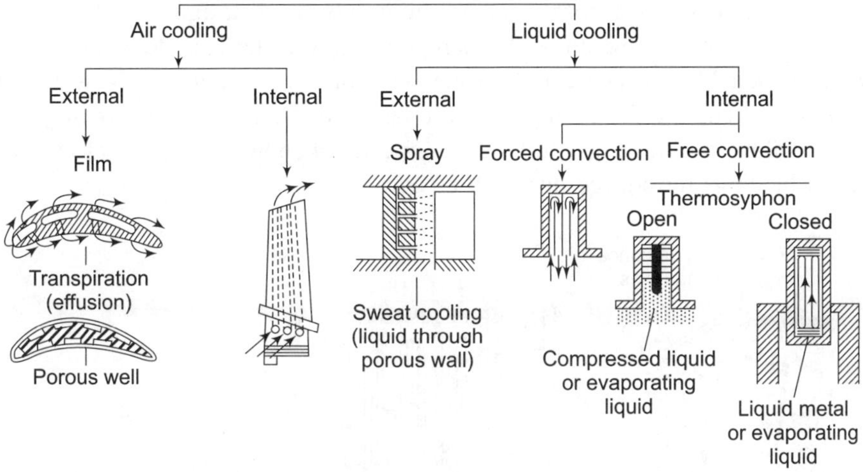

Figure 13.71 Various methods of blades cooling

not only does it remove heat from the wall more uniformly, but the effusing layer of air insulates the outer surface from the hot gas stream and so reduces the heat transfer to the blade resulting in reduced cooling mass flow rate. Among the three types of cooling, transpiration cooling is the most efficient and now film and transpiration are in current use in modern gas turbine. In combined cycle and co-generation plant, now closed loop steam cooling is being used in H-technology turbine (MS9000H) developed in 2002 by GE yielding thermal efficiency about 40% at ISO conditions.

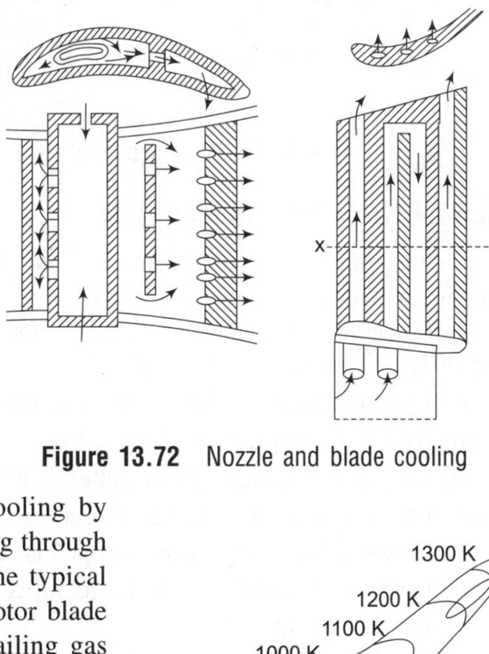

Figure 13.72 Nozzle and blade cooling

Figure 13.72 shows nozzle and blade cooling by working fluid air bled from compressor flowing through the internal passages. Figure 13.73 shows the typical temperature contours in a blade. The mean rotor blade temperature is about 350°C below the prevailing gas temperature after efficient blade cooling. ***As per current technology, turbine inlet temperature (TIT) is about 1700 K and safe blade temperature is 1123 K.*** **The coolant may be air, steam and water. The water as a coolant may be the best but still it has not been commercialised.**

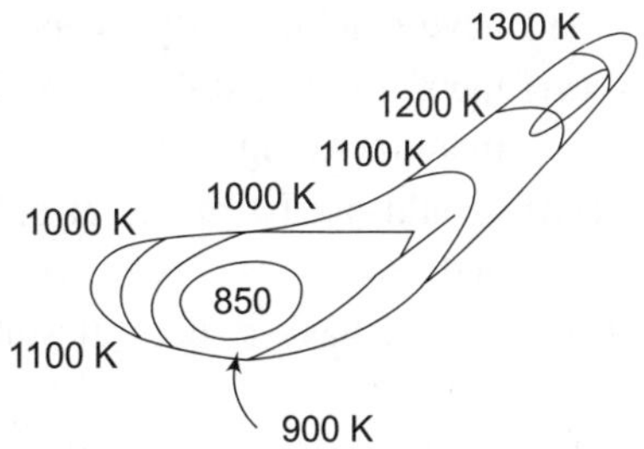

Figure 13.73 Temperature control across turbine blade

As much as 5% of the working fluid may be required from H.P. compressor delivery missing the combustion system to cool the first stage nozzles alone. In an advanced design, a further 2% total air cooling temperature requirement may be as high as 15% of the H.P. flow is required to cool the first stage rotor blades. The cooled blades last for 10,000 hours to 20,000 hours while uncooled may not survive for 10 hours. The total cooling air requirements may be as high as 20% of gas flow in modern gas turbines. Figure 13.74 shows the sectional view of cooled turbine rotor bladings introduced during 1980s.

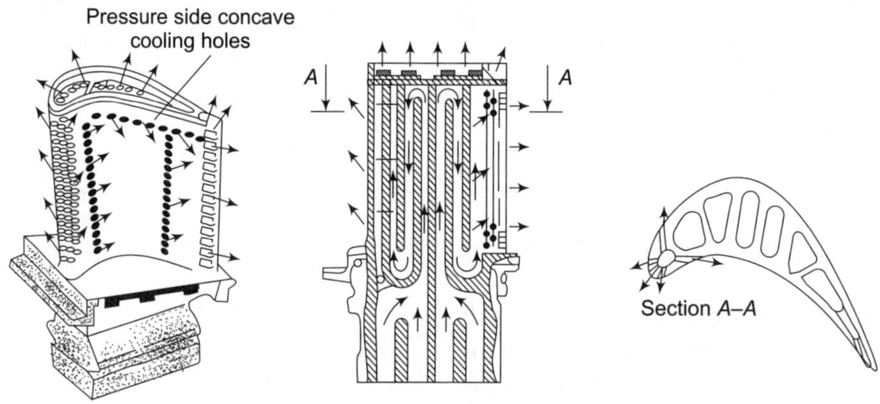

Figure 13.74 A sectional view of a cooling turbine rotor bladings

13.44.1 Gas Turbine Materials

In gas turbine plants, the components are subjected to hot gases and high pressure. As mentioned, blades are cooled. Blades are also subjected to high centrifugal stresses due to high relative speeds and thermal stresses. Blade materials should possess the following properties, otherwise they will fail:

(i) withstand high pressure, high temperature and high stresses generated,

(ii) have low creep rate for longer life,

(iii) have high resistance to oxidation, corrosion and erosion,

(iv) have resistance to fatigue failure,

(v) have machinability to achieve precise dimension and profile,

(vi) have good castability or forgeability characteristics depending on the method of manufacturing,

(vii) should not be brittle at ordinary temperature and not be plastic at high temperature, and

(viii) should have structural stability when exposed to varying temperature for avoiding failure.

It is to be noted that all these properties are not found in one material, so **alloys** are needed.

(A) Materials for Turbine Rotor Discs. Due to rotation of rotor and high gas temperature, the turbine rotor disc is subjected to centrifugal and thermal stress.

The disc hub stresses tend to cause tensile deformation in the disc. So, a material with low expansion coefficient will be suitable. In many cases, austenitic steels with 12 to 18% chromium, 8 to 12 per cent nickel and small percentage of tungsten, molybdenum and titanium are used for turbine rotor discs.

Nowadays, even rotor discs are being cooled by bled air from compressor. In such case, low expansive materials can be used, such as ferratic steels having higher creep strength at low temperature up to 600°C, the centre portion may be of austenitic steel.

(B) Materials for Turbine Rotor Blades. Though, gas turbine rotor blades are cooled by compressed air bled from compressor, still, they are subjected to high temperature and high stresses. The satisfactory materials for blades are the stainless steel alloys and 8-20% nickel chromium alloys, known as nimonic alloys. The important properties of nimonic alloys are high resistance to oxidation, scaling and deformation along with good creep and fatigue properties.

(C) Materials for Combustion Chamber. In the combustion chamber, the flame temperature is very high and the material has to withstand it. It is made of nimonic 75 alloys. The advantage of this alloy is that it has an excellent creep resistance, capacity to withstand heavy thermal shocks, and high resistance to oxidation.

(D) Materials for Compressor. For small gas turbine, centrifugal compressor is used. It is subjected to high centrifugal and thermal stresses. In general, aluminium alloys are used. Some allowances are provided to check thermal expansion.

In medium and large gas turbines, axial flow compressors are used. It is subjected to centrifugal and thermal stresses. In general, it is made of titanium alloys which are of low density, possess good strength at high temperature up to 500°C. It also has strong resistance to corrosion. The advantages of titanium alloys are: light in weight, good creep strength, and fatigue resistant.

13.45 OPERATION, MAINTENANCE AND TROUBLESHOOTING

13.45.1 Operation

(a) Starting. Starting sequence of any gas turbine from rest to its rated speed requires a certain order of events to be accomplished either manually or automatically. The major steps in sequence are cranking, ignition, acceleration and governing.

The following is typical starting sequence of a gas turbine.

1. Application of control power illuminates of a gas turbine.
2. Operate 'Reset switch' to reset malfunction circuits. By doing so, malfunction lights go off and all control devices assume the conditions for starting.
3. Operate "Start" switch to initiate starting sequence. By doing this, lube oil pump and cooling fan start. If there are separate switches for these, operate these.

4. When lube oil reaches a present pressure, the starter is energized and cranking of the engine begins.
5. With the cranking of starting of starter, the engines and exhaust ducts are purged of any combustible gases that might be present.
6. During the cranking cycle, the fuel boost pump is used and operated to increase fuel pressure.
7. As soon as the fuel pressure has reached a prescribed minimum value, fuel and ignition switches are turned on provided preset turbine speed has been reached.
8. The turbine accelerates due to combustion of fuel and assistance of cranking motor. At a preset value, say in the order of 70% of rated speed, the starter and ignition and cut-off automatically.
9. The turbine becomes self-sustaining and accelerates on its own to its governed speed till the governing system takes over the control.

(b) Shut-down. To stop the gas turbine, fuel supply should be turned off. This is accomplished by closing the fuel valve either manually or by deenergizing an electrically operated valve. In cases where sleeve bearings are used, circulation of lube oil to bearings after shut-down is necessary for cooling.

13.45.2 Maintenance

The types of maintenance which are done on the gas turbine is the same as that of steam turbine. From the experience of the most manufactures of gas turbine equipment, forced outages are frequently caused, at least in part by inadequate maintenance. The basic purpose of a preventive maintenance programme is to reduce forced outages. The following are the principal subsystems of the gas turbine for which maintenance instructions are presented by manufactures:

1. Turbine gear
2. Starting
3. Auxillary gear and main gear
4. Clutches and coupling
5. Gas turbine
6. Fuel system
7. Lube oil system
8. Pneumatic system
9. Overspeed protection
10. Fire protection system
11. Temperature control and monitoring system
12. Control equipment
13. Air conditioning system
14. Generator-exciter
15. Emergency power
16. Electrical controls
17. Motors
18. Related station equipment

Maintenance is carried out daily monthly, quarterly, semi-annually and annually. So far the sequence of overhaul is concerned, similar to that of steam turbines with some exceptions.

13.45.3 Troubleshooting

Modern gas turbines are usually equipped with a sophisticated protection system using microprocessor and computers which gives visual and audio alarm when any pre-established safe condition is violated. In all types of alarm, it is not necessary to shut-down the gas turbine. If an alarm condition is of sufficient duration and magnitude, the unit will trip and will shut-down automatically.

The following are the principal symptoms of gas turbine malfunctions and the most common causes of these malfunctions:

(A) Drop in compressor discharge pressure and subsequent drop in load:

1. Dirty compressor blades
2. Dirty intake screens
3. Loss of compressor blades
4. Damaged labyrinth seals

(B) Smoke or dark stack:

1. Burner nozzle dirty or worn.

Note: Puffs of smoke indicate that carbon is building up around the fuel nozzle and then passing through the turbine creating rapid wear of blades, vanes and shrouds.

2. Uneven distribution of fuel to combustion chambers.
3. Combustion chamber damaged or out of position.

(C) Spread in turbine discharge temperatures:

1. Bad thermocouples
2. Uneven fuel to burners or dirty nozzles
3. Combustion chamber damaged or out of position
4. Unlit burners
5. Damaged burner nozzles.

(D) High wheel space temperatures:

1. Cooling air lines plugged
2. Cooling air heat exchanger dirty, leakage water or loss of cooling water.
3. Bad thermocouples
4. Wheel space seals worn due to rubs by axial movement or rotor (worn thrust bearing), bowed shaft or casing out of round to open seal clearances
5. Cooling air supply not functioning properly.

(E) High turbine exhaust temperatures:

1. Loss of turbine blades or damaged inlet vanes
2. Bad thermocouples
3. Exhaust temperature controller out of adjustment
4. Increased blade tip clearances due to radial ribs
5. Dirty air compressor.

(F) High turbine exhaust pressure:

1. Turning vanes in turbines discharge duct damaged or missing
2. Discharge silencer damaged.

(G) Vibration:

1. Indicating instruments out of adjustment
2. Loose shaft couplings
3. Bowed turbine shaft
4. Broken or missing turbine blades
5. Damaged bearings
6. Shaft mal-alignment

(H) Loss of fuel pressure:

1. Fuel control valve out of adjustment
2. Fuel strainers dirty
3. Fuel pump or compressor damaged

(I) Light off failure:

1. Faulty spark plug
2. Combustion chamber cross-fire tubes out of place
3. Electrical control out of adjustment
4. Fuel proportion out of adjustment
5. Fuel atomizing air out of proportion
6. Burner nozzles dirty or worn
7. Combustion chamber damaged

(J) Machine 'Hunting':

1. Worn governor and control parts
2. Fluctuating fuel controllers
3. Fluctuating exhaust temperature controllers
4. Hydraulic control vanes leaking or strainers dirty

(K) Loss of oil pressure:

1. Filters dirty
2. Pump failure
3. Leakage in pump.

13.46 SUMMARY OF FORMULAE

(1) $\eta_{th} = 1 - \dfrac{1}{r_p^{(\gamma - 1)/\gamma}}$

(2) $\dfrac{n-1}{n} = \dfrac{\gamma - 1}{\gamma . \eta_{pc}}$ for compression

(3) $\dfrac{n-1}{n} = \dfrac{(\gamma - 1)\eta_{pt}}{\gamma}$ for expansion

(4) $AR = \dfrac{3600}{w_{net}}$

(5) $WR = \dfrac{w_{net}}{w_t}$

(6) $sfc = \dfrac{3600(\dot{m}_f / \dot{m}_a)}{w_{net}}$

(7) $r_{p_1} = r_{p_2} = \sqrt{r_p}$

(8) $w = \dfrac{Cr_1^2 - Cr_2^2}{2} + \dfrac{u_2^2 - u_1^2}{2} + \dfrac{C_2^2 - C_1^2}{2}$

(9) $w = (u_2 C_{u2} - u_1 C_{u1})$

(10) $(r_p)_{st} = \left[1 + \eta_{st} \dfrac{(\Delta T_0)_{stage}}{T_{01}}\right)^{\gamma/\gamma - 1}$

PROBLEMS

Things to Think About

13.1 Gas turbine plant is called prime mover of 21st century. Why?

13.2 What limits the increase in gas turbine output and plant efficiency?

13.3 If there is no control of gas turbine, what will happen? Explain.

13.4 Though, theoretically complex cycle system offers more power and efficiency, but still it is not commercialised. Why?

13.5 Is it possible to commercialise closed cycle in a big a way? Explain.

Short and Long Theoretical Questions

13.1 What will happen if there is no combustion chamber in the gas turbine plant?

13.2 Why is thermal efficiency of the gas turbine less, even though the turbine develops greater power? Classify the gas turbines.

13.3 List the parameters that affect the thermal efficiency of gas turbine plant. Discuss these effects in detail with performance curves.

13.4 Why is the thermal efficiency of gas turbine high at very high altitude?

13.5 Is the use of regenerator in a simple cycle gas turbine limited? Discuss.

13.6 Axial compressor is more preferred to centrifugal compressor in aviation. Give reasons.

13.7 What are the various methods which are used to improve the efficiency and output of a gas turbine? Give a neat sketch of each cycle and show the variation of efficiency and output with pressure ratio.

13.8 State merits and demerits of closed cycle gas turbine over open cycle.

13.9 Derive an expression for the optimum pressure ratio giving maximum cycle thermal efficiency of gas turbine cycle, if the compressor efficiency is η_c and turbine efficiency is η_t. The maximum cycle temperature is T_{max} and the minimum cycle temperature is T_{min}.

13.10 Derive an expression for the optimum pressure ratio giving the maximum specific output in simple cycle gas turbine.

13.11 Prove the following expression of thermal efficiency of gas turbine with the ideal regenerative and complete reheat, using reheat pressure for yielding the maximum output

$$\eta_{th} = \frac{1 - T_1(r_p)^{(\gamma-1)/2\gamma}[(r_p)^{(\gamma-1)/2\gamma} + 1]}{2\eta_c\, \eta_t . T_3}$$

where T_1 = compressor inlet temperature, T_3 = turbine inlet temperature, r_p = overall pressure ratio, η_c, η_t = compressor and turbine efficiencies.

13.12 Prove that the work done by the turbine per kg of gas is less by the following quantity, when pressure drops are compared to the case when they are neglected in an open cycle gas turbine using heat exchanger.

$$\frac{\gamma-1}{\gamma}\times\frac{c_p\,T_3\times\eta_t}{r_p^{(\gamma-1)/\gamma}.p_1}\left(\Delta p_1+\frac{\Delta_p}{r_p}\right)$$

where T_3 = turbine inlet temperature, η_t = turbine efficiency, r_p = pressure ratio, Δp_1 = pressure drop on gas side heat exchanges, Δp = combustion chamber pressure bar.

13.13 List the advantages and disadvantages of gas turbine over steam turbines and diesel or petrol engines.

13.14 Prove that the output of a simple gas turbine plant is positive only when the product of compressor and turbine efficiency is greater than $(T_1/T_2)\,(r_p)^{(\gamma-1)/\gamma}$ where r_p is pressure ratio, T_1 and T_3 are the compressor and turbine inlet temperatures.

13.15 Sketch diagrammatically the arrangement of an internal combustion turbine to work on an open cycle with two stages of compression and expansion of the working substance and regenerative after compression.

13.16 Describe the working principle of centrifugal compressor.

13.17 Discuss the mechanism of pressure rise achieved in centrifugal compressor.

13.18 Discuss the physical meaning of Euler's equation

$$\left[\frac{C_2^2-C_1^2}{2}+\frac{u_2^2-u_1^2}{2}+\frac{C_{r1}^2-C_{r2}^2}{2}\right].$$

13.19 Define slip factor of centrifugal compressor.

13.20 Discuss the working principle of axial flow compressor with the help of *T-s* representation.

13.21 What do you mean by pre-whirl in centrifugal compressor?

13.22 Discuss surging, choking and stalling phenomena.

13.23 Discuss the characteristics of centrifugal and axial flow compressors.

13.24 Discuss the operational problem of axial flow compressors.

13.25 Give a comparison between centrifugal and axial flow compressors.

13.26 With sketch discuss the various forms of combustion chambers for gas turbines.

13.27 Discuss the combustion process in combustion chamber.

13.28 Discuss the factor affecting the combustion chamber performance.

13.29 Discuss about gas turbine emission and its control.

13.30 Discuss the working principle of starting system of gas turbine plant.

13.31 Discuss the working principle of fuel system of gas turbine plant with sketch.

13.32 With sketches discuss working principle of various controls of gas turbine.

13.33 Discuss the hydro-mechanical speed governing system of gas turbine plant.

13.34 What are the protective devices.

13.35 Discuss the purpose of gas turbine control.

Numerical Questions

13.1 Air enters a compressor at 10°C and at atmospheric pressure and is compressed through a pressure ratio of 5. The temperature of gaseous mixture enters the turbine at 850°C and expands to atmospheric pressure. Determine the thermal efficiency of the cycle under following conditions: (a) The system operates on ideal Brayton cycle, (b) the isentropic efficiency of compressor and turbine are 0.85, (c) with conditions (b) a regenerator of 75% effectiveness is introduced, (d) with conditions (c) an intercooler which cools to 10°C is introduced at optimum pressure ratio. [***Ans.*** (a) 36.86% (b) 24.39% (c) 37.08% (d) 39.7%]

13.2 If the total pressure ratio is 15, the maximum cycle temperature is 1300°C and minimum 15°C compression and expansion are isentropic and regenerator are each 75% efficient find (a) the thermal efficiency of the cycle. If the isentropic efficiency of the turbine and compressor units are each 86, what is the (b) maximum thermal efficiency that may be expected from the turbine.

13.3 The air enters the compressor of a gas turbines at temperature 15°C and atmospheric pressure operating at pressure ratio of 5 : 1. The isentropic efficiency of compressor is 0.8. The inlet temperature to turbine is 1000°C. What must be the isentropic efficiency of turbine in order that the overall cycle efficiency may be 28%? Assume $c_p = 1$ for air and $c_p = 1.092$ kJ/kg K for gases and $R = 0.287$ for both air and gases. [**Hint:** Find γ air and γ gas]

13.4 A gas turbine is designed to operate under the following conditions:

Maximum temperature = 1400°C, $r_p = 15$, Inlet pressure and temperature = 1.05 bars and 15°C, Turbine isentropic efficiency = 86%, Compressor isentropic efficiency = 82%, Mechanical efficiency of the compressor and turbine = 99%. Efficiency of combustion = 98%.

Determine the improvement in plant thermal efficiency that would result from the addition of a heat exchanger of 75% effectiveness. Allow a pressure loss of 0.205 bar in the heat exchanger for air $c_p = 1.003$ and $c_p = 1.09$ for gas. Assume that in the heat exchanger, the mean specific heat of the exhaust products and air is the same.

13.5 The air enters the compressor of a gas turbine at 1 bar and 30°C and is compressed to 15 bars with an isentropic efficiency of 85%. The turbine inlet temperature is 1400°C. The isentropic efficiency of the turbine is 86%. Estimate the flow of air in kg/s for a net electrical power of 200 MW. Assume 2% of pressure loss in combustion chamber, 1% pressure loss in exhaust hood, mechanical efficiency of whole assembly 98% and alternator efficiency 98%.

[***Ans.*** 142 kg/s]

13.6 The minimum and maximum temperature limits in a gas turbine plant are 15°C and 878°C. The pressure limit is 1 bar and 4 bars. Determine: (a) optimum ratio for maximum work output and also, (b) for maximum cycle thermal efficiency assembly isentropic efficiency of compressor and turbine are 0.85 and 0.8 respectively. [***Ans.*** (a) 5.75, (b) 8.24]

13.7 A gas turbine plant operating on closed system uses 1.5 kg of air/s and operate between pressure ratio of 5 : 1. The air enters the compressor at 15°C and turbine at 730°C. Assuming an isentropic efficiency of compressor as 0.85 and that of turbine 0.87, mechanical efficiency of compressor and turbine 0.99 and combustion efficiency of 0.98, determine the shaft power of the plant and overall efficiency. [***Ans.*** 178.6 kW; 22.5%]

13.8 The following data relate to a gas turbine plant with free power turbine.

Total compressor ratio = 12, Total turbine inlet temperature = 1350 K, isentropic efficiency compressor (total to total) = 0.86, Isentropic efficiency of each turbine (total to total) = 0.89, Mechanical efficiency of each shaft = 0.99, Combustion chamber pressure loss = 6% of compressor delivery pressure. Exhaust pressure loss = 0.03 bar, Ambient conditions = 1 bar and 288 K. Calculate the specific work output, Specific fuel comsumption and cycle thermal efficiency. [***Ans.*** 277 kJ/kg, 0.2655 kg/kWh, 31.6%]

13.9 The following data relate to an open cycle single shaft gas turbine with reheat.

Total compressor pressure ratio = 30, Polytropic efficiency for compressor and turbine = 0.89, Both turbine inlet temperature = 1523 K, Pressure loss in first combustion = 2% of compressor delivery, Pressure loss in second combustor = 4% of H.P. turbine exhaust, Exhaust pressure = 1.02 bars, Power output = 240 MW, Inlet conditions = 288 K and 1.01 bars. Calculate the net work output, mass flow rate of air the fuel-air ratio and thermal efficiency.

[***Ans.*** 499.3 kJ/kg, 480.7 kg/s, 0.0342, 33.9%]

13.10 An open cycle gas turbine serving as a peak load generator develops 20 MW.

Following data relate to this turbine: Total compressor pressure ratio = 11.0, Compressor isentropic efficiency = 0.82, Combustion pressure loss = 0.4 bar, Combustion efficiency = 0.99, Turbine inlet temperature = 1150 K, Compressor turbine isentropic efficiency = 0.87, Power turbine isentropic efficiency = 0.89, Mechanical efficiency of each shaft = 0.98, inlet condition of air = 1 bar, 288 K.

Determine the air mass flow rate and sfc. [***Ans.*** 119.5 kg/s, 0.307 kg/kWh]

13.11 Calculate the specific fuel consumption, specific work mass flow rate of air and plant efficiency for a gas turbine generating 260 MW working on total compressor ratio of 20 and total turbine inlet temperature of 1400°C. Assume suitable component efficiencies and pressure losses.

13.12 Following particulars relate to a centrifugal compressor:

Inlet diameter of impeller = 61.4 cm, Outlet diameter of impeller = 12.3 cm, Speed = 5000 rpm, Velocity of flow = 61.6 m/s. Free air delivered = 1000 m^3/min. Pressure ratio = 1.33, Index of compression = 1.6.

Assuming that all pressure rise takes place in the impeller, find the angle at which air from impeller enters the casing breadth of the impeller blade at inlet and outlet.

[***Ans.*** $\beta_1 = 21°$, $\beta_2 = 17.2°$, $\alpha_3 = 26.6°$, $\beta_1 = 14.05$ cm, $b_2 = 6.04$ cm]

13.13 The following particulars relate to a double sided centrifugal compressor:

Speed of compressor = 15,000 rpm, Mass flow rate = 10 kg/s, Total head pressure at inlet = 1.12 bar, Total temperature at inlet = 295 K, Isentropic efficiency = 0.8, Eye root diameter = 15 cm, Eye tip diameter = 20 cm, Impeller tip diameter = 50 cm. Assuming a slip factor of 0.9, a power input of 1.04 and axial entry, calculate total head pressure ratio and compressor power. Also calculate angles at impeller root and tip, assuming axial component of initial velocity to remain constant. [***Ans.*** $r_p = 3.134$, power = 2585 kW, 31.1°, 50.5°]

13.14 A single-sided centrifugal compressor is to deliver 13.65 kg of air/s when operate at a pressure of 4 : 1 and a speed of 12,000 rpm. The total head inlet conditions may be taken as 283 K and 1.03 bars and an isentropic efficiency of 80%. Calculate the overall diameter of the impeller. If the Mach number is not to exceed unity at the impeller tip, and 50% of the losses are assumed to occur in the impeller, calculate the minimum depth of the diffuser. [***Ans.*** 69 cm 5.5 cm]

13.15 All the stages of an axial flow compressor have the following conditions at the mean diameter; degree of reaction is half, blade peripheral speed 200 m/s, angles of absolute velocity at inlet to and exit from the rotor measured from the axial direction 15° and 45° respectively. A stage efficiency of 88% and work done factor of 0.86 are to be taken as constant throughout the compressor. The combined efficiency of the intake and inlet guide vanes is 95% based on static conditions. If the ambient air pressure and temperature are 1 bar and 228 K respectively find, (a) the static temperature of the air at inlet to the first stage, and (b) the static pressure of the air at exit from the tenth stage. [***Ans.*** 277 K 1.195, 3.81 bar]

13.16 The first stage of an axial-flow compressor is to be designed for an axial velocity of 133 m/s, mass flow rate of air 22.7 kg/s. And for ambient condition of 1 bar and 10°C at the mean radius both moving and fixed blade sections are to be suitable for relative air angles of 50° at inlet and 30° at out let. A row of guide vanes giving an outlet air angle of 30° is to be fitted before the stage.

Draw the velocity triangles assuming the ratio of hub to tip diameter as 0.68, free vortex condition and ignoring end effects. Calculate (a) the compressor speed; (b) theoretical temperature rise across a stage, (c) the number of similar stages, which would theoretically be required to give a total head delivery pressure of 8.4 bars if the stage isentropic efficiency is 85%.

[***Ans.*** (a) 8250 r.p.m; (b) 16°C; (c) 13]

13.17 A single sided centrifugal compressor delivers 8.15 kg/s of air with a pressure ratio of 4.4 to 1 at 18000 r.p.m. at the entry to the eye, for which the internal diameter is 12.7 cm, is axial and the mean axial velocity at the eye section is 148 m/s with no pre-whirl. Static conditions at the eye section are 15°C and 1 bar respectively.

The slip factor is 0.94 and the isentropic efficiency of compression is 78%. Neglecting the losses, calculate (a) the rise in temperature during compression if the change in kinetic energy is negligible; (b) the tip speed and impeller tip diameter, (c) external diameter of the eye.

[***Ans.*** (a) 157°C, (b) 493 m/s, (c) 28.7 cm]

13.18 Design a centrifugal compressor of the following data.

$\dot{m} = 15$ kg/s, $r_p = 4.5$, $N = 16{,}000$ rpm, $p_{01} = 1.013$ bars, $T_{01} = 300$ K

13.19 Design an axial flow compressor, for the following data:

$\dot{m} = 500$ kg/s, $r_p = 30$, $N = 3000$ rpm

13.20 An axial flow compressor stage is to be designed for a stagnation temperature rise of 22 K and axial velocity of 155 m/s, which are constant from root to tip. The blade root, mean and tip velocitites are 160, 210 and 250 m/s. The work done factor is 0.92. Calculate the stage air angles at root, mean and tip and the degree of reaction at mean and free vortex design.

13.21 The following data relate to an axial flow compressor:

Overall pressure ratio = 4.0, Mass flow rate = 3 kg/s, Polytropic efficiency = 88%, Stagnation temperature rise per stage = 25 K, Absolute velocity approaching the last rotor = 165 m/s, Air angle approaching last stage measured from axial direction = 20°, Work done factor = 0.83, Velocity diagram = symmetrical, Mean diameter of the last stage rotor = 18 cm, Ambient conditions = 1.01 bars, 288 K.

Calculate the number of stages required, pressure ratio of the first and last stage, rotational speed and the length of last stage rotor blade at inlet to the stage.

[***Ans.*** 7, 1.273, 1.178, 414 rps, 1.325 cm]

13.22 The following data relate to the design of first stage of an axial flow compressor based on free vortex principle with no inlet guide vanes:

$N = 6000$ rpm, Stagnation temperature rise per stage = 20 K, Hub-tip ratio = 0.6, Work done factor = 0.93, Mach number relative to tip = 0.95, Isentropic efficiency of the stage = 0.89, Inlet velocity = 140 m/s, Ambient condition = 1.01 bars and 288 K, Calculate: (a) The mass flow rate entering the stage, (b) The tip radius and corresponding rotor inlet and outlet angles,

(c) The stage stagnation pressure, (d) The power required to drive compressor, (e) The rotor angle at the root section.

[*Ans.* (a) 65.55 kg/s, (b) 0.455 m. 64°, 56.5°, (c) 1.2335, (d) 1.318 MW, (e) 50.82° and 18.33°]

13.23 The following data relate to an axial flow compressor:
Mass flow rate = 50 kg/s, Inlet stagnation pressure and temperature = 100 kPa, 23°C, Outlet stagnation pressure = 500 kPa, Hub and tip diameter at inlet = 0.436 m and 0.728 m. The reaction at mean radius throughout = 0.5, Absolute air angles through all stages at stator exit = 28.8°, The speed of the rotor = 8000 rpm, Polytropic efficiency = 0.89, Axial velocity at mean radius = 1.05 × Average axial velocity. Calculate the number of similar stages.

[*Ans.* 14 stages]

Software Based Questions

13.1 Develop a software for the design of axial flow compressor flow power plant. List the design dimension for the following data:

$r_{p0} = 14, \quad \dot{m}_a = 600$ kg/s, $\quad N = 3000$ rpm, $\quad p_{01} = 1$ bar, $\quad T_{01} = 300$ K

13.2 Develop a software for the design of industrial centrifugal air compressor. List the design dimension for the following data:

$r_p = 300, \quad \dot{m}_a = 10$ kg/s, $\quad N = 16{,}000$ rpm, $\quad p_{01} = 1$ bar, $\quad T_{01} = 300$ K

13.3 Develop a software for the thermodynamic analysis of gas turbine plant considering simple cycle and other combinations such as intercooling, reheating and regeneration.

13.4 Develop a software for the thermal analysis of open gas turbine cycle covering all real aspects.

Design and Open Ended Question

13.1. Prepare a comprehensive report considering all major specifications of various models of gas turbine plants with their merits and demerits which may be helpful for the customers to select a particular product.

Multiple-Choice Questions

13.1 The thermal efficiency of a modern gas turbine plant ranges from

(a) 20 to 30% (b) 25 to 35% (c) 35 to 45% (d) 50 to 60%

13.2 The isentropic efficiency of an axial flow compressor ranges from

(a) 60 to 70% (b) 65 to 75% (c) 75 to 85% (d) 85 to 90%

13.3 The combustion efficiency of combustor ranges from

(a) 60 to 70% (b) 70 to 80% (c) 80 to 90% (d) 90 to 98%

13.4 Intercooling in gas turbine plant

(a) increases net work (b) decreases η_{th}

(c) increases η_{th} (d) increases net work and decreases η_{th}

13.5 Increasing turbine inlet temperature

(a) increases network (b) increasies η_{th}

(c) increases net work but decrease η_{th} (d) increases both network and efficiency

13.6. Decreasing compressor inlet temperature
(a) increases network
(b) increases η_{th}
(c) increases both network and η_{th}
(d) None of the above

13.7 Recuperation arrangement in gas turbine
(a) increases η_{th}
(b) increases network
(c) decreases η_{th}
(d) increases both η_{th} and network

13.8 Cooling of gas turbine blade is about
(a) to increase power out
(b) to increase η_{th}
(c) to safeguard against failure due to high thermal stresses and creep
(d) All of them.

13.9 Prime control of gas turbine
(a) speed
(b) pressure
(c) temperature
(d) None of above

13.10 The pressure ratio per stage in an axial flow compressor is about
(a) 1.1
(b) 2.2
(c) 3.2
(d) 4.2

13.11 In the present technology, a simple gas turbine with single casing can generate net output of about
(a) 100 MW
(b) 200 MW
(c) 300 MW
(d) 400 MW

13.12 The compressor ratio in a multistage axial flow compressor housed in one cylinder can generate pressure ratio up to about
(a) 20
(b) 30
(c) 40
(d) 50

Answers to Multiple-Choice Questions

13.1 (c) 13.2 (d) 13.3 (d) 13.4 (c) 13.5 (d) 13.6 (c) 13.7 (a) 13.8 (c)
13.9 (a) 13.10 (a) 13.11 (c) 13.12 (c)

CHAPTER

14

Combined, Co-generation and Mixed Cycle Power Plants

OVERVIEW

Energy is a fundamental requirement of human life, and thus a driving force of civilization, so it must be utilised very efficiently at all levels. Due to increased fuel prices and depletion of fossil fuels resources, high efficiency energy conversion systems are utmost desirable. There is a great potential for increased efficiency in energy converting devices by recovering waste heat to a greater extent. Development of combined cycle is the most efficient and effective effort in this direction. *Combined cycle power plants are gaining increasing acceptance as alternatives to conventional or nuclear steam cycle due to high thermal efficiency as high as 64% utilising natural gas as fuel.* This chapter deals with the analysis and performance of gas/steam combined cycle power plants. In India, about 4000 MW of combined cycle plant had been installed by 2000. Now more combined cycle power plants have been added. In Europe, USA and Japan, a large number of combined cycle and co-generation plants have been installed. The plant efficiencies vary from 47% to 64% as per pressure ratio, turbine inlet temperature, configuration, component efficiencies, etc. Extensive research is going on to obtain efficiency above 64%. For more details, References [46 to 70] may be consulted.

14.1 INTRODUCTION

The concept and development of combined cycle originated from the utilisation of waste heat of a thermal power plant. A combined cycle, as the name implies, is a *combination of two cycles operating at different temperatures, each of which could operate independently.* The heat rejected by the higher temperature cycle is recovered and used by a lower temperature cycle to produce additional power to realise an improved overall efficiency. For the combination, the separate cycles must operate on separate fluids. Combined cycles which have been commercial in service include diesel-steam, mercury-steam and gas turbine-steam turbine. Development work is going on for steam-organic fluid, gas-organic fluid, liquid metal-steam and MHD-steam.

The higher temperature cycle is called the ***topping cycle*** (or topper) and *the lower temperature cycle is the* **bottoming cycle (or bottomer)**. Topping cycle may include, Otto, **Brayton** and Rankine cycles. Bottoming cycle have all been **Rankine cycle**, in all the cases.

In the gas/steam combined cycle power plant (Fig. 14.2) the gas turbine working on Joule cycle is the topping cycle where steam cycle working on Rankine cycle is the bottoming cycle. *The gas/steam combined cycle efficiency ranges from 47% to 64% as compared to 35% to 40% of conventional steam or gas turbine power plant.*

The heat rejected by gas turbine is highly appreciable as the exhaust temperature varies from 450 to 650°C depending upon pressure ratio and turbine inlet temperature. This energy goes as a waste if it is rejected to atmosphere. This waste heat energy may be utilised to produce steam in a heat recovery steam generator (HRSG) and may be expanded in steam turbine to develop additional power based on Rankine cycle. This combination is known as gas/steam combined cycle power plant. The gas cycle and steam cycle may use separate generator or a single generator. *It is desirable that the exhaust temperature of gas turbine should be above **570°C** otherwise the steam cycle will be inefficient resulting in lower combined cycle plant efficiency.*

If the steam generated in HRSG is utilised for process heating the system is called co-generation, *which is defined as the simultaneous production of electrical power as well as heat energy from the same source.* The efficiency of co-generation ranges from **80 to 90%** due to utilisation of exhaust heat energy used as **heat power**.

A mixed cycle *utilizes the two working fluids in one power cycle (gas turbine) to improve the power output and thermal efficiency. Steam injected cycle is one of the examples of mixed cycle.* The efficiency of mixed cycle is slightly lower than combined cycle.

The most suitable fuel for combined cycle is **natural gas** and **LNG**. However, plants using liquid and coal based gaseous fuels are also established.

In the past twenty years, the development in power gas turbine is enormous. Previously, the power output from gas turbine was merely 1 MW to 30 MW but now simple open gas turbine is capable of developing 400 MW. Table 14.1 shows the specifications of *latest combined cycle* power plants manufactured by leading companies of the world. It depicts the total power, shared power, efficiency, number and type of gas turbines, and types of HRSG-used by various companies. Unfired HRSG is widely used in combined, mixed and co-generation cycles. The most suitable fuel for combined cycle is natural gas, LNG and (NG + H_2) but oil may also be used. In some cases, fired HRSG is used.

14.2 COMBINED GAS/STEAM, MIXED AND CO-GENERATION CYCLES

Figure 14.1 shows the classification of combined gas/steam, mixed and co-generation cycles.

Table 14.1 Specifications of combined cycle power plants [*Courtsey*: Gas Turbine World, 2000]

Model	*Year*	*Net plant output (MW)*	*Net plant efficiency (%)*	*Heat rate (kJ/kWh)*	*Electrical frequency*	*GT power (MW)*	*ST power (MW)*	*No. and Type of units*	*Comments (based on Gas turbine natural gas as fuel)*
ABB ALSTOM POWER									
KA 13 E2-2	1993	480	52.9	6805	50 Hz	318.6	167.0	2 × GT13E2	Dual pressure HRSG
KA 13 E2-3	1993	720	52.9	6805	50	477.9	249.0	3 × GTBE2	Dual pressure HRSG
KA 26–1	1996	393	57.9	6217	50	251.0	142.0	1 × GT26	Tripple pressure HRSG, single shaft
BHEL									
CC2942	1998	478	52.3	6883	50 Hz	308	177	2 × V94.2	Dual pressure
CC209FA	–	772	55.7	6461	50	503.4	279	2 × MS9001FA	Triple pressure
CC109 FA	–	383	554	6502	50	251.7	13.70	1 × MS9001FA	Triple pressure reheat
GE Power Systems									
S109FA	1994	390.8	56.7	6350	50 Hz	254.1	141.8	1 × MS9000FA	Triple pressure reheat
S209FA	1994	786.9	57.1	6305	50	508.2	289.2	2 × MS9001FA	Triple pressure reheat
S109H	1997	480.0	60.0	6000	50	–	–	1 × MSS9001H	Triple pressure reheat
9 HA.02	2020	1440.0	64.0	–	60	–	–	2 × MS 70001FB	Triple pressure reheat
Hitachi									
107 FA	1995	253.7	55.3	6510	60 Hz	164	89.7	1 × MS 7001FA	–
207 FA	1995	509.2	55.5	6490	60	328	181.2	1 × MS 7001FA	–
Mitsubishi heavy Ind. MPCIPI									
(M701G)	1997	484.4	58.0	6208	50	328.9	155.5	1 × M701G	–
MPCP2									
(M 701G)	1997	972.1	58.2	6187	50	657.8	314.3	2 × M701G	–
Siemens Westinghouse									
2V942A	1997	584.5	55.0	6550	50 Hz	367	230	2 × V942A	Dual pressure
ISV943A	1995	385.5	57.1	6300	50	–	–	1 × V943A	Triple pressure reheat
2V943A	1995	771.0	57.1	6300	50	512.0	275.0	2 × V943A	Triple pressure, reheat, single shaft
2W501G	1994	730.0	58	6210	60	490.0	253.0	2 × W502G	Triple pressure, reheat

14.3 EARLY HISTORY

The history of the combined cycle power plant goes back to the early part of this century. In 1925, Emmet described substantial development work on a mercury-steam plant. This presentation was based on the basic work of Charles Bradley of 1913.

The early development of the gas/steam turbine plant came somewhat later and was described by Seippel and Beseuter (1960). They proposed seven possible combinations of gas/steam combined cycle. In 1978, Wunsch gave classification of combined cycle plant.

By the 1979, the so called 'recuperative plant', higher level gas turbine, exhausting to a heat recovery steam generator (HRSG), which supplies steam to the lower (steam turbine) cycle with no supplementary heating of the exhaust had become well established, primarily by General Electric and Westinghouse in US and by Brown Boveri in Europe.

In 1997, Wood gave a list of some 40 such plants of USA in the range of 15–20 MW with one exception of 63 MW using supplementary firing.

Nowadays, gas/steam combined cycle are gaining popularity. Over 200 MW gas turbines, single units all over the world are working. Now research is going on for 400 MW single unit gas turbine. Specifications of some of the combined cycle plants are given in Table 14.1. Briesch et al. [46] and GE have advanced to achieve cycle efficiency greater than 64% [Ref [46] to [70] give an account of advancements made in the combined cycle].

14.4 COMBINED CYCLE POWER PLANTS IN INDIA

In india, NTPC initially planned three major combined cycle power plants totalling 1600 MW based on natural gas available from HBJ gas pipe line. This was a quite successful experience and NTPC continued to plan and build large size gas/steam combined cycle power plants by 2004. Tables 14.2 shows NTPC gas based projects.

Table 14.2 NTPC gas based combined gas/steam cycle power plants

Project	*Location*	*Total capacity (MW)*	*Units commissioned (MW)*	*Fuel source*
Anta	Baran, Rajasthan	413	3 × 88 *GT* + 1 × 149 *ST*	HBJ Pipeline
Auraiya	Etawah, UP	652	4 × 112 *GT* + 2 × 102 *ST*	HBJ Pipeline
Kawas	Surat, Gujrat	645	4 × 106 *GT* + 2 × 110 *ST*	South Basin Gas field
Dadri	Ghaziabad UP	814	4 × 131 *GT* + 2 × 146.5 *ST*	HBJ Pipeline
Jhanor Gandhar	Bharuch, Gujarat	648	3 × 131 *GT* + 1 × 255 *ST*	Gandhar Gas Field
Kanyan Kulan	Allapozha, Kerala	400 (3.50)	Under progress	Naptha from BPCL.

14.5 CONFIGURATIONS OF COMBINED CYCLE POWER PLANTS

A large number of configurations of unfired combined cycle is possible however, only the main configuration will be discussed and from that many other configurations may be evolved. It has been already discussed in the chapter on Gas Turbines that the efficiency/specific work may be increased by the addition of intercooling, reheating and recuperation arrangement in

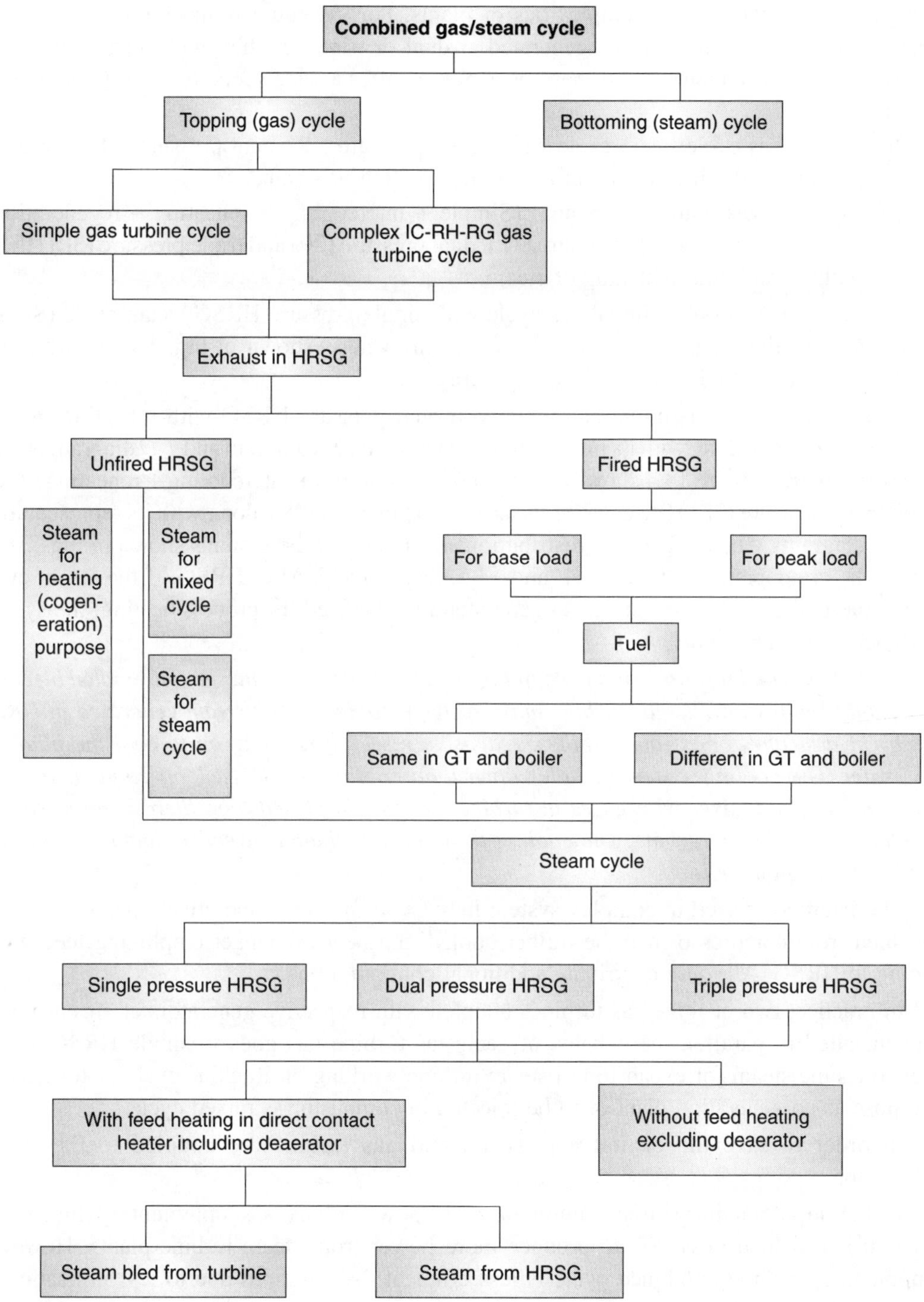

Figure 14.1 Classification of combined gas/steeam, mixed and co-generation cycles

simple cycle at the cost of complexities of plants. For the better utilisation of exhaust gas energy in HRSG, steam may be generated at dual or triple pressure with reheat instead of single pressure arrangement. All these possibilities offer a wide scope for a large number of configurations.

(a) Simple gas cycle/steam cycle with single pressure (IP), dual pressure (2P) and triple pressure (3P) HRSG without and with reheating of steam.

(b) Complex gas with two or more (Simple + intercooling + reheating + regeneration)/ steam cycle with single pressure (1P), dual pressure (2P) and triple pressure (3P) HRSG without and with reheating of steam.

Figure 14.2(a) shows a simple gas cycle with single pressure HRSG steam cycle (SGIP). The representation of processes of gas and steam cycle is shown in Fig. 14.2(b) while the *T-Q* diagram of HRSG is shown in Fig. 14.2(c).

Figure 14.3 shows a simple gas/cycle with dual pressure HRSG with reheating of dual pressure (SG2PR) along with its process representation on *T-s* diagram and *T-Q* diagram of dual pressure HRSG. Figure 14.4 shows the complex gas (simple + intercooling + reheating) cycle with triple pressure (3P) HRSG with reheating of steam (CG3PR) along with its representation on *T-s* diagrams. The temperature distributions in HRSG for the systems shown in Figs. 14.2. and 14.3. are depicted in Figs. 14.4. and 14.5. respectively. After HRSG, if the steam cycle is absent in above case, and steam so generated may be used for process heating, the system will act as *co-generation plant.*

In all these configurations as shown in Fig. 14.2(a), the gas turbines are air cooled bleeding air from compressor. The gas turbine may be steam cooled by the steam generated in HRSG or by bled steam from steam turbines. Extensive research is going on to cool the bladings by water. The cooling system includes open loop cooling or closed loop cooling. In open loop, the coolant mixes with gases in turbine whereas in closed loop there is no mixing of coolant and gases in turbine. The cooling techniques include convective, impingement, film and transpiration cooling.

The intercooler used in complex system may be surface or evaporative type. The cooling air bled from compressor may be further cooled in a heat exchanger employing feed water going to HRSG. All these contribute additional configurations.

In practice, two or three gas turbines coupled with respective generators at air entry side are installed in parallel, the exhaust of each gas turbine gas goes to single HRSG which generates the steam for expansion in steam turbine working on Rankine cycle or for heating purpose as co-generation plant or to be injected in combustor as mixed cycle.

In order to save the capital cost steam and gas turbine are coupled to the single alternator.

In the supplementary fired combined cycle power plant, a supplementary fuel firing system is provided in HRSG to produce more power from steam turbine plants. However, supplementary fired combined cycle is less efficient than unifired due to less utilisation of waste and supplementary heat and so it is seldom used.

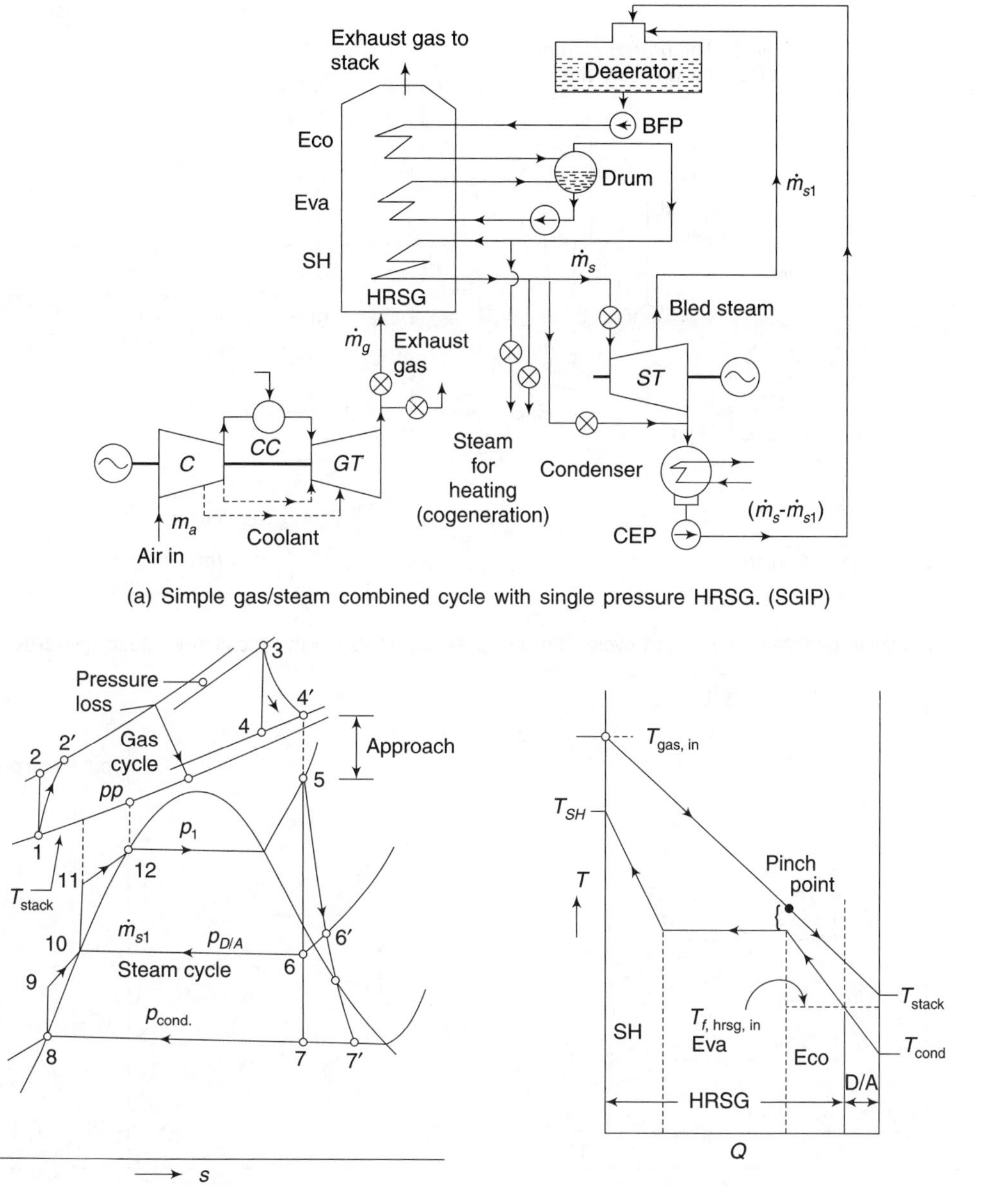

(a) Simple gas/steam combined cycle with single pressure HRSG. (SGIP)

(b) T-s representation of SGIP system

(c) T-Q representation single pressure HRSG (IP)

Figure 14.2

14.6 MIXED CYCLE

Since the bottoming cycles need extensive cost, so efforts are going on to disperse with it and may use mixed cycle whose performance is slightly inferior or comparable to combined

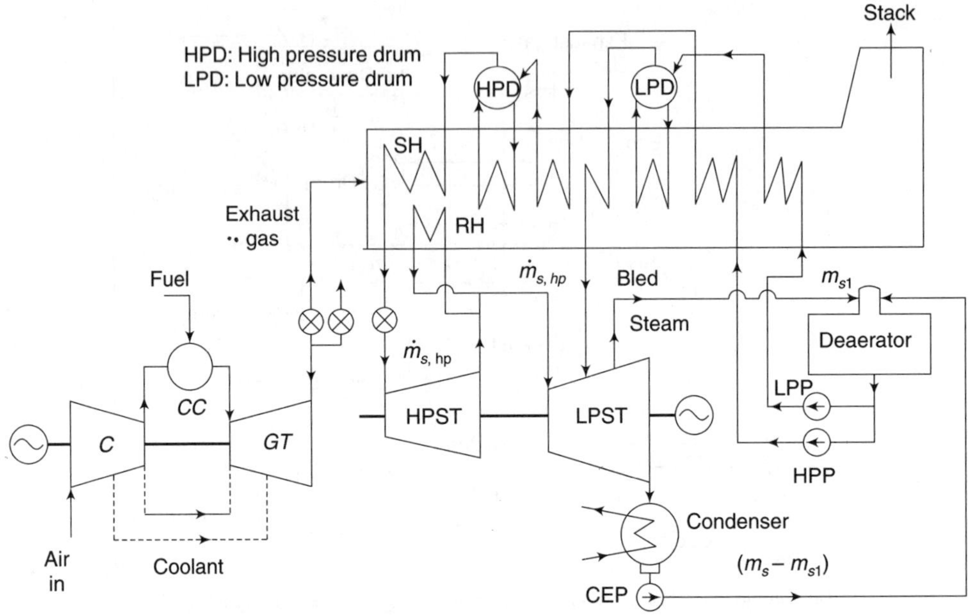

(a) Simple gas/steam combined cycle with dual pressure HRSG with reheating of steam **(SG2PR)**

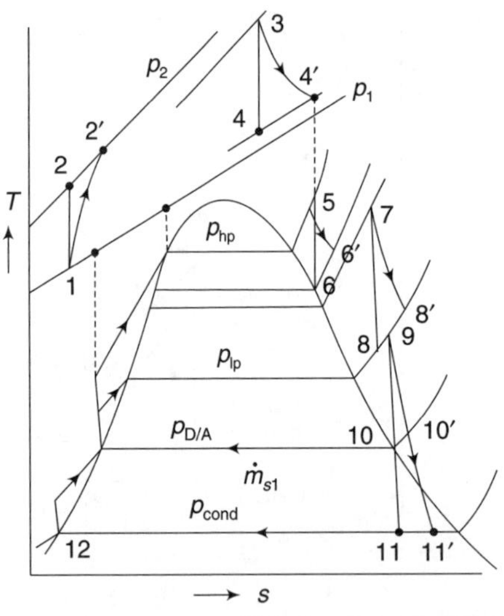

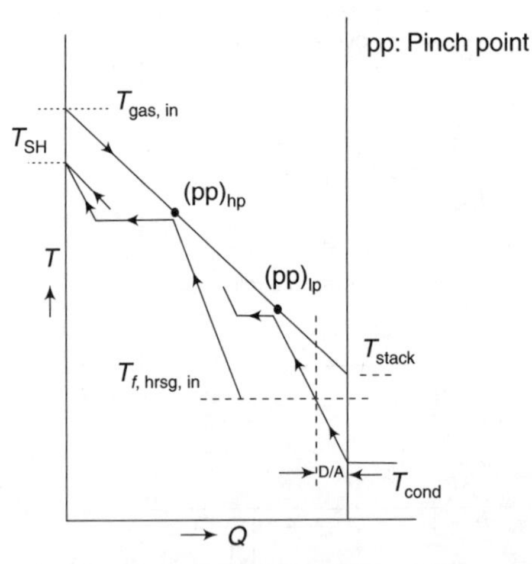

(b) T-s representation of SG2PR system

(c) T-Q representation of dual pressure HRSG with reheat (2PR)

Figure 14.3

cycle. In these cycles, major part of steam generated is fed to the combustion chamber, get heated upto turbine inlet temperature (TIT) and expands in gas turbine, however, the rest part of steam generated is used for bled cooling. Such system is either called steam injected

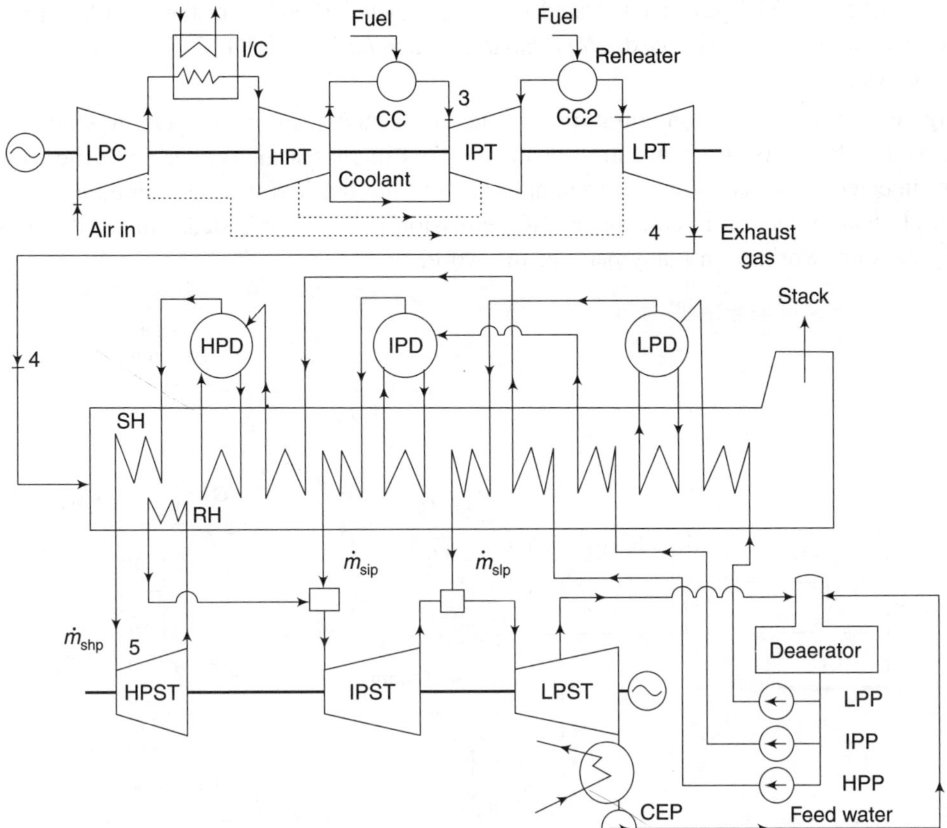

(a) Complex gas (Simple + Intercooling + Reheat) with triple pressure HRSG with reheating of steam **(CG3PR)**

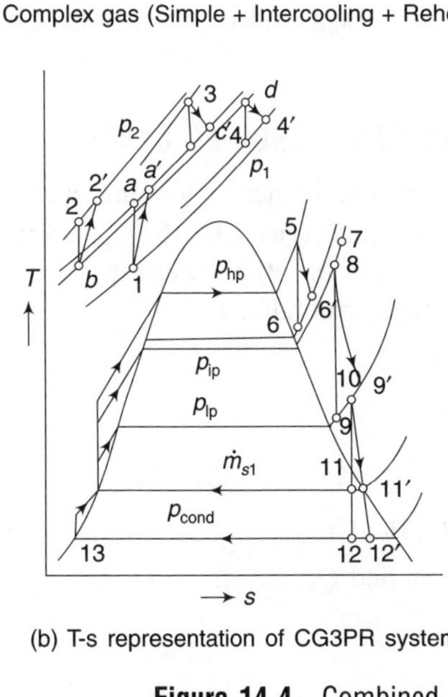

(b) T-s representation of CG3PR system

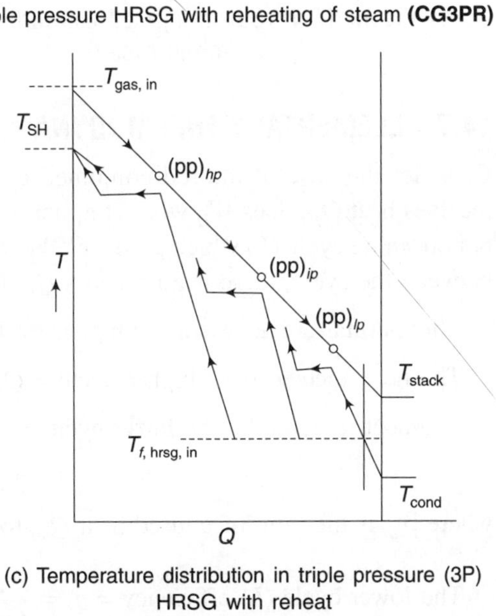

(c) Temperature distribution in triple pressure (3P) HRSG with reheat

Figure 14.4 Combined complex cycle with 3P HRSG with reheat

gas turbine cycle (SIGTC) or integrated gas steam cycle (IGSC). In other words, in mixed cycle. two fluid streams are used. *Humidified gas turbine cycle* (HGTC) is another example of mixed cycle.

Figure 14.5 shows the configuration of a steam injected gas turbine cycle in which steam generated is fed into the combustion chamber. This increases the output and efficiency of steam injected cycle as compared to simple gas turbine but less than combined cycle. As a result of steam injection in combustor, NO_x emission is less. Some steam injected plants of small sizes are working in many parts of the world.

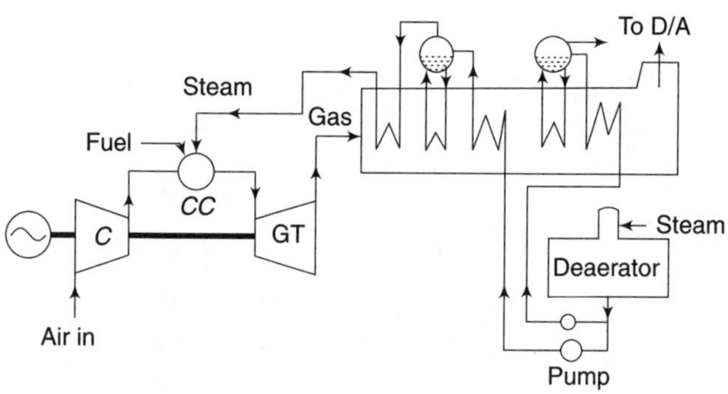

Figure 14.5 Steam injected gas turbine cycle (mixed cycle)

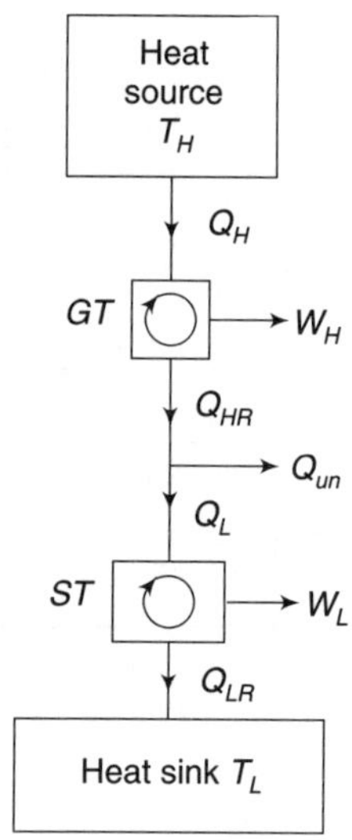

Figure 14.6 Unifired combined cycle

14.7 ELEMENTARY THERMODYNAMICS OF UNFIRED COMBINED CYCLE

Consider the case of unfired combined cycle (Fig. 14.6).The higher or topping cycle (H) receives heat Q_H, does W_H work at a thermal efficiency of η_H. It rejects heats Q_{HR} to a lower or bottoming cycle (L) which produces W_L work output at a thermal efficiency of η_L. However, between the cycle, there is a heat loss Q_{un} (unused) to the surroundings.

The thermal efficiency of the higher cycle = $\eta_H = W_H/Q_H$

The heat rejected from higher cycle = $Q_{HR} = Q_H(1 - \eta_H)$

The heat received by the lower cycle = $Q_L = Q_{HR} - Q_{un} = Q_H[1 - \eta_H] - Q_{un}$

$$= Q_H(1 - \eta_H) - \beta_{un} Q_H \quad ...(14.1)$$

where β_{un} is the ratio of unused heat Q_{un} to the heat supplied Q_H, i.e., $\beta_{un} = Q_{un}/Q_H$

The lower cycle (L) efficiency = $n_L = \dfrac{W_L}{Q_L}$

The thermal efficiency of combined cycle plant (η_{ccp}) is given by

$$\eta_{ccp} = \frac{W_H + W_L}{Q_H} = \frac{\eta_H Q_H + \eta_L \{Q_H(1-\eta_H) - \beta_{un} Q_H\}}{Q_H}$$

$$\eta_{ccp} = \eta_H + \eta_L - \eta_H \eta_L - \beta_{un} \cdot \eta_L \quad \text{...(14.2)}$$

The expression for combined cycle plant efficiency may also be written in terms of a "boiler efficiency", (η_B) of the heat exchange process between the two cycles (the heat recovery steam generator. HRSG for the gas/steam turbine plant.) Thus, η_B is defined as

$$\eta_B = \frac{Q_l}{Q_{HR}} = \frac{Q_{HR} - Q_{un}}{Q_{HR}} = 1 - \frac{Q_{un}/Q_R}{Q_{HR}/Q_R} = 1 - \frac{\beta_{un}}{1-\eta_H} \quad \text{...(14.3)}$$

The combined cycle plant efficiency (η_{ccp}) may be written as

$$\eta_{ccp} = \eta_H + (\eta_L Q_L)/Q_B = \eta_H + \eta_L \eta_B (Q_{HR}/Q_B)$$

$$= \eta_H + \eta_B \eta_L - \eta_B \eta_L \eta_H \quad \text{...(14.4)}$$

$$= \eta_H + (\eta_0)_L - \eta_H (\eta_0)_L \quad \text{...(14.5)}$$

where $(\eta_o)_L = \eta_B \eta_L$ is the overall efficiency of the lower plant, including the HRSG.

For illustrating this elementary thermodynamic analysis for a practical operating station, consider the Brown Boveri combined cycle plant which has the following data: Combined cycle output = 125 MW, Gas turbine output = 81 MW, Steam turbine output = 49 MW, Heat supplied based on 276 MW, LCV of natural gas, Gas turbine efficiency = 0.29, Temperature of gas turbine exhaust = 491°C, Temperature of HRSG exhaust = 90°C, Energy lost in the exhaust = 34 MW, Steam power plant efficiency = 32%

$$\eta_{un} = \frac{34}{276} = 0.12 \quad \text{and} \quad \eta_B = \frac{0.12}{(1-0.29)} = 0.83$$

Steam plant thermal efficiency = 0.32, so $(\eta_o)_L = 0.32 \times 0.83 = 0.27$

The combined plant efficiency = $\eta_{ccp} = \eta_H + \eta_L - \eta_H \eta_L - \beta_{UH} \cdot \eta_L$

$= 0.29 + 0.32 - 0.29 \times 0.32 - 0.12 \times 0.32 = 0.48 =$ **48%**

Also, $\eta_{ccp} = \eta_H + (\eta_o)_L - \eta_H (\eta_o)_L = 0.29 + 0.27 - 0.29 \times 0.27 = 0.48 =$ **48%**

The above results show that the combined cycle plant efficiency is much higher than individual topping and bottoming cycle plants.

If the individual cycle efficiency having higher value which is currently available

e.g., $\eta_H = 39\%$ and $\eta_L = 35\%$, and $\beta_{un} = 0.12$, then we have

$\eta_{ccp} = 0.39 + 0.39 - 0.35 \times 0.35 - 0.12 \times 0.35 = 56.15 =$ **56.15%**

14.8 THERMODYNAMIC ANALYSIS OF COMBINED CYCLE AND CO-GENERATION PLANTS

14.8.1 Ideal Cycle Analysis

In ideal cycle analysis, no pressure losses and aerodynamic losses are considered and specific heat of air and gases are the same and constant. There is no cooling of gas turbine bladings. Mass of fuel is neglected as heat input is due to heat transfer alone. Consider 1 kg/s of air entering the compressor. The analysis may be based on static or total values. Refer to Figs. 14.2 to 14.5

Gas properties: $c_{pa} = c_{pg} = 1.005$ kJ/kgK constant, $\Delta h = c_p \Delta T$ and $\dot{m}_a = \dot{m}_g$ as $\dot{m}_f = 0$

Compressor: The compressor work = $w_c = h_{out} - h_{in} = c_p (T_{out} - T_{in})$ ...(14.6)

Thus, for simple gas cycle (Fig. 14.2), the compressor work is given by

$$w_c = h_2 - h_1 = c_p (T_2 - T_1) \quad ...(14.7)$$

For complex gas cycle (Fig. 14.3) the compressor work is given by

$$w_c = (h_a - h_1) + (h_2 - h_b) = c_p [(T_a - T_1) + (T_2 - T_b)] \quad ...(14.8)$$

Intercooler: For perfect intercooling, $T_3 = T_1$

Combustion chamber: For a given turbine inlet temperature (TIT), the fuel flow rate per kg of air is found from the energy equation

$$q_A = \dot{m}_f \times (\text{LHV})_f = h_{out} - h_{in} = c_p (T_{out} - T_{in}) \quad ...(14.9)$$

Thus, for simple gas cycle (Fig. 14.2)

$$\dot{m}_f \times (\text{LHV})_f = (h_3 - h_2) = c_p (T_3 - T_2) \quad ...(14.10)$$

For complex gas cycle (Fig. 14.3), the fuel flow rate per kg of air is found as

$$q_A = \dot{m}_f (\text{LHV})_f = (h_3 - h_2) + (h_d - h_c) = c_p [(T_3 - T_2) + (T_d - T_c)] \quad ...(14.11)$$

For simplicity, $\dot{m}_f$ is neglected for calculating w_c and w_t

Gas Turbine: The turbine output = $w_t = (h_{out} - h_{in}) = c_p (T_{out} - T_{in})$; kJ/kg ...(14.12)

For simple gas cycle (Fig. 14.2), $w_{gt} = (h_3 - h_4) = c_p (T_3 - T_4)$

For complex gas cycle (Fig. 14.4), $w_{gt} = (h_3 - h_c) + (h_d - h_4)$ ...(14.13)

$$= c_p [(T_3 - T_c) + (T_d - T_4)] \quad ...(14.14)$$

HRSG: In general, it should be kept in mind that the temperature and pressure of steam depends upon the temperature of gas turbine exhaust entering HRSG. The steam cycle efficiency is satisfactory only when the gas turbine exhaust temperature is above 570°C. For single pressure HRSG (IP) system, the throttle pressure varies from 40 to 50 bars while in dual pressure (2P) system (Fig. 14.3) the HP throttle pressure vaies from 80 to 120 bars. In triple pressure HRSG system (Fig. 14.4), the HP throttle pressure varies from 120 to 166 bars. The throttle temperature is generally limited to 570°C, however in some cases it may be taken up to 600°C. It is to be noted that the approach temperature difference has to be at least 20°C.

The mass of steam generated per kg of air, $\dot{m}_s$ (Fig. 14.4) is expressed as

$$\dot{m}_s\,(\Delta h_{eco} + \Delta h_{eva} + \Delta h_{sup} + \Delta h_{reh}) = (h_{out} + h_{in})$$

where $\dot{m}_s$ is the mass of steam generated in HRSG per kg of gas.

For simple (SG2P) and complex gas cycle (CG3P) the mass of steam generated in HRSG is given by

$$\dot{m}_s\,(\Delta h_{eco} + \Delta h_{ev} + \Delta h_{sup} + \Delta h_{reh}) = (h_4 - h_{stack}) \quad \text{...(14.15)}$$

and

$$\dot{m}_s\,[\Delta h_{eco} + \Delta h_{ev} + \Delta h_{sup} + \Delta h_{reh}] = [h_4 - h_{stack}] \quad \text{...(14.16)}$$

where Δh = enthalpy rise in respective elements of HRSG.

The stack temperature is limited by the possibilities of condensation of combustion products in stack and thus its minimum value is 78°C. Generally, stack temperature ranges from 85 to 120°C. While calculating mass of steam ($\dot{m}_s$) generated, pinch point has to be satisfied.

Steam Turbine: Assuming isentropic expansion in the turbine, the work done by the steam turbine is given by for single pressure HRSG (Fig. 14.2)

$$w_{st} = \dot{m}_s\,(\Delta h)_{\text{isen},\ bb} + (\dot{m}_s - \dot{m}_{s1})\,(\Delta h)_{\text{isen},\ ab} \quad \text{...(14.17)}$$

where suffixes *bb* and *ab* stand for before bleeding of steam for deaerator and after bleeding of steam.

Condenser: Under ideal condition, the condensate temperature is equal to the saturation temperature corresponding to condenser pressure. The heat rejected by condenser is given by

$$q_R = (\dot{m}_s - \dot{m}_{s1})\,(h_{fg})_{\text{cond}};\ \text{kJ/kg of air} \quad \text{...(14.18)}$$

Feed Pump: The work input to the feed pump is given by

$$w_p = \dot{m}_{cs}\,(h_{fg} - h_{fi})_{\ p} = \dot{m}_{cs}\,v_{\text{fcon}}\,(p_{\text{boiler}} - p_{\text{cond}}) \quad \text{...(14.19)}$$

where $\dot{m}_{cs}$ = mass of condensed steam.

Overall Performance: $w_{ccp} = w_{gt,\ \text{net}} + w_{st,\ \text{net}} = (w_{gt} - w_c) + (w_{st} - w_p)$ kJ/kg of air ...(14.20)

$$\eta_{ccp} = w_{ccp}/q_A \text{ (First law of efficiency)} \quad \text{...(14.21)}$$

The second law of efficiency (exergetic efficiency) is given as

$$\eta_{ex} = \frac{w_{ccp}}{(e_x)_f} \quad \text{...(14.22)}$$

where $(e_x)_f$ = exergy content of fuel input.

Co-generation Performance Parameters: In general, mainly these parameters, namely, fuel utilisation efficiency (η_f), power to heat ratio (R_{ph}) and exergetic efficiency (η_{ex}) are used to predict the performance of a co-generation plant.

Fuel utilisation efficiency, $\eta_f = \dfrac{w_{gt,\,net} + q_P}{q_A}$

where q_P = process heat production = $(h_g)_{exit} - (h_g)_{stack}$ kJ/kg of air

Power to heat ratio, $R_{pH} = \dfrac{w_{gt,\,net}}{q_P}$

Exergetic Efficiency = $(w_{gt,\,net} + q_P)/(e_x)_f$

14.8.2 Actual Cycle Analysis

In the actual cycle, the fluid flow losses occur in every component of the combined cycle. The analysis of actual cycle is affected by introducing the concept of effectiveness, efficiency and pressure losses, etc.

Gas Properties: As mentioned above the most suitable fuel for combined cycle power plants is natural gas. The lower calorific value of natural gas is around 42 MJ/kg. The composition of natural gas is 94% CH_4, 3% C_2H_6, 1% N_2 on mass basis. All gaseous mixtures at any point are treated as the mixture of air, water vapour and combustion products with excess air in their varying proportions.

All non-reacting *gases may be arbitrarily assigned zero thermodynamical enthalpy at the ambient temperature regardless of chemical composition.* In practice, it is never constant and it is a function of temperature and very high pressure. The specific heats and enthalpies of air and gases have been obtained in the from of polynomial fits, as the functions of temperature

$$c_p(T) = a + bT + cT^2 + dT^3$$

$$h = \int_{T_a}^{T} c_p(T)\, dT \qquad ...(14.23)$$

For approximation, the average value of specific heat of gases may be taken which is much higher than that of air, i.e., $c_{pg} \approx$ 1.15 to 1.2 kJ/kgK.

Compressor: The compressor, which is axial flow in nature supplies compressed air to the combustion chamber/intercooler. Air is bled from the compressor for turbine blade cooling. For a compressor spool with specified inlet state and polytropic efficiency, the actual overall temperature is expressed as

$$\frac{T_2'}{T_1} = \left(\frac{p_2}{p_1}\right)^{\frac{n-1}{n}} \qquad ...(14.24)$$

The polytropic index, n is given by the relation $\dfrac{n-1}{n} = \dfrac{\gamma - 1}{\gamma \eta_{pc}}$ where η_{pc} may be taken as 90 to 91%. Thus, for polytropic stages (assuming large number of small stages)

Thus, $$\frac{T_{out}'}{T_{in}} = \left(\frac{p_{out}}{p_{in}}\right)^{\frac{n-1}{n}}$$

The number of stages for an axial flow compressor may be expressed as by assuming equal pressure in each stage (1.12 to 1.2) or by assuming equal temperature drop. Thus,

$$\text{The number of stages} = z = \frac{\ln r_p}{\ln(r_p)_{\text{stages}}} \qquad ...(14.25)$$

$$\text{Alternatively, } z = \frac{\Delta T'}{(\Delta T')_{\text{stage}}}$$

where $\Delta T'$ = overall actual temperature rise and $(\Delta T')_{\text{stage}}$ = temperature rise (20K–30K)

Using the concept of isentropic compressor efficiency, the overall exit temperature is expressed approximately as

$$\eta_c = \frac{T_{out} - T_{in}}{T'_{out} - T_{in}} \qquad ...(14.26)$$

where T'_{out} = actual temperature of compressed air.

It is worth to note that the concept of polytropic efficiency will give better representation of actual compression for any pressure ratio.

The mass and enthalpies balance for the compressor work yields as

$$\dot{m}_{in} = \dot{m}_{out} + \Sigma\, \dot{m}_{ci} \qquad ...(14.27)$$

$$w_c = m_{out}\, h_{out} - \Sigma\, m_{ci}\, h_{ci} - m_{in}\, h_{in} \qquad ...(14.28a)$$

Without considering cooling of turbine blading

$$w_c = \dot{m}_{out}\, h_{out} - \dot{m}_{in}\, h_{in} \qquad ...(14.28b)$$

where Σm_{ci} represents the total cooling flow requirements. After specifying the pressure at the desired state, the temperature and specific enthalpies may be determined at that state. After the turbines flow cooling, calculations have been performed, one may determine the mass flows extracted from each bleed port from the compressor. If the cooling of the turbine blading is neglected, then $\Sigma m_{ci} = 0$.

Intercooler: In intercooler, pressure loss occurs and the intercooling is imperfect. The actual temperature of air leaving the intercooler is related with effectiveness

$$\in_{ic} = \frac{\Delta T_{act}}{\Delta T_{\max}} = \frac{T_2 - T_3}{T_2 - T_1} \qquad ...(14.29)$$

If $\in_{ic}$ is known as the actual temperature leaving the intercooler may be calculated.

Combustion chamber: In combustor, the combustion does not take place under ideal conditions but is accompanied by fuel throttling loss, mainflow pressure drop loss, incomplete combustion loss, and other losses. For a given turbine inlet temperature (TIT), pressure drop and combustion efficiency, the fuel flow is found by the mass and energy balance.

$$\dot{m}_{out} = \dot{m}_{in} + \dot{m}_f \qquad ...(14.30)$$

$$\dot{m}_f\, \eta_{cc}\, (LCV)_f = \dot{m}_{out}\, h_{out} - \dot{m}_{in}\, \eta_{in} \qquad ...(14.31)$$

Turbine: The number of stages in the turbine depends on the expansion ratio which varies from 2 to 2.5. The maximum number of turbine stages is generally 4 to 5. The turbine is cooled by the air bled from compressor. However, blades may also be cooled by steam generated in HRSG or by fresh water (under development). The concept of polytropic efficiency is used to calculate the actual temperature. Alternatively, the concept of isentropic efficiency is used to calculate the actual exhaust temperature using efficiencies.

For cooled turbine stages the mass and energy balance may be written as

$$\dot{m}_{out} = \dot{m}_{in} + \Sigma m_{ci} \quad \text{...(14.32)}$$

$$w_{gt} = \dot{m}_{in}\, h_{in} + \Sigma \dot{m}_{ci}\, h_{ci} - \dot{m}_{out}\, h_{out} \quad \text{...(14.33)}$$

The actual temperature, $$\frac{T'_{out}}{T_{in}} = \left(\frac{p_{out}}{p_{in}}\right)^{\frac{n-1}{n}} \quad \text{...(14.34)}$$

where $$\frac{n-1}{n} = \frac{(\gamma - 1)\,\eta_{pt}}{\gamma}$$

Approximately, the concept of turbine isentropic efficiency may be used to calculate actual temperature at the exit of turbine.

$$\eta_t = \frac{h'_{out} - h_{in}}{h_{out} - h_{in}} = \frac{T_{out}' - T_{in}}{T_{out} - T_{in}} \quad \text{...(14.35)}$$

where $\Sigma \dot{m}_{ci}$ represents the sum of cooling flows to each blade row and are found from expression given in my book Steam & Gas Turbines, T, PPE.

An additional coolant leakage flow is measured as 0.4% of local gas flow. Generally, the air coolant required varies from 10% to 20% of entering air to the compressor depending on cooling medium techniques, turbine inlet temperature, compressor pressure ratio, etc.

HRSG: As discussed above, HRSG may be a *single pressure, dual pressure or triple pressure system.* It may also incorporate reheater unit. The inefficiency of HRSG is taken care by efficiency or effectiveness of HRSG. In triple pressure HRSG with reheat, the recovery of exhaust gas energy is the maximum at the cost of complexity.

The mass of steam generated per kg of air is expressed as

$$\dot{m}_s\,(\Delta h_{eco} + \Delta h_{eva} + \Delta h_{sup} + \Delta h_{reh}) = \dot{m}_g\,(h_{gin} - h_{gout}) \in_{hrsg} \quad \text{...(14.36)}$$

where $\in_{hrsg}$ = effectiveness of HRSG

Steam Turbine: The internal losses in steam turbine are taken care by introducing the concept of steam turbine isentropic efficiency. Thus, for system shown in Fig. 14.2.

$$\dot{m}_s\,(h_5 - h_6')_{bb} + (\dot{m}_s - \dot{m}_{s1})\,(h_6' - h_7')_{ab}\ ;\ \text{kJ/kg of air} \quad \text{...(14.37)}$$

where $\dot{m}_s$ is the mass flow rate per kg of air generated in HRSG and expanding through steam turbine and $\dot{m}_{s1}$ is the amount of bled steam for deaerator. Here, suffixes *bb* and *ab* represent before bleeding and after bleeding.

Condenser: In real situation, there is an undercooling of the condensate due to efficient heat transfer and pressure loss in condenser. The enthalpy of the condensate is calculated based on the degree of undercooling which generally varies from 4°C to 6°C. Thus, enthalpy of the condensate is calculated

$$q_{rej} = (\dot{m}_s - \dot{m}_{s1})(h_{ex,\ act} - h_{\text{fcond}})$$

$$= (\dot{m}_s - \dot{m}_{s1})(h_7' - h_{f8}) \quad \text{...(14.38)}$$

Feed pump: Due to inefficiency, more pump work has to be supplied

$$w_p = \dot{m}_s\,(h_{fo} - h_{f1})/\,\eta_p = \dot{m}_{cs}\,.\,v_{cs}\,(p_{hsrg} - p_{cond})/\eta_p \quad \text{...(14.39)}$$

where η_p = overall pump efficiency (say 80 to 90%)

Overall performance of combined cycle plant:

$$(w_{ccp})_{act} = (w_{gt} - w_c)_{act} + (w_{st} - w_p)_{act} \quad \text{...(14.40)}$$

$$(\eta_{ccp})_{act} = (w_{ccp})_{act}/q_{A_{act}} \quad \text{...(14.41)}$$

14.9 ADVANTAGES OF COMBINED CYCLE POWER GENERATION

The following are the advantages of combined cycle power plants.

(i) **High overall plant efficiency.** The overall plant efficiency of combined cycle varies from 50 to 64% depending upon many factors which are much more than conventional steam power plant (35 to 40%).

(ii) **Less investment costs.** Since 2/3 of the output is produced by gas turbine (GT) unit and only 1/3 by steam turbine (ST) plant, so the investment cost per kW is approximately 30% less than that of steam power plant.

(iii) **Greater operating flexibility.** Due to gas turbine plant and simple steam power plant, there is a greater operating flexibility in terms of start-up and shut-down. The start-up losses are less.

(iv) **Less cooling water requirement.** Since the steam power (bottoming) cycle contributes only about 1/3 of the total power output, so the cooling water requirement in condenser is greatly reduced to about 50% as compared to conventional steam power plant.

(v) **Phased installation.** The installation of gas turbine plant (topping cycle) is much quicker, so it can start generating electricity while the installation of steam plant (bottoming cycle) may be under construction. This makes it possible to adjust the growth in demand for energy in a grid. Later on, combined cycle may be operated with coal gasification if price of natural gas or oil is increased considerably.

(vi) **Simplicity in operation.** Due to gas turbine plant as topping cycle, the operation is very simple as in the era of state-of-art technology, the operation of combined cycle plant is fully automatic. Even with less experience staff, the plant may be run.

(vii) **Less pollution.** In general, combined cycle uses natural gas as the fuel and since the efficiency is high, so the pollution level is low. The low NO_x emission is one of the attractive features of a combined cycle plant. It is worth to note that gas fired combined cycle plants produce per kWh only about 40% of CO_2 produced by coal fired plant. Virtually, there is absence of fly ash.

(viii) Possibility of co-regeneration. The combined cycle power plant is most suitable for co-generation. A part or full quantity of steam generated in HRSG may be used for process heating or a part of steam may be bled from LP turbine for process heating.

14.10 EXERGY ANALYSIS OF COMBINED CYCLES

The exergy of a system is the maximum work obtainable as the system comes to equilibrium with the surrounding. The higher the value of exergy, more is the work obtainable from the system. The traditional first law cycle analysis based upon component performance characteristics coupled with energy balances can invariably lead to the correct final result. However, such analysis cannot locate and quantify the sources of losses which lead to that result. This is because the first law embodies no distinction between work and heat, for example, in adiabatically throttling a pressurized gas. *It is the second law which quantifies the quality of energy and losses in a system. Energies of two systems may be quantitatively equal, but qualitatively they may be different. Exergy is a measure of energy quality and second law (exergetic) efficiency is a measure of the perfectness of a thermal system.* While energy of a system in any process constant, for example, in throttling process a part of its energy is always destroyed.

The first law (i.e., thermal) efficiency is expressed as

$$\eta_t = \frac{W_{out}}{Q} = \frac{W_{out}}{\dot{m}_f \, \Delta H_{r\,n}} \qquad ...(14.42)$$

While second law (exergetic) efficiency,

$$\eta_{II} = \frac{W_{out}}{E_{x,\,in}} = \frac{W_{out}}{\dot{m}_f \, \Delta G_{r\,n}} \qquad ...(14.43)$$

where $\Delta H_{r\,n}$ is enthalpy of reaction while $\Delta G_{r\,n}$ is Gibbs function of reaction and $E_{x,\,\text{in}}$ is the energy input.

In finding out Gibbs function for fuels, the ratio (ϕ) of chemical energy, $e_{g,0}$ to the net calorific value, $(NCV)_0$ or $(LCV)_0$ plays an important role. Kotas [94] have given expression for various fuels.

The chemical exergy for gaseous fuels of a given composition is given by

$$e_{x0} = -\,\Delta h_0 + T_0\,\Delta s_0 + RT_0 \left[x_{02} \ln \frac{p_{02}}{p_0} - \Sigma x_k \ln \frac{p_k}{p_0} \right] \qquad ...(14.44)$$

where the subscript k refers to the components of products of composition.

Typical values for some industrial fuels are given in Table 14.3.

Table 14.3 Typical values of Φ for some industrial fuels

S.No.	*Fuel*	Φ
(i)	Coke	1.05
(ii)	Coal	1.06 – 1.10
(iii)	Wood	1.15 – 1.3
(iv)	Fuel oils and petrol	1.04 – 1.08
(v)	Natural gas	1.04 ± 0.5%
(vi)	Coal gas	1.0 ± 0.5%
(vii)	Blast furnace gas	0.98 ± 1%
(viii)	Hydrogen	0.985
(ix)	Carbon monoxide	0.973

For chemically reactive system, the specific exergy

$e_{x,0} = \Delta g_{r\,n,0}$ = the change in the standard Gibb's function of reaction (kJ/kg) and $\Delta h_{r\,n,o} = -(NCV)_o$ = the change in the enthalpy of reaction (kJ/kg).

We have $$\Phi = \frac{-\Delta G_{r\,n,o}}{-\Delta H_{r\,n,o}} \qquad ...(14.45)$$

and $$\Delta H_{r\,n,o} = -\dot{m}_f\,(NCV)_0,\ \text{kW} \qquad ...(14.46)$$

where $\dot{m}_f$ = fuel burning rate (kg/s)

Here, suffix $r\,n$ represents reaction and suffix 0 stands for final dead or equilibrium condition (i.e., atmospheric condition)

From Eqs. (14.45) and (14.46), the value of $\Delta G_{r\,n,0}$ can be evaluated.

Further, $\Delta G_{r\,n,0} = \Delta H_{r\,n,0} - T_0 . \Delta S_{r\,n,0}$

Here $T_0 . \Delta S_{r\,n,0} = \Omega_{rn}$ = Exergy loss in combustion or reaction

Thus, $$\Omega_{rn} = T_0\,\Delta S_{r\,n,o} = -\dot{m}_f\,(NCV)_0 + \Phi[\dot{m}_f \times (NCV)_0] = \dot{m}_f \times (NCV)_0\,(\Phi - 1) \qquad ...(14.47)$$

where Φ is given in Table 14.3 for various fuels.

The exergy of fuel = $(\Delta G_{m,0}) = \Phi\,\dot{m}_f\,(NCV)_{f,o}$ ...(14.48)

The specific exergy of a flow stream for a given pressure p_1 and temperature T_1 can be computed by the following equation

$$e_x\,(p_1, T_1) = (h_1 - h_0) - T_0\,(s_1 - s_0) \quad \text{kJ/kg} \qquad ...(14.49)$$

For an ideal gas, the term $(s_1 - s_0)$ can be expressed as

$$s_1 - s_0 = \int_0^1 c_p \left(\frac{dT}{T}\right) - R \int_0^1 \frac{dp}{p} = c_p \ln \frac{T_1}{T_0} - R \ln \frac{p_1}{p_0} \qquad ...(14.50)$$

The exergy of the stream is given by

$$E_x = \dot{m}\,[h_1 - h_0) - T_0\,(S_1 - S_0)];\quad \text{kW} \qquad ...(14.51)$$

where $\dot{m}$ = mass flow rate (kg/s).

In general, the specific exergy loss (Ω) of a stream given by

$$\Omega = W_{\max} - W_{act} = W_{rev} - W_{act}\ ;\ \text{kW} \qquad ...(14.52)$$

or

$$\Omega = T_0\,\Delta s \quad \text{kJ/kg} \qquad ...(14.53)$$

The exergy balance for a stream gives the exergy loss.

$$\Omega = e_{xin} - e_{xout}\ ;\ \text{kJ/kg} \qquad ...(14.54)$$

Specific exergy loss or loss of availability or irreversibility in various elements of combined cycle is given in Fig. 14.7, when working with 1 kg of air at inlet to the compressor.

(i) Compressor: The specific exergy loss in the compressor without bleeding of air for cooling is given by

$$\Omega_{\text{comp}} = T_0\,(s_2' - s_1) = T_0\left[c_p \ln\frac{T_2'}{T_1} - R_a \ln\frac{p_2}{p_1}\right] \quad \text{kJ/kg} \qquad ...(14.55)$$

where $T_2 = T_1\left(\dfrac{p_2}{p_1}\right)^{(\gamma-1)/\gamma}$ and $T_2' = T_1 + (T_2 - T_1)/\eta_c$

where η_c = isentropic compressor efficiency.

Alternatively, $T_2' = T_1\,(p_2/p_1)^{(n-1)/n} = T_1\,(p_2/p_1)^{(\gamma-1)/\gamma\,\eta_c}$

(ii) Combustor: The exergy loss in the combustion chamber is given by

$$\Omega_{\text{comb}} = T_0\,[(s_p)_3 - (s_r)_2] \quad \text{kJ/kg} \qquad ...(14.56)$$

where $(s_r)_2 = (s_a)_2 + (s_f)_2$

Here, the subscripts p, r, a and f stand for products, reactants, air and fuel respectively

Thus, $\Omega_{\text{comb}} = T_0\,[(s_p)_3 - (s_a)_2 - (s_f)_2]$

$$= T_0\left\{[(s_p)_3 - (s_p)_0] + (s_p)_0 - [(s_a)_2 - (s_a)_0] - (s_a)_0 - [(s_f)_2 - (s_f)_0] + (s_f)_0\right\}$$

But $(s_p)_0 - [(s_f)_0 + (s_a)_0] = (\Delta s_{rn})_0$

Thus, $\Omega_{\text{comb}} = T_0\,\{[(s_p)_3 - (s_p)_0] - [(s_a)_2 - (s_a)_0] + (\Delta s_m)_0\}$

or

$$\Omega_{\text{comb}} = T_0\,\{(1 + \dot{m}_f/\dot{m}_a)\left[c_{pg} \ln\frac{T_3}{T_0} - R_g \ln\frac{p_3}{p_0}\right] - \left[c_{pa} \ln\frac{T_2}{T_1} - R_a \ln\frac{p_2}{p_0}\right] + (\Delta s_{rn})_0\} \qquad ...(14.57)$$

where $T_0\,(\Delta s_{rn})_o$ is given by Eq. (14.57)

Alternatively, Ω_{comb} may be expressed by taking exergy balance

$$\Omega_{\text{comb}} = \dot{m}_f\left[\Delta g_{rn} + R_f\,T_a \ln\frac{p_f}{p_a}\right] - [(1 + \dot{m}_f/\dot{m}_a)\,(e_x)_3 - (e_x)_2] \qquad ...(14.58)$$

(iii) Gas Turbine. For uncooled gas turbine (GT), the exergy loss is given by

$$\Omega_{GT} = (1 + \dot{m}_f/\dot{m}_a)\, T_0\, (s_4' - s_3)$$

$$= (1 + \dot{m}_f/\dot{m}_a)\, T_0 \left[c_{pg} \ln \frac{T_4'}{T_3} - R_g \ln \frac{p_4}{p_3} \right] \text{ kJ/kg} \qquad \text{...(14.59)}$$

where $T_4 = T_3 \left(\frac{p_4}{p_3}\right)^{\gamma - 1)/\gamma}$ and $T_4' = T_3 - \eta_{gt}\,(T_3 - T_4)$

Alternatively, $T_4' = T_3 \left(\frac{p_4}{p_3}\right)^{\frac{n-1}{n}} = \left(\frac{p_4}{p_3}\right)^{(\gamma - 1)\eta_{gtp}/\gamma}$ where η_{gt} is polytropic efficiency of turbine.

(iv) HRSG: The exergy loss in the HRSG is given by considering the two streams, i.e., water/steam and gas streams.

$$\Omega_{\text{HRSG}} = T_0\, [\dot{m}_s\, (s_{hrsg,\, out} - s_{hrsg,\, in})_{s/w} + (1 + \dot{m}_f/\dot{m}_a)\, (s_{hrsg,\, out} - s_{hrsg,\, in})_{gas}]$$

$$= T_0 \left[\dot{m}_s\, (s_5 - s_{10}) + (1 + \dot{m}_f/\dot{m}_a) \left(c_{pg} \ln \frac{T_{stack}}{T_{hrsg,in}} - R_g \ln \frac{p_{stack}}{p_{hsrg,\, in}} \right) \right] \qquad \text{...(14.60)}$$

Here the feed water enters the HRSG at state 10, $\dot{m}_s$ is the mass of steam generated in HRSG per kg of gas

(v) Exhaust Gas: The exhaust gas exergy loss is given by

$$\Omega_{EXH} = \int_{T_{stack}}^{T_o} \left(1 - \frac{T_o}{T}\right) \delta Q_{rej}$$

$$= (1 + \dot{m}_f/\dot{m}_a) c_{pg} \left[(T_{stack} - T_o) - T_o \ln \frac{T_{stack}}{T_o} \right] \qquad \text{...(14.61)}$$

Sometimes, exhaust gas exergy loss in included with HRSG.

(vi) Steam Turbine: The exergy loss in the steam turbine is given by

$$\Omega_{ST} = \dot{m}_s T_0\, (s_{exit} - s_{in}) = \dot{m}_s\, T_0\, (s_7' - s_5) \qquad \text{...(14.62)}$$

(vii) Condenser. The exergy loss in the condenser by considering the two stream steam and cooling water.

$$\Omega_{cond} = \dot{m}_s\, T_0\, (s_{exit} - s_{in})_{steam} + \dot{m}_w\, T_0\, (s_{f,\,exit} - s_{f,\,in})_{cw}$$

$$= \dot{m}_s\, T_0\, (s_8 - s_7')_{steam} + \dot{m}_w\, T_0\, (s_{f,\,exit} - s_{f,\,in})_{cw} \qquad \text{...(14.63)}$$

(viii) Feed Pump: The exergy loss in the feed pump (FP) is given by

$$\Omega_{FP} = \dot{m}_s\, T_0\, (s_{exit} - s_{in}) = \dot{m}_s\, T_0\, (s_9' - s_8) \qquad \text{...(14.64)}$$

The total exergy loss, $\Omega = \Sigma\Omega_{\text{component}}$ ...(14.65)

Fuel exergy input, $(e_x)_f = -\,\Delta g_{rn,\,o} = -\,\Phi\Delta h_{rn,\,o}$ ...(14.66)

Total power generation, $w_{out} = w_{gt,\,net} + w_{st,\,net}$ ...(14.67)

Exergetic or second law efficiency $= \dfrac{w_{out}}{(e_x)_f}$ ...(14.68)

14.11 PERFORMANCE CURVES OF COMBINED AND STEAM INJECTED CYCLES

Figure 14.7 shows the variation of gas turbine steam injected and combined cycle efficiencies with specific cycle work at various TIT and r_p for simple gas/simple steam combined cycle. At all TIT, the cycle work decreases with an increase in cycle pressure ratio but cycle efficiency first increases and then decreases resulting in optimum pressure ratio in all three cases at any TIT. Further, it is obvious from the results of combined cycle that at any compressor pressure ratio (r_{pc}), the efficiency increases with TIT.

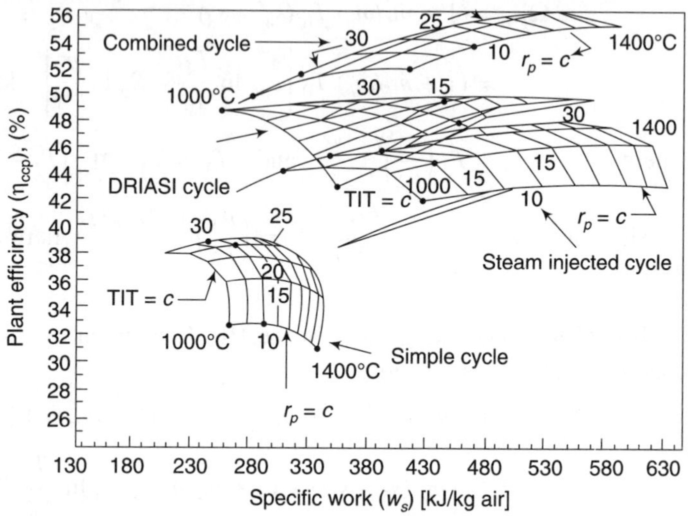

Figure 14.7 η_{ccp} vs. w_{ccp} of gas turbine, steam injected and combined cycles

Figure 14.8 shows the effect of gas turbine cycle efficiency, steam cycle efficiency and sizing ratio ($w_{st, net}/w_{gt, net}$) on the combined cycle efficiency. The higher plant efficiency is achieved at lower value of sizing ratio.

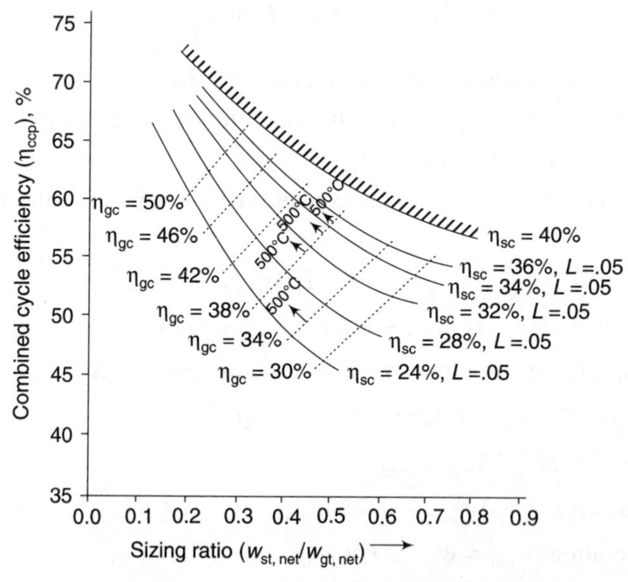

Figure 14.8 Combined cycle plant efficiency vs sizing ratio ($w_{st, net}/w_{gt, net}$)

Example 14.1 A combined cycle power plant takes in air at 1 bar and 15°C. The compressor pressure ratio is 13 and the gas turbine inlet temperature is 1400°C. The fuel used in natural gas.

The exhaust gas pressure of gas turbine is 1.08 bars. The stack temperature is 140°C. The HRSG is a single pressure system. The steam pressure and temperature are 50 bars and 500°C. The steam at 2 bars, is bled from the steam turbine for heating the feed water in the deaerator and the feed after the deaerator is pumped to the boiler. The condenser pressure is 0.05 bar. For ideal conditions calculate: (a) compressor work, (b) gas turbine work, (c) mass of fuel (natural gas having lower calorific value of 42 MJ/kg), (d) gas turbine cycle efficiency, (e) steam turbine work, (f) steam turbine cycle efficiency, (g) combined cycle specific work, (h) combined cycle power plant efficiency. Also calculate the above by assuming the following actual conditions: (i) compressor isentropic efficiency = 86%, (ii) combustion efficiency = 98%, (iii) gas turbine isentropic efficiency = 86%, (iv) HRSG effectiveness = 0.92, (v) steam turbine efficiency = 85%, (vi) mechanical efficiency for compressor and turbine each = 99%, (vii) generator efficiency = 98%, (viii) c_{pa} = 1.005 kJ/kgK, c_{pg} = 1.15 kJ/kgk K; (ix) overall pump efficiency = 90%. Neglect the pressure losses and cooling requirements by gas turbine bladings.

Solution Refer to Fig. 14.2.

For Ideal Conditions. The temperature at the end of compressor is given by

$$T_2 = T_1 \left(\frac{p_2}{p_1}\right)^{(\gamma - 1)/\gamma} = 288(13)^{0.4/1.4} = 288 \times 2.08 = 599.32 \text{ K.}$$

(a) Compress work, $w_c = h_2 - h_1 = c_p (T_2 - T_1) = 1.005(599.32 - 288)$
$= 312.87$ kJ/kg ***Ans.***

The gas turbine exhaust temperature is given by

$$T_4 = T_3 \left(\frac{p_4}{p_3}\right)^{(\gamma-1)/\gamma} = (1400 + 273)\left(\frac{1.08}{13}\right)^{0.4/1.4} = 821.84 \text{ K.}$$

(b) Gas turbine work, $w_{gt} = h_3 - h_4 = c_p (T_3 - T_4) = 1.005 (1673 - 821.84)$
$= 855.41$ kJ/kg ***Ans.***

Heat supplied, $q_A = h_3 - h_2 = c_p (T_3 - T_2) = 1.005 (1673 - 599.32) = 1079.04$ kJ/kg

(c) Mass of fuel required, $\dot{m}_f = \dfrac{q_A}{(LCV)_f} = \dfrac{1079.04}{42 \times 10^3} = 0.02569$ kg/kg of air ***Ans.***

(d) Gas turbine cycle efficiency, $\dfrac{w_{gt,net}}{q_A} = \dfrac{w_{gt} - w_c}{q_A} = \dfrac{855.41 - 312.87}{1079.04} = 50.27\%$

From Mollier diagram, the steam properties are: h_5 = 3440 kJ/kg, h_6 = 2650 kJ/kg, h_{f6} = 504.8 kJ/kg, h_7 = 2140 kJ/kg, $h_{f7} = h_{w7}$ = 137.8 kJ/kg

The mass of steam generated in HRSG ($\dot{m}_s$) in kg/kg of air is given by

$(h_4 - h_{stack}) = \dot{m}_s (h_5 - h_{f6})$ or $c_p (T_4 - T_{stack}) = \dot{m}_s (h_5 - h_{f6})$

$$\text{or } \dot{m}_s = \frac{c_p (T_4 - T_{stack})}{(h_5 - h_{f6})} = \frac{1.005(821.84 - 413)}{(3440 - 504.8)} = 0.1399 \text{ kg/kg of air}$$

(e) Steam turbine work, $w_{st} = \dot{m}_s (h_5 - h_6) + (\dot{m}_s - \dot{m}_{s1}) (h_6 - h_7)$

Energy balance on deaerator gives

$$m_{s1} = \frac{(h_{f6} - h_{f1})\dot{m}_s}{h_6 - h_{f7}} = \frac{(504.8 - 137.8) \times 0.1399}{(2650 - 137.8)} = 0.0204$$

$$\therefore \quad w_{st} = \dot{m}_s(h_5 - h_6) + (\dot{m}_s - \dot{m}_{s1}) (h_6 - h_7)$$

$$= 0.1399(3490 - 2650) + (0.1399 - 0.0204) (2650 - 2140) = 178.46 \text{ kJ/kg}$$

Feed pump work, $\dot{m}_{cv} . v_{comd} (p_{hsrg} - p_{cond})$

$= 0.001005 (50 - 0.05) \times 10^2 \times (0.1399 - 0.0204) = 0.594$ kJ/kg

(f) Steam turbine efficiency $= \dfrac{w_{st} - w_p}{q} = \dfrac{w_{st,\ net}}{q}$

$$= \frac{w_{st} - w_p}{\dot{m}_s (h_5 - h_{f6})} = \frac{178.46 - 0.594}{0.1399(3440 - 504.8)} = 43.31\% \qquad \textbf{\textit{Ans.}}$$

(g) The combined cycle power plant efficiency is given by

$$\eta_{ccp} = \frac{w_{gt,net} + w_{st,net}}{q_A} = \frac{533.54 + 178.46}{1079.04} = 65.98\% \qquad \textbf{\textit{Ans.}}$$

For Actual Conditions (Ignoring blade cooling)

From the concept of compressor isentropic efficiency, we have

$$\eta_c = \frac{h_2 - h_1}{h_2' - h_1} = \frac{T_2 - T_1}{T_2' - T_1}$$

or

$$T_2' = T_1 + \frac{(T_2 - T_1)}{\eta_c} = 288 + \frac{599.32 - 288}{0.86} = 650 \text{ K}$$

(a) Actual compressor work, $w_{ca} = \dfrac{(h_2' - h_1)}{\eta_m} = \dfrac{c_{pa}(T_2' - T_1)}{\eta_m}$

$$= \frac{1.005(650 - 288)}{0.98} = 371.23 \text{ kJ/kg of air} \qquad \textbf{\textit{Ans.}}$$

From the concept of gas turbine isentropic efficiency, we have.

$$\eta_{gt} = \frac{h_3 - h_4'}{h_3 - h_4} = \frac{T_3 - T_4'}{T_3 - T_4}$$

or $\quad T_4' = T_3 - \eta_{gt}(T_3 - T_4) = 1673 - 0.86\ (1673 - 821.84) = 941\ \text{K}$

Actual heat supplied, $q_{Aa} = (1 + \dot{m}_f/\dot{m}_a)\ h_3 - h_2' = (1 + \dot{m}_f/\dot{m}_a)\ c_{pg}\ T_3 - c_{pg}\ T_2'$

Thus, $q_{Aa} = (LCV)_f \dfrac{\dot{m}_f}{\dot{m}_a} \times \eta_{cc} = \left[\left(1 + \dfrac{\dot{m}_f}{\dot{m}_a}\right) c_{pg}\ T_3 - c_{pa}\ T_2'\right]$ gives $\dfrac{m_f}{m_a} = 0.0323$

(b) Actual gas turbine work $= (1 + \dot{m}_f/\dot{m}_a)\ (h_3 - h_4') = (1 + \dot{m}_f/\dot{m}_a)\ c_{pg}\ (T_3 - T_4')$

$$= (1 + 0.0323) \times 1.15\ (1673 - 941) = 868.99\ \text{kJ/kg} \qquad \textbf{\textit{Ans.}}$$

(c) Mass of fuel required, $\dfrac{\dot{m}_f}{\dot{m}_a} = \dfrac{c_{pg}\ T_3 - c_{pa}\ T_2'}{(LCV)_f \times \eta_{cc} - c_{pg}\ T_3}$

$$= \frac{1.15 \times 1673 - 1.005 \times 650}{42 \times 10^3 \times 0.98 - 1.15 \times 1673} = 0.0323\ \text{kg/kg of air} \qquad \textbf{\textit{Ans.}}$$

The actual heat supplied, $q_{Aa} = (1 + \dot{m}_f/\dot{m}_a) c_{pg}\ T_3 - c_{pg}\ T_2'$

$$= (1 + 0.0323) \times 1.15 \times 1673 - 1.005 \times 650 = 1332.84\ \text{kJ/kg}$$

(d) Actual gas turbine efficiency, $\eta_{gca} = \dfrac{w_{gt,\ net}}{q_{Aa}} = \dfrac{w_{ga} - w_{ca}}{q_{Aa}}$

$$= \frac{868.99 - 371.23}{1332.84} = 37.345\% \qquad \textbf{\textit{Ans.}}$$

The mass of steam generated ($\dot{m}_s$) in HRSG is given by

$$(1 + \dot{m}_f/\dot{m}_a)\ (h_4' - h_{stack})\ \in_{hrsg} = \dot{m}_s\ (h_5 - h_{f6})$$

or $\quad \dot{m}_s = \dfrac{(1 + 0.0323) \times 1.15(941 - 413) \times 0.92}{(3440 - 504.8)} = 0.1964$ kg/kg of air

(e) Actual steam turbine work, $w_{sta} = \dot{m}_s\ (h_s - h_6') + (\dot{m}_s - \dot{m}_{s1}')\ (h_6' - h_7')$

Here, $\dot{m}_{s1}$ is calculated by taking energy balance in deaerator

$$\dot{m}_{s1} = \frac{(h_{f6} - h_{f7})\dot{m}_s}{(h_6' - h_{f7})} = \frac{(504.8 - 137.8) \times 0.1964}{(2775 - 137.83)} = 0.02733\ \text{kg/kg of gas}$$

$$w_{sta} = 0.1964(3440 - 2775) + (0.1964 - 0.02733)\ (2775 - 2330) = 205.84\ \text{kJ/kg}$$

Actual feed pump $= \dfrac{\dot{m}_{cs} . \dot{v}_{cs}\ (p_{hrsg} - p_{cond})}{\eta_p} = 0.67/0.9 = 0.728$ kJ/kg

(f) Actual steam cycle efficiency, $\eta_{sca} = \dfrac{w_{sta} - w_{pa}}{\dot{m}_s\ (h_5 - h_{f6})}$

$$= \frac{205.84 - 0.728}{(3440 - 50.48) \times 0.1964} = 30.81\% \qquad \textbf{\textit{Ans.}}$$

(g) The combined cycle power plant efficiency is given by

$$\eta_{ccp} = \frac{w_{out}}{q_{Aa}} = \frac{w_{ccp}}{q_{Aa}} = \frac{(w_{gt,\, net} + w_{st,\, net})}{q_{Aa}} \times \eta_{gen}$$

$$= \frac{[(866.69 - 371.23) + (205.84 - 0.728)]}{1332.84} \times 0.98 = \frac{686.85}{1332.84} = 51.53\%$$

Obviously, the combined cycle power plant efficiency is much higher than individual gas or steam cycle efficiency. The gas turbine cooling has been neglected which will further alter the efficiency. CCP efficiency is less here due to low r_p and T_3.

Example 14.2 A combined cycle power plant takes in air at 1 bar and 27°C. The compressor pressure ratio is 8 and the gas turbine inlet temperature is 1200 K. The exhaust gas pressure is 1.08 bars. The stack temperature is 110°C. The HRSG is a single pressure system. The steam pressure and temperature entering the steam turbine is 45 bars and 450°C. The steam at 2 bars is bled from the system turbine for heating the feed water in the deaerator and the feed after deaerator is pumped to the boiler. The condenser pressure is 0.05 bar. Assume η_c, η_{cc}, ε_{hrsg}, η_{gt}, η_{st}, η_p and η_{gen} as 0.86, 0.98, 0.92, 0.86, 0.86, 0.99, 0.9 respectively. Take c_{pa} = 1.005 kJ/kg and c_{pg} = 1.15 kJ/kg and R_g = 0.289 kJ/kg. Neglect pressure losses and cooling requirements for gas turbine bladings. The fuel used is natural gas giving calorific value equal to 42 MJ/kg. Calculate first law of efficiency, second law efficiency and exergy losses in various elements of combined cycle. Take T_0 = 288 K (ISO)

Solution Refer to Fig. 14.2 for the system.

Compressor. The isentropic temperature at the exit of compression

$$T_2 = T_1\, (p_2/p_1)^{(\gamma-1)/\gamma} = 300(8)^{0.4/1.4} = 543.75 \text{ K}$$

The actual temperature at the exit of compressor

$$T_2' = T_1 + \frac{(T_2 - T_1)}{\eta_c} = 300 + \frac{543.75 - 300}{0.86} = 583.43 \text{ K}$$

The exergy losses in the compressor for 1 kg of air is

$$\Omega_{\text{comp}} = T_0\,(s_2' - s_1) = T_0\left[c_{pa} \ln \frac{T_2'}{T_1} - R_a \ln \frac{p_2}{p_1}\right]$$

$$= 288\left(1.005 \ln \frac{583.43}{300} - 0.287 \ln \frac{8}{1}\right) = 20.53 \text{ kJ/kg}$$

Combustor. For finding out exergy loss of combustion, the mass flow rate of fuel per kg of air ($\dot{m}_f/\dot{m}_a$) is to be found out.

$$\dot{m}_f \times LCV \times \eta_{cc} = \dot{m}_a\left[\left(1 + \frac{\dot{m}_f}{\dot{m}_a}\right) c_{pg}\, T_3 - c_{pg}\, T_2'\right]$$

or $$\frac{\dot{m}_f}{\dot{m}_a}[42000 \times 0.98 - 1.15 \times 1200] = 1.15 \times 1200 - 1.005 \times 583.43$$

or $$\frac{\dot{m}_f}{\dot{m}_a} = \left[\frac{1.15 \times 1200 - 1.005 \times 583.43}{42000 \times 0.98 - 1.15 \times 1200}\right] = 0.02 \text{ kg/kg of air}$$

The exergy of fuel, $e_{xf} = \dot{m}_f (LCV)_f \cdot \Phi = 0.02 \times 42000 \times 1.045 = 877.8$ kJ/kg

The exergy loss in combustion or reaction is given by

$$\Omega_{rn} = T_0 (\Delta s)_{\text{react, }0} = \dot{m}_f (LCV)_f \cdot (\Phi - 1) \text{ where } \phi = \text{exergy flux}$$

$$= 0.02 \times 42000 (1.045 - 1) = 37.8 \text{ kJ/kg}$$

The exergy loss in the combustor is

$$\Omega_{\text{combustor}} = T_0 [\{(s_p)_3 - (s_p)_0\} - \{(s_a)_2 - (s_a)_0\} + (\Delta s_{\text{react}})_0]$$

where $T_0 (\Delta s_{\text{react}})_0 = T_0 [(s_p)_0 - (s_a)_0] = 37.8$ kJ/kg

Thus, $$\Omega_{\text{combustor}} = T_0 \left[\left\{c_{pg} \ln \frac{T_3}{T_0} - R_g \ln \frac{p_3}{p_0}\right\} - c_{pa} \ln \frac{T_2'}{T_1} + R_a \ln \frac{p_2}{p_0}\right] + T_0 (\Delta s_{\text{react}})_0$$

$$= 288\left[\left\{1.15 \ln \frac{1200}{288} - 0.289 \ln \frac{8}{1}\right\} - 1.005 \ln \frac{583.75}{288} + 0.286 \ln \frac{8}{1}\right] + 37.8$$

$$= 288[1.6411 - 0.6 - 0.71 + 0.5947] + 37.8 = 304.43 \text{ kJ/kg}$$

Gas Turbine. The exergy loss for uncooled gas turbine is

$$\Omega_{GT} = \left(1 + \frac{\dot{m}_f}{\dot{m}_a}\right) T_0 (s_4' - s_3) = \left(1 + \frac{\dot{m}_f}{\dot{m}_a}\right) T_0 \left[c_{pg} \frac{T_4'}{T_3} - R_g \ln \frac{p_4}{p_3}\right]$$

But $$T_4 = T_3\left(\frac{p_4}{p_3}\right)^{(\gamma - 1)\gamma} = 1200\left(\frac{1.08}{8}\right)^{0.4/1.4} = 677.21 \text{ K}$$

And, $$T_4' = 1200 - 0.86(1200 - 677.21) = 750.4 \text{ K}$$

Hence, $$\Omega_{GT} = 1.03 \times 288 \left[1.15 \ln \frac{750.4}{1200} - 0.289 \ln \frac{1.08}{8}\right]$$

$$= 1.03 \times 288 [-0.5398 + 0.6] = 17.85 \text{ kJ/kg}$$

The gas turbine net power output is

$$w_{gt,\, net} = [(1 + \dot{m}_f/\dot{m}_a) \cdot c_{pg} (T_3 - T_4') - c_{pa} (T_2' - T_1)$$

$$= 1.03 \times 1.15 (1200 - 750.4) - 1.005 (583.43 - 300) = 247.67 \text{ kJ/kg}$$

HRSG. For finding out exergy loss in HRSG, the amount of steam generated per kg of gas is to be calculated.

Energy balance yields

$$\left(1+\frac{\dot{m}_f}{\dot{m}_a}\right)(h_4' - h_{\text{stack}})\,\varepsilon_{hrsg} = \dot{m}_s\,(h_5 - h_{f10})$$

or
$$\left(1+\frac{\dot{m}_f}{\dot{m}_a}\right)\cdot c_{pg}\,(T_4' - T_{\text{stack}}) \times \varepsilon_{hrsg} = \dot{m}_s\,(h_5 - h_{f10})$$

or
$$\dot{m}_s = \frac{1.03 \times 1.15(750.4 - 413) \times 0.92}{(3323.3 - 540.8)} = 0.1215 \text{ kg/kg of gas}$$

The exergy loss in HRSG is given by

$$\Omega_{hrsg} = T_0\,[\dot{m}_s\,(s_{hrsg,out} - s_{hrsg,in})_{s/w} + (1 + \dot{m}_f/\dot{m}_a)\,(s_{stack} - s_{hrsg,\,in})_{gas}]$$

or
$$\Omega_{hrsg} = 288\left[0.1215(6.8746 - 0.4556) + 1.03\left(c_{pg}\ln\frac{T_{\text{stack}}}{T_4'} - R_g \ln\frac{1.0}{1.08}\right)\right]$$

$$= 288\,[0.1215 \times 6.419 + 1.03]\left(1.15 \ln\frac{413}{750.4} - 0.286 \ln\frac{1.0}{1.08}\right)$$

$$= 288\,[0.7799 + 1.03\,(-\,0.6868 + 0.0222)]$$

$$= 27.475 \text{ kJ/kg of gas}$$

Exhaust gas. The exhaust gas exergy loss is given by

$$\Omega_{Exh} = \int_{T_{\text{stack}}}^{T_0}\left(1 - \frac{T_0}{T_{\text{stack}}}\right)\delta q_{rej} = (1 + \dot{m}_f/\dot{m}_a)\,c_{pg}\left[(T_{\text{stack}} - T_0) - T_0 \ln\frac{T_{\text{stack}}}{T_0}\right]$$

$$= 1.03 \times 1.15\left[(413 - 288) - 288 \ln\frac{413}{288}\right]$$

$$= 1.03 \times 1.15[125 - 103.81] = 25.099 \text{ kJ/kg}$$

Steam Turbine. The exergy loss in steam turbine is given by

$$\Omega_{st} = \dot{m}_s\,T_0\,(s_7' - s_5) = 0.1225(7.5 - 6.87) \times 288 = 22.22 \text{ kJ/kg of gas}$$

Steam turbine net output $= w_{st,\,net} = \dot{m}_s\,(h_5 - h_6') + (\dot{m}_s - \dot{m}_{s1})\,(h_6' - h_7') - w_p$

The energy balance at deaerator gives

$$\dot{m}_{s1} = \frac{(h_{f6} - h_{f7})\dot{m}_s}{h_6' - h_{f7}} = \frac{(504.8 - 137.38) \times 0.1225}{(2775 - 137.38)} = 0.01692 \text{ kg/kg of steam}$$

$$\therefore \qquad w_{st,\, net} = 0.1225(3380 - 2775) + (0.1225 - 0.01692)\ (2775 - 2330)]$$
$$= 74.11 \ + 0.10558 \times 445 = 121.09 \text{ kJ/kg}$$

Condenser. The exergy loss in the condenser is

$$\Omega_{\text{cond}} = \dot{m}_s\, T_0\, (s_{f8} - s_7') + \dot{m}_w\, T_0\, (s_{f,\, out} - s_{f,\, in})_{\text{water}}$$

But $\dot{m}_w$ (kg/kg of gas) is calculated from

$$\dot{m}_s\, (h_7' - h_{f8}) = \dot{m}_w\, c_{pw}\, \Delta T_w$$

or
$$\dot{m}_w = \frac{\dot{m}_s (h_7' - h_{f8})}{c_{pw} \times \Delta T_w} = \frac{0.1215(2330 - 137.83)}{4.18 \times 10} = 6.5281 \text{ kg/kg of gas}$$

Thus, $\Omega_{\text{cond}} = 0.1215 \times 288\ (0.4761 - 7.5) + 6.5281 \times 288(0.3672 - 0.2244)$

$$= -\ 245.78 + 268.47 = 22.686 \text{ kJ/kg}$$

Feed Pump. The exergy loss in the feed pump is

$$\Omega_{FP} = \dot{m}_s\, T_0\, (s_{fg} - s_{f8}) = 0.1215 \times 288(0.5 - 0.4761) = 0.8363 \text{ kJ/kg}$$

The exergy balance is tabulated in the following table.

Exergy input (kJ/kg)	*Power output (kJ/kg)*	*Components*	*Exergy losses (kJ/kg)*
877.8 (100%)	–	–	–
	$w_{\text{gt, net}}$ = 247.67 (23.21%) $w_{\text{st, net}}$ = 121.09 (13.79%)	Compressor	= 20.53 (2.33%)
		Combustor	= 304.43 (34.68%)
		GT	= 17.85 (2.03%)
		HRSG	= 27.475 (3.12%)
		Exhaust gas	= 25.59 (2.91%)
		ST	= 22.22 (2.53%)
		Condenser	= 22.68 (2.58%)
		Pump	= 0.8363 (0.95%)
		Unaccounted	= 877.8 – 368.76 – 411.61 = 97.43 (11.07%)
Input = 877.8	Output = 368.76		877.8(100%)

Note 1: In fact, exergy analysis is done for compressor and turbine assuming equal pressure ratio in each stage and for every stage exergy and exergy loss are calculated. The concept of polytropic efficiency is used.

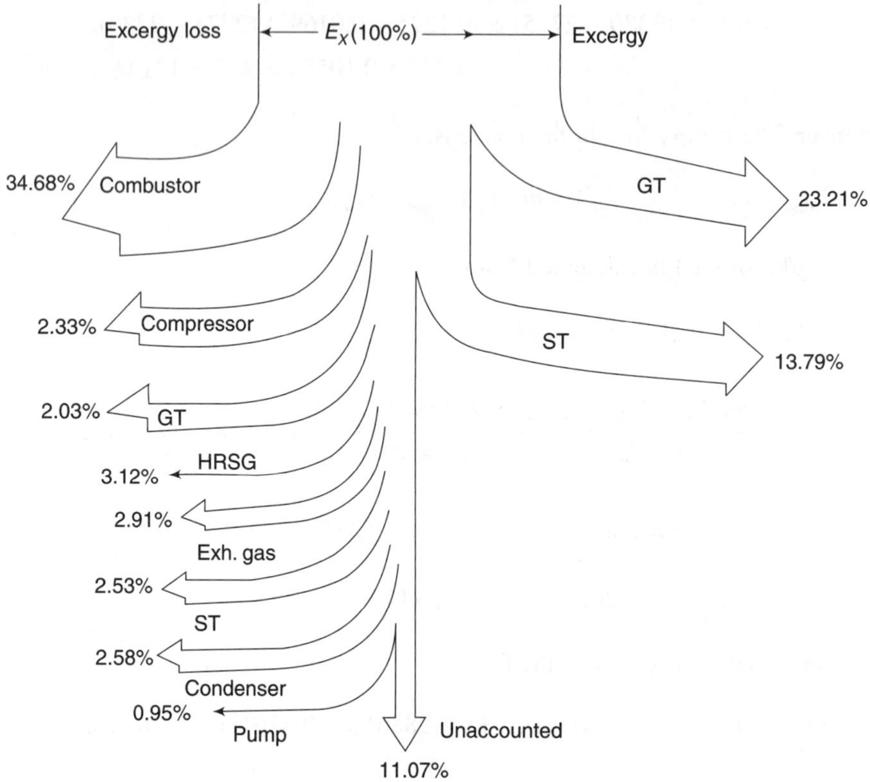

Figure 14.9 Exergy flow diagram based on problem 14.2

Note 2: If T_0 = 288 K (ISO condition) and p_0 = 1 bar are not given, the atmospheric condition whatever given are taken as T_0 and p_0.

$$\text{Second law efficiency, } \frac{w_{out}}{e_{xf}} = \frac{368.76}{877.8} = 42.0\% \text{ (low due to low } r_p \text{ and TIT)} \qquad \textbf{\textit{Ans.}}$$

The exergy flow diagram is shown in Fig. 14.9.

14.12 COAL BASED COMBINED CYCLE PLANTS

There are two dominant coal based combined cycle technologies. They are: (a) pressurized fluidized bed combustion (PFBC) system which may be either a bubbling fluidized bed or a circulating fluidized bed, (b) integrated gasification combined cycle (IGEC). The schematic diagram of these plants has been discussed in chapter 6.

14.13 REPOWERING

The conversion of older steam power plant into combined cycle units is generally known as **repowering**. Boiler is replaced by gas turbine unit and HRSG. For building, steam turbine, generator, condenser, water cooling system, and electrical equipment are used. By this way, the uneconomical steam power plants becomes economical.

14.14 OTHER TYPES OF COMBINED CYCLES

The following are the other possible combined cycle power plants.

(a) M.H.D.-steam power plant

(b) Thermionic-steam power plant

(c) Fuel-cells-gas turbine (hybrid) power plant

The first-two are still in the research stages while the third one, i.e., hybrid power plant has come up as prototype (40 MW unit) by 2007 in USA.

PROBLEMS

Things to Think About

14.1 Why the efficiency of combined cycle is higher?

14.2 What led to the development of combined cycle?

14.3 What parameters affect the performance of combined cycle?

Short and Long Theoretical Questions

14.1 Define combined cycle, co-generation plant and mixed cycle plant.

14.2 Derive an expression of combined cycle plant in terms of topping and bottoming cycle efficiencies and fraction of unutilized energy.

Numerical Question

14.1 Following data relate to combined cycle power plant based on simple gas and simple steam cycle-power output (electrical) = 240 MW, Total pressure ratio = 30, Total turbine inlet temperature = 1650 K, Pressure of steam is bled from turbine = 2 bars, Condenser pressure = 0.05 bar, Stack temperature = 140°C, Polytropic compression efficiency = 90%, Polytropic turbine efficiency = 91%, Mechanical efficiency = 98%, Alternator efficiency = 98%, Combustion efficiency = 99%. Pressure loss in combustor = 3% of entry pressure. Turbine exhaust pressure = 1.08 bars, Ambient condition = 1 bar, 255 K, Effectiveness of HRSG = 0.9, Steam turbine efficiency = 0.86, Overall pump efficiency = 90%. Calculate the net gas turbine specific work gas cycle efficiency, mass of fuel flow rate, air flow rate, steam turbine specific work, steam turbine efficiency, combined cycle specific work, combined cycle efficiency and exergy losses.

Software Based Question

14.1 Develop a software for predicting the performance of combined cycle power plant for various configurations.

Design and Open Ended Question

14.1 Prepare a comprehensive report on the design features of combined cycle of GE and Siemens and suggest various means to improve capacity and efficiency.

Multiple-Choice Questions

14.1 The efficiency of gas/steam combined cycle ranges from
(a) 40 to 50% (b) 50 to 60% (c) 60 to 70% (d) 70 to 80%

14.2 The efficiency of cogeneration plant ranges from
(a) 50 to 60% (b) 60 to 70% (c) 70 to 80% (d) 80 to 90%

14.3 The efficiency of combined cycle plant increase with
(a) increase of TIT (b) increase with r_p (c) decrease of CIT (d) All of them

14.4 The efficiency of combined cycle increase with
(a) increase of η_{gas} (b) increase of η_{steam}
(c) decrease in exhaust losses (d) All of them

Answers to Multiple-Choice Questions

14.1 (b) 14.2 (d) 14.3 (d) 14.4 (d)

CHAPTER

15

Energy Storage

OVERVIEW

The demand of electricity consumption is based on the consumer's demand which is highly fluctuating. *The objective of energy storage is to offset this adverse effect of fluctuating demand of electricity and to ensure a steady output from existing power plants.* This chapter deals with various types of energy storage systems used in practice.

15.1 NEED OF ENERGY STORAGE

The need for energy storage arises because the demand for electric energy in a utility system is characterized by hourly, daily and seasonal fluctuations based on consumer's demand, whereas the supply from system, in the majority of cases has a fixed capacity. As per practice, the capacity of the utility system is selected to correspond to the maximum demand plus a reasonable excess to take care of scheduled plant shut-downs for maintenance and unscheduled shut-down due to abnormal occurrences. As a result, the large expensive plants operate most of the time below capacity (except during peak demand), thus causing high operating and capital cost which necessitates the need of energy storage. The picture is more clouded, and the need for energy storage is greater, *if plants using renewable forms of energy such as solar and wind are used to generate electricity.* It is the output of these renewable energy plants that *fluctuates severely* because of *their input energy intermittence.*

The objective of energy storage, therefore, is to offset the disadvantages that result from the fluctuations in demand for electric energy by ensuring a steady high output from existing power plants. When the demand is lower than the capacity, energy is stored and when the demand is higher than the capacity, the stored energy is released which results in reliable, efficient and economical supply of electricity to the consumers.

The need for energy storage was not actually realized when the generating plants were cheap and the fuel supply plentiful. *Nature herself stored energy in the fuels in the form of chemical and nuclear energy.* The energy density in this natural storage is large. Fossil fuels, coals, oil and gas at 17 bars contain about 37 × 10^6 kJ/m^3, while natural uranium (0.071%, U-235) contains about 10^{14} kJ/m^3. With the depletion of fuels and their increasing costs, the need to conserve natural resources is a must and for this, the **following methods for energy management become necessary**.

(i) It is necessary to meet the peak load by *interconnecting power networks* that might have different power demands on them.

(ii) It is important to use *modern, state-of-art and more efficient power plants for base-load generation* and use older, less efficient plants for peak power generation.

(iii) Install smaller, *low and capital* cost, though not so efficient power plants, as power peaking such as special steam plants, small hydroelectric plants or gas turbine peaking plants.

(iv) Add energy storage system in the utility.

Now, it is an established fact that reliability and economy of electrical supply can best be achieved by having a mix of three types of power plants, namely, a base-load, a cycling (intermediate) plant and a peaking plant.

Base-load plants: They are used to provide a base electrical load to the grid. Such plants are usually large, efficient steam power plant or combined cycle plant powered by fossil or nuclear fuels. They operate continuously except for scheduled maintenance or forced outage and have annual power operating factor (POF) between 70 to 80 per cent ensuring low unit cost.

Intermediate and cycling plants: They usually are older, less efficient plants or new ones designed for cyclic operation such as repowered combined cycles. They operate primarily during hours of high load demand and have an annual POF between 25 to 50%.

Peaking plants: As per established practice, *peaking plants* are especially designed to provide relatively inexpensive power during peak demands period such as evening. They operate at low POF of 5 to 15 per cent.

The last action is the **energy storage** which will allow the plants to be designed for nearly constant load operation below peak demand, a process called *peak saving* which would thus reduce the high capital cost of the initial plants. It is attractive only when the overall costs (capital and operating) are less. The main drawback is that energy density of all storage systems is much lower than that of nature stored energy.

A typical weakly load curve of a utility with and without energy storage is shown in Fig. 15.1. In the event of large scale energy storage availability as illustrated in Fig. 15.1(b), a relatively efficient and base load generation could be increased and the excess beyond off-peak demand (lower shaded areas) could be used to charge the storage system. Discharge of the stored energy (upper shaded area) during periods of peak load demand would then reduce or replace fuel burning peaking plant capacity thus conserving fuel resources. This is called *peak saving* as mentioned above. Further, something of the daily load curve leads to reduced stresses of load following operation by the steam plant and consequently reduced maintenance costs.

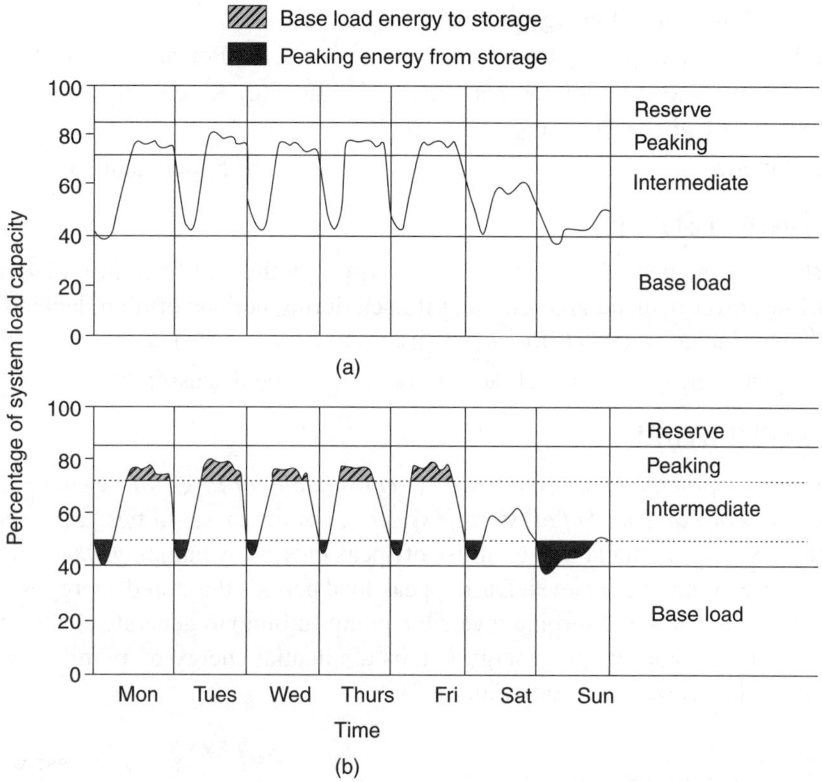

Figure 15.1 Typical weekly load curve of a power utility, (a) without storage (b) with storage

15.2 ENERGY STORAGE SYSTEMS

There are basically two generic approaches to energy storage in utility system, namely, (1) electrical storage and (2) thermal storage.

15.2.1 Electrical Storage

Figure 15.2 depicts the energy storage with constant thermal input either from the fossil or nuclear fuel. The primary electrical generating plant is continuously operated in a base load mode. The excess electricity production during the off-peak periods is $(ab + cd)$. Electrical storage is then used to hold this excess electricity for use during peak demand, period bc. Obviously, the total energy stored is greater than the total energy supplied because of conversion losses to and from storage. The following are the types of electrical storage system.

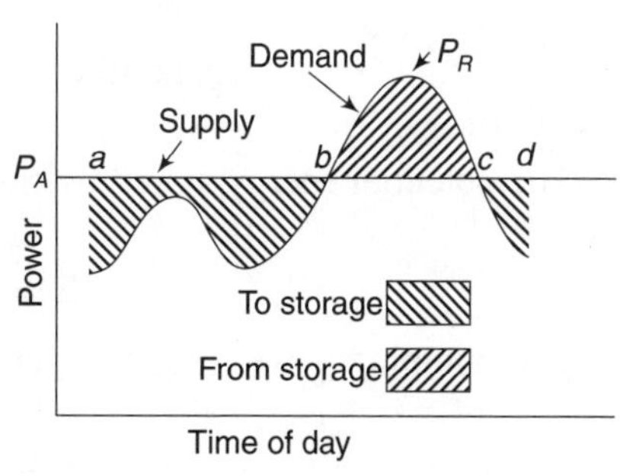

Figure 15.2 Energy storage with constant thermal input

(1) Electrical-mechanical energy storage
   (a) Potential, pumped hydro
   (b) Potential, compressed air
   (c) Potential, springs, torsion bars, mass elevation
   (d) Kinetic, flywheel
(2) Direct electrical energy storage
   (a) Batteries
   (b) Superconducting coils

15.2.2 Thermal Storage

In this system, all methods deal with storing energy in a thermal form in a material during periods of low power demand and releasing it back during periods of high demand. Thermal storage schemes includes: (i) *sensible heat,* (ii) *latent heat* and (iii) *chemical reaction.*

The energy storage systems which are in practice will be discussed here.

15.3 PUMPED HYDRO

Pumped hydro is a potential energy storage system suitable for large utility energy storage. It is the most developed and used of all storage systems. For decades, utilities have used pumped hydro storage as an economical way to utilise off-peak energy, by pumping water with the help of pump to a reservoir at higher level. During peak load periods the stored energy is discharged through pump which acts as a turbine reversible pump-turbine) to generate electricity to meet the peak demand. In other words, energy is thus a potential energy by pumping water from a low level to a high level reservoir (Fig. 15.3).

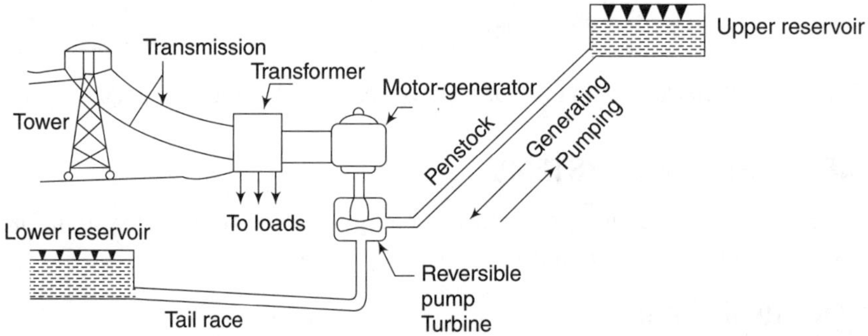

Figure 15.3 Pumped hydroelectric energy storage

The potential energy stored by raising mass of water (m) to an elevation (H) is expressed as

$$PE = \dot{m}gH \quad ...(15.1)$$

The pumping power (P_p) and generating power (P_T) are expressed as

$$P_p = \frac{\dot{Q}_p \, \rho \, g \, H_P}{\eta_p}; \text{ W} \quad ...(15.2)$$

and

$$P_T = \rho g H_T \dot{Q}_T \eta_T; \text{ W} \quad ...(15.3)$$

where $H_p = H + H_1$; $H_T = H - H_1$, H stand for the static head and H_1 stands for losses, Q_p is the pump discharge while Q_T is the turbine discharge and ρ is the density of water.

A sample calculation will reveal the facts about energy storage. If 1000 kg of water is raised through 100 m, it will store 1000 (kg) × 9.81 $(m/s)^2$ × 100 (m) = 9.81 × 10^5 J or 0.2725 kWh of energy. This shows that a large mass of water must be elevated to sufficient large height to store a large amount of energy. One or both of these reservoirs may be artificially excavated or may be natural river or lake.

It is worth to note that pump hydro system requires a suitable topography that will allow the design and construction or selection of two reservoirs with sufficient capacities, maximum available elevation difference, (*H*) and minimum horizontal distance (*L*). Although, high heads are desirable but some topographies do not offer high head and so the pump hydro-systems are classified as (a) ***above ground***, which includes the preferred high head and medium head, and (b) ***underground***.

15.3.1 Above Ground Pumped Hydro Storage System

Figure 15.4 shows a conventional above ground pumped hydro storage system. The upper and lower reservoirs are vast and difference in their elevation is also high. At the head of steel pipe, there is usually a surge tank and a valve house. The surge tank is built near the mouth of pressure tunnel to relieve the pipes of undue inertia pressure set up in the tunnel when the flow is checked following a reduction of load. The power house itself is located as close as possible to the lower reservoir into which the tail race discharges.

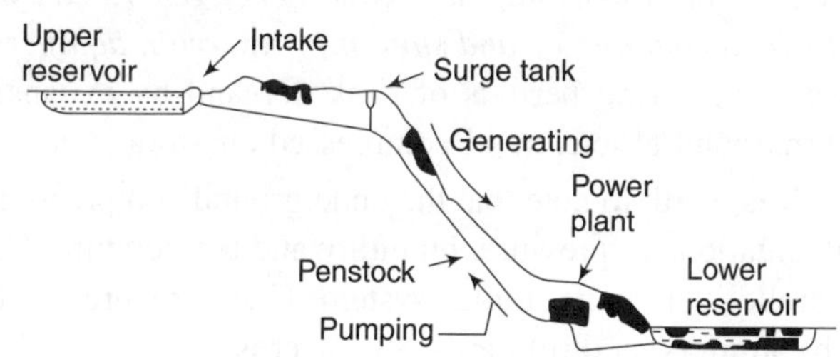

Figure 15.4 Conventional above ground pumped hydro-storage system

15.3.2 Underground Pumped Hydro Storage System

In the event of non-suitability of suitable topography for ground pumped hydro, underground pump hydro storage system is used. The upper reservoir may be at or near ground level. The lower reservoir is placed underground in natural caverns, old mines, or other underground cavities (Fig. 15.5). In all systems, a principal piece of equipment is a reversible pump-turbine or motor generator sets.

The **losses in pumped hydro system include** (i) motor and pump losses, (ii) flow losses during upflow (iii) seepage into ground, (iv) leakage of water from pipes and equipment, (v) evaporation during storage, (vi) turbine and generator losses and flow losses during downflow.

The performance of pumped hydro system is described by a parameter called ***turned around efficiency*** (η_{TA}). It is defined as

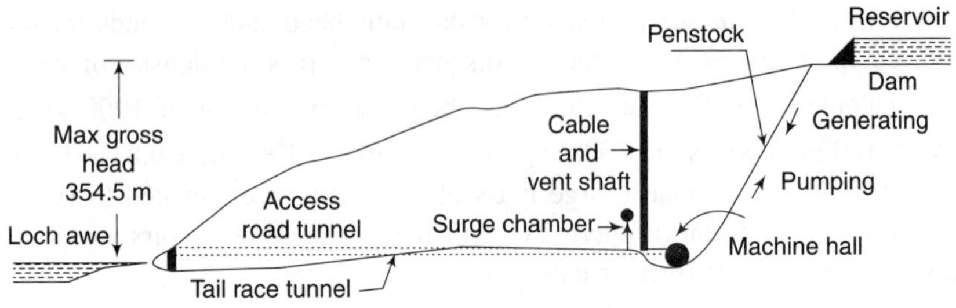

Figure 15.5 Schematic of an underground pumped hydro-storage system, built in Scotland

$$\eta_{TA} = \frac{\text{Total energy output}}{\text{Total energy input during a charge–discharge cycle}} \qquad ...(15.4)$$

For the most of plant, η_{TA} **is around 65%**.

15.4 COMPRESSED AIR ENERGY STORAGE (CAES)

Compressed air energy storage (CAES) is analogous to pumped hydro energy storage. During period of low demand, the excess energy generated by the base load plant is used to compress air by a compressor and store it in reservoir, aquifiers or caverns. The stored energy is then released during periods of peak demand by expansion of air through an air turbine. The turnaround efficiency of compressed air storage is around 65%.

It is worth to note that the underground compressed air reservoirs are subjected to repeated fluctuations in pressure, humidity and temperature. Usually, multiple reservoirs, operating in parallel, serve one storage system. There are three types of reservoir, namely, (a) salt caverns, (b) aquifers, and (c) hard-rock caverns.

Salt Caverns. These types of reservoirs may serve the purpose of storing compressed air as they have been used in the past to restore the petroleum products. In the design of salt caverns, the major concerns are cavern geometry, size and spacings, long-term creep and creep-rupture of rock salt, and air leakage.

Aquifers. These are naturally formed porous rock reservoirs. They have been used to store natural gas. Still the effects of different physical properties of air on the reservoir are still to be evaluated. Further, major concerns are, cyclic fatigue of porous rock, air-water interface movement and the generation and transport of fine particulate matter due to rocking environment.

Hard rock Caverns. They need water compensating surface reservoirs to maintain air pressure and thus are costly. The major concerns are effervescence of air in the water shaft called the *champagne effect*, hard rock properties under cyclic conditions are residual strength of hard rock after an initial failure.

The temperature of compressed air is given by

$$T_2 = T_1 \left(\frac{p_2}{p_1}\right)^{(n-1)/n} \qquad ...(15.5)$$

When the heat of compression may be retained in the compressed air or in another heat storage medium and then restored to the air before expanding through the turbine, it is called *adiabatic storage*. This results in high storage efficiency. If the heat of compression is allowed to dissipate, additional heat could be added by fuel combustion to retain the high storage efficiency but this results in extra expenses and more maintenance. Such a storage system is called *hybrid system*.

15.4.1 Adiabatic Storage System

Figure 15.6 shows a schematic diagram of simple adiabatic storage energy system. During off-peak hours, the electrical energy from the main power plant (not shown) is used by the motor-generator (MG) set operating in the motor mode to drive the compressor (*C*). In order to store the sensible thermal energy, the compressed air first passes through a packed bed (*P*) and then finally to an underground reservoir (*R*) for the storage. The constant pressure in the reservoir is achieved by displacing water to a pressure compensation pond that has a constant pressure above the reservoir. During peak hours, air from the reservoir flows through the packed bed (*P*) picking back sensible heat, then through the air turbine where the compressed air expands and drives the MG set in the generator mode. Clutches are used to separate the compressor during the peak periods and then turbines during off-peak periods.

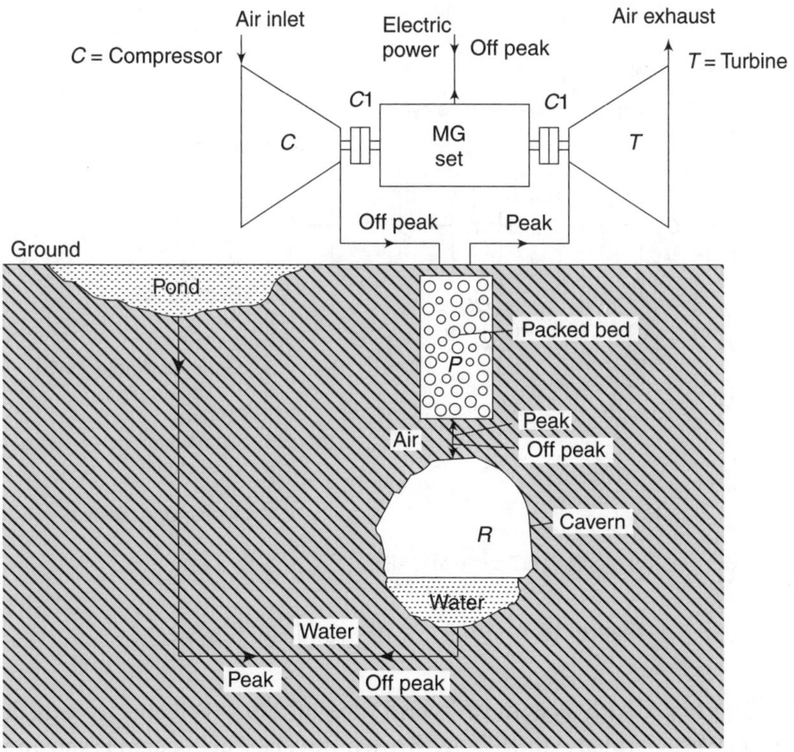

Figure 15.6 Adiabatic compressed air energy storage system

It is worth to note that the air reservoir volume of the cavern (V) greatly depends upon the storage pressure. Hence, for peak load capacity of 1500 MWh, the required volume will be 2×10^6 m^3 at 10 bar. For 100-bar storage pressure, the volume requirement will be 64,000 m^3. As a thumb rule, the packed-bed (*P*) volume is usually kept one-tenth of the reservoir volume (*R*).

15.4.2 Hybrid Air Energy Storage System (HAES)

Figure 15.7 shows the working principle of hybrid system. It is a simple gas turbine modified for CAES system. In hybrid system, fuel is burnt in the compressed air. During off-peak periods (at night), the MG set of gas turbine operates as motor to drive the compressor which is stored in a storage reservoir (cavern). During peak periods (obviously next day), air is taken from the reservoir, heated in a combustor by burning the fuel and then expanded in a turbine up to ambient pressure. The MG set now operates in the generator mode to generate electricity and supplies to the power grid.

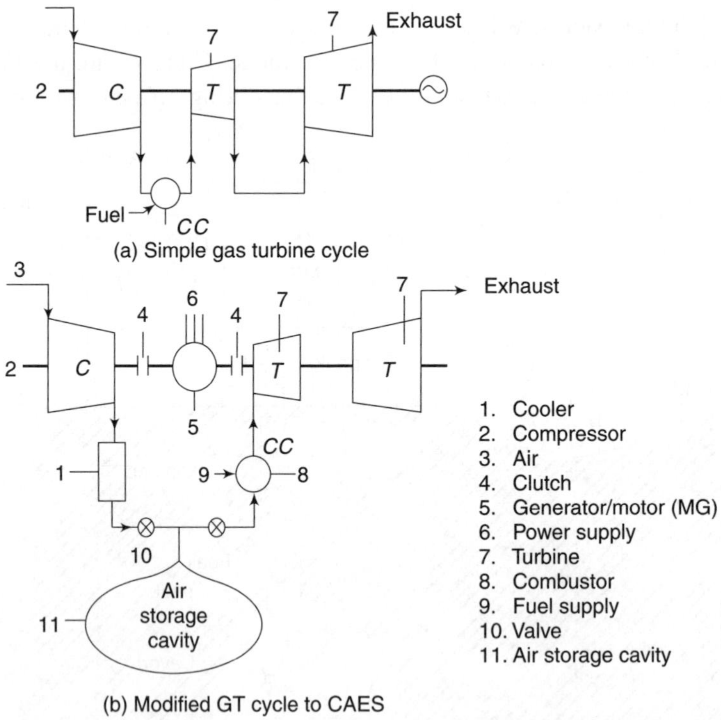

Figure 15.7 Principle of hybrid CAES system

Case Study: The first CAES system of hybrid type of 290 MW capacity was designed and built at Huntorf (West Germany) in 1978 by **Brown Boveri**. This system is shown in Fig. 15.8. It uses a 150 m high salt cavern with a total volume of 3×10^5 m^3 created by leaching a salt dome at a depth between 650 and 800 m below ground. The pressure is maintained at 50 to 70 bars. Storage occurs daily at about 8 hours and generation about 2 hours, so the charging ratio in 1:4. Another CAES hybrid system using rock reservoir was built at Illinois (USA).

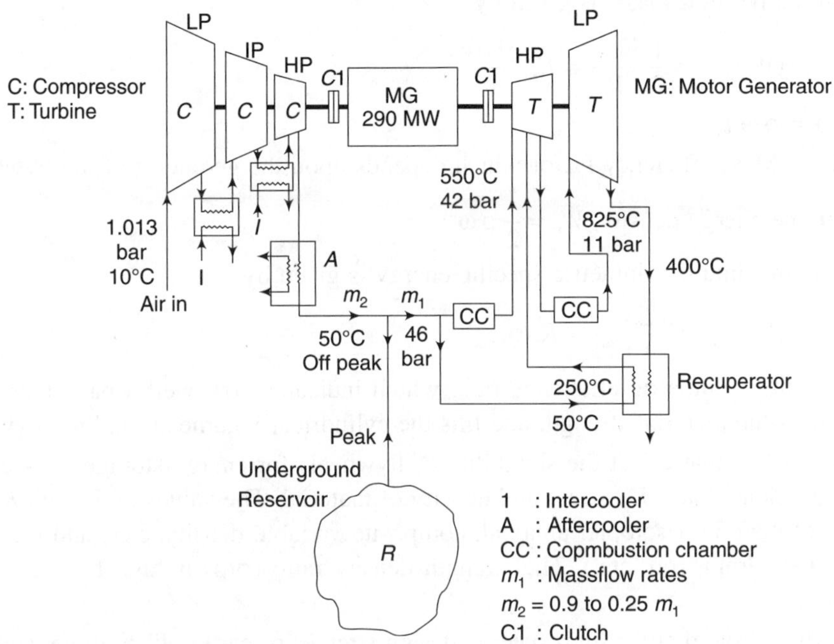

Figure 15.8 The Huntorf, Germany hybrid air storage gas turbine power plant

15.5 ENERGY STORAGE BY FLYWHEELS

It is a well-known fact that flywheels store off-peak energy as kinetic energy. They have been used extensively to smooth out power pulses from reciprocating engines. In this system, the flywheel rotor is physically connected to motor-generator set. In the charging mode, during off-peak periods, the motor adds energy to the flywheel. In the generation mode, during periods of peak demand, the flywheel rotor coasts driving the generator.

The energy stored in a flywheel is given by

$$KE = \frac{1}{2} I\omega^2 = \frac{1}{2} MR^2 (2\pi N)^2 = 2\pi^2 mR^2N^2 \quad ...(15.6)$$

where m is the mass of flywheel, R is radius of gyration and N is rps.

The energy absorbed or released by a flywheel between speeds N_1 and N_2 is

$$\Delta E = 2\pi^2 mR^2 (N_2^2 - N_1^2) = 4\pi^2 mR^2K_sN^2 \quad ...(15.7)$$

where $$K_s = \frac{N_2 - N_1}{N} = \frac{(N_2 - N_1)^2}{N_1 + N_2} = \text{Coefficient of speed fluctuation}$$

$$N = (N_1 + N_2)/2$$

K_s depends upon the desired closeness of speed regulations. For engines, K_s may vary between 0.005 and 0.2. In the design of flywheel, an important consideration is the stress level to which a rotating flywheel is subjected to. The theoretical maximum specific energy (energy stored per unit mass) is given by

$$(W_v)_{\max} = \frac{1}{2}\frac{\sigma_{\max}}{\rho} = K_m \frac{\sigma_{\max}}{\rho} \quad ...(15.8)$$

where $\sigma = \rho\,\omega^2 r^2$

K_m = Mass efficiency factor which depends upon the geometry of the flywheel.

The volume energy density, $W_v = \frac{1}{2}\rho\,\omega^2 r^2$ ...(15.9)

Thus, the maximum volumetric specific energy is given by

$$(W_v)_{\max} = \left(\frac{E}{V}\right)_{max} = K_v\,\sigma_{\max} \quad ...(15.10)$$

where K_v is the volumetric efficiency ratio which indicates how well a particular flywheel design utilises the material strength and fills the cylindrical volume around the flywheel.

The parameters that affect the suitability of flywheels for energy storage are—efficiency ratio, K_m, efficiency ratio K_v, stress and density of material. The values of K_m and K_v depend upon type of material (isotopic, uniaxial, composite variable density, etc.) and the shape of flywheel (disc, drum, rod, etc.). The strength density ratio (σ/ρ) is high for glass or silica fibres material.

It is worth to note that the high specific design stresses of carbon fibre/epoxy and Kevlar-fibre) epoxy allow storage of a large amount of energy in a relatively light flywheel. The turnaround efficiency (η_{TA}) of a flywheel energy storage system during the charge-store-discharge period will depend on the duration of the store period. Two main sources of losses in flywheel are windage bearing, and seal friction. *A NASA conceptual design for a flywheel electric energy storage system has nearly zero losses.*

It has been designed for a 10^4 kWh substation. The flywheel rotor, made of a filamentary anisotropic composite material to achieve high strength to mass ratio has inner and outer diameter and weighs of 4, 6, 1 and 3.05 m respectively. The weight is 333.4 tonnes and rotates at 1250 rpm. In order to eliminate rotor bearing friction, the rotor is magnetically levitated by permanent magnets that are incorporated in the stator with sets of teeth on both rims facing each other to maintain axial alignment. There are various other arrangements to remove losses such as hysteresis, eddy current and windage losses.

15.6 ELECTROCHEMICAL ENERGY STORAGE

It is the most traditional form of all energy storage devices for power system and may be classified into three categories, namely, ***batteries, secondary batteries*** and ***fuel cells***. The common feature of these devices is that devices are not limited by second law of thermodynamics. Figure 15.9 shows the working principle of batteries and fuel cells. Both consist of two electrode systems and an electrolyte-placed together in a special container and connected to an external source or load.

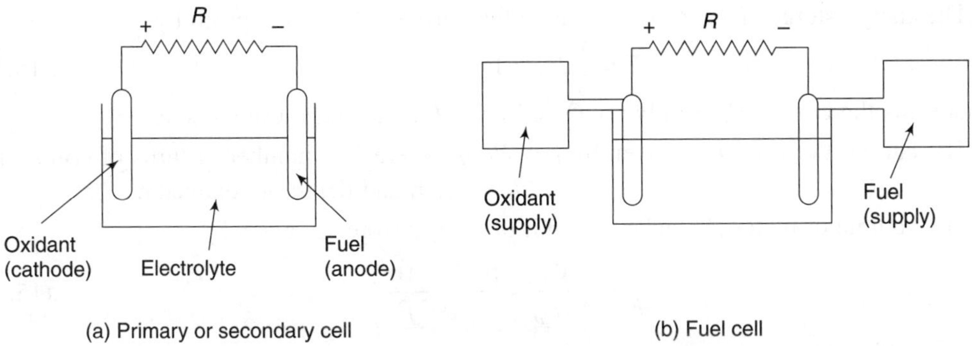

Figure 15.9 Principle of batteries and fuel cells

15.6.1 Battery Storage

The familiar lead-acid battery used in motor vehicles is a direct-current battery. It contains a number of voltaic cells that are connected. Each cell contains several lead plates, connected in parallel, made of grids that are filled with a spongy grey lead, Pb, and which forms the anode. The plate containing lead oxide, PbO_2, forms the cathode. All plates are immersed in a water solution of sulphuric acid (H_2SO_4) which acts as an electrolyte. The promising batteries system include (a) sodium-sulphur batteries, (b) lithium-chlorine and lithium-tellunide batteries, (c) zinc-chlorine batteries, (d) lithium-ion battery, etc.

15.6.2 Fuel Cells

The most common H_2–O_2 fuel cell was demonstrated in 1839. It has been discussed in the chapter 1 of this book.

15.6.3 Magnetic Energy Storage

It is well-known fact that electrical resistance of metals depends on temperature and in the neighbourhood of about zero temperature, the electrical resistance of mercury drops to zero. This is called *superconductivity*. This property is exhibited by all metals. The temperature below which a metal becomes a superconductor is called *critical* or *transition temperature*. All metals exhibit superconductivity behaviour in the cryogenic range (0–123 K). Electrical power transmission and distribution can best be accomplished by the use of high purity aluminium cables operated at liquid hydrogen temperature (20 K). The commercial success depends upon the economics of operation as achieving such temperature is costly affair.

The principle of superconducting magnetic energy storage is based on the fact that energy can be stored in a magnetic field associated, i.e., maintained below with the coil. If a coil is maintained in superconducting state (i.e., below critical temperature), then once it is charged, the current will not decay and the magnetic energy can be stored indefinitely. The stored energy can be released back to the network by discharging the coil.

The energy stored (E) is a coil in which the current (I) flows is given by

$$E = \frac{1}{2} L I^2; \text{ J} \qquad ...(15.11)$$

where L is the conductance in henry (V-s/A) and I is the current in ampere.

The current density (j) is given by $j = NI/ab$, where N = number of turns of coil, a and b = width and depth of conductor.

The volume of material needed per unit of energy stored is given by

$$\frac{V}{E} = \frac{V}{F.E_b} = \frac{0.33 \times 10^8}{FV^{2/3} j^2} \qquad ...(15.12)$$

where $F < 1$, $E_b = 3.028 \times 10^{-8}\ V^{5/2} j^2$ = energy stored in Brook's coil, j = current density limited to value between 50×10^6 and 100×10^6 A/m^2, V = volume of conductor in coil turn (m^3).

The above equation suggests that a huge structural mass is required to contain the magnetic field energy. At present, this field is under active research.

15.7 THERMAL ENERGY STORAGE

Thermal energy storage systems can operate at many desired temperature levels depending upon the use, choice of system and materials ranging from refrigeration temperature to 1250°C. They are widely used in many industrial applications, such as in the manufacturing of cement, iron and steel, glass, aluminium, paper, plastics and rubber and in food processing. There are two main types of thermal energy storage systems. They are:

(i) Sensible heat energy storage based on the heat capacity of storage medium.

(ii) Latent heat storage based on the energy associated with the change of phase for the storage medium (melting evaporation or structural change).

15.7.1 Sensible Heat Energy Storage

Sensible heat energy storage can be accomplished by raising the temperature of material, such as water, an organic liquid or a solid. The storage density (J/m^3) is given by $\rho.c.\Delta\theta$ where ρ is density (kg/m^3, c is heat capacity (kJ/kg.K) and $\Delta\theta$ is the temperature rise (K). The following types of sensible heat energy storage systems are used.

(a) Pressurized water storage (b) Organic liquid storage

(c) Packed solids beds (d) Fluidized solid beds

15.7.1.1 Pressurized Water Sensible Energy Storage System

Figure 15.10 shows the schematic of a pressurized water sensible energy storage system in a power plant in which the primary heat source is either a nuclear reactor or a fossil fuelled furnace. The base loaded portion of the plant is capable of supplying more steam than needed during periods of low demand, and the excess steam is bled at high pressure through turbine extraction. This extracted steam is fed to steel accumulators and mixed with water, thus producing saturated pressurized water. During the period of high demand, the accumulators are discharged through a small peaking turbine until a low specified pressure is reached in the accumulator resulting in low and varying steam temperature entering the peaking turbine. It is worth to note that this system involves steam condensing in water during accumulator charge and re-evaporating during discharge, the storage medium is the pressurized water in the accumulators and operates over a relatively wide temperature range. The typical values of a accumulator having high and low pressure are 20 bar and 2 bar corresponding to a saturation temperature of about 212°C and 120°C respectively.

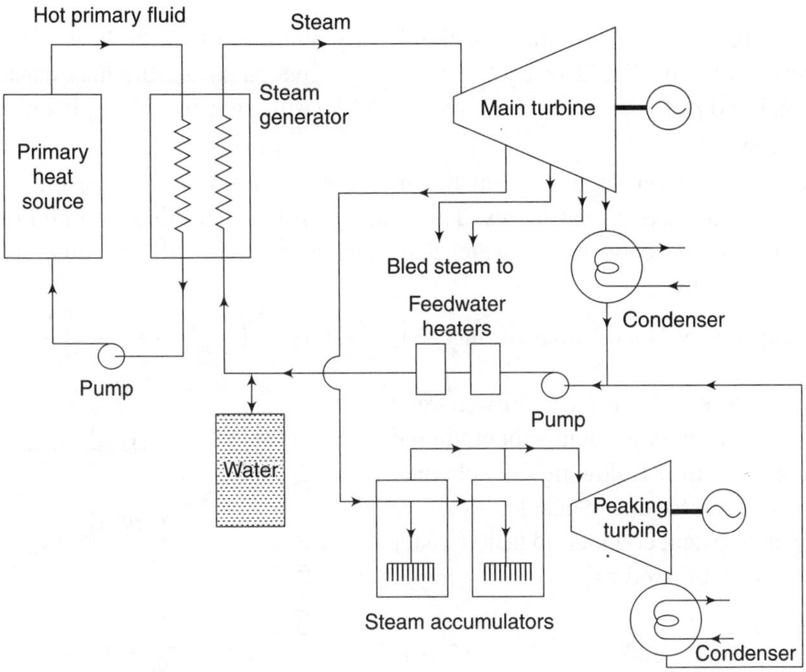

Figure 15.10 Pressurized water sensible energy storage system

The storage density of the thermal energy utilised in the peaking turbine per unit volume of the high-pressure saturated water is expressed as

$$\text{Storage density} = \frac{1}{v_{f,1}}(h_{f,1} - h_{f,2}) \qquad \text{...(15.13)}$$

where v_f and h_f are the specific volume and enthalpy of saturated water respectively. Here the subscripts 1 and 2 refer to the stored and emptied pressure respectively. The storage density is about 1 kWh/m^2. °C over the temperature range cited above.

The electric energy density obtained by the peaking turbine-generator depends upon two efficiencies, namely, *thermal turnaround efficiency (η_{TA}) and the peaking-turbine generator* efficiency *(η_{PT})*. The thermal turnaround efficiency is a complex function of the losses associated with sensible heat transfer to and from the steel cells, structural members of the accumulators, and interconnecting pipework, and time-dependent convective heat losses to the environment.

The losses associated with sensible heat transfer $(\delta Q_L)_{sh}$ is given by

$$(\delta Q_L)_{sh} = \frac{\text{Energy stored in the accumulator structure}}{\text{Energy stored in the contained water at a given pressure}}$$

$$= \frac{\pi D L t \rho_s c_s}{(\pi D^2/4)\, L \rho_f c_f} = 2\left(\frac{p}{\sigma}\right)\left(\frac{\rho_s c_s}{\rho_f c_f}\right) \quad \text{...(15.14)}$$

where p is the pressure, σ wall stress, ρ the density and c the specific heat. The subscripts s and f denote solid and liquid respectively. In the above equation, the heat capacities ratio is nearly equal and p/σ is around 0.003, and so the contribution of $(\delta Q_L)_{sh}$ is rather minimal and may be ignored.

The above discussion suggests that the major contributor to the thermal turnaround efficiency is the convective heat losses. They vary with time and depend upon overall heat transfer coefficient (U) between the water and outside environment. The time constant τ of the system is given by

$$\tau = \text{heat capacity} \times \text{Heat flow resistance} = \left(\frac{\pi D^2}{4} \rho_f c_f L\right)\left(\frac{1}{\pi D L U}\right) = \frac{D \rho_f c_f}{4U} \quad \text{...(15.15)}$$

If the temperature of the liquid at fully charged condition is T_1, its time-dependent temperature of $T(\theta)$ decreases with time following fully charged conditions at (time zero) due to heat losses above (Fig. 15.11) on transient conduction heat transfer analysis, $T(\theta)$ is expressed as

$$\frac{T(\theta) - T_1}{T_\infty - T_1} = 1 - e^{-\theta_s/\tau} \quad \text{...(15.16)}$$

where T_∞ is the environmental temperature. If the storage period is θ_s at which water temperature decreases to T_s, we have

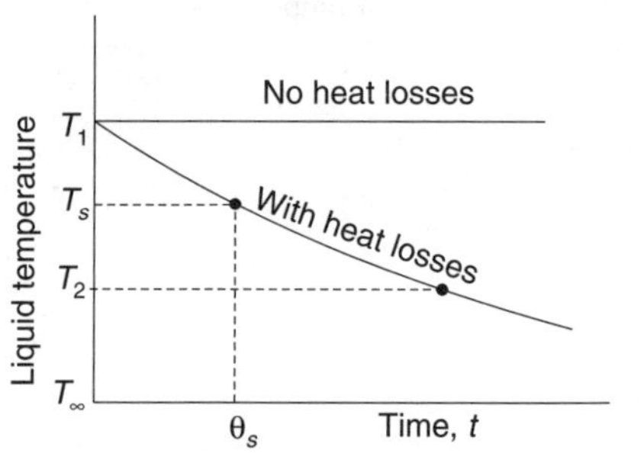

Figure 15.11 Storage liquid temperature-time change in sensible energy storage

$$\frac{T_s - T_1}{T_\infty - T_1} = 1 - e^{-\theta_s/\tau} \quad \text{...(15.6a)}$$

The thermal turnaround efficiency is expressed as the ratio of energy left in storage at T_s, after heat losses to the original energy stored. Thus,

$$\eta_{TA} = \frac{h_s - h_2}{h_1 - h_2} \quad \text{...(15.17)}$$

where h_1, h_2 and h_s are the compressed liquid enthalpies at T_1, T_2 and T_s respectively.

Assuming equal specific heats over the temperature range,

$$\eta_{TA} = \frac{T_s - T_2}{T_1 - T_2} \quad \text{...(15.18)}$$

From Eqs. (15.16) and (15.18), we have

$$\eta_{TA} = 1 - \frac{T_1 - T_\infty}{T_1 - T_2}(1 - e^{-\theta_s/\tau}) \quad ...(15.19)$$

The above equation suggests that η_{TA} is a strong function of the ratio θ_s/τ. The only important variable in τ is U.

It is worth to note that accumulators may be constructed above ground or underground, the choice should be based on the careful evaluation of efficiency, cost, operational problems and safety.

The efficiency of the peaking turbogenerator is low about 20 to 25 per cent due to variable inlet conditions, use of low temperature saturated steam, small size, absence of feed water heating and other factors. There are three main ways of operating the accumulators, such as *variable pressure, expansion and displacement accumulators.* The schematic diagram of each of them is shown in Figs. 15.12, 15.13 and 15.14 respectively.

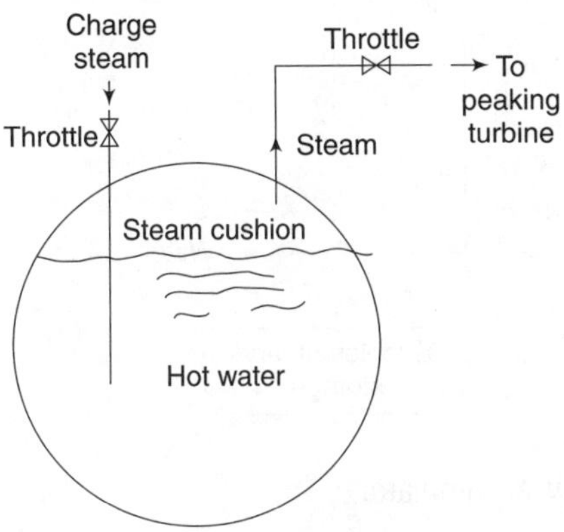

Figure 15.12 Variable Pressure Regulator

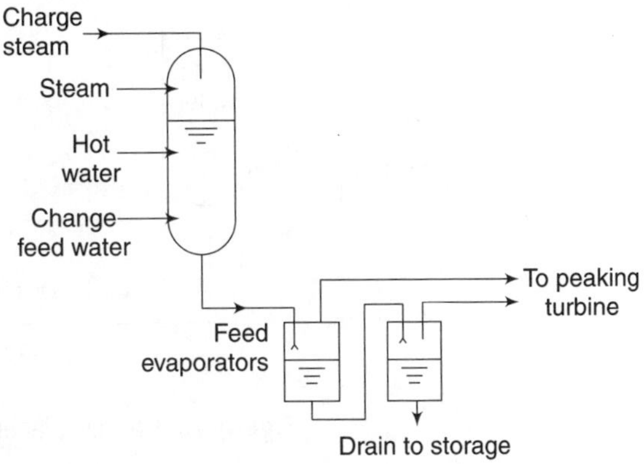

Figure 15.13 Expansion Accumulator

15.7.1.1.1 Variable Pressure Accumulator. Under the fully charged condition (Fig. 15.12), the accumulator is almost full of hot water with a small steam cushion as in the variable pressure mode. Charged steam is used to heat the water in the accumulator. The steam is taken from the steam space from top for peaking turbine and as the pressure in the cushion decreases, some water flashes into steam.

15.7.1.1.2 Expansion Accumulator. In this case also, under the fully charged conditions (Fig. 15.13), the accumulator is almost full of hot water with a small steam cushion on the top as in variable pressure mode. With the drawing of hot water from the bottom during discharge, some water flashes into steam resulting in the reduction of pressure and temperature of hot water and steam. It is possible to remove all the water to the flash evaporator with a reduction of pressure of about 30 per cent. The steam is withdrawn from the flash evaporator for peaking turbine and the water from the last flash evaporator is collected and stored.

15.7.1.1.3 Displacement Accumulator. During charge, steam is mixed with cold water taken from the bottom to maintain temperature at desired level and it is always completely full of water (Fig. 15.14). When fully charged, it contains hot water at desired temperature and when fully discharged, all the water becomes cold. Hot water is injected from the top during charge and removed from the top during discharge. A sharp temperature gradient separates the cold and hot water. During discharge one or more flash evaporators are used to generate steam for peaking turbine. The water drained from the evaporators and the condensate from the turbine is returned as cold water.

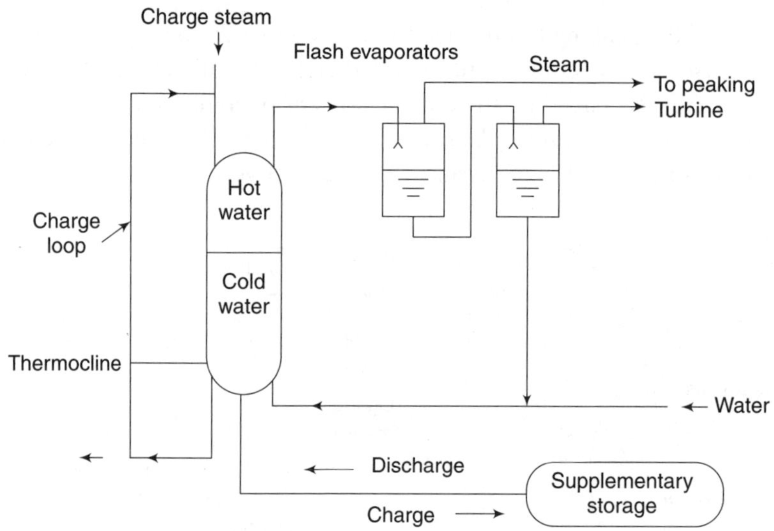

Figure 15.14 Displacement accumulator

15.7.2 Latent Heat Energy Storage

As the name implies, in this system, energy is stored in the form of latent heat caused by phase charge either by melting solid or vaporizing liquid. Energy release is achieved by reversing the process, i.e., solidifying the liquid or condensing the vapour. The storage density is equal to the product of latent heat of fusion or vaporization times the density of storage material. Obviously, it is greater than that in sensible heat storage as the latent heats are much longer than the specific heats of single phase material. There are various advantages of this system. It can operate essentially at constant temperature with low volume change during phase change. It may use variety of materials suiting the requirement of peaking turbine. The materials include fluoride salts, 70% NaF, 30% F_eI_2 eutectic salt, $ZnCl_2$, silicon, germanium, and sulfides of germanium. Still the system is in developing stage.

15.8 CHEMICAL REACTION ENERGY STORAGE

As the name implies, in this mode of energy storage, the heat of reaction of reversible chemical reaction is used to store thermal energy during endothermic reactions and to release it during exothermic reaction. This process also offers large energy storage density like latent heat energy storage. The following reaction was suggested by Schulten, et al., for long distance transmission of gas-cooled reactor thermal energy–

$$CO + 3H_2 \leftrightarrows CH_4 + H_2O \qquad ...(15.20)$$

Table 15.1 shows the various reversible reactions with the operating range and heats of reactions.

Table 15.1 Reversible chemical reaction for energy storage

Reaction	*Temperature range (K)*	*Heat of reaction of 298 K (kJ/g mole)*
$CO + 3H_2 \leftrightarrows CH_4 + H_2O$	700–1200	250.3
$2CO + 2H_2 \leftrightarrows CH_4 + CO_2$	700–1200	247.4
$C_6H_6 + 3H_2 \leftrightarrows C_6H_{12}$	500–750	207.2
$C_7H_6 + 3H_2 \leftrightarrows C_7H_{19}$	450–750	213.2
$C_{10}H_8 + 5H_2 \leftrightarrows C_{10}H_{18}$	450–700	314.0

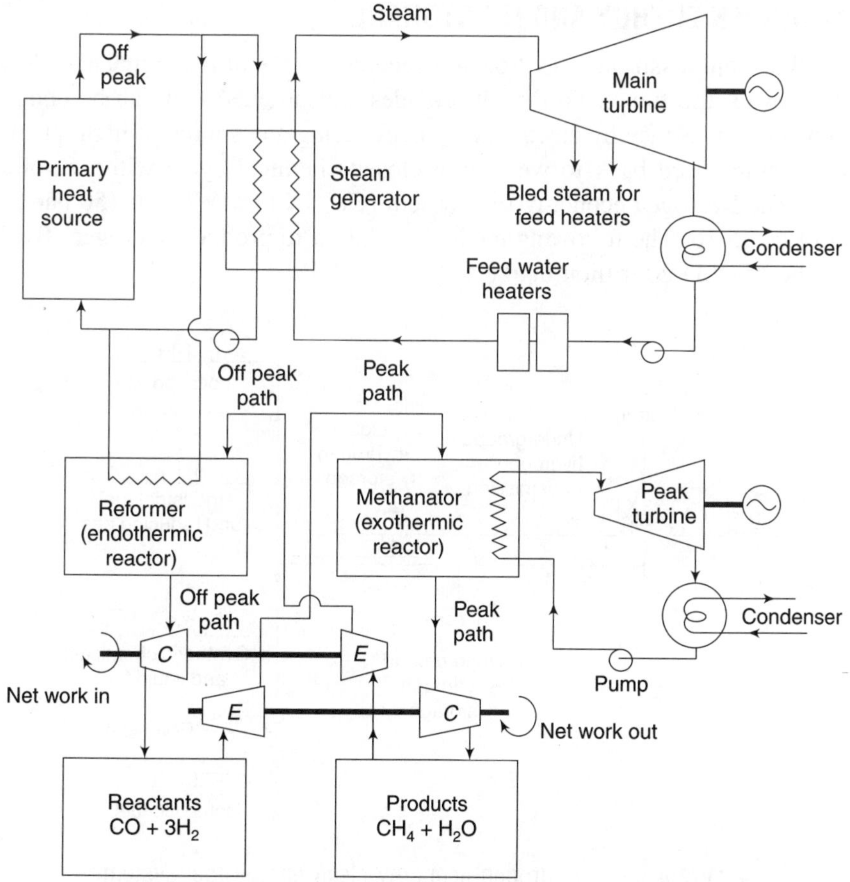

Figure 15.15 A schematic of chemical energy storage system used in power plant

Figure 15.15 shows the schematic of a power plant with a chemical energy storage system using the reaction given by Eq. (15.20). During the periods of low demand, some heat from the primary heat source is directed to the reformer (endothermic reactor) to convert the products $CH_4 + H_2O$ to the reactants $CO + 3H_2$ which are stored in a vessel at high pressure (about 70 bars) but at atmospheric temperature. During the periods of high demand, these reactants are fed to the methanator (exothermic reactor) where heat is generated to run a peak turbine (or generate more steam for the main turbine). The methanator converts the reactants into the product $CH_4 + H_2O$, which are stored in a separate vessel for later use in the reformer during period of low demand. The thermal turnaround efficiency of this system is very high, about 85 to 90% as per estimate. The losses in the system include heat losses to vessel and piping and pumping losses of the gases.

The size of the storage vessels, (generally two in number), even with operation near a pressure of 70 bars, which is compatible with current technology for gas storage, is still very large of the order of hundreds of thousands of cubic metres. The vessel may be steel tanks (costly), or underground caverns (cheaper) if suitable sites are available.

15.9 HYDROGEN ENERGY AND ITS STORAGE

Due to pollution from fossil fuels, hydrogen is regarded as the ultimate fuel and clean energy storage medium for the future. Further, it provides a clean energy. It can be obtained from water which is in abundance by electrolysis process using any source of high quality energy and it can be combusted back to water in a closed chemical cycle without polluting the environment. The hydrogen economy fuel system (CU = 140 MW/kg at 150 bar and 20°C) is shown in Fig. 15.16. The following methods are used to produce hydrogen. By 2050, H_2 energy may be of USD 20 trillion markets.

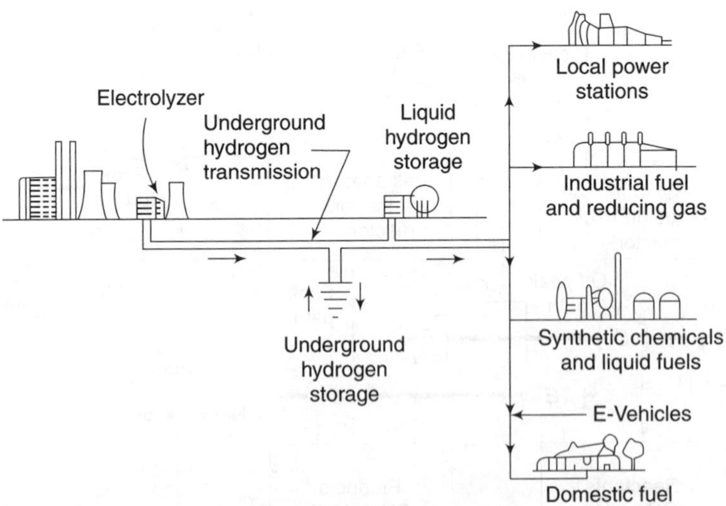

Figure 15.16 Hydrogen economy clean energy fuel system

(i) Electrolytic decomposition of water.
(ii) Catalytic steam reforming of natural gas (at present more prevalent)
(iii) Chemical reduction of coal (also prevalent)
(iv) Industrial photosynthesis
(v) Ultraviolet radiation
(vi) Partial oxidation of heavy oils
(vii) Thermal decomposition of water, utilizing thermochemical cycles.

Among all, electrolytic decomposition of water is gaining ground for producing clean H_2. It consists of two processes

Cathode : $H_2O + 2e^- \rightarrow H_2 + 2\,HO^-$

Anode: $2OH^- \rightarrow \frac{1}{2}O_2 + H_2O + 2e^-$

The overall reaction is

$$H_2O \rightarrow H_2 + \frac{1}{2}O_2 - 237 \text{ kJ/kg mole} \quad ...(15.21)$$

It is worth to note that hydrogen may be used as primary fuel for peak power generation producing water in the reaction. During off-peak hours, hydrogen may be produced by electrolysis of water with excess power available. Figure 15.17 shows the possible hydrogen based power utility concept.

The following are the **various methods of storing hydrogen.**

(i) Compressed gas, (ii) chemical compound, (iii) liquid and (iv) metallic hybrids.

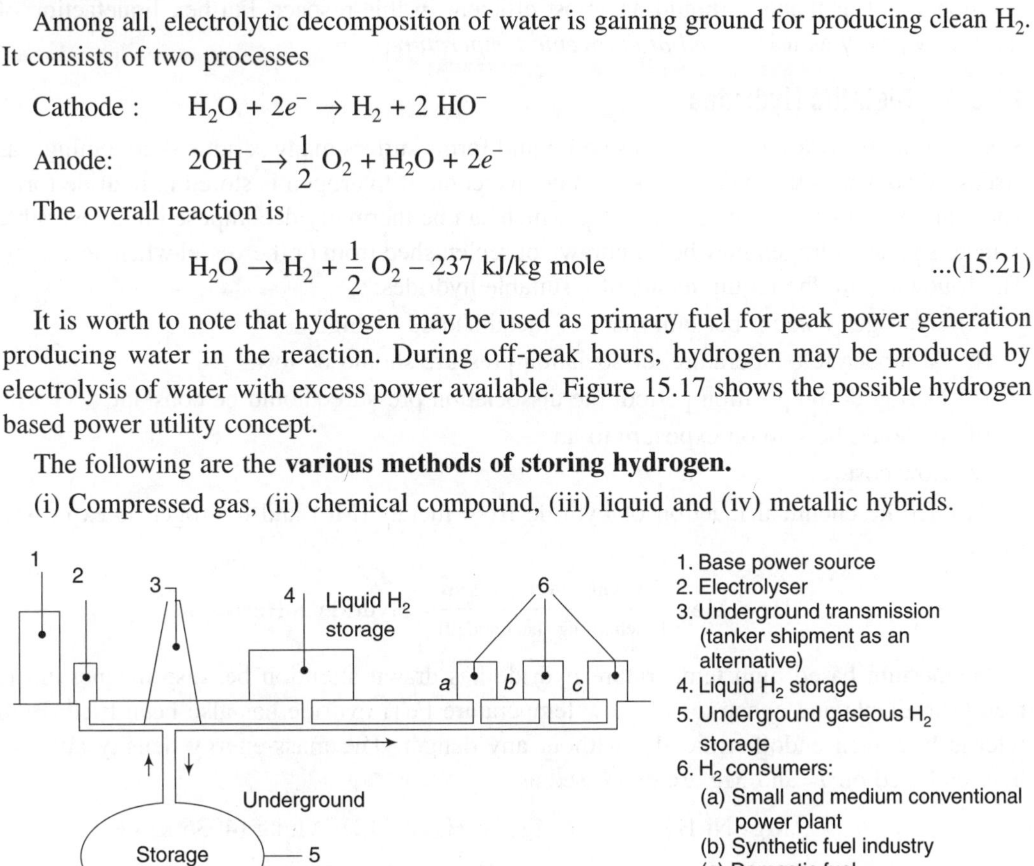

Figure 15.17 Power utility concept using hydrogen energy

15.9.1 Compressed Gas

Owing to the high cost of liquefaction, the hydrogen may be stored in the form of compressed gas in underground caverns, where it can be stored like natural gas. *The storing of hydrogen in the form of gas offers many disadvantages such as it needs large space, it is explosive and is not leak proof.*

15.9.2 Chemical Compound

Hydrogen in chemical combination with other elements like ethane and ammonia forming compound is more amenable to storage. But the energy release from compound is not efficient.

15.9.3 Liquid Hydrogen

Storage of hydrogen in the form of liquid is not so attractive because its density is low. Ammonia and methane as liquid are most efficient in this respect. Further, liquefaction of H_2 is very costly as it is *stored at cryogenic temperature.*

15.9.4 Metallic Hydrides

Since storing of hydrogen in gaseous and liquid forms suffers many serious shortcomings, as discussed above, so these drawbacks can be overcome if hydrogen is stored in hydride form. The main objective is to select a hydride which can be thermally decomposed in a reversible manner so that hydrogen may be withdrawn or replenished from or the vessel when necessary. The following are the requirements of a suitable hydrides:

(a) Hydrogen content per unit mass of metal should be high.

(b) At moderate temperature, dissociation pressure should be low.

(c) During decomposition period, the dissociation pressure should be constant

(d) It should be safe on exposure to air.

(e) Low cost.

Exothermic chemical reaction of hydride from metals (Me) and hydrogen is expressed as

$$H_2 + \text{Me} \xrightarrow[\text{Discharging (heat added)}]{\text{Charging (heat released)}} \text{Hydride} + \text{Heat}$$

Magnesium based high temperature hydride has drawn attention because magnesium is readily available and cheap metal. Low temperature FeTi hydride has also been tried. Both release hydrogen endothermically, without any danger. The mass-energy density (W_m) of hybrids based on T_2 and Mg are expressed as

$$Mg_2\,Ni\,H_4 \longrightarrow Mg_2\,Ni\,H_{0.3} \quad 1121 \text{ Wh/kg (4036 kJ/kg)}$$

$$Fe\,Ti\,H_{1.7} \longrightarrow Fe\,Ti\,H_{0.1} \quad 516 \text{ Wh/kg (1856 kJ/kg)}$$

Since different hydrides release different amount of heat so it can be used in a variety of applications such as heat pumps, central air conditioning and renewable energy systems.

Extensive research is going on to produce and store hydrogen energy.

Example 15.1 A simple compressed air energy system (CAES) supplies average air flow into a cavern of 65,000 m^3 at the rate of 8400 m^3/h. Air enters the compressor at 1 bar and 288 K and leaves at 90 bars. The isentropic compressor efficiency is 75% and the peaking turbine efficiency is 60%. The air is stored in the cavern at 90 bars and 288 K, calculate the following: (a) the compressed air temperature, (b) the storage time, (c) the total energy storage in MWh, and (d) the total energy delivered by the peaking turbine in MWh.

Solution Refer to Fig. 15.18.

$$T_2 = T_1\left(\frac{p_2}{p_1}\right)^{(\gamma-1)/\gamma} = 288(90)^{0.4/1.4} = 1043.05 \text{ K}$$

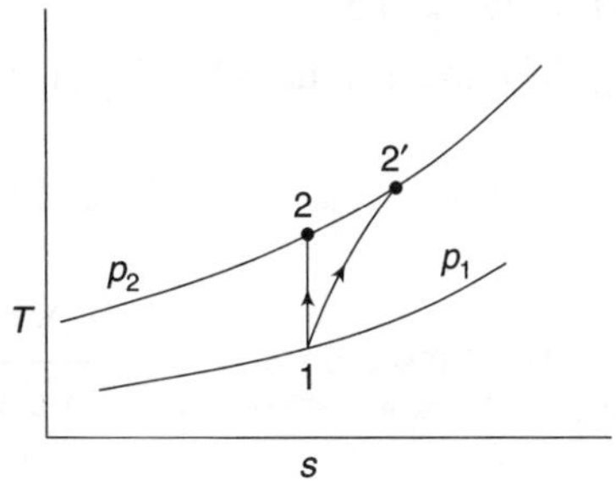

Figure 15.18 Example 15.1

The actual temperature at the exit of compressor is given by

$$\eta_c = \frac{T_2 - T_1}{T_2' - T_1} = 0.75$$

(a) or $T_2' = T_1 + \dfrac{T_2 - T_1}{0.75} = 288 + \dfrac{1043.05 - 288}{0.75}$

$= 1294.73$ K ***Ans.***

The specific volume of air at the condition of cavern, i.e., 90 bars and 288 K is given by

$$v = \frac{RT}{p} = \frac{0.287 \times 288}{90 \times 10^2} = 0.009184 \text{ m}^3\text{/kg} \qquad \textbf{\textit{Ans.}}$$

The mass flow rate of air into the cavern,

$$\dot{m} = \frac{\dot{V}}{v} = \frac{8400}{0.009184} = 914634.15 \text{ kg/h} = 254.06 \text{ kg/s}$$

The rate of energy storage is given by

$$\dot{E} = \dot{m}c_p(T_2' - T_1) = 254.06 \times 1.005(1294.73 - 288) \times 10^{-3} = 257.04 \text{ MW}$$

(b) Thus, the storage time, $\tau = \dfrac{V}{\dot{V}} = \dfrac{65000}{8400} = 7.738$ h ***Ans.***

(c) The total energy storage $= \dot{E} \times \tau = 257.04 \times 7.738 = 1988.97$ MWh ***Ans.***

The total energy delivered by peaking turbine = 1988.97 × 0.6 = 1193.34 MWh ***Ans.***

Example 15.2 A pressurized water sensible water heat storage system extracts the steam from the main turbine during off peak hours and feeds into steel accumulators of volume 18×10^4 m^3. The size of each steel accumulators is 4 m in diameter. These are well insulated yielding overall heat transfer coefficient (U) = 1.5 W/m^2 K. The maximum and minimum storage pressure are 20 bars and 2 bars. The ambient air temperature and specific heat of water may be taken as 293 K and 4.35 kJ/kg K respectively. Assume thermal turnaround efficiency as 96 per cent. If the peaking turbine efficiency is 20% calculate the following: (a) storage time, (b) total energy stored in the accumulators, and (c) total energy that can be delivered by the peaking turbine.

Solution From Steam Table, at 20 bars, $t_1 = 212.37°C$, $h_{f,1} = 908.5$ kJ/kg, $v_{f,1} = 0.0011766$ m^3/kg, At 2 bars, $t_2 = 120.23°C$, $h_{f,2} = 504.8$ kJ/kg, $v_{f,2} = 0.0010605$ m^3/kg. At 20°C, $h_f = 293$ kJ/kg.

The average density of water (ρ) is given by

$$\rho_f = \frac{1}{2}\left[\frac{1}{v_{f,1}} + \frac{1}{v_{f,2}}\right] = 896.43 \text{ m}^3/\text{kg}$$

The time constant, $\tau = \dfrac{D\,\rho_f\,c_f}{4U} = \dfrac{4 \times 896.43 \times 4.35 \times 10^3}{4 \times 1.5 \times 3600} = 722.12$ h. ***Ans.***

The thermal turnaround efficiency, $\eta_{TA} = 1 - \dfrac{T_1 - T_\infty}{T_1 - T_2}[1 - e^{\theta_s/\tau}]$

or
$$0.96 = 1 - \frac{212.37 - 20}{212.37 - 120.23}[1 - e^{\theta_s/\tau}]$$

or
$$e^{\theta_s/\tau} = 0.980841 \text{ or } \theta_s/\tau = 0.01931$$

(a) Storage time, $\theta_s = 0.01931 \times 722.12 = 13.94$ h ***Ans.***

The mass of water needed to be flashed

$$m_f = \frac{18 \times 10^4}{v_{f,1}} = \frac{18 \times 10^4}{0.0011766} = 1.52983 \times 10^8 \text{ kg}$$

(b) The total energy stored in the accumulator is given by

$$m_f\,(h_{f,1} - h_{f,2}) = 1.52983 \times 10^8\,(905.5 - 504.8) = 613 \times 10^8 \text{ kJ}$$

0.17027×10^8 kWh = 17027 MWh ***Ans.***

(c) The total energy delivered by the peaking turbine

$17027 \times 0.96 \times 0.2 = 3269.18$ MWh ***Ans.***

PROBLEMS

Things to Think About

15.1 What is the basic philosophy of energy storage?

15.2 What is the basic philosophy of pump hydro?

15.3 What limits the use of compressed air energy storage?

Short and Long Theoretical Questions

15.1 State the aim of energy storage in the utility system.

15.2 State suitable methods of energy management in utility system.

15.3 Explain the concept of mix of base-load plants intermediate plants and peaking plants for reliable and economical supply of electrical energy.

15.4 Explain the basics of energy storage schemes?

15.5 What are the various methods of energy storage schemes?

15.6 With the help of a neat sketch, discuss the working principle of the following energy storage systems: (a) Pumped hydro, (b) Compressed air energy storage, (c) Energy storage by flywheel, (d) Electrochemical energy storage, (e) Magnetic energy storage, (f) Thermal energy storage, (g) Chemical energy storage and (h) Hydrogen energy storage.

Numerical Questions

15.1 A pumped-hydro energy storage system with an elevation of 50 m has been chosen for a power grid which has a load pattern of 700 MW during 18 h and 1300 MW during 6 h for a day. Find (a) the power output of the power plant in MW that would meet the load demand with and without energy storage, and (b) volume of water in m^3 that must be pumped to meet storage demand. The electric generator efficiency is 0.8 and the density of water is 10^3 kg/m^3.

15.2 A simple compressed air energy storage system (CAES) supplies average air flow into a cavern of volume 7×10^3 m^3 at the rate of 8500 m^3/h. The air enters the compressor at 1 bar and 288 K and leaves at 110 bars. The isentropic efficiency of compressor is 80% and the peaking turbine efficiency is 60%. The air is stored in cavern at 110 bars and 30°C. Calculate: (a) the compressed air temperature, (b) the storage time, (c) the total energy storage in MWh, and (d) the total energy delivered by the peaking turbine.

15.3 A base load 2000 MW power plant is designed with a sensible thermal energy storage having accumulators for extracted steam in pressurized water. The thermal energy stored is designed to produce 8000 MWh daily. The size of the accumulators is 4 m in diameter each and are well insulated to produce U = 5 kJ/hm^2 K. The storage time is 18 hours. The maximum and minimum storage pressure of steam are 20 bars and 2 bars. Respectively. Determine (a) the following: (a) the thermal turnaround efficiency, and (b) the total accumulator volume.

Software Based Question

15.1 Develop a software for the design of various energy storage systems.

Design and Open Ended Question

15.1 Prepare a comprehensive report for various types of the existing energy system working in world over with their limitations.

Multiple-Choice Questions

15.1 The pumping power of pump hydro depends in

(a) discharge (b) head (c) efficiency (d) all of them

15.2 The turnaround efficiency of pump hydro system is around

(a) 55% (b) 65% (c) 75% (d) 95%

15.3 In hybrid air energy storage system, the pressure raised in storage tank is around

(a) 20 to 30 bars (b) 30 to 40 bars (c) 40 to 50 bars (d) 60 to 70 bars

Answers to Multiple-Choice Questions

15.1 (d) 15.2 (b) 15.3 (d)

CHAPTER

16

Major Electrical Equipment in Power Plants

OVERVIEW

It is essential to study the brief description of electrical equipment used in power plants by Mechanical Engineering students. This chapter deals with the working principle of major electrical equipment used in power plants. The major electrical equipment include generator (alternator), exciters, transformers, reactor, circuit breaker, switchgear and protective equipment, control board equipment, bus-bars, standby generators, etc.

16.1 LAYOUT OF ELECTRICAL EQUIPMENT

The layout of electrical equipment in the power plant plays a vital role. *It consists of the arrangement of the* ***generators, bus-bars, circuit breakers, transformers*** *and* ***controlling switchboards****. The generator (or alternator) is coupled to the turbine shaft.*

The controlling switchboard consists of instruments to indicate the behaviour of equipment at any time. The purpose of controlling instruments is to control the operation of the whole station and different types of relays to protect the station equipment against abnormal faults. The various types of layout of electrical equipment are as follows.

16.1.1 Single Bus-Bar System

Figure 16.1 shows a single bus-bar arrangement. This is suitable to a power station having number of generators and single bus-bar arrangement. The main advantage of this system is that a fault on one part of the bus-bar does not completely shuts down the whole station. The system comprises a circuit breaker and isolating switch between each generator and low voltage bus-bar, a circuit breaker and isolating switch on both of low voltage side and high voltage side of the step-up transformer and a circuit breaker and isolating switch to tie up the two bus-bar sections.

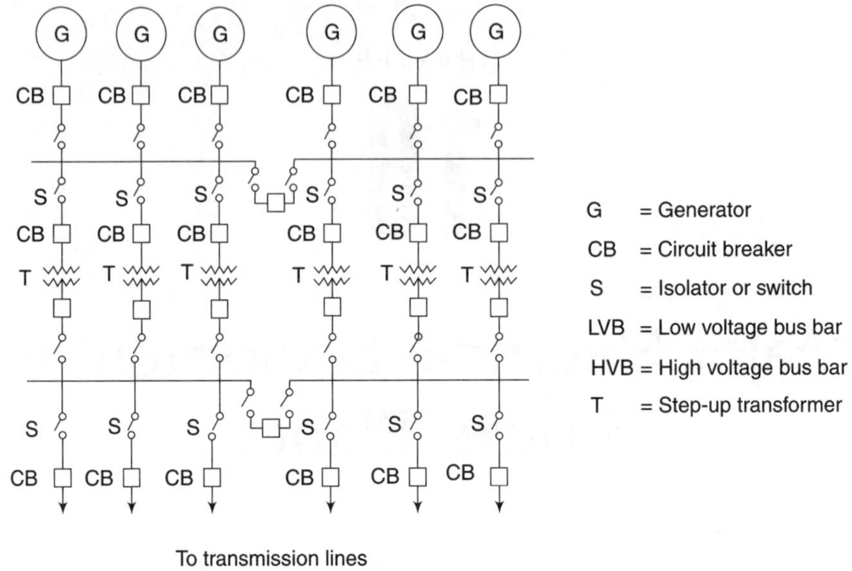

Figure 16.1 Single bus-bar arrangement

*In the **present technology**, the use of large number of circuit breakers is obsolete. Some modern power plants, where the unit system is used, they use **one generator** and **one transformer** as a unit connected direct to the **high voltage** bus-bars and there is only one set of **high-voltage switchgear**.*

16.1.2 Duplicate Bus-Bar System

Figure 16.2 shows the schematic diagram of a duplicate bus-bar system. The system comprises both the low-voltage and high-voltage bus-bar in duplicate. One of the two can be used as desired as a bus-bar coupling switch to transfer operation from one bus-bar to the other bus-bar.

The following are the advantages of this duplicate system:

(i) This makes possible to do repair work one bus-bar while other in operation without affecting power supply.

(ii) The repair of any bus-bar isolator is possible by switching out only that circuit in which repair work is to be done.

(iii) Quick restore of operation of the installation is possible even in the event of an occurrence of short-circuit on the working system of the bus-bar.

(iv) It is possible to repair the circuit breaker of any circuit without taking it out of service for an extended period.

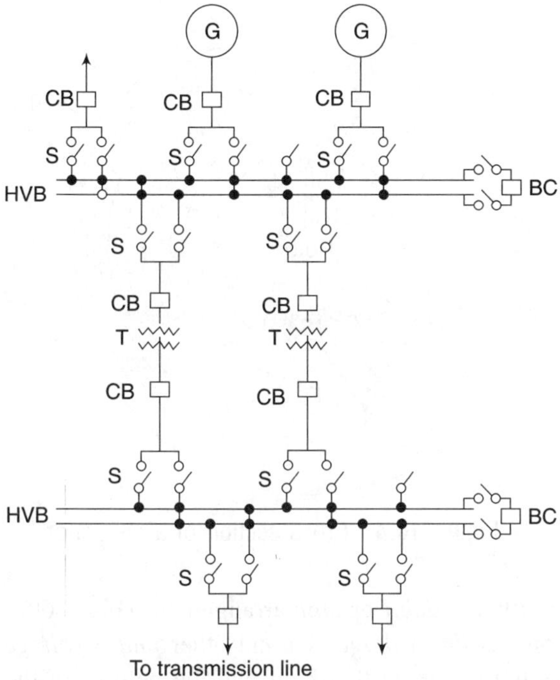

Figure 16.2 Duplicate bus-bar arrangement

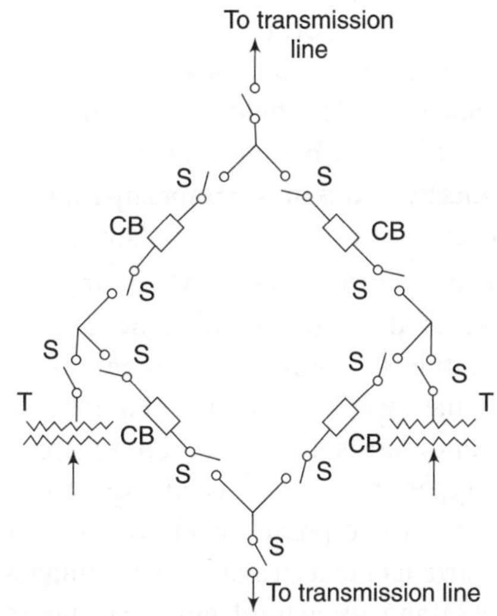

Figure 16.3 Ring and bridging bus-bar system

16.1.3 Ring and Bridging Bus-Bar System

Figure 16.3 shows the schematic of a ring and bridging bus-bar system. In this system, each line is served by two circuit breakers, therefore, any one of the latter can thus be repaired without disruption of the line. All the isolators are used for repair purpose only. The system offers high reliability of supply with the aid of a comparatively small number of circuit breakers as compared to double bus-bar system. *The other advantage is that there are always two parallel paths to the circuit and failure of one section does not interrupt the supply completely.* The main drawback includes the need to equip the circuits with a more involved relay protection and difficulty in making extensions. Due to this shortcoming, system is mainly suitable for a station where there is no plan for extension of installation.

This system finds applications for voltages of 110, 220 and 440 kV where two outgoing transmission lines are fed by two set-ups or auto-transformers.

16.2 GENERATORS

Electric generator generating a.c. (or alternator) coupled to the turbine shaft is the most important part of the power station as all other units in the power stations are installed to drive it. *The generators work on the principle of* ***electromagnetic induction***. It is to be noted that generator which generates d.c. is called d.c. generator or dynamo. Figure 16.4 shows the cross section of a generator (alternator). It essentially consists of a fixed stator and revolving rotor.

The stator core consists of a three phase winding in which alternating electromagnetic force (emf) is induced when the shaft of the rotor is revolved with the help of turbine (prime-mover). The rotor carrying field magnet provides the magnetic flux to the machine. The field is excited by a direct current brought into the field circuits by means of two rotor slip rings and set of brushes riding on them.

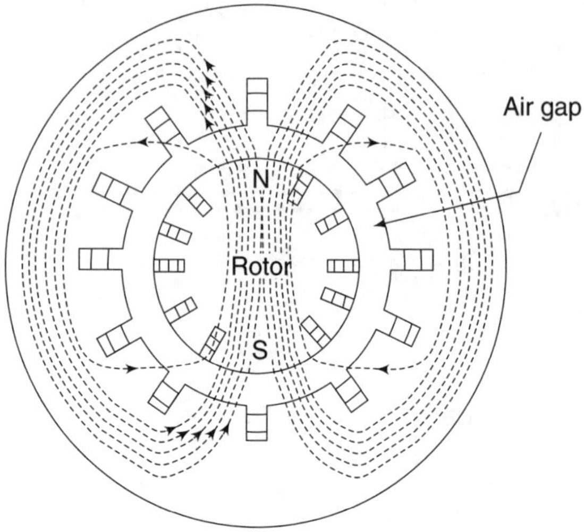

Figure 16.4 Cross section of a generator

The induced voltage in the single phase of the rotor winding depends upon the strength of the magnetic field, the speed of rotation and number of stator coils in series. There are two types of winding used in stator coil in a three-phase generator. It may be either a *delta* or *star* arrangement (Fig. 16.5). Star arrangement provides two voltages—one as *line voltage* (V) and other *phase voltage* ($V/\sqrt{3}$) and its neutral point can be connected to the earth which may be helpful in the design of protective system. In the case of delta system, in the event of a fault earthing one phase, the maximum voltage between two sound phases and earth is limited to phase voltage $V/\sqrt{3}$ and an earth fault on one line, a voltage of $\sqrt{3}\,V$ volts will be applied between the earth and other two healthy phases. This suggests that in the event of a fault in the stator winding, the stress on insulation will be higher in the case of a delta circuit.

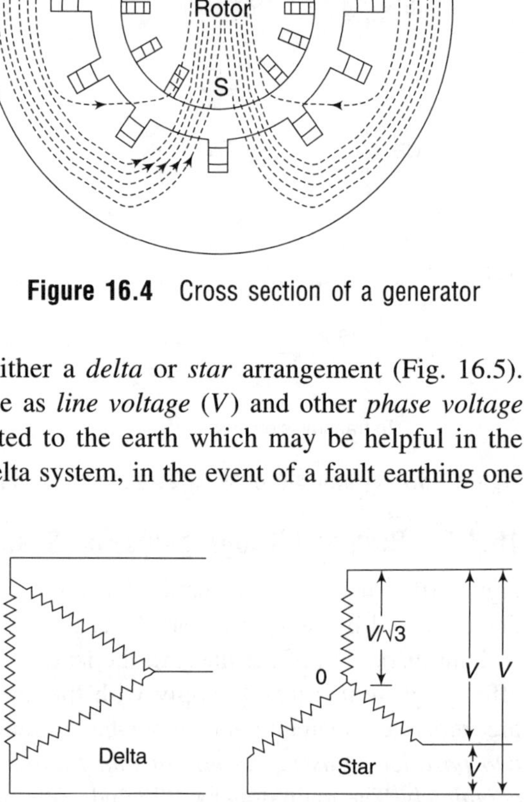

Figure 16.5 Delta and star arrangement of stator winding

As mentioned above, the main parts of the alternator are stator, rotor and exciter, the details of which are given below.

16.2.1 Stator

The main parts of the stator are *stator frame, stator core* and *stator windings.* The stator frame is circular in shape and is made of welded steel plate. The core is made of stampings of high permeability, low hysteresis and eddy current losses. By coating the lamination on each side with enamel, it is possible to reduce the eddy current losses. In order to cool the stator, the radial ducts are provided at regular intervals through the length of the core for efficient cooling of the stator.

As discussed above, the stator contains three groups of separate condenser conductors in which the voltage and current are generated. The conductors are arranged at 120° from each other and are connected in series to form three sets of coil called *phases*. The electrical power is generated in each of three phases. As the coils are displaced in space by 120 mechanical degrees, and there is only one magnetic field, the voltage and current produced in each of three phases are displaced in time by 120 electrical degrees. Such a power produced is known as *balanced three-phase power*.

16.2.2 Rotor

The design and construction of rotor is difficult as its weight is considerably high and rotates at fairly high speed (3000 or 3600 rpm). In order to accommodate the field windings to carry field current, a large number of deep slots are machined in the rotor. The length between the two bearings is limited to eight times the diameter of the rotor. *The approximate weight of the rotor of 120 MW is about 30 to 40 tonnes.*

In order to achieve efficient cooling of rotor, it is necessary to allow ample passage in the rotor, through which cooling fluid can be circulated freely.

16.2.3 Exciter

The field of the alternator requires magnetism. The permanent magnets are unsuitable for the field of alternator. Therefore, direct current is fed into the poles of alternator so that one end is a north pole and the other a south pole. This direct current is supplied by a separate generator which is mounted on the same shaft as the alternator. *Hence, all alternators have a separate exciter which generally produces high current and low voltage. Large machines use two exciters known as main exciter and pilot exciter.* The pilot exciter is a well-saturated constant voltage machine arranged to excite the main exciter.

Since the field strength depends on the d.c. exciting current in the field magnet coils, the exciting or field current is regulated accordingly to adjust the output voltage. In the large *machine, the field current may be as high as 400 amperes.* In order to regulate, the exciting current is obtained by connecting the regulating resistance in the field circuit of the exciter where comparatively low current has to be handled as shown in Fig. 16.6.

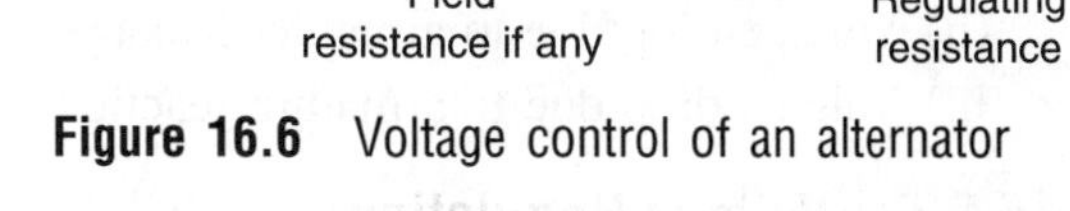

Figure 16.6 Voltage control of an alternator

It is worth to note that the terminal voltage of the alternator depends not only the exciting current but also on the load current and the power factor of the load. The voltage is directly proportional to the flux per pole. If the excitation of an alternator is adjusted to give normal voltage at no load and then a load is applied, the terminal voltage changes even though, the speed is kept constant. The voltage normally falls, but in certain circumstances, it may actually rise. These changes may be due to (i) resistance of each phase winding, (ii) the reactance of each phase windings and the armature reactors.

The exciters are generally built as a shunt wound d.c. machines and their ratings are 0.3 to 1% of the generator output. Exciters usually operate at a voltage between 1156 and 400 volts. For a new machine, it is necessary to apply the first excitation with a battery. Then later on the poles of the exciter have residual magnetism and thus require no external help. An exciter may be either directly coupled to the alternator shaft or it may be driven by a belt or chain drive. Directly coupled exciter is better.

In order to avoid commutation difficulties as well as commutator maintenance, *a.c. exciters* with silicon or germanium rectifiers have been developed for the largest turbo-alternator as well as for the smaller machines. The current from the static floor mounted rectifiers is fed to the rotor of the main machine through conventional slip rings. The a.c. exciter is mounted on the main generator shaft and may be designed to operate at an economically convenient frequency.

16.2.4 Power Generated

Figure 16.7 shows the components of power produced in the generator. The real power generated is kW. The kilovolt amps reactive (kVAR) component is the result of current that is out of phase with the generated voltage from the rotor field. The magnitude of kW is determined by the amount of mechanical power supplied by the turbine. The magnitude of kVAR is a function of the electrical system load to which the generator is connected and the amount of the current supplied to the field winding. The total generator output represented in kVA is given by

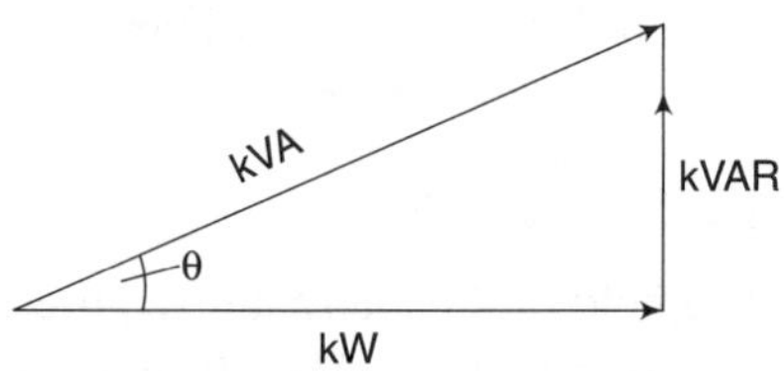

Figure 16.7 Components of power generated

$$kVA = [(kW)^2 + (kVAR)^2]^{1/2} \quad ...(16.1)$$

The power factor of a generator is defined as the ratio of kW to kVA and is the cosine of the angle of describing the angular difference between phase current and phase voltage.

16.2.5 Generator on Load

When load on generator varies, its terminal voltage also varies. This variation in terminal voltage is due to the following reasons:

(i) Voltage drop due to armature resistance
(ii) Voltage drop due to armature leakage reactants
(iii) Voltage drop due to armature reaction.

16.2.6 Voltage Regulation

When an alternator is subjected to a varying load, the voltage at the armature terminals varies to a certain extent, the amount of this variation determines the regulation of the machine. The voltage regulation is defined as the percentage rise in voltage when full load at the specified power factor is switched off, the excitation being adjusted initially to give a thermal voltage.

Thus,

$$\text{Percent of regulation 'up'} = \frac{E_0 - V}{V} \times 100 \qquad ...(16.2)$$

where E_0 = no load emf and V = terminal voltage

A normal alternator has a regulation of about 8 to 10 per cent at unity power factor, but the voltage rise is considerably increased at logging power factor. At 0.8 lagging power factor, the value lies between 25 and 35 per cent or even more.

16.2.7 Generator Cooling

Heat is produced in a generator as a result of resistive losses caused by current flow in the stator and field windings, stator core magnetic losses and windage losses. The heating effect which depends on the load and power factor is the limiting factor in the generator rating. Further, insulation gets damaged due to overheating. In order to prevent the temperature rise exceeding the maximum permissible value, every generator requires continuous cooling during operation.

16.2.7.1 Cooling Methods

There are two types of cooling methods, namely, (i) open system using air, (ii) closed system using air or hydrogen and (iii) hydrogen cooling.

16.2.7.1.1 Open System Using Air. In this system, the cold air is taken in from the atmosphere with the help of a fan and passed through the machine. The outcoming hot air is discharged to atmosphere or into the machine hall station warming. Since it is difficult to remove all the dust particles of air entering the machine so this system is rarely used.

16.2.7.1.2 Closed System Using Air. In this system, a given volume of air is circulated continuously through the generator. The discharged hot air cooled to about 40°C through a heat exchanger before it is recirculated through the generator. This system is universally used by large size generator (up to 100 MW unit)

16.2.7.1.3 Advantages and Disadvantages of Closed System over Open System

Advantages

(i) Quiter operation of the station.

(ii) No filter is required.

(iii) Less possibility of dust and impurities deposition in the machine passage.

(iv) Reduced amount of moisture in the humid atmosphere.

(v) Eaiser to combat a fire in an alternator since supply of fresh air is absent.

(vi) Appearance is quite good.

Disadvantages. It is a costly affair.

16.2.7.1.4 Hydrogen Cooling. The gaseous hydrogen is better cooling medium than air. In generator having capacity more than 100 MW, the windage losses are appreciable and increases with capacity. To reduce this loss, now hydrogen is universally used for cooling the generator in almost all thermal power plants having capacity more than 100 MW. **The following are the *advantages* of *closed loop hydrogen gas cooling*.**

Advantages

(i) The specific heat of hydrogen is 14.5 times that of air and the thermal conductivity of hydrogen is almost seven times that of air.

(ii) Less windage losses since the density of hydrogen is (1/14th) that of air.

(iii) Reduced noise level, due to lighter cooling medium and lower friction.

(iv) Reduced fire risk in the event of electrical breakdown as oxygen is not present to maintain combustion of the insulation material.

(v) Due to high heat transfer coefficient, the cooling surface required by H_2 cooling is considerably smaller.

(vi) Owing to lower windage losses and higher heat transfer coefficient, less cooling water is needed in the cooler (heat exchanger).

(vii) Less space is occupied by the hydrogen cooled machine.

(viii) The reliability of the insulation increases and its life span is prolonged.

(ix) The "corona" effects on the conductors in winding are less deleterious in hydrogen atmosphere than air. This results in further increase in the life of winding.

Disadvantages

(i) High cooling cost as hydrogen is costly.

(ii) High overall capital and maintenance cost.

(iii) There are chances of hydrogen leaking through the casing and for this reason the casing is made explosion proof against any pressure which can be developed by explosive mixture of hydrogen and air.

(iv) The major problem is the effectively sealing of the shaft glands at the alternator housing.

(v) A high degree of hydrogen purity is required ensuring high efficiency and avoiding explosions.

16.3 SWITCHGEAR

The switchgear constitutes all parts or equipment of the power plant *whose function is to receive and distribute electrical power*. It consists of assemblies of switching apparatus, protective and metering devices, intercooling bus-bar systems and relevant accessories. The functions of switchgear are as follows:

(i) The faults are localized and the faulty plant is automatically disconnected from system by the operation of protective equipment.

(ii) To break short-circuit to avoid dangerous conditions.

(iii) To facilitate redistribution of loads, inspection and maintenance of the system.

The following types of switchgear installations are used:

(i) Indoor switchgear, (ii) Unit type switchgear, and (iii) Outdoor switchgear.

16.3.1 Indoor Switchgear Installation

These are preferred for small capacity stations and in the voltage range of 20 kV to 110 kV. It is adopted for 110 kV station when the station surrounding the substation site is highly contaminated by wastes from chemical, cement and similar works whose chimney discharge dust, gases and vapours. This is also suitable to sites in region experiencing low temperature and those located near sea coasts. The different methods of arrangement of the switchgear installation are used as per the type of circuit breakers are used.

16.3.2 Unit Type Switchgear Installation

A unit switchgear is a switchboard consisting of cabinet obtained from manufacturer fully assembled and ready for installation. It houses the electrical equipment required for one main connection. Such units are constructed in the form of metal cabinets. Being housed in metal cabinet, they protect the operating personnel from coming into accidental contact with any one of the line parts and also protect the equipment from dust.

The advantages of unit type switchgear are: (i) quick erection of substation, (ii) their buildings can be smaller and simpler in design, (iii) the future extensions and rearrangements are possible and (iv) lower yearly cost.

Due to above advantages, the use of unit type switchgear is widespread in recent years.

16.3.3 Outdoor Switchgear Installation

They are preferred for the voltage range of 35 kV to 500 kV because they have the following advantages and disadvantages over indoor stations within this range.

Advantages

(i) The volume of construction work is smaller and therefore, the cost of switchgear installation is lower.

(ii) Danger of fault is less because the apparatus of adjoining connections can be spaced liberally without any noticeable increase in cost.

(iii) The time needed for the erection of substation is shorter.

(iv) All the equipment are within view and installation is comparatively easy to extend if need arises.

Disadvantages

(i) A much greater area is required by the substation.

(ii) Inconvenience is caused to maintenance worker due to abnormal working whether severe summer, winter and rainy season.

(iii) The effect of rapid fluctuation in ambient temperature and possibility of dust and dirt collection, the effect of rain and storm on the equipment of the installation require special design of equipment which can be used for outdoor installation.

Ignoring the disadvantages, outdoor installations are very widely used in high capacity power plants. It is worth to note that the main structural elements in outdoor substance are especially designed supports to which conductor systems are attached. The supports for overhead conductor are constructed of reinforced concrete or steel.

16.4 POWER TRANSFORMER

The function of transformer is to convert alternating current of one voltage to other desired voltage. More precisely, in ***electrical power engineering,*** the transformer converts electrical power from a generated voltage of about 11 kV (as determined by generator design limitations) to higher values of 132 kV, 220 kV, 400 kV and 765 kV thus permitting transmission of huge amounts of power to long distance to appropriate distribution points at tremendous saving at the cost of transmission lines as well as in power losses. A ***distribution points transformer*** is used to *reduce* these high voltages to a safe level of 400/230 volts for use in homes, offices, etc. When energy is transformed into a higher voltage, the transformer is called a *step-up transformer* but when the case is otherwise, it is called a *step-down transformer*. Most power transformers operate at constant voltage, i.e., if the power varies, the current varies while the voltage remains fairly constant.

16.4.1 Working Principle of a Transformer

A transformer works on the principle of *mutual inductance*, between two (and sometimes more) inductively coupled coils. It comprises two windings in close proximity as shown in Fig. 16.8 which are coupled by magnetic induction, without any conductive connection between the windings. One of the windings, called *primary,* is energised by sinusoidal voltage. The second winding, called *secondary,* feeds the load. The alternating current in the primary windings sets up an alternating flux (Δ) in the core. The secondary winding is linked by most of this flux, and emfs are induced in the two windings. The emf introduced in the secondary winding drives a current through load connecting to the winding. In other words, energy is transformed from the primary circuit to the secondary circuit through the medium of magnetic field.

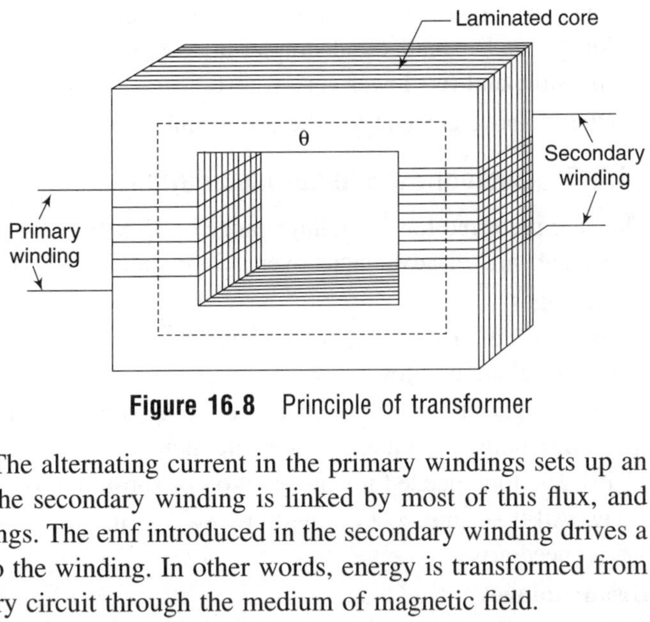

Figure 16.8 Principle of transformer

16.4.2 Main Elements of a Power Transformer

Construction. Figure 16.9 shows the main elements of a power transformer. The transformers are usually of oval shape in plant and are fabricated from steel plates of 5 to 12 mm thick. In order to provide more strengths, the tank bottom and covers are made of still heavier plate. The tank covers are bolted to the tanks to provide a sealed joint between them. The core and the windings arranged on it comprise an assembly which is placed within the tank.

The core of the transformer serves as the magnetic circuit for the flux set-up by the windings. In order to reduce the losses due to eddy currents, the core is assembled of laminations cut

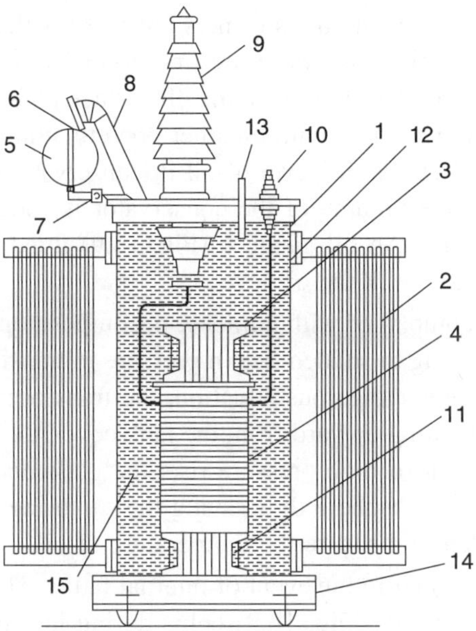

1. Tank
2. Radiators
3. Magnetic circuit (core)
4. Windings
5. Oil conservator
6. Oil gauge
7. Buchholz relay
8. Vent-pipe
9. Terminal bush
10. Terminal bush
11. Clamping bolts
12. Pipe underframe
13. Temperature indicating thermometer
14. Tank underframe
15. Transformer oil

Figure 16.9 Main elements of a power transformer

from special transformer steel sheets of 0.5 mm thick. Each lamination in the core is insulated from its neighbour in the stack by a thin coating of varnish or a sheet of thin paper. In order to obtain a rigid core structure, the assembled laminations are lightly pressed together by a series of clamping bolts.

The windings of transformer are made of electrolytic grade copper wire of rectangular or round section. The wire is insulated with the cable paper or cotton yarn. The cylinders of insulated materials insulate the windings from each other and from the core. For connection purpose, each winding has its ends brought up to porcelain terminal bushing 9 and 10 per load-out through the tank cover.

The tank contains the **transformer oil**—*a special grade of mineral oil.* The transformer oil *serves two purposes*. The first purpose is to serve *as a cooling medium* to transfer the heat generated in the windings and core steel to the air surrounding the transformer. The second purpose is to insulate the current carrying parts from each other and from the earthed tank. In order to increase the surface heat transfer area, **radiators are fitted** to the tank to cool efficiently the oil in transformer. The radiators are attached to the tank with a pipe connection 12 incorporating special type radiator valves.

Working. When the transformer is under operation, the oil in transformer undergoes temperature and volume changes. In order to take account of the volume change, it is essential to fill the tank partially leaving an air space under the cover free to communicate with atmospheric air. As the oil comes in contact with air containing oxygen and moisture, it gets oxidized and forms the sludge. This will result in deterioration of oil, mainly reduction of its dielectric strength.

In order to overcome these problems, the majority of the transformers are filled with an oil conservator 5 (Fig. 16.9). The main functions of oil conservator are first, to ensure that the transformer tank is always filled with oil, and secondary to reduce the oil surface maintained in contact with the air. The top of the conservator is filled with breather open to atmosphere which permits the air to be taken in or expelled out as the level of oil in conservator varies with changes in temperature. As a thumb rule, the volume of the conservator is about 8 to 10% of the volume of oil in the transformer tank. An oil gauge 6 (Fig. 16.9) fitted in the conservator in the form of gauge glass is used to check the oil level.

In the event of short-circuit, it is generally accompanied with immense decomposition of oil and the formation of large amount of gases resulting in sudden rise in pressure in transformer which may rupture the tank. In order to avoid this dangerous situation, the transformer are fitted with explosion vent pipe 8 (Fig. 16.9) extending upward from the tank cover, the upper end is shut off with a thin glass disk. In the event of high pressure rise, the glass disk gets ruptured and a small amount of oil will vent to atmosphere thus saving the rupturing of the tank. A thermometer (13) (Fig. 16.9) is provided to measure the temperature of oil.

Buchholz relay system (7) protects the transformer in the event of internal faults. The tank is supported by roller-mounted tank under frame (14) (Fig. 16.9) helps during installation, repair and maintenance.

Transformer Connections. As a common practice, power transformer are either single phase transformer banks of three each or 3-phase transformer. The 3-phase transformer has certain advantages, namely, it is cheaper, heaving less weight, occupies less space and it is more compact. This connection is widely used in 2-windings, 3-phase transformers are delta-star, star-delta, and delta-delta. The choice of delta-delta or delta-star in a particular power station is decided by the need of parallel operation with existing bank of transformer or the interconnection of the network.

16.5 CIRCUIT BREAKER

16.5.1 Function

The function of a circuit breaker is to break a circuit when various abnormal conditions arises and create a danger for the electrical equipment in an installation. The heaviest duty of the circuit breaker is to interrupt a short-circuit current which may reach a value of several tens of thousands and even more ($>10^5$ amperes) in order to quickly eliminate the sources of fault, the circuit breaker must open the circuit with least possible delay.

16.5.2 Principle of Circuit Breaker

The principle of circuit breaker is based on the concept of *rapid conversion of predetermined section of circuit from a conductor to insulator.* When the current carrying contacts are separated an arc (which contains an ionized gas) is produced between them. *This arc provides for the gradual change over from current carrying to voltage isolating states of contacts.* Therefore, it plays an important part in circuit interruption process. It is very essential to control the arc carefully as a good deal of energy in the form of heat is generated in it. The arc so produced may be extinguished by high resistance or low resistance methods.

16.5.3 Classification of Circuit Breakers

They are classified as follows:

(1) Low voltage circuit breaker
(2) High voltage circuit breaker
  (a) Oil circuit breaker
    (i) Bulk oil circuit breaker
    (ii) Low oil contact circuit breaker
  (b) Oil-less circuit breaker
    (i) Air blast circuit breaker
    (ii) Water circuit breaker
    (iii) Hard gas circuit breaker

High resistance method is generally used in d.c. circuit breakers. Low resistance method is used for a.c. circuit breakers.

16.5.3.1 Oil Circuit Breaker

Oil circuit breakers are invariably used in power plant. Their ratings lie between 25 MVA at 2.5 kV and 5000 MVA at 250 kV. The following are the *advantages* of oil as quenching medium of circuit breakers:

(i) It offers high dielectric strength
(ii) It absorbs arc energy while decomposing
(iii) The gas formed due to decomposition of oil offers good cooling properly
(iv) Surrounding oil in close proximity to the arc presents a large cooling surface
(v) It acts as an insulator between live part and the earth
(vi) Once the current is zero, it is the ability of cool oil of high dielectric strength to flow into the arc spaces

The following are the disadvantages:

(i) Oil is easily inflammable
(ii) It can cause explosion by mixing with air
(iii) It requires maintenance and periodic replacement.

Figure 16.10 shows the schematic of plain oil circuit breakers. It consists of strong weather tight earthed metal tank, containing oil up to a predetermined level, and an air cushion above the oil. The oil pressure in tank is solely due to the head of oil above the contact which are enclosed in the tank. When the contacts separate an arc is produced which vaporizes oil into the gas. The sudden formation of the gas by arc may dissipate heat generated as a result of arc.

The large quantity of oil required and long and inconsistent arcing times are the main drawbacks of this circuit breaker. As a result of these limitations, the plant oil circuit breaker are suitable for low current and low voltage operations, not exceeding 150 MVA at 11 kV.

The other type of oil circuit breaker is self blast oil circuit breaker in which the pressure of the gas bubbles set up in the oil by the arc is used to force fresh unionized oil into the arc path. **Now, *modern high speed circuit breaker is used.***

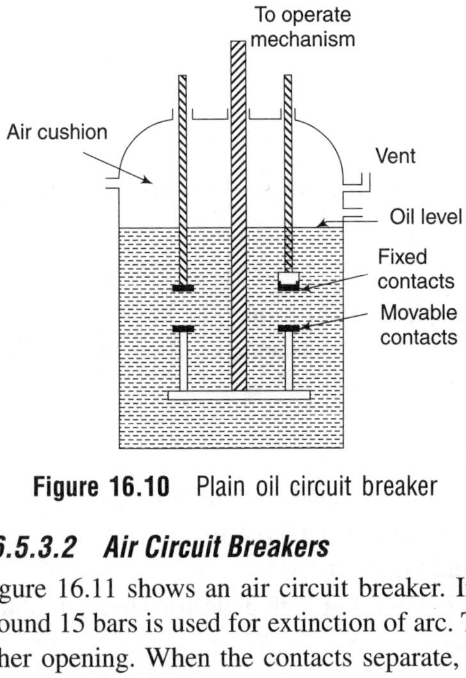

Figure 16.10 Plain oil circuit breaker

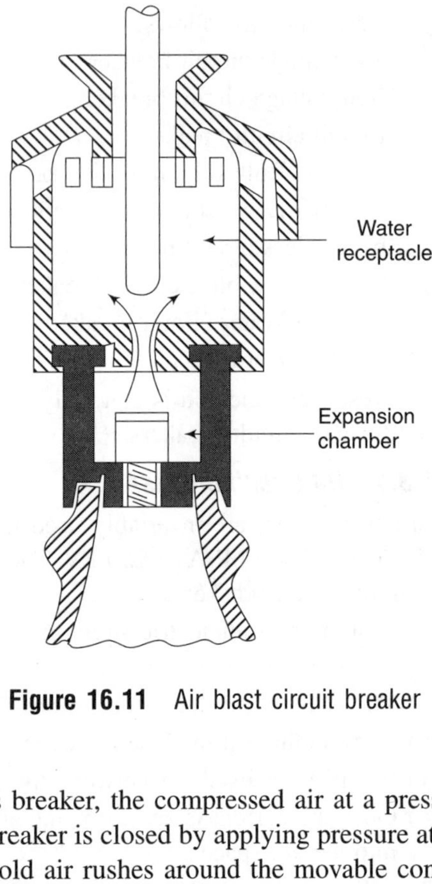

Figure 16.11 Air blast circuit breaker

16.5.3.2 Air Circuit Breakers

Figure 16.11 shows an air circuit breaker. In this breaker, the compressed air at a pressure around 15 bars is used for extinction of arc. The breaker is closed by applying pressure at the other opening. When the contacts separate, the cold air rushes around the movable contact and blows out the arc.

The following are the advantages of air circuit breaker:

(i) Fire hazards due to oil are eliminated.
(ii) The operation is quick.
(iii) The burning of contacts is less due to short duration of arc.
(iv) Suitable for frequent operation as cooling medium air flows constantly.

The following are the disadvantages:

(i) There is a chance of current choppings.
(ii) It is sensitive to restriking voltage change.
(iii) A compressor is constantly required to remain in operation.
(iv) The leakage of air at pipe fitting creates problem.

16.6 PROTECTION OF ELECTRICAL SYSTEM

Figure 16.12 shows a typical power system in which there is always a chance of malfunctioning of any system. *The functions of relays and circuit breakers is to prevent or limit the damage to the system due to faults or overloads and to isolate the faulty section from the remaining system.*

The speed with which relays and circuit breakers operate has a direct bearing on the following:

(i) The quality of service to the consumers
(ii) The stability of the system
(iii) The amount of power that may be transmitted without exceeding the stability limit
(iv) The damage done by a short circuit
(v) Safety to life and property

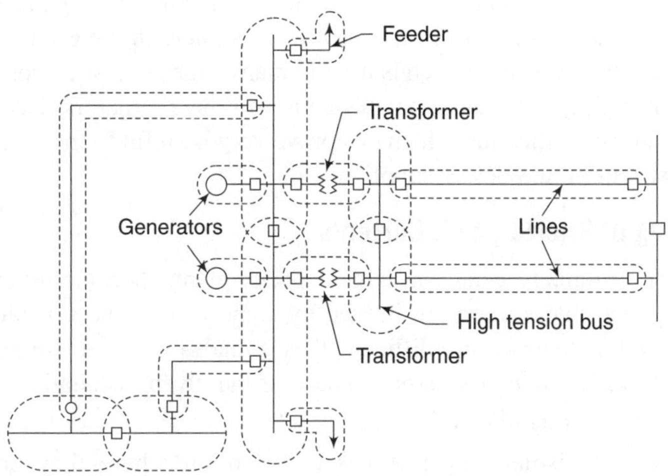

Figure 16.12 Typical power system showing zones of protection

16.6.1 Types of Relays

Relays have been developed for practically all types of system disorders. Generally, all relays comprise three elements, namely, (i) an actuating element, (ii) a movable element and (iii) a set of contacts.

The following types are the commonly used relays:

(a) Plunger type relays (b) The over-current (induction type) relays
(c) The current balance relays (d) The different relays

Among the four types, the current balance relays are mostly used. It operates on the principle of current balance in which two currents are compared on proportional basis (Fig. 16.13).

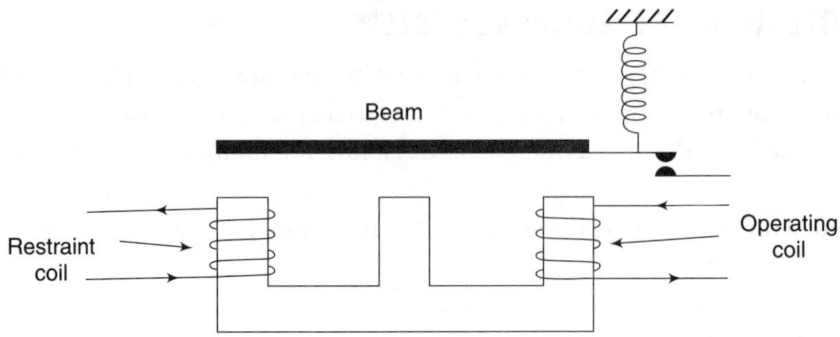

Figure 16.13 Current balance relay

16.7 SHORT CIRCUIT IN ELECTRICAL INSTALLATION AND LIMITING METHODS

A short circuit is any solid contact established between the differential phases of an electrical installation or any bridging of them by a very low resistance. In the event of short circuit, a short circuit current starts flowing which is usually many times excess of normal rated current in the current carrying parts circuit apparatus and even the alternators. Due to short circuit, the high current causes overheating which can prove very harmful to the electrical equipment leading to fire distributing the power supply.

16.7.1 Limiting of Short Circuit Currents

In electrical power system, especially in large capacity plant, short circuit current can attain such a high value that unless these are limited by some means, the selection of equipment capacity of withstanding them is very difficult. This is made possible by *interposing auxiliary inductive reactants, called reactors,* in each phase of the given installation.

The following are the functions of reactors:

(i) They localize or isolate the troubles at the point where they originate without communicating their distributing effects to the other parts of the network.

(ii) They limit the flow of current into short circuit with a view of protecting the equipment from overheating as well as from failure owing to destructive mechanical forces and also protect the system against shut-down.

(iii) They make possible to install the lower capacity circuit than needed.

Reactors may be located in the generator leads, between bus section in the low tension transformer leads or in out going low tension feeders.

16.8 EARTHING OF A POWER SYSTEM

It is utmost important to earth to neutrals of industrial power system. Earthing is necessary as it offers many advantages. They are as follows:

(i) Persisting arcing ground is eliminated

(ii) The over-voltage due to restriking is minimized

(iii) The ground faults can be located and isolated quickly.

(iv) Better sensitive protective apparatus may be used.

(v) Less maintenance, better safety, better reliability, and improved lighting protection are possible.

Methods of earthing are:

(a) Solid earthing (b) Resistance earthing

(b) Reactance earthing (d) Resonant earthing

The earthing of systems should be done at the neutral of the supply transformers and generators.

16.9 CONTROL ROOM

A control room of a power plant is supposed to be the nerve centre. The following controls are located in the control room.

(i) Circuit breakers, (ii) Load and voltage adjustment, (iii) Transformer tap changing, (iv) Emergency tripping of turbines, etc, (v) Various instruments, indicating load, voltage, frequency, power factor, etc. (vi) Synchronizing equipment, (vii) Voltage regulators, (viii) Relays, (ix) Integrating meters, etc. In the era of digital, each information appears on the computer screen.

16.10 VOLTAGE REGULATION

It is essential to maintain the supply voltage within narrow limits for the satisfactory operation of power system. In this connection, there are some legal requirements. In order to meet these requirements, the following methods are adopted.

(i) Voltage control by generator excitation

(ii) Tap changing transformer

(iii) Booster transformer

The voltage control by generator excitation is not suitable for long lines.

16.11 TRANSMISSION OF ELECTRICAL POWER

Because of economic reasons, three-phase circuits are generally used for the transmission of electrical power. The transmission lines may be classified as follows:

(i) Single line, (ii) Parallel lines, (iii) Radial lines, (iv) Ring system, and (iv) Network.

All forms of lines are in common use as per requirement. For a systems covering a large territory, the ring system of transmission is very important.

The electrical power may be transmitted by overload or underground conductors. Underground transmission with the exception of a few notable cases, is limited to voltage less than 45,000 volts.

PROBLEMS

Things to Think About

16.1 What is the basic philosophy of bus-bar?

16.2 On which principle a generator works?

16.3 Why there is a need of voltage regulation?

16.4 On what principle transmission lines are laid?

Short and Long Answer Theoretical Questions

16.1 List the major electrical machines/equipment used in power plants.

16.2 Give a layout of electrical equipment and discuss each of them.

16.3 Discuss the major construction detail of a generator (i.e., alternator)

16.4 Why is there a need of generator cooling?

16.5 Discuss the various methods of generator cooling.

16.6 What is the function of a switchgear?

16.7 What are the merits and demerits of various types of switchgear installations?

16.8 With the help of a sketch, discuss the working principles of a power transformer.

16.9 What is the function of a circuit breaker?

16.10 What is the principle of working of a circuit breaker?

16.11 Give a classification of circuit breaker.

16.12 With the help of sketches, discuss the working principle of various types of circuit breakers.

16.13 How a electrical system is protected?

16.14 What is the function of relay? Mention the type of relay. Discuss the current balance relay system.

16.15 What is short circuit? Discuss the method of limiting short circuit currents.

16.16 What is the function of reactors and give its location.

16.17 What are the advantages of earthing of a power system? What are the various methods of earthing?

16.18 What types of controls are located in a control room of power plant?

16.19 Why there is a need of voltage regulation? List the various methods of voltage control.

16.20 Classify the transmission lines.

Software Based Questions

16.1 Develop a software for the design of major electrical equipment.

16.2 Prepare a comprehensive report of all types of electrical equipment used in super power plant with their working principle, specification, merits and demerits.

Multiple-Choice Questions

16.1 In ring and bridge bar system, which one is true?
   (a) Each line is served by two circuit breakers.
   (b) Each line is served by one circuit breaker only.
   (c) Each line is suitable for 11 kV transmission line.
   (d) None of the above

16.2 The field current in the exciter used in large generator is around
   (a) 100 ampere (b) 200 ampere (c) 300 ampere (d) 400 ampere

16.3 The ratings of exciter ranges from
   (a) 0.1 to 0.5% of generator output (b) 0.3 to 1% of generator output
   (c) 0.5 to 1.5% of generator output (d) 1 to 2% of generator output

16.4 Generator is cooled by
   (a) water (b) light oil (c) heavy oil (d) hydrogen

16.5 The voltage of outdoor switchgear ranges from
   (a) 10 to 50 kV (b) 50 to 100 kV (c) 60 to 300 kV (d) 60 to 500 kV

16.6 The power transformer steps up from generating voltage of 11 kV to
   (a) 715 kV (b) 735 kV (c) 765 kV (d) 775 kV

16.7 The tank of the transformer contains
   (a) water for cooling (b) liquid hydrogen for cooling
   (c) liquid nitrogen for cooling (d) oil for cooling

Answers to Multiple-Choice Questions

16.1 (a) 16.2 (d) 16.3 (b) 16.4 (d) 16.5 (d) 16.6 (c) 16.7 (d)

CHAPTER

17

Air Pollution Caused by Power Generation and Its Control

OVERVIEW

The pollutants are contaminants in concentration high enough adversely affect environment and health. The problem of air pollution came into existence from the day the first fire was lit. Some specific definitions of air pollution are given below.

(i) Air pollution means the presence in the outdoor atmosphere of one or more contaminants such as dust, fumes, gas, mist, odour, smoke or vapour in quantities with characteristics and of duration such as to be injurious to human, plant or animal life or property or which unreasonably interfere with comfortable enjoyment of life and property.

–Engineer Joint Council, USA.

(ii) Air pollution is the excessive concentration of foreign matter in the air which adversely affects the well-being of the individual or causes damage to property.

–American Medical Association

(iii) Air pollution is the presence in ambient atmosphere of substances, generally resulting from the activity of man, in sufficient concentration, present for sufficient time and under circumstances which interfere significantly with the comfort, health or welfare of precious or with full use or enjoyment of property. ***–Indian Standards Institution, IS-4167, 1966***

The average major contributors to air pollution are motor vehicles (about 60%), industries (about 17%); fossil power plants (about 14%), rest (9%).

The power plant pollutants of most concern are as following.

(i) *From fossil power plants*

(a) Sulphur oxides (b) Nitrogen oxides

(c) Carbon oxides (d) Particular matter (smoke, dust and other solids)

(ii) ***From nuclear power plant***

(a) Radioactive release (b) Radioactive wastes

(c) Thermal pollution

In addition, pollutants such as CO_2, CO NO_x, SO_2 particulates, lead and hydrocarbons are contributed primarily by motor vehicles and it will not be covered here.

17.1 EFFECT OF AIR POLLUTION

Air pollution affects the human health, vegetation, materials and animals in various ways causing greenhouse effect, acid rain, health hazards, disturbed ecosystem, etc.

17.1.1 Greenhouse Effect

A phenomenon called the greenhouse effect is caused by the increasing amounts of CO_2 in the atmosphere. The primary source of CO_2 is the consumption of fossil fuels. The production of CO_2 is much more than the sinks (e.g., plants and sea) can absorb them and therefore the level of CO_2 in the atmosphere is increasing. The greenhouse effect as the name suggests is analogue to the phenomena used to keep the ambient condition of plants in a greenhouse. In the botanists greenhouse, the sun's rays while passing through the glass wall suffer scattering which changes (increase) its wavelength. These glass pans are opaque to this longer wavelength, so the radiant heat is retained. In earth's case, what are referred to as "greenhouse gases" which means primarily CO_2 act as the glass and the earth itself as the greenhouse. The sun's radiation can pass through the layer of CO_2 and heats up the earth but is stops the same heat being remitted by the earth. The net result of all this is that there is a danger of the earth gradually being heated up. If the CO_2 content in the atmosphere is doubled, which can occur by the middle of the next century, then it will result in increase in the present world temperature by about 3.6°C. As a result, all the glaciers will recede and the ice caps in the Antarctic and the Arctic will begin to melt which will increase sea level by a few metres and thus most of the cities on seashores may be submerged.

17.1.2 Acid Precipitation

Acid precipitation in the form of acid rain, snow, dry deposition and acid fog is the another effect of pollution. Large quantities of SO_2, and NO_2 emitted into the atmosphere through chimney react with moisture in the air and produce sulphuric acid and nitric acid which fall as acid precipitation on the earth's surface. The acidification of streams and lakes affects aquatic animals and plants leading to the destruction of ecosystem and also the important monuments.

17.1.3 Human Health

The human health is affected by air pollution as a result of contact between the pollutants and the body at the surface of skin and exposed membranes. The following are the effects of air pollution on the health:

(i) Eye, nose, throat and respiratory track irritation.

(ii) Gases such as H_2S, NH_3 and mercaptans cause odour nuisance even at low concentration.

(iii) Mortality and morbidity rate increase.

(iv) Pollen causes asthmatic attack to some.

(v) Chronic pulmonary diseases like bronchitis and asthma are aggravated by a high concentration of SO_2, NO_2, particulate matter (PM) below 10 μm in diameter and photochemical smog. PM 2.5 is very lower fist.

(vi) CO produces stress on those suffering from cardiovascular and pulmonary disease by combining with haemoglobin in the blood.

(vii) Hydrogen fluoride causes disease of the bone (osteofluorosis) and mottling of teeth.

(viii) Carcinogenic agents in the pollutants causes cancer.

(ix) Respiratory diseases like silicosis, asbestosis, etc., are caused by dust particles.

(x) Certain heavy metals such as lead may enter the body through lungs and cause poisoning.

It is, therefore, highly imperative that engineers and scientists of all disciplines consciously incorporate in their design sufficient constraints and safeguards to ensure that they do not contribute to atmosphere pollution. Over and above they must apply various methods to eliminate air pollution where it exists so that the natural environment is restored.

17.2 BASIC TYPES OF SYSTEMS FOR AIR QUALITY CONTROL

Systems that help solve air quality control problems associated with fossil fuel steam generator exhausts primarily deal with particulate collection and SO_2 emission clean-up. There are four basic types of systems:

(i) Flue gas desulfurization (FGD)

  (a) Wet-type scrubbers

  (b) Dry-type scrubbers

(ii) Particulate collection

  (a) Electrostatic precipitators (already discussed in Chap. 6)

  (b) Fabric filler (already discussed in Chap. 6)

Air pollution caused by gas turbine plants has already been discussed in Chapter 13 as Combustor.

17.3 FLUE-GAS DESULPHURIZATION (FGD) SYSTEM

Gas desulphurization can be accomplished by wet, dry or alkali scrubbing. They are discussed as separately follows.

17.3.1 The Wet-Gas Desulphurization System

The wet FGD system, also known as wet scrubber, is commonly based on low-cost lime-limestone in the form of an aqueous slurry. This slurry which is brought into intimate contact with flue gas by various techniques absorbs the SO_2 in it.

In modern version of scrubbing process, the flue gas is **scrubbed** with slurry that contains lime (CaO) and limestone ($CaCO_3$) as well as the salts, calcium sulphide ($CaSO_3 \cdot 2\,H_2O$) and calcium sulphate (in hydrate form or natural gypsum, $CaSO_4 \cdot 2H_2O$). The SO_2 in the flue gas reacts with slurry to form additional sulphide and sulphate salts which are recycled with addition of fresh lime or limestone. Probably the following chemical reaction takes place.

$$CaO + H_2O \rightarrow Ca(OH)_2$$

$$Ca(OH)_2 + CO \rightarrow CaCO_3 + H_2O$$

$$CaCO_3 + CO_2 + H_2O \rightarrow Ca(HCO_3)_2$$

$$Ca(HCO_3)_2 + SO_2 + H_2O \rightarrow CaSO_3 \cdot 2H_2O \downarrow + 2CO_2$$

$$CaSO_3 \cdot 2H_2O + \frac{1}{2}\,O_2 \rightarrow CaSO_4 \cdot 2H_2O \downarrow$$

A *spray tower* is employed downstream of the particulate removal system (i.e., electrostatic precipitator or fabric filler) as shown in Fig. 17.1. The flue gas is drawn into the spray tower by the main steam generator induced draft fan where it flows in counter current fashion to the limestone-slurry spray. Any spray droplets entrained by the gas is removed by a mist eliminator located at the upper exit of the spray tower. The gas may have to be slightly reheated before it enters the stack to improve atmospheric dispension.

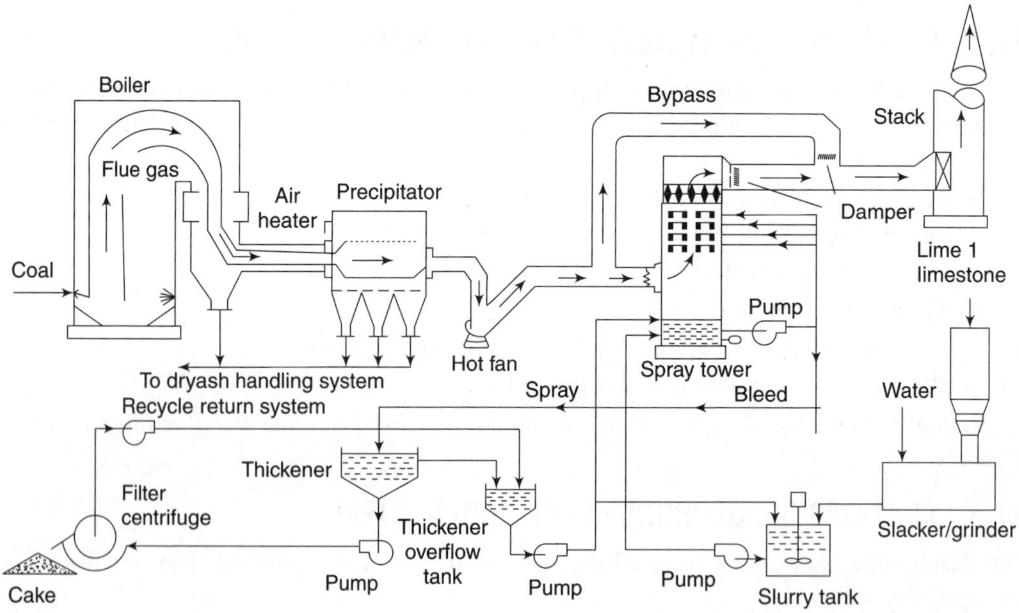

Figure 17.1 Wet flue gas desulphurization system with a spray tower

In the bottom of spray tower, the sprayed limestone slurry gets collected and is recirculated back to the spary nozzles by a pump. There is a system of feed and bleed in the system by which charges a fresh slurry, under pH control and discharges an equivalent amount from the circulating slurry. The preparation of fresh slurry is accomplished by mixing the limestone with water in a *slacker-grinder* and stirred in a slurry tank. There is dewatering system in the form of thickener and fillers or centrifuges which receives the bled slurry where water is removed from the calcium-sulphur salts. The reclaimed water is utilised in making fresh slurry. The wet FGD system is mostly used in power plants.

Advantages of wet scrubber.

(i) The SO_2 removal efficiency is high
(ii) The system offers good reliability
(iii) It requires low flue gas energy
(iv) It is capable of removing from the flue gases residual particulates that might have escaped the electrophile precipitator or fabric filler.

Disadvantages of this wet FGD system.

(i) It builds up scale in the spray tower and the possibility of plugging.
(ii) There is reheating of the flue gas.
(iii) The larger gas pressure drop requires higher fan power.
(iv) It offers higher capital and operating cost.
(v) It offers costly disposal of waste material from wet scrubbers which is waterlogged sludge.

17.3.2 The Dry Flue Gas Desulphurization System

It is also known as *dry scrubber*. It utilises an aqueous slurry of lime, CaO, to capture flue gas (SO_2) by forming calcium sulphides and sulphates in spray absorber. Such a dry scrubber is shown in Fig. 17.2 which is placed before fabric filter. The preparation of slurry in this case is different from wet scrubber. In dry scrubber system, the slurry is atomized usually by a centrifugal atomizer, into a fine spray in the spray absorber that promotes the chemical absorption of SO_2 and, because of the small spray particle size, is quickly dried by the hot flue gases themselves to a particulate suspension that is carried along with the desulphurized gas stream. The reaction particulates as well as those carried by the flue gases (fly ash) are then removed, mainly by fabric filter before the gas is drawn by the induced draft fan to the stack.

The dry scrubber system consists of a slurry generating system which is supposed to be its major component. In this, a '**slaker**' meters lime and water processed to remove inert impurities called ***grits***, which is disposed of. The cleaned lime slurry is pumped to the spray absorber as per requirement controlled by the amount of SO_2 in the flue gas.

It is worth to mention here that the particulates both coming in with the flue gas and generated in the FGD are collected from the spray absorbed and fabric filler hoppers and sent to a recycling silo for disposal or for recycling of a portion of it with slurry. Further, the recycled slurry is enriched by an alkaline material, such as CaO, MgO, K_2O or Na_2O.

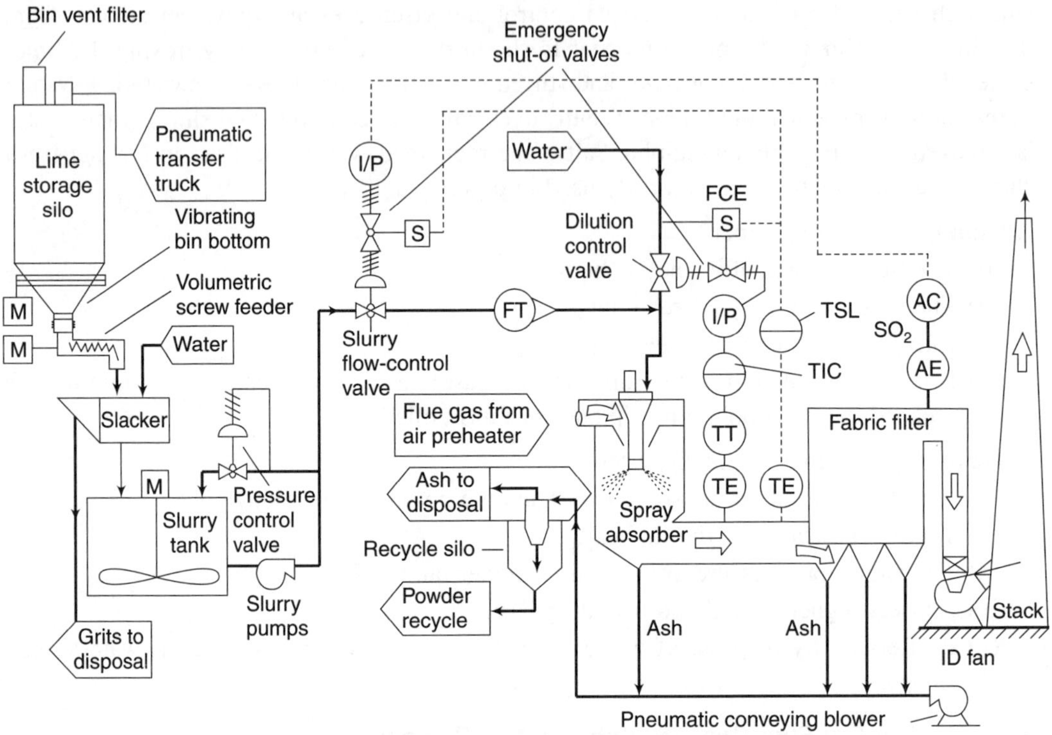

Figure 17.2 Dry flue gas desulphurization (FGD) system

The following are the *main advantages* of dry FGD system.

(i) It produces dry and powdery waste material which poses less costly disposal problem than the wet waste from the wet FGD system.

(ii) The system is simple.

The *main disadvantages* include the following:

(i) The efficiency of SO_2 removal is lower than that of the wet scrubber.

(ii) The spray absorber and slaker need careful design for better performance.

17.3.3 Single Alkali Scrubbing

It is an established fact that clear water solution of either sodium (usually in the form of sodium hydroxide, NaOH or sodium sulphide, $Na_2\ SO_3$) or ammonia are excellent absorbers of SO_2. The main advantage of alkali scrubing is that it avoids scaling and plugging problems of slurry scrubbing by using alkaline earth.

Figure 17.3 shows a well developed sodium scrubber known as *"Wellman-Lord SO_2 recovery process"* which is used by many power plants established in USA and Japan. The basic principle involved in this process is that it utilises a water solution of sodium sulphide (Na_2SO_3) for scrubbing and generates concentrated SO_2 (about 90%), in effect removing the SO_2 from other flue gases.

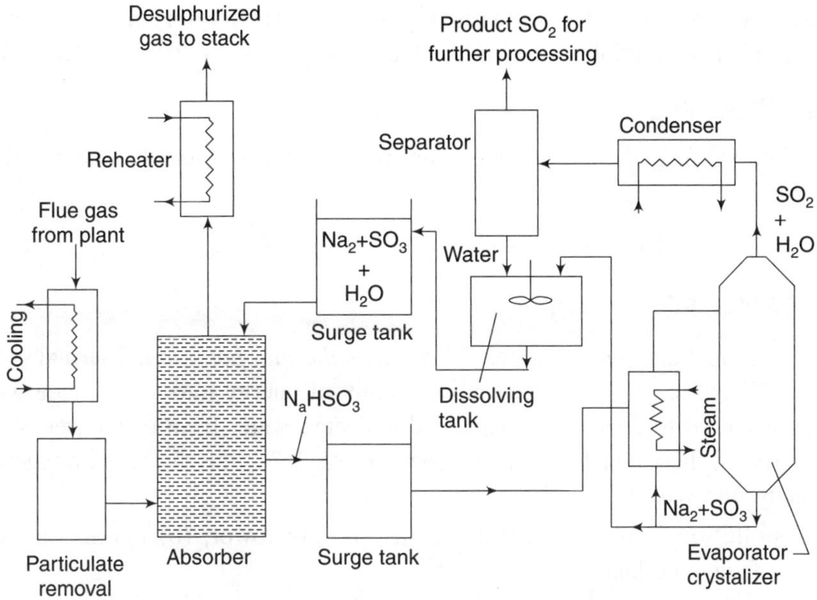

Figure 17.3 Schematic of the Wellman-Lord SO_2 recovery process

In this process, the flue gas from fossil power plants is first pretreated by cooling and removal of particulate matter either by electrostatic precipitator or fabric filler, prior to being sent to absorber, where the water solution of sodium sulphide absorbs the SO_2 in the pretreated flue gas and produces sodium bisulphate ($NaHSO_3$) involving the following reaction

$$SO_2 + Na_2SO_3 + H_2O \rightarrow 2NaHSO_3 \qquad ...(17.1)$$

The desulphurized gas is reheated before going to the stack in order to improve atmospheric dispension.

Sodium bisulphate ($NaHSO_3$) is fed to a forced circulation evaporator-crystallizer (i.e., the heat of system) via a surge tank which allows steady flow rates into the crystallizer despite gas flow and concentration fluctuations. With the applications of low pressure steam from turbine exhaust, the sulphide is regenerated in the form of a slurry involving the following reaction:

$$2NaHSO_3 \rightarrow Na_2SO_3 \downarrow + SO_2 \uparrow + H_2O \qquad ...(17.2)$$

H_2O is separated from the SO_2 in a condenser and recycled to a dissolving tank where the sulphide slurry (Na_2SO_3) is redissolved and sent back to the absorber via a solution surge tank (involving Na_2SO_3 + H_2O) that has same functions as discussed above.

It is worth to note that a small amount of the circulating solution oxidizes to nonregenerable sodium sulphate crystals that must be disposed of. This is possible by purging a small stream of solution and adding fresh sulphate.

SO_2 may be utilised either by producing liquid SO_2 or H_2 SO_4 on site or in a satellite plant or to produce elemental sulphur known as Claus process ($SO_2 + 2H_2S \rightarrow 3S + 2H_2O$.

17.4 NO_x REMOVAL

The removal of NO from the fossil power plant is under active research. One possible process is as follows:

$$NO + H_2S \rightarrow S + \frac{1}{2} N_2 + H_2O \qquad ...(17.3)$$

17.5 THERMAL POLLUTION

Discharge of thermal energy into water is known as *thermal pollution*. Thermal power plants invariably will have to discharge enormous amount of energy into water since water is one medium largely used to condense steam. If this heated water from condenser is discharged into lake or rivers, the water temperature goes up. At 35°C, the dissolved oxygen is so low that the aquatic life will die.

Thermal pollution is reduced by using (a) cooling tower, (b) cooling pond, and (c) construction of separate lake.

17.6 POLLUTION FROM NUCLEAR POWER PLANTS

As discussed earlier, the following are the risks from nuclear power plants (a) radioactivity release, (b) radioactivity waste, and (c) thermal pollution.

17.6.1 Radioactivity Release

This is the most dangerous and serious type of pollution. This is due to radioactive elements and fissionable products in reactor. The best way to abate is the radioactive shield around the reactor.

17.6.2 Radioactive Waste

Due to nuclear reaction in the reactor nuclear wastes, (mixtures of various Beta and Gamma emitting isotopes with various half lives) are produced which cannot be neutralized by any chemical method. If this nuclear waste is discharged in the atmosphere, air and water will be contaminated beyond the tolerable limits. The following disposal or storage methods are used

(a) **Storage tank.** The radioactive waste can be buried underground (very deep below the surface) in corrosion resistant tank located in isolated areas. After a long period, these will become stable isotopes.

(b) **Dilution.** After storing the radioactive waste for a short time, low energy wastes are diluted either in liquid or gaseous materials. After dilution, they are disposed of in sewer without causing hazard.

(c) **Sea Disposal.** This dilution can be used by adequately diluting the wastes and this method is used by the British.

(d) **Atmosphere Dilution.** This method is suitable for gaseous radioactive wastes. It is to be noted that the solid particles from the gaseous waste must be filtered out thoroughly since they are the most dangerous with higher half lives.

(e) **Absorption by the Soil.** This method is suitable for fission products. The radioactive particles are absorbed by the soil particles. However, this method is expensive.

(f) **Burrying in Sea.** Solid nuclear wastes can be stored in concrete blocks which are buried in the sea. Though, this method is expensive, but no further care is needed in this case.

17.7 HIGH EFFICIENCY ENGINES AND TURBINES (HEET) TECHNOLOGY WITH NEAR ZERO EMISSIONS

The U.S. Dept. of Energy's new High Efficiency Engines and Turbine (HEET) Program is currently focused on making technology available to produce a new class of gas turbine design in the 2015–2022 time frame that will be far superior to what is possible today (GT World 2002). DOE is aiming high in terms of fuel flexibility, efficiency, emissions and cost for coal derived syngas and natural gas fired plants.

(a) **Intermediate Goals.** Coal fueled combined cycle plant will be built with at least 50% efficiency (HHV) by 2008 for an installed cost of less than 1000 dollars per kW.

(b) **Long Term Goals.** At least 60% efficiency (HHV) on coal derived fuel (syngas) and 65% (LHV) on natural gas with near zero emissions will be achieved by 2020 through combined cycle.

(c) **H_2 Combustion Gas Turbine.** In WE-NET program of Japan extensive research since 2001 is going to develop hydrogen combustion turbine, using H_2 from electrolysis and O_2 from air liquifaction process. After combustion, water will be in the form of superheated steam having temperature more than 2200°C. Material is the main constraint. It will be clearly a clean energy power plant yielding 75 per cent efficiency.

PROBLEMS

Things to Think About

17.1 What is the basic reason of greenhouse effect?

17.2 What is the basic reason of acid rain ?

17.3 In what way the human health is affected by air pollution?

Short and Long Theoretical Questions

17.1 What do you mean by pollution?

17.2 Give various definitions of air pollution.

17.3 Discuss the effect of air pollution.

17.4 List the various types of pollutants from fossil based power plants.

17.5 List the various types of pollutants from nuclear power plants.

17.6 With the help of sketch discuss the various method of flue gas desulphurization (FGD) systems.

17.7 List the advantages and disadvantages of wet and dry FGD systems.

17.8 Discuss the single alkali scrubbing system.

17.9 What do you mean by thermal pollution? How are they taken care of.

17.10 What are the various methods of radioactive wastes ?

Software Based Question

17.1 Develop a software for the design of an efficient air pollution control unit to be used in a modern coal based power plant.

Design and Open Ended Questions

17.1 Prepare a comprehensive report on various types of air pollution controls used world over for coal, oil and gas based power plant.

17.2 Develop a technology to capture CO_2 in liquid form at a very cheap rate, so that all fossil based plant become environmentally friendly.

Multiple-Choice Questions

17.1 The contribution of motor vehicles towards air pollution is around

(a) 30% (b) 40% (c) 50% (d) 60%

17.2 The contribution of fossil fuel power plants towards are pollution is around

(a) 10% (b) 15% (c) 20% (d) 30%

17.3 The flue gas desulfurization system may be

(a) wet (b) dry (c) alkali scrubbing (d) any of the above

17.4 In the flue gas desulphurization system, the flue gas is scrubbed with

(a) slurry (b) SO_2 (c) NH_3 (d) H_2SO_4

Answers to Multiple-Choice Questions

17.1 (d) 17.2 (b) 17.3 (d) 17.4 (a)

CHAPTER 18

Non-Conventional/Renewable Energy Resources Power Plants

OVERVIEW

In the present setup, 70% of world's energy is met by fossil fuels which is creating global warming due to emission of greenhouse gases. To overcome the problem of global warming, non-conventional/renewable energy sources are the answer. They include:

(i) Solar energy (thermal and photovoltaic)

(ii) Wind energy

(iii) Biomass energy

(iv) Geothermal energy

(v) Ocean (tidal, wave and thermal) energy

(vi) Fuel cell

(vii) Hydrogen energy (already discussed in chapter 3 and chapter 15)

(viii) Magnetohydrodynamic power conversion system

(ix) Thermoelectric and thermionic power conversion system

(x) Small hydro resources (already discussed in hydroelectric plants, chapter 11). Resource (iv), (vi), (vii), (viii) and (ix) come under Non-conventional. Resources (i), (ii), (iii), (v) and (x) are **renewable forms** of energy. About 11% global energy came from renewables in 2019. *Powered by breakthroughs in cost, storage, etc., wind and solar will generate about 50% global energy by 2050.* Coal investment is expected to drop by 11%. **Renewals can increase global GDP by USD 98 trillion by 2050 creating 63 million new jobs**. Though renewal also come unler non-conventional.

18.1 SOLAR THERMAL ENERGY POWER PLANT

For decades to come, the sun will grow in importance as one of the most attractive alternatives to fossil fuels as a major source of energy to mankind. The solar radiation on the earth surface, though erratic due to cloud cover, discontinuous due to the night and seasons, and diffuse but it is enough to provide energy for industrial processes, commercial buildings, private homes, electricity generation and many other end uses. **The solar radiation constant is 1367 W/m^2 at the** *top of earth's surface.* The value of terrestrial radiation is much less than solar constant. Now, solar heating and cooling systems have been commercialized. Large size solar thermal power plants (100 MW) are in the developing stages. Solar power has low density per unit area, **say 500 W/m^2 to 1 kW/m^2**. Hence, in order to collect the solar energy, a large ground is to be covered by solar collector. There are two ways to use solar energy. They are **solar thermal** and **photovoltic solar cell**. *A solar* ***thermal energy collector*** *is an equipment in which solar energy is collected by absorbing radiation in an absorber and then transferring to a fluid which is transporting heat. There are mainly two types of* ***solar thermal collectors.***

Solar Thermal:

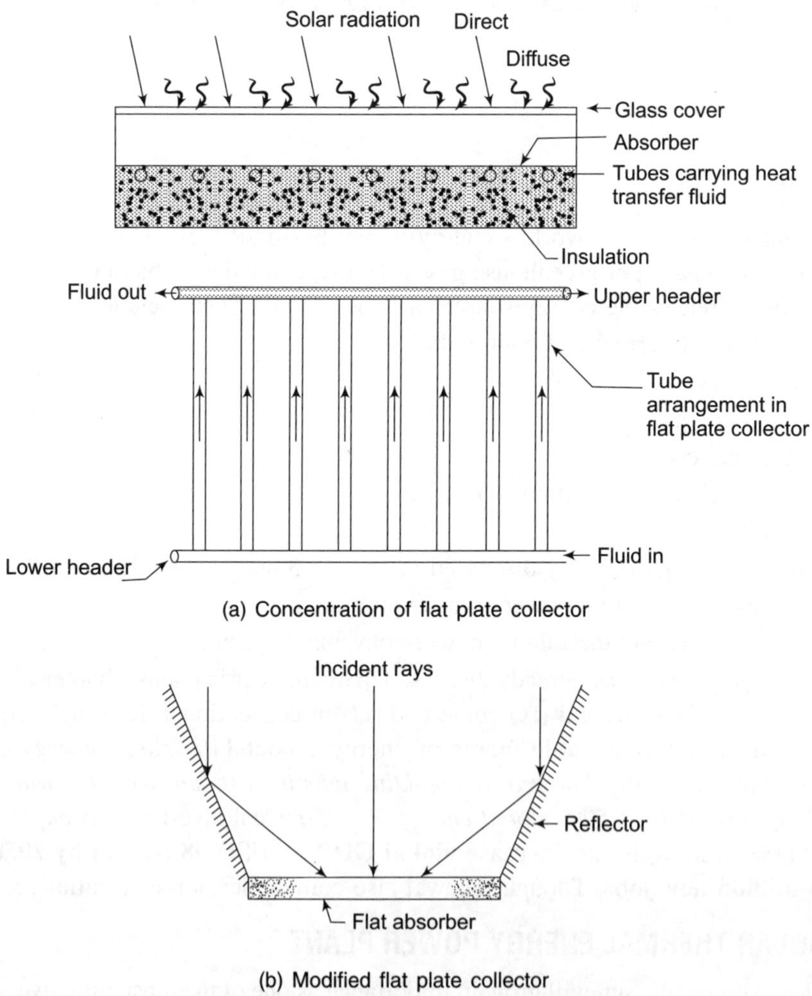

Figure 18.1 Solar flat plate collector

(i) **Non-concentrating type (flat plate collectors)**. (Fig. 18.1 a, b)

(ii) **Concentrating type (focus or non-focus type)** (Fig. 18.2a, b, c, d, e, f, g)

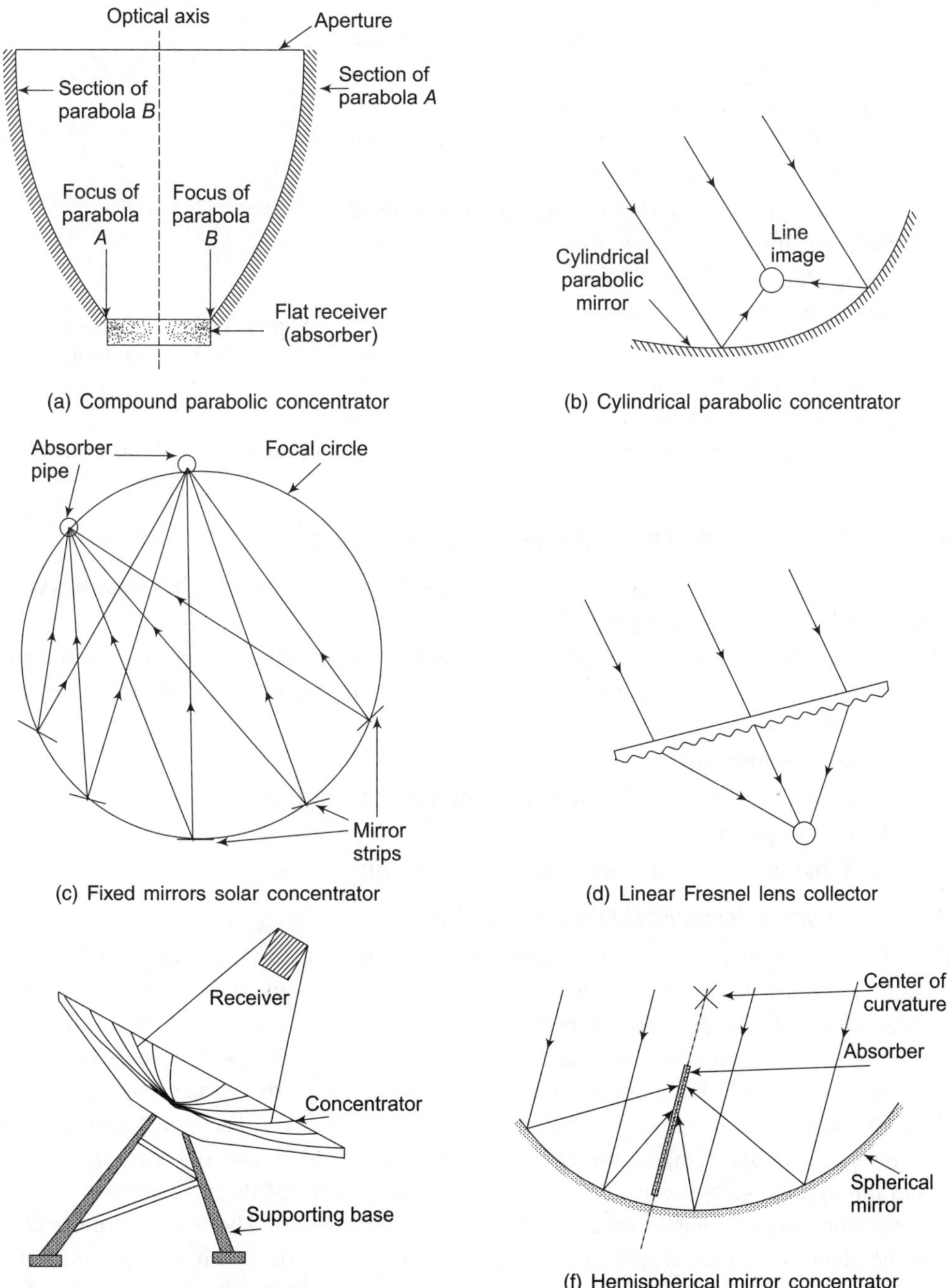

(a) Compound parabolic concentrator

(b) Cylindrical parabolic concentrator

(c) Fixed mirrors solar concentrator

(d) Linear Fresnel lens collector

(e) Parabolic disc collector

(f) Hemispherical mirror concentrator

Figure 18.2 Concentrating types solar collector

Parameters for Performance of Solar Collectors

The performance of a solar collector is assessed by three parameters, namely:

(i) Collector efficiency, (ii) concentration ratio and (iii) temperature range.

- **(a) Collector Efficiency** ($\eta_{collector}$): *It is defined as the ratio of the energy actually absorbed by collector to the energy incident on the collector.*
- **(b) Concentration Ratio (CR):** *It is ratio of the area of the aperture of the system to the area of receiver.*
- **(c) Temperature Range (TR):** *Temperature range is one in which the heat transport fluid is heated up by the collector.*

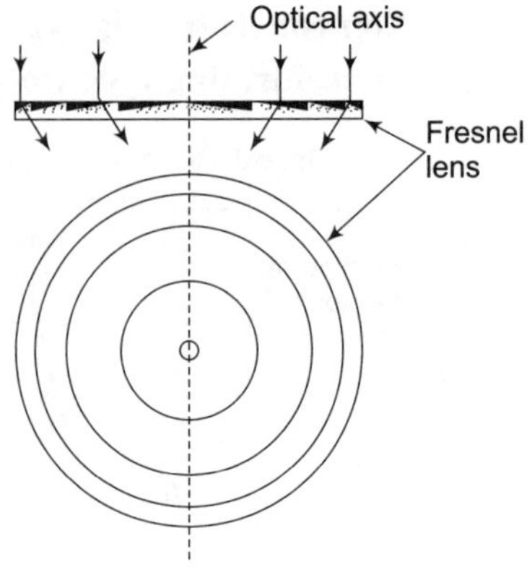

Figure 18.2(g) Circular Fresnel lens concentrator

It is worth to note that CR = 1 and TR is upto 100°C in flat plate collectors but line focus concentrations have CR ≈ 100 and RT = 150 to 300°C. The point focus concentrator has CR up to 1000 and TR = 500 to 1000°C.

18.1.1 Types of Solar Thermal Energy Power Plants

There are basically three types of solar thermal electric power systems. They are: **(a) central; (b) distributed and (c) solar pond.**

Several systems are being studied for both central and distributed approaches and many solar collector designs are being explored [4, 15, 31]. The major factors which determine a system are grouped into two areas:

(i) Types of subgroup

(a) Collector, (b) storage (c) energy conversion, and (d) hybrid.

(ii) Mode of operation

(a) Stand-alone, (b) fuel saver, and (c) repowering

18.1.1.1 Central Receiver Thermal Power Plant

Figure 18.3 shows an arrangement of a central receiver solar thermal power plant. It optically transmits radiant solar energy from a reflector known as **heliostat to a centrally located receiver on a tall tower (central receiver)** for converting water into steam. The central receiver is an insulated enclosure with a slot opening of about 1 × 4 m and with cooling tubes opposite the slot. The radiation through the slot can give rise to very high temperature (2000°C) and high heat transfer to the coolant (here feed water). The steam generated may attain the pressure and temperature about 90 bars and 510°C. A part of steam is allowed to expand through a conventional turbine and the rest is utilised for thermal storage, as storage is an essential requirement to take care variable solar radiation. The use of storage facility helps the plant to produce electrical output at or near rated power continuously. The working cycle is the Rankine cycle and the system is known as stand-alone solar power plant and most closely represents the manner in which conventional plants operate. *In one system, currently* *being designed; liquid sodium has been suggested as the coolant because it is able to cope with high heat flux in the receiver and steam is generated in a heat exchanger for expansion in turbine.*

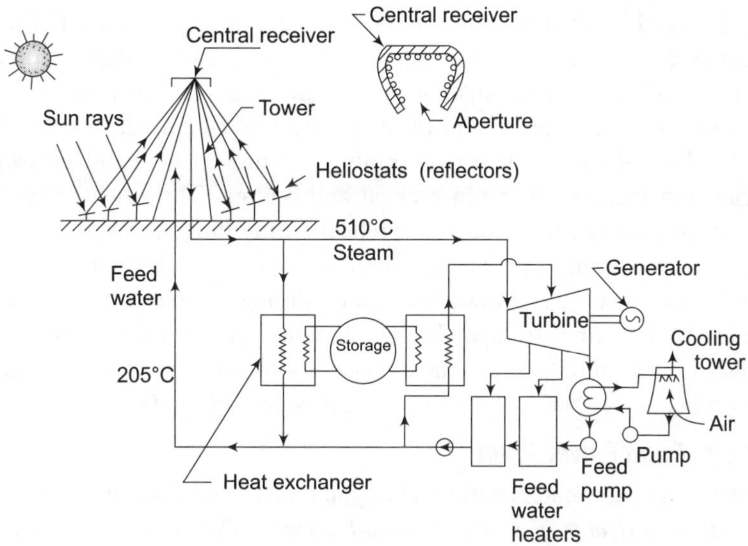

Figure 18.3 Central receiver solar thermal power plant

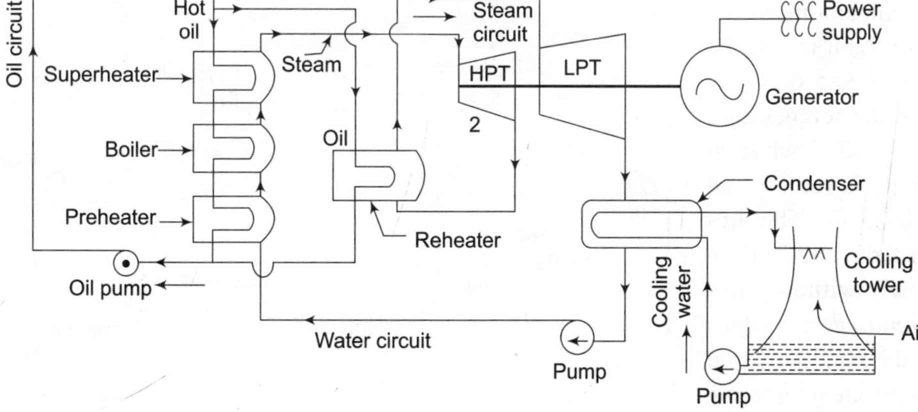

Figure 18.4 Distributed parabolic dish collector solar thermal power plant

18.1.1.2 Distributed Collector Solar Thermal Power Plant

It converts radiant solar energy to heat at each local reflector site known as distributed collector system. Figure 18.4 shows the use of a two axis-tracking paraboloidal dish collector coupled to a steam Rankine system. There are two options to absorb solar energy in collector. In one case, oil may be used to absorb solar energy. The heat absorbed in oil is used to generate steam in a heat exchanger. One heat exchanger is used as storage system. In other case, water may be used to absorb solar energy in collector and the steam so generated expands in turbine which produces power. The upper limit of temperature of steam attained is 540°C. In order to minimize the loss of superheat and to minimise transport heat and pressure losses, the superheat's function is separated from preheat and boiling. The essential requirement is to locate collectors providing superheat from preheat and boiling. The essential requirement is to locate collectors providing superheat at a short distance from the plant. A part of the steam generated by the solar energy is allowed to expand through a conventional steam turbine and the rest is utilised for thermal storage. The combination of thermal storage facility helps the plant to produce electrical output at or near a rated poor. The above mentioned power plants are now in working stage. Nano fluids may absorb solar energy better.

18.1.1.3 Solar Pond Power Plant

Figure 18.5 shows a solar pond electric power plant. Here, *the solar pond serves the purpose of a large flat plate collector as well as thermal storage. The pond consists of brine solution which is generally 1 m deep in which vertical gradient of salt concentration is maintained such that most concentrated and dense solutions are at the bottom which is blackened.* Solar radiation falling on the brine solution of pond penetrates through water up to the blackened bottom, where it gets absorbed and increased the temperature near the bottom. Beauty lies in the setup of minimum convection mixing current in the pond because the specific gravity at the bottom remains higher than that of top on account of concentration gradient in spite of high temperature at the bottom. As a result, the bottom layer of brine attains the temperature of 70 to 80°C while at the top, it remains at about 25°C. The hot brine from the bottom is slowly withdrawn in a *laminar flow pattern* from the pond and used to *evaporate an organic fluid in a heat exchanger*

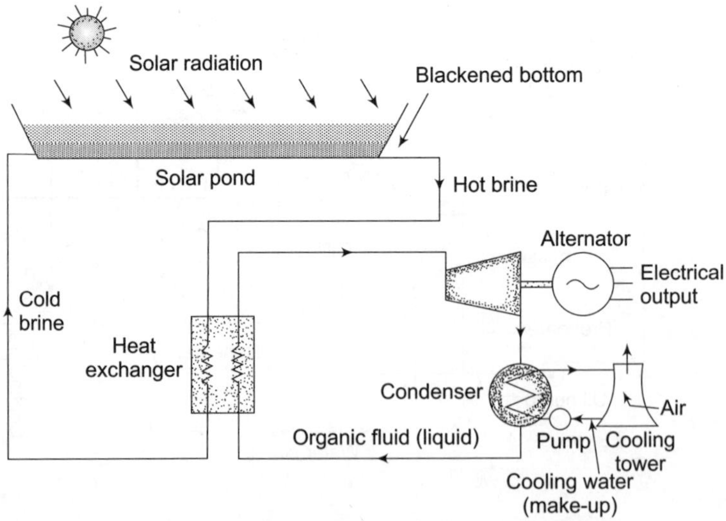

Figure 18.5 Solar pond power plant

and returned to the pond as a cold brine solution. The organic fluid vapour expands in a turbine working on a Rankine cycle and produces mechanical power which is converted into electrical energy through alternator coupled with turbine. *In Australia, there exists such power plant having a 2000 m² solar pond and generating 20 kW of electrical energy.*

18.1.2 Other Applications of Solar Thermal Energy

Over and above of generating electrical energy, solar thermal energy has many other applications. Some of these are given below:

(i) Solar water heating system (Fig. 18.6)
(ii) Solar passive space heating and cooling system
(iii) Solar industrial heating system
(iv) Solar refrigeration and air conditioning system
(v) Solar cooker
(vi) Solar furnace
(vii) Solar greenhouse
(viii) Solar desalination
(ix) Solar thermal water pump

Description of all of them is out of scope of the book. However; the schematic diagram of solar water heater is given in Fig. 18.6.

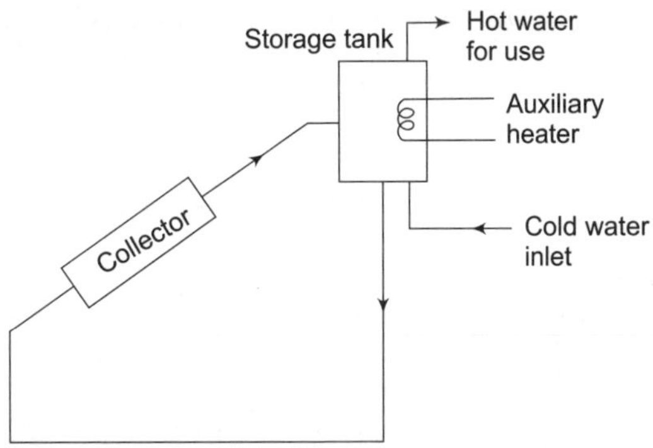

Figure 18.6 Solar water heater

18.1.3 Performance of Flat Plate Collectors

The performance of solar collector (Fig. 18.1a) is analysed by energy balance. It indicates the distribution of incident solar energy into useful energy gain and various heat losses. Under steady state condition, the useful heat delivered by a solar collector is equal to the heat energy absorbed by the metal surface minus the heat losses directly and indirectly to the surroundings. Thus,

$$Q_c = A \cdot [I_c \cdot (\tau \cdot \alpha)_c - U_c(T_{in} - T_a)]F_R \quad \text{Watt} \qquad \ldots(18.1)$$

where

Q_c = Useful heat output of the flat plate collector, W

A = Collector surface area, m²

I_c = Intensity of solar radiation on the collector surface, W/m²

τ = Transmissivity, a fraction of the solar energy reaching the surface that is absorbed.

α = Absorptivity, a fraction of solar energy reaching the surface that is absorbed.

$(\tau \cdot \alpha)_c$ = Effective product of transmissivity of the transparent cover and absorptivity of the absorber

U_c = Overall heat loss coefficient of collector, W/m²K

T_{in} = Collector fluid inlet temperature, °C
T_a = Ambient air temperature, *e*
F_R = Heat removal factor

The efficiency of solar collector is defined as the ratio of useful heat removal by the collector to the solar energy flux incident on the collector. Thus,

$$\eta_c = \frac{Q_c}{AI_c} = F_R\,(\tau_\alpha)_c - F_R\,U_c\left(\frac{T_{in} - T_a}{I_c}\right) \qquad \text{...(18.2)}$$

Figure 18.6(a) shows the variation of collector efficiency, η_c with term $(T_{in} - T_a)/I_c$ for different type of solar collectors. The highest collector is exhibited by parabolic trough concentrator and then evacuated tube collector and the poorest by flat plate collector.

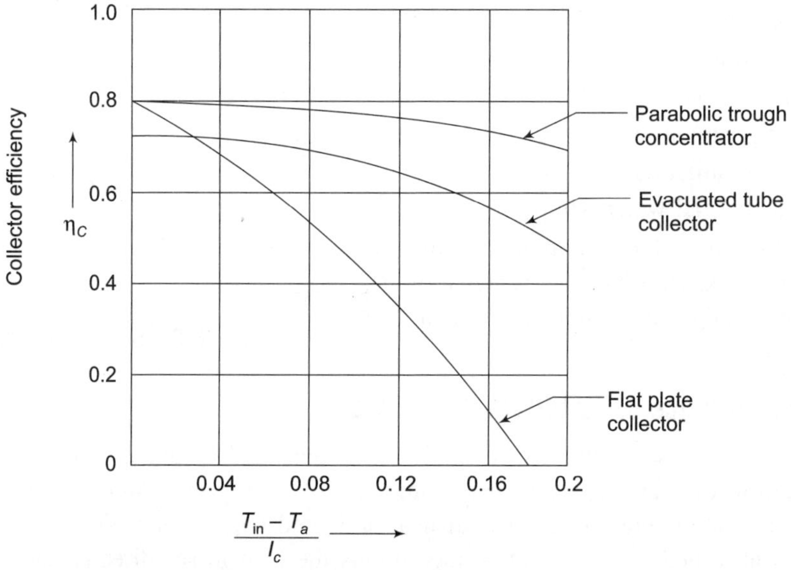

Figure 18.6(a) Efficiency curve for solar collectors

The outlet temperature of the fluid collector is given by

$$Q_c = \dot{m}c_p\,(T_{out} - T_{in}) \qquad \text{...(18.3)}$$

where Q_c = Useful heat output of collector, W
$\dot{m}$ = Mass flow rate of collector fluid, kg/s
c_p = Specific heat of collector fluid,
T_{out} = Outlet temperature of heated fluid, °C
T_{in} = Inlet tempesature of collector fluid, °C

The stagnation temperature, T_S of collector occurs when $Q_C = 0$, then $T_{in} = T_S$

$$T_s = T_{in} = T_a + I_c \cdot \frac{F_g \cdot (\tau \cdot \alpha)_c}{F_R \cdot U_c} \quad \text{...(18.4)}$$

When $(T_{in} - T_a)/I_c$ becomes zero, the collector efficiency is given by Eq. (18.2).

$$\eta_c = F_R(\tau \cdot \alpha)_c \quad \text{...(18.5)}$$

18.1.4 Performance of Concentrating Solar Collectors

The useful heat output of a concentrating solar collector is given by energy balance.

$$Q_c = F_R \cdot A_a[I_{be} \cdot \eta_{ope} - (U_c/C)(T_{in} - T_a)]; \quad W \quad \text{...(18.6)}$$

where A_a = Unshaded aperture area, m^2

F_R = Heat removal factor for solar concentrator

I_{be} = Intensity of beam solar radiation incident on the concentrator, W/m^2

η_{opte} = Optical efficiency of concentrating solar collector

U_c = Total heat loss coefficient of collector, W/m^2K

T_{in} = Collector inlet temperature, °C

T_a = Ambient temperature, °C

C = Correction factor.

The *optical efficiency,* η_{opte} *is defined as the ratio of the solar radiation absorbed by the absorber to the beam solar radiation on the concentrator.* It is expressed as

$$\eta_{opte} = \eta_{opt,\,o} \times C_{opt} = \rho\gamma\tau\alpha_a \cdot C_{opt} \quad \text{...(18.7)}$$

where $\eta_{opt,\,e}$ = Optical efficiency of collector at 0° incident angle of the beam

γ = Intercept factor

ρ = Mirror reflectivity

τ = Transmissivity of the transparent cover of the absorber

α_a = Absorptivity of absorber.

The efficiency of concentrating collector is expressed as

$$\eta_c = \frac{Q_c}{A_a \cdot I_{be}} \quad \text{...(18.8)}$$

The outlet enthalpy of heat transfer fluid in a concentrating collector with the phase change (vaporisation of fluid in the case of water, it will be steam) is expressed as

$$h_{out} = h_{in} + \frac{Q_c}{n_i}; \quad \text{J/kg} \quad \text{...(18.9)}$$

where h_{out} = Outlet enthalpy of heat transfer fluid, J/kg

h_{in} = Inlet enthalpy of heat transfer of fluid, J/kg

Q_c = Useful heat output of collector, W

$\dot{m}$ = Mass flow rate of heat transfer fluid, kg/s

Example 18.1 The following data relate to an evacuated tube collector:

The intensity of solar radiation on the collector surface = 700 W/m^2

The collector area = 4 m^2

The collector fluid inlet temperature = 30°C

The ambient air temperature = 27°C

The effective optical efficiency, $F_R\ (\tau_\alpha)_c = 0.78$

The effective heat transfer coefficient,

$$F_R \cdot U_c = 1.5 \text{ W/m}^2\text{K}$$

Mass flow rate of water = 0.03 kg/s

Specific heat of water = $c_p \approx 4.18 \times 10^3$ J/kgK

Calculate the following:

(a) The useful heat output of solar collector

(b) The outlet temperature of water

(c) The stagnation temperature

(d) The collector efficiency

Solution (a) The useful heat output of solar collector

$$Q_c = A\ F_R[I_c \cdot (\tau \cdot \alpha)_c - U_c(T_{\text{in}} - T_a)]$$

$$= 4 \times [700 \times 0.78 - 1.5(30 - 27)]$$

$$= 4[546 - 4.5] = 2166 \text{ W} \qquad \textit{Ans.}$$

(b) $$Q_c = \dot{m}\ c_p(T_{\text{out}} - T_{\text{in}})$$

or $$2166 = 0.03 \times 4.18 \times 10^3\ (T_{\text{out}} - 30)$$

or $$T_{\text{out}} = 30 + \frac{2166}{0.03 \times 4.18 \times 10^3} \approx 47.27°\text{C} \qquad \textit{Ans.}$$

(c) The stagnation temperature is given by

$$T_s = T_a + I_c \cdot \frac{F_R(\tau \cdot \alpha)_c}{F_R \cdot U_c}$$

or $$T_s = 27 + 700 \times \frac{0.78}{1.5} = 391°\text{C} \qquad \textit{Ans.}$$

(d) The collector efficiency is given by

$$\eta_c = F_R(\tau \cdot \alpha)_c - F_R U_c \cdot \frac{(T_{\text{in}} - T_a)}{I_c}$$

$$= 0.78 - 1.5 \times \frac{3}{700} = 0.7736 \qquad \textit{Ans.}$$

18.2 SOLAR PHOTOVOLTAIC CELLS POWER PLANT

Solar photovoltaic cells are semiconductor devices which can convert sunlight directly into electricity. A solar cell is basically an electrical current source, driven by a flux of radiation. Commercial photo cells may have efficiencies in the range of 15 to 20% and can produce d.c. electrical energy of 1 to 2 $kWh/m^2/day$ in ordinary sunshine. Typically, it produces a potential difference of about 0.5V and a current density of 1 kW/m^2. A typical commercial cell of 100 cm^2 area may produce a current of 2A. It has a life span in excess of about 20 years.

Applications

The major use of photovoltaic cells are:

- Space satellites
- Remote radio communication booster stations
- Marine warning lights
- Lighting, water pumping and remote area lighting and refrigeration of medicines.
- Solar energy powered vehicles, production of clean energy, H_2
- Battery charging
- Solar power plant attached to grid system without the need of battery.

18.2.1 Major Advantages and Disadvantages of Solar PV System Over Conventional Power System

Advantages

(i) *Direct conversion:* It converts solar energy directly into electrical energy without getting through thermal-mechanical link.

(ii) It has no moving parts.

(iii) It is reliable, modular, durable and generally maintenance free.

(iv) The systems are quiet, compatible with almost all environments.

(v) It can be located at the place of use (houses, hospitals, schools factories, etc.) and hence no distribution network is required.

(vi) It is a pollution free system, a green energy resource.

Disadvantages

(i) At present, its cost is comparable to fossile and so it is now economically competitive with other conventional power plants. Now, the cost is an advantage.

(ii) Its conversion efficiency is low (about 20%)

(iii) As solar radiation density is low, so large area of solar cells are required to generate sufficient useful power.

(iv) Due to intermittent nature of solar energy availability, some kind of electrical energy storage (say battery) is required, which makes the whole system expensive.

18.2.2 Solar Cell Construction

The basic structure of a solar cell is a typical *n*-on-*p* bulk silicon cell (Fig. 18.7). The bulk material is *p* type silicon with thickness varying from 100 to 350 microns depending upon the technology used in manufacturing. In general, a thin layer of *n* type silicon is formed at the top surface by *diffusing an impurity from V group (say phosphorus)* ***to get a-p-n junction***. The top active surface of the *n* layer has an ohmic contact with metallic gird structure to collect the current produced by the impinging photons of sunrays. It is worth to note that the metallic grid covers the minimum possible top surface area (say about 8% of total area) to ensure enough uncovered surface area for incoming solar energy photons. Similarly, the bottom inactive surface has an ohmic metallic contact over the entire area. These two metallic contacts on *p* and *n* layers respectively form the positive and negative terminals of the solar cells. Further, several enhancement features such as *anti-reflecting coating and textured finish are added to capture maximum photons.*

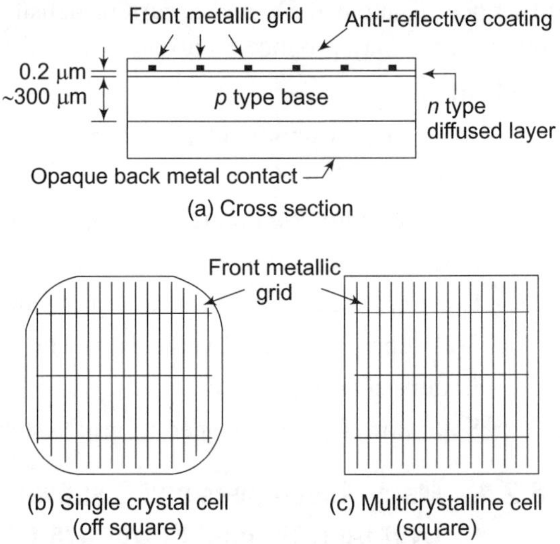

Figure 18.7 Construction of bulk silicon cell

A bare single cell cannot be used for outdoor energy generation by itself because of very small output. A workable voltage and reasonable power is possible to attain by interconnecting an appropriate number of cells which is known as **solar module**–*a basic building block of a PV system.* Most common commercial modules have a **series connection** *of 32 or 36 silicon cells to make it capable of charging a 12 V storage battery.*

Several solar modules are connected in series/parallel to increase the voltage/current ratings. *When connected in series, it gives the maximum power at the same current. While, when connected in parallel, it gives the maximum power at the same voltage. So connections are done as per requirement. A solar panel is a group of several modules connected in series, parallel combination in a frame that can be mounted on a structure.* Figure 18.8 show the construction of a solar PV module and panel.

On the basis of the type of junction structure solar PV cells are classified as (i) *pn* homo-junction cell, (ii) *pn* hetero-junction cell, (iii) *pn* multi-junction cell (iv) metal semiconductor, Schottky junction, and (v) *p-i-n* type intrinsic-*n*-type semiconductor junction.

Depending on the type of material used for fabrication of the solar PV cells, they are classified as (i) single crystal cell, (ii) multi-crystal cell, (iii) amorphous silicon cell (iv) gallium arsenide cell (v) copper indium diselanide cell, (iv) cadimum telluride cell, and (vii) organic PV cell.

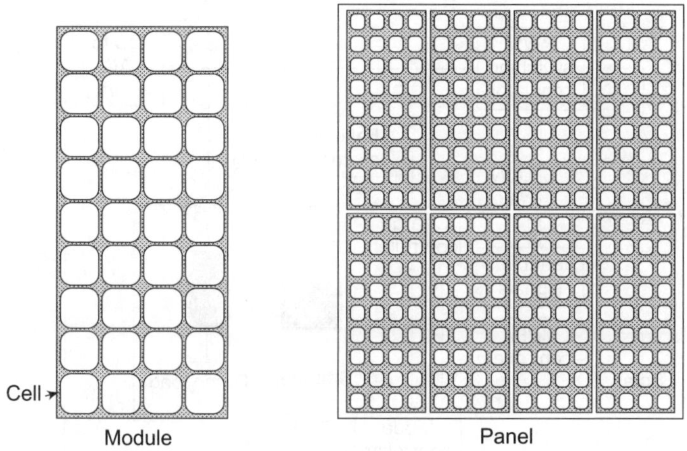

Figure 18.8 Solar PV module and panel

18.2.3 Solar PV System

Solar PV systems may be broadly classified into four categories, namely:

(i) Central PV power station system.
(ii) Distributed PV power system.
(iii) Hybrid PV system.
(iv) Space based solar power (a new concept).

18.2.4 Central PV Power Station System

Central PV power station system is similar to any other conventional central power station. The power generated is fed to the grid. They are installed in the capacity of a few MW range to meet the daytime peak load. Including both ground and roof mounted solar PV plants, India has installed solar power capacity of 39,083 MW as of 28 February 2021.

18.2.5 Distributed PV System

Distributed PV power system is the most unique in respect of use and success. It may be further classified as the following.

18.2.5.1 Standalone Solar PV System

This system is located at the load centre and dedicated to meet as the electrical loads of a village/community on a specific set of loads.

The need of energy storage becomes essential due to absence of sunrays in night and cloudy days. This system is the most suitable for remote and rural areas where there is no access of power from grid supply. Stand-alone system produces power in the range of 10 Wp to 100 kWp. Lithium-ion battery is the storage device.

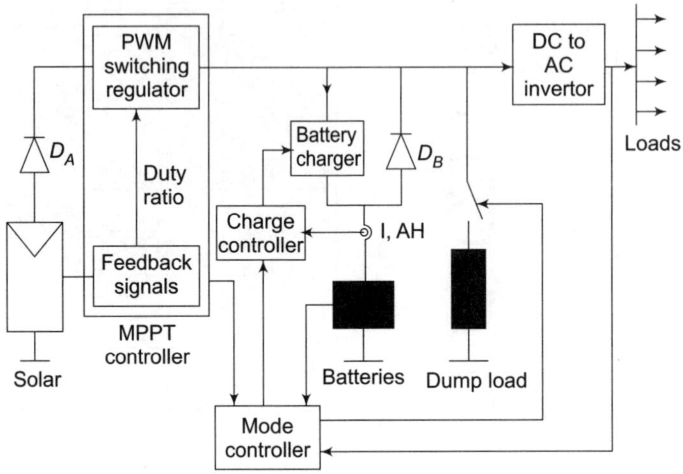

Figure 18.9 General stand-alone solar PV system

Figure 18.9 shows a schematic diagram of a stand-alone solar PV system. The MPPT senses the voltage and current output of the array and adjusts the operating point to extract the maximum power under the given situations. The d.c. output of the array after converting to a.c. is fed to loads. The excess output of the array is used to charge the battery. In the event of availability of excess power even after charging the battery, it may be shunted to dump heaters. When there is no sun energy, during night and cloudy day, the battery supplies the energy to the load through inverter. In order to prevent the battery from overcharging after the charger is opened, the battery is made to discharge diode D_B. *The function of the array diode (D_A) is to isolate the array from the battery to prevent the battery discharge through array during nights. Battery is costly.*

A central controller is used to control the whole system. This controller collects the system signals and keeps track of charge/discharge state of the battery, matches the generated output with load and supplies commanding signals to the charger and dump heaters for on-off operation.

18.2.5.2 Grid-interactive Solar PV System

Figure 18.10 shows a schematic diagram of a grid-intractive solar PV system. In this system, the excess power is fed to grid and so dump heaters are eliminated. Further, in the event of unavailability of sunshine, the supply of power is maintained from grid and thus battery is eliminated. There is a need of mechanism called MPPT controller synchronizing operation d.c. power obtained is first converted to a.c. with the use of inverter, harmonics are filtered and then filtered power is fed to the grid by adjusting voltage level in the system. Now, PV modules with inverter as an integral component have come up which provide utility-grade 60 Hz power directly from the model junction box.

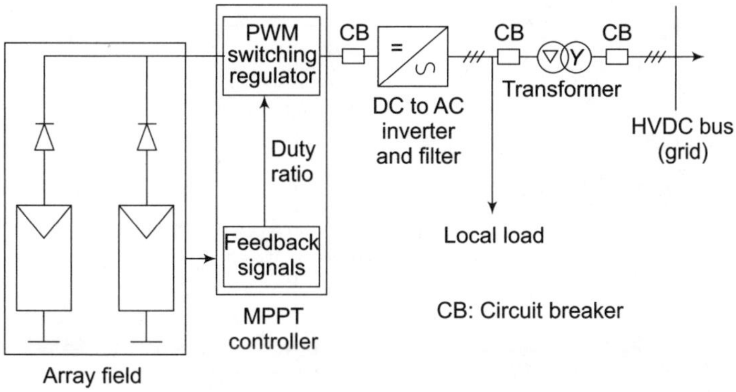

Figure 18.10 A grid-interactive solar PV system

(a) Hybrid Solar PV System. Since PV system was costly so in order to reduce the cost of generation a concept of **hybrid solar PV system** was used. In this system, the power supply was met with diesel/gasoline generator in conjunction with solar PV system.

(b) Space Based Solar Power (SBSP). A new concept and for this see chapter 1, Section. 1.16.9.3.

18.3 TIDAL WAVE ENERGY POWER PLANT

The tides in the oceans are caused by the joint action of the sun and the moon. The tides are rhythmic but not constant, nor do they occur on a regular schedule. The occurrence of the tides is due to a balance of forces, mainly the gravitational force of the moon and also that of the sun, both acting together with that of the earth to balance the centrifugal force on the water due to earth's rotation. This results in the rhythmic rise and fall of water, i.e., tides. They are characterized by two parameters, namely, **their schedule** and **range R.**

The orbit of the moon does not occur on a regular 24 h daily schedule, but it rotates around the earth every 24 h 50 minutes. This causes the variation of schedule from day-to-day basis. During this period, tides rise and fall twice which results in the formation of a tidal cycle that lasts 12 h and 25 minutes. This is called schedule.

The tidal range R is defined as

$$R = \text{Water elevation at high tide} - \text{Water elevation at low tide.}$$

Figure 18.11(a) shows the tidal cycle, the schedule and range. Point A indicates high tide point while point B indicates the low tide point. The average period of tidal cycle is 12 h 25 min while the average period of time for the water level to fall from A to B and then rise from B to C is each approximately equal to 6 h 12.5 min.

Figure 18.11(b) shows variation of range of tides during a lunar month. On the full or new moon time, the sun, the moon and the earth are approximately in one line, and as a result, the gravitational forces of the sun and the moon are enhanced, resulting in high tides. These high tides are called **spring tides**. Whereas near the first and third quarters of the moon, the sun and the moon are at right angle with respect to the earth, **neap** (low) tides occur. In the

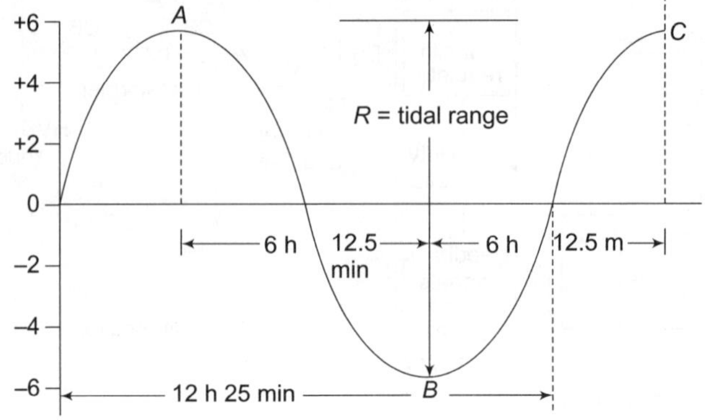

Figure 18.11(a) Tidal cycle showing schedule and range

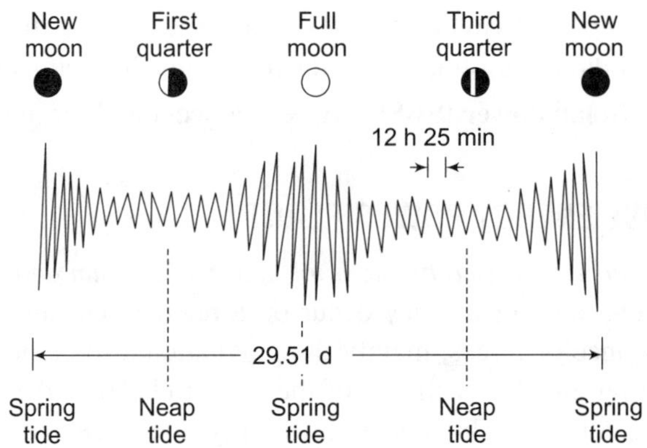

Figure 18.11(b) Variation of tides during a lunar month

neap tides, the tidal range is very small, because the high tides are lower, and low tides are higher. The tidal range is the maximum at the time of new and full moon known as **spring** tides, and the minimum during **neap tides** which occur during first and third quarter moon. During lunar month, a **typical mean range** may the calculated which is one-third of the spring range.

It is worth to note that in the design and operation of tidal power plant, the variations in the periodicity, monthly and seasonal tidal ranges along with the proper site are to be taken into consideration. Tidal ranges vary from one location to another. Since construction of tidal wave power plant is very costly, so the tidal range and site location must justify the cost involved.

The most suitable sites which are less available are **sea bay or estuary** where the width is small and tides are high. The tides along most coastlines are about one metre high, but at the site where plant is to be constructed is to have more than 10 m high tides. In these areas, the most effective tidal power plants are built. A dam or sluice gate is built across an ocean bay or estuary and a **row of hydraulic turbines** are installed. The incoming tides flow through these hydraulic turbines and the exit water fills the enclosed basin. The hydraulic turbines coupled with generators produce electricity. Once the basin is filled with water, the gates are closed and the turbines are shut down. Now, the turbine blades are reversed and the gates are opened through which the water in the basin passes through turbines and generates electricity.

As mentioned above, the building of tidal power plant involves the construction of a long barrier across a sea bay or estuary to create a large basin on the landward side of the bay. If the basin size is less, it has to be enlarged by digging. The barrier includes dykes, gate-controlled sluices, and the powerhouse consisting of row of hydraulic turbines, generators, transformers and other electrical equipment. Based on above discussion, the tidal wave power plants may have the following configurations.

(i) Single basin, single effect tidal power plant
(ii) Single basin, double effect tidal power plant
(iii) Linked basin, double basin single effect power plant.

18.3.1 Single Basin, Single Effect Tidal Power Plant

Figure 18.11(c) shows a scheme of a single basin, single effect tidal power plant. The site is selected at the narrow width of the ocean bay where a barrier has been built which separates the sea bay to the basin. Dyke and sluices have been built on both sides of the powerhouse. During the high tides, the basin is filled by keeping the sluices open and allowing the water to flow from sea to the basin. Then sluices are closed and water is allowed from the basin to flow through turbine to the sea bay which generates power.

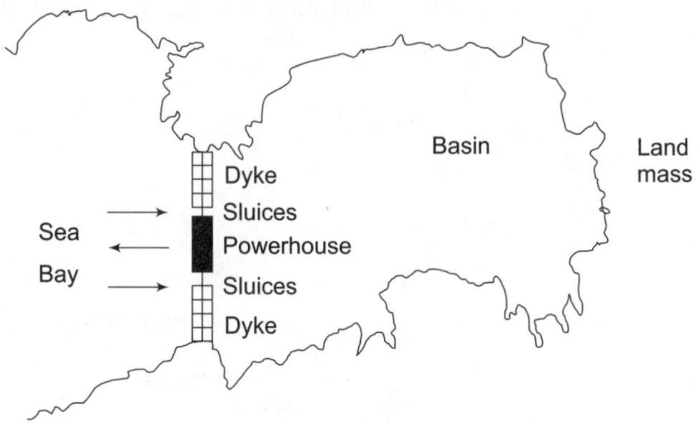

Figure 18.11(c) Scheme of a single basin, single effect power plant

18.3.2 Single Basin, Double Effect Tidal Power Plant

Figure 18.11(d) shows a scheme of a single basin, double effect tidal power plant. In this scheme, power is generated during high tides with water flowing from the sea bay to the basin through the row of turbines. The sluice gates are closed. During the low tides, water flows from the basin to the row of turbines which generate electricity. In this case, the turbine blades should be reversible with proper blade angles depending upon the direction of flow. The best way is the foundation of turbine should be such which should rotate turbine set by 180°.

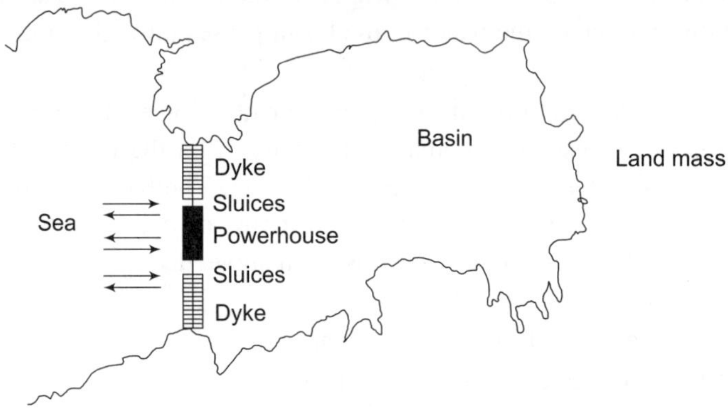

Figure 18.11(d) Scheme of a single basin, double effect power plant

18.3.3 Linked Basin-Double Basin, Single Effect Power Plant

Figure 18.11(e) shows a scheme of a linked basin-double basin, single effect power plant. These are two basins on the landward side with the powerhouse located between the barrier and land side. One basin acts as high basin and other low basin. During high tides, sluice gates are on the high barrier side to fill the basin. Power is generated by water flowing from the high basin to the low basin through the row of water turbines and water flows from the low basin to the sea bay during low tide. Generating low heads are available through which turbo-generators have to generate electricity.

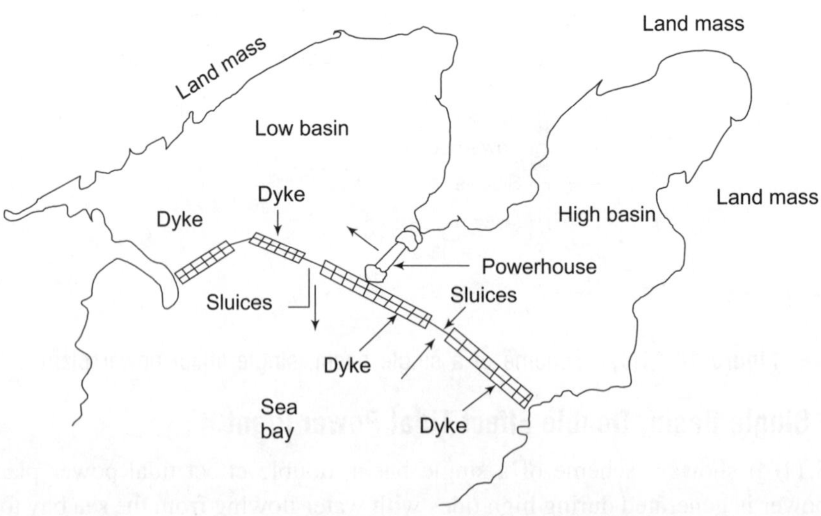

Figure 18.11(e) Scheme of a linked basin, double basin single effect power plant

Figure 18.11(f) shows a layout of a typical tidal wave powerhouse. Many tidal power plants have been installed world over. The most important ones are located in Bay of Fundy, Nova Scotia, Canada, Cook Inlet in Alaska and Maine, USA and other. The first tidal plant was established at La Rance, France on November 26, 1966, having 10 bulb turbines generating 10 MW each, totalling 240 MW. The other biggest tidal power plants are:

- Sihwa Lake Tidal Power Station, South Korea, 254 MW
- Swansea: Bay Tidal Lagoon, UK, 240 MW
- Meygen Tidal Energy Project, Scotland, 86 MW
- Annapolis Royal Generating Tidal Station Canada, 20 MW.

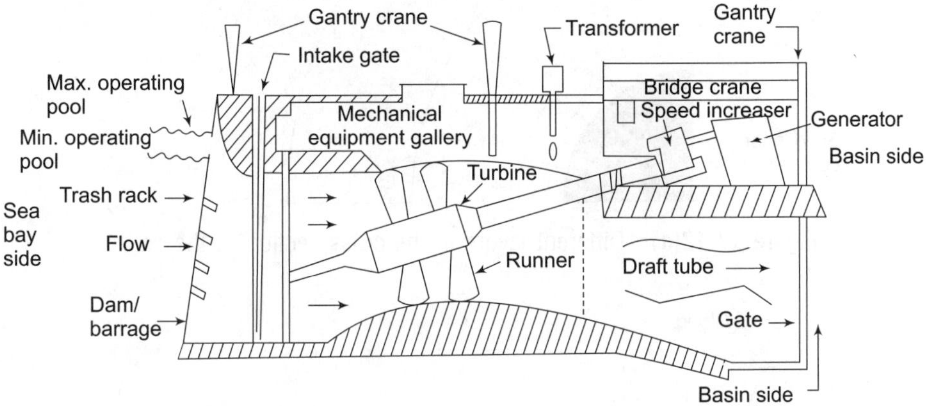

Figure 18.11(f) Layout of a typical tidal wave power plant

18.4 GEOTHERMAL POWER PLANT

The heat of the earth's interior is one of the most intensive energy available to man. Geothermal resources are derived from the distribution of temperature and thermal energy beneath the earth's surface. The natural earth has manifested itself for millions of year in the form of volcanos, lava flows, hot springs and geysers. Figure 18.12(a) shows the interior of the earth which is thought to consist of a central molten core surrounded by a region of a semifluid material called the **mantle**. The mantle zone starting from the lower boundary of the crust continues up to a depth of about 2900 km. It has been subdivided into upper and lower mantles, the boundary between the two layers being placed at 900-1000 km below the earth. The upper mantle has been divided into two layers of about 400 and 600 km thickness respectively. The **core is innermost concentric shell of the earth**. The core boundary begins at a depth of about 2900 km from the surface and it extends in the centre of the earth at about 6371 km. It is further divided into outer core and inner core. The outer core comprises the region from a depth of about 2900 km to 4580 km below the earth's surface and behaves like a liquid. The inner core with a thickness of around 1790 km is believed to be a **solid metallic body**.

The temperature of the crust increases with the depth of the rate of **about 30°C /km**. Below the crust, the mantle mass is called **magma**. It is in the process of cooling at the rate of about

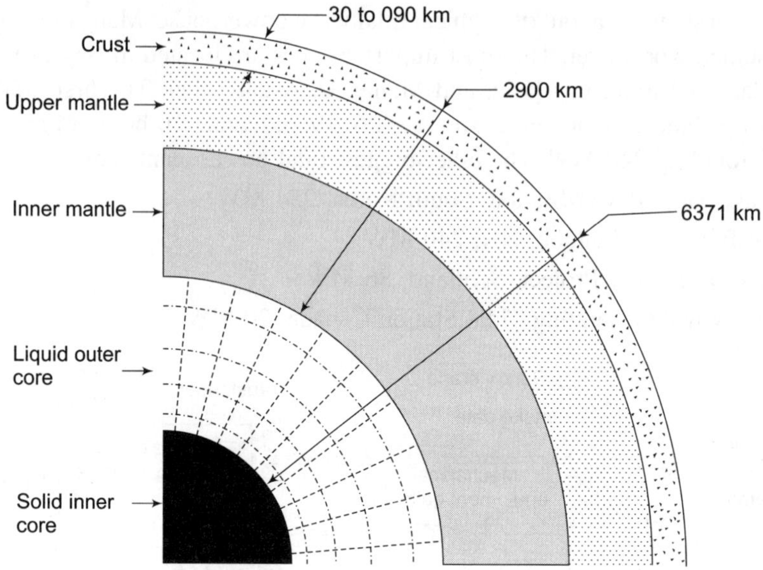

Figure 18.12(a) Different layers in the cross section of the earth

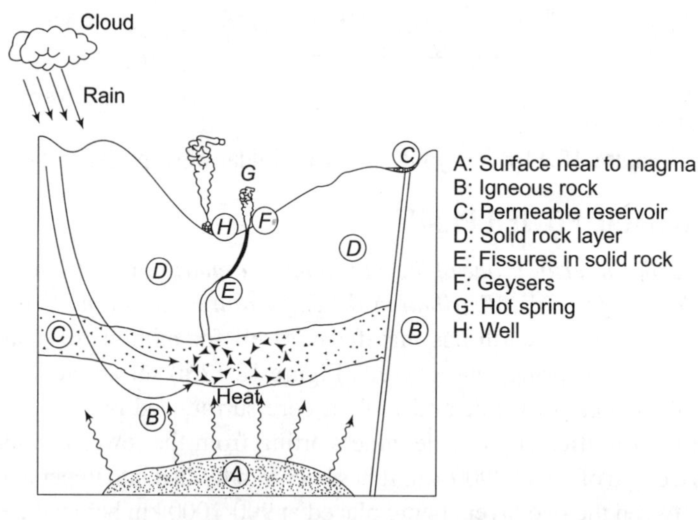

Figure 18.12(b) A typical geothermal field

0.063 W/m^2. Figure 18.12(b) shows a typical geothermal field. As a result of cooling process, the hot magma near the surface (A) solidifies into **igneous rock** (B) or volcanic rock. With the help of fissures in the rock, the groundwater that finds its way down to thick rock gets heated up by the heat of the rock or by mixing with hot gases and steam emanating from the magma. There is a porous and permeable reservoir (C) above the igneous rock. The heated water rises up and gets trapped in this reservoir as above this layer, there is a layer of solid rock (D) which acts as a cap for the reservoir. The solid rock has fissures (E) that act as a vent to the giant underground boiler. The vents act as the passage of geysers, fumaroles (F) or the hot spring (G) on the surface of the earth. A well (H) taps steam from the fissures for the use in a geothermal power plant. There are two types of geothermal steam, one is **magmatic steam** that originates from the magma itself and other is **meteoritic steam** with groundwater heated by the magma. Among the two, the latter one is the largest source of geothermal steam.

Like other thermal power plants, the geothermal plant involves production and use of steam expanding it through a turbine. The geothermal plant differs from the conventional fuel or nuclear plant in its method of steam production and the quality of its steam. In this plant, the steam is produced in nature's own boiler by the natural circulation of water coming into contact with hot rocks in the depth of the earth. Depending upon the conditions within the reservoir, the geothermal fluids may be in the form either of *slightly superheated dry steam or of pressurized hot water.* The condition of the fluid, in turn, controls the method of utilization and the potential environmental impact from its production. The requirement for the geothermal power plant are the following:

(a) Reservoir of high temperature (at least 180°C, preferably above 200°C)

(b) Depth of less than 3 km

(c) Natural fluids for transferring the heat to the surface and power plant

(d) Adequate reservoir volume (7 m^3)

(e) Sufficient reservoir permeable to ensure sustained delivery of fluids to wells at adequate rates

(f) No major unsolved problems

Depending upon the type of geothermal energy resources, three systems are prevalent for the utilisation of geothermal energy. These are as follows:

(i) Vapour dominated (dry steam) systems

(ii) Liquid-dominated (hot water) systems. It may be of two types—binary vapour system and flashed steam system

(iii) Dry (hot-rock) geothermal formation.

Each of these forms has a potential for exploitation. The vapour dominated system offers the best features for production of electricity. It is the form most sought after and the rarest. The hot water systems are many times more abundant than vapour dominated systems. The hot rocks comprise the bulk of the earth's surface and are the least exploited so far. *It has been estimated that if we merely cool the rocky crust under dry land 1°C (ignoring heat input from the sun and the magma core), there would be enough energy to supply the world's needs for 20,000 to 50,000 years.* Underground nuclear explosions may be used to fracture rocks but this has ecological disadvantages. A more promising approach is hydro-fracturing by injecting high pressure water [29]. Various ways are possible to extract the energy of hot rock. Either (i) or (ii) system may be used for power generation.

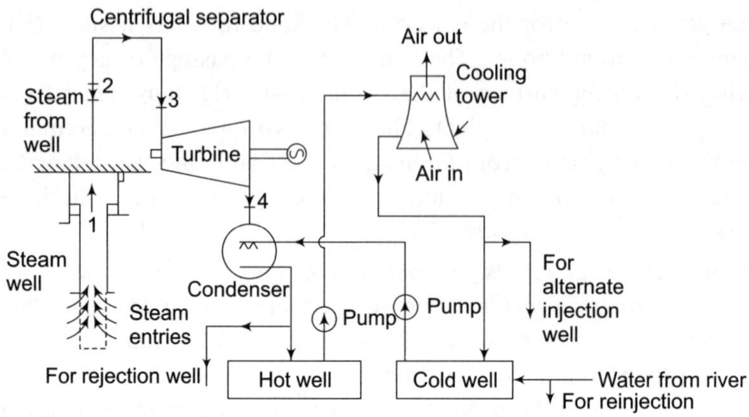

Figure 18.13(a) Geothermal power plant using dry steam working on Rankine cycle

18.4.1 Vapour Dominated System

Figure 18.13(a) shows a diagrammatic arrangement of geothermal power plant using dry steam from well. The inlet pressure is about 7.5 bars and temperature of about 180°C. Its process on T-s diagram is shown in Fig. 18.13(b).

Geothermal resources offer a clean energy and the cost per kWh is either less or comparable to the conventional system depending upon the type of the resources. In the present day of energy crisis due to enormous increase in the price of crude petroleum, this may prove to be of great relief. The total power production by geothermal energy in various countries like USA, Italy, New Zealand, Japan, Mexico, Soviet Union, Iceland, etc., amounts to be around 5000 MW, India has not entered into this field of energy production until now but with the formulation of separate wing of Non-conventional Energy Department of Government of India, there is a hope to enter in this area.

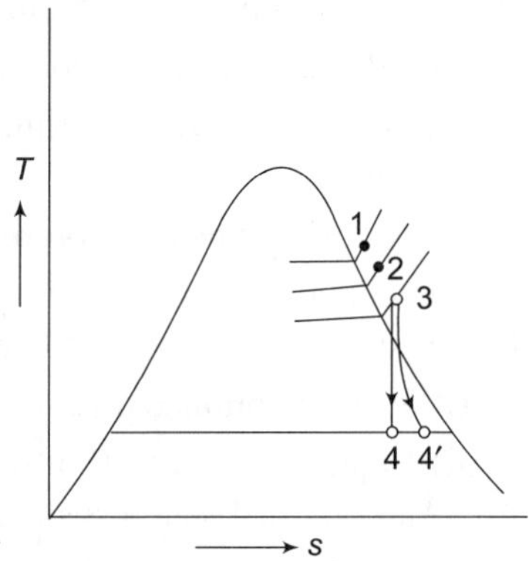

Figure 18.13(b) T-S diagram for vapour dominated system

18.4.2 Liquid Dominated Hot-water System

Figure 18.13(c) shows a closed system geothermal power plant using hot water as the main heat source and a secondary fluid driving turbo-generator. Sometimes, it is also called vapour-turbine cycle. Its process is shown in Fig. 18.13(d) on *T-s* diagram.

In this power plant, water is pumped to the surface by a deep well pump. To ensure that no steam is produced and no dissolved gases escape from the water, the water is kept above saturation pressure throughout the process. The water gives up its heat in a heat exchanger (boiler) to heat a liquid (isobutene) and superheat it. The cooled water is then injected into a nearby well, carrying all dissolved gases and most of the solids in solution with it. The

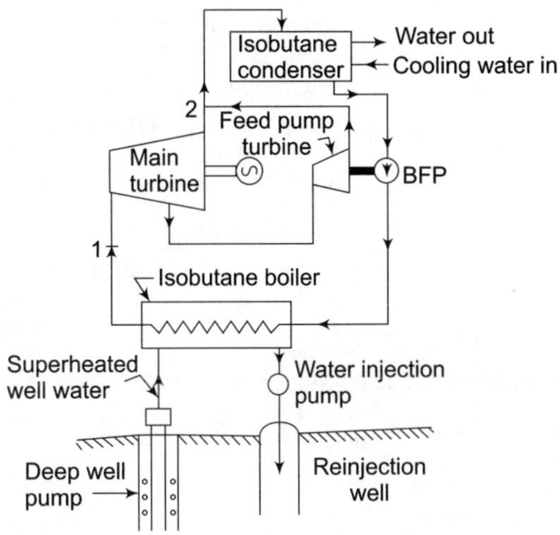

Figure 18.13(c) Geothermal vapour-turbine based on hot water cycle diagram working on Rankine cycle

water thus injected maintains the underground reservoir pressure and very likely may be reheated by the hot porous zone. In other words, the water becomes a means of transporting heat from the hot rocks beneath to the power plant above, in a continuous cycle.

Isobutane vapour, expanding through a turbine, yields its energy to drive the generator. The exhaust gas is condensed back to the liquid in a water-cooled condenser and pumped back to the isobutane boiler to repeat the cycle.

Since the dry steam availability is rare and hot water resources are abundant, the practical and economic advantages of this process may be great enough to turn the vast potential of geothermal energy into reality.

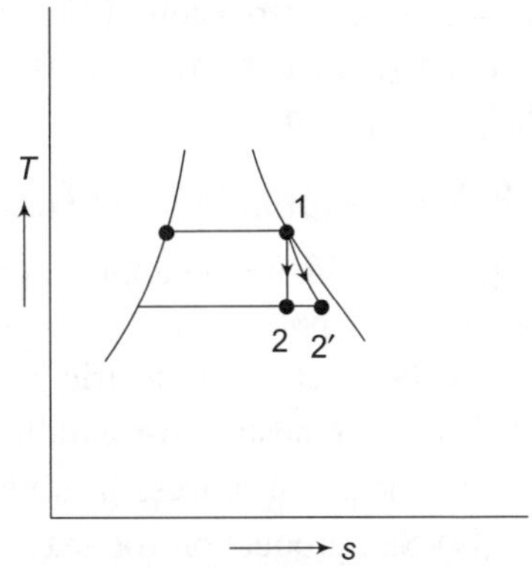

Figure 18.13(d) T-s diagram for binary system

18.4.3 Liquid Dominated-Flash Steam System

Figure 18.13(e) shows a schematic of a liquid dominated single flash steam system. Its process on *T-s* diagram is shown in Fig. 18.13(f). Hot water is drawn from the underground reservoir at state (1) and reaches at state (2) at low pressure due to throttling process (1-2). This throttling process results in a two-phase mixture of low quality at state (2). It passes through a flash separator at (3) where water is separated and is led to the underground as reinjection water at state (6). The flash separator acts as a throttling process. The dry saturated steam at state (4) at low prossure enters in steam turbine where it expands and develops electrical power as it is coupled to the generator. The extra hot steam at state (5) enters a direct contact (d.c.) condenser where it is condenses. The condensate is withdrawn by a condensate pump at state (7) and it is pumped at state (8). A part of this condensate

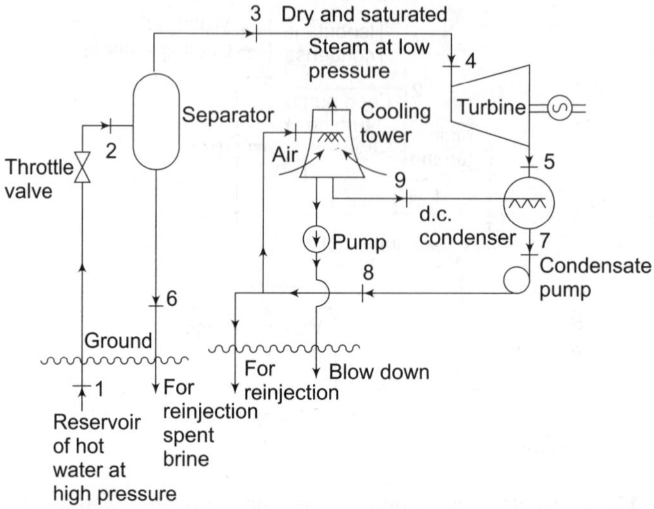

Figure 18.13(e) Liquid-dominated single flash steams power plant

is fed to the cooling tower for its cooling. The cooled condensate at state (9) is led to the d.c. condenser. The other part of condensate is used as reinjection water into the ground. A provision of blowdown of cooled water in cooling tower is sometimes used as blowdown with the help of a pump.

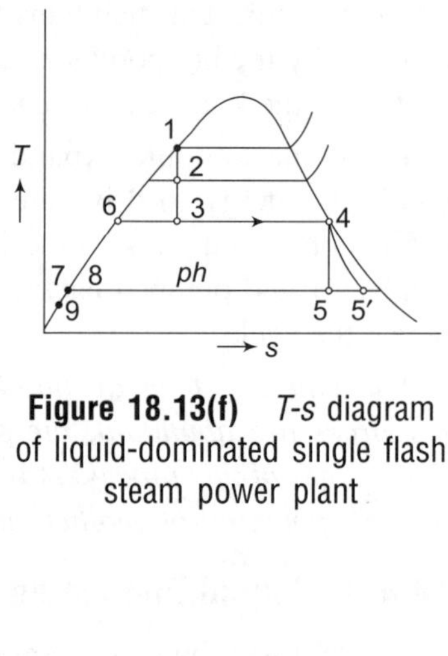

Figure 18.13(f) *T-s* diagram of liquid-dominated single flash steam power plant

18.4.4 Applications of Geothermal Energy

The following are the applications of geothermal energy without any pollution.

(i) Generation of electric power
(ii) Space heating for buildings
(iii) Industrial process heating
(iv) Salt production for sea
(v) Extraction manufacturing
(vi) Textile industry
(vii) Timber seasoning
(viii) Crop drying
(ix) Fish drying and fish meal production
(x) Production of dried milk
(xi) Mushroom farming
(xii) Food processing
(xiii) Gasification of coal

(xiv) Sewage heat treatment

(ixv) Tourism industries through hot springs. Many countries have a large number of hot springs. For example, Japan has 4000 hot springs. Many have bathing facilities. A large number of tourists visit every year to see these hot springs and bathing.

(xvi) Paper manufacturing

(xvii) Use in greenhouse to maintain required temperature.

18.5 WIND ENERGY POWER PLANT

Wind energy is the kinetic energy associated with movement of large masses of air and caused by uneven heating of atmosphere by the sun creating temperature, density and pressure differences. As per estimate, 1% of all solar radiation falling on the face of the earth is converted into kinetic energy of the atmosphere, 30% of which occurs in the lowest 1 km of elevation. *Thus, it is an indirect form of solar energy. It is cleans, cheap and a ecofriendly renewable source of energy but dispersed erractically and is location specific.*

Wind energy is harnessed as mechanical energy with the help of wind turbine from which electricity can be generated or pumps may be driven. **Moderate to high-speed wind typically from 5 to 25 m/s is considered favourable for most wind turbines.**

As per estimation, the total energy of the wind over the earth's surface is about 2×10^{10} kW. Clearly, to harness more than 5% is impractical. The concept of wind energy converter (also **known as wind mill**) is not new. It was used by the Chinese in 150 B.C. Wind energy converter utilizes the energy of the wind and converts it into mechanical or electrical energy. It must be recognised that wind does not blow continuously. A critical part of effective wind energy utilisation is **choosing an optimum site** as some parts of the world have significantly higher wind energy than others. It is important to evaluate the uses to which the wind energy may be put, especially, if no storage system is contemplated.

18.5.1 Origin of Winds

Parameters and Nature. *The origin of winds is caused by the uneven heating of earth's surface from the sun.* This produces widespread winds on **global basis and local basis**. *The wind speed varies with the height specially up to 100 m* (called **surface layer**).

(a) Global (or Planetary) Wind

These are two major factors which determine the speed and direction of wind on global basis.

(i) *Primary force for global winds is developed due to differential heating of earth equator and polar regions.* As a result, there is a neat gain of heat in tropical regions while there is a net loss of heat in polar regions. For balancing, earth's atmosphere has to transport heat towards polar region. Ocean currents act similarly.

(ii) *Spinning of the earth, about its axis produces* **Coriolis** force which moves air currents towards west. In the zone between 30°N and 30°S, heated air at the equator rises which is replaced by cooler air coming from north and south. This phenomenon is known as **Hadley circulation,** after its name of researchers. But due to coriolis force these winds deviate towards west which are also known as **trade winds**. In the zone between 30°N(S) and 30°N(S) western winds predominately similar to wave like circulation transferring cold air southwards and warms air northward. This phenomenon is known as **Rossby circulation**. Figure 18.14 shows global circulation of wind showing Hadley. Trade and Rossby circulation.

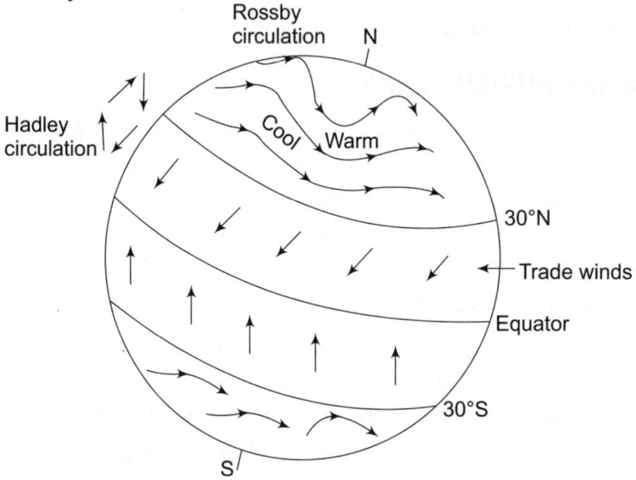

Figure 18.14 Global circulation of wind

(b) Factors Affecting Distribution of Winds

The following factors affect the distribution of energy on the surface of earth.

(i) Great mountain massifs
(ii) Surface roughness or friction of earth's surface
(iii) Hill, trees, buildings and similar obstructions
(iv) Climate disturbances
(v) Narrow mountains gaps

(c) Nature of Winds

In order to design and predict the performance of wind turbines, it is essential to know the behaviour and structure of the wind. The behaviour and structure of wind varies from site to site depending upon the general climate, physical geometry, surface condition of terrain, etc. Table 18.1 depicts the wind speed based on **Beaufort number**.

Table 18.1 Description of wind speed based on Beaufort number

Beaufort No.	*Wind speed*		*Wind description*	*Beaufort No.*	*Wind speed*		*Wind description*
	m/s	*Km/h*			*m/s*	*Km/h*	
0	0.0-0.4	0.0-1.6	Calm	7	14-17	51-63	Strong
1	0.4-1.8	1.6-6	Light	8	17-21	69-76	Gale
2	1.8-3.6	6-13	Light	9	21-25	76-88	Gale

Contd.

Table 18.1 Contd...

3	3.6-5.8	13-21	Light	10	25-29	88-103	Strong gale
4	5.8-8.5	21-31	Moderate	11	29-34	103-121	Strong gale
5	8.5-11	31-40	Fresh	12	34<	121<	Hurricane
6	11-14	40-51	Strong	–	–	–	–

Best sites are those for wind turbines where favourbale winds with speeds varying from 5 m/s to 25 m/s are available for most of the time (about 70% to 80% of whole day) such as sea coast.

Wind data are collected by all countries for meteorological services. *Wind speed is measured by* ***anemometer*** and wind direction is measured by ***wind vane*** **attached to a direction indicator. Wind velocity increases with height.**

18.5.2 Major Applications of Wind Power

In the recent past, major development has taken place in respect of wind turbine technology. Wind turbines have been built in power output range varying from 1 kW to a 5 MW to suit a wide range of applications. The major applications are grouped in three categories as given below.

(a) Applications Requiring Mechanical Power

(i) **Wind pumps.** For pumping water in remote areas, low power turbines are used to drive pumps. These turbines are also known as wind pumps. Simple and reliable reciprocating and centrifugal pumps are used.

(ii) **Heating.** By using paddle wheel in water run by wind turbine, there is a direct dissipation of mechanical power into heat. The available hot water is used for space heating or as such.

(iii) **Sea Transport.** Modern racing yachts, with a surface keel, harness lift forces and are much more efficient and can sail faster than wind. Large cargo ships employ efficient sails to boost power. In order to power propellers in ferries operating on short routes, wind turbines are installed on board.

(b) Applications as off Grid Electrical Power Source

(i) **Electrical Power for Domestic Applications.** Wind turbines having rotor diameter about 3 m may produce electrical power from 40 W to 1 kW which may be used for space heating, cooling and operating fan, light and small tools.

(ii) **Electrical Power for Navigation Signal, Remote Communication, Weather Stations and offshore Drilling Platform.** Wind turbine producing up to 50 kW of electrical power may be used in navigation signal, remote communication, weather stations and offshore oil drilling platforms.

(iii) **Electrical Power in Stand-alone Mode in the Range of 100–250 kW.** Wind turbines coupled with generator developing about 100 to 250 kW of electrical power may be used to supply electrical power to isolated populations, farm cooperatives commercial refrigeration, desalination and other small industries working as a stand-alone mode or supplying to a mini-grid for these applications.

(iv) Operating Pump to Lift water at the Top of Hill. In order to lift water, wind turbines (aero-generators) are installed on the top of a hill and electrical energy is transmitted to a pump located at a lower level.

(v) Applications as a Grid Connected Electrical Power. In order to supply power to a utility grid, large aero-generators in large arrays are installed. Such aero-generators known as **wind farms** are installed in open plains or offshore in shallow water to meet the demand.

18.5.3 Wind Turbine Aerodynamics

18.5.3.1 Energy Available in Wind

Power (P_0) available in wind having velocity C_0 of free wind of mass flow rate $\dot{m}$ in passing through an area A_1 is given by

$$P_0 = \frac{1}{2}\dot{m}\,C_0^2 \;;\quad \text{W}$$

where $\dot{m} = \rho\, A_1 C_0$ kg/s at inlet only as C_0 is available at area A_1

ρ = density of air, kg/m^3, A_1 = area swept by turbine rotor

Hence,

$$P_0 = \frac{1}{2}(\rho\, A_1)\, C_0^3 \;;\quad \text{W} \qquad ...(18.10)$$

or

$$\frac{P_0}{A_1} = \frac{1}{2}\rho\, C_0^3 \;;\quad \text{W/m}^2 \qquad ...(18.11)$$

From Eq. (18.11), *we conclude that power available in the wind is proportional to the cube of wind speed.* For a typical value of C_0 = 10 m/s and ρ at 15°C = 1.2 kg/m^3, the power available = 600 W/m^2.

In fact, $\dot{m} = \rho A_0 C_0 = \rho A_1 C_1 = \rho A_2 C_2$

Since $C_0 > C_1 > C_2$, so $A_2 > A_1 > A_0$ to maintain continuity equation.

Since A_1 is only known, so at stations 1, we have

$$\dot{m} = \rho A_1 C_1 \qquad ...(18.12)$$

18.5.3.2 Power Extraction from Wind Turbine

The purpose of wind turbine is to harness useful mechanical power which may be used either as mechanical or electrical power after conversion through alternator. The rotor of the turbine collects energy from the whole area swept by the rotor. Mr. Betz proposed a model for estimating wind power. This model assumes a smooth laminar flow with no perturbations. The rotor may be considered actuator disc across which there is a reduction of pressure on account of energy extraction by rotor. Figure 18.15 shows such a Betz model of expanding air stream tube through a turbine rotor disc.

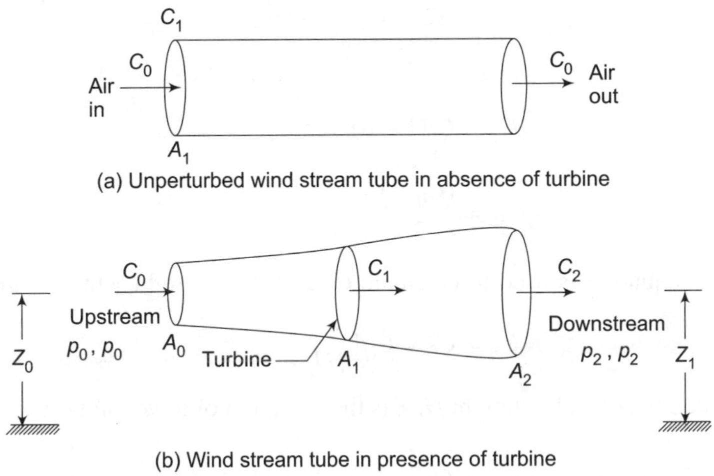

Figure 18.15 Flow through turbine blades as per Betz turbine

Let A_0, A_1 and A_2 = stream tube areas of coolant mass at upstream, turbine and down stream respectively.

C_0, C_1 and C_2 = wind speed at upstream, turbine and down streams respectively.

The force or thrust on rotor from Newton's 2nd law, $F = \dot{m}\,(C_0 - C_2)$ ...(18.13)

This force is applied by the air at uniform airflow speed at C_1, passing through actuator disc, hence the power extracted by turbine

$$P_T = F\,.\,C_1 = \dot{m}\,(C_0 - C_2)\,C_1 = P_0 - P_2 \qquad ...(18.14)$$

Further, the power extracted from wind by the turbine is also equal to the loss in K.E. per unit time. Hence,

$$P_W = P_T = \frac{1}{2}\,\dot{m}\,(C_0^2 - C_2^2) = P_0 - P_2 \qquad ...(18.15)$$

Since $P_T = P_w$

Hence, $$\dot{m}\,(C_0 - C_2)C_1 = \frac{1}{2}\,\dot{m}\,(C_0^2 - C_2^2) = \frac{1}{2}\,\dot{m}\,(C_0 - C_2)\,(C_0 + C_2)$$

or $$C_1 = \frac{C_0 + C_2}{2} \qquad ...(18.16)$$

In most hypothetical case, $C_2 \approx 0$ (which is not possible), so $C_1 = C_0/2$. Thus, as per linear momentum theory, *the air speed through the actuator disc cannot be less than half of the speed of air upstream.*

A term **interference or induction or perturbation factor (a)** *is introduced which is defined as the ratio of decrease in wind speed from free stream upto turbine to the free stream wind speed.*

$$a = \frac{(C_0 - C_1)}{C_0} \quad ...(18.17)$$

or $$C_1 = C_0(1 - a) \quad ...(18.18)$$

or $$a = \frac{(C_0 - C_2)}{2\ C_0} \quad ...(18.19)$$

An empirical equation is used to calculate C_0 at greater height which is given as

$$C_Z = C_0 = C_H\left(\frac{Z}{H}\right)^{\alpha}$$

where C_H is valocity at smaller height H, Z is the elevation of tower and α is power coefficient ≈ 0.14.

Hence, the power extracted by turbine by $\dot{m}$, C_1 and C_2 using Eq. (18.15), we get

$$P_T = \frac{1}{2}\ \dot{m}(C_0^2 - C_2^2) = \frac{1}{2}\ \rho A_1 C_1 (C_0 + C_2)\ (C_0 - C_2)$$

or $$P_T = \frac{1}{2}\ \rho A_1 C_0 (1 - a) \cdot 2aC_0 \cdot 2C_1$$

or $$P_T = \frac{1}{2}\ \rho A_1 (1 - a)\ 4a \cdot C_0 \cdot C_0(1 - a)$$

or $$P_T = 4a(1 - a)^2 \left[\frac{1}{2}\ \rho A_1 C_0^3\right] \quad ...(18.20)$$

or $$P_T = 4a(1 - a)^2 \cdot P_0 \quad ...(18.21)$$

Power Coefficient, C_P

Another term, Power Coefficient, C_P is defined as the ratio of power extracted by turbine to the power available in the wind at inlet. Hence,

$$C_P = \frac{P_T}{P_0} \quad ...(18.22)$$

or $$C_P = 4a(1 - a)^2 \quad ...(18.23)$$

There are two ways to find the condition of the maximum value of power coefficient (C_P). The one way is to differentiate C_P w.r.t. (a) and equal to zero, and the other is to plot graph between C_P and a.

First, differentiating C_P w.r.t. (a), we have

$$\frac{d(C_P)}{da} = 0 = \frac{d}{da}[4a\ (1 - a)^2]$$

or $$4(1 - a)^2 + 4a \times 2\ (1 - a)\ (-\ 1) = 0$$

or $$4(1 + a^2 - 2a) - 8a + 8a^2 = 0$$

or $$4 + 12a^2 - 16a = 0$$

or $$1 + 3a^2 - 4a = 0$$

or $$3a^2 - 3a - a + 1 = 0$$

or $$(3a - 1)(a - 1) = 0$$

or $$a = 1, \text{ and } a = \frac{1}{3}$$

The value of $a = 1$ is not possible, so the optimum condition of (a) for the maximum C_P is $\frac{1}{3}$.

Putting the value of (a) in Eq. (18.23), we have,

$$C_P = C_{P,\,\max} = 4 \times \frac{1}{3}\left(1 - \frac{1}{3}\right)^2 = \frac{16}{27} = \mathbf{0.593} \qquad \text{...(18.23a)}$$

The criteria for the maximum power is known as **Betz Criteria** which $\boldsymbol{a = \frac{1}{3}}$.

Figure 18.16 shows the variation of power coefficient (C_p) with interference factor (a). The ***following conclusions*** *are drawn from this curve.*

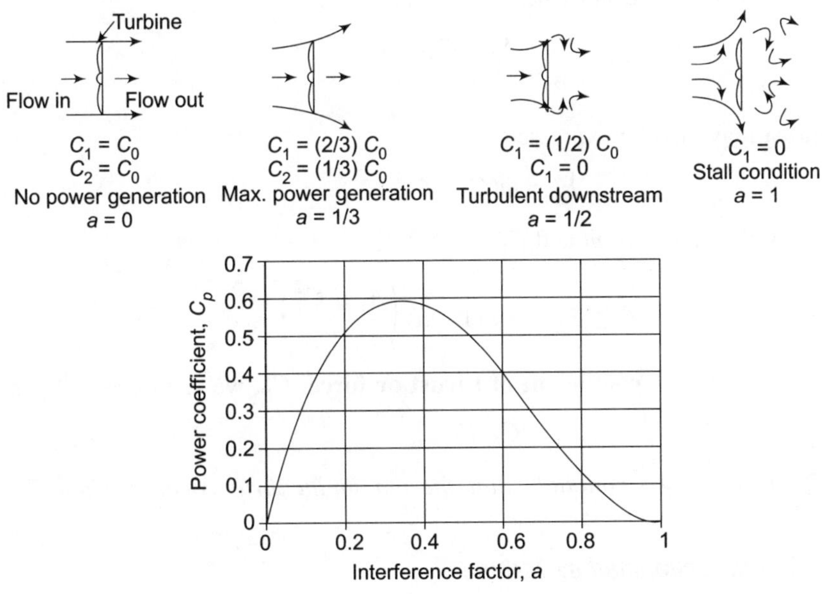

Figure 18.16 Variation of C_p with a

(i) *When no load is applied blades acts as free wheel.* Thus, $C_1 = C_0$, $C_2 = C_0$ and no power generation which gives $a = 0$.

(ii) *As load is applied, power increases, so C_p increases as C_1 decreases.* The maximum value of C_p, i.e., $C_{p\max}$ = **16/27 (i.e., 0.593) occurs at $\boldsymbol{a}$ = 1/3.**

Thus, $$C_1 = \frac{2}{3} C_0$$

and $$C_2 = \frac{1}{3} C_0$$

The criterion for the maximum power is known as **Betz criterion.**

(iii) When $C_2 = 0$, $a = 0.5$. Here, the simple model breaks down as no wind is predicted to be leaving downstream. In actual fact, this leads to equivalent to onset of **turbulence downstream**.

(iv) When $C_1 = 0$, a becomes unity and no power is extracted. This condition is called **stall states of bladings**.

18.5.3.3 Axial Thrust on Turbine

On horizontal machine turbine, a thrust acts along the turbine axis and therefore is known as axial thrust, F_A. Axial thrust is given by

$$F_A = \Delta p . A_1 = (p_0 - p_2) A_1 = (C_0^2 - C_2^2) \frac{\rho}{2} A_1$$

The maximum thrust occurs when $C_2 \approx 0$, i.e.,

$$F_{A,\,max} = \frac{A_1 \rho C_0^2}{2} \quad \text{...(18.24)}$$

Axial thrust may also be expressed as

$$F_A = \dot{m}(C_0 - C_2)$$

Substituting the values of $\dot{m}$ and C_2, we get

$$F_A = 4a(1 - a)\left(\frac{A_1 \rho C_0^2}{2}\right) \quad \text{...(18.25)}$$

By introducing a term **coefficient of thrust or force, C_F,** we can express F_A as

$$F_A = C_F . F_{A,\,max} \quad \text{...(18.26)}$$

From Eq. (18.26), *we conclude that the maximum thrust occurs when $C_2 = 0$, i.e., $C_F = 1$.*

18.5.3.4 Torque Developed by Turbine

In general, $T = F \times R$.

The maximum possible torque (T_{max}) on an ideal turbine rotor would occur if the maximum circumferential forces act at the tip of blade with radius R.

So, we have

$$T_{max} = F_{max} \times R = \frac{P_0}{C_0} \times R$$

Now, a new term **tip-speed ratio**, λ is defined as follows.

$$\lambda = \frac{\text{Speed of the tip of the rotor blade}}{\text{Speed of incoming air}}$$

or $$\lambda = \frac{R.\omega}{C_0} \qquad ...(18.27)$$

where ω is angular velocity of rotor.

For a working wind turbine producing a shaft torque, T is given by

$$T = C_T . T_{max} \qquad ...(18.28)$$

where C_T is called the **torque coefficient.**

Now, the power developed by the wind turbine in terms of tarque, T, we get

$$T.\omega = P_T$$

or $$C_T . T_{max} . \omega = C_P . P_0$$

Substituting the value of T_{max} in the above equation, we have

$$C_T . \frac{P_0}{C_0} \times R.\omega = C_P . P_0$$

or $$C_T . P_0 \, \lambda = C_P . P_0$$

or $$C_T = \frac{C_P}{\lambda} \qquad ...(18.29)$$

It is to be noted that both C_T and C_P are functions of tip-speed ratio, λ. As per Betz criterion, the maximum value of C_P can be only 0.593, so

or $$C_{T,\,max} = \frac{C_{P\,max}}{\lambda} \qquad ...(18.30)$$

Equation (18.30) suggests that the **wind turbines with higher tip-speed have low value of $C_{T,\,max}$ (i.e., low starting torque).**

18.5.3.5 Dynamic Matching for the Maximum Power Extraction

Though, as per Betz criterion, wind turbine can extract 59.3% of the available power of wind but this does not tell about dynamic rotational state at which this occurs.

There are two parameters which affect the performance:

(a) Tip-speed ratio (b) Blade setting or pitch angle, y.

(a) Tip-speed ratio. The wind turbine running at constant speed, the power extraction by the turbine will decrease under the following conditions:

(i) if the blades are so close together or rotating so rapidly that the blades run in turbulence created by blades; and

(ii) if the blades are so far apart or rotating slowly which does not allow to interact with blades.

From the above discussion, *it is concluded that for a particular wind speed, there exists and optimum turbine speed to produce the maximum output. This suggests that for achieving the maximum efficiency, it is important to match the rotational frequency of turbine to the corresponding wind speed.*

The tip-speed ratio at the maximum power extraction is given by

$$\lambda_{opt} = \frac{2\pi}{n}\left(\frac{R}{d}\right) \quad ...(18.31)$$

where n = number of blades

d = length of wind strongly perturbed the rotating blades

R = radius of rotor disk

In general, $$d = \frac{R}{2}$$

Hence, $$\lambda_{opt} = \frac{4\pi}{n} \quad ...(18.32)$$

Equation (18.32) suggests that for $n = 2$, $C_{p\max}$ occurs at $\lambda_{opt} = 2\pi$ whereas for $n = 4$, $C_{p\max}$ occurs λ at $\lambda_{opt} = \pi$.

Figure 18.17 shows the variation of power coefficient (C_p) with tip-speed ratio (λ) for various types of wind turbines.

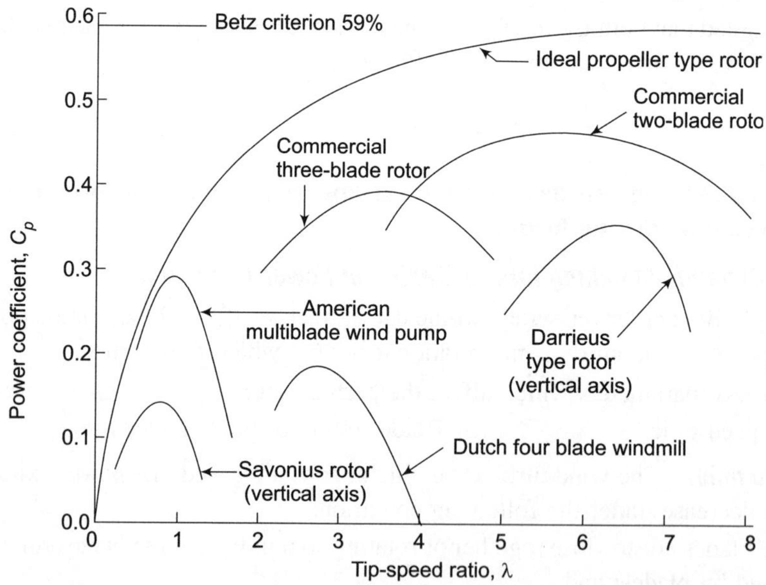

Figure 18.17 Variation of C_p with λ for various wind turbines

18.5.4 Wind Turbine Types and Their Construction

Basically, there are two types of wind turbines

(i) Horizontal axis wind turbine (HAWT)

(ii) Vertical axis wind turbine (VAWT)

In the case of HAWT, the axis of rotation is parallel to the air stream (i.e., horizontal) while in the case of VAWT, the axis of rotation is perpendicular to the air stream (i.e., vertical). Out of these two, the most common and used worldwide is HAWT. Commercial HAWTs are found in various sizes (small, medium, large and very large). The rated power ranges from **10 kW to 4 MW**. The rotor diameter ranges from **6.4 m to 130 m**. The rotor rpm ranges from **200 rpm (small) to 11 rpm** (large sizes). By 2024, Siemens will install a 14 MW horizontal axis wind turbine. The turbine uses 222 metre diameter rotor. Similarly, GE has also developed 14 MW wind turbine and will install in France.

18.5.4.1 Horizontal Axis Wind Turbine (HAWT)

Worldwide HAWTs have been working successfully in different sizes as per location and requirements.

Figure 18.18 shows the constructional details of the most common three-blades horizontal axis wind turbine. The control mechanism is shown in Fig 18.19.

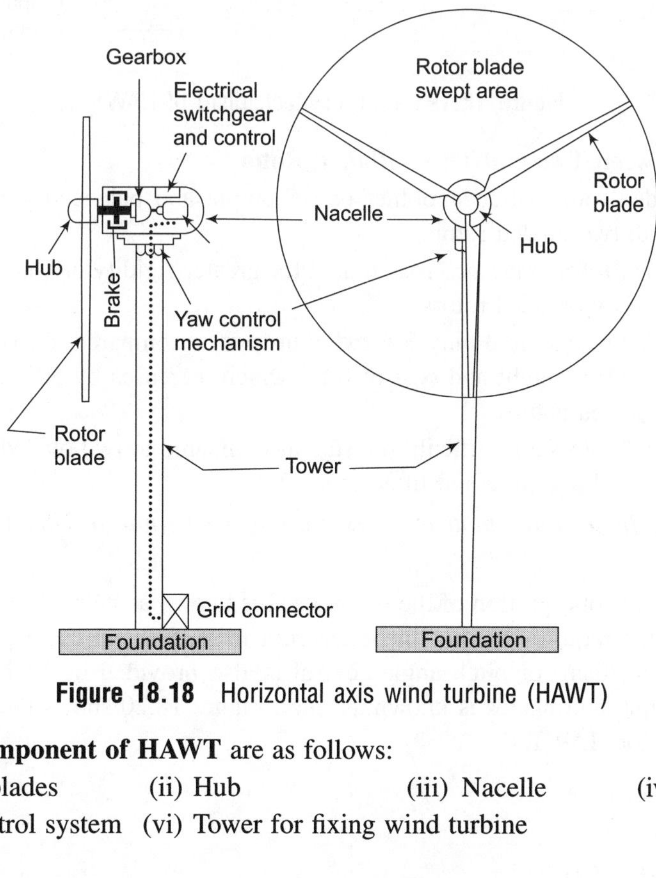

Figure 18.18 Horizontal axis wind turbine (HAWT)

The **main component of HAWT** are as follows:

(i) Turbine blades (ii) Hub (iii) Nacelle (iv) Yaw control

(v) Pitch control system (vi) Tower for fixing wind turbine

(i) Turbine Blades. In order to keep the weight of turbine blades light, these are made of **high-density wood** or **glass fibre and epoxy composites**. The cross section of the blades are of **air foil type and slightly twisted from the outer tip to the root to reduce the tendency to stall**. The blades are subjected to various forces such as centrifugal force, fatigue due to continuous vibrations, wind turbulence, gust, gravitational forces, direction change in the wind, etc. While designing the blades, all forces are to be taken care of.

Presently, wind turbines under operation are either two or three bladed. Being similar to propeller of old aeroplanes, they are also known as propellers but their speeds are much less than that of aeroplane propellers.

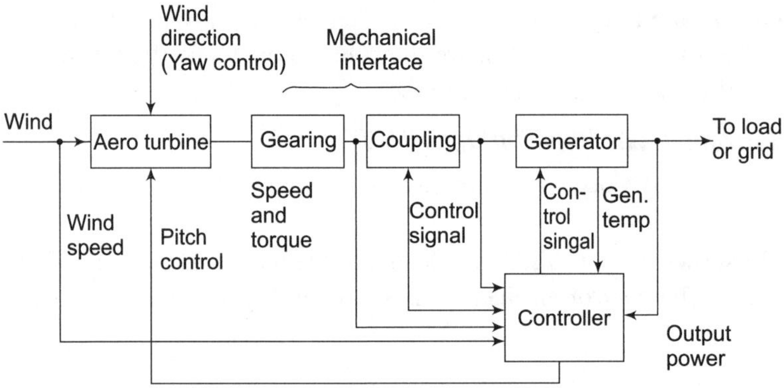

Figure 18.19 Control mechanism of HAWT

Comparison between Two and Three Bladed Rotors

(a) Three bladed rotors offer smoother power output and balanced gyroscopic force as compared to two bladed rotors.

(b) Three bladed rotors may be cross-linked for greater rigidity and no need of teeter as compared to two bladed rotors.

(c) Three bladed rotors yield only **5% extra output** as compared to two bladed rotors at the cost of extra weight and cost of rotor which increases by 50% and so the return is diminishing in nature.

(d) Three bladed rotors are difficult to install as compared to two bladed rotors which are assembled on the ground and lifted to shaft.

In Europe and India, three bladed rotors are preferred while in USA two bladed rotors are preferred.

(ii) Hub. The central solid portion of the rotors level is known as hub. All blades are attached to the hub and this requires blade attachment root in blades and corresponding recess in hub. Further, mechanism for pitch angle control is also provided in the hub. So basically, the assembly of hub and blades is known as rotor. Figure 18.20 shows the various types of rotors developed for HAWT.

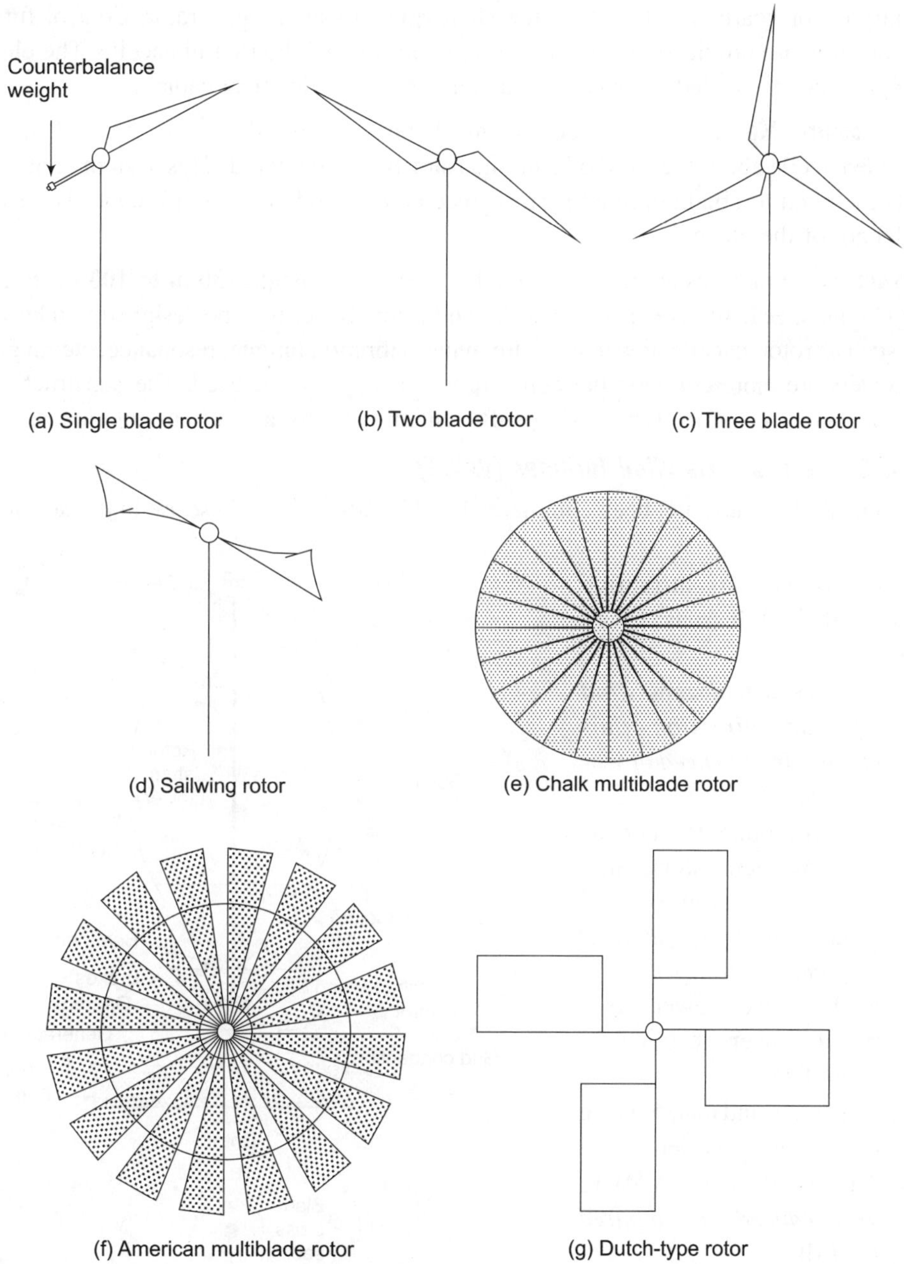

Figure 18.20 Various types of HAWT rotors

(iii) Nacelle. Nacelle is a housing system. The rotor is attached to nacelle, mounted at the top of tower. Nacelle houses rotor brakes, gearbox, generator and electrical switchgear and control. The purpose of brakes is to stop the rotors when power generation is not required.

The purpose of gearbox is to set up the shaft rpm to suit the generator. Control functions and protection are provided by switchgear and control back housed in nacelle. The electrical power so generated is fed to ground terminals through a electrical cable.

(iv) Yaw control Mechanism. Yaw control mechanism is provided at the base of nacelle to adjust the nacelle about the axis to bring the rotor facing the wind. This system continuously adjusts the pitch to obtain optimal performance and in modern system it controls only outer 20% length of the blade.

(vi) Tower. Wind turbines are to be mounted on **sufficient height (30 m to 100 m)** to receive enough wind speed. Tower supports nacelle and rotor. Tower is to be designed to take various loads such as rotor, nacelle, tower structure, wind, vibration fatigue, resonance, etc. In general, steel towers are more common but concrete tower may also be used. The construction may be either tubular or lattice type having sufficient strength to take all possible loads.

18.5.4.2 Vertical Axis Wind Turbines (VAWT)

At present VAWTs have not been commercialized in full scale as these are in the development stages.

There are various attractive features of VAWTs, as given below:

(a) VAWTs can accept *wind from any direction, thus eliminating the need of yaw control.*

(b) The components such as gearbox, generators, etc., may be located on the ground and hence there is *no need of heavy nacelle* at the top of tower. As a result, tower design gets simplified.

(c) Inspection and maintenance of system are easier.

(d) The overall cost of VAWTs *gets reduced as compared to HAWT.*

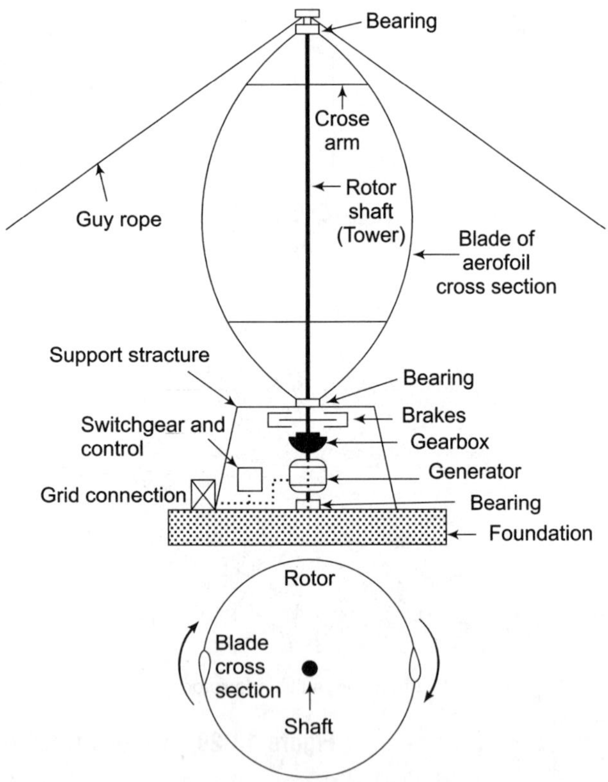

Figure 18.21 Vertical axis wind turbine (VAWT) with Darrieus type rotor

Figure 18.21 shows the construction details of a vertical axis wind turbine (VAWT). The main components include the following:

(i) Rotor shaft or tower (ii) blades (iii) support structure

(i) Rotor shaft (or Tower) *Rotor shaft or tower is a vertical hollow shaft which rotates freely about vertical axis between top and bottom bearings. It is installed above a small support structure.* As there is no load at the top, a very strong support structure is not required resulting in a simple design. The upper part of rotor shaft is supported by guy ropes for failure against bending. The height of rotor shaft or tower may vary from **30 m to 100 m** depending upon the size of turbine.

(ii) Blades. VAWT employs **two or three thin curved blades shaped like an eggbeats in profile.** These blades are curved in a form which minimizes the bending stress caused by centrifugal forces and that is why it is called **Troposkien profile**. So far the cross section of blades, it has airfoil shape with constant chord length. It is worthy to note that the pitch of blades cannot be changed and the diameter of the rotor is slightly less than rotor shaft or tower. *As a case study,* ***Canadian VAWT****, first very large Darrieus type turbine having a rotor height of* ***94 m with diameter as 65 m and a chord of 2.4 m generates 3.8 MW****.* The various types of rotors are shown in Fig. 18.22.

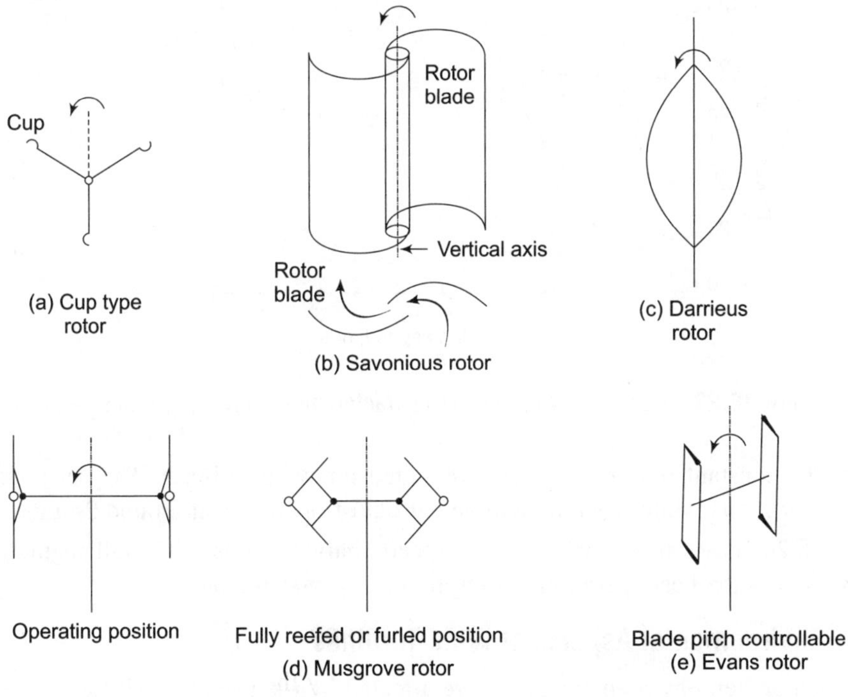

Figure 18.22 Various types of VAWT rotors

(iii) Support Structure. As mentioned earlier, the support structure needed for VAWT is simple and small. It is provided at the ground to support the weight of the rotor. In this structure, gearbox, generator, brakes, electrical switchgear and controls are housed at foundation which makes it simpler than HAWT system.

18.5.5 Wind Turbine Characteristics

The characteristics of wind turbine may be shown by exhibiting the power versus wind speed variation. Figure 18.23 shows such characteristics depicting four separate regions are given below.

(i) Low speed region (zero to cut-in speed)
(ii) Maximum power coefficient region
(iii) Constant power or constant turbine speed region.
(iv) Furling speed or cut out speed and above region.

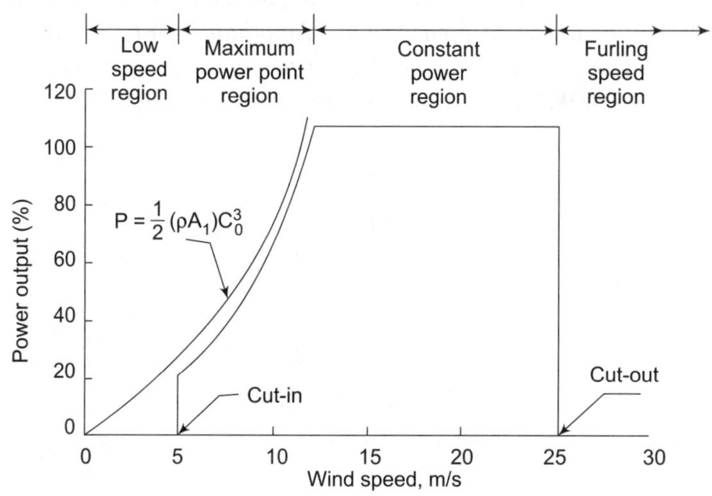

Figure 18.23 Power vs wind speed characteristics showing various regions

Obviously, constant power region is best suited for wind turbines. The cut-in starts at a wind speed of 5 m/s while machine will be cut out of operation at around 25 m/s.

Figure 18.24 shows power versus wind speed characteristics with stall regulation. This curve also shows the best operation during the peak power region.

18.5.6 Environmental Aspects of Wind Turbines

As discussed earlier, environment gets more affected by the use of fossil fuels based power plants and less in the case of renewable energy forms. Obviously, wind turbines are no exception in slightly affecting the environment. The following factors affect the environment created by the use of wind power plants.

(i) **Indirect energy use and emissions.** During the construction of wind turbines, some energy will be used which will pollute environment.

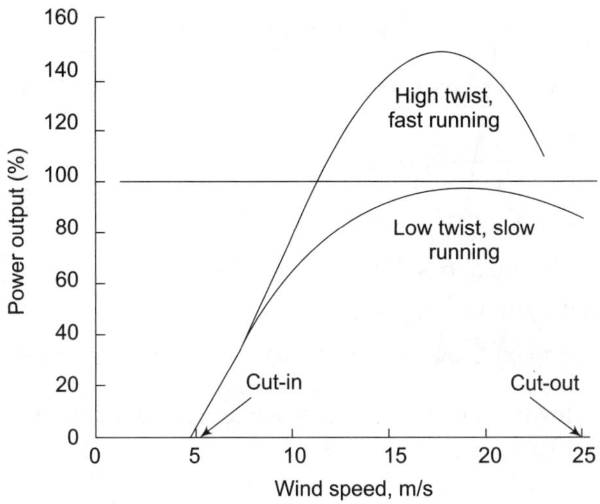

Figure 18.24 Power vs wind speed characteristics showing stall regulation

(ii) Bird life. There is always a danger of the lives of birds as they may get hit by tower or rotating rotor. Wind turbines also affect the resting and breeding pattern of birds.

(iii) Noise generation. An appreciable noise is created by the rotation of rotor and so it cannot be located near the inhabited areas. The reason behind acoustic noise by wind turbines are as follows:

(a) Mechanical noise due to movement of mechanical parts in the nacelle.

(b) Aerodynamic noise which is a function of wind speed.

A sound level of 35 dBA for residential area may be possible only if wind turbines are located at least 500 m away.

(iv) Visual impact. Visual impact of wind turbines is qualitative in nature.

(v) Telecommunication interference. Wind turbines offer an obstacle for incident electromagnetic waves.

(vi) Safety. When blades fail they may get thrown out which may hit population.

(vii) Effects on ecosystem. If it is used in large scale, wind speed may get reduced and cause stress to ecosystem.

Example 18.2 In a wind turbine, the speed of undisturbed wind is 15 m/s. The speed of the wind at the turbine rotor is 60% of inlet speed. The rotor diameter is 10 m and the density of air is 1.29 m^3/kg. Calculate (a) the power in the wind at the inlet, (b) the interference factor, (c) the coefficient of performance, (d) the electrical power developed by turbine if generator and mechanical efficiencies are 0.96 and 0.94 respectively, and (e) the power at the exit of wind turbine.

Solution Refer to Fig. 18.15. Let suffixes 0, 1 and 2 represent inlet, rotor and exit respectively.

(a) $P_0 = \frac{1}{2}\rho A_1 C_0^3 = \frac{1}{2} \times 1.29 \times \frac{\pi}{4} \times 10^2 \times 15^3$

$= 170{,}971.3 \text{ W} = 170.97 \text{ kW}$ ***Ans.***

(b) $a = \dfrac{C_0 - C_1}{C_0} = \dfrac{C_0(1-0.6)}{C_0} = 0.4$ ***Ans.***

(c) $C_P = 4a(1-a)^2 = 4 \times 0.4\,(1-0.4)^2 = 0.576$ ***Ans.***

(d) $P_T = P_1 = C_P \times P_0 = 0.576 \times 170.97 = 98.47 \text{ kW}$

$P_e = 98.47 \times 0.96 \times 0.94 = 88.86 \text{ kW}$ ***Ans.***

(e) $P_2 = P_0 - P_1 = 170.97 - 98.47 = 72.5 \text{ kW}$ ***Ans.***

Example 18.3 The following data relate to a propeller type wind turbine. Speed of free wind at a height of 10 m = 12 m/s.

Air density = 1.23 kg/m^3

Air angle coefficient, $\alpha = 0.14$

Height of tower = 90 m

Diameter of rotor = 70 m

The wind velocity at the turbine is 70% of inlet.

Generator efficiency = 0.95

Mechanical efficiency = 0.94

Calculate the following:

(a) The free wind speed at the inlet
(b) The total power available in the wind
(c) The power extracted by turbine and coefficient of performance
(d) The electrical power generated
(e) The axial thrust on the rotor

Solution Refer to Fig. 18.15. Data given

$C_H = 12$ m/s, $Z = 90$ m, $H = 10$ m, $D = 70$ m

$C_1 = 0.7\, C_0$, $\eta_g = 0.95$, $\eta_m = 0.94$, $\propto = 0.14$

From empirical equation

(a) $C_Z = C_0 = C_H\left(\dfrac{Z}{H}\right)^{\propto} = 12\left(\dfrac{90}{10}\right)^{0.14} = 16.32 \text{ m/s}$ ***Ans.***

(b) $P_0 = \frac{1}{2}\rho A_1 C_0^3 = \frac{1}{2} \times 1.23 \times \frac{\pi}{4}(70)^2 \times (16.32)^3 = 10287776.9 \text{ W} = 10.287 \text{ MW}$ ***Ans.***

(c) $a = (C_0 - C_1)/C_0 = (1-0.7) = 0.3$

$C_P = 4a(1-a)^2 = 4 \times 0.3\,(0.7)^2 = 0.588$ ***Ans.***

$P_T = P_1 = C_P . P_0 = 0.588 \times 10.287 = 6.048 \text{ MW}$ ***Ans.***

(d) $P_e = P_T \times \eta_g \times \eta_m = 6.048 \times 0.95 \times 0.94 = 5.4 \text{ MW}$ ***Ans.***

(e) $F_A = 4a\,(1 - a)\,(\rho A_1\, C_0^2/2)$

$$= 4 \times 0.588\,(1 - 0.588)\left(1.23 \times \frac{\pi}{4}(70)^2 \times \frac{(16.32)^2}{2}\right) = 6.1 \times 10^5 \text{ N} \qquad \textit{Ans.}$$

Example 18.4 In a wind turbine, the speed of the undisturbed wind is 14 m/s. The speed of the wind at turbine rotor is 70% of inlet undisturbed speed. The density of air is 1.23 kg/m^3. Calculate the following for an electrical output of 2 MW if generator and mechanical efficiencies are 0.95 each.

(a) The diameter of the rotor
(b) The axial thrust on the rotor
(c) The coefficient of performance

Solution Refer to Fig. 18.15

(a) The power developed by the turbine

$$P_T = P_1 = C_P \times P_0 = C_P\left(\frac{1}{2}\rho\, A_1\, C_0^3\right)$$

or $$\frac{2 \times 10^6}{0.95 \times 0.95} = 0.588\left(\frac{1}{2} \times 1.25 \times \frac{\pi}{4} D^2 \times (14)^3\right)$$

or $$D^2 = \frac{2 \times 10^6 \times 2 \times 4}{0.95 \times 0.95 \times 0.588 \times 1.25 \times \pi \times (14)^3}$$

or $$D^2 = \frac{16 \times 10^6}{5718.32} = 2798.02$$

or $$D = 52.89 \text{ m} \qquad \textit{Ans.}$$

(b) The axial thrust is given by

$$F_A = 4a(1 - a)\,(\rho A_1\, C_0^2/2)$$

or $$F_A = 4 \times 0.3 \times 0.7 \times 1.23 \times \frac{\pi}{4}(52.89)^2 \times (14)^2/2$$

$$= 2.22 \times 10^5 \text{ N} \qquad \textit{Ans.}$$

(c) The interference coefficient is

$$a = (C_0 - C_1)/C_0 = (1 - 0.7) = 0.3$$

The coefficient of performance

$$C_P = 4a\,(1 - a)^2 = 4 \times 0.3 \times (0.7)^2 = 0.588 \qquad \textit{Ans.}$$

18.6 BIOMASS ENERGY POWER PLANT

Since the existence of human race, men have used biomass, (or plant and animal materials) for various energy forms including food, shelter, clothing and fuel. *Even today, two-thirds of the world's population uses biomass for their heating and cooking.* Conversion of energy from biomass has enjoyed renewed public interest as the price of fossil fuels is increasing. *The main advantage of biomass is that it is renewable, can be produced in most regions of the world, has negligible amount of sulphur and has no major disposal problems.*

The methods of obtaining energy from biomass may be extremely simple or very complex. A family may chop some cords of wood and use it for room heating. A commercial enterprise may grow biomass on a large energy farm, harvest the material, convert it into methanol, mix it with gasoline and sell it as a transportation fuel. Whatever be the system, there are still three components of obtaining energy from biomass: (a) *growing the biomass,* (b) *transporting it to the conversion site and* (c) *conversion of it to a form of energy which man can use.*

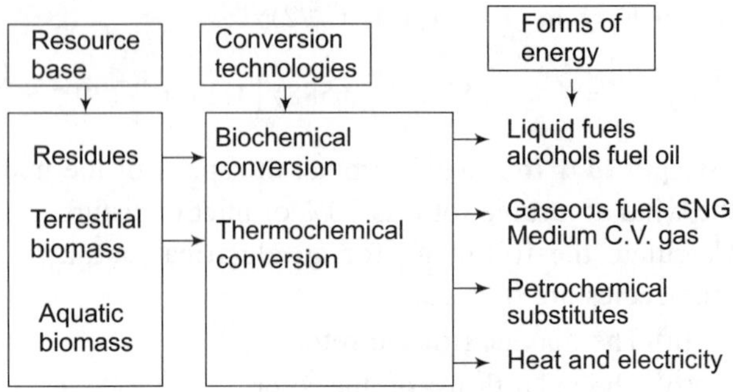

Figure 18.25 Conversion of energy from biomass in general

It is mainly the photosynthesis process occurring with the green plant in the presence of sunlight that is responsible for the growth of plant or biomass or total living material. Following are the types of biomass:

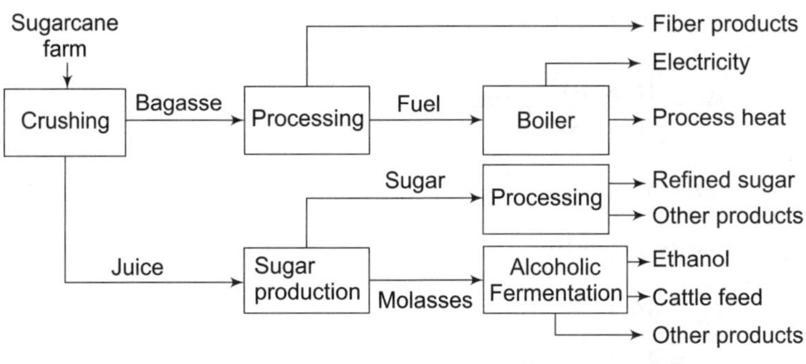

Figure 18.26 Conversion of energy from sugarcane

(i) Terrestrial energy crops (ii) aquatic energy crops, (iii) agricultural residues, (iv) forest residues, (v) animal manures, and (vi) standing vegetation.

Figure 18.25 shows the conversions of energy from biomass in general and Fig. 18.26 shows conversion of energy from sugarcane in particular. It can be converted to useful fuels by either biochemical or thermo-chemical processes; In the former, biological organisms are used to process the biomass, followed by initial chemical breakdown in some cases. In the thermo-chemical process, high temperature, catalysts and sometimes high pressure are used to react the biomass chemically. In addition, chemical process can be used to extract petrochemical substitutes from plant materials such as natural rubber or resin.

18.7 OCEAN THERMAL ENERGY CONVERSION (OTEC) POWER PLANT

An ocean thermal energy conversion (OTEC) power plant utilizes the temperature difference between warm surface water and cold deep water of the ocean, to drive a turbine. This concept was first used by do Arsonval in 1881. *It is a pollution free energy conversion system.*

Ocean covers about 71% of the earth's surface and receives the majority of the solar energy incident on the earth. In semitropical and tropical oceans, the available temperature difference (about 20°C) is sufficient for OTEC operation and is available for 24 hours a day. There is a vast potential for OTEC system which can be appreciated by the fact that a quantity of electric power equal to the 30 per cent of world's total energy requirement in 2010 could be obtained from the ocean in 10° latitude [1].

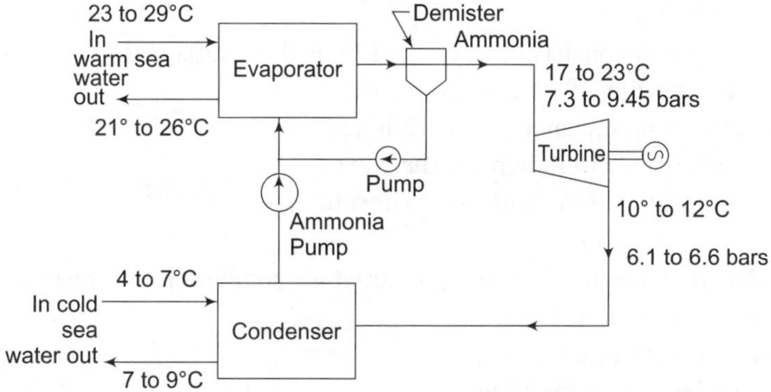

Figure 18.27 OTEC power plant based on closed Rankine cycle

Figure 18.27 shows a closed Rankine cycle OTEC system in which the working fluid ammonia is vaporized by heat exchange with warm water, drives a turbine generator and is condensed by heat exchange with cold water drawn from 700 to 1200 m depth. Still OTEC system is in the developing stage. Its conversion efficiency is very low (about 4%). However, OTEC project of 10 MW has been planned 40 km offshore of Tamil Nadu with the help of USA.

18.8 WAVE POWER PLANT*

Developments in harnessing the power of ocean waves have started recently. *The surface waves on the sea are in fact sets of fluid particles in circular motion with no set forward motion except when they break on a shore.* These waves are set up by winds acting on the surface of the sea and the effect can travel some 800 km from a storm centre before dying out. *The energy available on more stormy seas is about 50 to 75 kW per metre of wave front which is a very high figure but sometimes wave power is zero* [1]. Various energy converting devices have been tested but none have been found to be highly efficient over a wide range of wave heights. It is difficult to design a rugged device with a high efficiency at low wave heights,

---

* For Tidal Wave Power Plant, see Chapter 1, p 35.

where most power is required. *The ideal form of device is one which will have large waves ahead of it and calm water behind.* There are **at least five types of devices in developing stages which** may convert wave power into electrical power. These are:

(a) Salters Duck (b) Cockcroft Rafts (c) Russel Rectifier
(d) Pressure Change Device (e) Littoral Devices Employing Breaking Waves.

(a) ***Salters Duck.*** Figure 18.28 shows the principle of operation of Salters Duck. It consists of a pivoted raft forming a string of moored 'ducks' facing the oncoming waves. Strings 1 km long have been designed for output of 45 MW.

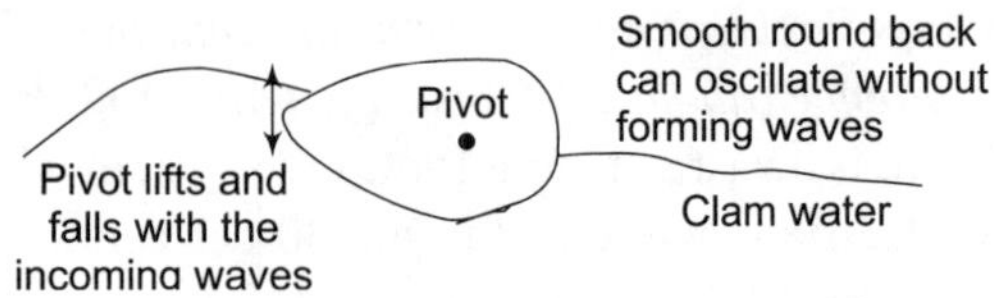

Figure 18.28 Salters duck

(b) ***Cockcroft Rafts.*** It consists of a series of hinged flat rafts (Fig. 28.29) which will tend to follow the profile of an advancing wave front. By extracting the energy at each hinge, the rafts can be made to flatten out the waves passing beneath them and to generate power.

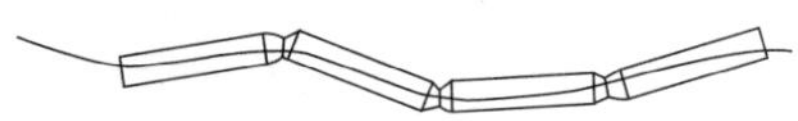

Figure 18.29 Cockcroft rafts

(c) ***Russel Rectifier.*** Device is a floating structure containing two reservoirs, one higher than the other. Non-return valves or vanes are fitted which allow the higher one to be filled by the wave peaks and the lower to empty into the troughs (Fig. 18.30). A low head water turbine is installed between two generating electric power for transmission to the shore.

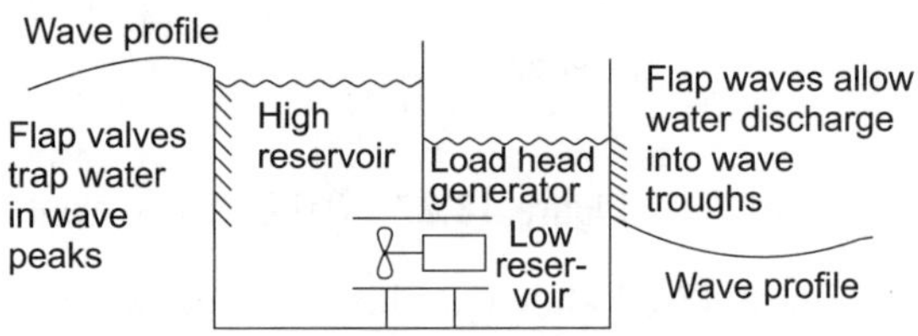

Figure 18.30 Russel rectifier

(d) ***Pressure Change Devices.*** Ocean waves cause pressure changes beneath the surface to water. These changes can be harnessed, either by means of hydraulic fluids in flexible containers, or by the compression of a gas such as air. Many devices are in the developing stage.

(e) ***Littoral Devices Employing Breaking Waves.*** In this arrangement, breaking waves are induced to throw water over a break water into a higher level reservoir. Where tidal movements are small and the water wind is usually in one prevailing direction, this system has many attractions.

(f) ***Wave Energy Module (WEM).*** Harold Hopfe, an American engineer has developed a practical way of harnessing the energy of ocean waves called 'Wave Energy Module' (WEM). This consists of two metal disks one floating on the surface of the ocean and the other anchored parallel to it beneath the waves. The two disks are connected by a piston filled with hydraulic fluid. The waves set off a relative motion of the disks which causes the piston to drive the fluid through a turbo-generator to produce electricity. Hopfe claims that a prototype of this device has generated about 8 watts of power from an 18 cm wave and a 36 m version of it could generate enough power for 40 homes for seaside community or a desalination plant in the US, the production cost would be about ₹1.5/kWh.

18.9 MAGNETO-HYDRO-DYNAMIC (MHD) POWER PLANT

Figure 18.31 shows a schematic diagram of MHD power plant. A jet of ionized gas flows through a magnetic field, so a high voltage direct current (above 2000V) is obtained from the electrodes placed in the gas stream. The high temperature (2760°C) is required to ionized the gas, imposes severe problems for materials. Its efficiency is also low because most of the power developed is consumed in magnetizing the magnet. The output is also low.

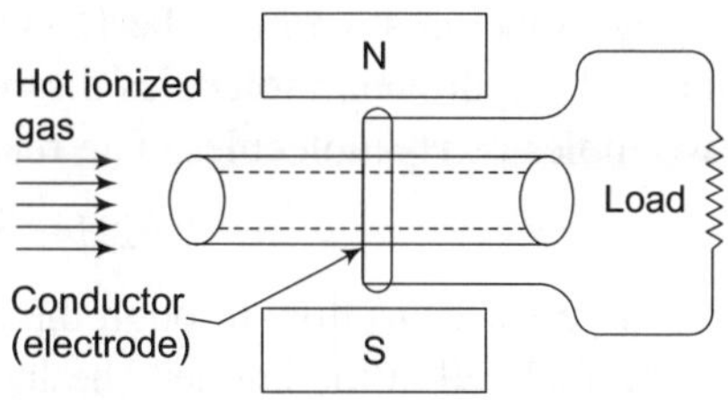

Figure 18.31 Schematic diagram of MHD power plant

18.10 ELECTRIC STORAGE BATTERIES POWER PLANT

Generally, two types of batteries are employed. They are the non-rechargeable zinc-silver oxide battery and the rechargeable nickel oxide-cadimum, silver oxide cadmium and lithium-ion batteries. They are in the developing stage and work on the principle of batteries. Now, electric cars, buses, motor cycles have come up using lithium-ion batteries world over. It is mostly pollution free.

18.11 FUEL CELLS POWER PLANT

A fuel cell is an electrochemical device which converts the chemical energy of conventional fuel, such as hydrogen, hydrocarbon, etc., directly into electrical energy. Broadly speaking, it consists of a *non-consumable anode, cathode and an electrolyte* (invariant), fuel and oxidant supply system and suitable control. In rural area, H_2 may be obtained from biomass whereas in urban areas it may be obtained from sewage waste, natural gas, and electrolysis. O_2 may be obtained from air liquifaction. Figure 18.32 shows a schematic diagram of H_2–O_2 fuel cell. The hydrogen gas serves as a fuel while O_2 serves as an oxidizer. Each gas is introduced at high pressure into separate chambers. H_2 is supplied to the anode side and oxygen to the cathode side. *Both electrodes are made*

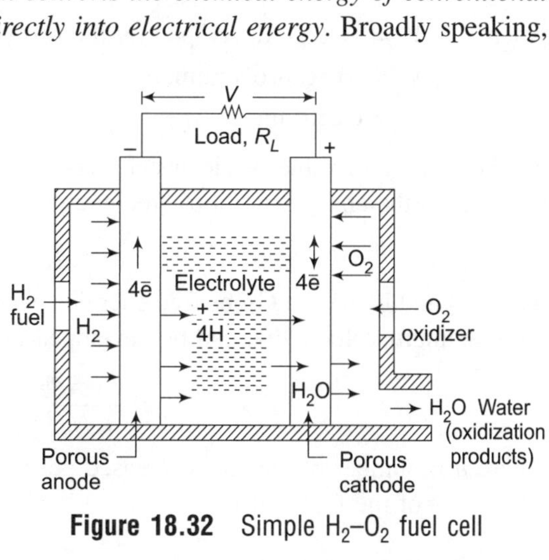

Figure 18.32 Simple H_2–O_2 fuel cell

of porous materials to permit passage of gas. The electrolyte is placed between the electrodes. The electrolyte may be solid or liquid. Porus ceramic (zirconia oxide) polymers and alkaline solution (potassium hydroxide) have been successfully used as electrolytes. When a solid electrolyte is used, its surface is first impregnated with a catalyst to facilitate the ion migration process. Impregnation allows the two sides of the electrolyte to be used as a electrodes. In the case of liquid electrolyte, separate electrodes are necessary. When the electrodes are connected to a resistive load, current flows.

The principle on which the H_2–O_2 fuel cell works is as follows: At the anode electrolyte interface, hydrogen, molecules dissociate, forming hydrogen ions and four free electrons per two moles of H_2 molecules. The reaction at the anode surface may be written as

$$2H_2 \rightarrow 4H^+ + 4e^- \qquad ...(18.33)$$

These electrons flow through the external circuit where, they do electrical work and return to the fuel cell at the anode. The hydrogen ions ($4H^+$) migrate through the electrolyte to its cathode interface, where they combine with returning electrons and with oxygen to form water. The reaction at the cathode is written as

$$4e^- + 4H^+ + O_2 \rightarrow 2H_2O \qquad ...(18.34)$$

The overall reaction for the fuel cell is just the sum of above two equations.

$$2H_2 + O_2 \rightarrow 2H_2O$$

In account of the chemical reaction, a quantity of energy representing the enthalpy of combustion of the fuel $\bar{h}_f^\circ$ is released. A part of this energy is available for conversion to electrical energy, the maximum amount is equal to the change in Gibbs function ($w_e = \bar{g}_1 - \bar{g}_2$). *The remaining part of the energy remains as heat. The electrical energy output per mole of* O_2 or per 2 mole of H_2 is expressed as

$$w_e = 4\ N_0.eV_t \qquad ...(18.35)$$

where V_r = voltage developed by the reversible cell

N_0 = Avogadro's number

e = electronic charge

The maximum value of electrical work (w_e) under reversible steady-state conditions will be equal to the change in Gibbs free energy.

$$w_e = \bar{g}_1 - \bar{g}_2 \qquad ...(18.36)$$

where $\bar{g}_1$ and $\bar{g}_2$ are the Gibbs free energies of the reactants and products respectively, Hence, the maximum voltage that can be developed in the reversible fuel cell is

$$V_r = \frac{\bar{g}_1 - \bar{g}_2}{n.N_0.e} \qquad ...(18.37)$$

where n = number of electrons released per molecule of O_2. The maximum ideal efficiency of the fuel cell is

$$\eta_{fc} = \frac{w_e}{(-\bar{h}_f^\circ)} = \frac{\bar{g}_1 - \bar{g}_2}{(-h_f^\circ)} = \frac{nN_0.\mathrm{eV}_t}{(-\bar{h}_f^\circ)} \quad ...(18.38)$$

It is to be noted that in the H_2–O_2 fuel cell, the enthalpy $\bar{h}_f^\circ$ (i.e., $\bar{h}_f^\circ$) is –286 kJ/gm mole. **This gives the efficiency of H_2–O_2 fuel cell equal to 83%.**

Fuel cells are not limited by the Carnot efficiency and theoretical efficiency and theoretical efficiency as high as can be expected of it. But high efficiencies are possible only at relatively light load and very low voltage (1 volt). The main advantages of the fuel cells are its simplicity, its high power to weight ratio, emission free and higher efficiency. The disadvantages are its high cost, need for equipment to control the flow of reactants to match the reaction rates, problems of selection of materials and short life, especially at high temperature.

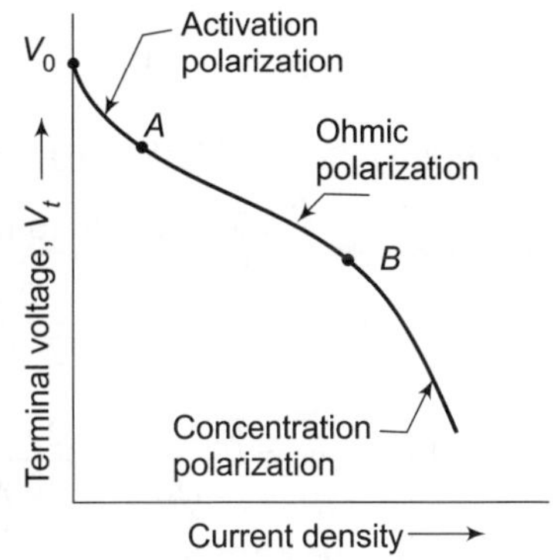

Figure 18.33 Polarization curve for a fuel

The performance of a fuel cell has been found to depend primarily on the current density (amperes per unit area) and temperature. Figure 18.33 shows a typical voltage current density for a fixed temperature. This is known as ***polarization curve***. Due to the interference at the electrodes, the measured no-load voltage (V_0) is less than that for the ideal reversible case, V_t.

It is obvious from Fig. 18.33 that the cell voltage drops rapidly with light loads. This is due to the reversible effects at the electrodes and is known as an *activation polarization.* The main reason for the linear drop in the operating range *AB* is the ohmic resistance in the electrolyte. Beyond the point *B* electrostatic effects, due to concentration gradient in the electrolyte, begins to limits the current, hence the term concentration polarization for this region of operation.

For an actual fuel cycle, the voltage (V_a) is less than (V_r). Hence, the efficiency is

$$\eta_0 = \frac{V_a\, n.N_0.e}{-\bar{h}_f^\circ} \quad ...(18.39)$$

With V_a = 0.6 volt, η_0 = 60% at a current density near 1075 amp/m^2.

These are the fuel cells which have been used in *Apollo system (long back).*

In Japan, Tyota Motors have developed a H_2–O_2 fuel cell for electric vehicles—cars and buses. Readymade H_2 gas cylinders at 70 bars is used and O_2 is taken from air. Electricity generated is fed to a battery which feeds electricity to an electric motor which drives the shaft. Some power plants are working based on H_2–O_2 fuel cells in USA (See Chapter 15).

Problem 18.5 A hydrogen-oxygen fuel cell operates at 298 K. Determine the voltage output of the cell, the efficiency and the electrical work output per mole of H_2 consumed and per mole of H_2O produced. Also find the heat transfer to the atmosphere.

Solution At 298 K from Gas Tables

$$\bar{g}_1 - \bar{g}_2 = 2{,}37{,}350 \text{ kJ/kgmole } H_2 \quad \text{and} \quad \bar{h}_f^\circ = -\ 286030 \text{ kJ/mole } H_2$$

According to the reaction equation, only two electrons are transferred per mole of H_2. Hence, the voltage developed in the reversible cell is

$$V_r = \frac{\bar{g}_1 - \bar{g}_2}{nN_0 . e.}$$

Here, e = (6.025 × 10^{23} mole/g mole) = 1.602 × 10^{-19} coulomb)

= 96,500 coulomb/g mole = 96,500 J volt-g-mole

$$\therefore \quad V_r = \frac{237350}{2 \times 96{,}500} = 1.229 \text{ volt} \qquad \textbf{\textit{Ans.}}$$

$$\text{Efficiency, } \frac{\bar{g}_1 - \bar{g}_2}{-(\bar{h}_f^\circ)} = \frac{237350}{286032} = 82.98\% \qquad \textbf{\textit{Ans.}}$$

The electrical work output per mole of fuel is calculated by

$$w_e = \bar{g}_1 - \bar{g}_2 = \Delta\bar{g} = 237350 \text{ kJ/kg mole } H_2$$

For each mole of fuel, there is one mole of H_2O produced, therefore, the heat transfer is

$$q = T\Delta s = \bar{h}_f^\circ + \Delta\bar{g} = -\ 286030 + 237350 = -\ 48680 \text{ J/mole} \qquad \textbf{\textit{Ans.}}$$

The negative sign shows the heat is removed from the cell.

18.11.1 Fuel Cell System Under Development

The development of fuel cells is taking place in many directions owing to its usefulness. Figure 18.34 shows a pictorial representation of various types of fuel cells which are under

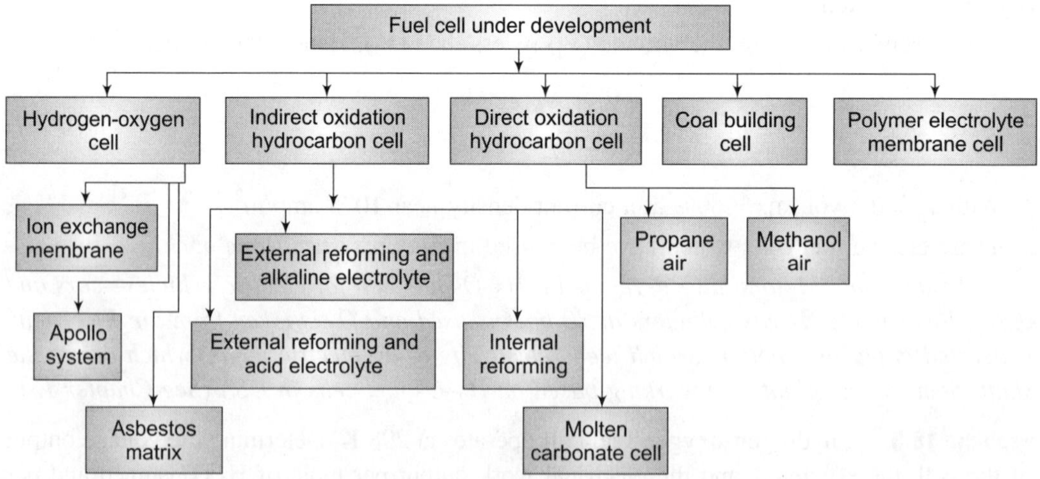

Figure 18.34 Fuel cells under development

development stage. Out of all fuel cells, hydrogen-oxygen fuel cell is the most developed cell system but now molten carbonate fuel cell and polymer electrolyte membrane fuel cell are promising ones.

18.11.2 Hydrogen-Oxygen Fuel Cells

The principle of working of H_2–O_2 fuel cell has been already discussed earlier. There are three types of highly developed fuel cells. They are:

(i) Ion exchange membrane cell of General Electric

(ii) Porous-nickel-electrode cell of Pratt.

(iii) Asbestos matrix cell of Allis Chalmers.

The main differences between these cells are due to the nature of the electrodes, the method of containing the electrolyte and the method of water removal.

18.11.2.1 Ion Exchange Membrane Cell

Figure 18.35 shows the working of such a cell. This cell was used in *Gemini V mission in* 1965. Hydrogen enters from one side and oxygen (or air) from the other side the membrane electrolyte and penetrate the thin porous electrodes to react with them near the surface of the membrane. From the hydrogen electrode (i.e., anode), electrons are emitted and are conducted by the external circuit to the cathode. The hydrogen ions travel through the membrane electrolyte to the other surface of the membrane when they combine with the returning electrodes and oxygen and by this way form the water on the oxygen (cathode) electrode. The water thus formed is drained away. The cell reactions are:

At anode: $2H_2 = 4H + 4e$

At cathode: $4H + 4e + O_2 = 2H_2O$

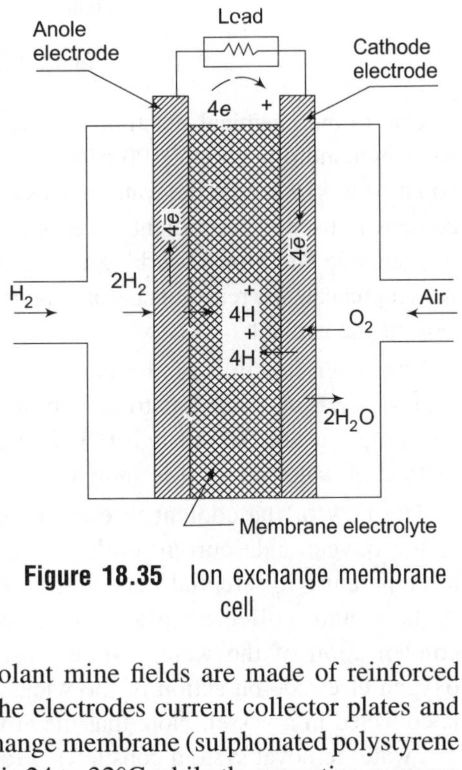

Figure 18.35 Ion exchange membrane cell

Figure 18.36 shows a schematic diagram of *Gemini cell.* The cell frame and hydrogen and coolant mine fields are made of reinforced plastic. The titanium made metallic parts include the electrodes current collector plates and coolant tubes. For electrolyte, a solid plastic ion exchange membrane (sulphonated polystyrene resin) is used. The operating temperature of the cell is 24 to 32°C while the operating pressure of the gas is about 1.5 bars. For large output, generally stack is formed by connecting 32 cells (17.5 cm × 20 cm each in size) in series. The stacks connected in parallel are contained in a single canister. Each canister, together with its accessories is called a stack with its forms an independent power supply. Two such sections, each producing 120 to 640 watts were employed in *Gemini spacecraft.*

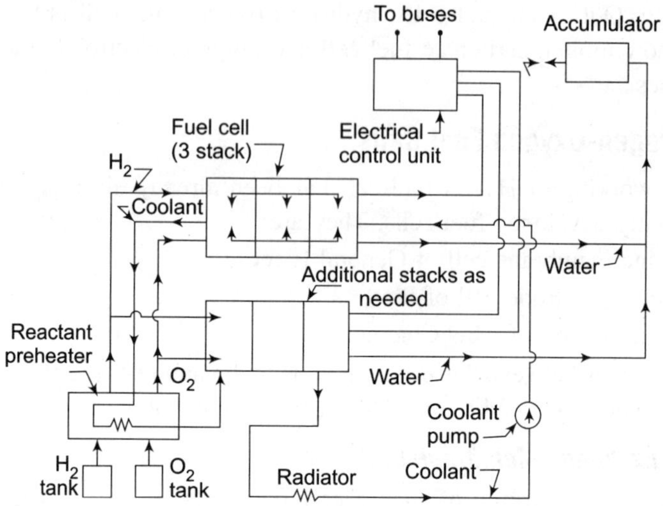

Figure 18.36 Gemini fuel cell

Due to the chemical reaction, heat is evolved in the fuel cell. To keep the cell in proper working condition about 2100 kJ/kWh (if the water formed is in saturated vapour state) and about 3140 kJ/kWh (if the water leaves as saturate liquid) has to be removed. Thus, a suitable coolant is used. The coolant used is glycol-water mixture. The coolant is pumped to the oxygen side and after absorbing heat from the fuel cell, it passes through a heat exchanger which preheats the reactants before entering the cell. A radiator is used to reject the remaining heat of the coolant.

The other function of the coolant is to remove the water formed in the cell. Nearly 0.21 kg/kWh of water has to be removed from the cell because the hydrogen and oxygen consumption are 0.0568 and 0.3442 kg/kWh, the method of water removal is shown in Fig. 18.37.

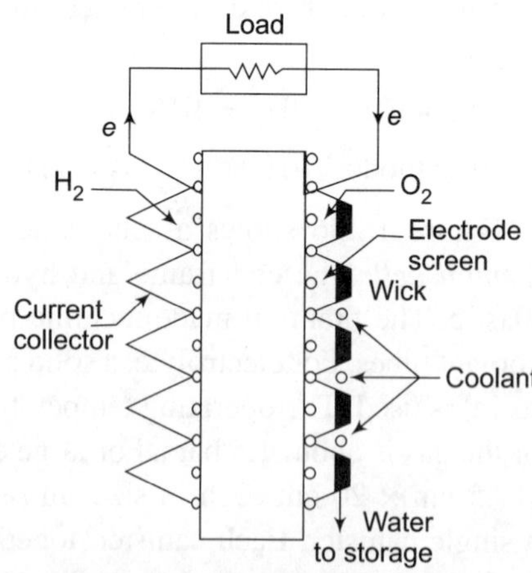

Figure 18.37 Water removal system of Gemini cell

The flow of the coolant through tubes attached to the oxygen side current collector plate, causes a temperature differential between the electrolyte-surface and collector plate. This causes the condensation of the water vapour formed at the oxygen electrode on action of the wicks; this water is collected in a vessel. Note that the auxiliary heat and water removal system consumes 3 to 4% of the gas power output of the system.

18.11.2.2 The Apollo Fuel Cell

This fuel cell was used in Apollo Lunar Mission in 1968. The electrodes are made of porous nickel. Concentrated KOH liquid serves as an electrolyte.

The cell operates between 200 and 250°C. The gases are supplied at 4 bars. To increase the power output, 31 cells are connected in series forming a single power unit called a *module*. Such three modules were placed at the Apollo spacecraft, each producing 563 to 1420 watts. Figure 18.38 shows a schematic diagram of Apollo fuel cell. Since the heat and water removal operation are carried out at the hydrogen electrode, hence an extra quantity of hydrogen than exactly required for the chemical reaction is constantly circulated past the hydrogen electrode. A separate coolant circuit is provided in the cell. With the help of the coolant, the inlet temperature of hydrogen is regulated, and thus it is possible to remove heat by transferring it to the excess hydrogen. Moreover, the water formed in the cell evaporates from the electrode's surface and is carried away by the hydrogen as stream. By varying the humidity of the incoming gas, the rate of this water removal is controlled. The excess hydrogen (moist and hot) leaves the fuel cell and passes through a fuel regenerator where it is partly cooled and then passes through a condenser where the water vapour is condensed. The hydrogen and condensate passes through a pump separator unit where liquid water (condensate) is removed from the gas due to the centrifugal action of the rotating separator. The hydrogen gas is brought back to the fuel cell.

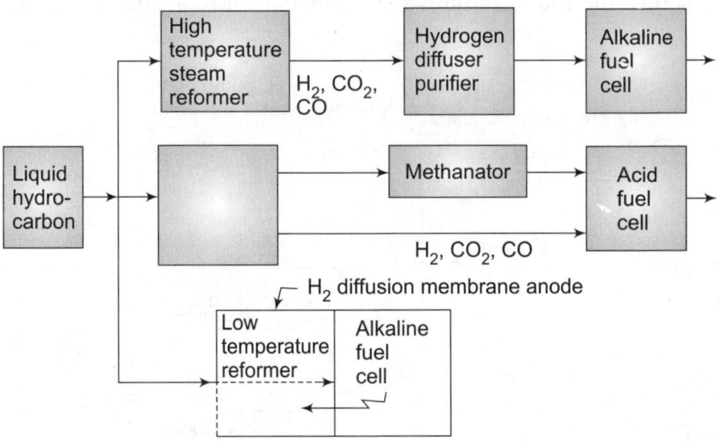

Figure 18.38 Apollo fuel cell

Note that the power consumed by the accessories such as pump and separator is about 10% of the gross output. But due to the high performance of nickel electrodes, the efficiency of the cell is the maximum among all. The total mass of the system is more or less same as Gemini cell.

18.11.2.3 Asbestos Matrix Cell

In this cell, the electrolyte is an asbestos matrix saturated with dilute KOH. The anode is of platinum-catalysed nickel and the cathode is of silver. The operating temperature of the cell varies from 90°C to 100°C. A 2 kW stack consists of 30-30 pairs of cell connected in series. Heat and water removal system are quite independent.

18.11.3 Direct Oxidation Hydrocarbon Fuel Cells

In this case, there is a direct electro-chemical oxidation of hydrocarbon fuels. There are many problems associated with this cell which are yet to be solved. Fuels which are electro-chemically more reactive are found to be uneconomical. For example, gasoline and kerosene would be highly desirable for mobile application but their reactivity is low. Many other hydrocarbons are under investigations.

18.11.4 Indirect Oxidation Fuel Cell Systems

This method consists of first converting the primary fuel into hydrogen and then fed to the cell. The conversion of primary fuel into hydrogen is achieved in a pre-reactor which may be steam reformer, a partial oxidizer or a catalytic cracker. Among various types of pre-reactor, steam reformers have been found to give higher efficiency and richer hydrogen yield.

In steam reformer, the hydrocarbon fuel mixed with a certain excess of steam is passed over a nickel-alumina catalyst at about 800°C. After this, the product is passed over an oxide catalyst at about 200°C. The final product normally contains 79.7% H_2, 20% CO_2 and 0.3% C.

Figure 18.39 shows the four alternative systems currently available for indirect oxidation fuel cells.

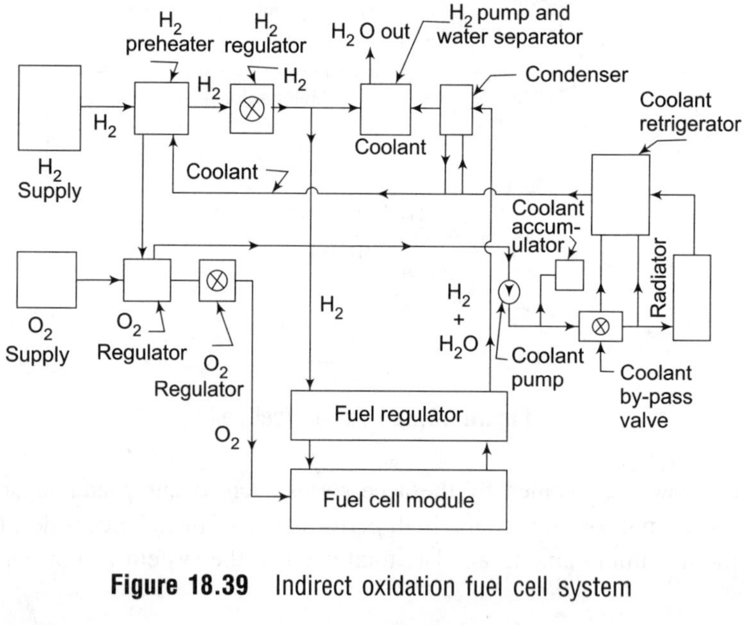

Figure 18.39 Indirect oxidation fuel cell system

18.11.4.1 External Reforming and Alkaline Electrolyte

This system is the most advanced at present. In this system, the impure hydrogen from the reformer is purified by passing it through palladium silver diffusers at temperature between 350°C to 600°C. Hydrogen alone diffuses through the alloy membranes and its directly supplied to the fuel cells. Platinum is the best suited for the electrocatalyst cells. KOH serves as electrolyte in American cynamide cells.

18.11.4.2 External Reforming and Acid Electrolyte

As the name implies, the impure hydrogen from the reformer is directly fed to fuel cells. This is possible due to the fact that the acid electrolyte (phosphoric or sulfuric) is not affected by CO_2. To protect the platinum anode catalyst from the harmful effect of CO_2, a methanator is placed between the reformer and fuel cell.

18.11.4.3 Internal Reforming Cell

This process is under development at Pratt and Whitney Aircraft and Leesona Mass Laboratories. In this system, process of reforming, purification and electrochemical oxidation of the fuel are all carried out at the fuel cell anode.

18.11.4.4 Molten Carbonate Fuel Cell

Figure 18.40 shows the schematic diagram of molten carbonate fuel cell. In this case, the electrolyte is molten carbonate which is generally, a mixture of lithium, potassium and sodium carbonates. The fuel used is liquid hydrocarbon (methane or kerosene). The cell operates at about 600°C or even more.

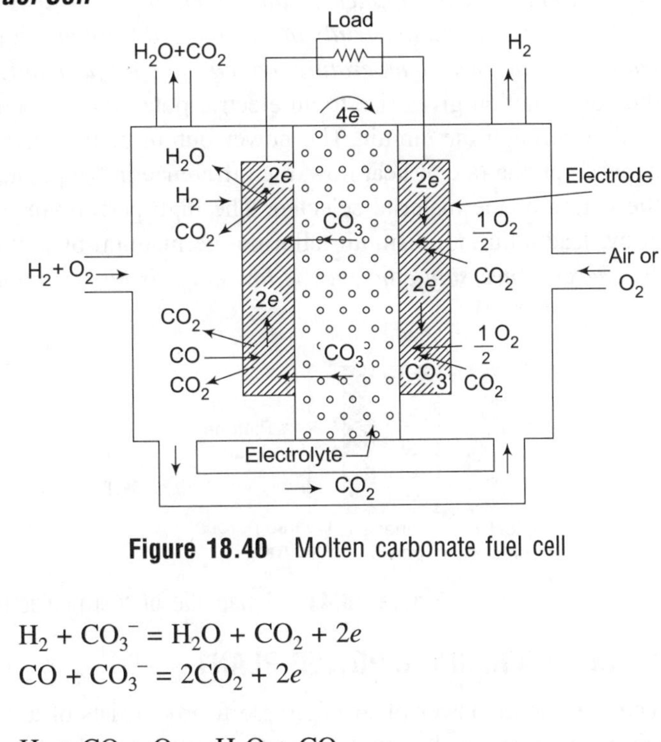

Figure 18.40 Molten carbonate fuel cell

Electrodes (metallic) are placed in direct contact with this solid electrolyte. The fuel is reacted inside the cell forming H_2 and CO. The reaction are as follows:

At fuel electrodes: $H_2 + CO_3^- = H_2O + CO_2 + 2e$

$CO + CO_3^- = 2CO_2 + 2e$

At oxygen electrode.

The overall reaction: $H_2 + CO + O_2 = H_2O + CO_2$

The cell has potential to give the maximum efficiency than any other cell but the main drawback is the corrosive effect of molten carbonate electrodes. *However, at present it is the most promising fuel cell.*

18.11.5 Biochemical Fuel Cells

The *biochemical fuel cell is a device which converts the chemical energy of fuel into electrical energy using enzymes as catalysts.* The enzymes undergo an oxidation-reduction cycle by sequentially releasing and taking up electrons. The released electrons can be tapped and transferred to an electrode in order to provide electrons current in the external circuit. This is under active research.

18.11.6 Hybrid (Fuel Cell and Gas Turbines) System

Now, Department of Energy **(DOE) of USA** has given fund to various power agencies to develop ***hybrid system* (fuel cells and gas turbine system) under distributed power concept using natural gas as fuel [64]**. The conversion efficiency is about 70%. This system is free from emission as there is no combustion. About 80% energy is produced from fuel cells and 20% by small gas turbine, which uses emitted gases from fuel cell at around 1000°C.

18.12 THERMOELECTRIC POWER PLANT

This plant works on the principle of Seebeck effect. *Two dissimilar materials are joined at one end and heat is supplied to this junction and so it is called* ***hot junction*** *(Fig. 18.41). At the other end, the materials are connected through an external load and this junction is maintained at lower temperature, called the* ***cold junction***. The flow of heat from the hot to the cold junction gives rise to an electric potential difference and causes an electric current to flow through the circuit. The power output of the thermoelectric device power for plant depends on the rate of heat flow, the difference in temperature between the two junctions and the nature of the material selected. The high performance materials are silicon-germanium alloy, lead-telluride, bismuth-telluride, germanium-bismuth-telluride. *Its output is very low. Its energy conversion efficiency is low but it is highly reliable.*

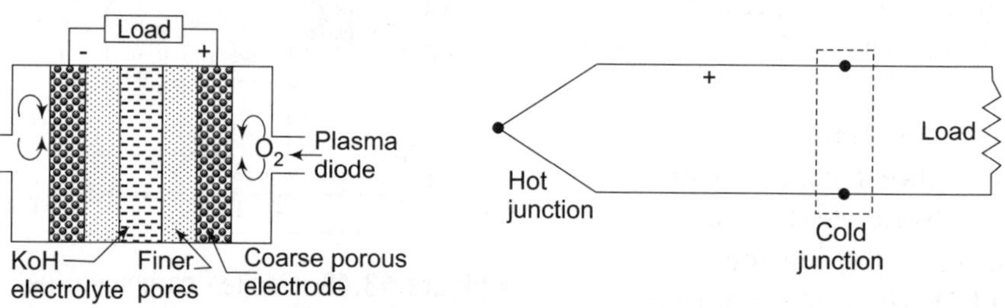

Figure 18.41 Schematic of thermoelectric power plant

18.13 THERMIONIC POWER PLANT

The thermionic power plant in simple form consists of a high temperature electron emitting surface (emitter) and a low temperature electron collecting surface (collector) separated by a small gap working on Peltier effect. Commonly the emitter temperature is 1200° to 1700°C, the collector temperature is 4000 to 900°C and the inter electrode gap is 0.125 to 0.25 mm. Due to high temperature of the emitter, sufficient energy is imparted to the electrons to escape from the surface and travel across the gap to the collector surface. If the collector and emitter are connected through an external load, a current will flow due to the potential difference between the two surfaces. The gap is filled by electrically conducting ionized gas (plasma) to neutralize the space charge. Its energy conversion efficiency is 20 per cent. It is a low voltage high current device (Fig. 18.42).

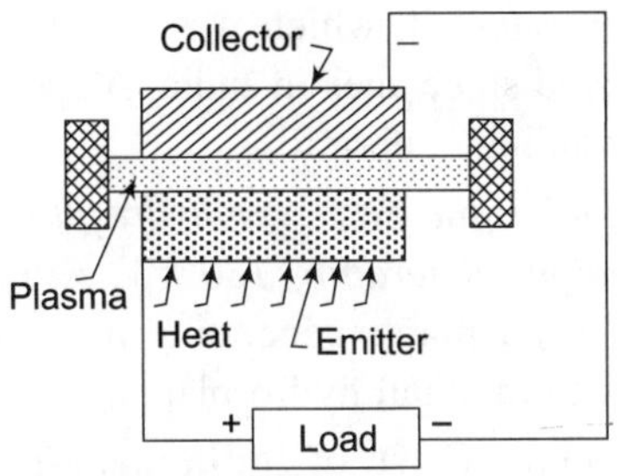

Figure 18.42 Thermionic power plant

18.14 ELECTRO-GAS-DYNAMIC (EGD) POWER PLANT

Figure 18.43 shows a schematic diagram of a simple EGO power plant. In this type, the thermal energy is converted into kinetic energy in the supersonic nozzle and ions are then injected into the air stream by Corona discharge between a pointed Corona electrode and a cylindrically shaped collector electrode. The air carries ions downstream to the collector against the action of an opposing electron field thus converting the source of kinetic energy of the gas into direct current electricity at a very high voltage. The thermal energy in hot compressed air is normally recovered by expansion through a turbine or generating steam and driving steam turbine. Conversion efficiency in this plant is slightly higher.

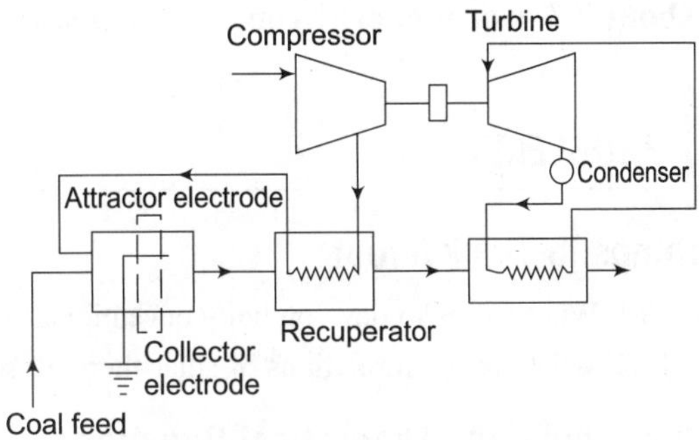

Figure 18.43 E.G.D. power plant

18.15 SMALL HYDROPOWER PLANT

Small hydropower has been included under renewal energy as it does not disturb the ecological balance like large hydropower plants. As per Central Electricity Authority of India and Bureau of Indian Standards, small hydropower plants are classified in the following categories based on capacity and head. It is a small version of large hydro.

(A) Depending on capacity of plant

- Micro plant is up to 100 kW
- Mini plant is from 101 to 1000 kW
- Small plant is from 1001 to 6000 kW

(B) Depending on head of plant

- Ultra low head: below 3 m
- Low head: less than 30 m
- Medium head: between 30 and 75 m
- High head: above 75 m

Small hydro is not a new invention but it is an age old practice in India and world over.

The watermill which was also known as Panchakki, Gharat, Water wheel, etc., has been employed since ages in India. As per one estimate, about 2,00,000 sites in India are suitable for installing watermills.

In *India, the first ever hydro plant for generating electricity was established in 1897 at Darjeeling generating 130 kW*. After independence of India, the focus was more on large hydropower plants generating more than 2100 MW, and out of which only about 500 MW comes from small hydro plants.

Recently, **Govt. of India has identified about 2000 sites for small hydro plants** which have generating capacity of **about 5000 MWe**. Another many sites having headless **than 3 m have** been identified whose generating capacity is about **1200 MWe**.

Small hydro is more reliable for 24 × 7 operation as compared to solar and wind power. **About 7% world energy come from small hydro in 2019**.

PROBLEMS

Things to Think About

18.1 What led us to develop non-conventional/renewable energy sources of energy?

18.2 What are the limitations of solar thermal, solar photovoltaic and wind power?

Short and Long Theoretical Questions

18.1 With the help of a simple sketch, discuss the working principle of central receiver and distributed system solar thermal power plant.

18.2 With the help of simple sketch, discuss the working principle of stand-alone and general grid interactive solar system.

18.3 With the help of a simple sketch, discuss the working principle of various types of wind turbine power plant.

18.4 With the help of a simple sketch, discuss the working principle of various types of geothermal power plants.

18.5 With help of a sketch, discuss the working principle of the following power plants:

(a) Ocean (b) fuel cell (c) biomass, (d) MHD (e) thermoelectric (f) small hydro-resources power plants.

Software Based Question

18.1 Develop a software to design various types of renewable energy resource power plants.

Design and Open Ended Questions

18.1 Prepare a comprehensive report of various types of wind power plant with their design features and specifications.

18.2 Prepare a comprehensive report about space based solar photovoltaic power plant.

Multiple-Choice Questions

18.1 What is the value of solar constant?

(a) 567 W/m^2 (b) 967 W/m^2 (c) 1367 W/m^2 (d) 1767 W/m^2

18.2 Flat plate collectors have CR and TR as follows

(a) 1,100°C (b) 1.3,150°C (c) 2,200°C (d) 2.5,250°C

18.3 The point focus concentrations have CR and TR as follows

(a) 1000, 1000°C (b) 1200, 1200°C (c) 1500, 1500°C (d) 2000, 2000°C

18.4 Suitable velocity range for wind turbine ranges from

(a) 1 to 2 m/s (b) 2 to 5 m/s (c) 5 to 25 m/s (d) 25 to 40 m/s

18.5 Number of blades in wind mills in around

(a) 3 (b) 6 (c) 36 (d) 72

18.6 The efficiency of modern solar P-V. cell ranges from

(a) 5 to 10% (b) 10 to 15% flat (c) 15% to 20% (d) 20% to 30%

18.7 Efficiency of flat plate solar thermal system for moderate outlet water temperature varies

(a) 10 to 20% (b) 20 to 30% flat (c) 30% to 40% (d) 40% to 50%

Answers to Multiple-Choice Questions

18.1 (c) 18.2 (a) 18.3 (a) 18.4 (c) 18.5 (a) 18.6 (c) 18.7 (d)

Appendices

APPENDIX A

SI Units

In 1960, the C.C.P.M. (Conference Generals des Poids et Measures) recommended the Universal adoption of units bases on metre (m), kilogram (kg), second (s), degree kelvin (K) and Candela (cd), to cover the entire system of science and engineering and give it the name "System International of Units" abbreviated SI.

Most of the countries including India has switched over to this unit.

The fundamental and supplementary units of the S.I. system are depicted in Table A1. The derived units from the base and supplementary units are present in Table A2. Table A3 shows multiples and sub-multiples.

Table A.1 Fundamental and supplementary S.I. units

Quantity	*Unit*		*Symbol*
		Fundamental Units	
length	metre		m
mass	kilogram		kg
time	second		s
temperature	degree kelvin		K
current	ampere		amp
luminous intensity	candela		cd
amount of substance	mole		mol
		Supplementary Units	
plane angle	radians		rad
solid angle	steradian		sr

Note: In addition to thermodynamic temperature scale degree Kelvin (K), the degree Celcius (°C) is also in use.

Table A.2 Derived units

Quantity	*Unit*	*Symbols*
area	square metre	m^2
volume	cubic metre	m^3
velocity	metre per second	m/s
acceleration	metre per second per second	m/s^2
density	kilogram per cubic metre	kg/m^3
force	newton	N, kgm/s^2
Pressure, stress	pascal or newton or per sq. metre	Pa; N/m^2
	metre	(IPa = IN/m^2
work, heat energy	joule	J; Nm
power	watt	W, J/s (1W = J/s)
specific entropy	joule per kilogram per degree kelvin	J/kgK
enthalpy	joule per kilogram	J/kg
rate of heat flow	watt	W
heat flux rate	watt per square metre	W/m^2
thermal conductivity	watt per metre per degree kelvin	W/m K
heat transfer coefficient	watt per sq. metre per degree kelvin	W/m^2K
radiation factor	watt per square metre per degree K to the power four	W/m^2K^4
frequency	hertz	Hz = c/s
dynamic viscosity	pascal-second or newton second per sq. metre	Pa's; Ns/m^2
kinematic viscosity	square metre per second	m^2/s

Note: In addition to pressure expressed in pascal or N/m^2, bar is also in commom use. 1 bar = 10^5 N/m^2, 1 atm pressure = 101,3 kN/m^2 = 1.013 bar.

Table A.3 Multiples and sub-multiples

tetra	T	10^{12}	milli	*m*	10^{-3}
giga	G	10^9	micro	μ	10^{-6}
mega	M	10^6	nano	*n*	10^{-9}
kilo	k	10^3	pico	*p*	10^{-12}
deci	d	10^{-1}	femto	*f*	10^{-15}
centi	c	10^{-2}	atto	*a*	10^{-18}

A.4 Conversion Factor

(i) Force	1N	$= 10^5$ dyne	$= 0.012\ kg_f$
	1 kg_f	= 9.81 N	
(ii) Pressure	$1N/m^2$	= 1 Pa	
	1 bar	$= 10^5\ N/m^2$	$= 10^5$ Pa $= 10^2$ kPa
	1 atm	= 1.01325 bar	$= 101.325\ kN/m^2$
		= 760 mm of Hg	$= 1.03\ kg_f/cm^2$
(iii) Energy	1 J	= 1 Nm	= 1 Ws $= 10^7$ erg
	1 kcal	= 4.1868 kJ	
(iv) Power	1 W	= 1 J/s	= 0.86 kcal/h
	1 h.p.	= 75 m.kgf/s	= 733.5 W = 0.7335 kW
(v) Heat Transfer Rate			
	$1\ W/m^2$	$= 0.86\ kcal/m^2h$	
		$= 0.317\ BTU/hft^2$	
1 tonne of refrigeration		= 50 kcal/min	= 3.517 kW = 200 BTU/min
(vi) Temperature	K	= °C + 273	$= \left(\frac{5}{9}\right)$ (°F + 460)
Temperature difference		$= \Delta T = \Delta t$	
(vii) Latent Heat	1 kJ/kg	= 0.2349 kcal/kg	
(viii) Thermal conductivity and Heat Transfer coefficient			
	1 W/mK	= 0.86 kcal/m hK = 0.5778 BTU/fhR	
	$1\ W/m^2K$	$= 0.0.8598\ kcal/m^2$ hK	
(ix) **Physical Constant**			
Speed of light		$= 2.988 \times 10^8$ m/s	
Universal gas constant		= 8314 J/kg mole K	

APPENDIX B

Table B.1 Comparison between different types of power plant

S.No.	Parameter	Diesel Engine	Gas Turbine	Steam Turbine Plant
1.	Usual power range	10 kW to 10 MW	100 kW to 260 MW	1 MW to 1000 MW
2.	Approx. capital cost	₹1100 – 1800/kW	₹800 – 1500/kW (aero) (aero) – 280/kW (heavy)	₹1400 – 2400/ kW (turbine only)
3.	Approx installed cost	–	₹1500 – 3000/kW	₹8000– to 12000/kW
4.	Size and weight/kW	Medium	Small	Large
5.	Specific installation requirements	Heavy foundations Vibration isolation Quietening enclosure	Intake and filters Intake and exhaust mufflers	Ash handling facility Copious supplies of water Exhaust de-pollution
6.	Typical plant life	10 to 20 years	15 to 30 years	25 to 35 years
7.	Inspection/overhaul periods	Annual	Several years	Annual
8.	Efficiency (typical)	35% to 40%	35% to 40%	30% to 40%
9.	Efficiency factor at half load.	1.00	0.8 to 0.85	0.91 to 0.96
10.	Fuel grade and type	High liquid (diesel grade)	High to medium liquid, natural gas	Medium to low Gas, liquid coal
11.	Staffing requirements	A few	None (extra to plant)	Many
12.	Lubrication needs	Frequent oil and filter	Oil topping up	Negligible
13.	Maintenance periods	Weekly and Monthly	Monthly and annually	Daily
14.	Site selection	No special site	No special site	Special site to meet water and pollution requirements

APPENDIX C

Table C.1 The properties of some gaseous fuels

Gases	*Chemical Formula*	*Molecular weight*	*Density (kg/m³)*	*Lower Calorific value (MJ/kg)*	*L.C. Value (MJ/m³)*
n–Butane	C_4H_{10}	58.17	2.458	45.71	112.41
Carbon monoxide	CO	28.01	1.185	10.1	11.97
Ethane	C_2H_6	30.09	1.272	47.48	60.43
Methane	CH_4	16.04	0.678	50.01	33.95
Propane	C_3H_8	44.14	1.865	46.35	86.42
Hydrogen	H_2	2.016	0.085	119.95	10.22

Table C.2 The properties of some solid fuels

Solid Fuels	*Density (No pores) kg/m³*	*Higher Calorific Value (MJ/kg)*
Charcoal	1400 to 1700	34.30
Coal	1250 to 1700	32 to 36
Coke	1750 to 2000	33.60
Peat	1150 to 1250	19.70
Wood	500 to 1100	17.30

Table C.3 The properties of some liquid fuels

Pure liquids	*Chemical formula*	*Molecular Weight*	*Boiling Point at 1 atm (K)*	*Density (kg/litre) at 298 K*	*Lower Calorific Value at 298 K (MJ/kg)*
Benzene	C_6H_6	78.11	353	0.879	40.14
n–Butane (LPG)	C_4H_{10}	58.17	273	0.575	45.34
n–octane	C_3H_{18}	114.23	399	0.703	44.43
n–Heptane	C_7H_{16}	100.21	371	0.684	44.56
Ethanol	CH_5OH	34.06	352	0.794	27.23
Methanol	CH_3OH	32.04	338	0.796	19.94
Propane (LPG)	C_3H_6	44.14	231	0.505	45.98

Liquid Fuels	*Density at 188 K (kg/filter)*	*Boiling range (K)*	*Kinematic viscosity (cSt)*	*Lower Calorific Value at 188K MJ/kg)*
Diesel	0.84	453 to 613	5	42.9
Fuel oil	0.8 to 0.97	453	1.4 to 7.0	53.1
Gasoline	0.72 to 0.76	301 to 473	0.5 to 0.65	41.9–44.0
Kerosine	0.73 to 0.83	–	2.1 to 3.0	51.7

APPENDIX D

D.1: Enthalpy of formation, Gibbs functions of formation, and absolute entropy of various substances at 25°C, 1 atm pressure

Substance	Molecular Weight M	$\bar{h}_f^0$ kJ/kg mole	$\bar{h}_f^0$ kJ/kg mole	$\bar{h}_0$ kJ/kg mole
CO(g)	28.001	−119529	−137150	197.653
CO_2(g)	44.001	−393522	−394374	213.795
H_2O(g)	18.015	−241827	−228583	188.883
H_2O(1)	18.025	−28538	−237178	70.049
CH_4(g)	16.025	−74873	−50751	186.256
C_2H_2(g)	26.038	+226731	+209234	200.958
C_2H_4(g)	28.054	+52283	+68207	219.548
C_2H_6(g)	30.070	−84667	−32777	229.602
C_3H_8(g)	44.097	−103847	−23316	270.019
C_4H_{10}(g)	58.124	−126148	+16914	310.227
C_8H_{18}(g)	114.23	−208447	+16859	466.835
C_8H_{18}(1)	114.23	−249952	+6940	360.896

D.2. Enthalpy of Combustion of some hydrocarbons at 25°C.

Hydrocarbon	Formula	Liquid H_2O in products (Negative of higher hearing value)		Vapour H_2O in products (Negative of lower heating value)	
		Liquid Hydrocarbons kJ/kg fuel	Gaseous Hydrocarbons kJ/kg fuel	Liquid Hydrocarbon kJ/kg fuel	Gaseous Hydrocarbon kJ/kg fuel
Paraffin family					
Methane	CH_4		−55496		−50010
Ethane	C_3H_6		−51875		−47484
Propane	C_3H_8	−49975	−50345	−45983	−46353
Butane	C_3H_{10}	−49130	−49500	−45344	−45714
Pentane	C_5H_{12}	−48643	−49011	−44983	−45351
Hexane	C_6H_{14}	−48308	−48676	−44033	−45101
Heptane	C_7H_{16}	−48071	−48436	−44557	−44922
Octane	C_8H_{18}	−47893	−48256	−44425	−44788
Decane	$C_{10}H_{22}$	−47641	−48000	−44239	−44598
Dodecane	$C_{12}H_{24}$	−47470	−47282	−44110	−44467

Contd.

Contd.

Olefin family					
Ethene	C_2H_4		−50296		−47158
Propene	C_3H_6		−48917		−45780
Butene	C_4H_8		−48453		−45316
Pentene	C_5H_{10}		−48034		−44996
Hexene	C_6H_{12}		−47937		−44800
Heptene	C_7H_{14}		−47800		−446612
Octene	C_8H_{16}		−47693		−44556
Nonene	C_9H_{18}		−47612		−44475
Decene	$C_{10}H_{20}$		−47547		−44410
Alkyl-benzene family					
Benzene	C_6H_6	−41831	−42266	−40141	−40576
Methyl-benzene	C_7H_8	−42473	−42847	−40527	−40937
Ethyl-benzene	C_8H_{10}	−42997	−43395	−40924	−41322
Propyl-benzene	C_9H_{12}	−43416	−43800	−41219	−41603
Butyl-benzene	$C_{10}H_{14}$	−43748	−44123	−41453	−41828

References

1. Anderson, J.H., et al., “Thermal Power from Sea Water” *Mechanical Engineering,* April 1966, pp. 41-46.
2. Bigg, J.V., “Design and Construction of Power Station Feed Heaters”, *Proceeding Ins. of Mech., Engrs.*, 1959.
3. Brown Boveri Review, Dec 1960, 1982.
4. Caputo R., *“An Initial Comparative Statement of Orbital and Terrestrial Central Power Systems”*, Final Report 77-44, Jet Propulsion Laboratory, CA, March 1977.
5. Church, E.F, *“Steam Turbines”*, McGraw Hill, New York.
6. Clemmer, A.B., and Lemezis, S., *“Selection and Design of closed Feed water Heaters, Trans”*. ASME, 1965.
7. Cox, H.R., *“Gas Turbine Principles and Practice”,* George Newness, London, 1955.
8. Cohen, H., Rogers, G.F.C., and Saravanamutto, H.I.H., *“Gas Turbine Theory,”* 4th ed., Longmans, London, 1946.
9. Dollin, F., *“Investigation of Steam Turbine Nozzle and Blading Efficiency”*, Proc, Instn., Mech. Engrs., 1940, p. 147.
10. Dollin, F., *“Design Problems in Development of Large Steam Turbines ins. Mech. Engrs.,* 1962.
11. Derby, *“The Jet Engine”* Rolls-Royce Ltd; 1969.
12. Dunham, J., and Came, P.M., *“Improvement to the Ainley – Mathieson Method of Turbine Performance Prediction,”* ASME Journal of Engg, for Power, 92, July 1970.
13. Eastop, T.D. and McConkey A., *“Applied Thermodynamics for Engineering Technologies”*, Longman, 1978.
14. Ecker, R., *“Current Trends in Steam Turbine Design”* Engineering, 24, Feb. 1961.
15. Fracia, C., *“Pilot Plants of Solar Steam Generating Stations”*, Solar Energy, Vol. 22, 1968, pp. 51–64.
16. Frass A.P., *“Trans ASME, Journal of Engg. for Power”*, Jan 1973, p. 53.
17. Gretizer, EM., *“Review – Axial Compressor Stall Phenomenon”*, *Trans, ASME Journal of Fluids Engg.*, Vol. 102, number 2 June 1980.
18. Hubbert, M.K., *“Resources and Man”*, San Francisco, Freeman, 1969.
19. Horlock, J.H., *“Losses and Efficiencies in Axial Flow Turbines”*, Int. J. Mech Science. 1960, 2 (No. 53), 228.
20. Horlock, J.H., *“Axial Flow Compressor, Axial Flow Turbines, Butterworths”*, London. 1966.

21. Harman, R.T.C., *"Gas Turbines Engineering – Applications, Cycles and Characteristics"*, Macmillan, 1981.
22. Hamed, A. and Baskharome, E., *"Analysis of the Three Dimensional Flow in a Turbine Scroll"*, *Trans, ASME, Journal of Fluids Engg.*, 1980.
23. *"Institution of Electrical Engineers Conference on Future Energy Concepts"*. London, 1979.
24. Kreider, J.F.. and Kreith F., *"Solar Energy Hand Book"*, McGraw Hill Book Company, 1981.
25. Kearton, W.J., *"Steam Turbines Theory and Practice"*, Pitman, 1962.
26. Lee, J.F., *"Theory and Design of Steam and Gas Turbines"*. McGraw Hill Book Company, N.Y.
27. *"Modern Power Station Practice"*, Vol. 3., Central Electricity and Generating Board, Pergamon Press, Oxford. 2nd ed., 1971.
28. Mollison D., et al., *"Wave Power Available in the North – East Atlantic"*, Nature. 263, 5574, 1976.
29. Paul Kruger and Carel Otte, *"Geothermal Energy"*, Standford University Press, Stanford, California, 1973.
30. Ram Kumar, R., et al., *"Solar Energy Conversion and Storage System for the Future"*. *IEE. Trans Power Apparatus and Systems*, Vol. PAS-94, 1975.
31. Seleuk, M.K., *"Survey of Several Central Receiver Solar Thermal Power Plant Design Concepts"*, *Internal Report, 900 – 419. Jet Propulsion Laboratory, Pasadena. CA., August* 1975.
32. Smith, A, *"Survey of Twisted Blading Development in Steam Turbines"*. Proc. Ins of Mech. Engrs., London. 1970.
33. Salisbury, J.K., *"Steam Turbines and their Cycles"*. Willey and Sons.
34. Shepherd. D.G., *"Principles of Turbo-machinery"*, Macmillan.
35. Shylakhin, P., *"Steam Turbines, Theory and Design"*. Foreign Language Publishing House, Moscow.
36. Stodola, A., *"Steam and Gas Turbines"*, Vols. I and II. Peter Smith.
37. *Sawyer's "Gas Turbines Engineering Hand Book"*, 1, 2, and 3 Volumes, Gas Turbines Publications, Inc.
38. Stenning, A.H., "Rotating Stall and Surge", *ASME, Journal of Fluids Engg,* Vol. 102, March 1980.
39. "Technical Advances in Gas Turbine Design" *Proc. Ins. of Mech. Engrs.* 183, part 3rd, 1969.
40. *"United Nations"*. World Energy Statistics, 1971.
41. Vincent, E.T., *"Theory and Design of Gas Turbines and Jet Engines"*. McGraw-Hill, 1960.
42. Wiesner, F.. "A Review of Slip Factor for Centrifugal Impellers". *Trans. ASIME, Journal of Engg for Power,* Series A. 1967.
43. Wang. W. "Direct Digital Control System" *I. Mech Engrs. Bulletin,* August 1982.
44. Yadav, R. and Yahaya S.M., "Flow Visualisation Studies and the Effect of Tongue Area on the Performance of Volute Casings of Centrifugal Machines". *Int. J. of Mechanical Science,* Vol. 22. Pergamon Press Ltd. 1980.
45. Yadav R. and Mishra S.S., *"Evaluation of Slip Factor of Centrifugal Impellers"* Inst. of Engineers (India). Vol. 53, 1973.

46. Briesch M.S., Bannister, R.L., Diakunchek, I.S. and Huber D.J., "A Combined Cycle Designed to Achieve Greater than 60 Per cent Efficiency, *Trans of the ASME. J. of Engg. For gas Turbine and Power,* Vol. 53, 1973.

47. Facchini, B, Fiaschi, D. and Manfrida, G., "Exergy Analysis of Combined Cycles Using Latest Generation Gas Turbines." *Trans of the ASME, J. of Engg for Gas Turbine and Power*, Vol, 122, 2000.

48. Horlock, J.H., "Limitations on Gas Turbine Performance Imposed by Large Turbine Cooling Flows." *Trans of the ASME, J. of Engg for Gas Turbine and Power,* Vol., 127, 2001.

49. Macchi, E. Consonni, S., Lozza G., and Chiesa, P., "An Assessment of Thermodynamic Performance of Mixed Gas-Steam Cycles : Part A – Intercooled and Steam – injected cycles," Part B – water injected and HAT cycles." *Trans of the ASME. J. of Engg. For Gas Turbines and Power,* Vol. 117. 1995.

50. Khodak, E.A. and Romakhova, G.A., "Thermodynamic Analysis of Air Cooled Gas Turbine Plants," *Trans of the ASME. J. of Engg for Gas Turbine and Power"*, Vol. 123, 2001.

51. Brandt, D.E., *"The Design and Development of An advanced Heavy-Duty Gas Turbines"*, *Trans of the ASME. J of Engg for Gas Turbines and Power.* Vol. 110 1988.

52. Horlock, J.H., Combined Power Plant-Past, Present and Future, *Trans of the ASME, J. of Engg. for Gas Turbines and Power*, Vol. 117. 1995.

53. Leis, D.M., Boss, N.J., and Melsert, M.P., "Medway; A High-efficiency Combined Cycle Power plant Design," *Trans of the ASME. J. of Engg. for Gas Turbine and Power",* Vol. 117, 1995.

54. Bolland. O. "A Comparative Evaluation of Advanced Combined cycle Alternatives" "*Trans of the ASME. J. of Engg. for Gas Turbine and Power,"* Vol. 113, 1991.

55. Lugand, P. and Parietti, "Combined Cycle Plants with Frame 9F Gas Turbine, *Trans. of the ASME, J. of Engg. for Gas Turbine and Power,"* Vol. 113, 1991.

56. Entenmann, DT., et al., *"Shop Test of the 5OIF-150 MW Combustion Turbine," Trans. of the ASME, J. of Engg. for Gas Turbine and Power,"* Vol. 113, 1991.

57. Bolland, O. and Stadaas, J.F., "Comparative Evaluation of Combined Cycles and Gas Turbine System with Water Injection and Reculperation" *Trans. of the ASME. J. of Engg. for Gas Turbine and Power,* Vol. 117, 1995.

58. Scalzo, A. J. Bannister R.H. Decorro M. and Howard, G.S. "Evolution of Westing House Heavy-Duty Power Generation and Industrial Combustion Turbines," *Trans of the ASME,* Vol. 118, 1996.

59. Korakianitis, T. and Svensson, K., "Off-design Performance of Various Gas-Turbine Cycle and Shaft Configurations, *Trans of the ASME. J. of Engg for Gas Turbines and Power,* Vol. 121, 1999.

60. Dechamps. P.J., "Advanced Combined Cycle Alternatives with the Latest Gas Turbines", *Trans of the ASME. J. of Engg for Gas Turbine and Power"* Vol. 120. 1998.

61. Chiesa P. and Consonni S., "Natural Gas Fired Combined cycle with low CO_2 Emissions", *Trans of the ASME J of Engg for Gas Turbines and Power*, Vol. 122. 2000.

62. Bannister, R.L., Cheruvu, N.S., Little, D.A. and McQuiggan, G., "Development Requirements for an Advanced Gas Turbine System", *Trans of the ASME. J. of Engg for Gas Turbines and Power,* Vol. 118. 1996.

63. Oh Si-Doeck, Pang Hye-Suu, Kim, Si-Moon and Kwak Ho-Young, "Exergy Analysis for Gas Turbine Co-generation System, *Trans of the ASME. J. of Engg. for Gas Turbine and Power,* Vol. 118, 1996.
64. *Gas Turbines World,* 99-,2000, 2001, 2002, Pequot Publishing Inc. USA.
65. Rice, I.J., P.E.., "Thermodynamic Evaluation of Gas Turbine Cogeneration Cycles : Past I – Heat Balance Method Analysis Part II – Complex Cycle Analysis," *Trans of ASME. J. of Engg : for Gas Turbines and Power"* Vol. 109, 1987.
66. El-Masri, M.A., "GASCAN–An Interactive Code for Thermal Analysis of Gas Turbine Systems," *Trans. of ASME. J. of Engg for Gas Turbines and Power,* Vol., 110, 1988.
67. El-Masri, M.A., "On-Thermodynamic of Gas Turbine Cycles", Part 1 : Second Law Analysis of Combined Cycles" *Trans. of ASME. J. of Engg for Gas Turbines and Power,* Vol. 107, 1985.
68. El-Masri, M.A., "On-Thermodynamics of Gas Turbine Cycles": Part 2 – A Model for Expansion in Cooled Turbines," Part 3-Thermodynamics Potential and Limitations of Cooled Reheat Gas-Turbine Combined Cycles," *Trans of ASME. J. of Engg for Gas Turbines and Power,* Vol. 108, 1986.
69. El-Masri, M.A., "A Modified, High-Efficiency Recuperated Gas Turbine Cycle," *Trans of ASME. J. of Engg for Gas Turbine; and Power,* Vol. 110, 1988.
70. Yadav, R. and Singh, L., "Comparative Performance of Gas/Steam Combined Cycle Power Plants, ASME, Int. Gas Turbine and Aero-engine Congress and Expositions, Houstan, Texas, June. 5–8, 1995.
71. Sanjay, Singh O. and Prasad, B.N., "Prediction of Performance of Simple Combined Gas/Steam Cycle and Co-generation Plants with Different Means of Cooling", *Proceeding of IECEC,* 2001, Paper No. IECEC - 2001 CT – 11.
72. Yadav, R; "Effect of Bottoming Cycle Alternatives on the Performance of Combined Cycle, *Proceeding of ASME Turbo. Expo.* 2003. June 16–19. 2003, Geogia Atlanta, USA; (GT 2003-38052)
73. Dixon, S.L., *Fluid Mechanics. Thermodynamics of Turbomachinery,* Pergamon Press, 2nd edn, 1975.
74. Yadav, R: Kumar Pradeep and Sarswati Samir, "Comparative Thennodyanmic Analysis of Combined and Steam Injected Gas Turbine Cycles". *Proceeding of IJPGC* 2003, June 165-19-2003, Georgia Atlanta, USA (IJPGC 2003. 40118).
75. Yadav. R; Srivastav Priyesh and Sarswati Samir, "Thermo-economic Analysis of Combined Cycles", *Proceeding of IJPGC* 2003, Jun-16–19 2003, Georgia, Atlanta USA (IJGPC 2003-40119).
76. Sanjay, Singh Onkar and Prasad, B.N.; "Thermodynamic Performance of Complex Gas Turbine Cycles" *Proceeding of IJGPC* 2002, June 24–26, 2002, Virginia, USA (IJPGC-2002-26109).
77. Sanjay, Singh Onkar and Prasad, B.N: "Thermodynamic Evaluation of Advanced Combined cycle Using Latest Gas Turbines, *"Proceeding of ASME/IGTI TURBO EXPO* 2003, June 16-19, 2003-38096).
78. Yadav. R. and Yadav, N.P., and Sarswati Samir, "Thermodynamic Analysis of Hydrogen Fuels Combustion Turbine of ICEE 2003," Dec. 11-13, 2003, Sanghai, China.
79. Yadav. R.L., Newby, R.A., and Yang W.C., 1999 "Final Report on the Development of Hydrogen Fueled Combustion Turbine Cycle for Power Generation". *ASME. J. of Engineering for Gas Turbine and Power,* 121, pp. 38-45.

80. Aschner, F.S., *Planning Fundamentals of Thermal Power Plants,* John Wiley, 1976.
81. Weisman, J. and Eckert, R: *Modem Power Plant Engineering,* Prentice-Hall of India, New Delhi, 1983
82. Li, K.W. and Priddy A.P.., *Power Plant System Design,* John Wiley, 1985.
83. Haywood, R.W., *Analysis of Engineering Cycles,* "Pergamon Press, Oxford, 1975.
84. Horlock, J.H., "Combined Heat and Power," Pergamon Press, Oxford, 1984.
85. Horlock, J.H., "Combined Cycle Plants, Pergamon Press, 1992.
86. Bannister, R.L., "Newby, R.A., and Diehrl, R.C., "Developing a Direct Coal Fired Combined Cycle, *Mechanical Engineering*, Dec. 1992. pp. 64–70.
87. Tuzson, J., "Status of Steam Injected Gas Turbines." *Trans, ASME. J. of Engg. for Gas Turbine and Power,* Vol. IIA. Oct 1992, pp. 682-686.
88. Bennett, B. and Bannister, R.L., "Pulverized Coal Power Plants: The next Logical Step; *Mechanical Engg.* Vol. 103, No. 12, 1981, pp. 18-24.
89. El-Wakil, M.M., *Power Plant Technology,* McGraw-Hill, New York, 1985.
90. Basu, P. and Haldar. P.K., *"Combustion of Single Carbon Particles in a Fast Fluidized. Bed of Fine Solids"*, "Fuel 68, 1989, pp. 1056-1063.
91. Basu, P. and Fraser, S.A., *Circulating Fluidized Bed Boiler: Design and Operation.* Butterworth-Heinamann, 1991.
92. Howard, J.R., Fluidized Bed Technology: Principles and Applications, Adam Hilger, 1989.
93. Kehlhofer, J.R., *Combined Cycle Gas and Steam Turbine Power Plants,* Fairmount Press, Lilburn, 1991.
94. Reznikov, L and Lipov Y.M., *Steam Boilers of Thermal Power Stations,* Mir Publishers, Moscow, 1983.
95. British Electricity and Generating Board (CEGB), *Modem Power Station Practice,* Third edition, Vol. 1 to 8, Pergamon Press, 1991.
96. Kotas, T.J., *"The Energy Method of Thermal Power Plants Analysis,"* Butterworths, 1985.
97. Nag, P.K., *Power Plant Engineering.* 3rd edn. Tata McGraw Hill Co Ltd. New Delhi. 2012.
98. Arora and Domkundwar, A Course in Power Plant Engineering, Dhanpat Rai & Sons. New Delhi.
99. Yadav, R., *"Internal Combustion Engines and Air Pollution",* 1st edn; 2002, Central Publishing House, Allahabad. (Now with IK Intl.)
100. Yadav, R., *"Fundamental Engineering Thermodynamics",* Central Publishing House Allahabad. (Now with IK Intl.)
101. Yadav, R., *"Applied Thermodynamics* Central Publishing House*",* Allahabad. (Now with IK Intl.)
102. Yadav, R., *"Steam & Gas Turbines and Power Plant Engineering",* Central Publishing House, Allahabad. (Now with IK Intl.)
103. Khan, B.H; *"Non-conventional Energy Resources",* 1st edition, Published by Tata McGraw Hill, New Delhi.
104. Kothari B.P.; Singal, K.C. and Rakesh Ranjan, *"Renewable Energy Resources and Emerging Technology"*. Published by Prentice Hall India, New Delhi.

105. D.J. Bennet and J.R. Thomson, *"The Elements of Nuclear Power"*, Longman, 1989.

106. Joel Weisman and Roy Eckart, *"Modern Power Plant Engineering"*, Printice-Hall of India, New Delhi, 1995.

107. Harry A. Sorensen, *"Energy Conversion Systems"*, John Wiley, 1983.

108. M.M. Dandekar and K.N. Sharma, *"Water Power Engineering"*, Vikas Publishing House, New Delhi, 1979.

109. R.K. Rajput, *"A Text Book of Power Plant Engineering"*, Laxmi Publications (P) Let. New Delhi, 1995.

110. Z. Hussain, *"Steam Turbines"*, Tata McGraw-Hill, N.Y. 1986.

111. Black and Veatch, *"Power Plant Engineering"*, Chapman & Hall, Inc. N.Y., Indian Edition, CBS Publishers & Distributors, New Delhi 1998.

112. British Electricity International, London (CEGB), *"Modern, Power Station Practice"*, 3rd edition Vol. 8, "Boilers and Ancillary Plant", Pergamon Press, 1991.

113. P. Basu and S.A. Frazer, *"Circulting Fluidised Bed Boilers"*, Butterworth, Heinemann, 1991.

114. M.J. Reznikov and Yu M. Lipov, *"Steam Boilers of Thermal Powar Stations"*, Mir Publishers, Moscow, 1985.

115. B. Bennett and R L Bannister, *"Pulvervised Coal Power Plants, the Next Logical Step"*, Mech Engg., Vol. 103, No 12, 1981, pp. 18-29.

116. S.P. Sharma and C. Mohan, *"Fuels and Combustion"*, Tata McGraw-Hill, New Delhi, 1984.

117. G.H. Lamb, *"Underground Coal Gasification"*, Noyes Data Corporation, New Jersey, USA, 1977.

118. J. Makansi, *"Combined Cycle Power Plants"*, Power, June 1990, pp. 91–126.

119. R.H. Finckh and H. Pfrost, *"Development Potential of Combined Cycles (GUP) Power Plant with and without Supplementry Firing"*, *Trans., ASME, J. of Engineering for Gas Turbines and Power,* Vol. 114, Oct. 1992, pp. 655-659.

120. T.J. Kotas, *"The Energy Method of Thermal Plant Analysis"*, Butterworths, 1985.

121. P.N. Modi and S.M. Seth, *"Hydraulics and Fluid Mechanics"*, Standard Book House, Delhi, 1985.

122. J. Jensen, *"Energy Storage"*, Newnes-Butterworths, London, 1980.

123. G. Beghi, *"Thermal Energy Storage"*, D. Reidel Publishing Co., Holland, 1981.

124. B.H. Khan, *"Non-Conventional Energy Resources"*, Tata McGraw-Hill, 3rd edition, New Delhi.

Index